THE EXCITEMENT AND FASCINATION OF SCIENCE:
Reflections by Eminent Scientists

Volume 4

THE EXCITEMENT AND FASCINATION OF SCIENCE

Reflections by Eminent Scientists

VOLUME 4 (1995)

Compiled by ROBERT H. HAYNES
York University
and WILLIAM KAUFMANN
Annual Reviews, Inc.

Reprinted from the following Annual Reviews:

Anthropology
Astronomy and Astrophysics
Biochemistry
Computer Science
Earth and Planetary Sciences
Entomology
Fluid Mechanics
Genetics
Immunology
Materials Science

Microbiology
Neuroscience
Nuclear and Particle Science
Nutrition
Pharmacology and Toxicology
Physical Chemistry
Physiology
Plant Physiology and Plant Molecular Biology
Sociology

ANNUAL REVIEWS INC.   4139 EL CAMINO WAY   P.O. BOX 10139   PALO ALTO, CALIFORNIA   94303-0139

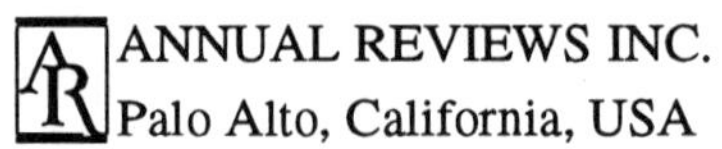

ANNUAL REVIEWS INC.
Palo Alto, California, USA

International Standard Book Number: 0-8243-2604-0

Library of Congress Cataloging-in-Publication Data
(Revised for volume 4)

The Excitement and Fascination of Science

Vol. 4 has subtitle: Reflections by eminent scientists
"Reprinted from the Annual review of anthropology," etc.—Vol. 4, t.p.
Includes bibliographies.
1. Science   2. Scientists—Biography.
Q171.E98
ISBN 0-8243-2604-0 (v. 4)

The paper used in this publication meets the minimum requirements of American National Standards for Information Sciences—Permanence of Paper for Printed Library Materials, ANZI Z39.48-1984

PRINTED AND BOUND IN THE UNITED STATES OF AMERICA

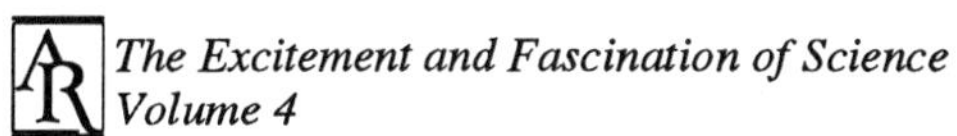

The Excitement and Fascination of Science
Volume 4

CONTENTS

(Note: These chapters have been repaged for this collection, but the source line of each article gives the original span.)

ANNUAL REVIEWS INC. is a nonprofit scientific publisher established to promote the advancement of the sciences. Beginning in 1932 with the *Annual Review of Biochemistry,* the Company has pursued as its principal function the publication of high-quality, reasonably priced *Annual Review* volumes. The volumes are organized by Editors and Editorial Committees who invite qualified authors to contribute critical articles reviewing significant developments within each major discipline. The Editor-in-Chief invites those interested in serving as future Editorial Committee members to communicate directly with him. Annual Reviews Inc. is administered by a Board of Directors, whose members serve without compensation.

FOREWORD

The first scientific memoir to be published by Annual Reviews (AR) appeared as a Prefatory Chapter in the *Annual Review of Physiology* for the year 1950. Its author was the distinguished physician and student of human metabolism Eugene F. DuBois of Cornell University Medical College. In his letter of invitation, Editor Victor E. Hall told DuBois that the motive for the request was "our desire to make the Review something more than a consideration in detail of the current advances in our science. Physiology is a form of human activity as well as an accumulation of knowledge." Over the next 41 years, some 425 such chapters have been published in 20 of the 26 AR series. Most editorial committees today maintain the tradition of inviting widely admired senior scientists to write reflectively on their lives and careers and/or on the evolution of their fields as seen from their personal perspectives. Many authors, editorial committee members, and readers have told us that these memoirs are often the articles they read first when a new Annual Review reaches their desks.

The genesis of the idea for including personal memoirs in AR series and the rationale for their compilation into anthologies were described by AR's founder, Dr. J. Murray Luck, in his Foreword to the first volume of *The Excitement and Fascination of Science* (1965):

> ... these essays are uniformly so fascinating, so rich in interesting bits of personal history, and so successful in conveying something of the excitement of science that each deserves a much wider audience than it has hitherto enjoyed .... Perhaps [they] will help to show that scientists are people, possessed of the same strengths and the same weaknesses as anybody else ... . [Furthermore,] we believe that this collection of essays will prove to be just as interesting to many who are outside of the sciences as to the insiders themselves.

In 1976, Dr. William C. Gibson, compiler of the second collection, observed in his Foreword:

> Thirteen years ago Annual Reviews Inc. published its first volume of autobiographical chapters by scientists who had made great contributions and who could recount engagingly their development as scientists. Such chronicles are all too rare, alas, for they should be the heady diet of secondary school and college students searching for fact, rather than fancy, on which to base the selection of a field of study and possibly a career... .
>
> The story of science is replete with instances of great personal decisions having been made on key, but minimal, evidence. Discoveries too may hang on the merest thread of chance. But, as Pasteur remarked, "Fortune favors the prepared mind." These essays can be instrumental in preparing young minds... .

In 1990, the third—and by far the most comprehensive—collection was published. It included for the first time both subject indexes and a bibliographical appendix that were cumulative for all of the first three volumes. That volume's compiler, Dr. Joshua Lederberg, contributed an illuminating introductory essay, "Reflections on Scientific Biography," in which he stated his criteria for inclusion of chapters in Volume 3. We have in general followed these criteria in making the selections for Volume 4. We urge all readers with an interest in life writing in the sciences to consult Dr. Lederberg's essay.

The Appendix to the present volume lists every prefatory chapter published by AR from 1950 through 1991. Some intriguing discoveries await readers who scan this list. For example, some of the outstanding chapters that have not, for one reason or another, been reprinted in this series of anthologies include:

Cosman, MP 1983. A feast for Aesculapius: historical diets for asthma and sexual pleasure. *Annu. Rev. Nutr.* 3: 1-33

Simpson, GG 1976. The compleat palaeontologist? *Annu. Rev. Earth Planet. Sci.* 4: 1-13

Schrimshaw, NS 1987. The phenomenon of famine. *Annu. Rev. Nutr.* 7: 1-21

Smith, CS 1986. On material structure and human history. *Annu. Rev. Mater. Sci.* 16: 1-11

Smith, EH 1976. The Comstocks and Cornell: in the people's service. *Annu. Rev. Entomol.* 21: 1-25

Readers who may be lured by the titles or authors of those memoirs that have not been reprinted and who do not otherwise have easy access to past AR volumes can obtain copies through document delivery services.

The publication of the present volume resulted from the very positive and gratifying reception accorded the first three volumes by individual readers and the science media. Among the comments published by reviewers were the following:

... These are the best of all scientific reflections... .

... these volumes contain a rich lode of ore for the future historian or philosopher of science to mine, as well as a bountiful source of interest and inspiration for the practicing scientist and teacher.

I ended up reading much more of this ... than I had intended ... . I can well imagine that these volumes might provide inspiration to many high school and college students... .

Nuclear physicist Emilio Segrè offers the best reasons that physicians and health professionals should read this collection: "My father once told me, 'You are living on Fermi's crumbs.' ... Virtually every chapter contains numerous personal triumphs, tragedies, insights, and experiences ... . Reading the entire volume is an enlightening educational process ...the reader will appreciate the scientific mind and the scientific method by reading this book ... This is a wonderfully rich collection. We recommend it."

The founder of the Institute for Scientific Information, Dr. Eugene Garfield, noted in his editorial column in the September 27, 1993 issue of *Current Contents* that "In recent years, autobiographies by scientists in book form have become more common and popular ..." and suggested that "short autobiographical accounts ought to become a standard practice for scientists ... ." We concur, and we believe there may well be an increase in the publication both of full-scale "lives" and of brief memoirs by leading scientists in the years ahead.

In this fourth volume of *The Excitement and Fascination of Science* we have included 52 engaging pieces. All are drawn from AR series published over the years 1988–1991 and are reprinted here without further editing. Taken collectively, the four volumes of *The Excitement and Fascination of Science* published to date constitute an important, indeed unique, historical record of the lives, views, and careers of many notable scientists. They provide a rich source of instructive and pleasurable reading for those who wish to become better acquainted with the men and women of science. Those interested in making cross-cultural comparisons of the experiences and motivating factors that shaped the careers of a diverse group of physical, biomedical, and social scientists will also find much useful information in these memoirs.

It is our impression that many undergraduate and even graduate students lack an accurate view of what sorts of people scientists actually are, or of what kinds of intellectual development and career paths successful scientists may experience. The popular image of the scientist, created traditionally by novelists, dramatists, filmmakers, journalists, and nonscientists generally, is more often evil, arrogant, inhuman, mad—or simply foolish—than is any real scientist of our acquaintance. At the other extreme of unreality, scientists are sometimes (though more rarely) depicted as heroes, saints, or saviors facing some Manichaean conflict or looming disaster.

Fortunately, teachers and professors can help their students bridge the enormous gulf between imaginative myth and benign reality by introducing them to memoirs like those brought together in *The Excitement and Fascination of Science.* The lives and attitudes of the authors presented in this volume reveal the deep humanity of scientists, and of science itself as a human endeavor.

Robert H. Haynes, *Distinguished Research Professor, York University, Toronto*

William Kaufmann, *Editor-in-Chief Emeritus, Annual Reviews Inc.*

Annu. Rev. Anthropol. 1991. 20:1–23

TOWARDS A ROOM WITH A VIEW:
A Personal Account of Contributions to Local Knowledge, Theory, and Research in Fieldwork and Comparative Studies

Jack Goody

St. John's College, University of Cambridge, Cambridge CB2 1TP, United Kingdom

KEY WORDS: history, Cambridge, kinship, literacy, LoDagaa

When the editors asked me to make a personal contribution to the *Annual Review of Anthropology,* I accepted mainly because my colleagues, Fortes and Leach, had already done so. I write here more to mark our joint achievement at what became the Department of Social Anthropology at Cambridge, and to review my own particular contribution, than to offer new insights or revelations about anthropologists.

Fortes arrived in Cambridge in 1950 as Professor, I returned there as a graduate student at the end of the same year, and Leach came in 1954. Our relationships were never entirely harmonious—what relationships are? But we collaborated from the beginning in a way that was helpful to each other and to others.[1] That collaboration began with seminars every Friday of term, which

[1] In her interesting piece on Meyer Fortes in the *American Ethologist,* Susan Drucker-Brown (2) describes her introduction to Cambridge and to the postgraduate seminar where Leach and Goody were always being rude to one another. That we argued vigorously is true, but such was the intellectual culture to which we were accustomed. We saw this not as rudeness but as frankness. We generally saw eye to eye on the way the Department should go, except that Fortes did not approve of the proposal to merge with Social and Political Sciences, which is where I thought the most favorable environment for social anthropology lay. However we remained close friends throughout our time in Cambridge.

0084-6570/91/1015-0001$02.00

GOODY

all those attached to the Department felt an obligation to attend—on the Oxford model, which in turn was based on the practice at the London School of Economics. There was literally a continuing clash and exchange of ideas. The seminar crystalized into the *Cambridge Papers in Social Anthropology*, to the first of which, *The Developmental Cycle in Domestic Groups* (1958), we each contributed. For that volume the central idea came from Fortes; but we circulated editorial responsiblity and each edited subsequent collections on a particular topic with a long theoretical introduction. We founded, too, a series, *Cambridge Studies in Social Anthropology,* which published some 70 volumes before changing its title—many by our staff and students, but including also works by authors from Europe, the Americas, and elsewhere.

Our backgrounds were very different. Fortes came from a family of Jewish migrants to South Africa, Leach from a lineage of mill-owners with interests in the Argentine, myself from a more modest background in the Home Counties. I had originally gone up to Cambridge on a scholarship to read for a degree in English literature and became interested in, among other things, aspects of the sociology of literature. After my first year at the University war broke out, and for the next six and a half years I was in the army. In the Near East one was brought face to face not only with a wide variety of humanity— Greeks, Turks, Egyptians, Palestinians, Jews—but also with the remnants of ancient civilizations. Directly I had an opportunity I began to read works such as Childe's *What Happened in History*? That opportunity came sooner than I anticipated, since I was captured by the German army at Tobruk in June 1942 and spent the next two and a half years in and out of prison camps in Italy and Germany. There I was also able to read fairly widely, following up various interests, but the situation in which I found myself also turned my attention to the variety of human behavior. Spending time with Italian peasants in the Abruzzi and with men of various nationalities, origins, and personalities in camp made me think about social relations in a more sociological and psychological frame. On my release I was particularly attracted to the work being done at the Tavistock Institute of Human Relations in London, which was concerned with returning prisoners of war.

When I returned to Cambridge in 1946 I completed a degree in English, then switched to the Faculty of Archaeology and Anthropology. After some years of inactivity in prison camp I was unwilling to think of spending more time in the rarified atmosphere of the University so, having completed a Diploma, I took a job in adult education of a more basic kind. Even when I decided to return to the social sciences it was not with the aim of spending my life working on "other cultures".

I went to West Africa not primarily because I wanted to become an Africanist but because I had become interested in the comparative study of human society more generally and intended to return with a perspective that would enable me better to look at European culture, possibly at those Italian peasants who had fascinated me when I was hiding among them, more likely the inhabitants of Britain itself with whose future I was deeply concerned in the aftermath of war. It is difficult to reconstruct the situation that faced many of my generation at that period. If I may put the matter a little more dramatically than an Englishman should, I saw post-war Britain under the Labor government of 1945 as the national outcome of many lives like my own: We had grown up between the two wars, with an interval of only 21 years from one terrifying destruction to the next, and lived our adolescence under the shadow of continental fascism, with its devastating oppression and annihiliation of man by man both in ideology and in practice. This period began with the Japanese attacks on China, the colonial wars of Italy, the civil war in Spain, and the expansion of Germany, against the background of widespread suppression and maltreatment in those countries. There followed for my generation some six-and-a-half years of life under arms, during which time all one could look forward to was post-war reconstruction, through the national government and through the United Nations.

That reconstruction obviously involved the dissolution of earlier empires, the whole process of decolonization that began with India in 1947 and that was envisaged, at least under Labor rule and to some extent by Conservative politicians as well, as gradually extending to the rest of the colonial territories. In a sense the deconstruction of the empire was part and parcel of the reconstruction of Britain. And although we were not fully aware of the consequences of decolonization, it was a heady prospect, indeed a heady actuality, too, for the major part of the process was over within some ten years, by 1960, the Year of Africa. The Gold Coast, where I chose to work, led the way and became the independent nation of Ghana in 1957 under the premiership of Kwame Nkrumah. It was this general background set against five years' residence in the country that led me to write about matters that touched on political theory and development.[2] Anthropologists had often tried to avoid administrators and problems of development. That seemed a churlish way of treating the new Africa that was coming into being so rapidly.

That new Africa was also demanding a history. Again earlier anthropologists had often steered away from historical considerations, partly because of the often rash speculations of their predecessors about the past, partly because of an alternative methodological approach associated with fieldwork, and partly for the practical reason that archives and records were officially closed for a 50-year period. The situation had changed in many respects. Particularly in France many anthropologists began to explore the historical dimension, and elsewhere, too, historians became interested in anthropological research

---

[2]For example, I wrote articles on "Consensus and dissent in Ghana" (1968), "Rice-Burning and the Green Revolution" (1980), as well as on general impediments to development in Africa as compared to Asia. Much else exists in the unpublished form of talks.

and approaches. The "structural-functional" paradigm was ready to be loosened in a number of ways.

My own choice of Ghana was largely personal. At Cambridge I was friendly with Joe Reindorf, later Attorney General, who started a PhD in history, finished a law degree at the same time, and then returned to Ghana to work with the firm of a relative, Victor Owusu. Owusu was closely identified with the Ashanti opposition to the Convention People's Party, although Joe's own sympathies lay elsewhere.[3] Since I had a family I did not want to work too far from Europe; in those days of boat travel, distance mattered. For this reason when I followed my colleagues to Oxford, after the spell in adult education, I was asked to talk to Meyer Fortes, the West Africanist, my first encounter with whom I have described elsewhere (17a).

In his review of social anthropology in West Africa, Keith Hart speaks of the structural-functional Cambridge school of Fortes and Goody (17). From the standpoint of ethnography and fieldwork, such a characterization is no doubt correct. But as he goes on to point out, the intellectual preoccupations in my research had other roots. Few students in the 1930s could avoid having to make some kind of resolution of their interests in two major figures, Marx and Freud. My contact with Marxism had by then been longstanding, but there were other influences. Through social anthropology I encountered Durkheim, through Harvard students of social relations I encountered Weber, as well as Parsons (with whom I later worked for a year when he was at Cambridge). But equally important to me in the early phases was Freud, as well as the interests I had developed in studying literature and history.

In literature I was strongly influenced by the "New Criticism" associated with F. R. Leavis and other contributers to *Scrutiny,* regarding not only the creative arts themselves but also their whole grounding in society. One further current that affected many of my generation was the logico-positivism of the Vienna circle and the subsequent transformations that appeared in the Anglo-Saxon world. Of particular importance for me was the work of the con-tributors to the *International Encyclopaedia of Unified Science* edited by 0. Neurath, R. Carnap, and C.W. Morris from Chicago (1938-), which included works by E. Nagel, L. Hogben, and many others.

Two other factors of a different kind played a part in defining my fields of interest. Over six years of war had left their mark: Life in a prison camp led me to wonder how people got on or came into conflict with one another; life in the army encouraged an interest in the military side of "other cultures". Evans-Pritchard's clinical analysis of feud in his lectures in Cambridge in 1946–1947 helped to place those years of experience in a wider context, and

---

[3]I had later the pleasure of acting briefly as his clerk in the reknowned Al-Hajji Baba case in Kumasi in 1957, where the Government tried to expel him from Ghana as an alien.

the threat of nuclear warfare continued to keep "conflict studies" in the center of one's vision. If Marx drew my attention to modes of production, the "New Criticism" and logical positivism to modes of communication and of thought, so did the times lead me to consider modes of destruction.

Largely at Fortes' prompting I went to work in a community in Northern Ghana. These were the LoDagaa (or Lobi as the British then called them, or the Dagari or Dagara as they were known to the French and now generally to themselves), a study of whom had been listed as a priority in Raymond Firth's report to the Colonial Social Science Research Council.[4] The LoDagaa inhabited an area that lay astride the Black Volta, which served as the boundary between the anglophone colony of the Gold Coast (later Ghana) and the francophone colony of Haute Volta (later Burkino Faso). I was thus required to become familiar with the early French writing about the region as well as with French anthropology more generally. That encounter enabled me to indulge my enduring attraction to France and at the same time to modify, in the course of time and in marginal ways, some of my Anglo-Saxon attitudes.

Fortes directed my attention to the LoDagaa partly because he had carried out research among the "patrilineal" Tallensi and the "matrilineal" Ashanti and was interested in those societies of which the Yakö of Nigeria, studied by Daryll Forde, was the paradigmatic case, in which named unilineal descent groups of a patrilineal and a matrilineal kind existed side by side—that is to say, everyone was a member of one of each set. I became involved in the topic of descent of necessity, for it was the subject of lively discussion among graduate students. I also took up other topics on which people were working at the time, such as the developmental cycle of domestic groups (8), the maintenance of social control in acephalous communities (7), the role of the mother's brother (9), and the general theme of incest itself (6). These topics drew me into comparative analyses that I later pursued with Esther Goody using material on the societies of northern Ghana for the purposes of con- trolled comparison (10, 11). Another general question that struck me was how communities defined themselves not only in opposition to one another but often in relation to particular activities, so that among the LoDagaa one's identification was with those to the east in one context and with those to the west in another. That is to say, in the Gold Coast at that time there were no "real" Lobi and Dagari, just a series of communities who defined their similarities and differences by means of these roughly directional terms—

---

[4]The Colonial Social Science Research Council, working under the inspired secretaryship of Sally Chilver, was responsible for administering the grants that enabled Lloyd Fallers, Jim and Paula Bohannan, Al and Grace Harris, and Bob Armstrong to work, first at Oxford, then in Africa. At the end of the war the Council promoted surveys of research needs in the social sciences in West Africa (undertaken by Raymond Firth), in East Africa (by Isaac Schapera), and in Sarawak (by Edmund Leach).

GOODY

Lo(bi) = west, matrilineal; and Dagaa(ri, ra, ti) east, patrilineal (to simplify a more complex situation).

I was interested in both the subtle differences in the importance of descent groups among and for the LoDagaa and in their systems of kinship, family and marriage at the domestic level. Indeed I attempted, in the spirit of the times, to look at the total range of the community's activities. When I came back after one year, between field trips, I wrote a general account for the Bachelor of Letters degree at Oxford, which was later published as *The Social Organisation of the LoWiili* (1956). I then returned to Africa to live in another area of the LoDagaa country where more emphasis was given to the matrilineal clans. Here was a case of "double descent" along the lines of the Yakö in that movable property (cattle, money, grain) was visualized as belonging to the matrilineal clan (or *belo*, species) and was inherited between, first, maternal siblings and then by sister's sons—that is, between uterine kinsfolk; whereas immovables (houses and land) were transmitted within the patrilineal clans, between, first, maternal siblings and then to sons—that is, between agnates. My major interest in this situation lay in the problematic of Malinowski and Fortes, relating to the different nature of interpersonal ties in patrilineal and matrilineal societies, depending upon the systems of authority, transmission, and organization. That is, my interest lay in an attempt to link the social and personality systems in Parsons's terms, or, in more general terms, the approaches of Marx and Freud.

My doctoral thesis, a much-revised version of which was published as *Death, Property, and the Ancestors* (Stanford, 1962), was intended to be an analysis of the religion of the LoDagaa. Having already written a general account of the social organization, on my second return I completed a historical and ethnographic outline of the region in order to supply a context for my more detailed studies.[5] The concentration on religion resulted partly from the fact that Fortes had written two major studies on the political and kinship systems of the Tallensi, which were sufficiently like those of the

[5]*An Ethnography of the Northern Territories of the Gold Coast, West of the White Volta* (1954). I have subsequently published, often in obscure places, a number of contributions to the ethnographic, historical, and linguistic study of the region. I would like to acknowledge the collaboration of a number of scholars working in the lively Institute of Africa Studies at the University of Ghana, Legon, at that time under the direction of Thomas Hodgkin—especially Kwame Arhin, Ivor Wilks, and Nehemiah Levtzion. At the same time I would pay tribute to my long-standing collaborators among the LoDagaa, especially S. W. D. K. Gandah, and among the Gonja, especially the late J. A. Braimah, who became paramount chief of his kingdom and wrote extensively about its history; with Braimah I edited *Salaga: The Struggle for Power* (1967). I have said little about these historical papers in this account, but they were important in providing a diachronic view of the region as well as in trying to allow for the influences, both stabilizing and disruptive, that colonial governments had on the peoples among whom earlier anthropologists were working, whether in Africa, the Americas, or Russia.

LoDagaa to make me direct my attention elsewhere. Like Evans-Pritchard, Fortes left the monograph on religion to the end, as was the usual progression in analysing field material at the time, *The Work of the Gods* following on from *We the Tikopia*. I chose to take another direction. My material on the religious life was rich, partly because I had attended so many funerals and sacrifices, partly because of an earlier interest in myth and ritual, partly because I knew people who would discuss these matters, and partly because the Bagre ceremony was being performed in the valley below my house during my first year. Not only could I attend the public aspects of this initiation, but I had the good fortune to meet a man who had become marginal to the society itself. He offered to recite his version of the Bagre so that I could write it down, and he then took the trouble to explain many of its details. It was unheard of at the time to recite before someone who was not a member of the association, and at first I was reluctant to publish. But local friends thought it would be better to do so as a way of recording the richness of their culture.

My first aim was to present a general analysis of LoDagaa religion. Having previously sketched out the social organization, I was obviously concerned with those aspects linked to relatively permanent roles and groupings (as with ancestors and descent groups, earth shrines and parishes). But I was also interested in the turnover displayed by cults such as those associated with medicine shrines, which migrated from one group to the next over a wide range of territory. Such cults were marked by a rise and fall, by birth and obsolescence, by the recognition of the God who failed as well as by the innovative capacities of those seeking new and different solutions to old and persistent problems. A quest was involved, one that was intellectual and problem-solving as well as emotional and theological. The structure of meaning to the actors had first to be elucidated, including the overt symbolism of ritual acts.

Although the aim of the thesis was to present a study of the whole domain of ritual and religion, I never got further than an analysis of funerals (and of the ancestors), the topic on which I had begun, taking it to represent the point of transition and of transmission between this world and the next. On this one aspect there seemed already too much to say, especially if one tried to take into account what previous writers in a variety of disciplines had contributed. In any case the interest the subject held for me ranged outside the sphere of religion narrowly conceived.

Among the problems with which I was concerned in discussing LoDagaa funerals were those I related to the work of Marx and Freud. There were other influences. The French school, including the work of Van Gennep and especially Hertz, was most relevant; some have seen this trend as dominant. Others have taken the analysis of the role of kin groups based on matrilineal and patrilineal descent to lie at the core. Typical of this first view is the

GOODY

comment of Jacques Lombard: "De même, J. Goodie (sic), reprenant certains thèmes de R-B (Radcliffe-Brown), a expliqué la fonction de rites mortuaires en les liant à la structure sociale et plus particulièrement au statut du défunt et de ses proches" (19). Lombard contrasts the work of Middleton, Turner, and myself with that of Douglas, Lienhardt, and Beidelman who wanted to "désociologise le religieux" following the studies of Evans-Pritchard and the volume edited by Forde entitled *African Worlds*.

The contrast is too stark. It seemed hardly possible to deal with a *rite de passage* without building upon Van Gennep's pioneering work, nor did it seem useful to discuss "symbolic" meanings without recourse to the classic but equally simple techniques developed by Radcliffe-Brown and Srinivas— that is, without being concerned with "action". While I dealt with the way social groups, including descent groups, emerged and participated so clearly in the funerals, I was also specifically interested in (*a*) differences in funeral practices that were not connected with sociological variables; and (*b*) meaning to the actors of the acts, verbal and gestural, in which they were engaged, as a counterpoise to the interpretations of the anthropologist. Indeed my hesitation to use a variety of hardy anthropological concepts, such as ritual, religion, the sacred, and the profane—except as vague sign-posts—was precisely because they were not based upon, nor did they reflect, indigenous categories, which were more complex and more shaded than such constructs allowed.

Above all, my central thrust was aimed in another direction, more closely linked to much of my later work. In looking at these aspects of religious action and belief, I reviewed earlier studies of ancestor worship, of funerals, and in particular of patterns of grief and mourning. How did the behavior of individuals in these situations correspond to the wider contexts of their lives? The question raised issues considered by Malinowski and Fortes when they examined matrilineal systems with the aim of specifying the typical patterns of tension and cathexis, of ties and cleavages, in interpersonal relations. It was a theme pursued by Malinowski when he argued that the "Oedipus complex" was not universal but was linked to a particular range of social institutions.

The central theme of the enterprise was parallel to one that developed in the study of witchcraft, in particular of witchcraft accusations. Evans-Pritchard's book *Magic, Witchcraft and Oracles among the Azande* (1937) had attempted to map out the logic of African witchcraft in opposition to the view of Levy-Bruhl that "the primitive mind" was illogical or alogical, unable to perceive contradiction, a theme that I pursued in connection with a very different line of research. What Evans-Pritchard did not do was look at witchcraft in terms of the relationships of accuser, witch, and victim. Starting with the work of Nadel, these relationships came to comprise an important topic in the two post-war decades, in Europe as well as Africa, as the work of

Favret and others has shown. Gluckman extended the analysis to the realm of gossip, and my own work on ancestor worship raised problems of a similar kind. Who were the ancestors seen as most likely to demand sacrifices from the living, and were these relationships linked to the control they had exercised and to the benefits received? The tensions that arose between those who gave and those who received were in turn connected with the incidence of grief, with the particular situations of the dead and the living, and with the guilt about a deceased parent for whom one may not have done enough or whom one had left on his or her own in old age. That connection was often made at the less explicit levels of the mind, and the general link with Freud is obvious.

I was concerned to elaborate and differentiate the relevant kinds of tension that marked relations between the generations. It was not simply a question of "splitting the Oedipus complex" (between roles) as Malinowski had argued for the matrilineal Trobriands, where mother's brother counterpoised father, but of splitting up the Oedipus complex, the intergenerational relationship of conflict, into its analytic components, some of which were concerned with sexual jealousy, others with the process of socialization, and others with authority more generally. This is where Marx was relevent. For an important element in preindustrial societies of this kind was the tension arising out of inheritance, the tension between the holder of property and the heirs. My field data came from the two adjacent communities I have mentioned, in one of which all male property was transmitted between fathers and sons while in the other immovables (land and houses) passed in that way and movables between a man and his sister's sons (in both cases after full sibling in the same farming group). In the latter case, that of a fully fledged double-descent system, these tensions were particularly in evidence; they were quite explicit and the splitting quite formal.

My enquiry aimed to examine differences in the funeral and ancestral ceremonies of the two adjacent commmunities and to try to explain at least some of these by reference to the differences in the transmission of property. Not all differences were so explained. I had no monocausal answer. Others I considered to be the result of intellectual exploration, yet others perhaps as the result of cultural drift. But certain central differences I did see related to social relationships, and these I discussed in a chapter entitled "The Merry Bells: Inter-generational Transmission and Its Conflicts," which was the prelude to a general analysis of property and inheritance, before I engaged upon an enquiry into the institutions of the LoDagaa themselves. Those 50 pages are for me among the most critical I have written. Beginning with Freud and Fromm, I went on to refer to Engels's discussion of production and reproduction in *The Origins of the Family* (1884). I attempted here, and later in *Production and Reproduction* (1977), to see kinship and the economy as

GOODY

distinct but related, and to do so at a more formal analytical level than those who had concentrated upon them separately. I saw that a key lay in the area of the intergenerational transmission of goods, including the instruments of livelihood, especially in the basic means of production at the domestic level in precapitalist (or non-industrial) societies, as well as in the authority relations associated with their management.

By attempting to deconstruct intergenerational relations in this way, tensions based on sexual rights and domestic authority could be given separate consideration from those based on different forms of property rights, each in turn being linked to the processes of production and reproduction. In addition I needed to distinguish the transmission of office (succession) from that of property (inheritance) since these might well diverge. The latter gave rise to the kinds of tension brought out in the speech of Henry the Fourth to his son, Prince Hal (the future Henry V), when the latter finds the king sleeping in his room and, thinking him dead, picks up the crown. The waking king reprimands his son:

> Thy wish was father, Harry, to that thought.
> I stay too long by thee, I weary thee.
> Dost thou hunger for mine empty chair
> That thou wilt needs invest thee with my honours
> Before thy hour be ripe?. . .
> Thou hast stol'n that which, after some few hours,
> Were thine without offence;. . .
> Thy life did manifest thou lov'dst me not,
> And thou wilt have me die assur'd of it. . .
> Then get thee gone, and dig my grave thyself;
> And bid the merry bells ring to thine ear
> That thou art crowned, not that I am dead.

(Henry IV, Pt. 2, iv, 2)

The notion of the "merry bells" was intrinsic to what I called, for short, the Prince Hal complex, which emerges in anticipatory acts of wish-fulfillment as well as in the mixed reactions to death itself, inducing feelings of guilt and appeasement.

Putting together these relationships, which featured conflict as well as cooperation, I presented a schematic table (Table 1).

Problems of succession to high office played a minimal part in LoDagaa life, for there was little or nothing that fell into that category. The last row of the table referred instead to enquiries among the Gonja of northern Ghana where I had been working with Esther Goody and where I was especially interested in one particular aspect of the political system. Instead of direct intergenerational succession, such as we find among modern monarchies,

Table 1 The Transfer of Rights Between Roles

Type of exclusive right	Authorized transfer (prescribed *propter mortem*)	Role relationships	Unauthorized transfer	Role relationships
property	inheritance	holder-heir	theft	holder-thief
sexual	levirate, etc	husband-levir, etc	adultery, incest, abduction, etc	cuckold-adulterer, etc
roles and office	succession	incumbent-successor	usurpation	ruler-rebel

each vacancy entailed a lateral shift of power between different segments of the ruling estate. While this rotational movement was sometimes phrased in terms of fraternal succession, the "brothers" were very distant "relatives," the kinship term referring to the members of other segments of the ruling group. The important rule was that, in an expression from Mampong in Asante, the elephant never sleeps in the same place twice. No segment should hold office for longer than the reign of one incumbent (12).

Rotational and next-of-kin systems resulted in different lines of tension and alliance. When the prince was no longer the next heir to his father, he had more to gain by keeping him on the throne than by his disappearance. The contrast lay not only in the implications or consequences of rotational as against direct succession, but also in the possible predisposing factors ("causal" factors) that pertained to the systems of political power. In rotational systems, power is distributed, for instance, among the members of a mass dynasty each of whom retains an interest in the holding (or holders) of high office. The confinement of those eligible for high office to close rather than to distant kin means a narrow rather than a mass dynasty. The different ways power is distributed may relate to, among other things, the control of the means of destruction (or the means of coercion, as Ernest Gellner has suggested) on which most early states and many modern ones are ultimately based. In West Africa a narrow dynasty tended to arise where the means of coercion were controlled by a professional group, as often with firearms when they reached the simpler societies, partly because of the investment and technical skills required when arms were imported. Whereas a mass dynasty tended to be associated with the equality of status and opportunity associated with a ruling class of armed horsemen. The ruling estate in Gonja constituted such a mass dynasty comprising roughly 20% of the population; power was rotated between the owners of the means of coercion, that is, between the chiefs with their horses and armed followers.

Interest in the nature of the ownership of the means of destruction ran

GOODY

parallel to interest in the nature and the ownership of the means of production and the way that this effected and was affected by the intergenerational distribution of resources. At the political level many anthropologists wrote of feudal and tributary systems in Africa as if they were similar to those of Europe or Asia (12a). The comparison seemed to me weak—not so much because of the distribution of power as such but because of the differences in productive systems (including craft production) within which power was exercised. The major states of Eurasia were characterized by the intensive cultivation of "advanced" agriculture typified by the use of the plough or irrigation while a more extensive hoe cultivation marked Africa (and other areas such as New Guinea), usually involving shifting, slash-and-burn farming. The extensive system of agriculture did not prevent the rise of the state in Africa, but that state had usually to be based on taxes on trade (or exports) and on raiding neighbors for slaves rather than on the internal accumulation of surplus through primary agricultural production, since any such surplus was usually limited in size, in transportability, and in exchange potential. Booty production and taxes on trade obviously affected the external relations between peoples, giving rise to the characteristic distribution of states separated by bands of "acephalous" peoples. Internally, too, that difference affected the nature of hierarchy in that under extensive systems the control of land as a factor of production gave no overwhelming advantage to the rulers. The differential transmission of resources was also affected. Under a more "advanced" agriculture, land was a scarce good and its control differentially distributed within the hierarchy and between particular families. In such a situation different strata would have their own strategies for passing down property and managing the estate, partly in order to preserve their hierarchical position. Indeed different families would have their own strategies of marriage, management, and heirship. In *Production and Reproduction* (1977), I tried to sketch out some of the general implications of this difference for kinship relations—for example, in encouraging in-marriage rather than out-marriage. If there are few resources to protect, then the marriage of chiefs to commoners may be a sound political strategy, extending the cross-cutting ties that ensue from unions between status groups. Such marriages frequently took place in Gonja, where the political benefits were overtly recognized.

This interest led me to ask why, for example, one found formal adoption in Eurasia but not in Africa, although fostering was common in parts of that continent, as Esther Goody discusses in her work on Gonja and West Africa generally (4, 5). One function of the widespread adoption that existed earlier in Eurasia was to provide an heir for the heirless, and often a person to continue the ancestral cult after death. These considerations became relevant when the transmission of property, especially property in land that differentiated positions in the hierarchy, became of importance. Similar considerations applied to the distribution of monogamy and polygyny as well as of other institutions such as filiacentric unions (the incoming son-in-law), for one aspect of all these practices was their role as mechanisms of heirship and continuity, of transmitting relatively scarce resources between generations, as well as of passing down the management of those resources. While none of these features (nor indeed any features of human society) are monolithically tied to any other variable, they do constitute a cluster of features that I called the "women's property complex". Under such hierarchical systems efforts were made to maintain the position of daughters as well as of sons, especially by means of endowment and/or inheritance. Such modes of transmission were significantly linked to advanced forms of agriculture under which dowry as distinct from bridewealth prevailed, for it was then that the maintenance of the position of daughters became an issue.

This distinction was discussed in a volume of Cambridge Papers, *Bridewealth and Dowry* (1973), written with S. J. Tambiah. This work entailed a very broad comparison between Africa and the major societies of Asia and Europe. My interest in European kinship had been partly fuelled by the examination of inheritance systems developed in the book on LoDagaa funerals. But I was also struck by the fact that in Western Europe, unlike Rome, India, and China (but like Islam and Judaism), adoption was not practiced after the 5th century AD until the 20th century, or the mid-19th in parts of the United States. It was this situation that constituted one of the starting points for *The Development of Family and Marriage in Europe* (1983). The absence of adoption and certain other mechanisms of heirship in Europe seemed related to the fact that it was the Church that effectively established itself as the heir when family continuity failed (and claimed a substantial share in other situations). One priestly writer of 5th-century France described the adopted as "children of perjury." They cheated God of the goods that He had first given to mankind and that should be returned to Him at death through his church. Some allowance was made for the children of one's own "blood", but others were improper recipients of the goods rightly belonging to the Church and which should be used for God's purposes. In this way Christian doctrine and the demands of an ecclesia were consistent with the realignment of strategies of heirship at the familial level, and by this and other means the Church rapidly became owner of a third of the available farming land of Europe.

The work on which I have recently been engaged, entitled *The Oriental, the Ancient and the Primitive* (1990), pursues this analysis among the major states of Asia, including the ancient and modern Near East. While one can differentiate Christian Europe from the rest of Eurasia on the basis of what the Church took and how it did so, certain main features of kinship systems in the major states of the Old World are sufficiently similar in a structural sense to

GOODY

cast serious doubt not only upon the way many anthropologists and sociologists have treated this aspect of "the uniqueness of the West," but also upon the way this same theme has been pursued by a number of European historians of the family (and especially by English ones). This difference with regard to the influence of an established ecclesia with monastic institutions is less clear under Buddhism, which may be why Japan, Tibet, and Sri Lanka share certain structural resemblances. It is true that the marriage age for both men and women was later in Europe and that in-living life-cycle servants were more common, but these features of domestic life in the West do not seem sufficient to account in a significant way for any predisposition it had towards the development of capitalism, as has sometimes been suggested or assumed. Indeed as we look at the rapid development of industrial capitalism in East Asia (and elsewhere in Asia) today, and at the increasing contrast with Africa, the uniqueness of the West seems less significant than the uniqueness of Eurasia. To this "uniqueness" Asia has contributed a great deal and has never been the stagnant oriental society (though all societies have been this at particular periods) of which European writers have written.

In this recent volume I tried to modify the notion that an unbridgeable gulf exists between the kinship systems of premodern Europe and Asia—a constant theme of European historians. But anthropologists, too, have been overly ready to compare Chinese and South Indian "kinship systems" with those of the Australian aborigines, or have singled out features in Ancient Egypt and Arabia to compare with those of Africa. Such comparisons take a restricted view of what constitutes a "system" of kinship and marriage. My thrust has been to look at the entire domestic domain, especially in relation to the economy, and to argue that neither Oriental nor Ancient systems were "primitive" in this sense but resembled the preindustrial societies of Europe in many significant respects.

This work on kinship inevitably raised matters that verge on demography, and it has often been historical demographers, such as the members of the Cambridge Group, who have discussed crucial issues about comparative family structures. Demographic themes have played a consistent part in West African ethnography; intensive census data had to be collected in any examination of the cycle of domestic groups discussed by Fortes (see 8), but both he and others such as Oppong made more systematic attempts to integrate their findings with wider demographic concerns. I sought a way of assessing the number of individuals who, under different demographic conditions, would be left without male or female heirs, a question that was of direct importance in the resort to certain strategies of heirship. There was also the possibility that the different kinship regimes of Africa and Eurasia might have varying effects on the number and especially the sex of children in a family. N. Addo and I collected a large sample of completed families in Ghana and found no evidence of any discrimination in favor of one sex, but rather a strategy of maximum numbers (12). This evidence was compared with other large samples of sibling groups where these were available, but we found less difference than anticipated (14, 15). However, the results were consistent with the fact that in Africa, as distinct from Asia, there was no evidence of any discrimination by sex in, for example, the survival of children. Other enquiries showed that people went as far to hospital for a daughter as they did for a son, and it was generally recognized that in a bridewealth system any domestic group needed equal numbers of both sexes to get the best results from marriage transactions.

I began this account by explaining how I was led to work on the ritual and religion of the LoDagaa partly because of my involvement with the Bagre association during whose rites a long myth is recited. This work became an important strand of my research from that day to this. I had been interested in myth and other genres, both in oral and written cultures, mainly because I had read some English medieval literature, to which I was directed partly by the interpretative notes to T. S. Eliot's "The Waste Land". In prisoner-of-war camp in Germany I had been fortunate to find a copy of Frazer's *The Golden Bough* in the library as well as E. K. Chambers's study of *The Medieval Stage* and other similar works much influenced by earlier anthropology. One of the first series of lectures I delivered in Cambridge in the early 1950s was on the subject of myth, before it had become so fashionable a topic with the publication of Lèvi-Strauss's major works. As the result of these interests I was never inclined simply to treat myths and legends in the Malinowskian fashion as "charters" of social situations, as many of his pupils tended to do (see, for example, 3 and 18). Just as in my ethnological work in the region I saw an element of "history" in legend and genealogy, so too I saw some intellectual quest in myth, with the important level of meaning located in the actor domain rather than purely in the deep structure posited by those who tended to see the surface meaning as "absurd" and fantastic. I wanted to understand the explicit as well as the implicit meaning to the actors, and hence I was interested in the words as words, in the text as text, or rather in the utterance as utterance.

The advent of the transistorized tape recorder made this possible to an altogether different extent than before. Of course, electronic recording had previously been available and was used in ethnographic research, mainly for recording music and song. That was the case for the important work of Milman Parry and A. B. Lord in Yugoslavia in the 1930s, when Parry set out to study some of the characteristic features of "oral" epic in order to determine the status of that great interstitial figure, Homer. Lord subsequently published

GOODY

a volume called *The Singer of Tales* (1960), which has rightly been the source of much subsequent inspiration, research, and debate.

But Parry and Lord were working with café singers in Europe, with secular song in a region where electricity was available and where the church maintained its hold over religious action and belief. It was a very different story in field situations where the recitals were ritual, often secret, and made only in formal conditions, and where the possibilities for mechanical recording were minimal. The change in technology had important effects on the study of oral discourse in the simpler societies, probably as important as many broad "theoretical" shifts, especially in the sphere of pragmatics. Its influence has been of a cumulative kind. For the first time it has enabled anthropologists to record long recitations like the Bagre in situ, in the actual conditions of performance—in other words, to create a reasonably accurate text from a ritual utterance, one that can be checked back against the tape. It has also enabled them to make repeated recordings over time and so to examine questions of continuity and change in these standard oral forms. In earlier times only one version was usually available, and even that was often a summary of a recitation written down from an "informant's" recollection of the original performance. Not only did the summary concentrate on the retellable, narrative elements, but the impression was also left that there existed a single authorized version. Any differences tended to be attributed to the mistakes of observers, or alternatively to structural factors. The new technology made it possible to examine the elements of individual intellectual exploration, of creative innovation, as Barth has done for the Ok of New Guinea (1) using comparative evidence, or as I have begun to do for the Bagre—more programmatically, perhaps, than in actuality. Of course such explorations occur within a set of constraints, but which constraits will be relevant in any specific case is difficult or impossible to predict.

The first version (published as *The Myth of the Bagre* in 1972) I wrote down from dictation; the process took a full ten days. Since then I have worked with my friend, S. W. D. K. Gandah, with whose help we have recorded, transcribed, translated, and annotated another four versions of the Black Bagre and ten of the White, a long and arduous program of work. One of these versions was published in 1981 as *Une Recitation du Bagré,* with French and English translations. Having originally thought the recitation highly standardized, we were suprised to find the range of variation. Together with Colin Duly, an attempt was made to analyse this range by means of the computer, the results of which were set out in an unpublished report for the Social Science Research Council (13). Our sample was small but nevertheless produced a far greater body of data than for any other long oral recitation to date. It enabled us to compare the performance of different speakers on the same occasion and of the same speaker on different occasions. The former varied more widely than the latter, indicating that no one learned precisely anyone else's version but always introduced variants of his own. The result was that suprisingly different recitations were to be found in nearby communities, though they were all recognized by the actors to be the same "Bagre." The dictated version was longer, richer, and more sophisticated linguistically and thematically than those we recorded electronically, owing something to the personal accomplishments of the speaker, but also to the nature of dictation, the pauses in which enabled him to elaborate to a greater extent than was possible in the rapid recitativo of the actual ritual performance. The dictated version was also significantly more theocentric in that greater attention was paid to the part played by the High God, although he is given much more prominence throughout than would appear from ordinary social interaction.

I have used the work on the Bagre in a number of ways, but chiefly to draw out some features that contrasted with written literary genres. That contrast has turned on the role of verbatim memorizing, which I find charateristic of societies with writing, the emerging division between creator and performer (both artists in their own way), and the tighter structure of written works. As a consequence I would see both the work of Homer in Greece and the Rig Veda in India as the products of early literate societies rather than purely oral cultures.

These studies of the effects of the introduction of writing on oral genres were published in *The Interface between the Written and the Oral* (1987). This work was part of a series of studies that began with a fortunate collaboration with Ian Watt, Professor of English at Stanford University, on "the consequences of literacy," the title of the article we published in 1963 (16). Watt and I had attended the same College at Cambridge where we were influenced by the work of Q. D. Leavis and others on the relationship among changes in the production of reading material, the nature of the audience, and the content and form of literary works (see, for example, 31 and 33). We had also been through a long period of warfare, and especially of imprisonment, where books were scarce or absent, that led us to pursue an interest in the influence of modes of communication on human societies, especially the introduction of writing. Watt provided much of the initial inspiration, but my later engagement was greater, for the project made possible an historical or even "evolutionary" approach to some of the problems of differentiating "simple" and "complex" societies, which were the staple of anthropological discussions and certainly of anthropological categories. Differences in cognitive processes could be looked at in a developmental way, which made me less willing to accept in its entirety the elusive relativisim of much anthropological discourse.

In our essay we attempted a broad assessment of the contribution that

GOODY

writing, and specifically alphabetic writing, had made to human cultures. Acknowledging the work of Havelock and of members of the Toronto School, we looked at the intellectual contributions of the Greeks against the background of their invention of a fully alphabetic script, especially in regard to ideas of "myth" and history, concepts of time and space, notions of democracy, categories of intellectual activity, and the notions of logic and contradiction that Levy-Bruhl had found absent from "primitive" societies. While accepting that Evans-Pritchard was correct to refute this proposition, we argued that "logic" in the limited philosophical sense, as well as the refinement of proof and contradiction, were critically dependent upon the use of writing.

Subsequently this research was continued in an edited volume, entitled *Literacy in Traditional Society* (1968), which examined instances of the impact of writing on mainly tribal societies. In the course of this later work I came to qualify the weight placed upon alphabetic writing in Greece, partly because the introduction of the alphabet itself was less clear cut than formerly suggested by Indo-European (as distinct from Semitic) scholars, and partly because a number of the features that had been seen as consequential on the introduction of that script appeared as embryonic in earlier forms of writing itself. The results of these enquiries into the cognitive effects of early writing systems appeared in the *Domestication of the Savage Mind* (1977) in which attention was drawn to the role of lists and tables in early written cultures as ways of organizing information in non-speech-like ways. That work led to a profitable collaboration in the field of "literacy" studies with psychologists Michael Cole and David Olsen. The former invited me to cooperate in his research into the uses and cognitive implications of the indigenous Vai script in Liberia, a test case in such investigations because learning this script took place separately from school education. We were fortunate to come across a body of writing by A. Sonie that amply illustrated the powerful role which writing could play in reorganizing information for the purposes of recall, logical presentation, and accountability.

The research proceeded in two directions. The *Interface* (1987) volume gathered together studies on genres and similar topics. Previous to that, in *The Logic of the Writing and the Organisation of Society* (1986), I had suggested ways the introduction of writing had influenced the domains of religion, the economy, politics, and law. At the same time I had tried to elucidate certain anthropological and sociological problems surrounding the broad difference between "primitive" and "advanced," simpler and more complex societies. One of the major pressures behind the studies on writing had been to get away from the simplistic dichotomies of much social science, in order to isolate some of the mechanisms behind sociocultural changes; one such set of mechanisms consisted of the mode of communication. But whereas scholars had paid much attention to the effects of language on social life, the forms and implications of writing had received comparatively little notice. That was a situation I tried to remedy.

The tape recorder has changed not only the analysis of myth and the procedures of the anthropologist. It has also produced an emphasis on discourse, on letting the actor speak for himself. Before the appearance of sound movies, ethnographic films had to invent a spoken script, a text. Now, as in the Granada television series *Disappearing Worlds,* the natives could speak for themselves. But the result was in some ways an impediment to research. Not only was there an overabundance of material, not only did investigators think less before they turned on the machine, but there was even a tendency, which remains strong, to consider a presentation of the actor's picture as the be-all and end-all of analytic endeavor. That is *verstehen* carried to an illogical extreme. Of course there is a lot to be said for recording as much as we can about "disappearing worlds" before they finally disappear. But this level of ethnography, of reportage ethnology, does not require any great expertise apart from a journalistic ability to get people to talk into the mike. And it is no substitute for theoretically oriented research or even for the analysis of research materials within the framework of a scholarly tradition. Understanding the meaning to the actor may be the ultimate goal for some literary studies; it hardly exhausts all levels of enquiry.

The remaining sector of my work relates both to the contrasting sociocultural situations in Africa and Eurasia and to the implications of writing. In reading analyses of cooking in Africa, I was struck by how they differed from my own experience, in terms both of the traditional patterns and of what was emerging in interaction with Europe and the rest of the world (*Cooking, Cuisine and Class,* 1982). Quantity apart, I found little evidence of different forms of cooking even within most of these African states where a hierarchy existed, no development of an *haute cuisine.* But as we have seen, different kinds of hierarchy were involved in Africa than in Eurasia where cultural activities and values were by and large held in common. Partly because of frequent intermarriage between the different groups in the hierarchy, there were few tendencies for the emergence of subcultures, except in those ethnically distinct and endogamous ruling classes found in some of the interlacustrine kingdoms of East Africa. Partly because of the nature of the hierarchy, partly because of the productive system, partly because of the absence, except under Islam, of writing, there was less tendency to elaborate a distinctive high cuisine of the sort we find in many of the major societies in Eurasia, with their considerable literatures on cooking and eating, and their institutionalization of the differentiated production and consumption of food.

My most recent book, *The Culture of Flowers* (in press), pursues what is on one level a similar theme. Visiting Bali and carrying out some limited

GOODY

fieldwork with Esther Goody in Gujerat, India, I was amazed at the use of cultivated flowers, especially for worship, for which they were deliberately produced and purchased. This "non-utilitarian" form of agriculture was virtually absent from Africa, where the usual form of offering to a shrine is blood sacrifice. A partial explanation of the difference is not hard to find. It concerns the lack of stress in Africa on intensive agriculture, although tobacco, onions, and rice were cultivated by such methods. When I looked at forms of ritual, at pictorial representations, or at the imagery of poetic forms, I found many references to trees, leaves, and roots, but little or nothing on flowers, wild or domesticated, except along the coastal regions of East Africa where influences from India and the Near East had played an obvious part. My suggested reasons for this state of affairs were several, and partly connected with the nature of agriculture. As a comparison I was led to look in greater depth at the major uses of flowers in Europe and Asia. But initially I concentrated upon the history of flower use since the Bronze Age, its development in the Near East, and especially Ancient Egypt, its culmination in Rome, and then the rapid decline under Christianity, where the use of flowers was connected with pagan worship, with luxury, as well as with a certain conception of God and a Biblical reluctance to represent His creatures. The effects of this ambivalence on the history of Europe were radical. After the collapse of Rome, that continent saw a dramatic decline in the use of flowers, which had hitherto so often decorated the altar, the priest, and the sacrificial animal itself. This decline was paralleled by the virtual disappearance of three-dimensional statuary. Europe also experienced a decline in botanical knowledge which, beyond a certain threshold of folk usage, was partially dependent upon the representation of plants, whether as separate images or in manuscripts. The relative backwardness of medieval Europe with respect to Asia is a significant comment on the effects on systems of knowledge, not only of the barbarian invasions but of Christianity itself.

Europe saw a gradual renaissance in the growth, the use, and the representation of flowers in graphic and literary works from the 12th century onwards. The repertoire greatly expanded with the expansion of Europe, with the demand of the aristocrats and bourgeoisie, and with the supply of new plant varieties by traders, missionaries, administrators and later by specialist plant hunters. Before flowers could play a role in cemeteries and in worship they had to overcome deeply established "puritanical" beliefs that tended to differentiate Protestant from Catholic regions. The continuing differences (and similarities) in the use of flowers in Europe and America, the expansion of their use with the expanding economy, and the world market in cut flowers made possible by the airplane (and before that by the train and truck) are the subjects of several chapters. That urban market was linked to the growth of the formal, written "Language of Flowers" that sprang up in Paris in the early part of the 19th century and soon became accepted as the rebirth of a vanished code of eastern origin. It spread throughout Europe, and the efforts made to adapt the language to the United States provide an interesting example of the domestication of the foreign mind.

I pursue the themes of specialist production, of the stratification of use, and of a perpetual ambivalence in the culture of flowers in India, China, and Japan, partly based on observations of the markets and rituals, of the houses and temples, in those countries. At the end I return to the theme of the ambivalence about representation in Africa itself, this time in respect not of God's creatures but of the High God himself, and often of other divinities as well. I argue that the reluctance to portray the Creator God displays a concept of divinity at odds with the characterization of African religion as animism and fetishism. The intellectual and cognitive problems of African thought appear closer to those of Western man than is often supposed, especially when we can point to precise reasons for some of the difference. At this level, too, it is necessary to reconsider assumptions about the uniqueness of the West.

I have tried to give some idea of the themes that have been important in my research. One constant regret has been the failure of sociocultural anthropology to become more cumulative. Part of this failure has to do with the methods favored over the past half century. Many anthropologists seem imprisoned by their rejection of any method other than fieldwork, being unprepared to see how their enquiries relate in a systematic way to those of others. Local knowledge is essential, but for many purposes it is a beginning rather than an end. We have come the full cycle since the 19th-century addiction to a version of the comparative method. In my view it is essential to integrate and test our observations and conclusions with material gathered by other scholars in this and related fields, using whatever methods we can, imperfect as these will always be. In these resources I include the Human Relations Area Files, which I have tried to use for this purpose. The rejection of techniques other than fieldwork and speculation tends to lead to a withdrawal into deep descriptions of "my people," combined on the one hand with unsystematic global statements and, on the other, with the search for new gods; as with medicine shrines, the old gods become obsolescent as they fail to meet impossible demands. Thus there are shifts—for example, from structuralism to post-modernism—on a basis analogous to the "global exchange" ("global replace") function in computer software rather than on the modification and development of existing programs.

I have said little on my work in changing social systems and in the history of languages and ethnography of Africa, even though these took up much of my time and energies in the 1960s when I was frequently in Ghana. That gap is partly because I have not completed as much in these areas as I wished. Nevertheless the bright hopes for the future and the harsher realities of the

GOODY

present have formed a constant theme of my other research. As with other colleagues, that research involved entering some of the no-go areas of earlier scholars and making use of the work of various fields, especially history.

This breach in disciplinary walls has its dangers, and it could be argued that the so-called crisis in social anthropology (though crises exist all the time and in all fields) has resulted in part from discarding the boundaries accepted by earlier workers in the field, including their tighter paradigms. There seems to me much truth in this claim. Anthropology has become diffuse in a number of ways, which means there is room for a range of approaches. For me the way ahead is not the way back but consists in supplementing the analysis of sociocultural "facts" or situations with a problem-driven approach. That means looking for evidence where it exists rather than confining oneself to arbitrary boundaries, useful as these have been for specific problems or topics.

I have drawn much satisfaction from working with scholars in other fields, with historians (especially those associated with *Past and Present, Annales E. S. C.,* and the Cambridge Group), with psychologists, and with those in the neighboring fields of literature and oriental studies. An anthropology that does not or cannot communicate with other disciplines stands in danger of placing an excessive value on its own conclusions. It is only too easy to draw back to the safety of the community one has studied (or even one's individual "informant" in some American traditions), about which one may know more than most of one's colleagues; or perhaps to seek the shelter of a hermeneutic approach to which only the writer holds the key. But knowledge exists in communication not only within arbitrarily defined fields (for that is essentially what we are dealing with) but also outside them. Fields such as cognitive anthropology seem sometimes to aim not to widen knowledge and provide a link with other students of cognitive processes but to protect and separate off a distinctive anthropological patch. Yes, we say, we anthropologists are also interested, but in our own way and on our own terms. That has always seemed to me the way to ruin. There is no anthropological truth, enlightenment, or even insight that is not related to the work of scholars in other fields.

Literature Cited

1. Barth, F. 1987. *Cosmologies in the Making: A Generative Approach to Cultural Variation in the Inner New Guinea.* Cambridge: Cambridge Univ. Press
2. Drucker-Brown, S. 1980. Notes towards a biography of Meyer Fortes. *Am. Ethnol.* 16:375–85
3. Fortes, M. 1945. *The Dynamics of Clanship among the Tallensi: Being the First Part of an Analysis of the Social Structure of a Trans-Volta Tribe.* London: Oxford Univ. Press
4. Goody, J. 1968. Consensus and dissent in Ghana. *Polit. Sci. Q. 83:337–52*
5. Goody, E. N. 1973. *Contexts of Kinship: An Essay in the Family Sociology of the Gonja of Northern Ghana.* Cambridge: Cambridge Univ. Press
6. Goody, E. N. 1982. *Parenthood and Social Reproduction: Fostering and Occupational Roles in West Africa.* Cambridge: Cambridge Univ. Press
7. Goody, J. 1956. A comparative approach to incest and adultery. *Br. J. Sociol.* 7:286–305

8. Goody, J. 1957. Fields of social control among the LoDagaa. *J. Roy. Anthropol. Soc.* 87:75–104

9. Goody, J. 1958. The fission of domestic groups among the LoDagaa. In *The Developmental Cycle in Domestic Groups,* ed. J. Goody. Cambridge: Cambridge Univ. Press

10. Goody, J. 1959. The mother's brother and the sister's son in West Africa. *J. Roy. Anthropol. Inst.* 89:61–88

11. Goody, J. 1962. *Death, Property and the Ancestors.* Stanford: Stanford Univ. Press

12. Goody, J., ed. 1966. *Succession to High Office.* Cambridge: Cambridge Univ. Press

12a. Goody, J. 1971. *Technology, Tradition, and the State in Africa.* Oxford: Oxford Univ. Press. Reprinted 1980, Cambridge Univ. Press

13. Goody, J. 1977. *The Domestication of the Savage Mind.* Cambridge: Cambridge Univ. Press

14. Goody, J. 1977. *Production and Reproduction.* Cambridge: Cambridge Univ. Press

15. Goody, J. 1980. Rice-Burning and the Green Revolution in Northern Ghana. *J. Dev. Stud.* 16:136–55

16. Goody, J. 1982. *Cooking, Cuisine, and Class.* Cambridge: Cambridge Univ. Press

17. Goody, J. 1983. *The Development of Family and Marriage in Europe.* Cambridge: Cambridge Univ. Press.

17a. Goody, J. 1983. Introduction. Memorial Issue for M. Fortes. *Cambridge Anthropol.* 2–3

18. Goody, J. 1987. *The Interface Between the Oral and the Written.* Cambridge: Cambridge Univ. Press

19. Goody, J. 1990. *The Oriental, the Ancient, and the Primitive.* Cambridge: Cambridge Univ. Press

20. Goody, J. 1991. *The Culture of Flowers.* Cambridge: Cambridge Univ. Press. In press

21. Goody, J. , Addo, N. 1977. *Siblings in Ghana.* Univ. Ghana Pop. Stud. Vol. 7. Legon, Accra: Univ. Ghana Press

22. Goody, J. , Duly, C. 1981. Studies in the use of computers in social anthropology. Rep. to the Soc. Sci. Res. Counc., London

23. Goody, J., Goody, E. 1966. Cross-cousin marriage in northern Ghana. *Man* 1:343–55

24. Goody, J., Goody, E. 1967. The circulation of women and children in northern Ghana. *Man* 2:226–48

25. Goody, J., Duly, C., Beeson, I., Harrison, G. 1981. On the absence of implicit sex-preferences in Ghana. *J. Biol. Sci.* 13:87–96

26. Goody, J., Duly, C., Beeson, I., Harrison, G. 1981. Implicit sex preferences: a comparative study. *J. Biol. Sci.* 13:455–66

27. Goody, J., Tambiah, S. J. 1973. *Bride wealth and Dowry.* Cambridge: Cambridge Univ. Press

28. Goody, J., Watt, I. P. 1963. The consequences of literacy. *Comp. Stud. Soc. Hist.* 5:304–45

29. Hart, K. 1985. The social anthropology of West Africa. *Annu. Rev. Anthropol.* 14:243–72

30. Leach, E. R. 1954. *Political Systems of Highland Burma: A Study of Kachin Social Structure.* London: Athlone Press

31. Leavis, Q. D. 1932. *Fiction and the Reading Public.* London: Chatto & Windus

32. Lombard, J. 1972. *L'Anthropologie brittanique contemporaine.* Paris: Press. Univers. de France

33. Watt, I. P. 1967. *The Rise of the Novel: Studies in Defoe, Richardson, and Fielding.* Berkeley: Univ. Calif. Press

Annu. Rev. Anthropol. 1990. 19:1–15

AN ANTHROPOLOGICAL ODYSSEY

Bernardo Bernardi

University of Rome "La Sapienza," Rome, Italy

Dedicated to Isaac Schapera with respect and affection

My anthropological experience may be described as a long and inspiring journey across the seas of the human condition. I have always been struck by what I call the paradox of human reality: Born to live, we are destined to die; ready to enjoy life, we must struggle to avoid discomfort and disease; conscious of individuality, we are still unavoidably bound from birth to a social group; claiming to be equal, we nevertheless differ both as individuals and groups.

Whenever it seemed I had a comprehensive view of human nature, I was led to a reorienting perspective. I regret none of the various phases of my long journey. Each has contributed to the evolution of my anthropological approach to reality. The reverse is also true: My anthropological perspective has often brought changes in my life's route. Indeed, the study of anthropology, which revealed the essence of the human condition, decisively led me to change my social and professional stations. I left the Catholic priesthood and its missionary activity, dedicating myself fully to the study of anthropology.

BASIC TRAINING

It was during the first stage of my long journey, my early missionary vocation, that my interest in anthropology was aroused. Latin literature, Greek literature, and history had been the main subjects in my secondary education. At the end of my teens I took the philosophical and theological training offered by a seminary for missionaries. Certain highlights from this experience may be of interest, especially to my fellow anthropologists concerned with the cultural background of Christian missionaries.

0084-6570/90/1015-0001$02.00

As prospective missionaries, we were not trained as ordinary priests—as mere ritual specialists versed in theological and biblical knowledge. We studied all that, but we were also instructed in how to approach peoples of various cultures. We were trained to meet others not as Christians, nor necessarily as religious men, but rather as fellow human beings. The emphasis in our training was practical, and manual labor was required of us every day. I owed much to that kind of education, especially during my field work. Even now, at an old age, I feel all its advantages.

My seminary training involved also the study of moral theology, the theology of human acts. I was thus encouraged early to ponder personal autonomy and responsibility, which prepared me to evaluate later, as an anthropologist, the impact of cultural context upon individual actions. I am convinced that study of the nature of the human act (a term implying a degree of consciousness and freedom that confers responsibility) should be a required element of anthropological training. It helped me to deepen both my inspection of human behavior and my understanding of the social institutions within which that behavior is conceived and realized.

For almost four years (1944–1947) I was attached to a special office of the Secretariat of State of the Vatican designed to conduct the Vatican's relations with the Allied Military Authority in Rome. During this extraordinary period I had the opportunity to meet people from all walks of life. As a reaction to a world of formalities and officialdom, of ambitions and jealousies, I felt strengthened in my determination to respect all persons equally, whatever their qualifications.

The Secretariat of State was then directed by the high officers of its two main sections (to the silent resentment and embarrassment of the ambassadors, who were forced to deal with officials not of their own rank). These officers were Domenico Tardini and Giovanni Battista Montini. Both bore a family name ending in the diminutive *"ini,"* so they were ironically described as "the two diminutives of the Secretariat of State." Both, on the contrary, were giants: Tardini became Cardinal Secretary of State under Pope John XXIII, and Montini succeeded John as Pope Paul VI.

HISTORY AND ANTHROPOLOGY: THE PROBLEM OF ORIGINS

The theoretical issue between history and anthropology engaged my intellectual interest from the very beginning of my anthropological training in the early 1940s. My classical education had been imbued with history. During my undergraduate years, the historicism of Benedetto Croce represented the liberal reaction against the rude authoritarianism of the collapsing fascist regime. We were convinced that there is no escape from history whenever you deal with the human condition, whatever your disciplinary orientation. As a consequence, I found congenial the historical approach to "primitive peoples." I also relished the debate between the early evolutionist school and the diffusionist school over the history of culture and society. I took the side of the diffusionistic school, attracted by the rigor of its methods. Both schools, in my view, sought the *historic* origin of human culture and society. During the 100 years when the two schools flourished—from 1859, the year of the first edition of Charles Darwin's *The Origin of Species*, to 1955, when the 12th volume of Wilhelm Schmidt's *Der Ursprung der Gottesidee* was posthumously published—an overwhelming literature arose on the theme of human origins. The debate's outcome, however, was disappointing; the origins of human culture remain uncertain.

When I entered the university, few students there showed interest in ethnology. (In Italy, ethnology is the equivalent of social and cultural anthropology, as distinct from physical anthropology.) The few who were interested, however, were extremely taken by methodological disputes. Personally, I had formed a high opinion of F. Gräbner and his attempt to define criteria of ethnological research as severe as those required by classical history. That rigor strengthened my sympathy for the German school of ethnology, including its Viennese version. I read, and even translated from German into Italian for my personal use, Wilhelm Schmidt's handbook of ethnological method. The construction of cultural cycles seemed to me a much more reliable way to trace the historical development of human culture and society than all the evolutionistic schemes. I considered such cycles to be broad generalizations about the styles of human culture, much the way the styles of art serve to periodize art history.

SCHMIDT AND THE ITALIAN CATHOLIC CONTEXT

The influence of Wilhelm Schmidt on the Italian Catholic intelligentsia of the early 1930s needs a special note.

Following the unification of Italy and the fall of the papal state (1870), Catholics were forbidden by the Church from participating in the political life of the new state. Their presence in the academic institutions of the state was also minimal. In addition, the positivism, idealism, and historicism dominant during the first decades of this century were not only anti-Catholic but also hostile toward such empirical disciplines as sociology, anthropology, and psychology. Against that trend, there came, in the early 1920s, a revival of Thomistic philosophy—"neo-scholasticism," with its Italian center at the recently established Catholic University of Milan. It was there that Schmidt found his main introduction to academic circles in Italy.

During the same period, Schmidt had been called to Rome in connection with the missionary exhibition of the holy year, 1925. Following its success, he was entrusted by Pope Pius XI with the establishment of an ethnographic museum, *Pontificio Museo Missionario Etnologico,* at the Lateran Palace in Rome, using material from the exhibition. Schmidt impressed his theories upon the museum, especially in an exhibit showing the succession of cultural cycles from the primeval through the secondary and tertiary cycles. A specialized library was added to the museum, which soon became one of the two best libraries of ethnology in Rome, the other being the library of the Pigorini Museum.

At these centers I did my reading, and in 1946 I wrote my first doctoral thesis in anthropology at the University of Rome "La Sapienza." Among the activities of the new museum, Schmidt inaugurated the production of an annual review, *Annali del Pontificio Museo Lateranense,* to provide space for missionary ethnographic reports as well as for essays by professional anthropologists. On the same lines he had founded his first journal, *Anthropos,* in 1904.

Schmidt, who had never been in the field, was succeeded by one of his best students, Michael Schulien, a German native of the Saar with some field experience in Mozambique. Schmidt returned to Austria, St. Gabriel, Mödling, wherefrom, on the eve of the *Anschluss,* he escaped to Switzerland in order to save himself and the *Anthropos* library from Nazi requisition. Both Schmidt and Schulien were fierce antagonists of Nazi racism, and they continued to be regarded by some Italian antifascists as a source of inspiration. Schulien, in fact, wrote a series of articles in *L'Osservatore Romano,* the Vatican newspaper, on the unity of mankind (29). These were widely read as a response to the fascist propaganda of the day. During that period the Vatican paper reached its largest circulation, for it was the only antifascist, or nonfascist, paper available to the Italians of those days. Anthropology had thus been brought into the heart of the political scene, and we young anthropologists felt more than a moral responsibility to reject the appalling false claim of a scientific basis for fascist racial laws. The situation reached its climax during the Nazi occupation of Rome. When they were assaulted by the Nazis, some Roman Jews—and most Italian politicians—found refuge in the numberless religious facilities in Rome, including the Lateran Palace where the ethnographic museum was lodged.

RAFFAELE PETTAZZONI

Schmidt's personal success and his school of thought were not so unanimously accepted outside Italian Catholic spheres. The strongest ideological opposition came from Raffaele Pettazzoni, the renowned historian of religions. Pettazzoni was the incumbent of the first chair of the history of religions at the University of Rome "La Sapienza." As a student, I greatly profited from his courses on the religion of the ancient Germans, and I recall some occasions when, because we were both from Bologna, I enjoyed special signs of his friendly kindness. He envisaged ethnology as an auxiliary discipline of the history of religions. Indeed, we owe to him the introduction of an academic course in ethnology at Rome University and the establishment of an ethnological institute initially named the *Istituto delle Civiltá Primitive* (Institute of Primitive Civilisations). Its name soon changed to *Istituto di Etnologia* (Institute of Ethnology) by Vinigi L. Grottanelli, this entity has recently been absorbed by the newly established *Dipartimento di Studi Glottoantropologici* (Department of Anthropology and Linguistics).

In my day there were two official courses of ethnology in Rome, one at Rome University "La Sapienza" taught by Alberto Carlo Blanc, a paleoethnologist; the other at the Urbaniana University of the Vatican by Renato Boccassino, who had attended Malinowski's famous seminars at the London School of Economics and had completed intensive fieldwork among the Acholi of Uganda. I followed both, with no great profit. My real teacher was Schulien. He did not hold a chair, refusing to stand publicly as a teacher while the Nazis were ravaging his own country and the world. We developed the habit of frequent encounters at the Lateran Museum, during which, patiently answering the queries of a curious young apprentice, he sought to elucidate the aims and problems of the *Kulturkreise* method.

The polemics between Schmidt and Pettazzoni had been constant and even bitter, from their early reviews of each other's works in the second decade of the century. Each in his own way was a scholar of exceptional value and consequence. When I first met Pettazzoni, the two had been battling for three decades. Our student loyalties were divided. Schmidt was acute in spotting the ethnological weaknesses of his opponent. He criticized particularly Pettazzoni's attribution of the idea of God among "primitives" to a mythical kind of thought *(mythische Denken)*, rather than to causal thought *(kausalen Denken)*. Pettazzoni was strengthened by his insistence upon a method of historical research and reconstruction that he claimed was not suggested by the idea of Schmidt's *Urmonotheismus*. He lived long enough (he died in 1959) to see the end of Schmidt's ethnological *Ur*-construction and *Kulturkreiselehre* (28; on Pettazzoni, see 16).

Strong criticism of Schmidt's historical approach came from another young Italian scholar, Ernesto De Martino, a brilliant student of Benedetto Croce. His work, though practically unknown outside of Italy, has had a tremendous impact on young Italian anthropologists. His first criticism of the cultural cycles of the German school placed him in the forefront (13), and his later research on tarantism, ritual mourning, and magic in Southern Italy gained him enduring consideration as the true innovator of Italian cultural anthropology (13, 26, 27).

When Malinowski's functionalism was first made known in Italy as a new method of research it excited considerable interest but no real acceptance. Pettazzoni recognized its ideological foundation and methodological utility, though—as he told me—he would stay a "historian." Indeed we students followed suit. Having a classical background, we held history to be the only correct approach to the analysis of human culture. As for the new functionalist propositions, we remained skeptical, regarding them as good only for the current phenomena of colonial encounter.

AFRICAN STUDIES AT CAPE TOWN

When, in the first days of February 1948, I landed in Cape Town, after a sea voyage of more than 20 days from Venice to Durban and three days on the railway from Durban to Cape Town, the chair of social anthropology at the University of Cape Town was held by Isaac Schapera. My missionary sponsors had expected that I would eventually teach in one of the secondary schools of Kenya. I was to have received my professional training at either Oxford or Cambridge, where I would earn a PhD in anthropology. However, because in 1948 the Oxbridge institutions were filled with ex-servicemen, I was diverted to Cape Town. Michael Schulien provided me with a letter of introduction to Isaac Schapera.

The climate and the surroundings of Cape Town and Cape Peninsula were congenial. Indeed, from that point of view, I thoroughly enjoyed my three years in South Africa. My arrival coincided with the high tide of the political campaign that led to the fall of Jan Smuts and the first victory of the Afrikaaner nationalists under Daniel Malan. I could feel the mounting tension as the existing practice of color-bar was intensified by the new policy of apartheid, at that moment a hazy concept filled with gloomy implications. The University of Cape Town was liberal, and there I met and befriended African students. (One of them, the Hon. Peter Okondo, is now a Minister of State in the Government of Kenya.) While discontent simmered and the causes for grievance increased day by day, we students were able to do little. We endured the situation created by the selfishness of most European South Africans.

Those three years in South Africa were really a determinant stage of my odyssey. For the first time I came into contact with what I think of as the true Africa. I had the good fortune to meet Isaac Schapera—a renowned Africanist and a great teacher. Never verbose, he was rather matter-of-fact. I always found his suggestions poignant and stimulating. Even today, whenever I reach London either alone or with my wife, I visit him as an old friend.

But Schapera, and all that was behind him ideologically and methodologically, meant a deep change in my approach to anthropology. Though an avowed follower and admirer of his two great teachers, Radcliffe-Brown and Malinowski, he did not unduly emphasize their rejection of history. It was as if he did not entirely share their opinion on that issue. In fact, in his comprehensive research on the Tswana, he succeeded in synthesizing the functionalist and historical perspectives, and in this sense his oeuvre has been exceptionally consistent—really, paradigmatic (on Schapera, see 14).

I was also fortunate to establish a lasting friendship with Meyer Fortes. When in 1949 he came for a semester to Cape Town to replace Schapera, who was visiting in the United States, I had already become familiar with all the tenets of functionalism. My former prejudices had been removed as I saw the possibility of a synthesis between the functionalist perspective and the historical approach, on the lines that Schapera had been pursuing without any formal or public ideological declaration. However, I recollect having occasional, deep discussions with Meyer on the value of history in anthropology.

Passage from my early stance as an anthropologist of cultural-historical extraction to a structural-functionalist of new convinction came with my first fieldwork, the result of shrewd guidance by Schapera. During that first semester in Cape Town, I had been studying Zezuru, a dialect of the Shona language, under the tutorship of a fellow PhD candidate, George Fortune, a distinguished Bantuist. Since I had read the scarce literature available on the Zezuru, Schapera suggested that I spend my term vacation among this people to obtain direct, even if brief, experience in the field. He helped me discard some ill-conceived proposals. Finally, I spent four weeks visiting, family by family, the five villages around the Catholic Mission of Musami, where I was lodging. My report was published in the *Communications of the School of African Studies* (1). My ethnographic contribution to the history of the Zezuru was fortified by three replications over the next 30 years (1, 11, 18 regrettably, I have not seen the results of Chavunduka's last survey, that of 1980).

After that first field work, my approach to anthropology became more practical and direct. I developed a capacity, if I may say so, to observe a social structure at work within a living community, be it the kinship system, the age system or any other, tempering my theoretical approach with the experience of sharing life with "others" and matching my interpretation of reality with theirs and with their behavior.

At the end of 1950 my dissertation on the age system of the Nilo-Hamitic peoples was finally approved, and I was awarded the PhD in African Studies by the University of Cape Town. My work had been intended as a preliminary evaluation of the existing literature, a task made difficult by the tremendously confusing ethnographic reports. Each author had used his own jargon or attributed different meanings to an apparently common language. I attempted to build a semantic concordance: *concordia discordantium*. The basic pattern of the age system seemed to emerge and make sense. I was convinced of its political significance in the structural context of the Nilo-Hamitic societies (3), a line of thought I was to pursue in my future, more general work on the same subject (7). I am happy to have witnessed an efflorescence of further important research on the same theme, the latest being that of Müller (23) and Tornay (34).

FIELD WORK IN KENYA

At the end of 1950, I went directly to Kenya from South Africa, having been summoned, as a teacher, to Kevote Secondary School among the Embu, where I stayed only for one term. During even so short a stay one could sense the general discontent mounting in the Central Province of Kenya, a simmering prelude to the Mau Mau rebellion.

At the end of the term I left for London, where I was to spend a full year at the Institute of Education, London University, for a Postgraduate Certificate in Education. I returned to Kenya at the end of 1953 when the State of Emergency had just been declared.

For the next six years I lived in Kenya. The East African experience served as a further stage of my odyssey. I lived through the Mau Mau resistance, sharing the heavy suffering of the Kikuyu and the Embu, whom I regularly visited, and of the Meru. To the latter I became particularly associated by my work, first as a teacher and afterwards, in the field, as an anthropologist. This connection I never discontinued, even when, in 1956, I was assigned to pastoral work among the Italian community in Nairobi. For a short period, at the beginning of 1954, I was the principal of the Nkubu secondary school among the Meru Imenti. I was soon relieved of my educational responsibility when, in mid-1954, having been awarded a fellowship from the Ford Foundation through the International African Institute, I became entirely engaged in my field research among the Tharaka.

Like the Imenti, the Tharaka were one of the nine Meru ethnic groups; but unlike the Imenti, they had until then been left to themselves, considered the most backward of all the Meru. I was to discover how unjust were the opinions that the first missionaries and colonial administrators had formed of the Tharaka. Secluded from the main trends of colonial presence, they were only marginally involved in the oppressive burdens of the Emergency, which rested heavily on all other Meru groups, especially the Imenti and the Chuka.

The Mugwe, a traditional institution of the Meru, was the main object of my research. He was a highly respected religious dignitary. His blessing was required to assure the welfare of the people and, in a special way, the success of raids by young warriors. He was to be protected from all intruders as "the queen of the bees." The existence of the Mugwe's office, as I have reported (3), had been kept secret from almost all previous Europeans, a fact that brought me both apprehension and excitement. When I realized that the Mugwe had been a living factor of Meru history and sensed that this social institution was doomed soon to end, I became motivated by a sense of service to the future generations of the Meru.

I confess that on disclosing the Mugwe's existence, I felt a certain intimate embarrassment, as of one encroaching on private grounds. The significance of secrets had always been strongly felt among the East African peoples. A Kikuyu proverb, which also helps us to understand the secret character of the Mau Mau organisation, goes as follows: *andu matari ndundu mahuragwo na njuguna imwe,* "a people without a secret is destroyed by a single stroke." Deeply conscious of this attitude, I feared at first that the Meru were critical and suspicious of my activities, but it soon became apparent that I enjoyed the general confidence of those among whom I worked.

I aimed for the most rigorous objectivity in reporting on my observation of the Meru social life. I avoided, as much as possible, any personal impingement on the ethnographic evidence. That is why my text is so frequently interspersed with the words of my informants. Following Monica Wilson's suggestion, I reported fully some of the most important texts recorded, providing direct evidence for the reader's evaluation. My Meru friends faulted only the generic implication of my study's subtitle, "The *Failing* Prophet." I have accepted their criticism, and in the edition now in press in Nairobi I have changed that subtitle to "A *Blessing* Prophet" (8).

Since my book appeared in 1959, the ethnography of the Mugwe has been the subject of several critical analyses, each inspired by a different theoretical approach. When the late Daryll Forde, the director of the International African Institute and the editor of the journal *Africa,* asked my opinion about Needham's symbolic and structural interpretation of "the left hand of the Mugwe," I said recognized its validity, as best expressed by George Dumézil's statement: *le système est dans les faits.* I have had no reason to change my opinion in the face of other polemic opinions (20, 21, 24, 33). The left hand of the Mugwe should, however, be analyzed together with all other aspects of his personality: The Mugwe also has a tail like an animal's, for example, and is compared to the queen of the bees. A global analysis would better respond to the conceptual categorization used by the Meru, who consider the Mugwe as one of their "men of wonder" *(antu ba kurigaria)* (3: 62–67, 72–76).

The Agwe I visited were the last of them all. No Mugwe is now at work in Meruland—none, in any case, relying on the traditional structure of Meru society, which has been markedly altered. Some have tried to adjust the conception of the Mugwe to the present, either for social prestige or as an adaptation to Christian ideals (see 6: 180–81); but mine was practically the last chance to record the Mugwe institution from living reality. The Mugwe is now a mere historical figure. I was greatly rewarded to note, after Independence on the 12th of December 1963, that in various parts of the District country clinics, coffee houses, roads, and other public facilities had been named after the Mugwe: a clear example of how popular exaltation creates myth out of history.

THE ITALIAN SITUATION

From 1959 to 1969, having been appointed Director of Education at the headquarters of the Consolata missionary Society in Turin, Northern Italy, I traveled extensively through Africa, America, and Europe, though I had no time for any professional anthropological activity.

In 1970, I was called to the chair of Cultural Anthropology in the Faculty of Political Sciences at the University of Bologna, thus beginning the last stage of my anthropological journey. Entering the Italian universities almost as an outsider, I was surprised by the weak position of anthropology and the jingoistic divisions among Italian anthropologists.

Two schools of thought had been formed, one upholding ethnology as the traditional study of exotic and "primitive" societies, the other promoting a new approach in terms of cultural anthropology, exclusively interested in the study of "complex" societies (primarily the study of peasant Italy). This new approach was largely supported by sociologists and by most folklorists. The two schools had a common interest in obtaining more space in the universities. In such a context my appointment to the Institute of Sociology at Bologna University had a special significance, my chair being the first one in Italy professing Cultural Anthropology. As its first incumbent, I proposed, of course, to work for a clarification of all the issues involved.

Conventionally I was considered an ethnologist, but my anthropological identity had grown more complex, because I had been deeply connected with the first Malinowskian generation of social anthropologists—Schapera, Evans-Pritchard, and Fortes. My appointment implicitly disavowed the nominalistic dichotomy setting cultural anthropology against ethnology, and thus my predicament reminded me of the early British situation described by Stocking (32) in his paper, "What's in a name?"

I am convinced that the terms ethnology and anthropology may legitimately be used to refer to the same field of study. Both are derived from essential attributes of the human condition—in one case from our being born as individuals *(anthropos)*, in the other from our belonging to a social community *(ethnos)*. Unfortunately, the development of so many anthropological schools of thought has caused an undiscriminating use of the two terms, now as if they were synonymous, now as if mutually exclusive.

On becoming more deeply involved with this situation I realized that our task was seriously to rethink anthropology to meet the needs of the Italian universities. With this in mind, in 1972 I convened a conference on the theme "Etnologia e Antropologia Culturale." I made a point of inviting Jack Goody to give an introductory paper on the state of social anthropology in Britain. My aim was to set the Italian situation against an international background and to show how anthropology, as a discipline, was on the move everywhere in the world. I sought to demonstrate that what matters is not where research is carried out, whether in an exotic or a domestic field, but rather whether it employs objective and reliable methods to produce evidence that furthers knowledge and theory (see 4).

At that time, Professor V. L. Grottanelli held the chair of Ethnology at the University of Rome "La Sapienza," whom I was to succeed in 1982. His fieldwork, first in East Africa and later in West Africa, made him an outstanding figure in Italian anthropology. Interested readers may refer to his paper (15) for more information on the Italian situation. Always refractory to pretentious novelty, and almost by right of aristocratic birth attracted by pedigrees (he loves to trace the source of Italian anthropology to Roman classics, like Julius Caesar, etc), Grottanelli has been reproached for neglecting the new trends of cultural anthropology. In 1984, a further paper on the Italian situation was published by G. R. Saunders, who apparently attempted to fill Grottanelli's gaps (27). Although he gathered a long bibliography of Italian works, Saunders was strongly biased towards the new Italian cultural anthropology; he missed the lively polemic that, from the antithesis between ethnology and cultural anthropology, had expanded on Marxist anthropology (see 25). Perhaps neither paper is wholly satisfactory, but from both one can sense how animated the debate among the Italian anthropologists had been. For readers of the Italian language, a recent book on the last one hundred years of Italian anthropology is also suggested (12).

The folklorist tradition, ethnological research, an innovative cultural anthropology, and finally the debate on the relationship between Marxism and anthropology have been the main trends of Italian anthropology over this period. Their outcomes include the introduction of a variety of academic disciplines, besides ethnology, forming a broad ethnoanthropological group (it may interest American readers that "ethnoanthropology" has a long currency in Italian anthropology: cf. 26:423); an emphasis on the mental and ideological implications of the concept of culture; an expanding research on cultural values; and a semiotic approach to cultural phenomena.

On entering the Italian arena I was impressed not so much by the development of a Marxist anthropology as by the emphasis on Gramsci, and if he had been the inspiring source of cultural anthropology. On reading his works, which I did from their first publication, I was impressed by his perceptive analysis of Italian events, but I had not noted any direct contribution to anthropology (4:232–33). Only through the writings of Alberto M. Cirese did such concepts as culture, hegemony, structure, and so on come to be conceived and discussed with direct reference to Gramsci (see 9, 10: 18–33).

Although I would not say that the Marxist approach has run its course, its fervor everywhere, and certainly in Italian anthropology, has subsided. Rarely has it engaged in either the discussion or radical criticism of ethnographic texts, and it has not led to any deconstructing revisionism. It failed, in any case, to inspire Italian anthropologists to undertake more intense fieldwork. Indeed, I think the weakest aspect of Italian cultural anthropology has been its neglect of direct, sustained research in the field—a weakness, I must hasten to state, that is beginning to be redressed by promising anthropologists of the youngest generation.

Anthropological disciplines are now (at the end of 1989) taught in many an Italian university by 29 professors and 66 associate professors. These figures refer to the whole system of Italian universities. One should include also a large number of junior and freelance professionals; these latter have formed a new society of freelance anthropologists to meet the apparently increasing demand for their profession. Since last year a reform has affected the Italian university system through the creation of a special *Ministero dell' Universitá e della Ricerca Scientifica e Tecnologica.* A law is currently under discussion in Parliament that aims to restore full academic autonomy to each university institution. If such a law passes, one may dream of a new kind of anthropology department in Italy, assembling (on the model of the best French *laboratoires*) Africanists, Americanists, Asiatists, Oceanists, and Europeanists who, sharing their field experience, would together contribute to anthropological theory.

CONTINUITY IN DIVERSITY

At the end of my anthropological journey (22) I am convinced that the essential aim of anthropology is a deeper understanding of the human condition, expressed by the social formations constantly changing under the pressure of each individual's urge to form a personal and ethnic identity. I believe fieldwork to be the only way to acquire reliable evidence about the determinant elements of the human condition. If fieldwork has grown to be the traditional form of anthropological initiation, it is because it affords the only means of observing the human condition as a concrete reality. Only in people's behavior are the patterns of social life revealed in the making and interconnection.

With these premises, I am always puzzled by the claim that only the exotic can provide significant research in anthropology. Anthropologists are too frequently and perhaps unconsciously inclined to attach, as Ioan Lewis puts it, a "deep metaphysical significance to field work in *exotic* settings" (19:1; italics added). We owe to Lewis's ingenuity a refined and witty analysis of fieldwork, "the anthropologist's Muse"; but I do not share his belief that to-day "The ultimate test of the professional anthropologist's standing remains the successful completion of a major piece of field research in *exotic* surroundings" (19:1). Emphasis on the exotic was indeed correct in the past, when world communications were poor and many peoples, unknown to the Western world and thus considered "out of history," were assigned to the invented category of the so-called "primitives." Anthropologists set themselves the task of exploring the real situation of these societies, and their works contributed to a changed perspective. The image of the world is now deeply altered, having developed into that of a great village wherein mankind shares more and more a common way of life and a common destiny.

The essence of the human condition can never be exotic, being the same (despite its surprising variety of expressions) wherever a human being is found. I am thus impelled by my classical bias to cite the old saying of Terence: *Homo sum; humani nihil a me alienum puto* ("being a man, nothing human is exotic to me").

To Heraclitus is attributed the observation *ethos anthropo daimon:* The demon in our individuality drives us where we would not; our own unavoidable ethnocentrism can cause us to despise and even commit violence upon other peoples. In such a predicament, a thoughtful anthropological approach may be useful, if not decisive, in restoring our judgment and recovering the human solidarity of Terence's aphorism.

Such contradictory situations result from the structure of human personality. The fact that the essential components of that structure are at the same time individual and social constitutes one of the paradoxes of the human condition, affecting us all. Though we share some patterns of behavior with other individuals, we still differ, both as individuals and as social groups. Our individual differences and changing social interests alter our ideologies, our knowledge, and our social norms, however firmly these may have been established by tradition. Contrary to old theories, even myths do change. Change is the basic propeller of the development of human culture and society.

The century now nearing its end has been characterized by such great technological achievements as nuclear fission and the exploration of the cosmos. However, I personally assign greater historical consequence to a different accomplishment: We opened up our concept of history to include those peoples who, less than 50 years ago, were thought to be "out of history." They are now a part of the network of human nations, conscious of their own cultural inheritance, and respectfully accepted by others. This

BERNARDI

achievement, like the so-called discovery of the New World and its peoples, has altered our perception of reality, modifying our "common knowledge" and lofty ideologies alike.

It is the business of the anthropologist to explore the human condition, with its variety of ideologies and social institutions. Such a task requires alertness, sensitivity, and the ability to decode observed facts in order to further knowledge without letting preestablished or ethnocentric theories overshadow reality. May these qualities of discernment and perception always be used for the benefit of humanity.

Literature Cited

1. Bernardi, B. 1950. *The Social Structure of the Kraal among the Zezuru in Musami, Southern Rhodesia.* University of Cape Town. *Commun. Sch. Afr. Stud.* (NS, No. 23)
2. Bernardi, B. 1952. The age-system of the Nilo-hamitic peoples. *Africa* 22(4): 316–32
3. Bernardi, B. 1959. *The Mugwe. A Failing Prophet.* London: Oxford Univ. Press for Int. Afr. Inst.
4. Bernardi, B. 1973. *Etnologia e antropologia culturale.* Milano: Franco Angeli Editore
5. Bernardi, B. 1974. *Uomo Cultura Societá. Introduzione agli studi etno-antropologici.* Milano: Franco Angeli Editore (1988: 10th reprint)
6. Bernardi, B. 1983. *Il Mugwe. Un Profeta che scompare.* Milano: Franco Angeli Editore
7. Bernardi, B. 1985. *Age Class Systems. Social Institutions and Polities Based on Age.* Transl. D. I. Kertzer. London: Cambridge Univ. Press
8. Bernardi, B. 1989. The *Mugwe, A Blessing Prophet.* Nairobi: Gideon Were Press
9. Cirese, A. M. 1972. *Cultura egemonica e culture subalterne.* Palermo: Palumbo
10. Cirese, A. M. 1977. *Oggetti, segni, musei. Sulle tradizioni contadine.* Torino: Einaudi
11. Chavunduka, G. L. 1970. *Social Change in a Shona Ward.* Dept. Sociol., Univ. Rhodesia, Occas. Pap. No. 4
12. Clemente, P., et al. 1985. *L'antropologia italiana. Un secolo di storia.* Bari: Laterza
13. De Martino, E. 1941. *Naturalismo e storicismo nell'etnologia.* Bari: Laterza
14. Fortes, M., Patterson, S., eds. 1975. *Studies in African Social Anthropology.* London: Academic
15. Gallini, C., ed. 1986. Ernesto de Martino. *La ricerca e i suoi percorsi. Ric. Folklor.* 13, Aprile
16. Gandini, M. 1969. *Raffaele Pettazzoni e gli studi storico-religiosi in Italia.* Bologna: Forni
17. Grottanelli, V. L. 1977. Ethnology and/or cultural anthropology in Italy: traditions and developments. *Curr. Anthropol.* 18:593–614
18. Kingsley-Garbett, G. 1960. *Growth and Change in a Shona Ward.* University Coll. Rhodesia and Nyasaland. Occas. Pap. No. 1, Dept. Afr. Stud., Salisbury
19. Lewis, I. M. 1986. *Religion in Context. Cults and Charisma.* Cambridge: Cambridge Univ. Press
20. Makarius, L. 1970. Du "roi magique" au "roi divin". *Annal. Econ. Soc. Civil.* 3:668–98
21. Makarius, R., Makarius, L. 1968. Le symbolisme de la main gauche. *Homme Soc.* 9:195–212
22. Marazzi, A., ed. 1989. *Antropologia. Tendenze contemporanee. Scritti in onore di Bernardo Bernardi.* Milano: Hoepli
23. Müller, H. K. 1989. *Changing Generations: Dynamics of Generation and Age-sets in Southeastern Sudan (Toposa) and Northwestern Kenya (Turkana).* Saarbrücken: Verlag Breitenbach
24. Needham, R. 1960. The left hand of the Mugwe: an analytical note on the structure of Meru symbolism. *Africa* 30(1):20–33
25. Remotti, F. 1986. *Antenati e antagonisti. Consensi e dissensi in antropologia culturale.* Bologna: Il Mulino

26. Sangren, P. S. 1988. Rhetoric and the authority of ethnography: "postmodernism" and the social reproduction of texts. *Curr. Anthropol.* 29(3):405–35
27. Saunders, G. R. 1984. Contemporary Italian cultural anthropology. *Annu. Rev. Anthropol.* 13:447–66
28. Schmidt, W. 1951. In der Wissenschaft nur Wissenschaft. *Anthropos* 46:611–14
29. Schulien, M. 1943. *L'unitá del genere umano alla luce delle ultime risultanze antropologiche, linguistiche ed etnologiche.* Milano: Vita e Pensiero
30. Signorelli, A. 1986. Lo storico etnografo. Ernesto De Martino nella ricerca sul campo. *Ric. Folklor.* 13:5–14
31. Solinas, P. 1985. "Idealismo, marxismo, strutturalismo." See Ref. 12, pp. 207–64
32. Stocking, G. W. 1971. What's in a name? The origin of the Royal Anthropological Institute (1837–1871). *Man* 6(3):369–90
33. Tcherkézoff, S. 1983. *Le roi nyamwezi. La droite et la gauche.* Paris: Éditions de la Maison des Sciences de l'Homme; Cambridge: Cambridge Univ. Press
34. Tornay, S. 1989. *Un système générationnel. Les Nyangatom de l'Ethiopie du sud-ouest et les peuples apparentés.* PhD thesis. Univ. Paris X: Lab. Ethnol. Sociol. Comparative

Elizabeth Colson

Annu. Rev. Anthropol. 1989. 18:1–16

OVERVIEW

Elizabeth Colson

Department of Anthropology, University of California, Berkeley; and Refugee Studies Programme, Queen Elizabeth House, Oxford University

FORGING AN APPROACH

The last 50 years has seen anthropology become part of the intellectual baggage of American and western European thought. The media, novelists, literary critics, historians, politicians and political scientists, development economists, and religious leaders speak of culture, cite ethnographic examples, and bolster their arguments by invoking the names of anthropologists. Social workers and educators report their research as based on "the ethnographic method" and "the holistic approach." At a time when so many nonanthropologists think they know what anthropology is, it is disconcerting to realize how little many of us in fact have in common, and the extent to which we have forged our own approaches through questions that have seemed to us worth pursuing.

I see certain consistent assumptions informing whatever I have written as an anthropologist during the last 50 years, whether it has been on contemporary American society, including the bureaucracy of development agencies (11), American Indian reservations (4), the political order of an Australian city, or various aspects of the life of the Tonga-speaking people of the southern Zambian Plateau and the Gwembe Valley. Whether I have written about rituals or residence patterns (7), familial relationships (5), social controls (5, 10), shifts in drinking behavior (13), or the consequences of forced resettlement (9, 12, 40), I have assumed the transiency of social forms and a higher degree of freedom of action than many anthropologists appear to grant to whoever it is they define as "others." With his persistent probing about why I asked certain kinds of questions and formulated my research as I did, Asraf Ghani recently helped to put all this into perspective during three days of intense discussion.

0084-6570/89/1015-0001$02.00

In the first place my central focus is on how people behave and the relevance of that behavior for their own further actions and the activities of others. I assume that they are pursuing goals and have choices, and that these are limited both by their own skills and by the fact that they compete with others in a social universe where access to power is unequal. People act in an opportunistic fashion, which does not mean that they have no regard for others or that they lack generosity. But they are not bound to a code that tells them what must be done. They act and then explain their actions, to themselves and others, in universalistic terms to justify what was in fact contingent. Since the conditions under which choices are made shift from one moment to the next, all action in a sense is exceptional and can be justified as such. The universalistic terms are what James (26:6) has called "an archive" that constitutes "a lasting base of past reference and future validation." They provide continuity because they are an assurance that a normative order exists. While these assumptions derive in part from my training in anthropology they have been reinforced by the people among whom I have carried out research. It would be instructive to examine the extent to which our anthropological theory reflects the assumptions of the people with whom we have worked. The humans I have known have been skeptical, empirical, and prepared to take what they regarded as acceptable risks. At the same time they have wished to be regarded with respect, and to think of themselves as worthy of respect, and so they have explained both themselves and others in terms that make mutual interaction possible, understandable, and, it is hoped, justifiable.

Despite the hypotheses of linguistic philosophers and the more extreme arguments of cultural relativists, I think it possible to communicate across a barrier of different experiences and to understand about as much of a code as is understood by native users of that code. Indeed, under the probing questions of outsiders, native users may make explicit to themselves for the first time matters that have never previously been publically discussed and in so doing discover how much their interpretations differ. We can communicate across barriers of experience because we share pretty basic desires and fears; the fact that we need contact with others to survive provides common experience of the problems of social interaction. If we can never penetrate completely the thinking of someone from another society, neither can their nearest and dearest. People respond to what they believe others are thinking; it may be fortunate that they do not *know* what the other thinks. Social life depends on face saving and the tacit agreement that understanding is mutual.

Anthropological data come primarily from the observation of living people who are getting on with life and making decisions that have implications for the future. Our discipline is not about some reified entity called culture. We do not have anything like a holistic approach and it would not do us much good if we had. For one thing, good research is based on a sense of what is important, and this means selectivity. For another, whatever it is we do observe, it is not an integrated culture or an integrated social system. Of course it may be more comforting to talk in terms of cultures rather than people, for while culture can be destroyed—and it is painful to think of lost art, lost music, or lost social conventions—culture cannot be hurt. People can be, and horribly so.

One way to hurt people is to destroy their familiar environment so that they can no longer achieve what they value, a message driven home by anthropologists who have worked with the organization Cultural Survival. Pain can likewise be inflicted by physical torture, which has become the resort of many contemporary regimes, sometimes tutored by American officials. This is now the background experience of many who once hosted anthropologists, some of whom have been tortured while others have become torturers. In the past, and probably for good reasons, including the protection of human subjects, we downplayed the violence, cruelty, and unhappiness existing in the areas where we worked. One reason may have been the belief that such actions were momentary departures from cultural norms that generated long-term harmony, but whatever the reason, by doing so, we falsified the record.

If anthropology has a rationale beyond the description of a particular moment in the history of a given community, then it lies in trying to understand the range, bitter to sweet, of human experience. Thus there are good theoretical as well as ethical reasons for looking at what is happening now to those who once provided us with hospitality and accepted our request to be taught what they knew. It is not enough to turn anthropology into a critique of its own past or to define it as the study of the structuring of myth, art, or other forms of discourse. Torture, too, is a form of discourse. I am not arguing here against the relevance of studying myth and other forms of art. Kapferer (27), after all, has demonstrated how Sri Lankan and Australian myth and ritual are invoked and transformed to create environments wherein ethnic violence, including murder, becomes acceptable.

If anthropology is more than a game or an academic discipline that provides employment to anthropologists, then at least some of its research effort should be focused on policy environments that lead to the impoverishment both of other countries and of our own people, and create the political chaos that currently has driven 10 million refugees into flight across international borders while an unknown number are displaced and homeless within their own countries. Many people are trying to flee killing poverty across the barriers set up by governments now fearful of their own prosperity and the reaction of their citizens to the stranger. Still others are uprooted because big development projects are financially profitable to a powerful minority, and multinational companies can take advantage of the situation. The political and economic choices of people everywhere are formulated within an international community based on the interplay of nation states and multinational corporations (16).

At various times and places it has been fashionable to deny that anthropologists may legitimately advise programs that work to bring about change. Such a role, it is claimed, attacks the cultural integrity of a community and may lead to new gross inequities as people become dependent upon relationships with the world economy. Applied anthropology, action anthropology, development anthropology, and practicing anthropology—whatever it is called—has also been characterized as "mere application" that leads to no new theoretical advances. I can understand skepticism about the impact that anthropologists can have on the implementation of programs based on nonanthropological priorities (11). But the second claim, that anthropology is advanced only to the degree that its practitioners remain detached from world events, has never seemed to me to have a reasoned basis. Theory, after all, should inform practice; when theory fails, practice has a chance to inform theory.

Sorting over data in academia does not automatically produce new theory; in fact, it can be a dead end. What counts is quality of thought, and this is likely to be stimulated by new experience, whether through the acquisition of new data or through brainstorming with people who, while pursuing different questions, query whatever the anthropologist says. I was therefore delighted when in 1988 the president of the American Anthropological Association, Roy Rappaport, created a number of panels to examine how anthropologists can help to tackle current issues in both industrial and nonindustrial regions of the world. This action encourages anthropologists to involve themselves in an examination of contemporary life.

One of my own long-term concerns (9, 12, 40) has been with the consequences of forced resettlement. Once resettled, people can no longer rely upon survival strategies worked out in their old environments. These prior habitats may have been destroyed by large-scale irrigation schemes, urban clearance, or warfare; they may have been placed out of bounds by political disorder; or they may now be inaccessible for other reasons. Though some people move as communities, they must deal with the loss of the old setting and the outmoding of skills. How does all this affect their relationships with one another, their trust in political leaders, their explanations of misfortune, and their belief in an orderly universe? Who emerges empowered? How long does it take to reestablish confidence in one's ability to cope with an environment that must be learned afresh? What measures by external agencies help or hinder adjustment? Thayer Scudder, who has collaborated with me in the longitudinal study of the Gwembe people displaced by the creation of Kariba Hydroelectric Dam and Kariba Lake, has looked at resettlement schemes throughout the tropics and is now able to predict the response over time to displacement (37, 38, 40). That work has had some impact. Recently the World Bank has issued guidelines to be followed when it agrees to finance major projects that will result in the displacement of large numbers of people. These guidelines require the same attention to the social effects as to the technical problems of construction (3). Many of the same issues are relevant to the work of the Refugee Studies Programme in Queen Elizabeth House, Oxford University, founded by Barbara Harrell-Bond, an anthropologist who has studied at first hand the impact on refugees of relief measures fielded by international organizations and the multitude of voluntary organizations specializing in humanitarian assistance that are now endemic to the so-called Third World (23). We are now at a stage where it is possible to see the commonalities among the experiences of those who flee as refugees, the victims of large-scale forced resettlement carried out by governments in the interest of economic gain, those who find their communities destroyed by urban renewal projects, and the old who are forced from their homes into care-centers by the logic of public health systems (22, 34). It is possible to analyze their experience as a process in time, with reactions appearing in a predictable sequence as "displaced people strive to maintain whatever power they can" (21 : 14) in an attempt "to recreate a social world which would offer the highest degree of safety and control" (29 : 244–45).

ACADEMIC STIMULUS

My current work is a culmination of years of exposure to teachers, colleagues, and those who have educated me during the course of field research. I have also had to do a good deal of thinking about the environments within which I have worked.

In the late 1930s when I began graduate work in anthropology at the University of Minnesota, the world was in the midst of a depression and fearful of the outbreak of further war in Europe, where the Spanish Civil War had just ended. Refugees arriving from Nazi Germany were a reminder of how easy it was to become vulnerable to an oppressive political regime. The economic chaos of the depression and growing recognition of the savageries being carried out in Germany and other European countries undermined belief in the inevitability of progress and the superiority of Western civilization.

Like most anthropology students, I had chosen the subject for the love of it, not because I expected that it would lead to a job. (A joke at the time was that one needed a PhD to find a job as a dishwasher.) It is difficult to remember, however, what it was that made us love anthropology, other than the fact that it seemed to provide a powerful critique of the world as we knew it— particularly of the social rules that confined us. But we could also believe that anthropology had an important role in encouraging tolerance of differences and in combating racial and ethnic prejudices. It was also an exciting time to be a student. The discipline was being transformed as anthropologists experimented with participant observation in a great many different contexts, which gave them a lively awareness that the field dealt with the very stuff of human life. A few anthropologists were doing research on the social life of American cities and towns, providing a critique of accepted views on the nature of American society. A number were looking at the influences of the reservation system on the formation of 20th century American Indian culture. Others, inspired by Freud and the neo-Freudians, were engrossed with questions of the interplay between personality and culture.

Students in that era had an advantage over those entering the field today because anthropologists were few and the theoretical issues that divided them were matters of common discussion. The number of monographs and journal articles appearing annually was small enough so that we could be expected at least to look at them. This still left time for us to read widely in other fields. At Minnesota (1935–1940) and Radcliffe (1940–1941, 1943–1944), David Mandelbaum and Clyde Kluckhohn encouraged me to read in the psychoanalytic and social psychology literature since I then planned to work on personality and culture, using life histories as a data base. I discovered Moreno, later known for the psycho-drama but then examining social dynamics through methods that led on into sociometry. That work was a predecessor of both small group dynamics and network analysis.

Equally important for my own development was the field theory of Kurt Lewin, who conceived of human behavior as a response to current forces operating within a given environment rather than as the precipitate of the past. Coser (14:15) has suggested that Lewin was strongly influenced by his own experiences as a refugee, since refugees "are compelled to cultivate a detached form of knowledge that will help them along paths not marked by traditional signposts. They are structurally compelled to use analytical reasoning, where for the insiders custom and habit are sufficient guides to conduct." Nevertheless, such an approach was appealing to a midwesterner who shared the typical American belief in the importance of the individual and the possibility of exercising choice in a world that was not predetermined. Trying to think in Lewin's terms, however, encouraged skepticism about current theories of "culture." These either dealt with culture as a thing in itself (whether it was seen as an integrated whole or a compound of shreds and patches) or, following Sapir, reduced culture to the realm of the individual. Skepticism about the utility of "culture" as an analytic concept was not lessened when I worked as Clyde Kluckhohn's research assistant while he and William Kelly wrote their critique of culture theories (28) for a volume edited by Ralph Linton entitled *The Science of Man in the World Crisis*. If culture was problematic, then so was Herskovits's concept of acculturation (24) and Malinowski's views on culture change (31): It was never clear what interacted with what.

Later, I found new problems with the treatment of cultures as things in themselves when I found anthropological formulations of culture to be similar to arguments used in South Africa to justify the color bar with its denial of civil rights to Africans and to entrench European domination. This was evidence of the power exercised by conceptual frameworks, for the belief in cultures as enduring entities was a common heritage from Herder and other German romantics of the early 19th century. That same belief had empowered the ethnic nationalisms of Europe. Wilson Wallis at Minnesota earlier had aroused my interest in the intellectual ancestry of the discipline. He stressed the influence of Locke, Hume, Ferguson, and other moral philosophers of the 17th and 18th centuries upon the development of a discipline that examined the interplay between the givens of biology and the freedom to create displayed by human inventiveness (44).

In the early 1940s Kluckhohn was already involved with Talcott Parsons and Henry Murray in the interchanges that led to the creation of the Social Relations group at Harvard after World War II. By that time I was gone, but it was Talcott Parsons who introduced me to Max Weber and the vocabulary of social action (35). His structural-functionalism was stimulating but too rigid for the direction in which I was developing. Durkheim's work, especially his concern for a collective morality, was more difficult to come to terms with, and I have found generations of American students responding in much the same fashion. It was Balzac's novels that gave me an inkling of the massive questioning of order, including the categories of time and space, in revolutionary France, which provided the background to Durkheim's concerns. Since then, the struggle of new African governments to create a national consciousness and give their regimes legitimacy has made the Durkheimian legacy more relevant.

THE IMPACT OF FIELD WORK

My first field work was with Native Americans, the Pomo of northern California in the summers of 1939, 1940, and 1941. My supervisors were Burt and Ethel Aginsky, who had founded the New York University Field Laboratory for Research in the Social Sciences. Field schools were a characteristic of the period, but the Aginskys may have been unique in recruiting students from sociology, psychology, and literature as well as anthropology and encouraging us to examine the interplay among townspeople, ranchers, migrant agricultural workers, and Pomo. It was a dynamic situation with strong overtones of caste. The older Pomo told of being hunted down for internment at Round Valley and returning to find their lands preempted by whites. Since then their lives had been dominated by the multiracial society, and that domination was strongly reflected in the life histories of the Pomo women with whom I spent much of my time during those summers (8).

Dissertation research at Neah Bay, Washington followed (1941–1942). During the course of that year, the town doubled in size as construction workers arrived to build a naval base and soldiers came to guard the seafront against a possible Japanese invasion. My field data inevitably reflected a period in the history of the town that all knew to be ephemeral, though it would have consequences for the future. This experience convinced me of the importance of grounding ethnographic description in the historical moment, whatever one does in analysis. Dates are as important in ethnography as they are in history.

It was obvious that people within the town were interacting on some basis of understanding although they came from different regions and shared no common past. It was problematic that they shared the same culture. The Makah also disagreed about much that had happened in the past, and I began to doubt that there had ever been a homogeneous entity that could be called Makah culture. History, as they handled it, provided convincing demonstration of the validity of Malinowski's dictum that myth should be seen as a social charter reflecting the interests of its tellers. People disagreed about historical events in the same way that they disagreed about relative social status. They said there had been such a thing as a Makah culture but the details of that culture were in dispute. They did agree that to be a Makah had little to do with either language or culture. I came to suspect that disputes might be at least as important as consensus in producing social cohesion (4).

Many Makah themselves linked their existence as a people to their continued control of their land. They had never been subject to forcible exile, although their children might have been taken away to boarding schools, and they had been able to retain some semblance of economic independence. Their confidence in themselves and their refusal to be subordinated to white Americans was a revelation after experience among California Indians, who had known exile and thoroughgoing disruption. Also revealing was the Makah fear that the government would use the war as an excuse to remove them. They insisted that they could remain a viable community only on their own terrain, where those who chose to live elsewhere could later find a refuge in adversity. They were prepared to use the courts to fight off efforts to displace them.

How vulnerable we all are to arbitrary action was demonstrated in 1942 when, as one response to American hysteria after Pearl Harbor, the US government removed all Japanese and Japanese-Americans from the west coast into so-called "war-relocation camps." There was no pretense of fair trial. Instead people were stigmatized, segregated, and denied their civil rights. I had the chance to join a research team in Poston, one of the camps, where I worked under Alexander Leighton, a psychiatrist with anthropological training, and Edward Spicer, a social anthropologist trained by Radcliffe-Brown. In studying the dynamics of camp life, the former emphasized the forces playing upon individuals, while the latter described the creation of social roles that were part of an interactive system. Again we were looking at a situation that had only just come into existence, that changed before our eyes, and that we hoped would not last (30).

Poston displayed many of the characteristics now being reported from camps set up for refugees in Hong Kong (2, 29), on the Thai-border (20, 25, 29, 36, 41), in Sudan (23), and in Zambia (21). In describing camp life observers invoke Goffman's formulation of the total institution (19). Staff and inmates become progressively more polarized. Each stereotypes the other. Homogenization is the general aim of the procedures in the interest of administrative control; but factionalism is the order of the day, since those who once held authority try to maintain some kind of status and control over their own actions. Irrespective of the cultural traditions of those who administer or those who are administered, the social order and culture of the camp are created as people respond to what are basically the same kinds of forces operating within a similar environment.

Recently anthropologists like myself who carried out research in "war-relocation" camps have been criticized on the grounds that we provided legitimation for the internment and the losses suffered by those interned. I can only say that we regarded the internment as a gross violation of civil rights. But I thought then as I think now that witnesses were needed and that anthropologists had skills suited to that task.

By the time the war ended, I was eager for further field work, this time outside the United States. It was exhilarating to arrive in Northern Rhodesia at the Rhodes-Livingstone Institute in 1946 to discover scholars who shared my approach and interests. The Institute had been more or less on hold during the war years but was now ready to start new research. The proposal submitted by Max Gluckman to the Colonial Social Science Research Council called for the investigation of how involvement in the market economy affected rural African communities that were either exporting labor or growing cash crops. We were asked to look at people who were moving about, making choices, adjusting to changing circumstances.

By then labor migration had long been seen as relevant to the research of anthropologists working in southern Africa. Such studies had been carried out at least as early as the 1930s when two things became apparent: that people were moving between rural areas and employment centers in response to economic forces acting throughout the region, and that rural areas were being transformed by the absence of men and the dependence on cash. Research had also begun in the receiving areas where new urban societies were being born. Godfrey Wilson, the first director of the Rhodes-Livingstone Institute, worked in Broken Hill (now Kabwe) at the end of the 1930s (46). He reported that a significant number of African workers were now permanently resident in town and dependent on wage labor for a livelihood. They and their families needed adequate housing and other services. The government of Northern Rhodesia was unwilling to concede that the towns were generating new communities, new needs, and new aspirations and that the Africans in towns should be treated as town residents. This conflicted with the official policy, which maintained that Africans were "tribesmen" who should be encouraged to retain their own cultures and also continue to support themselves in the "tribal" areas.

Gluckman, who had replaced Wilson as director, was a student of Radcliffe-Brown; but as a reader of Marx and Freud, he saw conflict as intrinsic to society and social organization as dependent upon economic factors. As a South African he was well aware of the color bar and the power relationships that maintained it, but he also observed that cooperation existed across the color bar that could not be explained simply in terms of raw power or by the role of ideology (17, 18). From his observation of the ceremony of the opening of a bridge in Zululand in the 1930s, he had developed what was later described by Van Velsen (43:xxv) as "a situational approach." In this, as in Lewin's field theory, actors make choices within a field where multiple forces are operating, and their behavior shifts as they encounter different situations. Mitchell (33:70) has summed up the approach succinctly in his recent reassessment of his work in the 1950s on labor migration and urbanism, carried out at the Rhodes-Livingstone Institute: "*behaving* . . . is *situational* and not an immutable characteristic of actors."

THE BEGINNINGS OF THE MANCHESTER SCHOOL

Those of us who worked with the Rhodes-Livingstone Institute in the 1940s and 1950s were fortunate in having contracts that gave us contact with the people of a region through several years. This led to the emphasis upon time as an important methodological tool. Later, in the 1960s, when travel became easier and repeated visits to the same region became feasible, the recognition of the importance of extended observation led to the advocacy of long-term field research and longitudinal studies (15, 39). It was observation over a period of several years that made it possible for Van Velsen (43) to use the extended case method and Turner (42) to develop what he termed "processual analysis." It is instructive, however, that each method also focused upon conflict, for it was conflict that connected the various episodes through which a set of actors assessed and reassessed their commitments. Conflict likewise made possible the examination of "the variations in actual behaviour, the confusion and conflicting events which occur within the same frame of values (43:xxv)." Barnes, also affiliated with the Rhodes-Livingstone Institute in the late 1940s, went on to develop the concept of social networks, where the emphasis lay upon the multiplicity of relationships maintained at any one time with individuals who might or might not otherwise be associated (1). Network analysis was vital to much later urban research (32, 33).

Spurred by Gluckman's enthusiasm and love of field data, we engaged in close-grained analysis of case material and saw ourselves as moving towards a more dynamic view of social action and one better adapted to deal with complex situations than the structuralism of some of our colleagues.

In the analysis of the African material, we recognized that we were looking at people caught up in a colonial system whose influence was pervasive. This was a basic datum. We certainly indicated that in both rural and urban areas people resented the economic and political domination and the gross incquities of the system. At the time, however, we let much that they said go unreported because it was dangerous to them, and sometimes to the continuation of our research, to do otherwise. Nor did writing a polemic about the inequities of the colonial system then seem a contribution to ethnography. That we failed to provide descriptions of the working of the colonial administration adequate to the needs of later readers is due to our assumption that since administrative practices were everywhere similar, readers who knew anything about Africa could fill in the details. Today, of course, readers cannot. This is one of the dilemmas of the ethnographer: Just how much needs to be spelled out for readers distant in time and space? Those who described urban situations were more likely to include administrators and employers as prominent actors in the social dynamics of the rapidly evolving society, and they also included the trade unions and political parties that were organizing African demands for better living and a role in government. Their studies now seem less dated than the monographs based on work in rural areas.

At the Institute and later at Manchester University I contributed to the general pool of ideas my own concerns about how people define themselves, a refusal to see individuals as overpowered by roles, a view of disputes as having integrative functions, a desire to discover what kept conflict from getting out of hand, and the concept of cross-cutting ties as basic to the maintenance of community. I do not claim that these were unique contributions, for when people are working on close collegial terms a certain degree of homogeneity develops. Where I differed from some of the others was in a skepticism about the value of reifying a method or approach under a label or of promoting a particular approach as useful for examining all questions. I thought we needed a more eclectic toolkit, which may be a characteristic feminine attitude. The contention that social arrangements are contingent was also a reflection of the way Pomo, Makah, and Tonga women explained the social environment in which they lived. What they thought had to be taken seriously, since they were actors in their own right. Information from males ought not to be privileged over their testimony.

FORCED MIGRATION AS A RESEARCH FOCUS

In the 1940s I carried out research among the Plateau Tonga who lived near the railway line in what is now Zambia. They had already lost a good deal of land to European settlers and were afraid they might again be forced out. On the other hand, many were themselves immigrants drawn by the opportunities for cash cropping provided by the proximity to the railway line and markets. Many of them, or their parents, were immigrants from Tonga-speaking villages in Gwembe Valley to the east, where the majority of the population lived in large villages near the Zambezi River. Nevertheless, Plateau villagers spoke of the Gwembe Valley as the epitome of backwardness, where people followed ancient custom. This they contrasted to their own access to education and adoption of European styles. Gwembe men, moreover, were forced to become labor migrants by the lack of economic opportunities in their home region. Forced mobility and a clinging to old custom were both considered undesirable, since each conflicted with fundamental values. Plateau villagers stressed the right of people to make their own decisions. They did not regard custom as sacrosanct and were prepared to jettison whatever prevented them from experimenting with new possibilities. The right of mobility was cherished; to be restricted was to be treated like a slave. Before I left in 1950 a few of the villagers in my sample had moved to a completely unfamiliar area 150 miles away where they could get land that would allow them to expand farming operations. The attempt of colonial administrators to restrict settlement in the towns, which were seen as providing amenities and opportunities lacking in rural areas, caused resentment against the system. Migration leading to settlement in urban areas or distant agricultural regions was an experience men and women were willing to risk, whatever their emotional ties with home.

Why, then, is forced migration such a traumatic experience? This question has dominated much of my research over the past 30 years, a period during which Thayer Scudder and I have been examining the impacts on Gwembe District and its people of the building of Kariba Dam, a project that flooded much of the middle Zambezi valley and forced the resettlement of many of its people.

Gwembe villagers had moved in numbers to the Plateau in the 1920s and 1930s, and more would have gone if land had been available. Some had settled in cities to the south. Yet in 1956 people were adamant that being *forced* to move was something they could not face (6, 9). Like the Makah, they argued that a particular place gave meaning to their own identity.

We saw in the aftermath of the resettlement that the stress had been severe even though villages were moved as units and an attempt was made to keep neighborhoods together. Because people had to face an unfamiliar physical environment at a time when they had also lost confidence in themselves, their political leaders, and their divinities, they became immediately disoriented and experienced great insecurity. Political refugees who have chosen flight against great odds to reach sanctuary are reported to feel empowered by the knowledge that they have won through and demonstrated their own fortitude and intelligence (29). None of this was apparent in Gwembe, where people saw resettlement as a failure that demonstrated their loss of control over their destinies. They began to recover only after they had mapped their new regions and when opportunities associated with the opening of the Kariba Lake fisheries and the control of tsetse fly restored their view of themselves as independent actors.

Since 1956 we have returned many times to Gwembe. I have also worked with students who have carried out research among refugee and other migrant populations elsewhere in the world (34) and read widely in the literature on forced migration and its consequences. I have concluded that humans are in fact migratory animals in the sense that they treasure possibilities for mobility but at the same time see any attempt to force mobility upon them as an infringement upon their personal space and upon their sense of integrity. Also important to the sense of self is the existence of a place of refuge, a home community. When forcible removal involves also the destruction or consequent inaccessibility of that refuge, people suffer immensely, even those who left voluntarily but with the expectation that they could always return. Those who have worked with refugees report that even those long resettled report grief over the loss of home.

Since so many parts of Africa, Asia, and Latin America are now in turmoil and people are being forced away from their home territories, this sense of loss must now be endemic, permeating contemporary definitions of self and the construction of reality. If anthropologists continue to define themselves as ethnographers, they are going to have to take cognizance of this. This century is characterized by displacement. Many who now make decisions have been formed under the stress of displacement. We need to understand the phenomenon's causes and effects. A start has been made, with the development of such centers as the Refugee Studies Programme at Oxford (involved in comparative work on uprooting and resettlement) and the creation of various more specialized centers in the United States and Canada.

A SUMMING UP

Although an anthropologist's life involves a great many activities besides ethnographic research and writing—teaching and service to professional and other organizations being common to most of us—our most meaningful experiences are probably associated with field research, just as the experience of enlightenment is most commonly associated with those rare moments when disparate bits and pieces of information fall into a pattern. For those of us who have been engaged in long-term studies, it is field work—and the mulling over the field data—that provides a good deal of our own personal sense of continuity. We live between two worlds and feel somewhat detached from both, but each gains meaning through its contrast with the other. But if field research becomes a process over time, those who pursue it have to develop methods that can make accumulated data useful in the testing of hypotheses— something more than private memorabilia. This necessity lay behind the formation of the Linkages group in 1987, which initially brought together Douglas White, Lilian Brudener White, Scarlet Epstein, Conrad Kottak, Nancy Gonzalez, Michael Burton, Thayer Scudder, and myself. We are using the Gwembe material accumulated over the past 39 years (I first visited the region in 1949, and the most recent material was added by Jonathan Habarad, who worked in Gwembe in 1987–1988) as an experimental base for longitudinal analysis. We aim to systematize it in such a way that it will be susceptible to reworking by other anthropologists who want to test the findings or use the data to pursue their own questions (45).

It is good to know that something more may come of our many months of association with Gwembe men and women.

ACKNOWLEDGMENT

My thanks to Norman Buchignani, who commented on an earlier draft.

Literature Cited

1. Barnes, J. 1954. Class and committees in a Norwegian island parish. *Hum. Relat.* 7:39–58
2. Bousquet, G. 1987. Living in a state of limbo: a case study of Vietnamese refugees in Hong Kong camps. See Ref. 34, pp. 34–53
3. Cernea, M. 1988. *Involuntary Resettlement in Development Projects: Policy Guidelines in World Bank-Financed Project.* Washington, DC: World Bank
4. Colson, E. 1953. *The Makah Indians: A Study of an Indian Tribe in Modern American Society.* Minneapolis, Minn: Univ. Minnesota Press
5. Colson, E. 1958. *Marriage and the Family among the Plateau Tonga of Northern Rhodesia.* Manchester: Manchester Univ. Press
6. Colson, E. 1960. *Social Organization of the Gwembe Tonga.* Manchester: Manchester Univ. Press

7. Colson, E. 1962. *The Plateau Tonga of Northern Rhodesia: Social and Religious Studies*. Manchester: Manchester Univ. Press
8. Colson, E. 1967. *Life Histories of Pomo Women*. Berkeley, Calif: Archaeol. Facil., Univ. California
9. Colson, E. 1971. *The Social Consequences of Resettlement: The Impact of the Kariba Resettlement upon the Gwembe Tonga*. Manchester: Manchester Univ. Press
10. Colson, E. 1973. *Tradition and Contract: The Problem of Order*. Chicago: Aldine
11. Colson, E. 1982. *Planned Change: The Creation of a New Community*. Berkeley, Calif: Inst. Int. Stud., Univ. California
12. Colson, E. 1987. Introduction: migrants and their hosts. See Ref. 34, pp. 1–16
13. Colson, E., Scudder, T. 1988. *For Prayer and Profit: The Ritual, Economic, and Social Importance of Beer in Gwembe District, Zambia, 1950–1982*. Stanford, Calif: Stanford Univ. Press
14. Coser, L. 1984. *Refugee Scholars in America, Their Impact and Their Experiences*. New Haven, Conn: Yale Univ. Press
15. Foster, G., Scudder, T., Colson, E., Kemper, R., ed. 1978. *Long-Term Field Research in Social Anthropology*. New York: Academic
16. Gellner, E. *Nations and Nationalism*. Oxford: Basil Blackwell
17. Gluckman, M. 1985. *Customs and Conflict in Africa*. Oxford: Basil Blackwell
18. Gluckman, M. 1958. *An Analysis of a Social Situation in Modern Zululand*. Rhodes-Livingstone Pap. 14. Manchester: Manchester Univ. Press. (Originally published 1940)
19. Goffman, E. 1961. *Asylums: Essays on the Social Situation of Mental Patients and Other Inmates*. London: Penguin
20. Habarad, J. 1986. Five villages: culture and resources among Lao Iu Mien. *Kroeber Anthropol. Soc. Pap.* 65–66:83–100
21. Hansen, A. 1982. Self-settled rural refugees in Africa: the case of Angolans in Zambian villages. See Ref. 22, pp. 13–35
22. Hansen, A., Oliver-Smith, A. 1982. *Involuntary Migration and Resettlement: The Problems and Responses of Dislocated People*. Boulder, Colo: Westview
23. Harrell-Bond, B. 1986. *Imposing Aid: Emergency Assistance to Refugees*. Oxford: Oxford Univ. Press
24. Herskovits, M. 1939. *Acculturation: the Study of Culture Contact*. New York: Augustim
25. Hitchcox, L. 1988. *Vietnamese Refugees in Transition*. Oxford: Oxford Univ.
26. James, W. 1988. *The Listening Ebony: Moral Knowledge, Religion, and Power among the Uduk of the Sudan*. Oxford: Clarendon Press
27. Kapferer, B. 1988. *Legends of People, Myths of State: Violence, Intolerance, and Political Culture in Sri Lanka and Australia*. Washington DC: Smithsonian Inst. Press
28. Kluckhohn, C., Kelly, W. 1945. The concept of culture. In *The Science of Man in the World Crisis*, ed. R. Linton, pp. 78–106. New York: Columbia Univ. Press
29. Knudsen, J. 1988. *Vietnamese Survivors: Processes Involved in Refugee Coping and Adaptation*. Bergen: Dept. Soc. Anthropol., Univ. Bergen
30. Leighton, A. 1945. *The Governing of Men*. Princeton, NJ: Princeton Univ. Press
31. Malinowski, M. 1945. *Dynamics of Culture Change: An Inquiry into Race Relations in Africa*. New Haven, Conn: Yale Univ. Press
32. Mitchell, J., ed. 1969. *Social Networks in Urban Situations: Analysis of Personal Relationships in Central African Towns*. Manchester: Manchester Univ. Press
33. Mitchell, J. 1987. *Cities, Society, and Social Perception: A Central African Perspective*. Oxford: Clarendon Press
34. Morgan, S., Colson, E., ed. 1987. *People in Upheaval*. Staten Island, NY: Cent. Migration Stud.
35. Parsons, T. 1939. *The Structure of Social Action*. Glencoe, Ill: The Free Press
36. Reynell, J. 1988. *Political Pawns: Refugees on the Thai-Kampuchean Border*. Oxford: Refugee Studies Programme
37. Scudder, T. 1973. The human ecology of big dam projects: river basin development and resettlement. *Annu. Rev. Anthropol.* 12:45–55
38. Scudder, T. 1985. A sociological framework for the analysis of new land settlement. In *Putting People First: Sociological Variables in Rural Development*, ed. M. Cernea, pp. 121–53. New York: Oxford Univ. Press
39. Scudder, T., Colson E. 1978. Long-term field research in Gwembe Valley, Zambia. See Ref. 15, pp. 227–54
40. Scudder, T., Colson, E. 1982. From welfare to development: a conceptual framework for the analysis of dislocated people. See Ref. 22, pp. 267–87

41. Tapp, N. 1988. The reformation of culture: Hmong refugees from Laos. *J. Refugee Stud.* 1(1):20–37
42. Turner, V. 1957. *Schism and Continuity in an African Society*. Manchester: Manchester Univ. Press
43. Van Velsen, J. 1964. *The Politics of Kinship*. Manchester: Manchester Univ. Press
44. Wallis, W. 1937. *Culture and Progress*. New York: Whittlesey House, McGraw Hill
45. White, D. 1989. Linkages: designing and implementing optimal scientific experiments for evaluating socio-economic development strategies. In *Anthropological Research: Process and Application*, ed. J. Poggie. In press
46. Wilson, G. 1941–1942. *An Essay on the Economics of Detribalization in Northern Rhodesia*. Rhodes-Livingstone Pap. 5 & 6. Livingstone: Rhodes-Livingstone Inst.

Sol Tax

Ann. Rev. Anthropol. 1988. 17:1–21
Copyright © 1988 by Annual Reviews Inc. All rights reserved

PRIDE AND PUZZLEMENT: A Retro-introspective Record of 60 Years of Anthropology

Sol Tax

Department of Anthropology, University of Chicago, Chicago, Illinois 60637

I have been so lucky, in my personal and professional life alike, that I am happy to risk telling as candidly as I can whatever bears on my career. It was in the summer of 1928 that, while supervising a playground in Milwaukee, I read Marett's *Anthropology*. When I returned to Madison to find that the University of Wisconsin had hired its first anthropologist, Ralph Linton, I took his course and after several lectures decided to change my major. Linton was encouraging: "It's a good field," he said. "There are only about 50 anthropologists in the United States." The four fields (physical anthropology, archaeology, cultural anthropology, and linguistics) were then undivided, and I immediately began to prepare myself with courses in geology, evolution, comparative anatomy, embryology, neurology, psychology, and sociology. All the courses in anthropology were taught by Linton (who had just returned from fieldwork in Madagascar) and by Charlotte Gower, a University of Chicago PhD whose thesis had been in Caribbean archaeology and who was fresh from fieldwork in rural Italy. None of us either inside or outside of our small community of anthropologists had any doubts. The "four fields" have since grown tenfold, and my first question is whether it is the same anthropology.

Mainly in Britain and the Americas was "anthropology" the master term. In most of Europe it meant physical anthropology; ethnology had its own terms, and archaeology and linguistics seemed also to be isolated, but in fact a "general anthropology" seems to have been recognized. After the First World War "social anthropology" began to be used, at least by Radcliffe-Brown and Malinowski and their students, who distinguished their work from ethnology in terms of its being nonhistorical and for a time threatened to secede. Partly because linguistics became interesting to many of them, this movement slowed just as countervailing forces, especially in North America, were strengthened by government funding for "anthropology" as traditionally recognized. In the summer of 1952, 80 scholars gathered at the Wenner-Gren Foundation in New York for two full weeks of discussion of all aspects of anthropology (7, 41). Although East Europeans were unable to join this first postwar gathering, it was self-consciously international, and it brought into the closest face-to-face contact scholars of every field of anthropology. It made a great difference; Carlos Monge, who studied the effects of very high altitude on Indians in the Peruvian Andes, said to me one day in the course of it, "I have been a biologist; but no more. Now I am an anthropologist."

Reviewing the conference (29), I noted that nonanthropologists had taken for granted or admired the breadth of our subject, but some of the anthropologists had found it problematic. When I attacked the problem historically, I found that anthropology as I knew it had had its beginnings in 1839 in Paris with the founding of a network of societies embracing the four fields and concluded that

> except possibly in Britain . . . anthropology is becoming more rather than less integrated. The danger is not in increasing numbers, or adding specialisms, or permitting free access to our ranks; but rather the contrary is true. If anthropology in the United States disintegrates, it will most likely be because increasing professionalization leads to restrictive definitions of anthropology or anthropologists and to the counting out of some activities—scientific, applied, or action-oriented. An essential value in the case of anthropology is its historic reaching for new ideas, tools, and subject matters that may further the study of man.

These words might have seemed wishful thinking on the part of one as committed as I was to anthropology. I need to ask myself how I might have become so committed—60 years ago—and why I still am. It happens that from early childhood I had Walter Mitty dreams of greatness that became increasingly sociopolitical and worldwide. By the time I was in high school I was impatient to *do* things, and I thought of law as a career leading to politics and of political science and economics as prelaw majors. But I had also adopted values against which these disciplines would be tested and—I feared—found wanting. In any case, in my first college years I was much more interested in campus political activities, and in writing and playacting, than in required courses, which I quietly neglected. That a sudden return to disciplined study coincided with the turn to a positive, named career— whatever it might be—is understandable. But I came to credit the outlook of anthropology as critically coinciding with philosophical values that, also, I had adopted in my third year of high school. Specifically, I recall one evening reviewing in my mind the different conclusions of great thinkers on such large questions as whether there are knowable absolutes as opposed to conditional truths. It dawned on me then that the very diversity of views among the wisest of men settled the question in favor of relativism, and I have never wavered from that view.

Fieldwork

When I came to the Department of Anthropology of the University of Chicago in the autumn of 1931, I had had a four-month experience with Alonzo Pond's Logan Museum (Beloit College) archaeological expedition in Algeria (15), another two months with George Grant McCurdy's American School of Prehistoric Research in Europe (both in 1930), and then a summer's field school under Ruth Benedict among the Mescalero and Chiricahua Apache in New Mexico. Having learned basic four-field anthropology from Linton and Gower, I felt well prepared.

A. R. Radcliffe-Brown arrived in the department at the same time, and even before I had taken his course in kinship he suggested that I do fieldwork among the Fox because they had an interesting variant of the Omaha type of kinship system. Even the phrase was new to me, and I quickly went to Lewis H. Morgan's *Systems of Consanguinity and Affinity* to learn about Omaha and Crow types of kinship. In the next three years I studied in depth the history and theory of the study of kinship systems; spent the summer of 1932 with the Fox (Mesquakies) in Iowa and that of 1933 visiting related tribes in Wisconsin, Michigan, Minnesota, and Ontario (none of which had Omaha or Crow systems) for comparison; and, as a research assistant, collected data on the Indian tribes of California and the Northwest Coast. By the summer of 1934, when I returned to the Mesquakies for last-minute checking of data and ideas to write the part of my thesis that later was published as "The Social Organization of the Fox Indians" (18), I had in hand the part that was eventually published as "From Lafitau to Radcliffe-Brown" (30). That fall I worked on the theoretical portion, eventually published as "Some Problems of Social Organization" (17). Robert Redfield had invited me to join his Carnegie Institution of Washington project to do ethnography among the Indians of the highlands of Guatemala, and I finished writing only two or three days before my wife and I were to embark from New Orleans for Guatemala. The faculty was miraculously cooperative in reading and examining me over the thesis, and we were off.

Redfield had suggested that we learn Spanish and read Schulze-Jena's newly published book on the Quiché, and that was almost all we did in preparation for our first Guatemalan fieldwork. We were scheduled to spend eight months in the field and four months back home each year to put together what we had learned, and this we did for three years (1934–1937). Then I had a short field season in the winter of 1938, and then—together now with our newborn daughter—we had six months in 1939 and a full season of fieldwork in 1940–1941 (see R. A. Rubinstein, *Fieldwork: The Correspondence of Robert Redfield and Sol Tax,* in preparation). Our fieldwork consisted of fairly intensive work in Chichicastenango (Quiché) and in the towns bordering Lake Atitlán, particularly Panajachel (Cakchiquel). Part of this time the Redfields lived in nearby Agua Escondida, a Ladino town, and our contact was close. I worked particularly on the world view and the economy of the Indians (19–22, 26, 28, 32, 39). I also had considerable success in helping Antonio Goubaud Carrera and Juan de Dios Rosales to become, in the early 1940s, the first native Guatemalan professional anthropologists. I worked with both of them during the war years on a nutritional survey of rural Guatemala and in the same period (1942–1944) taught in Mexico City and gave students field training among Maya-speaking Indians in Chiapas.

Professional Activities

We returned to Chicago for permanent residency in the spring of 1943. It was wartime. Between trips to Guatemala I wrote and filled in for absent faculty by teaching undergraduates in Robert Maynard Hutchins's famous college. I was a research associate in the Department of Anthropology and met weekly with the faculty. I was also a tenured scientist of the Carnegie Institution of Washington, but when it abandoned its Division of Historical Research it was time to think of a university position. Redfield and I were too close in our interests for me to think of the University of Chicago; but there was no hurry. I was writing and enjoying contacts in the department and the college. I presume that it was at one of our department meetings that we discussed the coming postwar influx of mature students, and the idea came to me of developing a graduate curriculum after the pattern I admired in the college, using a syllabus with selected readings and with lectures by specialists, classroom discussions of issues, and written examinations to measure progress. I was encouraged to develop these ideas, and eventually there was a plan for three year-long courses: "Human Origins," "Peoples of the World," and "Culture, Society, and the Individual." In 1944 I became associate professor, and Robert Braidwood and I worked out the first course, which was followed by Fred Eggan's work on the second and a cooperative effort with the Department of Sociology and the Committee on Human Development accomplishing the third. In 1948 I became full professor and had a five-year term as associate dean of the Division of the Social Sciences, also chairing its Curriculum Committee. These were years of intense interaction with many of my 150 colleagues of every social science discipline. It is probably no accident that these were the years both of the development of action anthropology and of my work toward the racial integration of the Hyde Park community in which the University of Chicago is located (33). These same years also saw publication of the three Americanist volumes (23–25) and of *Heritage of Conquest* (26) and the completion of *An Appraisal of Anthropology Today* (41) and *Penny Capitalism* (28).

The professional activities that occupied my time over the following three decades might be grouped into three large, overlapping categories:

1. Research, teaching, writing and lecturing, and administration.

2. Editing and arranging publication of journals and of books and advising on such matters; organizing conferences on scientific topics and national and international academic congresses; and serving state, national, and international bodies such as UNESCO, the Smithsonian Institution, the Illinois State Museum, the National Science Foundation, the National Institutes of Mental Health, the Office of Education, and (once) the White House.

3. Pursuit of a variety of intellectual problems and practical topics to which anthropological knowledge could be applied, such as (*a*) the (wartime/peacetime) draft (36); (*b*) community development and race relations in Chicago, Mexico and Guatemala, and Bangladesh; (*c*) expansion of the possibilities open to North American Indians through education and the modification of non-Indian institutions and attitudes; (*d*) worldwide population problems; (*e*) concepts of mankind-as-a-whole and means of introducing a worldwide second language; (*f*) ways of living in space colonies, involving futuristic experiments with internationally mixed communities in isolated areas of the world connected only electronically to sister-communities and to their home nations; and (*g*) the idea (38) of substituting for powerful nation-states powerless "localities" of 250,000 people under a democratic constitution ensuring open information and movement but not the possibility of forming alliances (38).

These extracurricular activities represent varying amounts of time and energy over the years. As examples I choose only one from each of two sets: editorial and publication matters and concerns of American Indians.

My editorial activities from the beginning have been associated with organization and community development. In Milwaukee as I passed my twelfth birthday there was a Newsboys' Republic, modeled after the United States government, with which I became associated, winding up as editor of the citywide *Newsboys' World*. The *World* had a circulation of 5,000, and I was on a par with high-school editors at their state convention in my junior year. The innovation I remember was a front cover with a colored cartoon. As an undergraduate in the Milwaukee State Teachers' College, one semester I organized a forum on comparative religions and with two friends published a *Forum Free Press* to carry on the discussions. The next year, at the University of Wisconsin, we founded a Liberal Club and published the *Student Independent*, for sale on campus corners in competition with the staid *Daily Cardinal*. Meanwhile, I was active in the new Hillel Foundation, editing its *Bulletin* and serving as president.

As a graduate student at Chicago, and while doing fieldwork in Guatemala, I dropped everything but anthropological research. Not until the early 1950s would I again be editing a journal.

TAX

My first contact with book publishers was in 1942–1943 with the then new Fondo de Cultura Económica in Mexico, helping them to get started in anthropology with translations of Linton's *The Study of Man* and Redfield's *The Folk Culture of Yucatan*. Then in 1944, after conversation with Alfonso Caso on how the voluminous data of ethnographic fieldwork might be published, I originated (with the University of Chicago Library) the microfilm series that, after 395 numbers, is still being edited by N. A. McQuown. In 1948, when we needed out-of-print books for students, I negotiated with Jeremiah Kaplan's fledgling Free Press an agreement to reprint books such as Radcliffe-Brown's *The Andaman Islanders* with a guarantee by our bookstore to keep them on order until the initial publication costs were covered. This taught me something about the economics of books. Then it was easy, after the Viking Fund seminar in 1949 that produced *Heritage of Conquest*, to see that book subsidies should be used to buy at wholesale—and for good uses—the books produced. At the same time, when in 1949 I was asked to edit the proceedings of the 29th International Congress of Americanists and told that there was insufficient money to pay for the paperback books promised to members, I learned that one can make good books from selected congress papers and produce hardbound copies for profits that enable a publisher to provide the customary free copies. I do not know if this innovation suggested that I be appointed to the university's Board of Publications (which after a year or two I also chaired), but of course that brought me close to the University of Chicago Press, from which I continued to learn.

When in 1952–1953 the Wenner-Gren Foundation needed to publish quickly the two large volumes from its "Anthropology Today" symposium, I knew how to be helpful to both sides. By 1960, when appointed editor of the Viking Fund Publications in Anthropology, I was prepared to arrange with publishers for a commercial hardcover edition that would pay for the free paperbacks needed for distribution by the Foundation, and we published 21 volumes in both forms. By happenstance the idea could be extended in 1962 to a worldwide audience. The Voice of America asked me to develop a series of lectures taped by 20 anthropologists for an international audience and also made available in pamphlet form. We were permitted to publish the lectures, and Aldine made of them a successful college text (35). Meanwhile, in at least Italy, India, and Colombia, the pamphlets were translated and published as books locally.

With the postwar inflation of the early 1950s, the American Anthropological Association found itself with too little money to publish more than a fraction of what the growing profession was writing, and the thinness of the quarterly journal and the reduced number of issues in its series of memoirs became a serious problem for members, whose numbers declined. I was asked to assume responsibility, with three or four associates; I accepted the challenge and publicly promised (27) to double in four years the capacity of the journal and the number of memoirs. I did so in three years and turned over a thriving journal to a new editor. Doubtless at least partly because of this success I was elected president of the Association two years later. This certainly also led directly to my involvement in the founding of *Current Anthropology*.

The Wenner-Gren Foundation had in 1955 launched a biennial yearbook, but a year later Paul Fejos, the president, approached me to say that the burden on the Foundation had proven too great and the yearbook would be continued only if I would carry it forward outside. Eventually I agreed, provided that I could first discover what the world's anthropologists would find most useful. After a series of conferences in the United States and Western Europe the decision was made to have an open-ended international journal reflecting the changing discipline. Next, through consultation with colleagues in all parts of the world, the form and content of the journal were determined, and in the autumn of 1959 a "pre-issue" was presented that attracted subscriptions from some 3,000 scholars. The first issue was dated January 1960. Editorial and policy decisions were to be made by all "Associates" through "Reply Letters" accompanying each issue. *Current Anthropology* thus became equally a journal and a community, now 23 years old and in its third (editorial) generation.

Out of *Current Anthropology* in 1971 came LARG, the Library-Anthropology Resource Group. To help resolve problems of keeping up with the ever-increasing quantity of reference materials, I asked advice of Jan Wepsich, the Slavic bibliographer of the University of Chicago Library. We decided to call a meeting of anthropologists and librarians from the several universities in the Chicago area to discuss the subject, and this gathering produced the idea of developing reference materials cooperatively. Since then a changing group of us has met monthly and produced three books (6, 9, 13), with a fourth (a biographical dictionary of anthropologists by C. Winters) in preparation. We have worked happily together without funds except as royalties now repay some out-of-pocket costs.

The Ninth International Congress of Anthropological and Ethnological Sciences, in Chicago in 1973, was built on all of these experiences. Responsibility for this congress came to me at the preceding congress in Tokyo in 1968, and as plans for it developed it became apparent that we would need a great deal of money to support them. In some countries governments pay the costs of such congresses, and indeed eventually we received some helpful US government grants for special projects. But we were proposing the largest use of simultaneous translation ever attempted (five languages for six days), the production and mailing of thousands of prepared papers all over the world for advance reading as background for the translated discussions, and, if possible, payment of transportation and living costs in Chicago for scholars unable otherwise to come from far places.

TAX

Therefore the congress was designed to make good books, and a publisher advanced $200,000 in potential royalties to make this possible. We produced the 91-volume World Anthropology series, which is expected at least to repay the advance. Sufficient funds were provided for a congress of close to 4,000 participants from every part of the world. The only disappointment was that the volume describing the congress (42), which was to be sent free to all members, was denied publication by the Berlin firm that in the meantime had purchased the original publisher.

Native Americans

My first contact with American Indians was with the Mescalero and Chiricahua Apache people in New Mexico, where in the summer of 1931, as earlier noted, I was part of Ruth Benedict's Laboratory of Anthropology field party. It was a pure-science learning experience, though Blanchard (1) cites my correspondence to show that I encouraged at least one Indian to think of independent reservation development. Again, during three summers (1932 and 1934 with the Mesquakies of Iowa and 1933 visiting mainly Ojibwa communities and then settling for some weeks with the Menominee), I studied kinship systems and remained relatively "pure." In Guatemala and Mexico from 1934 to 1944 I worked on social and community problems mainly to understand the social structure and the economy but also (particularly in Chiapas) to advise colleagues working in their important development programs. It was not until the summer of 1948, however, that six graduate students getting their first fieldwork experiences under my direction—on the Mesquakie settlement near Tama, Iowa, where I had lived 14 years before—urged me to permit them to work on problems of the people as well as on the anthropological problems they had brought with them. Hitler, the War, and the Bomb had all played a part in turning me back to my earlier interest in social action and in the philosophical issues involved in the use of anthropological theory to benefit the people among whom we worked. Thus began what we have since called "action anthropology" (11, 37, 45).

The six students who were doing research on the Mesquakies and, with my permission, also trying to help their newfound friends were warned that they would be facing problems of values. Soon they found themselves divided over what they would like to see as the Indians' future, some favoring an inevitable (but, ideally, happy) assimilation and others hoping for preservation of at least some traditional values. When they returned to the university in the autumn that question remained until, in the course of discussion of substantive issues, it was one evening revealed to all that this question was not ours but the Mesquakies'—that we might only help them to see alternatives. This was perhaps the real beginning of action anthropology; the rest has been legitimizing the relationship of the concepts behind the two words. There followed excellent students and continuing seminars, with participation from other disciplines, and in 1955 a foundation grant made it possible for Robert Rietz (of the "first six") to return to the Fox for four years to help develop several programs (4; C. Tjerendsen, privately published, 1980).

In the autumn of 1949 I was asked to meet with John Provinse, then Assistant Commissioner of Indian Affairs, and Galen Weaver, director of race relations of the National Council of Churches, who were worried about the fate of the Affiliated Tribes on the Fort Berthold Reservation, soon to be flooded by the opening of the new Garrison Dam. Did congressional provisions for the Indians appear to be satisfactory? I agreed to go to Fort Berthold and took with me one of our students, Robert Merrill, a theoretical chemist turned anthropologist interested in economic development. After a few days there I left Merrill to make a thorough study; I had already seen the complex divisions within the reservation community. Merrill's full report was excellent, resulting in the appointment of Robert Rietz to a position on the ground. Rietz's work there for four years was remarkable in that what he did—in most difficult circumstances—was much appreciated by the Indian community, the Bureau of Indian Affairs, and the missionaries; indeed, his greatest problem turned out to be avoiding moving "up" to higher positions in the B.I.A.!

The next summer I accepted an invitation to teach at the University of California at Berkeley, and on our return through Navajo and Hopi country I came to a realization that changed my thinking about American Indians' future. One of our students (John Connelly), a schoolteacher among the Hopi, introduced us to the people of the Mesas. To me the Hopi had always symbolized traditionalism among isolated Indians; we "knew" them well from the work of Fred and Dorothy Eggan, and it had always seemed to me that nobody would ever say, as they said of the Mesquakies, that *these* people were about to disappear into the melting pot. Now, seeing them for the first time in their isolated habitat, I suddenly recognized that they were no different from the Mesquakies or from any of the eastern Indians with whom I had visited. People in Iowa told the Mesquakies that they were temporary; nobody seemed to be saying that to the Hopi. But the Hopi seemed to me no more permanently Indian than the Mesquakies—or the Penobscot or the Iroquois. And I had to conclude that we had no right to count out any of the peoples that are with us today—that existing notions about acculturation and assimilation were not only harmful but quite possibly mistaken. During the next year examination of the data and discussion with our students convinced me of the validity of this simple shift in perception, and three of us prepared papers, given in March of 1952 at the annual meeting of the Central States Anthropological Society, arguing as clearly as we could that the assimilation of our Indian communities was not inevitable. So fixed were Americans on the notion of the melting pot, however, that the first 45 minutes of the discussion that followed was at cross-purposes. Our colleagues took us to be saying that acculturation was proceeding more slowly than we had assumed, and it took that long before they realized that we were saying, rather, that it might *never* occur. The misunderstanding by sophisticated anthropologists was itself a significant cultural datum.

A month or two after this meeting I attended a conference in Washington sponsored by the Association on American Indian Affairs. The topic was assimilation, and the conference was intended to persuade Congress and the B.I.A. that they should slow down the increasingly destructive withdrawal of federal services. For the occasion tribal leaders were brought from Western reservations. The meeting was introduced with the argument that the Indian people would of course eventually assimilate but meanwhile needed help. Then, one after another were introduced the Indian leaders, and each in turn repeated this theme in his own way. When they had finished, I rose and said simply that there was no scientific evidence that the Indian tribes would indeed assimilate and to say that they would expressed the damaging value judgment that those who did so could be called progressive and the others backward. Then each of the Indians rose and recanted vehemently, and a motion was made to change the conference topic from "assimilation" to "integration." This happened so quickly and definitively that I had to conclude that I had been the boy who said that the Emperor wore no clothes.

The melting-pot theory with respect to European immigrants has been weakened since the 1930s by a sentimental third-generation "return"; but Indian *peoples,* in contrast to immigrants from Europe and some individual Indians, have rarely doubted their survival. Indeed, Indians sometimes express a belief that it is only their continuing presence that has protected all of us from destruction. It puzzles me that we anthropologists—even though most of us are immigrants who have ourselves melted into the pot—could have fallen in any degree into what has been an error hurtful to our Indian friends. It is true that acculturation was rapidly changing at least the externals of our ethnographic data-base, and museum-affiliated anthropologists may have thought more than we do today of anthropology as the study of man's *works.* They themselves did not confuse artifacts with people; but the emphasis on the one against the anonymity of the other may well have set in motion the idea that the people were disappearing. If so, perhaps it is no accident that a correction had to begin with anthropologists who study the interrelations of people, requiring understanding of their ideas and emotions. Such a correction would become most necessary when anthropologists came to try not only to understand but to help to improve.

I have seen at least one Indian carrying a copy of the United States Constitution with every mention of Indians marked heavily for reference. Indians understand that they had some protection against their neighbors when the Crown was across the ocean that was lost with the Revolution. The Constitution, enforced in federal courts, became their only protection, providing as it did for the making of treaties that remain today the symbol of their special status. The struggle has been to protect these rights against a Bureau of Indian Affairs that might still carry traditions from its time in the War Department and a powerful Congress whose constituents often compete for resources with tribes and communities. There have rarely if ever been "good times" for Indians in relation to government, but the short-lived F.D.R.-Ickes-Collier "New Deal for Indians" had a more liberal spirit than existed before or after. It added to reservation resources in some cases and improved education and health by taking more account of local preferences; but tribal constitutions embodying majority rule rather than consensus and vesting Washington with veto powers vitiated the effort and in many cases induced what among the Mesquakies we saw as "structural paralysis." Congress, meanwhile, had its collective eye on the contrary plan to "solve the Indian problem" by forcing their assimilation into the general society. The way for this had been prepared back in 1946 by the establishment of the Indian Claims Commission, which would "once-and-for-all" compensate with cash for all the chicanery by which whites had obtained tribal lands. The stage was set for the withdrawal of services to Indian tribes and the termination of their special relations to the federal government as expressed in 1953 in House Concurrent Resolution 108 and the program of relocation of Indians to cities, followed in 1954 by Public Law 280, which transferred much jurisdiction over Indians to the states.

I had at most a vague awareness of all this when at the 1952 conference I spoke against the notion that assimilation was inevitable. (The Association had doubtless organized the meeting to counter the congressional pressures for termination of which I was ignorant; evidently an anthropologist at the grass-roots level may never learn what's up!) In any case, I soon became a vocal opponent of "termination," especially since the close-by Menominee were among the first tribes to be cast off (31). In October of 1957 I spoke at the annual convention of the National Congress of American Indians, in Claremore, Oklahoma, putting together what I thought was nationally appropriate policy concerning Indians, and it was so well received that I felt I had finally learned what tribal leaders themselves might want. Meanwhile, since 1953 Indians had been pouring into Chicago under the federal relocation program, and the American Friends Service Committee had been helping local Indians to establish a center to receive the new arrivals. Hundreds—eventually thousands—of individuals and families came from reservations, far and near, on one-way railroad tickets. Most had been recruited because they were having difficulties on the reservation, so it is not surprising that they had difficulties in the city. The federal relocation office in Chicago met their trains, provided funds for some weeks, and found them housing and jobs. Happily settled or not, the Indians were then on their own. Not surprisingly, most of them were unhappy with their housing or jobs or both and lonely in the big city. With new ones arriving, the relocation office could do nothing to help, and soon there were problems for the city's welfare agencies, which, because of the government's rules regarding privacy, could not learn even when or where new people were arriving. Nor was the Indian Center apprised of any facts, but it became a repository for the problems. Many—eventually most—of the relocatees returned to their reservations, but the Center persisted, and my students and I came to know its people well.

The Indian Center depended for funds mostly on grants and employed a variety of tradional Indian techniques that did not compromise Indian values. But it was also an institution without Indian precedent—involving people of very different cultures and even traditions of inimical relations in a strange environment. It was a new challenge for those selected for leadership, and the problems seemed insurmountable until in 1954 Robert Rietz tapered off his work with the Fox Project and became the nondirective director. Rietz helped the Indians to shape, finance, and manage the Center in as Indian a way as was possible in our complex society. In Indian fashion, too, when in 1971 Rietz died and a dissident faction took control, the ousted people quickly established a new and successful organization, the Native American Committee. Its leaders eventually also founded Native American Education Services, Inc., which became NAES College, accredited and operating in Chicago and on several reservations.

In 1956, meanwhile, Galen Weaver, who had brought us to the Fort Berthold problems, called our attention to the plight of young reservation Indians in college, who were not only lonely but embarrassed because they could not answer questions about Indian affairs. Fred Gearing and I suggested a summer-session workshop, and Colorado College agreed to house it (with academic credit provided by the University of Chicago). That summer 25 Indian students from as many tribes began—as it turned out—to make history. They not only produced a book but also discovered together their own interests and problems. The first week of the 1961 workshop was held at the American Indian Chicago Conference, where alumni of previous workshops also organized the American Indian Youth Council, important in the turbulent 1960s and still active. Eventually the workshop moved to the University of Colorado under D'Arcy McNickle's American Indian Development (on whose board I sat), and alumni themselves later developed similar workshops elsewhere.

A detailed map of the 1950 locations of Indian communities in the United States and Canada (43), distributed at the Chicago conference, showed clearly that, except for those removed to Oklahoma, almost all Indian communities still lived on parts of their aboriginal lands. It countered the myth of Indian disappearance and now provides a benchmark for understanding the movement since then of Indians to urban areas. The conference (8), with its more than 800 leaders in national Indian affairs, at least three-quarters of them Indians from all regions, religions, historical perspectives, and political persuasions, was a national media event that attracted international attention, exciting and genuine. It also appears to have been a turning point in modern American Indian history—the beginning of the end, perhaps, of the myth of the disappearance of Indians. It was probably the first time that Indians had ever been asked to express in public their collective hopes for their future. I am proud to have had a part in this, and most proud of the way we anthropologists and other friends of Indians permitted them to do what really became *their* own thing. To me it was the ultimate success—and on a large scale, too—of the philosophy and methods of action anthropology.

When we sent the *Declaration of Indian Purpose* produced by the conference to our Indian mailing list of some 4,000 names, many Indians who had attended were disappointed not to have received it; apparently not all those who came to Chicago had been on that list. On the evidence offered by this response that on the reservations mail—at least ours!—was appreciated, the Carnegie Corporation gave us funds to experiment with encouraging literacy through such mailings. For five years we then published *Indian Voices,* written mostly by Indian readers. We also had an exceedingly fruitful field project in eastern Oklahoma, where the Cherokee, who in the late 19th century had had excellent academies in which even Greek and Latin were taught, now found themselves among the least literate of people (44, 46). This was our most difficult action program but perhaps the most rewarding, as we helped these Cherokee regain some of their lost independence.

In the 1970s, when the United States Senate investigated American Indian education, Indians gave testimony in Washington, and it appears, as I read the record, that only two of them "spoke up." One was a young Mesquakie and the other a Cherokee from eastern Oklahoma. How this happened I do not know, but it seems more than a coincidence that, out of the scores of Indian communities with problems and grievances, the only Indians to question the status quo should be from two tribes that had felt our influence. Also: in September of 1968 an attorney hired by the Mesquakies telephoned to ask me to testify in federal court in Cedar Rapids, Iowa, in a suit brought against the Bureau of Indian Affairs for closing the Mesquakie day school and even selling its furniture. He said the Indians wanted "only one witness: Sol Tax." Years had passed since my last contact, but I left a conference in Princeton to fly to Cedar Rapids. In the courthouse I found the entire tribal council waiting while a settlement wholly favoring the Indians was being worked out in the judge's chambers. There was no trial, then, and we had time to reminisce;

they said that this victory had been made possible by the unity that we had always urged upon them.

There is another story that is perhaps worth telling. In *Science* (November 30, 1951) appeared a letter signed by five anthropologists, including me, who had studied the peyote religions of different Indian tribes. The letter strongly protested the campaign of "propagandists [who] argued that Peyotists are simply addicted to a narcotic . . . which they use orgiastically" and showed that the "Native American Church" was "a legitimate religious organization [using] peyote . . . sacramentally." In the summer of 1954 I was invited to attend the national convention of the Native American Church, for which the Mesquakie church was to be host. With the issue of legality in the public eye, it occurred to me that a documentary film of the entire convention might someday prove to the courts that the church deserved protection. The film, I thought, would have to include both the mundane business sessions and the drumming and praying around the sacred fire in the ceremonial tepee over the whole of Saturday night. Otis Imboden, a young filmmaker on campus, eagerly accepted my invitation to come along, and the filmmakers at the University of Iowa agreed to bring all the necessary equipment to Tama as soon as I telephoned them that the Indians had consented. The rest of the story, as I told it to colleagues 10 months later (34), follows:

All Thursday afternoon and Friday morning we were part of the business of the convention. . . . On Thursday I explained at length and carefully the possible importance to them of the film, and the unusual good fortune that made it possible at no cost. There were questions and discussions, and a night to sleep on it. We were optimistic. The next morning the discussion resumed. Again I made explanations and answered questions. I promised that they would help edit the film and would have to approve it before its use. I said it was up to them. Then followed speech after speech, some for me and some against me. It became clear that everybody thoroughly understood that the film, perhaps shown as evidence in court, could someday establish them as a legitimate religion and peyote as the sacrament they felt it to be; otherwise the church seemed to them in danger. The rub came in the prospect of filming their sacred ceremony. The ritual itself would be inevitably disturbed by technical problems, but perhaps more important they could not picture themselves engaged in the very personal matter of prayer in front of a camera. As one after another expressed their views, pro and con, the tension heightened. To defile a single ritual to save the church became the stated issue, and none of them tried to avoid it. Not a person argued that perhaps the church was not in as great danger as they thought; there was no suggestion of distrust of me; they seemed to accept the dilemma as posed, as though they were acting out a Greek tragedy. I sat in front with the president and his wife, facing the assembly. Fascinated, I listened to the speeches, and gradually the realization came that they were choosing their integrity over their existence; that although these were the more politically oriented members of the church, they could not sacrifice a longed-for and sacred night of prayer. When everybody else had spoken, the president spoke, and said if the others wished to have the movie made he had no objections; but he begged then to be excused from the ceremony. Of course this ended the movie, and the sense of the meeting was clear. It was over, and then the realization seemed to come over the Indians that I must be hurt; for all my good and unselfish intentions, and high hopes, and hard work—my reward, and Otis's, was a clear rebuff. They had suffered through their dilemma, and had made the painful choice that should have relieved their tension. But they realized now that their peace with themselves had been bought at our expense, and they began speeches painfully to make amends.

They were wrong, of course. As their decision was being made I understood that what I had proposed was akin to asking a man to deliver his wife to a lecherous creditor to save the family from ruin. Now, therefore, I arose to speak, and could with genuine sincerity apologize for having brought so painful an issue to them. I had meant to be a friend, but had hurt them. I agreed with their decision. I would be a poor friend indeed if I resented their deciding an issue for their own good.

Relief was great; the euphoria was instantly restored; and it was evident then and in the days that followed that they were more genuinely grateful to me than any Indians had ever been to me for any material or moral help, and felt closer rapport with me.

What I learned that day about the Peyotists' view of their ceremony, about the nature of group discussion in an Indian assembly faced with a real issue, and about the sensitivity of Indians to a situation of aggression against an individual could come in other ways. But never I believe so convincingly. In anthropology we can't prove interpretations of the behavior we see; but in this incident I was so overwhelmingly convinced—and so by the way was my young friend Otis—as to remove the doubt to quite another level.

In concluding this account of my relations to North American Indians, I must say that I am always embarrassed to be thought knowledgeable of their ethnology and history. True, for four or five years as a student, I absorbed a great deal of specific knowledge about many groups on some topics and always tried to "keep up," but since I did not teach courses in the subject and did so many other things, this was a losing battle. What I *do* have is a sense of what Indian people feel about themselves and about us. I find that I cannot treat them, as once I could, as subjects of study. Indians and I are friends, respecting each other as equals who are different.

Many Indians, then, have given me friendship that I want to return. Is that enough to explain these nearly 60 years of trying to be helpful? There is satisfaction and pleasure in any chance to help a friend, and from childhood I sought such opportunities. The associated emotion I knew was anger at injustice—the wrong schoolmate punished, the child beaten by a bully. But I grew up to see that, bad as were the individual acts of injustice that bred in me flashes of anger and dismay, it was *social* injustice that deeply pained the heart and bred lasting resentment. From my first summer at Mescalero, with people giving us firsthand accounts of remembered lives of independence, I could feel the rancor of their loss. And as I came to understand that it was so with all the Native American peoples that I learned to name, and to see the administrative follies that constrained them, I could not but share their dismay. What little I might do to understand those bonds and perhaps to loosen them I had to do. But I soon learned that ultimate success or failure on my part for them, or theirs for themselves, was no measure of the pleasure of the effort, and so I have kept on supporting their long struggle.

Explanations

Since it is the business of an anthropologist to add to and interpret the world's store of anthropological knowledge and to teach it to others, I need now to explain how I permitted myself to engage in the apparently extraneous activities of which those just described are only a sample. Clearly, the first answer is that I have rarely done any of these things alone. But that is no answer; either enlisting others in a cooperative venture or managing helpers takes time and energy. I presume that the list is long because I have always found it difficult to resist what I hope will be opportunities. Since it appears that I have been a willing victim, the question arises how, with a full load of teaching, administration, and research, I managed so many additional things. Two personal characteristics may help explain this.

The first is an almost lifelong reluctance to do only one thing at a time. As a child given routine tasks such as stuffing envelopes for an advertising campaign for my father's business, I took to choosing a subject about which to daydream while my hands did the work. Later, as a newspaper carrier on a regular route I easily delivered my papers with my mind far away. When, still later, I was a telegraph-company messenger I often wondered how I managed to make six or eight or a dozen miscellaneous pickups and deliveries on the busy streets of downtown Milwaukee and return to the office with all the proper receipts and messages in my cap and without any remembrance of the journey. Even today I feel let down when, unable to find at least one other thing, I have to do the single one that is necessary. Foolish as this habit is, it may account for the relative ease with which I have combined (for example) publication with conferences—even congresses. A related tendency is willingness to let things happen or grow—a hospitality to other people's ideas, desires, and needs. Another obvious tendency—shared with most people—is a desire to improve institutions, and in the course of time I have put whatever inventive skills I have to work on institutions with which I have had dealings.

The second characteristic I should mention is a deliberate indecisiveness that comes in part, I suppose, from my generally relativistic philosophy but more specifically from my immersion in action anthropology. At its best, it is "Look before you leap"—learning as much as possible before acting, playing a waiting game when the choices are poor and the consequences of a mistake perhaps irreversible. This all seems wise. But it can also be interpreted as reversing the proverbial expression to read "Never do today what can safely be put off until tomorrow," with the word "safely" always to be questioned. For better or worse, the practice has become habitual with me. It grew out of trying to satisfy the needs and wants of persons who were not sure what they wanted; "trial and error" takes times and patience. But I think that I began to use it also in my personal life, and, noting that I myself was often not sure what I wanted, I became more confident in delaying actions on behalf of others. When working with people, it became my peculiar virtue to be able to accept and quickly internalize and recombine the ideas of others. I recall an extreme case in which, after a lecture in New York to public health workers who had to take into account cultural differences, a number of nondisputatious comments began by referring to something I had said. I was surprised by several that I did not recognize as mine and wished I *had* said! The experience caused me to wonder if indeed I had misunderstood myself. I was reminded of the story of the Chinese philosopher who dreamt that he was a butterfly and ever after wondered if he might not be a butterfly dreaming he was a man.

I have also been influenced by the philosophy and method implied in the way North American Indian communities operate. The distinction between war-party (or hunting-party) behavior and the way decisions are taken and policy made in community affairs is striking. When immediate action appears necessary, followers by definition are committed to a leader they have chosen. I have rarely, if ever, been a leader in that sense. Political leaders of an Indian community, often called "peace chiefs," are successful only if they satisfy the perceived needs and wishes of the people they serve. It is that sort of model that includes anthropologists interested in helping communities long suspicious of officious outsiders who threatened their valued ways and their independence. If I did not actually learn from Indians how to operate helpfully without exercising undue personal influence, they strongly reinforced propensities already there. "Ideas" in profusion come to me, and I express them quickly and enthusiastically and may even appear to be promoting them; but I am saved by my correspondingly quick willingness to accept their rejection happily. Indeed, my "ideas" seem only to suggest others better suited to the situation. Although prone to seek answers to problems posed, I have never been a "planner" in the sense of one who sets goals and then determines means to achieve them. We explicitly rejected this means-end model of thought and action in the Fox Project (3, 10), and though I had doubtless assumed it as part of our common definition of intelligent behavior, I doubt if it had commonly governed my own thought and action even before.

It occurs to me that although I have lived through history's greatest changes in technological knowledge, they have not destroyed what I take to be a conviction that continuities are at least equally important. I have mentioned continuities in anthropology and of beliefs and motives in my own life. I recall also that I continually emphasized the continuities in the Guatemalan Indian villages and, later, stressed the persistence of North American Indian peoples and cultures. So I come back to fieldwork and to the stuff of anthropology, where I would like to leave the reader. But I cannot without first anticipating what colleagues might wonder: Do I have no regrets? Perhaps I am expected to ask at least about my priorities, but that sort of question implies that I would on principle violate another that I hold dear. I must act at any time according to the situation as I understand it; so clearly I am not the kind of person to set up priorities in advance. But perhaps the question is whether I regret having spent so much time in some ways that I had no time for other things that I might have done. Yes, of course, set aside was my schoolboy ideal of pure science as a means for solving the world's major problems. Even in my first fieldwork in Guatemala I recognized that devoting time to applied science could slow the building of the structure of knowledge that might make application feasible (21), but eventually—as I dealt with human problems themselves—I found myself incapable of giving up present smaller goods for possibly larger ones beyond some horizon. Ironically, however, I fear that my two major contributions to pure science have not served anthropology as theoretically they should have, and I cannot have too many regrets at having sacrificed to "good works" other such abortive successes.

In Guatemala, in happy association with Redfield, I came to understand a kind of society that I described in two complementary books (28, 40) and an article (20). Doubtless it is my fault, but it turns out also that in the end anthropology lost its curiosity about the kinds of types this work explored. *Penny Capitalism* became very influential in economics as presenting what was the first quantitative study of an economy like that conceived but never seen in Adam Smith's Europe, but without capital firms or technology that would permit economic growth (14). I wrote in the preface, "My work falls short of an ideal because I had no model. Here is a pattern from which others may depart." But I should not have been surprised when my colleague the Nobel laureate T. W. Schultz asked me in recent years why it was that other anthropologists had not made similar studies elsewhere. Presumably the answer is that we all like to pursue new problems and methods. But there is also the question whether I would have served science better had I myself pursued the method elsewhere instead of spending my energy on the activities I have described.

The other case is perhaps more interesting because it concerns my first professional work, on the most traditional of anthropological subjects: kinship. My PhD thesis (16) included what I thought was a major breakthrough in studies of kinship terminology. Having noted the sporadic distributions of Omaha and Crow systems (17) I required a theory more widely applicable than any then extant theory of kinship. In working it out I hit upon the sort of structural equivalences (which I called "rules") that Floyd Lounsbury would rediscover as "transformational analysis" some 30 years later. I saw utterly no reference to this until Alan Coult (2) discussed my "rules" in a paper that began "It is commonplace in the history of ideas that certain concepts are too advanced for their time and thereby fail to be accepted, although at a later time their rediscovery is accepted with great enthusiasm." By then I was far too busy with other matters to ask his views on my general theory (and he died soon after), and when, 15 years later, Scheffler (12) made use of the "rules" I had developed, again there was no mention of the theory. Indeed, I would never have heard a word about it had not Greenberg in 1980 called my 1937 article "perhaps the wisest article on kinship ever written" (4).

It happens that in the examination of my PhD thesis back in 1934, the psychologically oriented political scientist Harold Lasswell shocked the anthropologists present by putting to me the ad hominem question "How did you become so antitheoretical?" Since I considered myself a theorist objecting to some of the bad theories that preceded my own, I found this a difficult question to answer, but I never forgot it. Only now do I realize what he might have meant. My explanation of how kinship systems operate in small societies requires intimate knowledge of the intrafamily relations within which the "rules" that I had noticed and systematized were observed; and my explanation involved the ways in which in such societies solutions to problems by individuals can become institutionalized changes. It occurs to me now that if, when Charles Darwin returned to England after his voyage on the *Beagle*, there had not been a belief in Creation, he would have tried to explain the variations in plants and animals he noted and described the processes of adaptation to different environments, showing how individuals with characteristics valuable in changing environments were likely to survive and reproduce more than others. Without the alternative of the biblical Creation, he might very well never have needed the term "evolution" to explain what he saw. Would one then have described his explanation as "a theory"? By analogy, to work through my explanations of changing kinship systems would have required detailed study of variations in kinship behavior in the variety of circumstances to which they could be related in living societies all over the world—a formidable task indeed.

I must, then, forgive us all for bypassing the problem, and I cannot feel myself seriously at fault for choosing problems that people themselves— rather than theoretical anthropologists—feel. It is with the most significant of these human problems that I conclude these comments: the general malaise of most of our people. We feel responsibility for both the increasing economic interdependence of all the nations of the world and the danger to the entire species of any major war with nuclear weapons. It is only the most recent eight millenia of evolution that have brought us to this pass. Every anthropologist knows that the millions still living in small communities—some even in hunting-gathering kinship societies—show that basically humans want more than anything to control their lives and destinies. There are signs that many even in complex urban societies would choose to use our new technologies to help form smaller communities that together we could control,

TAX

thus beginning a new evolution to replace the course subverted in the Neolithic when villages became subject to nation-states. There is a peaceful and legal model for converting the localities in which we live and work into voluntary communities bound together electronically (38). It happens that in the United States the model size of localities is the congressional district. Thus localities do not cross state lines, and statistics are provided in the census. Each has its representatives in the state legislature as well as in Congress. Any group of residents and organizations can form an educational-philanthropic community open to all and seek support from the existing bodies that collect their taxes. Beginning perhaps in the states of the Northeast, Midwest, and Far West, scores could organize within weeks or months and provide examples for others. The model would spread most quickly in Europe, perhaps, but some such voluntary communities would certainly appear in Asia and Africa as well, everywhere adapted to local conditions. A fresh start now—with our new technologies and our need for community as well as survival—might well permit an evolution fast enough to dissolve the warring nation-states before they dissolve us all.

Literature Cited

1. Blanchard, D. 1979. Beyond empathy: the emergence of an action anthropology in the life and career of Sol Tax. In *Currents in Anthropology,* ed. R. Hinshaw, pp. 419–43. The Hague: Mouton
2. Coult, A. 1967. Lineage solidarity, transformational analysis, and the meaning of kinship terminologies. *Man* 2:26–47
3. Diesing, P. 1960. A method of social problem solving. See Ref. 4, pp. 182–97
4. Gearing, F., Netting, R. M., Peattie, L., eds. 1960. *Documentary History of the Fox Project.* Microfilm Coll. Manuscripts on Cult. Anthropol. 394. Enl. ed.
5. Greenberg, J., ed. 1980. *On Linguistic Anthropology: Essays in Honor of Harry Hoijer.* Malibu: Undena
6. Grollig, F. A., Tax, S., eds. 1982. *Serial Publications in Anthropology.* South Salem, NY: Redgrave. 2nd ed.
7. Kroeber, A. L., ed. 1953. *Anthropology Today.* Chicago: Univ. Chicago Press
8. Lurie, N. O. 1961. The voice of the American Indian: report on the American Indian Chicago Conference. *Curr. Anthropol.* 11:478–500
9. Mann, T. 1988. *Biographical Directory of Anthropologists.* New York: Garland. In press
10. Peattie, L. 1960. The failure of the means-ends scheme in action anthropology. See Ref. 4, pp. 300–4
11. Rubinstein, R. A. 1986. Reflections on action anthropology: some developmental dynamics of an anthropological tradition. *Hum. Org.* 45:270–79
12. Scheffler, H. W. 1982. Theory and method in social anthropology: on the structure of systems of kin classification. *Am. Ethnol.* 9:167–84
13. Smith, M., Damien, Y. 1981. *Anthropological Bibliographies: A Selected Guide.* South Salem, NY: Redgrave
14. Solow, R. M. 1970. *Growth Theory: An Exposition.* New York: Oxford Univ. Press
15. Tarabulski, M. 1986. *Reliving the Past: Alonzo Pond and the 1930 Logan African Expedition.* Videotape, Centre Prod., Inc., 1800 30th St., Suite 207, Boulder, CO 80301
16. Tax, S. 1935. *Primitive Social Organization with Some Description of the Social Organization of the Fox Indians.* Microfilm Coll. Manuscripts on Cult. Anthropol. 393
17. Tax, S. 1937. Some problems of social organization. In *Social Anthropology of North American Tribes,* ed. F. Eggan, pp. 3–35. Chicago: Univ. Chicago Press
18. Tax, S. 1937. The social organization of the Fox Indians. In *Social Anthropology of North American Tribes,* ed. F. Eggan, pp. 243–85. Chicago: Univ. Chicago Press
19. Tax. S. 1937. The municipios of the

midwestern highlands of Guatemala. *Am. Anthropol.* 39:423–44

20. Tax, S. 1941. World view and social relations in Indian Guatemala. *Am. Anthropol.* 39:423–42

21. Tax, S. 1942. Ethnic relations in Guatemala. *América Indígena* 2:43–48

22. Tax, S. 1949. Folk tales in Chichicastenango: an unsolved puzzle. *J. Am. Folklore* 62:125–35

23. Tax, S., ed. 1951. *The Civilizations of Ancient America.* Chicago: Univ. Chicago Press

24. Tax, S., ed. 1952. *Acculturation in the Americas.* Chicago: Univ. Chicago Press

25. Tax, S., ed. 1952. *Indian Tribes of Aboriginal America.* Chicago: Univ. Chicago Press

26. Tax, S., ed. 1952. *Heritage of Conquest: The Ethnology of Middle America.* Glencoe, Ill: Free Press

27. Tax, S. 1953. Editorial. *Am. Anthropol.* 55:1–3

28. Tax, S. 1953. *Penny Capitalism: A Guatemalan Indian Economy.* Washington, DC: Smithsonian Inst. Press

29. Tax, S. 1955. The integration of anthropology. In *Yearbook of Anthropology,* ed. W. L. Thomas, Jr.. New York: Wenner-Gren Found. Anthropol. Res.

30. Tax, S. 1955. From Lafitau to Radcliffe-Brown: a short history of the study of social organization. In *Social Anthropology of North American Tribes,* ed. F. Eggan, pp. 445–81. Chicago: Univ. Chicago Press. Enl. ed.

31. Tax, S. 1957. Termination vs. the needs of a positive program for American Indians. *Congr. Rec.* 103(116), July 3

32. Tax, S. 1958. Changing consumption in Indian Guatemala. In *Consumer Behavior,* ed. L. H. Clark, pp. 227–38. New York: Harper

33. Tax, S. 1959. Residential integration: a Chicago case. *Hum. Org.* 18:22–27

34. Tax S. 1960. Learning through action. See Ref. 4, pp. 304–7

35. Tax, S., ed. 1964. *Horizons of Anthropology.* Chicago: Aldine

36. Tax, S., ed. 1968. *The Draft: A Handbook of Facts and Alternatives.* Chicago: Univ. Chicago Press

37. Tax, S., ed. 1971. *April Is This Afternoon: Redfield-Tax Correspondence.* Microfilm Coll. Manuscripts on Cult. Anthropol. 330

38. Tax, S. 1977. Anthropology for the world of the future. *Hum. Org.* 36:225–34

39. Tax, S. 1979. Autobiography of Santiago Yach. In *Currents in Anthropology,* ed. R. Hinshaw, pp. 1–68. The Hague: Mouton

40. Tax, S. n.d. *Practical Animism.* Microfilm Coll. Manuscripts on Cult. Anthropol. 392

41. Tax, S., et al, eds. 1953. *An Appraisal of Anthropology Today.* Chicago: Univ. Chicago Press

42. Tax, S., Neuberger, G., eds. 1976. *Proceedings of the Ninth International Congress of Anthropological and Ethnological Sciences, Chicago 1973.* Microfilm Coll. Manuscripts on Cult. Anthropol. 395

43. Tax, S., Stanley, S., Thomas, R., MacLachlan, B. 1960. *The North American Indians.* Chicago: Dept. Anthropol.

44. Tax, S., Thomas, R. 1969. Education "for" American Indians: threat or promise? *Florida FL Rep.* 7:1

45. Van Willigen, J. 1986. *Applied Anthropology.* South Hadley, Mass: Bergin and Garvey

46. Wahrhaftig, A. 1968. The tribal Cherokee population of eastern Oklahoma. *Curr. Anthropol.* 9:510–18

ASTRONOMY AND ASTROPHYSICS

V. Ginzburg

Annu. Rev. Astron. Astrophys. 1990. 28: 1–36

NOTES OF AN AMATEUR ASTROPHYSICIST

Vitaly L. Ginzburg

P. N. Lebedev Physical Institute of the Academy of Sciences USSR, Moscow, USSR

KEY WORDS: radio astronomy, synchrotron radio emission, cosmic-ray astrophysics, gamma-ray astronomy.

INTRODUCTION

I was surprised to receive the kind invitation to write the Prefatory Chapter for this volume of the *Annual Review of Astronomy and Astrophysics*, in that I do not consider myself a "real" astronomer. Rather, I am an amateur astrophysicist or, possibly better to say, amateur astronomer. In any case, I'm somewhere on the periphery of the astronomical community. For this and other reasons I did not think that one day I would be asked to write such a paper. On receiving the invitation, however, I immediately decided to make an attempt because I have always liked the idea of publishing in review volumes autobiographical papers or other nonstandard reviews including autobiographical elements.

The content of scientific knowledge does not, of course, depend on who has established some facts or how they have carried out observations and measurements. However, it is human beings, with their tastes, passions, and fates, that produce this knowledge, and it is much more difficult to understand a person than to decipher the spectrum of a star. At the same time, isn't it a natural temptation to desire insight into the inward life of a "comrade-in-battle" whom you meet at conferences and/or on journal pages? This problem is solved to some extent by obituaries and by other posthumous publications, but they can in no way substitute for auto-biographical narration.

Below I shall try to outline the contributions due to my astrophysical attempts [which has been partially done already in (1, 2)]. I shall also dwell

0066–4146/90/0915–0001$02.00

on my biography, which I hope will be of interest to my colleagues in the West and help them better understand the living conditions in my country, especially during those now already remote years when my generation of physicists and astronomers was brought up.

AUTOBIOGRAPHY I

I was born on October 4, 1916, in Moscow, where I have lived all my life (with the exception of two war years that I spent in Kazan'). My father was an engineer engaged in water purification and held several patents. He married for the first time in 1914 at the age of 51. My mother, a physician, was then 28. I was the only child in the family. My mother died of typhoid fever in 1920, and I have almost no memories of her. After she died, her younger sister came to live with us and was as good a mother to me as she could be. It was a very hard time—first World War I and then the Civil War. Moscow became the capital and was in general in a privileged position, but all the same there was little food and diseases raged. My memory is generally poor or, at any rate, has a high threshold. Above this threshold I recall one scene that I witnessed not far from our home in the center of Moscow in about 1920: a cart carrying coffins, a carter dragging himself along, and arms and legs sticking out of the coffins. Another reminiscence, not so ghastly, but still typical, also comes to mind: We managed to buy fresh meat somewhere, but it appeared to be dog meat; and under normal conditions, dog meat was never used as food in Russia. Nevertheless, my family suffered much less from what are called practical difficulties, I think, than did most people in the country at that time—we had a roof above our heads, Moscow was not occupied by troops, and there was no real hunger. What I had in excess was loneliness. It was aggravated by the fact that I did not go to elementary school and went only to the fourth form at age 11. I do not remember why this was so. The schools, like almost everything in the country at that time, were being subjected to all kinds of reorganization, and my parents probably thought it better for me to get my education at home. I believe this was legal at that time. (I don't know of such cases now, except for sick children.) It was undoubtedly a mistake, since when I finally did go to school, I found that it was not so bad. My school was a former gymnasium where many of the old teachers worked. But bad luck is bad luck. When in 1931 I finished at this seven-year school, it was decided, somewhere by somebody, that 7 years were quite enough, and thus the remainder of the secondary schooling was liquidated. (Secondary school has been different at different times; some time later, they changed their minds and returned to the ten-year school, as it is today.) My total school education, however, amounted to only four years.

After seven years of school one was supposed to enter a factory-plant school, where one was simultaneously to advance in education and to train for a worker's qualifications. But I did not take this route and instead became a laboratory assistant at an X-ray diffraction laboratory at one of the high schools. There I mainly communicated with two other laboratory workers who were three years my senior and took a great interest in physics and invention. (By the way, both my fellow lab workers later became good physicists.) I did not learn much but was imbued with something more important than knowledge—enthusiasm and an interest in work. In 1933, for the first time in many years, one could become a student at Moscow State University through open competitions at the entrance exams, and I decided to attempt to enter the physical faculty. It took me three months to "cover" the three-year course corresponding to the eighth, ninth, and tenth years of school. I passed the entrance exams but was not admitted— preference was given to those with better biographical particulars (i.e. more proletarian social origins and occupations of parents). But there was no special discrimination (for instance, related to the fact I am a Jew)— the results of my exams were not brilliant. I decided not to wait until the next year, left my work, entered a correspondence course at the university, again studied almost without assistance, and finally, in 1934, at the age of 18, started attending lectures and studying normally "like other students."

AMATEUR ASTROPHYSICIST

Why all these details? My aim is only to warn people against this route because I am deeply convinced that one shouldn't follow my example if given the choice. Of course, the school curriculum can be mastered in much less than ten years, but for this one generally must pay a high price. In my case, this price was a deficient background in Russian grammar, which to some extent has hampered my ability to write in this fairly rich language. Mastering a language requires exercise, exercise, and more exercise. I hated cramming and did not do it without pressure. The same also applies to mathematics. To understand something, one need only solve a dozen or so problems. But at school, I would have solved ten, if not a hundred, times more such problems, which would have provided the necessary automatism. All this is obviously so clear that further explanations are not needed here, and, in any case, I have already written about it in more detail elsewhere (3). However, I shall give a less trivial example. First-year students of the direct department (i.e. not including students taking correspondence courses) were taught astronomy, and my schoolmates recalled it with pleasure. However, I somehow contrived to become a second-year student without passing exams in astronomy or in chemistry. For this reason and because of my lack of corresponding school knowledge, even on becoming a professor of physics I remained illiterate in astronomy and chemistry. In both cases, I'm sure, I have paid a high price. As far as chemistry is concerned, I am still paying the price even now, because since 1964 I have been engaged in research on high-temperature superconductivity, which is today the center of gravity of my work (see e.g. 4), and this research appears to be closely connected with some chemical problems of which I am ignorant. As for astronomy, I'll begin with a curious illustration. Many times, especially when I took a great interest in the "new astronomy" with its quasars, pulsars, etc., I have spoken in lectures or with friends about various astronomical discoveries, about the radio, X-ray, and gamma-ray "sky." But the ordinary stellar sky is unfamiliar to me, and when asked what star or constellation is it, I must honestly say that I don't know. And if I have called this situation funny rather than shameful, it is because I do not consider myself to be a professional astronomer. I'm somewhat ashamed nonetheless, but such is my life.

When in 1946, at the age of 30, I wrote my first paper in astronomy, I had already authored many papers in physics, and an even greater number were waiting their turn. I had neither spare time nor extra strength, and life was hard. How could I learn the map of the stellar sky, remember it, and get used to it?

This may seem strange, but the facts show that a lack of elementary knowledge in one or another field—I mean now astronomy and physics—is not yet an obstacle for obtaining interesting and important results in these fields. Examples are especially numerous among mathematicians engaged in solving (and very successfully) various physical problems in spite of their ignorance of physics as a whole, to say nothing of numerous details. Analogously, quite a number of physicists, myself included, have written papers in astronomy that are of interest and were published without any allowances in the astronomical literature in spite of a very poor general astronomical background of the authors.

This is my personal opinion, and although I haven't had a chance to discuss it with anyone, it would be of interest to know what other people think. This is the first reason why I have entered upon this discourse here. The second is my desire to explain why I consider myself to be an amateur astrophysicist, as can be seen from the title of this article. Finally, and of most interest to me, I am curious as to how my work as astronomer would have been influenced if I had a "normal" astronomical education, i.e. the same as any astronomer who chose this profession as a university student, if not earlier. Unfortunately, it is extremely difficult, indeed almost imposs-ible, to answer such a question. Alas, one cannot start one's life anew. The best of all imaginable ways of obtaining the answer would be to trace the life of identical twins developing under different conditions. But given the absence of not only a twin brother but even analogous examples of other people, one can only make assumptions. I do not, of course, mean a fairly trivial general assertion concerning the benefit of being well educated and well informed. I refer to quite concrete hypotheses, results, and even discoveries for which I've asked myself whether I could have been the author of some of them. Sometimes the answer was negative, whereas in certain cases I'm sure I would have immediately given a correct answer if asked, or if at least acquainted with the corresponding astronomical material, I would have asked myself this or that question. But can a man who has never heard of neutron stars ask why they can rotate rapidly, possess a huge magnetic field, and be superfluid in some part of them? In short, and this is of course common knowledge, once a question is formulated, the work is sometimes half done. And in order to formulate a question concerning astronomical objects or effects, one must be acquainted with them.

Overall, I'm sure that it was a mistake that when I was first entering the field of astronomy, I restricted myself only to the material that was directly connected with what I was working on [at first it was the Sun; see (5)]. I should have eliminated my astronomical illiteracy in spite of all the obstacles that I have already mentioned. I was only 30 years old then, and of course I could, if I had realized this, have postponed some $n+1$ paper in physics and instead studied a textbook in astronomy. Today, as in the past, some physicists and engineers who are far from astronomy make attempts, say, to measure cosmic gamma-ray emission or to detect gravi-tational waves. It seems to them that they may remain unacquainted with astronomy as a whole. They are wrong. I know, of course, that such advice usually is not taken, but all the same I try to convince people that extending their astronomical horizons and advancing in their general level of knowl-edge is justified even within an exclusively pragmatic approach.

AUTOBIOGRAPHY II

I shall, however, return to my biography. From 1934 to 1938 I studied quite conscientiously at the physical faculty of Moscow State University. This was a flourishing period for the physical faculty (one that ended with the beginning of the Second World War and the evacuation from Moscow). My sympathy was from the very beginning with L. I. Mandelstam and his school. (Representatives of this school were I. E. Tamm, G. S. Landsberg, and others, including A. A. Andronov, although by that time he had already moved to Gorky and I only made his acquaintance later). The name of L. I. Mandelstam (1879–1944) is little known in the West, although he made great scientific achievements in optics and radio physics. (Suffice it to say that Mandelstam and Landsberg simultaneously and, of course, absolutely independently of C. V. Raman discovered the effect named after Raman; in the USSR, however, we prefer the term "combinational light scattering.") Mandelstam delivered lecture courses on various topics that had a wide audience.[1] These lectures, as well as subsequent discussions and occasional seminars, were a perfect school of true physics.

A third- or fourth-year student (the whole course consisted of five years, after which a student obtained the degree of "diploma") had to make a precise choice of his speciality. The Department of Theoretical Physics was headed by I. E. Tamm (1895–1971) and seemed most attractive to me. But my mathematical aptitude is rather modest, whereas in theoretical physics mathematics is considered to be not simply important (this is undoubted) but very essential. Therefore I chose experimental optics (the department was headed by Landsberg) and started working under the guidance of S. M. Levi. We tried to study the spectrum of canal rays, but in hindsight it is now clear that our experimental potentialities were not adequate to this difficult task. Nonetheless I received my degree. More interesting were my contacts with Levi. He was a Jew from Lithuania who had worked for a long time in Berlin, if I am not mistaken, in R. Landenburg's laboratory. The advent of fascism to power impelled him to leave for Moscow. Jumping ahead, I shall mention that later (in 1937 or 1938) Levi was dismissed from the physical faculty but, fortunately, not imprisoned. He could move abroad and found himself in the USA. In the 1960s I visited the USA three times (1965, 1967, 1969) and tried to find Levi. Miss Helen Dukas, the former secretary of Einstein, and some other people tried to help me, but all in vain. Later, when back in Moscow, I learned Levi's address when I complained to an acquaintance of mine of the bad luck in my attempts to find Levi in the USA. It turned out that this man had for a long time exchanged New Year's greetings with Levi. Such tangles are, of course, not surprising. It is more interesting that as far back as the 1930s Levi (and obviously many others) understood clearly the sense and the possible role of induced emission. Levi told me straight-forwardly, "Create an overpopulation at higher atomic levels and you will obtain an amplifier; the whole trouble is that it is difficult to create a

---

[1] These lectures have been published, although I do not know whether they have been translated into English. In such cases, I shall henceforth not refer to the Russian editions, which are practically inaccessible to the readers of the present volume.

substantial overpopulation of levels." As is now known, it is not so difficult to create an overpopulation, and, more importantly, by using mirrors the optical pathlength can be extended so that we obtain a laser. Why lasers were not created as far back as the 1920s, I do not understand. Much becomes obvious in hindsight. Perhaps there were some obstacles or simply nobody thought of using mirrors. This idea did not occur to me either, and here I definitely cannot explain this by my lack of knowledge. But the grains dropped by Levi have produced shoots. (I do not mention the general influence upon me of this pleasant and educated man.) As mentioned below, at the beginning of the Second World War, I was engaged in work on radio wave propagation. In particular, I paid attention to the decisive role of induced emission for propagation of radio waves in some ranges of the Earth's atmosphere (6). O_2 molecules possess magnetic moment, and therefore in the Earth's magnetic field the O_2 molecule levels (the lower electron levels) split. But the difference between the magnetic sublevels is $\hbar\omega \sim ehH/mc \sim 10^{-19}$ erg. Therefore, at a temperature $T \sim 100$–300 K, the energy is $kT \gg \hbar\omega$ and the sublevels are filled almost equally. As a result, waves with frequencies corresponding to the energy difference between sublevels propagate by means of a continuous alternation of absorption of a wave (radio photon) from a lower sublevel and an induced emission of the same photon from the upper sublevel. If the induced emission is ignored in this situation, the resulting absorption will be $kT/\hbar\omega \sim 10^5$ times stronger than the actual absorption. It is now, of course, a well-known fact that it is necessary in astrophysics that allowance be made for induced emission. To the best of my knowledge, my paper (6) was the pioneering application of this to the Earth's atmosphere. However, radio wave propagation in the ionosphere and atmosphere has long been beyond the scope of my interests, and I shall not dare estimate the role of the paper (6). This role, I think, was negligible because the resulting effect is small and the paper (6) itself most likely remained unnoticed. In 1942, people had other troubles.

On graduating from the university in 1938, I became a postgraduate student,[2] but not everything went smoothly. The war was coming, and deferments for postgraduates were rescinded. I was called up and remember receiving a document in which I was called an "espirant" [because, I think, the word "aspirant" (a postgraduate student) was associated with the then-popular language Esperanto]. The physical faculty still managed to grant deferments to its postgraduate students, but in 1938 this happened for the last time. I do not doubt that had I been then inducted into the

---

[2] This is a three-year period intended for preparation of a candidate dissertation, which generally corresponds to a PhD dissertation.

GINZBURG

Army, I wouldn't be writing this paper now. Only a few of my university fellows who entered the army survived the war. The situation on the whole, however, was more complicated. Three times I quite accidentally failed to get in the army; two of the three times I tried to volunteer. All of this, as well as much else that I mention only casually, is not without interest, but I'm afraid that I have already paid too much attention to my biography and further details would be out of place. (Some of these details, however, will become clear from what follows.)

While awaiting the summons (or more precisely the call-up "to present myself with things") I did not rush to the black-walled windowless room where I conducted measurements. Instead, I tried to somehow explain the strange angular dependence of the canal-ray radiation that we were attempting to find. (The indications of strange angular dependence found in the old literature were erroneous, as I am now sure.) It occurred to me that the electromagnetic field of a moving charge may play the same role as a photon flux and, in particular, induce radiation. This idea is erroneous, since a charge field is not equivalent to a free field. But in those times the situation was less clear. At any rate, when in the fall of 1938 (if I am not mistaken, this historical event in my life happened on September 13) I asked Tamm a corresponding question, he got interested, advised me to look through the literature, and wished me well. I soon found out that the difference between the methods used in classical and quantum electrodynamics gave rise to misunderstandings and obscurities that I partially managed to clarify, and that what was important was not the application of a complicated mathematical technique but in understanding the formulation of the problem. This is how I became a theoretician and left experimentation for life. Obtaining on my own these (maybe modest) results, I realized that I could fruitfully work in theoretical physics. This had a great effect on me, and my new life began. In this respect I am much obliged to Tamm and his attitude toward people. I do not write about Tamm in more detail here, since the interested reader may refer to my reminiscences about him (7). Here I only emphasize how important a friendly atmosphere and mere kindness are for many (although not all) beginners [for more details, see (7)]. Neither do I dwell on the content of my first three papers devoted to quantum electrodynamics (published in 1939). The corresponding references can be found in my book (8) and paper (9), where the subject itself[3] is also considered, of course.

In 1939 S. I. Vavilov and P. A. Cherenkov discovered the phenomenon that is now known as the Vavilov-Cherenkov effect or radiation. [True, in

---

[3] Here and below I try to refer to more recent and accessible literature, especially in what does not concern astronomy.

the West it is most often referred to as Cherenkov radiation, but many Soviet authors (I among them) who know well the history of this discovery use only the term Vavilov-Cherenkov (VC) radiation.] The classical theory of the effect was formulated in 1937 by I. M. Frank and Tamm, who showed that a charge uniformly moving in a medium at a "super-light" velocity $v > c/n$ (where n is the refractive index of the medium) must radiate. This theory was already within my memory in 1939, and it was thus very natural that while studying quantum electrodynamics at this time, I constructed the quantum theory of VC radiation as well as considered this effect in an anisotropic medium. Since then, the study of radiation of uniformly moving sources (including, besides the VC effect, the transition radiation and the Doppler effect) has become one of my favorite fields. All of these effects seem beautiful to me. Of course, the definition of "beautiful" is always subjective. For example, L. D. Landau (1908–1968) was almost absolutely indifferent to the above-mentioned effects. By the way, I regard L. D. Landau, as well as Tamm, to be one of my teachers. Like Tamm, Landau was an outstanding personality [see my contribution in the book of reminiscences about Landau (10)]. Therefore, it may not be out of place to narrate, as an illustration of the above-said, how we first met (and, in fact, got acquainted) on scientific grounds. At one of the joint seminars [for more details, see (10, 11)] Tamm mentioned the quantum theory of VC radiation formulated by me, to which Landau immediately made a remark to the effect that this is absolutely unnecessary, since the effect is classical. As was usual in such cases, Landau's criticism had serious grounds—the quantum corrections in the theory of VC radiation are of the order of $\hbar\omega/mc^2$ (ω is the radiation frequency, and m is the mass of a radiating particle) and are small because at high frequencies ω, the refractive index $n(\omega)$ tends to unity and VC radiation disappears. However, the quantum theory of VC radiation is in fact not of narrow methodical interest only, since it also substantially clarifies the situation when applied to the Doppler effect in a medium [for a discussion of this and practically all my papers devoted to the theory of uniformly moving sources, see (8, 11)].

The papers mentioned above were written within about a year and a half and were the basis for my candidate dissertation, defended at Moscow State University in May 1940. A candidate dissertation and the corresponding candidate degree in the USSR correspond to the PhD in the West. Of course, the level of candidate dissertations is sometimes very low, but in physics and at good institutes and universities, the level of candidate dissertations seems to me to be rather high. We also have another degree— the Doctor of Sciences (Dr. Sc.). In this case, and again in the proper places, a Doctor's thesis is usually defended by an author of dozens of papers, quite an independent research worker not younger than 30 to 40. The Doctor's degree gives one the right to hold a professor's post, although there are not a few professors in our country without the Doctor's degree. At that time (1940) the USSR Academy of Sciences had special vacancies for those who were to prepare a Doctor's dissertation within three years. There was one such vacancy in the Department of Theoretical Physics at the P. N. Lebedev Physical Institute of the USSR Academy of Sciences (FIAN), which I filled on September 1, 1940. Ever since, I have been working in this department, which was founded by Tamm in 1934. (Until mid-1941 he had also been head of the theoretical department at the physical faculty of Moscow State University.) Before entering FIAN, I had already been engaged in the theory of particles with higher spin states. This work continued for many years (partially in collaboration with Tamm and young colleagues). I shall restrict myself to mentioning only my most recent papers (12, 13) that touch upon this problem.

As is well known, Germany invaded the USSR on June 22, 1941. Several cities had already been under bombing attacks for many hours when, at about midday, Molotov announced over the radio that war had broken out. I clearly remember listening to his speech with my two-year-old daughter on my lap. I also remember Stalin's speech on July 3, when this dictator, who had poured the country with blood and tears, called his audience "brothers and sisters" for the first and last time. I was a "private untrained" and therefore was not mobilized. But for people like me and, it seems to me, for all those who were not mobilized, a people's volunteer corps was organized. Soon a great many of these people were killed or captured near Moscow. Of course, I immediately joined the volunteer corps and spent the whole first day in a school building, where we were "formed," and in the evening left with an order to come "with things" at the first call. But soon there came a decision to evacuate the Academy of Sciences, and, if I am not mistaken, exactly one month after the war broke out, my elderly father (he was 78 years old), aunt, wife, and I left for Kazan'. (My only daughter had been evacuated with her grandmother some time before.) Before our departure, Moscow underwent only one air attack by the Germans. When we heard the sound of alert, my wife and I happened to be near a metro station, where we had to spend that night. In the morning I saw only traces of the attack. I shall note that in Kazan' I had to join the "labor front"—we dug trenches not far from the town, which fortunately were not used. I had no deferment ("reserve") because in those times it was intended only for people of higher standing. But having had no military training I was of no value to the army, and in any event I was not called up. While waiting for the summons, I wished to finish my Doctor's dissertation as soon as possible, and in May 1942 I defended my thesis on the theory of higher spin particles. I have no reason for considering my thesis to be weak, and indeed it was highly estimated. Under peaceful conditions, however, I would not have hurried to get my Doctor's degree. But in the situation described, and taking into account the fact that I changed my subject of research (see below), my haste was justified. By the way, the defense of a thesis did not have any influence, as far as I know, upon one's being called to military service, at least in those hard times—by that time, the Germans had reached the Volga. Furthermore, I was invited to volunteer to become a paratrooper, and I tried to do so but was rejected because of the state of my health by virtue of an anecdotic set of circumstances. In Kazan' our institute occupied a floor in the wing of the Kazan' University building, and there were no more than 100–200 people in the institute. We very often were sent to fulfill different kinds of jobs, and I remember unloading a barge of logs on the Volga. Everybody worked, including Tamm, Landsberg, and others who then seemed to me to be very old although they were no more than fifty. As for me, I carried big logs "on crate"—this is a setup in which straps are put on like a rucksack and a "ground" is then fixed on one's back, on which a log or something else is put. Such a contrivance was widely used in Russia for carrying bricks and other heavy objects. I was not at all mighty (180 cm tall and weighing about 60 kg at that time). Using this contrivance, however, I carried rather heavy short logs that two men could hardly put on the "ground." But the load was apparently too heavy, and one time I found blood in my mouth—a small vessel must have broken. Tuberculosis was suspected and I was sent to the dispensary, where they found some "petrificated foci," after which I was registered. Therefore I was regarded as unfit for the landing force. I am not going to play the hypocrite and say that I regret this. But at that time, I felt that it was better to die in a battle than to find myself under German occupation and then in a death camp.

Life in Kazan' was not a honey, of course. My family lived in a small room in the university hostel where the temperature in winter fell below 0°C. My father could not endure such a life and died in mid-1942. The Academy was supplied with some food and thus we did not actually starve, but we did feel hungry all the time. I remember a canteen where for a corresponding recompense waitresses served pea porridge. People came in with a tin stuck into a briefcase, filled the tin with porridge or soup, and then carried away the briefcase as if it were filled with papers.

At the end of 1943 the members of FIAN began returning to Moscow as victory approached. Below I narrate another important chapter of my biography, and thus now I shall only note that on returning to Moscow the theoretical department started growing. For several years I was Tamm's deputy, and after his death in 1971 I had to become head of the department. I say "had to become," for I did not want this position, but in the Academy it was necessary "for the welfare of the department staff" that its head be a man of position. In 1971 there were only two full members of the Academy (academicians) in the department: A. D. Sakharov and myself. Sakharov had already become a dissident and was not suitable as a chief, so I had to be head of the department for 17 years. A new rule has recently been introduced obliging all members of the Academy beyond the age of 70 to resign their administrative posts with the right to become a councillor (adviser) without a decrease in salary. Of course, to be head of the department that now numbers nearly 60 workers, including three full and four corresponding members of the USSR Academy of Sciences (the same number as in the rest of our great institute), required time and strength. But I do not complain, for our department is known for its friendly terms and very kind-hearted atmosphere, and during the lifetime of the department (54 years!) we have not had a single serious conflict.

Since 1943 I have been engaged in the theory of superconductivity and have written a number of papers in this field; the most well-known among them (14) (in collaboration with Landau) was published in 1950. I have also been concerned with the theory of ferroelectrics, superfluidity, and many other things, including astrophysics, which will be a special subject here. But all this work came some time later. In spite of all the warnings, the coming war was not openly regarded as inevitable in the USSR. Therefore, to the best of my knowledge, no direct preparation for the war was conducted by the Academy. In any event, when the war broke out on June 22, 1941, physicists were in general out of it. Suffice it to recall that even I. V. Kurchatov set himself to the problem of demagnetization of ships against magnetic mines, and it was only at the end of 1942 that he was appointed to head the nuclear program. As has already been mentioned, I was occupied in an absolutely abstract analysis of equations for higher spin particles. As the war began, we all started searching for applications of our abilities closer to practice, if not to defense. Accidentally I was given advice to concern myself with radio-wave propagation in the ionosphere, which was supposedly an important problem for defense. I followed this advice and spent much time on this subject.

SIDE ASSOCIATIONS

There is a special type of person who usually has "side associations." I belong to this type, in that I am constantly developing various associations that require comment.

I have mentioned above the title "academician," the use of which in the

West has always surprised me and aroused an unpleasant feeling. I first experienced this feeling many years ago, when in the list of participants of some congress I came across the names of Prof. P. Dirac, Prof. N. Bohr, and Academician X. Why a special title for a Soviet scientist? Does it not suffice to be a professor, especially in the same row with Bohr and Dirac?

In the USSR Academy of Sciences there exist two "degrees" (titles): full member (or academician) and corresponding member. Unfortunately, both these titles often are used in the Soviet literature. I think that this is the influence of the prerevolutionary habit not to forget to mention the title of count or duke, if not grand duke. Strong, also, was the German influence, whereby a colleague is thanked as, say, Herr Geheimrat Professor Doctor X. The progressive tendency in our country is now to omit titles, say, in acknowledgments (acknowledge X, Y, or Z but not professor, doctor, or academician X, Y, or Z). In the West, as far as I know, such a style has already become conventional. In any case, I am a resolute opponent to the use of the academician title in the English literature. However, one cannot totally abandon titles in the USSR, because otherwise some authority will not answer the telephone, among other things.

Since I have already touched upon the subject, it seems reasonable to say some more about the USSR Academy of Sciences. (We have numerous other academies as well.) This academy numbers about 800–900 members, of which about one third are academicians and the rest are corresponding members. Such a structure was created before the revolution of 1917. Merely for the title, an academician now receives 500 roubles a month (without any duties), and a corresponding member receives half that. These sums are rather large if one takes into account that my salary as head of a department or councillor is also 500 roubles, while a young candidate of sciences gets 150–200 roubles a month. Members of the Academy have some other privileges as well; for instance, an academician may call a car with a driver for several hours a day. This may greatly surprise some readers, but one should remember that the salary of 500 roubles a month is often much less than the salary of a man of equal qualifications in the West. As for the use of a car, some time ago the members of the Academy were fewer in number than now, the majority of them were elderly people (the average age of academicians in 1985 was still 69.9), and there were very few private cars.

I am convinced that in principle the only correct thing to do is to abolish all material privileges for members of the Academy. But in our society this is evidently a long way off,

Academic privileges have, naturally, a very pernicious effect upon the elections: Not only scientists, but also designers (engineers), high-ranking officials, etc., try to join the Academy. Despite this, the USSR Academy

GINZBURG

of Sciences, especially in the exact sciences, is on the whole a focus of the most qualified people of this country. Besides, the Academy does not act exclusively on the basis of arbitrariness but is still guided by a charter that provides for a competitive ballot and not "one place–one candidate" type elections. However, until recently the charter has, in essence (although not formally), been repeatedly violated. (For example, special vacancies were allotted for candidates welcome to the authorities.) But under the conditions of the deep changes now taking place in my country, I hope that the charter will be made more precise and, most importantly, observed. The Academy and all of Soviet science have, however, problems that are no less important. One of them is the necessity to fight against fantastic bureaucratism. It would be out of place to dwell on this subject here, the more so as I can refer the interested reader to my paper (15) that elucidates this problem [for those who wish to obtain insight into the state of Soviet society in mid-1988, I advise them to get acquainted with the whole of the volume containing (15)].

Another "association" arises in connection with the above-mentioned fact that I am the author of a large number of works and a still larger number of papers. By "work," I mean some scientific results in the form of a paper or papers that I have registered in a special list. Long ago, when I was defending my theses, by some formal requirement I had to start such a list, and I have since continued it voluntarily because it is convenient to do so for references. The list now numbers 325 items, and one item often includes several papers devoted to one problem. A number of notes and reviews, popular scientific papers, etc., are not included, nor will the present paper figure in the list, although it was more difficult for me to write this one than a whole number of scientific papers. The total number of my publications probably approaches one thousand, and the number of my "works" is also large. What does this prodigality signify? Is it good or bad? I have had to face such or similar questions, and opinions in this connection are very diverse. For example, I have been reproached for publishing papers very easily. There also exists an opinion that authors who publish many papers do so for the sake of glory, for increasing the number of references to their papers, etc.

It is quite obvious that a large number of publications or even "works" cannot in itself be put to one's credit. One essential work may, of course, be more valuable than a thousand weak papers. So in estimating the contribution of one or another author, the decisive role is undoubtedly played by the content of his publications. All other conditions being equal, the number of publications characterizes basically the style of the activities and tastes of the author. Personally, I write rather easily if, of course, I have an idea of what I am going to write about. Moreover, I do not feel satisfied until the paper is written; of course, dissatisfaction often remains afterward, but nonetheless the very process of writing is, at least for me (and I think for many people), an important element of the work itself. With some exceptions, a paper is written not in the way of a duty but as a result of some interest. And if the author himself is interested, why not share this with others and send a paper or a note to press? As for those who write little, there exist different cases. Some merely have nothing to write about; others find it difficult to write or believe that they have already reached such a level that publication of a not very important result, a popular paper, etc., will add nothing to their scientific reputation and may even arouse ironical smiles. I am well aware of the fact that many of my publications only give rise to criticism. But I have no fear for the "clamours of Boeotians," for I do not want to be led by snobs. Sometimes there are mistakes and I regret having sent a paper to press, but idlers alone never make mistakes. Finally, tastes and opinions are diverse [a very typical example is given in (3)]. The author must heed his inner voice and not try to please everybody. If a paper is not interesting to someone, he can choose not to read it. This, by the way, consoles me with respect to this paper, in spite of the fact that I realize that some may not like it.

The value of the works and their quality is an important question in scientific activities. For young people who are only beginning their scientific life it is even of vital importance, but irrespective of age and position, one wants to hear a just opinion. An unambiguous and reliable way of judging the merits of works probably does not exist. In the West, one common way of estimating papers is based on the citation index. This method is seldom used in the USSR and is in general unpopular, but I do not think that there is enough reason for such a negative attitude. The citation index may, of course, absolutely distort the picture. For instance, an erroneous but sensational paper may give rise to a lot of references. Another source of errors is that only the first author's name[4] is taken into account. Finally, papers published in Russian are not as frequently cited as those appearing in English in well-known journals. But information from the citation index is nonetheless interesting and significant. True, I do not look in the index. (The library of our institute does not have it, and

---

[4] By the way, I, along with the majority of my colleagues in the USSR, use only an alphabetical ordering of names for my publications. (I have written only two or three papers in which this order was more or less accidentally violated.) One should bear in mind, however, that the Russian alphabet differs from the Latin one. In my case it is particularly important that the letter Γ (G) is the fourth in the Russian alphabet, and to V or W there corresponds the third letter of the Russian alphabet (B). Also, the Russian Φ (F) is toward the end of the alphabet. Therefore, when a paper is translated from Russian into English, the alphabetical ordering often appears to be violated.

if I am not mistaken it is available only in two places in the whole of Moscow.) But once in a Russian publication I came across a listing (16) of the 249 most frequently cited scientists in the world, based on the overall number of references to their papers in 1961–75, i.e. within 15 years. In this list, L. D. Landau was in second place (18,888 references), and five other Soviet authors were mentioned (four physicists and one chemist), of which I had the most citations (6834 references, or an average of 456 references a year); overall, I was in 66th place. In subsequent years the number of references to my papers has become somewhat smaller, which is natural. [In recent years the papers that I have published have been less in number and maybe also in quality; I do note, however, that some of my old papers, especially (14), are now most frequently cited in the text without any mention in the list of references.] Unfortunately, I do not know which place is taken by references to my astronomical papers (including cosmic-ray astrophysics).

In spite of all the limitations of the conclusions that can be drawn from the citation index, it would nonetheless be of interest to have such information with respect to astronomical themes. (True, it is not always easy to decide to what branch of science this or that paper should be ascribed.)

AUTOBIOGRAPHY III

In 1985 the Soviet Union entered a new epoch, of which one of the most important and distinctive features is *glasnost* (openness). We now (especially since 1987) read books and articles in magazines and news-papers whose publication was unthinkable some time ago. Possessing, to say nothing of spreading, such literature used to lead to arrest and many years of imprisonment. *Glasnost* also allows me to dwell here on the part of my biography of which I could not write or speak publicly of before. The ignorance of some foreign colleagues with respect to life in the Soviet Union has often amazed us, especially since quite a lot of materials fairly sufficient to remove such "illiteracy" for anyone who can read have already been published. But not everybody is fond of reading, and for astronomers a concrete example concerning an "amateur" astronomer, but still one of their colleagues, may be of some additional interest.

When the war ended in 1945, physicists in the USSR were already held in great esteem. The existing institutes grew rapidly, and new scientific institutions and high schools appeared. As far back as the 1930s, there was at Gorky University a strong group of physicists and mathematicians, the most outstanding of which was A. A. Andronov. This group decided to organize a special radio physical faculty. Since they were short of their own experts in this field, they invited three professors from Moscow (from FIAN) for whom it was to become their "part-time" job, i.e. they were expected to remain working in Moscow but at the same time come to Gorky from time to time to deliver lectures. I was one of these three and was put in charge of organizing and heading the department of radio-wave propagation because, as I have already mentioned, I had been occupied with precisely this problem since the beginning of the war and had already published a number of papers. I remember, at the end of 1945, arriving in Gorky by train (it is a night's journey) in far from comfortable conditions and hiring a fellow to carry my suitcase by sledge (we were walking) from the station to the central part of town, where the university and hotel were located. At the beginning, if I am not mistaken, there was only one colleague (M. M. Kobrin) in "my department" and one strange student, who later left the university. But soon after there appeared capable students, followed by postgraduates. Many of those who graduated from the department have long since obtained their Doctor's degree, and the department continues to exist. That I headed this department under hard postwar conditions now seems to me to have been a venturesome enterprise. But I was 29 then, and as a doctor of sciences had the right to be at the head of a department. I was willing to teach youngsters, but in Moscow it was difficult for me to realize this ambition. I think I would have left Gorky in a couple of years, as did my colleagues from Moscow, had fate not willed things differently. I met there my present (second) wife Nina Ermakova, and in 1946 we got married. I would not, of course, even mention here my family life if it were not for some special circumstances— namely, that Nina was, in fact, in exile in Gorky and had no right to move to Moscow. In short, the story (which has even been mentioned in some publications) is this. Her father was arrested in 1938 and died in Saratov's prison in 1942. (He was in the same prison as the well-known geneticist N. I. Vavilov, who died there during the same period.) A student of the mathematical faculty of Moscow State University, Nina was arrested in July 1944 and along with her fellows (some of them, as she herself, were the children of repressed parents) was accused of plotting an attempt on Stalin's life—Stalin was supposedly to be shot from the window of her flat in the Arbat. But the "scriptwriters" from the KGB had never taken the trouble before the arrest to check everything, and only later was it established that the windows of the room where Nina lived with her mother did not face the Arbat. For all the nonsense of the accusations brought by the KGB, the investigators somehow tried not to make easily refutable assertions. In any event, the accusation of terror was remitted, and there remained "only" the accusation of counterrevolutionary group anti-Soviet activities (Articles 58.10 and 58.11 of the then Criminal Code). She spent

9 months in prison and in March 1945, without any trial and by the decision of the so-called special assembly, was sentenced to three years' confinement in a camp. This was a very "short" term for a person convicted under Article 58. Maybe it was for this reason that when an amnesty was announced on the occasion of the end of the war, even prisoners charged under Article 58 were given amnesty if their terms were three years or less. (The majority of those charged under Article 58 had longer terms and were not included in the amnesty.) Thus, in September 1945 Nina was released, but without the right of residing in a number of large cities.[5] She had an aunt in Gorky, and therefore she chose Gorky, but nonetheless she could get registered only on the other bank of the river Volga in the village Bor. Despite this, she managed (and this was nontrivial—kind people helped her) to enter the polytechnical institute in Gorky, from which she graduated in 1947. Until 1949 she lived illegally in my room, but at the end of 1949 she was registered in Gorky by Andronov's application. (This happened after a large accident on the river Volga on October 29, 1949, in which a ship carrying people from Gorky to Bor was wrecked; Nina was among some dozen or so out of about 250 passengers who survived.) I, naturally, handed in an application each year (it could not be done more frequently) with a request to allow my wife to move to Moscow, but all in vain. It was only in 1953, after Stalin's death, that a new amnesty was announced and Nina could leave for Moscow at last. In 1956 she, as well as all her fellows from the "anti-Soviet group," was completely rehabilitated—that is, the accusation was recognized as fully groundless. To characterize the rehabilitation process, I will say that an inspector came to Nina's mother's place to draw up a statement, in the presence of witnesses, that the windows of the room did not face the Arbat.

In addition to the reasons that I have already mentioned, I narrated this story also to explain why I taught students and conducted research work in Gorky for so long. I had students, postgraduates, and colleagues there, and much research was conducted. Therefore, after 1953 I continued, although more and more seldom, to visit Gorky. I headed the department, I believe until 1961, and the last two times (in 1980 and at the end of 1983) I went to Gorky it was mainly to see Sakharov, who, by irony of fate (at least I feel so) was also exiled to Gorky.

In 1942 I became a candidate member of the CPSU (Communist Party of the Soviet Union) and in 1944 a CPSU member. The war was in full swing then, and today, when we know what was Stalin's true face, it is still hard to believe that many millions of people, including myself, were absolutely blind about him. Even when I learned the "criminal" story of

---

[5] In order to live in a given place, Soviet citizens must be registered there by the police and get visas in their passport. Amnestied people were given passports with limitations for visas.

my wife and her fellows, I did not think yet that everything came from the "great leader." My eyes were opened wide only after N. S. Khrushchev's report on February 25, 1956, at the 20th Congress of the CPSU.

If I continued writing my biography in further detail, it would, of course, overstep the admissible page limits of the present paper. Therefore, I shall restrict myself only to a few more remarks. By unwritten rule, CPSU members were not, of course, supposed to marry "counterrevolutionaries." In addition, on October 4, 1947, there appeared in the Literary Gazette an article in which a certain physicist D. Ivanenko (he wrote denunciations of many others as well, including Tamm) accused me of "servility" and "cosmopolitism," i.e. admiration for bourgeois science. This gave me much trouble, and still more was to be faced. I think it would have cost me my head if it were not for I. V. Kurchatov, who invited Tamm in 1947 to take part in nuclear research. Tamm, in turn, enlisted me, Sakharov, and a number of other workers of our department. Soon I made an important proposal; another was made by Sakharov. Unfortunately, even now, at a time when the USSR and the USA exchange military observers and the Secretary of Defense of the USA has recently examined the new Soviet armaments, these 40-year-old works still remain classified. Therefore, so as not to get into new troubles (I have had enough of them in my life), I shall only say the following. Tamm, Sakharov, and some others left for far-off lands sometime in 1948. As one that aroused suspicion (the wife, you know, was in exile), I remained in Moscow, but a sentry was posted at the door of my office. I knew no real secrets and was not at all interested in them, but in 1950, on the initiative of Sakharov and Tamm, research work in the field of controlled thermonuclear synthesis was begun. At the first stage this was, of course, a physics problem, and I was also engaged in it and obtained some results. By 1952 (or late 1951) this work was considered so important and secret that I was altogether removed from it. Maybe it was for this reason that after this problem was declassified (this was due to Kurchatov, who made his well-known report on the thermonuclear problem in England in 1956), I decided in 1962 to publish my old reports (17).

As is well known, in 1952 and early 1953 the situation in my country became still more heated. Stalin went absolutely mad, and vivid evidence of this was the notorious "case of the physicians." New monstrous repressions were coming. Fortunately, on March 5, 1953, "the greatest man of genius of all times" was finally gathered to his fathers. There followed many rapid changes in the country, the "case of the physicians" was disavowed and they were released, the next amnesty was announced (I have mentioned it above), and Beria—one of the closest bloody assistants of Stalin—was shot. It was noted that the Academy of Sciences of the USSR had not called elections since 1946, and late in 1953 the elections were finally held. I was then elected a corresponding member. At the end of the year I was awarded the order of Lenin (the highest civil order in this country) and the Stalin Prize of 1st degree (later this prize was renamed the State Prize, and three degrees were abolished and reduced to only one) for the above-mentioned "closed" works. In general, by 1954, from a half-disgraced man I had become a person. Since then, my life has been generally normal. From 1962 to 1970 I even could rather often go abroad, sometimes with my wife (this was a rare thing for Soviet citizens, but an exception was made for full members of the Academy; I was elected in 1966). Since 1971, true, I have gotten into new troubles, but these were not a threat to my life and, therefore, I shall say only a few words about them. First, I was almost never allowed to go abroad. For instance, I could not deliver my Darwin Lecture myself. [It was delivered at a meeting of the Royal Astronomical Society from a text that I had submitted beforehand (18).] Only after several indignant letters sent by me "to the very top" did I manage to go to several international conferences, but each time with such nervous strain that I would not wish it on my enemy. Only since 1987 have I been going abroad without extreme difficulties, although, as is quite a usual thing with us, with a great many bureaucratic impediments (see 15). Second, complications arose owing to the fact that in 1969, A. D. Sakharov again started working in our department. I refused to sign any letters condemning his activities, and generally the attitude toward him was quite loyal in the department, even when in 1980 he was exiled to Gorky. On our initiative he remained a worker in our department, and, with permission from the administrators, his colleagues from the department went to see him there.[6] When, at the very end of 1986, Sakharov could finally return to Moscow, he appeared in the department on the day of his arrival and was heartily greeted. He is working in our department now. Being head of the department, I had, of course, to deal with "the Sakharov case." I think that I have done my best and I have no reason to reproach myself, but on the whole it is not up to me to judge.

In general, as can be seen from what has been said, it was only in 1987 that I got out (is it forever?) of "external" difficulties. In return, by virtue of what is called the "law of conservation," so to speak, my "internal" difficulties increased. The point is that beginning at the age of about 65, it has become more and more difficult for me to work fruitfully, although I try to overcome this obstacle (see 3).

In concluding this section, I would like to answer one, obviously natural question: What was the influence of my severe trials upon my scientific

---

[6] About this period, see, in particular, my article (written 17 December 1989, after A. D. Sakharov's death, on behalf of the editor) in the literary magazine *Znamja* (1990, no. 2, p. 3) (noted added in proof).

activities? A simple answer suggests itself—under better conditions I would have managed to do more. But, frankly speaking, I am not sure of that. During all of my "scientific life" (since 1938) I could almost without interruption be engaged in anything I wanted and not worry about earning my bread, even though it was not always with butter. This was first. Second, and this is also typical of many of my colleagues, science and scientific activities occupied a predominant place in our lives; they were simultaneously work, hobby, and rest (and even a narcotic). I think that if I had lived under better conditions, I would have probably been happier and have rested and seen more. But the integral of my scientific activity, if I may say so, most probably would not be larger than it is.

RADIO ASTRONOMY[7]

As has already been mentioned, in the middle of 1941 I turned to the study of radio-wave propagation in the ionosphere and generally to plasma physics. This activity is reflected in the monograph (19) as well as in (8). It was, in fact, the ionosphere that was the starting point for my astronomical and, more precisely, radio astronomical studies.

The well-known Soviet physicists and radio specialists L. I. Mandelstam and N. D. Papaleksi contemplated the problem of Moon radiolocation (radar) long before the Second World War. In 1944, stimulated by the progress in radiolocation, Papaleksi returned to this idea by considering, naturally, radiolocation of the planets and the Sun. In this connection, at the end of 1945 or the beginning of 1946, he asked me to clarify the conditions of radio-wave reflection from the Sun. As it turned out, this was a typical ionosphere problem, and I had all the formulas at hand. The results of the calculations did not seem to be very optimistic, since for a large set of parameters, such as electron concentration and temperature in the corona and chromosphere (which then remained unknown in many respects), radio waves should be strongly absorbed in the corona or chromosphere and not even reach the level of reflection. [The reflection due to inhomogeneities was not considered; the "point" $n^2(\omega) = 1 - \omega_p^2/\omega^2 = 0$ was, roughly speaking, taken as the level of reflection.] But this was immediately followed by a more interesting conclusion: The source of solar radio emission must not be the photosphere, but rather the chromosphere and, for longer waves, the corona. At the same time, the corona was already assumed to be heated up to hundreds of thousands or even a million

---

[7] In this and the following sections I have used material from, and sometimes even the exact texts of, my papers (1, 2). Since these papers are easily accessible, I refer those (probably very few) interested in the details to them. A number of references omitted here can also be found in these papers.

degrees. Thus, even under equilibrium conditions, i.e. in the absence of perturbations and sporadic processes, the temperature of emission from the corona at waves longer than about a meter must reach approximately 10^6 K at a photospheric temperature $T_{ph} \approx 6000$ K. All this was presented in my first astronomical paper (5). In the same year, analogous conclusions were published by Martyn (20) and Shklovsky (21). [I can only say that my paper (5) was submitted for publication on March 27, 1946, whereas the dates of submission of the other two papers are not indicated; they appeared, respectively, in *Nature* on November 2, 1946, and in the November–December 1946 volume of *Astronomicheskii Zhurnal*.] In the calculations, I used absolutely clear and reliable formulas known from the theory of ionospheric wave propagation (19). Martyn did not present the formulas, but he evidently acted in the same way. As for Shklovsky, he believed (21) that the absorption due to "free-free" transitions and the absorption due to electron-proton collisions should be taken into account separately and then summed up, whereas, in reality, these are one and the same mechanism (19). However, this circumstance was of no importance, since the parameters of the corona were not exactly known at that time.

The existence of thermal radio emission of the corona at $T \sim 10^6$ K was confirmed in the paper by Pawsey published immediately after Martyn's paper (20), in which such emission was shown to play the role of a lower limit that is reached as soon as the sporadic component of solar radio emission becomes sufficiently weak.

A weak point of radio astronomy of that period was a low angular resolution that prevented investigations of the Sun even for regions of sizes of arcminutes—it is difficult to believe this today when the angular resolution of radio interferometers far surpasses that of the best optical telescopes. In this connection, Papaleksi suggested measuring solar radio emission during the total solar eclipse of May 20, 1947, with the help of a 1.5-m wavelength antenna that was installed on board a ship and had a wide directivity pattern of several degrees. These measurements were successful [see the literature cited in the book containing (1)] and appeared to be the first of their kind. Whereas the intensity of optical emission from the Sun during a total eclipse decreases by several orders of magnitude, at a wavelength of 1.5 m the intensity during the eclipse decreased no more than by 60%. Thus, meter-wave radio emission was proven to come from the corona, which remains uncovered by the Moon even during a total optical eclipse.

In 1947 I took part in the Brazil expedition of the USSR Academy of Sciences, on board the ship *Griboyedov*, in which radio observations of the Sun were conducted. It seems that I was included on the staff of the expedition in recognition of my early work in the field of radio astronomy, which was only then beginning in the USSR. I did not take part, however, in the measurements themselves—they were carried out on board the ship while the main part of the expedition made its way to Brazil to take optical measurements, which were, unfortunately, unsuccessful because of bad weather. This main part of the expedition also included a small ionospheric group headed by Ya. L. Al'pert. (I was in this group; weather could not, certainly, prevent ionospheric measurements.)

The above-mentioned activities drew me further into radio astronomy, and I became for a while almost a professional radio astronomer. (I tried to acquaint myself with all the available material, methods of measurement, etc.). As a result, I wrote two reviews of radio astronomy (22, 23), among the first in the world literature. (I cannot, however, vouch for this.)[8] Now that 40 years have passed, it is difficult for me to judge the value of these papers, and I do not feel like analyzing them in detail. I shall only mention the suggestion made by me in (23) that the radio-wave diffraction on the Moon edge should be used to increase angular resolution of details on the Sun during eclipses. This question I considered in further detail in my 1950 paper with G. G. Getmantsev [see the reference in (1)], where we were already thinking more of discrete sources of cosmic radio emission than of the Sun. The diffraction of radio emission on the Moon edge has since been widely used, and therefore I need only add here that I also discussed in (23) the possibility of enhancing the angular resolution still more by observing a source located on the line that joins the Moon center with the point of observation. (Here we are evidently dealing with the Arago-Poisson light spot.)[9] The nonspherical shape of the Moon and the necessity of having a source on or very near to the indicated line strongly hamper such observations, of course, and I do not know of any attempts to exploit this method. But maybe one should nonetheless analyze such a possibility in more detail, not only for the Moon but also for the planets

[8] How far I still remained from astronomy as a whole, in spite of what has been said, can be seen from my note (24) in the volume dedicated to the 80th birthday of J. Oort. There, I recount that on the way back from Brazil, the participants of our expedition were lucky enough to visit Leiden, although quite accidentally. And while there, instead of taking the opportunity of meeting Oort and generally taking part in the discussion of astronomical problems, I rushed to the Kamerling Onnes Cryogenic Laboratory because I was then most interested in low-temperature physics.

[9] As is known, as an objection to the wave theory of light, Poisson pointed out a consequence of this theory that seemed to him quite absurd: On the axis of the geometrical shadow of a round nontransparent screen a light spot must be observed. (The source is considered to be pointlike and is positioned behind the screen on the axis perpendicular to its plane.) The experiment conducted by Arago immediately after this confirmed the existence of the light spot. For an opaque sphere as screen, the conditions for observing a central peak are facilitated, as compared with for a flat screen.

GINZBURG

and their satellites and artificial screens (both flat and spherical). Perhaps this has already been done and I do not know about it, for I have not looked through the corresponding literature.

If I tried to dwell on my subsequent papers in so much detail, it would take too much space. My work in astrophysics has been rather sporadic and chaotic, and the portion closest to radio astronomy may somewhat conditionally be divided into three main trends:

1. Twinkling of radio sources in and beyond the ionosphere, oscillations of the intensity of solar emission, the use of polarization measurements, and satellite measurements.

2. The theory of sporadic emission from the Sun. V. V. Zheleznyakov and I began studying this range of questions in 1958 (25). A number of other papers followed, but it would be out of place to refer to them here, since the corresponding results (with references) are fully given in the book by Zheleznyakov (26).

3. The theory of cosmic synchrotron radio emission, and its connection with the problems of the cosmic-ray origin and with high-energy astrophysics as a whole. This was the subject of my main astrophysical interest, and I still work in this field, although not so much as before. From the point of view of historical information that I can offer, these problems also play the most important role. Therefore, they are discussed in the next section [for more details, see (1, 2)].

I shall now dare to mention some more of my papers close to radio astronomy. The origin of radio galaxies and, specifically, the question of the energy source of their radio emission were not immediately clear. For instance, one can mention the hypothesis of colliding galaxies, which proved, as a rule, incapable of explaining the mechanism of energy output in radio galaxies. Another suggested idea was that of a sharp increase in the number of supernova flares in radio galaxies, an idea that has always been rather groundless. Therefore, in my opinion, my paper (27) was not without value, since it stated that the required energy output and cosmic-ray acceleration in radio galaxies are, in principle, easily provided by gravitational energy—in particular, in the process of star formation. I pointed out that "it seems more attractive to associate the galactic flares not with supernova flares, but with another large-scale mechanism, for example, gravitational instability of a galaxy or of its central part." This trend of thought continued to a certain extent in later papers devoted to quasars [see (28) and the literature cited therein, as well as (29)]. I would also like to mention paper (71) about heating of intergalactic gas.

The discovery of pulsars gave rise to the temptation to clarify the mechanism of their radio emission. Zheleznyakov and I (and partially V.

V. Zaytsev) published several articles on this subject, the last being the review (30) published in these volumes. But the problem proved to be much more complicated than it at first seemed. (It happens sometimes that a problem turns out to be particularly sophisticated when this is not suspected in advance.) For this reason I decided long ago to leave this field, and I am sure I was right to do so. Only representatives of the younger generation (or even generations) appeared to be able to gain a real understanding (though not to the bottom) of the very interesting but complicated and many-sided range of questions concerning pulsars, including the mechanism of their radiation and the whole theory of pulsar magnetospheres (see 31).

SYNCHROTRON RADIO EMISSION, COSMIC-RAY ASTROPHYSICS, AND GAMMA-RAY ASTRONOMY

Now I turn to cosmic-ray astrophysics and gamma-ray astronomy, which are very important areas of astronomy in which I have taken part. My activity in these fields started with the theory of synchrotron radio emission.

It is reasonable to comment first on the terminology, which has not yet become conventional. Cosmic rays are now usually understood to be charged particles (protons, other nuclei, electrons, etc.) of cosmic origin that possess high energy (say, kinetic energy $E_k \gtrsim 100$ MeV, but this limit is rather conditional). The whole range of questions connected with the origin and role of cosmic rays in space may be called cosmic-ray astrophysics, although it is frequently referred to as the problem of cosmic-ray origin. However, the origin (in the literal sense of the word) of cosmic rays is only a part of cosmic-ray astrophysics. The term high-energy astrophysics is also used and includes, besides cosmic-ray astrophysics, gamma-ray astronomy, X-ray astronomy, and high-energy neutrino astrophysics. Many authors, I among them, have written much on these subjects. I refer to the reviews (32–34) and particularly to the Proceedings of the International Cosmic-Ray Conferences (ICRC); the 20th ICRC was held in 1987 (35). [Paper (36) is my introductory talk at this conference.] Therefore I shall only briefly comment on the early stages in the development of the theory of cosmic synchrotron radiation and the appearance of cosmic-ray astrophysics [for more details, see (1, 2)].

Approximately in 1947–49 it became quite clear that comparatively long-wave, nonsolar cosmic radio emission (including, in particular, the very first radio astronomical measurements made by K. Jansky at a wavelength of about 15 m) possesses an effective temperature T_{eff} exceeding 10^4 K. Therefore, it was impossible to interpret such radio emission as thermal radiation from interstellar gas, since this gas, generally speaking, has a temperature $T \lesssim 10^4$ K, and in any case radiation with $T \gg 10^4$ K cannot be explained by thermal radiation of gas. Thus, one had to assume the existence of some nonthermal source analogous, for example, to the sporadic sources of nonthermal solar radio emission. This was how the "radio star hypothesis" arose in a quite natural way. According to this idea, some stars are anomalously powerful radio sources responsible for the nonthermal, cosmic radio emission with its continuous spectrum and diffuse directional distribution (37–39). The radio star hypothesis, however, faced many difficulties, mainly involving assumptions (sometimes arbitrary and unrealistic) concerning the hypothetical radio stars. An alternative soon appeared that with time became stronger and stronger— the synchrotron hypothesis of the origin of nonthermal radio emission, which in the end proved to be valid.

The competition or struggle between these two hypotheses took several years. From the physical point of view, synchrotron radiation had been known and understood for many years (40), and in the 1940s it was especially widely discussed in the physical literature in connection with the analysis of synchrotrons. But it was only in 1950 that the first papers (41, 42) appeared in which the synchrotron mechanism was considered as related to cosmic radio emission: Alfvén & Herlofson (41) discussed radiation from radio stars and Keipenheuer (42) that from interstellar space. It seems rather strange to me that these papers appeared in a journal of physics, and only then in the form of short letters. In any case, as far as I know, these papers did not attract the attention of astronomers. However, I at once believed that the synchrotron mechanism was responsible for nonthermal cosmic radio emission. I do not attribute this to any keen insight on my part, but rather to the above-mentioned fact that I was closer to physics and far from classical astronomy. In this situation the synchrotron mechanism seemed clear and realistic, whereas hypothetical, strange radio stars remained purely speculative. I immediately verified the calculations in these two papers, but, if I am not mistaken, I did not add anything essential. [I have not now compared all the expressions and estimates from (41), (42), and my own paper (43), submitted for publication on October 31, 1950, because it does not seem essential here, especially in that I have never claimed priority.] But the fact remains that my paper (43) was the first, and for some time the only one, that responded to the proposals of Kiepenheuer and of Alfvén & Herlofson to use the synchrotron mechanism in astronomy. The reaction of astronomers was probably quite the opposite, i.e. that the synchrotron mechanism seemed mysterious and speculative, whereas radio stars, although posing riddles, were more acceptable, for what kinds of stars cannot exist? In this respect, I. S.

Shklovsky was not an exception. He not only developed the radio star hypothesis (39), but also positively denied the synchrotron hypothesis ["which for a number of reasons does not seem to us to be acceptable" (39)]. This quotation from Shklovsky is presented in more detail in (2). I dare make this remark here only because it is just Shklovsky's paper (39) that has been repeatedly cited in the world literature as the first and principal discussion of the application of the synchrotron hypothesis. A similar error is often spread concerning the question of exploiting polarization of synchrotron radiation as a criterion for establishing the validity of a synchrotron origin of cosmic radiation [for more details, see (1, 2)]. After the appearance of a very important paper by Pikelner (44), who emphasized that the interstellar magnetic field exists in the entire galactic volume, Shklovsky (45) realized that his earlier objection (39) to the efficiency of the synchrotron mechanism had been groundless. For reasons that will become clear from the section to follow, I do not here dwell in more detail on Shklovsky's paper (45) and his subsequent papers [see (1, 2)].

Judging from the proceedings of the 1955 Manchester IAU symposium on radio astronomy (46), which included a paper on the radio star hypothesis while my paper sent to the symposium was not even published, astronomical "public opinion" in 1955 was still on the side of the radio star model. However, by the next IAU radio astronomy symposium, in Paris in 1958, the synchrotron mechanism was unconditionally accepted as the dominant production mechanism of nonthermal cosmic radio emission. (I do not mean, of course, the radiation coming from the solar atmosphere and generally from relatively dense regions.) This time my report "Radio Astronomy and the Origin of Cosmic Rays" was included in the proceedings of the symposium (47), although I myself was not able to participate.

The establishment of a connection between radio astronomy and cosmic rays has led, as a matter of fact, to the appearance of a new field of research in astronomy—namely, cosmic-ray astrophysics (now usually included in the term high-energy astrophysics). Although cosmic rays were recognized earlier (prior to 1950–53) to be cosmic objects, they were investigated only on the Earth (in the atmosphere and on its boundaries) exclusively for the purposes of high-energy physics. As is well known, the cosmic-ray results have led to exceedingly important discoveries in physics [$\mu^{\pm}$-leptons, $\pi^{\pm}$-mesons, and some other particles were discovered; see the sources cited in (2)]. But since cosmic rays are highly isotropic, studying them near the Earth is similar to making a spectral analysis of the light of all stars taken together. It is clear that under such conditions, without additional information about celestial bodies, astronomy could not have developed. The reception of synchrotron radio emission from cosmic rays (more precisely,

GINZBURG

from their electron component) has sharply changed this situation. It has become clear that cosmic rays are a universal phenomenon: They are present in interstellar space, in supernova remnants, and in galaxies. (Radio galaxies and then quasars have been discovered.) Radio astronomical data, together with the information on primary cosmic rays near the Earth and with the available astronomical concepts, have promoted further advances. I note that as far back as 1934, Baade & Zwicky (48) associated cosmic-ray generation and neutron star formation with supernova flares. In 1949, in considering cosmic rays as a gas of charged particles, Fermi (49) discussed a possible mechanism of their acceleration. These ideas, together with radio astronomical data, have been the basis for constructing the galactic model of the origin of the main part of cosmic rays observed near the Earth. Since 1953, I have supported only the galactic model with halo. The halo problem aroused many discussions (see e.g. 50), which seem to me to be the results of misunderstanding [see (50, 51)]. It is impossible to write here about this and many other problems of cosmic-ray astrophysics in further detail, and I once again refer the reader to the literature [see (32–36, 51)]. True, it is necessary to lay special emphasis on the principal achievement of recent years or, more precisely, of the last two decades. I refer to gamma-ray astronomy and, more specifically, to the reception of gamma-ray emission due to π^0-meson decay, with the energy of photons $E_\gamma > 30$–50 MeV. π^0-mesons are produced, in turn, during collisions of protons and nuclei [which make up nearly 99% (by number of particles) of cosmic rays] with gas nuclei in the interstellar medium. Therefore, identification of gamma-ray photons generated by π^0-meson decay provides information on the proton-nuclear component of cosmic rays far from the Earth. In this respect, gamma-ray astronomy plays the same role that radio astronomy plays in the study of the electron (or, more precisely, electron-positron) component of cosmic rays. Thus, only through the combination of radio and gamma-ray astronomical methods can we, in principle, obtain rather complete information on cosmic rays far from the Earth. Unfortunately, gamma-ray astronomy is not yet sufficiently developed. The reason is that it is necessary to use special satellites (gamma-ray observatories) to detect photons in the range 30 MeV $< E_\gamma < (1$–$5)$ GeV, but not a single gamma-ray observatory has been in operation, at the time of this writing (1989), since the very successful satellite *COS-B* finished its work in 1982.

The problems to be solved by gamma-ray astronomy cover the whole energy range from $E_\gamma \gtrsim 10^5$ eV to $E_\gamma \gtrsim 10^{16}$ eV [gamma-ray bursts, gamma-ray lines, diffuse gamma-ray emission, radiation from discrete sources—in particular, gamma-ray emission from SN 1987A (52)]. Here, too, I restrict myself only to references (34–36, 53).

The proton-nuclear component of cosmic rays generates neutrinos as well, and the potentialities of high-energy neutrino astrophysics are very large. [Since I have participated in a single work (54) in this field, I shall refer only to it and the review contained in (34).] The prospects for the development of cosmic-ray astrophysics and the whole of high-energy astrophysics have been repeatedly discussed, by me among others [most recently in (36)].

Summarizing, I can say that within the memory of my generation, astronomy has undergone a profound transformation—from optical to "all-wave"; in the course of this process, cosmic-ray astrophysics has been added, and in the near future it will also include high-energy neutrino astrophysics. Along with physics, I was lucky to plunge rather early into astronomical problems as well.

A FEW WORDS ABOUT PRIORITY

Questions of priority have long played an important role in the scientific community. Suffice it to recall how much time and strength the great Newton spent on priority questions (see e.g. 55) even in the period when he had already obtained general recognition. His was, of course, another era, but in not so remote times and even now the priority passions have flared and do flare up, although people have learned to hide them. At the same time, a great increase in the number of scientific publications and of various conferences has created additional difficulties.

On August 3, 1987, I gave an introductory talk at the 20th International Cosmic Ray Conference in Moscow; in the talk itself I did not mention the names of the authors for the majority of the papers that I touched upon. That is why I gave some explanations that were absent in the published version (36). For this reason and for purposes of further discussion, I briefly dwell on these explanations here.

First, the gleam of names in a talk (and, say, on overhead transparencies) hinders discussion of the material itself in that it diverts attention from the content. Second, papers and authors now are so numerous that it is impossible to mention everybody, while selection often brings displeasure and offense. In this connection I showed a transparency with the following two phrases:

"Priority questions are a dirty business. Priority mania or supersensitivity is a disease."

I then advised not to attach much importance to the absence of references. In most cases, this omission is explained not by evil intent but rather by the abundance of literature, by ignorance, by the desire to refer only to

GINZBURG

the most recent review, etc. It is only a negligible minority of authors who avoid citations deliberately and act from ill-intentioned motives. Such people and such actions, with some exceptions, do not deserve attention. As I understood, my transparency and comments were taken exactly as I expected, i.e. as advice given half-jokingly, inspired by my experiences of many years in the fields of physics and astrophysics. By the way, two weeks after (on August 21, 1987) I made a remark in the same spirit at the 18th International Conference on Low-Temperature Physics but in quite different conditions and atmosphere, due to which some people misunderstood me [see (4)]. But, of course, my way of thinking is exactly as I have tried to explain it above. My point is not that I am somehow indifferent to the question of priority. Yes, I do notice when my papers are cited and when not, but I never lay claims to authors who do not refer to my works. Besides feeling that such claims (which are, unfortunately, not rare) do not seem to be very decent, I think that in most cases, as I have already mentioned, the reasons for which a citation is sometimes omitted are not ill intentioned.

I decided to touch upon this question of priority for two reasons. First, it seems to me that it deserves attention because some people are deeply agitated about it and get upset. Therefore, one should not pretend that such a problem does not exist. Second, from time to time I happen to be, to this or that extent, involved in priority disputes or some collisions. I may state that I have never tried to prove or defend my priority. In all cases my indignation was provoked by the tactless behavior of my opponents. It is, of course, up to the reader to decide whether or not I am sincere—a man may be sophisticated and often play the hypocrite or distort, maybe unconsciously (subconsciously?), the truth. But I also have the right to express my opinion, which is what I have done above.

I would not do this, however, if I did not think it necessary to somehow explain here my relations with I. S. Shklovsky (1916–1985). It has been well known in my country and possibly abroad that since the end of 1967 I was on bad terms with Shklovsky, that we did not speak to nor greet each other. I do not know about the West, but in our country such a fantastic form of relations is sometimes encountered. By the way, it does not necessarily show some special enmity. In this case it simply happened, and neither of us wanted afterward to break the ice of estrangement. In the opinion of some colleagues, the break-off of my relations with Shklovsky was due to my priority claims. In my view, this is absolutely wrong. Fortunately, I need not prove this assertion here, since it is quite clear from papers (1, 2) written while Shklovsky was alive.

Of course, he made use of my papers and took into account my results, but I did the same with respect to his papers. True, we had different manners of citation, but this question is of secondary importance and therefore is not the point here.

We were of the same age and studied almost simultaneously at Moscow State University, but we made a closer acquaintance only during the Soviet expedition to Brazil, aimed at observing the total solar eclipse of May 20, 1947, in which we both participated. From then our relations, if not very close, were quite friendly up to 1966. By the way, all those "claims" that I could, in principle, lay to Shklovsky refer to the pre-1966 period [see references in (1, 2)]. But I did not pay attention to his tactlessness and attached no importance to it. But from 1966 on, beginning with the IAU conference on radio astronomy in Nordwijk, Shklovsky was ill-disposed toward me for some reason, of which I can only guess. This had no consequences until the fall of 1967, when Shklovsky published his paper (56) in which he associated powerful X-ray radiation from a number of sources with binary systems, in which one of the components is a neutron star onto which plasma accretes.Meanwhile, all of this had, as a matter of fact, been said at a round-table discussion held during the Nordwijk conference the year before (50). The results of the discussion were reported at the conference by G. Burbidge, who was the chairman of this discussion [see (50, p. 463)]. I was present at the discussion, and so was Shklovsky, who, as far as I remember, did not utter a word. In his paper (56) he did not refer to this discussion, which was published some time later (50). The history of the interpretation of powerful X-ray radiation has been described in more detail by Burbidge (57), who also narrated what I have said above. I spoke at the round-table discussion and after Burbidge's report [see (50)] as well, but I have not published anything on this subject and have no special priority claims. But as far back as 1966, Ya. B. Zel'dovich, together with I. D. Novikov, had published a paper (58) containing the idea of accretion onto a neutron star. According to Zel'dovich, Shklovsky knew about it but made no reference to it in (56). Zel'dovich was generally very sensitive to priority questions, and with respect to Shklovsky's paper (56) he was indignant and told me about this. I, in turn, told him about the Nordwijk round-table discussion. Eventually, we wrote (of course, with both our signatures) a corresponding letter to Shklovsky, and I added there the examples presented in (1, 2). I have made no attempts now to find the copy of this letter among my papers. The letter was not distributed—it was sent to Shklovsky only. However, his close colleagues read this letter. Zel'dovich and I were led by the naive idea that Shklovsky's eyes would be opened, but nothing of the sort happened. He wrote an answer, but I was not shown it because I left for England for three months and when I returned, Shklovsky either was ill or fell ill soon thereafter, and thus his friends and coworkers decided to "damp down" this matter.

After that I did not communicate with Shklovsky, but he remained on diplomatic terms with Zel'dovich. I think that in sending this letter, Zel'dovich and I made a mistake, the more so as it did not change in any way Shklovsky's position. In particular, in his paper written in 1982 on the occasion of the 20th anniversary of X-ray astronomy, and even in spite of the paper by Burbidge (57), Shklovsky insisted on the priority of his paper (56).

Shklovsky was a very capable man and had a number of actual achievements in the field of astrophysics. As regards his unfortunate statements concerning radio astronomy and X-ray astronomy, they were harmful, first of all, to himself; I will not take the liberty of interpreting his motives. In his last years, Shklovsky wrote autobiographical short stories. At the end of 1987 I was given a volume of these short stories, entitled *Echelon*. I found this book very interesting and hope it will someday be published. Influenced by Shklovsky's short stories, I wrote a rather long letter (addressed to L. S. Marochnik) in which I narrated the story of my relations with Shklovsky and the cause of our quarrel. This letter was read by Shklovsky's former coworkers from the Institute for Space Research, and I asked one of them (B. V. Komberg) to keep this letter and, if possible, other documents. Historians of astronomy (or psychology?) may one day be interested in these documents.

There remains only to answer the following question: Was this worth writing? Shklovsky is unfortunately not with us now, and so my position is rather delicate—he cannot object to me, and how can I avoid suspicions of ignoble posthumous settling a score with him? I understand this, and at first hesitated, but I still decided not to delete this section. First, all my remarks are completely based on already published materials, which are cited in (1, 2) and above. Thus, those interested may check everything. Second, some unpublished material (including the letter—Shklovsky's reply—that I have not read) is with his former coworkers, and I will only be too glad if they now publish everything that they may care to, including their own comments. Finally, as I write this paper (in fall 1988) I am already 72 years old, and I am not at all sure I shall live to see it published. Under such conditions it seems to me simply preposterous to write not with the goal of clarifying the truth but from some other considerations. I hope nobody will doubt that.

CONCLUDING REMARKS

In this paper I have not intended to enumerate all my papers that refer to astronomy. The main references, however, were made, directly or indirectly, when I referred the reader to papers (1, 2) and others. In view of the fact that this paper is to be published in the *Annual Review of Astronomy and Astrophysics*, it seems not out of place to refer separately to my papers (59, 60), which were published in this series and devoted to the theory of synchrotron radiation. I shall also mention papers (61, 62) on X-ray and gamma-ray astronomy. In (63) I discussed the problem of superfluidity and superconductivity in astrophysics.

Besides these, I shall dwell only on the papers connected with the general theory of relativity (GTR). GTR is, of course, a physical theory, but it is so closely connected with astronomy that it can be referred to as astrophysics as well. This, in particular, is what I did in my popular book (64).

In general, I did not work hard on GTR and do not belong to the so-called relativists. I shall mention the study of the collapse of a magnetic star (29, 65) which led to the conclusion that the magnetic moment of the star tends to zero as the Schwarzschild radius is reached. This was the first indication of an effect that later was described by the phrase "a black hole has no hair." How large the magnetic field of collapsing stars can be was pointed out in (29), several years prior to the discovery of pulsars and, therefore, neutron stars.

The application of the equivalence principle, the fundamental principle of GTR, in taking into account the quantum effects and, specifically, zero-point fluctuations of different fields is discussed in (66). Finally, several of my papers were devoted to experimental verification of GTR [my last review on this subject is (67); see also (64)] and to the limits of its applicability. [I refer only to (68, 69).] The latter problem is associated with the question of limitations on certain physical concepts when applying them to astronomy, a question that has attracted the attention of some astronomers. My Darwin Lecture (18) was devoted to this subject. Today I am of the same opinion as that expressed in this lecture—namely, that one can give no definite guarantees, but it is most probable and realistic that if some deviations from the known physical laws and, specifically, from Einstein's original, classical GTR can be expected, they may happen only at very early stages of cosmological evolution. [I mean the observable, expanding Universe, which, from the point of view of some inflationary cosmological models, may appear to be part of a larger or even infinite system—see the review (70).] True, I am not going to touch upon black holes that are deep below the horizon (i.e. for a Schwarzschild black hole for $r \ll r_g = 2GM/c^2$), nor upon mini–black holes, cosmic strings, etc.

One of the possible ways in which GTR might be violated is connected with the existence (purely hypothetical) of a fundamental length $l_f \gg l_g = (G\hbar/c^3)^{1/2} \sim 10^{-33}$ cm [for more details, see (69)]. At later stages of cosmological evolution (for example, for the redshift parameter $z < 10^3$), there is absolutely no reason to expect that the well-known laws of macroscopic physics do not apply. Of course, this is on the whole a large and complicated problem, and it is impossible to consider it here more thoroughly.

In concluding, I would like to ask this final question: Have I contributed anything essential to astronomy? This is undoubtedly up to other people to decide, but the author himself also has the right to have his own opinion, even if it is erroneous. For instance, I believe that I have made a noticeable contribution to the theory of superconductivity, but with respect to astrophysics I do not have any definite opinion. I have done some things, of course, but how important are they, and what role have they played? It would be interesting to have the answer, but unfortunately I do not know a single work that considers in detail and reliably assesses the history of "new astronomy"—i.e. the development of astronomy over the last 40–50 years. On the contrary, I have come across rather many papers and even booklets or books containing errors concerning this history. Absurd assertions, adapted by repetition, roam from one such paper or book to another. It is surprising how credulous some authors who elucidate history are not to consult all the original literature and check their assertions. I hope that the situation will be clarified with time and believe that the prefatory chapters published in the Annual Reviews will make their contribution to this process. I hope the present paper is not an exception, although I have not tried [with some exceptions; see (1, 2)] to take on the role of historian—that is, to compare, verify, and assess the results and assertions of various authors. Besides, against my own expectations and original aim, this paper proved to be devoted not so much to astrophysics as to autobiographical and other remarks. But maybe this is not so bad.

ACKNOWLEDGMENT

I would like to warmly thank Kip Thorne and Keith Dodson for reading and commenting on the manuscript.

Literature Cited

1. Ginzburg, V. L. 1984. In *The Early Years of Radio Astronomy*, ed. W. Sullivan, pp. 289–302. Cambridge: Univ. Press
2. Ginzburg, V. L. 1985. In *The Early History of Cosmic Ray Studies*, ed. Y. Sekido, H. Elliot, pp. 411–26. Dordrecht: Reidel
3. Ginzburg, V. L. 1986. *Priroda* 1986(10): 80–94 (In Russian)
4. Ginzburg, V. L. 1988. *Prog. Low-Temp. Phys.* 12: 1–44
5. Ginzburg, V. L. 1946. *C. R. (Dokl.) Acad. Sci. URSS* 52: 487–90
6. Ginzburg, V. L. 1942. *C. R. (Dokl.) Acad. Sci. URSS* 36: 8–13
7. Ginzburg, V. L. 1987. In *Reminiscences About I. E. Tamm*, p. 190. Moscow: Nauka (In Russian)
8. Ginzburg, V. L. 1989. *Applications of Electrodynamics in Theoretical Physics and Astrophysics*. New York: Gordon & Breach

9. Ginzburg, V. L. 1984. *Sov. Phys. Usp.* 26: 713
10. Ginzburg, V. L. 1989. In *Reminiscences About L. D. Landau.* Oxford: Pergamon. In press
11. Ginzburg, V. L. 1986. In *The Lesson of Quantum Theory,* p. 113. Elsevier [Unfortunately, my paper was cut down by the editors (evidently for want of space). The complete version is published in Russian in 1986. *Tr. Fiz. Inst. P. N. Lebedev Akad. Nauk SSSR* 176: 3. This volume is being translated into English by Nova Science Publ., New York)
12. Ginzburg, V. L., Man'ko, V. I. 1976. *Sov. J. Part. Nucl.* 7: 1
13. Ginzburg, V. L. 1987. In *Quantum Field Theory and Quantum Statistics,* 2: 15–33. Bristol, Engl: A. Hilger
14. Ginzburg, V. L., Landau, L. D. 1950. *Zh. Eksp. Teor. Fiz.* 20: 1064–82
15. Ginzburg, V. L. 1988. In *Ynogo ne dano* (*One Way*), pp. 135–53. Moscow: Progress Publ. (In Russian)
16. Garfield, E. 1977. *Current Contents,* 1977(49): 5–15
17. Ginzburg, V. L. 1962. *Tr. FIAN* (*Proc. P. N. Lebedev Phys. Inst.*) 18: 55–104
18. Ginzburg, V. L. 1975. *Q. J. R. Astron. Soc.* 16: 265–85
19. Ginzburg, V. L. 1970. *Propagation of Electromagnetic Waves in Plasmas.* Oxford: Pergamon
20. Martyn, D. F. 1946. *Nature* 158: 632
21. Shklovsky, I. S. 1946. *Astron. Zh.* 23: 333
22. Ginzburg, V. L. 1947. *Usp. Fiz. Nauk* 32: 26
23. Ginzburg, V. L. 1948. *Usp. Fiz. Nauk* 34: 13
24. Ginzburg, V. L. 1980. In *Oort and the Universe,* p. 129. Dordrecht: Reidel
25. Ginzburg, V. L., Zhelezhnyakov, V. V. 1958. *Astron. Zh.* 35: 694 (*Sov. Astron. AJ* 2: 653)
26. Zhelezhnyakov, V. V. 1970. *Radio Emission of the Sun and Planets.* Oxford: Pergamon
27. Ginzburg, V. L. 1961. *Astron. Zh.* 38: 380
28. Ginzburg, V. L., Ozernoy, L. M. 1977. *Astrophys. Space Sci.* 50: 23 (see also 1970. *Astrophys. Space Sci.* 9: 116)
29. Ginzburg, V. L. 1964. *Dokl. Akad. Nauk SSSR* 156: 43–46 (*Sov. Phys. Dokl.* 9: 329)
30. Ginzburg, V. L., Zheleznyakov, V. V. 1975. *Annu. Rev. Astron. Astrophys.* 13: 511–35
31. Beskin, V. S., Gurevich, A. V., Istomin, Ya. N. 1986. *Sov. Phys. Usp.* 29: 946 (see also 1988. *Astrophys. Space Sci.* 146: 205)
32. Ginzburg, V. L., Syrovatskii, S. I. 1964. *The Origin of Cosmic Rays.* Oxford: Pergamon
33. Ginzburg, V. L., Ptuskin, V. S. 1976. *Rev. Mod. Phys.* 48: 161, 675
34. Ginzburg, V. L., ed. 1990. *Astrofizika kosmicheskych luchey* (*Cosmic Ray Astrophysics*). Moscow: Nauka. 2nd ed. (expected to be translated into English by North-Holland Publ., Amsterdam)
35. 1987. *Proceedings of International Cosmic Ray Conference* (*ICRC*), 20th, Moscow
36. Ginzburg, V. L. See Ref. 35, 7: 7 (see also 1988. *Usp. Fiz. Nauk* 155: 185; 1988. *Sov. Phys. Usp.* 31: 491–510)
37. Ryle, M. 1949. *Proc. Phys. Soc. London Sect. A* 62: 491
38. Unsöld, A. 1949. *Z. Astrophys.* 26: 176
39. Shklovsky, I. S. 1952. *Astron. Zh.* 29: 418
40. Schott, G. A. 1912. *Electromagnetic Radiation.* Cambridge: Univ. Press
41. Alfvén, H. A., Herlofson, N. 1950. *Phys. Rev.* 78: 616
42. Kipenheuer, K. O. 1950. *Phys. Rev.* 79: 738
43. Ginzburg, V. L. 1951. *Dokl. Akad. Nauk SSSR* 76: 377
44. Pikelner, S. B. 1953. *Dokl. Akad. Nauk SSSR* 88: 229
45. Shklovsky, I. S. 1953. *Astron. Zh.* 30: 15
46. 1957. *Radio Astronomy, IAU Symp. No. 4.* Cambridge: Univ. Press
47. 1959. *Paris Symposium on Radioastronomy, IAU Symp. No. 9.* Stanford, Calif: Stanford Univ. Press
48. Baade, W., Zwicky, F. 1934. *Phys. Rev.* 46: 76 (also 1934. *Proc. Natl. Acad. Sci. USA* 20: 259)
49. Fermi, E. 1949. *Phys. Rev.* 75: 1169
50. 1967. *Radio Astronomy and the Galactic System, IAU Symp. No. 31.* Academic
51. Ginzburg, V. L. 1989. In *Essays on Particles and Fields,* p. 103. Bangalore: Indian Acad. Sci.
52. Beresinsky, V. S., Ginzburg, V. L. 1987. *Nature* 329: 807
53. Dogiel, V. A., Ginzburg, V. L. 1989. *Space Sci. Rev.* 49: 311–83
54. Beresinsky, V. S., Ginzburg, V. L. 1981. *MNRAS* 194: 3
55. Westfall, R. 1982. *Never at Rest: A Biography of Isaac Newton.* Cambridge: Univ. Press
56. Shklovsky, I. S. 1967. *Ap. J. Lett.* 148: L1
57. Burbidge, G. 1972. *Comments Astrophys. Space Sci.* 4(4): 105
58. Novikov, I. D., Zel'dovich, Ya. B. 1966. *Nuovo Cimento Suppl.* 4: 810

59. Ginzburg, V. L., Syrovatskii, S. I. 1965. *Annu. Rev. Astron. Astrophys.* 3: 297–350

60. Ginzburg, V. L., Syrovatskii, S. I. 1969. *Annu. Rev. Astron. Astrophys.* 7: 375–420

61. Ginzburg, V. L., Syrovatskii, S. I. 1965. *Space Sci. Rev.* 4: 267–312 (see also 1964. *Dokl. Akad. Sci. SSSR* 154: 557)

62. Ginzburg, V. L. 1967. *Sov. Phys. Usp.* 9: 543–50

63. Ginzburg, V. L. 1969. *Sov. Phys. Usp.* 12: 241–51 (see also 1971. *Physica* 55: 207–12)

64. Ginzburg, V. L. 1985. *Physics and Astrophysics (A Selection of Key Problems)*. Oxford: Pergamon

65. Ginzburg, V. L., Ozernoy, L. M. 1965. *Sov. Phys. JETP* 20: 689–96

66. Ginzburg, V. L., Frolov, V. P. 1987. *Usp. Fiz. Nauk* 153: 633–74 (Engl. transl., *Sov. Phys. Usp.*)

67. Ginzburg, V. L. 1980. *Sov. Phys. Usp.* 29: 514–27

68. Ginzburg, V. L., Kirzhnits, D. A., Lyubushin, A. A. 1971. *Sov. Phys. JETP* 33: 242–46

69. Ginzburg, V. L., Mukhanov, V. F., Frolov, V. P. 1988. *Zh. Eksp. Teor. Fiz.* 94: 1 (Engl. transl., *Sov. Phys. JETP.* 1988. 67: 649)

70. Barrow, J. D. 1988. *Q. J. R. Astron. Soc.* 29: 101

71. Ginzburg, V. L., Ozernoy, L. M. 1966. *Astron. J.* 9: 726

Lyman Spitzer, Jr.

Annu. Rev. Astron. Astrophys. 1989. 27:1–17

DREAMS, STARS, AND ELECTRONS

Lyman Spitzer, Jr.

Princeton University Observatory, Peyton Hall, Princeton,
New Jersey 08544

"What were you doing today, Daddy?" my children often asked me. I would reply, "I was thinking what I would do if I were an electron." Much of my professional work has been based on an ability to visualize a physical system and how it operates and thus to predict its behavior without benefit of mathematics—though mathematics provides a much needed check. This particular facility has led me to work on a variety of physical processes important in astrophysics.

Another characteristic that has greatly affected my research is an attraction to spectacular and difficult projects that strike me as important. Three major such projects, which provided long-term goals for my career, concerned the theory of star formation, a large general-purpose space telescope, and generation of useful power from fusion energy. The objective of the first was to understand the sequence of complex processes occurring; those of the last two involved engineering as well as scientific problems.

This paper deals with my various research activities, especially those that now seem relevant to further work in astronomy. The personal characteristics described above are mentioned because they provide the unifying strands in what might otherwise seem a scattered, somewhat miscellaneous professional career. Some personal items relevant to my scientific life are included also in the subsequent text. In the words of Pooh-Bah (1), perhaps these will give verisimilitude to an otherwise bald and unconvincing narrative.

Early Years

As a boy I enjoyed much of my school work, but a serious interest in science developed only after entering Phillips Academy, Andover, in 1929, at age 15, for two precollege years. There I was introduced to physics by Frederick Boyce, who had a unique capability for making this topic both clear and exciting. It was a wonderful experience to find that by understanding a few simple principles one could master an entire area of physics. In response to a request from several of us, Mr. Boyce gave an astronomy course, supplementing the books by Jeans and Eddington that I had been reading and leaving me permanently fascinated with astrophysics.

At the end of my stay at Andover I became involved in my mind with the first of the big projects that have appealed to me. This concerned the development of a global transportation system, based on the electromagnetic propulsion of levitated cars in evacuated tubes. Others have suggested such ideas. My concept was to integrate intercity travel with local transportation, so that one could get into a car on a high floor of an office building, dial the local or long distance number desired, and get out some minutes later at the desired floor in one's own city or in some remote center. I spent so much time considering various details of such a system that my parents, seriously concerned for my sanity, extracted a promise from me to stop thinking about this enterprise!

From Andover I went on to Yale, where astronomy (as well as the global transportation system) was put aside while majoring in physics. Under Leigh Page I took a general survey course in theoretical physics and found a keen aesthetic pleasure from his elegantly organized mathematical presentations. A limitation of my own in such areas became painfully evident during my final year at Yale, when I took a final oral examination on a physics thesis I had written for some independent work. Said one of my examiners, "Your second equation, Mr. Spitzer, states that the power P equals ir^2. Could you derive this for us, please?" Swiftly going through the elementary steps, I suddenly realized that I had inadvertently written ir^2 instead of the familiar i^2r, and that all the subsequent equations in my thesis were incorrect! My mind tends to make inversions of this sort, and while I have learned to ferret out most of them, they still dog my work from time to time.

In graduate school, first at Cambridge, England, (as a Henry Fellow) in 1935–36 and then at Princeton, I returned to astrophysics, partly because I was so enchanted by the beautiful theories that S. Chandrasekhar had presented at a series of informal evening seminars in his Trinity College rooms, and partly because I was so impressed by the physical lucidity of several contemporary papers by Bengt Strömgren. Studying under Henry Norris Russell was also a unique experience; his enormous knowledge, physical insight, and unfailing enthusiasm made him a very stimulating mentor. The lectures by Edward Condon were particularly helpful in extending my own insight into various aspects of the quantum theory.

My thesis topic, suggested by Russell, was an analysis of high-dispersion spectra that Walter Adams had obtained for three cool supergiant stars. I proposed a fountain model to explain the observed dependence of radial velocity on excitation potential. While later work by others disproved this model, the research was educational for me.

During my postdoctoral year at Harvard (as a National Research Council Fellow) in 1938–39, I found tremendously stimulating the scientific discussions among the active scientists whom Harlow Shapley, Donald Menzel, and Bart Bok had assembled at the Observatory. In particular, the free-ranging exchange of views at Shapley's "hollow square" discussion groups and at the evening seminars organized by Bok at his home gave all of us many exciting, important problems to work on. As pointed out below, my subsequent research in stellar dynamics and in star formation and interstellar matter grew out of these discussions.

In this active ferment of astronomical ideas Martin Schwarzschild was an imposing figure. He greatly impressed me by his insight, originality, and forcefulness. I felt strongly that he would be an ideal colleague if events should make it possible for us to be in the same institution.

Yale and Princeton

A brief discussion of my years at these two institutions will provide a backdrop for a discussion of my research work there. At Yale, where I spent two and a half years before shifting to war research in New York and Washington, and another year and a half afterward, I was alone much of the time as far as astrophysics was concerned. This did not worry me, since the impetus I had received at Princeton and Harvard would suffice for some years, and informal get-togethers of astronomers in the region from Washington to Cambridge were frequent.

My initial appointment in 1939 was as an instructor in the physics department, which gave me a welcome chance to broaden my background in this basic field, as well as to continue some research I had begun at Harvard. Under the influence of Henry Margenau I published a few papers on the collisional broadening of spectral lines, giving relatively exact solutions in certain idealized cases. When I returned to Yale after four years of underwater sound research, I had a joint appointment in physics and astronomy and concentrated my research on problems related to interstellar matter.

In 1947 I was offered a professorship at Princeton and also the directorship of the Observatory, succeeding Russell. I soon accepted; ever since my graduate student days I had felt this would be an ideal position. For this offer I am greatly indebted to three of this country's greatest astronomers—to Russell for his support, to Chandrasekhar for not accepting the earlier offer to him of Russell's professorship, and to Shapley for informing Princeton of my long-standing interest in this particular post and then serving as a helpful intermediary between me and the Princeton administration. In addition to his widespread other activities, Shapley ran sort of an informal employment bureau for young astronomers. Following the tradition established by Russell, I held the directorship (together with the departmental chairmanship) for some three decades; my brilliant colleague Jerry Ostriker succeeded me in 1979, three years before my retirement from the faculty (compulsory then at age 68).

One of my most important achievements at Princeton was to persuade Martin Schwarzschild to join me there. We worked closely and effectively together, his wisdom in practical affairs supplementing my willingness to get absorbed occasionally with administrative details. Our joint objective, when we both moved to Princeton in 1947, was to build a significant graduate program in theoretical astrophysics, with some emphasis also on observations. In support of this latter purpose the Princeton administration agreed to send Schwarzschild and me to the western US (in practice, usually the Mt. Wilson Observatory) for observing, each of us in alternate years for one semester. This arrangement certainly encouraged observational work by the two of us and our students. After 15 years, the shift of our observational research plans to space telescopes forced us to drop, with great reluctance, these regular observing trips to the West Coast.

Interstellar Matter

On arriving at Harvard in 1938 I found widespread discussion of Bethe's now classic explanation of stellar energy generation (half then in published form, half in press), showing that nuclear fusion provided the basic energy source for stars. This conclusion had far-reaching consequences for the age and evolutionary history of stars. From active discussions during that year a simple picture of stars and galaxies began to take shape in my mind.

A basic element of this picture as I then envisioned it was the obvious fact that the highly luminous O and B stars, which are seen in spiral galaxies, must exhaust their hydrogen fuel in much less time than the age of the Universe. That such stars are not seen in elliptical galaxies or in the globular clusters of our Galaxy suggests that these systems are composed of old or primordial stars, formed early in the life of the Universe. It seemed reasonable to attribute the formation of the young stars in spiral galaxies to the interstellar gas, which is a characteristic feature of spiral systems; there were very few traces of gas and dust in elliptical galaxies or globular clusters. Moreover, the absence of pervasive gas in these latter systems is theoretically not unexpected, since the random velocities of atoms are limited by thermal processes and tend to be much less than the random velocities required to support the stars in the observed elliptical or spherical systems.

This broad though tentative picture, which identified the interstellar clouds in spiral galaxies as the birthplace of massive early-type stars, strongly appealed to me and led me to embark on a program of interstellar matter studies, with the goal of understanding how new stars are formed. The essential first step in this program was to determine the physical conditions in the interstellar gas and dust observed in our Galaxy. Hence during my first year at Yale, I computed such items as the mean free path of electrons, positive ions, neutral atoms, and dust grains in interstellar space; I showed that these grains would tend to be negatively charged because of the higher random velocities of electrons as compared with positive ions. An important inference of these early studies was that under normal conditions, the grains would not drift much through the gas. Another conclusion was that different regions of the gas would normally tend toward pressure equilibrium with each other.

In a draft of a paper presenting these detailed results I included an exposition of the broad picture that is described above and that provided the motivation for the work. Several older astronomers advised me not to publish this speculative discussion, which one of them described as "metaphysical." I followed their advice, though with some reluctance. Over subsequent years this picture has become generally accepted, thanks in large part to the detailed observations of elliptical and spiral systems by Walter Baade (2) and others. Quite possibly my published results (3) on the kinetic properties of the interstellar gas would not have received serious attention if they had appeared as part of a grandiose and controversial package.

On returning to Yale after the war, I resumed the analysis of physical conditions in the interstellar gas. In particular, I began a lengthy calculation of the kinetic temperature of this gas, based on the assumption of radiative equilibrium (i.e. that this temperature adjusts itself so that the gains of kinetic energy, per unit of time and volume, just equal the corresponding losses by radiation). For planetary nebulae many of the relevant reactions had been analyzed earlier by Menzel and his collaborators, but for much of the interstellar gas additional physical processes are important; the density, radiation flux, and state of ionization can be quite different in these two environments. Strömgren had pointed out (4) several years earlier that interstellar H atoms in regions of average density or higher must normally be almost entirely neutral or almost entirely ionized; these two types of regions are now generally designated (3) as H I and H II, respectively.

The actual calculations of temperature were made (5) for somewhat idealized atoms, but they sufficed to show a large difference of temperature between H I and H II regions. In H II zones, a strong heating source is provided by the photoionization of neutral H atoms, at a rate set by the number of proton-electron recombinations per unit time and volume. The electrons ejected in the photoionization process carry off kinetic energy, which goes into heating the gas, maintaining a temperature of some 10^4 K despite rapid radiation by collisionally excited heavy ions such as C^+ and O^+. In H I regions the ionizing photons are absent as a result of absorption by the neutral H atoms. Hence this energy source is missing; other such sources are weaker by several orders of magnitude. However, the radiation rates at a given density and temperature are not so very different between the two regions. As a result, the H I regions are necessarily much colder, with T in the general neighborhood of 100 K. In pressure equilibrium the H I region will have a particle density roughly 100 times greater than that of an H II zone, with its temperature of about 10^4 K.

More recent calculations (6, 7) are, of course, much more detailed, taking into account the individual energy levels of the abundant atoms and allowing for important additional processes, such as radiative transfer of ionizing photons and enhanced photoelectric efficiency of dust grains for short wavelengths and small grain radius. For H I regions of low density ($n_H < 1$ cm^{-3}), these more realistic models give significantly higher temperatures than were obtained earlier. The important conditions for possible instability of all these equilibrium temperatures were thoroughly explored by George Field (8), our former graduate student and colleague.

Another physical quantity of some importance in interstellar space is the electrical conductivity, and I resolved to compute this more accurately for fully ionized gas than had been done previously. This work, which spread over several years, was part of my effort in plasma physics and is discussed in the last section below.

At about this time a new observational result was obtained in collaboration with Paul Routly, based on observations made at the Mt. Wilson Observatory over several years. High-resolution spectra of some 20 O and B stars were obtained (9), showing components of the interstellar Na D lines, which could be compared with similar data on the K and H lines of Ca II previously obtained by Adams (10). Quantitative analysis of these data indicated (9) that the ratio of neutral Na to Ca$^+$ along lines of sight to various clouds varied strongly with cloud velocity; this ratio decreased by about an order of magnitude when the velocity (measured in the local standard of rest) increased from less than 10 to more than 20 km s^{-1}. A later survey, including some 60 stars, gave very similar results (11). These observations may be understood (12) if one assumes that the Ca atoms tend preferentially to be "locked up" in the dust grains (an effect termed "depletion"), but that in high-velocity clouds the grains have evaporated by some mechanism during the process of cloud acceleration, producing a Na-Ca ratio more in accord with cosmic abundance. This assumption was subsequently confirmed by measures of other elements, especially Si and Fe, with the *Copernicus* satellite (see below); these data also indicated that depletion tends to disappear at the higher cloud velocities (13).

A brief research effort combining observations with quantum mechanical theory grew out of a luncheon discussion with Jesse Greenstein, who pointed out the existence of a blue continuous emission in the spectra of planetary nebulae. The upshot was a joint paper (14) that computed for the first time the precise transition probability and the spectral distribution of two-photon emission by hydrogen atoms in the metastable 2s state; our paper used these theoretical results in a preliminary interpretation of the limited astronomical data.

Two physical effects that may be important for interstellar clouds were discussed in joint papers with visiting astronomers at Princeton. The acceleration of clouds by O-type stars was discussed in a paper with Jan Oort (15), where we pointed out that on the side of the cloud facing the star the ionized, heated gas would stream back toward the star, accelerating the cloud by a rocket effect. The detailed theory agrees (16) with the observed brightness distribution of the luminous rims between dark interstellar clouds and H II regions. This process of cloud acceleration may account for an appreciable fraction of the kinetic energies in the observed H I diffuse clouds.

A paper with Leon Mestel (17) discussed how cloud collapse would be affected by the presence of a magnetic field. An important process in this context is the systematic motion of the plasma, including positive ions, electrons, and magnetic lines of force, through the gas of neutral H and He atoms. This process is known in the laboratory as "ambipolar diffusion" (drift of positive and negative ions at the same rate through neutrals) and is sometimes called "plasma drift." Especially in view of the very large cross section (10^{-14} cm^2) for encounters between positive ions and H atoms at low temperature (18), this mechanism is most likely to be important in the later stage of cloud collapse, when the density and opacity are relatively high and the fractional ionization is consequently very low.

My emphasis on the tendency toward pressure equilibrium (12) led to a speculative suggestion (19) concerning the interstellar gas far from the galactic plane. Observations of the interstellar K line in stars at different distance z from the galactic plane indicated (20) that for some of the clouds absorbing this line, z must be at least 1000 pc. To prevent such clouds from rapidly expanding and disappearing, the pressure of a surrounding medium seems required, and if such a medium consists of hydrogen and extends up to $z \approx 1000$ pc its temperature must be of order 10^6 K. I called this a "coronal gas." While subsequent ultraviolet observations from the *Copernicus* and *IUE* satellites, and observations of soft X-rays also, have confirmed the existence of a hot gas within the galactic disk, whether such a gas extends up to high z is still not quite certain (21, 22).

Over the years I took an active interest in the polarization of starlight by magnetically aligned grains and collaborated with other scientists in a number of analyses. Perhaps the most noteworthy of the papers resulting was one with R. V. Jones, a solid-state physicist at Harvard; we pointed out (23) that alignment by dissipative processes, apparently the most effective mechanism (24), would be much enhanced if the Fe atoms in grains were assumed to be distributed in clumps, leading to ferromagnetic or superparamagnetic grains with a much greater energy dissipation rate for rotation in a magnetic field. The increased rotation rate produced by "spin-up" (25) also increases the rate of rotational energy dissipation and may make grain alignment possible even with conventional values for grain paramagnetism. On the other hand, Fe atoms must be present in grains, and their aggregating in clumps seems not unlikely.

One may ask what happened to the initial objective of this interstellar research. How was this knowledge used in theories of star formation? During a dozen or so years after I moved to Princeton I wrote a number of papers, applying to star formation theories the results that I had obtained on the properties of the interstellar gas. At the time I found some of these results rather exciting, but in retrospect they seem to be rather small steps in a long journey. To analyze the condensation of turbulent gas in a rotating galaxy, in the presence of such complicating factors as magnetic fields, cosmic rays, and a full spectrum of electromagnetic radiation, is a very complex undertaking. My own preliminary efforts in this direction (26) led to interesting idealized models and have been of some use as stepping stones for later researchers.

In my opinion there have been two special consequences from my research on these interstellar problems. First was a book—in two successive versions (7)—on this topic that summarized our knowledge of the relevant physical principles. My own research concerned only a fraction of the material presented, but writing the book provided excellent training for me in many subjects. While the text is selective rather than exhaustive, it seems to provide a useful introduction to the physical laws and mechanisms operating in the interstellar gas.

A second consequence was the interstellar research with the *Copernicus* satellite. This program, which led to a substantial jump in our understanding of the interstellar gas, is discussed below in connection with my space astronomy activities.

Stellar Dynamics

My involvement in the dynamics of star systems began one evening in the fall of 1938 in the home of Bart and Priscilla Bok. The occasion was an informal discussion of various research problems, including in particular the equilibrium of a globular cluster. In such clusters random encounters between stars tend to produce a Maxwellian distribution of velocities in a time much shorter than the cluster age. Thus, one would expect these clusters to approach isothermal spheres. But theory shows that such a sphere must have an infinite mass, which makes it a difficult model for an actual cluster to approach. What happens?

After pondering this problem for a while I finally came up with the correct answer, which now seems rather obvious. As random encounters tend to establish a Maxwellian velocity distribution, stars in the tail of this distribution will acquire a velocity greater than the escape velocity (which on the average is only twice the rms random velocity) and promptly leave the cluster. I called this process "evaporation" and sent off a paper (27) discussing an approximate theory of this effect. Not until several years later did I discover that the great Russian astrophysicist V. A. Ambartsumian had published a similar theory (28). At least our results agreed! These theories were based on the totally unrealistic but tractable model of a uniform cluster. To obtain a better approximation would have involved mathematical complications, and I left this more general problem untouched for some 30 years.

A decade later, Martin Schwarzschild pointed out to me the observed increase of velocity dispersion with average age for stars in the galactic disk. It occurred to us that random encounters between stars and massive interstellar clouds might explain this result. My work on the electrical conductivity of a fully ionized gas, which was under way at that time, gave me a good training in the theory of encounters between particles subject to mutual inverse-square forces. Our analysis of encounters between stars and clouds suggested (29) that these could account for the observed stellar velocities if cloud masses were as great as $10^6 \ M_\odot$. Detailed numerical simulations (30) have given recent support to this view. The mass required seems high for clouds in the solar neighborhood (31), but extended cloud complexes, such as spiral arms, can under some conditions produce the same acceleration of stars (29) without any systematic velocity of their own other than differential galactic rotation.

A few years later Oort raised informally the question of why so few open clusters could be found with ages exceeding some 10^8 yr. I considered this problem and computed (32) the rate at which a cluster in the galactic disk could be disrupted by encounters with passing clouds, whose transient tidal force must tend to increase the kinetic energy of the cluster stars. To give a simple result I introduced the "impulsive approximation," in which the random stellar motions during the passage of the cloud were ignored. A separate, more exact computation for an idealized cluster in a parabolic gravitational potential made it possible to estimate the error of the impulsive approximation. The results of all this analysis showed that the "large clouds" in the solar neighborhood, evident from optical extinction data and subsequently identified (31) as "giant molecular clouds," would provide the greatest effect and could give maximum ages consistent with those observed for some open clusters. Since that time, more complete analyses have been made (33); the resultant change in predicted disruption times is comparable with the uncertainty resulting from incomplete information on giant molecular clouds.

The discovery of quasars with their very high luminosities led me to consider the evolution of galactic nuclei. Research in collaboration with W. C. Saslaw led to the development of an idealized model (34) that included evaporation of stars from the nucleus (as from a globular cluster) and also direct collision of stars with each other; analysis of this latter process was made possible by a number of drastic physical simplifications. The model calculations showed that the gas released in collisions falls to the center, releasing energy and forming new stars. While the computed maximum luminosity is some 10^{45} erg s^{-1}, comparable with that of quasars, it is not clear whether nuclei have actually reached the compactness required for this process.

Because of my work on this topic, I was invited to attend a small scientific conference on galactic nuclei, held in 1970 by the Pontifical Academy of Sciences. In preparation for this meeting I devoted some further thought to the evolution of spherical star systems. Analysis of equilibrium in a system containing a small fraction of its mass in stars appreciably heavier than the average led to the discovery of the mass stratification instability (35). The heavier stars will lose energy by equipartition with the lighter ones and will sink toward the center of the system. If the fraction of the cluster mass in the heavier stars exceeds a certain small value (about 3×10^{-2} if each heavy star has a mass three times that of a light star), no equilibrium is possible when the heavy stars concentrate toward the center; instead, according to the virial theorem, the self-attraction of these stars will lead to an increase rather than a decrease of their rms random velocity, and the tendency toward equipartition will lead to a continuing contraction of the heavy-star subsystem.

Since the theory leading to this instability was somewhat idealized, I resolved to study the actual evolution of a spherical system by numerical simulation, taking advantage of the fast electronic computers available at

Princeton University. A Monte Carlo method suitable for this purpose, which had been designed by Michel Hénon (36), computed collisional perturbations of orbital elements for each of 1000 representative stars. It seemed to me that a method making more direct use of the velocity diffusion coefficients, which I had already employed in the plasma conductivity research, would have certain advantages. In collaboration with six Princeton students in succession (M. H. Hart, S. L. Shapiro, T. X. Thuan, R. A. Chevalier, J. M. Shull, and R. D. Mathieu), such a Monte Carlo method was designed and gradually extended to explore a wide variety of problems. The emphasis was on globular clusters, since these are known to have relaxation times shorter than their ages.

The results obtained in this research, together with important work by many other theorists, have been summarized in my recent book (37). In addition to verifying the presence of evaporation and of the mass-stratification instability, and computing the relevant rates, the detailed dynamical study of cluster evolution has indicated the dominant importance of the gravothermal instability (38, 39) in leading to collapse of a cluster core. While the nature of postcollapse evolution is still being explored, the evolution until core collapse now seems rather well understood, and in this sense the cluster research begun in 1938 has now reached a satisfying conclusion.

Space Astronomy

As World War II drew near its end, I was approached by a friend on the staff of the RAND Project, an Air Force "think tank." He told me that his group was carrying out a secret study of a possible large artificial satellite, to circle the Earth a few hundred miles up. "Would you be interested," he asked me, "in writing a chapter on how such a satellite might be useful in astronomy?" With my long and ardent background in science fiction, I found this invitation an exciting one and accepted with great enthusiasm. I spent some time analyzing a number of possible research programs for telescopes of different sizes above the atmosphere and prepared a brief description of these.

While no orbiting astronomical telescope resulted from the RAND study, my work on space astronomy convinced me of the paramount scientific importance that a large general-purpose optical telescope in orbit would have. The wartime success of the large German V-2 rockets made the launching of large instruments into Earth orbit seem not too many years in the future. In my thinking, personal association with the development and operation of such a large orbital telescope gradually became a major professional goal. In 1947, when I was urging Martin Schwarzschild to resign from Columbia University and to join me at Princeton, he asked me how likely I was to stay there. I replied that a leadership position with a large space telescope project was one of the very few attractions that might be able to lure me away from Princeton.

It was several years before my fascination with space telescopes led to any actual research in this field. While still at Yale I tried to organize a small program in solar ultraviolet spectroscopy. My attempts to persuade Leo Goldberg to join me in such an effort turned out to be a helpful contribution to astronomy, since the offer to him of a Yale professorship may have been a factor in his promotion to the directorship at Michigan. In the early 1950s Martin's research in solar convection led him into balloon astronomy, thanks to an informal suggestion from James Van Allen (see below). High-resolution imagery above most of the atmosphere did not require orbital altitudes; at about 80,000 ft atmospheric refraction produces very little widening of a stellar image. Accordingly, Project Stratoscope was begun; a 12-inch balloon-launched solar telescope obtained beautifully sharp images in 1957 and 1959, and a 36-inch achieved similar success with planets and galaxies some dozen years later. Since I was convinced that Stratoscope was an important step toward the large space telescope, I encouraged Martin in this effective program by all the pathways that I could think of; in particular, I helped to muster financial support initially and suggested various effects that the unfamiliar environment might conceivably have on the various items of equipment.

The launch of the USSR sputniks, starting in 1957, gave tremendous impetus to the nascent US space program; the launch into orbit of an intermediate-size telescope began to appear possible. An agency of the Air Force asked me if Princeton might wish to carry out some preliminary work on the problems and possibilities of such an instrument. I gave an affirmative answer, of course. A contract was arranged, and a small group of us explored scientific possibilities for interstellar matter research at wavelengths between 1000 and 3000 Å (40) and also possible engineering components for an astronomical satellite (41). In 1959–60 the newly formed NASA (National Aeronautics and Space Administration) gradually took over the support of this program; after the usual combination of technical difficulties, funding shortages, and spacecraft delays, the *Copernicus* satellite was launched into orbit. For me, perhaps the high point of my career was that day in August 1972 when our ultraviolet spectrometer was turned on from the Goddard Space Flight Center, several days after the launch, and was found to be operating properly. I am sure that similar emotions were experienced by the outstanding group of astronomers who shared this historic enterprise with me—Jerry Drake, Kurt Dressler, Ed Jenkins, Don Morton, Don York, and especially Jack Rogerson, who was with this program almost from the start and who became Co-

Principal Investigator; his keen physical insight, his unfailing accuracy, and his dedication to this work were key ingredients in the program's success.

As expected, the combination of high spectral resolution, accurate spectrophotometry, and sensitivity in the far ultraviolet opened up a new chapter in the study of interstellar gas (42, 43). Interstellar H_2 molecules were observed in the spectra of most early-type stars, confirming the ubiquity of this interstellar component and giving information on the kinetic temperature and density within interstellar clouds. The ratio of D to H was measured, indicating (through the theory of element formation during the Big Bang) a material density too small to close the Universe. Observed absorption features of highly ionized atoms such as O^{+5} showed the presence of a hot interstellar gas with T of order roughly 10^6 K in the galactic disk. The relative abundances of many different elements were measured accurately, indicating how the depletion (presumably resulting from condensation on grains) varies from element to element and from cloud to cloud. In addition to the Princeton group, some 170 astronomers in many different countries made effective use of the *Copernicus* telescope-spectrometer for a wide variety of programs. The instrument operated effectively for almost nine years.

The general progress of the US space program after 1957 aroused interest in a possible general-purpose telescope, with a mirror in the 3-m class, providing a unique capability for diffraction-limited imagery and for observations in the ultraviolet and infrared. Because of its high cost and its dedication primarily to use by guest observers, this great instrument required widespread support by astronomers before it could possibly become a reality. A group of us formed a committee (44) under the auspices of the National Academy of Sciences and organized a series of small discussion meetings on various scientific programs for which the Space Telescope (ST) might be particularly helpful. Astronomers unfamiliar with space research were invited, and their participation helped to persuade them that ST would be a very powerful tool. At the beginning there was widespread doubt as to the practicability of such ideas. One astronomer with whom I was discussing a projected orbital telescope remarked to me, "Lyman, you're young. You'll live to see it fail." After the scientific successes obtained with other, though smaller, instruments, these arguments faded away, and by the 1970s most astronomers were enthusiastic supporters of ST.

Almost as difficult as getting support from astronomers was to get Congressional votes for financial support of ST. Fortunately, success with the former group was very helpful with the latter. John Bahcall and I collaborated in urging astronomers to express their support of ST by contacts with Congress as well as with NASA (44). After several years of such efforts, full Congressional approval for this enterprise was finally obtained in 1977. Since then, the program has moved forward on a broad front, despite several delays for technical reasons—especially the 32-month hiatus in space shuttle flights because of the *Challenger* disaster. The launch of the Hubble Space Telescope (HST), as this great observatory has been appropriately named, is now expected in late 1989 or in 1990.

For a variety of reasons my own personal involvement with HST since the program was given a go-ahead by Congress has been very much less than I had hoped for some 40 years ago. However, as an advisor to the Marshall Space Flight Center and as chairman of the AURA (Association of Universities for Research in Astronomy) oversight committee for the Space Telescope Science Institute, I have kept in reasonable touch with what is going on, and it is good to see the program approaching its goals.

I look forward to participating with Bob O'Dell in the analysis of some HST data, especially spectrophotometric observations of interstellar absorption lines. As compared with *Copernicus*, the high-resolution spectrograph on HST has much greater spectral resolution (10^5 as compared with about 2×10^4) and an enormously more rapid photon-gathering capability (up by some two orders of magnitude for interstellar line observations). With these capabilities one can explore the properties of separate clouds, if these have different radial velocities, and can measure in each the relative abundances of different atomic species. Thus we can hope to explore how cloud properties depend both on velocity and on distance from the galactic plane.

Plasma Physics

An important topic that forms part of both plasma physics and stellar dynamics is the set of effects produced by two-body random encounters between particles whose mutual force varies as the inverse square of their separation. These encounters modify the velocity distribution functions for the different types of particles present, tending to make these functions approach the Maxwellian form and to establish equipartition of kinetic energy between particles of different masses.

I became exposed to this type of process in connection with my work on the evaporation of stars from globular clusters (27). To determine the rate at which two groups of stars approach equipartition of energy with each other, I first computed ΔE_A, the change of kinetic energy of a star of group A (and mass m_A) in a single encounter with a star of group B (mass m_B). Multiplying ΔE_A by the rate of such encounters and averaging over all the orbital parameters then gave $\langle dE_A/dt \rangle$, the mean energy exchange rate. A central feature was the assumption of separate Maxwellian velocity distributions for each group. The calculation was trivial but is mentioned here because the resulting formula for $\langle dE_A/dt \rangle$ is so frequently used. It is curious how sometimes the simplest results, which seem in retrospect transparently obvious, are among the most widely quoted.

Some 10 years later, as part of the program to determine the physical properties of the interstellar gas, I became interested in another problem involving similar encounters, this time among electrons and positive ions. My objective was to determine the electrical conductivity of a fully ionized gas, using methods more appropriate to this particular problem than those developed (45) for collisions with mostly neutral particles. In encounters between charged particles, distant collisions producing relatively small velocity changes are dominant, and large changes in velocity are produced by the cumulative effects of many small random changes, as in diffusion or Brownian motion. A recent review paper by Chandrasekhar (46) had provided a detailed discussion of such processes and had discussed the Fokker-Planck equation, which gives the resulting rate of change of the velocity distribution function; this paper provided a basis for the determination of electrical conductivity.

To derive and justify the complex integro-differential equations to be used constituted a PhD thesis for a Yale physics student (R. S. Cohen). To solve these equations numerically provided a year's research for a Princeton graduate student (Routly). These computations were repeated a few years later, including a term that I had previously assumed would vanish, and were extended to include thermal conductivity also (47). Detailed agreement of these results with those obtained by later researchers indicates that the computations are essentially correct.

The behavior of an ionized gas, or plasma, in the presence of a magnetic field was explored in pioneering research by Hannes Alfvén, who had visited Princeton in late 1948 and whose papers and subsequent book (48) were eagerly read and discussed by Martin and me. We both did our best to become familiar with the new field of magnetohydrodynamics so that we could apply it in our respective fields of interest.

An opportunity for such an application came to me in the spring of 1951, when the Argentines announced that they had successfully achieved a controlled thermonuclear reaction in the laboratory. In view of the nearly unlimited supply of nuclear energy available in the deuterium of the oceans, this announcement (which was subsequently shown to be incorrect) raised intense interest, together with considerable skepticism. My wife and I left for a week's ski trip the day after this announcement, and during the long rides on the Aspen single-chair lift I had ample time to consider how a hot thermonuclear plasma might be confined without contact with any material walls. Thomas Cowling was a Visiting Professor in Princeton during that term, and his beautifully systematic lectures on magnetohydrodynamics provided the training needed to consider magnetic confinement of a hot plasma. Such confinement seemed to be a promising possibility, and back in Princeton I worked out (mostly in bed during the middle of the night) a magnetic configuration that I hoped would be effective. The Atomic Energy Commission (AEC) provided funds for a year of theoretical research by a small group that I assembled; the results seemed to confirm earlier hopes, and we embarked on a small experimental program. The dream of a virtually limitless power supply for mankind became for me a primary driving force.

For reasons that I now find quite understandable, the AEC was unwilling to support a major experimental program in controlled fusion at Princeton unless some distinguished experimental physicist was willing to head the effort at least for a few years. Fortunately I was able to persuade James Van Allen to come in this capacity. In addition to his major contributions to our experimental plasma program, he suggested that Martin make use of balloon-launched instruments (one of Van Allen's specialties) for observing the Sun, and he persuaded me to prepare a simple text on plasma physics; the material I put together was designed for our own plasma research group but became the basis for my published plasma physics tract (49), which fortunately appeared at the right time to satisfy a genuine need.

This is not the place to review the progress and problems of the Princeton Plasma Physics Laboratory, as our controlled fusion program was called after its declassification in 1958; this history has been related elsewhere (50, 51). I withdrew from the Laboratory in 1966, after 15 years of mostly half-time participation in the effort. The large staff (now about 1000 persons) engaged on this program have achieved many successes. The theoretical group have made outstanding contributions not only in basic plasma physics but also in plasma astrophysics. On the experimental side, the heating and confinement of a plasma in a magnetic field have been steadily improved and are approaching the levels required for a full-scale power-producing thermonuclear reactor. Of course, as the program has developed over the years my initial concepts and suggestions have formed only a small part of the growing enterprise, as with the Space Telescope also.

In retrospect, I have indeed been fortunate to be engaged in such varied and exciting research and to have had stimulating contacts with so many brilliant teachers, colleagues, and students. While one's early dreams are rarely fulfilled in detail, working toward these goals and taking an active part in fascinating, worthwhile efforts has brought its own rewards.

Literature Cited

1. Gilbert, W. S. 1921. *The Mikado*, Act II
2. Baade, W. 1944. *Ap. J.* 100: 137
3. Spitzer, L. 1941. *Ap. J.* 93: 369
4. Strömgren, B. 1939. *Ap. J.* 89: 526
5. Spitzer, L., Savedoff, M. P. 1950. *Ap. J.* 111: 593
6. Osterbrock, D. E. 1974. *Astrophysics of Gaseous Nebulae.* San Francisco: Freeman
7. Spitzer, L. 1968. *Diffuse Matter in Space.* New York: Wiley-Interscience; 1978. *Physical Processes in the Interstellar Medium.* New York: Wiley
8. Field, G. B. 1965. *Ap. J.* 142: 531
9. Routly, P. McR., Spitzer, L. 1952. *Ap. J.* 115: 227
10. Adams, W. S. 1949. *Ap. J.* 109: 354
11. Siluk, R. S., Silk, J. 1974. *Ap. J.* 192: 51
12. Spitzer, L. 1954. *Ap. J.* 120: 1
13. Spitzer, L. 1976. *Comments Astrophys.* 6: 177
14. Spitzer, L., Greenstein, J. L. 1951. *Ap. J.* 114: 407
15. Oort, J. H., Spitzer, L. 1955. *Ap. J.* 121: 6
16. Pottasch, S. R. 1958. *Bull. Astron. Inst. Neth.* 14: 29
17. Mestel, L., Spitzer, L. 1956. *MNRAS* 116: 503
18. Osterbrock, D. E. 1961. *Ap. J.* 134: 270
19. Spitzer, L. 1956. *Ap. J.* 124: 20
20. Münch, G., Zirin, H. 1961. *Ap. J.* 133: 11
21. Savage, B. D. 1987. In *Interstellar Processes*, ed. D. J. Hollenbach, H. A. Thronson Jr., p. 123. Dordrecht: Reidel
22. Jenkins, E. B. 1987. In *Exploring the Universe with the IUE Satellite*, ed. Y. Kondo, p. 531. Dordrecht: Reidel
23. Jones, R. V., Spitzer, L. 1967. *Ap. J.* 147: 943
24. Davis, L., Greenstein, J. L. 1951. *Ap. J.* 114: 206
25. Purcell, E. M. 1979. *Ap. J.* 231: 404
26. Spitzer, L. 1968. In *Nebulae and Interstellar Matter (Stars and Stellar Systems*, Vol. 7), ed. B. M. Middlehurst, L. H. Aller, p. 1. Chicago: Univ. Chicago Press
27. Spitzer, L. 1940. *MNRAS* 100: 396
28. Ambartsumian, V. A. 1938. *Ann. Leningrad State Univ. No. 22 (Astron. Ser., Iss. 4)*, p. 19; 1985. In *Dynamics of Star Clusters, IAU Symp. No. 113*, ed. J.
29. Spitzer, L., Schwarzschild, M. 1951. *Ap. J.* 114: 385; 1953. *Ap. J.* 118: 106
30. Villumsen, J. V. 1983. *Ap. J.* 274: 632
31. Blitz, L. 1980. In *Giant Molecular Clouds in the Galaxy*, ed. P. M. Solomon, M. G. Edmunds, p. 1. New York: Pergamon
32. Spitzer, L. 1958. *Ap. J.* 127: 17
33. Wielen, R. 1985. In *Dynamics of Star Clusters, IAU Symp. No. 113*, ed. J. Goodman, P. Hut, p. 449. Dordrecht: Reidel
34. Spitzer, L., Saslaw, W. C. 1966. *Ap. J.* 143: 400
35. Spitzer, L. 1969. *Ap. J. Lett.* 158: L139
36. Hénon, M. 1971. *Astrophys. Space Sci.* 13: 284, 14: 151
37. Spitzer, L. 1987. *Dynamical Evolution of Globular Clusters.* Princeton, NJ: Princeton Univ. Press
38. Lynden-Bell, D., Wood, R. 1968. *MNRAS* 138: 495
39. Lynden-Bell, D., Eggleton, P. P. 1980. *MNRAS* 191: 483
40. Spitzer, L., Zabriskie, F. R. 1959. *Publ. Astron. Soc. Pac.* 71: 412; 1988. *Publ. Astron. Soc. Pac.* 100: 509
41. Spitzer, L. 1960. *Astron. J.* 65: 242
42. Spitzer, L., Jenkins, E. B. 1975. *Annu. Rev. Astron. Astrophys.* 13: 133
43. Spitzer, L. 1982. *Searching Between the Stars.* New Haven, Conn: Yale Univ. Press
44. Spitzer, L. 1979. *Q. J. R. Astron. Soc.* 20: 29
45. Chapman, S., Cowling, T. G. 1939. *The Mathematical Theory of Non-Uniform Gases.* Cambridge: Univ. Press
46. Chandrasekhar, S. 1943. *Rev. Mod. Phys.* 15: 1
47. Spitzer, L., Härm, R. 1953. *Phys. Rev.* 89: 977
48. Alfvén, H. 1950. *Cosmical Electrodynamics.* Oxford: Clarendon
49. Spitzer, L. 1956, 1962. *Physics of Fully Ionized Gases.* New York: Wiley-Interscience. 1st, 2nd ed.
50. Bishop, A. S. 1958. *Project Sherwood.* Reading, Mass: Addison-Wesley
51. Bromberg, J. L. 1982. *Fusion: Science, Politics and the Invention of a New Energy Source.* Cambridge, Mass: MIT Press

Ann. Rev. Astron. Astrophys. 1988. 26: 1–9

A MORPHOLOGICAL LIFE

W. W. Morgan

Yerkes Observatory, University of Chicago, Williams Bay,
Wisconsin 53191

"Partial answers are the only answers."
                                    James Gunn

My first encounter with the astronomical world was in 1920 or 1921, when
I was a student in a small private high school attached to a tiny college
(Marvin College) in a small town (Fredericktown) in southeast Missouri.
My Latin teacher, Miss Alice Witherspoon, overheard me talking about
the stars and arranged for me to look at the Moon through a small
telescope on a theodolite. This view, after all of the intervening years,
remains the most exciting visual experience of my life.

I was born on January 3, 1906, in a small town that has since vanished
from the map of Tennessee (Bethesda). My father, William Thomas
Morgan, and mother, Mary Wilson Morgan, became home missionaries
in the early years of the twentieth century. They served in the states of
Kentucky and Tennessee: In the former they sought to bring the Gospel
to families engaged in bloody feuds with other families.

Later, my father held a series of pastorates in the Southern Methodist
Church; the last two were in Colorado Springs, Colorado, and Poplar
Bluff, Missouri. Because of an illness at the age of six, I did not attend
school regularly until I entered the fourth grade in Colorado Springs at
the age of nine. These rapid changes in location, together with the presence
in the home of an unstable father, resulted in a stressful life that was only
relieved by my saintlike mother, who taught my sister and me and who
made it possible through her care for us to have a bearable life.

The situation changed drastically for the better in the fall of 1919, when
my mother, my sister, and I settled in Fredericktown, Missouri. My first
two years of high school were spent there; these were the crucial ones for
my orientation toward my future life.

In addition to her astronomical introduction, my Latin teacher pre-

0066–4146/88/0915–0001$02.00

sented that language as a living thing, and helped me to realize that a "dead" language can possess a vibrant, living form. My continuing preoccupation with morphology probably began with this experience; it has continued, with a progressively deepening effort, during the ensuing 66 years.

Also, in the Fredericktown years, there was my first encounter with the Elizabethan play *Doctor Faustus*, by Christopher Marlowe, written shortly before the author was murdered in 1593 at the age of 29. Marlowe's portrayal of Faustus as a person who refuses to accept any limits in his desire to push out into the Unknown has been the model for my inner life since the age of 15.

The final two years of high school were spent attending Central High School in Washington, DC; I graduated there in the spring of 1923. In the fall of the same year, I enrolled at Washington and Lee University in Lexington, Virginia. My interest in astronomy continued, but I had no idea of becoming a professional astronomer; I had, in fact, decided to specialize in English, in preparation for a career of teaching in that field.

My three years at Washington and Lee were spent on a general course program that included mathematics, English, Latin, and science. My only astronomical activity consisted in convincing my physics professor to buy a small astronomical refractor, with which I observed superficially stars and sunspots; no astronomy courses were available.

As a result of a series of chance occurrences in the summer of 1926, I was offered an assistantship at the Yerkes Observatory of the University of Chicago, at a salary of $1200 for the year; I accepted the offer and arrived in Williams Bay, Wisconsin, on September 6, 1926. Four days after this is being written, on September 6, 1987, I shall have completed 61 years of work at the Yerkes Observatory.

I was allowed to take sufficient courses during the 1926–27 school year to permit me to receive the BS degree from the University of Chicago in June 1927. I enrolled immediately thereafter in the graduate program at the University of Chicago at the Yerkes Observatory. The 1929–30 school year was spent on the University campus in Chicago.

My thesis research work was under the supervision of Otto Struve, one of the greatest astronomers of the twentieth century. The thesis was concerned with certain categories of the peculiar stellar spectra near type A: silicon stars, manganese stars, europium stars, etc. For some of these categories, only a single example was known. From the large collection of spectrograms obtained in earlier years by various observers on radial velocity programs at Yerkes, I found it possible to discover a considerable number of new examples of these peculiarities; and it was shown that such stars are not as uncommon as had been thought. My thesis was published in three papers in the *Astrophysical Journal*, and I received the PhD degree in December 1931. I subsequently accepted an offer of an Instructorship in the University of Chicago to begin on July 1, 1932.

The person who began that appointment of Instructor on that date had certain unusual characteristics: The astronomy courses at Yerkes were informal at that time, and the year spent on the Chicago campus did not include any formal astronomy courses, since none were given. Preparations for the final PhD examination had been carried out by private reading, such as F. R. Moulton's *Celestial Mechanics*. This served me well on my examination; the future Astronomer Royal, Harold Spencer Jones, was present and asked me for a development out of that volume. But such a preparation, which did result in a respectable examination, was hardly the optimum conditioning for an astronomical research career; as a result, I learned on the job.

Since youth, I have been entranced with what I later discovered Husserl has called the "thing itself." In my field of astronomy, it could be a star (that is, a *particular* star), a star cluster, a galaxy, or a cluster of galaxies. When the mind turns in other directions, the "thing itself" could be a tree of special beauty or interest, a chickadee, or a chipping sparrow. And it must be said that the mental state of this observer toward all of these subjects would be remarkably similar. *The "thing itself," whether breathing or not, growing or not, has a fundamental property or existence far deeper than any conceptual construction.* And how does this bear on the classification of stars, galaxies, and clusters of galaxies? It induces a very deep feeling of respect for those things *for themselves*, as independent structures of Nature.

It is of supreme importance to the creation of our operational morphologies that we use the "things themselves" as our picture-language for stellar spectra, and direct photographs for galaxies and clusters of galaxies. The derived morphologies must be "true" to the information revealed in the "things" (or specimens), *not* from superposition of formal constructions on the "things." After the specimens have been arranged in the order indicated by the *observed* forms, we have brought into existence morphological types of lasting value and richness.

In what follows, I comment briefly on morphological aspects of four different areas of past activity: the MK classification system for stellar spectra, the discovery of the spiral arms of the Milky Way galaxy, the invention of the *UBV* system of brightness of stars, and the Yerkes system of classification of the optical forms of galaxies.

The first two-dimensional spectral classification of stars over the entire range of spectral class A–M was carried out by Walter S. Adams and his associates at Mount Wilson (1914–17). The second dimension (luminosity) they determined by calibration of the spectroscopic line ratios in terms of absolute magnitude from trigonometric parallaxes. For this reason, the resulting absolute magnitudes can be affected by serious errors in the calibrating process itself, and this reduces the degree of discrimination possible from the spectroscopic evidence.

The unsatisfactory nature of this situation became obvious to me in the mid 1930s, and I decided to create a two-dimensional spectral classification that would be determined completely from the appearance of spectral lines, bands, and blends. It was to be autonomous; that is, it would be defined completely by the spectra of standard stars, without having to appeal to any theoretical picture. The classification process would then be a judgment of "like" or "not like" for an unclassified spectrum in comparison with the spectra in the array of standard stars.

The problem of terminology was solved in a 1938 paper with the introduction of a family of "luminosity classes" to specify the vertical dimension:

Ia—the brightest supergiants,
Ib—somewhat fainter supergiants,
II—bright giants,
III—normal giants,
IV—subgiants,
V—main-sequence stars.

The complete MKK type would then consist of the horizontal type, followed by one of the above luminosity classes, such as α Cygni A2 Ia, α Persei F5 Ib, β Geminorum K0 III, β Aquilae G8 IV, and the Sun G2 V. These classes are to be defined by an assemblage of standard stellar spectra, which, in effect, define the MK system itself, *without any calibration whatsoever being necessary for the classification process itself.*

Not long after this time, I was joined by Dr. Philip Keenan as co-investigator on this program. It must be emphasized that the resulting quality of the 1943 *Atlas of Stellar Spectra* is due, in large measure, to the gifts and deep knowledge of this distinguished stellar spectroscopist, especially in the case of stellar spectra later than the Sun. I also wish to acknowledge the help of Miss Edith Kellman for photographic work during this period; she prepared a large number of the sets of photographic prints that became, in effect, the *Atlas of Stellar Spectra* in 1943.

A limited edition of the *Atlas* was obvious, in these circumstances; the situation was worsened because of the Second World War, which resulted in the US Government's refusal to allow any copies of the *Atlas* to be sent to Europe "because it would give aid and comfort to the enemy." I made a serious mistake in not holding back a significant number of copies for shipment after the war. By that time, Miss Kellman had changed her profession to school teacher.

I introduced the term *the MK process* at a conference at Toronto in 1983. By this I implied a procedure similar to that which brought the MK system into existence—and which is capable of developing future systems that are complementary to the MK system. The latter system is incomplete; in its most satisfactory mode, it can be used properly only for stars of Baade's Population I. It will turn out to be even more incomplete in time, as more distant—and fainter—stars are observed in our Galaxy and in other galaxies. And it should be used with caution in the far ultraviolet and in the infrared spectral regions. It is an exciting task for the future to devise the most *discriminating* classification systems for the latter regions— systems that will complement the MK system in the blue and reveal beautiful new discriminations for stellar astronomy and astrophysics.

It was the inspired work of Walter Baade of the Mount Wilson Observatory between 1940–50 that established the crucial part played by the blue supergiant stars in defining the spiral arms of galaxies. I had considered the possibility of using the 1943 *MKK Atlas* for a more detailed investigation of the arms, but it was the clarity and beauty of Baade's presentation that determined this course.

Between 1947 and 1949, Dr. J. J. Nassau and I obtained a series of objective-prism spectrum plates at Cleveland that led to the discovery of around 900 O- and early B-type stars of high luminosity around the observable part of the Galactic equator. A number of these stars were found to be located in H II regions from observations by Donald Osterbrock and Stewart Sharpless, two Yerkes students.

This particular group of stars was then put on the observing program at Yerkes to obtain slit spectrograms for the purpose of determining their MKK spectral classification, which permitted an absolute magnitude and distance to be derived.

When the distances were plotted, it could be seen that the small H II regions surrounding the central supergiant stars defined two armlike sections that resembled closely the spiral arms in the Andromeda Nebula, M31. And this comparison could be made because the unselfish Dr. Baade had sent me a magnificent original plate taken of the spiral arms in M31 with the 100-inch telescope; Dr. Baade had told me earlier that as soon as he got a good plate he would send it to me!

In 1953, A. E. Whitford, A. D. Code, and I published a map of the Galactic spiral structure obtained from Yerkes spectroscopic distances of 27 large-scale clusterings of high-luminosity O, B, and A stars. The two spiral arms announced earlier were strongly confirmed, and the suspected third arm, located between the Sun and the Galactic center at a distance of around 1400 pc, was established.

The *UBV* system of magnitudes and colors was invented by me during the period 1947–51. During this same period, I was involved in improving the discrimination of the *MKK Atlas* of 1943; this was accomplished by (*a*) making minor changes in the system itself and (*b*) revising the types of some of the standard stars. The revised system was then labeled the MK system. This system was published in the *UBV* paper of 1953, coauthored with Harold Johnson. I considered the new *UBV* system primarily as an adjunct to the new MK system of spectral classification.

There are interesting overtones in the two-color diagrams for $(B-V)$, $(U-B)$ as illustrated in the above paper: (*a*) The white dwarfs are separated from the standard main sequence; (*b*) the yellow giants tend to separate off from the normal main sequence; and (*c*) the reddening path for the O stars and early B Ia supergiants occupies a narrow, well-defined area on the diagram. These last two characteristics could be of value for determining distances of faint, distant stars in our Galaxy and in other neighboring galaxies.

The above can be considered as characteristic of applications of the *UBV* system to problems of galactic structure and astrophysics. The very complicatedness of the relationships turns out to be a blessing in disguise. The possibilities of discrimination furnished by the variation in behavior of different kinds of "specimens"—*stars, star clusters, galaxies, and galaxy clusters, with all but "stars" being treated as composite sources*—have been, as yet, hardly touched.

The classification of the optical forms of galaxies is a very different problem from that of the classification of stellar spectra. The ordered appearance of the spectrograms contrasts sharply with the disorder of the forms of many galaxies; and even in the case of galaxies possessing a degree of order, we find them in fantastic variety. This clearly requires a differing point of view toward the classification problem: (*a*) A satisfactory frame of reference has to be brought into existence for the purpose listed below, and (*b*) we can never expect to obtain the clean, detailed matching of specimen form to standards that we take for granted in the case of stellar spectra.

Thirty years ago, N. U. Mayall and I suggested a new spectral classification of giant galaxies. The spectral type would apply to (*a*) the galaxy as a whole, when there is little concentration of light toward the nuclear region; and (*b*) to the nuclear region, when it dominates the system.

The Yerkes MK system was used. There result five different categories of types of giant galaxies from classification in the spectral region $\lambda\lambda3850$–4100 Å:

The A systems Seven giant galaxies are classified as A systems. Their range in Hubble type is Irr, Sbc, and Sc.

The AF systems The seven giant galaxies in this category are classified in the range F0–F2 in the spectral range $\lambda\lambda3850$–4100; at $\lambda4340$ Å, their types are near F8. All are spirals of Hubble type Sc.

The F systems Four systems are classified in this group; they have slightly later spectral types than those in the preceding group. They include M33 (Sc) and M51 (Sc).

The FG systems Six galaxies are listed here; all but one are of Hubble type Sb.

The K systems Twenty-three galaxies are listed here, including M31 (Sb) and M81 (Sb). In this group of giant galaxies are the giant ellipticals and many galaxies classified as Sa and Sb spirals in Hubble's system. The presence in the spectra of well-marked absorption of CN indicates that G8–K3 giants are a major source of the light in the blue-violet region.

This five-part spectral classification of giant galaxies shows a clear correlation between the relative size of the amorphous central regions as compared with their main bodies, and composite spectral type, as determined in the wavelength range $\lambda\lambda3850$–4100 Å.

What we now need for emphasis on stellar population in giant galaxies is a form classification that will have a single criterion: the relative size of the amorphous central region as compared with the size of the main body itself. This is, of course, one of the two criteria of the classical Hubble system, but, as Sandage has shown in his *Hubble Atlas of Galaxies* (1961), the classification of giant galaxies in the Sa, Sb, Sc sequence is based primarily on the arms.

A system of form classification that is basically one dimensional was developed in the light of the Morgan-Mayall *Spectral Classification of Galaxies* described above. The fundamental dimension is denoted by the small letters a, af, f, fg, g, gk, and k, ordered by increasing relative size of the nuclear bulge of each galaxy that is in the giant category.

This is a form classification, but it also relates to the five spectral groups outlined above. That is, the mean blue-violet spectral type of giant galaxies in any one of the five spectral groups will, in general, have similar form types. Spectral group AF will have, in general, form types near a, af, and f; spectral group K will have, in general, form types near gk and k. In addition to the form type, a family type (S, B, E, I) is added. Two catalogs of these types were prepared and published in *Publications of the Astro-*

nomical Society of the Pacific in 1958 and 1959. A later state of this classification was published by Osterbrock and me in 1969.

An unusual situation developed in the course of a 1964 study of the galaxies identified with radio sources, by T. A. Matthews, Maarten Schmidt, and me: Many of the strongest radio sources were found to have an elliptical-like inner region, surrounded by an external envelope. They belong to the "D systems" of my classification. The largest of these D systems are classified as "cD galaxies"; they are observed in clusters of galaxies, centrally located and far larger than any of the commonly observed galaxies. These cD galaxies are not always strong radio sources, but wherever they appear, they dominate their surroundings.

The study of these giant structures is still in its infancy; their optical morphology requires reexamination on plates of high resolution, to discover whether a finer classification of their optical forms might be possible.

The "natural groups" in my spectral classification are not concerned with the MK system. They deal with spectra of very low dispersion, where only a few spectral features may be visible. In some such cases, the few visible features are sufficient to segregate groups of stars located in narrow areas of the HR diagram—and therefore of similar absolute magnitude. This procedure is rich in possibilities for future use in studies of galactic structure.

The statement by James Gunn[1] at the head of this paper, "Partial answers are the only answers," suggests beautiful new methodologies. When we add to it, "What can be said about *this* question *now*?" we move toward a deeper reality in our morphologies.

And now, what have we been doing in the above discussion? *We have been attempting to discriminate* between "things" ("specimens"), from their spectra or optical forms. *This process of discrimination carries something in common with the Gunn statement. The process of discrimination seems never to be carried so far that it cannot be carried further; that is, the discrimination process can never be considered to be complete.* But this does not at all imply that we do not gain by continuing the discrimination process; the further we go, the more we suspect the existence of worlds unexplored.

The great American poet T. S. Eliot was once asked by a student whether there could be things of beauty in his verse that he himself was not aware of. His reply was "yes, of course." So much of the creative process takes place in the subconscious, in an area of incomplete communication; and

---

[1] I am indebted to Dr. Gunn for permission to quote his statement.

poetry is far from being the only discipline where such things can happen. It is in such uncharted areas of mental space where some of the deepest science has its origins, and where the revolutionary philosophy of Wittgenstein's *Philosophical Investigations* must have labored before birth.

ACKNOWLEDGMENTS

I am deeply indebted to a number of persons for crucial discussions, plates, and collaboration, among them H. Abt, V. Ambartsumian, W. Baade, R. Garrison, J. Gunn, Al Harper, R. Hildebrand, L. Hobbs, H. L. Johnson, P. Keenan, R. Kron, D. Mihalas, J. Nassau, D. Osterbrock, B. Stromgren, O. Struve, and P. Vandervoort.

In addition, I owe an unpayable debt to the University of Chicago and to the Yerkes Observatory for having furnished facilities and an internal climate that made possible the creative activity that I longed to pursue and am still pursuing.

I also wish to thank Elizabeth Cox for preparing and typing the manuscript, as well as Richard Dreiser for preparing photographic materials.

BIOCHEMISTRY

Annu. Rev. Biochem. 1991. 60:1–37

50 YEARS OF BIOLOGICAL RESEARCH—FROM OXIDATIVE PHOSPHORYLATION TO ENERGY REQUIRING TRANSPORT REGULATION

Herman M. Kalckar *

Department of Biochemistry, Harvard University, Cambridge, Massachusetts 02138

KEY WORDS: ATP, molecular enzyme defects in infants, galactose chemotactic mutants, energized down-regulation.

CONTENTS

* The Editorial Committee was saddened to learn that Dr. Kalckar died at the age of 83 on May 17, 1991.

0066-4154/91/0701-0001$02.00

My School and Family

This autobiography, requested by the Editorial Committee of the *Annual Review of Biochemistry*, has, I hope, mobilized my wit sufficiently to make the accidental reader dwell for more than a few moments and to raise curiosity about my intellectual pursuits in Copenhagen, where I grew up.

Copenhagen at that time was often referred to as "The Athens of the North." Although I was an erratic pupil in school and occasionally felt miserable, I have many interesting and rewarding memories from the old-fashioned school I attended, which, according to modern standards, especially for athletic activities, must be considered very modest. Its name was "Østre Borgerdyd Skole," which literally meant "the school for civic merits in the Eastern (Østre) borough of Copenhagen." I suppose that "merits" were meant to be applied to a larger part of the world than that quiet, attractive part of the Danish capital in which the school was located (the three Kalckar sons, of which I was the middle one, lived in easy walking distance from the school).

The Athenian flavor of "Borgerdyd Skolen" could not help making itself felt. The retired "Rector" (headmaster), J. L. Heiberg, was a world-renowned scholar in Greek, having translated Archimedes from Latin to Greek with great precision before the original Greek inscriptions were discovered. Our physics teacher, H. C. Christiansen, from the heather-covered, austere region of Western Jutland, was a formidable and passionately devoted teacher in mathematical physics. He had written a very concentrated, and in many ways very fine, textbook in physics, in which calculus played a prominent role. We were a little scared of him, because his temper would flare up if we tried to cut corners in our preparations and tried instead to memorize the text and formulas from his book. Christiansen insisted in no uncertain terms that physics and calculus, vector analysis, etc can only be grasped by using pencil and paper over and over again in order to find the right polarity in the fine network of a creative scientific argument. I dare say that we got the Ørsted-Ampere rules right.

I have allocated so much space to this alert teacher because I happened later to choose biological research and teaching as my occupation. Scientific arguments in pursuit of "truth," or I prefer the term "high fidelity," require a special kind of alertness.

Although our education in biology in school was largely systematic and somewhat static, it was brightened greatly by some extraordinary demonstrations in human physiology by a world-famous "Jutlandian" zoo-physiologist, August Krogh, professor at the University of Copenhagen and Nobel Laureate of 1923 for his description of capillary blood flow and its regulation. Krogh was, to my knowledge, the only physiologist in the 1920s who took an active interest in introducing the principles of human physiology to Danish high school boys. Krogh had constructed the microtonometer for measuring oxygen tension in small samples of blood as well as a respiratory spirometer for metabolic measurements. The spirometer and the ergometer bicycle were very popular at these school demonstrations.

Although I was perhaps the most restless of three, very different, brothers, the relatively modest Kalckar home where I grew up was an oasis for me. Here most of my interest in the humanistic disciplines developed and thrived. My mother, Bertha Rosalie, born Melchior, read French (Flaubert, Proust, etc) and German (Goethe, Heine, Lessing, etc) and spoke both languages. English came home at a slower rate (see below). Both my parents insisted that I treat the Danish language with care, reverence, and love. We belonged to a middle-class, Jewish-Danish family—Danish for several generations. We were mainly free-thinkers; however, my mother was attached to the Jewish faith and thought that it behooved us to acquire some religious education under the auspices of the main synagogue in Copenhagen and learn a minimum of Hebrew. My father, Ludvig Kalckar, took interest in reading Jewish and Yiddish humorists. All of us, however, responded most to Danish humor: cartoons, limericks, prose, and especially the theater.

My father, born in 1860, attended the world premiere of Ibsen's "A Doll's House" at the Royal Theater in November 1879. He wrote an enthusiastic report about the event. Although he was a businessman by necessity rather than by choice, my father carried on his profession as a consultant and broker most conscientiously. During World War I, he could easily have earned 10 to 20 times more than he did, had he not stubbornly refused to deal with stocks. He said in his quiet, unassuming way that he did not care to earn money on sunken ships and lost lives, and he "meant business." He always insisted, in his shy fashion, that this was a personal issue, and that he enjoyed seeing his relatives and friends socially, speculators and nonspeculators alike. So in spite of my father's late evening hours in his office, we did not grow rich. My mother was very patient and understood his principles. During the war she helped the International Red Cross in its relief efforts for prisoners of war and their families.

In contrast to its experience in World War II, Denmark was rather lucky during and after the World War I, almost too lucky in terms of gluttony, be it profiteering at the stock market or in the wild export of milk products.

We were very interested in the history of Denmark's relations with its neighbors, especially Germany. Germany under the Kaiser was not felt to be a trustworthy neighbor. Like so many other Danes, neither of my parents liked the Kaiser. My father had an extensive library on the history of Slesvig and Holstein, the disputed border provinces annexed by Bismarck in 1864. In 1918–1920, the popular, nationalistic Danish King Christian X wanted both provinces returned to Denmark. My Francophile mother fully supported this idea. In contrast, my father felt that restraint was more important than blind nationalism. He supported the return of North Slesvig, but felt unable to make a judgment in regard to the southern provinces. Fortunately, Denmark was blessed with a democratic statesman, Danish by birth and by sentiment, yet very familiar with the populations of both North and South Slesvig. His name was Hans Peter Hansen-Nørremølle, and he and his equally wise friend, H. V. Claussen, approached people unfailingly as human beings, regardless of whether they were from North or South Slesvig.

The verdict of their fair and patient inquiry was that Denmark should concentrate its efforts on North Slesvig, where the pro-Danes were in the majority, and not on South Slesvig, where the majority of the population remained pro-German. The liberal Danish government at that time respected this genuinely democratic recommendation, although it was not universally popular at the time. In general, the liberal Danes at that time showed great and truly democratic foresight. Imagine the abysmal conditions in Denmark under the Nazi occupation 1940–1945 if it had annexed the Slesvig-Holstein provinces. The real Nazi terror in Denmark started only three years after the invasion. I may be wrong in my analysis of these factors; after all why should the brave Norwegians have been exposed over five years to Nazi Terror? Both countries were lucky to have their staunch Kings at that time, Christian X and the heroic Haakon VII of Norway.

My Two Great Teachers

Before discussing my debut in research on biological phosphorylations in 1935, it seems fitting to pay my special respects to my two great teachers, Ejnar Lundsgaard and Fritz Lipmann. Fritz Lipmann's memorial words about Ejnar Lundsgaard in 1968 are so beautifully chosen that I have selected excerpts from them:

> Word has come from Copenhagen of the death, during the last days of 1968, of Ejnar Lundsgaard. In these times of rush and short recall, I am particularly keen to refresh the memory of the scientific community about his truly great discovery which is now slipping into the background. Personally and scientifically, I encountered Lundsgaard during my period of maturation and I owe him much. When I moved to Copenhagen in 1932 and stayed until 1939, we saw each other a great deal. In the fall of 1967, his friends and colleagues went to Copenhagen to celebrate the 40th anniversary of the discovery of what he called the a-lactacid contraction of iodoacetate-poisoned muscle. When this startling news reached us in Meyerhof's laboratory, it was very upsetting to our group, which looked upon glycolytic lactic acid as the link between metabolic energy generation and muscle contraction.
>
> The interest in iodinated organic compounds which might have metabolic actions similar to thyroxin induced Lundsgaard to study iodinated acetic acid. Its injection into animals, however, yielded a rather unexpected effect: for a few minutes the animal behaved quite normally, but suddenly it turned over and its muscles became rigid. Such rigor was dependent on prior muscle activity, since denervated or curarized muscles did not respond.

But when rigor developed, the expected burst of lactic acid was missing. Lundsgaard elegantly solved the puzzle. He showed that: (i) iodoacetate inhibited glycolysis; and (ii) poisoned muscle performed a limited number of normal contractions at the expense of dephosphorylation of creatine phosphate, then newly found in need of a function. The rigor mortis-like condition developed when the limited supply of creatine phosphate was exhausted.

Lundsgaard had discovered that the muscle machine can be driven by phosphate-bond energy, and he shrewdly realized that this type of energy was "nearer," as he expressed it, to the conversion of metabolic energy into mechanical energy than lactic acid. He was right, because it soon developed that the glycolytic reaction is a feeder of phosphate-bond energy and not of acid. On the way, Lundsgaard also provided an enormously useful tool for studying enzyme mechanisms. Iodoacetate has become one of the standard reagents for SH-blocking in enzymes. Thus, iodoacetate inhibits glycolysis because it blocks the functional SH in phosphoglyceraldehyde dehydrogenase.

In the middle thirties, Lundsgaard became professor of physiology at Copenhagen University and trained many biochemists and physicians. Herman Kalckar was one of his graduate students. And, even though I did not formally work with him, I consider myself his pupil. My subsequent work was profoundly influenced by his discoveries, which changed our concepts of metabolic energy transformation.

First Descriptions of Oxidative Phosphorylations in Tissue

During my studies in physiology and pathology at the University of Copenhagen, my interest was captivated by a few revolutionary developments in cell physiology that forced us all to reinterpret the role of lactic acid in mammalian cells. Otto Warburg found that tumor cells were unable to suppress their high generation of lactic acid formation even if they were supplied with excess oxygen. Ejnar Lundsgaard discovered that muscle contractions could occur without the formation of lactic acid, and Fritz Lipmann was able to interpret the Pasteur reaction. Lundsgaard found that frog muscles that have been exposed to iodoacetate are only able to perform a limited number of twitches; the phosphocreatine splits into phosphate and creatine. Lipmann was the first to believe in Lundsgaard's discovery as well as his interpretation. A few years later, in 1933, Lipmann was invited to Copenhagen. I was indeed lucky to have the good fortune of having Lundsgaard and Lipmann become my mentors.

In 1934 Lundsgaard became chief of the physiology department. His time was therefore more restricted, and so Fritz Lipmann became my foremost mentor. Lipmann encouraged me to study the newer literature on carbohydrate and phosphate metabolism in isolated mammalian tissue extracts or tissue particle preparations. For example, Hans Krebs seemed to have had much luck in disclosing different types of respiratory metabolism in tissue particle preparations from a variety of organs, especially pigeon breast muscle. About the same time, he formulated the tricarboxylic cycle, later called the Krebs tricarboxylic cycle.

For a variety of reasons, I elected to study the metabolism of phosphate and carbohydrate in extracts of kidney cortex. One might argue that pigeon breast muscle, bristling with enzymes catalyzing the Krebs cycle, would be superior. In order to obtain this tissue, however, I would have had to train myself to cut off the heads of pigeons, and I greatly preferred to avoid killing animals, probably from a lack of courage. In any case I was lucky to obtain fresh mammalian kidney from the perfusion experiments on anesthetized cats and rabbits, conducted at least once a week in the physiology department.

My studies on phosphorylation and respiration in kidney cortex extracts turned out to be rewarding. In the presence of oxygen (aeration in Warburg manometer vessels), inorganic phosphate was captured and esterified to various substrates such as glucose or glycerol. A special gift of 5'adenylic acid (5'AMP) from the late Dr. Pawel Ostern of Lwow, Poland gave me additional valuable insight. The kidney extracts converted the 5'AMP to adenosine triphosphate (ATP). When phosphatase activity was arrested by the addition of sodium fluoride, the P/O ratio approached 1. Under anaerobic conditions, no phosphorylation was detectable,[1] even in preparations well "spiked" with fluoride. I also noted that addition of dicarboxylic acids, like fumarate or succinate, stimulated oxygen consumption as well as phosphorylation. I never managed to get P/O ratios higher than 1.5, however, and then only if succinate was added to the cortex preparation (1).

Oxidative phosphorylation remained at about the same order of magnitude when I merely added fumarate or succinate to the extract, i.e. in the absence of glucose. This observation seemed very puzzling, and it induced me to pursue this particular aspect. I found that fumarate (or malate) in the absence of glucose not only stimulated respiration, but also gave rise to an accumulation of a phosphoric ester with properties different from those of hexose phosphate or glycerophosphate. What was the nature of this phosphoric ester? In pursuing this question, I once more received encouragement and advice from Lipmann.

In 1934–1935, work by Gustav Embden opened up new perspectives regarding phosphoric esters and their role in glycolysis; this new panorama was further expanded with the discovery of *phosphoenolpyruvic acid* (PEP) by Lohmann & Meyerhof. It was customary at that time to use the stability of phosphoric esters in acid or alkali as a first criterion of the nature of the different esters. The properties of the phosphoric ester formed in my fumarate experiments did not correspond to those of ATP. Since PEP was able to serve as a phosphate donor, and even able to catalyze the conversion of ADP to ATP, PEP attracted my special interest. PEP was known to be chemically

---

[1]The awareness that respiration stabilizes P-ester levels was expressed in 1932 not only by Lundsgaard but independently by Engelhardt in his work on nucleoated avian erothrocytes, to which he added redox dyes.

Figure 1 V. A. Belitser, H. M. Kalckar, June 1960, Kiev

dephosphorylated rapidly at room temperature in alkaline iodine and even more rapidly by mercuric chloride. I used these highly specific criteria and found that the ester formed from fumarate or malate fulfilled all of these criteria. PEP seemed indeed to be formed in the cortex preparations from oxidation of malate in the presence of inorganic phosphate.

This type of generation of PEP could scarcely have originated from hexose phosphate, since the presence of sodium fluoride barred not only phosphatase activity but, as shown by Lohmann & Meyerhof, enolase activity as well; in this way the conversion of 2-phosphoglycerate to PEP was arrested. I therefore believed that with fumarate or malate in the experiments, PEP must have originated from one of the dicarboxylic acids, including oxaloacetate, all members of the Szent-Gyorgyi-Krebs di-tricarboxylic acid pathway. I published my new observations in *Nature* (3).

Lipmann's response to my discovery of PEP generation from dicarboxylic acids in respiring kidney cortex dispersions was unreservedly positive. As I mentioned, in spite of my repeated efforts, I never succeeded in obtaining P/O ratios significantly above 1.5. The further development in this direction we owe to V. A. Belitser in the USSR; in an extensive paper on oxidative phosphorylation he showed that pigeon breast muscle and rabbit heart muscle, in the presence of oxygen, are able to bring about phosphorylation of creatine to phosphocreatine. Belitser & Tsybakova showed that in the presence of arsenous acid and fluoride, the oxidation of succinate to fumarate did in fact generate P/O ratios as high as 3–4. The paper appeared shortly before the outbreak of World War II in 1939 (2). Much later, in June 1960, I had the privilege to meet Belitser in Kiev (see photo, p. 7), when the National Academy of Sciences sent me to the USSR Academy (Akademi NAYK) at the beginning of the scholarly exchange between the two academies.

In 1938, I was an active participant at the International Congress of Physiology in Zürich. I presented some of my recent findings on the oxidative phosphorylation of glycerol and the generation of PEP from dicarboxylic acids (3). My text and communication were presented in German, since I did not yet trust my ability to lecture in English. However, Professor R. A. Peters from the University of Oxford, who was chairman of the session, and Dorothy and Joseph Needham, University of Cambridge, were actively interested in some of the metabolic aspects of my topic, and they asked questions. I therefore had to switch to English, and enjoyed a very reassuring dialogue. I can still see and hear the very tall Joseph Needham rising and introducing himself to the audience: "Joseph Needham, Caius ("Kee-ees") College, Cambridge—can I ask a few questions?" My English apparently sufficed.

Plans for Leaving Denmark

While I was trying to finish my PhD thesis, we lost my younger brother Fritz, just six months after his exciting journey to California in 1937. He died suddenly in a "status epilepticus," which, according to his California friends, he had been close to a couple of times in 1937. His fatal attack happened barely a year before the effective treatment of epilepsy became available in the United States and Denmark. My mother never got over this loss. The Bohr family, who also loved Fritz dearly, rendered deep personal comfort. We were touched by Niels Bohr's speech at the modest cremation ceremony and by his request to bring with him young Aage Bohr, who had enjoyed a warm friendship with Fritz. This all happened in January 1938. My PhD thesis became dedicated to the memory of Fritz Kalckar.

Although I admittedly was depressed during 1938, I had much encouragement and inspiration as well. Lundsgaard and Lipmann were formidable mentors and generous personalities. Funding for my research came from three sources: the University of Copenhagen, financed by the Ministry of Education; the Carlsberg Foundation; and the Rockefeller Foundation.

My personal salary for teaching, during the last years as assistant professor, came from the University of Copenhagen. The medical students who came to my afternoon lectures in physiology numbered 30 to 40, only about 20–30% of those who showed up at the professor's morning lectures. Yet I felt that I had the cream of the crop of medical students and that we generated a sophisticated dialogue. Our texts were Lundsgaard's excellent textbook interwoven occasionally with the British classic, "Starling's."

In the late 1930s the defense for a "Dr. Med." degree was a highly ceremonial event, whether at Danish, Swedish, or Norwegian universities. In my case, the ceremony took place in the old original section of "København" under the auspices of the almost 500-year-old university, in the middle of January 1939. My formal defense of my thesis was not as formidable as Professor Lundsgaard's concentrated and stylish opposition, but once he was over the critique of the language (mainly my casual Danish text and some of the tables), the scientific dialogue proceeded very well indeed. Lundsgaard agreed with Lipmann that the thesis revealed several new and important features regarding a relatively new field, metabolic biology, barely 20 years old.

During 1938 I came to know a most interesting scholarly chemist and biochemist, Kaj Linderstrøm-Lang at the Carlsberg Laboratory. Lang attracted scores of bright young biochemists, especially from Great Britain and the United States. Among the latter I came to know one of Lang's favorite American scholars, Rollin Hotchkiss from the Rockefeller Institute. By 1937 he and Lang had already developed protein chemistry along fundamentally new lines.

Through Lipmann, Lang had developed a strong interest in the bioenergetics of protein synthesis and wanted to learn more about phosphorylation systems. Lang fully agreed with Lipmann that the postulate that peptidases catalyzed synthesis as well as hydrolysis of peptides in proteins was not the answer to the problem of cellular protein synthesis. I was mainly a listener to these fascinating discussions and cherished the ideas that transphosphorylations and ATP might well be involved in cellular protein synthesis. Lipmann's visionary ideas began to take shape in 1938 via his bold experiment in which he generated ATP from his "homemade" acetylphosphate, catalyzed by a bacterial enzyme. This was performed in early 1939, after I had left Denmark, and only six months before the outbreak of World War II. It was Lipmann's last performance in Denmark during his last months there.

In 1938, Dr. Warren Weaver, Director of the Rockefeller Foundation, visited Copenhagen. My candidacy for a research fellowship was being discussed. I was appointed a Rockefeller Research Fellow in January 1939, shortly after my defense of my PhD degree at the University of Copenhagen. Now came the important problem: Which laboratory in the United States would stimulate and widen my horizon in general biology and bioenergetics? Linderstrøm-Lang was very familiar with Cal Tech's Departments of Chemistry, with Linus Pauling and Charles Coryell, and Biology, with the great geneticists T. H. Morgan, A. Sturtevant, and Calvin Bridges. Bridges had once shown my brother the *Drosophila* laboratory. Lang also knew the biochemist there, Henry Borsook, who was interested in thermodynamics and the bioenergetics of peptide formation. Borsook still operated on the old-fashioned notion of reversal of peptidase activity in order to reclaim polypeptides and knew as yet nothing about phosphorylations. I followed Lang's recommendation and decided on Cal Tech.

California Institute of Technology and Writing a Review on ATP

In January 1939 I visited London for the first time. My purpose was to visit Dr. Robert Robison, an outstanding biochemist and the Editor of the *Biochemical Journal*. He was instrumental in publishing my fumarate-PEP paper (4) in the journal. Although I had decided to spend most of the year 1939–1940 at Cal Tech's Biology Department, I wanted to make a brief stop at Washington University in St. Louis, where Carl and Gerty Cori worked and where in 1936 they had discovered the first phosphorolytic fission with isolation of glucose-1-phosphate, generated from glycogen and catalyzed by a muscle enzyme, which they now called phosphorylase. I arrived in the bleak and sooty city of St. Louis in February 1939, and was well received by the Coris. Unexpectedly, they were actively interested in my articles on oxidative phosphorylation but were unable to repeat the experiments. Gerty wanted me to make a few demonstrations, but Carl cut it short and simply asked a few questions, and so did I. I saw that the Coris used the old-fashioned Meyerhof extract technique, using test tubes. I explained to them that kidney cortex extracts show a very high oxygen consumption, and in order to see oxidative phosphorylation I always aerated the cortex extracts by shaking them in Warburg vessels.

A bright young researcher in the Cori department, Dr. Sidney Colowick, witnessed my discussion with the Coris. He was asked to compare my technique with theirs. Colowick soon confirmed that the simple shaking technique was indeed needed in order to observe oxidative phosphorylation in the cortex preparations. The Coris wrote to me with the good news and with their best "thanks." This response may well have prepared the way for Carl Cori's later generous invitation to come to their lab when my first-year fellowship was running out.

On March 1, 1939 I arrived in Pasadena. I was received by a young student who soon became my good friend, David Bonner, originally from Salt Lake City. My other friend there was Max Delbrück, whom I had met briefly in Copenhagen at Bohr's Institute. I should later share many interesting reunions with Delbrück.

Thanks to the Bonner brothers, James and the younger David, I managed to find a bungalow shortly after the arrival of my wife Vibeke. We all agreed that the Cal Tech faculty club, "Atheneum," was nice but too expensive as a domicile for a research fellow. The Bonners were also most helpful in introducing me to a superb auto mechanic, José Plannes (of Basque origin); José was THE auto mechanic all the young Cal Tech scholars turned to then. José helped me superbly too. A Ford coupe, vintage 1934, was for sale and I bought it. It held up, not only on desert roads but in the mountains as well. After I received my driver's license, I had the nerve in early April to drive Dr. Hugo Theorell from Stockholm the whole way up to Mt. Wilson observatory, which he wanted to see. The snow was just beginning to melt.

Through Theorell I met Linus Pauling, who later came to play a very positive role in my career. I gave seminars and began to write an extensive review, much inspired by one of Dr. Linus Pauling's former coworkers, Dr. Charles Coryell, who was very interested in the energetics of phosphoryl donors like phosphocreatine and ATP. Pauling's ideas about resonance had greatly influenced Coryell's thinking, and I incorporated some of this into the review. I did not know then about the intriguing review that Lipmann had started, for which he was collecting thermal data. Lipmann's review was to have a great impact in biochemical circles.

I was lucky in the summer of 1939 to be enrolled in a combined lab and lecture course taught by an unusual microbiologist, Dr. Cornelius B. van Niel, who, perhaps more than anybody else I had then encountered, sensed what he called "The Unity of Life." van Niel was Dutch and a graduate of Delft University, Biotechnological School, a highly distinguished school. There he wrote his PhD thesis about a new type of sugar fermentation catalyzed by certain bacteria that generated propionic acid and carbon dioxide. van Niel reminded me that a Danish cheese expert, Orla Jensen, had first sniffed the existence of such a type of fermentation when he found out that the gas in the "eyes" of Swiss cheese was pure carbon dioxide. I had not paid attention to these matters while in Denmark. Perhaps worse, I was ignorant of some important work at the Carlsberg Lab. O. Winge had made great basic contributions in yeast genetics by describing the existence of diploid yeast as well as monoploid. Clearly my fixation on the four "big L's" had made me overlook the great tradition of Danish genetics at that time, founded by Winge's great teacher and researcher, Wilhelm Johannsen.

Although the van Niel Lab was very near the Pacific coast (in Pacific Grove, California), we discussed rather sparingly the role of the microorganisms in the ocean, except algae and the chemosynthetic microorganisms. Photosynthesis was, of course, broadly covered and various postulates critically discussed. These discussions inspired me to expand my biochemistry review that I had started at Cal Tech.

The autumn of 1939 was unusually hot, even for Pasadena, in stark contrast to the climate of Pacific Grove. Fortunately, the hospitable Physics Professor,

C. C. Lauritsen, and his wife invited us over to their swimming pool. November was sunny and pleasant, as was December 1939. My wife and I were invited to a real traditional American Christmas dinner at the Morgan's (T. H. and Lilian) in the stylish old house built and designed by the founder of The Kerckhoff Institute for the Biological Sciences. The founder was of course Dr. Thomas Hunt Morgan himself, and the multidisciplined institute had largely served as my scientific home base. Morgan asked about Denmark and more specifically about Niels Bohr.

In early 1940 I said farewell to my friends at Cal Tech, and Linus Pauling generously offered to get my extensive review on bioenergetics published in *Chemical Review*, provided I could manage to send him a finished script by the end of 1940.

We sold our car and left California in January 1940 and travelled more or less comfortably by train to the Midwest. I sat up most of the night on the train to St. Louis writing page after page of the review for Linus Pauling. After a brief couple of days in colorful New Orleans, we continued north, arriving late February 1940 in the damp and smoky St. Louis. It was a depressing city, but there I had the benefit of working in a lab that had developed a new type of enzymology, which now had become my main interest.

I had partly lost my enthusiasm for pursuing P/O ratios in various tissue particle preparations. Yet Carl Cori and his able coworkers had managed to obtain good P/O ratios in heart tissue dispersions, keeping in mind my warning against using unshaken test tubes for trying to obtain P/O ratios in tissue dispersions. Admittedly, by now I would rather have learned enzymology "a la Cori." Nevertheless, Carl persuaded me first to participate in collaborative work with him and Sidney Colowick on P/O ratios. Later, Sidney and I tried to delve into enzymology on our own. Thus began some of my happiest months in research.

While writing these lines I realized that W. A. Engelhardt and M. N. Lyubimova discovered 50 years ago that myosin was endowed with ATP-ase activity (5). In my big composition entitled "The Nature of Energetic Coupling on Biological Synthesis" (6), I took the liberty to make a speculative sketch from the Engelhardt discovery on the possibility that phosphorylation and dephosphorylation of myosin are "synonymous" with contraction and relaxation of myosin.

More important than my speculative efforts of 1939 on the ATP-ase report, however, was Engelhardt's experimental design, which was based on a hypothesis that expressed much the same ideas, albeit with experimental data. The interesting article by Engelhardt came to light in the *Yale Journal of Biology & Medicine* 15:21–38 (1942–1943), thanks to Paul Talalay's translation from the Russian [Engelhardt, W. A. 1941. *Adv. Contemp. Biochem.* (USSR) 14:172–90; cf. Kalckar, H. M. 1969. *Biological Phosphorylations,*

Development of Concepts, pp. 444–87. Engelwood Cliffs, NJ: Prentice Hall)]. In this masterful article, Engelhardt studied the changes of extensibility of isolated myosin fibers from muscle preparations upon the addition of ATP. A torsion balance with a reflecting mirror was used for recording. Engelhardt found that addition of ATP to freshly made myosin fibers brought about a specific increase in the extensibility of the myosin fiber, with a splitting of ATP to ADP. Engelhardt wisely refrained from interpreting whether this mechano-chemical effect was a contraction or a relaxation of the myosin fiber. In any case, this work opened the way for new discoveries in the life sciences.

Resumption of Laboratory Projects

Only 25 years old, Dr. Sidney Colowick had achieved wisdom as well as a warm sense of humor and brilliant analytical insight. Colowick and I were interested in studying the enzymes of transphosphorylations. Our first attempt centered around a further fractionation and purification of yeast hexokinase. We soon found that the semipurified hexokinase preparations catalyzed the transfer of only one mole of P from ATP per hexose phosphate formed, and ATP was converted to ADP; in contrast, in crude hexokinase preparations, ATP was converted to 5'AMP, and 2 moles of hexose phosphate were formed.

Although we did not have the luck to find the suspected factor in yeast preparations, we did have luck testing extracts from skeletal muscle. Through a simple and effective purification of muscle extracts (merely by acidification with HCl and heating the acidified extract to 90°C for one to two minutes, then neutralizing partly to pH 6.5 and spinning), we obtained a clear supernatant (containing only 1% of the original protein content). This fraction turned out to contain the complementary activity. In other words, the mixing of the fractionated yeast preparation and the heat-resistant muscle fraction brought about the catalysis of formation of two hexose phosphates per ATP consumed, and the appearance of one 5'AMP. We called the active complementary fraction from muscle, "myokinase" (7).

At the suggestion of Dr. Marvin Johnson, University of Wisconsin, we pursued the following research problem: Does myokinase catalyze an "ADP dismutation" in which two ADP molecules are converted to one ATP and one 5'AMP? In order to prove this option, we needed to expand our analytical methods. Meanwhile Pearl Harbor forced the United States into the War, and Sidney Colowick was "drafted" for defense-related research in early 1942.

I was therefore left alone to find analytic methods to prove the existence of an enzymic "ADP dismutation." For this difficult task I used fractionation by means of barium acetate and analytical determinations of ammonia and of pentose as well as phosphorus (7a). The specific 5'AMP deaminase (described by Gerhard Schmidt in 1928) served well for this purpose. The analytical data supported the ADP dismutation scheme. The enzyme was later called "adenylate kinase." In 1946 I was able to reconfirm this enzymic ADP dismutase by UV spectrophotometric shifts.

In 1943 I received an invitation from Dr. Oliver H. Lowry to join a research institute of biochemistry, partly supported by the city of New York, partly from private funds. The division was headed by Oliver Lowry. The section on nutrition was headed by Dr. Otto A. Bessey (an old associate of C. G. King, President of the Nutrition Foundation). The names of the Committee of Overseers I scarcely remember, except that I do indeed remember Dr. Michael Heidelberger, then at Columbia University Medical College.

The biochemistry lab was very well equipped. For me the number one attraction was a Beckman UV spectrophotometer, which soon became my "Stradivarius." Playing this instrument, I came to develop a new field, the enzymatic synthesis of purine nucleosides (8). One of my favorite "tunes" was the metabolism of inosine, a hypoxanthine riboside. Why did I select this odd nucleoside? Because I soon found out that addition of xanthine oxidase gave rise to a highly conspicuous spectral change from 2480 to 2900 Å, since xanthine oxidase converted the hypoxanthine via xanthine to uric acid. I soon suspected that I was essentially dealing with a genuine phosphorylase reaction (8).

In pursuing this interesting problem, Oliver Lowry, a master analyst in the field of phosphate metabolism, came to play a leading role. He and his coworker, Jeanne Lopez, first tried to develop a better method for analyzing phosphocreatine. They succeeded and, in addition, via the same method (which operates at pH 4), they demonstrated that phophorolysis of inosine and guanosine gave rise to the formation of an acid-labile phosphoric ester, ribose-1-phosphate (9). Incubating this labile ester with hypoxanthine or guanine in the presence of the "new" phosphorylase gave rise to the resynthesis of inosine or guanosine. Thirty to 40 years later, it was found that inborn defects in the phosphorolytic fission of inosine to hypoxanthine and phosphate interfere in the conversion of T-cells to B-cells.

Reunion with Linderstrøm-Lang in New York

Contact with Denmark until 1943 had been possible provided the letters reported about "neutral" events, since all the letters were opened and censored by the German occupation army. Denmark was now, as were so many other European countries, in the hands of the SS army and the Gestapo (joined by a small but vicious band of Danish Nazis). I received a laconic message that my mother had died in her home near the end of August 1943. It was very disturbing news, so much more so because I did not know what happened to my older brother or to my brave friends who had actively resisted the Nazi infestation in Denmark. I learned before my return that my brother Emil had been saved in Sweden.

In late 1945 the great Kaj Linderstrøm-Lang came to New York, invited by Warren Weaver, director of the Rockefeller Foundation. After a joint meeting, I had the rich opportunity to spend time alone with Lang. He told me in his low-keyed way about his encounters with the Gestapo and their brutalities, which led to the loss of many lives, including members of his own family. He knew about my mother's death, and he expressed that moving, warm empathy that so characterized him. Then we came to talk about the future. He had followed my scientific projects from my bioenergetics paper in early 1941 in the *Chemical Review* to my most recent paper on nucleoside phosphorolysis, which had just appeared as a full paper in the *Federation Proceedings, 1945*. He expressed the hope that I eventually would return to Copenhagen, but at present he agreed with me that I should continue to pursue my highly promising career in the United States.

Curiously enough, during our joint meeting (without Warren Weaver's knowledge), a very zealous assistant manager for fellowships of The Rockefeller Foundation expressed his personal feelings in a letter addressed to me that a former Rockefeller Foundation Fellow (even though the fellowship had terminated in early 1941) still ought to return to his homeland, i.e. Denmark. Fortunately, Ollie Lowry, Lang, and Weaver found this message under my circumstances absurd. I did not let myself be "bamboozled" by such a schoolmasterish message. My future decisions were to be influenced by other forces.

In 1945 I went to Cold Spring Harbor for Max Delbrück's first laboratory course in bacteriophage genetics. We were only four or five "students," among them my friend Rollin Hotchkiss from the Rockefeller Institute. We were all working happily more than ten hours a day reading plaques (lysed bacterial hosts, by various *Escherichia coli* strains). *E. coli* strains that were resistant to a phage mutant would be read on the plate as a "carpet" without any plaques. Delbrück was with us in the lab until well after midnight.

Late in 1945 there arrived an important letter from Ejnar Lundsgaard in Copenhagen with a generous offer to expand my research position at the University of Copenhagen. Keeping Lang's words in mind, I responded warmly in favor of returning to Denmark.

Return to the University of Copenhagen

I returned to Denmark on a nice Norwegian ship, after first anchoring in Bergen, an ancient city, which probably was not right then in the mood to call itself an old "Hansa City" after all its bitter fight against five years of Nazi tyranny. After Bergen, we finally reached Copenhagen in April 1946. Among the survivors of the dark years was my older brother Emil (rescued by friends in Sweden), his wife, and their two children, Jørgen (10 years) and Birgit (7 years). I shall not dwell anymore on the loss of my mother and her brother.

My scientific life was rapidly restored, thanks to help from many quarters. First of all, my mentor of 12 years ago, Professor Ejnar Lundsgaard, saw to it that a sizable unit of his institute was set aside for my research. He initiated a promotion in my rank at the University and found starting funds for the continuation of my research. I had managed in advance to obtain some American financial support from The Rockefeller Foundation, and thanks to Dr. Thomas Jukes, from Lederle Laboratories. This together with support from the Carlsberg Foundation enabled me to purchase equipment for my lab, which I called the "Cytofysiologisk Institute."

In 1947, I was lucky enough to be approached by a young student, Hans Klenow ("studiosis magistrarum," as he was assigned, since he came from the graduate school, University of Copenhagen): Klenow was interested to learn and participate in research projects. Another bright "stud. mag.," Niels Ole Kjeldgaard, joined us. As a starting point in the new lab, we investigated the intriguing enzymological puzzles of the activities observed in two oxidases: xanthine oxidase and xanthopterin oxidase. This project soon disclosed quite a number of interesting observations, which we monitored not only by spectrophotometry but also by fluorometry. We isolated xanthine and pterine oxidase from cows' milk by using a milk centrifuge of Danish make (10). Our preference was to collect the enzyme from the whey fraction and purify the enzyme by ammonium sulfate fractionation.

Most folic acid preparations exerted a marked inhibitory effect on xanthine and xanthopterine oxidase, due apparently to an inhibitory factor in most pteroyl glutamate acid preparations (PGA). Much of this was resolved when my friend Tom Jukes sent us a PGA preparation of particularly high purity that seemed free of most of the inhibitory effect on the two oxidases. In addition, Oliver Lowry identified the oxidase inhibitor to be 2-amino-4-hydroxy-6-formylpteridine (i.e. a 6-aldehyde) as the strong inhibitor of the oxidases. This finding was further supported by our observations that incubation with xanthine oxidase catalyzed the 6-aldehyde group to a 6-carboxyl group, a derivative that did not exert any inhibition of the oxidases.

Aside from its value as an exercise in biochemical methods, I cannot say that this project deserved to be further pursued, and soon Klenow as well as Kjeldgaard showed their own great abilities in independent research. I was longing to resume work on the nucleoside phosphorylase project, and was lucky to be approached by a brilliant young PhD postdoctoral biochemist on a National Institutes of Health fellowship. His name was Morris E. Friedkin.

Young American Biologists Choose Copenhagen Laboratories

Young Dr. Morris Friedkin from Lehninger's laboratory at the University of Chicago wanted to come to the University of Copenhagen to spend a year with me on his NIH postdoctoral Fellowship, 1949–1950. This association was to become one of the culminations in my own career. Friedkin had followed my work on inosine and guanosine phosphorolysis from 1947, and he was also aware of the Manson-Lampen abstract on phosphorolysis of deoxynucleosides from thymus preparations.

Friedkin and I preferred first to monitor the phosphorolysis of guanosine-deoxyriboside (DR) by enzymatic oxidation, observing the spectral shift as I had described in 1947. The enzymatic phosphorolysis of guanine-DR yielded a highly acid-labile phosphoric ester, which reacted with hypoxanthine in the presence of liver nucleoside phosphorylase to form hypoxanthine-DR. The highly acid labile ester was rapidly dephosphorylated at pH 4 at room temperature and subsequently lost its ability to generate purine-DR (11).

Upon enzymatic phosphorolysis (at pH 8) of guanine-DR, Friedkin then isolated a deoxyribosyl-1-phosphate as the crystalline salt of the organic base, cyclohexylamine. The salt contained no inorganic phosphate unless subjected to hydrolysis at pH 4. Hypoxanthine-DR was synthesized when the cyclohexamide salt of DR-1-phosphate was incubated with hypoxanthine in the presence of nucleoside phosphorylase. The formation of hypoxanthine-DR was shown by spectrophotometric methods. It indicated that the enzymatic synthesis of hypoxanthine-DR was greatly favored. The same was observed in the case of deoxy-guanosine. This is illustrated in Figure 2 (12). Friedkin and I were fortunate to have the excellent Danish microbiologist E. Hoff-Jørgensen join us. He showed that the microbiologic assay (with *Thermobacterium acidophilus* R 26) confirmed our enzymatic data for deoxy-guanosine (13).

On my visit to Chicago in early 1950 I was invited to a meeting on bacteriophage with Max Delbrück and Salvadore Luria. They were running the meeting at a hectic tempo. I was introduced to a shy young man from Indiana University. His name was James Dewey Watson. Both Delbrück and Luria recommended to young Watson that he use his upcoming fellowship to visit Dr. Ole Maaløe and me in Copenhagen, since we were supposedly attuned to phage biology and biochemistry. The research topic had actually been initiated by an able virologist, Alfred Hershey. Hershey's game plan was to label T-phages via their *E. coli* host with ^{32}P and ^{35}S to see whether the two isotopes were transmitted to the progeny. ^{32}P was supposed to represent DNA and ^{35}S, protein. It turned out that 40–50% of either isotope was transmitted to the progeny.

When I learned that Maaløe and Watson would also examine various T-phages by labeling them with ^{32}P, I wondered whether I could not enrich their program by providing them with ^{14}C-labeled adenine.

Thanks to the generous help from Professor Alexander Todd (head of the University Chemistry Laboratory, Cambridge) and his young coworker, Dr. Malcolm Clark, I was able to provide our lab with highly labeled [8-^{14}C]–adenine. Actually, Clark and I worked together, and half a year before

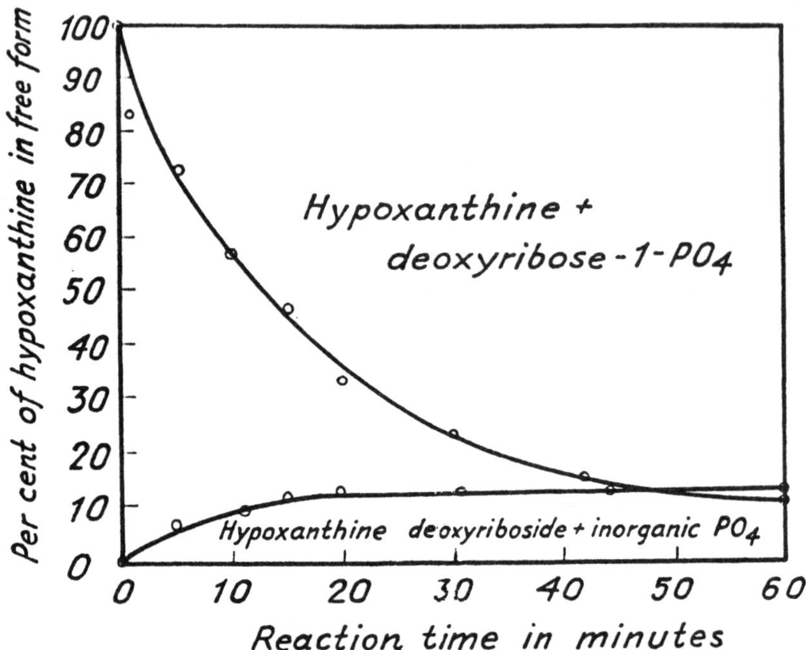

Figure 2 Equilibrium between deoxyribose-1-phosphate plus hypoxanthine and hypoxanthine deoxyriboside plus phosphate. Abscissa: time. Ordinate: per cent hypoxanthine in free form.

Watson arrived we had [8-^{14}C] adenine ready for him and Maaløe. Todd let us publish alone and helped us to get our findings published in the British *Journal of the Chemical Society*, which he found needed some useful publications in the field of "radio-chemistry" (14).

A hard-working and deeply original American microbiologist, Walter McNutt, wanted to come to our Copenhagen Institute in 1950. Within a few months, McNutt found that dialyzed enzyme preparations from *Lactobacillus helveticus* catalyzed a seemingly related reaction of fission of deoxynucleosides (DR), albeit in the total absence of phosphate. Moreover, this type of DR transfer, as McNutt showed, operated between pyrimidine and purine DRs. This bacterial enzyme was classified as a *trans-N*-glycosidase (15).

The ^{14}C-labeled adenine was mobilized to try to solve a dispute, the enzymic interaction of deoxyribosides between adenine-DR and hypoxanthine-DR. The *L. helveticus* was the enzyme source with which McNutt had demonstrated the existence of *trans-N*-glycosidases between purines and pyrimidines. In the case of adenine-DR reacting with hypoxanthine or vice versa, however, two options were plausible, and in this case the use of a ^{14}C adenine turned out to be critical. The enzyme might catalyze 1. a transamination or 2. a transglycosidation.

1. Transamination would catalyze the following reaction: ^{14}C adenine + hypoxanthine-DR $<= = =>$ ^{14}C hypoxanthine + adenine-DR.
2. Trans-N-glycosidase would catalyze the reaction: ^{14}C adenine + hypoxanthine-DR $<= = =>$ hypoxanthine + ^{14}C adenine-DR.

The experimental outcome supported Reaction 2, i.e. the transglycosidase reaction (16).

The second occasion for which our [8-^{14}C]adenine was put to use was when Watson and Maaløe were studying the transmission of nucleic acid to the progeny of a T-phage in a purine requiring *E. coli* host. It was found that the progeny of the T-phages had recovered 40–50% of the ^{32}P label as well as 40–50% of the ^{14}C adenine label (17).

Otto Warburg in Denmark

In the spring of 1952, Kaj Linderstrøm-Lang at the Carlsberg Laboratory in Copenhagen invited Otto Warburg as a special guest lecturer at Carlsberg. After all, thanks to the great Warburg, Denmark had been fortunate enough to claim two great biologists, Albert Fischer, a pioneering tissue culturist, and Fritz Lipmann, a leading genius in biochemistry, both of whom he inspired. Besides, Hans Klenow and I had been Otto Warburg's personal guests in Berlin-Dahlen in 1950. Warburg was pleased to accept the invitation from the Carlsberg Laboratory in the spring of 1952. He gave two lectures, concentrating on his work on crystalline metabolic enzymes. Lang was impressed by the depth and style of the Warburg lectures, which barely consumed one hour. (Fortunately, photosynthesis was not one of the topics selected, since Warburg was disinclined to discuss quantum yields less than perfect, namely one!, i.e. 1.0!)

After delivering a second lecture on the kinetics of the crystalline enolase for a wider circle of Danish scholars, Warburg expressed a warm interest in seeing more of Denmark. There was, of course, a variety of choices, but to my mind the superior one was Elsinore and Hamlet's Castle. My good friend from Pacific Grove (a fellow student in C. B. van Niel's second microbiology course in 1939), Andrew Benson, happened to be a visitor in Copenhagen, and he joined Warburg and myself on the little excursion to Elsinore.

The excursion turned out better than anticipated. It was a crystal-clear, windy, and sunny day. Warburg seemed to like the scenery very much. In fact he seemed moved to open up and ask me, "Why do I encounter so much alienation from British and American biochemists?" I tried to answer this difficult question by quoting his wise and warm admirer, Hans H. Weber, who, like so many of us, treasured the monumental Warburgian prose, but occasionally worried about some of his opinionated footnotes. "Be honest with me," Warburg said, "give me one example." "O.k.," I said. "Why did you call the late Sir Frederic Gowland Hopkins a Romantic?" "Did I really?"

KALCKAR

Warburg said. "Yes, in one of your footnotes," I answered. Warburg assured me that he had always been an admirer of Gowland Hopkins, "The generous 'Dean' of British Biochemistry" and the former President of the Royal Society, who undoubtedly had voted for Warburg as a Foreign Associate in 1934. In any case, Warburg's open confession was remarkable; perhaps Hamlet's castle played its part. I believe that I transmitted this Warburgian adventure to Carl and Gerty Cori, old friends of Warburg. It might also interest other believers in the Warburgian enzymological approach, such as Arthur Kornberg and Eugene Kennedy.

The Galactose Pathway and UDP-Glucose

By 1951, Leloir and coworkers in Buenos Aires had already described the complex nature of the biochemical conversion of glucose phosphate to galactose phosphate. This seemingly so simple positional "swap" of a hydroxyl of the C-4 group of glucose with the hydrogen on the same carbon, thus forming galactose, was found to require at least one coenzyme, UDPGlc or UDPGal. We suggested that such nucleotides might be formed by an interaction between galactose-1-phosphate (Gal-1-P) and UDPGlc, and by a transfer between Glc-1-P and UTP, as summarized in Table 1. In 1952 we initiated our studies on the complex galactose metabolism in eukaryotic cells. These studies took place in our research unit of cell physiology at the University of Copenhagen. Along with Agnete Munch-Petersen and visiting research fellows, among them Evelyn Smith-Mills from Scotland, we conducted studies on the enzymatic interactions of UTP, UDPGlc, Glc-1-P, and Gal-1-P. The enzymes involved were prepared from the same type of yeast as the one used

Table 1 The galactose pathway towards glucose metabolites

Reaction	Enzyme	EC number
1. Gal + ATP → Gal-1-P + ADP	Galactokinase	2.7.1.6
2. Gal-1-P + UDPGlc ↔ UDPGal + Glc-1-P	P-Gal-uridylyl transferase	2.7.7.12
3. UDPGal ↔ UDPGlc	UDPGal-4-epimerase	5.1.3.2
Sum:		
Gal + ATP → Glc-1-P + ADP		
UDP Gal formation from glucose		
4. Glc + ATP → Glc-6-P + ADP	Hexokinase	2.7.1.1
5. Glc-6-P ↔ Glc-1-P	Phosphoglucomutase	5.4.2.2
6. Glc-1-P + UTP ↔ UDPGlc + PPi	PP-uridylyl transferase	2.7.7.9
7. UDPGlc ↔ UDPGal[a]	UDPGal-4-epimerase	5.1.3.2
Sum:		
Glc + ATP + UTP → UDPGal + ADP + PPi		

[a] UDPGal is donor of galactosyl units to various low- or high-molecular-weight acceptors; the transfer is catalyzed by various galactosyl transferases.
Sources: (20, 21)

by the Leloir group, i.e. the yeast *Saccharomyces fragilis* preinduced by galactose. We were able to detect two types of "uridyl transferase" in yeast:

1. Gal-1-P + UDPG ↔ Glc-1-P + UDPGal (Gal-1-P uridyl transferase) (Ref. 18)

2. Glc-1-P + UTP ↔ UDPGlc + PP (Pyro P uridyl transferase) (Ref. 19)

Description of a Human Enzyme Defect

In 1953 my contact with biochemical research in the United States was renewed in, let me stress, a very creative way via "The Visiting Scientists Program" at the National Institutes of Health in Bethesda. Thanks to Dr. Bernard L. Horecker, chief of research at the "National Institute of Metabolic Diseases," as it was called in 1953, I received a generous invitation to conduct basic medical research in a field of biochemistry of my own choice. In 1954, I had not only Horecker's encouragement and support but also that of Dr. Hans De Witt Stetten, incoming Institute director. Stetten, as well as Horecker, soon became my trusted friends during these years.

At NIH I was able to pursue further research on the enzymology of the UDP-glucose system, thanks to an enlightened and active group of biochemists and pharmacologists there. I joined forces with Drs. Jack L. Strominger and Elizabeth S. Maxwell. For the first opus we had Dr. Julius Axelrod (from the National Institute of Mental Health) as an additional participant. The joint papers described a novel dehydrogenase pathway from liver by which UDP-glucose (UDPG), through a two-step dehydrogenation of the free 6-hydroxyl group of the glucosyl moiety, could be converted to UDP-glucuronic acid ("active glucoronic acid"). Here I merely emphasize a few aspects of this reaction in connection with our ongoing studies of galactose metabolism. UDP-glucose dehydrogenase turned out to be highly specific for UDP-glucose; it required NAD as a coenzyme and was inhibited strongly by NADH. In my further studies on galactose metabolism, I used this specific and sensitive spectrophotometric enzyme test for UDPGlc, catalyzing the reduction of 2 moles of NAD per UDPGlc oxidized (i.e. UDPGlc + 2 NAD → UDP-glucuronic acid + 2 NADH). In human hemolysates, however, UDPGlc was only consumed if Gal-1-P also was present.

Several research groups had clearly demonstrated the presence of active galactokinase in hemolysates from normal as well as galactosemic subjects, i.e. incubation of galactose and ATP brought about a marked accumulation of Gal-1-P. The view on galactosemia in 1952 seemed to center around defects in the so-called "Galactowaldenase," now named "epimerase," which supposedly was needed to convert Gal-1-P to Glc-1-P. The latter was sketched in an oversimplified way,

Gal-1-P$\leftrightarrow$Glc-1-P,

without taking into account whether UDPGlc was involved or not.

My own view on the question of the defective step in galactosemia differed decisively from this formulation, and for several reasons. Some of my reservations arose from numerous reported clinical observations on galactosemic infants. The development of toxic symptoms, such as cataracts of the lens or mental retardation, did not develop in galactosemic infants, provided they were administered, as early as possible, a lactose-free diet. Soybean formulas, containing sucrose instead of lactose, serve as a telling example. In short, galactosemic infants kept on a galactose-free diet seem to have an excellent chance of normal development.

It was known that the central nervous system, and especially the mammalian brain, contains large amounts of galactolipids and that almost half of the galactose is deposited after birth, i.e. converted from UDPG to UDPGal. Yet, normal brain function is actually better preserved in galactosemic infants kept on galactose-free diets. A defect of the enzyme Gal-1-P uridyl transferase would not interfere with the synthesis of galactolipids, even on a galactose-free diet. In this case a normally functioning 4-epimerase would secure a normal conversion from glucose via UDPGlc to UDPGal and hence an unhindered formation of galactosyl compounds from glucose (Reactions 4–7, Table 1).

The disease symptoms observable in galactosemic infants fed lactose may rather be due to a "traffic jam," i.e. an accumulation of Gal-1-P that is unable to be converted to UDPGal due to the absence of the enzyme Gal-1-P uridyl transferase. Our first priority, therefore, was to compare the levels of the Gal-1-P uridyl transferase enzyme levels to hemolysates from galactosemic and nongalactosemic subjects.

Although galactosemia is a rare disease, it was possible for us, thanks to the skill and efforts of Dr. Kurt Isselbacher (with additional help from Dr. Josephine Kety and Dr. Elisabeth P. Anderson), to acquire blood samples from eight subjects with galactosemia (and more than a dozen subjects who were either normal or afflicted with unrelated diseases). We found that the galactosemic patients specifically lacked Gal-1-P uridyl transferase (20, 21) but did have normal levels of the other uridyl transferase, which reacts with pyrophosphate.[2]

We wanted to ascertain that the type of galactosemia that we had been studying was an exclusive defect in the enzyme Gal-1-P uridyl transferase, and did not affect UDPGal-4-epimerase. The absence of the latter in hemoly-

---

[2]This was further corroborated by Dr. Kurt Isselbacher in tests on white blood cells and on a few trace amounts of liver biopsies.

Table 2 Quantitative comparison of activities of enzymes of galactose metabolism in three classes of subjects

Classes of subjects	Average activities, μM conversion/hour/gram cells (37°C) (lysates of erythrocytes)			
	Galactokinase	Gal-1-P uridyl transferase	UDP-Gal-4- epimerase	G-1-P, PP-uridyl transferase
Nongalactosemics	0·10	4·8	0·32	>10·0
Galactosemics	0·08	<0·02	0·35	>10·0
Galactosemic carriers		2·9		

sates from galactosemic as well as normal subjects was clarified through the following sequence of events in our laboratory, all within the same lucky year of 1956. Dr. Elizabeth Maxwell in our group was able to purify UDPGal-4-epimerase (abbreviated: epimerase) from calf liver. It turned out that the purified epimerase required NAD for its activity and was strongly inhibited by NADH. The absence of epimerase activity in normal as well as in galactosemic hemolysates in earlier experiments was due to an unanticipated need for addition of NAD to the hemolysates in order to overcome the inhibition of the NADH present in all human hemolysates (22).

In 1957 Dr. H. N. Kirkman joined our group. Kirkman was an MD interested in biochemistry. As a pediatrician, he was also interested in the field of inborn errors of metabolism in humans. He developed improved methods for studying the kinetics of the enzyme Gal-1-P uridyl transferase. This enabled him to demonstrate partial enzyme defects in galactosemic carriers in siblings or parents of galactosemic subjects (totaling 15 carriers). Table 2 summarizes these records as well as records from nongalactosemic families (23, 24[3]).

Liz Maxwell and I decided that the chemical name "Waldenase" should not be used anymore in the context of the enzymatic reversible conversion of UDP-galactose to UDP-glucose. The enzyme catalyzing this reaction should be called "UDPGal-UDPG epimerase" (25).

Gal *Mutants in* E. coli

In 1957 Dr. Kiyoshi Kurahashi and I were lucky enough to obtain a valuable gift from Dr. E. M. Lederberg: some samples from the Morse-Lederberg collection of *E. coli* galactose mutants. Kurahashi and an able French microbiologist, Dr. Huguette de Robichon-Szulmajster (from Gif sur Yvette), were

---

[3]Reports from Great Britain and Switzerland about a human hereditary UDP-Gal epimerase defect that does not respond favorably to a galactose-free diet were recorded around 1970.

able to classify the *Gal* mutants and to study their physiology just before an international biochemistry meeting in Tokyo, Japan. Some of the *Gal* mutants were highly defective in the enzyme Gal-1-P uridyl transferase, the type we had singled out as the defective enzyme in the children afflicted with the "inborn error, galactosemia." Indeed, the *E. coli* with a defective Gal-1-P uridyl transferase enzyme, if grown in media containing galactose, turned out to lag in rate of growth. The effect of galactose on this mutant was especially noteworthy when the growth rate of the mutant was monitored in a minimum medium, such as ammonia mineral glycerol medium, with or without galactose. In the presence of galactose, growth was very slow and was soon arrested (26).

Japan—India—Israel

At the International Symposium on Enzyme Research in Tokyo, September–October 1957, I had the opportunity to present our joint work (H.M.K. Huguette de Robichon-Szulmajster, and Kiyoshi Kurahashi) on mutants of Gal-1-P uridylyl transferase in man and microorganism (46). After the symposium I flew to India, where Professor J. B. S. Haldane, now a resident of India, had invited me to stay for a week at the Indian Statistical Institute.

Professor Haldane received me at the airport and brought me safely out of Calcutta to the Statistical Institute. Haldane was greatly interested in our work on hereditary galactosemia. His own concern about galactose evolved around the consumption of the so-called wrinkled peas, which are stuffed with a galactose-containing trisaccharide, stachyose (containing fructose, glucose, and galactose). Could this type of undeveloped pea (which Mendel had discovered to be recessive) prove to be toxic? What would happen if stachyose peas were offered to a galactosemic subject? Later the verdict came from a friend, Professor G. Semenza in Zürich: Because the human intestine is unable to hydrolyze stachyose, galactose would not be released.

Haldane had a broad view of the diversities of evolution. One of his favorite illustrations was Alkaptonuria. "All insects have Alkaptonuria. They have phenolic acid around which they use to tar and blacken their cuticle. Alkaptonuria in man merely results in black ear cartilages."

After my visit with Haldane, I stopped at several places in mid-India (Sarnath, Delhi, and Bombay). I was invited to Thrombay by Professor Bhaba, head of the Atomic station. Bhaba was a friend of my brother, Fritz Kalckar, from the Niels Bohr Institute in Copenhagen.

Finally, my last invitation was to The Weizmann Institute of Sciences with Drs. Ephraim Katchalski and Michael Sela. They were then both heavily engaged in work on a novel type of immunology. They were making synthetic polypeptides, conferring them with specific antigenic markers, like DNP, from which, after injection into rabbits, they could harvest specific antibodies to DNP. My friend, Dr. Edgar Haber, happened to be visiting Michael Sela at the same time.

Johns Hopkins University Years

In 1958 I was invited by Dr. William D. McElroy, Scientific Director of The McCollum Pratt Laboratories of Biochemistry at Johns Hopkins University, to assume a full professorship. Among the advanced graduate students, I had the great luck to have Michael Yarmolinsky and Elke Jordan join me, and we were all enthusiastic to have Dr. K. Kurahashi join us too.

We developed methods to analyze the levels of the enzymes of the "galactose sequence." We were puzzled by the fact that the galacto kinase-negative mutant (*Gal-2* mutant) invariably also showed high, "constitutive," levels of the next enzyme of the Gal sequence, i.e. Gal-1-P uridyl transferase ("Gal-3"). In contrast, if galactokinase was operating normally in the *E. coli,* the Gal-1-P uridyltransferase levels were low and depended on induction from outside, i.e. by addition of galactose to the growth medium, in order to increase the levels of both enzymes (27, 28). This involved response by the *E. coli* Gal sequence continued to occupy my mind. Ten years later, it eventually awakened my great interest in galactose chemotaxis (discussed below).

In April 1959 I received a warm personal congratulation from Johns Hopkins University President Milton Eisenhower. I had been elected to be a Member of the US National Academy of Sciences (NAS) in Washington, D.C. Curiously enough, I was the only biochemist elected. It was said that the Academy President, Detlev Bronk, an outstanding biophysicist in his own right, was hesitant to crowd the Academy with too many biochemists. I definitely felt that Carl and Gerty Cori had convinced Detlev Bronk that I was a deserving choice. Perhaps my academicianship in the Royal Danish Academy of 1948 was a contributing factor in my favor.

Be that as it may, a call was extended for a young academician to participate in a scientific exchange between the NAS and the USSR Academy NAYK. I volunteered immediately and was accepted. Hans Stetten at NIH encouraged me to tell the Russians part of our hereditary galactosemia story. Dr. Bentley Glass at Hopkins (also elected an NAS member) supported this idea. Thanks to the skill and culture of my friend Dr. Simon Black, this became a reality. Black had taught himself Russian, and when in Russia travelled without a guide on modest third-class trains with native Russians. I arrived in Moscow with his Russian slides ($2\frac{1}{2}'' \times 3\frac{1}{4}''$) in my pocket for my lecture there.

My Solo Visit to the USSR

On June 1, 1959, I flew to Paris and from there flew with "Aeroflot" directly to Moscow; the jet was not more than half filled. In Moscow I was met by a young Scandinavian-looking student from Akademy NAYK, Michkael Kritsky, who turned out to be an interesting and attractive travel companion. Kritsky was a 24-year-old graduate student in biochemistry at one of the academy institutes in Moscow under Prof. Belozorsky, a highly regarded biochemist in the field of nucleic acids, whom I met later.

At the Bach Institute, I had the opportunity to meet several distinguished and active biochemists in my own field. Among them was Wladimir A. Engelhardt, a grand old friend and scholar, whom I had recently seen at Johns Hopkins University during his visit to the Biology Department there. Engelhardt was studying membrane ATP-ases and cation transport. In a few months he was due to move to a new academy institute. His wife, M. N. Lyubimova, was studying ATP-ases in various filamentous proteins— spermosin, mimosin (from mimosa plants), and filaments in trypanosomes; all these filamentous proteins seem to contract when exposed to ATP. My lecture, entitled "Nasledstvennaya Galacktotsemia," was to take place in Academician A. L. Oparin's division. Oparin presided and Dr. V. Gorkin translated paragraph by paragraph into Russian. In this way almost two hours elapsed, in spite of the Russian slides, which I duly presented. The audience included junior as well as senior biologists and biochemists (Engelhardt, among the latter). The topic of inborn errors of metabolism in man seemed to interest the audience greatly.

I had arranged for a 20-minute question period, if necessary. Five minutes turned out to suffice, since only four questions were raised, either philosophical or physiological. I then decided to give a brief summary about the X-linked Glc-6-P dehydrogenase defects. The audience was surprised to learn about the high incidence in their own country of Favism (especially among oriental Jews). This is a sex-linked hereditary Glc-6-P dehydrogenase defect in red blood cells that leads to hemolytic anemia.

During this visit to Moscow, I had the pleasure to meet the great biochemist and polymer chemist, Ephraim K. Katchalski, from Israel. Our subsequent friendship resulted in several invitations to visit The Weizmann Institute of Sciences in Israel. I did not participate actively in the polymer symposium in Moscow. I learned, however, that this discipline was considered important in the USSR, and that during the Krushchev regime at the time, Jewish scientists were held in high esteem, several of whom, such as N. N. Semenov, occupied leading positions.

My guide Kritsky and I then travelled by train to Leningrad (the night express was called "Red Arrow" and was comfortable). The light nights in Leningrad encouraged us to take long evening walks near the Neva. The gracious city had a spirit of European urbanity. At the Hermitage there was an exhibition of the ancient Crimean Gold. Yet, for me, the unforgettable sight there was of the French paintings, culminating with the magnificent painting by Henri Matisse, "Les Danseuses."

From Leningrad a jet carried us nonstop to Kiev. My main interest there was to meet V. A. Belitser, who, as I have mentioned, in 1939 first described in detail a reliable system that generated oxidative phosphorylation. I later received permission from the USSR reprint office to have his 1939 article with Tsibakova translated to English and reprinted in my monograph of 1969. My immediate problem in 1959, however, was to find Belitser. I did find him. He was most cordial and seemed to enjoy very much to meet me and to be photographed with me (see photo of both of us on p. 7). We spent about half an hour together. I then travelled on to Armenia (see Ref. 46 for my further trip to Armenia and Yerevan).

The year 1959 was, in a way, with the U-2 incident, a tragic year. President Eisenhower and his wise advisor, George Kistiakovsky of Harvard, were constantly fed lies by the generals in the "Defense"—nay! "War"—Department. Later, back in Moscow at the International Congress of Biochemistry in 1961, the mood was mixed. In all the turmoil, a few of us were lucky enough to hear Marshall Nirenberg disclose publicly the first breaking of the genetic code. UUU stands for phenylalanine. My friend and gentle humorist, the biochemist, Rollin Hotchkiss, coined the moving event in 1961 "the U-3 incident."

Massachusetts General Hospital

In 1960 I received a distinguished invitation that pleased me very much from Dr. Paul C. Zamecnik, chairman of The Huntington Laboratories at Massachusetts General Hospital (MGH) and Professor of Medicine at Harvard Medical School (HMS), and the new chairman of The Department of Biological Chemistry at HMS, Dr. Eugene P. Kennedy. Professor Carl Cori, who had become an adviser to MGH for the last couple of years, added his voice in strong support. The invitation was to fill the position left open since Fritz Lipmann had accepted the Professorship at the Rockefeller Institute in 1957. Dr. Oliver Cope, chairman and Professor of Surgery at MGH, and Dr. George Packer Berry, Dean of Harvard Medical School, had carefully looked into the possibility of financing and restoring a new professorship at MGH. Moreover, thanks to Dr. Walter Bauer, Professor of Medicine at HMS and MGH, and his friendship with Sir Henry Dale, chairman of The Wellcome Trust in London, further generous support of the new professorship was offered from London. I was pleased and honored to accept this new professorship.

In Boston, I met two old friends from the Carlsberg Lab in Copenhagen, Mahlon Hoagland and Paul Plesner, and their guest, a most exciting, beautiful lady with serene poise, who also came from Copenhagen. Thirty years ago I had met and admired Agnete Fridericia. Agnete had lost her husband, Svend Laursen, in 1960. Her friends and colleagues respected and admired more than ever Agnete's courage in the face of the agony, and her ability to get through it with her two young children. Agnete's and my cheerful conversations during the 1960s, most of it in Danish, changed my life. We married in 1968.

BACTERIAL MUTANTS AFFECTING THE TOPOGRAPHY OF THE CELL WALL
In starting my research at MGH, I was lucky enough, through the encouragement of Dr. Roger Jeanloz, head of the Carbohydrate Unit at MGH, to have Dr. A. M. Rapin join our unit in early 1962. Moreover, Dr. T. A. Sundararajan from India also joined my unit at the same time. Already this joint work opened up the discovery of a new type of *E. coli* Gal mutant, which we called *Gal 23*. Because *Gal 23* contained both K. T. and E. enzymes, we wondered whether the new Gal mutant might be unable to synthesize UDPG (i.e. UDPG pyrophosphorylase). We were able to substantiate this idea. We found that induction with galactose (or D-fucose) in minimum medium or in broth failed to increase the low levels of UDPG synthetase (29).

At the same time and completely independently, Fukasawa, Jokura & Kurahashi in Tokyo also isolated an *E. coli* mutant unable to synthesise UDP-glucose. Dr. Hiroshi Nikaido, who had recently joined our lab at MGH, became a guiding spirit in the study of bacterial cell walls. Nikaido analyzed the lipopolysaccharides in *Salmonella enterides,* the so-called O-antigens. From his collaboration with Fukasawa on the UDP-galactose-epimerase-less *Salmonella* mutants, and adopting the concept of a core and side chains and the data from the epimerase-less mutants, Nikaido invoked a theory in which galactose forms a link between the core of the side chains. Missing the galactose link elicited the lack of incorporation of a number of core sugars, as well as those of the side chain (i.e. mannose, rhamnose, tyvelose, etc) (30). Clearly, Nikaido was a great observer. I was fortunate to have him visit our lab and to be able to obtain funds for him as an independent researcher in our department at MGH.

LEARNING PROTEIN BIOCHEMISTRY AND BIOPHYSICS IN AN UNANTICIPATED WAY   In 1958, E. Maxwell, R. Szulmajster, and I had observed that a highly purified yeast enzyme, UDP galactose epimerase with NAD(H) bound to the enzyme, showed a conspicuously intense blue fluorescence. In our lab at MGH, Robert A. Darrow and his coworkers found that purified epimerase from yeast is asymmetrical in the sense that one dimer enzyme protein carries only one bound NAD, not two bound NAD (31, 32). At about the same time, Creighton & Yanofsky revealed a related type of asymmetry in tryptophan synthetase. Later, when Dr. Othmar Gabriel joined our lab, he found a related type of asymmetry in *E. coli* TDPG oxidoreductase (33).

We had studied the effects of NAD and NADH on the conformation of alcohol dehydrogenase at the Carlsberg Laboratory, where we found that the addition of NAD or NADH exerted a tightening effect on the apo-enzyme in terms of changes in optical rotatory dispersion (ORD) and circular dichroism (CD). In our joint work, Professor Martin Ottesen had recommended additional studies of conformational changes of the apo-enzyme, such as tritium exchange methods combined with sephedex fractionation.

A. U. Bertland studied the changes in fluorescence parameters as well as changes in ORD and CD. He found that if the purified yeast epimerase was incubated with 5' UMP and D-galactose or 5' UMP and L-arabinose for 2 to 3 hours, a 30-fold fluorescence enhancement (based on the mol. fluorescence of free NADH) ensued. The "superfluorescent" epimerase lost more than 90% of the catalytic activity (34). A quantum yield of 3% for the fluorescence of free NADH and an enhancement factor of 30 for the epimerase-bound [NADH] would amount to a quantum yield of nearly 90% for the bound NAD (35). The fluorescence of the bound [NADH] showed high polarization, not only through the excitation band (3400 Å) to the emission band (4400 Å), but also from the aromatic region (2800 Å). Since the *S. fragiles* epimerase contains 40 tyrosines but only 3–4 tryptophans per dimer (see summary by Kalckar et al 1969), the energy transfer from the tyrosine rings to the reduced nicotine amide ring is remarkably high.

In the discussion at the Copenhagen Symposium (cf. Kalckar et al. 1969), Dr. Howard Schachman had noted that one of the slides carried vector signs and wondered whether we were also following the polarization of fluorescence. I was pleased to answer his question affirmatively. Bertland had indeed been studying the polarization of fluorescence and did find polarization of fluorescence from the aromatic band as well as from the excitation and emission bands (36). The weird device by which Bertland managed to obtain catalytically active epimerase (NAD) with quenched fluorescence was also described at the Symposium. Prolonged storage at high dilution (0.1–0.3 mg protein per ml buffer) would suffice to accomplish the transformation. I should add that epimerase reduced by sodium borohydride did not respond in this way but remained generally unresponsive as NADH, which remained catalytically inactive. As mentioned above, the reductive inhibition of yeast epimerase by $NaBH_4$ is only short-lived. Bertland, Y. Seyama, and I found, however, that the further addition of 5'UMP would transform it into a long-lived, reductive inhibition (37).

A HOME-SPUN TRANSITION TO GAL CHEMOTAXIS   In 1964, Henry C. P. Wu approached me regarding graduate work to obtain a PhD degree, which he eventually achieved in 1969. In 1966, another brilliant young biochemist, Dr. Winfried Boos from the University of Konstanz, Bodensee, also joined our laboratory.

We again concentrated our efforts on trying to understand the regulation of the Gal operon. To make a long story short, *GalK,i* releases galactose from its cellular UDPGal and seems unable to recapture it from the medium. In contrast, *GalK,c* seems to possess a galactose "scavenger mechanism" sufficiently effective to maintain the induced state of *GalK,c*.

Winfried Boos then developed biochemical methods to isolate the Gal scavenger from either of the GalK strains. By a skillful use of polyacrylamide gel electrophoresis of cell extracts, obtained by osmotic shock devices, he found that the fractionated shock fluid from *GalK,c* showed a specific high-affinity Gal-binding protein, whereas shock fluid from *GalK,i* did not. The Gal-binding protein was considered a periplasmic protein, residing in the space between the outer and inner membranes (38).

Shortly afterwards, while listening to a fascinating lecture by Julius Adler from the University of Wisconsin on galactose chemotaxis in various *E. coli* mutants, I became excited by certain parallels. Adler had found a galactose chemotactic receptor located in the periplasmic space of *E. coli*. Chemotaxis was then only partly characterized, but it was suspected to be related to one of the most primitive sensory responses known in nature.

I decided that Boos and Adler should exchange the Gal mutants that had scored positive (+) or negative (−) in their respective tests. For what purpose? I had exhilarated myself with the notion that the *GalK,c* strain would score positive in Adler's chemotactic test and that the *GalK,i* strain would score negative in the chemotactic test. Unfortunately, due to my habitual type of "daydreaming," as Boos used to call it, my monomaniacal mumbling, "soliloquizing," was obviously not deciphered by either Adler or Boos. Finally, after a good deal of cajolery, both friends began to understand me. They did finally exchange their bacterial mutants for executing their respective tests. The finding was in fact that only the strain with high-affinity Gal binders *(GalK,c)* scored positive in Gal chemotaxis, and this was also reflected in Boos's tests of Gal binding versus Gal chemotaxis. The receptors in Gal chemotaxis seemed indeed identical with the Gal binder.

As Boos mused in his generous and thoughtful chapter in the "Festschrift" to my seventy-fifth birthday, "I feel obliged to tell you on this occasion that it was really Herman Kalckar who first perceived the idea that a transport-related binding protein is chemoreceptor in *E. coli*, even though the published record may suggest otherwise." In an article in *Scientific American*, Adler (39) also recalled our fruitful encounter of 1970.

Since one of my favorite pupils, Dr. Paul Plesner, recently challenged me to try to express some of my philosophical thoughts regarding my long trot along the trail of biochemistry, I shall try to construct a philosophical parable to the story that I have just finished. The grand old British biochemist, Sir Frederic Gowland Hopkins, ventured to answer a question posed to him through the philosophy of Alfred North Whitehead, "Has the modern biochemist, in analyzing the organism into parts, so departed from reality that his studies no longer have biological meaning?" In a dialogue with himself, Hopkins wrote: "So long as his analysis involves the isolation of events, and not merely of substances, he is not in danger of such departure." Boos, Wu, Adler, and I actually managed in our own way, somehow, to make a transition from biochemical analysis to cellular events, like endogenous induction and sugar chemotaxis. Our case, that a transport-related protein might also be used as a chemoreceptor, could only be postulated on account of the recognized specificity of mutational steps, selecting mutants by two widely different criteria. Adler and his group had been seeking mutants in chemotaxis, while we were looking for mutants in galactose metabolism. Now both fields came together. This makes it feasible to "isolate" two different types of events and thus to disclose one of the capricious ways of the economy of Nature.

My old friend, the late Max Delbrück, who had always teased me about my mumbling, seemed to have enjoyed the mutual Gal taxis adventure, and he invited me in 1970 to give a lecture at Cal Tech, Biology Division. This happened also to be the first "Jean Weigle Lecture," which was later published in *Science* (40).

TUMOR BIOLOGY AND HEXOSE TRANSPORT REGULATION   Some time during my stay at MGH and Harvard Medical School, I became captivated by a new important field that had originated through observations on nutrient transport into tumor cell cultures. It was possible to transform a fibroblast culture into a sarcoma culture by exposure to a virus. This was first explored on amino acid transport by Foster & Pardee in 1969 and by Isselbacher in 1972 on glucose as well as on amino acids.

My own interest in tumor biology was aroused in 1973, and I was lucky to be invited to visit Dr. S. Hakomori at the University of Washington, Seattle, who had initiated active studies on tumor biology. We studied the biology of glucose and galactose transport in nontransformed and polyoma-transformed hamster cell cultures. We found enhancement of transport, of glucose as well as galactose, in the transformed cultures (41), and also morphological changes in the transformed cultures. These changes may well have represented the loss of a surface protein, as described in transformed cells by R. Hynes the same year. The surface protein that is present in normal cultures Hynes named "fibronectin."

The same year my interest in general regulation of hexose transport was alerted by some exciting observations by Harold Amos and his coworkers in the Microbiology Department at Harvard Medical School. Their studies focused on regulation of glucose transport into primary chick embryo fibroblasts. They found persistently that depriving normal chick embryo fibroblasts (CEF) of glucose brought about a dramatic enhancement of the rate of influx of glucose, as compared with tests on glucose-fed cultures. As Dr. Amos was happy about my enthusiasm, and he was as always concerned with the future of his young coworkers, he asked Dr. C. W. Christopher if he would be interested to join my unit at MGH. At that time, although I was retired from Harvard, I was fortunate to have a unit in Dr. Zamecnik's department and to be well funded. Moreover, I had my able coworker, Donna B. Ullrey, in my unit, and so in 1975 we had the pleasure of having Christopher join us.

Hexose uptake by hamster fibroblast cell cultures was increased 5–10-fold by either substituting D-fructose for glucose or by completely omitting D-glucose from the culture medium for 24 hours. Conversely, when cyclohex-imide was present for 24 hours in media containing glucose, up to 20-fold decreases in hexose uptake were observed. These decreases in uptake activity were only observed over a narrow range of cycloheximide (CHX) concentrations, however. After extended exposure to low concentration of c CHX (0.05 to 10 μg/ml), the uptake by the fed cells decreased in parallel with inhibition of protein synthesis, whereas at high concentrations of CHX (>50 μg/ml), uptake was increased. The effect of CHX on the uptake mechanism was found to be biphasic with respect to the concentration of CHX (42). Cells deprived of glucose and maintained in the presence of cycloheximide did not show decreases in uptake activity. In separate experiments, the high uptake rates of glucose-starved cells could be decreased by the addition of glucose to the glucose-free medium. The reversal was complete in 6 to 8 hours. The analog of glucose, 2-deoxy-D-glucose, did not promote the time-dependent decrease, thus suggesting that the 6-phosphoester of glucose is not an inhibitor of transport. In addition, when cycloheximide was added at the same time as glucose, there was no decrease in uptake for at least 12 hours. It was proposed that the turnover of components of hexose uptake systems could account for part of the control of hexose transport.

Japan 1977

Once more, in September–October 1977, my wife and I were the guests of the hospitable Japan Science Council. Dr. Yousuko Seyama, who had been in my laboratory in 1972–1973, was our host in Tokyo, and Dr. Kiyoshi Kurahashi was our host in Osaka. This was also the year we visited Hiroshima. The joint Japanese-American Atom Bomb Committee (to whom I was introduced by a letter from Dr. Alexander Leaf, MGH) invited me to summarize my article on stromtium fallout and my article in *Nature* about collection of children's teeth, a study that was organized by the Dental School of Washington University in St. Louis (see Ref. 46).

A unique contribution to peace was initiated by the president of the University of Hiroshima, Dr. Orata Osada, with the book, *Children of the*

A-Bomb—A Testament of the Surviving Boys and Girls of Hiroshima. Dr. Osada, in spite of his own radiation sickness, had managed to collect hundreds of letters from the schoolchildren of Hiroshima. One of the translators of this unique book, Dr. Jean Clark Dan, a marine biologist and a mother of five children, was an old friend of mine and sent me the English translation by her and her friend Ruth S. Morgan. The testaments of the schoolchildren, from grammar school to junior high school, are completely free of any propaganda and only tell what they experienced. It was finally possible, after several years, to get this book distributed through a Canadian publishing house. No US publisher, as far as I know, was willing to distribute the book.

After a few more stops in southern Japan we returned to Tokyo where our friends helped us get ready for our trip to China.

China 1977

Thanks to the help of Dr. Joseph Needham and to his friendship with Kuo Mo Jo, a prominent Chinese historian, I was invited to Academia Sinica in Peking and Shanghai for two weeks in late November 1977. I lectured in both places.

Just prior to our arrival, the outskirts of Peking had been hit by an earthquake. The epicenter was in Tiensin. A couple of days later, I was invited by the director, Dr. Pei Shih Shang, to visit the Biophysics Institute of Academia Sinica, Peking. Dr. Shang, 77 years old, was a member of the standing committee of the party central committee. He did not speak English, but since he had been in Freiburg, Germany in 1927, he still spoke German, and I had not forgotten my German from Denmark, so our conversations in German went quite well. In view of the prevalence of earthquakes in China one would expect that the Biophysics Institute would have had some seismographic monitors. They did not have such luxuries, however. They used pigeons as monitors, a method they called "bio-seismography." I was told that the feet of pigeons have some sensitive "corpuscles"; these sensors are called the "Herbst sensors." They are supposed to transmit mechanical vibrations into increased spike frequency. I am not sure that I followed the pigeon "tale."

Obviously the large staff had to deal with a lot of biophysical topics such as radioactive isotopes and crystallography especially of insulin crystals, which they received from Shanghai Institute of Biochemistry. The equipment was protected against the late November chill, but the director and the senior or junior scholars had merely their old overcoats on.

After a week in Peking we flew to Shanghai where I visited the Shanghai Institute of Biochemistry, Academia Sinica. The director, Wang Ying-lai, was born in 1907 in Fukian Province, China. After graduating from Nanking University, he earned his PhD at Cambridge University in 1941. He remained in Cambridge for four years of postdoctoral research on cytochromes with

David Keilin at the Molteno Institute. After his return to China, he published in the 1950s procedures for the isolation of succinic dehydrogenase, which received world-wide use. In 1958 he was appointed director of the newly founded Institute of Biochemistry of the Academia Sinica in Shanghai. It was under his administration that the Institute undertook the major effort in basic research that culminated in the total synthesis of insulin. Wang Ying-lai has represented the Chinese biochemists internationally and was a visitor in the United States in 1975.

With the changes in administration that accompanied the Cultural Revolution, Wang was appointed the senior Vice-Chairman of the Revolutionary Committee of the Shanghai Institute of Biochemistry. At the time fundamental studies on peptide synthesis included research on the structure-function relationships of insulin and methods of synthesis of oxytocin, vasopressin, and glucagon. Selected publications were credited to groups rather than to individual authors: *The Total Synthesis of Crystalline Insulin*, Institute of Biochemistry (Shanghai), Institute of Organic Chemistry (Shanghai), and Department of Chemistry.

I was lucky to attend a biochemistry meeting in which Professor Wang requested one of his coworkers, Dr. Pan Xiu Jia, to report on an unusual hormone project undertaken by the biochemistry lab on the behalf of the fresh-water fish farmers. Dr. Pan gave her report in Chinese, which was then translated into English. The first step was to sneak one D-amino-acid, D-alanine, into a nonapeptide of L-amino acids of a hypothalamic hormone luteinizing releasing hormone (LRH). Dr. Pan brought this unique mosaic hormone analogue to the Shanghai cooperative fresh-water fish farm in order to study the possible spawning response of the silver carp at the temperature of the farm, 25°. The injection of a few micrograms of the mosaic D-analogue per kilogram carp elicited a dramatic spawning response that lasted for weeks, since no antibodies against the mosaic hormone were formed. The D–amino acid protected the hormone against hydrolysis.

I also paid a visit to the Shanghai Institute for Physiology and Acupuncture. The Institute tried to map the neural paths of acupuncture response. I found a visit to the Da Hua Hospital most interesting. There I observed the following: Female 20–30 years old—acupuncture for appendectomy: 2–3 needles in right ear (120 vibrations per minute) + 1 needle in right foot (200 vibrations per minute). Male 60–70 years old—acupuncture for gastrectomy: pretreatment with luminal, 4 needles in right ear, and 2 + 2 needles in each foot.

I stayed for 30–45 minutes during these operations, which were surprisingly peaceful. Afterwards I was the only one of ten foreign observers who volunteered to have the acupuncture needle inserted in my hand between the thumb and index finger; with the vibrations it felt as if my hand "grew" bigger and bigger within 10 minutes. This method is used by pregnant women giving birth, according to Ms. Wang, our intelligent guide, who had had it during the birth of her child.

Before we left China, Agnete and I were invited by a lady superintendent of a junior high school in Peking to attend a class. The teacher was a young, spirited lady who taught vector analysis. The usual standard pictures of Mao, Lenin, and Marx were on the wall, but the blackboard was bristling with mathematical formulas. The rapport between students and teacher seemed excellent.

A few days later we left China for Tokyo, and, with the help of our Japanese friends, from there on to United States.

Boston University

After my second retirement from MGH, my friend and colleague, Dr. John M. Buchanan of MIT, approached his colleague, Dr. Norman Lichtin, Chairman of the Chemistry Department at Boston University Graduate School, who invited me as visiting professor. Thanks to the skill and energy of my assistant, Donna Ullrey, we performed innovative research in biochemistry there for more than ten years.

As mentioned above, feeding D-glucose to Chinese hamster lung fibroblast down-regulates its own transport system. This effect depends on glucose metabolism, since a mutant that lacks phosphoglucose isomerase depends specifically on glucose metabolism (43). In collaboration with Dr. Paul Plesner, we found that the mutant was unable to keep its ATP at normal levels in glucose media, whereas mannose as a hexose source was a competent energy source (44). Later, Ullrey and I found that the rare sugar D-allose is much more effective than glucose in controlling the glucose transport system (45); it is also phosphorylated (46).

Summary

In 1930 adenosine triphosphate appeared in the literature from W. A. Engelhardt's work on avian erythrocytes. This was an early example of oxidative phosphorylation in intact cells, and it required methylene blue and oxygen. Both Belitser and I realized that the use of Warburg manometers for aeration was critical in order to generate oxidative phosphorylation of glucose in tissue preparations. Test tube techniques did not work.

In 1956 we were able to describe a human type of diabetes called "galactose diabetes," in which consumption of human or cows' milk provokes mental retardation. Replacement of human or cows' milk products with "vegetable milk" formula in early infancy can prevent retardation. We determined that the disease results from a defect of galactose-one-phosphate uridylyltransferase, a hereditary enzyme. This type of enzyme defect, if discovered and treated in early infancy, is a benign molecular disease.

Regulation of transport systems in mammalian cell cultures are frequently complex energized systems. Perhaps my greatest surprise in this regard was the mere fact that an all-cis "odd" hexose-D-allose turned out to be a highly intense down-regulator of the hexose transport system. Additions of inhibitors of oxidative phosphorylation (such as oligomycin or di-nitrophenol) arrested the allose-mediated down-regulation. We have reason to suspect that the strong down-regulator is a phosphorylated form of D-allose.

Thus ends my story about oxidative energized biological phosphorylation systems.

ACKNOWLEDGMENTS

I thank Dr. Rollin D. Hotchkiss, Mrs. Agnete Kalckar, Dr. Eugene Kennedy, and Donna B. Ullrey.

Literature Cited

1. Kalckar, H. M. 1937. *Enzymologia* 2:47–53
2. Belitser, V. A., Tsibakova, E. T. 1939. *Biokhimya* 4:516–34
3. Kalckar, H. M. 1938. *Nature* 142:871
4. Kalckar, H. M. 1939. *Biochem. J.* 33:531–41
5. Engelhardt, W. A., Lyubimova, M. N. 1941. *Nature* 144:668–72
6. Kalckar, H. M. 1941. *Chem. Rev.* 28:71–178
7. Colowick, S. P., Kalckar, H. M. 1943. *J. Biol. Chem.* 148:117–26
7a. Kalckar, H. M. 1943. *J. Biol. Chem.* 148:127–37
8. Kalckar, H. M. 1945. *Fed. Proc.* 4:248–52
9. Lowry, O. H., Lopez, J. 1946. *J. Biol. Chem.* 162:421–28
10. Kalckar, H. M., Klenow, H., Kjeldgaard, N. O. 1950. *Biochem. Biophys. Acta* 5:575–85
11. Friedkin, M., Kalckar, H. M. 1950. *J. Biol. Chem.* 184:437–48
12. Friedkin, M. 1950. *J. Biol. Chem.* 184:449–59
13. Hoff-Jørgensen, E., Friedkin, M., Kalckar, H. M. 1950. *J. Biol. Chem.* 184:461–64
14. Clark, V. M., Kalckar, H. M. 1950. *J. Chem. Soc.*, pp. 1029–30
15. McNutt, W. S. 1950. *Biochem. J.* 50:384–97
16. Kalckar, H. M., McNutt, W. S., Hoff-Jørgensen, E. 1952. *Biochem. J.* 50:397–400
17. Watson, J. D., Maaløe, O. 1953. *Biochim. Biophys. Acta* 10:432–42
18. Munch-Petersen, A., Kalckar, H. M., Cutolo, E., Smith, E. B. 1952. *Nature* 172:1936–39
19. Kalckar, H. M., Braganca, B., Munch-Petersen, A. 1952. *Nature* 172:2–4
20. Kalckar, H. M., Anderson, E. P., Isselbacher, K. J. 1956. *Biochim. Biophys. Acta* 20:262–68
21. Kalckar, H. M. 1985. *BioEssays* 3:134–37
22. Maxwell, E. S. 1957. *J. Biol. Chem.* 229:139–51
23. Kirkman, H. N., Kalckar, H. M. 1958. *Ann. NY Acad. Sci.* 75:274–78
24. Kirkman, H. N., Bynum, E. 1959. *Ann. Hum. Genet.* 29:117–26
25. Kalckar, H. M., Maxwell, E. S. 1958. *Physiol. Rev.* 38:77–84
26. Kalckar, H. M., de Robichon-Szulmaister, H., Kurahashi, K. 1958. *Proc. Int. Symp. Enzymol. Chem., Tokyo-Kyoto-Maruzen*, pp. 52–56
27. Kalckar, H. M., Kurahashi, K., Jordan, E. 1959. *Proc. Natl. Acad. Sci. USA* 45:1775–86
28. Jordan, E., Yarmolinsky, M. B., Kalckar, H. M. 1962. *Proc. Natl. Acad. Sci. USA* 48:32–40
29. Sundararajan, T. A., Rapin, A. M., Kalckar, H. M. 1962. *Proc. Natl. Acad. Sci. USA* 48:2187–93
30. Nikaido, H. 1962. In *The Specificity of Cell Surfaces*, ed. B. D. Davis, L. Warren, pp. 67–71. Englewood Cliffs, NJ: Prentice-Hall
31. Darrow, R. A., Creveling, C. R. 1964. *J. Biol. Chem.* 239:362–63
32. Darrow, R. A., Rodstrom, R. 1968. *Biochemistry* 7:1645–54

33. Wang, S. F., Gabriel, O. 1969. *J. Biol. Chem.* 244:3430
34. Gabriel, O., Kalckar, H. M., Darrow, R. A. 1975. In *Subunit Enzymes*, ed. K. E. Ebner, pp. 85–135. New York: Dekker
35. Bhaduri, A., Christensen, A., Kalckar, H. M. 1965. *Biochem. Biophys. Res. Commun.* 21:632–37
36. Kalckar, H. M., Bertland, A. U., Johansen, J. T., Ottesen, M. 1969. *The Role of Nucleotides for the Function and Conformation of Enzymes. First Benzon Symp. 1968–1969*, ed. H. M. Kalckar, H. Klenow, A. Munch-Petersen, M. Ottesen, J. Hess-Thaysen, pp. 247–75. Copenhagen: Munksgaard
37. Bertland, A. U., Seyama, Y., Kalckar, H. M. 1971. *Biochemistry* 10:1545–51
38. Boos, W. 1969. *Eur. J. Biochem.* 10:66–73
39. Adler, J. 1976. *Sci. Am.* 234:40–42
40. Kalckar, H. M. 1971. *Science* 174:557–65
41. Kalckar, H. M., Ullrey, D. B., Kijomota, S., Hakomori, S. 1973. *Proc. Natl. Acad. Sci. USA* 70:839–43
42. Christopher, C. W., Colby, M., Ullrey, D. B. 1976. *J. Cell Physiol.* 89:684–92
43. Ullrey, D. B., Franchi, A., Pouyssegur, J., Kalckar, H. M. 1982. *Proc. Natl. Acad. Sci. USA* 79:3777–79
44. Plesner, P., Ullrey, D. B., Kalckar, H. M. 1985. *Proc. Natl. Acad. Sci. USA* 82:2761–63
45. Ullrey, D. B., Kalckar, H. M. 1986. *Proc. Natl. Acad. Sci. USA* 83:5858–60
46. Kalckar, H. M. 1990. *Compr. Biochem.: Autobiographical Notes from a Nomadic Biochemist.* 37:101–76

Annu. Rev. Biochem. 1990. 59:1–27

HOW TO SUCCEED IN RESEARCH WITHOUT BEING A GENIUS

Oliver H. Lowry

Department of Pharmacology, Washington University School of Medicine, St. Louis, Missouri 63110

KEY WORDS: quantitative histochemistry, rapid rise of biochemistry, biomedical categories, Linderstrøm-Lang, enzymatic cycling.

CONTENTS

0066-4154/90/0701-0001$02.00

GROWING UP

Ever since receiving the invitation to write this prefatory chapter, I have been wondering "Why me?" After reviewing the list of previous authors of this chapter, I was even more puzzled until I realized that this may have been a move to show that it is not necessary to be a genius to contribute to science.

I grew up in a very religious family with ancestors on both sides of the American Revolution. Several ancestors were preachers. One exhorted our soldiers in the 1812 War to "fight with the sword of the Lord and of Gideon." Another was John Rankin, a prominent pre–Civil War abolitionist who had a price on his head in Kentucky (which he ignored in his travels). He operated a very successful station on the underground railroad at the Ohio-Kentucky border where he passed 2000 slaves North without a single loss (Harriet Beecher Stowe wrote the part about Eliza crossing the ice from his house on the Ohio River). But as far as I know, none of my ancestors was a scholar or a doctor.

My father was the son of a carpenter who was killed during a barn raising, leaving an impoverished family, held together by a determined mother with a reputation for uncommon sense and a great respect for education. She also had a wholesome level of skepticism expressed by "what *they* say is a lie, and what *they all* say is a lie and a half."

At age 19, my father started teaching in a one-room country school and began a program of self-education. He managed in his early 20s to get a job teaching physics in the Chicago school system, where he introduced the first physics laboratory in the city. He was a master of the Socratic method. Whenever as a child I asked him a "why" question, he would always respond by asking me a series of questions to show me that I, myself, could figure out the answer. His use of this technique, I believe, had an important influence on my eventual attitude toward scientific problems. I recall a specific reinforcement of this attitude as a graduate student: I needed to know certain physical properties of a particular compound. I knew that my thesis advisor would not know the answer—the answer was probably not in the literature—but I could go into the laboratory and in a short time determine the answer. This reinforcement of my father's teaching, and the confidence it gave me, may have been the most important lesson of my graduate training.

My father went on to become a school principal, then a district superintendent, and finally Acting Superintendent of all the Chicago public schools. Not being much of a politician, he never became the permanent Superintendent, but ended his career as superintendent of all the Chicago high schools.

During most of his career, he was continuing his program of self-education. He arranged for an individualized degree program with Northwestern University, which he carried out as a district superintendent by studying on the "Elevated" going from school to school. He would split each textbook into segments that would fit into his pocket. After getting a bachelor's degree this way, he started on a PhD program with a thesis designed to test for innate musical ability among his public schoolchildren. Unfortunately, after years of testing and documentation, his thesis material was accidentally discarded, and he never found time to start over.

All of his children were provided an opportunity (on his teaching salary) to obtain advanced degrees: my sister an M.S. in mathematics; my three brothers, respectively, a law degree, an advanced engineering degree, and a PhD in organic chemistry (this last with postdoctoral training under Willstaetter in Munich).

As the youngest child, I felt I had to live up to much of what my admired siblings accomplished. I never aspired to the law, but conceived of combining chemistry and engineering to emulate my two oldest brothers. Unfortunately, my youngest brother was an outstanding athlete and extremely popular, and I was neither. These qualities, being much more important during school years than scholastic achievement, gave me a feeling of inferiority that undoubtedly did all kinds of bad things to my psyche.

One thing this probably did was make me determined to excel at something. My determination may have been reinforced by learning that I scored only 100 on a high school intelligence test. I gathered that 100 was not really a sign of brilliance. This in turn may have been reinforced at some point by my father expressing his opinion that when choosing a career, persons with mediocre talent should not attempt to master a broad comprehensive field, but instead should specialize in some narrow aspect of a field where they might hope to become truly expert. This I have in fact done, although I doubt it was done consciously. And whether the high school IQ score was accurate or not, my father's idea seems to have worked for me.

My father was convinced that public schools were better than private ones for a number of logical reasons. I am sure his children would have gone to public schools anyway, not only for financial reasons, but because it would not be fitting for a prominent public school teacher to send his children to a private school.

At any rate, I went exclusively to public primary and secondary schools and have never regretted it. Most of my teachers were good, and some were outstanding. I remember an exceptionally good physics teacher saying (circa 1925) "I do not know why there should only be 92 elements, perhaps additional ones will be discovered some day." The large class sizes (40 was the norm) did not seem to be much of a disadvantage. Perhaps this discouraged spoon feeding. Standards were high.

I had skipped ahead a number of grades in elementary school, which was easy in those days. But I had not skipped ahead socially, so I stayed out of school now and then. One semester after finishing elementary school was spent working on an uncle's farm. A year after high school was spent half as an ordinary seaman on a freight boat to the Philippines and Korea, the other half working on another uncle's ranch in Nebraska. These were distinctly maturing experiences, particularly the shift at age 16 from a sheltered religious home environment to that of a tramp ships' forecastle (learning a whole new range of adjectives).

Graduate School

As mentioned above, my first inclination was to combine my brothers' vocations, and thus I enrolled at Northwestern in chemical engineering. But then I spent my sophomore year in Germany at the University of Frieburg with a schoolmate who was a premed (how this came about is not particularly relevant). My companion was so enthusiastic about medicine that I decided I wanted a piece of the action. He suggested that perhaps I should go into biochemistry. He said that so little was known about biochemistry that anything you found out would be new (which was not far from wrong in 1929!). So when we came back, I switched to a chemistry major, and two years later entered the University of Chicago as a graduate student in "physiological chemistry."

My thesis advisor was Frederick Koch, who together with Thomas Gallagher, was trying to isolate the male sex hormone from enormous volumes of urine (every male who came on the premises had to contribute). They assumed that the potency in biological units per mg would be as great as that of the estrogens that Doisy had already isolated. I remember the day (in 1936?) when it was announced that Butenandt had isolated testosterone, and that a unit of activity was much larger in mass than expected, i.e. their preparations were purer than they thought. Whereupon they looked at their best preparations and, in fact, found crystals of the hormone! This was a very blue day in the department.

As a graduate student, I did not have sense enough to pick a thesis project in the field of my advisor. Koch was very tolerant and let me choose for myself. I picked a subject that had something to do with ketone body metabolism, a subject no one in the department knew anything about or had much interest in. After floundering around for a time with some very naive in vitro experiments, I ended up concentrating on the development of a micro method for measuring ketone bodies in one ml of normal blood. This involved the construction of a very complicated homemade multichambered glass distillation apparatus, which permitted delivery from a single volume of blood extract, first acetone itself plus acetone from the degradation of acetoacetic acid, and second, the acetone from oxidation of β-hydroxybutyric acid. The acetone in the two fractions was determined by an iodometric titrimetric procedure I had modified to increase the sensitivity 10- or 20-fold and decrease the blank more than 20-fold. The procedure worked well in my hands and provided the first reliable values for normal blood levels in the rat. But no one in his or her right mind would ever have used the method, and it was never published.

I believe this atypical graduate program increased my self-reliance and self-confidence and may have been better in the long run than a program designed and monitored by a conscientious thesis advisor. Although in one sense my graduate school research was wasted, my thesis subject did get me hooked on micro methods. I continued to be fascinated all my life with ways to increase analytical sensitivity. This turned out to be for me the specialization that my father recommended for people of limited ability. I had the good sense to recognize that biological analytical methods, micro or macro, were of little value unless they were designed to meet specific needs. Consequently, in most cases my methods were published only in the methods section of papers in which they had been used.

During the second year at the University of Chicago, the Dean asked if I would be interested in working for an M.D. along with a PhD. He pointed out that I already had taken many of the preclinical courses, that he was willing to back-date my admission to medical school, and the quarter system made it easy to squeeze four academic years into three calendar years. M.D.-PhD programs were rare in those days; Chicago was one of the few universities that made such programs feasible. My family was supportive because in the depths of a depression (which this was), an M.D. looked like good insurance. So at a commencement five years after my matriculation, I received two diplomas. When President Hutchens of the "Great Books" fame handed me the second diploma, he asked if he hadn't seen me somewhere before. Although I have never practiced medicine and would not claim that medical training greatly changed my life, I still feel lucky to have received this educational dividend. It has certainly added to my enjoyment of biomedical research, broadened my perspective about living systems, and been good for my ego.

Another dividend of my University of Chicago experience was meeting Baird Hastings and working briefly in his laboratory in Billings Hospital. His attitude about research was that it was an exciting game. There was competition, but it was between friends who were all working for the same goals. One should therefore rejoice in the success of the other fellow. This helped restore my somewhat idealistic concept of research: that the scientific edifice is so grand and so important, that adding even one sound brick to the growing structure is a worthy achievement.

HARVARD

Baird and I hit it off well together, and after graduation I wanted very much to work in his department at Harvard, where he had since moved to succeed Otto Folin. Postdoctoral fellowships were almost non-existent in those days, but the Rockefeller Foundation offered a few, and Baird suggested I apply for one.

I made two alternative proposals, one of which I will describe since it illustrates considerable naiveté and my micro method hang-up. I proposed to confirm directly and measure the relationship between mass and energy, which was still somewhat theoretical. I calculated that if I built a closed glass apparatus containing a liter of bromine and an equivalent amount of sodium that were so situated that the two elements could be made to react slowly enough to dissipate the heat without disaster, that the weight change should be measurable (a few micrograms).

Not surprisingly, the Rockefeller Foundation was not enchanted with this idea nor with the other proposal on a subject that I have forgotten. (As far as I know, no one since then has ever tried to weigh directly a decrease in mass from a large dissipation of chemical energy.)

Fortunately, Baird found money ($2000 per year!) for a job as subinstructor, which I heard about a month or two before graduation, and which would start immediately thereafter. (My luck continued, although even during the Depression, $2000 per year was not easy to live on, particularly since I was married by this time.)

The research plan was for me to continue one of Baird's basic interests, that of electrolyte metabolism (which involved measurements of Cl^-, Na^+, K^+, Mg^{2+}, and Ca^{2+}), and I offered to develop micro methods that would extend the investigation to milligram-size tissue samples. Perhaps the most useful applications were the studies of electrolyte changes in the myocardium as the result of ischemia (1–3), and in the heart, skeletal muscle, liver, brain, and kidney as the result of aging (4–6). The ischemia study was made in collaboration with Herman Blumgart, the cardiologist. The hypoxia experiments were made with Otto Krayer, Chairman of the Pharmacology Department, who was an expert with the heart-lung apparatus. The aging studies were made in collaboration with Clive McKay of Cornell University, who was able to double the life span of rats by drastically restricting their food intake. (Unfortunately, restricted food intake also delayed their maturation and did not prove to be of much value to rats once they were fully grown.)

A spin-off from the aging study was the development of micro methods for measuring collagen and elastin, which proved useful to a few other investigators (7). [Dorothy Gilligan and I measured these in everything from a rat's aorta to an elephant's ligamentum nuchea (7).]

Advancement at Harvard in those days was rather slow, and hearsay is that this may still be the case. For many years after he became Chairman of Pharmacology, with a worldwide reputation, Otto Krayer was still Associate Professor. It was not until I had been at Harvard for four years and was about to leave that I finally worked my way up to full Instructor. It was therefore not too difficult for my good friend, Otto Bessey, to persuade me to join him at the brand new Public Health Research Institute in New York City where he was to be the Head of the Department of Physiology and Nutrition.

Carlsberg Laboratory

One thing that Baird did for me while I was still at Harvard, and for which I am especially grateful, was to arrange a fellowship from the Commonwealth Fund that permitted me to work for five months with Kai Linderstrøm-Lang at the Carlsberg laboratory in Copenhagen. This was one of the most rewarding experiences of my life. Lang became one of my two scientific idols (the other being Baird himself). World War II began four days after I arrived with wife and baby. Because of the war, fellows from other European countries had to stay home, so the three American fellows (the other two were Paul Zamecnik and Chris Anfinsen!) had almost the full attention of Lang and his colleague Heinz Holter.

Lang was the most talented human being I have ever known. In addition to being a superb investigator (physical biochemist), he played the violin beautifully, sang delightfully, and was a self-taught artist who painted incredibly fine works of art. To top it off, he was intrigued by micro analytical methods and had invented and developed a whole scheme of quantitative histochemistry together with the appropriate devices. The constriction pipette, for example, was invented by Milton Levy while he was a fellow in Lang's laboratory (8).

If I was attracted to micro methods before I went to Copenhagen, I was an incorrigible addict by the time I left.

THE PUBLIC HEALTH RESEARCH INSTITUTE OF NEW YORK CITY (PHRI)

One of the reasons why Otto Bessey, who was a nutrition expert, wanted me to join him in New York was his belief that the studies he envisaged of the biochemical effects of nutritional deficiencies would require new microchemical methods. This belief was reinforced by the fact that the new institute had just opened when the attack on Pearl Harbor occurred. We decided that one of the most useful things we could do for the war effort was to devise a battery of practical blood and urine tests to screen for nutritional deficiency in the general public.

LOWRY

Otto Bessey and I shared equally in the research and the credit from the very beginning. Later, we were joined by Helen Burch, who became a key participant, particularly in the nutritional studies.

Urine tests (for thiamine and riboflavin) were not particularly micro, but the blood methods had to be quite sensitive, and those we devised permitted assay of the plasma from a single, 0.1 ml blood sample (from finger or ear lobe) for vitamin A, carotene, ascorbic acid, iron, total protein, and alkaline phosphatase, this last an index of vitamin D deficiency. We used these methods in a number of New York City high schools, from poor and rich neighborhoods, and on an international study in Newfoundland made before and after flour enrichment. The methods were also widely used by others for studies throughout the United States, and immediately after the war on nutritionally jeopardized populations in Europe. In one instance, when the methods were applied to a large sample of Munich residents, the ascorbic acid levels seemed unreasonably high, considering the acute shortage of fresh fruits and vegetables. Upon further investigation, it was discovered that large quantities of potatoes were being smuggled in from the countryside. Potatoes are an excellent source of vitamin C if they are boiled with their skins on, as was the local practice.

One of our own wartime studies still has considerable nutritional relevance. We collaborated in an elaborate study of ascorbic acid nutrition conducted by the Royal Canadian Air Force with "volunteer" personnel. For eight months, groups were maintained on diets supplying from 8 to 78 mg of ascorbic acid per day. At the end of this period we were invited to measure ascorbic acid in the plasma and in the buffy coat (white cells plus platelets). Measurements were made just before and during realimentation with large amounts of ascorbic acid (9). The results showed that with an average ascorbic acid intake of 23 mg per day, the buffy coat ascorbic acid is maintained at only about half the level attained with 78 mg per day, which in turn is about 90% of that attainable by realimentation with 2000 mg per day for four days. The data on retention during realimentation indicated a maximum body storage capacity of almost 4 g.

Another study, which also concerned vitamin C, was made with four genuine volunteers from our own staff (10). This was an assessment of the effects of ingesting for 90 days what at that time seemed like an excessively large intake of this vitamin: 1000 mg per day in divided doses. These volunteers had been receiving an estimated 75 to 100 mg per day from their regular well-balanced diets. The plasma ascorbic acid level rose an average of 50% during the first day where it stayed for the rest of the time; the buffy coat vitamin level (a good measure of body stores) did not change significantly at any time, and 80% of the 1000 mg intake was promptly excreted in the urine. No adverse symptoms were detected. Thus, vitamin C intakes that are much above what can be obtained on a good diet are promptly eliminated. This may well be why the enormous doses some enthusiasts recommend (up to 10,000 mg per day) usually do little harm (or good).

We also devised during the war an alkaline phosphatase method, which is still widely used (11). It was based on a study by King & Delory (12) of a wide variety of potential phosphatase substrates, and without our knowledge had already been introduced by Ohmori (13). So we received more credit than we deserved. The substrate, p-nitrophenyl phosphate, was originally obtained from Eastman Kodak, but they subsequently discontinued it. One day I happened to sit next to Dan Broida on the train (sic) coming back from a FASEB meeting in Atlantic City. I asked if his small, versatile company might like to make p-nitrophenyl phosphate for general use. He agreed and later gave this idea partial credit for getting Sigma Chemical Co. started.

A more famous method that also came out of the PHRI days was our protein procedure, which employs the Folin phenol reagent (14) and is merely a modification of the original 1922 method of Wu (15). We needed a quick and easy method for measuring antigen antibody precipitates from small amounts of plasma of nutritionally deficient rats. We tried the method that had been used for a similar purpose by Pressman (16), and by Heidelberger & MacPherson (17), but could not help tinkering with it, particularly in regard to the Cu^{2+} requirement that was first recognized by Herriot (18).

In the complete absence of Cu^{2+}, color development reflects only the content of tyrosine and tryptophan. The addition of Cu^{2+} gives a major increase in color owing to reaction with some of the peptide bonds themselves. When no Cu^{2+} is added, adventitious Cu^{2+} contamination gives partial, erratic color development, which had given the method a bad reputation.

After moving to St. Louis, we continued to use our modified method without publishing the details, but passed them on to whoever wanted them. This included Earl Sutherland, then in Carl Cori's department. He complained of being tired of referring to "an unpublished method of Lowry." So we finally got down to making a thorough study of the procedure: its limitations and virtues, and the results it gave with different proteins and tissues in comparison with the Kjeldahl method (an analytical headache). The first submission to the *Journal of Biological Chemistry* was returned for drastic shortening. This shortening may have improved the paper, but forced us to omit some details that perhaps would have lessened the plethora of subsequent papers by others describing improvements and precautions.

It may be worth commenting on why this paper, which really was not very original, came to be used so widely in spite of its inherent limitations. I believe this was because most biochemists had to measure proteins; the method was simple, sensitive, and reproducible; and it was used early by two outstanding biochemists who happened to be my friends, Earl Sutherland and Arthur Kornberg.

Another method we developed at PHRI was a colorimetric procedure for measuring inorganic phosphate (P_i) under conditions mild enough not to significantly hydrolyze the more unstable organic phosphates (19). We had already experimented a great deal with modifications of methods to measure P_i with acid molybdate reagents, all of which depend on the fact that phosphomolybdate is easier to reduce (to a blue compound) than molybdate alone. The factors that affect the rate of color development with P_i (as well as with molybdate itself) are molybdate concentration, pH, temperature, and the type and concentration of reducing agent. All but the last also affect the rate of hydrolysis of labile organic phosphates. Herman Kalckar, who was in the Department at that time, was working with ribose-1-phosphate generated by nucleoside phosphorylase (20). This phosphate is too unstable to permit P_i measurement by the classic Fiske & Subbarow method (21) and other modifications thereof. I bet Herman that we could work out a molybdate method to do this. We won the bet, but after more work than expected. We raised the pH from below 1 to 4 and substituted a stronger reducing agent, ascorbic acid.

Along with necessary work in developing and applying specific analytical methods at PHRI, we did a modest amount of work on instrument adaptation. When the Beckman DU spectrophotometer came out, we were among the first to get one and were particularly impressed by it because Otto and I had grown up with visual colorimeters (ugh). Because the standard cells required a wasteful 3 ml of solution, we promptly had a local company (Pyrocell) make special microcuvettes and an adapter that permitted us to use as little as 30 μl of solution without reducing the light path, giving a 100-fold increase in sensitivity (22).

We also had been introduced to fluorimetry because others had found this offered the best modality for measuring riboflavin and the riboflavin coenzymes, as well as thiamine (after conversion to thiochrome). We were delighted with the extreme inherent sensitivity of fluorescence measurement, but unhappy with the low sensitivity of available commercial fluorometers. We therefore replaced the simple phototube of a commercial instrument with a photomultiplier tube and made other modifications to reduce light leaks that were giving intolerably high blanks. The result was a 1000-fold increase in useful sensitivity (23). On the basis of this prototype, we persuaded the Ferrand Optical Company to manufacture a similar instrument, which proved eminently satisfactory, and which has gone through many model changes since then.

QUANTITATIVE HISTOCHEMISTRY

In 1947, I was invited to become Head of the Department of Pharmacology at Washington University in St. Louis. This was quite a gamble on the part of the university. I had never had a real course in pharmacology, nor had I done any research that was even marginally pharmacological. Moreover, my two predecessors, Carl Cori and Herbert Gasser, were both Nobel Laureates, and there was no sign that I would get to Sweden except as a tourist. At any rate, I was terribly flattered and of course accepted.

This permitted me to return to a deep interest in quantitative histochemistry, which I had acquired in Baird Hastings' Department at Harvard and had been further fostered by exposure to Linderstrøm-Lang and Holter in Copenhagen. I, therefore, immediately applied for support from the Committee on Growth of the American Cancer Society for a study of the "Quantitative Histochemistry of the Nervous System." It was obvious that if any part of the body required a histochemical approach, it was the brain, because it is such an incredible mixture of different kinds of cells. Generous support was soon forthcoming and has continued ever since, even though our direct applications to cancer research have been minimal.

The original histochemical approach of Linderstrøm-Lang was to analyze alternate histological sections for the substance of interest and to stain intervening sections to permit quantification of the cell types present. Correlations between cell type and substance were then looked for. This worked well with the tissues that Lang had examined, in which only a few cell types were present, and where the cell proportions changed gradually over a considerable distance. This did not seem appropriate for brain, where more cell types are present and changes can be abrupt, even occurring within a single section. I therefore proposed to make freeze-dried sections, which could be examined at room temperature and from which small identified portions could be dissected out, weighed, and analyzed. This was a modification of a procedure that Chris Anfinsen and I had developed for retina in Baird Hasting's department six or seven years earlier (24), and which Chris had applied to good advantage (25, 26).

I was gambling that substances to be measured, particularly enzymes and metabolites, would withstand freeze-drying and subsequent brief exposure to room air and temperature. Fortunately, stability was not much of a problem, although a few enzymes and such easily oxidized substances as NADH and NADPH did not tolerate more than a few hours in room air. On the other hand, after freeze drying, all components of the sections appeared to be stable indefinitely under vacuum at $-70°C$.

A major advantage of the use of freeze-dried sections, instead of fresh sections, was the preservation of metabolically labile substances at the levels that existed in vivo at the time of freezing.

Instrumentation

To exploit the analytical possibilities with freeze-dried material required appropriate special tools and ultimately much increased analytical sensitivity. The first requirement was to measure the size of the samples dissected out of the dry sections. The easiest way to do this proved to be by weight, and the simplest imaginable analytical balance proved to be the best. This is merely a quartz fiber of appropriate thickness and length mounted horizontally, like a fishpole, with one free end on which the sample is placed (27). The displacement of the tip is measured on the scale of an eyepiece micrometer of a horizontal dissecting microscope. For larger samples (0.1 μg or more), a small pan of very thin glass or quartz is affixed to the fiber tip. For smaller samples, no pan is needed, since surface forces ensure adherence. The most sensitive quartz fiber "fishpole" balance (made by Takahiko Kato for weighing nuclei of large individual neurons) could weigh 0.1 nanogram samples to 2% (a dry erythrocyte weighs about 0.03 nanograms). The fiber for this balance was about 3 mm long, had a thickness of 0.3 μm, and the tip drooped 0.6 mm under the weight of the fiber itself.

This type of balance is a simplification of an earlier balance that was inspired by studying with Linderstrøm-Lang (28), but actually a quartz fishpole balance was used in 1915 by Bazzoni (29) to prove that musk loses weight in giving off its odor. This fact had been challenged because the loss is so small as to easily escape detection.

To achieve high sensitivity usually requires reducing the analytical volume; otherwise the concentration of the substance measured becomes too low for precision. We have found the best solution with analytical volumes less than 5 μl is to work under oil in small wells drilled in a Teflon block (30). Volumes in the 0.05 to 0.5 μl range are quite manageable.

The clear choice for manipulating small volumes of liquid is the Lang-Levy constriction pipette, which as mentioned earlier was invented by Milton Levy when he was a postdoctoral fellow in Linderstrøm-Lang's laboratory. These pipettes had been shown to be capable of precise delivery in the 1 μl range. Necessity forced us to explore smaller pipettes. It proved possible to make constriction pipettes out of quartz tubing down to 0.000,2 μl volume that still had a precision of 2%. However, we have rarely required those smaller than 0.01 μl.

Unlimited Sensitivity

In our earlier attempts to achieve high sensitivity, we used a variety of colorimetric and fluorometric methods, choosing those with the highest absorption coefficients or fluorescence. Later, one development made it easier to design sensitive methods for a wide variety of enzymes and metabolites, and a second development made it possible to increase sensitivity almost without limit. The first improvement was to take advantage of the fact that NADH and NADPH are fluorescent, and that as Kaplan et al showed (31), NAD^+ and $NADP^+$ can be converted into highly fluorescent compounds wth strong alkali. This improvement made it possible to measure with high sensitivity any enzyme, or the substrate of any enzyme, that directly or with the aid of auxiliary enzymes can oxidize NADH or NADPH, or reduce NAD^+ or $NADP^+$. The technique used is strictly analogous to what had already been done spectrophotometrically by others following the initial lead of Negelein & Haas (32). The difference is that the fluorescence measurements are about 100 times more sensitive than those based on light absorption. Paul Greengard had already had the idea of using pyridine nucleotide fluorescence in this way, and had applied it to the measurement of a number of tissue metabolites (33).

If the reaction is in the direction to produce NADH or NADPH, the fluorescence is measured directly. If the reaction is in the direction to produce NAD^+ or $NADP^+$, the excess NADH or NADPH are first destroyed with acid to which both nucleotides are very sensitive, and then strong alkali is added and heated to produce the highly fluorescent products described. There are few substances of metabolic interest that could not be measured with the aid of an enzyme sequence terminating in a pyridine nucleotide reaction. The versatility of this approach improved as more purified enzymes became commercially available.

What finally gave us all the sensitivity we could use was enzymatic cycling. This technique is an exploitation of enzyme systems to amplify the pyridine nucleotides generated by the specific enzyme reactions just described (34). (Janet Passonneau joined the laboratory about this time and was a key figure in most of the work for the next 10 years.)

The following is an example of an enzymatic cycling amplifier system and its use. The problem is to measure metabolite A or enzyme a:

$$A \longrightarrow B \longrightarrow C \longrightarrow D$$

$$NAD(P)^+ \qquad NAD(P)H$$

After the specific reaction, whether a timed reaction to measure an enzyme a or a stoichiometric reaction to measure the metabolite, the excess pyridine nucleotide used to drive the specific step is destroyed with alkali (as in this case), or with acid if the pyridine nucleotide reaction is $NAD(P)H \rightarrow NAD(P)^+$. In either case, the pyridine nucleotide formed is used to catalyze a two-enzyme cyclic reaction, which alternatively oxidizes and reduces the nucleotide, thereby yielding one molecule of the product of each enzyme for each turn of the cycle:

$$\text{6-P-gluconate} \quad NADPH \quad \alpha\text{-ketoglutarate} + NH_3$$
$$\text{glucose-6-P} \quad NADP^+ \quad \text{glutamate}$$

After a sufficient number of cycles (which can be 25,000 per hour or more), the reaction is stopped, usually with heat; and one of the products is measured (again with an enzyme reaction that yields NADPH or NADH):

$$6\text{-P-gluconate} + NADP^+ \rightarrow \text{ribulose-5-P} + CO_2 + NADPH$$

The yield from 2×10^{-14} mol of initial nucleotide when amplified 25,000 times gives a fluorescence signal that is easily read in a final volume of 1 ml.

Somewhat greater amplification can be achieved by cycling longer [As a stunt, 400,000-fold amplification of NADP was obtained with a three-day incubation (35).] More practical is to simply repeat the cycling step: In the NADP cycling example given, after the indicator step reaction, the excess $NADP^+$ is destroyed with alkali and heat, and the NADPH is further amplified as needed. Two serial 25,000-fold cycles would yield 625,000,000-fold amplification, or sufficient sensitivity to measure about 10^{-18} mol of original sample, i.e. less than a million molecules.

One could imagine further amplification by triple cycling. This we have never tried for several reasons: (*a*) We have never needed more amplification; (*b*) we recognized some difficult problems; and (*c*) we lacked the courage. The biggest problem, even with the degree of sensitivity that can easily be obtained with double cycling, is analytical noise. Only rarely can the concentration of the reagent blank at the initial specific step (i.e. before any amplification) be kept below 10^{-8} M. A good rule of thumb for reasonable precision in any assay is to keep samples at least equivalent to the overall blank. A 10^{-8} M solution contains 10^{-14} mol of the solute in 1 μl and 10^{-18} mol in 0.1 nl.

Some historical perspective on enzymatic cycling may be in order. Although we have substantially refined and exploited this powerful tool, we did not invent it. Warburg et al (36) were the first to use the cycling principle for measuring NADP ("TPN"!) with a system containing glucose-6-phosphate and the "old yellow enzyme." Although they obtained 330 cycles in 10 min, because the signal was O_2 consumption measured manometrically the sensitivity was not great.

Jandorf et al (37) used the cycling principle to measure NAD in a system containing the enzymes needed to convert fructose-1,6-bisphosphate to glycerolphosphate and phosphoglycerate with the release of CO_2 from bicarbonate buffer. The NAD functioned to alternatively oxidize glyceraldehyde-3-phosphate and reduce dihydroxyacetone-phosphate. The cycling rate was about 1300 per hour. By the use of this system in the Cartesian diver (an analytical exploitation due to Linderstrøm-Lang), Anfinsen was able to measure with precision as little as 2×10^{-12} mol of NAD (38). More recently, Glock & McLean (39) obtained 30- to 50-fold enzymatic cycling of NAD or NADP with cytochrome c and either alcohol dehydrogenase or glucose-6-

phosphate dehydrogenase. Useful enzymatic cycling systems for compounds other than NAD or NADP have been devised by other investigators as well as ourselves (40).

ENZYMES AND METABOLITES

The glycolytic pathway consists of a long series of enzymes, which in average brain differ 100-fold in potential activity. And yet in a steady-state situation, the net flux through every enzyme step must be the same (ignoring side reactions). For example, during peak glycolytic flux in mouse brain, 50% of potential aldolase activity is used, but only 0.5% of that of phosphoglycerate kinase.

In each case, the kinetic properties of the respective enzymes, together with the steady-state levels of their substrates and products (plus the concentrations of any other effectors) must yield the same net velocities. Obviously, a full understanding of a metabolic system involves a great deal more than knowing just the levels of the enzymes concerned.

All our early histochemical studies concerned only the tissue distribution of enzymes, particularly enzymes of energy metabolism. This is because it takes much less sensitivity to measure the activity of an enzyme than the concentration in a tissue of its substrate or product. Brain lactate dehydrogenase can produce in vitro several moles of lactate per kg per hour, whereas the brain lactate concentration is normally only about one mmole per kg. But, as suggested, there is good reason to measure both the enzyme and its metabolites. The enzyme measurement indicates the capacity to carry out the metabolic reaction, whereas the levels of substrate and product of that enzyme, taken together with the flux, can indicate its actual function under the conditions of observation. Therefore, with the major increase in sensitivity at our disposal due to the substitution of fluorometry for spectrophotometry, Janet Passonneau and I decided to measure the levels in whole brain of all the intermediates of the glycolytic pathway plus ATP and phosphocreatine under control conditions, and during the sixfold increase in glycolysis that results from total ischemia (41).

Mice were decapitated and the heads frozen at intervals from 3 seconds to 10 minutes. Mice were used because the small head size minimizes artifacts from the delay in freezing the deeper portions.

During the first few seconds, fructose-6-phosphate fell and fructose-1,6-bisphosphate rose dramatically together with the other metabolites below it in the pathway. These results clearly indicated a control point at the phosphofructokinase (PFK) step, which was activated by the consequences of the lack of oxygen. [Cori et al (42) had previously concluded that PFK is a control step in muscle.] The changes in the other metabolites indicated the absence of any other important control step between glucose-6-phosphate and lactate. However, the first step in glucose metabolism was clearly also a control point, but the data did not permit distinction between control by hexokinase from control by glucose transport into the cells.

A companion in vitro study was made of the maximal activities and kinetics of all the enzymes of the glycolytic pathway in mouse brain homogenates under conditions simulating the pH, ionic strength, and temperature of brain (43).

Putting together the data on enzyme capacities and their kinetic properties, with the differences in the levels of the substrates and products of these enzymes under two different glycolytic fluxes, permitted a much better picture of the logistics of this important pathway. It helped to explain, for example, why some enzyme levels (expressed in terms of their maximum capacities) had to be much higher than those of others.

A full discussion of these results would go beyond the present purpose. However, I would submit that assessment of metabolite levels and their changes under various circumstances of interest can be a very informative approach, which has been rather underutilized.

PHOSPHOFRUCTOKINASE

Although study of biological problems requiring high analytical sensitivity has constituted the theme of most of our research, we have had several major distractions. One of these involved phosphofructokinase (PFK). Our first encounter with this remarkable enzyme was simply concerned with setting up optimal, reproducible, stable conditions for measuring it in brain. As usual, we tested different buffers, and to our surprise found not only that a phosphate buffer was by far the best, but also that without phosphate, activity was very low, and accelerated remarkably during the assay, as it was being followed in the spectrophotometer. We had stumbled onto what was later designated an allosteric phenomenon, and weren't smart enough to realize it. In the absence of P_i, the reaction was severely inhibited by ATP, as Lardy & Parks had discovered (44); but as the reaction proceeded, ADP accumulated and probably some of the fructose bisphosphate generated was not removed fast enough. These two PFK products, both deinhibitors of PFK (45), were probably sufficient to overcome the ATP inhibition. In any event, we did not pursue this exceptional opportunity, and about this time or soon after, Pardee discovered (1956) the feedback inhibition of aspartate carbamoyl transferase by CTP, probably the first clear-cut example of allosterism (46).

We did, however, go back to PFK later, after observing its dramatic activation in brain during ischemia. We were able to report that the kinetic properties of PFK made it perfectly suited for controlling glucose metabolism according to need (45). PFK is inhibited by ATP, which falls in ischemia, and this inhibition is overcome by fructose-6-phosphate, fructose-1,6-bisphosphate, ADP, AMP, P_i, and NH_4^+, all of which usually increase during ischemia.

Soon afterwards, we discovered that citrate is another potent inhibitor of PFK, and that its action is synergistic with ATP (47). This discovery was rather exciting, because it meant that the citrate cycle can feed back to control the glycolytic pathway. Recently we learned that Neifakh et al in 1953 had already reported in the Russian literature that citrate is a PFK inhibitor (48).

We have come to regard PFK as "the most complicated enzyme alive." In addition to the effectors already mentioned, Uyeda & Racker found that phosphocreatine and 3-phosphoglycerate are negative effectors (49), Krzanowski & Matschinsky reported that 2-phosphoglycerate, 2,3-bisphosphoglycerate, and phosphopyruvate are all potent negative effectors (50), Mansour & Mansour showed that cyclic AMP is a positive effector (51), and Van Schaftingen et al (52) found that a previously unknown metabolite, fructose-2,6-bisphosphate, is probably the most potent positive effector of all (52). Many of these effectors interact in a synergistic way, probably indicating a multiplicity of allosteric sites (53).

APPLICATIONS OF QUANTITATIVE HISTOCHEMISTRY

As mentioned earlier, our original purpose in trying to extend the quantitative histochemical approach of Linderstrøm-Lang & Holter was to determine the composition of different parts of the brain. However, practically every organ and tissue is heterogeneous, not only in regard to the types of cells present, but often in regard to the composition of any given cell type. Let me briefly review some of the aspects of this heterogeneity that have been explored by ourselves and others.

Linderstrøm-Lang & Holter, using their original approach, made some classical studies of the gastric and intestinal mucosas, and found, for example, that gastric chief cells are the source of pepsin (54). This approach has been used by Alfred Pope in several studies of the different layers of the cerebral cortex (55). David Glick has made a wide range of important applications of his own adaptations of the Linderstrøm-Lang approach (56). Of particular importance are his comprehensive studies of the different layers of the adrenal gland. Giacobini has been especially active in studies of the metabolism of single neurons, making use of an ultrasensitive adaptation of the Cartesian diver (57).

Nervous System

Investigations of the nervous system from our laboratory started with measurements of metabolic enzyme levels in 0.1 to 1 μg samples of specific layers of such structures as the cerebellum (58), hippocampus (59), and retina (60), and finally progressed to the 1- or 2-nanogram level with large single neurons (61) and even their nuclei (62). One study was made with Janet Passonneau on the effect of ischemia on ATP, phosphocreatine, glucose, and glycogen in single neurons from the spinal cord anterior horn and the dorsal root ganglia (63). This study involved measurements at the 10^{-15} mole level.

One of the most impressive brain histochemical studies was made by Takahiko Kato (64). He dissected out eight different types of neuron cell bodies from freeze-dried sections, with dry weights ranging from 0.2 to 10 ng, and analyzed them individually for one of seven different enzymes of the glycolytic pathway and citrate cycle. As an add-on (65), he measured the distribution of nine enzymes between nucleus and cytoplasm of individual dorsal root ganglion cell bodies.

Kidney

Each kidney nephron consists of a chain of structures with very different functions and with very different enzyme and metabolite compositions. This makes the kidney an ideal candidate for quantitative histochemical exploitation. I believe the first kidney study along this line was reported in 1956 by W. Peter McCann, a postdoctoral student in this laboratory (66). This was soon followed by a renal paper by Dubach & Recant from the Department of Medicine (67) and one by Kissane from the Department of Pathology (68). Somewhat later Dr. Helen Burch began a long series of very fruitful investigations of quantitative renal histochemistry, which continued for almost 15 years until her death in 1987 at 80 years of age. Meanwhile, this approach to renal biochemistry and pathology spread outside this institution, first to the University of Illinois Medical School in Chicago through the interest of Bonting & Kark (69), and then on a larger scale to Switzerland and Germany, mainly through the influence of Dr. Dubach, who is now Professor of Medicine in Basel.

Skeletal Muscle

For many years, biochemists treated skeletal muscle as though it were a homogeneous tissue. However, when enzyme-staining methods were applied, this was found to be far from true. Credit for the first quantitative enzyme measurements of single muscle fibers goes to James Nelson, then in the Department of Pathology at Washington University. He found large differences in the levels of glycogen phosphorylase among fibers from the same muscle (70). In 1975, a Swedish group began to apply quantitative enzyme methods to individual freeze-dried muscle fibers. Instead of making sections of the frozen muscle, Essén et al (71) freeze-dried a portion of the muscle and then dissected out segments of intact fibers several mm long. When a little later, we could not resist joining in a medical school–wide muscle program, we adopted this fiber isolation procedure. It proved to be quite feasible to analyze single fiber segments 2 or 3 mm long for many different enzymes and/or metabolites. For any particular assay, samples weighing 10 to 20 ng (10 to 50 μm in length) were simply cut off one end of the fiber, and the rest of the fiber returned to cold storage under vacuum for future use. This made possible direct comparisons between the levels of many different enzymes within the same fiber. The advantage of this was apparent when it was found that among fibers from a given muscle, the ratios between an enzyme of glycogenolysis and one of the citrate cycle might vary 30-fold or more (72). Similarly, it was possible to compare metabolite levels in single fibers from stimulated muscle with the relevant enzymes of the same fibers (e.g. malate with malate dehydrogenase) (73).

MAMMALIAN OVA

To me, one of the most satisfying applications of our microchemical methodology has been to the study of individual ova—first mouse ova and very recently human ova. In 1974, Elizabeth Barbehenn, then a graduate student, and Raymond Wales, a visitor from Monash University in Australia, with major experience in culturing mouse ova, decided to tackle an interesting puzzle: why fertilized mouse ova, before the eight-cell stage, cannot grow with glucose as the sole carbon source, but can do so if pyruvate or lactate is substituted.

The experiments were simple: ova from superovulated mice were starved for 60 min (that is, placed in medium with no carbon source), and then re-fed for 15 min with glucose or pyruvate or both. Ova were freeze dried before and after starvation and after refeeding, and then individually analyzed for glucose-6-phosphate, fructose-6-phosphate, fructose-1,6-bisphosphate, citrate, or malate. The metabolite results clearly showed that before the eight-cell stage, there was a block at the phosphofructokinase (PFK) step (74). The mechanism appears to be that the level of ATP (i.e. a potent PFK inhibitor) is high, and that the level of fructose-6-phosphate is too low to overcome the inhibition. The low fructose-6-phosphate is attributable to a low level of hexokinase, which in competition with highly active glycogen synthase cannot maintain an adequate level of the glucose-6-phosphate:fructose-6-phosphate equilibrium mixture. By the eight-cell and morula stages, there is a sufficient increase in fructose-6-phosphate to overcome the block. The fructose-6-phosphate increase in turn is probably due to the rise in hexokinase known to occur at this time. The biological importance of all this appears to be that with both glucose and pyruvate available, pyruvate satisfies the energy requirements, and glucose is diverted into glycogen to build up a reserve for the implantation process.

This study only required about 1000 ova from 50 mice, and would probably have required ova from many thousands of mice with more conventional methods.

The human ova studies were initiated 15 years later. The impetus came from two sources. Two daughters-in-law and a neighbor's daughter were attending in vitro fertilization clinics without success. During this same period, I reviewed a grant application from Henry Leese from the University of York for funds to support metabolic studies of human ova obtained from the famous Edwards and Steptoe Clinic. (Leese was using the highly sensitive microchemical techniques developed at Harvard Medical School by Claude Lechene.)

I was faced with an ethical dilemma. On the one hand, the success rate of in vitro fertilization was (and is) exceedingly poor, possibly owing in part to the fact that the in vitro incubation media were designed for optimal growth of mouse ova. The reason for this design choice is that practically nothing was known about the metabolism or growth requirements of human ova. We had the tools to at least find out if there are major metabolic differences between the two species, and felt almost obligated to apply these tools. On the other hand, the idea for us to get involved had come from my reviewing a privileged grant proposal.

We solved the dilemma, as far as our consciences were concerned, by writing to Dr. Leese describing the situation and stating that we were going ahead, but would keep him in touch with what we were planning and doing, and would share our results with him before they were published. I half expected an angry letter in return. Instead, I received a most cordial response and a welcome into this field, which only he and Claude Lechene (besides ourselves) had the tools plus the inclination to investigate.

Problems with Federal Support of Research on Human Ova

But solution of our own ethical problem did not mean that an ethical problem of a different sort might not be raised by others. With start-up funds contributed by our own Department and discarded human ova plus normal mouse ova, both kindly made available by Dr. Ronald Strickler of Washington University from the in vitro fertilization clinic under his control, we were soon able to compare metabolic enzyme levels in ova from the two species. Because of the severe limitation in numbers of available human ova, we modified our methodology to permit each ovum to be assayed for as many as 8 or 10 enzymes, or for 4 or 5 enzymes plus as many metabolites. Data were obtained for 17 enzymes of 8 metabolic pathways that demonstrated some dramatic species differences. For example, enzymes of fatty acid metabolism were as much as 15-fold higher relative to size in human than in mouse ova (75). A variety of data, including the levels of ATP and phosphocreatine, indicated that the limitation of our assays to discarded human ova did not invalidate the results.

With these data, we applied to NIH for funds, and after a little backing and filling, received approval with a very high priority score.

And then the trouble began. It was ruled (as I was told) that before funding, the application "had to go through the Ethics Committee." It was later revealed that there was no Ethics Committee and had been none for eight years! This was in the spring of 1988. Because of our high priority rating and what would appear to be a negligible ethical problem, the NIH decided to use this as a test case to clear the way for other research proposals concerning preimplantation human embryos. Subsequently, the Department of Health and Human Services agreed to appoint an ethics committee, but as of the fall of 1989, there is still no action, and the outcome for the near future in the present political-judicial climate seems dim. Fortunately, nonfederal funds have been granted for two years, and we hope to achieve something useful for in vitro fertilization before those funds run out. (My father never told me that hyperspecialization might get me into trouble.)

2-DEOXYGLUCOSE TO MEASURE GLUCOSE METABOLISM

In 1977, Sokoloff et al introduced the use of ^{14}C-2-deoxyglucose to measure the regional glucose metabolism of brain (76). This radioautographic method is based on the fact that although 2-deoxyglucose (DG) is phosphorylated by hexokinase in parallel with glucose, the 2-deoxyglucose-6-phosphate (DG6P) that is formed cannot be further metabolized along the glycolytic pathway. It therefore accumulates as an index of glucose metabolism. This method has been widely used and has yielded very valuable results. However, it has one important disadvantage. Because the radioautograph cannot distinguish ^{14}C-DG from ^{14}C-DG6P, it has been necessary to wait 30 to 45 minutes after DG injection for the brain DG to largely dissipate before preparing the brain for the radioautographic procedure. This limits the method to studies of long-term events, whereas many brain events of interest take place on a time scale of a few minutes or less.

When we recently found that DG and DG6P could be separately measured enzymatically with NADP$^+$ as cofactor, we realized it would be possible to use the principle introduced by Sokoloff et al to assess brain glucose metabolism on a time scale of a minute or two (77). Moreover, we could employ enzymatic cycling to give the sensitivity needed to study very small brain regions, even down to the level of single neurons.

So far, we have been mainly perfecting the analytical procedures and exploring the new use of DG with whole mouse brain and brain slices incubated in vitro. The methods depend on the fact that DG6P is oxidized by glucose-6-phosphate dehydrogenase, but at a rate 2000 times slower than with glucose-6-phosphate itself. This rate difference can introduce problems, for example, with enzyme impurities. However, these and other problems are manageable, and we are confident that this method of assessment of rapid changes in brain glucose metabolism will prove a useful additional way to study the quantitative histochemistry of brain.

BIOCHEMISTRY: 1932–1990. A PERSONAL VIEW

I have enjoyed almost 60 years of participation in this greatest game on earth and hope to continue a while longer. I feel much the same way about biochemical research that my mother, a fine artist, felt about painting. She said an artist ought not to complain about the poor financial rewards, because the pleasure in making the painting is reward enough. I tried to promote this idea around the laboratory on pay days: "You have all this fun and get paid too!" (But somehow this was never much of a substitute for better pay.)

One of the things one is supposed to acquire with age is wisdom. So I assume my final duty in writing this chapter should be to think of wise things to pass on to future generations. Unfortunately, in spite of much thought, I have come up with very few words of wisdom. So let me instead simply touch on three topics that seem to me particularly impressive concerning the biochemical achievements of the past 58 years.

The Rapid Rise of Biochemistry Since 1932

The changes in biochemistry in 58 years have been astounding. In 1932 many of the vitamins had not been identified, nor had their structures been determined. The accepted structure of cholesterol was incorrect. Only a few of the hormones had been isolated. Relatively few enzymes had been purified and only one (urease) crystalized. "Yeast" and "thymus" nucleic acid had not yet become RNA and DNA, and their functions were completely unknown, but for sure they had nothing to do with genetics or protein synthesis. The members of the Embden-Meyerhof pathway had been identified, but the citrate cycle was still being worked out, and the pentose pathway was unknown. ATP was known to have something to do with muscle contraction, but its broader function had not been realized.

Biochemical progress subsequent to 1932 was remarkably rapid considering the relatively few investigators, that most of them carried heavy teaching loads, and that the amount of financial support was minimal. In 1937–1941, Baird Hastings' whole department at Harvard had one modest outside research grant, two technicians, and I believe $2000 per year from the school for supplies and equipment.

World War II interrupted much of the pure research. However, applied biochemical research, which was supported quite well with federal funds, stimulated the development of improved tools and techniques that paid off well after the war.

The war also convinced influential persons in and out of government that biomedical research, biochemistry included, was a good investment. In consequence, a program of federal research support was instigated that soon expanded to a size no one would have believed possible. (Private foundations also joined in, with the American Cancer Society taking the lead.)

More research required more researchers. The war had turned the consensus around in regard to the intellectual potential of the average citizen. In 1932 it was generally held that only a minority of the population could benefit by a college education, and of these only rare individuals had the special talent needed to do worthwhile research. Both these views proved fallacious (although this elitist attitude is unfortunately not completely dead).

We now have enormously more investigators than in 1932. The increase is in all categories: brilliant, good, and poor. I doubt if the ratios between these categories are much different than in 1932, and the increase in output of high-quality research, by anyone's yardstick, has been sensational.

Biological Aids to Biochemical Research

It is remarkable how much this phenomenal progress in biochemical research has been dependent on the use of natural tools offered by biology itself. Before my day, bioassays were of necessity used to follow the purification of vitamins and hormones. These bioassays usually required measurements with whole animals, and progress was slow and tedious. Later on, bioassays with isolated organs were introduced for rapidly acting substances ranging from epinephrine to prostaglandins and atriopeptins. This type of assay reached its highest sophistication with Vane's organ cascades. Microbiologists made great use of bioassays in the isolation of growth factors for bacteria. And a spin-off was to turn this around and use growth or acid production of bacterial cultures to measure the levels of specific substances in tissue extracts.

I have already stressed the importance of enzymes for measuring other enzymes and their metabolites and cofactors, as well as in participating in enzymatic amplifier systems. But it is not just the convenience and sensitivity that are important. What is really invaluable is the specificity conferred by the use of enzymes as reagents. Whole branches of biochemical investigation would slow down to a snail's pace if it were not for the use of enzymes to cleave proteins and nucleic acids at specific sites, as well as to add specific fragments according to plan. Antibodies, both monoclonal and polyclonal, have similarly proved to be powerful biological tools for biochemical research.

To generalize, the very nature of biological systems to carry out innumerable, highly coordinated, synthetic, degradative, and identification functions requires machinery that the skillful experimenter can turn around to unravel the systems themselves. Biology supplies the keys to unlock its own secrets.

The Revolution in Biomedical Categories

In 1932 and for many years thereafter, preclinical medical departments were strictly segregated not only with respect to teaching, but with few exceptions with respect to research as well. Each subject was designated as a separate "discipline," which aptly indicated the strict party lines then existing. Washington University Medical School broke ground when it appointed Carl Cori in 1931 to be Head of Pharmacology, but it was several years before he was accepted into the American Society for Pharmacology and Experimental Therapeutics. Similarly, Cori had to make somewhat of a fuss to get me into ASPET when I took his place in 1947.

One of the most pleasant transitions (even revolutions) in the biomedical world has been the blurring of party lines, first in research, and more gradually in teaching. I hope I am not being a chauvinist by pointing out that this transition was due to a gradual realization that biochemistry in fact pervades every biomedical discipline. How can a cytologist do research or teach without considering the biochemical nature of the cells, or a physiologist investigate secretion or nerve transmission without taking account of the biochemical elements involved? And so on.

When the term "molecular biology" was first introduced, I thought it was somewhat silly, since biochemists had been studying the molecules of biological systems since the late 1800s. I now realize it was a face-saving device for physiologists and biophysicists who had discovered biochemistry and wanted to apply it without seeming to cave in.

In any event, party lines have largely come down, much to the advantage of biomedical research and teaching.

ACKNOWLEDGMENTS

I have had the good fortune to work with a great many fine research people: collaborators, assistants, and trainees. Those with whom I worked directly for a significant period include: at Harvard, Baird Hastings, Tatiana Hull, Andrea Brown, Thomas Hunter, Chris Anfinsen, William Wallace, and Dorothy Gilligan; at the Public Health Research Institute, Otto Bessey, Helen Burch, Herman Kalckar, Robert Shank, Jeanne Lopez, Elizabeth Crawford, Mary Jane Brock, Elizabeth Davis, and Ruth Love; at Washington University, Mary Buell, Janet Passonneau, Catherine Smith, Susamma Berger, James Ferrendelli, Mei-Ling Wu, Ben Senturia, Charles Lowry, Franz Matschinsky, Helen Graham, Vera Yip, Madelon Price, Barbara Cole, Patti Nemeth, Jan Henriksson, Stanley Salmons, Kenneth Kaiser, John Holloszy, John Ivy, Harold Teutsch, David McDougal, Thomas Woolsey, Lewis Farr, Rose

Randall, Katherine Leiner, Nira Roberts, Joyce Kapphahn, Demoy Schulz, Shirley Kahana, Martha Rock, Mary Ann Reynolds, Francis Hasselberger, Joseph Brown, Charles Lewis, Joyce Carter, Maggie Chi, Carol Hintz, Mary Ellen Pusateri, Polly Passonneau, Ronald Hellendahl, Jill Manchester, Wayne Albers, Jack Strominger, Eli Robins, Roger Hornbrook, Robert Kuhlman, Mark Stewart, Bruce Breckenridge, Wolff Kirsch, David Gatfield, Stanley Nelson, Bertil Diamant, Richard Young, Wallace Tourtellote, Marie Fleming, Nelson Goldberg, Lucy King, Philip Needleman, Jaroslava Folbergrovà, Frederick Kauffman, Matti Harkonen, Carl Rovainen, Clinton Corder, Takahiko Kato, Lawrence Austen, Thomas Duffy, Norman Curthoys, Elizabeth Barbehenn, Adolph Cohen, Raymond Wales, Allen Blackshaw, Luis Glaser, Harry Orr, Eugene Butcher, Michael McDaniel, Donald Godfrey, William Outlaw, Stephen Felder, Julie Kimmey, Robert Narins, Berlin Hsieh, Ronald Hellendahl, Joseph Knorr, Marvin Natowicz, Mildred Yang, Dalton Dietrich, Dierdra McKee, Beverly Norris, Catherine Henry, Douglas Young, Evan Dich, Vicki Yang, Robert Rust, Jean Bastin, and Bernard Ferrier.

From this list I would like to single out Philip Needleman, who was first a postdoctoral trainee in the Pharmacology Department, then a departmental member, and finally my boss as Department Head. He was exceedingly generous to his predecessor in regard to space and support, long past the normal retirement age. (In turn, I promised never to tell him how to run the Department.)

I would also like to acknowledge the generous support from the American Cancer Society, the National Institutes of Health, the Muscular Dystrophy Association, the National Science Foundation, the Nutrition Foundation, and Williams-Waterman Fund.

Literature Cited

1. Hastings, A. B., Blumgart, H. L., Lowry, O. H., Gilligan, D. R. 1939. *Trans. Assoc. Am. Physicians* 54:237–42
2. Lowry, O. H., Hastings, A. B. 1942. *Am. J. Physiol.* 136:474–85
3. Lowry, O. H., Krayer, O., Hastings, A. B., Tucker, R. P. 1942. *Proc. Soc. Exp. Biol. Med.* 49:670–74
4. Lowry, O. H., Hastings, A. B., Hull, T. Z., Brown, A. N. 1942. *J. Biol. Chem.* 143:271–80
5. Lowry, O. H., McCay, C. M., Hastings, A. B., Brown, A. N. 1942. *J. Biol. Chem.* 143:281–84
6. Lowry, O. H., Hastings, A. B., McCay, C. M., Brown, A. N. 1946. *J. Gerontol.* 1:345–57
7. Lowry, O. H., Gilligan, D. R., Katersky, E. M. 1941. *J. Biol. Chem.* 139:795–804
8. Levy, M. 1936. *C. R. Trav. Lab. Carlsberg, Ser. Chim.* 21:101–10
9. Lowry, O. H., Bessey, O. A., Brock, M. J., Lopez, J. A. 1946. *J. Biol. Chem.* 166:111–19
10. Lowry, O. H., Bessey, O. A., Burch, H. B. 1952. *Proc. Soc. Exp. Biol. Med.* 80:361–62
11. Bessey, O. A., Lowry, O. H., Brock, M. J. 1946. *J. Biol. Chem.* 164:321–29
12. King, E. J., Delory, G. E. 1939. *Biochem. J.* 33:1185–90
13. Ohmori, Y. 1937. *Enzymologia* 4:217–31
14. Lowry, O. H., Rosebrough, N. J., Farr,

A. L., Randall, R. J. 1951. *J. Biol. Chem.* 193:265–75

15. Wu, H. 1922. *J. Biol. Chem.* 51:33–39
16. Pressman, D. 1943. *Ind. Eng. Chem. Anal. Ed.* 15:357–59
17. Heidelberger, M., MacPherson, C. F. C. 1943. *Science* 97:405–6
18. Herriot, R. M. 1935. *J. Gen. Physiol.* 19:283–99
19. Lowry, O. H., Lopez, J. A. 1946. *J. Biol. Chem.* 162:421–28
20. Kalckar, H. M. 1945. *J. Biol. Chem.* 158:723–24
21. Fiske, C. H., Subbarow, Y. 1925. *J. Biol. Chem.* 66:375–400
22. Lowry, O. H., Bessey, O. A. 1946. *J. Biol. Chem.* 163:633–39
23. Lowry, O. H. 1948. *J. Biol. Chem.* 173:677–82
24. Anfinsen, C. B., Lowry, O. H., Hastings, A. B. 1942. *J. Cell Comp. Physiol.* 20:231–37
25. Anfinsen, C. B. 1944. *J. Biol. Chem.* 152:267–78
26. Anfinsen, C. B. 1944. *J. Biol. Chem.* 152:279–84
27. Lowry, O. H. 1953. *J. Histochem. Cytochem.* 1:420–28
28. Lowry, O. H. 1941. *J. Biol. Chem.* 140:183–89
29. Bazzoni, C. B. 1915. *J. Franklin Inst.* 180:463
30. Matschinsky, F. M., Passonneau, J. V., Lowry, O. H. 1968. *J. Histochem. Cytochem.* 16:29–39
31. Kaplan, N. O., Colowick, S. P., Barnes, C. C. 1951. *J. Biol. Chem.* 191:461–72
32. Negelein, E., Haas, E. 1985. *Biochem. Z.* 282:206–20
33. Greengard, P. 1956. *Nature* 178:632–34
34. Lowry, O. H., Passonneau, J. V., Schulz, D. W., Rock, M. K. 1961. *J. Biol. Chem.* 236:2746–55
35. Chi, M. M.-Y., Lowry, C. V., Lowry, O. H. 1978. *Anal. Biochem.* 89:119–29
36. Warburg, O., Christian, W., Giese, A. 1935. *Biochem. Z.* 282:157–205
37. Jandorf, B., Klemperer, F. W., Hastings, A. B. 1941. *J. Biol. Chem.* 138:311–20
38. Anfinsen, C. B. 1944. *J. Biol. Chem.* 152:285–91
39. Glock, G. E., McLean, P. 1955. *Biochem. J.* 61:381–88
40. Lowry, O. H. 1980. *Mol. Cell. Biochem.* 32:135–46
41. Lowry, O. H., Passonneau, J. V., Hasselberger, F. X., Schulz, D. W. 1964. *J. Biol. Chem.* 239:18–30
42. Cori, C. F. 1942. *A Symposium on Respiratory Enzymes*, p. 175. Madison, Wis: Univ. Wis. Press
43. Lowry, O. H., Passonneau, J. V. 1964. *J. Biol. Chem.* 239:31–42
44. Lardy, H. A., Parks, R. E. Jr. 1956. *Enzymes: Units of Biological Structure and Function*, p. 584. New York: Academic
45. Passonneau, J. V., Lowry, O. H. 1962. *Biochem. Biophys. Res. Commun.* 7:10–15
46. Yates, R. A., Pardee, A. B. 1956. *J. Biol. Chem.* 221:757–70
47. Passonneau, J. V., Lowry, O. H. 1963. *Biochem. Biophys. Res. Commun.* 13:372–79
48. Neifakh, S. A., Melnikova, M. P., Mozhajko, F. 1953. *Dokl. Akad. Nauk Uzb.* 91:557
49. Uyeda, K., Racker, E. 1965. *J. Biol. Chem.* 240:4682–88
50. Krzanowski, J., Matschinsky, F. M. 1969. *Biochem. Biophys. Res. Commun.* 34:816–23
51. Mansour, T. E., Mansour, J. M. 1962. *J. Biol. Chem.* 237:629–34
52. Van Schaftingen, E., Jett, M. F., Hue, L., Hers, H.-G. 1980. *Biochem. J.* 192:897–901
53. Lowry, O. H., Passonneau, J. V. 1966. *J. Biol. Chem.* 241:2268–79
54. Glick, D., Holter, H., Linderstrøm-Lang, K., Søeberg Ohlesen, A. 1935. *C. R. Trav. Lab. Carlsberg Ser. Chim.* 20(11):
55. Pope, A. 1967. *Arch. Neurol.* 16:351–56
56. Glick, D. 1949. *Techniques of Histo- and Cytochemistry.* New York: Insterscience
57. Giacobini, E. 1969. *J. Histochem. Cytochem.* 17:139–55
58. Buell, M. V., Lowry, O. H., Roberts, N. R., Chang, M-L. W., Kapphahn, J. I. 1958. *J. Biol. Chem.* 232:979–93
59. Lowry, O. H., Roberts, N. R., Leiner, K. Y., Wu, M-L., Farr, A. L., Albers, R. W. 1954. *J. Biol. Chem.* 207:39–49
60. Lowry, O. H., Roberts, N. R., Schulz, D. W., Clow, J. E., Clark, J. R. 1961. *J. Biol. Chem.* 236:2813–20
61. Lowry, O. H. 1957. *Metabolism of the Nervous System*, pp. 323–28. New York: Pergamon
62. Lowry, O. H. 1963. *Harvey Lect., Ser.* 58:1–19
63. Passonneau, J. V., Lowry, O. H. 1971. *Recent Advances in Quantitative Histo- and Cytochemistry*, pp. 198–212. Bern/Stuttgart/Vienna: Hans Huber
64. Kato, T., Lowry, O. H. 1973. *J. Neurochem.* 20:151–63
65. Kato, T., Lowry, O. H. 1973. *J. Biol. Chem.* 248:2044–48

66. McCann, W. P. 1956. *Am. J. Physiol.* 185:372–76
67. Dubach, U. C., Recant, L. 1960. *J. Clin. Invest.* 39:1364–70
68. Kissane, J. M. 1961. *J. Histochem. Cytochem.* 9:578–84
69. Bonting, S. L., Pollak, V. E., Muehrcke, R. C., Kark, R. M. 1958. *Science* 127:1342–43
70. Nelson, J. S., Tashiro, K. 1973. *J. Neuropathol. Exp. Neurol.* 32:371–79
71. Essén, B., Jansson, E., Henriksson, J., Taylor, A. W., Saltin, B. 1975. *Acta Physiol. Scand.* 95:153–65
72. Lowry, C. V., Kimmey, J. S., Felder, S., Chi, M. M.-Y., Kaiser, K. K., et al. 1978. *J. Biol. Chem.* 253:8269–77
73. Hintz, C. S., Chi, M. M.-Y., Fell, R. D., Ivy, J. L., Kaiser, K. K., et al. 1982. *Am. J. Physiol.* 242:C218–28
74. Barbehenn, E. K., Wales, R. G., Lowry, O. H. 1974. *Proc. Natl. Acad. Sci. USA* 71:1056–60
75. Chi, M. M.-Y., Manchester, J. K., Yang, V. C., Curato, A. D., Strickler, R. C., Lowry, O. H. 1988. *Biol. Reprod.* 39:295–307
76. Sokoloff, L., Reivich, M., Kennedy, C., Des Rosiers, M. H., Patlak, C. S., et al. 1977. *J. Neurochem.* 28:897–916
77. Chi, M. M.-Y., Pusateri, M. E., Carter, J. G., Norris, B. J., McDougal, D. B. Jr., Lowry, O. H. 1987. *Anal. Biochem.* 161:508–13

Annu. Rev. Biochem. 1989. 58:1–30

NEVER A DULL ENZYME

Arthur Kornberg

Department of Biochemistry, Stanford University School of Medicine, Stanford, California 94305

CONTENTS

GROWING UP WITHOUT SCIENCE (1918–1942)

The current generation of scientists may be suprised to know that I had no formal research training. I was well started in a career of clinical medicine until World War II placed me in the National Institutes of Health (NIH) where I soon became an eager investigator of rat nutrition. Three years later, in 1945, I responded to the lure of enzymes and have remained faithful to them ever since.

Science was unknown in my family and circle of friends. Once, in 1947, when I was in the biochemistry department of Washington University in St. Louis, working under the guidance of Carl and Gerty Cori, Gerty told me that Carl had collected beetles and butterflies in his youth, and then asked: "Arthur, what did you collect?" "Matchbook covers," was my sheepish response. What else? They were the dominant flora in the Brooklyn streets where I played and in the subways where my father often risked being trampled when he stooped to add one more to my collection.

My early education in grade school and Abraham Lincoln High School in Brooklyn was distinguished only by "skipping" a few grades and finishing three years ahead of schedule. I recall nothing inspirational from teachers or courses except encouragement to get good marks. I remember the glow of my chemistry teacher when I received a grade of 100 in New York State Regents examination. It was the first time, in more than twenty years of teaching, a student of his had gotten a perfect grade. Once when I boasted about this to my wife, Sylvy, she remarked that she too had gotten 100, not only in chemistry, but also in algebra and geometry.

I chose the cachet of City College in uptown Manhattan over nearby Brooklyn College, even though commuting from Bath Beach (near Coney Island) meant three hours a day in crowded subways. Competition among a large body of bright and highly motivated students was fierce in all subjects. I carried over my high school interest in chemistry, but the prospects for employment in college teaching or industry were dismal. For lack of graduate studies or research laboratories at City College then, these possibilities barely existed. At age 19 in 1937, with a Bachelor of Science degree, and no jobs to be had in the depths of the Great Depression, I welcomed the haven that medical school would provide for four more years.

Throughout college I worked evenings, weekends, and school holidays as a salesman in men's furnishings stores. This left little time for study or sleep and none for leisure. With these earnings, a New York State Regents Scholarship of $100 a year, no college tuition, and frugal living, I saved enough to see myself through the first half of medical school at the University of Rochester.

I enjoyed medical school and the training to become a doctor. Among my courses, biochemistry seemed rather dull. The descriptive emphasis on the constituents of tissues, blood, and urine reflected biochemistry in the United States in the 1930s. The dynamism of cellular energy exchanges and macromolecules was still unknown, and the importance of enzymes had not penetrated my course or textbook. By contrast, anatomy and physiology presented integrated and awesome structures and functions. The aberrations presented in pathology and bacteriology were absorbing, as were the responsibilities to diagnose and treat patients during the clinical years.

Did I as a medical student consider a career in research? Not really. I expected to practice internal medicine, preferably in an academic setting; the idea of spending a significant fraction of my future days in the laboratory had no appeal. The medical school of the University of Rochester granted some students fellowships to take a year out for research. I had hoped but failed to get such an award from any of the departments. In those years, ethnic and religious barriers were formidable, even within the enlightened circle of academic science.

I did some research on my own, which grew out of curiosity about jaundice. I had noticed a slightly yellow discoloration of the whites of my eyes, and found that my blood bilirubin level was elevated and my tolerance to injected bilirubin reduced. I made similar measurements on as many medical students and patients as I could. I collected samples at odd moments and did the analyses on a borrowed bench, late at night and on weekends. The report I published (1) called attention to the frequent occurrence of high bilirubin levels and reduced capacity to eliminate bilirubin, now recognized as signs of the benign familial trait called Gilbert's Disease.

Looking back, I realized that I enjoyed collecting data. I kept on collecting bilirubin measurements during my internship year and started setting up to do more analyses in the small sickbay of a Navy ship soon after I joined it. A lucky consequence was that the publication of my student work on jaundice attracted attention and led to my transfer from sea duty to do research at the NIH, a rare assignment at that time.

JOINING THE VITAMIN HUNTERS (1942–1945)

The Nutrition Laboratory at NIH to which I was assigned in the fall of 1942 as a commissioned officer in the U. S. Public Health Service had been started by Joseph Goldberger (1874–1929). He was among the first to recognize that a vitamin deficiency can cause an epidemic disease, and in tracking the missing vitamin in the diets of pellagra patients, he emerged as one of the greatest of the vitamin hunters. W. H. (Henry) Sebrell, whom he had trained, was now chief of the laboratory and my senior boss. The laboratory had moved in 1938 from downtown Washington to suburban Bethesda, Maryland, but some of Goldberger's animal caretakers, kitchen staff, and diet notebooks, as well as his aura were still around.

My initial project as a nutritionist was to find out why rats fed a purified ("synthetic") diet containing a sulfa drug developed a severe blood disorder in a few weeks and died. A stock animal ration or inclusion of a yeast or liver supplement in the purified diet was effective in preventing and curing the disease (2). After other vitamin hunters (3) with the use of a microbial assay had succeeded in isolating folic acid and made it available to us, we could show that an induced deficiency of this vitamin was responsible for the sulfa drug effect.

It seemed clear that sulfa drugs, as analogues of para-aminobenzoic acid (PABA), a component of folic acid, were preventing bacteria from synthesizing this essential constituent and thus preventing their growth. We also knew that many animals rely on their intestinal bacteria for an adequate supply of folic acid and other vitamins, including vitamin K. I was therefore puzzled by a report (4) that PABA prevented the sulfa drug from producing a vitamin K deficiency even when given by injection. This result was taken to mean that the sulfa drug was not exerting its toxic effect on the intestinal bacteria, but somewhere else in the body.

I repeated the experiments with PABA injections and sulfa drugs, and developed a method to measure the amounts of vitamin K and PABA in the intestinal contents and feces. Ample quantities of vitamin K were produced by the intestinal bacteria of rats on a purified diet and this production was eliminated by sulfa drugs. As for rats injected with PABA, high levels of this substance accumulated in the intestinal contents, amounts sufficient to offset the action of the sulfa drug taken in the diet. These findings were reported in my first contribution to the *Journal of Biological Chemistry* (5), one of a very few biochemical papers from the NIH.

With the isolation of folic acid, it was apparent that virtually all the vitamins had been discovered. But we did not understand what most of the vitamins did in the body. How was folic acid serving in the growth of blood cells? What clues did the structure of folic acid offer to understanding its precise metabolic function? Could this understanding explain why sulfa drugs kill bacteria but not animal cells?

The answers to these questions, as well as to similar questions about the functions of the other vitamins, would be answered in the next two decades by enzymology. Just as the microbe hunters, who led the way in the first two decades of this century, were succeeded in the 1920s and 1930s by the vitamin hunters, so the latter would be overrun in the next two decades by the enzyme hunters.

I had come to nutrition in its twilight, decades late for the excitement and adventures of the early vitamin hunters who had solved the riddles of diseases that had plagued the world for centuries. My envy of their exploits impelled me to search for a new frontier. The discoveries of each of the vitamins—nicotinic acid, riboflavin, and thiamine in intermediary metabolism, and folic acid in nucleotide biosynthesis—became part of my heritage as I went on to learn about their biochemical functions. The rush to biochemistry depopulated the ranks of nutrition. How tragic that diet remains to this day as controversial as politics and the science of nutrition is in disarray.

FROM RATS TO ENZYMES (1945–1947)

By 1945, with the war over, I had become bored with feeding rats variations of purified diets. I was excited reading for the first time about enzymes, coenzymes, and ATP, in papers by Otto Warburg, Otto Meyerhof, Carl Cori, Herman Kalckar, and Fritz Lipmann. I had learned nothing about these things or people in medical school. While at NIH, I was startled and fascinated by a seminar in which Edward Tatum described his and George Beadle's work with *Neurospora* mutants and their one gene-one enzyme hypothesis. I knew even less about genetics than about biochemistry.

Fortunately, I was able to persuade Dr. Sebrell to let me quit my nutritional work and go to a laboratory where I could learn about ATP and enzymes. Immediately, I apprenticed myself to Bernard Horecker, a friend at NIH, who had been studying effects of DDT on cockroaches and was returning to the subject of his doctoral dissertation, the cytochromes of cellular respiration. Bernie introduced me to succinoxidase, cytochrome c, and the Beckman Model DU spectrometer. The unsolved problem of oxidative phosphorylation seemed to me to be the most important thing to do in biochemistry.

While still in uniform, I spent the year 1946 with Severo Ochoa at New York University Medical School; it was one of the happiest and most exhilarating in my life. Never had my learning curve been so sharply exponential and sustained. And in the few waking hours outside the laboratory, Sylvy and I discovered the theater, music, and museums that are the heartthrob of New York. Despite my being a native of Brooklyn and having attended City College in uptown Manhattan, and despite Sylvy's many visits from Rochester where she grew up and studied biochemistry, we were strangers to the city.

My mission from Ochoa was to purify heart muscle aconitase. This was my first solo stab at enzyme purification. We expected to resolve the activity into two enzymes to account for the successive subtraction and readdition of a water molecule that converts citric to isocitric acid. Despite repeated failure (aconitase proved to be one enzyme), this immersion in enzymology was intoxicating. Aside from the fascination of seeing an enzyme in action, the pace of the experimental work was breathtaking. By coupling aconitase action to isocitrate dehydrogenase, spectrophotometric assays could be performed in a few minutes, and many ideas could be tested and discarded in the course of a day. Late evenings were occupied preparing a series of protocols for the following day. What a contrast with the tedious pace of nutritional experiments on rats.

In my work on aconitase, I learned the philosophy and practice of enzyme purification. To attain the goal of a pure protein, the notebook record of an enzyme purification should withstand the scrutiny of an auditor or bank examiner. Not that I ever regarded the enterprise as a business or banking operation. Rather, it often seemed like the ascent of an uncharted mountain: the logistics resembled supplying successively higher base camps; protein fatalities and confusing contaminants resembled the adventure of unexpected storms and hardships. Gratifying views along the way fed the anticipation of what would be seen from the top. The ultimate reward of a pure enzyme was tantamount to the unobstructed and commanding view from the summit. Beyond the grand vista and thrill of being there first, there was no need for descent, but rather the prospect of ascending even more inviting mountains, each with the promise of even grander views.

I was luckier in my second attempt at enzyme purification when I joined Ochoa and Alan Mehler, his first graduate student, in purifying the liver malic enzyme, the enzyme that converts malic to lactic acid (6). Mehler was already on the scene when I arrived in Ochoa's lab and became my indefatigable and devoted tutor. Having always been the youngest in my class, it was a shock to find that I was so far behind someone four years my junior.

To let me pursue my training and the problem of aerobic phosphorylation, the NIH extended my stay with Ochoa to a full year, and allowed me another six months in the laboratory of Carl and Gerty Cori at the Washington University Medical School in St. Louis. Right after the war, the Cori laboratory was the mecca of enzymology. There I joined a young Swedish visitor, Olov Lindberg, who was investigating a striking observation made six years earlier by Ochoa when he worked in the Cori laboratory. Liver particles metabolizing pyruvic and related acids produced inorganic pyrophosphate (PP), a compound previously unknown as a cellular constituent. We began by ruling out the possibility that PP was released from an unstable form of ATP, but then found little else to guide us.

Later, while trying to enhance the levels of respiration and coupled aerobic phosphorylation by kidney particles, we observed a strong stimulation by NAD, and discovered that the effect could be traced to AMP generated by its hydrolysis.

$$\text{Nicotinamide-ribose-P-P-ribose-adenine} + H_2O \rightarrow \text{Nicotinamide-ribose-P} + \text{AMP} \qquad 1.$$
$$\qquad\qquad (\text{NAD}) \qquad\qquad\qquad\qquad\qquad (\text{NicRP})$$

The AMP produced by NAD cleavage stimulated the reaction because it served as an acceptor of inorganic phosphate to form ATP. This mundane result marked the end of my search for the source of ATP in aerobic phosphorylation. The search had been doomed from the start because I was committed to finding discrete soluble enzymes that linked the synthesis of

ATP to respiration. As the late Albert Lehninger recognized a few years later, these enzymes are firmly embedded in mitochondria.

MY ROOKIE YEAR (1948)

Nineteen forty-eight, the year I set up my own biochemistry lab, was a great year for me. When I returned to the NIH, my former laboratory space in the Nutrition Division (in Building 4) was occupied. Just about then, one of the frequent organizational convulsions in the Industrial Hygiene Division (in Building 2) threatened Bernie Horecker and Leon Heppel, a close friend and medical school classmate, with a transfer to Cincinnati. Fortunately, Henry Sebrell agreed to let me start an Enzyme Section that would include the three of us in a few laboratory rooms in Building 3. (Considering the present mammoth size of the NIH, covering 300 acres and employing 13,000 people, it is hard to believe that in 1947 there were only six small buildings and that the research emphasis was still on infectious disease, dominated by a small corps of commissioned medical officers.)

I continued the work on the rabbit kidney enzyme that Lindberg and I had discovered in St. Louis and established that it cleaves NAD at the pyrophosphate linkage (7). However, the enzyme was firmly attached to tissue particles and there was little hope of obtaining it in pure form. At the suggestion of the late Sidney Colowick and Oliver Lowry, I looked for and found a similar enzyme activity in potatoes, from which it could readily be extracted in a free, soluble form (8).

The purified enzyme cleaved not only NAD, but all nucleotides with a pyrophosphate bond. I called the enzyme *nucleotide pyrophosphatase*. By using the enzyme to cleave NADP, I could show that the position of the extra phosphate, then unknown, was part of the AMP moiety on carbon 2 of the ribose. Best of all, having isolated NicRP from NAD cleavage, I wondered whether it might serve in the synthesis of NAD. It did! Enzymes purified from yeast and liver condensed NicRP and ATP to produce not only NAD, but PP as well, the first clue to the origin of PP after years of speculation. The reaction was readily reversible and could support a vigorous exchange of PP with ATP (9).

$$\text{NicRP} + \text{PPPRA} \rightleftarrows \text{NicRPPRA} + \text{PP} \qquad\qquad 2.$$
$$\quad\ \ (\text{ATP}) \qquad\qquad (\text{NAD})$$

This mechanism immediately led us to the discovery of the enzyme that synthesizes flavin adenine dinucleotide (FAD) from riboflavin phosphate and ATP (10). In the ensuing years, the mechanism of nucleotidyl transfer from a nucleoside triphosphate for the biosynthesis of coenzymes was discovered again and again in the biosynthesis of proteins, lipids, carbohydrates, and nucleic acids. A variety of phosphoric, carboxylic, and sulfuric acids (XO^-) accept a nucleotidyl group from a nucleoside triphosphate (PPPRN) to generate an activated form of XO^- with the release of PP.

$$XO^- + PPPRN \rightleftarrows XO-PRN + PP \qquad\qquad 3.$$
$$\downarrow$$
$$2\ Pi$$

Hydrolysis of PP by a strong and ubiquitous inorganic pyrophosphatase drives these reversible condensations toward biosynthesis (11).

What a wondrous enzyme, the humble potato pyrophosphatase! It helped solve an aspect of NADP structure, set up the discovery of coenzyme biosynthesis, and with it a major theme in biochemistry, and then led me on to the enzymes that assemble DNA, genes, and chromosomes.

OROTIC ACID IS ON THE MAIN TRACK (1953–1955)

In 1955, two years after the historic Watson and Crick reports (12) of the double helix and its implications for replication, I found an enzyme that synthesizes DNA chains from simple building blocks. Based on this chronology, it is commonly assumed that the Watson-Crick discovery spurred me to search for the enzymes of replication. But that is not the way it happened. In 1953, DNA was far from the center of my interests. The significance of the double helix did not intrude on my work until 1956, when the enzyme that assembles the nucleotide building blocks into a DNA chain was already in hand.

My interest in the replication of DNA, the focus of my research for the past 33 years, developed primarily from a fascination with enzymes. Having found an enzyme that incorporates a nucleotide into a coenzyme, I began, around 1950, to wonder about enzymes that might assemble the many nucleotides that make up the chains of nucleic acids, particularly RNA. But first we had to know the building blocks of the nucleic acids. It was not at all obvious in 1950 what they might be. Was the backbone assembled first and were the bases attached later? Was each link added to the chain as a single nucleotide? If so, was the phosphate in each component nucleotide initially attached to carbon number three or five, or to either one randomly or in a cyclic form to both?

In anticipating what the building block might be, I was influenced by what I had learned from the biosynthesis of coenzymes. I also felt that in searching for the form of the nucleotide that might serve as a building block for RNA and DNA, it would help to know how a nucleotide itself is built from simpler molecules, and thus what its nascent form might be. Inasmuch as Jack Buchanan and Bob Greenberg were already pursuing purine biosynthesis, I decided to go after the pyrimidines.

During a brief interlude, I acted on the hunch that biosynthesis of the phosphodiester bond, accessible in phospholipids, might offer a model for building the backbone of nucleic acids. In exploratory experiments with $[^{32}P]$-α-glycerophosphate and $[^{14}C]$-phosphoryl choline, I could find no evidence for their condensation to form the diester (glycerophosphoryl choline), but I did stumble on the formation of phosphatidic acid and phosphatidyl choline in the cell-free extract (13). I worked out the enzymatic synthesis of phosphatidic acid (14), the key precursor of phospholipids, but still was eager to get away from greasy molecules and return to pyrimidines and the aqueous phase. In the future, I would not rely on intuition about model systems, but would head toward an objective directly.

Osamu Hayaishi came as a postdoctoral fellow in 1950 experienced in the use of soil bacterial enrichment cultures. Among the huge variety of species in soil, at least one can be found that will respond to virtually every natural organic compound and use it as a source of carbon and energy. Believing too that reversibility of metabolic pathways might provide clues to biosynthesis, we examined the breakdown of uracil and thymine in extracts of bacteria isolated from soil by aerobic enrichment on these pyrimidines. Uracil and thymine were converted to the corresponding barbiturates, not at all promising as biosynthetic precursors (15). But the next year, during my first visit to California, H. A. Barker helped me find an anaerobe in San Francisco Bay mud that consumed orotic acid. Back at NIH, with the participation of Irving Lieberman, who had been a student of Barker, studies of this organism identified as metabolic products dihydroorotic acid and carbamyl aspartate (16), which later proved to be intermediates in the biosynthesis of orotic acid.

Orotic acid was known from intact cell studies to be a precursor of nucleic acid pyrimidines, but it was uncertain whether it was on the main track or connected to it by a spur. With orotic acid tagged in its carboxyl group, the release of CO_2 to form uracil might lead us to the enzyme that took orotic "up" to nucleic acid. CO_2 release by extracts from yeast or liver was terribly feeble, yet showed a tantalizing requirement for ATP and ribose 5P. One happy day, instead of using extracts of either yeast or liver, I combined them. The reaction was explosive, hundreds of times greater than before, one of those rare moments in a scientific lifetime.

The enzyme abundant in liver extracts transferred a PP group from ATP to carbon 1 of ribose 5P to produce the novel phosphoribosyl pyrophosphate (PRPP) (17), later recognized as the key precursor of purine nucleotides, histidine, tryptophan, and NAD. The enzyme in yeast extracts (actually two enzymes) formed orotidine 5P, which then was decarboxylated to UMP (18), the direct precursor of all the nucleic acid pyrimidines (Figure 1).

The transfer of pyrophosphate to ribose 5P entails an attack on the middle phosphate of ATP, as Gobind Khorana showed during one of his whirlwind and productive visits to my lab (19). (Other examples of this unusual reaction are the synthesis of thiamine PP, and guanosine tetraphosphate.) PRPP synthetase remains one of my favorite enzymes. As I wrote in the 1975 Festschrift for Ochoa (*Reflections on Biochemistry*, Pergamon Press): "Most of us anticipated that ribosyl activation for nucleotide biosynthesis would use the same device of phosphorylation, so well known for glucose. But the novelty of pyrophosphorylation used by this enzyme (coupled with elimination of inorganic pyrophosphate upon subsequent condensations) established my unalloyed awe for the ingenuity and fitness of an enzyme."

Knowing that PRPP enables a free pyrimidine (orotic acid) to be converted directly to a nucleotide, we sought and found enzymes that used PRPP to convert free purines (adenine, hypoxanthine, guanine) directly to nucleotides (20). Yet, I also knew from Buchanan's and Greenberg's studies (21, 22) that a purine ring is assembled from the very outset attached to ribose phosphate (later shown to be derived from PRPP). These facts, coupled with the knowledge that nucleotides can be formed from nucleosides by kinases, made it clear to me that cells have alternate pathways to the biosynthesis of nucleotides: salvage of preformed bases and nucleosides, and de novo routes from smaller molecules (e.g. sugar phosphates, amino acids, ammonia, one-carbon units). We have since realized that the role of salvage pathways can be as vital as the de novo pathways even under normal conditions when the de novo routes are not blocked by mutation, drugs, disease, or excessive traffic (23).

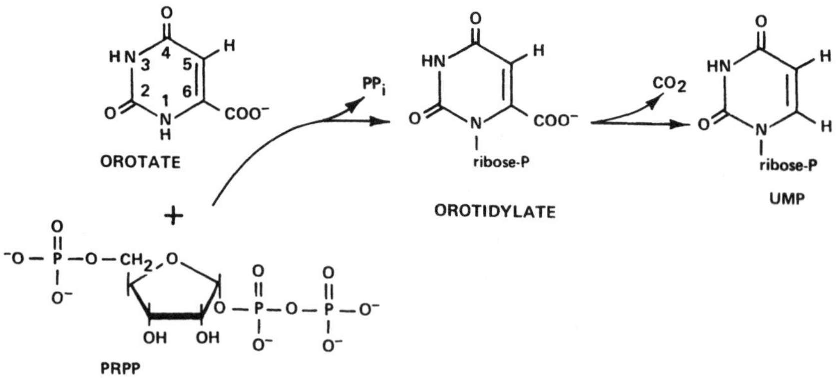

Figure 1   Condensation of orotic acid with PRPP produces the nucelotide, orotidylate (orotidine 5P), which upon decarboxylation generates uridine 5P (UMP).

DISCOVERY OF DNA POLYMERASE (1955–1959)

Having learned how the likely nucleotide building blocks of nucleic acids are synthesized and activated in cells, it seemed natural that in 1954 I would look for the enzymes that assemble them into RNA and DNA. Such an attempt might have been considered by some as audacious. Synthesis of starch and fat, once regarded as impossible outside the living cell, had been achieved with enzymes in the test tube. But, the monotonous array of sugar units in starch or the acetic acid units in fat was a far cry from the assembly of DNA, thousands of times larger and genetically precise.

Yet, I was only following the classical biochemical traditions practiced by my teachers. It always seemed to me that a biochemist devoted to enzymes could, if persistent, reconstitute any metabolic event in the test tube as well as the cell does it. In fact better! Without the constraints under which an intact cell must operate, the biochemist can manipulate the concentrations of substrates and enzymes and arrange the medium around them to favor the reaction of his choice.

I have adhered to the rule that all chemical reactions in the cell proceed through the catalysis and control of enzymes. Once, in a seminar on the enzymes that degrade orotic acid (16), I realized that my audience in the Washington University chemistry department was drifting away. In a last-ditch attempt to gain their attention, I pronounced loudly that every chemical event in the cell depends on the action of an enzyme. At that point, Joseph Kennedy, the brilliant young chairman, awoke: "Do you mean to tell us that something as simple as the hydration of carbon dioxide (to form bicarbonate) needs an enzyme?" The Lord had delivered him into my hands. "Yes, Joe, cells have an enzyme, called carbonic anhydrase. It enhances the rate of that reaction more than a million-fold."

By 1954, the rapidly growing *Escherichia coli* cell had become a favored object of biochemical and genetic studies, and for me had replaced yeast and animal tissues as the preferred source of enzymes. To explore the synthesis of RNA, Uri Littauer, a postdoctoral fellow, and I prepared [^{14}C-adenine]-ATP and maintained it as ATP with a regenerating system. Upon incubation with an *E. coli* extract, a small but significant amount of the radioactivity was incorporated into an acid-insoluble form, presumably RNA, and we proceeded eagerly to purify the activity responsible.

I also pursued the synthesis of DNA. Here, I had the invaluable help of Morris Friedkin, who had synthesized ^{14}C-thymidine and was studying its uptake into the DNA of rabbit bone marrow or onion root tip cells. Disinclined to work with cell-free extracts, he generously saved the spent reaction fluid from which I recovered radioactive thymidine to use in trials with extracts of *E. coli*.

The results were mixed. Very little thymidine was incorporated into the acid-insoluble form indicative of DNA, only about 50 cpm out of the million with which we started. On the other hand, 5–10% of the thymidine was converted to novel soluble forms that resembled the phosphorylated states of the nucleotide building blocks, possibly better precursors than thymidine for DNA synthesis.

At this juncture, Herman Kalckar on a visit to St. Louis brought us the startling and unsettling news that Ochoa and Marianne Grunberg-Manago, a postdoctoral fellow, had just discovered the enzymatic synthesis of RNA. It was for them a totally unexpected finding made while exploring aerobic phosphorylation in extracts of *Azotobacter vinelandii*. They observed an exchange of phosphate into ADP and the reversible conversion of ADP (or other nucleoside diphosphates) into RNA-like chains (24) and they named the enzyme polynucleotide phosphorylase.

On the strength of this new information, we shifted to using ADP rather than ATP in our studies with *E. coli*. The rate and extent of reaction were far greater and we readily purified the enzyme involved (25). We had made a classic blunder. Accounting for a phenomenon does not insure that it is the only or the best explanation of it. In this instance, we were diverted from the discovery of RNA polymerase, which depends on ATP. By switching to ADP, we tracked the synthetic activity of polynucleotide phosphorylase and missed the key enzyme for gene transcription.

Ten months passed before I repeated the experiment of converting radioactive thymidine to an acid-insoluble form. Once again, only a tiny amount of this presumed precursor was converted. But several things were different. For one, the radioactivity of the thymidine happened to be three times as great and so the results seemed more impressive. For another, believing I had lost out on the synthesis of RNA, the synthesis of DNA became a more precious goal. Finally, I exposed the product this time to pancreatic DNase and found that it became acid-soluble, a strong indication that it was DNA.

Even before I calculated the DNase results, I stopped to tell Bob Lehman about them. Although his postdoctoral problem was well started, he was eager to switch to DNA synthesis. Progress was rapid. Bob soon found that thymidine phosphate was a far better precursor than thymidine and later showed that thymidine triphosphate was much better still. With improvements in the assay of DNA synthesis by these crude extracts, our goal was to purify the enzyme that assembled nucleotides into a DNA chain, the enzyme we would name DNA polymerase (26, 27).

The most complex and revealing insights into the reaction would come from exploring the function of the DNA that I had included in the reaction mixture in my earliest attempt to incorporate thymidine into DNA. Some assume that DNA was included to serve as a template and that its primer role emerged many years later. Not so. I added DNA expecting that it would serve as a primer for growth of a DNA chain, because I was influenced by the Cori work on the growth of carbohydrate chains by glycogen phosphorylase. I never thought that I would discover a phenomenon utterly unprecedented in biochemistry: an absolute dependence of an enzyme for instruction by its substrate serving as a template.

I had added DNA for another reason. Nuclease action in the extracts was rampant, and I wanted a pool of DNA to surround the newly incorporated thymidine and protect at least some of it. Only later did Lehman and I learn with elation that the added DNA fulfilled two other essential roles. It indeed served as a template and also as a source of the missing nucleotides. The DNA was cleaved by DNases in the extract to nucleotides. These were converted by ATP and five kinases in the extract to the di- and triphosphates of the A, G, C, and T deoxyribonucleotides, which were then still unknown.

Maurice Bessman, Steve Zimmerman, and Julius Adler joined Bob Lehman, Sylvy, Ernie Simms (my research assistant), and me and occasionally one or two others, all in a small laboratory, only about 20 by 20 feet. Crowded and excited, we shared ideas, reagents, and data. The sum of our efforts was far greater than if we had been diluted into a larger room or separated by walls.

CREATION OF LIFE IN THE TEST TUBE (1960–1967)

With purified DNA polymerase, we could show that the DNA product reflected the base composition of the template and the frequencies of the 16 possible dinucleotides. The "nearest-neighbor" sequence method, which we devised to determine the dinucleotide frequencies, also revealed that the two strands of the double helix have opposite polarities, a structural feature that had not been experimentally demonstrated up to that time (28).

We also made the unexpected discovery that the enzyme, in the apparent absence of any template would, after a considerable delay, make DNA-like polymers of simple composition (29, 30): the alternating copolymers polydA·dT and polydG·dC and the homopolymer pairs of polydA with polydT and of polydG with polydC. These polymers, once made, proved to be superior templates and have been widely used in DNA chemistry and biology. Generation of the polymers de novo could be ascribed to the reiterative replication of short sequences in the immeasurably small amounts of DNA that contaminate a polymerase preparation (31, 32).

For more than 10 years, I had to find excuses at the end of every seminar to explain why the DNA product had no biologic activity. If the template had been copied accurately, why were we unsuccessful in all our attempts to multiply the transforming factor activity of DNA from *Pneumococcus, Hemophilus,* and *Bacillus* species? Finally, with the arrival of ligase in 1967, a crucial test could be made. [The enzyme had been discovered that year in five laboratories: those of Martin Gellert, Charles Richardson, and Jerard Hurwitz, in Lehman's next door, and in mine by Nicholas Cozzarelli.] Mehran Goulian and I could replicate the single-stranded circle of phage ϕX174 with DNA polymerase and then seal the complementary product with ligase. The circular product was isolated and then replicated to produce a circular copy of the original viral strand, which could be assayed for infectivity in *E. coli* (33). We found the completely synthetic viral strand to be as infectious as that of the phage DNA with which we started!

After so many years of trying, we had finally done it. We had gotten DNA polymerase to asemble a 5000-nucleotide DNA chain with the identical form, composition, and genetic activity of DNA from a natural virus. All the enzyme needed was the four common building blocks: A, G, T, and C. At that moment, it seemed there were no major impediments to the synthesis of DNA, genes, and chromosomes. The way was open to create novel DNA and genes by manipulating the building blocks and their templates.

In a very small way, we were observers of something akin to what those at Alamogordo on a July day in 1945 witnessed in the explosive force of the atomic nucleus. Harnessing the enzymic powers of the cellular nucleus had neither the dramatic staging of light and sound nor the stunningly apparent global consequences. Yet, this demonstration of our power with enzymes that build and link DNA chains would soon help others forge a different revolution, the engineering of genes and modification of species.

A hundred newspaper and television reporters and photographers came to a press conference called by the Stanford News Bureau on December 14, 1967, because of many inquiries about the paper we had just published in that month's issue of the *Proceedings of the National Academy of Sciences* (34). The title was: "Enzymatic Synthesis of DNA, XXIV. Synthesis of Infectious Phage ϕX174 DNA." To the editors who sent the newsmen, it seemed that a virus had been synthesized and life created in the test tube.

At the news conference, I tried to explain why the definition of life and a living molecule is so elusive. Afterwards, I overheard a reporter on the telephone to his office: "It's not what we expected. They haven't made a virus. It's only a molecule, a short chain of DNA. They've been making DNA in the test tube for 12 years." Hairy little monsters had not been created in the test tube. Yet, the story rated banner headlines worldwide and a newspaper article on January 4, 1968, was titled: "Creation of Life Rates Best of Science Stories in 1967." In smaller type: "Human Heart Transplant Second."

PROOFREADING AND EDITING BY A REPLICATING ENZYME (1967–1971)

Knowing that DNA polymerase synthesized a chain in the 5' to 3' direction, it made no sense to me then that the enzyme degraded the very 3'-end of the chain it would normally be extending. In the absence of the nucleotide building blocks needed for synthesis, nucleotide units were cleaved slowly and serially from the 3' end of a DNA chain. Then a simple fact about the nuclease gave us our best clue. Douglas Brutlag observed that the degrading activity was far more potent on a single strand of DNA than on the usual double-stranded form. This preference became extreme when the temperature of the reaction was lowered, presumably because the ends of duplex DNA are less frayed at lower temperatures.

Why should a loose primer end be a substrate for degradation by a synthesizing enzyme? We prepared a variety of duplex DNAs in which a few residues at the primer end of a chain were not matched to the other strand. The mismatched residues were removed immediately, after which the others were removed far more slowly. When deoxynucleoside triphosphates were supplied to permit chain extension, the mismatched residues were still removed quickly, but now the subjacent nucleotides remained intact and were extended by synthesis (35).

Thus, the enzyme removes all mismatched units, permitting fresh units to be added to the growing chain end only when it is correctly matched to the template chain. We could infer that if the synthesizing enzyme were to make a rare mistake during elongation of a chain, such as inserting a C opposite an A (estimated to happen once in 10,000 times), it would remove the mismatched C before proceeding with extension of the chain. This astonishing proofreading ability of the enzyme, coupled with its fine discrimination in the initial choice of correct building blocks during synthesis, reduces errors in the overall process of replication to one in 10 million.

Having finally made sense of why an activity that degrades DNA is part of the very enzyme that makes it, we were unprepared for the paradoxical observation by Lehman (36) that the nuclease action of DNA polymerase on double-stranded DNA was enhanced tenfold when all four building blocks required for synthesis (i.e. A, T, G, and C) were present. How could synthesis be enhancing degradation? After all, we had observed earlier that synthesis extends the primer end of a chain and thereby protects it from nuclease action.

The solution came from Edward Reich and his colleagues (37) and from Murray Deutscher in my laboratory (38), showing that nuclease activity in DNA polymerase persists even with the 3'-end blocked by an analogue or phosphate. A separate domain in the enzyme removes nucleotides from the 5'-end of a chain.

Now we could explain how the four building blocks, and the synthesis they make possible, enhance nuclease action by DNA polymerase (39). By removal of DNA from the 5'-end at a nick, a stretch of template becomes exposed for pairing with a substrate nucleotide and further synthesis. In its synthetic progress along the template, the polymerase is brought up to a 5'-end of the chain, which it then degrades. It was immediately obvious how this "nick translation" by polymerase could be useful in the repair of lesions in DNA (40), and as we recognized some years later, could perform an essential step in replication by removing the RNA that initiates the start of a DNA chain.

DNA POLYMERASE UNDER INDICTMENT (1970–1972)

DNA polymerase was called a "red herring" and charged by *Nature New Biology* in a series of editorials with masquerading as a replication enzyme (41). The replicative role of DNA polymerase was questioned because of the Cairns mutant of *E. coli* (42), which appeared to lack the enzyme and yet grew and multiplied at a normal rate. In addition to the apparent dispensability of DNA polymerase for cell multiplication and its more estimable quali- fications as a repair enzyme, genes were being discovered (designated *dnaA, dnaB, dnaC*, etc) that strongly implicated many other proteins as essential for a replication process far more complex than had been imagined.

The rising skepticism about the importance of DNA polymerase was fanned by the *Nature New Biology* vendetta. Not only was the enzyme attacked, but the basic mechanism, the building blocks, and the assays of DNA synthesis were judged to have prevented the discovery of the true DNA-replicating enzymes. At this juncture, my middle son, Tom, entered the fray. (I was at the Molecular Biology Laboratory in Cambridge, England, in the second half of a sabbatical year devoted to membranes.) An injury to his hand prevented him from continuing his career as a cellist at the Juilliard School, and he was disturbed by disparaging comments about DNA polymerase in his biology course at Columbia College where he was also a full-time student. Despite lack of laboratory experience, he found, within a few weeks, a DNA polymerase in *E. coli* cells, distinct from the one I had discovered. The new activity, named DNA polymerase II (pol II) (43), was clearly different from the "classic" DNA polymerase (pol I) and from still another, DNA polymerase III (pol III) (44), which he discovered in the course of purifying pol II. Subsequently, he and Malcolm Gefter located the gene for pol III and showed that conditionally lethal mutations in this gene blocked DNA replica- tion (45). Pol III, in a far more elaborate form, was to gain recognition as the central enzyme of DNA replication in *E. coli*.

All three polymerases, although differing significantly in structure, proved to be virtually identical in their mechanisms of DNA synthesis, proofreading, and use of the same building blocks. The maligned polymerase (pol I) became the prototype for all DNA polymerases in plants, animals, and viruses, as well as in *E. coli*. The gloomy prophecies of *Nature New Biology* soon dis- appeared, as did the magazine itself.

HOW DNA CHAINS ARE STARTED (1971–1975)

Despite the excitement over the synthesis of a chain of infectious viral DNA, I had felt a certain uneasiness. One of the inferences drawn from the replication of a single-stranded, circular template was that DNA polymerase I could start a new chain. Yet we were never able to find direct proof of this. Moreover, we had observed that replication of the circular template was far more efficient if a small amount of boiled *E. coli* extract was present. Although it seemed unlikely that a random fragment of DNA in the extract would match the viral DNA template accurately enough to serve as a primer, this possibility became a reality. DNA polymerase removed the unmatched regions of the fragment by proofreading at the 3' end; with generous editing at the 5' end, no trace of the fragment remained in the synthetic product.

We were left with the question of how DNA chains are started, how a single-stranded, circular viral DNA is converted to the duplex form upon entering the cell, how nascent chains are initiated in the replication of virtually all chromosomes. Indeed, Reiji Okazaki had shown earlier (46) that chains are started not just once, at the beginning of the chromosome, but repeatedly in staccato fashion during the progress of replication.

After several years of unproductive attempts, I recognized a basic flaw in our work, how hopeless it was to answer the question about chain starts with the DNA we were using as template and primer. A tenet I was taught and to which I had faithfully adhered is that one must purify an enzyme to understand what it does. An aphorism attributed to me, but actually due to Efraim Racker, is: "Don't waste clean thinking on dirty enzymes." Another basic tenet of enzymology, too obvious, it would seem, to mention, is that one must provide the enzyme, clean or not, with a pure substrate.

The blunder we had made for too many years was accepting the DNA extracted from bacterial and animal cells as an adequate substrate for the enzymes of replication. The huge chromosomes are very fragile and the mechanical forces of flow and mixing used during isolation are violent enough to reduce the DNA to a heterogeneous collection of damaged fragments. In short, we were giving our relatively clean enzyme a very "dirty" substrate.

When I finally recognized the futility of searching for the replication enzymes with bacterial DNA, let alone animal DNA, I also realized that we had been ignoring a proper DNA substrate, the chromosome of a tiny bacteriophage. I recalled belatedly the virtues of the intact, clean, phage chromosome that four years earlier had served us in demonstrating the synthesis of infectious DNA by DNA polymerase I. As small, single-stranded circles, we could actually view them with an electron microscope and verify that in a purified sample, they were intact, homogeneous, and uncontaminated by fragments of the bacterial host DNA. We also knew that immediately upon entering the cell, the phage DNA is converted by bacterial enzymes to a double-stranded circle, an event we could easily assess. Probing how a new phage circle is started and completed might illuminate the intricate enzymatic machinery the cell uses to replicate its own chromosome.

We could use the chromosome of either of the two classes of small phages, the icosahedral ϕX174 or the filamentous M13. Luckily, as events later proved, I chose to work on M13 and within a week or two switched the efforts of my entire group to the various stages of its life cycle. I have sometimes felt wistful reflecting on the boldness of that move and the exciting events in the weeks that followed.

Being preoccupied with the initial event in M13 replication enabled me to connect three otherwise unrelated facts and arrive at an idea as to how a DNA chain might get started. For one, RNA polymerase, unlike DNA polymerase, can start chains. Furthermore, DNA polymerase, while routinely excluding ribonucleotides in assembling a DNA chain, does accept an RNA chain end matched to a DNA template as a primer for extension in DNA synthesis. Finally, DNA polymerase I has an editing function, which can remove something foreign, like thymine dimers and RNA from the start of a DNA chain, and replace it with proper DNA.

Might RNA polymerase make a short piece of RNA on single-stranded M13, which DNA polymerase could use to start a DNA chain? Then, when the enzyme had come full circle in copying the available template, its editing system would erase the RNA and synthesize DNA in its place. We could test this hypothesis by using rifampicin to inhibit RNA polymerase in vivo (47). When Doug Brutlag did so, the M13 circle was not replicated! We went on to show with cell extracts and partially purified enzymes that RNA polymerase initiates DNA replication by forming a primer RNA for covalent attachment of the deoxyribonucleotide that starts the new DNA chain (48).

There was one discordant note. Rifampicin did not prevent the conversion of the phage ϕX174 single-stranded circle to its duplex form, nor did it interrupt the ongoing replication of the *E. coli* chromosome, despite the repeated initiations of DNA strands presumed to be occurring at the growing fork. Either RNA priming was of limited significance, or another mode of RNA synthesis, independent of RNA polymerase, existed. As we probed the initiation on ϕX174 circles, it became clear that this virus, instead of relying on RNA polymerase, exploits the extraordinarily complex assembly of initiation proteins that the cell uses for replicating its own chromosome.

PRIMOSOMES, HELICASES, AND REPLISOMES (1972–)

After a 10-year drought of discoveries of new enzymes, they now came in a torrent, and in 1972 a bright and boistrous group, led by Randy Schekman and

Bill Wickner, was there to collect and sort them. Upon fractionating the components responsible for conversion of the single-stranded circle of ϕX174 to the duplex form, we could separate them into two groups, one that primed the start of a chain and the other that extended it. Then, as we tried to purify each of these fractions, they splintered into many separate components. The joy of uncovering the trails to so many novel proteins soon gave way to the discouragement of being unable to track down any one of them. We judged there might be as many as eight different proteins, all scarce, that were needed to make the tiny bit of RNA that primed the synthesis of a DNA chain; the DNA polymerase seemed just as complex. What were all these proteins and what was each of them doing?

A major assist came from New Haven sewage via Nigel Godson, who discovered in it a new phage (G4) that resembled ϕX174 (49). Replication of G4 DNA required only only three of the eight fractions needed by ϕX174 (50). One of the fractions was purified easily because it withstood heating. It was the single-strand binding protein (SSB) (51), which coats the DNA, except at a region that forms a duplex, hairpinlike structure and is used by the second fraction as a template to synthesize a stretch of primer RNA. This protein, known to complement the deficiency of cell extracts of dnaG mutants, was named primase (52). The third fraction was DNA polymerase III in a complex, unstable form (53) with many auxiliary units that clamp the polymerase to the template and enable it to replicate great lengths of DNA with astonishing speed and accuracy. Replication of G4 DNA thus provided us with our best assays for purifying each of these components: SSB, primase, and the super polymerase we now called DNA polymerase III holoenzyme.

We could now return to ϕX174 and begin to explain the molecular operations of the multiprotein assembly (called a primosome) (54) that starts a DNA chain. The image of a locomotive seemed helpful for a time in accounting for primosome actions. The engine, protein n' (55) (its gene unknown to this day), powered by ATP energy, is a helicase (56, 57); it unzippers the DNA duplex and is equipped with a cowcatcher to remove SSB in its path. Another protein, dnaB, is both helicase (58, 59) and engineer (60), using ATP to locate or shape a section of DNA track upon which primase will find it possible to lay down a short stretch of RNA, which will then attract DNA polymerase to start a DNA chain. The primosome is translocated on DNA in only one direction, the one that keeps it at the advancing fork of a replicating chromosome (61).

Another awesome property of the primosome is the persistence of its attachment to the ϕX174 DNA circle even after the duplex circle has been completed (62). The attached primosome directs the next replication event and is used over and over again as part of a stamping machine to generate the many duplex forms needed for transcription into messages for the 10 proteins encoded by the phage DNA.

To multiply duplex circles (Figure 2), the phage employs a single gene (gene A) of its own and relies on host proteins to do the rest. The isolated gene A protein breaks the viral strand backbone at a particular diester bond and becomes covalently attached to the 5'-end of the break (63). In so doing, it makes the 3'-end available as a primer for DNA polymerase III holoenzyme replication. For lack of exposed template at this nick, the phage exploits the host rep protein, a helicase, to unzipper the duplex (64, 65). This was the first occasion in which a helicase was seen in its role of opening of a duplex for an advancing replication fork. Coordinate helicase and polymerase actions generate single-stranded viral circles repeatedly by a mechanism called rolling-circle replication (66). Each completed circle is then replicated to form a duplex in the same way as the one originally injected by the phage.

Based on these insights gained from the proteins involved in the replication of the small phages, we now propose that progress of replication of the host chromosome (Figure 3) (67) depends on helicases, SSB, and topoisomerases to prepare the templates for the continuous synthesis of a leading strand and the discontinuous synthesis of the other. We regard the polymerase III holoenzyme as a rather loose assembly of many auxiliary units attached

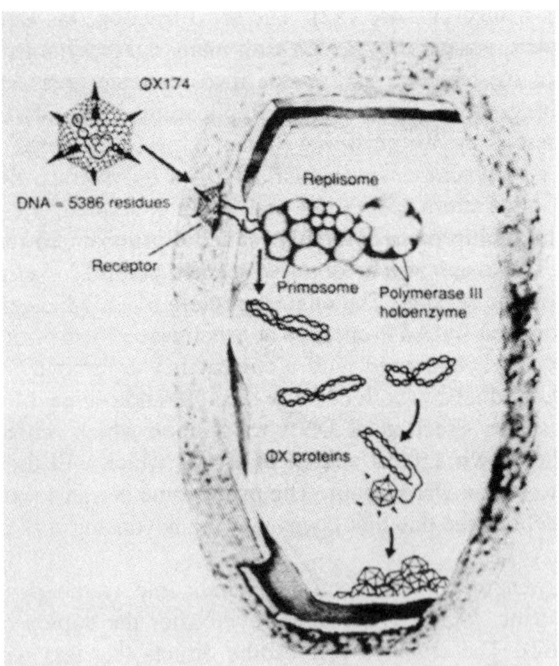

Figure 2   Life cycle of phage φX174 in *E. coli*. Conversion of the viral circle to a duplex is followed by multiplication of the duplexes by a rolling-circle mechanism.

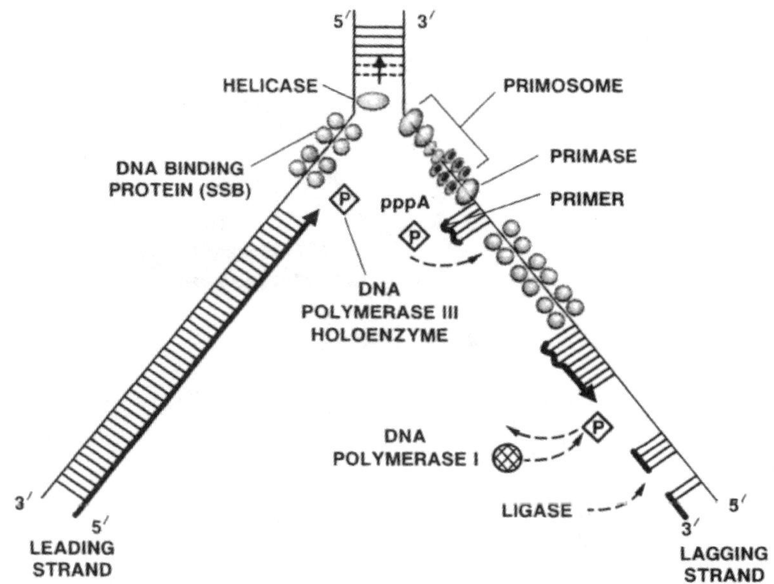

Figure 3   Scheme for enzymes operating at a replication fork in *E. coli:* continuous synthesis of a leading strand and RNA-primed discontinuous synthesis of the other strand.

asymmetrically to a pair of polymerase cores. One arm appears suited to the highly processive synthesis of the leading strand and the other arm to the discontinuous synthesis of the other strand (68, 69).

We now wonder whether essentially concurrent replication of both strands might be achieved were priming of nascent fragments of discontinuous synthesis integrated with continuous strand synthesis. A replisome comprising the holoenzyme, primosome, and helicases might periodically generate a loop in the lagging strand template to place it in the same orientation as the leading strand at the fork (Figure 4) (67).

We had been aware that the primosome and the primase in it moved in opposite directions on a DNA chain. With additional knowledge, the difficulty grew. The isolated dnaB component was found to move in the direction of the primosome (58), whereas the n' component by itself moved in the opposite (chain elongation) direction (57). A smoldering annoyance had become a serious paradox. How could integral parts of the primosome locomotive move in opposite directions on a fixed track of DNA? The analogy would have to be scrapped. Instead, it now makes more sense to invert the relative movements of the primosome and DNA and regard the primosome machine as fixed in place with the DNA chains being drawn through it (Figure 4) (57). This view also offers an attractive mechanism for coordinating the

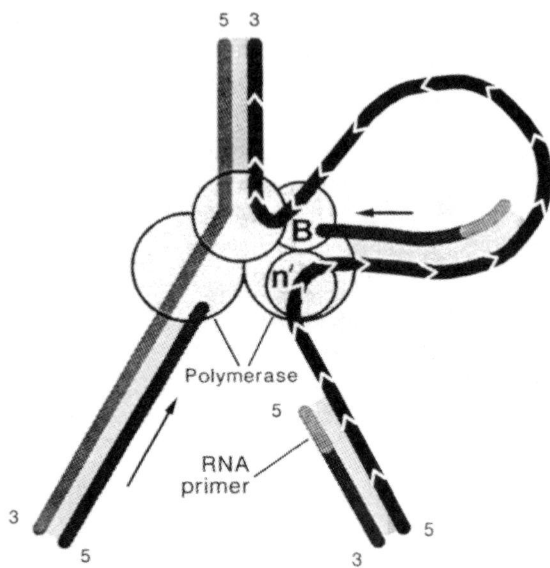

Figure 4 Hypothetical view of the replisome as a fixed machine through which a loop of DNA is pulled to provide for essentially concurrent replication of both strands. As the DNA is pulled through the primosome, the n' protein is translocated on the DNA in the direction of chain elongation and the B protein in the opposite direction.

helicase, priming, and replication actions at the advancing fork of a chromosome.

INITIATION OF CHROMOSOMES (1979–)

An aspect of replication that has long intrigued me is how an increase in *E. coli* cell mass triggers the initiation of replication that commits the cell to start a new cycle. What is the biochemistry of the replication switch, which in *E. coli* regulates the cell cycle and in eukaryotes responds to signals that turn the embryonic cell to a quiescent adult, or the quiescent cell to proliferation? With the cloning of the highly conserved, unique, 245–base pair chromosomal origin (*oriC*) in plasmids, we were afforded a substrate with which to seek the comparably complex, conserved multiprotein system that uses *oriC* to initiate a cycle of replication. After 10 man-years of fruitless effort to obtain a cell-free initiation system, Bob Fuller and Jon Kaguni were the ones who finally succeeded (70). Two apparently illogical maneuvers were essential. One was the inclusion of a hydrophilic polymer (e.g. polyethylene glycol) at high levels which, as we later realized, acted by a "macromolecular crowding" effect, which concentrates the numerous proteins and DNA into a small volume. The other was subjecting an inert lysate to a refined ammonium sulfate fractionation, a trick that had worked for me 30 years earlier in the discovery of the yeast enzyme that converts NAD to NADP (71). In the active fraction the numerous required proteins are concentrated and potent inhibitors are excluded.

The proteins that we found responsible for initiating replication at *oriC* include *initiation proteins* (particularly dnaA protein) that recognize super-coiled *oriC*, alter its structural conformation, and lead to its further opening by dnaB helicase action, *specificity proteins* that suppress potential origins elsewhere on the chromosome, and the *replication proteins* that prime and elongate chains on the opened plasmid and propel two forks in opposite directions (72, 73).

The motif we are finding in the mechanism for initiation of the *E. coli* chromosome (Figure 5) seems to apply to a wide variety of bacteria and their plasmids and phages (74). Control of dnaA protein by ATP-ADP cycling (75) and by binding to the acidic headgroups of a fluid membrane (76), and activation of an inert origin in an overly relaxed supercoil by nearby transcription (77) are among the factors already discovered that influence initiation, and likely more will be found as we explore this crucial event in the cell cycle.

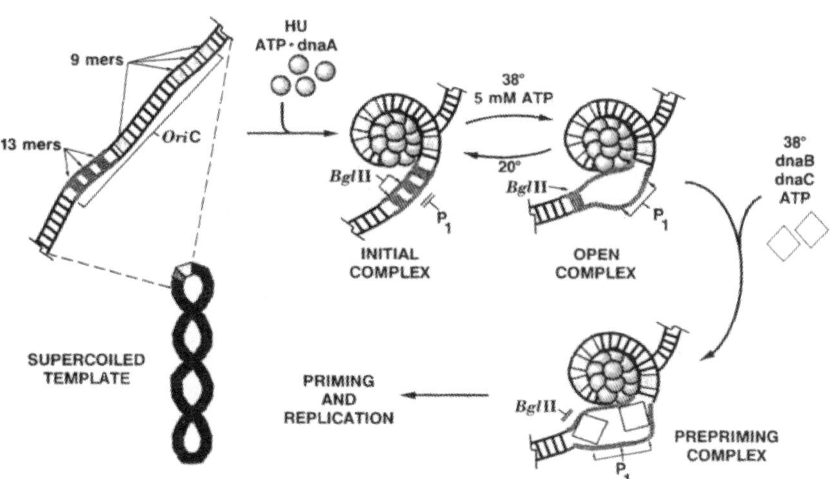

Figure 5 Scheme for early events in initiation of replication at the origin of the *E. coli* chromosome operating in a plasmid.

SPORULATION AND GERMINATION (1962–1970)

Those familiar with my research are aware of my intense concentration on a single subject—the enzymatic synthesis of DNA—and the blinders I have worn to maintain this focus. Nearly forgotten now by us all are the eight years, in the midst of the DNA work, when half of my research effort was devoted to an arcane subject, the development of spores (78).

During my tenure as chairman of a department of microbiology (1953–1959), I had become interested again in spores as agents of disease: anthrax, tetanus, botulism. I constantly remembered with deep anguish the ghastly *Clostridium perfringens* (gas gangrene) spore that killed my mother within a day of a "routine" gall bladder operation in 1939. Now I could look beyond the "bad" spores to the vast array of innocent species whose mysterious biology and biochemistry fascinated me.

How is a spore made and how does its chemical organization endow it with astonishing abilities: dormancy and resistance to extremes of heat, desiccation, disinfectants, and ultraviolet rays lethal to the cell? How does a spore, after years of hibernation, respond instantly to a substance that signals conditions are right for growth into a new cell? I believed then, and still do, that this knowledge would contribute in a major way toward understanding the embryonic development of animals and their response to environmental stresses.

During the eight-year period in which a succession of students and postdoctoral fellows (including Pieter Bonsen, Pierre Chambon, Murray Deutscher, Arturo Falaschi, David Nelson, Tuneko Okazaki, Peter Setlow, Jim Spudich, Henrique Tono, and Jim Vary) worked on spores, we published 26 papers and still were making little progress toward answering the global questions that had attracted me.

I abandoned the spore work when I came to realize how much more complex the problems were than I had imagined and that hardly anyone else in the world seemed to care. The little research on spores, then and even now, was largely of a practical nature: how to destroy spores in food canning or how to use them as pesticides on crops.

Beyond the discouragement and loneliness of working on a tough problem in an unfashionable area was the distraction of having the other half of my research group of eight or so engaged in the more glamorous and productive work on DNA replication. Because of my own ambivalence, I offered no resistance on occasions when a member of the sporulation group defected to the replication team. Finally, after this eight-year siege, I too gave up. Eventually only Peter Setlow has continued a biochemical interest in spores.

Progress in science depends on how vigorously the field is cultivated. In contrast with sporulation, interest in cancer is enormous. Hundreds of laboratories worldwide attract many thousands of scientists, including the brightest, to unravel the processes responsible for malignant growth. Yet studies of sporulation are also deserving of resources and talents. Sporulation, dormancy, and germination are fundamental processes in nature, more accessible to incisive examination, and, if better understood, might yield as much information relevant to the cancer process as some of the massive programs on tumor-bearing animals.

SCIENTIST, TEACHER, AUTHOR, CHAIRMAN: IN WHAT ORDER?

In May 1988, when my former and present students and colleagues gathered in San Francisco for a gala 70th birthday party, I thought it would be fun to select from the 30 or more enzymes I had worked with, the 10 that I favored most. I was surprised to find that 6 of the 10 were discovered in the brief period from 1948 to 1955: nucleotide pyrophosphatase (8), NAD synthetase (9), phosphatidic acid synthetase (14), PRPP synthetase (17), polyphosphate synthetase (79), and DNA polymerase (26). That left only 4 to be selected from more than 20 enzymes that appeared in the next 30+ years. Inasmuch as some of the enzymes omitted from the top 10 are far more deserving of selection than some of the chosen ones, it is clear that the basis for the choice of the first 6 was largely sentimental. I was most attached to those enzymes that came during the time of my life when I collected the data myself, from conception to delivery.

In my marriage to enzymes, I have found a level of complexity that suits me. I feel ill at ease grappling with the operations of a cell, let alone those of a multicellular creature. I also feel inadequate in probing the fine chemistry of small molecules. Becoming familiar with the personality of an enzyme performing in a major synthetic pathway is just right. To gain this intimacy, the enzyme must first be purified and I have never felt unrewarded for any effort expended this way.

I once shocked the Dean of the Washington University School of Medicine by telling him that my prime interest as Chairman of the Department of Microbiology was to do and foster research rather than teach. It has never been otherwise. Experiments are far more consuming and fulfilling for me than any form of teaching. Still, I have enjoyed a rather modest amount of formal lecture and laboratory instruction and have done it conscientiously. For the student, didactic teaching fails without the infusion of scientific skepticism and a fervor for new knowledge, and these things are naturally conveyed by someone dedicated to research. For me, some 10 lectures a year freshen my awareness of basic subjects, and on one occasion the preparation of a laboratory exercise on DNA opened a major avenue for my experimental work.

The most rewarding teaching for me has been in the intimate, daily contact with graduate and postdoctoral students. Well over a hundred of them spent from two to five years in my laboratory and were exposed to my tastes and my obsession with the use of time. I felt closest to those who shared my devotion to enzymes and my concern with the productive use of our most precious resource: each of the hours and days that so quickly stretch into the few years of a creative life. I recall in 1948 relating to Sidney Colowick and Ollie Lowry (both senior to me in age and experience) my failure in purifying an enzyme by a certain procedure. "I wasted a whole afternoon trying that," I said. Colowick turned to Lowry and said with mock gravity: "Imagine, Ollie, he wasted a *whole* afternoon."

Imagination or hard work? At either extreme—speculating about complex phenomena or doggedly collecting data—success may come on occasion and draw acclaim. But the most consistent approach for acquiring a biochemical understanding of nature lies in between. The novel is yet to be written that captures the creative and artistic essence of scientific discoveries and dispels images of the scientist as dreamer, walking in the woods awaiting a flash of insight or of the scientist as engineeer, at an instrument panel executing a precisely planned experiment. Some intermediate ground, hard work with a touch of fantasy, is what I have sought for myself and my students.

If asked to name varieties of mental torture, most scientists would place writing near the top of the list. As a result, scientific papers are usually put off or dashed off and demean the quality and value of the work they describe. Writing a paper is an integral part of the research and surely deserves the small fraction, say five percent, of the time spent finding the thing worth reporting. Yet, I feel uneasy seeing students and colleagues writing at their desks during "working hours" rather than busy at the laboratory bench. Whereas taking time to prepare a scientific report is unavoidable, writing a book always seemed an unconscionable abdication from research until I wrote one.

Writing *DNA Synthesis*, a 400-page book (80), was a surprise in many ways. First, the effort was far greater than I imagined. Very little from lecture notes and reprints could be lifted and placed in the right context and still remain readable. I was also surprised by the pleasure I found in reworking and polishing sentences and paragraphs for brevity and clarity, a satisfaction I had never found in crossword puzzles or other word games. Best of all, I could present my work, views, and excitement about the enzymology of DNA replication to an unexpectedly wide audience. The book—adopted as a text for some courses—became the reference source for writers of reviews of DNA replication and for authors of textbooks of biology and biochemistry.

The sequel to *DNA Synthesis,* entitled *DNA Replication* (23), came out in 1980. Twice the size of its predecessor, it was really a new book in scope and organization as well as in expanded contents. Despite its inflated size, it was a better book and found a wider readership. However, progress in this field is so rapid that revisions are needed annually. As an experiment in publishing, I assembled a 273-page "1982 Supplement to DNA Replication" (67), which extended the life of its parent. The publishers objected to "1982" in the title and correctly saw it as an advertisement of obsolescence. There have been no further supplements.

Having regarded teaching and book writing as deviant activities for a dedicated scientist, then surely the administrative work of a departmental chairman should be beyond the pale. Yet, I served as chairman for more than 20 years and never found it a serious intrusion on my time or attention. On the contrary, the benefits of creating and maintaining a collegial and stimulating scientific circle were well worth the investment I made. With excellent administrative assistance and the eager participation of my faculty colleagues, direction of departmental activities took no more time than being a con scientious member of the department.

Involvement in medical school and university affairs is a far different matter. I never found the skills and patience to function at these levels. For me, the most burdensome feature of being a departmental chairman was the obligated service on the Executive Committee of the Medical School, pre-occupied with budgets, promotions, interdepartmental feuds, and salaries. In 6 years at Washington University and 10 at Stanford, I cannot recall a deliberate discussion of science or educational policy. No wonder I had no interest in being the dean of a medical or graduate school on occasions when this possibility was raised.

Increasingly conspicuous in current scientific life are the extramural administrative and educational activities, which, with the attendant travel, may consume half the time of prominent members of a science faculty. Lectures and visiting professorships, scientific meetings and society councils, government panels and advisory boards, consultantships in industry—all are prestigious, diverting, less demanding than research, and terribly tempting. I have done less than most, but have been unable to resist participating, particularly in writing essays (81–83), testifying for federal support of research and training, and most recently in the founding and development of a biotechnology enterprise (the DNAX Research Institute, Inc., later acquired by the Schering-Plough Corp.) with the mission of applying the techniques of molecular and cellular biology to the therapy of diseases of the immune system.

All these nonresearch activities, in and out of the university, fail to give me a deep sense of personal achievement. In research, it is up to me to select a corner of the giant jigsaw puzzle of nature and then find and fit a missing piece. When after false starts and fumbling, a piece falls into place and provides clues for more, I take pleasure in having done something creative. By contrast, in my other activities, which are just as personal, all I do, it seems, is try to behave in a commonsensical, fair, and responsible way, as anyone else would. With research so dominant over my teaching, writing, and administrative activities, in sharply descending order of importance to me, I sometimes wonder whether, valued for their contributions to science, this order might be inverted.

Beginning with administration, consider the creation and management of the Stanford Biochemistry Department. It was started in St. Louis as the Microbiology Department. From there, Paul Berg, Bob Lehman, Dave Hogness, Dale Kaiser, and Mel Cohn moved with me to Stanford in 1959 to be joined by Buzz Baldwin and a few years later by George Stark and Lubert Stryer. In polls of peers, the Department has been accorded a top rating for many years, and is regarded as a major source of discoveries basic to recombinant DNA and the genetic engineering revolution. More than 500 people trained in the department now staff and direct departments of biochemistry and molecular biology all over the world. The organization and development of this notable faculty and its preservation, largely intact, against strong and attractive centrifugal forces, I would have to admit is a unique achievement.

As for writing, the monographs on DNA replication, with more than 40,000 copies sold, have made it easier for others to enter and work in this field. More than offering a readable account of a forbiddingly specialized area of biochemistry, these books have helped revive an appreciation that enzymology provides a direct route toward solving biologic problems and creates reagents for the analysis and synthesis of a great variety of compounds for all branches of biologic science.

With regard to teaching, assumption of credit for the success of a student has always puzzled me. There simply are no controls in these experiments. How do I know, given a motivated, gifted student, whether I have been a help or hindrance? Nevertheless, having involved myself in the daily scientific lives of my students, I may have guided some of them in directions that attract me and thereby diverted them from a career in biology or chemistry to the love and pursuit of biochemistry and enzymes. These progeny now include illustrious figures in science who have spread this gospel to a widening circle of "grandstudents" and "great-grandstudents."

Finally, even were I forced to agree that my activities in administration, writing, and teaching had a singular quality, I would have to concede further that my discoveries in science did not. Very likely, they would have been made by others soon after. Yet in the last analysis, I will argue that for me it was the research that mattered most, because all my attitudes and activities were shaped by it.

Acknowledgments

To the gifted students, postdoctoral fellows, and research associates (notably LeRoy Bertsch), including many whose work I have not cited here, I am deeply grateful. I also thank the NIH and NSF for 35 years of generous and uninterrupted research support. In preparing this memoir, I have borrowed extensively from *For the Love of Enzymes: The Odyssey of a Biochemist* (Harvard University Press, 1989) and I am indebted to the publishers for their permission to do so. Finally, I dedicate this autobiographical account to the memory of Sylvy for 43 years of love and devotion, for unwavering support for me to start and sustain a career in science, and for a happy family life with our three sons: Roger, Tom, and Ken.

Literature Cited

1. Kornberg, A. 1942. *J. Clin. Invest.* 21:299–308
2. Kornberg, A., Daft, F. S., Sebrell, W. H. 1943. *Science* 98:20–22
3. Hutchings, B. L., Stokstad, E. L. R., Bohonos, N., Slobodkin, N. H. 1944. *Science* 99:371
4. Black, S., Overman, R. S., Elvehjem, C. A., Link, K. P. 1942. *J. Biol. Chem.* 145:137–43
5. Kornberg, A., Daft, F. S., Sebrell, W. H. 1944. *J. Biol. Chem.* 155:193–200
6. Ochoa, S., Mehler, A. H., Kornberg, A. 1947. *J. Biol. Chem.* 167:871–72
7. Kornberg, A., Lindberg, O. 1948. *J. Biol. Chem.* 176:665–77
8. Kornberg, A., Pricer, W. E. 1950. *J. Biol. Chem.* 182:763–77
9. Kornberg, A. 1950. *J. Biol. Chem.* 182:779–93
10. Schrecker, A. W., Kornberg, A. 1950. *J. Biol. Chem.* 182:795–803
11. Kornberg, A. 1957. *Adv. Enzymol.* 18:191–240
12. Watson, J. D., Crick, F. H. C. 1953. *Nature* 171:737–38, 964–67
13. Kornberg, A., Pricer, W. E. 1952. *J. Am. Chem. Soc.* 74:1617
14. Kornberg, A., Pricer, W. E. 1953. *J. Biol. Chem.* 204:345–57
15. Hayaishi, O., Kornberg, A. 1952. *J. Biol. Chem.* 197:717–32
16. Lieberman, I., Kornberg, A. 1954. *J. Biol. Chem.* 207:911–24
17. Kornberg, A., Lieberman, I., Simms, E. 1955. *J. Biol. Chem.* 215:389–402
18. Lieberman, I., Kornberg, A., Simms, E. 1955. *J. Biol. Chem.* 215:403–15
19. Khorana, H. G., Fernandes, J. F., Kornberg, A. 1958. *J. Biol. Chem.* 230:941–48
20. Kornberg, A., Lieberman, I., Simms, E. 1955. *J. Biol. Chem.* 215:417–27
21. Buchanan, J. M., Wilson, D. W. 1953. *Fed. Proc.* 12:646
22. Greenberg, G. R. 1953. *Fed. Proc.* 12:651
23. Kornberg, A. 1980. *DNA Replication.* San Francisco: Freeman. 724 pp.
24. Grunberg-Manago, M., Ochoa, S. 1955. *J. Am. Chem. Soc.* 77:3165–66
25. Littauer, U. Z., Kornberg, A. 1957. *J. Biol. Chem.* 226:1077–92
26. Kornberg, A., Lehman, I. R., Bessman, M., Simms, E. 1956. *Biochem. Biophys. Acta* 21:197–98
27. Kornberg, A. 1957. *The Chemical Basis of Heredity.* ed. W. D. McElroy, B. Glass, pp. 579–608. Baltimore: Johns Hopkins
28. Josse, J., Kaiser, A. D., Kornberg, A. 1961. *J. Biol. Chem.* 236:864–75
29. Schachman, H. K., Adler, J., Radding, C. M., Lehman, I. R., Kornberg, A. 1960. *J. Biol. Chem.* 235:3242–49
30. Radding, C. M., Josse, J., Kornberg, A. 1962. *J. Biol. Chem.* 237:2869–76
31. Radding, C. M., Kornberg, A. 1962. *J. Biol. Chem.* 237:2877–82
32. Kornberg, A., Bertsch, L., Jackson, J. F., Khorana, H. G. 1964. *Proc. Natl. Acad. Sci. USA* 51:315–23
33. Goulian, M., Kornberg, A. 1967. *Proc. Natl. Acad. Sci. USA* 58:1723–30
34. Goulian, M., Kornberg, A., Sinsheim-

er, R. L. 1967. *Proc. Natl. Acad. Sci. USA* 58:2321–28

35. Brutlag, D., Kornberg, A. 1972. *J. Biol. Chem.* 247:241–48
36. Lehman, I. R. 1967. *Annu. Rev. Biochem.* 36:645–68
37. Klett, R. P., Cerami, A., Reich, E. 1968. *Proc. Natl. Acad. Sci. USA* 60:943–50
38. Deutscher, M. P., Kornberg, A. 1969. *J. Biol. Chem.* 244:3029–37
39. Kornberg, A. 1969. *Science* 163:1410–18
40. Kelly, R. B., Atkinson, M. R., Huberman, J. A., Kornberg, A. 1969. *Nature* 224:495–501
41. Editorials. 1971. *Nature New Biol.* 229:65–66; 230:258; 233:97–98
42. De Lucia, P., Cairns, J. 1969. *Nature* 224:1164–66
43. Kornberg, T., Gefter, M. L. 1970. *Biochem. Biophys. Res. Commun.* 40:1348–55
44. Kornberg, T., Gefter, M. L. 1971. *Proc. Natl. Acad. Sci. USA* 68:761–64
45. Gefter, M. L., Hirota, Y., Kornberg, T., Wechsler, J. A., Barnoux, C. 1971. *Proc. Natl. Acad. Sci. USA* 68:3150–53
46. Sugimoto, K., Okazaki, T., Okazaki, R. 1968. *Proc. Natl. Acad. Sci. USA* 60:1356–62
47. Brutlag, D., Schekman, R., Kornberg, A. 1971. *Proc. Natl. Acad. Sci. USA* 68:2826–29
48. Wickner, W., Brutlag, D., Schekman, R., Kornberg, A. 1972. *Proc. Natl. Acad. Sci. USA* 69:965–69
49. Godson, G. M. 1974. *Virology* 58:272–89
50. Zechel, K., Bouché, J.-P., Kornberg, A. 1975. *J. Biol. Chem.* 250:4684–89
51. Weiner, J. H., Bertsch, L. L., Kornberg, A. 1975. *J. Biol. Chem.* 250:1972–80
52. Bouché, J.-P., Rowen, L., Kornberg, A. 1978. *J. Biol. Chem.* 253:765–69
53. Wickner, W., Kornberg, A. 1973. *Proc. Natl. Acad. Sci. USA* 70:3679–83
54. Arai, K., Low, R., Kobori, J., Shlomai, J., Kornberg, A. 1981. *J. Biol. Chem.* 256:5273–80
55. Shlomai, J., Kornberg, A. 1980. *Proc. Natl. Acad. Sci. USA* 77:799–803
56. Lee, M. S., Marians, K. J. 1987. *Proc. Natl. Acad. Sci. USA* 84:8345–49
57. Lasken, R. S., Kornberg, A. 1988. *J. Biol. Chem.* 263:5512–18
58. LeBowitz, J. H., McMacken, R. 1986. *J. Biol. Chem.* 261:4738–78
59. Baker, T. A., Funnell, B. E., Kornberg, A. 1987. *J. Biol. Chem.* 262:6877–85
60. Arai, K., Kornberg, A. 1981. *J. Biol. Chem.* 256:5267–72
61. Arai, K., Kornberg, A. 1981. *Proc. Natl. Acad. Sci. USA* 78:69–73
62. Low, R. L., Arai, K., Kornberg, A. 1981. *Proc. Natl. Acad, Sci. USA* 78:1436–40
63. Eisenberg, S., Kornberg, A. 1979. *J. Biol. Chem.* 254:5328–32
64. Scott, J. F., Kornberg, A. 1978. *J. Biol. Chem.* 253:3292–97
65. Kornberg, A., Scott, J. F., Bertsch, L. L. 1978. *J. Biol. Chem.* 253:3298–304
66. Gilbert, W., Dressler, D. 1968. *Cold Spring Harbor Symp. Quant. Biol.* 33:473–84
67. Kornberg, A. 1982. *1982 Supplement to DNA Replication.* San Francisco: Freeman. 273 pp
68. Maki, H., Maki, S., Kornberg, A. 1988. *J. Biol. Chem.* 263:6570–78
69. McHenry, C. S. 1988. *Annu. Rev. Biochem.* 57:519–50
70. Fuller, R. S., Kaguni, J. M., Kornberg, A. 1981. *Proc. Natl. Acad. Sci. USA* 78:7370–74
71. Kornberg, A. 1950. *J. Biol. Chem.* 182:805–13
72. Funnell, B. E., Baker, T. A., Kornberg, A. 1987. *J. Biol. Chem.* 262:10327–34
73. Bramhill, D., Kornberg, A. 1988. *Cell* 52:743–55
74. Bramhill, D., Kornberg, A. 1988. *Cell.* In press
75. Sekimizu, K., Bramhill, D., Kornberg, A. 1988. *J. Biol. Chem.* 263:7124–30
76. Yung, B. Y., Kornberg, A. 1988. *Proc. Natl. Acad. Sci. USA* 54:915–18
77. Baker, T. A., Kornberg, A. 1988. *Cell* 55:113–23
78. Kornberg, A., Spudich, J. A., Nelson, D. L., Deutscher, M. P. 1968. *Annu. Rev. Biochem.* 37:51–78
79. Kornberg, A., Kornberg, S. R., Simms, E. 1956. *Biochem. Biophys. Acta* 20:215–27
80. Kornberg, A. 1974. *DNA Synthesis.* San Francisco: Freeman. 399 pp
81. Kornberg, A. 1976. *New Engl. J. Med.* 294:1212–16
82. Kornberg, A. 1984. In *Medicine, Science and Society,* ed. K. J. Isselbacher, pp. 6–17. New York: Wiley
83. Kornberg, A. 1987. *Biochemistry* 26:6888–91

Ann. Rev. Biochem. 1988. 57:1–28

SEQUENCES, SEQUENCES, AND SEQUENCES

Frederick Sanger

Retired from Medical Research Council Laboratory of Molecular Biology, Hills Road, Cambridge CB2 2QH, England

CONTENTS

INTRODUCTION

These prefatory chapters are usually accounts of biochemists' experiences in research, teaching, and administration. In my case the last two are easily dealt with as I have done hardly any and have indeed actively tried to avoid both teaching and administrative work. This was partly because I thought I would be no good at them, but also out of selfishness. I do not enjoy them, whereas I find research most enjoyable and rewarding.

Of the three main activities involved in scientific research, thinking, talking, and doing, I much prefer the last and am probably best at it. I am all right at the thinking, but not much good at the talking. "Doing" for a scientist implies doing experiments, and I managed to work in the laboratory as my main occupation from 1940, when I started as a PhD student with Albert Neuberger in Cambridge, until I retired in 1983. Unlike most of my scientific colleagues, I was not academically brilliant. I never won scholarships and would probably not have been able to attend Cambridge University if my parents had not been fairly rich; however, when it came to research where experiments were of paramount importance and fairly narrow specialization was helpful, I managed to hold my own even with the most academically outstanding.

This article is a personal account of my research career, which has been concerned primarily with the development of sequencing methods. Most of the significant work has been summarized in a number of reviews and articles. In these there was, of necessity, a good deal of simplification and omission of detail, both for reasons of space and, sometimes, to make a good and logical story. With the passage of time even I find myself accepting such simplified accounts, with no detail, no failures recorded, and very little reference to the colleagues who have helped me. This article is perhaps an appropriate place to clarify things somewhat and to describe certain lesser known aspects of the work, some of which have never appeared in print.

When I arrived at the University of Cambridge as an undergraduate, I had to decide which three scientific subjects I should take. I had chosen two-and-a-half subjects and was looking through the list of "half" subjects when I noticed one I had not heard of before: "Biochemistry, supervisor Ernest Baldwin." The idea that biology could be explained in terms of chemistry seemed an exciting one, and this was amply confirmed when I met the enthusiastic Dr. Baldwin. My first real academic success was when I stayed on at Cambridge for an extra year (1939) to take the advanced course in Biochemistry. I left Cambridge after taking the final examination with no high hopes of having achieved a good result, so I was very surprised to read in the newspaper two weeks later that I had earned a first class degree. This raised the possibility of my becoming a PhD student and I wrote to the Biochemistry Department at Cambridge enquiring if there was any prospect of this. As I received no reply I decided to visit Cambridge to investigate and found that plenty of research work was going on in spite of the war and that there were several people who would be glad to have a PhD student (especially one who, like me, did not need any money). Furthermore, I was a conscientious objector and had been given exemption from military service. Much of the work in the department centered on enzymes and intermediary metabolism, and the main tools were Warburg manometers and methylene blue tubes. During the advanced course I had been supervised by E. Friedmann, for whom I had developed a great respect. He was a refugee from Germany and a chemist of the old school, and in biochemical studies he believed in isolation and structure and did not have much use for what he called "making nonsense with methylene blue tubes." I think it was partly his influence that made me choose to work with N. W. (Bill) Pirie, who was the protein expert in the lab. Bill's chief interest at that time was in making edible protein from grass. His method of dealing with PhD students was the fairly effective one of "throwing them in at the deep end" and letting them find out things for themselves. He presented me with a large bucket of frozen grass extract and suggested that I investigate it. Unfortunately by the time it had thawed Bill had left the lab for a new job, but not before I had assimilated some biochemical wisdom from him and enjoyed his caustic comments about the shortcomings of other members of the lab. I then needed another supervisor and was fortunate in that Albert Neuberger, who was a postdoctoral fellow in the laboratory, was prepared to take me on. I regard Albert as my main teacher. The most important thing he taught me, both by instruction and by example, was how to do research. I shall always be grateful for his kindness and patience. He also had an extremely wide knowledge of biochemistry, which I admired and used but could never emulate. The subject of my thesis was the "Metabolism of lysine": nothing very profound, but through it I gained much experience in amino acid chemistry, a good introduction to "sequences."

The Department of Biochemistry was an interesting and exciting place in which to work. It had been founded largely for F. G. Hopkins, and most of the staff members were his disciples, but also leaders in their own fields. Biochemistry was then a new subject, and there was plenty going on and a lot of enthusiasm. It seems to me that the quality of research in a laboratory is very much dependent on the atmosphere of the place. I remember being impressed with the friendliness and enthusiasm of everyone with whom I came into contact. The researchers seemed to be interested not only in what they themselves were doing, but also in what I was doing, and they were forever talking about biochemistry with no time wasted on "small talk." I liked this sort of environment and found it stimulating, and I have been fortunate in that both the laboratories in which I have worked have had this kind of atmosphere. The Laboratory of Molecular Biology, to which I moved in 1962, we built up ourselves and it could scarcely have failed to generate excitement and enthusiasm, but that initial impetus seemed to survive at least until I retired, and there were none of the major personal frictions that can have such an adverse effect on the research output of a laboratory.

INSULIN

When I had completed my PhD, in 1943, and Neuberger had left the lab, I was given the opportunity to work with Professor A. C. Chibnall and his group supported by an MRC (Medical Research Council) grant. Chibnall had just been appointed Professor of Biochemistry in succession to Hopkins and was about to move to Cambridge from Imperial College in London with a number of collaborators. Their main research interest at that time was the amino acid analysis of proteins, especially of insulin. Insulin had been chosen both for its medical importance and because it was one of the few proteins that could be bought in a pure form. It was unlikely to have been chosen because of its small size (M_r 6000), as this was not known then, but it was of course very fortunate for our later work that it was so small. At that time amino acid analysis was a painstaking undertaking, and in many cases it was difficult to get accurate results. The analyses obtained by Chibnall's group were probably the best published at that time.

My work on insulin was done about 40 years ago and, because it is associated with the beginning of protein sequencing, has already been described, with varying degrees of accuracy, in a number of reviews and books. Some (e.g. Ref. 1), give a rather romantic impression of me, all alone, resolving to determine the structure of insulin against the advice of my colleagues who believed that proteins were some ill-defined amorphous mixture that could not be studied by chemical methods. This impression, however, is entirely wrong. Although there had previously been papers suggesting this pessimistic view of proteins, it was not one held by those with whom I associated. I think we all followed the lead of Emil Fischer and his school, which depended entirely on a chemical approach to proteins. If anything, the errors were in the opposite direction, ascribing more stoichiometry to proteins than was actually the case. In Britain the chief proponent of stoichiometry was the X-ray crystallographer W. T. Astbury. On the basis of a few spots on an X-ray film he proposed detailed structures for the folding of protein chains, and these were for a while widely accepted. The theory of Bergmann & Niemann (la) was particularly popular. They suggested that residues of a particular amino acid were arranged at regular intervals along polypeptide chains and that the content of each amino acid could be expressed by the formula $2^m \times 3^n$. A useful outcome of this theory was the stimulation of interest in amino acid analysis. People would express their results in this form, and, because of the lack of accuracy of the methods and as most small numbers fit the formula, this was not too difficult. Chibnall's was probably the first group to have results accurate enough to cause the theory to be questioned (2).

So the environment in which I worked was entirely favorable, and my initial experiments were logical continuations of the group's work, and in fact were suggested to me by Chibnall. Had it not been for him I would probably have continued working on lysine metabolism. Chibnall et al (2) had found that the number of free amino groups in insulin was considerably greater than could be accounted for by the lysine content. They suggested that the excess was due to free α-amino groups, indicating that the chains were relatively short and therefore particularly suitable for chemical studies. Chibnall suggested that I try to obtain a quantitative estimation and identification of the amino acids on which these free α-amino groups were located, and that is how my work on insulin started. In fact, the work on insulin could be considered to have started somewhat earlier than that as Neuberger had done some preliminary experiments on the end groups, and probably his interest and experience were passed on to me. There was no question of my setting out to determine the complete sequence of insulin, then or for a considerable time thereafter. Probably the only time that I felt any real pressure to do so was in the final stages when only the disulfide bridges and one amide group remained to be located.

According to reviews that I and others have written, I then developed the use of fluorodinitrobenzene (FDNB) as a general reagent for determining end groups in proteins (3). Although this is so, on looking through my old notebooks I find it was by no means that simple. Several methods had already been suggested for labeling free amino groups, although in only one case was an N-terminal residue identified. In 1935 Jensen & Evans (4) had detected phenylalanine at the N-terminus of insulin, using phenylisocyanate, which is a somewhat less reactive analogue of the Edman reagent (phenylisothiocyanate) (5). This phenylalanine residue was, I think, the first amino acid to be located in any protein. For a general method what we needed was a reagent that would react under mild conditions with amino groups to form a derivative that was stable to acid hydrolysis. It was also important to have a relatively simple method for the separation and identification of the substituted amino acids. For this purpose we were anxious to use the newly discovered partition chromatography (6). It was a much better fractionation procedure than any that had been used previously. It worked well with the acetyl amino acids (6) and was in fact being used in the group by G. R. Tristram.

Besides the isocyanates the main possibilities were sulfonyl chlorides (7) or substituted benzenes (8). Benzene sulfonyl chloride had been used and was readily available, but the products did not seem to have very suitable solubility characteristics. The first reagent I tried was methane sulfonyl chloride. The main reason for this choice was that I expected the methane sulfonyl amino acids to behave similarly to the acetyl derivatives on partition chromatograms. This they did, but I was unable to get any clear results and we next turned to the substituted benzenes. 2,4 dinitrochlorobenzene had been studied by Abderhalden & Stix in 1923 (8); however it only reacted at high temperatures, where there was some hydrolysis of peptide bonds. I used it to make some dinitrophenyl (DNP) amino acids, and found they could be fractionated well on silica gel partition chromatograms. One great advantage was that the DNP derivatives were colored yellow. At that time chromatography really was "chroma"tography: there were no really reliable fraction collectors so that it was important to be able to see the bands on the columns. Clearly the chloro compound was not reactive enough, but the corresponding fluoro or nitro compounds could be expected to be more reactive. I made a small amount of the 1,2,4 trinitrobenzene and Chibnall discovered that Dr. B. C. Saunders, who worked in the nearby Chemical Laboratories, had a supply of the fluoro compound, and he kindly let us have some. Both reacted satisfactorily in the cold. We had a good supply of the FDNB and so it was used in further work. The first experiments were, however, somewhat discouraging. I tried out the reaction of FDNB with glycine, but when I applied the product to a column, instead of the expected single band of DNP-glycine, I obtained two bands in significant amounts. It seemed that the second product was diDNP-glycine, with two DNP groups on the one N atom. I had considerable difficulties getting over this problem, but it turned out that glycine was an exception: other amino acids did not give so much of the second product and a change to bicarbonate from phosphate buffer almost completely eliminated the effect.

When I applied the method to insulin, DNP-phenylalanine and DNP-glycine were identified (3). There was one residue of each per M_r 6000 (although at the time we thought the M_r was 12,000). This suggested that there were two types of chain. These were joined by disulfide cross links of the cystine residues, which could be broken by performic acid oxidation (9), and it was possible to isolate two fractions that corresponded to the two chains (10).

At this time (1947) I had an opportunity to visit the laboratory of A. Tiselius in Uppsala, Sweden. Tiselius was one of the best-known names in proteins as he had invented the analytical electrophoresis method that bore his name and was very widely used. I had shown that the cross-linking disulfide bonds could be broken by oxidation and was trying to find a way to fractionate the chains. There were really no good methods for products of that size, but Tiselius's colleagues were working on a chromatographic system based on adsorption on charcoal which looked rather hopeful. It was suggested that I spend some time working on this in Uppsala. It seemed an exciting opportunity to me, especially as England was still in the postwar period of austerity and Sweden was, comparatively, a land of plenty.

The experiments I carried out with the charcoal columns were not very satisfactory. Detection of the bands coming out of the bottom of the columns was by a rather sophisticated method depending on refractive index: one had to look through an eyepiece observing "fringes" whose positions depended on refractive index. The trouble was the fringes would disappear when the refractive index changed and only reappear in different positions when the refactive index settled down again. One thus got rather few points on the plot. The custom in the lab was to have skilled technical assistants do the practical work, and there was one technician who was considered particularly good at reading these columns, so I got her to do some readings for me. A frontal analysis method was used, so that each product should show up as a step on the plot. The technician did several experiments, and one time the plot had four sharp steps. I was quite excited by this because at the time we thought the molecular weight of insulin was 12,000 and that there were four chains. I showed this to Tiselius, and he immediately suggested we send a letter to *Nature* under both his and my name and include all the preliminary work on oxidation of the chains that I had already done in Cambridge. I was rather shocked as he had not really contributed anything and I had never seen him working in the lab. I managed to persuade him that I was not quite ready to publish the oxidation work, but being his guest and very much his junior I did not protest too hard about publishing the one promising result we had. The resultant paper (11) is fortunately one of my lesser known ones, and it is the only one of which I am ashamed. There are of course only two chains in insulin. I think the four sharp steps were due to the expert technician's way of plotting her readings; if the fringes disappeared she would assume there was no change and only record a higher reading when they reappeared, thus producing artificial steps. Tiselius was really a very kind and charming person, and I am certainly grateful for all the support he gave me both then and later, but this experience did make me realize how lucky I was to have worked with Chibnall, who had allowed me to publish my paper on the amino groups of insulin by myself even though he had actually initiated the work and could justifiably have put his name on it. One bonus from my visit to Sweden was meeting R. L. M. (Dick) Synge, who was working there and who introduced me to zone ionophoresis on starch, which was the forerunner of other more successful ionophoreses such as paper or acrylamide gel.

Complete acid hydrolysis of DNP insulin yielded DNP amino acids, but if the conditions of hydrolysis were reduced partial hydrolysis resulted and DNP peptides were obtained. By studying the DNP peptides from insulin I was able to identify short sequences of four to five amino acids near the N-terminus (12). The fact that only two sequences were obtained made it fairly certain that there were only two types of chains, and, what was probably more important, showed that specific sequences could be determined in proteins by partial hydrolysis, thus eliminating any doubts that they were pure stoichiometric compounds that could be studied by the methods of organic chemistry. In most subsequent work, both on proteins and nucleic acids, partial hydrolysis has been the main general approach to all sequencing, and it is only relatively recently in DNA sequencing that it has been largely superseded. Although my first paper on the N-terminal residues (3) has been more widely quoted (13), I was myself more excited about this peptide paper (12), and indeed in retrospect the results seem to me to be of greater significance. It was the first time that sequences were located in a protein molecule, and we could draw important conclusions about stoichiometry. Also it was done on my own initiative, whereas the first paper relied more on Chibnall's stimulus.

Apart from my two teachers, Neuberger and Chibnall, my first actual collaborator was Rodney Porter. Although a little older than I was, he was officially my PhD student. This was because he had been in the army during the war, while I had been working in the lab. Rodney was not the sort of person one could supervise, but he was very good as a collaborator and I think I learned more from him than he did from me. Probably the main thing he taught me was a certain lighthearted attitude toward research, which makes it more enjoyable and, provided standards of complete integrity and honesty are maintained, can do no harm. He joined me just after the development of the DNP method, when we were extending it to peptides and to other proteins. Rodney had just developed an interest in γ-globulin and wanted to study its end groups. I warned him that it was a complex mixture and that it was not worth trying to determine its end groups. However, this did not stop him, and to everybody's surprise he got a sensible, simple result; and from then on he was inseparable from γ-globulin, to the great benefit of science.

During this period we were consigned to a lab in the basement next to the experimental rats. In spite of these rather unsavory surroundings, it was probably the most enjoyable lab I have worked in. We shared it with Kenneth Bailey and his PhD student, S. V. (Sam) Perry, who had also been in the army and had been at the University of Liverpool with Porter. Thus Bailey and myself, two reserved and serious-minded scientists, were thrown together with these two wild boys fresh from the army. It certainly kept us awake. Both had vocal tendencies: Perry's favorite tune was "Mighty mountain" rendered after the manner of Paul Robeson, and Porter's was "Nobody knows the trouble I've seen." It was a good atmosphere and at the same time we accomplished quite a lot. The only serious black spot was when a flask exploded in front of Rodney's eye and damaged the cornea. Although the sight was not completely lost, I think he never had good vision in that eye.

Although the DNP method was an improvement on previous ones and important results were obtained using it, there were certain disadvantages, the most serious being that the DNP amino acids were not completely stable in acid and a correction factor had to be used. This factor differed with the various amino acids and for some (e.g. glycine) was fairly large; in fact in the case of hemoglobin this problem led to an error in the estimation of the free amino groups. I was therefore still interested in trying to develop a better method. The sulfonamide (SO_2NH) bond is considerably more stable than the DNP, so that it seemed a sulfonyl chloride might be an improvement, and I was anxious to try a more highly colored material to increase the sensitivity. The reagent on which I spent quite a lot of time was dimethylaminophenyl-azo-benzene sulfonyl chloride (or helianthyl chloride). The helianthyl amino acids gave beautiful red bands on the chromatograms and appeared to be stable to acid, but we did not have much success when we tried this reagent with proteins. There were probably some undesirable side reactions with the azo group. There were also complaints from other occupants of the lab when their biological preparations all became bright red. Eventually we gave it up, but more success was obtained later by Hartley & Gray (14) with another sulfonyl chloride, diamino naphthalene sulfonyl chloride (dansyl chloride). The dansyl amino acids were satisfactorily stable and their strong fluorescence made them easy to detect and assay at low concentrations.

The DNP method was fairly widely used after its development, but is now completely superseded by other methods, especially the Edman procedure (5), which has the advantage that it can be used not only for determining N-terminal residues but also for extensive sequential degradations of polypeptide chains. It is interesting that the Edman method was developed about the same time as the DNP method but at first was not used much. The main reason for this was probably the color of the DNP derivatives, which made them much easier to handle at a time when reliable fraction collectors and micro methods of assay were not available.

For sequencing purposes the DNP method was very limited as it could only be used for a few residues at the N-terminus, and the more general approach of subjecting a whole chain to partial hydrolysis seemed necessary to extend the sequence. The main technical problem was fractionation of the complex mixture of peptides produced. Indeed in almost all sequencing work fractionation has been a crucial factor, and progress in sequencing has often depended on progress in fractionation methods. With the insulin chains we were again lucky in that the method of paper chromatography had just been developed by Martin and his colleagues (15), and had been used to determine the pentapeptide sequence of gramicidin S (16). The fractionation of small peptides was far superior to anything that had been achieved previously, and it seemed something of a miracle at the time to see these hitherto intractable products separated from one another on a simple sheet of filter paper. Using this method we were able to isolate small peptides from acid hydrolysates, some of which could be fitted together to deduce larger sequences (17, 18). However, with acid hydrolysates alone we could not obtain the complete sequence of either chain, and what we clearly needed were longer peptides. A more specific method of hydrolysis was required and the obvious choice was proteolytic enzymes. Hitherto we had been reluctant to use enzymes because it was rather generally thought that they might synthesize bonds as well as hydrolyze them, and thus sequences could be rearranged. Enzyme reactions were known to be reversible, and a good deal of work was being done on the synthetic activity of proteolytic enzymes, since it was considered that this might be involved in protein biosynthesis. In spite of these warnings, we decided to risk it and to try using proteolytic enzymes, and we were able to obtain the complete sequence of both chains (19, 20). Knowing that we already had much of the sequence from the acid hydrolysis results and that there were no anomalies between the two methods convinced us that there had been no rearrangement. In all subsequent work proteolytic hydrolysis has been the method of choice, and in fact rearrangements have been shown to occur during acid hydrolysis (21).

The work on the sequencing of the two chains owes much to my two collaborators—Tuppy and Thompson. Hans Tuppy was from Vienna and was one of the hardest workers I have known. Just after the war there were only limited facilities for research in Austria, and he was delighted to come to a well-equipped lab where there was no limitation on the amount of work he could do. We were at the time collecting peptides from acid hydrolysates of the chains of insulin. He took the longer phenylalanyl chain, and I worked, among other things, on the glycyl chain. Much of the work required the use of paper chromatography; the chromatography room was at the other end of the basement corridor, and a common sight at that time was Tuppy walking at full speed along that corridor bearing chromatograms. He never ran, but to keep up with him anyone else had to run. He only stayed a year, but by the end of that year the phenylalanyl chain was practically finished and the glycyl chain had to wait for my next collaborator, E. O. P. (Ted) Thompson from Australia. I have worked with several Australians and have always found them easy to collaborate with. They seem to have a certain practical "no nonsense" view of research that takes the attitude, "If there's an experiment to be done, get on and do it." Ted was no exception and we had great fun working together. Among other things we finished up the glycyl chain, which I think really was harder than the longer phenylalanyl chain, and his persistence in tracking down the final amide group was ultimately rewarded.

Once we had determined the complete sequence of the two chains, the only remaining problem was the arrangement of the disulfide bridges. This proved particularly difficult and probably involved as many man-hours, and more frustrations, than the rest of the work put together. The idea was to hydrolyze the intact insulin, isolate cystine peptides, oxidize them, and see which cysteic acid peptides were produced. The results were, however, very equivocal, and it eventually transpired that there was an interchange reaction taking place at the disulfide bonds during acid hydrolysis (21a). One bond between the two chains could be identified using enzymic digestion, but the other two occurred in a Cys-Cys sequence in the glycyl chain, and no enzyme would split the two half-cystine residues. Clearly, in this respect insulin had been a bad protein to choose; hardly any other protein would have given us this problem. It was quite a long time before we found conditions for avoiding this interchange reaction during acid hydrolysis and were able to complete the structure (22).

Insulin was the first protein to be sequenced (22, 23), but its chains were relatively short, and it is doubtful if the methods used would have succeeded with larger, more normal, chains. Further fractionation methods were needed and the most important was that introduced by Moore & Stein (24, 25): ion-exchange chromatograms to separate the large peptides from proteolytic digests. They also used ion-exchange columns to develop an accurate method of amino acid analysis, and this, though considerably modified and improved in detail, is still the method of choice today. Using these techniques they were able to determine the complete sequence of ribonuclease (120 amino acids) (25), and subsequent work on protein sequence has more closely followed their techniques than the less quantitative methods employed for insulin.

THE LEAN YEARS

After completion of the insulin sequence around 1955, there followed a number of "lean years" when there were no major successes. I think these periods occur in most people's research careers and can be depressing and sometimes lead to disillusion. I have found that the best antidote is to keep looking ahead. When an experiment is a complete failure it is best not to spend too much time worrying about it but rather to get on with planning and becoming involved in the next one. This is always exciting and you soon forget your troubles. One disadvantage of this attitude, though, is that it makes the writing of an article of this sort difficult, because if you are always thinking about the future you tend to forget the past.

Toward the end of the insulin work people would ask me what I would do after I finished insulin. My usual reply was that knowing the structure was only a beginning and that the next thing was to find out what it meant and how the insulin worked. I have always been rather unsuccessful in predicting the future, especially in my own research, and although I did try to do some experiments toward this end, my interest began to go in other directions, and perhaps the sequencing bug had already taken hold of me.

During these "lean years" there were a number of events that are worth recording as they had an important effect on the subsequent work on nucleic acids, although they may not have led to publishable results. The first was initiated by the visit of Chris Anfinsen on sabbatical leave to our laboratory in 1954. Chris had been using, and was very enthusiastic about, the new technique of labeling with radioactive isotopes, and in discussion with him I caught some of his enthusiasm. Previously I had assumed that isotopes were in the realm of the physicists and that the apparatus and techniques would be beyond my means. But I learned that this was not the case and that already a number of radioactive substrates were available. Together we labeled a rat with ^{35}S in an attempt to prepare radioactive insulin. The experiment was not a success, but nevertheless I became a complete convert to the use of radioactive isotopes and to the autoradiographic technique, which was a very powerful and simple tool especially when used in conjunction with paper fractionation methods (paper chromatography and paper ionophoresis). After the introduction of ion-exchange chromatography for fractionating peptides (24), paper methods became less fashionable and were largely replaced by columns and fraction collectors, which were more accurate and reproducible. I was, however, never really converted to columns and retained a certain affection for the paper methods. It seemed to me much more exciting to develop a two-dimensional paper separation and to see all the spots spread out before one than to collect innumerable fractions from a column, analyze them all, and end up with what was usually the expected clear, clean data. There was more, if less clean, data on the paper, and hence more opportunity for unexpected and exciting findings. Although my preference for paper methods was to some extent sentimental, I think it was scientifically justified if only because of its subsequent successful application to RNA.

Somewhat more successful than the ^{35}S-labeling were later experiments with ^{32}P. Some proteins, notably ovalbumin (26), could be labeled with [^{32}P]-phosphate, and some enzymes, such as trypsin and chymotrypsin (27, 28), could be labeled in their active centers with [^{32}P]-diisopropyl phospho-fluoridate (DFP). In this way just one or two residues in a whole protein were labeled, and if a partial hydrolysate were fractionated on a two-dimensional paper system and autoradiographed a relatively simple picture was obtained. I was particularly interested in developing methods for deducing the sequence of peptides without resorting to the standard procedure of carrying out complete amino acid analyses, which were somewhat time-consuming and tedious. The ^{32}P-labeled proteins were particularly suitable for trying this type of approach. For instance, information could be obtained by comparing autoradiographs of ionophoreses of partial acid hydrolysates of ^{32}P-labeled proteins. In this way it was shown that the sequence around the reactive serine in elastase was the same as that in trypsin and chymotrypsin (27). Each labeled residue gave a specific pattern of bands or "fingerprint," and techniques were worked out for identifying the sequences by autoradiographic methods without carrying out an amino acid analysis. The main rationale behind this approach was that the radioactive peptides were usually heavily contaminated with nonradioactive substances, but they were radiochemically pure and thus the autoradiographs gave a simpler and more interpretable fingerprint. The relationships of the different bands in a fingerprint could be determined by subjecting them to a further partial hydrolysis. Information about charged amino acids was obtained from the ionophoretic behavior of the peptide bands, and about neutral amino acids from chromatographic behavior. Treatment with phenylisothiocyanate or with periodate gave information about the free amino groups. Using this type of method it was often possible to determine a sequence of four or five amino acids around a specific labeled residue (27–31). Particularly rewarding at this time was a conversation I had with Cesar Milstein, who was working at the other end of the corridor on phosphoglucomutase. This led to a collaboration in which we labeled its active center with ^{32}P and determined a sequence of five amino acids (30). Although this was our only collaboration, my continued association with him has meant a great deal to me.

Even more ambitious but less successful was a means I devised for determining the complete sequence of a protein by autoradiographic methods. The idea was to prepare 20 samples of the protein, each labeled with a different amino acid. Ovalbumin was chosen to study as it was relatively easy to label in isolated oviduct tissue. The samples were then digested with a proteolytic enzyme and the digests fractionated in parallel by paper electrophoresis and an autoradiograph prepared. Assuming that a peptide was pure, its amino acid composition could be deduced from the darkness (radioactivity) of its band in the separate channels. The whole paper was then to be subjected to the Edman procedure and washed to remove the phenyl-thiohydrantoin, and another autoradiograph was to be prepared. The N-terminal residue of each peptide should then be identified as the band missing or weaker in the second autoradiograph. By doing further Edman degradation on the paper, further sequences of the individual peptides were to be determined. The idea was certainly rather wild, and it did not show much sign of working. However, if it had worked it would have been exciting and would have been an alternative to the standard partial hydrolysis approach, and it might have been quicker and simpler to carry out. We spent quite a lot of time working on this idea and the question arises whether this was really justified. At that time I was in the fortunate position of having a permanent research appointment with the (British) Medical Research Council, and was not under the usual obligation of having to produce a regular output of publishable material, with the result that I could afford to attack problems that were more "way out" and longer term: in fact, as few others could adopt this approach, I felt under some obligation to do so. This type of work also suited my temperament. I like the idea of doing something that nobody else is doing rather than racing to be the first to complete a project, and I prefer to concentrate more on the practical work itself than on its ultimate outcome. Although this work was unsuccessful, the proposed method bears some resemblance to the DNA sequencing methods we finally developed, and so probably our efforts with proteins were worthwhile, and they give some indication of how our ideas about sequencing were developing. My only regret about this work is that it seemed rather unfair to my excellent assistant, Biddy Segall, who devoted so much time and skill to this project but whose name never appeared on a paper.

During most of the work on insulin I was supported by a Beit Fellowship; this lasted seven years (a Junior Beit, a fourth year, and a Senior Beit), and I was certainly very grateful for this continued support, which allowed me to devote all my time to research in the Biochemistry Laboratory. Thereafter I was supported by the Medical Research Council—initially as an external member of the scientific staff. In 1962 I moved to the new MRC Laboratory of Molecular Biology where I joined forces with the group led by Max Perutz from the Cavendish (Physics) Laboratory; this included, among others, Francis Crick, Sydney Brenner, and John Smith. The new laboratory was divided into three more or less self-governing divisions, with Max as Chairman. One division (the Division of Protein Chemistry, which later became the Division of Protein and Nucleic Acid Chemistry) was allotted to my group. There was much more space than I had ever had previously, which meant that I could expand considerably. As I wanted to continue doing experimental work, I had no ambition to have a large team working under my direction, but this gave me a chance to bring together a number of people with similar interests in protein chemistry to make a viable group—a few collaborating directly with me and others either working independently or with a more senior member of the group. Initially the senior members were Ieuan Harris and Brian Hartley, who had been fairly closely associated with me in the Biochemistry Department, and each had his own group. We were later joined by Cesar Milstein. The laboratory rapidly acquired a good reputation so that we were able to attract plenty of good postdoctorals and PhD students, which greatly helped in maintaining the enthusiasm and momentum of the lab. At first there was plenty of room and everyone had a large bay to himself, which seemed a great luxury; however this did not last long and soon the laboratory was as overcrowded as most. In fact the work seemed to go better in the overcrowded conditions.

I think the move to the new laboratory was probably an influence in my conversion to nucleic acid sequencing. Previously I had not had much interest in nucleic acids. I used to go to Gordon Conferences on Protein and Nucleic Acids when the two subjects were bracketed together, and would sit through the nucleic acid talks waiting to get back to proteins. However, with people like Francis Crick around, it was difficult to ignore nucleic acids or to fail to realize the importance of sequencing them. An even more seminal influence was John Smith, who was the nucleic acid expert in the new laboratory and who was extremely helpful to me, so that I could turn to him for advice in this new field. He might well have felt I was "muscling in" on his area, but he did not, and was always ready to give me the benefit of his long experience.

RNA

Initially the sequencing of nucleic acids seemed much more difficult than that of proteins, and little progress had been made until the early 1960s. This was partly because of the lack of pure small substrates on which to develop methods, and partly because of their composition. For nucleic acids with only the four monomers, the interpretation of results would be expected to be more difficult than for proteins with their 20 amino acids, and larger degradation products would have to be isolated to give significant overlaps for sequence deduction. On the other hand, only having four components would make the final analyses much easier. To begin with the difficulties of interpretation were predominant, but as the techniques have improved and larger molecules have been studied the question of analysis has become more important, so that today nucleic acid sequencing is much quicker and simpler than protein sequencing. Whereas for proteins much of the time is spent in identifying and estimating the amino acid end products, it has now been possible to evolve methods for DNA in which this final analytical step is completely eliminated.

The first nucleic acid to be sequenced was in fact the first small RNA to be purified. This was the alanine transfer RNA, which was purified by Holley and his collaborators (32). They used sequencing methods that were very similar to those in use with proteins: partial hydrolysis with enzymes, and fractionation of the products on ion exchange columns. They introduced the use of the enzyme ribonuclease T_1, which, because of its unique specificity for splitting after G^1 residues, has been used as the main degradative method in most subsequent work.

Our own early efforts in RNA sequencing were directed toward the development of more rapid and simpler fractionation techniques and were influenced by our recent experience with ^{32}P-labeled proteins. Since every nucleotide contains a phosphorous atom, RNA is particularly suitable for ^{32}P-labeling in vivo, making very sensitive detection by autoradiography possible. The use of columns for fractionation was somewhat laborious, and it seemed that the simplicity and resolving power of two-dimensional techniques would offer considerable advantages for separating the complex mixtures of small nucleotides that one would obtain from partial degradation of larger RNAs. In general, oligonucleotides did not separate well by simple paper techniques (partition chromatography or ionophoresis), but we were able to develop a two-dimensional method that used ionophoresis on cellulose acetate followed by ionophoresis on ion exchange paper, and this proved considerably better than any of the previous methods (33).

I am sometimes asked such questions as "What was the most exciting moment of your scientific career?" or "What did it feel like when you discovered. . . .?" Such questions are usually prompted by the popular idea

---

[1]Where there is no possible ambiguity in the text the terms A, G, C are used for both the ribo and deoxyribonucleotides; otherwise rA, rG, rC, U are used for the ribo and dA, dG, dC, T for the deoxyribonucleotides.

that scientific progress depends on sudden breakthroughs or moments of inspiration (cf Archimedes jumping out of his bath, or Watson and Crick discovering the double helix). For me, however, progress does not seem to go like that, and I find it hard to remember moments of sudden exhilaration. Although there were such moments, they were probably more often associated with small and gradual advances; and these were more numerous and therefore gave more enjoyment than would have been the case if there had been sudden big leaps forward. However, I do have a clear recollection of one occasion when Bart Barrell, who usually developed the day's autoradiographs first thing in the morning, came into my lab brandishing a beautiful sheet of film with clear, round, well-separated spots on it. This was certainly exciting after the streaky, unresolved pictures we had been getting before.

Figure 1 *(left)* shows a diagram of an autoradiograph of a ribonuclease T_1 digest on ribosomal RNA. In such a digest all products have G at their 3' ends; there are three dinucleotides and nine trinucleotides, and most of these occupy unique positions so that in subsequent experiments 10 oligonucleotides could be identified simply by their position. This avoided a good deal of final analysis, and was a first step toward our ambition of eliminating the analytical step in nucleic acid sequencing. Furthermore, the composition of larger nucleotides could be determined from the "graticules," shown in the righthand side drawing. All fragments contain one G. The residue that has the greatest effect on the mobility is U, so that the pattern is divided into separate, distinct groups of spots; all oligonucleotides with no U are in the fastest

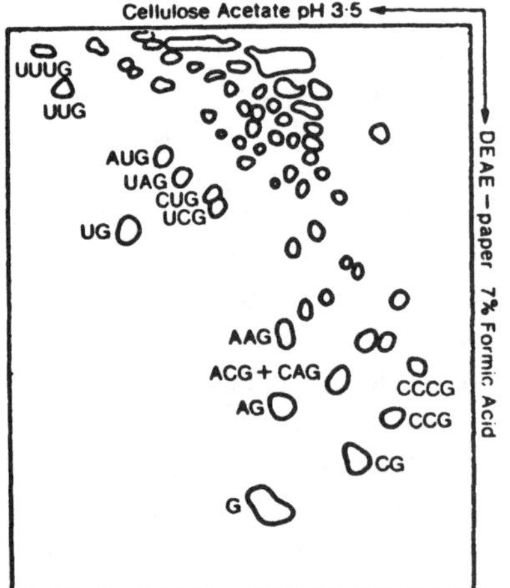

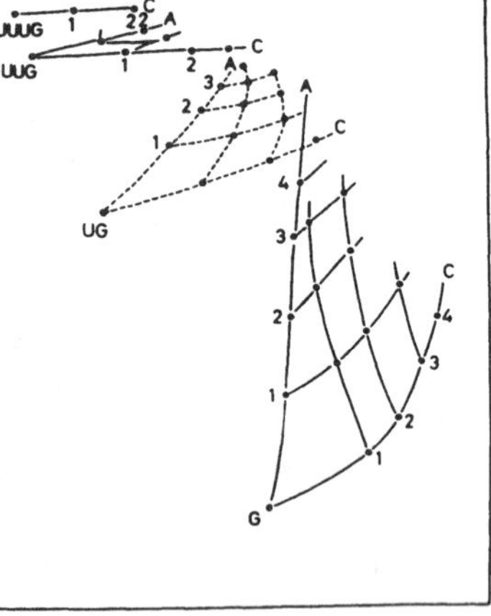

Figure 1 *Left:* Drawing of autoradiograph of two-dimensional separation of ribonuclease T_1 digest of ribosomal RNA. *Right:* Graticules showing the relation of the composition of an oligonucleotide to its position on the autoradiograph.

moving group, those with one U in the next, and so on. Within the groups it is possible to draw a graph or graticule in which one axis gives the number of C residues and the other the number of A residues.

Following this general approach we developed a number of methods for studying the isolated nucleotides. This was in a way an extension of the work on ^{32}P-labeled peptides, though much easier for RNA with only the four components to be considered. One such method that was used with oligonucleotides from pancreatic or T_1 ribonuclease digests was to subject them to partial digestion with a 5' exonuclease and to run the products on ionophoresis on DEAE-cellulose paper at pH 1.9. Sequential degradation from the 5' end will give a mixture of fragments all with the same 3' end but differing in their 5' ends. On the ionophoresis these will be arranged in size order. From the relative positions of two adjacent bands it was possible to identify the nature of the nucleotide by which they differed. Figure 2 shows an example of this approach with the oligonucleotide GAAGC from a pancreatic ribonuclease digest. We defined a value M as the distance between two adjacent oligonucleotides, x and y, divided by the distance of x (the larger) from the origin. If the two differ by an A the M value is 0.5–1.0, whereas in the case of G it is 1.2–3.1. In Figure 2 a is the unchanged material and b the

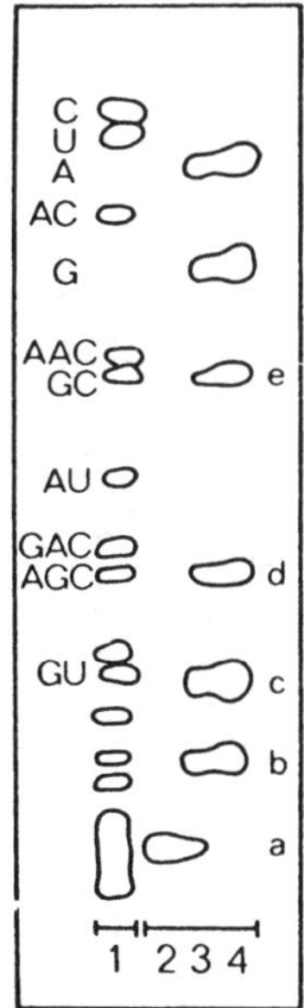

Figure 2 Sequence determination by partial digestion with 5' exonuclease of the oligonucleotide GAAAGC. 1. A pancreatic digest of ribosomal RNA used as marker. 2. Untreated GAAAGC. 3. and 4. GAAAGC digested with 5' exonuclease for 30 and 60 min, respectively.

first degradation product. The M value is 1.9 and therefore the first residue must be G. The M value for *b* and *c* is 0.55, so the second residue is A, and so is the third. Bands *d* and *e* are recognized as AGC and GC by their unique position relative to known markers.

Another useful method was the "wandering spot" technique, which was also sometimes known as "walking the graticule." This was essentially a two-dimensional extension of the above technique and was used mainly with DNA. It was first worked out by Victor Ling to sequence large depurination products (34). A two-dimensional system was devised in which the fragments from an exonuclease digest were arranged according to size, so that each spot differed from the one next to it by a single nucleotide. The system was also arranged to that the relative positions of two neighboring spots depended on the nucleotide by which they differed, as shown in the inset in Figure 3. The method was extended for use with more complex digests, but it was not possible to distinguish A and G with absolute certainty. Figure 3 shows the application of the method to a dodecamer that contained only C, T, and A. The diagram on the left shows a 3' exonuclease digest from which the sequence TTACCATT can be read, and that on the right a 5' exonuclease digest giving the sequence CTTATTAC. Combining the results gives the full sequence CTTATTACCATT.

A suitable substrate for trying out our new RNA sequencing strategies was the 5S ribosomal RNA of 120 residues, and this was completely sequenced mainly by G. G. Brownlee; at that time it was the largest RNA to have been

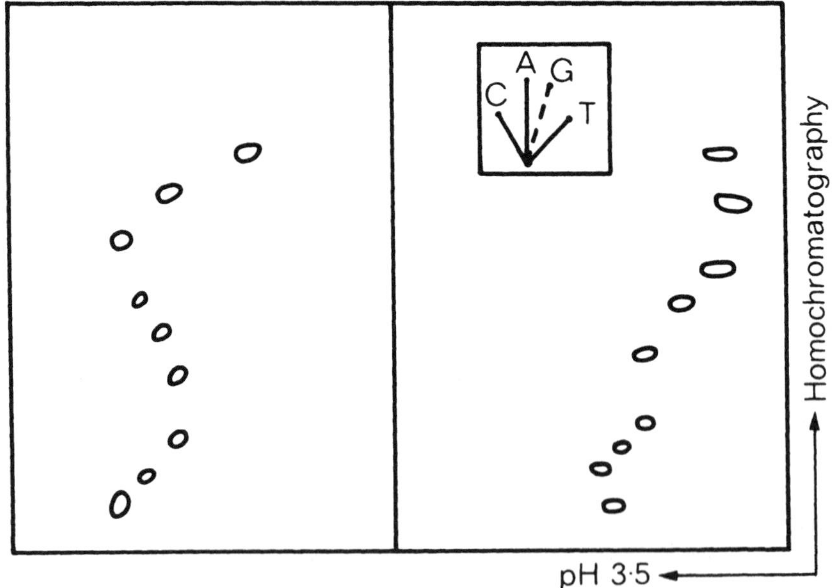

Figure 3 The "wandering spot" method applied to the oligonucleotide CTTATTACCATT.

sequenced (35). We were also able to determine sequences in the RNA bacteriophage R17 that were coding for a part of the head protein whose amino acid sequence was known (36, 37). When we started sequencing RNA one of the main objectives was to try to "break the genetic code." However this was done by entirely different in vitro translation techniques before the sequencing methods were sufficiently advanced. Nevertheless the results with R17 were a useful confirmation of its correctness.

During the work on nucleic acids I had more collaborators than previously and I have been extremely fortunate in the people who have worked with me. Many have been American postdoctorals. We used to get rather many applications, and it was always difficult to choose from such a glittering group. The main problem is that it is not only academic brilliance you are looking for. For a successful collaboration one of the most important things is to have someone you like and can get on well with, and references are often not helpful in this respect. When I was considering John Donelson, one of the referees had added a postscript saying "I think you will enjoy his sense of humour." This proved to be the case and I found the comment helpful. For successful collaborative work there must be complete understanding and trust, and a sharing of ideas and resources. If frictions arise then usually the work suffers. Usually we have had a good atmosphere in the group, but at one time when we were starting our work on DNA we had a relatively large number of people and I noticed that a certain amount of rivalry and bad feeling seemed to be building up. This worried me rather as I did not know how to deal with it, never having had the problem before, so I largely "acted ignorant" and took no notice of it. I think this was probably the best policy; by assuming that they were all friends they were shamed into recognizing the stupidity of their behavior.

The two people to whom I owe most during the nucleic acid work are my two personal assistants, B. G. (Bart) Barrell and Alan Coulson. At about the time I was changing from proteins to RNA I was looking for a new assistant and there were very few applications. The only serious possibility was B. G. Barrell, a young fellow fresh from school. He did not have the academic qualifications we were looking for, but he had plenty of enthusiasm. He came for an interview; I took to him and decided to risk it. He already knew a surprising amount about molecular biology, which he had clearly found out for himself—probably at the expense of his more orthodox school work. He came from Gloucestershire, which was also my own home county, and spoke with a nice thick Gloucestershire accent. He soon became a first-class sequencer and subsequently took charge of his own group. He gradually took over more and more of the nucleic acid sequencing work, especially the actual determinations, while I concentrated more on methods. This meant that I needed a new assistant, and I was fortunate to find Alan Coulson, who was my main collaborator in the later DNA work, and who in his own quiet way contributed to every aspect of it. Hitherto I had preferred working with the more outgoing, extroverted types whose personalities were in marked contrast to my own. I suppose I admired them and found them stimulating. With Alan it was different; we were both quiet by nature. Perhaps as I became more mature I felt less need for external stimulation, though there were always plenty of noisy people around in the lab.

DNA

The first attack on DNA in our group was made in the mid-1960s by Kenneth Murray, who developed some two-dimensional systems for fractionating and analyzing small oligodeoxyribonucleotides (38). Small sequences could be obtained, but at that time there was no way of locating them in the chains. The main problem with DNA was the very large size: the smallest pure DNAs that were available were the genomes of the single-stranded bacteriophages (such as øX174, which will be referred to as øX) of about 5000 nucleotides, and these were rather large for testing out methods. Another difficulty was the absence of suitable degradative enzymes. The sequencing of RNA depended very much on the use of ribonuclease T_1 with its specificity for one residue (G). No enzyme with an analogous specificity exists for DNA. The restriction enzymes, which have since played a very important part in DNA sequencing, had not yet been developed. We did however, around 1973, carry out some experiments using similar techniques to those employed for RNA and were able to determine a few sequences of about 50 residues (39–42). However, the methods were both slow and laborious, and it seemed that if we were really going to be able to attack the vast sequences of genetic materials, then an entirely new approach was needed. As an alternative to partial hydrolysis, some work had been done on the possibility of using copying techniques for sequencing, particularly by C. Weissmann and his colleagues (43). The RNA bacteriophage $Q\beta$ contains an RNA polymerase that copies its own RNA, and they had developed elegant pulse-labeling techniques for the RNA and for deducing sequences. An obvious enzyme to choose for copying DNA was DNA polymerase, which had already been used by Wu & Kaiser in 1969 (44) to determine the sequence of the "sticky ends" of bacteriophage λ. This dodecanucleotide sequence was in fact the first bit of DNA to be sequenced (45).

An elegant approach to specific digestion of DNA, which could be combined with a copying procedure, was suggested by Berg, Fancher, & Chamberlin in 1963 (46). In normal conditions the only substrates for DNA polymerase are the deoxyribonucleoside triphosphates, but if magnesium is replaced by manganese in the medium, ribonucleoside triphosphates can be used. Thus, for instance, if incubation is with one ribonucleoside triphosphate (say rCTP) and three deoxyribonucleoside triphosphates, a DNA chain is built up with all the dCs replaced by rCs. These bonds are labile to alkali or ribonuclease, and so it was possible to prepare a C-specific digest, and similar digests to split at other residues. The method looked very promising, and we did a good deal of work on it. DNA polymerase requires a single-stranded DNA as template and a primer that is another piece of single-stranded DNA that base-pairs with a specific region on the template. As the template we could use the single-stranded DNA of bacteriophage f1. Our intention was to use as primer an oligodeoxyribonucleotide having a sequence that could be predicted from the known amino acid sequence of the coat protein of the bacteriophage. The main question was how to make such a primer. At the time Khorana and his colleagues were developing methods for synthesizing oligodeoxyribonucleotides, but the methods were rather slow and specialized and we had no experience. Fortunately when attending a meeting at about that time I met Hans Kössel from the University of Freiburg, who had worked in Khorana's lab. I found that he had the same idea and was thinking of synthesizing the same primer, so we decided to collaborate. It took him and D. Fischer more than a year of hard work to make the octanucleotide primer. If we had waited a year or two restriction enzymes would have been available and we could have used a restriction enzyme fragment as primer. The ribo-substitution method worked reasonably well and we were able to de-termine a sequence of about 80 nucleotides (39, 40). Our plan was to make another primer with a sequence corresponding to the 3' end of the one we had determined and get another 80 residues, and so gradually work round the genome. This would obviously still be a slow process with many fragments to be analyzed, as well as primers to be synthesized. Fortunately, the research developed in a different direction.

In the above experiments we wanted to obtain very highly labeled DNA, and therefore used the radioactive substrate at high specific activity and in low concentrations. In such experiments (using ^{32}P-ATP) it was frequently observed that the DNA products formed ended at a position before an A would have been incorporated. Presumably the enzyme was running out of ATP. This suggested a new approach to DNA sequencing. If one could produce a mixture of chains all having the same 5' end (corresponding to the 5' end of the primer) and finishing at the 3' end at positions corresponding to the A residues, determination of the relative sizes of all these chains should give a measure for the relative positions of the A residues, and this, combined with similar data for the other three nucleotides, is all one needs for the complete sequence determination. Various methods were studied for fractionating the products of synthesis according to size, and one of these was found to be satisfactory. In fact it proved to be much more efficient than we had ever imagined would be possible. This was electrophoresis on acrylamide gel. If carried out in suitable denaturing conditions, oligonucleotides up to over 300 residues long could be separated according to size, the smaller ones migrating more rapidly on the gel than the larger ones, with each product clearly separated from its neighbors, which differed in size by only one nucleotide.

This new approach to DNA sequencing was I think the best idea I have ever had, being original and ultimately successful, so I have attempted to describe its development in some detail, but on reading it through I must confess that I am by no means certain that it really did happen like that, I certainly do not remember having the idea, whereas I do remember doing some preliminary experiments and discussing it with Alan Coulson and John Donelson, I have a feeling that the above account may have originated to some extent from my attempts to explain the method in a simple way when giving lectures, and that subsequent frequent repetition resulted in its being established as part of my "official," but perhaps not actual, memory.

The method as first conceived, using low concentrations of the substrates, was not very satisfactory, but we did not wish to give up an approach that seemed so promising and, after introducing various modifications of the conditions, two analogous techniques (which we called the "plus and minus" method [47]) were developed and proved to be a much more rapid way of sequencing DNA than anything that had gone before. Most of the sequence of øX was determined using this method. However, a much more efficient and reliable method soon superseded this. For the approach to work we needed to be able to prepare a mixture in which the various end products (all with the same residue at the 3' end) were present in about equal amounts. In the plus and minus method this was not the case, and the sizes of the products were distributed over a rather narrow range so that only relatively short sequences could be determined from one incubation. Another way of achieving the same effect was suggested by the work of Kornberg and his colleagues (48) with analogues of the normal DNA polymerase substrates, which act as chain-terminating inhibitors. One of these was 2',3'-dideoxythymidine triphosphate (dideoxy TTP), identical to the normal TTP but lacking a 3' hydroxyl group. It is a substrate for DNA polymerase, though not such a good one as TTP. Once incorporated the chain no longer has a 3' hydroxyl group and so cannot be extended. If a reaction with DNA polymerase is carried out using the four normal nucleoside triphosphates and a suitable concentration of dideoxy TTP, one would expect on long incubation to end up with a mixture of products all with dideoxy T at their 3' ends. The main problem initially was that these analogues were not available. However, at a meeting in Germany I discovered that K. Geider had made some dideoxy TTP, and he very kindly let me have some. The first experiment we did with it gave a beautiful autoradiograph with sharp bands of equal intensities extending over a long sequence. Clearly this would be much better than the plus and minus method. The problem then was to obtain the other three dideoxy NTPs, which had never been synthesized. I was told by a representative of one of the drug companies that his company was going to make them. After waiting about a year, I met him again and he casually told me that his company had decided not to, so Alan Coulson and I decided to make them ourselves. We had no experience, but we had some expert chemists in the lab to advise us, and quite enjoyed this different type of activity. Eventually we succeeded in preparing all four and the method worked well (49). It was used first to complete the øX sequence. In this case the single-stranded phage was used as template, and restriction enzyme fragments from the double-stranded (replicative) form of øX were used as primers. We naturally wanted to extend the method to other and larger DNAs; however there were two major problems. The first was the eternal one of fractionation, which has always played such an important part in sequencing studies. As the methods became more speedy the preparation of restriction fragments for primers became more and more of a limiting factor. This had been done by acrylamide gel electrophoresis. With øX reasonably pure products could be obtained from a single run, but with larger DNAs this was clearly going to be more of a problem. The other difficulty was that the method requires single-stranded DNA as a template. This was fine for øX whose normal form is single-stranded, but most DNAs are double-stranded, and it is difficult if not impossible to prepare the single strands. Various means were tried to overcome this problem, but the best was cloning the DNA in a single-stranded phage vector. Not only is this a means of obtaining single-stranded DNA, but it also solves the fractionation problem as fractionating is by cloning, which is really the ultimate method of purification and can be applied to a mixture of any complexity. The most efficient method was devised by Messing and his colleagues (50, 51) and is now the quickest method for sequencing DNA of almost any size. The vector used is the double-stranded form of the single-stranded bacteriophage M13, modified to contain a site into which fragments of the DNA to be sequenced can be inserted. The mixture of recombinants is then plated out and grown. This is essentially the fractionation method. Each plaque is derived from a single molecule and is therefore pure. The phage from the plaque is then grown and the single-stranded DNA isolated from it. This is used as the template for the dideoxy sequencing, using as a primer a synthetic oligonucleotide that is complementary to a part of the M13 vector that flanks the insert in such a way that the insert is copied by the DNA polymerase. In this way the same synthetic primer, which can be prepared pure in relatively large amounts, is utilized to prime the different recombinants.

The method is normally used as a random procedure. Originally restriction enzyme digests were used for inserting into the vector, but this was subsequently found to be unnecessary and better results are now obtained by random cleavage, usually by sonication (52).

I suppose the dideoxy method can be regarded as the climax of my research career and the fulfilment of an ambition that had gradually been forming as I became more and more involved in sequencing. It is of course very exciting and gratifying to read of the method now being used in many laboratories and of vast regions of the genomes of both simple and higher organisms becoming exposed in the form of sequences, and these are helping in the understanding of some of the fundamental problems of life. But I think I have derived even more pleasure from the development of the work—seeing the method gradually improving until we were able to read a sequence straight from an autoradiograph. Before this reading became automated, people complained that it was a tedious process, but to me it was always a delight, having in the back of my mind the way we used to do sequences one residue at a time by painstaking partial hydrolysis, fractionation, and analysis. At one stage in the work I would take the autoradiographs home with me and look forward to the pleasure of reading them in the peace of the evening.

At about the same time that the plus and minus method came out, another rapid sequencing technique was developed by Maxam & Gilbert (53, 53a). This was somewhat similar to our method in that it gave an autoradiograph of an ionophoresis from which a sequence could be read off. However, whereas we obtained the mixtures of fragments all terminating at the same nucleotide by a copying procedure using DNA polymerase, they used partial degradation by chemical methods of fragments labeled at their 5' ends. The method worked well and was widely used in preference to the plus and minus method, and in the United States probably in preference to the dideoxy technique until the introduction of the M13 cloning greatly increased the applicability of the dideoxy method. I think it would be fair to say that at present, for a long sequence, the dideoxy-M13 approach is preferable, whereas if one is interested only in a short specific sequence it is probably easier to set up the chemical method.

While the two techniques are to some extent complementary and one must welcome any scientific progress, I cannot pretend that I was altogether overjoyed by the appearance of a competitive method. However, this did not generate any sort of a "rat race," and I do not think it affected our subsequent work at all. I was by no means satisfied with the plus and minus method and certainly the dideoxy method would have been developed in any case. It may be wondered why I did not also try a similar chemical method. I had indeed considered similar ideas but suspect I rejected them because of lack of specificity. Throughout the work on proteins and RNA it had usually been found that enzymic methods were more successful than chemical ones because of their greater specificity. I was thus rather prejudiced in favor of the use of enzymes and assumed that for a rapid degradative sequencing method to work the reactions would have to be specific. In fact those used by Maxam & Gilbert are not specific, but it turns out that this does not matter and may even result in stronger data than would be obtained by completely specific reactions.

In this account I have dealt almost entirely with our work on methods for sequencing and have said little about the results obtained. This is partly for reasons of space, but also because I consider that my own personal contributions have been to the methods. In the course of the work we have determined many sequences and obtained significant results, which have been reported elsewhere, but in these I have usually been part of a team. I quite enjoyed doing the somewhat routine DNA sequencing work where it was possible to produce results without too much effort, and I probably found this a form of relaxation compared with the more original, exacting, variable, and often frustrating, work on development of methods. With the larger DNA sequences it is necessary to have someone collect and analyze the data, and this was usually done by Bart Barrell, who is something of a wizard with sequences. An exception was when we were working on bacteriophage λ (54), which was sequenced mainly by Alan Coulson and me. At first the routine work of collecting the data was good fun, but I think both of us were happy when the sequence was finally completed. Filling up the last gaps in a sequence can be both slow and frustrating. The main high spots of these sequencing results were two unexpected findings—the overlapping genes in bacteriophage øX174 (55, 56), and the altered genetic code in mitochondrial DNA (57). This work is adequately described in other articles (57–61).

Table 1 summarizes the progress made in sequencing, and lists the various landmarks discussed in this article, starting from Jensen & Evans's positioning of a single amino acid in insulin in 1935 and ending with the recent enormous acceleration in DNA sequencing. At present the number of sequences in the data banks is in the millions, and is still increasing rapidly.

Although I have dealt only with sequencing methods in this article, my interests have not been quite so narrow, and from time to time I have ventured into other fields, although usually unsuccessfully. The most successful diversion was the discovery of the initiator tRNA, formyl-methionyl tRNA (62). This was largely the work of Kjeld Marcker, who came as a visiting worker to the lab at about the time we were first trying to develop methods on RNA. The idea was to get some sequence information near the amino acid binding sites of tRNAs by labeling them with radioactive amino acids. Partial hydrolysis should then yield oligonucleotides of various lengths joined to an amino acid, and methods could be designed for sequencing them. Methionine was chosen to try out the method because it could be labeled at high specific activity with

Table 1 The progress in sequencing

Year	Protein	RNA	DNA	Number of residues	Ref.
1935	Insulin			1	4
1945	Insulin			2	3
1947	Gramicidin S			5	16
1949	Insulin			9	12
1955	Insulin			51	22
1960	Ribonuclease			120	25
1965		tRNA$_{Ala}$		75	32
1967		5S RNA		120	35
1968			Bacteriophage λ	12	45
1978			Bacteriophage ϕX 174	5,386	61
1981			Mitochondria	16,569	58
1982			Bacteriophage λ	48,502	54
1984			Epstein-Barr virus	172,282	64

^{35}S. To see if the coupling of the methionine to the tRNA had worked, the product was digested with pancreatic ribonuclease and subjected to ionophoresis. The expected adenosyl ester of methionine was identified, but there was another, unexpected, spot present. This could easily have been dismissed as an artifact, but that did not satisfy Marcker. It turned out to be the adenosyl ester of N-formyl methionine, and further experiments enabled him to show that there were two methionyl tRNAs in *Escherichia coli*: one, tRNA$_m^{Met}$, which incorporates methionine within protein chains, and the other, tRNA$_f^{Met}$, which incorporates it only in the N-terminus and is the chain initiator (63).

Unlike many scientists, I decided to retire and give up research when I reached the age of 65. This surprised my colleagues, and to some extent myself also. I had not thought about retirement until I suddenly realized that in a few years I would be 65 and would be entitled to stop work and do some of the things I had always wanted to do and had never had time for. The possibility seemed surprisingly attractive, especially as our work had reached a climax with the DNA sequencing method and I rather felt that to continue would be something of an anticlimax. The decision was I think a wise one—not only because I have greatly enjoyed the new life-style, but also because the aging process was not improving my performance in the laboratory and I think that if I had gone on working I would have found it frustrating and have felt guilty at occupying space that could have been available to a younger person. For more than 40 years I had had wonderful opportunities for research, and had been given the chance to fulfill some of my wildest dreams.

Literature Cited

1. Loewy, A. G., Siekevitz, P. 1969. *Cell Structure and Function*, pp. 193–94. New York: Holt, Rinehart & Winston
1a. Bergmann, M., Niemann, C. 1938. *J. Biol. Chem.* 122:577–96
2. Chibnall, A. C. 1942. *Proc. R. Soc. London Ser. B* 131:136–60
3. Sanger, F. 1945. *Biochem. J.* 39:507–15
4. Jensen, H., Evans, E. A. 1935. *J. Biol. Chem.* 108:1–12
5. Edman, P. 1950. *Acta Chem. Scand.* 4:283–93
6. Gordon, A. H., Martin, A. J. P., Synge, R. L. M. 1943. *Biochem. J.* 37:79–86
7. Gurin, S., Clarke, H. T. 1934. *J. Biol. Chem.* 107:395–99
8. Abderhalden, E., Stix, W. 1923. *Hoppe-Seyler's Z.* 129:143–56
9. Sanger, F. 1947. *Nature* 160:295–96
10. Sanger, F. 1949. *Biochem. J.* 44:126–28
11. Tiselius, A., Sanger, F. 1947. *Nature* 160:433–34
12. Sanger, F. 1949. *Biochem. J.* 45:563–74
13. Sanger, F. 1985. *Curr. Contents* 28(12):23
14. Gray, W. R., Hartley, B. S. 1963. *Biochem. J.* 89:59
15. Consden, R., Gordon, A. H., Martin, A. J. P. 1944. *Biochem. J.* 38:224–32
16. Consden, R., Gordon, A. H., Martin, A. J. P., Synge, R. L. M. 1947. *Biochem. J.* 41:596–602
17. Sanger, F., Tuppy, H. 1951. *Biochem. J.* 49:463–81
18. Sanger, F., Thompson, E. O. P. 1953. *Biochem. J.* 53:353–66
19. Sanger, F., Tuppy, H. 1951. *Biochem. J.* 49:481–90
20. Sanger, F., Thompson, E. O. P. 1953. *Biochem. J.* 53:366–74
21. Sanger, F., Thompson, E. O. P. 1952. *Biochim. Biophys. Acta* 9:225–26
21a. Ryle, A. P., Sanger, F. 1955. *Biochem. J.* 60:535–40
22. Ryle, A. P., Sanger, F., Smith, L. F., Kitai, R. 1955. *Biochem. J.* 60:541–56
23. Sanger, F. 1959. *Les Prix Nobel en 1958*, pp. 134–46. Reprinted in *Science* 129:1340–45
24. Spackman, D. H., Stein, W. H., Moore, S. 1958. *Anal. Chem.* 30:1190–1206
25. Hirs, C. H. W., Moore, S., Stein, W. H. 1960. *J. Biol. Chem.* 235:633–47
26. Sanger, F., Hocquard, E. 1962. *Biochim. Biophys. Acta* 62:606–7
27. Hartley, B. S., Naughton, M. A., Sanger, F. 1959. *Biochim. Biophys. Acta* 34:243–44
28. Naughton, M. A., Sanger, F., Hartley, B. S., Shaw, D. C. 1960. *Biochem. J.* 77:149–63
29. Sanger, F., Shaw, D. C. 1960. *Nature* 187:872–73
30. Milstein, C., Sanger, F. 1961. *Biochem. J.* 79:456–69
31. Sanger, F. 1963. *Proc. Chem. Soc. (UK)*, March 1963:76–83
32. Holley, R. W., Agpar, J., Everett, G. A., Madison, J. T., Marquisee, M., et al. 1965. *Science* 147:1462–65
33. Sanger, F., Brownlee, G. G., Barrell, B. G. 1964. *J. Mol. Biol.* 13:373–98
34. Ling, V. 1972. *J. Mol. Biol.* 64:87–102
35. Brownlee, G. G., Sanger, F., Barrell, B. G. 1968. *J. Mol. Biol.* 34:379–412
36. Adams, J. M., Jeppesen, P. G. N., Sanger, F., Barrell, B. G. 1969. *Nature* 223:1009–14
37. Jeppesen, P. G. N., Barrell, B. G., Sanger, F., Coulson, A. R. 1972. *Biochem. J.* 128:993–1006
38. Murray, K. 1970. *Biochem. J.* 118:831–41
39. Sanger, F., Donelson, J. E., Coulson, A. R., Kössel, H., Fischer, D. 1973. *Proc. Natl. Acad. Sci. USA* 70:1209–13
40. Sanger, F., Donelson, J. E., Coulson, A. R., Kössel, H., Fischer, D. 1974. *J. Mol. Biol.* 90:315–33
41. Ziff, E. B., Sedat, J. W., Galibert, F. 1973. *Nature New Biol.* 241:34–37
42. Robertson, H. D., Barrell, B. G., Weith, H. L., Donelson, J. E. 1973. *Nature New Biol.* 241:38–40
43. Billeter, M. A., Dahlberg, J. E., Goodman, H. M., Hindley, J., Weissmann, C. 1969. *Nature* 224:1083–86
44. Wu, R., Kaiser, A. D. 1968. *J. Mol. Biol.* 35:523–27
45. Wu, R., Taylor, E. 1971. *J. Mol. Biol.* 57:491–511
46. Berg, P., Fancher, H., Chamberlin, M. 1963. In *Symposium on Informational Macromolecules*, ed. H. Vogel, V. Bryson, J. O. Lampen, pp. 467–83. New York/London: Academic
47. Sanger, F., Coulson, A. R. 1975. *J. Mol. Biol.* 94:441–48
48. Atkinson, M. R., Deutscher, M. P., Kornberg, A., Russell, A. F., Moffatt, J. G. 1969. *Biochemistry* 8:4897–904
49. Sanger, F., Nicklen, S., Coulson, A. R. 1977. *Proc. Natl. Acad. Sci. USA* 74:5463–67
50. Gronenborn, B., Messing, J. 1978. *Nature* 272:375–77
51. Messing, J., Vierira, J. 1982. *Gene* 19:269–76
52. Deininger, P. 1983. *Anal. Biochem.* 129:216–23
53. Maxam, A. M., Gilbert, W. 1977. *Proc. Natl. Acad. Sci. USA* 74:560–64

SANGER

53a. Maxam, A. M., Gilbert, W. 1980. *Methods Enzymol.* 65(P. 1):499–560

54. Sanger, F., Coulson, A. R., Hong, G. F., Hill, D. F., Petersen, G. B. 1982. *J. Mol. Biol.* 162:729–73

55. Barrell, B. G., Air, G. M., Hutchison, C. A. III. 1976. *Nature* 264:34–41

56. Smith, M., Brown, N. L., Air, G. M., Barrell, B. G., Coulson, A. R., et al. 1977. *Nature* 265:702–5

57. Barrell, B. G., Anderson, S., Bankier, A. T., de Bruijn, M., Chen, E., et al. 1980. *Proc. Natl. Acad. Sci. USA* 77:3164–66

58. Anderson, S., Bankier, A. T., Barrell, B. G., de Bruijn, M. H. L., Coulson, A. R., et al. 1981. *Nature* 290:457–65

59. Anderson, S., de Bruijn, M. H. L., Coulson, A. R., Eperon, I. C., Sanger, F., et al. 1982. *J. Mol. Biol.* 156:683–717

60. Sanger, F. 1981. *Les Prix Nobel en 1980*, pp. 143–59. Reprinted in *Biosci. Rep.* 1:3–18; *Science* 214:1205–10

61. Sanger, F., Coulson, A. R., Friedmann, T., Air, G. M., Barrell, B. G., et al. 1978. *J. Mol. Biol.* 125:225–46

62. Marcker, K., Sanger, F. 1964. *J. Mol. Biol.* 8:835–40

63. Clark, B. F. C., Marcker, K. A. 1966. *J. Mol. Biol.* 17:394–406

64. Baer, R., Bankier, A. T., Biggin, M. D., Deininger, P. L., Farrell, P. J., et al. 1984. *Nature* 310:207–11

COMPUTER SCIENCE

Annu. Rev. Comput. Sci. 1990. 4:1–12

REFLECTIONS ON COMPUTER SCIENCE

John E. Hopcroft

Computer Science Department, Cornell University, Ithaca,
New York 14853

CONTENTS

INTRODUCTION

In 1986, Annual Reviews Inc added the *Annual Review of Computer Science* to its distinguished series of state-of-the-art reviews. This was a significant event for the discipline because the establishment of a new series by Annual Reviews serves as acknowledgment of the maturity and stature of a field.

I have been actively involved in computer science research and education since 1964. Over the years, I have felt the excitement generated by the acquisition of new knowledge in computer science and witnessed the phenomenal growth as the field developed into a vigorous and independent discipline. However, recognition by those outside the field lagged considerably behind growth. I am happy to see that this situation has been rectified. I view the publication of the *Annual Review of Computer Science* as an indication of an increased awareness of the importance of computer science by the scientific community at large. Thus, I am extremely pleased to have been given the opportunity to write the prefatory chapter for this fourth volume of the new series.

As one of the early contributors to the field of computer science, I can describe the development of computer science into a discipline, and speculate on future areas of growth. In reviewing the history of computer science, one would normally mention such important evolutionary factors as the beginning of artificial intelligence research, Project MAC, Illiac, the role of DARPA, and the influence of major research institutions. However, here I present a slightly different historical view. I omit these well-documented events in favor of other, less well-known developments. We often overlook the fact that what contributes most to the expansion of a discipline is not the major events but the total of all of the advances that take place within it, both major and minor. Here I also concentrate on computer science developments in which I was personally involved.

THE PAST

The modern era of computing began in 1946 with the development of ENIAC, the Electrical Numerical Integrator and Computer, built at the Moore School of Electrical Engineering at the University of Pennsylvania. ENIAC, the first electronic digital computer, was capable of performing 5000 arithmetic operations per second. Thus ENIAC possessed less computing power than the hand-held calculators used today by high school students; but for its time, ENIAC performed at extraordinary speed and heralded the scientific age of computing. While many researchers and scientists were attracted to computing by exciting computing devices like ENIAC, their energies were channeled primarily toward applying computing instruments to scientific problems rather than toward developing a science of computation.

The science of computation traces its beginnings to the work of Alonzo Church and Alan Turing. It is interesting to note that Turing's paper "On Computable Numbers with an Application to the Entscheidungsproblem," in which he introduced the concept of the Turing machine, was published in 1936, predating ENIAC by ten years. While the roots of computer science go back to 1936, computer science as an academic discipline did not begin until 1962, when the first independent computer science departments were created at Stanford University and Purdue University.

Early computer scientists came to the field with training primarily in electrical engineering or mathematics. Thus it was not surprising that the early discipline placed heavy emphasis on digital switching theory and mathematical logic. Back then, computer science research centered on the design of Boolean switching circuits and the implementation of finite-state machines. In fact, IEEE's Foundations of Computer Science Conference was initially called the Switching Circuit Theory and Logical Design Conference. Work on switching circuits concentrated on the minimization of the number of components needed to realize Boolean functions in electronic circuits. Work on the implementation of finite-state machines was concerned with the state assignment problem. David Huffman was the key contributor in this area. One of the major intellectual advances of the time was his formulation of the abstraction of the internal state, an idea taken for granted today.

This early work tended to take place in electrical engineering departments; consequently it was viewed as "computer engineering" rather than "computer science." When academic departments of computer science were established, they incorporated the mathematical concept of finite automata but left the circuit realization material to electrical engineering departments. Two books of collected papers were influential in determining the content of early courses in computer science: *Automata Studies*, edited by C. E. Shannon and J. McCarthy, and *Sequential Machines: Selected Papers*, edited by E. F. Moore. Shannon & McCarthy's book contained a paper by Warren McCulloch and Walter Pitts that characterized the behavior of neural activities. Their work led to the regular expression notation, a notation that was subsequently proved equivalent in expressive power to the finite-automaton formalism. Moore's book contained a survey paper on finite automata by Michael Rabin and Dana Scott, which gave a more mathematical treatment to automata theory than found in previous works on finite-state machines. Unlike Huffman's work, which focused on the circuit realization of the machine, their paper focused on the sets of strings that were accepted.

Two other important advances of the time were context-free grammars and Backus-Naur Form (BNF). The context-free grammar was a formal linguistic model developed by Noam Chomsky to capture the syntax of natural language; BNF was a formalism developed by John Backus and Peter Naur to express the syntax of programming languages, such as Algol 60, that were then being developed. The context-free grammar and the BNF formalism were shown to be equivalent. This result, together with the equivalence of regular expression notation and sets accepted by finite automata, placed a coherent structure on automata theory and languages.

Another major influence on academic computer science stemmed from work on Turing machines and computability. In 1964, Juris Hartmanis and Richard Stearns published their paper on "The Computational Complexity of Algorithms" in the *Transactions of the American Mathematical Society*. Although many papers on complexity had been published previously, this paper tied complexity to the atomic steps of the computation and to the actual number of tape cells used by a Turing machine. Hartmanis & Stearns showed that, with very concrete measures of complexity, modest increases in resources allowed new functions to be computed that could not have been computed using the original resources. This idea sparked enormous interest and led to the study of the intrinsic complexity of problems, an area that would expand significantly over the next 20 years. In fact, it was from this paper that the subdiscipline of computational complexity received its name.

This was the state of the field of computer science in 1964 when I obtained my PhD in Electrical Engineering from Stanford University. Following my graduation, Edward McCluskey at Princeton University hired me to help build a computer science program within their Electrical Engineering Department. My first assignment was to teach a graduate course in theoretical computer science. Since no textbooks were available on this topic, I had to start from scratch. I gathered material on what I considered the most important concepts in the area. In class we discussed the work of McCulloch & Pitts, Backus & Naur, Rabin & Scott, Hartmanis & Stearns, Chomsky, and Turing. Subsequently Jeffrey Ullman and I developed the notes from this course into the text *Formal Languages and their Relation to Automata*, which would shape the way theoretical computer science was taught for the next two decades.

The systems' side of academic computer science lagged behind the theoretical side. Research in computing systems was driven by the need for compilers and operating systems. In the area of compilers, work often focused on the lexical and parsing aspects of compilation. New programming languages were designed; control mechanisms, block structure, parameter passing, and their impact on implementations were explored. Operating systems focused on resource management. It was not until much later, when distributed systems came along with notions of concurrency and synchronization, that operating systems began to flourish as a rich discipline.

Techniques for writing programs and proving them correct began to appear around this time. Much of this work used well-defined mathematical tasks as sample problems—for example, computing the greatest common divisor of two integers, or sorting a set of integers. These tasks possessed a common property: they had very precise specifications. Thus, programming became a science of how to take a formal specification and convert it into a correct program, what today is called programming in the small.

An important next step in the development of computer science was the publication of the first two volumes of the *Art of Computer Programming*, what was to be a seven-volume set by Donald Knuth. Knuth had intended to write a seven-chapter book on programming, but after the publisher received the first chapter, it was clear that what they had agreed to publish as a book would be more on the order of a small encyclopedia. Fortunately the publisher did not abandon the project. Prior to this time, no organized, comprehensive work had been written on computer algorithms. Knuth's books substantially reduced the entrance cost for researchers in this area and opened up the area of algorithms to scientific inquiry. Thus the domain of computer science was broadened to include the study of algorithms.

During 1970, I spent my sabbatic at Stanford University, sharing an office with Robert Tarjan. Together we worked on various graph algorithms: planarity, triconnectivity, depth-first search, and graph isomorphism. Initially we sought a nonlinear lower bound on the complexity of determining if a graph was triconnected. In the course of our attempts, we discovered linear-time algorithms for testing the planarity and triconnectivity of graphs. Our successes led to the realization that courses in algorithms could be taught on a theoretical basis. *The Design and Analysis of Computer Algorithms* by Aho, Hopcroft, and Ullman was a product of that realization.

In 1970, Stephen Cook at the University of Toronto proved what at first appeared to be an obscure and highly technical theorem; namely, that the problem of whether or not a nondeterministic polynomial time–bounded Turing machine accepts a string could be transformed into the problem of determining if a Boolean formula was satisfiable. This meant that the problem of determining whether a Boolean formula was satisfiable was at least as difficult computationally as any problem that could be solved by a nondeterministic algorithm in time polynomial in the size of the problem description—i.e. any problem in the class that today is known as nondeterministic polynomial time (NP). He presented his results in the paper "On the Complexity of Theorem-Proving Procedures" at the 1971 ACM SIGACT Symposium on Theory of Computing. At first, researchers did not know what to make of this curious result. But a year later, Richard Karp at Berkeley presented his now renowned paper on "Reducibility among Combinatorial Problems" at an IBM symposium at Yorktown Heights. In this paper, Karp showed not only that the problem of satisfiability of Boolean formulas had this property (which today we call NP-completeness), but also that almost all packing, matching, routing, and scheduling problems had it. Thus the notion of classifying problems as complete for a class, such as polynomial time (P) or nondeterministic polynomial time (NP), entered every computer scientist's working vocabulary. Within a year or two, a dozen complexity classes (e.g. PSPACE, polylog time, and exponential time) had been defined, each with its own set of complete problems. Problems were then classified by the complexity class for which they were complete. Mathematical theories, such as Presburger arithmetic and the theory of reals, were examined to determine their intrinsic complexity, and tools such as relativization were developed.

The exciting work in computational complexity soon attracted many new scientists to computer science, in particular logicians and mathematicians. When these scientists realized the depth of the contributions being made in this new field, they accepted it as a separate and distinct area of inquiry. This was a first step in the process of gaining recognition for computer science as an independent discipline.

During the years that followed, rapid advances were made in a number of areas of computer science, among them randomized algorithms, factorization, cryptography, and polynomial-time algorithms for linear programming. After 1975, an explosion of new subfields developed: relational data bases, semantics, distributed systems, concurrent systems, computational geometry, and VLSI design. By this time, computer scientists could no longer be considered generalists. The field had grown so large that an individual computer scientist could no longer teach every course that needed to be taught. The depth and the variety of the results obtained during the discipline's formative years had led computer scientists to specialize within specific subdisciplines. Computer science had to undergo this process to achieve maturity as a discipline.

Another important step in the growth of computer science occurred during the late 1970s. Jerome Feldman and William Sutherland edited a report to the National Science Foundation entitled "Rejuvenating Experimental Computer Science." The Feldman Report, as it came to be called, documented the need for an infusion of research funds to build up the infrastructure of academic computer science departments. Unlike Stanford, MIT, and CMU, which had been well funded by DARPA, most of the remaining institutions in the country were equipment impoverished; they did not have the resources necessary to carry out broad research programs. To help alleviate this problem, the National Science Foundation responded by creating the Coordinated Experimental Research (CER) Program. This program provided large, five-year block grants to approximately 20 institutions. These funds allowed many computer science departments to acquire equipment necessary to move them beyond the realm of single-investigator projects. Within five years, the CER Program transformed what had been primarily paper-and-pencil research efforts into experimental projects across a wide spectrum of research areas. Many consider the CER Program to have been one of NSF's most successful programs.

By 1985, computer science had firmly established itself and had begun to assert leadership at a national level. No other discipline has grown so much in so short a period. Now, approximately 25 years after the first computer science departments were created, 125 such computer science departments across the nation produce over 500 PhDs per year. In addition, there are over 40 major research journals devoted entirely to computer science or computer engineering; these journals publish more than 300 issues annually. Computer science has had a major intellectual impact on the academic environment and contributed to the reshaping of such other disciplines as mathematics, linguistics, and cognitive psychology.

THE FUTURE

Phenomenal progress has been made since the science of computation began approximately 50 years ago. Back then few people could have imagined the state of computing today and the ease with which computers have permeated all aspects of business, science, and education. The same holds true for academic computer science; significant advances have been achieved that have changed the very nature of the discipline. With this in mind, I present some speculations on what the future holds.

I believe that computers will cause an intellectual revolution for mankind. During the next 50–100 years we will witness research advances that will change in a fundamental manner our understanding of knowledge, enhancing our ability to acquire and utilize it. I believe computer science has brought us to the threshold of an information revolution whose consequences will be more profound than the scientific and industrial revolutions combined. But rather than speculate on the distant future, I confine my remarks here to trends in computing that are already visible. In the near term, computer science will be influenced and altered by several forces: parallelism, man-machine interfaces, knowledge fusion, and a trend towards more expansive computer applications. I believe that these factors will have a great impact on research in computer science.

Parallelism

Already numerous parallel architectures have been developed: the single-instruction multiple-data (SIMD) architecture used in the Connection Machine, the message-passing architecture used in the Hypercube, and the shared-memory architectures used in the Butterfly and Encore machines. The driving force behind parallelism is the potential for significant increases in computing power.

Ralph Gomory, senior vice president and chief scientist at IBM, has described computing in terms of marks (those binary quantities that indicate the presence or absence of a signal) and the ease with which marks are made, read, and erased. He has noted that the phenomenon of doubling computer power every three years resulted from the development of technology that made marks smaller and smaller. In the 1950s a vacuum tube was used to record a single mark; later the transistor was used, then the integrated circuit, and, more recently, very-large-scale integration. Over time, the mark has decreased in size; now it is only a few microns long. Consequently the distance that marks are transmitted has decreased, as has the time required to read or write them. There is no end in sight to the drive to decrease the size of the mark and thereby increase the speed of computation.

However, it is becoming more and more expensive to decrease the size of the mark. As costs increase, it becomes more effective to read and write many marks at once—i.e. to build parallel machines. In the next decade much of our thinking about computation will undoubtedly change from the single-instruction-stream model developed by von Neumann to a parallel model. Reflecting this change, computer scientists will place more and more emphasis on languages and architectures for exploiting parallel computation. Research will focus on the connection between processors and memory, the amount of parallelism that can be extracted from a sequential program, and the linguistic concepts for expressing computations that are carried out on a parallel machine. The switch from the von Neumann model of computation to a parallel paradigm is likely to be rapid. It will take place within the next 10–15 years.

In the long run, massively parallel machines will change fundamentally the way we view computation. For example, problems may have to be reformulated to better utilize parallel technology. Rather than simulating fluid flow by solving differential equations on a highly parallel machine, the problem may be restructured for a cellular-automata model, where each cell will have an internal state and communicate only with neighboring cells. A massively parallel machine may carry out the simulation by assigning one processor to each cell. Such a change in the computational model will produce a significant change in the way scientists view physics. One might be able to prove global properties of material or fluid flow from the local properties of a cell and the interconnection patterns among cells. This is not as farfetched as it may appear; at some level of granularity, the world is discrete, and differential equations are only an approximation to this discrete world.

The Man-Machine Interface

Another area that will play a key role in the future of computer science is computer programming, in essence the man-machine interface. While initially computers were programmed using machine code, the first major advance over machine code was assembly language. Although it involved almost a one-to-one translation of machine code, assembly language's ability to use mnemonics for operations made it a major advance. In addition, programmers were no longer required to allocate physical locations for instructions and storage in memory; instead they worked within the space of logical addresses and left the task of making physical assignments to the computer. The next important advance in programming was the development of high-level languages, such as FORTRAN. High-level languages made it possible to write arithmetic expressions in a familiar notation. This, along with subsequent advances in control structures, block structures, data abstractions, etc, tremendously improved programming productivity. The sequence of languages that followed—Algol, Pascal, etc—exhibited minor advances compared to the fundamental advance of FORTRAN.

Programming is a challenging intellectual task. As such it is a major barrier to the full use of computers in everyday life. To overcome this barrier, it may be worthwhile to learn from the industrial revolution. The industrial revolution produced fundamental changes in production methodologies, and it had a profound effect on quality of life and on technology. By breaking up the work done by skilled artisans into simple, well-defined tasks, laborers with minimal technical skill could produce the same high-quality products as artisans. The skill level for computer programming today appears to be moving in the opposite direction, requiring higher and higher skill levels. Unless we reverse this trend and make computers accessible to the average worker, many of our advances will be of little benefit to society as a whole. Procedures for accessing computers are too complex for the average citizen. Current man-machine interfaces must be altered to allow those with less skill to interact easily with computers, at all levels. We have seen the beginning of this kind of process with systems like the Macintosh computer and with spreadsheets. It is likely that spreadsheet technology has expanded the use of computers more than any other invention; it may provide us with the key to make computer systems accessible to large numbers of people. Just as during the industrial revolution, this increased accessibility will open up new and important areas for growth. Thus the computer revolution is only now beginning.

Knowledge Fusion

A deeper understanding of the process of knowledge fusion will be achieved in the near future, and this will have a significant impact on computer science. Humans instantaneously comprehend images; when we see a tree, we immediately recognize it as one. Since trees come in an infinite variety of shapes, it is obvious that our recognition does not depend upon geometrical pattern-matching. Moreover, we recognize trees we have never seen before. What makes a tree a tree is not its shape and texture alone, but something more intrinsic. Our recognition of the tree involves the fusion of many bits of information, some that support the decision for treeness and some that do not. By some as yet unknown technique, humans immediately combine various sensory data to arrive at a correct decision. This combination algorithm must be far from the bottom-up methods of object recognition used in computer vision systems today, which identify features such as edges and corners and try to combine these features into more complex patterns. Somehow the human computer possesses top-down information about what is most likely in view. (For example, if one is outdoors, an object is more likely to be a tree than if one is indoors.) The human computer simultaneously processes bottom-up information about what it is seeing and top-down information about what it expects to see; then it fuses these two computations. The fusion probably occurs at many different levels rather than at a single interface. Our recognition processes are rarely fooled; and when they are, the error is usually only momentary. For example, when looking out a train window, we may believe that our train is moving when a train on an adjacent track starts to move. However, since additional information does not jibe with this belief our brain immediately corrects its impression. No computer program yet written has capabilities so advanced, but I foresee developments like this in the near future.

Applications

Computer usage was at first limited to scientific computations and the management of large, similarly structured data; for example, computers were used to calculate trajectories and track payroll files, insurance records, and airline reservations. Today there are applications on the horizon whose sophistication far exceeds these early applications. A single earth-scanning satellite may transmit 16,000 photographs a day, each covering a 150×150 square mile area; merely cataloging this data requires considerable computer automation. Imagine a computer program that will scan these images and report the occurrence of anything atypical, such as a major earthquake or a reactor meltdown.

An application that promises to improve manufacturing productivity dramatically is electronic prototyping, which allows us to substitute computer models for physical ones. Using electronic prototyping, one can design an object such as a multifingered gripper and test its design by programming the gripper to manipulate an object or perform an assembly. Since the gripper is only a computer file, it can be edited easily to change its size, shape, or finger configuration. This ability to explore a wide design space before building an object will have a tremendous impact on the variety and quality of products. Standardized design descriptions will lead to highly automated production with greatly reduced costs.

The concept of electronic prototyping illustrates the importance of the information content of a physical object or a design. It is in the management and transformation of information that computers will have their greatest impact. For example writing, a tool for storing and communicating ideas, has been by and large a linear medium; but we are now seeing other forms of writing, like HYPERTEXT, where the order in which we wish to access information can be restructured. The computer gives us the opportunity to interact actively with written work, including pictures and 3-dimensional models. The computer allows us to interact with what might be called artificial realities. The ability to represent knowledge in procedural form may become as important as the ability to record knowledge in written form.

The pervasiveness of computer applications is bound to have a profound impact on computer science, expanding future research beyond the narrow confines we know today.

CONCLUSIONS

Computer scientists can look back with pride on the last 50 years, the years during which computer science evolved from an interesting area of research into a mature and independent discipline. Many changes took place during this time, many research avenues were explored, and much knowledge was gained. It was an exciting period, but the next 50 years will be even more exciting for computer science.

The course that computer science will follow in the near future seems clear. Parallelism will be a driving force for change in the field; advances in high-level languages will reduce the complexity of programs and allow more people to interact with computers; computer capability will increase substantially as advances in knowledge fusion occur; and computer applications will continue to expand across a broad spectrum of areas.

My long-term vision for computer science is not as clear. I believe that revolutionary advances in computing principles lie just out of sight. During the next century, we will begin to unravel these mysteries.

In the 1600s Newton revolutionized our understanding of the physical world by formulating the most elementary laws of motion. Today, less than 350 years later, we have developed a deep understanding of the nature of matter, quantum mechanics, and the creation of the universe. Computing is on the brink of a similarly Newtonian development. Fundamental computational principles lie on the horizon that are capable of vastly changing the way we process information. I predict exciting times ahead for computer science.

Literature Cited

Aho, A. V., Hopcroft, J. E., Ullman, J. D. 1974. *The Design and Analysis of Computer Algorithms.* Reading: Addison-Wesley. 470 pp.

Cook, S. A. 1971. The complexity of theorem proving procedures. *Proc. 3rd ACM Symp. on Theory of Computing,* pp. 151–58

Feldman, J. A., Sutherland, W. R. 1979. Rejuvenating experimental computer science. *Commun. ACM* 22: 497–502

Hartmanis, J., Stearns, R. E. 1965. On the computational complexity of algorithms. *Trans. AMS* 117: 285–306

Hopcroft, J. E., Ullman, J. D. 1969. *Formal Languages and their Relation to Automata.* Reading: Addison-Wesley. 242 pp.

Karp, R. M. 1972. Reducibility among combinatorial problems. In *Complexity of Computer Computations,* ed. R. E. Miller, J. W. Thatcher, pp. 85–104. New York: Plenum

Knuth, D. E. 1973. *The Art of Computer Programming,* Vol. 1. Reading: Addison-Wesley. 634 pp. 2nd ed.

McCulloch, W. S., Pitts, W. 1943. A logical calculus of the ideas immanent in nervous activity. *Bull. Math. Biophys.* 5: 115–33

Moore, E. F., ed. 1964. *Sequential Machines: Selected Papers.* Reading: Addison-Wesley. 266 pp.

Rabin, M. O., Scott, D. 1959. Finite automata and their decision problems. *IBM J. Res. Devel.* 3(2): 114–25

Shannon, C. E., McCarthy, J., eds. 1956. *Automata Studies.* Princeton: Princeton Univ. Press. 285 pp.

Turing, A. M. 1936–1937. On computable numbers, with an application to the Entscheidungsproblem. *Proc. London Math. Soc.* 42: 230–65

Annu. Rev. Earth Planet. Sci. 1991. 19:1–16

SOME CHICAGO GEORECOLLECTIONS

Julian R. Goldsmith

Department of the Geophysical Sciences, University of Chicago, 5734 South Ellis Avenue, Chicago, Illinois 60637

KEY WORDS: isotope fractionation, Harold Urey, geochemistry, University of Chicago, interdisciplinary research

All thoughtless people who have lived long enough probably feel that their life span embraced a period of great change, of progress, and perhaps even of revolution. As one fortunate enough at more than three score and ten to be still personally active in the laboratory, I can look back at what I used to be paid for doing and observe how my day-to-day work conjoins with the views of those probing the Earth since the 1930s. I shall attempt to look back without too much reinventing or tampering with history, and shall also attempt to give my views of things then and now, limiting myself to that area of our science best known to me.

I arrived as an undergraduate at the University of Chicago in 1936, one year behind my age group because of an overly protective mother who delayed my entrance into first grade, ensuring an unfortunate midyear schedule throughout school. I came to the University of Chicago not so much because of any reputation for quality, but rather for its convenience to my suburban home in Oak Park. Oak Park and River Forest Township High School was a good school (we were all told), but, sadly, I was able to neglect certain aspects of my education, particularly in science. I came to the university with too little mathematics and with no chemistry or physics. Having grown up on the prairie, I had an interest in nature, but more toward the biological than the physical sciences. My most eye-opening subject in high school was a course called "Social Problems"; not only was the subject of sociology unknown in most secondary schools, but in the then stuffy village of Oak Park, who had ever heard of a social problem? I was tempted to major in it, but then again I might also have become a professional photographer, for photography was an ardent hobby since fourth grade. My freshman advisor at Chicago was Bill Krumbein, who coincidentally was an instructor in the Department of Geology, and when I was seeking an elective fourth subject in my first year, he suggested the introductory course in geology. Never having heard of geology, I took it, perhaps in part because of the lack of prerequisites. If I had taken chemistry first, as I should have done to be better prepared for geology, would I have ever heard of geology?

I had never thought about the origin of the things that are "there"— the mountains, the rocks, the minerals (what were they?), even the soil— although I avidly read the family set of the Book of Knowledge, which linked shrinking wrinkled apples with the origin of mountains. I became interested, and my involvement was fostered and promoted by the inspirational teaching of J Harlen Bretz, to whom I was exposed in my second quarter of the first year. I was hooked, but it became immediately obvious that I had real deficiencies to make up, so I girded my loins and sailed into mathematics and science. My third year in college almost did me in, and at one point I had run myself down to the point that I just got sick and went home to bed for a week, but wonder of wonders, I *liked* the subjects!

Sometime during this period I became aware of a quiet guy who had a title that differed from the others: Norman Levi Bowen, Charles L. Hutchinson Distinguished Service Professor. NLB had come to the university shortly after I did, in 1937, as a replacement for the retiring Albert Johannsen. Johannsen was an outstanding petrographer, and in addition to having a love affair with igneous rocks, he also was an authority on dime novels, postage stamps, first editions of nineteenth-century illustrated books, coins, genealogy and autographs, to mention a few of his interests. Words of erudition, wisdom, and humor stand out in and at the beginning of each of the chapters of his four-volume compendium, *A Descriptive Petrography of the Igneous Rocks*. The difference in interests, however, between Joh and Bowen was striking—petrography versus petrogenesis. I had become aware of the research going on in Bowen's "hot lab" (before "hot" meant radioactive), in which the students, mostly Canadians, were doing things not done by the other students. By this time I had decided that I was interested in the more quantitative aspects of Earth science, and experimental petrology filled the bill.

In the last half of the 1930s, the Department of Geology at Chicago was highly respected and, if one uses attractiveness to students as a criterion of quality, had produced more PhD degrees in geology than any other institution. This lead continued until the early 1960s, when it was probably lost to Columbia. Reputations of institutions and departments, however, persist *at least* a generation after the fact, and although in the 1930s the geology department at Chicago was considered one of the elite, and received a large boost when Bowen arrived, in hindsight I feel that it had benefited from that lag and was overrated. Chicago *was* a world leader when T. C. Chamberlin left the presidency of the University of Wisconsin to build a department at the newly created University of Chicago in 1892. William Rainey Harper, the first president of the university, was an enormously persuasive man, who convinced Chamberlin not only to leave the presidency of a major university, but to come as chairman in a non-existent department at an embryonic university. Remarkable how one so far removed from science (a Hebrew scholar and nominal theologian) was insightful enough to *select*, in addition to *acquire*, outstanding men of science such as Chamberlin. Chamberlin became deeply involved in inter-departmental cooperation with F. R. Moulton of astronomy, and the two developed the long-lived and important planetesimal hypothesis of the origin of the solar system.

Even if the aura of the "early days" of the Chicago department persisted into the 1930s, it had become, I feel, quite conservative and, in spite of the firebrand Bretz and the shot in the arm by Bowen, rather stodgy. Although we were thoroughly drilled in Chamberlin's concepts of multiple working hypotheses, little time was spent in courses in the discussion of alternate theories, at least with respect to the older, or more grandiose, aspects of geology. Perhaps this was just an expression of the general American conservatism as exemplified by the ruling concept of permanence of continents and ocean basins and, unlike European teachings, rejection of any consideration of continental drift. In sharp contrast to this attitude, however, was the Bretzian catastrophism of the channeled scablands, which was ridiculed by a large part of the geologic community. The story of the ultimate exoneration and appreciation of the renegade Bretz, who used observations and perceptions that violated the sacred cow of uniformitarianism, need not be told here, but the importance of his work wasn't fully felt until Martian surface features indicated the validity of his views on other planets! Bretz was a forceful, incisive person, whose total effort as a teacher was to make students think and to look at all sides of a problem, and he taught me that little in science is cut-and-dried.

My interests, however, drew me closer to Bowen, and by the time my graduate work began, we had developed something of a teacher-pupil bond, fostered by my respect for him and aided by his appreciation of my willingness to approach him directly and discuss things on a one-to-one basis; he was bothered by the fact that most students held back in awe. A more reasonable picture of Bowen can be illustrated by an action that took place when the war interrupted my graduate research work in the "hot lab." An enlightened Selective Service Board had called me in and told me to bend every effort in the next month or two to find employment in a situation in which my background would be used to aid the war effort, other than as a soldier. I was hired by the Corning Glass Works and prepared to leave Chicago. One day, while working in the lab before leaving, I suddenly felt my arms pinned to my sides from behind as I was lifted and rotated 90°—and released. Looking up from the floor, I saw a grinning Bowen, holding a card from the Office of the Registrar and reading "Drop Goldsmith."

At Corning I became involved in the surface chemistry of silicates, especially 96.5% SiO_2 glass (Vycor). We were using it in a ceramic process to make insulators and other parts for a mysterious factory in Oak Ridge, Tennessee. It turns out that these were used in the Calutron for isotopic separation of uranium, from which atomic bombs were made, but at that time what went on was a very well kept secret. I became involved with the role of water in silicates and in OH-bonding to unsatisfied silicon atoms, all of which has carried over to current interests in silicate behavior. In 1946 I returned to Chicago to complete my PhD, which was granted in June 1947. The war had created much more than just a hiatus in the workings of the university, the faculty, and in the lives of most students. During the war Bowen had gone back to the Geophysical Laboratory, Carnegie Institution of Washington (1942–44), for war work, from which he returned to Chicago as departmental chairman. Bowen broke the long-lasting financial drought in the department: The expense and equipment budget went up by a factor of 30, and staff changes took place. Carey Croneis had gone to Beloit College as president; Krumbein formally resigned; R. T. Chamberlin retired, with Bretz also retiring shortly afterward; Marvin Weller came to Chicago, as did Tom Barth, Walter Newhouse, Lee Horberg, N. A. Riley, and Robert Balk; and in 1947 I was taken on as a Research Associate, as were Hans Ramberg and Kalervo Rankama. Barth then was the prime mover in recruiting several other "Scandahoovians," as they were called: Frans Wickman, who left after one year to become the director of the Riksmuseets (Swedish National Museum), Brunjoff Bruun, and a few others in the area of analytical chemistry and spectroscopy, but they were essentially transients. Bowen shook the place up a great deal, and the university administration respected him and demonstrated that respect by increasing the departmental budget. After two years as chairman he resigned from that position and in 1946 returned once again to the Geophysical Laboratory, where he retired in 1952. Unfortunately (or fortunately?) for me, he left Chicago before I finished my thesis research, and although Tom Barth became my titular advisor, I was really on my own.

I inherited Bowen's laboratory and shortly afterward was able to install hydrothermal equipment, based on the designs of Frank Tuttle. My first graduate student, Irving Friedman, had been associated with Frank at the Naval Research Laboratory during the war. The laboratory, as set up by NLB, consisted of one platinum-wound quenching furnace and one Pt-wound melting or preparation furnace, plus a commercially available Tagliabue temperature controller, a White single potentiometer, and a reflecting wall galvanometer. Only one temperature at a time could be measured. This, plus an old analytical balance, which had been Johannsen's, was *it*. Yet in the late 1930s, it was the only working high-temperature petrological lab in the USA outside of the Geophysical Laboratory! Experimental petrology was pretty well confined to Washington, DC, and then Chicago; the University of Michigan had also obtained a furnace or two, but with few exceptions it never seemed to catch on there.

But this is all prehistory—setting the stage, as it were. The changes that took place right after the war were enormous. The University of Chicago was a truly inspiring place to be—the scientists here, including those brought here during the war for the "Metallurgical Project" [better known as the Manhattan Project (the self-sustaining nuclear reaction)], chiefly in physics and chemistry, were outstanding. I suppose I still mourn the premature death of Enrico Fermi, a man to whom *any* scientist could talk on *any* subject and come away enriched. I shall limit myself to a quick rundown on the postwar revolution in geochemistry and associated geophysics that took place, the seeds of which were planted by the presence of rather few people—Harold Urey, Harrison Brown, Bill Libby, Mark Inghram, Subrahmanyan Chandrasekhar, Andrew Lawson—and fostered by the presence of many others, including Joe and Maria Mayer, Sam Allison, Gerard Kuiper, Cyril Smith, Charlie Barrett, Tony Turkevich, Hans Suess, Clyde Hutchison, George Reed, and others. Curiously, Willie Zachariasen, who became one of my dearest friends, although potent in crystallography and an outstanding mineralogist, interacted little with the other Earth scientists. Urey fathered isotopic geochemistry, and Libby carbon-14 dating; Mark Inghram developed a mass spectrometer that made it possible to work with small samples, opening the field up to the Earth scientists; double beta decay was pioneered by Inghram and John Reynolds, and also a search for extinct short-lived nuclides was carried out by Jerry Wasserburg and Richard Hayden. The cross-disciplinary interactions were the guide to future developments. No one was afraid to talk to anyone else in a different field, and this was particularly important for the young people. It was at this time that new and innovative techniques were introduced by Inghram and others into other fields such as geochemistry, astronomy, and cosmology. Harrison Brown, along with Harold Urey, gave a major boost to the work in meteorites and cosmic abundances begun by V. M. Goldschmidt in Norway and W. D. Harkins at Chicago prior to 1920. G. N. Lewis stated, presumably referring to the odd-even rule, "It was Harkins who first called attention to the striking connection between the atomic weights of the elements and their abundance, not only in the earth's crust, but in the meteors."

When Bowen left Chicago, W. H. Newhouse became chairman of the Department of Geology. Walter Newhouse was an idealist and a reformer, whose mission was to eliminate the trivial that he felt cluttered up much of geology. He was a completely dedicated, totally honest and forthright man whose directness and attitudes angered some elements of the department. In addition to his fierce determination to modernize and eliminate traditionalism, personal problems became intertwined with his science and administration, and unpleasant conflicts developed within the department, ultimately leading to the departure of at least three faculty members. Perhaps unrelated to this was the fact that Harold Urey, needing more space than available to him in the chemistry department, wanted to move his laboratory to the basement of Rosenwald Hall, the principal home of the Department of Geology. For reasons not clear, possibly because of Newhouse's concern with domination from an "outside" source, he was turned down and ended up in the Research Institutes building, several blocks away. Urey was resented by some as an outsider. At the time I certainly had no basis to fault Newhouse's decision, for Harold (he *insisted* that I call him that, though the fact that he was the same age as my father and his Nobel Laureate status made it *most* difficult at first) wanted me to work with him on diamonds in meteorites. I declined, feeling I had to do "my own thing." By that time many, if not most, of Urey's laboratory colleagues were students, or recent PhDs, from the Department of Geology. Heinz Lowenstam, associate professor of geology, brought his knowledge of carbonate-depositing organisms to the laboratory and has consistently applied mineralogy and geochemistry to organisms and fossils. Sol Silverman was the first student in the department to work with Urey on oxygen in silicates; Harmon Craig, Jerry Wasserburg, Cesare Emiliani, and Irving Friedman all played important and varied roles.

Things were moving fast. In rather few years, new ideas, tools, and outlooks were developed that produced a quiet revolution in the chemistry of the Earth and planets, the implications of which remained for the most part obscure to classical geologists. Even George Kennedy, *far* from a "classical geologist," asked me (sometime later) if I thought isotopic analysis of Earth materials would really prove valuable. Although students in the Department of Geology at Chicago participated and contributed significantly to the revolution, as I have indicated, it extended beyond the department into other disciplinary areas, and the sharpness of focus provided by hindsight gives me a better perspective not only of its importance but of the resistance provided by most formal departments of Earth science. Harold Urey was a remarkable man, and with this same hindsight I would go so far as to say that his importance to Earth science has been matched by very few. His interest in the Earth and planets began with extension of his earlier Nobel Prize work on hydrogen isotopes to oxygen isotopic fractionation as a function of temperature and thus to paleotemperatures. His first published work involving isotopic fractionation in nature and leading to paleotemperatures was in 1932 (with G. M. Murphy) on N and O isotopes in air, Chilean nitrates, coal, and magnetite. In 1934, with S. H. Manian and W. Bleakney, he measured $^{18}O/^{16}O$ ratios in meteorites and igneous rocks. As an outgrowth of discussions with Harrison Brown on compositions of meteorites, he then became interested in the Earth and its origin, and rather soon thereafter he became fascinated with the Moon. He had a large picture of it in his office and would waylay anyone, especially visitors, and expound on it. His concept of the origin of the Moon, as a primitive object, turned out to be wrong, yet the energy (and the authority) with which his case was pursued greatly stimulated research and, as incorrect ideas at times do, proved beneficial. He rather soon expanded his interests to the planetary bodies, and disagreements with Gerard Kuiper, of the Department of Astronomy, led to a (apparently unidirectional) degree of ill feeling that to my knowledge was unique with Urey. His excitement and enthusiasm produced an ongoing series of seminars in the Department of Geology, in which one could witness the workings of his mind, for he thought out loud and did his calculating at the blackboard from the top of his head. He "learned" about geology during this period and in fact reinvented important concepts that were not part of his formal background! The series was ongoing because he would continuously change or modify his views and reschedule something for the next day! It was a wonderful sight to behold. And now? Has any MBA made a study of the number of jobs created in the areas of isotopic analyses and interpretation?

The phrase "revolution in Earth science" is generally thought of, and rightfully so, as the geodynamic or plate tectonic revolution, which related diverse aspects of Earth science to a major theme. How long is the time constant that distinguishes revolution from evolution? In my opinion another revolution had been taking place, but with a longer time constant, in geochemistry, beginning with V. M. Goldschmidt and relating Earth science with the physical chemistry of the cosmos. When I was a student, Bowen (arguably the world's leading petrologist) taught petrology without really considering pressure, except in terms of gas or fluid pressures. He *knew* about the role of pressure as a variable, but it did not enter into his petrogenetic scheme, in magmatic differentiation. These were the days when crustal concerns were dominant, and little or no thought was given to the mantle. The major issue in petrology was the granite controversy. Arguments flew on the matter of granitization vs magmatism, on the issue of solid diffusion in crustal genesis, and other esoterica. I think it fair to say that the first real scientific thinking about rocks developed at that time, largely through the efforts of Bowen. Percy Bridgman at Harvard and J. Johnston and L. H. Adams at the Geophysical Laboratory were the only ones doing or who had done high-pressure research, but this was unusual. Bowen did, however, develop the concept of a petrogenetic grid, which of course involved pressure, particularly as applied to metamorphism: This came when he developed a graduate course in metamorphic petrology at Chicago. Metamorphic petrology was more or less primitive, and my notes in the course are quite thin compared with those in igneous petrology. Granulites, for example, have really prospered—they were uncommon rocks then and known to but a few Scandinavian geologists. They have since somehow undergone an enormous volumetric increase in the Earth. The enormous budding of interest and development in metamorphic processes and rocks didn't really begin until after the war, promoted by laboratory studies at elevated pressures. The development of hydrothermal apparatuses and of higher pressure piston-cylinder and other devices took place, at first with George Morey and then O. F. Tuttle of the Geophysical Laboratory. Mention should be made also of the pioneering use of X-ray diffraction in opposed-anvil devices, including diamond cell designs pioneered by John Jamieson and Andrew Lawson. Lawson was (a very young) chairman of the Department of Physics, doing solid-state physics at high pressures. John did his PhD research with Lawson, although his degree was granted in geology. As a young faculty member, I too used Lawson's high-pressure laboratory and benefited from the insight of a physicist (perhaps tempered by being the grandson of the geologist A. C. Lawson of Berkeley). Jamieson was the first to determine the equilibrium relations of polymorphs (calcite-aragonite) by measuring relative solubilities in a high-pressure cell.

Little use had been made of thermodynamic calculations, in large part because few thermochemical data were available. A. L. Day and E. T. Allen's derivation of the melting curves of the plagioclase feldspars from ideal solution theory, and Bowen's elaboration of this, plus his location of the solidus curve prior to World War I, were about it. To my knowledge, Hans Ramberg was the first, using a combined field and theoretical approach, to seriously apply thermodynamics to rock-forming processes, and his 1951 paper with George Devore on element partitioning of Fe^{++} and Mg^{++} between phases as a function of temperature (still referenced in 1990) is a major conceptual advance in geothermometry. This was added to the pioneering work at the Geophysical Laboratory, including that of Bowen, Leason Adams, George Morey, Roy Goranson, George Tunell, and others on homogeneous and heterogeneous equilibria and the creation of the acid-solution calorimetric laboratory by F. C. Kracek, T. G. Sahama, and K. J. Neuvonen in the late 1940s, based on the pioneering work of K. K. Kelley. The calorimetry of silicates was first carried out in the Geophysical Laboratory by W. P. White earlier in the century, and solution calorimetry of geologically important substances carried out by K. K. Kelley at the Berkeley, California, laboratory of the US Bureau of Mines. Hans Ramberg also set up an acid-solution calorimetric laboratory at Chicago, and Dick Robie, who did his PhD work in calorimetry with Ramberg and in low-temperature calorimetry with J. W. Stout in chemistry at the Research Institutes, went to the US Geological Survey and there developed a world-renowned laboratory. These and other developments at the Geophysical Laboratory, including high-pressure apparatuses so important to experimental mineralogy and petrology, have been well covered by Yoder[1]; I am attempting to recount here, for the most part, the happenings that took place essentially on my home turf. Incidentally, and while mentioning names, Hat Yoder received his SB degree at Chicago in 1941, was tutored in meteorology by Carl-Gustaf Rossby, commissioned in the Navy, returned in 1946 for one quarter in the "hot lab" with Bowen, and then (after being advised by Bowen that high-pressure facilities would not be available at Chicago) missed all the excitement by going off to MIT. He ultimately became director of the Geophysical Laboratory.

Shortly after the war our department became alerted to the possible availability of German scientists; Tom Barth, in particular, knew Fritz Laves. In 1947 I had been granted one of the early Office of Naval Research (ONR) contracts, for the study of order-disorder phenomena in silicates, and arrangements were made with the US Navy to bring Laves to Chicago as a "Paperclip Specialist," the code name for German scientists who were willingly brought here by the navy without the knowledge of immigration and customs. Laves was put under my care (he would now be called an "illegal"), under the aegis of my ONR contract. I felt foolish, for he was a mature, internationally known and respected crystallographer and mineralogist, a student of V. M. Goldschmidt, yet he was my ward and was paid by my contract. Ushering him, after the fact, through immigration and customs was quite an experience: The officials didn't like it one bit!

[1] See Yoder, H. S. Jr. 1989. Scientific highlights of the Geophysical Laboratory, 1905–1989. In *Annual Report of the Director, Geophysical Laboratory*, pp. 143–97. Washington, DC: Carnegie Inst. Washington.

Although I held the title of research associate, I was in fact a faculty member, embedded in the university budget, and had students and gave classes. These were the days before research associates were equated with "postdocs," before the concept of "soft money" was part of the system, and I don't recall ever hearing of "summer salaries." My contract paid all of the stipends of Irving Friedman, Gunnar Kullerud (who did his research in my laboratory, but got his ScD at Oslo!), and Ursula Chaisson (now Ursula Marvin); Fritz Laves and Tom Barth were also involved. Kalervo Rankama and Hans Ramberg were research associates at that time, and Frans-Erik Wickman's status was that of a fellow for one year, because he had accepted the position of director of the Riksmuseets in Stockholm. My collaboration with Laves, which lasted until 1954, when he took Paul Niggli's post at the ETH in Zurich, was a fruitful and highly rewarding one. I cannot say too many kind things about Laves and how he treated this tenderfoot as an equal. Incidentally, the ONR experience itself was rewarding, for their support of areas of science not directly related to the navy's mission was generally recognized as a model for the way government money should be used to support research, with no strings attached. The ONR became the organization that showed the way for the National Science Foundation, first funded in 1951.

In the "early days" of ONR support, the naval officers in charge didn't quite know how to rank civilians. The admiral in charge of the Chicago branch office apparently considered me (with a PhD) to be the equivalent of a moderately high-ranking officer, but it was amusing to watch the way these contacts were (softly) handled. My contract officer was a metallurgist and quite interested in what happened to my metal (mostly stellite) pressure vessels in the experimental work, and much less so about what happened *inside* the "bombs."

A second personal fruitful collaboration was with Donald Graf, of the Illinois Geological Survey, who for some years came to Chicago one day a week. Oiva Joensuu, our wonderful optical spectroscopist, who did first-rate chemical analyses with now obsolete apparatuses, called him "Mr. Friday." We were both interested in carbonates, and got together after Keith Chave, then a student of Heinz Lowenstam's, introduced me to the problems of Mg in calcite. Keith had no intention of looking into the laboratory determination of equilibrium phase relations. I get the impression that much of this work in the 1950s has held up, and the concept of protodolomite, or poorly ordered Ca-rich dolomites, is ensconced in the lore of sedimentary petrology. My venture in carbonates was made more entertaining by the willingness of carbonates to undergo reactions, including order-disorder equilibria, that were kinetically difficult or impossible in the laboratory with the more recalcitrant feldspars.

At that time the research of chemists, physicists, and astrophysicists was certainly of more concern to those in the Earth sciences than vice versa. At some point between the days of the Chamberlin-Moulton association and circa 1940, geology seems to have lost its dignity. Urey's interest in and major contributions to Earth science were part of a process that helped make "geology" an active component of modern science. The issue of the "softer" nature of the geological sciences has been around for a long time, and the perception of geology as an easier way to satisfy science requirements than the more rigorous physics or chemistry has been long known, and remains so today. Although when I was a student this matter did not seem to be openly used as a discriminatory factor within the faculty of the Division of the Physical Sciences (Mathematics, Statistics, Chemistry, Physics, Geology, Astronomy), it *was* there, however, although Bowen did his part to reduce it. After the war several factors came into play to help alleviate the situation: (*a*) the presence of a group of outstanding faculty of different disciplines who had been involved with the "Project" and thus had *lived* with close interaction and cooperation for some time; (*b*) the formation of interdisciplinary research institutes incorporating scientists from several departments in a new building; (*c*) the presence of a group of mature, serious students whose graduate careers had been delayed by the war; and (*d*) the intermingling of students from geology, chemistry, physics, and astrophysics in the laboratories of Harold Urey, Harrison Brown, Mark Inghram, and others.

In addition to those people from this period whose names have already been mentioned, to give an idea of the products of what I could call this golden age, I shall list, at random, the following people who were at Chicago as faculty, students, or research associates. The cutoff in time that separates the names listed from those who followed them is both arbitrary and ill defined, but it does not get too deep into the 1950s. Omissions can thus be blamed on their youth and/or my weakness of memory. To keep the length of this piece down, I do not identify their fields or accomplishments, but many may be known to you:

Sam Epstein	Robert Nanz
Clair Patterson	Stanley Miller
Toshiko Kuki Mayeda	Jack S. Kahn
Edward Goldberg	James Arnold
George Tilton	Johannes Geiss
Peter Eberhardt	Anthony Turkevich
Fred Begemann	John A. S. Adams
Truman Kohman	Thomas Sugihara
Giovanni Boato	Edward Martell

George Reed	Ernest Nickel
Ernest Anderson	D. C. Hess
A. D. Suttle	Sherry Rowland
Peter Baertschi	Ernst Schumacher
Jacob Bigeleisen	H. B. Wiik
George Wetherill	Denis Shaw
Edward Olsen	Robert Ginsburg
H. Hamaguchi	Bertram Donn
H. Kigoshi	William Chupka
Haro Von Butlar	Charles McKinney
John McCrea	Edward Chao
Ray Siever	Leon Atlas
Meyer Rubin	Ernest Ehlers

This list and the preceding names include seven Day Medalists of the Geological Society of America, fourteen members of the National Academy of Sciences, three Nobel Prize winners, one Vetlesen Prize winner, and one recipient of the Crafoord Prize. Out of this environment, the following people personally influenced my scientific career the most:

1. J Harlen Bretz, whose full name was really Harley Bretz. The J (no period!) was added, and the spelling of Harley changed, for the sake of dignity—and that from the least dignified person I had met until that time!

2. Norman L. Bowen, my mentor and role model. His humor seemed not apparent to those who didn't know him well. His nickname, "Ham," did not originate as a play on words (ham bone), but rather was the result of neighborhood kids hearing and misinterpreting his Welsh father pronounce, with amusement, the shortened form of "Harmon," as "Hahm," after NLB's Sunday school report card came back as Harmon Bowen.

3. Tom F. W. Barth, who got me concerned with order-disorder phenomena and other crystal-structure matters. His work, with E. Posnjak, on "variate atom equipoints" helped open my eyes. Barth really didn't like V. M. Goldschmidt, perhaps out of jealousy, and once told me that Goldschmidt had never really done anything important. For a time, to avoid invidious comparison, he would introduce me and spell g-o-l-d-S-M-I-T-H.

4. Frans-Erik Wickman, who in one short year inspired me with his clear thinking in crystal chemistry and other structural matters.

5. Hans Ramberg, who arrived on the scene at the same time that Wickman and I did. Considered unconventional (to say the least!) by some, and certainly controversial, his originality and insight shone through,

and he was the first to show me the value, in mineralogy and petrology, of practical thermodynamics. In my opinion Ramberg stands as a major figure in the modernization of petrology, and hopefully his contributions will not be forgotten before being fully appreciated.

6. Fritz Henning Emil Paul Berndt Laves, a cultured, kind friend, who unselfishly shared a great deal of science with me and also showed me a view of life and history and its interaction with science in Europe through the eyes of one who had come from Nazi Germany.

The postwar developments directly influenced at least *two* other institutions over and above any induced by the normal dispersion of students and research associates. The first blooming of Caltech resulted when Robert A. Millikan was brought from Chicago in 1921 as president. Prior to his arrival, it was a small local institution, of little reputation. In his autobiography, Millikan says, "The institution was indeed a very weak institution, with practically no endowment, but with three buildings on campus...." The Earth sciences at Caltech got another great shot in the arm after World War II when Harrison Brown, Heinz Lowenstam, Clair Patterson, Sam Epstein, Charles McKinney, and (shortly afterward) G. J. Wasserburg all went from Chicago to Pasadena. I was personally most strongly affected by Sam's departure, for we had planned a joint study of oxygen isotopic equilibrium at elevated temperatures and pressures, a field of research that thus lost the opportunity to get started ahead of its time. Thirty years elapsed before I took up isotopic work, and then not with Sam, but with his first graduate student, Bob Clayton, and with Tosh Mayeda, who *didn't* leave for Caltech and is still energetically here. We have, with colleagues, been working on oxygen isotopic fractionation equilibria at high pressures between a variety of phases, first with water as the exchange medium, and now with $CaCO_3$ and directly with CO_2. Reactions are made possible in the dry systems by a large rate enhancement at pressures greater than those used (1–2 kbars) in earlier experimental work. In addition to fractionation factors useful for geothermometry, we are looking at diffusion and reaction mechanisms.

The second California migration was to San Diego, when in 1955 Harmon Craig and Hans Suess joined the faculty there along with Walter Elsasser from Johns Hopkins to become the first three people brought to the University of California at San Diego by Roger Revelle to build a new way of life. Harold Urey joined them after retiring from Chicago in 1958, and afterward Harmon and company brought Stanley Miller, Jim Arnold, Joe and Maria Mayer, Walter Kohn, and others from Chicago.

The cultural difference that I sensed as a student (and later) between the different sciences was perhaps as real as C. P. Snow's "two cultural" distinction between scientists and humanists. Urey and his colleagues in several departments helped, in my opinion, in reducing the cultural differences not only between Earth scientists and those in the "harder" sciences, but also between the various types of Earth scientists. I hasten to add that the concepts of plate tectonics also played an important and even pivotal role in producing convergence of the subcultures (and I learned that although apples and the Earth are approximately spherical, they have little else in common, wrinkles notwithstanding). As a student and young faculty member, I was made aware of the fact that I was not a "field geologist," or that I did not do field work. A certain amount of disdain existed between the experimentalists and the khaki-pants-and-scuffed-boots group, although perhaps neither side was free of a concealed touch of envy and even respect. There are obviously things that cannot be resolved by field work alone, just as there are matters that would never be considered by an experimentalist or theoretician isolated from the outside world. Today, field work and theoretical/experimental work, both of which are essential, are not the exclusive domains of disparate groups, but a good geologist may be competent in or at least conversant with a variety of fields; not only is there a greater understanding between disciplines, but there is also increased mutual respect. Pioneering work on the "geochemistry" of the solar system and beyond by Robert Clayton, Edward Anders, Jerry Wasserburg, and Lawrence Grossman—work that bears on the origin of meteorites and planetary bodies—has had an impact far beyond the Earth science community. New frontiers have continued to develop. The intrusion of fluid dynamics into geology was first applied to postglacial rebound, and then to mantle convection and convection in magma chambers. One might say it has helped stir up geology.

Allow me to recount a history of internal unification that, although local, I feel to be of some significance to the Earth science community. Meteorology was a latecomer to Chicago, beginning in 1940 with the arrival of Horace Byers, who was instrumental in setting up an institute within the Department of Physics. In 1941 Carl Gustaf Rossby came to Chicago as its first director; he had started a Department of Meteorology at MIT in 1928 (the first in the US) and in 1939 went to Washington as assistant chief of the US Weather Bureau. During World War II meteorological cadets were trained for the Air Force, and in 1944 the institute became the Department of Meteorology. In 1960–61 I was the associate dean of the Division of the Physical Sciences, and Willie Zachariasen was the dean. At that time the Department of Geology remained relatively isolated on a campus containing a variety of people in other departments also concerned with some aspect of the Earth and planets. This seemed more and more to be an artificial fragmentation of talent. At this time, fluid dynamics, applicable to the atmosphere, hydrosphere, and the solid Earth, was traditionally treated independently in separate departments or even institutions, with little or no interchange of ideas. Attempts to shape cooperative science usually fail, but inducing the intermingling of independent scientists with the attendant interchange of thoughts and inspirations can be very fruitful. Willie Zach wholeheartedly agreed with me that a condensation of sorts would be worthwhile, perhaps in part because of a dean's desire to reduce the number of areas of responsibility under his wing and thus simplify the administration of the division. He appointed me chairman of a committee to investigate the matter, and our 1960 recommendation, with the unanimous consent of the faculties of the Departments of Geology and of Meteorology, resulted in a merger of the two in 1961; at the same time joint appointees from the Department of Chemistry and the Research Institutes (Clayton, Anders) and the Department of Astronomy (Joe Chamberlain) joined the new department. Bill Reid became a joint appointee with the Department of Mathematics shortly thereafter. Why shouldn't fluid dynamicists talk to paleontologists? William McNeil, historian, expressed amazement, saying that it was probably the first time in recorded history that two departments had, by choice, given up their individual autonomy. The institute concept, so successful during and after the war in bringing and keeping people together, was extended to a unified department, an educational as well as research unit.

The formalization of the new department took much less effort than the choice of a name! Name selection became tangled with distinction. The term Earth Sciences, which embraced more of the faculty than most other designations (exclusive of meteoritic and planetary types), was looked down upon by several, perhaps self-consciously for reasons already mentioned (Earth is a dirty word). The least troublesome name turned out to be "The Department of the Geophysical Sciences," soon abbreviated to DoGS. This name was even approved of by paleontologists and geochemists, perhaps seasoned by the name of the Geophysical Laboratory, long respected, even if it might have been more aptly named "The Geochemical Laboratory." The "new" department, now 30 years old, has prospered and grown. Times have indeed changed, for the (unwarranted) complaints we received 30 years ago from former students (for the most part geologists) that we had deserted field studies and the traditional subjects are no longer heard. It seems to us that the study of the Earth and its environment is best handled by all concerned with it, and that rigid, old-fashioned subdivisions served no purpose other than to stifle knowledge and obstruct free interchange of ideas and information. A larger view has evolved during my lifetime, one in which I feel my university has played an important role. Not only is there a kinship between various types who deal in the workings and history of the Earth, but with those who look to the planets and to the stars beyond. A more recent interest in Earth science by a host of physicists, chemists, and computer modelers appears to have been promoted by fear—a belated concern with the health of the planet. Many outside of the traditional Earth science community are now involved in the physics and chemistry of environmental changes. As one who sees little cause for optimism in the behavior of people in the world at large, including their diminishing interest in science at a time when its importance to the world is growing, it is heartening to note that great progress has been made in the extension of interests and the broadening of vistas of geology into areas formerly inhabited for the most part by scientists in other disciplines, as well as by an increasing interest of physicists, chemists, and applied mathematicians in the Earth sciences.

ACKNOWLEDGMENTS

I am indebted to the old (University of) Chicago Mafia for unknowingly providing the inspiration for this work. I also want to thank a variety of friends and colleagues for filling in memory lapses, improving expression, corrections, and admonishments. These include Bob Clayton, Harmon Craig, Sam Epstein, Clyde Hutchison, Mark Inghram, Tosh Mayeda, Frank Richter, Tony Turkevich, Jerry Wasserburg, and Hat Yoder.

James A. Van Allen

Annu. Rev. Earth Planet. Sci. 1990. 18:1–26

WHAT IS A SPACE SCIENTIST? AN AUTOBIOGRAPHICAL EXAMPLE

James A. Van Allen

Department of Physics and Astronomy, University of Iowa, Iowa City, Iowa 52242

INTRODUCTION

Space science is not a professional discipline in the usual sense of that term as exemplified by the traditional terms astronomy, geology, physics, chemistry, and biology. Rather, it is a loosely defined mixture of all of these fields plus an exotic and expensive operational style. The distinctive features of space science are the use of rocket vehicles for propelling scientific equipment through and beyond the appreciable atmosphere of the Earth; the rigorous mechanical, electrical, and thermal requirements of such equipment; and (usually) the remote control of the equipment and the radio transmission of data from distant points in space to an investigator at a ground laboratory. Space science is primarily observational and interpretative; it is directed toward the investigation of natural conditions and natural phenomena. But it can be and sometimes is experimental in the sense that artificial conditions are created and the consequences observed. Most space science has been and will continue to be conducted by unmanned, automated, commandable spacecraft. But some is conducted by human flight crews performing direct hands-on manipulation of equipment. The latter mode of operation is of dubious efficacy and, in any case, will probably be the technique of choice only in specialized subfields involving preliminary laboratory-type experiments under free-fall or low-g conditions.

The personal and professional backgrounds of space scientists are diverse, as is commonly the case in new and interdisciplinary fields. In accepting the invitation of Editor Wetherill of the *Annual Review of Earth*

0084–6597/90/0515–0001$02.00

and Planetary Sciences to write an autobiographical account of my career as a space scientist, I did so with a full realization of the diversity and individualism of those who belong to the fraternal order of space scientists. My account is a personal one and does not include references to primary sources, as would a proper scholarly paper. Some of this account is abridged from my monograph *Origins of Magnetospheric Physics* (Smithsonian Institution Press, 1983), but most of it is not.

PARENTAGE, BOYHOOD, AND EARLY EDUCATION

I was born to Alfred Morris and Alma Olney Van Allen on the 7th of September 1914, the second of their four sons, in Mount Pleasant (population then about 3000), Iowa, the county seat of Henry County. My mother grew up near Eddyville, Iowa, on a small farm that her father had inherited from his father, who had moved from Ohio to Iowa in the mid-1840s. My paternal grandfather, George Clinton Van Allen, was one of 11 children of Cornelius and Lory Ann Van Allen, the former a shipbuilder in Pillar Point, New York, at the eastern end of Lake Ontario. He attended Wesleyan University in Connecticut for two years and later studied law and became proficient in land titles and surveying. He passed through Mount Pleasant in 1862 as a member of the survey party that was laying out the route of the Burlington and Missouri River Railroad, which later became part of the Chicago, Burlington and Quincy (now the Burlington Northern) Railroad. (Several of my prized possessions are the magnetic compass and the drafting instruments that he used.) In late 1862, with his wife of five years, Jennie, he settled in Mount Pleasant, built a small house, and established a law office. My father, an only child, was born in Mount Pleasant in 1869. He attended the local public schools and Iowa Wesleyan College, and then studied law at the University of Iowa in Iowa City, receiving an LLB degree in 1892. He joined his father as a practicing lawyer and continued to practice law for the remainder of his life.

My boyhood activities were centered within our closely knit family, which had a strong resemblance to earlier pioneer families.

The virtues of frugality, hard work, and devotion to education were enforced rigorously and on a daily basis, especially by my father. My mother exemplified the pioneer qualities of affection and nurture for her husband and their children and of comprehensive self-reliance: cooking all meals from scratch, baking delicious bread twice a week, washing clothes with a washboard and tub, maintaining a meticulous standard of household cleanliness, canning large quantities of fruits and vegetables, and, most important of all, ministering to her children through health and frequent sickness during the many epidemics of those days. Before her marriage, she had taught in one-room country schools near Eddyville and had attended the Iowa Wesleyan Academy for two years.

My first clear recollection is waving a tremulous farewell to her as I set off on foot to kindergarten a few days after my fourth birthday. Two months later, my older brother and I went with my father to the public square in Mount Pleasant to witness the celebration of the armistice of World War I by a horde of raucous and exuberant people of all ages. The culmination of this celebration was the burning of a huge straw-filled effigy of Kaiser Wilhelm.

I enjoyed school work greatly under the guidance of devoted teachers, most of whom were unmarried woman who had gone into teaching as a durable profession. Our father read to my brothers and me for about an hour after supper nearly every evening—from the *Book of Knowledge*, *An Illustrated History of the Civil War*, the *National Geographic* magazine, and, occasionally, from the *Atlantic Monthly*. Then he shooed us off to our respective corners to do our homework for two or three hours. Our chores varied with the seasons. We raised a large flock of chickens year-round. In the summer we planted and cultivated a one-acre vegetable garden and a large apple orchard, and in the winter we split wood for the cook stove, shoveled snow, ran errands, fired the furnace, and tried to keep warm. We had a car but seldom used it, even during the summer. During the winter, the car was set up on wooden blocks in the barn to "save the tires." For the most part, we walked everywhere.

I was intensely interested in mechanical and electrical devices. *Popular Mechanics* and *Popular Science* were my favorite magazines. I built elementary electrical motors, primitive (crystal) radios, and other devices described therein. Two highlights were the construction of a Tesla coil that produced, to my mother's horror, foot-long electrical discharges and caused my hair to stand on end; and the complete disassembly and reassembly of those mysterious "black boxes"—the engine and planetary transmission of an ancient Model T Ford that my older brother and I had bought for $25 (later recovered on resale).

In high school my favorite subjects were mathematics (including solid geometry), Latin, grammar, and manual training (woodworking). As a senior in 1930–31, I had my first course in physics, with many opportunities for laboratory work, a memorable experience. During the same year I edited the senior annual *The Target*. I graduated from Mount Pleasant High School in June 1931 as class valedictorian. My valedictory oration was entitled "Pax Romana—Pax Americana," based on my study of Roman history in school and on my father's tutelage. The thesis of this oration was that America, by virtue of its economic, cultural, and military stength, would dominate world affairs and enforce world peace for a limited period of history but would then lose its influence because of its preoccupation with "bread and the circus games."

COLLEGE AND GRADUATE WORK

Throughout my boyhood, there was never any doubt that my three brothers and I would go to college and have an opportunity to "amount to something." The matter was not subject to discussion. In the autumn of 1931, following in the footsteps of my father, mother, and older brother George, I entered Iowa Wesleyan College in Mount Pleasant. The tuition was $45 per semester, and I lived at home. The academic work was demanding, and I took all the courses offered there in physics, chemistry, and mathematics (four years of each), a summer field course in geology, and the one available course in astronomy (using Moulton's 1933 *Astronomy*, which I still have), the only formal course in astronomy that I ever took. Professor Thomas Poulter in physics and Professor Delbert Wobbe in chemistry were my principal inspirations. Each was the one-man faculty of his respective department. I wavered between choosing physics or chemistry as my major but decided on physics after Poulter offered me a part-time student assistantship. I worked in his high-pressure research laboratory and learned to blow glass, to run a metal-turning lathe and a milling machine, and to braze, silver-solder, and weld. More importantly, I came to have an almost worshipful regard for his mechanical ingenuity, his intuitive use of physics and chemistry as a way of life, and his devotion to experimental research. Poulter was in the process of preparing for his role as chief scientist of the Second Byrd Antarctic Expedition, a part of the Second International Polar Year. Following my freshman year I became a part of those preparations. I helped build a simple seismograph and was entrusted with checking out a field magnetometer on loan from the Department of Terrestrial Magnetism of the Carnegie Institution of Washington (DTM/CIW), one of the most beautiful instruments that I have ever seen. In the autumn of 1932 I used this instrument to make precision measurements of the geomagnetic field at three ad hoc locations in Henry County. The measurements involved also the determination of latitude and longitude by observation of the Sun with the theodolite on the magnetometer. All of this was done by carefully following the third edition of Daniel L. Hazard's *Directions for Magnetic Measurements* (US Department of Commerce, Serial Number 166, 1930). I copied my field notes onto clean forms and mailed them proudly to John A. Fleming, then director of DTM/CIW, as a modest contribution to the world survey that was underway. I received a prompt acknowledgment from him that concluded by making it clear that only raw field notes could be accepted as valid. I then sent him those, thereby learning a durable lesson about the sanctity of raw data.

My other introduction to geophysical research was serving as an observer of meteor trails during the Perseid shower of August 1932. Arrangements for the observations were worked out between Poulter and astronomy professor C. C. Wylie of the University of Iowa; sky "reticles" devised and built by Poulter from welding rods were used to carry out the observations. These 6-ft-long conical devices with an eye-ring at the vertex and a coordinate system of radial and circular rods at the other end were mounted on fixed stands. One was located in my backyard in Mount Pleasant and the other in Iowa City, 50 miles to the north. The conical fields of view were positioned so that they included a common volume of the atmosphere spanning the estimated altitude range of meteoric luminosity. During the early morning hours of 22 August, Raymond Crilley manned the Iowa City reticle, and I the Mount Pleasant one, using accurate watches for coordination. Each of us observed about 20 bright meteor trails. Of these, Wylie identified seven as identical cases. He later published the calculated altitudes of the beginning and end points of each of these trails. At the time, I had the impression that this was the first successful attempt to make such measurements, and the impression provided part of the thrill of making them. Later, I learned that my impression was not true. During the ensuing Antarctic expedition Poulter used this system to obtain one of the world's most comprehensive sets of observations of meteor trails. Also, he made extensive use of the DTM/CIW magnetometer and the seismograph that I had helped construct. The 1935 graduation ceremony at Iowa Wesleyan College included a public parade honoring Poulter and Admiral Richard E. Byrd. The latter gave the commencement address. I graduated summa cum laude and was the first student to walk across the platform. Poulter moved forward to congratulate me, but I was so flustered that I scurried past him, clutching my diploma.

During the summer of 1934, I went by automobile to California with my mother, father, and two of my brothers to visit prospective graduate schools in the west. Two of my most pleasurable recollections were visits to the laboratories of Jesse Du Mond at Caltech and Paul Kirkpatrick at Stanford. My eyes popped at the elegance and scope of their laboratories, and I was deeply grateful for the careful explanations of their research that they gave me, a young kid who had dropped in uninvited. But in the end I followed my family's tradition of attending the University of Iowa. In 1935, the faculty of its Department of Physics numbered five: George W. Stewart (head of the department since 1909), John A. Eldridge, Edward

P. T. Tyndall, Claude J. Lapp, and Alexander Ellett. The latest addition occurred in 1928 with Ellett's arrival. My assigned advisor was Tyndall, a warm-hearted and spirited individual with a PhD from Cornell University. My central preoccupation was with introductory graduate-level courses based on Slater and Frank's *Introduction to Theoretical Physics*, Abraham and Becker's *Classical Electricity and Magnetism*, and Pauling and Wilson's *Introduction to Quantum Mechanics*; on instructors' original lectures on classical mechanics, statistics, and partial differential equations; and on lectures and laboratory studies in atomic physics. I found the work to be rigorous and demanding.

I was eager to start research, and soon after my arrival Tyndall introduced me to the art of growing large single crystals of spectroscopically pure zinc and of measuring their physical properties. I completed an MS degree in June 1936 with an original experimental thesis, "A Sensitive Apparatus for Determining Young's Modulus at Small Tensional Strains." By that time Ellett, who formerly worked with atomic beams, was actively converting his research interests to the new field of experimental nuclear physics. I decided to join in this work. Together with Robert Huntoon, a more senior graduate student, and others, I helped build a copy of the famous Cockroft-Walton high-voltage power supply and accelerator. Our capacitors were made of plates of window glass on which we glued aluminum foil; the rectifiers and the accelerator tube used glass cylinders from a local company that supplied them to service stations for the then prevalent model of gasoline pumps. Everything was improvisation. Central elements of the measuring equipment were an ionization chamber and a Dunning-type pulse amplifier with a voltage gain of about one million, built with vacuum tubes of course and a nightmare to shield adequately against pickup of AC ripple and coronal discharges, of which we had a plethora. Because of the absence of air conditioning or any effective humidity control, operation during the summer was impossible. But on a good day in the autumn of 1938, we finally got an ion beam of a few microamperes with an accelerating potential of 400 kV. My objective was to measure the absolute cross section of the reaction

$$H^2 + H^2 \rightarrow H^1 + H^3$$

over as great a range of bombarding energy as possible. The novel feature of my experiment was the use of a gaseous (i.e. infinitesimally thin) target, which involved the controlled flow of deuterium gas through the custom-built reaction chamber. After several months of fixing leaks in the vacuum system, replacing burnt-out filaments in the rectifiers, repairing damage from high-voltage spark-overs, etc, etc, I finally got everything to work at the same time. With the help of a fellow graduate student, I then made a continuous run of 40 hr, being unwilling to turn off anything because of the well-founded expectation that many weeks might be required to restore full operation. However, with good luck, I was able to make a confirmatory run two weeks later. These two runs provided the basis of my PhD dissertation, which, with Ellett's approval, I then wrote up under the title "Absolute Cross-Section for the Nuclear Disintegration $H^2 + H^2 \rightarrow H^1 + H^3$ and its Dependence on Bombarding Energy" (50–380 keV). I defended my work successfully before the examining committee and received the degree in June 1939.

Following an oral paper that I gave at the spring 1939 American Physical Society meeting, Hans Bethe expressed a keen interest in the results but found that the trend of my curve of cross section vs. bombarding energy was impossible to believe at the lower energies because of basic quantum-mechanical theory. This criticism was unsettling to put it mildly. Ellett and I went over the entire matter critically and eventually realized that my method of measuring the beam current through the reaction chamber was faulty. I had collected the ion beam in a Faraday cup *after* it had passed through the chamber and had measured the charge collected per unit time there. I failed to take account of the partial neutralization of the beam by charge exchange in the target gas, an effect of increasing importance at the lower energies. As a result, the measured current was too low and the calculated cross section was correspondingly too large. A follow-on experiment by Stanley Atkinson, using the same apparatus, established the magnitude of this effect and corrected my results.

Many years later, the cross section of the deuteron-deuteron reaction at much lower energies became a matter of importance in the development of equipment for the current major effort to achieve controlled fusion in the laboratory.

DEPARTMENT OF TERRESTRIAL MAGNETISM OF THE CARNEGIE INSTITUTION OF WASHINGTON

Concurrently with the early nuclear physics work at Iowa, Merle Tuve, Lawrence Hafstad, and Odd Dahl had built a Van de Graaff (electrostatic) power supply and an ion accelerator tube at DTM and had succeeded in getting a stable beam at bombarding energies up to 1 MeV. The principal emphasis of their early work, under the urging of theoretician Gregory Breit, was the careful measurement of the proton-proton scattering cross section, then regarded as one of the most fundamental problems in nuclear physics. Norman Heydenberg, also one of Ellett's former students at Iowa, was one of Tuve's principal collaborators. In the spring of

1939 Ellett recommended me to Tuve, and I received a Carnegie Research Fellowship to work at DTM.

Earlier, in late 1938, Otto Hahn and Fritz Strassman in Germany had discovered nuclear fission. The DTM laboratory was converted to confirmatory experiments, which were successful. More importantly, Richard Roberts discovered the delayed emission of neutrons from fission products. This discovery provided the basis for the control of nuclear fission in all subsequently developed nuclear power plants.

My own work at DTM during 1939–40 was the measurement of the absolute cross section for photodisintegration of the deuteron by 6.2-MeV gamma rays from protons on fluorine. This was done in collaboration with Nicholas Smith, another Carnegie fellow, formerly at the University of Chicago. In addition, Norman Ramsey, yet another Carnegie fellow, and I measured neutron-proton cross sections using a small proportional counter that I had devised for observing the recoil protons.

Of much greater importance to my future career was my crossing of the culture gap at DTM from nuclear physics to the department's traditional research in geomagnetism, cosmic rays, auroral physics, and ionospheric physics. I was impressed especially by the work of Scott Forbush and Harry Vestine. Also, there were occasional visits by Sydney Chapman and Julius Bartels, who were then completing their great two-volume treatise *Geomagnetism*. As a result, my interest in low-energy nuclear physics dwindled, and I resolved to make geomagnetism, cosmic rays, and solar-terrestrial physics my fields of research—at some unidentified future date.

PROXIMITY FUZES

By late 1939 the war in Europe was already several months old, and Tuve foresaw the inevitable involvement of the United States. He abandoned experimental work and turned his remarkable talents to the problem of what scientists in the United States should be doing to help remedy the desperately inadequate quality of our military establishment. He made intensive inquiries, especially among high-ranking naval officers, and returned to DTM with a vivid impression of the ineffectiveness of anti-aircraft guns and with full knowledge of the embryonic British work on proximity fuzes for eliminating the range error of time-fuzed projectiles. He seized on this as *the* matter to which he would devote his own staff and, by recruitment, other physicists and engineers of kindred inclination— including Ellett from Iowa and Charles Lauritsen, his son Thomas, and William Fowler from Caltech. As a Carnegie fellow, I was apart from these early efforts, but by the summer of 1940 I asked to become a part of this enterprise and was appointed to a staff position in Section T (for Tuve) of the National Defense Research Council (NDRC) of the newly created Office of Scientific Research and Development, headed by Vannevar Bush.

I worked first on a photoelectric proximity fuze and succeeded in solving the basic problem of making a circuit such that the fuze would have equal sensitivity over a large range of ambient light levels. My circuit gave an output approximately proportional to the logarithm of the current from a photoelectric cell by using a fundamental characteristic of a vacuum-tube diode. My demonstration of a breadboard of this circuit to Charlie Lauritsen and Willy Fowler showed that I got the same size pulse by waving my hand in front of a photocell when illuminated by full sunlight as I got in a darkened room. Their exuberant response not only made my day, but it also propelled the photoelectric fuze into the realm of serious consideration.

But soon thereafter, I was transferred to work on the radio proximity fuze. Dick Roberts had built a simple self-excited rf oscillator operating at about 70 MHz after the fashion of the one that the British called an autodyne circuit. In brief, the plate current of the one-tube oscillator with a short antenna was affected by the reflected signal from a nearby conductor. The basic scheme was that the transient pulse as a fuze passed an aircraft could be amplified so as to trigger a gaseous tube (thyratron) to fire the detonator of the projectile. This device became the focus of a truly huge development.

For the first time in my life I worked under conditions in which urgency was the motto, multiple approaches to a problem were fostered, money was no object, and the first approximation to a solution was the prime objective. As Tuve put it, " I don't want you to waste your time saving money."

THE APPLIED PHYSICS LABORATORY OF JOHNS HOPKINS UNIVERSITY

The radio proximity fuze group soon outgrew the capacity of DTM, and thus Tuve negotiated an arrangement with Johns Hopkins University (JHU) such that JHU would assume contractual oversight of the project. In early 1942 JHU rented a large Chevrolet garage in Silver Spring, Maryland, and established the Applied Physics Laboratory (APL). Along with other members of the group, I was transferred to APL/JHU in April 1942, thereby qualifying as a plank-owner, as that term is used in the navy for a member of the crew who places a new ship on commission.

My own work was principally on developing what was termed a rugged vacuum tube, i.e. one that would survive acceleration of some 20,000 g as it was propelled through the barrel of a 5″/38 navy gun. The starting point was the miniature vacuum tubes that had been developed for use in electronic hearing aids by the Raytheon and Sylvania companies. I worked principally with tube engineer Ross Wood of Raytheon in the trial-and-error process of remedying the numerous shortcomings of the early tubes. I conducted field tests of each batch of tubes by putting them in a small cylinder that was mounted in a projectile. These projectiles were then fired vertically by a converted 10-pounder gun at a test site in southern Maryland along the Potomac River. We recovered the projectiles with a posthole digger and returned the tubes to the laboratory for detailed scrutiny. (In July 1942, I was commissioned a deputy sheriff of Montgomery County in order to legally carry a loaded revolver for coping with hypothetical hijackers on our daily expeditions to and from the test site.) I would then report the results to Ross by phone or, if we had important conclusions, by personal visit by train to Newton, Massachusetts, where he operated a pilot line. On most of these trips I would return to Silver Spring with a batch of improved tubes. One of the most nagging problems was the breakage of the fine filaments. I reasoned that distortion of the structure that supported the filaments was the cause of the failure. In a moment of inspiration I sketched out a scheme for a minute coil spring (wrapped around a mandrel), to the free end of which one end of the filament would be welded. My hope was that the spring would maintain nearly constant tension on the filament during acceleration in the barrel of the gun, and also that the tension would be such as to tune microphonics outside of the frequency pass-band of the amplifier. Wood executed this idea using the skills of the women who built these tubes with the aid of microscopes. The scheme worked and became an essential feature of the millions of tubes that were manufactured during the three subsequent years of World War II.

By late autumn 1942, the first of the Section T radio proximity fuzes were coming off the production line. Realistic and extensive testing at the Dahlgren Proving Ground over the Potomac River ("airbursts" as the projectile approached the water) and past an aircraft suspended between two towers at Jack Workman's test facility near Socorro, New Mexico, had been conducted. Despite numerous duds and premature bursts, it was estimated that the effectiveness of naval antiaircraft fire would be increased by a factor of order five if the proximity fuzes were substituted for the time fuzes then in use throughout the fleet.

In early November 1942, the Naval Bureau of Ordnance (Bu Ord) determined that the fuzes were ready for issue to the Pacific Fleet. Neil Dilley, Robert Peterson, and I were given spot commissions as United States Naval Reserve line officers with the rank of lieutenant junior grade.

Our job was to assist Commander William S. ("Deke") Parsons, United States Navy (USN), principal liason officer from Bu Ord during the development work, in introducing this new fuze to gunnery officers of combatant ships in the South Pacific. Parsons (later the weaponier on the *Enola Gay*, which dropped the first atomic bomb at Hiroshima) flew ahead to an unrevealed location in the Pacific theater. Dilley, Peterson, and I oversaw the loading of the first secret issue of some 5000 carefully counted, proximity-fuzed (also called VT fuzes to disguise their nature) 5"/38 projectiles into the hold of a troop ship at Mare Island near San Francisco. Within a week of receiving our commissions, signed personally by Frank Knox, Secretary of the Navy, we were at sea en route to a secret destination. The ship traveled without escort. I was able to keep track of our progress in latitude by elementary celestial observations and in longitude by the progressive change in mean time between sunrise and sunset and the occasional one-hour changes in ship's time.

About two weeks later we arrived in Nouméa, New Caledonia, headquarters of the Commander of the South Pacific Fleet (COMSOPAC). Parsons had already laid the groundwork and assigned us to various segments of the fleet. I was assigned as assistant gunnery officer on the staff of Rear Admiral Willis A. Lee, a task group commander of Task Force 38 (commanded by Admiral William F. Halsey) and Task Force 58 (commanded by Admiral Raymond A. Spruance). Admiral Lee was also type commander of battleships in the Pacific Fleet (COMBATPAC) with headquarters on the USS *Washington*. I arrived on the *Washington* only about two weeks after her celebrated role in the major engagement with a Japanese task force in the strait between Tulagi and Guadalcanal, thereafter called iron-bottom bay. Lee was the informal president of the Navy "gun club" and was acknowledged to be one of the leading gunnery officers of the US Navy. He was thoroughly familiar, both theoretically and practically, with the fundamental ineffectiveness of antiaircraft weapons and of the often fatal fallacy of supposing that an attacking aircraft could be stopped by "filling the air with shrapnel." He was deeply impressed by my briefings on the VT fuzes and immediately recognized their potential in quantitative terms. I gave him a clear statement on the necessity of a clear field of fire (not over our own ships), of the expectation of at least 15% duds and premature bursts (which posed no hazard to the firing ship), and of the airbursts that occurred as projectiles approached the sea at the end of flight. Also, I informed him of the then prevailing doctrine that despite the potential effectiveness of proximity-fuzed projectiles for shore bombardment, such usage was forbidden on the security ground that duds might be recovered by the enemy and either duplicated by them or used as a basis for countermeasures, i.e. "jamming" by radio transmitters so as to cause premature bursts. He endorsed my written description of the properties of the new ammunition and immediately ordered a pro rata distribution of the available supply to all combatant ships of his task group. My job was to effect this distribution and to brief gunnery officers and commanding officers on their proper use. I encountered a wide range of understanding and lack of understanding of the range-error problem and varying degrees of acceptance. The toughest operational problem was the restriction on firing over other ships of the task group under the complex conditions of actual air attack.

After eight months of sea duty on the *Washington* and other ships I was ordered back to Bu Ord to serve as liasion officer with APL/JHU and to read and summarize combat reports from ships using the VT fuze against attacking aircraft. Finding such desk work onerous, I requested transfer back to the Pacific Fleet to help remedy the grave shortcomings of the fuzes—most notably, the large percentage of duds that were occurring as the useful shelf life of their batteries expired during their long period of transport, usually at elevated temperature conditions, by cargo ships from the states to combatant ships. I then made contact again with Admiral Lee on the *Washington* and with Commander Lloyd Muston, COM-SOPAC staff gunnery officer, in Nouméa and engaged in setting up rebat-terying stations at ammunition depots at Nouméa, Espiritu Santo, Tulagi, Guadalcanal, and Manus Island, and on ammunition barges at Eniwetok Atoll, Kwajalein, and Ulithi. I also had temporary duty on a succession of destroyers to instruct gunnery officers and conduct tests of the fuzes. And I made frequent reports to Bu Ord on the status of the work and (usually urgent) requests for fresh batteries, tools, and equipment—by air transport, if possible, to try to maintain the fleet's supply of workable fuzes. During this period I was on the *Washington* as assistant staff gunnery officer during the Battle of the Philippines Sea, in which the ship suc-cessfully defended herself against kamikaze attack. In March 1945 I returned to duty at Bu Ord and as liaison officer at APL/JHU until my transfer to the inactive reserve as a lieutenant commander in March 1946, after the end of World War II hostilities.

The period 1940–45 was a part of my life totally foreign to my previous aspiration to become an academic physicist. But I lost no energy grieving over the turn of events. On the contrary, I plunged into "the war effort" with the patriotic fervor of those days and with the exhilaration of applying my knowledge of physics and mathematics and my laboratory skills to solving difficult problems of practical importance and national urgency. My service as a naval officer was, far and away, the most broadening experience of my lifetime. I had considerable responsibility in the real world of life-or-death, and for the first time, I dealt with a vertical cross section of the human race on a one-to-one basis, from apprentice seamen to admirals. I was deeply impressed by every such relationship, by the code of honor of the navy, and by the validity of military protocol. I gained a profound respect for the raw power and grandeur of the sea and a corresponding respect for seamen. Much of my boyhood reading was in that vein. As a high-school senior, I had hoped for an appointment to the US Naval Academy, and our US congressman, a close friend and former classmate of my father's in college and law school, nominated me subject to passing the academic and physical examinations. But I failed the latter. Eleven years later I received a spot commission as a lieutenant junior grade in the Naval Reserve under the relaxed wartime standards.

Among other things that I learned in the navy by close observation of my peers and superiors was how to make a sound decision when the basis for a decision was diffuse, inadequate, and bewildering. This lesson has served me well. Another strong and durable impression was the great gap between the life of a bureaucrat in Washington and the real situation on a combatant ship.

HIGH-ALTITUDE RESEARCH

While still on terminal leave from the navy, I was rehired as a physicist at APL and encouraged to organize a research group to engage in high-altitude research based on the prospectively available opportunity to conduct experiments with captured and refurbished German V-2 rockets. I had earned my spurs by my wartime work, and Tuve gave me a free hand and ample financing to develop this field as I saw fit. My interest in geophysics stemmed from Tom Poulter's work and from my association with the " old-line" geophysicists at DTM. This line of scientific interest and my laboratory experience in nuclear physics and with rugged electronic devices and high-performance ordnance combined to lay the groundwork for my future research career. I was eager to attack problems of the primary cosmic radiation, the ionosphere, and geomagnetism by rocket techniques, which promised direct observation of many phenomena that had been previously a matter of conjecture, albeit sophisticated conjecture.

I gathered together a spirited group of like-minded individuals—Robert Peterson, Lorence Fraser, Howard Tatel, Clyde Holliday, John Hopfield, and several others—and got to work. Parallel efforts were underway at several other military or quasi-military laboratories. Of these, the group at the Naval Research Laboratory (NRL), inspired by Ed Hulburt, their long-time leader in atmospheric and ionospheric physics, was the most noteworthy. We adopted NRL as our principal competitor and sometime collaborator.

The opportunity to use V-2's for scientific work was provided by the Army Ordnance Department by virtue of the foresight and broad vision of Colonel Holger N. Toftoy.

Under the leadership of Ernst Krause of NRL, a small and highly informal group of prospective participants in this effort was assembled to maximize the scientific work and to allocate flight opportunities in an equitable manner. I was a member of this group, which called itself the V-2 Rocket Panel. We had no formal organization, no official authority, and no budget. Nonetheless, we oversaw, in effect, the entire national effort in this field for over a decade. Krause was the original chairman, but he left the NRL in 1946 and I was chosen to succeed him, continuing thereafter as chairman until the effective termination of our functions in 1958.

In 1946, with the support of Merle Tuve, I initiated and supervised the development of a high-performance American sounding rocket, the Aerobee, to be used exclusively for scientific purposes. This rocket soon joined the V-2 as a basic vehicle for high-altitude research. During the period 1946–51, payloads of scientific instruments were carried by 48 V-2's and 30 Aerobees.

The emphasis of our APL work was in the fields of cosmic rays, solar ultraviolet, high-altitude photography, atmospheric ozone, and ionospheric current systems. The site for most of the launchings was the White Sands Proving Ground (later renamed White Sands Missile Range) near Las Cruces, New Mexico. But in 1949 and 1950, I organized successful Aerobee-firing expeditions on the USS *Norton Sound* to the equatorial Pacific and the Gulf of Alaska, respectively. (As of 17 January 1985, a total of 1037 Aerobees had been fired for a wide variety of investigations in atmospheric physics, cosmic rays, geomagnetism, astronomy, and other fields.)

The national effort in high-altitude research during those early free-wheeling and spirited days was characterized by many failures and many noteworthy successes. Substantial advances in knowledge were achieved in atmospheric structure, ionospheric physics, cosmic rays, high-altitude photography of large areas of the cloud cover and surface of the Earth, geomagnetism, and the ultraviolet and X-ray spectra of the Sun.

The V-2 Rocket Panel [later renamed the Upper Atmosphere Rocket Research Panel (UARRP) and, still later, the Rocket and Satellite Research Panel] presided over the entire effort. Beginning in the mid-1950s, the panel spawned one of the important components of our national participation in the 1957–58 International Geophysical Year (IGY). Its members became influential in the planning of the IGY, actively promoted the adoption of scientific satellites of the Earth as an element of the IGY program, and laid the foundations for the scientific program of the

National Aeronautics and Space Administration, a major agency of the federal government created in 1958.

RETURN TO THE UNIVERSITY OF IOWA

In 1950, and despite the flourishing of our high-altitude work, the new director of APL, R. C. Gibson, split my assignment so as to include supervision of the residual proximity fuze group. I was competent to provide such supervision but had no interest in pursuing further developmental work on fuzes. I did the job but interpreted the split assignment as foreshadowing the termination of academic-style research in geophysics at APL. A few months later I received a telephone call from Professor Tyndall, my former research mentor at the Universtity of Iowa. He informed me that Louis A. Turner had resigned as head of the Department of Physics after four years and that he (Tyndall) had suggested me as a possible successor. I was thrilled by this prospect and soon thereafter made a short visit to Iowa City for interviews and a departmental colloquium. Several weeks dragged on after I returned to Silver Spring, with no news. I finally received a letter from Tyndall advising me that they had offered the position to the individual who was their first choice and were awaiting his response. Another few weeks of suspense came to an end when Tyndall called to offer me the position, which would also carry the rank of full professor. At that time Abigail, my wife of five years, had been west of the Mississippi only once and considered Iowa to be terra incognito from the cultural point of view. Nonetheless, she agreed to support my decision whatever it might be. I then accepted the offer but told Tyndall that I would need six months to wind up my obligations at APL. This was agreed.

On a very cold first of January 1951, my wife and I with our two young daughters arrived in Iowa City in our old station wagon, pulling an even older trailer containing most of our earthly possessions. We plowed through the snow to move into a "barracks apartment," one of a cluster of small metal-sheathed buildings that had been erected during the war as temporary quarters for naval cadets and other personnel associated with the university. The sole source of heat was a cast-iron stove that was fed fuel oil from an external 55-gal drum by gravity flow through a small copper tube. The small living room could be made comfortably warm, but the remainder of the apartment presented a challenging problem in heat transfer. However, the monthly rent was only $35.

I entered my new duties with enthusiasm and dedication. I had no research budget, but the department had an excellent machine shop and two skilled instrument makers, as well as a large stock of more-or-less obsolete but still usable electrical instruments.

With the help of George W. Stewart I got a small but very important grant from the private Research Corporation as seed money and started research on cosmic rays using balloon-borne equipment; I also recruited several able graduate students as collaborators. Soon thereafter, I wrote a proposal to the US Office of Naval Research (ONR) for measuring the primary cosmic-ray intensity at high latitudes above the appreciable atmosphere, using small military-surplus rockets carried to an altitude of about 50,000 ft by a balloon and launched from that starting point to reach a summit altitude of some 250,000 ft. By this inexpensive technique, I hoped to resume high-altitude research on a low budget. The proposal was accepted. Support by the ONR has continued without a break for the subsequent 38 years and has provided the base for all of my research during this period.

In the summer of 1952, two of my students, Leslie Meredith and Gary Strein, our lab technician Lee Blodgett, and I made our first rockoon (rocket-balloon combination) expedition to measure the cosmic-ray intensity above the atmosphere in the Arctic. We traveled on the Coast Guard icebreaker USCGC *Eastwind*, whose primary mission was the resupply of the weather station at Alert at the shore of the Arctic Ocean on northeastern Ellesmere Island. We released balloon-borne Deacon rockets from the heliocopter deck whenever we could persuade the captain to steam downwind for an hour while we inflated and released the balloon under zero relative wind conditions. After several failures we diagnosed and cured the problem and got a succession of successful flights to altitudes of about 200,000 ft at locations off the coast of Greenland—the first research rocket flights ever made at such high geomagnetic latitudes. All of the instrumentation, including the telemetry transmitters and nose cones, were built in our own shop; a single Geiger-Mueller (G-M) tube was the radiation sensor.

We reported our results with pride to the UARRP and set to work to refine the instrumentation using ionization chambers and scintillation counters as well as G-M tubes.

During succeeding summers, Arctic rockoon expeditions on various ships were the heart of our work. These were led by Melvin Gottlieb and Frank McDonald. The 1953 expedition yielded a remarkable new finding—namely, the first direct detection of the electrons that, we surmised, were the primaries for producing auroral luminosity.

For a 15-month period in 1953–54, my family and I took a leave from the University of Iowa to join Lyman Spitzer at Princeton University in an experimental program to investigate the confinement of hot plasma by a magnetic field in a twisted figure-eight-shaped tube that he called a stellerator—so named with the hopeful prospect of providing a demonstration of controlled thermonuclear fusion in deuterium and eventually in a mixture of deuterium and tritium. All of this recalled my PhD thesis. I built and operated a crude model of such a machine, called the Model B-1 stellerator. (Model A was a previously built smaller device of table-top size.) I demonstrated the validity of Spitzer's rotational transform of magnetic fields in the twisted toroid with a miniature electron gun and fluorescent screen and got plasma confinement times of a few milliseconds in a hydrogen plasma. The difficulty of making a much larger machine of this nature and, as it appeared to me, the remote prospect for achieving self-sustained fusion on a reasonable time scale convinced me to return to Iowa and resume my high-altitude research, which was already yielding significant original results. This I did in August 1954.

Plans for the International Geophysical Year were by then being formulated, and my colleagues and I on the UARRP were eager to add investigations with rocket-borne equipment to these plans. I proposed to continue the rockoon program with further auroral, cosmic-ray, and magnetic-field measurements in the Arctic, equatorial latitudes, and the Antarctic. This program was funded by the National Science Foundation within its special IGY program. Our field work culminated in 1957 with two shipboard expeditions that I led. The first was aboard the USS *Plymouth Rock* from Norfolk to northwestern Greenland; the second, aboard the large icebreaker USS *Glacier* from Boston via the Panama Canal to the Central Pacific and thence to Antarctica. Of the 30 rockoon flights that Laurence Cahill and I attempted during these two expeditions, which ranged from 79°N to 75°S latitude within a four-month period, 20 were successful in yielding high-altitude cosmic-ray, auroral particle, and magnetic-field data.

EARLY SATELLITE WORK

In 1956 I made a formal proposal to the IGY directorate for a simple but globally comprehensive cosmic-ray investigation using one of the early US Earth satellites as a powerful follow-on to my previous and planned work with rockets. In addition, I proposed study of the auroral primary radiation, also on a global basis, if and when orbits of sufficiently high inclination were available.

The cosmic-ray proposal was given a favorable rating, placed in the pool of experiments to be conducted by one of the early IGY satellites, and funded by the National Science Foundation. The development of the instrument was in the capable hands of George H. Ludwig, a former Air Force pilot and a graduate student at Iowa. He introduced many novel features, including the use of then new transistors throughout the electronics and the design and construction of a miniature magnetic tape recorder.

Following the Soviet's successful flights of the first Earth satellite *Sputnik I* and then *Sputnik II*, the Army's rocket vehicle Jupiter C was adopted as a US alternative to the planned but faltering Vanguard vehicle for placing an early US payload into Earth orbit. By virtue of preparedness and good fortune, the Iowa cosmic-ray instrument was selected as the principal element of the payload of the first flight of a four-stage Jupiter C, launched on 31 January 1958 (1 February GMT).

Both the vehicle and our instrument worked. The data from the single Geiger-Mueller tube on *Explorer I* (as the payload was called) yielded the discovery of the radiation belt of the Earth—a huge region of space populated by energetic charged particles (principally electrons and protons), trapped within the external geomagnetic field. The attempted launch of *Explorer II* was a vehicular failure, but the launch of *Explorer III* in 26 March 1958, with an augmented version of the Iowa instrument, was successful. The *Explorer III* data provided massive confirmation of our earlier discovery and clarified many features of the earlier body of data.

Soon thereafter we were invited to provide radiation-detecting instruments for two satellites that were to observe the effects of several nuclear bombs to be detonated after delivery to high altitudes by rockets. On a time scale of less than three months, Carl McIlwain, George Ludwig, and I designed and built the radiation packages for these satellites—using much smaller and more discriminating detectors, chosen for the first time with knowledge of the existence of the natural radiation belts and the enormous intensity of charged particles therein.

Explorer IV was launched successfully on 26 July 1958. Our apparatus operated as planned and provided the principal body of observations of the artificial radiation belts that were produced by the three high-altitude nuclear bursts—called Argus I, II, and III. The back-up launch of our apparatus on *Explorer V* was a vehicular failure. Analysis of our *Explorer IV* data on the natural radiation belt as well as on the artificial radiation belts from the Argus bursts propelled the entire subject to a new level of understanding and broad scientific interest.

The first Soviet confirmation of the existence of natural radiation belts came from *Sputnik III*, launched in May 1958.

Late in 1958, the Iowa group supplied radiation detectors on two missions, *Pioneer I* and *Pioneer III*, that were intended to impact the Moon. The lunar objective was not achieved, but our data established the large-scale structure and radial dimensions of the region containing geomagnetically trapped radiation. Another lunar flight, *Pioneer IV* (also unsuccessful in reaching the Moon), carried our apparatus in early 1959 and provided a valuable body of confirmatory data.

SPACE RESEARCH UNDER THE AUSPICES OF THE NATIONAL AERONAUTICS AND SPACE ADMINISTRATION AND THE OFFICE OF NAVAL RESEARCH

Our early satellite work was done under the auspices of the IGY/National Science Foundation program, the Office of Naval Research, the Army Ballistic Missile Agency, the Jet Propulsion Laboratory, and the Air Force.

The creation of the National Aeronautics and Space Administration (NASA) as a new, civilian agency of the federal government was formalized by President Eisenhower's signature on the enabling legislation on 10 October 1958. As NASA got itself organized, it moved toward becoming the central national agency for the planning and support of space science and applications. Nonethless, a substantial effort in these areas continued under the auspices of the military departments of the Department of Defense, our principal relationship therein being with the Office of Naval Research.

The creation of NASA led to a dramatic change in space research in the United States. Whereas previously it had been performed by only a small cadre of individuals who might well be described as members of the UARRP and their immediate associates, principally in military and quasi-military laboratories, it then assumed national scope and became "civilianized." In anticipation of the NASA legislation, the National Academy of Sciences established the Space Science Board (SSB) in the early summer of 1958 to advise the federal government on the conduct of scientific research in space. This board was chaired during its important and most influential period by Lloyd Berkner. I was an original member of the SSB and served from 1958 to 1970 and again from 1980 to 1983. A major planning study was conducted under my chairmanship during a two-month period in the summer of 1962 at the University of Iowa. This study yielded a classical document and became the prototype for subsequent summer studies by the SSB. Space science in the United States benefited greatly from the close relationship and mutual respect between Berkner and James Webb, the second administrator of NASA and an especially effective one during the period of great growth of the agency. Indeed, members of the SSB had the heady, but only partially true, perception that they were writing the national scientific program in space in the form of well-considered advice from the interested segment of the scientific community.

Space exploration was transformed from being an arcane field with only a handful of participants to an activity of high national visibility. Members of Congress vied for membership on freshly created space committees.

VAN ALLEN

Senator Lyndon B. Johnson and Congressman John W. McCormack led the legislative drive to put the United States in space on a scale adequate to restore national pride and international prestige following what was perceived as national humiliation by the early successes of the Soviet Union, previously thought by most Americans to be a technologically backward nation. The political scene culminated in the hesitant but eventually dramatic decision by President John F. Kennedy to undertake the landing of a man on the Moon and his safe return to the Earth. His formal public announcement of the Apollo project was made on 25 May 1961. The politics of this decision has been discussed voluminously by many, many others, and I have nothing to add.

In parallel with the Apollo project, programs of space science and the numerous practical applications of space technology were also flourishing. These had much less public visibility but in the long run have proven to be of far greater importance and durability.

Our research at Iowa centered on expanding our knowledge of the energetic particle population of the Earths's external magnetic field and the multifold physics therof. In 1959 Thomas Gold suggested the term " magnetosphere" for the region around the Earth in which the geomagnetic field has a controlling influence on the motion of charged particles, and the term "magnetospheric physics" was widely adopted. Magnetosphere joined the already established list of "spheres"—atmosphere, ionosphere, mesosphere, thermosphere, etc—as a geophysical term.

Much earlier, the presence of thermal and quasi-thermal plasma (ionized gas) in the Earth's magnetic field had been established by Owen Storey and others in the interpretation of "whistlers," a low-frequency electromagnetic phenomenon resulting from lightning strokes.

In situ measurements of this plasma became a central objective of magnetospheric physics using Earth satellites. Another central objective was the investigation of the physics of the aurorae, geomagnetic storms, and the ring current of the Earth. Also, space technology opened up new fields of investigation of cosmic rays, energetic particles from the Sun, solar X-ray flares, and the detailed nature of the interplanetary medium.

The Iowa group played an imprtant role in these developments. Louis Frank made a marked advance in our understanding of the plasma physics of the magnetosphere by developing and flying, first on the *NASA/Orbiting Geophysical Observatory II* in 1965, a low-energy proton electron differential energy analyzer (LEPEDEA) that was sensitive to particles having energies as low as 1 keV.

With the support of the Office of Naval Research and later of NASA we developed and built complete satellites with full complements of scientific instrumentation, thus becoming the first university to succeed in this comprehensive undertaking. Indeed, at one point in time, we had built and flown more satellites than the combined number built by all foreign nations, excepting the Soviet Union. The Injun series of Iowa satellites comprised *Injun I* (launched 29 June 1961), *Injun II* (not placed in orbit because of a launch vehicle failure on 24 January 1962), *Injun III* (launched 12 December 1962), *Injun IV* (launched 21 November 1964), and *Injun V* (launched 8 August 1968). These were placed in low-altitude, high-inclination orbits and had investigation of the aurorae as one of their primary objectives. A notable advance was made by Donald Gurnett in devising and successfully flying a VLF (very low frequency) radio receiver on *Injun III*. A large variety of plasma/wave phenomena were observed, and this new field of investigation began to assume an important role in magnetospheric physics.

We also provided, under increasingly competitive circumstances, instruments as part of the scientific payload of NASA spacecraft: *OGO I, II, III*, and *IV*; *Explorers VII, XII*, and *XIV*; and *IMP's* (Interplanetary Monitoring Platforms) *-D, -E*, and *-F* (*Explorers 33, 35*, and *34*). *IMP-D* was placed into a very eccentric orbit of Earth with apogee beyond the Moon's orbit, and *IMP-E* was injected into a durable orbit around the Moon. Both of these spacecraft were exceedingly fruitful in studying solar X rays and solar energetic particles and in exploring the outer fringes of the magnetosphere—especially the magnetotail, which had been discovered by other groups using *Explorer VI* and studied further by us and others with *Explorer XII* and *Explorer XIV* in very eccentric orbits.

The latest of the Universtiy of Iowa's small satellites was *Hawkeye I*, which was placed in an eccentric, 90° inclination orbit in 1974 and continued to operate properly until its reentry into the atmosphere nearly four years later. It yielded important results on the configuration of the bow shock and magnetopause and on the topology of the geomagnetic field at large radial distances over the northern polar cap and in the vicinity of the polar cusp—a special feature of central importance in the entry of solar plasma into the magnetosphere.

Other Earth satellite missions in which the University of Iowa has had an important role have included *IMP-G, IMP-I, IMP-H, IMP-J, UK-4* (*United Kingdom-4*), S^3 (*Small Standard Satellite*), *DE-1* (*Dynamics Explorer I*), and *ISEE-1, -2*, and *-3* (*International Sun Earth Explorers 1, 2*, and *3*). Frank's imaging camera on *DE-1* has provided a large and classical album of global images of the aurorae and other faint light features of the Earth under lighting conditions previously thought to make such images impossible. One of my great regrets has been that Sydney Chapman did not live to see these, inasmuch as he often expressed such an aspiration when he lectured on auroral physics at the University of Iowa in 1963. Meanwhile, the investigations of Gurnett and his colleagues with their VLF receivers on *ISEE-3* have been exceedingly fruitful, as have the plasma measurements on *ISEE-1* and *-2* by Frank et al. The imaginative scheme of Robert Farquhar at the Goddard Space Flight Center for diverting *ISEE-3* from its original station at the Ll Lagrangian libration point of the Earth-Sun system to an orbit enabling it to fly by Comet Giacobini-Zinner, the spacecraft being then renamed *ICE* (*International Comet Explorer*), resulted in the first in situ observation of dust and plasma physical phenomena associated with a comet (September 1985).

PLANETARY EXPLORATION

As early as 1960 and in parallel with the activities just sketched, one of my driving aspirations was to push on with magnetospheric studies of the other planets.

The emphasis at NASA in the early 1960s was on manned lunar flights. But the Jet Propulsion Laboratory (JPL) and other groups had already made extensive general studies of the ballistics of flight to other planets—especially Venus and Mars. The interest in Mars was driven by the desire for geological studies of its surface and, perhaps more importantly, by the desire to search for any form of biological activity there. Also, Mars and Venus were ballistically much easier to reach than was Mercury or the outer planets. The first planetary target to be adopted by JPL/NASA was Venus. I proposed a simple radiation detector for the first mission with the purpose of searching for the existence of a Venusian radiation belt and the consequent inferences on the magnetization of this planet, then completely unknown. My instrument was selected and was incorporated into the payload of *Mariner I*, an early in-flight failure, and of *Mariner II*, launched successfully on 27 August 1962. The cruise phase was quite successful, yielding, most importantly, the first continuous measurements of the solar wind by Conway Snyder and Marcia Neugebauer and of the interplanetary magnetic field by Paul Coleman et al, as well as the detection of numerous solar energetic particle events by my apparatus and a companion instrument of Hugh Anderson and Victor Neher. *Mariner II* passed by Venus on 14 December 1962 at a radial miss distance of 41,000 km. In our measurements there was not the slightest indication of the presence of the planet, thereby implying an upper limit on its magnetic moment as 0.18 that of the Earth, its "sister" planet. A casual, and perhaps even correct, interpretation of this result is that Venus is simply rotating too slowly (period 243 days) to drive an internal self-excited dynamo

I had an improved model of the radiation instrument on *Mariner III* (launch failure) and *Mariner IV* (launched successfully of 28 November 1964), which made the first-ever encounter with Mars on 15 July 1965. The interplanetary data yielded a nearly continuous record of solar X-ray flares and of the presence of energetic solar particles, including the discovery of energetic solar electrons and, in stereoscopic combination with data from *Explorer 35* in lunar orbit, the first determination of the altitude in the Sun's atmosphere at which 2–10 Å X rays are emitted. At Mars, as at Venus, we got a null result and inferred an upper limit on Mars' magnetic moment as 0.001 that of the Earth.

Meanwhile, as a member of the Space Science Board and, from 1966–70, of the Lunar and Planetary Missions Board, I had adopted the role of being a special and unremitting advocate of missions to the outer planets—especially Jupiter. The first fruits of these efforts were the adoption by NASA of two missions to Jupiter—later called *Pioneer 10* and *Pioneer 11*—with emphasis on energetic particle and magnetic-field measurements. A special motivation for this emphasis was the radio-astronomical evidence that Jupiter has a huge radiation belt whose population of relativistic electrons emits the observed synchrotron radiation in the decimetric wavelength range. The *Pioneer 10/11* project was managed by the Ames Research Center of NASA with a keen concern for optimizing the scientific yield of the mission. The spacecraft were built by the TRW Company. My proposal for an energetic particle instrument was accepted, after some reduction in its scope, during the vigorous national competition for payload space. *Pioneer 10* was launched successfully on 3 March 1972 and *Pioneer 11* on 6 April 1973. For the subsequent 17 years, the inflight data from these two spacecraft have been a central part of my research life and that of several of my students and associates. *Pioneer 10* made the first-ever encounter (December 1973) with Jupiter and yielded a large body of new knowledge, most especially on its magnetosphere. *Pioneer 11* encountered Jupiter a year later along a different trajectory and confirmed and substantially expanded the earlier findings. *Pioneer 10* has continued on an escape trajectory out of the solar system and, at the date of writing, is about 47 AU from the Sun (nearly 7 billion km)—the most remote man-made object in the Universe. It is still working well and providing daily data on cosmic rays and the interplanetary medium in the outer heliosphere. After its close encounter with Jupiter, *Pioneer 11* made the first-ever encounter with Saturn in September 1979, discovering its magnetosphere and yielding a rich body of new information on the planet itself and its system of rings and satellites. This spacecraft is now also on a solar-system-escape trajectory at a current distance of over 28 AU, and it too is transmitting data on a daily basis.

In the late 1960s and early 1970s a Grand Tour of the Outer Planets was being advocated by the Jet Propulsion Laboratory, in particular, and by other planetary enthusiasts who were advising NASA on new programs. JPL had shown that the forthcoming configuration of the outer planets Jupiter, Saturn, Uranus, and Neptune (a once-in-179-year phenomenon) would make it ballistically feasible to have a single spacecraft fly by all four of these remote planets. The Grand Tour, as such, was a budgetary casualty of late 1970. Soon, thereafter, I was asked by JPL to chair a Science Working Group to develop a more modest-sounding mission, tentatively called MJS (Mariner/Jupiter Saturn). The two-spacecraft mission that we developed was eventually approved and came to life in 1974. It was later renamed Voyager. Although the term Grand Tour was now eschewed in polite conversation, it did not escape our attention that the configuration of the outer planets was independent of budgetary-political considerations in the White House and the Congress.

The successes of the two Pioneer missions produced a greatly enhanced interest in the Voyager missions, as well as in ground-based study of the outer planets. Competition for payload space brought forth a wealth of proposals of new and sophisticated instruments and eventual selection of an excellent complement.

Both Voyagers were launched successfully in the late summer of 1977. Each flew by Jupiter and Saturn and provided great advances, most notably in high-resolution imaging of the atmospheres of the planets, of their satellites, and of their rings and in our understanding of the plasma physics of their magnetospheres. Since its Saturn encounter, *Voyager 1* is on a solar-system-escape trajectory, but *Voyager 2* made the first-ever encounter with Uranus in early 1986 and is now approaching Neptune for a 25 August 1989 encounter—thus prospectively achieving the objectives of the Grand Tour as visualized at the outset of this program.

I have had no part in the execution of the Voyager program but have been a guilty bystander, so to speak, and one of its enthusiastic fans.

In 1976–77, I chaired still another JPL/Ames Research Center science working group, called JOP/SWG [Jupiter Orbiter with (Atmospheric) Probe/Science Working Group]. Our purpose was to develop a follow-on Jupiter mission of more advanced capability than the Pioneers and the Voyagers. This program would have a deep atmospheric entry probe and an orbiter having a useful lifetime of at least two years, in contrast to the limited period (days to weeks) of nearby observation available on a fly-by.

The mission, renamed *Galileo*, has suffered a plethora of delays—financial, political, and technical—principally as the result of the inadequacies and defaults of the shuttle launching system, which had been adopted by NASA in the late 1970s and 1980s. The launch of *Galileo* is now scheduled for October 1989, but many uncertainties remain. Also, because of the less than originally planned capability of the shuttle, it has been necessary to adopt an ingenious but very long flight path to Jupiter, requiring over 6 years vs. the 20-month flights of *Pioneers 10/11*. I recognize that the probability of my own survival to 1995 is substantially less than unity.

Nonetheless, I still hope to function in my interpretative role as an Interdisciplinary Scientist beginning after *Galileo*'s scheduled entry into orbit around Jupiter in that year.

CONCLUDING COMMENTS

In the period 1951–85, I served as head of the Department of Physics (which became the Department of Physics and Astronomy in 1969) of the University of Iowa. My formal teaching involved a full gamut of courses: General Physics, General Astronomy, Electricity and Magnetism, Introduction to Modern Physics, Radio Astronomy, Intermediate Mechanics, and a specialized course in Solar-Terrestrial Physics. Perhaps my favorite was General Astronomy, an introductory but rigorous course on the solar system, with laboratory, which I taught for 17 years.

My closest working relationships with students involved ones at the graduate level. The following are those who finished advanced degrees under my guidance. The first date in the parentheses after each name is the date of an MS degree, the second of a PhD degree. Some students who did their MS work with me later earned a PhD elsewhere or under another advisor at Iowa. Others terminated their graduate work at the MS level. Every one was a collaborator in the fullest sense of the word, a fact that is amply represented in authorship or coauthorship of published work.

JOHARI BIN ADNAN	(1983, —)
SIXTEN INGVAR ÅKERSTEN	(1969, —)
HUGH RIDDELL ANDERSON	(1958, —)
THOMAS PEYTON ARMSTRONG	(1964, —)
DANIEL N. BAKER	(1973, 1974)
KENNETH E. BUTTREY	(1955, —)
LAURENCE JAMES CAHILL, JR.	(1956, 1959)
CHARLES P. CATALANO	(— , 1971)
JAMES R. CESSNA	(1965, —)
PHILLIP CHANG	(1962, —)
TSAN-FU CHEN	(1973, 1978)
JOHN D. CRAVEN	(1964, —)
JOSÉ M. da COSTA	(1971, —)
JERRY F. DRAKE	(1967, 1970)
ROBERT A. ELLIS, JR.	(— , 1954)
R. WALKER FILLIUS	(1963, 1965)
HERBERT R. FLINDT	(1968, —)
LOUIS A. FRANK	(1961, 1964)
JOHN W. FREEMAN	(1961, 1963)
THEODORE A. FRITZ	(1964, 1967)
SISTER JEAN GIBSON, O.S.B.	(— , 1969)

CYNTHIA LEE GROSSKREUTZ	(1982, —)
DONALD A. GURNETT	(— , 1965)
ROLLIN CHARLES HARDING	(1966, —)
H. KENT HILLS	(1964, —)
WILLIAM G. INNANEN	(— , 1972)
ROBERT CHANDLER JOHNSON	(1957, —)
JOSEPH E. KASPER	(1955, —)
STAMATIOS MIKE KRIMIGIS	(1963, 1965)
CURTIS D. LAUGHLIN	(1960, —)
WEI CHING LIN	(1961, 1963)
THOMAS A. LOFTUS	(1969, —)
GEORGE HARRY LUDWIG	(1959, 1960)
CARL E. McILWAIN, JR.	(1956, 1960)
LESLIE H. MEREDITH	(1952, 1954)
RAYMOND F. MISSERT	(1955, 1957)
STEVEN R. MOSIER	(1967, —)
MICHAEL O'CONNOR	(1968, —)
MELVIN N. OLIVEN	(1966, 1970)
MARK E. PESSES	(1976, 1979)
GUIDO PIZZELLA	(— , 1962)
ROBERT C. PLACIOUS	(1953, —)
RICHARD LOUIS RAIRDEN	(1981, —)
BRUCE A. RANDALL	(1969, 1972)
JOANNA M. RANKIN	(— , 1970)
ERNEST C. RAY	(1953, 1956)
NICOLAOS A. SAFLEKOS	(— , 1975)
EMMANUEL T. SARRIS	(— , 1973)
MELVIN SCHWARTZ	(— , 1958)
DAVIS D. SENTMAN	(— , 1976)
STANLEY D. SHAWHAN	(1965, —)
HAROLD E. TAYLOR	(— , 1966)
JAMES DENNIS THISSEL	(1963, —)
MICHELLE F. THOMSEN	(1974, 1977)
WILLIAM R. WEBBER	(1955, 1957)
JOEL M. WEISBERG	(1975, 1978)
CHARLES D. WENDE	(1966, 1968)
MICHAEL JAMES WIEMER	(1964, —)
SAIYED MASOOD ZAKI	(1964, —)

In addition, I have benefited from the highly competent efforts of an uncounted number of members of our technical staff, of whom the following are representative: William A. Whelpley, Roger F. Randall, Robert B. Brechwald, Evelyn D. Robison, John E. Rogers, Joseph G. Sentinella, Donald C. Enemark, W. Lee Shope, Edmund Freund, and Robert Markee.

Finally, and most importantly, I am indebted to my wife of 44 years, Abigail, and our five children, who have provided the circumstances under which sustained and intensive professional work has seemed worthwhile.

Kiyoo Wadati

Ann. Rev. Earth Planet. Sci. 1989. 17:1–12

BORN IN A COUNTRY OF EARTHQUAKES

Kiyoo Wadati

1-8 Naitomachi, Shinjuku-ku, Tokyo 160, Japan

1.   NATURAL PHENOMENA—MY LIFELONG FRIENDS

My country, Japan, is favored with a mild climate and with sufficient precipitation. We enjoy this gift of nature heartily, even though we also suffer from violent natural phenomena: earthquakes, typhoons, and volcanic eruptions. I was born in 1902 in Nagoya, during a raging typhoon. I think I was fortunate to have been born in this wonderful country that exhibits such a wide range of natural phenomena.

When I was six years old and sitting on the veranda of my grandfather's home in Hamamatsu, I suddenly felt strong earthquake shocks. I was surprised by these and rushed to the garden, where the family had already gathered. They looked anxious. Fortunately, the shock was not violent enough to cause damage to the house, but still it impressed me with the terror of earthquakes. The magnitude of this earthquake was 7.0; 41 victims died in the epicentral region, which was about 150 km distant from our home.

I would also like to tell about a volcanic eruption. When I was a small boy in elementary school in Osaka, one morning in the winter of 1914 I was surprised to find the school grounds and roofs entirely covered by white powder. I was told that this was ash from a large eruption of Volcano Sakurajima, which is far away, about 550 km from Osaka. It seemed very strange to me to think that those ashes had come flying from such a distant place.

During my school and college days, I used to spend my whole summer vacation at the seashore of Hamana Bay near Hamamatsu. There I had an opportunity to become intimate with marine life and to observe closely various kinds of natural phenomena in the air and sea, such as tidal motion, sea waves, typhoons, and storm surges. Perhaps these experiences remained in my mind and latently influenced me throughout my life.

On October 1, 1917, a violent typhoon struck our house in Tokyo. The strong wind destroyed the window glass, and the storm rushed into my room. This typhoon was the strongest to have attacked Tokyo in recent years. Unusually heavy flooding, originating from storm surges, caused heavy damage, especially in low coastal areas.

The most impressive natural disaster that I ever experienced was the great Kwanto earthquake of September 1, 1923. I was a student of physics at Tokyo University at that time. As I sat at a desk in a second-floor room of my wooden house, I suddenly felt very strong earthquake shocks accompanied by awful earth sounds. Then the mud began to fall from the wall, and the bookshelf at my side fell down. Fortunately, my house was damaged only slightly and my family was safe. Our house was in a high section of Tokyo, where the earthquake shocks were not so strong as those in the downtown area. There, a terrible fire added to the damage directly caused by the strong shocks. Consequently, there was a great loss of life. For several days after the earthquake and fire, I walked about the ruins under the leadership of professors investigating the effects of the shocks and fire. Looking back, this opportunity may have been important to my becoming a geophysicist.

At that time, Professor A. Imamura, taking the place of Professor F. Omori, lectured on seismology at Tokyo University. In an examination he asked us a question concerning the possibility of practical earthquake prediction. I answered that this might be attained if we try to identify premonitory events by combining precise seismological observations with the measurement of minute crustal movements. Now, it is with feelings of emotion that I realize that since then, 60 years have passed, and that now research on earthquake prediction is being seriously carried out in many countries, directed toward or actually realizing its practical use.

2. ON DEEP-FOCUS EARTHQUAKES

After I graduated from the university I went to the Japan Central Meteorological Observatory in Tokyo, which at that time dealt not only with meteorology but also with seismology, oceanography, terrestrial magnetism, and other geophysical subjects. I became a member of the seismological section and engaged in observation and research on earthquakes.

After about one month, on May 23, 1925, a strong earthquake occurred in the northern part of Hyogo Prefecture, and 40 persons died. Many aftershocks were observed afterwards, and there occurred among them an unusually big earthquake. This earthquake was publicly mentioned as the largest aftershock, but with no other special comment. I was attracted to the fact that this earthquake showed a special wave propagation feature, areas of high seismic intensity extending far from the epicenter.

On looking back now, I realize that this was a memorable earthquake. Not only did I feel it with my body, but I also noticed its characteristic nature, and this led me to investigate deep-focus earthquakes. (By the way, the focal depth of this earthquake was 400 km and the magnitude was 6.9.)

My routine job at that time was mainly to reduce the seismological reports sent from local observing stations and determine the positions of the foci of the respective earthquakes. While occupied in this work, it seemed to me that the current way of doing this was unsatisfactory, because insufficient attention was paid to the depth of focus. At that time, earthquakes were generally considered to occur at depths no greater than 60 km. Accordingly, when deep earthquakes actually occurred, such as at depths of 300 km or more, we had much trouble locating the positions of their epicenters.

In those days, we could not keep accurate time. Furthermore, it was sometimes difficult to know the accuracy of the phase determinations reported by the observing stations. In addition, these stations were confined to a long chain of islands, and we had almost no observations from oceanic areas. Finally, at that time we did not have sufficiently accurate values for seismic wave velocities in the crust and upper mantle, especially for some parts of the Japanese Islands. Such being the case, the results obtained in those days concerning the locations of earthquake foci were not always accurate, particularly in the case of deep-focus earthquakes.

In the end, it was the precise investigation of seismic wave propagation that made me gain confidence in the existence of deep-focus earthquakes. But actually, what gave me an impressive early hint was the irregular distribution of seismic intensities, a phenomenon that had been known among us for a long time. That is, seismic shocks were felt by persons in irregularly distributed regions, sometimes at places far from the epicenter. These regions were usually along the Pacific coast of North Japan.

Two early papers that strongly attracted my attention were those of Dr. K. Hasegawa and of Prof. F. Omori. The former reported that shocks of an earthquake that occurred in the northern part of the Japan Sea were felt only by persons on the Pacific side of Japan. The latter, entitled "Seismograms Showing No Preliminary Tremors," described seismograms of an earthquake that clearly showed large P-waves but no S-waves. Needless to say, these properties are exactly those of deep-focus earthquakes as we know them today.

Well, numerous earthquakes occur in and near the Japanese Islands, and these earthquakes are mostly shallow ones, with foci at depths less than 60 km. In addition, there also occur some earthquakes at any depth down to about 600 km. These facts, now generally well known, were my conclusions 60 years ago from my investigations of the abundant seismological data of the previous years in this region.

I first published a paper in Japanese on these results of my research and then a series of papers entitled "Shallow and Deep Earthquakes" in 1927, 1929, and 1931. These papers attracted the attention of geophysicists in several countries, and many investigations were then made concerning deep-focus earthquakes (for instance, Dr. F. J. Scrase's paper entitled "The Characteristics of a Deep Focus Earthquake: A Study of the Disturbance of Feb. 20, 1931").

3. DEEP-FOCUS EARTHQUAKES AND DEVELOPMENT OF SEISMOLOGY

While making a map showing the epicenters of deep-focus earthquakes, I noticed that the epicenters of these earthquakes were located in zones. In the region of Japan and its neighborhood there are two deep-focus earthquake zones that meet at nearly right angles with each other. One traverses the Japanese main island, and the other extends northeastward from the northern part of the Japan Sea. Also, there is another deep earthquake zone along the southwestern islands of Japan.

By examining precisely the distribution of the foci of these deep-focus earthquakes, as well as those of intermediate depth, I found that these earthquake foci are distributed quite regularly within the Earth's interior. Thus, isopleth lines for the earthquake foci could be drawn, and these showed that the foci of deep and intermediate earthquakes are confined to a surface or thin layer in the interior of the Earth.

My paper concerning this problem was published in 1935, entitled "On the Activity of Deep-Focus Earthquakes in the Japan Islands and Neighborhood." In this paper, it is also noted that these isopleth lines and the distribution of active volcanoes have a close connection with each other— that is, the volcanic belts coincide with the isopleth lines of 120–200 km for the foci of intermediate-depth earthquakes.

As time passed, the theory of plate tectonics has been developed and the problem mentioned above has come to be considered anew. Especially in discussion of island-arc tectonics, this distribution of earthquake foci is closely related to the subduction of the oceanic plate or the sinking lithosphere. I feel honored that the name Wadati-Benioff zone is sometimes used in this connection, and am much impressed by the recent studies on subduction zones and the concept of "comparative subductology."

These deep-earthquake zones are now observed in various parts of the world, but the subduction angles, the focus of the deepest earthquake, and the magnitudes of the related large shallow earthquakes are different in different regions. The recent development of plate tectonics theory has led to many remarkable results concerning the structure and activity in the crust and upper mantle. Certainly, seismology has played an important role in this, together with all fields of Earth science. Studies of the mechanism of earthquake occurrence, the propagation of seismic waves, and other phenomena concerning earthquakes are contributing to the recent progress in solid-earth science, but still the precise spatial distribution of earthquake foci is, I think, the fundamental matter.

When I first studied deep and intermediate earthquakes, it was difficult to obtain the precise spatial distribution of their foci, even in the region of the Japanese Islands, as the observational data were not very accurate. But gradually the observations have improved, and now the foci of deep and intermediate earthquakes are being observed clearly on both the upper and lower sides of the downgoing thrust sheet.

Now, the seismic waves propagating from the foci of deep earthquakes usually show very elegant features on the seismograph. Phases of P, S, and their reflected and refracted waves can clearly be observed. By studying these waves, we can obtain the velocity variation and also the effect of absorption in the interior of the Earth. I sometimes noticed that the wave amplitudes, especially for S-waves, of some deep earthquakes were observed to be much smaller than expected theoretically from the focal mechanism of these earthquakes, and I considered this to be the result of the wave passing through an especially soft layer in the Earth's interior. Many studies have been made of the attenuation of seismic wave amplitudes caused by internal friction in the upper mantle. In the Japanese island-arc region, for example, a detailed study was reported by Prof. T. Utsu in his paper "Seismological Evidence for Anomalous Structure of the Island Arcs with Special Reference to the Japanese Region" (*Rev. Geophys. Space Phys.*, 9: 839–90, 1971).

I hope for future developments in the study of this part of the upper mantle that is closely related to volcanic activity. As is well known, earthquakes generally take place in the neighborhood of volcanic regions, but if we look closely, shallow tectonic earthquakes very seldom occur at places very near the active volcanoes, whereas intermediate earthquakes occasionally do.

In Japan, it is very important to predict volcanic eruptions and also to promote countermeasures against those disasters. One must realize that at present there occur many great disasters caused by volcanic eruptions in various parts of the world. A concerted effort of researchers in various branches of science is required because in most cases, areas of active volcanoes are densely populated.

As for the magnitude of earthquakes, although its definition is fixed and is generally used for each earthquake, it is sometimes difficult to determine the magnitude of earthquakes from observations made at small epicentral distances, especially for earthquakes of deep origin and for those occurring in island-arc regions. I have tried often to find some formula to determine the magnitude of earthquakes in Japan, but I did not succeed because seismic waves propagate in such a complicated way in this region. I was only able to obtain the variation wave amplitude as a function of epicentral distance for earthquakes of various focal depths, and for those in different areas. It seems to me that there still remains room for further study of the magnitudes of intermediate and deep earthquakes.

4. PREDICTION OF EARTHQUAKE AND VOLCANIC ERUPTION

Looking back at the development of seismology, it may be pointed out that there have been two major schools. One school developed in countries of high seismicity, and the other in countries of low seismicity. In the former, the subject of studies has been local earthquakes and studies related to the prevention of earthquake damage, focused on earthquake motions and related phenomena, and eventually on earthquake mechanisms. In the latter, along with studies of earthquakes, more emphasis has been placed on studies focused on the internal structure of the Earth.

I was born in a country of high seismicity and have studied earthquakes, feeling these shocks frequently with my body. So I have been mostly interested in such subjects as seismic motion, seismicity, magnitude, upheaval and subsidence of the ground, tsunamis, landslides, and groundwater levels, bearing in mind the problem of earthquake prediction.

The study of earthquake prediction has been enthusiastically carried on since the early days of seismology, and researchers in Japan have discussed effective ways for the prevention of disasters. By their incessant effort, the importance of this project was clearly recognized, and since 1965 the Japanese government has appropriated a budget for it. Research on earthquake prediction has been pushed forward as a national project, aimed toward its practical use. As this project advances, various new seismological observational facilities have been established year after year.

Furthermore, in 1976 the Headquarters for the Promotion of Earthquake Prediction was established by the Japanese Cabinet, and in June

1978, the Large-Scale Earthquake Countermeasure Act was concluded to take effective actions to intensify the system for disaster prevention in case a major earthquake prediction is made. This Act was made on the premise that the occurrence of a large-scale earthquake as large as magnitude 8 class will be predicted.

As mentioned above, prediction at present is mainly aimed toward earthquakes in the magnitude 8 class, but it is rather important to also predict such destructive earthquakes in the magnitude 7 class, which occur more frequently in the densely populated areas. We are now endeavoring to establish a prediction system for those earthquakes.

To carry out the work of earthquake prediction, a complete observational network is indispensable, especially in Japan, and it is important to extend the observational network to the sea area. Thus, at present two systems of submarine seismographs are located off the coasts of the Boso and Omaezaki Peninsulas. No formal earthquake warning has yet been issued since this system was established, so it is difficult to discuss its practical usefulness. Still, we are delighted to have reached this stage. Earthquake prediction is now being developed in many countries according to the characteristic seismic activity of these regions. Naturally, it is desirable that this work be advanced further by close international cooperation.

Now I wish to take up a problem concerning the prediction of volcanic eruptions. Needless to say, Japan has suffered incessantly from disasters caused by volcanic eruptions. The disasters caused by recent activity of the Mount St. Helens and Nevado del Ruiz volcanoes are well known. In Japan, fairly strong eruptions have occurred recently on the islands of Miyake and Ohsima, causing lava flows and considerable damage. For some active volcanoes we have geophysical observing stations for monitoring volcanic activity. In some cases these are near the crater, but it is still hard to understand precisely volcanic movements and to issue a warning of an outburst of the volcanic activity. Recently, however, research on volcanic prediction has been strengthened, and it is expected to make great progress, along with that on earthquakes.

In our country, and also in some other countries of the world, volcanic areas are in most cases densely populated or crowded by sightseers. Protection against volcanic disasters has become an urgent problem. There are two types of volcanic activity: one for which active movement is frequent and provides eruptions now and then, and the other kind that produces great sudden eruptions. These eruptions occur after a long silence. Because of this difference, the sociological consequences of a formal warning will also be different, and it seems to me that further studies are required by both natural and social scientists.

Although the relationship between the occurrence of earthquake and volcanic activity has been discussed for a long time, still more precise research on this problem is needed, as well as on the relationship between the eruptions of two volcanoes located near one another.

5.   TSUNAMI AND STORM SURGES

Since ancient times, Japan has suffered frequently from disasters caused by "tsunamis." In particular, great earthquakes that occur off the Pacific coast of the Japanese Islands usually cause very heavy damage to the coastal areas, even though the shocks are not always as violent there.

The Sanriku coast, which faces the Pacific, is well known by the tragic damage caused there twice, by earthquakes on June 15, 1896, and March 3, 1933. Along this coast of Liassic topography, there occurred various types of tsunamis in its different bays and harbors. In the former case, the wave height of the tsunami was 38.2 m in Ryori-Bay and 24.4 m in Yoshihama-Bay, and the number of deaths totalled about 22,000. In the latter case, the scale of the tsunami was nearly the same, but the disaster caused was not so heavy because of the advanced development of counter-measures.

Considerable tsunami damage has also been produced along the Pacific coast of Japan by large earthquakes originating at far remote places in the Pacific Ocean. For example, those coastal areas were attacked by tsunami waves on May 23, 1960, caused by the great earthquake of magnitude 8.5 that occurred off the coast of Chile. The tsunami propagated across the Pacific Ocean and arrived at the Japanese coast with long waves of 3–6 m height, and 140 lives were lost. After this event the Pacific Tsunami Warning System was established with its center in Hawaii and with the co-operation of the countries concerned. It is one example of an international system for prevention of natural disasters, and consideration is now being given to consolidate it with a similar system to monitor other global phenomena, such as air and marine pollution.

In Japan, storm surges caused by typhoons are very important natural phenomena, and we frequently suffer from these abnormally high tides. Typhoons usually give rise to heavy disasters by strong wind and rain, but the largest disasters sometimes are caused by the storm surges. The two typhoons of Muroto and Isewan raised terrible storm surges in Osaka Bay on September 21, 1934, and in Ise Bay on September 26, 1959, respectively. The height of sea level caused by these typhoons was 3 m or more and caused heavy damage to the cities of Osaka and Nagoya and their surroundings. In the latter case, about 5000 lives were lost by this typhoon, mostly due to the storm surge.

The causes and mechanisms of tsunamis and storm surges are different from each other, but the means for prevention of disasters due to these phenomena seem to be very similar. In Japan, cities are mostly located on low land, sometimes lower than the mean sea level, along the coast or inside bays. So it is a very serious problem to defend the coast against the attack of these phenomena, and therefore to issue the tsunami warning immediately after the occurrence of big earthquakes or to forecast accurately the passage of a typhoon.

6. DEVELOPMENT OF A SCIENCE FOR DISASTER PREVENTION

In view of the heavy damage caused by the Muroto typhoon, a new research institute for disaster prevention was established in Osaka in 1937, and I worked there for five years. At first, I did research on typhoons and earthquakes, but my main research made there was concerned with the phenomena of ground subsidence and air pollution.

The disaster of the lowlands of West Osaka was, of course, caused by the high storm surge of the Muroto typhoon, but the reason its damage was so heavy was ground subsidence that had been already occurring there for many years.

Ground subsidence had already been observed in the lowlands of Tokyo, where it reached even 20 cm yr^{-1} at some places. It was considered that this might have some relation to the occurrence of great earthquakes, and this made people feel uneasy.

There were many discussions about the cause of ground subsidence but no theory was established in scientific circles. In Osaka, I first tried to make a special instrument to record the daily vertical movement of the ground and also that of the underground water level, and I started observing at a suitable place in the subsiding area.

By investigating ground subsidence and underground water level data, I concluded that subsidence is caused by the marked lowering of underground water level—that is, a decrease of underground water pressure. Also, it was found that this subsidence was nothing but a contraction effect of the surface soft layer. The amount of subsidence correlated closely with the decrease of underground water pressure. It was concluded that this remarkable decrease of underground water pressure was caused by too much use of the groundwater, primarily by manufacturing plants. Accordingly, it can be regarded as a phenomenon caused by human activity and consequently quite different from the crustal movement related to the occurrence of earthquakes.

My papers on this problem were published in 1939–42, but in those days when the war was close at hand, my suggestions about the control of the use of underground water were not listened to by the authorities. But after more than 10 years, these arguments were generally accepted because the ground subsidence stopped during and after the war, since almost no pumping of the underground water had been made owing to the annihilation of industrial activities.

Special waterworks for industrial use were established in Osaka in 1951 and successively in Tokyo and in other cities. Now such marked ground subsidence in big cities and other districts has nearly stopped.

By the way, some years ago there occurred remarkable ground subsidence in the plain of Niigata. It resulted from heavy pumping of deep underground water for the purpose of obtaining natural gas. In this case underground water contains dissolved natural gas, and the ratio of water to gas is nearly 1 : 1. The depth of the underground water is about 600 m, and it caused remarkable ground subsidence around there that amounted to even more than 40 cm yr^{-1}. The mechanism of this phenomenon is regarded to be similar to that of the ground subsidence generally experienced in Tokyo and other places, although the depth of underground water was different.

The other main subject I studied in Osaka concerned the air pollution in this area. Before, Osaka was sometimes called the "Manchester of the Orient," and the citizens were proud of this name as a symbol of the flourishing industry. But actually they knew its harmful influence on health. I thought this air pollution over a great city should be scientifically studied as an important environmental problem, and I made routine observations and research on the phenomenon, particularly on the nature of floating particles.

In the course of this study, I was most impressed that the cause of the large-scale stoppage of electric service that occasionally occurred in the morning was found to be a harmful effect of the dirty dense fog that covered this area before dawn on the transmission lines.

Looking back upon my work in the Research Institute of Osaka, I see that I started to study disasters caused by natural phenomena, but as time went by I turned to research primarily on disasters brought about by human activities. Certainly, disasters occurring nowadays are caused by the combined effects of natural and artificial causes. The latter have come to play a dominant role.

When I worked in those days in Osaka, I believed that my duty was to investigate natural disasters, and so to my regret I did not carry out research on disasters caused mainly by human activities as fully as I might have. My full recognition of the importance of environmental problems came only later.

7. PROGRESS OF GEOSCIENCES AND THE CREATION OF AN IDEAL EARTH

I was appointed Director of the Central Meteorological Observatory of Japan in 1947 and worked to establish a new organization for meteorological service as well as investigations in oceanography, seismology, geomagnetism, and to promote scientific research in Earth sciences. In 1950, I had an opportunity to travel to the United States and Canada for three months to study the weather services and geophysical research carried out in governmental agencies, institutes, and universities.

This was my first travel abroad, and I was deeply impressed by seeing with my own eyes the splendid advancement of research in various fields of geophysics, and particularly by the newly developed instruments for geophysical observations. Thereafter, I attended various kinds of international geophysical meetings and had a good opportunity to become personally acquainted with the geophysicists of the world.

It is unnecessary to say that recently remarkable progress is being made in various fields of geophysics and geochemistry, such as environmental problems, plate tectonics, prediction of earthquakes and volcanic eruptions, and observations by satellites. These are now mostly made on a global and planetary scale. Recent geophysical research is being advanced by the international cooperation of many branches of science and by consideration of effects and variations of phenomena far distant in space and time. On the other hand, practical applications sometimes require that investigations be made on a very local scale.

In 1974–75, at the age of 72, I joined the Japanese Antarctic Research Expedition Team and experienced for the first time the glorious world of snow and ice of the South Pole region. I was deeply impressed by the beauty of nature and thought of my good fortune that I was born in a country of earthquakes and could study geophysics.

I am now mainly working on the prevention of natural disasters and environmental problems. The environment surrounding us is of both natural and artificial origin, and these are intertwined in a complicated way. Recently the artificial side has become predominant, particularly in densely populated districts. It is not an exaggeration that the Earth is changing its original state, and nature will be destroyed if things go as they are going now. Actually, it may be impossible to stop this present trend of gradual worsening of environmental conditions. But we must at least endeavor to create a new ideal Earth, searching for the best way to human happiness today and in the future. In the quest for this goal, research in geophysics will naturally play an important part, and its advancement is strongly anticipated.

Lastly, I think that we should be always be modest while carrying out this serious task, as our knowledge about the Earth is not yet enough.

Robert P. Sharp

Ann. Rev. Earth Planet. Sci. 1988. 16: 1–19

EARTH SCIENCE FIELD WORK: ROLE AND STATUS

Robert P. Sharp

Division of Geological and Planetary Sciences, California Institute of Technology, Pasadena, California 91125

Introduction

This essay contends that the outstanding contributions currently being made to the earth sciences by theoretical and laboratory endeavors increase rather than decrease the need for sound field observations. The presentation is strongly prejudiced in favor of field studies and, accordingly, invites critical examination by skeptical minds.

Many earth science problems being investigated have their source in field studies. Samples of materials worthy of analysis by sophisticated laboratory techniques and apparatuses are selected on the basis of field studies. Furthermore, the field is where the results of theory and laboratory experimentation are tested for conformity to nature and the truth. Field investigators of all types are as sorely needed now as at any time in the past. Their role in the earth sciences merits respect and recognition.

A statement supporting field studies may seem to be championing the obvious, but Francis Pettijohn's (1984) outspoken memoirs amply demonstrate that field activities have not always been respected. Many earth scientists would profess regard for, if not devotion to, field activities, and the US Geological Survey, the greatest assemblage of earth science talent ever, is strongly field oriented. Yet many field geologists feel their discipline is on the decline, that their efforts and products are looked down upon, that time spent in field work is seen as less productive than time spent in the laboratory or before a computer, that greater and more spectacular advances are made by laboratory experimenters or theoreticians than by field workers, and that field geologists may be an endangered species.

An unusually large number of exciting earth science developments have recently come from experimental and theoretical work. Much of the glamor in the earth sciences now seems to be associated with such endeavors, stimulated and supported by rapid developments in instrumentation and generous research grants. But close inspection shows that many of these fruitful advances have depended upon complementary and cooperative field investigations.

Field and laboratory work are mutually supportive, not adversative. Most field geologists need laboratory analyses to support their field work, and most laboratory investigators depend upon field data to bring reality to their studies. Few field geologists now make a career solely from field mapping, as Tom Dibblee (Steller 1986) has done so spectacularly. By the same token, few theoreticians or experimentalists proceed without reliance upon field data.

Field work may seem at times routine, unproductive, or even boring, but the same can be said of much laboratory work. One can labor long in either arena without rewarding results. Still, the chance of turning up something exciting is as great in the field as in the laboratory.

Most broadly, the term *geology* encompasses the study of planet Earth (Gary et al 1974, p. 293). In this sense, it includes the disciplines of geophysics, geochemistry, and geobiology (paleontology), as well as other subdisciplines involving the solid earth. A common modern practice, however, is to use the term *earth science* as embracing all these disciplines, including geology. This usage is followed here.

The terms *field geology* and *field work* are used in their broadest sense to include the observation, study, and investigation of natural materials, features, phenomena, and processes in their natural setting by any of a wide spectrum of procedures, techniques, and instruments. This broad concept of field work is developed further in a following section. A more classical concept is based on a geologist walking out contacts between rock units and transferring data to a base map. Geological mapping, however, is only one kind of field work.

Historical Perspective

My candidate for the greatest North American geologist would be G. K. Gilbert (Pyne 1980). Gilbert was above all a field geologist. His two monumental contributions, the *Lake Bonneville* monograph (Gilbert 1890) and the *Report on Geology of the Henry Mountains* (Utah) (Gilbert 1877), are products of field work. This is not surprising, because during the 50 years of his professional career, 1869–1918, geological studies were primarily field oriented, and western United States, where Gilbert did the bulk of his work, was geologically unexplored.

In some respects, Gilbert was before his time in setting up flume experiments to study the behavior of alluvial wastes derived from California gold placers (Gilbert 1914). From this study he formulated some of the basic laws governing fluvial transport of coarse rock debris. Gilbert (1893) was even more adventuresome in peering through an astronomical telescope at the moon's surface. He was one of the first to look at lunar features through the eyes of a field geologist. As a result, he postulated an impact origin for lunar craters, anticipating by many years conclusions drawn by Ralph Baldwin (1949) and Harold Urey (1951). Gilbert even resorted to throwing steel ball bearings into soft mud to simulate impact features. Through such activities, he pointed the way for modern earth scientists who combine field activity and laboratory experiments (Baker & Pyne 1978).

Outside his administrative duties for the US Geological Survey, Gilbert probably spent 80 to 85% of his research time and effort on field work and in preparing field data for publication. For many earth scientists today, laboratory effort is more likely to predominate. Modern instrumentation, techniques, and data-processing facilities make laboratory work especially productive and rewarding. As a result, fewer present-day earth scientists who devote the major part of their time and effort to field activities are regarded as outstanding by their peers, in contrast to Gilbert's time.

Types of Field Work

One's concept of field work naturally depends on special needs and interests. Near one end of the spectrum is the classical field geologist who wants to lay hands upon the earth's rocks, minerals, and fossils in their natural setting and to observe natural processes in action on the earth's surface. Such a person uses mostly eyes, feet, training, and experience in conducting field work and is likely to end up making a map showing relationships between geological units. Near the other end of the spectrum are those who employ highly sophisticated instruments, apparatuses, and techniques to learn about the physical properties, behavior, and relationships of masses composing the earth that cannot be ascertained by direct visual observation. The objective of such studies often lies within the earth rather than on its surface. Between these extremes is a wide variety of tasks, such as collecting specimens, measuring stream velocity, operating tiltmeters, and detailing stratigraphic sections.

Most exploratory geophysical procedures are forms of remote sensing, which does not at first thought seem like field work. Seismology, magnetometry, and gravimetry are examples. When Vening-Meinesz (1948) boarded a Netherlands submarine planning to operate a gravity meter over ocean basins, he actually embarked on a field program. If a magnetometer is used by someone walking over the ground, its measurements would probably be accepted as a product of field work. Why not regard measurements made by the same magnetometer flown over the area in an aircraft as the product of field work? If a geologist charters a plane for one day to fly over an area being mapped, has that day been devoted to something other than field work? Most of us would think not. The same geologist may make liberal use of aerial photos for mapping without thinking of them as a product of remote sensing.

A wide array of geophysical techniques is used for probing the earth, and if one, such as magnetometry, qualifies as field work, why not similarly regard the others? Geophysical techniques are used to observe and record characteristics of the earth in its natural state, whether the instruments are carried by hand, car, boat, submarine, airplane, spacecraft, or donkey. In most instances, field data not available in any other way are gathered by techniques of remote sensing. The fact that a remote-sensing technique, such as shuttle imaging radar in North Africa (Elachi et al 1982), benefits from subsequent ground studies (McCauley et al 1982) does not make it any less a form of field work. To map mantle tomography by use of earthquake waves, without ever leaving the laboratory, is in the broadest sense also a type of field work.

Remote sensing has the virtue of providing integrated views of large-scale relationships. Its needs have stimulated the development of techniques, instruments, and procedures that make possible more effective scientific observation of our Earth. It seems high time to recognize that remote sensors are engaged in an important form of field activity.

Essentially everyone accepts geological field mapping as classical field work. Two easily identified types of mapping are exploratory and directed. Exploratory mapping is carried on primarily to discover what exists within areas of unknown terrane. It is widely practiced in government and industry, and by some individuals. Exploratory mapping raises more questions than it answers, because by reconnaissance it turns up many new and unexpected findings. Directed mapping is normally conducted to solve specific problems or to support other field activity. It is normally more detailed but of more limited scope, both geographically and intellectually, than exploratory mapping, and it is designed to produce answers. To a purist, geological mapping of either type is the most basic form of field work.

Direct observations of geological phenomena in action, such as floods, surging glaciers, and volcanic eruptions, are productive and exciting—even hazardous—types of field work. Underground mapping in mines is a specialized activity largely of commercial interests. Although field geologists tend to rely on natural exposures, modern earth-moving tools, such as backhoes, are now often employed by Quaternary geologists, in place of shovels, to make artificial exposures at critical sites.

A widely exercised type of field activity is what might be called "show-and-tell." This is a procedure in which persons who have conducted field studies of relationships in a specific location show them to others and present explanations and interpretations. Show-and-tell field trips, usually sponsored by organizations, are an effective way of sharing knowledge.

Field work involves the observation, visual or instrumental, of natural materials and processes of the earth in their natural setting. Sometimes the only workable means are by the techniques of remote sensing. Defined this broadly, field work is done far more extensively by earth scientists than they or others may realize.

Field Work as Related to Research

Field activity introduces reality into earth science research. Nature can be a harsh critic, destroying elegant theoretical models with a few hard, cold facts. It makes sense to base such models on as much salient, sound field data as possible. No matter how attractive, a theoretical product remains incomplete until shown to be compatible with field relationships. Successful theoreticians respect and value field data and do not hesitate to enlist the cooperation of field workers in obtaining more. Essentially the same can be said for laboratory experimentation.

Much modern analytical equipment is so productive that one has to guard against letting the satisfaction of doing laboratory procedures overwhelm the significance of the analyses. Time and resources can be wasted analyzing specimens that do not merit the effort and expense. Since the earth sciences deal with complex, messy systems, a large number of relatively imprecise data are often more useful than a few highly precise values. There is little point in measuring the width of a city street to fractions of a millimeter with a micrometer. Field work is relatively inexpensive compared to most laboratory procedures, so it makes sense to invest in a thorough field study before launching an expensive analytical program. Contrariwise, in some situations, a few blind laboratory analyses may be required to establish the need for detailed field work.

A good field map is a necessity for many earth science research projects (US Geological Survey 1987, Reinhardt & Miller 1987). Such a map may have already been made, or it can be custom made, perhaps by someone other than the principal investigator. The problem being investigated may even have been identified in the first place by mapping. Geological knowledge of the area will help any research program, no matter how specialized or localized, to avoid later surprises.

Not all earth scientists need be adept at field mapping, but awareness of its value is desirable, as has been demonstrated many times. Consider the Heart Mountain overthrust of Wyoming, which for many years after its initial description (Hewitt 1920) was regarded as a rooted thrust produced by compression in the crust. Years of detailed field work by William G. Pierce (1941, 1957, 1960, 1963, 1973, 1979) were required to demonstrate that the Heart Mountain structure is a detachment thrust, essentially a huge slide, originating in the northeast corner of Yellowstone Park. The field relationships are striking, but a program of devoted mapping over a large region was required to demonstrate their true meaning.

Some geological settings allow us to observe and monitor experiments being conducted by nature at real scales in natural environments. Tracking natural experiments has yielded good results for rapidly acting processes and agents, such as glaciers, rivers, volcanism, wind, tectonism, and shoreline activities. Working with such situations requires patience, sometimes for years, before useful results are obtained, and even then it can be frustrating. Nature has a habit of changing parameters and variables indiscriminately at inopportune moments. Because of their duration and attending uncertainties, many natural experiments are not suitable subjects for graduate student research; they lie more in the domain of established professionals with a stable base.

Pedagogical Value of Field Work

The first geological field experience of many people is likely to be a show-and-tell field trip (Figure 1). Such trips are effective for stimulating interest among nonprofessionals and attracting the attention of other scientists to the opportunities and satisfactions of working on geological problems.

Show-and-tell can generate interest in the earth sciences without a large investment of time and effort on the part of nongeologists. Something seen in its natural state arouses more enthusiasm than it does through the medium of written or photographic representation, and it is certainly better retained in memory. Through stimulation by show-and-tell experiences, physicists, chemists, and planetary scientists, among others, have actually become professional earth scientists. It is a rewarding experience to have a radio astronomer wax enthusiastic about geological field phenomena after participating in some show-and-tell experiences. This has happened more than once. The show-and-tell procedure sows seeds widely, and one never knows what may sprout from some unexpected spot.

As a purely pedagogical procedure, show-and-tell has limitations in not demanding enough from the audience (students). Its educational function can be enhanced by presenting problems and challenges that require audience participation. The emphasis can be on show, with students providing the tell.

Educational insitutions could use field experiences more effectively in elementary geology courses by designing self-guided field excursions to replace some of the indoor exercises of the usual physical geology laboratory. This works better in some environments than others, but even the polished slabs of granite and gneiss in the local village bank front, the glacial erratic in the quad fronting the library, the cut behind the bookstore, or the gully and stream through the arboretum can, with imagination and thought, be used as a basis for field observation and interpretation by students. Send students into the field with written descriptive guides and thought-provoking questions. Don't lead them by the hand. It is also probably more effective to send them back into the field alone, with a corrected exercise in hand, than to conduct them en masse for a review of relationships.

Participation in a full-fledged field project is a still more effective means of generating a commitment to the earth sciences, on the part not only of students but of scientists from other disciplines. The British seem to practice this with particular success. Take, for example, Gerald Seligman's (1941) Jungfraujoch research program in glaciology, which enlisted the talents of Nobel laureate-to-be Max Perutz (Perutz & Seligman 1939), among others. Similar results attended a program of field research on the Austerdalsbrea in Norway, which led to major glaciological contributions from British physicists John Nye (1952, 1960, 1963) of Bristol and John Glen (1955, 1956) of Birmingham.

Geological field mapping may be regarded by the uninitiated as a simple-minded task of putting lines on maps. In truth, it is a first-class pedagogical discipline requiring keen observation, synthesis, and interpretation. Learning to arrive at workable conclusions, often on the basis of insufficient evidence, is part of the art of doing both geology and field mapping. Field mapping demands decisions; otherwise the map remains blank. Any reasonably intelligent person can be trained to do geological mapping, but as with many other pursuits, really good field mappers seem to be born, not made. The knack is not given to everyone, and those who have it deserve to be nurtured and respected.

Nature is a perverse ego humbler, and she exercises that trait freely in field geology. She delights in throwing spitball curves that send the overconfident neophyte, and often the hardened, experienced field mapper, back to the dugout muttering to themselves.

Except in the simplest areas, mapping involves a steady flux of surprises and complications. It is a detective game, solving ancient crimes committed by nature, with the clues now obscured. The experience is good for students. One of the main goals of a college education should be to establish a discipline of self-education. Few places are better suited to do that than the field. Nature is a stern teacher.

Figure 1 (*above* and *facing*)   Sharp and students during a field excursion east of the Sierra Nevada in the early 1980s devoted to inspection of glacial features.

Special Projects Related to Field Work

At least two long-range communal projects could be initiated by field geologists for the benefit of future generations of earth scientists.

THE CENTURY PROJECT   A central committee of retired field workers could be formed under the auspices of one of our national earth science organizations. This committee would design, select, and implement a number of field experimental and observational projects aimed at determining the rates at which geological changes take place over a period of 100 years.

Initiators of the program would obviously derive no direct benefit from it, hence the emphasis on older, retired people, who presumably have arrived, at least intellectually, and can afford to be generous with their time and effort. Raising funds for the project would be the responsibility of the central committee, which would also be charged with seeing that monitoring was maintained. The committee would have to be a continuing body with a slowly changing membership, managed by the sponsoring organization.

Suggestions for experiments should be solicited nationwide, and a selection of the most desirable, that could be adequately financed, would be made by the committee or a panel operating under its auspices. The central committee would assign responsibility for installation and periodic monitoring of each experiment, probably most often to the originators of the accepted proposals. For each experiment, the responsible entity would be charged with seeing that the baton of maintenance and monitoring was passed along to assure continuity to the completion of the project. Periodic reviews of each experiment by the central committee should be made to assure continuity.

Reports could be prepared periodically by experimenters, and the central committee could issue summaries through publication or other means, with ample credit to individuals. In 100 years, a final report on each experiment would be published under auspices of the central committee with full credit to all who participated.

Monitoring already going on in fields such as volcanology, glaciology, and neotectonics would not be duplicated. The types of experiments considered, all in natural settings, might include sedimentation rates for deposits in various environments, weathering and erosion rates, slow mass-movements, ground-ice wasting, dune migration, clay and caliche accumulation in soils, groundwater deposition, and shoreline modifications. There is no dearth of things that might be done. The task would be to select feasible experiments, likely to produce meaningful results, that could be maintained and monitored at reasonable cost and effort over a century.

A CENTER FOR FIELD TRIP GUIDEBOOKS   Unfortunately, much excellent field information is never published except in road logs and guides for field trips associated with a one-time event. The amount of valuable information packed away in such media is huge, but it is not easily accessible.

Field trip guides are issued by a number of widely dispersed sponsoring organizations; only a limited number are part of established serial publications. Guidebooks are commonly printed in small numbers, as special publications, and rapidly become unavailable. Few libraries have more than a sampling of field guides, and those they do have are a headache because of classification.

The Guidebooks Committee of the Geosciences Information Society has compiled a splendid catalog of geological field trip guidebooks of North America, up to 1980, and a limited listing of library holdings. A fourth edition of their catalog has recently been published by the American Geological Institute (1986). This is a commendable accomplishment, but more is needed.

Someone with library skills, organizational sense, and entrepreneurial spirit could render a major service to the earth sciences, hopefully at a profit, by establishing a Center for Field Trip Guidebooks. The Center should search out and assemble single copies of every field guide that can be located. Frequently, updated lists of these holdings should be distributed.

Through arrangements with authors and publishers, permission should be obtained allowing the Center to reproduce and sell copies of the guides at a price designed to support the Center's operation and provide a profit. If some tax-exempt organization undertook the service, it could be operated more economically on a no-gain basis and initial outlays might be underwritten by grants.

Personal Experiences

Field geology can be done almost anywhere at any time, with no more equipment than a notebook and pencil, by anyone with reasonable training and experience. Furthermore, it is usually worth doing. The following personal experiences are offered to illustrate the point.

In the summer of 1937, while finishing field work on a PhD thesis on geology of the Ruby–East Humboldt Mountains of northeastern Nevada (Sharp 1939), I began getting letters from Ian Campbell at the California Institute of Technology exploring possibilities of my joining a geological boat expedition through the Grand Canyon in October and November. The purpose of the expedition was to study exposures of the Archean igneous-metamorphic complex within the inner gorges, which are hard to reach except from river level. The expedition was to be jointly sponsored by Caltech and the Carnegie Institution.

That arrangement came about because John C. Merriam, famed vertebrate paleontologist at the University of California (Berkeley), had become Director of the Carnegie Institution of Washington. He wished to see the geology of the Grand Canyon thoroughly studied. Aside from early explorations, largely by Powell (1875), Gilbert (1875), Dutton (1882), and Walcott (1883, 1890, 1894, 1895), modern work had focused primarily on the Paleozoic (McKee 1934, 1938, Wheeler & Kerr 1936) and Proterozoic rocks (Van Gundy 1934, Hinds 1935). The only published product of comprehensive field mapping in the canyon was Levi Noble's (1914) excellent bulletin on the Shinumo quadrangle, which lies west of that part of the national park normally visited by tourists.

Solomon-like, Merriam apportioned the Paleozoic section to E. D. McKee, then ranger naturalist of the National Park Service, later of the US Geological Survey; the Proterozoic sedimentary and volcanic rocks to N. E. A. Hinds of the University of California (Berkeley); and the Archean complex to Ian Campbell and John H. Maxson of Caltech. Merriam supported these investigators with financial grants and facilitated publication of results in Carnegie Institution monographs (McKee 1934, 1938, McKee & Resser 1945, Hinds 1935).

Thus, I found myself in earliest October 1937 at Lee's Ferry on the Colorado River in the company of three experienced boatmen and three senior geologists—Campbell, Maxson, and J. T. Stark of Northwestern University. We were to board three wooden Stone-Galloway river boats for a two-month voyage of 280 miles through the Grand Canyon into Lake Mead, then filling behind Boulder (Hoover) Dam. McKee later joined the party at the foot of the old Bass Trail, partway along our route.

In 1937, the Colorado River was not a tourist's run; probably less than 100 people had made the trip through the canyon. Only one professional geologist other than Powell was known to have preceded us on such a voyage: Raymond C. Moore (1925), Professor of Geology at the University of Kansas and Kansas State Geologist, was a member of the 1923 Birdseye expedition (LaRue 1925) sponsored by the US Geological Survey to locate and evaluate dam sites.

Our expedition was truly exploratory with a promise of scientific discovery. I was very much the junior member of the group, and why Campbell took me rather than an experienced igneous-metamorphic petrologist, I'll never know. We traveled as fast as conditions, mostly rapids, permitted in reaches through Paleozoic and Proterozoic rocks and more slowly within the inner-gorge Archean exposures.

I asked Campbell if he had any specific geological chores in mind for me, and he decided I should keep track of pegmatite bodies in the Archean terrane. This and other activities did not fully occupy my time, so I cast around for something to do on my own. The Paleozoic belonged to McKee, the Proterozoic to Hinds, and the Archean to Campbell and Maxson, so I had to find something in between to avoid stepping on toes. Two great uncomformities are spectacularly exposed in the canyon walls: the younger one at the base of the Paleozoic beds where they rest upon truncated Proterozoic strata or the Archean complex, and the older one separating Proterozoic beds from truncated Archean rocks.

These appeared to fill the bill nicely. The near-horizontal pre-Paleozoic surface is exposed in cross section in the canyon walls along the wandering course of the river and its tributaries, so one gets a reasonable view of its three-dimensional relief. The pre-Proterozoic surface is preserved only in tilted fault blocks scattered throughout the region, so it is not so completely exposed. The study had to be improvised without a literature search or preconceived ideas. I had a Brunton compass, geological hammer, pencil, and notebook. Modern US Geological Survey topographic maps were available for only two or three quadrangles along the entire extent of the canyon, but river profile maps from the Birdseye work were helpful.

My improvised project meant some inconvenience for others of the expedition. Most of their work was conducted from daytime stops along the river, rather than from established camps. When a stop was made, if at all feasible I took off up a tributary canyon to get a look at one of my unconformities. As a result, when the rest of the party was prepared to move on, Sharp was often 500 to 1000 feet above on the Cambrian-Archean contact and out of touch. Nonetheless, Campbell was remarkably patient and supportive in letting me pursue my project.

The unconformities were produced by uplift and subsequent terrestrial erosion and weathering over hundreds of millions of years. Both of them retained remnants of the regolithic mantle produced on and within the underlying rocks. This weathering and erosion had occurred on a landscape devoid of vegetative cover under oxidizing conditions. The ancient erosion surfaces were each subsequently slowly invaded by a shallow sea, and the regolith had clearly been reworked into the basal layers of the initial marine sediments (Sharp 1940a). Striking sea cliffs were cut into residual knobs on the pre-Paleozoic surface by waves of the encroaching Cambrian sea. Looking at cross-section exposures of such cliffs in the canyon walls, one can almost hear the roar of the Cambrian surf hurling itself against the cliff and retreating to gather strength for its next attack. The knobs, with their sea cliffs flanked by outward-thinning tongues of coarse, reworked debris, were eventually submerged by the ever-deepening water, and ultimately buried by finer seafloor deposits.

In one locality, a cuesta on the pre-Paleozoic surface, created by sub-aerial erosion of tilted layers of Proterozoic quartzite, had been undercut by waves, generating a huge rockslide that spread outward over the seafloor across earlier beds of fine sediment. The slide contorted these beds into convolutions and whorls and created a tongue of coarse, angular quartzite breccia (Sharp 1940b).

The opportunity to work with these relationships was a stimulating scientific adventure for a neophyte geologist. The Grand Canyon voyage alone would have been a great adventure through magnificent scenery. Thanks to the ease with which one can do field geology, it was an enriching intellectual experience for me.

In late 1940 and early 1941, as a young faculty member at the University of Illinois, I received communications from Walter A. Wood, Director of Exploration and Field Research at the American Geographical Society in New York City, telling of explorations he and others had been conducting for years in the remote Mt. Steele and Wolf Creek (later renamed Steele Creek) area in the ice-bound St. Elias Range along the Yukon-Alaska border. Wood, trained as a geodesist in Switzerland, was using modern Swiss techniques to set up a triangulation network in this unsurveyed region.

Would I be interested in joining the expedition as a geologist, doing whatever I wished in the way of field work? My early wanderings in California's Sierra Nevada and mapping of glacial features in the Ruby–East Humboldt Range of Nevada (Sharp 1938) had whetted my interest in glaciers. The invitation to become familiar with active ice bodies was irresistible, and the summer's experience led to subsequent glaciological work extending over twenty years.

Although the Mt. Steele project allowed greater preparation than the Grand Canyon exercise, I was working entirely by myself with minimal equipment; the only special items were a Swedish increment borer for coring trees and a small ROTC tripod and plane table for mapping. No base maps were available, but Wood's triangulation network proved useful. The only paths were game trails made by Dall sheep and grizzly bears. The bears repeatedly devastated our caches of canned goods, except for an early version of the Army C-ration, which they disdainfully rejected.

It was a productive and educational summer of new experiences with glaciers (Sharp 1947), bedrock (Sharp 1943), frozen ground and ground ice (Sharp 1942c), debris flows (Sharp 1942a), and patterned ground (Sharp 1942b). A lot can be learned in two and a half months doing field work in a virgin area. I didn't have the background, however, to realize that the stagnant condition of the lower few kilometers of Wolf Creek (Steele) glacier showed it had undergone an earlier episode of surging. Twenty-five years later, in 1965–66, the glacier surged again, rejuvenating its stagnant lower reach (Post 1969, p. 230).

During World War II, as part of an Army–Air Force intelligence unit dealing with arctic, desert, and tropic environments, I was in the Aleutians during the summer of 1945 working on survival problems for the Eleventh Air Force. In the course of this duty, a visit to Shemya, the third western-most island of the Aleutian chain, brought me to a firm-minded general who said, "I don't believe all this junk you fellows write about survival in this area. Let's put you over on Agattu for a few days and see how you do." So a small speedboat deposited me on the east coast of Agattu with a sleeping bag and the meager bailout kit of a single-seater fighter pilot. They kindly gave me a cup of coffee before putting me ashore.

Even in summer, the Aleutians are cool and damp. I had matches, but the Aleutian tundra provides little fuel, and shoreline driftwood is meager and wet. A small cooking fire was possible, but not a warming fire. I would have appreciated the pilot's parachute, but that was not available because I had not bailed out. Some shelter from light rains and mist was obtained beneath a stream-cut bank capped by tundra.

Food proved to be no problem. The bailout kit contained a line and fishhooks. Within the first day, I caught enough small arctic char out of a stream to feed myself and a friendly gray fox for days. Sea urchins were plentiful along shore, and their roe is extremely nutritious, but it proved too rich for my stomach without blander foods. Chitons and small mussels were abundant, and edible plants could be found within the complex of tundra vegetation.

Once settled in and tired of watching puffins and whales, I explored coastal cliffs in search of a cave for shelter. Instead, I soon found extensive, modestly inclined exposures of thinly bedded, light-colored, siliceous sedimentary rocks some 600 meters thick (Sharp 1946). I was flabbergasted. The Aleutians are known to be an island arc, a chain of volcanic cones. What were sedimentary rocks like these doing on Agattu?

A literature search later revealed that other small accumulations of sedimentary rocks were known in the islands, and more may have been discovered since. The Agattu rocks proved to be porcellanites, presumably formed by silicification of fine-grained pyroclastics, so they were at home in a volcanic province, although they reflected an unusual environment of accumulation.

I was elated by the discovery. This was the one occasion during the war when I could do a bit of geology, albeit solely with notebook and pencil. An isotopic geochemist might have found Agattu uninteresting and forbidding. To a field geologist, it was fascinating.

When plans were formulated for experiments to be carried aboard spacecraft Mariners 3 and 4, during a flyby of Mars in 1965, Caltech physicist R. B. Leighton, designer and principal experimenter for the television camera, realized that pictures of the martian surface needed to be studied by a geologist, so he drafted Bruce Murray of Caltech's geology division. Murray recognized that the pictures would show primarily landscape features, so he drafted geomorphologist Sharp, and we three constituted the TV scientific team. Craters and faults proved to be the two primary forms shown on the 22 photos taken by Mariner 4. (Mariner 3 had shroud troubles and never got out of Earth orbit.)

This early work led to further participation in the Mariner 6 and 7 flybys and the Mariner 9 orbiter, which produced many good photos of martian surface features. Extensive experience in field work on Earth was an invaluable background for interpreting the martian terrains.

I have been fortunate to participate as a colleague in modern geochemical stable isotope research (Epstein & Sharp 1959) and in planetary science projects (Sharp 1973), largely because as a classical, old-time field geologist, I could bring something useful to the investigations.

Conclusion

Are field geologists an endangered species? Only if they themselves think so and behave accordingly.

As the earth sciences use more sophisticated laboratory procedures and theoretical models, the need for good field data increases rather than decreases. The relationship is symbiotic: The field and laboratory need each other, and as one prospers, the other benefits. Not everyone need be a field geologist, but the earth sciences will advance more effectively if workers use and have respect for field data and those who produce it. The earth sciences will always need people who are skilled at making field observations.

Literature Cited

American Geological Institute. 1986. *Union List of Geologic Field Trip Guidebooks of North America.* 200 pp. 4th ed.

Baker, V. C., Pyne, S. 1978. G. K. Gilbert and modern geomorphology. *Am. J. Sci.* 278: 97–123

Baldwin, R. B. 1949. *The Face of the Moon.* Chicago: Univ. Chicago Press. 238 pp.

Dutton, C. E. 1882. Tertiary history of the Grand Canyon district. *US Geol. Surv. Monogr. 2.* 264 pp.

Elachi, C., Brown, W. E., Cimino, J. B., Dixon, T., Evans, D. L., et al. 1982. Shuttle imaging radar experiment. *Science* 218: 996–1003

Epstein, S., Sharp, R. P. 1959. Oxygen isotope variations in the Malaspina and Saskatchewan glaciers. *J. Geol.* 67: 88–102

Gary, M., McAfee, R., Wolf, C. L., eds. 1974. *Glossary of Geology.* Washington, DC: Am. Geol. Inst. 805 pp.

Gilbert, G. K. 1875. Report on the geology of portions of Nevada, Utah, California, and Arizona examined in the years 1871 and 1872. *US Geogr. Geol. Surv. W. of the 100th Meridian Rep.* 3: 17–187

Gilbert, G. K. 1877. *Report on the Geology of the Henry Mountains.* Washington, DC: US Geogr. Geol. Surv. Rocky Mtn. Reg. 2nd ed. (1880). 160 pp., 170 pp.

Gilbert, G. K. 1890. Lake Bonneville. *US Geol. Surv. Monogr. 1.* 438 pp.

Gilbert, G. K. 1893. The moon's face; a study of the origin of its features. *Philos. Soc. Washington Bull.* 12: 241–92

Gilbert, G. K. 1914. The transportation of

debris by running water. *US Geol. Surv. Prof. Pap. 86.* 263 pp.

Glen, J. W. 1955. The creep of polycrystalline ice. *Proc. R. Soc. London Ser. A* 228: 519–38

Glen, J. W. 1956. Measurement of the deformation of ice in a tunnel at the foot of an ice fall. *J. Glaciol.* 2: 735–45

Hewitt, D. F. 1920. The Heart Mountain overthrust, Wyoming. *J. Geol.* 28: 536–57

Hinds, N. E. A. 1935. Ep-Archean and Ep-Algonkian intervals in western North America. *Carnegie Inst. Washington Publ. 463.* 52 pp.

LaRue, E. C. 1925. Water power and flood control of Colorado River below Green River, Utah. *US Geol. Surv. Water Supply Pap. 556.* 171 pp.

McCauley, J. F., Schaber, G. G., Breed, C. S., Grolier, M. J., Haynes, C. V., et al. 1982. Subsurface valleys and geoarcheology of the eastern Sahara revealed by shuttle radar. *Science* 218: 1004–20

McKee, E. D. 1934. The Coconino sandstone—its history and origin. *Carnegie Inst. Washington Publ. 440,* 7: 77–115

McKee, E. D. 1938. The environment and history of the Toroweap and Kaibab formations of northern Arizona and southern Utah. *Carnegie Inst. Washington Publ. 492.* 268 pp.

McKee, E. D., Resser, C. E. 1945. Cambrian history of the Grand Canyon region. *Carnegie Inst. Washington Publ. 563,* pp. 3–168

Moore, R. C. 1925. Geologic report on the inner gorge of the Grand Canyon of Colorado River. *US Geol. Surv. Water Supply Pap. 556,* pp. 125–71

Noble, L. F. 1914. The Shinumo quadrangle, Grand Canyon district, Arizona. *US Geol. Surv. Bull. 549.* 100 pp.

Nye, J. F. 1952. The mechanics of glacier flow. *J. Glaciol.* 2: 82–93

Nye, J. F. 1960. The response of glaciers and ice sheets to seasonal and climatic changes. *Proc. R. Soc. London Ser. A* 256: 559–84

Nye, J. F. 1963. On the theory of the advance and retreat of glaciers. *Geophys. J. R. Astron. Soc.* 7: 431–56

Perutz, M. F., Seligman, G. 1939. A crystallographic investigation of glacier structure and the mechanism of glacier flow. *Proc. R. Soc. London Ser. A* 172: 335–60

Pettijohn, F. J. 1984. *Memoirs of an Unrepentant Field Geologist: A Candid Profile of Some Geologists and Their Science.* Chicago: Univ. Chicago Press. 260 pp.

Pierce, W. G. 1941. Heart Mountain and South Fork thrusts, Park County, Wyoming. *Am. Assoc. Pet. Geol. Bull.* 25: 2021–45

Pierce, W. G. 1957. Heart Mountain and South Fork detachment thrusts of Wyoming. *Am. Assoc. Pet. Geol. Bull.* 41: 591–626

Pierce, W. G. 1960. The "break-away" point of the Heart Mountain detachment fault in northwestern Wyoming. *US Geol. Surv. Prof. Pap. 400-B,* pp. 236–37

Pierce, W. G. 1963. Reef Creek detachment fault, northwestern Wyoming. *Geol. Soc. Am. Bull.* 74: 1225–36

Pierce, W. G. 1973. Crandall Conglomerate, an unusual stream deposit and its relation to Heart Mountain faulting. *Geol. Soc. Am. Bull.* 84: 2631–44

Pierce, W. G. 1979. Clastic dikes of Heart Mountain fault breccia, northwestern Wyoming, and their significance. *US Geol. Surv. Prof. Pap. 1133.* 25 pp.

Post, A. 1969. Distribution of surging glaciers in western North America. *J. Glaciol.* 8: 229–40

Powell, J. W. 1875. *Exploration of the Colorado River and Its Canyons.* New York: Dover. 400 pp.

Pyne, S. J. 1980. *Grove Karl Gilbert: A Great Engine of Research.* Austin: Univ. Tex. Press. 306 pp.

Reinhardt, J., Miller, D. M. 1987. Cogeomap: a new era in cooperative geologic mapping. *US Geol. Surv. Circ. 1003.* 12 pp.

Seligman, G. 1941. The structure of a temperate glacier. *Geogr. J.* 97: 295–315

Sharp, R. P. 1938. Pleistocene glaciation in the Ruby–East Humboldt Range, northeastern Nevada. *J. Geomorph.* 1: 296–323

Sharp, R. P. 1939. Basin-range structure of the Ruby–East Humboldt Range, northeastern Nevada. *Geol. Soc. Am. Bull.* 50: 881–919

Sharp, R. P. 1940a. Ep-Archean and Ep-Algonkian erosion surfaces, Grand Canyon, Arizona. *Geol. Soc. Am. Bull.* 51: 1235–69

Sharp, R. P. 1940b. A Cambrian slide breccia, Grand Canyon, Arizona. *Am. J. Sci.* 238: 668–72

Sharp, R. P. 1942a. Mudflow levees. *J. Geomorph.* 5: 222–27

Sharp, R. P. 1942b. Soil structures in the St. Elias Range, Yukon Territory. *J. Geomorph.* 5: 274–301

Sharp, R. P. 1942c. Ground-ice mounds in tundra. *Geogr. Rev.* 32: 417–23

Sharp, R. P. 1943. Geology of the Wolf Creek area, St. Elias Range, Yukon Territory, Canada. *Geol. Soc. Am. Bull.* 54: 625–49

Sharp, R. P. 1946. Note on the geology of Agattu, an Aleutian Island. *J. Geol.* 54: 193–99

Sharp, R. P. 1947. The Wolf Creek glaciers,

St. Elias Range, Yukon Territory. *Geogr. Rev.* 37: 26–52

Sharp, R. P. 1973. Mars: fretted and chaotic terrains. *J. Geophys. Res.* 78: 4073–83

Stellar, D. L. 1986. Thomas Wilson Dibblee, Jr. *Calif. Geol.* 39: 202–6

Urey, H. C. 1951. The origin and development of the earth and other terrestrial planets. *Geochim. Cosmochim. Acta* 1: 209–77

US Geological Survey. 1987. National Geologic Mapping Program. *US Geol. Surv. Circ. 1020.* 29 pp.

Van Gundy, C. E. 1934. Some observations on the Unkar group of the Grand Canyon Algonkian. *Grand Canyon Nat. Notes* 9: 338–49

Vening-Meinesz, F. A. 1948. *Gravity Expeditions at Sea 1923–1938.* Delft: Publ. Neth. Geod. Comm. 4. 233 pp.

Walcott, C. D. 1883. Pre-Carboniferous strata in the Grand Canyon of the Colorado, Arizona. *Am. J. Sci.* 26: 437–42

Walcott, C. D. 1890. Study of a line of displacement in the Grand Canyon of the Colorado, in northern Arizona. *Geol. Soc. Am. Bull.* 1: 49–64

Walcott, C. D. 1894. Pre-Cambrian igneous rocks of the Unkar terrane, Grand Canyon of the Colorado, Arizona. *US Geol. Surv. Ann. Rep. 14*, pt. 2, pp. 497–524

Walcott, C. D. 1895. Algonkian rocks of the Grand Canyon of the Colorado. *J. Geol.* 3: 312–30

Wheeler, R. B., Kerr, A. R. 1936. Preliminary report on the Tonto Group of the Grand Canyon, Arizona. *Grand Canyon Nat. Hist. Assoc. Bull.* 5: 1–16

Photo of near-life-size caricature of Bob Sharp in the field, complete with staff and Filson jacket, as created in 1986 by Sharon Martens, emphasizing a characteristic gesture being imitated by the roadrunner, one of Bob's favorite birds.

ENTOMOLOGY

Sir Boris P. Uvarov

Annu. Rev. Entomol. 1990. 35:1–24

SIR BORIS UVAROV (1889–1970): THE FATHER OF ACRIDOLOGY

N. Waloff

Department of Pure and Applied Biology, Imperial College at Silwood Park, Ascot, Berkshire, SL5 7PY, England

G. B. Popov

Formerly of Anti-Locust Research Centre and Centre for Overseas Pest Research, London, and later Staff Member of FAO Locusts and Migratory Pests Emergency Operation Group

Introduction

It is now 20 years since Sir Boris Uvarov KCMG, FRS died on March 18, 1970, at the age of 82, and just over a century since he was born. For more than half a century he spearheaded *acridology,* a term he coined to cover all aspects of investigations on locusts and grasshoppers and of measures of their control.

Not only was he one of the great entomologists of the century and a world authority with an encyclopedic knowledge of locusts and grasshoppers, he also possessed a wide grasp of biological subjects and remarkable powers of organization, together with a clear vision of the necessity for international cooperation in control of migrant pests.

Uvarov, or B. P., as he was known to many, was a man of astonishing vitality, who was able to combine his academic expertise with practical ability. His skill and above all his scientific integrity enabled him to persuade and at times to argue relentlessly with those in power to produce the necessary resources for the development of acridology. His attitude towards his subject was dominated by two basic principles: First, that any rational economic control of an insect pest can only be achieved through profound knowledge of

0066-4170/90/0101-0001$02.00

its biology—he was angered by those who set boundaries in studies of economic entomology, for to him all investigations on Orthoptera formed a continuum. He thus stimulated work in taxonomy, biogeography, ecology, ethology, and control measures, and he influenced physiologists by persuading them to use locusts as their experimental animals. Second, he saw clearly that the vast scale and geographical complexity of the distribution of the migrant species of locusts and grasshoppers necessitated international coordination and cooperation in any control measures applied to them. He was fond of saying that insects do not recognize any artificial, political frontiers.

In writing this article we have been fortunate to have available such outstanding sources of information as the "In Memoriam" (3a) issued by the Anti-Locust Research Centre, under the auspices of its Director, Professor P. T. Haskell CMG.; the memorials of B. P. Uvarov written by Professor Sir Vincent Wigglesworth (75) and by Professor G. I. Bei-Bienko (4); and the erudite reviews by Professors M. P. Pener (32) and D. K. McE. Kevan (24). We have freely used the information summarized by them.

Also, in an attempt to assess Uvarov's impact on the development of acridology and on its present state, we have consulted a number of eminent entomologists who knew him personally, or who have been concerned with locusts and grasshoppers. We have listed their names in "Acknowledgments" and to them we owe immense gratitude for their generous help.

Youth and Education

Boris Petrovich Uvarov was born on November 5, 1889, in Uralsk in southeastern Russia. His father was a State Bank Manager. Both his parents were keen on outdoor activities and encouraged the love of nature in their three sons, of whom B. P. was the youngest. Among Sir Boris's earliest memories were the family's frequent outings in summer to the unspoilt countryside. The boys loaded a horse-drawn wagon with tents and other paraphernalia, among which the family cat was included. The weekends were spent camping, shooting, and fishing. From an early age Uvarov developed an interest in natural history and in entomology in particular. His interests were greatly enhanced by a gift from his father of six volumes of Brehm's *Tierleben*.

Uvarov attended a local school in Uralsk from 1895 to 1902. His interest in natural history grew, and he became a keen collector of insects, for which he was awarded several school prizes. This interest was encouraged by S. M. Zhuravlev, a teacher in the Agricultural School near Uralsk, whom B. P. remembered with affection all his life.

In 1904 Uvarov started his university studies in the School of Mining at Ekaterinoslav (now Dnepropetrovsk), but in 1906 he transferred to the Faculty of Biology of the University of St. Petersburg. He was greatly influenced by the lectures of Shimkevitch, Wagner, and Palladin but was even more stimulated by the informal meetings of the Russian Entomological Society, where students mixed freely with the most eminent entomologists. A close circle of friends also attended meetings of the Students Biological Society, usually in the apartments of D. N. Borodin, A. A. Lyubishev, and others. There they enjoyed lively discussions on biological problems and on the results of current investigations. Students were also encouraged to visit and work on taxonomic problems in the entomology section of the Zoological Museum of the Academy of Sciences. All his life Uvarov remained indebted to the stimulating environment of his years as a student.

Two sets of circumstances influenced B. P.'s particular interest in Orthoptera, namely, the richness and diversity of the insects in his native Uralsk province, and the publication in 1905 of Jacobson & Bianka's volume on Orthoptera of the Russian empire. His own diploma paper, published in 1910 (45), was on the orthopteran fauna of the Uralsk province. This taxonomic work included many observations on habitats, i.e. it had an ecological approach to taxonomy, which was only incipient in entomology of the time. In 1910 Uvarov was awarded a first class degree and married Anna Fedorovna (née Prodanjuk).

The diploma paper was not the first of his publications. In 1909, together with D. N. Borodin, he published observations on the flora of middle Emba and an essay on Lake Chalkar, and in early 1910 papers on the lepidopteran fauna of the Trans-Ural-Kenghiz Steppe and botanical-geographical notes on Inder.

Early Life in Russia

Immediately after graduation Uvarov took up a post as entomologist at the Murgab Crown Cotton Estate in Trancaucasia, but in 1911 he joined the Department of Agriculture in St. Petersburg. From there he was sent to Stavropol province to investigate the biology and methods of control of the migratory locust, *Locusta migratoria*. At the age of 23 he became the director of the Entomological Bureau of Stavropol, where he spent three years between 1912 and 1915 developing a control strategy for *L. migratoria* and the Moroccan locust, *Dociostaurus maroccanus,* as well as for other agricultural pests. It was then that he made the classic field observations which led him to formulate his famous phase theory; at the time his conclusions were so revolutionary that he withheld their publication for a number of years (see Phase Theory).

By 1915 Uvarov was firmly established in his career, and he was asked to organize plant-protection stations in Transcaucasia. From 1916 to 1920 he held the post of director of the Bureau of Plant Protection in Tiflis, Georgia.

The bureau served a wide territory, covering present-day Georgia and Armenia.

During this period Uvarov published many papers on the Orthoptera of Caucasus and wrote reports on the activity of the bureau and a book on agricultural entomology, dealing with insects of Georgia. Simultaneously he accumulated data for an important paper on the geographical distribution of orthopteran insects in Caucasus and western Asia, later published in London (46). In this paper he emphasized the importance of arid conditions in the origin of the peculiar fauna of the desert subregion of the Palaearctic stretching from central and western Asia to north Africa. In addition to these activities, in 1919 and 1920 he became the keeper of entomology at the State Museum of Georgia and a reader in entomology in the State University of Tiflis.

In the turbulent years of 1918–1920 in the Transcaucasia, and with the rise of Georgian nationalism, Uvarov's position became difficult and he found himself without most of his paid employment. Often he had to supplement his income by selling home-made pies in the market square of Tiflis, before proceeding to the museum or to give a lecture at the university. It was fortunate that at that time young Patrick A. Buxton was among the British troops in Georgia; he provided a link with London. In 1920 Uvarov received an invitation from G. A. K. Marshall (later Sir Guy Marshall) to join the Imperial Institute (later the Commonwealth Institute) of Entomology. With his wife and their small son, B. P. left for London, which became his home for the next 50 years, until his death. He did not visit Russia again until 1968, when as an honored guest of the Academy of Sciences USSR he attended the XIIIth International Congress of Entomology, held in Moscow. There he was received with great warmth and this gave him much pleasure.

Life in England

This major part of Uvarov's life falls into three periods—the first ten years 1920–1930, the years from 1930 until his retirement in 1959, and the post-retirement period.

The first ten years are characterized by a prodigious output of publications. Of the 465 publications listed in his bibliography (75), exactly a third appeared between 1920 and 1930. Uvarov's official duties at the Imperial Bureau of Entomology (under the directorship of Sir Guy Marshall and later S. A. Neave) were to identify insects sent there from all parts of the Commonwealth. With his remarkable powers of concentration, he also found time for much taxonomic work at the British Museum of Natural History and to prepare a number of outstanding publications. Among the latter were his classic paper, formulating his theory of locust phases (48), and his book, *Locusts and Grasshoppers,* published both in English and in Russian (53, 56), in which he summarized the available data on Acridoidea. It became the bible of Orthopterologists for several decades and found its way to the shelves of many nonspecialist entomologists. He produced simultaneously two outstanding reviews, *Insect Nutrition and Metabolism* (57) and the classic *Insects and Climate* (58), which inspired much research then and in subsequent years.

In this period of 10 years, 27 of his 156 publications appeared in the USSR. Some of the major ones, "Acrididae of the European part of USSR and Western Siberia" (51), "Acrididae of Central Asia" (52), and the keys to the orthopteran orders of insects (55) laid the foundations of Russian acridology, as has been acknowledged by eminent Russian orthopterologists [Bei-Bienko, (4) and in letters to us by Mishchenko and Stebaev].

It is impossible to overestimate Uvarov's influence in his crucial early years in Great Britain, when he continually advocated more financial support for locust research and set in train an important new line in entomology. When he was still an early immigrant, without a perfect command of English, his close friendship with Francis Hemming CMG, CBE, was particularly valuable. Hemming willingly helped him with presentation of papers. They had many fruitful discussions and worked closely together until the start of the war, when Hemming was given the post of Principal Assitant Secretary, War Cabinet Offices.

In the late 1920s serious locust plagues were widespread in Africa and southwest Asia. In 1929 the Committe on Civil Research asked the Commonwealth Institute to take part in investigations into bionomics, biogeography, and periodicity of swarming locusts. Uvarov's duties included the organization and supervision of this work. By 1930, it had been formalized into a Committee on Locust Control of the Economic Advisory Council, with Francis Hemming as its Secretary (64).

A small unit of workers on locusts, initially consisting only of B. P. and his assistant (as she then was) Zena Waloff, and later a few additional members of staff, was located in one of the towers of the British Museum, Natural History. This unit was soon recognized and (unofficially) known as the International Unit of Locust Research. Uvarov's drive and initiative to achieve coordination in research in acridology and international cooperation in control of locusts led to the famous series of International Locust Conferences held in Rome (1931), Paris (1932), London (1934), Cairo (1936), and Brussels (1938). Plans were formulated for international organizations to be set up in Africa to control the African migratory and the red locusts, but these schemes were interrupted by the outbreak of war.

When the war started, investigations on acridology came to a temporary halt, but Uvarov's advisory and organizational work greatly expanded when he became the locust control adviser to the War Cabinet. At that time, countries in the Middle East and East Africa were suffering simultaneous plagues from the desert, red, and the African migratory locusts. As protection of strategic food supplies became vital, the organizations—the Middle East Anti-Locust Unit and the East African Anti-Locust Directorate—were created. Both organizations were inspired by Uvarov and, in an overall sense, were guided by him. It was largely through his efforts that the Middle East Supply Centre in Cairo was given the responsibility of organizing operations of locust control throughout North Africa and the Arabian peninsula and coordinating anti-locust campaigns on a paramilitary basis from Morocco to India. Although it is impossible to evaluate exactly how effective these efforts were, for the first time large scale operations clearly demonstrated the practicability of international action in combatting an insect pest which hitherto was considered to be uncontrollable. In 1945 the London Centre became established as an independent institution, known as the Anti-Locust Research Centre (ALRC) under the Colonial Office and later under the Ministry of Overseas Development. Uvarov then became the director of the Centre, a position he kept until his retirement in 1959. Wigglesworth (75) said of this time, "During the next fourteen years the Centre developed into the foremost laboratory in the world for research on locusts." Simultaneously it continued to further international cooperation in locust control and coordinated and instigated much extramural research on acridids in universities and other scientific institutions in Great Britain and abroad.

Uvarov also excelled in dissemination of information on Orthoptera and was able to create a coherence and a sense of purpose in the "World of acridology" not only of entomologists, but also in other participants in the work on locusts and grasshoppers, who were scattered over many countries and several continents.

Information, which flowed so freely from ALRC, or the Centre, was supported by a unique library, founded on Uvarov's own collection of acridological literature. It became the largest specialist library on Orthoptera in the World, containing over 30,000 publications as well as unpublished manuscripts and theses. Also, all those who worked on locusts and grasshoppers were kept up to date by the circulation of *Acridological Abstracts* and of copies of *Current Research on Orthoptera* which gave the names, addresses, and special interests of investigators on acridids.

As the work and influence of the Centre expanded, two journals were launched, the *Anti-Locust Memoirs* in 1946, which presented biogeographical and cartographical studies and the *Anti-Locust Bulletins* in 1948, which contained accounts of both laboratory and field studies. Shorter papers were also published as *Occasional Reports*. Up to Uvarov's death in 1970 eleven *Memoirs* and 48 *Bulletins* had been published.

A major task of the Centre was to ensure a regular flow of information on the current situation of the three main species of locusts, *Schistocerca gregaria*, *Locusta migratoria*, and *Nomadacris septemfasciata*. From the time when the work on locusts was initiated at the Imperial Institute of Entomology, reports were received from all the British territories in tropical Africa and western Asia. These were coordinated, analyzed and the data regularly plotted on monthly maps. Over the years, these data accumulated into archives now housed in the Overseas Development Natural Resources Institute (ODNRI)—the successor of ALRC in Chatham. These archives have formed the basis of many biogeographical studies and made possible the formulation of monthly summaries and forecasts of the locust situation. Uvarov saw that special coordination of forecasting and planning of control was necessary, especially for the desert locust, whose invasion area spreads over 50 countries. He interested the Food and Agricultural Organization (FAO), which by 1953 undertook coordination of control campaigns against the desert locust. By 1958 the practice of issuing monthly situation summaries and forecasts was formally recognized by the establishment of the International Desert Locust Information Service (IDLIS, later DLIS), by agreement between the United Kingdom and the FAO; DLIS was always an integral part of ALRC, but it was partly financed by the FAO.

Among his numerous activities Uvarov played a major part as a consultant to the FAO Technical Advisory Committee, particularly as the chairman of an FAO panel of experts on the long-term policy for desert locust control. He also helped to initiate the United Nations Development Program (UNDP/FAO Desert Locust Project), in which more than 40 countries participated.

Throughout the years of his directorship, numerous governments and international organizations sought Uvarov's advice on locust and grasshopper problems, so he travelled widely to many countries in Africa, the Middle East, and southwest Asia as well as in Canada, the USA and Australia. He also enjoyed his visits to workers engaged in research or control projects in the field, unhesitatingly accepting their fairly rough camp life. Their feelings toward him were of affection and respect; many in the field referred to him as "Uncle Boris," though few did to his face.

After retiring from the directorship of the Centre in 1959, he remained there as a consultant and spent much of his time writing a new book, *Grasshoppers and Locusts,* which provided a masterly synthesis of the enormous amount of new information that had accumulated since the publication of his *Locusts and Grasshoppers* in 1928. The first volume of 481 pp. was published in 1966, and he worked on the second volume until his death in 1970. B. P.'s colleagues, Zena Waloff, R. F. Chapman, and N. D. Jago, edited the text after his death and added some supplementary material, but they did not write the additional chapters that he had planned. The second volume of 631 pp. appeared in 1977. In reviewing it, J. S. Kennedy referred to it as the "Acridologist's Vademecum."

Uvarov received many awards and public honors in his lifetime. He was elected Fellow of the Royal Society in 1950 and received The Order of Companion of St. Michael and St. George in 1943 and Knighthood of the same order in 1961. Among his other honors were an honorary Doctorate of Science of the University in Madrid in 1935 and of Commandeur de l'Ordre Royal de Lion in 1948. He served as President of the Royal Entomological Society of London from 1959 to 1961 and received Honorary Fellowships of the entomological societies of London, of Russia, France, the Netherlands, Egypt, and India. More than 80 species and subspecies and eleven genera of Orthoptera have been named after him, among them *Uvarovium* Dirsh and *Uvaroviola* Bei-Bienko.

Taxonomy

Given his vast interests and prodigious output, it is not easy to decide in which branch of acridology, or even in the wider field of pure and applied entomology, Uvarov's contribution was the greatest. He himself regarded taxonomy as his first love, specializing in orthopteroid insects with an emphasis on the old world Acridoidea. This love for systematics began with the publication of his study of the Orthoptera of Ural in 1910 and extended for over half a century. In the later part of his life other and more pressing occupations prevented him from devoting more than a token of his time to taxonomy, but he pursued this work almost as a form of relaxation and often commented that "it is good for the soul." Over 230 of his 431 publications are on the taxonomy and general systematics of orthopteroid insects, but true to his wide interests, those of a born naturalist, most of his publications contain some data on zoogeography, biology, and ecology.

His working philosophy was to produce quick taxonomic guidance where none was previously available, rather than to undertake major revisionary studies on a limited number of groups. The quality of Uvarov's taxonomic work compares favorably with that of most of his contemporaries, not all of whom recognized, as he did, the importance of examining type material. He abhorred verbosity and his style of writing is concise; his keys are clear and unambiguous; his diagnostic figures, which he regarded as essential, were rough and simple, but adequate.

Although he himself made little use of the more modern investigations into diagnostic characters, such as those of internal genital structures, he encouraged others to do so. He had a sharp, discerning eye and was remarkably quick at identification of taxa and recognition of good diagnostic characters, but he was not one to describe new genera or species lightly and there are still some specimens in the British Museum that bear his labels "sp.n."

Although his work was not altogether faultless and occasionally he did make errors in identification, the greater part of his taxonomic work has withstood the test of time; most of the genera (241 out of 284) and the 900 species described by him are still valid.

Uvarov's theory of locust phases stemmed partly from his taxonomic research and was first published in taxonomic context of his classic 1921 revision of *Locusta*. Among his more significant taxonomic publications were the revision of *Dociostaurus* (47), old world Cyrtacanthacridinae (49, 50), *Thisoecetrus* (61), Trinchini (62), *Sphodromerus* and its allies (63) and *Caloptenopsis* and its allies (65). Much of the rest of his taxonomic work took the form of accounts of particular collections with descriptions of new taxa in them. Most of these papers were quite short. A significant exception to this was his long-delayed account of Malcolm Burr's collection of Acrididae of Angola and Northern Rhodesia (66); this is 217 pages long. Two early papers in Russian are also important: one on the Acrididae of European Russia and Western Siberia (51) and the other on the Acrididae of Central Asia (52). His studies on the Orthopteran fauna of the Caucasus, though almost completed, were never published, but the original manuscript is deposited at the Overseas Development Natural Resources Institute, Chatham.

His last contribution to taxonomy was his final chapter in Volume I of *Grasshoppers and Locusts* (70). There he presented an up-to-date systematic suprageneric arrangement based on, but slightly modified from, researches of V. M. Dirsh. To a great extent, this arrangement seems to be generally accepted even by the more conservative systematists.

Closely related to his taxonomic research was his major involvement in curating a large section of the Orthopteran collection in the British Museum, Natural History. The size, arrangement, and the scientific reliability of this collection owes much to Uvarov's energetic curatorial work over a period of some 30 years. Uvarov was never a member of the Museum staff, but his contributions to the scientific value of the collection throughout the 1920s and 1930s were of crucial importance to its development. He added much material that he collected personally and more by exchanges with colleagues in other countries. He also encouraged potential collectors among the members of the locust control organizations and such renowned explorers as Bertram Thomas, St. John Philby, and Wilfred Thesiger.

Uvarov's impact on the taxonomy of Orthoptera will long continue, not only through his own publications, but also through many of those whom he helped, encouraged, and advised, among them such outstanding taxonomists as V. M. Dirsh, N. D. Jago, D. Keith Kevan, and L. L. Mishchenko. Moreover, he initiated full-time systematic research as part of the program of the Anti-Locust Research Centre; this had a profound influence on the progress of taxonomic acridology, particularly in Africa. It is highly gratifying that this research has been maintained by the successors of ALRC, first by the Centre for Overseas Pest Research (COPR) and now by the Overseas Development Natural Resources Institute (ODNRI).

Phase Theory

The mystery of sudden appearances and disappearances of the devastating swarms of locusts, which was a puzzle at the beginning of this century, was clarified by Uvarov's field observations in the Northern Caucasus between 1911 and 1914. From these he concluded that the two allegedly distinct species, the solitary *Locusta danica* and the swarming, migratory *Locusta migratoria*, were in fact two forms of a single species. The two forms differed in their behavior, coloration, and morphology but were able to transform from one to another and to form a continuous series of intermediates. Uvarov's conclusions were principally based on the appearance of *danica* hoppers from eggs laid by *migratoria* and were strengthened by parallel observations by Plotnikov in central Asia and by Faure on the brown locust in South Africa. However, it was not till 1921 that he proposed his momentous theory of locust phases within his taxonomic revision of the genus *Locusta* L. (= *Pachytylus* Fieb.), wherein he stated that *L. danica* and *L. migratoria* were in fact two forms of the same species in phase *solitaria* and phase *gregaria*. Further, he put forward the hypothesis that the periodical outbreaks of locust plagues were associated with phase transformation of *solitaria* to *gregaria*, which formed bands of hoppers and swarms of adults that emigrated from their breeding grounds. He also postulated that phase transformation was governed by environmental factors and that swarming could be prevented by changes in conditions of the breeding grounds by cultivation and agricultural practices.

Uvarov elaborated these concepts in *Locusts and Grasshoppers* (56) when he described phase changes in several species of acridids. By that time Faure (11) and Plotnikov (33) had confirmed their earlier observations on the effect of crowding (i.e. density) on phase change in *Locustana pardalina* and *Locusta migratoria*, and evidence of phase transformation in the desert locust, *Schistocerca gregaria*, was presented by Johnston (19) and in *S. paranensis* (now *S. piceifrons*) in Central America by Dampf (6). In addition, by examining museum material Uvarov found evidence of phase polymorphism in the red locust *Nomadacris septemfasciata* and the Moroccan locust, *Dociostaurus maroccanus*. This was later confirmed for these species and for the Madagascar race of the Migratory Locust, *Locusta migratoria capito*, the South American *Schistocerca cancellata*, and the Australian Plague locust, *Chortoicetes terminifera*. Uvarov also pointed out that the amplitudes of fluctuations in phase characters vary greatly from species to species, i.e. they may be striking or slight, but they invariably form a continuous series within a species.

Uvarov's theory of phase transformation and its relationship to locust outbreaks has provided a tremendous stimulus to investigations on all aspects of the locust problem and has led to the discovery of permanent outbreak areas of the red and migratory locusts in Africa.

The theory was developed and elaborated in a number of publications, among the most outstanding of which was the joint paper with Zolotarevsky (73) on phase nomenclature, where the terms *congregans* and *dissocians* were first introduced. In a later paper (59) he discussed the importance of effects of fluctuating rainfall and unstable environment on rapid multiplication, concentration of individuals, and phase transformation. He also suggested that hoppers forced into close association become "attuned" to their proximity and "strive" to remain in a crowd—in other words, they become habituated to one another. He expressed these views later, in a Paris colloquium (68), as well as his belief that phase transformation does not depend on increase in numbers alone, but also on sensory reactions of individuals to one another, usually in a highly unstable environment. He added that the changes in external characters were merely outward expressions of phase differences in behavior and physiology, which arose from aggregation of individuals at high densities. The title of his 1966 and 1977 volumes was deliberately changed from that of his earlier classic *Locusts and Grasshoppers* to *Grasshoppers and Locusts,* to emphasize the fact that locusts are grasshoppers which are capable of gregarious behavior.

The present day concepts of phase are well summarized by Pener (32). Typically, locust species show density dependent changes in behavior, coloration, morphology, physiology, ecological responses, and ultimately in their geographical distribution. When crowded as nymphs or adults they develop the characters of the gregarious phase, and when they are isolated, of the solitary phase. These characters may shift in either direction in ontogeny or in successive generations, and phase transformation is reversible in any developmental stage of an individual.

Phase change, however, is not the cause of locust plagues, for it follows and does not precede changes in density. Also it may occur in some populations which may not be large enough to give rise to plagues.

For gregarization to occur, the first prerequisite is a set of conditions that bring the insects into close proximity, with resulting mutual encounters that lead to habituation to being touched, which then lead to aggregation and formation of groups. This process, already recorded in the field by Kennedy (23), was studied in the laboratory by Ellis (9) and Gillett (13). Such processes seen in an outbreak of the desert locust in southern Sahara have been described in detail by Roffey & Popov (40). Yet one other important behavioral change, in addition to gregarization, that occurs with phase change is in flight behavior: solitarious adults flying by night and the gregarious individuals in swarms flying by day. This behavior, coupled with gregarious stimulation in swarms and the differences in wind and temperature conditions between day and night, explains the marked differences in the distribution of populations during outbreaks and recessions.

Many of the differences between the phases were summarized by Uvarov in

Grasshoppers and Locusts (70) and his Table on the effects of increased density on locusts is reproduced here in Table 1. To this, we add a few sentences from Pener's (32) review of endocrines and phase. Pener concluded that while hormones are involved in the regulation of locust phase polymorphism, there is no clear evidence that endocrine factors constitute a physiological prima causa in phase transformation. He emphasized that the complexity of these processes has not always been fully appreciated and that while some experimental treatments may induce considerable shifts in some phase characters, others may remain unaffected. He also considers that the widely held view that the "Juvenile hormone controls locust phase polymorphism" is an oversimplification.

Throughout his life Uvarov stressed the importance of phenotypic variation in phase polymorphism, and the role of genetic variation still remains largely unexplored. Genetic factors may be involved in phase polymorphism, as was indicated by the early experiments of Gunn & Hunter-Jones (14) as well as by those of other investigators (2, 29). It has been suggested that either genetic inheritance, or extrachromosomal inheritance through the cytoplasm of the egg, or both, may be implicated.

It is now realized that phase characters, where they occur, are species specific and show a wide variation between the species (1). At the lower end of the scale are some typical grasshoppers such as *Cyrtacanthacris tatarica* or *Kraussaria angulifera* that exhibit no density effects except for development of melanic forms in the hopper stage. Between them and the true locusts, like the desert and the migratory locusts, there are many intermediate gradations. Some pyrgomorphid species, such as *Zonocerus variegatus* and *Phymateus* spp. aggregate as hoppers, but scatter on reaching the adult stage although *Zonocerus* aggregates later, at the time of egglaying. Yet others, like *Oedaleus senegalensis* and *Aiolopus simulatrix,* on crowding may form mobile hopper bands and swarm as adults, developing melanic forms in their hopper stages. They do not, however, show any detectable morphological changes. This capacity was termed by Pasquier (30) an "aptitude," different species being more or less "gregariapte." Uvorov (70) proposed the English equivalent "gregarisable," but this term did not gain usage. The number of "gregariapte" species is probably greater than is known today. For instance, following the deforestation in Mato Grosso, Brazil, there appeared previously unrecorded swarms of *Rhammatocerus* species. Conversely, with the changes of environment associated with land usage in United States, the former plagues of the Rocky Mountain locust, *Melanoplus spretus* are now no more.

A discovery as momentous as the phase theory—surely no longer a theory, but an established fact—inevitably found numerous adherents and critics. Thus Key (25, 26) and Key & Day (27) maintain that phases must be essentially defined by morphological criteria and proposed the term

Table 1 Main effects of increased density on locusts. (> Increase; < Decrease; = No effect.) (From Uvarov 1966)

	Locusta	Schistocerca	Nomadacris	Locustana	Dociostaurus	Chortoicetes
Hatchlings						
Size and weight	>	>	>			
Water content	=	>	>			
Food reserves	>	>	>			
Vitality	>	>	>			
Hoppers						
Melanin	>	>	>	>		
Insectorubin	>	>	>	?		
Respiratory metabolism	>	>				
Activity	>	>				
Rate of development	>	>	>			
Instar number	=	<	<			
Adults						
Size ♂	<	<	<	>	>	>
Size ♀	<	<	<	>	>	>
Activity	>	>		>		
Maturation rate	<	>	>			
Ovariole number	<	<	<			
Eggs per pod	<	<	<	>	>	<
Pods per ♀	<	<	<			
Total eggs	<	<	<			
Viable eggs	<	<	<			

"kentromorphic phases" (kentron = stimulus). In his letter to us, dated 24.3.88 Key still defends his views. This proposal, however, did not find favor with Uvarov (70) who reiterated his view that the principal difference between phases lies in their bionomics and that changes in morphology are the result of some deep physiological differences between the phases. Uvarov considered that the term "kentromorphic" detracts from this fundamental concept. Some other criticisms concern not so much phase polymorphism per se, as its role in plague dynamics, and they are referred to in the next section.

Ecology, Zoogeography, and Applied Biogeography

Throughout his life Uvarov had a lively interest in ecology. Between 1912 and 1920 he worked in the field on the Moroccan locust and the migratory locust in northern Caucasus, and in 1931 and 1932 he paid two visits to the Middle East for the study of distribution and of further ecological characteristics of *Dociostaurus maroccanus* in Turkey, Syria, and Iraq. In later days, he much enjoyed his visits as a consultant and adviser to field investigators in East Africa and in many other parts of the world. However brief his visits, he invariably brought his sweep net and tried to find time for collecting Orthoptera and noting their habitats.

Many of his taxonomic papers contain valuable ecological data and discussions, but perhaps one of his major contributions to ecology was the publication of *Insects and Climate* (58). In this monumental work he reviewed 1150 articles in eleven languages (74). Few major changes or additions have been made to the conceptual framework provided by this work. Virtually for the first time, the many ways in which climate can affect not only Orthoptera but all types of insects in all their stages were listed and brought to light. This seminal review stimulated much interest and research.

In *Insects and Climate* Uvarov put forward his view that weather factors are the prime agents in controlling populations, and he questioned the idea that all populations are in stable equilibrium with nature. Throughout the 1950s many acrimonious disputes occurred between population ecologists as to whether natural control of populations is by density-dependent or by density-independent processes. Although Uvarov did not enter these polemics in public, his name was associated with the so-called "climatic school" wherein his true sympathies lay. These arguments are now in the past, and modern ecologists have arrived at more balanced views (see 16).

Another impact on population ecology made by Uvarov was through his phase theory of locusts, which highlighted the role of behavior in population dynamics by focusing on responses of individuals to high densities. This theory also illuminated effects of density on the qualitative changes that may occur in the individuals that make up a population.

Yet another paper that had a considerable impact on the thinking of a number of insect ecologists concerned the aridity factor in the ecology of locusts and grasshoppers (67). Uvarov stressed the three requirements of acridids, i.e. their need for plant cover, for food and shelter, and for bare ground for oviposition. Mosaics of vegetation and bare soil thus provide areas of high carrying capacity for populations of many acridid species. On similar lines, Uvarov argued that acridids can become important pests in the zones of contact between climatic and vegetational belts, and he pointed out that many such sites arise from human activities, such as deforestation, burning, and overgrazing. The importance of changes caused by agricultural practice, especially in the tropics, on populations of insect pests were further amplified by him in his publication on the problems of insect ecology in developing countries (69) as well as in his address to the XIIIth International Congress of Entomology (71). There he discussed the significance of changes in ecological conditions, either man-made or historically longer-term ones, for the dynamics of populations of acridid species, which may be affected in opposite ways. His examples included the virtual disappearance of the Rocky Mountain locust, *Melanoplus spretus*, and conversely, the great increase in abundance in recent years of the Senegal grasshopper, *Oedaleus senegalensis*.

Among Uvarov's earlier contributions of wide zoogeographical interest was a series of papers on the origin and composition of the orthopteran fauna of the Palaearctic and of the montane Orthoptera of that region (51, 52, 54). In a later paper (60), on the composition and origin of the orthopteran fauna of the great "Eremian Desert," extending through the Sahara to western and central Asia, he developed his concept of "life-forms" in Acridoidea. The term "life-form" was introduced and explored by Raunkiaer (37) and is usually used by plant ecologists as the basis of ecological classification. In his second volume of *Grasshoppers and Locusts* (72), Uvarov has a fascinating chapter on life forms, ecofaunas, and life zones, where acridid species are grouped into Terricoles, Aquaticoles, Arboricoles, Herbicoles, and their intermediate forms, on the basis of morphological, physiological and behavioral characteristics, which evolved in response to environmental pressures.

An aspect of zoogeography in which Uvarov and his colleagues have been outstanding pioneers can be called "applied biogeography," since the results of their investigations are, up to the present day, used in planning of strategies of control.

Biogeographical studies on the desert locust, the red locust, and the African migratory locust began in 1929, when Uvarov organized the collection of information and its systematic and cartographic analysis. After formation of the ALRC, a biogeographical division was set up, headed by Zena Waloff. Its aim was to analyze and plot maps of geographical distribution, breeding sites, and migratory routes of locusts and to try to relate these to climatic, vegetational, and topographic conditions. From these analyses and from field investigations of O. B. Lean and B. N. Zolotarevsky on *Locusta,* it became apparent that the plagues of the African migratory locust and of the red locust started in relatively restricted ecological regions. These smaller regions, within the enormous total territory occupied by these species, became known as their "outbreak areas," and the vast areas occupied by them at the times of plagues as their "invasion areas."

A more complex situation was found in the desert locust, whose vast geographical distribution extends from North Africa eastwards to the Indian subcontinent. Its outbreak areas are less permanent and may arise in various locations of its area of distribution when environmental conditions, mainly climatic and vegetational, favor breeding and the subsequent multiplication in numbers, sometimes over several generations. Uvarov was the first to recognize the interdependence of the widely separated breeding areas, since they are linked by the seasonal migrations of locusts.

In view of this complexity, Uvarov established, first, an information service, with its monthly summaries of the locust situation and forecasts of future developments based on the analyses of past situations, and subsequently, of the Desert Locust Information Service (DLIS), already referred to. One of the main problems of DLIS was the insufficiency of reports from the desert locust area, which spreads over more than 50 countries and has many virtually uninhabited tracts. In building up this service every conceivable source of information was tapped: consular offices, native chiefs, missionaries, district commissioners, explorers, captains of ships at sea, to all of whom simple instructions were given (later this took the form of a small locust handbook) for collecting and reporting useful information. Simultaneously, sponsored field investigations by British entomologists attempted to locate the source areas, and similar investigations were soon taken up by entomologists from many other countries concerned with the locust problems. As the result of laboratory and field research on the desert locust, sufficient knowledge of its breeding and migration patterns accrued to allow forecasting of events, useful in the planning of control operations.

Past and Present Trends in Research and Development of Acridology

The postwar years were justly regarded by many as the golden years of research on locusts. The last 20 years of Uvarov's life saw the publication of some 7000 papers, or about one per day; they were the theme of his two volumes of *Grasshoppers and Locusts.* It is almost presumptious to enumerate the highlights among them, but some of those more familiar to us spring to mind. Among them is the classical work of Weis Fogh, and his and Jensen's work on the flight dynamics of locusts. Rainey's hypothesis on the downwind displacement of locust swarms and their concentration in areas of convergence of air masses is fundamental to understanding and forecasting of swarm movements, as are his subsequent studies on meteorology and locust migrations (e.g. 36). The development of ultra-low-volume spraying which stemmed from the work of John Sayer of the Desert Locust Control Organization is significant, as are Z. Waloff's studies on biogeography and flight behavior of locusts, Ellis's investigations on aggregation of locust hoppers, and Norris's work on reproduction in locusts. Studies on population dynamics of the Moroccan locust by Dempster, of the migratory locust by Farrow, and of the British grasshoppers by Richards and N. Waloff, are noteworthy, as are Greathead's studies on the natural enemies of acridids, Dirsh's monumental work on acridoid taxonomy, and the publication of Johnston's *Annotated Catalogue of the African Grasshoppers* (20, 21).

After Uvarov's retirement, the stimulus of his continued presence was such that initially the Centre increased its scope of work, notably by creating its Field Research Division in 1963. After Uvarov's death the pace began to slacken when in 1971 ALRC was transformed into the Centre for Overseas Pest Research (COPR) with a greatly expanded program of investigations into other pest species, inevitably at the expense of research on acridids. In the 1980s COPR had undergone two other transformations, first into the Tropical Development and Research Institute in 1983 and then into the Overseas Development Natural Resources Institute in 1987. The declared policy of the Overseas Development Administration was that all work on acridology should be phased out by the mid-1980s. However, some outstanding research was done in this period, and in a modest way taxonomic and advisory work is still pursued.

Among the notable investigations at COPR were those of Ewer and McCaffery on the water uptake, development, and survival of acridid eggs. Taxonomic studies by Jago and his colleagues (15, 18), which included hybridization, elucidated the systematic position of the neotropic species of *Schistocerca*. Meanwhile, in the world at large, contemporary work by Key and by White in Australia on speciation in morabine grasshoppers led to important developments in the application of cytology to systematics, a method subsequently used by Hewitt in France and Baccetti in Italy. Further advances that led to the definition of several new families were made by Amedegnato (3) and Descamps (7) in the hitherto neglected orthopteroid fauna of Central and South America.

In the field, among the most important studies were those on *Zonocerus variegatus*, conducted by COPR and the University of Ibadan, under the general supervision of R. F. Chapman. These resulted in over a score of papers on different aspects of the biology, behavior, and population ecology of the grasshopper, with recommendations for its control.

After the severe drought in the Sahelian belt in Africa in 1974–1975, rains brought massive outbreaks of grasshoppers, spearheaded by *Oedaleus senegalensis*. This initiated two research and control projects: the first by COPR and OCLALAV (5, 12, 34, 35) on ecology, taxonomy, and seasonal migrations including radar observations. The second, by PRIFAS (Projet de Recherche Interdisciplinaire Francais sur les Acridiens de Sahel), resulted in many publications and in a biomodel of *O. senegalensis* (28).

A new development in control strategy was the use of remote sensing imagery in monitoring, which narrowed down the ground areas to be surveyed. Biogeographical and ecological studies during the long recession period of the desert locust (31, 35) indicated that some parts within its vast recession area, notably those associated with major relief features, show a particularly high incidence of occurrence and breeding of the locust. Monitoring for rainfall and ecological conditions suitable for the building up of populations was initially carried out by aerial surveys but is now increasingly done by satellite imagery. Such a method has been applied with considerable success in Australia for monitoring of the dynamics of *Chortoicetes terminifera* (17, 43), where the all-important phenology of the locust's foodplants is discernible from satellite imagery.

Among the most interesting developments during this period was radar entomology. The pioneering attempt to monitor locust flight in the Sahara, sponsored by ALRC in 1968, was a spectacular success. The radar showed that locusts could actively select their height of flight and maintain apparently purposeful orientation and that wing-beat frequencies, in some cases, help to identify the species (Schaefer 41). Studies in 1973, 1975, and 1978 provided further information on flight behavior and established that migrations in terms of hundreds of kilometers occurred quite regularly, not only in locusts, but also in many species of grasshoppers (38, 39, 42). Later, these studies were extended to the Australian plague locust, *C. terminifera*. Combinations of ground-based and airborne radars have also been used in investigations on migratory insect pests, other than acridids.

In Canada and the United States the problem of range grasshoppers persists, and Canada now issues regular forecast bulletins. For environmental considerations, biological control by the protozoan *Nosema locustae* baits is increasingly favoured.

In the USSR taxonomic work on acridids is continued in Leningrad by Mishchenko and Gorokhov, while studies on the ecology and zonal distribution of Acridoidea, inspired by works of Uvarov and developed by Bei-Bienko, continue under the supervision of Stebaev at the Novosibirsk Research Centre.

A number of computer models, based on several systems, have been developed during this period; among them are those by de Boer in the

Netherlands, which simulates desert locust migration as determined by wind trajectories at different heights, and by PRIFAS for *Oedaleus senegalensis*. PRIFAS needs a special mention, for it has evolved into what is now the most important center of acridological research in the Old World. It is part of the International Cooperation Center for Agricultural Research and Development (CIRAD) at Montpellier, France. PRIFAS pursues all the more topical lines of locust research, but its speciality is modelling. Having developed biomodels for the Senegalese grasshopper and the Madagascan and the African migratory locusts (8), scientists at the Center are now facing their greatest challenge in developing a model for the desert locust; this is done with participation of G. B. Popov and support from FAO.

Research and control of locusts have also been actively pursued in Australia. Among the notable recent contributions from CSIRO are Farrow's ecological studies on *Locusta migratoria migratorioides* (10). This tropical migratory locust presented no problems until the 1960s, when areas of grass, poor scrub, and woodland in the Central Highlands of Queensland were cleared for development of pasture lands. In the 1970s, these changes in land use, together with high rainfall, favored multiplication and gregarization of the locust and resulted in a severe outbreak.

Economically the most important locust in Australia is *Chortoicetes terminifera*. Past studies on this species by Key and subsequent studies by D. P. Clark in collaboration with (the then) Anti-Locust Research Centre established its patterns of migration. In more recent years, members of a federal organization, the Australian Plague Locust Commission (under the directorship of P. M. Symmons, formerly of COPR) have made further advances in the understanding of biogeography, flight behavior, and the importance of phenology of the food-plants to the dynamics of the plague locust. They have established a monthly forecasting service which contributes to the success of their control measures.

Space does not allow for enumeration of other recent advances in acridology, except for the outstanding publications of COPR, namely, Uvarov's second volume of *Grasshoppers and Locusts* in 1977, the *Desert Locust Forecasting Manual* edited by D. Pedgley in 1981 (31), and *The Locust and Grasshopper Agricultural Manual* (44) published in 1982 and initiated on Uvarov's inspiration in the mid-1960s.

Appreciation

Uvarov was not only a great entomologist, he was also a great personality, as all who came into contact with him recognized. With his far-reaching mind, his energy and enthusiasm, he was able to build up a monumental knowledge of locusts and grasshoppers and to found an institute (ALRC) which in his time became the foremost center in the world for research on locusts and guidance to their control.

His approach to the locust problem was eminently practical; he always emphasized the necessity of international cooperation in control of acridids. In his time other research organizations also developed which in a sense were extensions of ALRC into field work (e.g. the Desert Locust Survey; International African Migratory Locust Control Organization or OICMA). Uvarov's intellectual stimulus and force of personality held the expanding world of acridology together. B. P. translated a personal vision into a concrete achievement. He was able to achieve much, because of his unquestionable scientific integrity—a phrase recurrent in memorials dedicated to him and in many letters written to us by orthopterologists.

Acridology was Uvarov's life. Although he regarded himself primarily as a taxonomist, many biologists will remember him for his phase theory of locusts. He had a great breadth of biological knowledge and a broad grasp of ecological principles. The only area of study that did not interest him was genetics, though this might have added to the understanding of the phase change and variation of the widespread species. All his life he advocated ecological methods in control of acridids, keeping the use of insecticides to a minimum. By now, his views are well supported by the numerous examples of changes in the pest status of grasshoppers and locusts (e.g. 10) that have followed on changes in land use, but it remains doubtful that manipulation of the environment is practicable in control of such wide-spread species as the desert locust.

After Uvarov died, the running down of the locust organizations gradually set in, for there was no substitute for his organizing genius. Moreover, since most of the African countries gained independence, control measures have fragmented, a process to which almost 30 years of recessions from locust plagues contributed. It is to be hoped that the recent massive outbreak of the desert locust, with the unprecedented flights of its swarms across the Atlantic on at least two occasions, will stimulate new work in acridology. It is regrettable that the successors of ALRC have given up the training of new locust personnel. Now that experienced field acridologists are needed, FAO is relying on those who have retired or are nearing retirement. New generations of acridologists will be thankful to Uvarov for his monumental volumes of *Grasshoppers and Locusts,* which will continue to be a source of information for many years to come.

In spite of the wide international recognition of his achievements and the scientific and civic honors bestowed on him, B. P. remained an unassuming and an approachable person. He was rarely satisfied, but his highly critical attitude was tempered by his dry wit and a sense of fun. He was always helpful and understanding to young people and remained receptive to new ideas and findings to the end. He also established an easy rapport with the heterogeneous collection of locust field officers, who regarded him as a living legend.

Although in many ways he remained very Russian, he had total and unswerving loyalty to his adopted country. When asked what was it that he liked about England, he instantly replied, "the decency of the common people".

B. P. never suffered fools gladly and was a hard taskmaster, but all those who have been through his school will always remember him with gratitude, respect, and affection.

ACKNOWLEDGMENTS

We thank Zena Waloff for her comments and advice; though best qualified to write this article, unfortunately she was not well enough to do so.

Also, we would like to express our sincere gratitude to all those who have replied to our letters and to those who have generously provided us with information and encouragement. Limitations of space do not allow for individual thanks, but below is a list of their names in alphabetical order: C. Ashall OBE, R. E. Blackith, H. D. Brown, Valerie Brown, R. F. Chapman, J. P. Dempster, Peggy E. Ellis, R. A. Farrow, A. V. Gorokhov, D. Greathead, P. T. Haskell CMG, C. F. Hemming, J. Hewitt, N. D. Jago, R. J. V. Joyce, J. S. Kennedy F.R.S., D. K. McE. Kevan, K. H. L. Key, M. Launois, W. Loher, L. L. Mishchenko, M. P. Pener, J. Phipps, D. R. Ragge, R. C. Rainey FRS, J. Roffey, J. Roy, Gurdas Singh, R. Skaf, I. V. Stebaev and M. V. Stolyarov, P. M. Symmons, M. V. Venkatesh. Finally, we thank ODNRI for the permission to reproduce the table from *Grasshoppers and Locusts*.

Literature Cited

1. Albrecht, F. O. 1967. *Polymorphisme phasaire et Biologie des Acridiens Migrateurs*. Paris: Masson. 194 pp.
2. Albrecht, F. O., Verdier, M., Blackith, R. E. 1959. Maternal control of ovariole number in the progeny of migratory locust. *Nature* 184:103–4
3. Amedegnato, C. 1974. Les genres d'acridiens neotropicaux, leur classification par familles, sous-familles et tribus. *Acrida* 3:193–203
3a. Anti-Locust Res. Cent. 1970. "In Memoriam" Sir Boris Uvarov, KCMG, FRS, 1888–1970. London: ALRC Publ.
4. Bei-Bienko, G. I. 1970. Sir Boris Uvarov (1888–1970) and his contribution to science and practice. *Entomol. Obozr.* 49:915–22 (In Russian)

5. Cheke, R. A., Fishpool, L. D. C., Forrest, G. A. 1980. *Oedaleus senegalensis* (Krauss) (Orthoptera:Acrididae:Oedipodinae). An account of the 1977 outbreak in West Africa and notes on eclosion under laboratory conditions. *Acrida* 9:107–32
6. Dampf, A. 1926. Der Färbungswechsel bei den Wanderheuschreckenlarven ein biologisches Rätsel. Verh. III *Int. Entomol. Kongr. Zurich* 2:276–90 (In German)
7. Descamps, M. 1976. La faune dendrophile neotropicale. 1. Revue des Proctdabinae Orth. Acrididae. *Acrida* 5:63–167
8. Duranton, J. F., Launois, M., Launois-Luong, M. H., Lecoq, M. 1983. De

l'acridologie à l'ecologie operationnelle. *Pour Sci.* 63 (Jan. 1983).

9. Ellis, P. E. 1962. The behaviour of locusts in relation to phases and species. *Colloq. Int. CNRS* 114:123–43

10. Farrow, R. A. 1987. Effect of changing land use on outbreaks of tropical migratory locust. *Locusta migratoria migratorioides* (R. & F.). *Insect Sci. Appl.* 8(4/5/6):996–75

11. Faure, J. C. 1923. *The Life History of The Brown Locust* (Locustana pardalina (Walker)). *Bull. Fac. Agric. Transv. Univ. Coll.*, No. 4. 30 pp.

12. Fishpool, L. D. C., Popov, G. B. 1984. The grasshopper faunas of Mali, Niger, Benin and Togo savannas. *Bull. Inst. Fondam. Afr. Noire Ser. A.* 43:275–410

13. Gillett, S. D. 1973. Social determinants of aggregation behaviour in adults of the desert locust. *Anim. Behav.* 20:526–33

14. Gunn, D. L., Hunter-Jones, P. 1952. Laboratory experiments in phase differences in locusts. *Anti-Locust Bull.* 12:1–29

15. Harvey, A. W. 1981. A reclassification of the *Schistocerca americana* complex (Orthoptera:Acrididae). *Acrida* 10:61–77

16. Huffaker, C. B., Berryman, A. A., Laing, J. E. 1984. Natural control of insect populations. In *Ecological Entomology*, ed. C. B. Huffaker, R. L. Rabb, pp. 359–98. New York: Wiley

17. Hunter, D. M. 1981. Forecasting migrations of the Australian Plague Locust. *A.P.L.C. Ann. Rep. Res. Suppl.* 1979–1980:62–64

18. Jago, N. D., Antoniou, A., Scott, P. 1979. Laboratory evidence showing the separate species of *Schistocerca gregaria, americana* and *cancellata* (Acrididae, Cyrtacanthacridinae). *Syst. Entomol.* 4:133–42

19. Johnston, H. B. 1926. A further contribution to our knowledge of the bionomics and control of the migratory locust *Schistocerca gregaria* Forsk. (*peregrina* Oliv.) in the Sudan. *Bull. Wellcome Trop. Res. Labs. (Entomol.)* No. 22. 14 pp.

20. Johnston, H. B. 1956. *Annotated Catalogue of African Grasshoppers.* London: Cambridge Univ. Press for ALRC xxii + 883 pp.

21. Johnston, H. B. 1968. *Annotated Catalogue of African Grasshoppers.* Suppl. London: Cambridge Univ. Press for ALRC xiv + 448 pp.

22. Deleted in proof

23. Kennedy, J. S. 1939. The behaviour of the Desert Locust (*Schistocerca gregaria*) (Forsk.) (Orthopt.) in an outbreak centre. *Trans. R. Entomol. Soc. London* 89:385–542

24. Kevan, D. K. McE. 1971. The contribution of the late Sir Boris P. Uvarov (1888–1970) to the systematics of Orthoptera—An Appreciation. *The Entomologist* 104:324–30

25. Key, K. H. L. 1950. A critique on the phase theory of locusts. *Q. Rev. Biol.* 25:363–407

26. Key, K. H. L. 1957. Kentromorphic phases in three species of Phasmatodea. *Aust. J. Zool.* 5:247–84

27. Key, K. H. L., Day, M. F. 1954. The physiological mechanism of colour change in the grasshopper *Kosciuscola tristis* Sjöst (Orthoptera:Acrididae). *Aust. J. Zool.* 2:340–63

28. Launois, M. 1983. Modelisation d'*Oedaleus senegalensis* (Krauss) Sautirian Ravageur des Cultures vivrières du Sahel. *Rapport Projet TCP/RAF/128/CLS.* Rome: FAO., UN

29. Papillon, M. 1960. Études preliminaire de la repercussion du groupement des parents sur les larves nouveau-nées de *Schistocerca gregaris* Forsk. *Bull. Biol.* 93:203–63

30. Pasquier, R. 1952. Quelques propositions de terminologie acridologique. Première note. Terminologie concernant le comportement et l'aspect des Acrididae gregariaptes. *Ann. Inst. Agric. Ser. Rech. Exp. Agric. Algér.* 6:1–16

31. Pedgley, D., ed. 1981. *Desert Locust Forecasting Manual,* Vol. 1, 268 pp. Vol. 2, 142 pp. London: COPR

32. Pener, M. P. 1983. Endocrine aspects of phase polymorphism in locusts. In *Endocrinology of Insects,* ed. R. G. H. Downer, H. Laufer, pp. 379–94. New York: Liss

33. Plotnikov, V. 1927. *Locusta (Pachytylus) migratoria* L. and *L. danica* as independent forms and their derivatives. *Uzb. Opytn. Stn. Zashch. Rast.* 33 pp. (In Russian with German summary)

34. Popov, G. B. 1988. Sahelian grasshoppers. *ODNRI Bull.* No. 5. 87 pp.

35. Popov, G. B., Wood, T. G., Haggis, M. J. 1984. Insect pests of the Sahara. In *Key Environments: Sahara Desert,* ed. J. L. Cloudsley-Thompson, pp. Oxford: Pergamon

36. Rainey, R. C. 1963. Meteorology and the migration of desert locusts. *Wld. Met. Org.* No. 54. x + 115. Also, *ALRC Mem.* No. 7

37. Raunkiaer, C. 1934. *The Life Forms of Plants and Statistical Plant Geography.* Oxford: Clarendon

38. Reynolds, D. R., Riley, J. R. 1988. A migration of grasshoppers particularly

Diabolocatantops axillaris (Thunberg) (Orthoptera:Acrididae) in West African Sahel. *Bull. Entomol. Res.* 78:251–71

39. Riley, J. R., Reynolds, D. R. 1983. A long-range migration of grasshoppers observed in the Sahelian zone of Mali observed on two radars. *J Anim. Ecol.* 52:167–83

40. Roffey, J., Popov, G. B. 1968. Environmental and behavioral processes in a Desert Locust outbreak. *Nature* 219: 446–50

41. Schaefer, G. W. 1972. Radar detection of individual locusts and swarms. In *Proc. Int. Study Conf. Curr. Future Problems of Acridology, London 1970,* ed. C. F. Hemming, T. H. C. Taylor pp. 379–80. London: COPR

42. Schaefer, G. W. 1976. Radar observations of insect flight. *Symp. R. Entomol. Soc. London* 7:157–97

43. Symmons, P. M., McCulloch, L. 1980. Persistence and migration of *Chortoicetes terminifera* (Walker) (Orthoptera:Acrididae) in Australia. *Bull. Entomol. Res.* 70:197–201

44. *The Locust and Grasshopper Agricultural Manual.* 1982. London: COPR. 690 pp.

45. Uvarov, B. P. 1910. Contribution to the orthopteran fauna of the Ural province. *Tr. Russk. Entomol. Obshch.* 39:359–90 (In Russian)

46. Uvarov, B. P. 1921. The geographical distribution of orthopterous insects in the Caucasus and in Western Asia. *Proc. Zool. Soc. London* 447–72

47. Uvarov, B. P. 1921. A preliminary revision of the genus *Dociostaurus* Fieb. *Bull. Entomol. Res.* 11:397–407

48. Uvarov, B. P. 1921. A revision of genus *Locusta* (= *Pachytylus* Fieb.) with a new theory as to the periodicity and migrations of locusts. *Bull. Entomol. Res.* 12:135–63

49. Uvarov, B. P. 1923. A revision of the Old World Cyrtacanthacrinae (Orthoptera, Acrididae) I, II, & III *Ann. Mag. Nat. Hist.* 11:130–44, 345–67, 473–90

50. Uvarov, B. P. 1924. A revision of the Old World Cyrtacanthacrinae (Orthoptera, Acrididae) IV, V. *Ann. Mag. Nat. Hist.* 13:1–19, 14:96–113

51. Uvarov, B. P. 1925. Acrididae of the European part of U.S.S.R. and the Western Siberia. *Novaya Derev.* 119 pp. (In Russian)

52. Uvarov, B. P. 1927. Acrididae of Central Asia. *Tr. Uzb. Opytn. Stn. Zashch. Rast.* 215 pp. (In Russian)

53. Uvarov, B. P. 1927. *Locusts and Grasshoppers.* Moscow: Chief Cotton Committee. 306 pp. (In Russian)

54. Uvarov, B. P. 1928. Orthoptera of the mountains of palaearetic region. *Mém. Soc. Biogéogr.* 2:135–41

55. Uvarov, B. P. 1928. Blattodea (68–70), Mantodea (70–72), Phasmodea (114), Dermaptera (120–22). In *Key for Determination of Insects,* ed. I. N. Filipiev. Moscow: Gos. Inst. Opyt. Agron. 943 pp. (In Russian)

56. Uvarov, B. P. 1928. *Locusts and Grasshoppers: A Handbook for Their Study and Control.* London: Imperial Bur. Entomol. 352 pp.

57. Uvarov, B. P. 1929. Insect nutrition and metabolism: a summary of literature. *Trans. R. Entomol. Soc. London* 76: 255–343

58. Uvarov, B. P. 1931. Insects and Climate. *Trans. R. Entomol. Soc. London* 79:1–247

59. Uvarov, B. P. 1937. Biological and ecological basis of locust phases and their practical application. *4th Int. Locust Conf., Cairo 1936.* Appendix 7. 16 pp.

60. Uvarov, B. P. 1938. Ecological and biogeographical relations of Eremian Acrididae. *Mém. Soc. Biogéogr.* 6:231–73

61. Uvarov, B. P. 1939. A preliminary revision of the palaearctic species and subspecies of *Thisoicetrus* Br. W. (Orthoptera, Acrididae). *Novit. Zool.* 41:377–82

62. Uvarov, B. P. 1943. The tribe Trinchini of the subfamily Pamphaginae and the interrelations of the acridid subfamilies (Orthoptera). *Trans. R. Entomol. Soc. London* 93:1–72

63. Uvarov, B. P. 1943. A revision of the genera *Sphodromerus, Metromerus* and *Sphodronotus* (Orthoptera, Acrididae) *Proc. Linn. Soc. London* 154:69–85

64. Uvarov, B. P. 1951. Locust research and control, 1929–1950. *Colonial Res. Publ.* No. 10. 67 pp. London: HMSO

65. Uvarov, B. P. 1951. The genus *Caloptenopsis* 1. Bolivar and its allies (Orthoptera, Acrididae). *Eos, Madr.* Tomo extraord. (1950) 385–414

66. Uvarov, B. P. 1953. Grasshoppers (Orthoptera, Acrididae) of Angola and Northern Rhodesia collected by Dr. Malcolm Burr in 1927–1928. *Publçoes Cult. Co. Diam. Angola* No. 21. 217 pp.

67. Uvarov, B. P. 1957. The aridity factor in the ecology of locusts and grasshoppers of the Old World. In *Arid Zone Research VIII. Human and animal ecology. Reviews of research,* pp. 164–98 Paris: Unesco

68. Uvarov, B. P. 1962. Allocution de la séance de clôture. *Colloq. Int. CNRS* 114:319–28

69. Uvarov, B. P. 1964. Problems of insect

ecology in developing countries. *J. Appl. Ecol.* 1:159–68

70. Uvarov, B. P. 1966. *Grasshoppers and Locusts: A Handbook of General Acridology,* Vol. 1. Published for ALRC. London: Cambridge Univ. Press. 481 pp.

71. Uvarov, B. P. 1969. Current and future problems of acridology. *Entomol. Obozr.* 48:233–40 (In Russian). Also in 1977. *Grasshoppers and Locusts,* pp. 524–31. London: COPR

72. Uvarov, B. P. 1977. *Grasshoppers and Locusts: a Handbook of General Acridology. II.* London: COPR. 613 pp.

73. Uvarov, B. P., Zolotarevsky, B. N. 1929. Phases of locusts and their interrelations. *Bull. Entomol. Res.* 20: 261–65

74. Wellington, W. G., Trimble, R. M. 1984. Weather. In *Ecological Entomology,* ed. C. B. Huffaker, R. L. Rabb, pp. 399–425. New York: Wiley

75. Wigglesworth, V. B. 1971. Boris Petrovich Uvarov 1889–1970. *Biogr. Mem. Fellows R. Soc.* 17:713–40

FLUID MECHANICS

Osborne Reynolds at the age of $\sim$24 (ca. 1866).

Annu. Rev. Fluid Mech. 1990. 22 : 1–11

NOTE ON THE HISTORY OF THE REYNOLDS NUMBER

N. Rott

Department of Aeronautics and Astronautics, Stanford University, Stanford, California 94305

1. NAME ORIGINS

In 1908, Arnold Sommerfeld presented a paper on hydrodynamic stability at the 4th International Congress of Mathematicians in Rome (Sommerfeld 1908). In the equation known today as the Orr-Sommerfeld equation, he introduced a number R as "eine reine Zahl, die wir die Reynolds'sche Zahl nennen wollen." (freely translated: "R is a pure number; we will call it the Reynolds number.") The terminology introduced by Sommerfeld has not changed ever since, and the use of the expression "Reynolds number" has spread into all branches of fluid mechanics.

Actually Sommerfeld's work is not foremost in one's mind when one thinks of the direct continuation of the ideas set forth by Osborne Reynolds in 1883 (Reynolds 1883). This is probably the reason that Sommerfeld's use of the expression "Reynolds number" was generally forgotten before von Kármán (1954) drew attention to it in his book *Aerodynamics: Selected Topics in the Light of Their Historical Development.* He referred there to work by Sommerfeld from the year 1908 but did not mention the title and the place of publication. Von Kármán returned to this subject in his paper published in the Albert Betz anniversary issue of the *Zeitschrift für Flugwissenschaften* (von Kármán 1956). Unfortunately the reference there is to Sommerfeld's (1904) paper on the Reynolds theory of lubrication. In that work the inertial terms are neglected, and the notion of the Reynolds number is neither needed nor used. While it is not difficult to reconstruct the facts by the use of von Kármán's book, it may be useful, nevertheless, to have these sources collected and recollected here again.

I also wish to correct and to explain an incorrect statement that I made on this subject in an earlier article in this series (Rott 1985). There I stated

0066–4189/90/0115–0001$02.00

that the expression "Reynolds number" was introduced by Prandtl. This remark was based on personal recollections of conversations with Jakob Ackeret. Given the fallibility of memories that are not supported by tangible evidence, this remark should never have been made. Thanks are due to numerous colleagues and friends who drew my attention to the facts uncovered by von Kármán. However, I have decided to investigate, after having made the allegations, the actual role of Prandtl in the history of the Reynolds number. This has led also to studies of the early history of the notion and notation before Sommerfeld and Prandtl. A summary of my findings follows.

Many people in fluid mechanics, if asked to guess, would be inclined to attribute the expression "Reynolds number" to Blasius. His work is the first that is fully devoted to the extraction from experiments of a function that, according to Reynolds' similarity law, depends only on the Reynolds number. The reader consulting the originals will find that Blasius actually uses the expression "Reynolds number," but only twice in the first publication of his results (Blasius 1912, pp. 640, 642). In the subsequent extended VDI report, he uses the expressions again in the corresponding passages (Blasius 1913, pp. 7, 9).

Prandtl's role in the history of the Reynolds number, as revealed in his collected works, begins with his early paper on the Reynolds analogy (Prandtl 1910). There he introduces "die in der Hydrodynamik bekannte Reynolds'sche Zahl" but calls it later simply ζ. The words "known in hydrodynamics" refer to the general field. Thus it cannot be ascertained whether Prandtl has chosen his words following Sommerfeld or some other source known to him, or whether he made this choice independently. On the other hand, Prandtl's influence on Blasius is highly probable.

In his encyclopedia article on "Flüssigkeitsbewegung," Prandtl (1913) writes more deliberately: "Die vorstehende Grösse, eine dimensionslose Zahl, wird nach dem Entdecker dieses Ähnlichkeitsgesetzes, Osborne Reynolds, die Reynolds'sche Zahl genannt." ("The forementioned quantity, a nondimensional number, is named after the discoverer of this similarity, Osborne Reynolds, [and is called] the Reynolds number".) I am indebted to Professor Itiro Tani for pointing out this passage (in Volume 3, p. 1445 of the collected works), as well as for other information used in this note.

A study of Prandtl's collected works also shows the crucial role that he and his coworkers in Göttingen played in the application of hydrodynamic similarity to the drag problem. In the years around 1912, an international dispute erupted over the drag of spheres, with Prandtl and Eiffel (who made the famous drop experiments from his tower) as the main protagonists. In 1913, Lord Rayleigh noted briefly that this discussion should be considered in the light of hydrodynamic similarity (Rayleigh 1913). He credited Stokes before Reynolds with the discovery of this notion. Stokes indeed noted the similarity properties of his basic equations of viscous flow together with their derivation in 1850. The resolution of the drag problem, however, called for the specific similarity law established by Reynolds (1883) for the pressure drop in pipes, which is also applicable—*mutatis mutandis*—to the drag problem. Interestingly, it was Lord Rayleigh who made what is almost certainly the first reference ever to Reynolds' similarity law, in the introduction to his paper "On the question of the stability of the flow of fluids" (Rayleigh 1892). (This is discussed later.)

The resolution of the sphere drag problem was completed by Prandtl in his paper of 1914, where he introduced a new concept that complemented his boundary-layer theory—namely, the idea of the transition of the boundary layer from laminar to turbulent at a *critical* Reynolds number (Prandtl 1914). Prandtl made here no more explanatory comments on the usage of this word. After his paper it became common knowledge that the drag coefficient depends on the Reynolds number, a term that became a household word in aerodynamics and aeronautics. The general acceptance of the term *and* of the notion came much later in hydraulic engineering, in spite of the fact that a textbook on hydraulics by von Mises appeared in 1914 that fully exploited Reynolds similarity (albeit without using the term "Reynolds number"). The memory of this book (von Mises 1914), which is now largely forgotten, is kept alive in Rouse & Ince's *History of Hydraulics* (Rouse & Ince 1957).

After World War I, a treatise on similarity and its use for model experiments appeared in the *Jahrbuch der Schiffbautechnischen Gesellschaft* by Moritz Weber, Professor of Naval Architecture in Berlin (Weber 1919). As noted by Rouse & Ince, Weber not only put the Reynolds number to use but also introduced the Froude number and a new number involving capillarity, which later was named for him. This was apparently the beginning of a new era in the use of names for nondimensional numbers.

Reference has been made here repeatedly to the seminal paper of Reynolds (1883). There he gave, by dimensional analysis, unprecedented applications that led to specific results: He introduced the notion of the critical Reynolds number and established the similarity law for the pressure drop in pipes. The early history of these ideas is discussed in the following sections.

2.   REYNOLDS' FLOW-VISUALIZATION EXPERIMENTS

Reynolds gave a visual demonstration of the transition from laminar to turbulent flow in a pipe, using an experimental setup that is still popular today. Figure 1, which is reproduced from Reynolds (1883), shows an artist's concept of the original device. An elevated platform permits the use of a siphon that is high enough to reach the critical velocities; the valve at the exit is manipulated by a lever that extends to the platform. Water is drawn from a glass-walled box into a glass tube, together with a filament of dye. The tube has a trumpet-shaped inlet. Readings of the water level, accurate to 1/100 of an inch, were used to determine the velocity.

The dye filaments at transition are reproduced in many texts, using Reynolds' original drawings; but as his apparatus is still in existence at the University of Manchester, modern photographs can also be obtained. A series is shown in Van Dyke's (1982) *Album of Fluid Motion*, with a telling comment: "Modern traffic in the streets of Manchester made the critical Reynolds number lower than the value 13,000 found by Reynolds."

Cautioning remarks were already voiced by Reynolds himself concerning the importance attached to the actual value of this critical number.

Figure 1 Artist's concept of Reynolds' flow-visualization experiment.

It was clear to him that he needed a carefully shaped inlet to avoid eddies created at the entrance of his tube. He also wrote (Reynolds 1883, p. 955): "The fact . . . that this relation has only been obtained by the utmost care to reduce the external disturbances in the water to a minimum must not be lost sight of." This is a reminder repeated from p. 943, where he notes (we revert to present-day terminology) that turbulent flow has been observed at much lower Reynolds numbers; he continues: "This showed that the steady motion was unstable for large disturbances long before the critical velocity was reached, a fact which agreed with the full-blown manner in which the eddies appeared."

This discussion admits the interpretation that, by more careful experiments, an upper critical Reynolds number could be found that is the stability limit for small disturbances. Reynolds' experiments were repeated a quarter-century later by V. Walfrid Ekman, who visited Manchester to use Reynolds' original equipment. By smoothing the wooden trumpet-shaped inlet, he was able to reach critical Reynolds numbers up to 44,000 (and later more) but found strong scatter approaching these high values. He conjectured (Ekman 1910) that the flow is stable for small disturbances. In a footnote on the subject of the existence of the upper critical Reynolds number, Ekman notes: "It appears that Reynolds himself was somewhat doubtful on this point. It is not easy to find out from his paper what his final opinion really was."

The modern point of view, as formulated by Drazin & Reid (1981, p. 219), is that investigations in this century "have led to the belief that Poiseuille flow in a circular pipe is stable with respect to axisymmetric disturbances. There is also increasing evidence that it is stable with respect to non-axisymmetric disturbances. . . ." An infinite upper critical Reynolds number explains the experimental results. However, a full explanation has also to account for the stability in the varying environment of an inlet, where the Poiseuille profile is not yet fully developed.

In summary, the simplest and most successful visualization of flow transition devised by Reynolds is not, at the same time, the simplest demonstration from the theoretical point of view.

Reynolds first presented his experimental results in a form that did not take full advantage of the clarification of the concepts that he had obtained by dimensional analysis. (A close account is given in Lamb 1932.) For instance, Reynolds wrote for his upper critical velocity the expression $U = P/BD$, where D is the diameter, $P = v/v_0$ is the kinematic viscosity v of water divided by its value v_0 ($= 0.01779$ cm^2 s^{-1}) at 0°C, and P is a function of the temperature as measured by Poiseuille. Finally, B is the parameter for which the critical value is determined by experiment. Reynolds found that B is about 43.79 in seconds per square meter units.

The Reynolds number is the reciprocal of the product Bv_0; this gives 12,830.

In early classic (pre-Reynolds) hydraulics, measurements of the temperature dependence of the viscosity in water were often combined with studies of the flow resistance in pipes. Reynolds actually used the function P to extend the scope of his experiments: Critical velocities obtained for 5°C and 22°C were compared and found to have the ratio of about 1.4, in accordance with Poiseuille's formula for P. This complemented the experiments with different tube diameters of 1, 1/2, and 1/4 inch.

3. THE CRITICAL REYNOLDS NUMBER

In his original investigations Reynolds (1883, p. 946) came to the conclusion that "there must be another critical velocity, at which previously existing eddies would die out, and the motion become steady as the water proceeded along the tube. This conclusion has been verified."

Reynolds determined this critical velocity by measuring the pressure drop in a 5-ft section at the end of a 16-ft pipe. He found the pressure loss to be proportional to the first power of the velocity for low speed but varying with a higher power beyond a "lower critical velocity," which occurred for the value $B = 278$ s m^{-2} of his parameter defined above. This parameter value corresponds to a "lower critical" Reynolds number of 2020. (Later, this was simply called *the* critical Reynolds number.) The supercritical pressure-loss dependence on the velocity was found to follow a power law with the exponent 1.723.

Reynolds returned to the discussion of these results in 1895, when he first presented critical values by using explicitly the Reynolds number proper (Reynolds 1895). He called it K and based it (as an engineer always would, reading from a drawing or a caliper) on the pipe diameter. He quoted for the critical K of transition a value between 1900 and 2000, based on his own experience.

Reynolds' purpose in his 1895 paper was to calculate K, or at least to find a lower bound for K, by identifying dissipation with what is today called turbulent production. He thus created the foundation of the main branch of modern turbulence theory by writing down the "Reynolds-averaged" equations of motion. The history of these ideas is beyond the scope of this note.

Reynolds had already observed that plugs of laminar and turbulent flow alternate in a pipe near the critical Reynolds number, causing what is today called "intermittency" of the flow. For this reason Reynolds, as well as experimenters after him, preferred to give an interval for the critical Reynolds number instead of an "exact" value; it was found practically impossible to make an inlet sufficiently "rough" so that a weakly super-critical flow would start without any laminar spots.

Intermittency is easily demonstrated by letting a horizontal jet emerge from a near-critical pipe flow. The jet oscillates—i.e. it reaches different distances as laminar and turbulent flow alternate at the exit. As explained by Julius Rotta (1956), this occurs because for the same mass flow, laminar and turbulent parts do not have the same impulse. The laminar part has the higher impulse and thus moves farther horizontally when it emerges as a jet.

Oscillations are strongly accentuated if they are coupled to changes of the mass flow. This always happens when the flow resistance in the near-critical pipe, where laminar and turbulent flow plugs alternate, is a significant part of the overall pressure drop in the system. Experiments in such systems are not suitable for the determination of the critical Reynolds number. To assure the constancy of the mass flux, Rotta experimented with a (low-pressure) airflow regulated by a sonic throat. Velocity and intermittency were determined by hot-wire measurements. Rotta's observations, supported by theoretical arguments, showed that the supercritical turbulent plugs grew mostly at their front end. He proposed to define the critical Reynolds number by the constancy of the intermittency factor. However, as the growth speed became very slow approaching the critical Reynolds number, reliable measurements would have required a tube of excessive length. Rotta measured a 2% excess of the plug growth speed over the mean speed at a Reynolds number of 2300, from which he extrapolated to a putative critical Reynolds number of 2000. (This happens to be the value first proposed by Reynolds.)

For pipes without a streamlined inlet, the critical value quoted in most contemporary textbooks is 2300.

4. THE POWER LAW

Reynolds determined the critical Reynolds number by the change of the dependence of the pressure loss on velocity. As already noted, he found for turbulent flow a velocity power law with the exponent 1.723, valid for his experiments over a range of 1 to 50 (Reynolds 1883, p. 975). He actually wrote down (on the same page) the similarity law for the pressure drop in its full generality, namely (in present-day notation)

$$p(L) - p(0) = \frac{1}{2}\rho U^2 \frac{L}{D} f(R),$$

$$R = \frac{\rho U D}{\mu} \equiv \frac{U D}{v},$$

and then considered the special function $f(R) = cR^{-n}$.

Representation of experimental results by a power law has led, during its long history, to the discovery of analytic relations at best and to useful local approximations at least. Dimensional analysis extends the scope and power of this tool immensely. The experimentalist then is guided by knowing that the exponents obtained for the different physical variables that enter the similarity parameter have to fulfill certain relations.

Reynolds was the first to use the power law in this sense. Many laws with odd exponents were established before him without consideration of dimensionality or similarity. We can understand why Reynolds did not see the necessity of quoting them.

Lord Rayleigh (1892) pointed to the two limits of Reynolds similarity where exact explicit laws follow: The classical slow-flow result ($n = 1$) is independent of density, while the limiting velocity-square law ($n = 0$) for turbulent flow at high speeds is independent of viscosity. Rayleigh considered the latter case as a further illustration of his principle that inviscid solutions and solutions obtained in the limit of vanishing viscosity can be fundamentally different. He proceeded to show that inviscid stability calculations for simple channel flows do not lead to unstable solutions.

The next discussion of Reynolds' paper of 1883 (but with no mention of his work of 1895) appeared in 1897. That it was "next" is conjectured because the author, G. H. Knibbs of the University of Sydney, apparently was very conscientious in searching the literature, and he quotes only Lord Rayleigh's paper dealing with Reynolds' work. Knibbs (1897) could not accept Rayleigh's point of view and, moreover, he rejected Reynolds' similarity principle. To analyze such errors is futile. With some good will, one can say that Knibbs attempted (following older ideas) to establish formulas both for open channels and for pipes, using a common principle. When both viscosity and gravity play a role, then two empirical exponents are required for the most general power law, and the guidance given by Reynolds' similarity is lost. Knibbs described the history of the many contributions to the power law and took issue with Reynolds for ignoring them.

We have already sided with Reynolds and adopt now an idea of Blasius for the selection of references: Only papers are quoted in which the power-law dependence is stated both for the velocity U and for the diameter D, in a way that is compatible with Reynolds similarity. According to the formula that states the similarity law (see above), the difference of the exponents of U and of D has to be 3. Then, only one author is left on Knibbs' reference list, namely the German hydraulic engineer Gotthilf Heinrich Ludwig Hagen (1797–1884), of Hagen-Poiseuille fame.

Knibbs was aware of the compatibility of Hagen's empirical results with Reynolds' theory, but this was for him only an isolated case. Hagen obtained his results without knowing about similarity; he found that the best fit to the averaged results of his measurements with three different tube diameters was obtained with the exponent $n = 1/4$, the same as today's accepted value. Actually, Hagen gave credit to the German engineer Reinhard Woltman (1746–1822) for first proposing the exponent 1.75 for the velocity, in work dating back to 1790. The fact that a power less than 2 describes pressure losses in pipes was already observed earlier, by Pierre Louis Georges Du Buat (1734–1809), one of the great hydraulicists of eighteenth-century France. For more historical material, the reader may want to consult the eminently readable book of Rouse & Ince (1957).

Hagen's measurements were published in 1854 (Hagen 1854); in the same paper he made an observation for which he is probably best known. Knibbs quoted the whole passage in its original form: It is the first description of the transition between laminar and turbulent flow in a pipe. Hagen used sawdust ("Sägespähne") as a means of flow visualization; later (Hagen 1869), he recommended the use of filings of dark amber. Without the help of similarity laws, however, no general conclusions could be drawn.

After the 1897 paper of Knibbs, it took 15 more years before Blasius reintroduced Reynolds similarity to the power law for pipes. In the meantime, however, important experiments were conducted with new means, and Hagen's feat was repeated on a grand scale. Pressure drop in (smooth) pipes was measured by two graduate students at Cornell University, August V. Saph and Ernest W. Schoder, Jun., working at the Hydraulic Laboratory (founded in 1899). Their results were published in the *Transactions of the American Society of Civil Engineers* (Saph & Schoder 1903). They fitted their data, both for velocity and diameter, with $n = 1/4$ but did not consider the dependence on viscosity. Blasius only had to introduce the kinematic viscosity of water to obtain the nondimensional constant that multiplies the power law. (Actually he also made experiments of his own and surveyed data from many other sources. Reynolds' own experiments did not turn out to be useful for the nondimensional constant.)

The paper of Saph & Schoder (1903) was published together with a written discussion of their work by leading hydraulicists. One of them, A. Flamant of France, pointed out that he had already obtained the formula of Saph & Schoder in 1892. This lead proved to be interesting. Flamant's influential textbook *Hydraulique* first appeared in 1891. In the second edition (Flamant 1900), he gave a "new formula" that he obtained by inspecting existing results for accuracy and convenience; it agreed with the results of Saph & Schoder. Flamant also mentioned Reynolds' results in his book: He quoted Reynolds' power-law exponent 1.723 but not the law of similarity.

Finally, around 1910, the acceptance of Reynolds similarity began to spread. In 1911, von Kármán exhorted certain authors who were measuring pressure loss in pipes for different fluids to be mindful of Reynolds similarity (von Kármán 1911). Blasius wrote in the extended version of his paper (Blasius 1913, p. 5) that the Reynolds law "has not penetrated, as of today, into the pertinent fields of engineering" (in the original: "ist in die einschlägigen Gebiete der Ingenieurwissenschaften bis heute noch nicht eingedrungen"). His paper became influential in leading to a change, and it was widely used in aeronautical and mechanical engineering.

The combination of the power law with similarity considerations proved to be a valuable tool for further developments in the early stages of turbulence theory and has led to results of lasting importance for engineering applications. The connection between the wall stress, the dynamic pressure, and the Reynolds number, as given by Blasius' formula for pipe flow, is directly applicable to the turbulent boundary layer on a flat plate and can be extended to a multitude of interesting cases.

A higher level of sophistication in the application of these tools was reached with the derivation of the 1/7-power law for the turbulent velocity profile. However, this also signaled the demise of the power-law era: Both discoverers of the 1/7-power law, Prandtl and von Kármán, moved on to the logarithmic velocity distribution. (According to a personal communication from Ackeret, Prandtl never believed in the deep physical significance of simple fractions as exponents: He *wanted* to find a logarithm.)

Actually, neither Hagen nor Reynolds felt particularly committed to the use of the power law. Hagen, who was also a practical engineer and builder of public works, discussed in 1869 other types of formulas (Hagen 1869). For Reynolds, the power law was mostly a convenient tool for the determination of transition. He had continued interest in the critical Reynolds number and inspired other researchers to measure it. He remained active until 1905, when he retired from the position to which he was appointed in 1868: Professor of Engineering at the University of Manchester.

5. EPILOGUE

Reynolds begins his 1883 paper by stating: "The results of this investigation have both a practical and a philosophical aspect." This could be a useful quotation for instructors who teach a first course in fluid mechanics to juniors in an engineering college. There is a yearly battle going on for students' minds; history might help to convince them that the use of the Reynolds number as an independent variable is an application of a basic truth, and not just a useful convention for a handy diagram.

ACKNOWLEDGMENTS

Vital information provided by Professor Itiro Tani, as well as help and encouragement given by Professors Milton Van Dyke, John F. Kennedy, Arthur F. Messiter, Frederick S. Sherman, and John V. Wehausen, are thankfully acknowledged.

Literature Cited

Blasius, H. 1912. Das Aehnlichkeitsgesetz bei Reibungsvorgängen. *VDI-Z.* 56: 639–43

Blasius, H. 1913. Das Aehnlichkeitsgesetz bei Reibungsvorgängen in Flüssigkeiten. *VDI Mitt. Forschungsarb. No. 131.* 39 pp.

Drazin, P. G., Reid, W. H. 1981. *Hydrodynamic Stability.* Cambridge: Univ. Press

Ekman, V. W. 1910. On the change from steady to turbulent motion of liquids. *Ark. Mat. Astron. Fys.* 6(12): 1–16

Flamant, A. 1900. *Hydraulique.* Paris: Béranger. 2nd ed.

Hagen, G. 1854. Ueber den Einfluss der Temperatur auf die Bewegung des Wassers in Röhren. *Math. Abh. Akad. Wiss. Berlin,* pp. 17–98

Hagen, G. 1869. Ueber die Bewegung des Wassers in cylindrischen, nahe horizontalen Leitungen. *Math. Abh. Akad. Wiss. Berlin,* pp. 1–29

Knibbs, G. H. 1897. On the steady flow of water in uniform pipes and channels. *J. R. Soc. N.S.W.* 31: 314–55

Lamb, H. 1932. *Hydrodynamics.* Cambridge: Univ. Press. 6th ed. Reprinted, 1945, by Dover (New York)

Prandtl, L. 1910. Eine Beziehung zwischen Wärmeaustausch und Strömungswiderstand der Flüssigkeiten. *Phys. Z.* 11: 1072–78. See Prandtl 1961, 2: 583–96

Prandtl, L. 1913. Flüssigkeitsbewegung. In *Handwörterbuch der Naturwissenschaften,* 4: 101–40. See Prandtl 1961, 3: 1421–85

Prandtl, L. 1914. Der Luftwiderstand von Kugeln. *Nachr. Ges. Wiss. Göttingen,* pp. 177–90. See Prandtl 1961, 2: 597–608

Prandtl, L. 1961. *Gesammelte Abhandlungen,* ed. W. Tollmien, H. Schlichting, H. Görtler, Vols. 1–3. Berlin: Springer-Verlag

Rayleigh, Lord. 1892. On the question of the stability of the flow of fluids. *Philos. Mag.* 34: 177–80. Reprinted, 1920, in *Scientific Papers,* 3: 575–84. Cambridge: Univ. Press

Rayleigh, Lord. 1913. Sur la résistance des sphères dans l'air en mouvement. *C. R.* 156: 109. Reprinted, 1920, in *Scientific Papers,* 6: 136. Cambridge: Univ. Press

Reynolds, O. 1883. An experimental investigation of the circumstances which determine whether the motion of water shall be direct or sinuous, and of the law of resistance in parallel channels. *Philos. Trans.* 174: 935–82

Reynolds, O. 1895. On the dynamical theory of incompressible viscous fluids and the determination of the criterion. *Philos. Trans.* 186: 123–64

Rott, N. 1985. Jakob Ackeret and the history of the Mach number. *Annu. Rev. Fluid Mech.* 17: 1–9

Rotta, J. 1956. Experimenteller Beitrag zur Entstehung turbulenter Strömung im Rohr. *Ing.-Arch.* 24: 258–81

Rouse, H., Ince, S. 1957. *History of Hydraulics.* Iowa City: State Univ. Iowa. Reprinted, 1963, by Dover (New York)

Saph, A. V., Schoder, E. W. 1903. An experimental study of the resistance to the flow of water in pipes. *Trans. ASCE* 51: 253–330

Sommerfeld, A. 1904. Zur hydrodynamischen theorie der Schmiermittelreibung. *Z. Math. Phys.* 50: 97–155

Sommerfeld, A. 1908. Ein Beitrag zur hydrodynamischen Erklärung der turbulenten Flüssigkeitsbewegung. *Int. Congr. Math., 4th, Rome,* 3: 116–24

Van Dyke, M. 1982. *An Album of Fluid Motion.* Stanford, Calif: Parabolic. 176 pp.

von Kármán, Th. 1911. Über die Turbulenzreibung verschiedener Flüssigkeiten. *Phys. Z.* 12: 283–84. Reprinted, 1956, in *Collected Works,* 1: 321–23. Butterworth

von Kármán, Th. 1954. *Aerodynamics: Selected Topics in the Light of Their Historical Development.* Ithaca, NY: Cornell Univ. Press

von Kármán, Th. 1956. Dimensionslose Grössen in Grenzgebieten der Aerodynamik. *Z. Flugwiss.* 4: 3–5

von Mises, R. 1914. *Elemente der technischen Hydromechanik.* Leipzig: B. G. Teubner

GEORGE W. BEADLE

Annu. Rev. Genet. 1990. 24:1–4

G. W. BEADLE

Adrian M. Srb

Section of Genetics and Development, Cornell University, Ithaca, New York 14853

George Wells Beadle was born in 1903 and died in 1989. During his lifetime, the science of genetics changed dramatically. Beadle contributed definitively to that change. Perhaps more than that of any other investigator, his work prepared the way for the transition from neoclassical to molecular genetics. His contributions were recognized with a Nobel Prize and many other high honors.

Beadle did his work with three among the relatively few organisms that offer well-developed facilities for genetic research, and learned these facilities in excellent environments for scientific accomplishment. His Ph.D. research was with maize. The studies were carried out under the direction of Professor R. A. Emerson, and in the company of some remarkable fellow students at Cornell, including Barbara McClintock, Marcus Rhoades, and Charles Burnham. In postdoctoral work, Beadle turned to Drosophila—his major studies with that organism being carried out in collaboration with Boris Ephrussi. The evolution of Beadle's interests, which in part parallels the changing course of genetic science at that time, can be seen in the circumstance that Emerson was a magnificent formal geneticist, primarily interested in inheritance patterns, mapping and variations on the classical Mendelian ratios. Ephrussi, in contrast, while a highly competent geneticist, might better be described as a developmentalist who was adept at using genetic materials and techniques to address the problems that interested him.

The third experimental organism with which Beadle worked, and the one with which he made his definitive contributions, was Neurospora. Indeed, largely due to Beadle and his coworkers, that organism has become one of the better established and most useful objects of genetic research. Until Beadle, Neurospora was relatively obscure in the scientific world. The mycologist B. O. Dodge had found Neurospora to have intriguing possibilities for fungal

G. W. BEADLE

genetic studies but his vision of the possibilities was different than emerged in Beadle's work.

When I first came to Stanford in 1941 to do graduate work with Beadle, who had come there from Harvard as a professor in 1937, the transition of his work with Drosophila to that with Neurospora was nearly complete. A refrigerator or two held bottles containing Drosophila pupae in storage, but in the laboratory Neurospora mutants were starting to emerge for study. In an informal seminar, or discussion with his students and coworkers, I remember Beadle saying, in words as nearly as I can recall, that he was "tired of hearing other biologists say that geneticists only work with trivial attributes of organisms, like kernel color in maize or eye color in Drosophila." Clearly Beadle felt challenged to do something about such a criticism, justified or not.

Beadle had in mind to study genetic control of the steps in metabolic pathways. What was truly revolutionary in his plan was a practicable, although by modern standards not highly efficient, scheme for obtaining a large number of appropriate *conditional mutants* for study. The mutants that were sought, and in fact found, were nutritional mutants, each representing a genetically determined block at a particular step in a metabolic pathway. The standard, or wild-type, Neurospora has simple nutritional requirements, little more than a few salts and sugar being needed. From this *minimal medium* the organism makes all the biochemicals necessary for its existence—amino acids, nucleic acids, vitamins, complex structural components, and so on. By manipulating nutritional conditions, Beadle's scheme permitted detection, preservation, and study of mutant Neurosporas representing alteration of genes that control steps in essential biosyntheses.

In essence, after treatment of experimental material with mutagens, appropriate mutants can be preserved on media supplemented with essential metabolites, and detected by transfer to the minimal medium for the unmutated organism. Systematic studies with particular nutritional substances, or the absence of them, refine the relationship of the mutant gene and its unmutated allele to particular biochemical reactions. One can view the Neurospora nutritional mutants as special instances of *conditional lethals* now so useful and widely employed in molecular genetics and other research on many organisms.

Insight into Beadle's innovative research with Neurospora emerges from his immediately precedent work with Drosophila. Shortly after Beadle arrived at Stanford, E. L. Tatum, had come as a Research Associate to work on the biochemical genetics of eye pigmentation in Drosophila. The work was an extension of investigations carried out by Beadle and Ephrussi in Paris.

Those studies had utilized different eye color mutants. Reciprocal transplants of imaginal discs were made between larvae of different genotype. In particular instances, the occurrence of wild-type eye pigmentation in a derivative adult Drosophila revealed the presence of diffusible precursor substances that had accumulated behind genetic blocks in the mutants. In spite of the elegance of these sophisticated complementation experiments, the experimental system had clear limitations. The chemistry of the substances involved was extremely difficult to resolve. More importantly, the system as such could not be adapted to a broad attack on problems of gene action.

At this point the association with Tatum, the biochemist in the team, was particularly fortunate. A microbiologist as well as biochemist, Tatum was well aware of a growing body of literature that showed a remarkable diversity of growth requirements among microorganisms, some them quite closely related. A number of findings showed that in nature many of these organisms satisfied their growth factor requirements by some form of symbiotic relationship between strains or species. The similarity between such a situation and that shown in the transplant experiments is fairly obvious—at least it was so to Beadle and Tatum. And out of such a background, the Nobel prize-winning work with Neurospora was generated.

The various details of the Neurospora studies, and their extensions, have been often and well reviewed, and have become part of standard text book fare. They need not be summarized here. What needs to be said is that the volume and excellence of the work are in part due to Beadle's ability to attract outstanding associates. In the group that assembled early on at Stanford were, for example, Norman Horowitz, David Bonner, Herschel Mitchell. and Mary Houlahan Mitchell. Outstanding post doctorals like Francis Ryan were also drawn to Stanford, as was a full complement of graduate students.

Beadle was well aware that in the long run his work with Neurospora would benefit from information about the organism in addition to that derived from experiments on the biochemical mutants per se. At one point he enticed Barbara McClintock to come to Stanford to have a look at the chromosomes of Neurospora. He fostered some detailed studies of heterocaryosis. And he persuaded his graduate students to do at least minimal mapping of the mutants with which they worked. Indeed, the fact that Neurospora became a *major experimental object* in the broad sense of the words is largely due to him and his generosity with strains of Neurospora that he had developed.

The year 1946 was the beginning of the end of Beadle's experimental career. At that time he moved to Cal Tech as successor to T. H. Morgan as head of the Department of Biology. Highly successful in that position, he accepted the presidency of the University of Chicago in 1961, where he served with distinction until retirement in 1968.

When Beadle retired from the presidency of the University of Chicago, he came full circle and began again to do experiments with maize. These had to do with the origin of corn, a problem that had interested his first scientific mentor, R. A. Emerson. Energetic and incisive as ever, Beadle attacked this problem. He became an enthusiastic proponent of the theory that the important crop plant maize (corn) derives from teosinte. At the last seminar he ever gave at Cornell, on a visit made after his retirement, he described how he had disposed of arguments that teosinte was too useless a food source to tempt primitive native peoples to take it through its first stages as a crop plant. Beadle told how he had shown that the very tough kernels of teosinte could be popped like popcorn, thus making them edible, and had shown in addition the possibility of gathering enough of the kernels to make them a feasible source of food. Being Beadle, of course he had done genetic and cytogenetic studies, some in his graduate student days, some after retirement, that bore on the problem. He inevitably knew all the relevant studies by others, as reported in the literature, even to archaeological and cultural anthropological information.

Beadle was not only a great scientist and highly successful academic administrator. He was also an attractive and interesting human being. Born and raised on a farm near Wahoo. Nebraska, he was the kind of person whose fancy was tickled by his derivation from a community with an amusing name. He was also pleased to have come from a small place that in addition had nurtured such notables as Wahoo Sam Crawford, a famous old-time professional baseball hero; Howard Hanson, American composer and Director of the Eastman School of Music; and Darryl Zanuck, a historic director of films. He often expressed gratitude to the local schoolteacher who had inspired and encouraged him to go to college at the University of Nebraska. He was similarly grateful to F. D. Keim, a professor of agronomy at the university. Keim, an unusually perceptive talent scout for young men of scientific promise, persuaded Beadle to go to Cornell for graduate study.

Beadle's tastes were rather simple. He once told me that for pleasure-reading his favorite author was H. H. Munro. Beadle liked activities that involved physical exertion. In all these activities he showed his basic competiveness. He was a ferocious if not highly skilled tennis player. He invited his graduate students to compete with him in a now, and even then, outmoded athletic event called the standing broad jump. An enthusiastic gardener, he frequently wanted to place a small bet as to whether some other gardening friend could produce a larger pumpkin or grow earlier edible sweet corn. The very few bets he lost were cheerfully paid.

Sewall Wright

Annu. Rev. Genet. 1989. 23:1–18

SEWALL WRIGHT'S CONTRIBUTIONS TO PHYSIOLOGICAL GENETICS AND TO INBREEDING THEORY AND PRACTICE

Elizabeth S. Russell

The Jackson Laboratory, 600 Main Street, Bar Harbor, Maine 04609

CONTENTS

INTRODUCTION

Throughout Sewall Wright's long and varied scientific careers, stretching from graduate days at the Bussey Institute of Harvard, through ten years collecting and analyzing data on inbred guinea pigs at the USDA in Beltsville,

29 years of teaching and research at the University of Chicago, and finally on to 33 years of fruitful thinking and writing at the University of Wisconsin, he consistently made fundamental conceptual and practical contributions.

Sewall Wright's work on evolution is widely known from his long controversy with Oxford's R. A. Fisher, and from his four volume summarization, *Evolution and Population Genetics* (44–49), brought together during his years at the University of Wisconsin. We already have an excellent biography (7) that emphasizes Wright's work on evolution. In the course of writing this biography, Provine spent many informal and highly profitable hours consulting directly with Wright, which brings to his presentation a welcome ring of truth.

I was delighted when I was asked to prepare, for the *Annual Review of Genetics,* a "short biography of the first half of Sewall Wright's life." I wanted to accept the challenge, but hesitated. What could I add to Provine's excellent account? How could a scientific biography be limited to the first half of anyone's life? This objection is especially pertinent with Wright since, once he got into an interesting area, he never let it go! Pigmentation genetics and analysis of results of inbreeding, both of which he began to study very early, were both subjects of papers written more than 50 years later! I decided to concentrate on Wright's times at the Bussey and the USDA, and the first part of his time at the University of Chicago, and to reminisce about Wright as a teacher, concentrating on the two fields of mammalian genetics that I know best: physiological genetics and characterization and use of inbred strains.

Before I plunge in, I want to express my deep gratitude to Wright's nephew, Christopher Wright, who loaned me a series of fascinating letters, recently rediscovered, written by Wright between 1912 and 1933, mostly to his mother (13).

This is also an appropriate place to make some generalizations about Wright's personality and character.

1. Although Wright did not enter readily into everyday chitchat, he loved discussions and maintained numerous deep friendships throughout his life, most of which began from scientific connections. Some (C. C. Little, Harrison Hunt) dated from the Bussey. For a couple of years while at the USDA, he shared an apartment with Paul Popenoe, then editor of the *Journal of Heredity,* and developed firm connections with experts in livestock breeding. While at Chicago he greatly enjoyed discussions with biology faculty companions during long walks on the Indiana dunes. One important feature of his work on evolution was his friendship, including field trips, with Dobzhansky. His best and very much appreciated friend late in life was Jim Crow.

2. Wright was a very conscientious, hard-working man, who kept excellent records. His special love was trying to convert those records into forms

that could be analyzed mathematically. He also threw himself whole heartedly into analyzing research problems of all the many scientists who came to him for help.

3. Wright was an instinctual mathematician, having learned the basics from his father while an undergraduate at Lombard College. When tackling a new problem, he always went back to first principles. Perhaps because of this approach, he stuck firmly to his conclusions. Provine quotes him as saying, "I acquired some facility in translating questions into mathematical symbolism and solving as best I could." Provine continues, "He was not primarily interested in pure mathematics, but in devising ways to analyze problems quantitatively, and became adept at teaching himself new mathematics" (7).

   While I was his graduate student at Chicago, he often became completely engrossed, in the midst of a course lecture, in deriving a new formula. Usually none of his students could follow his derivation, nor even see most of what he had written on the blackboard.

4. Wright was a great believer in Occam's razor, or the principle of parsimony: Always cut assumptions to the bare minimum, adopting the simplest hypothesis concordant with the facts. And, please, "Don't go the long way around Robin Hood's barn."

5. At least up to 1940, Wright prided himself on reading and digesting all publications in genetics, and presented neat summaries to his students, usually referring to a stack of notes on 3x5 cards.

THE BUSSEY INSTITUTE

Wright's training as a mammalian geneticist started soon after the rediscovery of Mendel's laws and the reorganization (1909) of the Bussey Institute of Harvard University to be a center for graduate studies in genetics. Its director, W. E. Castle, was an early leader of mammalian genetics, and his associate, E. M. East, was an excellent botanical geneticist. Together they provided a very stimulating environment for a group of excellent graduate students, including C. C. Little, E. C. MacDowell, and J. A. Detlefson, working on mammals, and E. W. Sinnott and R. A. Emerson, working with plants. All of these became real scientific leaders. There were excellent seminars and discussions made especially lively by the fact that Castle and East, both ardent believers in Mendelism, did not completely agree on other genetic questions. Students were encouraged to contribute to the discussions.

When Wright arrived in 1912, Castle and students had already demonstrated Mendelian inheritance of several distinct coat-color differences in rabbits, mice and guinea pigs. In 1911 new and different stocks from South

America had been added to the guinea pig colony. By 1912, Castle himself was deeply involved in selection experiments on spotting patterns in hooded rats. Clarence Cook Little, who had begun to work with Castle even before the Bussey became a genetics institute, was by 1912 enthusiastically working on multiple factor inheritance of cancer susceptibility in mice. Detlefson, who was to leave soon, was in charge of the guinea pig colony. In 1912, the guinea pig colony contained new, not yet understood color variations, brought in from Peru. Castle immediately assigned to Wright responsibility for maintenance of the guinea pig colony, with freedom to analyze all genetic differences. Guinea pigs, with small litters and a long gestation period, are not the easiest mammals for genetic analysis, but Wright managed to make critical tests to assign all the new variability to the proper allelic series. Each of these new recessive diluting mutants could have been like "blue dilution" in the rabbit, or pink-eyed dilution in the mouse, or members of the albino series, or could belong to a new, previously unknown series. At this time, only one multiple allelic series had been described. One important contribution made by Wright at the Bussey was his demonstration that the guinea pig albino series was a second multiple allelic series that contained at least four distinct alleles.

Wright was also much interested in problems of selection and inbreeding. The importance of Johannsen's pure self-fertilized lines of Phaseolus beans, and of the work of G. H. Shull and E. M. East on inbred and F_1 hybrid corn, were becoming recognized, but there was considerable doubt among animal breeders about the possibility and utility of brother-sister inbreeding for development of animal stocks. As I've heard informally from his co-students C. C. Little and H. D. Fish, Wright's mathematical mind led him to contribute very effectively to local discussions of inbreeding. Wright was interested both in continuing to work on guinea pig coat-color genetics and in studying mammalian inbreeding and inbred lines.

Early in 1915, Dr. George Rommel, Chief of the USDA Bureau of Animal Industry, came to Castle seeking a geneticist who could tackle analysis of a very large body of data already accumulated at the USDA during nine years of inbreeding guinea pigs. Before 1906, non-inbred guinea pigs were maintained at USDA to supply research needs of government workers. Rommel, with considerable astuteness, started 35 brother-sister mated inbred lines in 1906, to test feasibility of inbreeding animals. While some of these lines died out, others survived, though their characteristics had never been studied. Castle felt that Wright's diligence and mathematical approach were ideal for analysis of this data, and highly recommended Wright for the position. Wright accepted Dr. Rommel's job offer and in the summer of 1915 packed up representatives of the special guinea pig coat-color stocks and took off for the USDA in Washington and Beltsville.

PERSONAL LIFE AT THE USDA

Before considering Wright's scientific work at the USDA, I want to talk a bit about developments in his personal life during those ten years. He arrived from academia as a bachelor, first living in a rented room, then sharing an apartment with Paul Popenoe. After a time he joined the Cosmos Club and found it a good place for lively discussions. He maintained scientific friendship contacts by attending and presenting papers at AAAS and American Naturalist meetings. This activity included getting to know members of the new Drosophila genetic group, especially Sturtevant and Morgan.

In the summer of 1920, he spent all of his 30-day annual vacation at the Station for Experimental Evolution at Cold Spring Harbor. C. C. Little was there, and Wright's good friends Phineas and Anna Rachel Whiting, and a young woman, Louise Williams, who had obtained a master's degree in Zoology with Harold Fish, one of Wright's friends at the Bussey. Miss Williams was now an instructor at Smith College, but had come to Cold Spring Harbor for the summer to take care of Fish's rabbit colony. The first evening, when Wright and Miss Williams met at dinner, Louise told him she wished he had not been "just another geneticist," but rather the new animal caretaker who was supposed to clean her rabbit cages. He volunteered to help her next day, a good start to a deepening friendship that included conversations, evening strolls, and flower gathering. She was somewhat lame from a congenital hip dislocation, but this did not keep her from walking, somewhat slowly, for miles. "She had an attractive gaiety and a good sense of humor, and was sympathetic with anyone in trouble . . . I was so much in love with her at the end of my vacation . . . that I proposed on the last evening" (7). They were married at her home in Ohio, in September, 1921. After the wedding they had a canoe-trip honeymoon above the falls on the Potomac. During his last four years at the USDA, he was a happily married man, with two sons (Dick and Bob) born by 1925. Their third child was a daughter (Betty) born in Chicago in 1929. Louise Wright was a truly lovely woman, always welcoming to students. Sewall Wright, as usual, by following his instincts, had made an excellent choice.

WORK ACCOMPLISHMENTS AT THE USDA

Wright found the atmosphere at the USDA very different from the Bussey. For quite a while he was very busy with both scientific and not-so-scientific assignments. In addition to animal caretakers, he worked with livestock breeders, and developed quite an interest in breeds of horses, cattle, pigs, and sheep and went as a part of his job to the yearly International Livestock

Show. He also received and answered many inquiries from farmers. He liked and appreciated his boss, Dr. Rommel, chief of the Bureau of Animal Industry, who helped Wright to get new metal animal cages. One early job that amused Wright (and me) was working out the market value of a "lot of very fancy cattle," which were quarantined for a year because they got hoof and mouth disease. The government had figured these cattle to be worth $70,000, the cattlemen, $186,000. Wright worked out a scientific scheme of evaluation that took their ancestry into account, "a modification of Galton's law. I don't know if the USDA will use my figures. Their problem probably really calls for a diplomat rather than a student of heredity" (13).

However odd these "real world" assignments may seem, Wright's ten years at the USDA were extremely productive. His contributions to methodology for quantitative genetic improvement of livestock breeds made him greatly revered by both theoretical and practical agriculturalists. He was called on frequently to solve quantitative analytic problems for the whole Agriculture Division; perhaps his most famous assistance there was his study, involving use of path coefficients, of "Corn and Hog Correlations" (37). By 1921 members of the Agriculture Division had persuaded him to offer courses in genetics and statistics for government workers. In the early years, he wrote two series of papers for the *Journal of Heredity*, these providing a simple, concise background for a wide audience. The first such series, covering ten species, was on color inheritance in mammals (16–26). Later he published five papers on Systems of Mating (27–31), which became the "bible" on genetic improvement of livestock. This series involved early use of the concept of path coefficients, tracing all pathways of influence between two individuals, and evaluating the contribution of each path to the total variance.

I am limiting my detailed scientific discussion of his work at the USDA to the developing area of physiological genetics, and to his studies analyzing and characterizing guinea pig inbred strains.

Studies of Coat-Color Genetics

Wright found it hard to prepare his Bussey thesis for publication, approximately one and a half years after its original composition. He found many points on which his opinion had changed, and it was difficult to make alterations so that the new text (15) harmonized with his new views (13). He kept collecting more data on the same and new genetic color combinations. Soon after arriving at USDA, he developed a series of grades for each of the color types, by matching each guinea pig's color(s) to a series of standard skins that marked barely perceptible increases in color intensity. He assigned genotypes through a combination of color-grade and progeny test. Many of the guinea pigs were tricolored, with patches of black-sepia, red-yellow, and white. This pattern was particularly useful since the observer could grade both sepia and yellow spots on the same individual, which obviously had only one genotype. Sepia grades went from 3 to 14 (intense black), red/yellow grades from 1 to 13 (intense red), and pink-eyed sepia from one (barely distinguishable from white) to 12 (light sepia, almost identical to dark-eyed sepia grade 5). He really had no direct measurement of the amount of melanin deposited in the hair. The only pigment granules he actually saw were in a few histological sections prepared in collaboration with Harrison Hunt. Using the color-grading methodology, he wrote three interim papers (20, 35, 36), a collaborative paper (6), and a big summary paper (39). These papers dealt almost entirely with the effects of five allelic series affecting quality and intensity of pigment over the whole body. He did not study the agouti series, which puts a red-yellow band on a sepia hair shaft. In the guinea pig, the extension locus affects quality of pigment, with two alleles controlling production of eumelanin (dominant E-) vs phaeomelanin (e/e) on the whole body, and a third allele (e^P/e^P) producing both colors in different areas (spots or brindling).

The multiple allelic albino series (C, c^k, c^d, c^r, c^a) affects intensity of both black-sepia and red-yellow. The C allele is always top dominant, and the c^a the lowest recessive. The effects of c^k are somewhat different on sepia and yellow, and the red-eyed c^r gene substitution does not reduce sepia pigment intensity in $E/-c^r/c^r$, but completely eliminates all color in $e/e\ c^r/c^r$. This very different ordering of albino allele affects in combination with sepia vs yellow is very unusual.

Two allelic series, B/b and P/p, alter eumelanic E-pigment, but have no effect on phaeomelanin.

One series, F/f, affects only red-yellow pigment, f/f reducing red to yellow. The only place where f/f affects eumelanin is in combination with pink-eyed dilution. $E/-\ C/-\ f/f\ p/p$ is very pale cream.

In the discussion section of his summary paper (39), he speaks of tyrosine as a possible chromogen, and tyrosinase as the probable product of the albino locus. He develops a "paper scheme" that suggests how the gene series *might* interact, but limits his conclusions as follows: "First, the immediate product (I) of action of genes of the C series has different thresholds of effectiveness depending on whether an accessory substance (II) (product of E-), necessary for any sort of sepia or brown, is present or not and whether or not II, if present, is modified by the P-series; second, that above these thresholds there is competition between the precursor of yellow (I alone) and the precursor of sepia and brown (I–II). Factors B and F must be assigned effects following, or otherwise without effect upon, the above threshold and competition effects" (40).

Early in these studies, Wright said, "coat patterns in guinea pigs, and doubtless in other animals, must be determined by a complex of causes of very diverse kinds (19).

Wright concluded that products of different genes must interact with each other. Certain genes must act before or after others, and two different gene products may compete for the same chromogen. When this was written, only Wright and Goldschmidt (4) were really thinking about what genes actually do.

Effects of Inbreeding on the Original Brother-Sister Inbred Lines

In 1906, a total of 35 lines were set up. Twenty-three of these survived through enough successive generations of full-sib matings to be called "lines," or "inbred families." Seventeen of these were still doing well in 1915. By that time, it had been clearly established that guinea pigs *could* be inbred! However, no one had carefully analyzed the effects of this inbreeding, nor studied the characteristics established. Rommel's original idea and Wright's special qualities formed a very happy combination. Wright found excellent records on over 30,000 guinea pigs from 23 families maintained over ten years in a single colony, with good information on date of birth, maternal age at birth, litter size, mortality at birth and between birth and weaning, birth weights, and number of litters. These were all available to a geneticist who loved to look for correlations and patterns. Quotes from a series of letters in spring, 1916, show his attitude: "My work is getting very interesting; every correlation determined suggests the desirability of finding another. I find all kinds of relationships between rate of growth and size of litters both complex and hard to interpret. It is exceedingly difficult to distinguish cause from effect, and from effects of a common cause" (13).

The system even contained a control outbred stock, with no matings closer than between second cousins. In this colony, there were strong environmental influences, both annual and seasonal, that affected both inbred and outbred stocks.

During 13 years of inbreeding, from 1906 to 1920, "there was an average decline in vigor in all characteristics," "most marked in frequency and size of litters" and "ability to raise large litters has fallen off much more than ability to raise small litters" (33).

Experimental inoculation with tuberculosis had shown that, on the average, the inbreds were inferior to their outbred counterparts in disease resistance (32, 50).

There was both a decrease in vigor during inbreeding and a differentiation among families as to the most affected elements of vigor. "Each family came to be characterized by a particular combination of traits, usually involving strength in some respects with weakness in others." "Crosses between different inbred families resulted in marked improvement over both parental stocks in every respect, due allowance having been made for the effects of size of litter on the other characters" (34).

Differentiation among Inbred Families

Even more interesting (to me) than the decline in vigor was the differentiation Wright observed among families. Each family produced a variety of colors in early generations, but "as time went on, different colors automatically became fixed in different lines" (34). Intensity of color and amount of white in spotting patterns also differed markedly between stocks. "The 23 families . . . have automatically become so differentiated from each other that a new litter could easily be recognized at a glance as belonging to its particular family" (34).

Many of the distinguishing traits of a particular family were not seen in all individuals within that family, since incidence was affected by environmental as well as genetic influences. Wright's method of path coefficients, tracing all possible connections between individuals, provided a means to evaluate the contributions of heredity and environment to expression of each observed character. Variability of expression persisted even after genetic homozygosity had been achieved, but there was no longer any correlation between expression in parent and in offspring.

Wright's analysis of tricolor coat patterns (26) provides an early, neat example using his "threshold" concept and his new method of path coefficients. He measured the amount of white in the color patterns of tricolored guinea pigs from three inbred families and the outbred control stock B, and attributed observed variance in amount of white to contributions from heredity (H^2), environment common to littermates (E^2) and "accidents of development" (D^2). Wright assumed that each area of a guinea pig's coat required at least a minimum "threshold" level of influences favoring color before pigment could be formed. The "sum of favorable and unfavorable influences could be arranged in a graded series," roughly fitted to a normal bell-shaped curve, whose mean was genetically fixed, surrounded by plus and minus environmental variations. Pigment was deposited only where the curve covered areas above threshold level. In the three inbred families, individual guinea pigs varied in proportion of white, but the overall mean for family 39 was 20% (left skew), for family 35 was near 50%, and for family 13, 85% (right skew). The distribution curve for the outbred stock B looked very much like that for family 35, with approximately 50% white. For the two families with skewed distributions, Wright determined medians, and upper and lower quartile values, then transformed the scale so that the distribution approximated a normal curve, whose variance and standard deviation he determined. Using these values, he calculated H^2, E^2, and D^2 for each stock. In the three inbred families, H^2 was practically zero, since there was essentially no genetic variation within strain. In family 35, with over 50% white in tricolors, less than 3% could be attributed to heredity, 5% to environment common to littermates, and 92% to accidents of development. In the very similar looking outbred stock B, the causes of variation were very different: 42% of the variance in stock B could be attributed to the large amount of hereditary difference within the stock, and 58% to accidents of development. My summary is: identical appearance does not necessarily mean genetic identity, and genetic identity does not guarantee identical appearance.

A later study of family 35 (38) showed correlation between amount of white and increasing maternal age. The coats of offspring of 3-month old mothers averaged 60% white, while offspring of 21- to 46-month old mothers averaged 73% white. Since 3-month females had usually achieved only one half of their adult weight, Wright felt the correlation between proportion of white with advancing age of the mother probably implied some kind of growth competition between mother and offspring.

Genetic Monsters

One type of trait that Wright found that differed among inbred families was the tendency to produce low but characteristic numbers of some specific developmental anomaly. The defects were not just culminations of the deleterious effects of inbreeding, since the affected strains were usually among the healthiest. The first "monstrosity" observed, otocephaly, was very rare, with 82 cases among almost 40,000 guinea pigs. There was a genetic basis for susceptibility to the trait, since 50 of these cases appeared among the 3253 individuals in family 13. Otocephaly varied from slight reduction of space between the ears (grade 1) to complete absence of the head (grade 12). "In grades 6 to 12 the primary feature . . . was progressive arrest of the brain" (49). Except for the head defect, the rest of the body was near normal in size, plump, and well developed. "The genetic basis determines an individual, not a maternal character . . . By elimination it is concluded that the main factor is probably chance delay, or other irregularity, in implantation, acting at a particular moment in ontogeny, to produce the resulting temporary arrest in development" (50). Otocephaly seems to have been a threshold character in family 13.

Guinea pigs normally have only three toes on each hind foot. The relatively mild condition, atavistic polydactyly, or return to the ancestral rodent type with four toes per hind foot, was found in a fair proportion of individuals in several different lines. Within one line, incidence of four toes was determined by environmental influences, but between lines, differences in susceptibility appeared to be controlled by 3 or 4 pairs of genes (41).

Quite independently, a different form of polydactyly appeared, dependent upon presence of a single semi-dominant gene *(Px)* (40, 41). Heterozygous

Px/px individuals often developed thumbs as well as little toes. Homozygous *Px/Px* individuals developed paddle-shaped feet, each with "7–12 undifferentiated digits, and other severe developmental anomalies."

Advances in Understanding of Inbred Lines

Wright's careful studies and development of mathematical methods for analyzing characteristics of inbred strains led to increased interest in genetic differences, reduced prejudice against inbred lines, and encouraged their use by both agriculturists and biomedical researchers. His studies of vital statistics confirmed that vigor declined during inbreeding, but showed differences between strains in the particular nature of deleterious effects. Usually F_1 hybrids between strains were much more vigorous than were pure lines.

The importance of using particular inbred strains depends upon genetically influenced characters differing between strains. Wright found many cases of such differentiation, and developed excellent methods for studying characteristics affected by internal and external environmental factors as well as heredity. His own classification of types of characters fixed in inbred strains (49) has been very helpful to me in thinking about the power of inbreeding, and I hope it may also be useful to other researchers:

1. Complete fixation: all individuals within a strain appear identical, as in quality of coat color or nature of histocompatibility genes.
2. Fixation of average grade: "There are characters affected by nongenetic factors in addition to genetic ones" where "the former are of the nature of irregularities of development unrelated to tangible external conditions." One can think of myriad examples; he cites proportion of white in tricolors.
3. Fixation of ranking: Although weight is much influenced by tangible environmental factors, each strain appears to have a fixed degree of genetic susceptibility, so that ranking, but not values, remains constant under varying conditions.
4. Fixation in reaction to particular conditions. Determination of vigor, or disease resistance, or many other characters depends on susceptibility to different insults in different strains. In these cases, what is fixed is a specific susceptibility, not the final character.

Physiological Genetic Concepts

From the very beginning, Wright saw that genes acted within cells, and that they must interact with other genes and with environmental influences. His first studies of action of coat-color genes demonstrated this kind of interdependence. Later he deduced that pathways of interdependence were important in expression of variable characters in inbred strains. In these cases it was usually difficult to sort out effects of particular genes, but analysis of inbred strain characteristics broadened the range of processes on which genetic factors could be seen to have effects. The genetic monsters, otocephaly and single-gene-induced polydactyly, both seemed to involve gene-induced changes in the rate of development at critical times and in particular places in early embryonic development.

In early 1925, Dr. Frank R. Lillie, head of the University of Chicago Zoology Department, offered Wright a position as associate professor, with responsibility for research and graduate courses in genetics. By this time, the Wrights were planning to build a home in the Washington area and research at the USDA was progressing very well. They weren't at all sure that they wanted to move. Wright wrote to Lillie of his concerns, including need for guinea-pig housing and for ample time for writing and research. Then, in February, 1925, Wright learned that President Coolidge, attempting to cut expenditures, had dismissed E. D. Ball, director at research at the USDA. Wright approved highly of Ball, who had benefited his research program extensively. As soon as Wright learned of Ball's dismissal, he wrote to Lillie, showing much more interest in the Chicago offer, and before long it was agreed that Wright would move to the University of Chicago in January, 1926.

Wright looked forward to interacting with embryologists at Chicago, and to increasing his attack on theoretical, rather than practical, problems.

THE UNIVERSITY OF CHICAGO ZOOLOGY DEPARTMENT

Analysis of development was a central interest in the Chicago Zoology Department. Its first chairman, C. O. Whitman, and its second chairman, Frank R. Lillie, were both embryologists. Charles M. Child proposed and promoted the concept that dominant apical to subordinate basal physiological gradients controlled development and differentiation in planaria. Carl R. Moore analyzed mammalian endocrinological development and differentiation. For seven years after Wright arrived, B. J. Willier, a leading amphibian embryologist, was at Chicago, and they became close friends. After Willier left, Victor Hamburger and Paul Weiss, both experimental embryologists, joined the Chicago department. All of these individuals were lively, driving people, devoted to their own research topics at a time when developmental biology was advancing rapidly, with much interest in Spemann's induction phenomena. These developmentalists thought of genes as influencing superficial characters and, at least at first, did not seem too eager to think about gene action in early development.

Other research fields were represented on the faculty. W. C. Allee was a fine ecologist with special interest in animal aggregations. He managed to be very active in spite of being confined to a wheelchair. Alfred Emerson was an entomologist, ecologist, zoogeographer, and expert on termites. H. H. Newman, who organized the undergraduate biology courses, was interested in one phase of genetics, study of characteristics of identical versus fraternal twins, reared together or apart. Dr. H. H. Strandskov, who had been Newman's student, particularly in human genetics, became a member of the faculty after 1931.

All of the zoology courses, and some of the research, took place in the stone Gothic-style Hull building on the main campus. This was two short blocks from the Whitman Laboratory, where both staff and students in genetics and ecology, plus some of those in endocrinology, had laboratory space. They all worked well together. Between 1925 and 1936, there was relatively little contact between zoology and the nearby Billings Laboratories of The University of Chicago Medical School.

Wright's office, on the ground floor of Whitman, was right next to the free-standing guinea-pig house. It was always interesting to hear the chirping voices of the guinea pigs when Wright rattled his keys, preparatory to entering their quarters.

Wright's Development at Chicago

During his first ten years at Chicago, Wright's career advanced rapidly. He was more than ever asked to present important papers, including one for the major session on Physiological Genetics at the International Genetics Congress in 1932, in Ithaca, New York. He was president of that Congress. He became a very strong proponent of Physiological Genetics, and came more and more to introduce his ideas with graphs and diagrams, showing first "the relation of factors," environmental and genetic, "to any developmental process of a vertebrate, occurring at a particular time and place." Usually this was followed by a "diagram illustrating the chain of processes relating to the immediate physiological action of a gene to characters at different levels."

Wright's accomplishments and abilities came to be very widely appreciated during his first decade at Chicago. He was elected to the American Philosophical Society in 1932, to the National Academy of Sciences in 1934, and served as vice-president (1933) and president (1934) of the Genetics Society of America. He served on many editorial boards and committees, and continued to receive even more requests for scientific advice, and through all of this, he continued to record data on his guinea pigs and to write papers!

In 1935 Harvard University made a very attractive offer to Wright, and the University of Chicago responded by offering him even more. The zoology faculty unanimously wrote him a letter, urging him to stay at Chicago. I remember his talking with his graduate students about the problem, trying to make up his mind. The school atmosphere and lifestyle of Chicago pleased both Wrights, and they decided to stay. Everyone was delighted.

WRIGHT AS A TEACHER

Wright took very seriously his responsibility for teaching all of Chicago's graduate genetics courses. I feel my courses with Wright were invaluable. Besides Fundamental Genetics, where I was teaching assistant rather than student, I had *two* biometry courses, each presented formally, although there were only two or three students. I couldn't possibly take satisfactory notes on all that he presented, but I absorbed a great deal about probability, normal curves, threshold characters, transformations of scale, variance, means, standard deviations, heritability, and even path coefficients, though I still can't really use that method effectively. The two most exciting courses, which drew a larger attendance, were Physiological Genetics and Evolution. In Physiological Genetics he discussed the development of ideas and concepts, and reviewed practically all recent genetic publications, putting them into appropriate perspective. He discussed what was known at that time (1934) about the chemistry of chromosomes, and the basic problem of the nature of the gene and its duplication. At that stage, he leaned toward the protein component as the site of gene specificity.

The evolution course was also exciting. His long controversy with R. A. Fisher about the processes of evolution, and the nature and significance of dominance, had just begun. I agree with Paul Scott (personal communication) that we all learned from that controversy: "Treat your opponent with respect, but state your own viewpoint firmly."

The majority of Wright's graduate students tackled research problems related to Wright's studies of guinea pigs, and used animals from his research colony, but he left us pretty much on our own in developing methodology. His first doctoral student, John Paul Scott, analyzed the embryonic development of the polydactylous monsters. He did a very credible job, first producing a "Normentafel" of guinea-pig embryology, then comparing "pollex" monsters with normal littermates. Beginning at 18½ days, almost all organs of the "pollex" embryos showed a much higher than normal rate of mitoses, with abnormal morphogenesis, and by 26 days almost all of these embryos were dead. Scott called these 8 days the critical period. I hope he won't mind my saying this, but aside from concern with genetics, the most apparent carryover from his thesis to later research was use of the term "critical period."

Herman Chase studied the distribution of sepia, yellow, and white areas on tricolor guinea pigs with highly variable total amounts of white. Certain regions were much more apt than others to be colored. Genetic factors influenced the amount of white on the body as a whole, and non-genetic factors acted locally. The intermediate allele e^p was not completely dominant over e, nor was S completely dominant over s (1, 2).

William L. Russell studied reactions of frozen sections of guinea pig hair follicles and of pigment cells in the basal epidermis to exposure to buffered solutions of 3,4-dioxyphenylalanine, or dopa. For each genotype tested, the graded responses in hair follicles agreed very well with the intensity of pigment in yellow hairs of corresponding e/e genotypes, but not at all well with intensity of sepia (9). Especially noticeable was the complete absence of dopa reaction in both dark sepia $E/\text{-} \ c^r c^r$ and white $e/e \ c^r/c^r$. The reactions in epidermal pigment cells corresponded with colors in the eyes.

Benson Ginsburg studied dopa reactions using extracts from guinea pig skin, and obtained results corresponding with yellow hair pigment-color grades (3).

Attempts to extract and measure concentrations in solutions of pigment from hair of different genotypes were made by myself (8) and Gertrude Heidenthal (5). Both of us were successful with solutions of yellow pigment in dilute alkali. Sepia and brown melanins were very insoluble. Both of us dissolved hair keratin with 6N hydrocloric acid, and obtained a precipitate of eumelanin. I tried weighing this, and oxidizing it with potassium permanganate. Heidenthal was more successful in making a melanin solution by long boiling in strong alkali. Conversion curves were developed for evaluating the relative amounts of pigment corresponding to the color grades and to hair genotypes.

Although Wright's students carried out experiments largely on their own, the genotypes and color grades of our guinea pigs had been assigned by Wright. The writing up of our results involved very long, fair, but piercing sessions with Wright.

In spite of his busyness, Wright enjoyed having students go with him to lectures and seminars in other departments. I particularly enjoyed going with him to hear Maud Slye tell about cancer in her mouse colony. She kept complete pedigrees, but never inbred her animals and found many different kinds of cancer irregularly distributed in her colony. She believed the tendency to all kinds of cancer was inherited as one simple recessive that might be transmitted hidden for many generations. Her lecture was a good demonstration of the need for background genetic control, and for testing your assumptions.

All of Wright's students learned a great deal about genetics, and continued to do genetic research after leaving Chicago. Scott and Ginsburg worked on genetics in social behavior in dogs. The rest of us worked with mice. Chase continued with pigment and with skin and hair. William Russell became an expert on induced mutation rates. My physiological genetic research came to center on anemias and other hematological characters.

THE WANING OF THE GUINEA PIG

Although Wright continued his research with guinea pigs until he left Chicago in 1955, his central interest came to be evolution in natural populations, particularly Drosophila. More examples, and more varied types of mutations were found in mice, and linkage maps were developed. The characterization and supply of a great variety of inbred mouse strains and mutants was also important in providing tools for research. Obviously, it was less expensive to work with mice than with guinea pigs.

Some, but not all, of his five favorite guinea pig inbred families were maintained at NIH after he moved to Wisconsin in 1955. This loss of inbred families seems very regrettable to me, especially as the early guinea pig studies of inbreeding and of gene-actions contributed so much to the development of mammalian genetics. We can not even fall back on the excuse that there was no published record of guinea pig genetics. As I worked on this review, I thought how useful the dark sepia $E/\text{-}\ c^r c^r$ vs. the white $e/e\ c^{cr}$ guinea pigs might be for today's molecular study of enzymes and enzyme reactions in pigmentation. I very much doubt if any one anywhere can find a live c^r/c^r guinea pig!

I would like to end by expressing the great debt of The Jackson Laboratory to Sewall Wright for his careful study of inbreeding and for his example and his training of many Jackson Laboratory staff members. Earl L. Green, the second director of the Jackson Laboratory, was a postdoctoral student of Wright, and four staff members (J. Paul Scott, William L. Russell, Elizabeth S. Russell, and Willys K. Silvers) were Wright's doctoral students. In 1954, Wright gave the important closing paper, *Summary of Patterns of Mammalian Gene-Action* (43) at The Jackson Laboratory's 25th Anniversary Symposium, and he was our special honored guest at the Laboratory's 50th Anniversary Symposium in 1979.

Wright was a wonderful man, and a truly excellent pioneer geneticist.

Literature Cited

1. Chase, H. B. 1939a. Studies on the tri-color pattern of the guinea-pig I. The relations between different areas of the coat in respect to the presence of color. *Genetics* 24:610–21
2. Chase, H. B. 1939b. Studies on the tri-color pattern of the guinea-pig II. The distribution of black and yellow as affected by white-spotting and by imperfect dominance in the tortoise shell series of alleles. *Genetics* 24:622–43
3. Ginsburg, B. 1944. The effects of the major genes controlling coat colors in the guinea-pig on the dopa oxidase activity of skin extracts. *Genetics* 29:176–98
4. Goldschmidt, R. 1938. *Physiological Genetics*. New York: McGraw-Hill. 375 pp.
5. Heidenthal, G. 1939. A colorimetric study of genic effect on guinea pig coat color. *Genetics* 25:197–214
6. Hunt, H. R., Wright S. 1918. Pigmentation in guinea-pig hair. *J. Hered.* 9:178–81
7. Provine, W. B. 1986. *Sewall Wright and Evolutionary Biology*. Chicago: Univ. Chicago Press
8. Russell, E. S. 1939. A quantitative study of genic effects on guinea pig coat color. *Genetics* 24:332–55.

9. Russell, W. L. 1939. Investigation of the physiological genetics of hair and skin color in the guinea pig by means of the copa reaction. *Genetics* 24:645–67.

10. Scott, J. P. 1937. The embryology of the guinea pig I. A table of normal development. *Am. J. Anat.* 60:397–432

11. Scott, J. P. 1937. The embryology of the guinea pig III. The development of the polydactylous monster. A case of growth acceleration at a particular period by a semi-dominant gene. *J. Exp. Zool.* 77:123–57

12. Scott, J. P. 1938. The embryology of the guinea-pig II. The development of the polydactylous monster. A new teras produced by the genes PxPx. *J. Morphol.* 62:299–321

13. Wright, C. 1989. Loan of rediscovered letters from S. Wright to family, 1912–1933

14. Wright, S. 1915. The albino series of allelomorphis in guinea-pig. *Am. Nat.* 49:140–48

15. Wright, S. 1916. An intensive study of the inheritance of color and other coat characters in guinea-pigs, with especial reference to graded variation. *Carn. Inst. Wash. Publ.* 241:59–121

16. Wright, S. 1917. Color inheritance in mammals. *J. Hered.* 8:224–35

17. Wright, S. 1917. Color inheritance in mammals II. The mouse. *J. Hered.* 8:373–78

18. Wright, S. 1917. Color inheritance in mammals III. The rat. *J. Hered.* 8:426–30

19. Wright, S. 1917. Color inheritance in mammals IV. The rabbit. *J. Hered.* 8:473–75

20. Wright, S. 1917. Color inheritance in mammals V. The guinea-pig. *J. Hered.* 8:476–80

21. Wright, S. 1917. Color inheritance in mammals VI. Cattle. *J. Hered.* 8:561–64

22. Wright, S. 1917. Color inheritance in mammals VII. The horse. *J. Hered.* 8:561–64

23. Wright, S. 1918. Color inheritance in mammals IX. The dog. *J. Hered.* 9:89–80

24. Wright, S. 1918. Color inheritance in mammals X. The cat. *J. Hered.* 9:178–81

25. Wright, S. 1918. Color inheritance in mammals XI. Man. *J. Hered.* 9:227–40

26. Wright, S. 1920. The relative importance of heredity and environment in determining the piebald pattern of guinea-pigs. *Proc. Natl. Acad. Sci. USA* 6:320–32

27. Wright, S. 1921. Systems of mating I. The biometric relation between parent and offspring. *Genetics* 6:111–23

28. Wright, S. 1921. Systems of mating II. The effects of inbreeding on the genetic composition of a population. *Genetics* 6:124–43

29. Wright, S. 1921. Systems of mating III. Assortative mating based on somatic resemblance. *Genetics* 6:144–61

30. Wright, S. 1921. Systems of mating IV. The effects of selection. *Genetics* 6:162–66

31. Wright, S. 1921. Systems of mating V. General considerations. *Genetics* 6:168–78

32. Wright, S. 1922. The effects of inbreeding and crossbreeding of guinea-pigs I. Decline in vigor. *USDA Bull. #1090*, pp. 1–36

33. Wright, S. 1922. The effects of inbreeding and crossbreeding on guinea-pigs II. Differentiation among inbred families. *USDA Bull. #1090*, pp. 37–63

34. Wright, S. 1922. The effects of inbreeding and crossbreeding on guinea-pigs III. Crosses between highly inbred families. *USDA Bull. #1121*

35. Wright, S. 1923. Two new color factors in the guinea-pig. *Am. Nat.* 57:42–51

36. Wright, S. 1925. The factors of the albino series of guinea-pigs and their effects on black and yellow pigmentation. *Genetics* 10:223–60

37. Wright, S. 1925. Corn and hog correlations. *USDA Bull. #1300*

38. Wright, S. 1926. Effects of age of parents on characteristics of the guinea-pig. *Am. Nat.* 60:552–69

39. Wright, S. 1927. The effect in combination of the major color-factors of the guinea-pig. *Genetics* 12:530–69

40. Wright, S. 1934. The results of crosses between inbred strans of guinea-pigs differing in number of digits. *Genetics* 19:537–51

41. Wright, S. 1934. Polydactylous guinea pigs: Two types respectively heterozygous and homozygous in the same mutant gene. *J. Hered.* 25:359–62

42. Wright, S. 1934. Physiological and evolutionary theories of dominance. *Am. Nat.* 68:25–53

43. Wright, S. 1954. Summary of patterns of mammalian gene action. *J. Nat. Cancer Inst.* 15:837–51

44. Wright, S. 1968. *Evolution and the Genetics of Populations, Vol. 1. Genetic and Biometric Foundations.* Chicago: Univ. Chicago Press

45. Wright, S. 1969. *Evolution and the Genetics of Populations, Vol. 2. Genetic and Biometric Foundations.* Chicago: Univ. Chicago Press

46. Wright, S. 1977. *Evolution and the Genetics of Populations, Vol. 3. Genetic and Biometric Foundations.* Chicago: Univ. Chicago Press

47. Wright, S. 1977. *Evolution and the Genetics of Populations, Vol. 4. Genetic and Biometric Foundations.* Chicago: Univ. Chicago Press

48. Wright, S., Chase, H. B. 1936. On the genetics of the spotted pattern of the guinea-pig. *Genetics* 21:758–78

49. Wright, S., Eaton, O. N. 1929. The persistence of differentiation among inbred families of guinea-pigs. *USDA Techn. Bull. #103*

50. Wright, S., Lewis, P. 1921. Factors in the resistance of guinea-pigs to tuberculosis with especial regard to inbreeding and heredity. *Am. Nat.* 55:20–50

51. Wright, S., Loeb, L. 1927. Transplantation and individuality differentials in inbred families of guinea-pigs. *Am. J. Pathol.* 3:251–85

James F. Crow

Ann. Rev. Genet. 1987. 21:1–22

POPULATION GENETICS HISTORY: A PERSONAL VIEW

James F. Crow

Genetics Department, University of Wisconsin, Madison, Wisconsin 53706

CONTENTS

INTRODUCTION

Space limitations dictate that this review be selective, and I have restricted it to the mainline, classical theory. The choice of subjects is arbitrary; they are topics that I think are interesting and historically important.

It is customary, even de rigueur, to point out the great contributions of Haldane, Fisher, and Wright. Indeed, they dominated the field for thirty years and converted it into a new scientific discipline, with mathematical theory, broad generalizations, and quantitative predictions. This review is dedicated to the proposition that the three pioneers constructed a remarkable foundation,

0066-4197/87/1215-0001$02.00

but that the edifice itself is still under construction and the foundation, for all its strength, needs some shoring up.

Various aspects of population genetics have been frequently reviewed in this and related publications (17, 19, 24, 46–50, 53, 80, 81, 97). Provine (75) has written a history of the early years.

MENDEL, HARDY, AND WEINBERG

The beginnings of genetics and of population genetics are one. Both started with Mendel (60); and of course both were unrecognized until the rediscovery in 1900.

In his classic paper Mendel considered the consequences of repeated self-fertilization, a natural line of inquiry since his peas were normally self-fertilized. Mendel showed that heterozygosity is reduced by half each generation and gave formulas for genotypic frequencies in successive generations, starting with an F_1 population derived from two homozygous strains.

Curiously, Mendel did not consider the consequences of random mating, again perhaps because of the breeding habit of his peas. The first to solve this problem, and receive credit for it, was the distinguished British mathematician G. H. Hardy (31). Although the principle is trivially simple, it is nevertheless the foundation for theoretical population genetics. It has two aspects: (a) If there are no genetically determined fitness differences, no migration, no mutation, and no random fluctuations, the allele frequencies do not change from generation to generation. (b) With random mating, the array of diploid genotypes is given by the binomial expansion of the square of the gametic array.

The first aspect is a truism: if no factors are present that change allele frequencies, they don't change. But there is the useful corollary that genotype frequencies can change while allele frequencies do not, as with inbreeding and assortative mating. The second principle is the one that made possible the rapid development of diploid population genetics theory. Zygotes are constructed anew each sexual generation, and when multiple loci are considered, essentially every zygote is unique. Allele frequencies, on the contrary, are relatively stable. This permits the great simplification of regarding allele frequencies as the fundamental quantities for evolutionary change. Since the binomial square principle (or multinomial if there are multiple alleles) relates zygotes to gametes, it does not matter whether the gametic frequencies are genes, linked clusters, chromosomes, or entire gametes. An enormous body of data supports the idea that most gene loci in most diploid populations conform to this principle well enough to make it one of great utility.

It is not surprising that a principle as simple as this has been rediscovered several times. Special cases were published soon after the rediscovery of

Mendelism, and doubtless many geneticists understood the principle without thinking it worth publishing. I once asked Sewall Wright when he had first heard of the Hardy law. His answer was that he had never thought otherwise; he had used the idea before he ever read the Hardy paper.

One other person should be specifically mentioned, Wilhelm Weinberg (82). Weinberg, a German physician, published the binomial square principle the same year as Hardy (86). But, in distinction to Hardy, who did no further work in population genetics, Weinberg went on to many other discoveries. Most important for population genetics were his extension of the principle to multiple alleles and multiple loci, his attempts to reconcile Mendelian inheritance with quantitative traits, his determining the correlation between relatives, and his taking environment into account (87, 88). Like Mendel's, his work went unrecognized for many years. His papers were hard for British and American geneticists to understand, only partly because they were written in German. By the time his work was recognized in the English-speaking world, it had been superseded, especially by Fisher. Weinberg also made basic contributions to the methodology of human genetics. There is at least some justice, for his name is enshrined in genetic history; the binomial square principle is now called the Hardy–Weinberg law.

EARLY THEORISTS, 1912–1918

Before 1918 British geneticists were embroiled in an acrimonious dispute as to whether Mendelism and biometrical genetics were compatible (75). Curiously, and fortunately, the dispute did not carry across the Atlantic. Most early American geneticists—East, Shull, Castle, Wright, Sturtevant, Muller—assumed from the beginning that continuously varying traits were determined by the cumulative effects of Mendelian factors whose individual effects were too small, or too obscured by environmental influences, to be measured.

Most of the early theoretical work in the United States dealt with the effects of inbreeding. The early leader was H. S. Jennings (32–34; see also 11). The pattern, followed by him and others at the time, was to work out arithmetically the genotype frequencies for a few generations and look for a pattern. The danger of such a procedure is obvious. The most egregious faux pas was made by Raymond Pearl (73). Starting with two homozygotes, AA and aa, he correctly calculated that the F_1 are all Aa, and with sib-mating, the F_2 and F_3 are 50% Aa. He then assumed that if two successive generations remained unchanged so would all subsequent generations; ergo, there is no cumulative increase in homozygosity with repeated sib-mating. This conclusion is contrary to the most elementary intuition, and Pearl himself soon corrected his results (74). Nevertheless, such extrapolation was the methodology of the time, and much of the early knowledge of the effects of systematic inbreeding systems and simple selection was gained in this way. This work pattern came to an abrupt halt when Wright (89) used his correlation (path-analysis) method to produce a simple algorithm to compute the inbreeding coefficient and the decreased heterozygosity for any pedigree or mating system, however complex.

Extension to multiple alleles usually involves no difficulties, but the theory becomes enormously more complicated when more than one locus is considered. Even two loci introduce complications. Both Jennings (34) and Weinberg (87) showed that the approach to gametic equilibrium for two loci is not immediate, as is zygotic equilibrium, but is approached asymptotically. After pages of painful struggling Jennings (34) concluded by saying, "The present writer would find it a relief if some one else would deal thoroughly with the laborious problem of the effects of inbreeding on two pairs of linked factors." The challenge was taken up by Robbins (78), who used methods that, although well known to mathematicians, were not part of the knowledge of geneticists at the time. Robbins formulated a measure of linkage disequilibrium, essentially the same as that used today. It was not, however, until much later that Bennett (1) solved the problem for an arbitrary number of loci. Regardless of the number of loci and for any amount of recombination greater than zero, the frequency of a composite gametic type approaches the product of the constituent allele frequencies. For two loci, the rate of approach is equal to the recombination rate, but for more than two the rate is more complicated. These results apply to an infinite population with no selection; in a finite population or with selection, linkage disequilibrium may be generated.

Jennings (33) also considered simple selection. A more detailed study was made in England at the instigation of R. C. Punnett. He enlisted the help of a mathematician, H. T. J. Norton, who worked out formulas for selection of arbitrary intensity over many generations. Norton prepared tables that listed the number of generations required to change the genotype frequencies from one specified frequency to another. The results appeared in Punnett's textbook on mimicry published in 1915 (77).

Thus, by 1918 the simpler problems of random mating, inbreeding, and selection had been solved.

THE THREE GIANTS

From the end of World War I until the 1950s, Haldane, Fisher, and Wright completely dominated the field. Other contributions were miniscule by comparison, and no doubt the prominence and greatness of these three were inhibiting.

Each of the three had other interests. Wright spent the major part of his time on physiological genetics, especially of pigments, in guinea pigs. Fisher almost single-handedly laid the foundations for modern statistics. Haldane's multifarious activities ranged from politics to fiction writing to astronomy, and included many areas of biology. To a substantial degree their work overlapped. Each was concerned with generality; physics was the paradigm. Yet each had his own style, his own way of looking at problems, and his own view of what was most important. Full-length biographies of all three are available (2, 3, 76); for their personal side, see also (8–10).

J. B. S. Haldane

Haldane started by continuing the studies of Norton. He asked how rapidly the genetic composition of a population would change, given the relative fitnesses of different genotypes. Many of his results were qualitatively predictable, but Haldane quantified them. For example, one would expect an inverse relationship between rate of change and selection intensity, but Haldane demonstrated that for weak selection the number of generations required for any specified change in the composition of the population is proportional to the reciprocal of the selection intensity. He also considered strong selection and showed that the replacement of a light-colored variety of the peppered moth by a more heavily pigmented one as industrial smoke permeated the British landscape required selection intensities of about 50%.

Haldane also showed, for a continuously variable trait, the relationship between the selection differential (the difference between the mean of the selected group and the population average) and the selection intensity (the proportion selected). He noted that most people overestimate the differential for intense selection; for example, saving 1% of the population produces a differential only 51% greater than saving 10%. This relationship has been of great value in livestock breeding by helping the breeder decide how to optimize selection for multiple objectives in species with limited reproductive potential.

One of my favorite Haldane papers (26) introduced the idea of representing a population by a point in an *n*-dimensional cube, in which the edges represent allele frequencies. Selection moves the population along a trajectory to an equilibrium. Of particular interest are alleles that are favorable in some combinations but unfavorable in others. In a simple two-locus example, used by Haldane, genotypes *aa bb* and *A- B-* are favored by selection, while *A- bb* and *aa B-* are less fit. The situation is illustrated in Figure 1. (The axes are reversed in Haldane's book. His papers are models of clear exposition; but there are frequent errors, I suspect because he worked very rapidly and relied too heavily on an extraordinary, but not quite perfect, memory.)

There is an unstable internal equilibrium, but from any other starting point

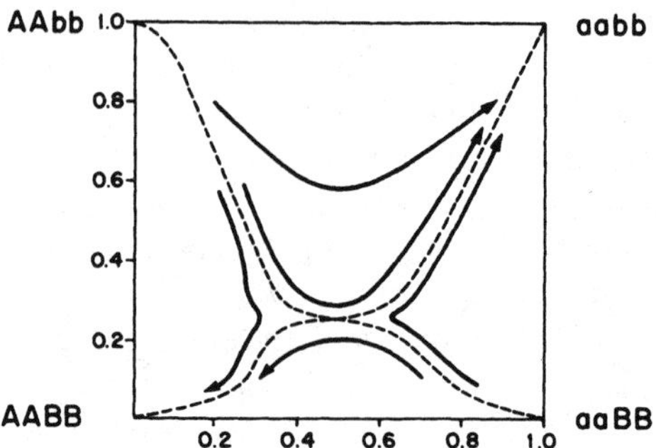

Figure 1 The trajectories of gene frequency change. The abscissa is the frequency of allele *a*, the ordinate that of *b*. The relative fitnesses of the four phenotypes are: *A- B-*, 1; *aa B-*, 1 − *s*; *A-bb*, 1 − 4*s*; *aa bb*, 1 + 11*s*. The borders between the four zones are indicated by dotted lines. There is an unstable equilibrium where these lines meet. Modified from Haldane (27).

(and from this one if there is a displacement for any reason) the population will go to fixation of one of the two favored genotypes. The important point is that if *A- B-* is less fit than *aa bb* and the initial population is mainly *A- B-*, the population cannot get to the more fit situation where *aa bb* predominates. Mass selection of loosely linked genes does not guarantee increased population fitness, a point given great emphasis by Sewall Wright in his shifting balance theory. Curiously, Haldane never included mean population fitness as another dimension in his graph; otherwise he would have anticipated Wright's peaks-and-valleys metaphor.

Haldane's early work is summarized in an appendix to his book, *The Causes of Evolution* (27), from which references to individual papers can be obtained.

Haldane's most original papers, I believe, were those on the effect of variation in fitness (genetic load) and the cumulative cost of an allelic substitution (cost of selection). In his genetic load paper (29) Haldane demonstrated that the effect of recurrent mutation on population fitness does not depend on the deleteriousness of the individual mutations, but only on their rate of occurrence (and on their interaction). For independently acting recessive mutations, the load is simply the gametic mutation rate and for partially or completely dominant mutations, twice this value.

The Haldane load principle was later discovered independently by Muller (67). Haldane also showed that the load involved in maintaining an overdominant locus (the segregation load) is much larger than the mutation load for a classical locus with intermediate or complete dominance. This observation seemed to offer a way to distinguish between the importance of loci of the two types, leading to a controversy in the 1960s that I have reviewed elsewhere (12). The main difficulty with load theory in setting a realistic limit on the total mutation rate or number of segregating overdominant loci is that truncation selection, by permitting multiple mutants to be eliminated by one genetic death, renders the principle much less restrictive (15, 44, 61, 84). It may be possible to use load theory for human mutation risk assessment and at the same time eliminate some of its faults by measuring the "mutation component" (13).

Haldane's genetic load is a measure of the loss of fitness due to static variation—recurrent mutation, heterosis, epistasis, and environmental mismatches. It is the cost of maintaining the status quo. The second Haldane principle, the cost of selection, is dynamic; it is the cost of changing (30). This principle offers at least a partial quantification of the fact, painfully apparent to the breeder, that there cannot be simultaneous strong selection for several independent traits. The cost of a favorable gene substitution has the property, similar to that of the mutation load, of being independent of the magnitude of the change. Haldane showed that the cumulative loss of fitness, compared to that of the population after the favorable allele has been fixed, is a function of its initial frequency. Haldane suggested that, as a typical example, to substitute one allele every 300 generations would require a reproductive excess of about 10%. This, he thought, would place some sort of limit on the rate of evolution. As with the load principle, Haldane did not really come to grips with multi-locus interactions. (For a review of load and cost principles, see 5.)

R. A. Fisher

Fisher augmented his already great mathematical and statistical skills with a remarkable creative faculty. His work has an elegance not found in the writings of either Haldane or Wright. On each rereading of *The Genetical Theory of Natural Selection* (22) I am again impressed by his clever insights, often tossed off as throw-away lines in his book. Fisher's first great contribution was reconciling Mendelian inheritance with biometry, showing that the correlations between physical measurements of close relatives were well in accord with Mendelism (20), which the early biometricians had denied. Weinberg had already realized that deviations from perfect correlation with the proportion of shared genes could be caused either by dominance and epistasis or by environmental effects. Fisher went further and showed that dominance contributes to the sib correlation but not to that of parent and offspring. In principle, this permits dominance and environmental effects to be separated. Fisher undoubtedly erred, however, in attributing the greater sib than parental correlation in height entirely to dominance rather than to the greater environmental similarity of sibs. Although Fisher's 1918 paper is a masterpiece and one of the foundation stones for quantitative genetics, it was an anticlimax in so far as its main purpose was concerned. By the time the paper appeared—it was unfavorably reviewed and its publication was delayed more than two years—Mendelism had already been widely accepted as a sufficient explantion of metrical traits, especially in the United States, where it had never been seriously questioned. But the mathematical foundations of genetic analysis of quantitative traits were laid.

Fisher placed great emphasis on the variance-conserving property of Mendelism. He noted that, with blending inheritance and random mating, variance is reduced by half each generation, thus requiring (as Darwin realized qualitatively) an enormous input of new variability. With Mendelism, on the contrary, a small amount of mutation is sufficient to maintain a large standing variance in all but very small or inbred populations. The problem is the reverse of what Darwin thought. Instead of a problem in accounting for the amount of observed variability, there is the opposite problem of deciding the relative importance of several factors, each of which is sufficient. Fisher also argued forcefully for evolutionary gradualism, noting that the larger the effect of a mutant gene the greater its chance of being highly deleterious and therefore of no long-time evolutionary consequence. In this belief in Darwinian gradualism—that large evolutionary changes were the accumulation of small changes, and not a single saltation—Fisher was joined by Haldane and Wright. On this point their agreement was emphatic.

Fisher (21) showed for the first time the principle, now regarded as obvious, that selection in favor of the heterozygote leads to a stable gene-frequency equilibrium. He also showed how to include the accidents of gene-sampling that occur during reproduction in finite populations, although, contrary to Wright, he placed little emphasis on this.

Haldane's writings for the most part ignored epistasis. Although his views were not necessarily so simple, his mathematical work on evolution dealt mainly with the successive substitution of favorable mutations. Fisher, in contrast, specifically took dominance and epistasis into account. His great contributions were realizing that natural or artificial selection for any character acts on the additive, or genic, component of the genetic variance of that character, and showing how to extract this component by various measures, such as correlations between relatives. In nature, where the all-important character is fitness, this is summarized by his Fundamental Theorem of Natural Selection: "The rate of increase in fitness of any organism at any time is equal to its genetic variance in fitness at that time." By genetic variance he means additive or genic variance. Fisher made it clear that this is a measure of fitness increase due to gene-frequency change relative to conditions at the time. Most such change is offset by overcrowding and changes in the environment, and there may be deviations from the theorem caused by changes in heterozygosity, linkage relations, and age structure, and by frequency-dependent fitnesses. Some of these factors have been introduced explicitly into the theorem by Kimura (38). Kimura's treatment has been widely regarded as clarifying and extending Fisher's theorem. But, characteristically, Fisher regarded his original statement as complete and not in need of either clarification or extension.

To me, this idea initiated by Fisher is a remarkable synthesis. It is obvious that the rate of evolutionary change depends on the amount of variability, but it is not obvious that the appropriate measure of variability is the mean square deviation, or variance (a name that Fisher himself coined in his 1918 paper). It is also not obvious that the additive component of the variance, extracted by least squares estimation, yields that variance component that is responsive to selection.

Leigh (47, 48) has written an insightful review of Fisher's evolutionary theory.

Some Extensions of Fisher's Fundamental Theorem

There are three interesting extensions of Fischer's theorem. The first, due to Robertson (79), states that the rate of change of a character correlated with fitness is the additive genetic correlation of this trait with fitness. It permits predictions from selection on traits other than fitness itself from knowledge of the correlation of such traits with fitness (16).

The second extension is a remarkable property discovered by Kimura (40; 14, p. 217 ff). Fisher (20) had shown that although dominance variance does not contribute to parent–offspring correlations, epistatic variance does. When there is epistasis, selection generates linkage disequilibrium. Kimura showed, however, that the degree of disequilibrium soon reaches a nearly stable amount as gene frequencies slowly change. When this stage is reached the measure of disequilibrium is opposite in sign to the epistatic variance, and these two factors approximately cancel, leaving only the genic or additive variance as the measure of fitness change. Natural selection produces just enough linkage disequilibrium to balance the epistatic variance. Thus, in long-continued directional selection, natural selection manages to select on the basis of average effects of genes, not on their interactions. The most important exception to this rule is strong epistasis combined with close linkage. In this case the linked cluster behaves almost like a single gene. Intermediate values of linkage and epistasis require a more detailed treatment.

The third extension is a Haldane-like idea. The major defects in the Haldane cost of selection as a rate-limiting factor in evolution are that it is measured as a deviation from ultimate fitness and that it fails to include properly the effects of epistasis. It has seemed to me that there is another way of determining the extent to which genetic variability limits evolutionary progress (6). This equation is like the Haldane cost because it sums the effect over several generations and like Fisher's theorem because it is measured in

CROW

terms of variances rather than means. We ask for the amount of variance required to increase the mean fitness by an amount S. Using the Fisher theorem, and letting w stand for fitness and V_g for genic variance, we can write an approximate formula for the change in mean fitness in one generation:

$$\Delta \bar{w} = V_g, \text{ and therefore } S = \Sigma \Delta \bar{w} = \Sigma V_g.$$

So to change the mean fitness by amount S requires a total variance of this same amount, regardless of the time required. The change can be rapid, with the variance used up quickly, or slow if it is distributed over many generations.

Assume that 10^4 loci are evolving at a rate roughly that of hemoglobins, or 2.5×10^{-6} per generation. Assume further that each substitution confers a selective advantage of 0.01, and by the principle above involves a total genic variance of 0.01. The total genic variance required to evolve at this rate is $10^4 \times 2.5 \times 10^{-6} \times 10^{-2} = 2.5 \times 10^{-4}$. This is a very small amount. The genic variance for viability alone in *Drosophila* is about 0.01 (65). There appears to be more than enough additive variance to account for observed evolution rates. Although the example I have given is one in which alleles are independently substituted, the principle still holds if there are interactions.

Sewall Wright

Sewall Wright's shifting-balance theory grew out of his studies on guinea pig coat color, his observations of the effects of inbreeding, and his analysis of the history of domestic livestock. Coat-color genetics convinced him that epistatic interactions are common and complex. Gene combinations often give results that would not have been predicted from the single gene effects, and that could render selection toward a particular color ineffective. His studies of inbreeding showed that the different strains acquired different characteristics, as a result of random fixation of alleles that were segregating in the foundation stock. Finally, he studied the history of domestic animals, especially Shorthorn cattle, for which extensive records were available. He astutely noticed that the overall improvement of the breed took place, not by selection within herds, but between herds. A particular herd would turn out to have exceptionally fine animals, seemingly largely by chance rather than by selection within the herd. From such herds bulls were exported and by repeated backcrossing other herds were "graded up" to the quality of the imported bulls.

Wright was especially impressed by epistatic combinations such as those I mentioned earlier in which unfavorable individual components lead to an improvement in combination. Yet there is no way to reach this endpoint by mass selection. Wright thought of this in his now-famous peaks-and-valleys metaphor. For two loci, each with two alleles, we can think of each allele frequency combination as a point in a unit square in which the scales on the sides are the allele frequencies at the two loci. This square is the same as that on which Haldane plotted the trajectories shown in Figure 1. If we take this unit square and erect at each point an ordinate proportional to the mean fitness of a randomly mating population with this combination of allele frequencies we have a Wrightian fitness surface. [Provine (76) has argued that the meaning of Wright's abscissas is fuzzy; and indeed Wright's early writings are not explicit on this point. It is clear, however, that the most transparent and useful interpretation is the one I have just given, in which the abscissas are allele frequencies, as in Haldane's diagram, and more recently Wright has adopted this interpretation. Wright tells me that although he read Haldane's paper, he had thought of the peaks-and-valleys metaphor quite independently.]

The highest peak occurs at the upper right, with a lower one at the lower left. There is a saddle point where the zone borders meet. From the trajectories we see that a population near the lower left corner tends to move toward that corner. Unless the gene frequencies somehow jump past the boundary into the upper right zone, the population cannot reach the higher fitness peak at the upper right corner. To Wright the essential dilemma is how a population can evolve from one harmonious gene combination to a better one, when intermediate combinations are discordant.

One possibility is for the genes to be closely linked, but nature cannot count on favorable combinations happening often to be linked. The best solution, according to Wright, is a population structure that permits a certain amount of local differentiation. If there are many subpopulations there is a chance that one of them will randomly drift into a high frequency of a favorable gene combination. This population will then grow more rapidly than the others and send out migrants, in the manner of bulls being exported from outstanding herds of cattle (or perhaps African bees moving through Central America). In this way Wright visualizes the gradual upgrading of the species as a whole. Then, after subsequent subdivision, the process can be repeated.

Fisher and Wright therefore arrive at opposite conclusions regarding the kind of population in which evolutionary progress is optimized. For Fisher, dominance and epistasis are facts of life, but decrease the rate of response to environmental changes, since natural selection acts on the additive component. Natural selection is most effective in a large panmictic population where variability is great and random noise is minimal. If the environment remains constant, the population is not far from the optimum combination of allele frequencies at any one time. But the environment will not remain constant, if for no other reason than that competing species continue to change. If the environment changes, a large population has sufficient variability to adapt quickly to the new conditions. Fisher thought that environmental changes are so ubiquitous that, as he once said, Wright's peaks and valleys are more like the undulating wave crests and troughs of an ocean than a mountainous landscape. He believed that a population rarely, if ever, finds itself in a position where *no* allele frequency change could increase its fitness.

For Wright, gene interaction is not simply a nuisance impeding the rate of allele frequency change; rather, it is a part of evolutionary creativity. In his view, epistatic hangups are common enough to constitute a major barrier to progress under mass selection. Only some means by which favorable combinations can be put together will solve this dilemma. Wright's theory calls for a rather delicate balance of migration rates, selection differences, and local deme sizes. The between-group selection that occurs when a favorable combination arises must not permit much mixing of groups, lest recombination destroy the very combinations that are the basis for the one group's superiority. But Wright does not argue that his process happens regularly or often, only that important evolutionary advances depend on such a situation.

There the argument stood in the 1930s, and there it stands today. Fisher held to his views until his death in 1962. Wright's views have remained essentially unchanged; compare his 1931 paper (90) with one of his latest (97). In his later years Wright wrote four volumes summarizing his, and much other, work (93–96). He was 97 in December 1986, and as recently as March 1987 sent a long manuscript to Provine defending his fitness surface diagrams.

Who Was Right?

We must remember that Fisher was mainly interested in how an organism can increase its fitness, or rather how it can maintain its fitness position on the treadmill of a continuously worsening environment. He didn't especially care what biological structures or processes this involved; the important trait is fitness, however it is achieved. Wright, on the contrary, is interested in evolutionary progress or novelty, not simply fitness adjustment. They may both be right.

My own view is that the large area of agreement between Fisher and Wright has had insufficient emphasis. They were in complete agreement about the importance of polygenic inheritance and individually small effects. They agreed that a great deal of natural selection consists of keeping up with changes in the environment. Both thought that much of evolution depends on quantitative traits, and that such traits usually have an intermediate optimum.

It was clear to both Wright and Fisher that in a panmictic sexual population, selection acts on the additive component of the genetic variance (genic selection). An asexual population, in which selection can act on the total genetic variance (genotypic selection), could well have an important advantage. Yet both regarded the ubiquity of sexual reproduction with its attendant Mendelian inheritance as an argument that Mendelism is of great evolutionary benefit. One could describe Wright's model, with its combination of intragroup random drift followed by intergroup (genotypic) selection, as a way of bringing some of the benefits of asexual reproduction to a sexual species.

Recent research has demonstrated that genes with very small effects tend to have incomplete dominance and little epistasis; in particular this is true for those affecting fitness (7, 66). Polygenic inheritance has the evolutionarily useful property of having a large potential variance with a small standing variance. Selection toward an intermediate phenotype, especially when the fitness function is concave, maintains substantial variance on the primary scale with relatively little on the fitness scale. If the optimum value changes because of a changed environment, the population has a large additive variance by which to move the mean to the new optimum. That this mechanism exists in nature and is sufficient for much of fitness evolution is, I think, clear. It is possible that the extreme epistasis that Wright found for guinea pig coat colors is not typical of genes, probably mostly regulatory, having minor effects on fitness. I regard the ubiquity of sexual reproduction as an argument that a great deal of evolution depends on genes that are roughly additive. Wright and Fisher agreed that much of evolution, such things as keeping up with changing environments and evolutionary fine-tuning, is of this type.

If there is to be evolutionary fine-tuning, two conditions are required (52). One is a certain amount of continuity; if a little bit of something causes a phenotypic change in a certain direction, a little more should ordinarily continue in the same direction. The other is a certain independence; it must be possible to adjust one character without adversely affecting others. In these regards, organisms are not like computers where an error in one part often upsets the whole thing. Carrying computer analogies into biology has, I believe, mistakenly led some to doubt the sufficiency of mutation and natural selection as a mechanism for evolution (62). Clearly nonindependence and discontinuities occur, but perhaps those sexual organisms that have survived the evolutionary struggle are those in which these impediments have not been overwhelming. Those species that have opted for an asexual mode of reproduction may be preserving a particularly happy gene combination that would be broken up by recombination.

So, it seems to me, we know that most adaptive changes, adjustment to environmental changes, and fine-tuning occur under the Fisher model. One must not forget some coadapted complexes, locked into an inversion or very tightly linked, and often polymorphic, as Fisher himself emphasized. But in most sexual species this must be a minor part of fitness adaptation. The question remains whether evolutionary novelty requires something more, as Wright believes. As long ago as 1932, Haldane (28) said:

> It would seem that we must envisage the possibility that there are two rather different types of evolution. The first, primrily studied by the paleontologist, is that of dominant species in a fairly stable environment. Such species change slowly by the gradual spread of genes, each with a relatively slight effect. . . . The second type is characteristic of species whose members exist in quite small and nearly or quite isolated groups. Such a group may undergo a cytological change or a change in several genes at once. Such changes, while they must ultimately stand the test of natural selection, are not themselves due to natural selection.

It's a very Wrightian statement.

Wright, although agreeing that the Fisher mechanism may account for most fitness change, believes that the evolution of novel processes and structures requires something more. He notes that major evolutionary innovations have left a weak fossil record, as if they occurred in populations that were sparse— thus favoring local differentiation. A strong argument in favor of the Wright view is that groups with various rapid evolutionary innovations, such as the mammals, have also had a large number of cytogenetic changes, such as translocations. If such cytogenetic changes occurred in a large panmictic population, there would be selection against individuals heterozygous for the arrangement. The result argues for the importance of some sort of random process in the evolution of novel types.

It may be that this is too big a question to be answered any time soon. We need to know not only what is the optimum structure, but what is the actual structure. This will depend on detailed case-by-case analysis. But we can count on more and better information about actual population structures and about the extent of gene interaction. One difficulty is that Wright's arguments become verbal, rather than mathematical, just at the point where the theory becomes controversial. We can surely expect modern developments in mathematics and computer technology to lead to a theory that is deeper, more rigorous, more powerful, and more predictive. Then perhaps we shall have an adequate test of Wright's appealing theory.

TWO NEWCOMERS IN THE 1950s

Haldane, Fisher, and Wright dominated the field of population genetics for the first half of the twentieth century. But in the early 1950s things began to change. Some new names appeared on the scene.

Gustave Malécot was a distinguished French mathematician who brought a new level of mathematical power to population genetics. He published his work in journals that, although not always obscure, were not ordinarily read by population geneticists; furthermore, he wrote in French. The combination of linguistic and mathematical difficulties made his work inaccessible, even to those who were aware of it. His thesis, published in 1939 (54), follows the pattern of Fisher's 1918 analysis of correlation between relatives. It is notable for attention to assumptions and details that Fisher glossed over.

I first became acquainted with his work through a short 63-page book, *Les Mathématiques de l'Hérédité*, published in 1948 (56; for an English translation see 58). I still remember the excitement it brought. It had a clarity and logic that I had not seen before in problems of relationship and inbreeding. Although I was already familiar with Malécot's simpler results because of Wright's work, his concept of identity by descent permitted a clear and elegant derivation of all the standard formulas. With his formulation it was easy to introduce mutation into the process, thus measuring the consequences of finite populations over long times. He also introduced the coefficient of kinship *(parenté)*, the probability that two alleles, one drawn at random from each of two individuals, are identical by descent. For the study of natural populations this probability is more suitable than Wright's earlier coefficient of relationship (89). The coefficient of kinship has permitted more realistic models of migration, isolation by distance, and the relationship between kinship and distance. It has been a most useful supplement to Wright's procedures based on correlation analysis and has been widely used (59, 71).

Malécot introduced the idea of population transformations as a Markov chain, which brought in the tools of modern probability analysis (55, 57). He was also influential in developing ways of studying population structure. In particular he found a simple relation between geographical distance and kinship. This formulation has been useful in the study of human populations, especially in the work of Morton and his colleagues (64).

The concept of identity by descent and a probabilistic approach to problems of inbreeding and relationship were independently introduced by Cotterman (4) in an unpublished thesis. By the time it became known, however, the concept had already been presented by Malécot. Furthermore, Cotterman did not consider geographical structure, the subject in which Malécot's methods have been particularly useful.

I may have had some influence in bringing Malécot's work to the attention of English-speaking geneticists by introducing his notion of identity by descent to a summer statistical conference at Iowa State University in 1952 and suggesting him as a speaker for the 1955 Cold Spring Harbor Symposium. Newton Morton, who was associated with me at the time we both first encountered Malécot's work, has been particularly effective in bringing his contributions to the attention of human population geneticists.

Motoo Kimura also first gained recognition in the English-speaking world in the early 1950s. His first paper published in *Genetics* (35) was a thorough analysis of random drift created by variation in selection coefficients, a problem on which Wright had also worked. I saw the manuscript and was greatly excited by it, as was Wright. Kimura had found a transformation that converted the equation into a form similar to that for heat conduction, and therefore it could be readily solved. Soon after, he showed the distribution through time of allele frequencies in a finite population (36). Wright (92) had shown earlier that his gene frequency distributions could be expressed in terms of the Kolmogorov forward equation. He had solved only the equilibrium distribution. Kimura then proceeded to work out the distribution through time from any arbitrary starting point. The work of Malécot, and especially of Kimura, attracted attention of mathematicians such as Samuel Karlin, and the result has been a much higher level of mathematical sophistication in the field. Another mathematician who became interested in population genetics about this time was Moran (63).

There is not space to review Kimura's voluminous contributions; some of the early work has been summarized (43). But I will emphasize one part. Kimura was especially creative in using the Kolmogorov backward equation. The forward equation gives the distribution of allele frequencies at some future time, given the starting values. The backward equation gives the distribution of starting values for a specified end result. With appropriate identification and boundary conditions, Kimura was able to compute, with greater accuracy than before, the probability of ultimate fixation of a mutant (37). Although it attracted little interest at the time, this computation turned out to be beautifully preadapted for the study of molecular evolution.

MOLECULAR EVOLUTION

The rich stochastic theory of population genetics has until recently suffered from the fact that it deals with allele frequencies, whereas the observations are on phenotypes, often complex ones. Molecular studies, first at the protein level and then direct DNA determinations, have totally changed this imbalance. The data applicable to this theory have been made available. The subject has changed from one that is rich in theory and poor in data to one that is almost the opposite.

The discoveries that there is much more DNA than is required for the expected number of proteins and that the bulk of the DNA has no obvious function have called into question the primacy of selection in determining molecular evolution and polymorphism. The first measures of isozyme variability made it appear that a large fraction of loci were polymorphic and that average heterozygosity was high, perhaps 40% or more (51). This observation immediately raised the possibility that a large segregation load was involved. Several people suggested that one way out of this apparent dilemma was provided by the hypothesis that much of the variability was essentially neutral. However, increasing data (72) have shown that the early extrapolations were too high and that the average level of heterozygosity per locus is in the vicinity of 5 to 10%, reasonable on a mutation-selection balance model.

The first suggestion that considerable molecular evolution was caused by mutation-driven neutral substitutions came from Freese (23) and Sueoka (83). Both were impressed by the large differences in base ratios in different bacterial species despite similar amino acid makeup, and independently suggested that biased mutation rates and neutrality, together with synonymy of the code, could account for the differences. Recent support of this idea has come from the more extreme base-ratio differences in noncoding and synonymous nucleotides (68).

The random drift hypothesis was first suggested by Motoo Kimura (41). His first paper used the Haldane cost principle to argue that the total amount of DNA substitution was so great as to produce an unacceptable fitness reduction. Quite independently the neutral idea was suggested by King & Jukes (45). In the years since that time the argument has been strongly defended by Kimura. In the process the theory has been greatly refined (42).

The key quantity determining the rate of molecular evolution is the probability of ultimate fixation of a new mutant. The first person to work on this problem was Haldane (25), who, using a method suggested by Fisher (21), showed that the probability of fixation of a new mutant with selective advantage s is approximately $2s$. This was later refined by Fisher (22) and Wright (90) to give the value $2s/(1 - e^{-4Ns})$, where N is the population number.

The formula was further generalized by Wright (91), Malécot (57), and finally by Kimura (37, 39), for any level of dominance, any initial frequency, and fluctuating selection coefficients, and taking into account the likely difference between census population number, N, and effective population number, N_e. For a partially or fully dominant gene the probability is $[1 - \exp(-2N_e s/N)]/[1 - \exp(-4N_e s)]$ or, when s is small and $N_e s$ is one or more, approximately $2s(N_e/N)$. For a recessive the value is $\sqrt{(2s/\pi N)}$. For a neutral allele ($s = 0$), the value becomes $1/2N$, as would be expected.

Kimura's work is much more than a refinement of the formula. He was the first to perceive that the relevant quantity for molecular evolution is the probability of fixation, not the time required for fixation of an individual mutant. For, if we view the process over a time scale that is long relative to the time required for an individual fixation, we need only count the number of substitutions that happen during the period; the time required for each is irrelevant.

The neutral theory has generated a great deal of controversy and a plethora of theoretical and experimental work. This is not the place to attempt a review. It is abundantly clear, however, that evolution, as observed at the nucleotide level, is following rules different from those of morphological and physiological evolution. The neutral hypothesis has opened up a whole new area for theoretical and experimental work. The connection between molecular, fitness, and morphological evolution remains to be clarified.

SOME TRENDS IN POPULATION GENETICS

Population genetics theory in the past has been dominated by attempts to find the broadest generalizations. This viewpoint especially characterized the theories of Wright, Fisher, and Haldane. Much of the experimental work, when it was related to theory at all, had similar broad aims; as a result it was often inconclusive. Recent trends, both in theory and experiment, have been toward a more detailed, case-by-case analysis.

One trend is toward a steady improvement in rigor. Mathematicians such as Karlin, Moran, Watterson, Ewens, and Nagylaki, following the trends started by Malécot and Kimura, have changed the character of the field. The study of inheritance and evolution of quantitative traits is still based largely on methods of Wright and Fisher. Some improvement in rigor was introduced by Malécot (54), but there is much more to be done. Characters with intermediate optima are of great importance, but the role of mutation and selection in determining phenotypic variance is not fully understood. According to one model, the fitness is reduced in proportion to the square root of the mutation rate; according to another it is proportional to the mutation rate. A major step toward a resolution of this seeming paradox was provided by Turelli (85), who showed that the two models represented opposite ends of a continuum of necessary assumptions about mutation rates. Another area where work at the very foundations is needed is in geographical structure. Still another is the accuracy of diffusion approximations, especially as the allele frequency approaches zero or one.

Another kind of work that seems particularly important to me is a careful assessment of the accuracy of the simple formulas now in standard use. The idealized models under which the formulas were derived are never exactly true. The question is: how robust are these formulas to departures from the idealized assumptions? The leader in studying this is Nagylaki (69, 70). His work has been reassuring in showing that standard equations, derived from the simplest assumptions (e.g. discrete generations, multiplicative fertility), are excellent approximations to a more realistic situation. More importantly, he has shown how to determine the magnitude of error involved.

Finally, there is a whole new theory being developed for the newer findings of molecular biology and their bearing on evolution. One example is the study of multigene families. Another is research on transposable elements. These are currently active fields of investigation as the theory is developed to fit the rapid growth of new facts.

FINAL REMARKS

My concluding remarks are very similar to those expressed by Ewens (18, p. 33). The "golden age" of population genetics was the period when Haldane, Fisher, and Wright were producing their great work. They reconciled biometry with genetics, quantified the approach to evolution, and created a totally new science. It was arguably the most successful mathematical theory in biology.

The period around 1960 marked the end of the beginning. Like the early theories in physics, population genetics in the golden age was rough and ready. By modern mathematical standards it lacks precision and rigor. We can expect that the foundations will be solidified. A likely result is that this will prove that what we think we know is correct, and that is good in itself, but there will be surprises. There are also new developments in the theory, carrying it in new directions. Some of this will deal with special items—mitochondrial evolution, meiotic drive, multigene families, transposons, viruses. These need to be integrated into the overall theory, with the objective of a deeper understanding of the relative importance of these compared to classical processes as evolutionary forces.

Molecular biology has enormously enriched the field. So has the neutral theory, by bringing in a revolutionary new concept. No longer can one treat molecular evolution without stochastic considerations.

The connections between molecular, fitness, physiological, and morphological evolution are now obscure. A major task for the future is to fit these together. As experimental techniques improve, I think we can also count on a closer coordination of experiment with theory than has characterized the field in the past.

In his 1932 book (27), Haldane closed his mathematical appendix with this statement: "The permeation of biology by mathematics is only beginning, but unless the history of science is an inadequate guide, it will continue, and the investigations here summarized represent the beginning of a new branch of applied mathematics." He was right.

ACKNOWLEDGMENTS

For many fruitful discussions over the years and for interesting ideas, several of which have been incorporated into this article, I am indebted to Carter Denniston and Tom Nagylaki. This is contribution number 2931 from the Laboratory of Genetics, University of Wisconsin.

Literature Cited

1. Bennett, J. H. 1954. On the theory of random mating. *Ann. Eugen.* 18:311–17
2. Box, J. F. 1978. *R. A. Fisher: The Life of a Scientist.* New York: Wiley. 512 pp.
3. Clark, R. W. 1968. *JBS: The Life and Work of J. B. S. Haldane.* New York: Coward-McCann. 326 pp.
4. Cotterman, C. W. 1940. *A calculus for statistico-genetics.* PhD thesis, Ohio State Univ., Columbus, Ohio
5. Crow, J. F. 1970. Genetic loads and the cost of natural selection. In *Mathematical Topics in Population Genetics,* ed. K. Kojima, pp. 128–77. Berlin: Springer-Verlag
6. Crow, J. F. 1976. The theory of neutral and weakly selected genes. *Fed. Proc.* 35:2083–86
7. Crow, J. F. 1979. Minor viability mutants in *Drosophila. Genetics* 92:s165–72
8. Crow, J. F. 1982. Sewall Wright, the scientist and the man. *Persp. Biol. Med.* 25:279–94
9. Crow, J. F. 1984. The founders of population genetics. In *Human Population Genetics: The Pittsburgh Symposium,* ed. A. Chakravarti, pp. 177–94. New York: Van Nostrand
10. Crow, J. F. 1985. J. B. S. Haldane—an appreciation. *J. Genet.* 64:3–5
11. Crow, J. F. 1987. Seventy years ago in *Genetics:* H. S. Jennings and inbreeding theory. *Genetics* 115:389–91
12. Crow, J. F. 1987. Muller, Dobzhansky, and overdominance. *J. Hist. Biol.* In press
13. Crow, J. F., Denniston, C. 1981. The mutation component of genetic damage. *Science* 212:888–93
14. Crow, J. F., Kimura, M. 1970. *An Introduction to Population Genetics Theory.* New York: Harper & Row. 591 pp. Reprinted 1977. Minneapolis: Burgess
15. Crow, J. F., Kimura, M. 1979. Efficiency of truncation selection. *Proc. Natl. Acad. Sci. USA* 76:396–99
16. Crow, J. F., Nagylaki, T. 1976. The rate of change of a character correlated with fitness. *Am. Nat.* 110:207–13
17. Ewens, W. J. 1977. Population genetics theory in relation to the neutralist–selectionist controversy. *Adv. Hum. Genet.* 8:67–134
18. Ewens, W. J. 1979. *Mathematical Population Genetics.* Berlin: Springer-Verlag. 325 pp.
19. Felsenstein, J. 1976. The theoretical population genetics of variable selection and migration. *Ann. Rev. Genet.* 10:253–80
20. Fisher, R. A. 1918. The correlation between relatives on the supposition of Mendelian inheritance. *Trans. R. Soc. Edinburgh* 52:399–433
21. Fisher, R. A. 1922. On the dominance ratio. *Proc. R. Soc. Edinburgh* 42:321–41
22. Fisher, R. A. 1930. *The Genetical Theory of Natural Selection.* Oxford: Clarendon. 272 pp. Revised ed. 1958. New York: Dover. 291 pp.
23. Freese, E. 1962. On the evolution of base composition of DNA. *J. Theor. Biol.* 3:82–101
24. Gillespie, J. H. 1986. Rates of molecular evolution. *Ann. Rev. Ecol. Syst.* 17:637–65
25. Haldane, J. B. S. 1927. A mathematical theory of natural and artificial selection. Part V. Selection and mutation. *Proc. Cambridge Philos. Soc.* 23:838–44
26. Haldane, J. B. S. 1930. A mathematical theory of natural and artificial selection. VIII. Metastable populations. *Proc. Cambridge Philos. Soc.* 27:137–42
27. Haldane, J. B. S. 1932. *The Causes of Evolution.* New York: Harper. 235 pp. Reprinted 1966. Ithaca, NY: Cornell Univ. Press
28. Haldane, J. B. S. 1932. Can evolution be explained in terms of known genetical facts? *Proc. 6th Int. Congr. Genet.* 1:185–89
29. Haldane, J. B. S. 1937. The effect of variation on fitness. *Am. Nat.* 71:337–49
30. Haldane, J. B. S. 1957. The cost of natural selection. *J. Genet.* 55:511–24
31. Hardy, G. H. 1908. Mendelian proportions in a mixed population. *Science* 28:49–50
32. Jennings. H. S. 1914. Formulae for the results of inbreeding. *Am. Nat.* 48:693–96
33. Jennings, H. S. 1916. Numerical results of diverse systems of breeding. *Genetics* 1:53–89
34. Jennings, H. S. 1917. The numerical results of diverse systems of breeding, with respect to two pairs of factors, linked or independent, with special relation to the effects of linkage. *Genetics* 2:97–154
35. Kimura, M. 1954. Process leading to quasi-fixation of genes in natural populations due to random fluctuations of selection intensities. *Genetics* 39:280–95

36. Kimura, M. 1955. Solution of a process of random genetic drift with a continuous model. *Proc. Natl. Acad. Sci. USA* 41:144–50

37. Kimura, M. 1957. Some problems of stochastic processes in genetics. *Ann. Math. Stat.* 28:882–901

38. Kimura, M. 1958. On the change of population fitness by natural selection. *Heredity* 12:145–67

39. Kimura, M. 1962. On the probability of fixation of mutant genes in a population. *Genetics* 47:713–19

40. Kimura, M. 1965. Attainment of quasi linkage equilibrium when gene frequencies are changing by natural selection. *Genetics* 52:875–90

41. Kimura, M. 1968. Evolutionary rate at the molecular level. *Nature* 217:624–26

42. Kimura, M. 1983. *The Neutral Theory of Molecular Evolution.* Cambridge: Cambridge Univ. Press. 367 pp.

43. Kimura, M., Ohta, T. 1971. *Theoretical Aspects of Population Genetics.* Princeton: Princeton Univ. Press. 219 pp.

44. King, J. L. 1967. Continuously distributed factors affecting fitness. *Genetics* 55:483–92

45. King, J. L., Jukes, T. H. 1969. Non-Darwinian evolution. *Science* 164:788–92

46. Krieber, M., Rose, M. R. 1986. Molecular aspects of the species barrier. *Ann. Rev. Ecol. Syst.* 17:465–85

47. Leigh, E. G. 1986. Ronald Fisher and the development of evolutionary theory. I. The role of selection. *Oxford Surv. Evol. Biol.* 3:187–223

48. Leigh, E. G. 1987. Ronald Fisher and the development of evolutionary theory. II. Influences of new variation on evolutionary process. *Oxford Surv. Evol. Biol.* In press

49. Lewontin, R. 1967. Population genetics. *Ann. Rev. Genet.* 1:37–70

50. Lewontin, R. 1973. Population genetics. *Ann. Rev. Genet.* 7:1–17

51. Lewontin, R. 1974. *The Genetic Basis of Evolutionary Change.* New York: Columbia Univ. Press. 346 pp.

52. Lewontin, R. 1978. Adaptation. *Sci. Am.* 239(3):213–31

53. Lewontin, R. 1985. Population genetics. *Ann. Rev. Genet.* 19:81–102

54. Malécot, G. 1939. *Théorie Mathématique de l'Hérédité Mendelienne Généralisé.* Paris: Guilhot. 100 pp.

55. Malécot, G. 1944. Sur un probleme de probabilities en chaine que pose la genetique. *C. R. Acad. Sci.* 219:379–81

56. Malécot, G. 1948. *Les Mathématiques de l'Hérédité.* Paris: Masson. 63 pp.

57. Malécot, G. 1952. Les processes stochastiques et la methode des fonctions generatrices ou characteristiques. *Pub. Inst. Stat. Univ. Paris* 1(3):1–16

58. Malécot, G. 1969. *The Mathematics of Heredity.* Transl. D. M. Yermanos. San Francisco: Freeman. 88 pp.

59. Maruyama, T. 1977. *Stochastic Problems in Population Genetics.* Berlin: Springer-Verlag. 245 pp.

60. Mendel, G. 1866. Versuche über Pflanzen Hybriden. *Verh. Naturforsch. Ver. Brünn* 4:3–47

61. Milkman, R. D. 1967. Heterosis as a major cause of heterozygosity in nature. *Genetics* 55:493–95

62. Moorhead, P. S., Kaplan, M. M., eds. *Mathematical Challenges to the Neo-Darwinian Theory of Evolution.* Philadelphia: Wistar Inst. Press

63. Moran, P. A. P. 1962. *The Statistical Process of Evolution Theory.* Oxford: Clarendon. 200 pp.

64. Morton, N. 1969. Human population structure. *Ann. Rev. Genet.* 3:53–74

65. Mukai, T., Cardellino, R. A., Watanabe, T. K., Crow, J. F. 1974. The genetic variance and its components in a local population of *Drosophila melanogaster. Genetics* 78:1195–208

66. Mukai, T., Chigusa, S. I., Mettler, L. E., Crow, J. F. 1972. Mutation rate and dominance of genes affecting viability in *Drosophila melanogaster. Genetics* 72:335–55

67. Muller, H. J. 1950. Our load of mutations. *Am. J. Human Genet.* 2:111–76

68. Muto, A., Osawa, S. 1987. The guanine and cytosine content of genomic DNA and bacterial evolution. *Proc. Natl. Acad. Sci. USA* 84:166–69

69. Nagylaki, T. 1977. *Selection in One- and Two-Locus Systems.* Berlin: Springer-Verlag. 208 pp.

70. Nagylaki, T. 1987. Evolution under fertility and viability selection. *Genetics* 115:367–75

71. Nei, M. 1975. *Molecular Population Genetics and Evolution.* Amsterdam: North-Holland. 288 pp.

72. Nevo, E., Beiles, A., Ben-Shlomo, R. 1984. The evolutionary significance of genetic diversity: Ecological, demographic and life history correlates. *Lecture Notes in Biomathematics,* 53:1–213. Berlin: Springer-Verlag

73. Pearl, R. 1913. A contribution towards an analysis of the problem of inbreeding. *Am. Nat.* 47:557–614

74. Pearl, R. 1914. On the results of inbreeding a Mendelian population: a

correction and extension of previous conclusions. *Am. Nat.* 48:57–62

75. Provine, W. B. 1971. *The Origins of Theoretical Population Genetics*. Chicago: Univ. Chicago Press. 201 pp.

76. Provine, W. B. 1986. *Sewall Wright and Evolutionary Biology*. Chicago: Univ. Chicago Press. 545 pp.

77. Punnett, R. C. 1915. *Mimicry in Butterflies*. Cambridge: Cambridge Univ. Press. 188 pp.

78. Robbins, R. 1918. Some applications of mathematics to breeding problems III. *Genetics* 3:375–89

79. Robertson, A. 1966. A mathematical model of the culling process in dairy cattle. *Anim. Prod.* 8:95–108

80. Slatkin, M. 1985. Gene flow in natural populations. *Ann. Rev. Ecol. Syst.* 16:393–430

81. Spiess, E. 1968. Experimental population genetics. *Ann. Rev. Genet.* 2:165–208

82. Stern, C. 1962. Wilhelm Weinberg. *Genetics* 44:1–5

83. Sueoka, N. 1962. On the genetic basis of variation and heterogeneity of DNA base composition. *Proc. Natl. Acad. Sci. USA* 48:582–92

84. Sved, J. A., Reed, T. E., Bodmer, W. F. 1967. The number of balanced polymorphisms that can be maintained in a natural population. *Genetics* 55:469–81

85. Turelli, M. 1984. Heritable genetic variation via mutation-selection balance: Lerch's zeta meets the abdominal bristle. *Theor. Pop. Biol.* 25:138–93

86. Weinberg, W. 1908. Über den Nachweis der Vererbung beim Menschen. *Jahresh. Ver. Vaterl. Naturkd. Wuerttemb.* 64:368–82

87. Weinberg, W. 1909. Über Vererbungsgesetze beim Menschen. *Z. Indukt. Abstamm. Vererbungsl.* 1:277–330

88. Weinberg, W. 1910. Weitere Beiträge zur Theorie der Vererbung. *Arch. Rass. Gesellschaftsbiol.* 7:35–49

89. Wright, S. 1922. Coefficients of inbreeding and relationship. *Am. Nat.* 56:330–38

90. Wright, S. 1931. Evolution in Mendelian populations. *Genetics* 16:97–159

91. Wright, S. 1942. Statistical genetics and evolution. *Bull. Am. Math. Soc.* 48:223–46

92. Wright, S. 1945. The differential equation of the distribution of gene frequencies. *Proc. Natl. Acad. Sci. USA* 31:382–89

93. Wright, S. 1968. *Evolution and the Genetics of Populations. 1. Genetic and Biometric Foundations*. Chicago: Univ. Chicago Press. 469 pp.

94. Wright, S. 1969. *Evolution and the Genetics of Populations. 2. The Theory of Gene Frequencies*. Chicago: Univ. Chicago Press. 511 pp.

95. Wright, S. 1977. *Evolution and the Genetics of Populations. 3. Experimental Results and Evolutionary Deductions*. Chicago: Univ. Chicago Press. 613 pp.

96. Wright, S. 1978. *Evolution and the Genetics of Populations. 4. Variability Within and Among Natural Populations*. Chicago: Univ. Chicago Press. 580 pp.

97. Wright, S. 1982. The shifting balance theory and macroevolution. *Ann. Rev. Genet.* 16:1–19

Annu. Rev. Immunol. 1991. 9:1–26

465

WHEN ALL IS SAID AND DONE . . .

Baruj Benacerraf

Department of Pathology, Harvard Medical School, and
Dana-Farber Cancer Institute, 44 Binney Street, Boston,
Massachusetts 02115

KEY WORDS: training, Ir genes, MHC restriction, antigen processing, reticulo-
endothelial system

Abstract

Dr. Benacerraf describes his experience in training immunologists for a
period of over 30 years. He then analyzes the circumstances in which
selected discoveries were made in his laboratory and the events and reasoning that led to these findings. The following topics are discussed: The
phagocytic activity of the reticulo-endothelial system. The pathogenesis of
immune complex diseases. The impact of activated macrophages on tumor
rejection. The evidence that T-cell immunity is specific for internal protein
determinants generated by antigen processing. The maturation of the
antibody response in terms of antibody affinity. The discovery of Ir gene
control of specific immune responses and the role of MHC molecules as
presenters of peptide antigens to T cells. The generation of alloreactivity.

Lastly, he addresses the administration of science and, particularly, the
problems related to the transfer of technology to industrial partners.

INTRODUCTION

My professional life[1] has focused on three activities which by necessity are
interrelated: the training of immunologists and the teaching of immun-

---

[1] In 1985, at the request of Goran Moller, I wrote an autobiographical contribution entitled
"Reminiscences" (1), in which I described the significant events in my life as a scientist and
a businessman, and chose as a major scientific theme the description of those discoveries
which, because of accidents of fate or interpretation, I failed to make, rather than those to
which I was fortunate to contribute. There is still much ground left to cover here.

ology, basic research, and the administration of a department and of an Institute. I propose therefore to discuss here:

1. My experience in training immunologists for a period of over 30 years, documented by a chronological list of all the trainees with whom I had the pleasure and privilege to interact in the laboratory, and their present positions.
2. The circumstances in which selected discoveries were made in our laboratory, with an emphasis on the contributions of my junior associates and the events and reasoning that led to these findings.
3. My involvement in professional scientific societies, my experience as an administrator, particularly in immunology in a medical school, and my thoughts and conclusions concerning: technology transfer, relationships with industry, and the need for clear ethical norms of behavior that preserve scientific integrity in the midst of valuable industrial relationships; and the future development of immunology.

TRAINING AND TEACHING

It is not fortuitous that I chose to write first about training: It has been my most rewarding and enriching experience as a scientist. In fact, my strong feelings in this respect and my thoughts and conclusions concerning the training of young scientists were the theme of my presidential address to the American Association of Immunologists in 1974 (2); these have not changed since.

My experience with the training of younger scientists began in 1956 as assistant professor of pathology at New York University, to which I had just been recruited by Lewis Thomas. Jeannette Thorbecke was my first postdoctoral fellow. Beginner's luck! Jeannette, who made a brilliant career, completed her term as president of the American Association of Immunologists in 1990.

I am tempted to describe individually and in detail the many accomplishments of the exceptional group of men and women it was my good fortune to be associated with over the 30 years during which I have headed my laboratory, but the space allotted me here would not suffice to do them justice. Instead I list them in Table 1, for the record—their names, the dates of their stay in my laboratory, and their present positions.

Some statistics concerning the group of immunologists who trained in my laboratory are informative. Fifty-four graduate trainees (about 64%) earned MD degrees, whereas 31 were awarded PhD degrees. Of the MDs 8 were MD-PhD students. Of the 84 who trained with me, 76 (over 90%) had successful careers in research and teaching of immunology. The remaining 10% went into the practice of medicine or accepted industrial positions. Over these more than 30 years, 38 of my trainees reached the rank of professor or equivalent rank as chief of an independent laboratory. If one takes into account the time required to reach high rank in academia, the numbers are even more significant and impressive. Thus, if we focus on the early period up to 1976, 36 out of 54 (66%) reached the top academic rank. Two from this group, Bill Paul and Lloyd Old were elected to the National Academy of Sciences (USA), and two, Bill Paul and Jeannette Thorbecke, served as presidents of the American Association of Immunologists. Abroad, François Kourilsky, who founded the Institut d'Immunologie de Marseille-Luminy, was recently appointed Directeur du Centre National de la Recherche Scientifique, the largest French government agency responsible for the support of Science. Pierre Vassalli received last year the major Swiss research prize. I am very proud of the achievements of this select group.

It is also noteworthy that 37 of the 87 young immunologists who worked in my laboratory (i.e. 42%) came from abroad. My laboratory has been an international enterprise with many foreign workers. To illustrate this interesting point, I have presented in Table 2 an analysis of the 16 countries of origin of my young associates.

The reader may be interested to know, as an anticlimax to this record and as an illustration of the uncertainty of funding criteria, that the latest competing renewal of my training grant in Immunology from the National Cancer Institute was downgraded, with the criticism that the program proposed had the "characteristics of an apprenticeship system which can be as successful as the mentors and trainees can make it."

In addition to my highly rewarding experience with graduate students and fellows, a set of close and fruitful collaborations with established scientists was important to my work over the years and significantly influenced my evolution as an immunologist. It is a pleasure to acknowledge their contributions. I would like to record in chronological order:

Guido Biozzi, Institut Curie, Paris, retired.
Phillip Gell, University of Birmingham, retired.
Zoltan Ovary, New York University Medical School.
Peter Miescher, University of Geneva.
Robert McCluskey, Massachusetts General Hospital.
James Howard, Wellcome Laboratories.
Paul Maurer, Jefferson Medical School.
Norman Cooper, New York University Medical School.
Gerald Edelman, Rockefeller University.
Emil Unanue, Washington University School of Medicine.
Ron Schwartz, National Institute of Allergy and Infectious Diseases, NIH.

Table 1 Fellows and graduate students

Name	Period	Present position
G. J. Thorbecke	1957–60	Prof. Pathology, New York University
M. M. Sebestyen	1957–69	Prof. Pediatrics, University of Vienna
S. Schlossman	1957–59	Prof. Medicine, Harvard Medical School and Dana-Faber Cancer Institute
F. Gorstein	1959–61	Prof. Pathology, New York University
F. Miller	1960–62	Prof. and Chair. Dept. Pathology SUNY at Stony Brook
L. J. Old	1959–62	Prof. Memorial Sloan Kettering
A. Ojeda	1961–63	Madrid, Spain
B. B. Levine	1962–65	Prof. Medicine, New York University
C. Jenkin	1960–62	Prof. Microbiology, University of Adelaide, Australia
F. Kantor	1961–63	Prof. Medicine, Yale University
R. Binaghi	1963–65	College de France, Paris, retired
K. J. Bloch	1962–65	Prof. Medicine, Harvard Medical School and Mass. Gen. Hospital
V. Nussenzweig	1962–66	Prof. Pathology, New York University
R. Nussenzweig	1963–65	Prof. Parasitology, New York University
C. Merryman	1963–65	Assoc. Prof. Biochemistry, Jefferson Medical School
A. Berken	1965–67	Clin. Prof. Medicine, SUNY at Stony Brook
F. M. Kourilsky	1962–64	Directeur, Centre National de la Recherche Scientifique, Paris
T. Brunner	1965	Institut Ludwig, Lausanne, Switzerland
W. E. Paul	1964–68	Chief, Laboratory of Immunology, National Institute of Allergy and Infectious Disease
Z. Trnka	1964–66	Basle Institute for Immunology, Basle, Switzerland
M. E. Lamm	1965–67	Prof. and Chair. Dept. Pathology, Western Reserve University
K. F. Oettgen	1964–66	Prof. Medicine, Memorial Sloan Kettering
J. Foerster	1966–68	Prof. Medicine, University of Winnipeg
G. W. Siskind	1965–67	Prof. Medicine, Cornell University
P. Frei	1965	Hopital Nestlé, Lausanne, Switzerland
I. Green	1964–68	Senior Scientist, National Institutes of Health
P. Vassalli	1965–67	Prof. Pathology, University of Geneva, Switzerland
H. Spiegelberg	1966	Scripps Clinic and Research Foundation, La Jolla
S. Cohen	1966–68	Prof. and Chair, Dept. Pathology, Hahnemann Univ. Medical School
T. Yoshida	1967–70	Merck Laboratories, Japan
J. P. Lamelin	1967–70	Centre Anti-Cancereux, Lyon, France
C. W. Pierce	1968–70	Prof. Pathology, Washington University
D. Hurez	1967	Universite de Rennes, France
L. Ellman	1968–70	Assoc. Prof. Medicine, Harvard Medical School
A. Amerding	1972–74	Germany
Y. Stupp	1968–70	Weizmann Institute, Israel
D. H. Katz	1968–72	President, Medical Biological Institute, La Jolla
H. Bluestein	1970–72	Assoc. Prof. Medicine, University of California, San Diego
R. M. Williams	1970–74	Prof. Medicine, Northwestern University
W. J. Martin	1969–71	Australia
E. K. Dunham	1972–74	Private Practice of Allergy
P. Debré	1973–75	Prof. Medicine, Paris
J. Kapp	1972–75	Assoc. Prof. Pathology, Washington University
E. Martz	1971–74	Assoc. Prof. Biology, University of Massachusetts

Table 1 (*continued*)

Name	Period	Present position
T. Hamaoka	1972–74	Prof. Osaka University, Japan
M. E. Dorf	1972–76	Prof. Pathology, Harvard Medical School
Z. Eshar	1974–76	Weizmann Institute, Israel
M. Pierres	1975–77	Centre d'Immunologie, Marseille-Luminy, France
A. Pierres	1975–77	Centre d'Immunologie, Marseille-Luminy, France
S. Burakoff	1973–77	Prof. Pediatrics, Harvard Medical School and Dana-Farber Cancer Institute
N. K. V. Cheung	1975–77	Memorial Sloan Kettering Inst.
R. N. Germain	1973–78	Senior Scientist, National Institutes of Health
W. E. Bullock	1974–76	Prof. Microbiology, Oregon Health Science University
A. Soyano	1974–76	IVIC Caracas, Venezuela
M. I. Greene	1976–79	Prof. Pathology, University of Pennsylvania
C. R. Waltenbaugh	1975–78	Assoc. Prof. Microbiology, Northwestern University
Shyr-Te Ju	1976–78	Assoc. Prof., Boston University
J. Théze	1976–78	Directeur, Institut Pasteur, Lyon
F. Lemmonier	1976–79	Centre d'Immunologie, Marseille-Luminy, France
A. Dessein	1977–79	Centre d'Immunologie, Marseille-Luminy, France
P. Billings	1976–78	Harvard School of Public Health
T. J. Kipps	1976–78	Scripps Clinic & Res. Foundation
B. A. Bach	1977–79	
L. A. Sherman	1977–79	Assoc. Member, Scripps Clinic and Research Foundation, La Jolla
D. Sherr	1978–80	Assoc. Prof. Pathology, Harvard Medical School
J. Bromberg	1980–81	Dept. of Pathology, University of Pennsylvania
L. L. Perry	1977–80	Asst. Prof. Microbiology, Emory University
N. Gualde	1980–82	Ecole de Medecine de Bordeaux, France
R. Finberg	1977–80	Assoc. Prof. Medicine, Harvard Medical School and Dana-Farber Cancer Institute
J. Z. Weinberger	1977–79	Dept. of Medicine, Columbia University
O. Weinberger	1977–79	Dept. of Physiology, Columbia University
M. M. Dietz	1979–81	Private Practice, Rheumatology
R. H. Zubler	1979–81	Dept. of Medicine, University of Geneva, Switzerland
M. E. Sunday	1979–81	Asst. Prof. Pathology, Harvard Medical School
M. S. Sy	1979–82	Assoc. Prof. Pathology, Harvard Medical School
N. Letvin	1979–82	Assoc. Prof. Medicine, Harvard Medical School
G. T. Nepom	1980–82	Virginia Mason Research Center, Seattle
R. B. Whitaker	1980–82	Assist. Prof. Biology, Bates College
K. L. Rock	1981–85	Assoc. Prof. Pathology, Harvard Medical School
C. F. Scott	1982–84	Private Industry
K. T. Hayglass	1983–86	Assist. Prof. Immunology, University of Manitoba
E. T. H. Yeh	1983–85	Asst. Prof. Medicine, Harvard Medical School and Massachusetts General Hospital
L. D. Falo	1985–88	House Officer, Massachusetts General Hospital
H. Reiser	1986–88	Assist. Prof. Pathology, Harvard Medical School
M. Michalek	1988–	Research fellow, Dana-Farber Cancer Institute
L. Dang	1989–	MD-PhD student, Harvard Medical School
Laurent Vidard	1990–	Graduate student, University of Paris

The list is not yet closed, since there are still three young immunologists in training in my laboratory

Table 2 Countries of origin of my 37 foreign post-doctoral fellows

Australia	2	Hong Kong	1
Argentina	1	Hungary	1
Brazil	2	Israel	2
Canada	3	Japan	2
Czechoslovakia	1	Spain	1
France	11	Switzerland	5
Germany	2	Taiwan	1
Holland	1	Venezuela	1

FOLLOWING THE TRAIL OF SCIENTIFIC DISCOVERY

When interviewing postdoctoral fellows, I have often asked myself what motivates a career in research, a particularly difficult and unpredictable profession, subject to the hazards of creativity and the vagaries of competitive funding? I have concluded that productive scientists are driven by an obsessive curiosity to uncover nature's hidden secrets and to reach the elusive goal of understanding the world around them, rather than by any ambition for power or glory.

After more than 40 years in research and over 600 publications, I have learned that discoveries are determined primarily by chance observations and are conditioned by past experience and advances in technology. To illustrate the basis for these conclusions I shall describe the circumstances in which many of the contributions from my laboratory were made over the years and point out the often invisible thread between them.

The Phagocytic Activity of the Reticulo-endothelial System (1950–1956)

From 1950–1956 I worked in Paris in the laboratory of Bernard Halpern (known for his contributions to the discovery of the antihistamines). Teamed with Guido Biozzi I studied a basic physiological problem, i.e. the development of methodology to measure the clearance of particulate matter from the blood by the phagocytes of the reticulo-endothelial system (RES). Using these techniques, we derived the equations that describe the blood clearance of phagocytizable particles by the liver Kuppfer cells and spleen macrophages (3). In the course of this study several incidental observations set the stage for later studies by ourselves and others. In particular, we noted that experimental infections of mice with live BCG vaccine caused a marked stimulation of the phagocytic system, a phenomenon illustrated by considerable hyperplasia and rapid blood clearance of carbon particles (4). We then demonstrated that activated macrophages were most efficient in combating infection with Gram negative organisms.

We also observed that denatured proteins, insoluble at their isoelectric points, were avidly taken up and digested by macrophages such as liver Kuppfer cells, so that even minute doses could be cleared from the portal circulation in a single passage (5). As a consequence of this finding, we introduced a new test to measure liver blood flow, based on the clearance of heat denatured radiolabelled human serum albumin (denHSA) (6). To my great surprise, I recently learned from radiologists that radiolabelled denHSA is still used in this context.

These studies aroused my lifelong interest in and respect for the macrophage both as a critical defense mechanism and as an important cell of the immune system. The studies also alerted me to the importance of denaturation in determining the fate and treatment of proteins.

Immune Complex Diseases (1957–1960)

When I joined Lewis Thomas's department of pathology at New York University in 1956 and sought grant support, my choice of an experimental problem was determined by my recent experience with the reticulo-endothelial system and my interest in immunopathological mechanisms. Robert McCluskey, a young pathologist with a strong commitment to the study of kidney diseases, occupied the adjacent laboratory. Based on my experience in studying the properties of particles cleared by the RES, I reasoned that the logical fate of soluble immune complexes formed in the circulation should be their phagocytosis by the RES and that the lesions of serum sickness could result from the abnormal deposition of phlogistic immune complexes in kidney glomeruli and arterial endothelia. After demonstrating the clearance of antigen-antibody complexes from the blood by the RES in mice, with Martha Sebestyen (7), I initiated a collaborative project with Bob McCluskey (a) to generate the renal and arterial lesions of serum sickness in mice and rats, by the intravenous administration of preformed immune complexes of ovalbumin and rabbit anti-ovalbumin dissolved in antigen excess, and (b) to identify the deposited complexes in the lesions by the recent Coons technique of immunofluorescence. This approach was successful in reproducing the acute lesions of serum sickness, which could be observed when the animals were sacrificed 24–48 hours after the intravenous injection of the immune complexes. We further demonstrated that the lesions resolved spontaneously if further injury was not inflicted (8–11).

The Effect of Activated Macrophages on Tumor Rejection (1958–1961)

Motivated by a long-standing interest in cancer, Lloyd Old graduated from the University of California Medical School at San Francisco and immediately accepted a position in the division of cancer chemotherapy in the Memorial Sloan Kettering's Rye laboratories. Prompted by a strong interest in the immunological approach to the treatment of tumors, Old called me at New York University to explore the possibility of working together on this project. I was intrigued, challenged, and much impressed by Lloyd's enthusiasm. Because of my earlier experience with the spectacular macrophage activation that resulted from BCG infection in mice, I proposed to explore whether such a treatment would induce the rejection of experimental tumors in mice. We were delighted to observe that mice infected with BCG previous to tumor inoculation showed marked increase in survival after challenge with sarcoma 180 as well as significantly delayed mortality following implantation of Ehrlich ascites tumor intraperitonealy (12). Moreover, a study of the phagocytic activity of mice injected with syngeneic transplantable tumors revealed a considerable spontaneous activation of the RES (13).

Related to these observations, at the time, was our finding that activation of the macrophages such as could be induced by zymosan injection or BCG infection was associated with heightened susceptibility to endotoxin toxicity (14). This result pointed to a critical role of macrophages as the primary target of endotoxin toxicity and the probable source of the mediators of the toxicity phenomena. It is very much to the credit of Lloyd Old that he pursued this trail with tenacity and skill to its ultimate destination; in the process he discovered and characterized tumor necrosis factor (TNF). This macrophage-derived cytokine is now known to be a major mediator of the antitumor effect in mice treated with BCG or endotoxin (15).

Cellular Immune Reactions. First Evidence That Cellular Immunity is Directed to Hidden Internal Determinants Requiring Processing by Antigen Presenting Cells. Critical Role of Macrophages in the Initiation of Specific Immune Responses (1958–1962)

In the late 1950s Philip Gell joined my laboratory as a visiting professor, and for several years he spent nearly six months of each year with us. By that time, following my earlier involvement with the study of anaphylactic reactivity and the Arthus reaction, in Kabat's laboratory, I had become very interested in the pathogenic mechanism of cellular immunity reactions, exemplified then by delayed hypersensitivity. Since Philip had similar interests, we decided to initiate a common project to compare the immunological specificity of cellular immunity with that of humoral antibodies. We felt that the hapten-protein conjugate systems, introduced by Landsteiner and investigated by Chase, were appropriate for such a study. Philip introduced me to the chemistry of several hapten-protein conjugate systems, particularly those involving the trinitrophenyl and the benzenearsonate haptens.

Our experiments were highly informative and indicated fundamental and previously unsuspected differences between the specificities of antibodies produced and cellular hypersensitivity manifested by Hartley guinea pigs immunized with hapten-protein conjugates. Strictly hapten-specific delayed sensitivity could not be demonstrated, whereas antihapten antibodies were readily produced. The specificity of cellular immune reaction was always, at least in part, directed to and conditioned by the carrier protein molecule. Similarly, the specificity of contact reactivity was directed to hapten conjugates of autologous proteins (16, 17). Later, with Zoltan Ovary, we extended the requirement for carrier recognition to eliciting secondary antihapten antibody responses (18).

Essentially, we had rediscovered the critical role of the carrier protein in the initiation of immune responses, an idea contributed originally by Landsteiner (19), and we extended our study to the development of cellular immunity. However, because our experiments were carried out with random-bred guinea-pigs, we failed to appreciate at the time the restricting contribution of the major histocompatibility complex to the specificity of these reactions (20, 21).

Our study of the specificity of cellular immune responses revealed another important and as yet unsuspected character of these reactions, i.e. that cellular immune responses, later identified as T-cell immunity, are directed to hidden determinants in native proteins which for exposure require unfolding and denaturing of the molecule. In contrast the epitopes interactive with antibodies are only expressed by the native folded molecule (22). I do not recall the precise reason why we investigated the specificity of delayed sensitivity reactions of guinea-pigs immunized with native proteins using denatured molecules, although my earlier observation of the avid uptake of denatured proteins by phagocytes may have motivated this study. These experiments were the first to indicate that protein antigens must undergo some form of processing to generate the determinants that initiate cellular immune responses, a theme I return to below. The preferred candidate for such a processing step was our old friend the macrophage. Evidence for the critical role of macrophages in the initiation of immune responses was provided by a study with Frei and Thorbecke. We demonstrated that whereas the intravenous injection of bovine serum albumin (BSA) to rabbits was uniformly immunogenic, BSA preparations from which their readily phagocytizable fraction had been removed by previous biological filtration in a rabbit were no longer immunogenic, but rather tolerogenic (23). This result may be related to the more recent experiments on the relative abilities of macrophages/ dendritic cells vs B cells to initiate T-cell responses, rather than to tolerize these cells.

Correlated with the discovery that protein denaturation and unfolding are required to generate molecules immunogenic for T cells are other observations I made with McCluskey and Miller. The significance of these is much clearer now that the importance of processing of autologous antigens for the induction of clonal deletion in the thymus has been documented. We noted that guinea pigs, unresponsive to their own gamma globulins could nevertheless be readily sensitized to their autologous gamma globulin if it had been previously denatured in vitro (24, 25). In retrospect thymic tolerance is only induced to the epitope generated by natural processing. We are always potentially responsive to our autologous proteins, denatured or modified by other routes; this process may be an important pathogenic mechanism of autoimmunity.

Identification of $\gamma1$ and $\gamma2$ Guinea Pig and Mouse Immunoglobulins. Distinct Biological Activity of These Two Classes of Antibody (1963–1964)

Pursuing my interest in anaphylaxis, which originated in Kabat's laboratory, I initiated a collaborative project in 1962 with my good friend Zoltan Ovary to study the biological properties of the two classes of guinea-pig 7S antibodies.

This study was motivated by a chance observation of my colleague and former fellow Jeannette Thorbecke, who excitedly showed me an immunoelectrophoresis of purified 7S anti-DNP antibodies I had prepared. This immunoelectrophoresis indicated that the antibodies could be separated into two populations with distinct electrophoretic mobility and different antigenic determinants.

Kurt Bloch and François Kourilsky carried out most of the experiments that established that the ability of guinea-pig 7S antibodies to sensitize for anaphylaxis was the property of $\gamma1$ antibodies, whereas the fixation of complement and the lysis of cells were exclusively the properties of $\gamma2$ antibodies (26–28). These results revealed, for the first time, that different biological properties should be assigned to distinct immunoglobulin isotypes, and the results led to the later demonstration that the heavy chains and particularly the Fc fragments contained sequences responsible for their biological properties (29).

With Ruth Nussenzweig and Carmen Merryman, we soon extended these findings to mouse antibodies and similarly separated 7S murine antibodies into $\gamma1$ responsible for anaphylaxis and $\gamma2$ complement fixing immunoglobulins (30).

The FC Receptor (1966)

Building on my long interest in macrophages and our recent finding of the distinct biological activities of different classes of immunoglobulins, I undertook to identify the antibodies described by Boyden as having cyto-philic properties for macrophages. These experiments which resulted in the original description of the Fc receptor (31) were carried out by Arthur Berken and my wife, Annette. We investigated the ability of different classes of murine anti-sheep red cell antibodies to form rosettes with live pulmonary macrophages in tissue culture chambers at room temperature. The ability to adhere to macrophages was clearly the property of 7S $\gamma2$ rather than $\gamma1$ antibodies and resided in the Fc fragment of the immuno-globulins. Moreover, we documented that the stage of adherence to live macrophages, which was stable at room temperature, was followed by immediate phagocytosis, visible under the microscope, if the preparation was heated to 37°C. The ability to adhere to macrophages was later observed also with IgM antibodies. My friend James Howard, who was a visiting professor at the time, and I made a productive study of the properties of the macrophage's Fc receptor (32).

Maturation of the Antibody Response and the Increase in Antibody Affinity with Time After Immunization (1967–1969)

During our work with antihapten antibodies, which are appropriate can-didates for the measurement of affinity, I became acutely aware of and curious about the phenomenon of increase in antibody affinity with time following immunization. With Bill Paul and Gregory Siskind we noted that this phenomenon was very much dependent on the initial immunizing dose. Above a minimum threshold amount, the increase in affinity as well as the rate of increase were found to be inversely related to the immunizing dose of antigen. Low immunizing doses resulted in the rapid production of the highest affinity antibodies. Based on the clonal selection theory of Burnet (33) for which we provided experimental evidence (detailed in the next section), we reasoned that the increases of antibody affinity reflected the continuous selection by antigen of those cells with the higher affinity antibody receptors (34). Under conditions of continuous antigen catabolism and decreasing concentration, the B cells with the highest affinity receptors would be the only ones expected to bind antigen. Moreover the antibody they would secrete should be immunosuppressive for cells with antibody receptors of lower affinity. I might note in passing that it was not until more than a decade later that we came to understand that these early findings were actually based on the efficient and specific uptake of antigen by the B-cell immunoglobulin receptors and the MHC-restricted presentation of the processed peptide to the helper T cell.

In a review with Gregory Siskind, we documented the evidence for cell selection by antigen in the immune response and came to the conclusion, in 1969, that somatic mutation of immunoglobulin genes in differentiated B cells was most probably an important component of this efficient process (35). I was gratified in the mid-1980s when the molecular biological evidence of somatic mutation of immunoglobulin genes in B cells was developed.

Experimental Evidence for Burnet's Clonal Selection Theory (1967)

In 1967 the experimental evidence concerning the clonal selection theory was controversial. Contrasting with the elegant original data of Nossal (36), a detailed study by Attardi, Lennox, and Cohn (37) reported a large number of double antibody-producing cells in rabbits immunized with two phages. As a result, immunologists developed considerable doubts about cellular clonality. We felt that the experimental approaches previously used were open to errors and that such errors would result most often in the identification of double producers. Moreover, technologies by then available used a combination of immunofluorescence and historadiography to identify the specificity of antibodies produced by individual plasma cells in animals immunized with hapten-protein conjugates. With Ira Green and Pierre Vassalli, we were delighted to find that the antibody-producing cells of animals immunized with DNP-BSA or DNP-BGG produced antibodies specific for either the hapten or the carrier molecule, but never reactive with both, in spite of the presence of both determinants on the immunizing antigen (38, 39). At about the same time Makela came to the identical conclusion by a different approach (40).

At the first Cold Spring Harbor Symposium in Immunology in 1967, Sir Macfarlane Burnet expressed his satisfaction that his theory had been definitely proven. The stage was set for the development of the useful clonal antibody technology by Kohler & Milstein (41).

Discovery of Ir Gene Control of Specific Immune Responses

THE INITIAL EXPERIMENTS (1963–1967)   The discovery in 1963 of Ir gene control of specific immune responses in guinea pigs was the result of a chance observation made possible by advances in technology. The availability of synthetic polymers and copolymers of L amino acids provided the immunologists with antigens that were immunogenic for T cells and yet possessed restricted heterogeneity and a limited number of determinants. Poly-L-lysine(PLL) was available in the laboratory because of Bernie Levine's interest in allergy to penicillin and because PLL, presumed nonimmunogenic, was an ideal carrier to conjugate with penicillin and to test for clinical hypersensitivity.

I decided to immunize Hartley guinea pig with DNP-PLL conjugates in the expectation that the anti-DNP antibody produced would be of restricted heterogeneity. This was not the case, but I was intrigued by the finding, which we made with Fred Kantor and Antonio Ojeda, that DNP-PLL was a strong antigen for about 40% of the random bred guinea pigs and was not immunogenic for the remaining animals (42). I had the intuition that these differences were probably determined by autosomal dominant genes. I thus initiated an analysis, with Bernie Levine, of the genetic control of the response to DNP-PLL, first in Hartley guinea pigs and later in the two inbred strains of guinea pigs, 2 and 13, then available from the NIH. The two NIH inbred strains turned out to be, respectively, responders and nonresponders to DNP-PLL. Breeding guinea pigs is a long and tedious process, as they have a long gestation period and small litters. Both Bernie Levine and I became acutely hypersensitive to guinea pigs, a small price to pay for the success of this project (43, 44). We concluded that the responses to both DNP-PLL and to the copolymer of L-glutamic acid and L-lysine (GL) were under the control of the same dominant autosomal gene and that the gene was not concerned with degradation of the antigens.

An early analysis, carried out with Ira Green and Bill Paul, of the process controlled by the DNP-PLL gene, revealed that the gene was responsible for the recognition of PLL as a carrier molecule and as an initiator of immune responses to any hapten it bore. In addition, we found that the process controlled by the gene was intimately involved with the elicitation of specific T-cell responsiveness (45). At the Cold Spring Harbor Symposium in Immunology, in 1967, we concluded that "Most probably, the PLL gene controls an initial process required for the subsequent induction of the immune response, and the stimulation of specifically sensitized cells. A limited number of such genes, specific for the processing of different antigens with different amino acid sequences, may indeed exist. Their activity could only be detected when polyamino acids of relatively simple structure are used" (46).

LINKAGE OF IR GENES TO THE MAJOR HISTOCOMPATIBILITY COMPLEX (MHC) (1969–1971)   The availability of inbred strains of mice, generated by transplantation geneticists, led to the rapid mapping of Ir genes in this species.

The linkage of murine Ir genes to H-2, observed by McDevitt & Chinitz in 1969 (47), should be considered the major advance opening the way to understanding the identity of Ir genes and MHC molecules and the biological function of MHC molecules as presenters of peptide antigens to the T cell receptor.

We immediately extended this critical finding to guinea pigs and the systems we had been studying and thus demonstrated the linkage of guinea-pig Ir genes to MHC (48, 49). We verified that responsiveness of random bred animals was determined by and could be predicted upon the possession of distinct histocompatibility specificities shared with the inbred responder strains (50).

MHC RESTRICTION OF ANTIGEN PRESENTATION AND T-B CELL INTERACTION (1973–1974)   The next major advance in our understanding of Ir gene function stemmed from the demonstration by several laboratories of the restriction imposed by the MHC on antigen presentation to T cells and on T-B cell interactions essential for the generation of immune responses.

These discoveries were also very much the result of accidental observations. Investigating the proliferative responses in vitro of primed (2×13)F1 T cells from animals immunized with the antigens under Ir gene control, Rosenthal and Shevach observed that parental macrophages only from responder strains could present antigen to immune F1 T cells (20, 51). Moreover, anti–class II MHC antibodies were able to block the specific antigen presentation as well as macrophage–T cell interaction.

Similarly, in our own laboratory, while attempting to transfer immune responsiveness to irradiated F1 mice with primed parental B and T cells, Katz, Hamaoka and I were amazed to discover that class-II MHC molecules restricted the T and B cell interactions required for secondary antibody responses (52, 53). In a more recent study with Abbas & Rock (54), which demonstrated the very efficient role of specific B cells as antigen presenters to primed T cells, we concluded that MHC restriction of antigen presentation is also the basis of the MHC restriction of T-B cell interaction.

Independent of the observations of Rosenthal and Shevach and of our own contributions, Zinkernagel and Doherty explored the specificity of cytolytic T cell activity in mice and noted that virus-specific cytolytic T cells could only lyse virally infected targets that shared class-I MHC specificities with the killer cells (21).

It took some time, however, for all of us to realize that these various observations reflected the same fundamental MHC dependence of antigen presentation, and the effects of allelic variations on such presentation, that are the basis of Ir gene phenomenology.

IR GENES CODE FOR MHC MOLECULES. EVIDENCE FROM COMPLEMENTING IR GENES (1975)   The definitive evidence that MHC molecules are Ir gene products resulted from: (*a*) the structural and genetic analysis of class-II murine MHC molecules which revealed the two chain structure of the two major classes of murine Ia molecules (55) and identified the genes in I-A and I-E that code for these chains; and (*b*) the demonstration by Martin Dorf and me that the complementing Ir genes needed for the response of mice to the terpolymer of L-glutamic acid, L-lysine and L-phenylalanine (GLPhe) mapped in the same manner as the genes for the alpha and beta chains of the I-E antigen (56, 57).

HYPOTHESIS DESCRIBING THE SPECIFIC INTERACTION BETWEEN PROCESSED PEPTIDE ANTIGENS AND MHC MOLECULES REQUIRED FOR INTERACTION WITH THE SPECIFIC T CELL RECEPTOR (1978)   By 1978, 15 years had elapsed since the initial description of specific Ir genes, and 9 years since the demonstration of their linkage to MHC. Enough information had been generated to permit a theoretical treatment of antigen processing and presentation. Combining the insight from our earlier experiments and Unanue's data (which demonstrated the essential role played by macrophages in antigen processing and presentation) with the recently discovered role of the MHC molecule in antigen presentation, these together led me to an inescapable conclusion, i.e. that MHC molecules needed to bind processed antigen specifically prior to the interaction of the MHC-antigen complex with the T cell receptor (58). A similar model was independently offered by Rosenthal at the same time (59). We both proposed that this interaction was the basis of Ir gene phenomena and also of determinant selection for T cells.

Although Kenneth Rock and I provided biological evidence, based on the phenomenon of antigen competition, in support of our hypothesis of the specific interaction between processed antigen and MHC molecules (60), our ideas were initially received with considerable skepticism on the part of MHC geneticists such as Jan Klein. Eventually, direct molecular evidence for such specific peptide-MHC interaction was provided, by Unanue and associates (61) in 1985, supplemented by crystallographic analysis of the structure of human HLA molecules, a few years later (62).

The Generation of Alloreactivity (1976–1978)
The Expansion of T Cells Specific for Autologous Class-II MHC Antigens Following Interaction with Thymic Adherent Cells (1984)

The commitment of large numbers of T lymphocytes to reactivity with MHC allogeneic specificities without the need of previous immunization, as manifested by the MLR reaction, has puzzled immunologists for many years. It was difficult to reconcile such alloreactivity with the clonal reactivity of individual T lymphocytes to foreign protein antigens in association with self-MHC.

We formulated the hypothesis that alloreactivity is the inevitable consequence of the selective expansion in the thymus of T cells committed to reactivity with self-MHC antigens and of the deletion of those clones specific for autologous proteins in the context of self-MHC antigens. Accordingly, with Steven Burakoff, Ron Germain, François Lemonnier, and Bob Finberg, we undertook to investigate the fine specificity and cross-reactivity of cytolytic T cells. When mouse spleen cells were stimulated with xenogeneic, allogeneic, or TNP-modified syngeneic lymphoid cells, the strongest cytolytic responses were induced by alloantigens. Moreover, mouse cytolytic T cells generated to rat cells exhibited extensive cross-reactive lysis of selected murine strains (63, 64).

In other experiments we observed that immunization of mice with Sendai virus resulted in the generation of effector cells that lysed unmodified allogeneic target cells of selected haplotypes as well as Sendai virus–infected syngeneic cells. Furthermore, the same clones of cells exhibited both forms of reactivities. We concluded that MHC alloantigens mimic foreign epitopes presented in the context of autologous MHC antigens (65) and that alloreactivity is indeed the by-product of the commitment of T lymphocytes to recognize foreign peptides in association with self-MHC molecules.

To explore the commitment of T cells to reactivity with self-MHC molecules in the thymus, we investigated, with Ken Rock and Ed Yeh, whether autologous thymocytes reactive with autologous MHC antigens could be selectively expanded in vitro when cultured with autologous thymic or splenic adherent cells. We found that such thymocytes, reactive with autologous MHC, could be readily expanded (66) and hybridized with thymoma BW5147. A major fraction (60%) of the T-cell hybrids reactive with autologous MHC antigens exhibited alloreactivity with selected individual haplotypes, and a smaller number (10–25%) with several haplotypes (67).

The GAT and GT Suppressor T Cell Systems (1974–1979)

It is a pleasant stage in the evolution of scientific problems when all fundamental issues have been resolved. In the ten previous sections, I discussed contributions we made to fields that have reached the stage of general agreement in the immunological community on the facts and interpretations of the original phenomena.

Other immunological problems, however, not less important, have not been as fortunate nor been resolved definitively. One such topic, to the phenomenology of which my laboratory contributed a number of papers in the 1970s, concerns the suppressor T cell system.

Of late, this problem has become unpopular with immunologists because of the difficulties encountered: (*a*) In characterizing the active molecules responsible for these phenomena, and (*b*) In identifying the mechanism of specific suppression at the level of T or B cells.

I must claim my share of the blame for failing to put into the study of this problem the energy or determination needed to bring it to some satisfactory conclusion. In part this was due to the weight of administrative responsibilities of the last ten years, which has forced me to cut my research program by more than half. Under these circumstances, I have elected to focus the activity of my laboratory in the last decade on the timely issues of antigen processing and presentation, and on T-cell activation, which were easier to address profitably and definitively than was the study of suppressor T cells and their factors.

I have not lost interest in suppressor T cell phenomena which, I feel, are some of the most important unresolved issues in immunology. Moreover, I am convinced that, despite the absence of molecular and genetic data accounting for specific immunological suppression, the wealth of important and fascinating phenomenology involving suppressor T cells cannot be brushed aside as nonexistent or insignificant. The problem deserves to be reexamined in the light of our understanding of modern molecular and cellular immunology.

I have decided, therefore, to review briefly my own laboratory's experience with a fascinating suppressor system concerned with the response of mice of selected H-2 haplotypes to certain random copolymers of L-amino acids such as GAT, GA, and GT. I hoped that I could come to a new assessment of the data, their meaning and significance, and that I might be able to propose a novel experimental approach to offer some solution to this problem.

In 1974, motivated by Gershon's suggestion that Ir gene–controlled unresponsiveness might be the result of suppressor T cell activity, I initiated a study with Kapp and Pierce of the response of inbred mouse strains to GAT. We observed that the lack of response of $H-2^s$ and $H-2^q$ mice to this antigen could be explained by the development of GAT-specific suppressor T cells. Following immunization with the copolymer, the mice produced GAT-specific suppressor T cells that were able to suppress the anti-GAT antibody response of these haplotypes to GAT coupled to an immunogenic carrier, methylated bovine serum albumin (MBSA) (68, 69). In later experiments we demonstrated that GAT-specific suppression could be transmitted by a suppressor factor extracted from lysed spleen or thymus cells from $H-2^s$ or $H-2^q$ mice immunized with GAT. The factor, which bore I-region determinants, could be specifically absorbed and eluted from GAT columns (70, 71).

These findings raised the issue whether H-2 controlled unresponsiveness was always associated with the generation of antigen-specific suppressor T cells. The answer to this question was unequivocal. With Patrice Debré we carried out experiments with the GT copolymer that did not induce antibody responses in any of the inbred mouse strains tested. Mice of the H-$2^{d,f,s}$ haplotypes developed GT-specific suppressor T cells in response to GT, whereas mice of the H-$2^{a,b,q}$ haplotypes did not. Moreover, the generation of GT-specific suppressors behaved as a dominant trait since H-$2^{a/d}$ or H-$2^{q/s}$ mice developed GT-specific suppressor T cells (72). The genes controlling the development of GT-specific suppressor T cells mapped in the I region and behaved as complementing genes in both the *cis* and *trans* configuration (73). As in the case of GAT-specific suppressors, lysis of GT-suppressor T cells yielded a GT-specific factor capable of suppressing GT-MBSA responses (74). In more recent experiments with this system, Vidovic et al investigated the ability of GT to initiate T cell–proliferative responses rather than antibody production in inbred strains of mice; they observed that some of the suppressor haplotypes we had identified, i.e. H-$2^{s,f,d}$, were capable of GT-specific proliferative T cell responses (75).

The results from this series of experiments indicate that the generation of T suppressors bears all the characteristics of having been determined by the H-2 haplotype of the antigen-presenting cell and therefore should require antigen presentation by MHC molecules to T suppressor cells as is the case for conventional helper T cells.

In other experiments with Ron Germain we noted that GAT-pulsed macrophages from nonresponder strains could stimulate primary in vitro responses by unprimed (responder × nonresponder) F1 T cells, but not by GAT-primed F1 T cells. Furthermore, this inability of GAT-primed (b × q) F1 T cells to respond to GAT-pulsed H-2^q macrophages was the result of specific suppression of that particular MHC-restricted response; this effect was not observed if the cells had been challenged with responder H-2^b GAT-pulsed macrophages (76). In addition, we pretreated with small doses of cyclophosphamide previous to priming the F1 animals with GAT (a protocol that we have shown does away with the ability of the mice to generate suppressor T cells in the GAT and GT systems—77). This pretreatment restored the ability of the GAT-primed (H-$2^{b×q}$) F1 T cells to respond to GAT-pulsed H-2^q macrophages.

I interpret these data to indicate that GAT and GT suppressor strains have an easily tolerized immature population of suppressor T cells of these specificities, whose activation is dependent upon a cyclophosphamide-sensitive cell. Because of the findings of Ron Germain and Michel Pierres (78), who demonstrated that spleen cells from mice of responder haplotypes, depleted of adherent cells, develop GAT-specific suppressor T cells in vitro, I suggest that conditions must exist in the nonresponder haplotype strains whereby these antigens bypass appropriate macrophage presentation. Instead, perhaps, presentation is mediated primarily by specialized macrophages or by GAT or GT specific B cells (known to be a very cyclophosphamide-sensitive population), incapable of delivering the secondary signal(s) required for activation of helper T cells. Such conditions should result in the induction of suppressor inducer T cells with GAT or GT specificities.

I have chosen here not to comment on the puzzling issues of the mechanism of action of suppressor T cell factors, the origin of the Ia determinants they bear, or the nature of their antigen specificity, since it is clear that these issues will not be resolved until we have stable clones of suppressor T cells, with easily assayed functional activity in vitro.

THE ADMINISTRATION OF SCIENCE

The need to exercise some control over my environment and to generate the resources to support my research and that of my younger associates prompted me to accept the increasingly onerous administrative responsibility, successively, as chief of the Laboratory of Immunology at NIAID for 2 years, chairman of the Department of Pathology at Harvard Medical School for 20 years, and president of the Dana-Farber Cancer Institute for the last 10 years. I have also served as an officer and president of our scientific societies, AAI, FASEB, and IUIS. I have been pleased to see the importance and world recognition of immunology grow almost exponentially during my lifetime as a scientist and an administrator.

From this varied experience I have gathered thoughts and reached conclusions that might be appropriately discussed here. At first I was concerned that immunology, a discipline to which I was committed, had no administrative home in universities and medical schools, and I shared a sense of insecurity with other immunologists in this respect. For a while I discussed with others the wisdom of creating departments of immunology. I soon realized that the trend in modern science is in the reverse direction. It is increasingly difficult to differentiate between the continuum of biochemistry, molecular biology, and cell biology, sciences to which immunological techniques and approaches have made and continue to make important contributions. In addition the unique relevance of immunology to medicine has created strong ties to pathology, and the clinical disciplines. As a consequence, immunologists have found congenial homes in both basic sciences and clinical departments.

Accordingly, at Harvard Medical School I promoted the development of immunology in many departments besides my own Department of Pathology, through the creation of a Committee for Immunology, responsible for teaching and graduate training. The consequence of this policy has been the development in our institution of a thriving community of immunologists located in both clinical and basic science departments, who foster and benefit from their close interactions. It has been a source of great satisfaction to me that the best and most well attended seminars in the last few years at Harvard Medical School have been those sponsored by the Committee for Immunology.

In addition to the positive developments in immunology with which we are all familiar, the dynamic growth of the biological sciences and the dramatic success of immunology in the last 40 years that span my experience as a scientist have had some unavoidable and unpleasant consequences. When I entered science, and for many years afterward, the conduct of research and the keeping of records were informal and left to individual initiative. The uniform manner of record keeping in official notebooks with numbered pages, now required, was not then in effect. Everyone was assumed to be responsible for his or her own data. As a consequence the laboratories had the pleasant aspect of an entrepreneurial atmosphere. In the last 15 years, under the pressure of rare but highly publicized cases of scientific misconduct, on the one hand, and of the increased awareness of the potential commercial value of inventions on the other hand, very strict and legalistic record keeping has become prevalent in our laboratories.

The emergence of biotechnology as a highly profitable activity focused the interest of investors and drug companies on research laboratories and their scientists in academic institutions. Investors and pharmaceutical companies have become aware of the potential benefits for them that derive from government regulations vesting ownership of research discoveries on the grantee institutions. These policies urge the academic institutions to accelerate the transfer of basic scientific information, generated with the support of public funds, into applicable technological advances for the benefit of mankind. As a consequence of this generous and enlightened governmental policy, an enormous wealth of scientific data, obtained with public support, has become the property of academic institutions. Investors may profit from this wealth of scientific information if they choose to fund research programs designed to develop new agents based on the existing science in these institutions. In return the investors can derive exclusive licensing rights for the products generated with their support.

I do not object to the unavoidable increase of the interaction between academic institutions and industry, which contributes significantly to the research mission of academic institutions. I agree wholeheartedly that our research efforts must extend beyond the limits of pure science to the generation of biotechnologies and their products that are beneficial for mankind. Such a task is best accomplished in close partnership with industry which has the expertise in the generation of practical applications, and which can contribute the required resources. Moreover, at the time when the financial contribution of the government to biomedical research appears to have reached a plateau, under the pressure of other needs and of the federal deficit, academic institutions desperately need the financial contribution of industry to basic research. Such support of basic research by industry can best be negotiated as part of the licensing agreements granted industrial partners.

Two major and serious problems have nevertheless arisen with the escalating relationships of academia with industry. These concern: (*a*) potential conflicts of interest for the faculty involved, and (*b*) the possibility that the basic goals and values of the academic institution's may be adversely affected by the financial attractiveness of these relationships. The increase in industrial relationships has been so rapid and recent that only now are rules and ethical norms of behavior to govern these relationships being debated. A Harvard Medical School committee, on which I served, has labored for nearly a year with these issues. Precise recommendations to the Dean concern new rules governing potential conflicts of interest. It is easily understandable that industrial partners and investors, eager to secure the cooperation of the most gifted researchers as well as the best expert advice, have attracted academic scientists with highly profitable consulting agreements and sometimes have offered them equity positions to secure their services. In an era when greed is not only condoned but sometimes encouraged, if such agreements were not scrutinized very carefully and seriously monitored by appropriate institutional authorities, the temptation to cross the line of propriety would be too great to resist for more than a few.

Strict reporting rules for such agreements are essential, as is a monitoring organization responsible to the highest level of the institution. Moreover, in my opinion, those responsible for the decisions as to the acceptable types of institutional relationships with industry and of personal agreements must themselves be untainted by any form of relationships with industry. Finally, these policies and their implementation need to be seriously scrutinized by the trustees who have, in the last analysis, the fiduciary responsibility for the institution and its faculty.

The concerns voiced here, which have been the result of my experience in managing a large research institution for the past ten years, should not be interpreted to mean that I oppose scientists benefiting financially from their discoveries, if these are marketable. Such benefits, however, are more appropriately derived through the academic institution which has the only legal right to the ownership of the technology, as well as the responsibility for the research. Thus, academic scientists at our institution, and in many others I know, receive for the discoveries they have made an important share of the royalties and licensing fees that are paid by the industrial partner to the institution.

This policy does not, of course, cover the most controversial of issues, i.e. the equity position offered some academic scientists as part of their consulting agreements. It is my view that such equity positions should not be barred, but when they exist, the institution must be careful to ensure that the scientist is neither a grantee of that company nor an administrator in the academic institution whose decision might affect the relationship of the institution with the company. These strict rules, I think, are the essential prerequisites to mutually successful partnership with industry, protecting our integrity as independent scientists.

THE FUTURE

I would like to end with my assessment of the future of immunology. This is a sensitive topic for one who has just focused on the contributions of his laboratory. I must be careful to refrain from Burnet's attitude at the first International Congress of Immunology in Washington in 1971, when he stated that all the important work in immunology had already been done and all major issues resolved. Obviously this was not the case, since many discoveries and fundamental contributions, which have greatly clarified the field, have been made in the last 19 years.

I firmly believe that major problems still exist to be resolved concerning: antigen processing, thymocyte differentiation, tolerance induction, specific T cell suppression, lymphokines, and signal transduction—all of these are clearly appropriate for immunologists. Perhaps an even greater challenge is posed by trying to generate an adequate account of how all the cellular and molecular events are integrated in a coherent whole, i.e. how the system is regulated and how the many effector modalities integrate for an effective immune response to diverse challenges.

At the same time, by 1990, enough fundamental knowledge has been acquired about the immune system, its genes, molecules, and cells that very real opportunities exist in the application of this information to the generation of new therapeutic agents active on a variety of diseases, starting with cancer, AIDS, autoimmune diseases, transplantation, and allergy.

ACKNOWLEDGMENTS

I would like to thank Steven Burakoff, Martin Dorf, Kenneth Rock, and particularly Ron Germain for their valuable editing contribution.

Literature Cited

1. Benacerraf, B. 1985. Reminiscences. *Immunol. Rev.* 84: 7–27
2. Benacerraf, B. 1974. The training experience. Presidential address. *J. Immunol.* 113: 431–37
3. Biozzi, G., Benacerraf, B., Halpern, B. N. 1953. Quantitative study of the granulopectic activity of the reticuloendothelial system. II. A study of the kinetics of the granulopectic activity of the RES in relation to the dose of carbon injected. Relationship between the weight of the organs and their activity. *Br. J. Exp. Pathol.* 34: 441–57
4. Biozzi, G., Benacerraf, B., Grumbach, F., Halpern, B. N., Levaditi, I., Rist, N. 1954. Etude de l'activité granulopexique du systéme réticuloendothelial au cours de l'infection tuberculeuse experimentale de la souris. *Ann. Inst. Pasteur* 87: 291–300
5. Benacerraf, B., Biozzi, G., Halpern, B. N., Stiffel, C., Mouton, D. 1957. Phagocytosis of heat-denatured human serum albumin labelled with ^{131}I and its use as a means of investigating liver blood flow. *Br. J. Exp. Pathol.* 38: 35–48
6. Biozzi, G., Benacerraf, B., Halpern, B. N., Stiffel, C., Hillemand, B. 1958. Exploration of the phagocytic function of the reticuloendothelial system with heat denatured human serum albumin labeled with I^{131} and application to the measurement of liver blood flow in normal man and in some pathologic conditions. *J. Lab. Clin. Med.* 51: 230–39
7. Benacerraf, B., Sebestyen, M., Cooper, N. S. 1959. The clearance of antigen-antibody complexes from the blood by the reticuloendothelial system. *J. Immunol.* 82: 131–37
8. Benacerraf, B., McCluskey, R. T., Patras, D. 1959. Localization of colloidal substances in vascular endothelium. A mechanism of tissue damage. I. Factors causing the pathologic deposition of colloidal carbon. *Am. J. Pathol.* 35: 75–91
9. McCluskey, R. T., Benacerraf, B. 1959. Localization of colloidal substances in vascular endothelium. A mechanism of tissue damage. II. Experimental serum sickness with acute glomerulonephritis induced passively in mice by antigen-antibody complexes in antigen excess. *Am. J. Pathol.* 35: 275–94
10. McCluskey, R. T., Benacerraf, B., Potter, J. L., Miller, F. 1960. The pathologic effects of intravenously administered soluble antigen-antibody complexes. I. Passive serum sickness in mice. *J. Exp. Med.* III: 181–94
11. Benacerraf, B., Potter, J. L., McCluskey, R. T., Miller, F. 1960. The pathologic effects of intravenously administered antigen-antibody complexes. II. Acute glomerulonephritis in rats. *J. Exp. Med.* III: 195–200
12. Old, L. J., Clarke, D. A., Benacerraf, B. 1959. Effect of bacillus Calmette-Guerin (BCG) infection on transplanted tumors in the mouse. *Nature* 184: 291–92
13. Old, L. J., Benacerraf, B., Clarke, D. A., Carswell, E. A., Stockert, E. 1961. The role of the reticuloendothelial system in the host reaction to neoplasia. *Cancer Res.* 21: 1281–1300
14. Benacerraf, B., Thorbecke, G. J., Jacoby, D. 1959. Effect of zymosan on endotoxin toxicity in mice. *Proc. Soc. Exp. Biol. Med.* 100: 796–99
15. Old, L. J. 1985. Tumor necrosis factor. *Science* 230: 630–32
16. Benacerraf, B., Gell, P. G. H. 1959. Studies on hypersensitivity. I. Delayed and Arthus-type skin reactivity to protein conjugates in guinea pigs. *Immunology* II: 53–63
17. Gell, P. G. H., Benacerraf, B. 1961. Studies on hypersensitivity. IV. The relationship between contact and delayed sensitivity: A study on the specificity of cellular immune reactions. *J. Exp. Med.* 113: 571–85
18. Ovary, Z., Benacerraf, B. 1963. Immunological specificity of the secondary response with dinitrophenylated proteins. *Proc. Soc. Exp. Biol. Med.* 114: 72–76
19. Landsteiner, K. 1936. *The Specificity of Serological Reactions.* Springfield, Ill: Thomas. 178 pp.
20. Rosenthal, A. S., Shevach, E. M. 1973. Function of macrophages in antigen recognition by guinea pig lymphocytes: I. Requirement for histocompatible

lymphocytes and macrophages. *J. Exp. Med.* 138: 1194–1212

21. Zinkernagel, R. M., Doherty, P. C. 1975. H-2 compatibility requirement for T cell mediated lysis of target cells infected with choriomeningitis virus: Different cytotoxic cell specificities are associated with structures coded for in H-2-K or H-2D. *J. Exp. Med.* 141: 1427–36

22. Gell, P. G. H., Benacerraf, B. 1959. Studies on hypersensitivity. II. Delayed hypersensitivity to denatured proteins in guinea pigs. *Immunology* II: 64–70

23. Frei, P. C., Benacerraf, B., Thorbecke, G. J. 1965. Phagocytosis of the antigen, a crucial step in the induction of the primary response. *Proc. Natl. Acad. Sci. USA* 53: 20–23

24. McCluskey, R. T., Miller, F., Benacerraf, B. 1962. Sensitization to denatured autologous gamma globulin. *J. Exp. Med.* 115: 253–73

25. Benacerraf, B. 1965. The antigenicity of altered autologous proteins. A mechanism of autoimmune reactions. *Ann. N.Y. Acad. Sci.* 124: 126–32

26. Benacerraf, B., Ovary, Z., Bloch, K. J., Franklin, E. C. 1963. Properties of guinea pig 7S antibodies. I. Electrophoretic separation of two types of guinea pig 7S antibodies. *J. Exp. Med.* 117: 937–49

27. Ovary, Z., Benacerraf, B., Bloch, K. J. 1963. Properties of guinea pig 7S antibodies. II. Identification of antibodies involved in passive cutaneous and systemic anaphylaxis. *J. Exp. Med.* 117: 951–63

28. Bloch, K. J., Kourilsky, F. M., Ovary, Z., Benacerraf, B. 1963. Properties of guinea pig 7S antibodies. III. Identification of antibodies involved in complement fixation and hemolysis. *J. Exp. Med.* 117: 965–81

29. Ovary, Z., Tarenta, A. 1963. Passive cutaneous anaphylaxis and antibody fragments. *Science* 140: 193–95

30. Nussenzweig, R. S., Merryman, C., Benacerraf, B. 1964. Electrophoretic separation and properties of mouse antihapten antibodies involved in passive cutaneous anaphylaxis and passive hemolysis. *J. Exp. Med.* 120: 315–28

31. Berken, A., Benacerraf, B. 1966. Properties of antibodies cytophilic for macrophages. *J. Exp. Med.* 123: 119–44

32. Howard, J., Benacerraf, B. 1966. Properties of macrophage receptors for cytophilic antibodies. *Br. J. Exp. Pathol.* 47: 193–200

33. Burnet, F. M. 1959. The clonal selection theory of acquired immunity. Cambridge: Cambridge Univ. Press

34. Paul, W. E., Siskind, G. W., Benacerraf, B., Ovary, Z. 1967. Secondary antibody responses in haptenic systems. Cell selection by antigen. *J. Immunol.* 99: 760–70

35. Siskind, G. W., Benacerraf, B. 1969. Cell selection by antigen in the immune response. *Adv. Immunol.* 10: 1–50

36. Nossal, G. J. V. 1962. Cellular genetics of immune responses. *Adv. Immunol.* 2: 163–204

37. Attardi, G., Cohn, M., Horibata, K., Lennox, E. S. 1964. Antibody formation of rabbit lymph nodes. I. Single cell responses to several antigens. *J. Immunol.* 92: 335–45

38. Green, I., Vassalli, P., Nussenzweig, V., Benacerraf, B. 1967. Specificity of the antibodies produced by single cells following immunization with antigens bearing two types of antigenic determinants. *J. Exp. Med.* 125: 511–20

39. Green, I., Vassalli, P., Benacerraf, B. 1967. Cellular localization of anti-DNP-PLL and anticonveyor albumin antibodies in genetic nonresponder guinea pigs immunized with DNP-PLL albumin complexes. *J. Exp. Med.* 125: 527–36

40. Makela, O. 1967. The specificity of antibodies produced by single cells. *Cold Spring Harbor Symp. Quant. Biol.* 32: 423–30

41. Kohler, G., Milstein, C. 1975. Continuous cultures of fused cells secreting antibody of predefined specificity. *Nature* 256: 495–97

42. Kantor, F. S., Ojeda, A., Benacerraf, B. 1963. Studies on artificial antigens. I. Antigenicity of DNP-polylysine and DNP copolymer of lysine and glutamic acid in guinea pigs. *J. Exp. Med.* 117: 55–69

43. Levine, B. B., Ojeda, A., Benacerraf, B. 1963. Studies on artificial antigens. III. The genetic control of the immune response to hapten poly-L-lysine conjugates in guinea pigs. *J. Exp. Med.* 118: 953–57

44. Levine, B. B., Benacerraf, B. 1965. Genetic control in guinea pigs of immune response to conjugates of haptens and Poly-L-Lysine. *Science* 147: 517–18

45. Green, I., Paul, W. E., Benacerraf, B. 1966. The behavior of hapten-poly-L-lysine conjugates as complete antigens in genetic responder and as haptens in nonresponder guinea pigs. *J. Exp. Med.* 123: 859–79

46. Benacerraf, B., Green, I., Paul, W. E. 1967. The immune response of guinea pigs to hapten-poly-L-lysine conjugates as an example of the genetic control of

the recognition of antigenicity. *Cold Spring Harbor Symp. Quant. Biol.* 32: 569–75

47. McDevitt, H. O., Chinitz, A. 1969. Genetic control of antibody response: Relationship between immune response and histocompatibility (H-2) type. *Science* 163: 1207–9

48. Ellman, L., Green, I., Martin, W. J., Benacerraf, B. 1970. Linkage between the PLL gene and the locus controlling the major histocompatibility antigens in Strain 2 guinea pigs. *Proc. Natl. Acad. Sci. USA* 66: 322–28

49. Bluestein, H. G., Ellman, L., Green, I., Benacerraf, B. 1971. Specific immune response genes of the guinea pig. III. Linkage of the GA and GT immune response genes to histocompatibility genotypes in inbred guinea pigs. *J. Exp. Med.* 143: 1529–37

50. Bluestein, H. G., Green, I., Benacerraf, B. 1971. Specific immune response genes of the guinea pig. IV. Demonstration in random bred guinea pigs that responsiveness to a copolymer of L-glutamic acid and L-tyrosine is predicated upon the possession of a distinct Strain 13 histocompatibility specificity. *J. Exp. Med.* 134: 1538–44

51. Shevach, E. M., Rosenthal, A. S. 1973. Function of macrophages in antigen recognition of guinea pig lymphocytes: II. Role of macrophages in the regulation of genetic control of the immune response. *J. Exp. Med.* 138: 1213–29

52. Katz, D. H., Hamaoka, T., Benacerraf, B. 1973. Cell interactions between histoincompatible T and B lymphocytes. II. Failure of physiologic cooperative interactions between T and B lymphocytes from allogeneic donor strains in humoral response to hapten-protein conjugates. *J. Exp. Med.* 137: 1405–18

53. Katz, D. H., Hamaoka, T., Dorf, M. E., Benacerraf, B. 1973. Cell interactions between histoincompatible T and B lymphocytes. The H-2 gene complex determines successful physiologic lymphocyte interactions. *Proc. Natl. Acad. Sci. USA* 70: 2624–28

54. Rock, K. L., Benacerraf, B., Abbas, A. K. 1984. Antigen-presentation by hapten-specific B lymphocytes. I. Role of surface immunoglobulin receptors. *J. Exp. Med.* 160: 1102–13

55. Cullen, S. E., Freed, J. H., Nathanson, S. G. 1976. Structural and serological properties of murine Ia alloantigens. *Transplant. Rev.* 30: 236–70

56. Dorf, M. E., Stimpfling, J. H., Benacerraf, B. 1975. Requirement for two H-2 complex Ir genes for the immune response to the GLPhe terpolymer. *J. Exp. Med.* 141: 1459–63

57. Dorf, M. E., Benacerraf, B. 1975. Complementation of H-2 linked Ir genes in the mouse. *Proc. Natl. Acad. Sci. USA* 72: 3671–75

58. Benacerraf, B. 1978. Hypothesis to relate the sensitivity of T lymphocytes and the activity of I region specific Ir genes in macrophages and B lymphocytes. *J. Immunol.* 120: 1809–12

59. Rosenthal, A. S., Barcinski, A. M., Blake, T. J. 1977. Determinant selection is a macrophage dependent immune response gene function. *Nature* 267: 156–58

60. Rock, K. L., Benacerraf, B. 1983. Inhibition of antigen-specific T lymphocyte activation by structurally related Ir gene controlled polymers: Evidence of specific competition for accessory cell antigen-presentation. *J. Exp. Med.* 157: 1618–34

61. Babbitt, B. P., Allen, P. M., Matsueda, G., Haber, E., Unanue, E. R. 1985. Binding of immunogenic peptides to Ia histocompatibility molecules. *Nature* 317: 359–61

62. Bjorkman, P. J., Saper, M. A., Samraoui, B., Bennett, W. S., Strominger, J. L., Wiley, D. C. 1987. Structure of the human class I histocompatibility antigen, HLAA-A2. *Nature* 329: 506–18

63. Lemonnier, F., Burakoff, S., Germain, R., Benacerraf, B. 1977. Cytolytic T lymphocytes specific for allogeneic stimulator cells cross-react with chemically modified syngeneic cells. *Proc. Natl. Acad. Sci. USA* 74: 1229–33

64. Burakoff, S. J., Ratnofsky, S. E., Benacerraf, B. 1977. Mouse cytolytic T lymphocytes induced by xenogeneic rat stimulator cells exhibit specificity for H-2 complex alloantigens. *Proc. Natl. Acad. Sci. USA* 74: 4572–76

65. Finberg, R., Burakoff, S. J., Cantor, H., Benacerraf, B. 1978. The biologic significance of alloreactivity. T cells stimulated by sendai virus coated syngeneic cells specifically lyse allogeneic target cells. *Proc. Natl. Acad. Sci. USA* 75: 5145–49

66. Rock, K. L., Benacerraf, B. 1984. Thymic T cells are driven to expand upon interaction with self-class II MHC gene products on accessory cells. *Proc. Natl. Acad. Sci. USA* 81: 1221–24

67. Yeh, E. T. H., Benacerraf, B., Rock, K. L. 1984. Analysis of thymocyte MHC specificity with thymocyte hybridomas. *J. Exp. Med.* 160: 799–813

68. Kapp, J. A., Pierce, C. W., Benacerraf, B. 1974. Genetic control of immune

responses *in vitro*. III. Tolerogenic properties of the terpolymer L-glutamic acid[60]-L-alanine[30]-L-tyrosine[10] (GAT) for spleen cells from nonresponder (H-2[s] and H-2[q]) mice. *J. Exp. Med.* 140: 172–84

69. Kapp, J. A., Pierce, C. W., Schlossman, S., Benacerraf, B. 1974. Genetic control of immune responses *in vitro*. V. Stimulation of suppressor T cells in nonresponder mice by terpolymer L-glutamic acid[60]-L-tyrosine[10] (GAT)[1]. *J. Exp. Med.* 140: 648–59

70. Kapp, J. A., Pierce, C. W., Benacerraf, B. 1977. Immunosuppressive factor(s) extracted from lymphoid cells of nonresponder mice primed with L-glutamic acid[60]-L-alanine[30]-L-tyrosine[10] (GAT). II. Cellular source and effect on responder and nonresponder mice. *J. Exp. Med.* 145: 828–38

71. Théze, J., Kapp, J. A., Benacerraf, B. 1977. Immunosuppressive factor(s) from lymphoid cells of nonresponder mice primed with L-glutamic acid[60]-L-alanine[30]-L-tyrosine[10] (GAT). III. Immunochemical properties of the GAT-specific suppressive factor. *J. Exp. Med.* 145: 829–56

72. Debré, P., Kapp, J. A., Dorf, M. E., Benacerraf, B. 1975. Genetic control of specific immune suppression. II. H-2 linked dominant genetic control of immune suppression by the random copolymer L-Glutamic Acid[50]-L-Tyrosine[50] (GT). *J. Exp. Med.* 142: 1447–53

73. Debré, P., Waltenbaugh, C., Dorf, M. E., Benacerraf, B. 1976. Genetic control of specific immune suppression. III. Mapping of H-2 complex complementing genes controlling immune suppression by the random copolymer L-glutamic acid[50]-L-tyrosine[50] (GT). *J. Exp. Med.* 144: 272–76

74. Théze, J., Waltenbaugh, C., Germain, R. N., Benacerraf, B. 1977. Immunosuppressive factor(s) specific for L-glutamic acid[50]-L-tyrosine[50]. IV. *In vitro* activity and immunochemical properties. *Eur. J. Immunol.* 7: 86–92

75. Vidovic, D., Klein, J., Nagy, Z. A. 1985. Recessive T cell response to Poly-(Glu50TYR50) possibly caused by self tolerance. *J. Immunol.* 134: 3563–68

76. Germain, R. N., Benacerraf, B. 1978. The involvement of suppressor T cells in Ir gene regulations of secondary antibody responses of primed (responder × nonresponder)F$_1$ mice to macrophage-bound L-glutamic acid[60]-L-alanine[30]-L-tyrosine[10] (GAT). *J. Exp. Med.* 148: 1324–37

77. Debré, P., Waltenbaugh, C., Dorf, M. E., Benacerraf, B. 1976. Genetic control of specific immune suppression. IV. Responsiveness to the random copolymer L-glutamic acid[50]-L-tyrosine[50] (GT) induced in BALB/c mice by cyclophosphamide. *J. Exp. Med.* 144: 277–81

78. Pierres, M., Germain, R. N. 1978. Antigen-specific T cell mediated suppression. IV. The role of macrophages in the generation of L-glutamic acid[60]-L-alanine[30]-L-tyrosine[10] (GAT) specific suppressor T cells in responder mouse strains. *J. Immunol.* 121: 1306–14

Annu. Rev. Immunol. 1990. 8:1–21

FROM PROTEIN SYNTHESIS TO ANTIBODY FORMATION AND CELLULAR IMMUNITY: A Personal View

Brigitte A. Askonas[1]

National Institute for Medical Research, Mill Hill, London NW7 1AA, England

KEY WORDS: CTL, virus infection, macrophages, antigen presentation, antibody formation.

INTRODUCTION

It has been exciting to be part of the immunological world for nearly 35 years and to experience the progress in the understanding of the immune system during my professional life, yet many challenging and unanswered problems are still left to pursue. I am writing this chapter for two reasons; first, to try to convey how little was known about immunology in the mid-fifties, and how one's projects tended to develop; second, to express my appreciation of the people I have shared this excitement with: my mentors, who influenced me profoundly, and my numerous associates, collaborators and PhD students who made most important contributions to our projects and became long-term friends. Their presence in the laboratory made research fun, and their successes have been a great pleasure to me. I am sorry that I shall not be able to mention all of them here. I am equally grateful to several associates on technical appointments, who provided essential continuity in the lab, for their loyalty, hard work, and contributions to our projects (Marion Perryman, Jackie Hunter, Brian Wright, Dianne Millikan, and Pat Taylor). Without them the laboratory would

[1] Present address: Institute of Molecular Medicine, Molecular Immunology Group, John Radcliffe Hospital, Headington, Oxford OX3 9DU, England.

have collapsed. And of course I should like to thank the Medical Research Council for its long-term support.

My involvement in Immunology spans three intertwining phases: (i) antibody formation and the life-style of B cells; (ii) macrophages as antigen-presenting cells and as mediators of adjuvanticity or immunosuppression, and (iii) T-cell immunity in virus infection and viral recognition patterns. My original intent was to study the biochemistry of antibody formation and B-cell triggering. However, gradually I became totally fascinated by the immune system as a whole and by the cells that defend us against infection, including the macrophages, the scavenger cells. Eventually I moved away from my original aims and ended up concentrating mostly on cellular aspects of the immune system and our cell-mediated defense against virus infections.

THE BEGINNINGS

My parents encouraged me in every way and always gave me the strongest support. My scientific training at McGill University, Montreal, and then my PhD at Cambridge University were entirely biochemical. At the beginning, unlike so many budding young scientists, I did not know what I would like to do with my life, even midway through University. I liked languages, art, literature, nature, and biological books. I chose to study biochemistry, however, because of one person. This was David Landsborough Thomson, Professor of Biochemistry and Dean of Science at McGill University. A graduate of Cambridge University, he was the most brilliant lecturer I have ever heard speak, with a marvellous sense of humor. Even steroid hormones and metabolic cycles came to life. Dean Thomson had an original cartoon by Thurber above his desk: The psychiatrist with a large rabbit head and floppy rabbit ears is asking his patient: "You said a moment ago that people look like rabbits to you. Now what do you mean by that, Mrs. Sprague?" It was my unrestrained laughter at this marvellous cartoon that got me my first job at the newly founded Allan Memorial Institute of Psychiatry (associated with McGill). Dr. Karl Stern had come to interview prospective biochemistry graduates. He was a musician, a scholar, a psychiatrist, and a most witty person. I learned an enormous amount about life and psychiatry at the Allan Memorial and made lifelong friendships with other researchers there. We were meant to study biochemical changes in mental disease, as Dr. Stern thought that in many cases an organic imbalance must exist. 1944 was a bit early to tackle these problems! Fortunately I realized this as well as the fact that I needed more research training. Dean Thomson arranged for me to do postgraduate work toward a PhD at the School of Biochemistry in Cambridge in 1949.

It was a marvellous stroke of luck to join this outstanding and friendly department where even PhD students were taken seriously as scientists. There was continuous scientific interchange in the department. Two particularly distinguished scientists, Margaret Stephenson and Dorothy Needham, who were among the first four women to be elected to the Royal Society, were members of the department. This taught me that good science gets recognition regardless of the sex of the scientist. Prof. Malcolm Dixon was my PhD supervisor. I am most grateful to him for putting me on my scientific feet by suggesting a project and then leaving me to it, and for his scientific generosity. My project was to try to use cold organic solvents to purify muscle enzymes (as newly used for serum protein fractionation). In those days salt precipitation and pH changes were the available methods. Almost any experiment you did was novel; science moved far more slowly, and one had a good chance to contribute.

For unknown reasons creatinephosphokinase did particularly well in propanol, and I spent a lot of time examining the kinetics and metal activation of the purified enzyme and the importance of its sulphhydryl groups. It was a wonderful three years. I learned a lot about protein and enzyme research from many distinguished members of the department, and I emerged with some confidence that I might be able to do independent research.

HOW I WAS INTRODUCED TO IMMUNOLOGY

When I joined the National Institute for Medical Research (NIMR), Mill Hill, in 1952, Tommy Work and Peter Campbell were focusing on problems of protein synthesis, i.e. the biosynthesis of milk proteins in goats. I am most grateful for their patience and for their teaching me that research has to be organized carefully and that tidy laboratory notes are absolutely essential. The question then being asked (which will seem strange now, 35 years later) was whether polypeptide chains were biosynthesized in one piece or as smaller peptides, which then joined together. We hoped to find the answer by giving animals, in this case goats, intravenously very short pulses of radiolabeled amino acids, then newly available. The biosynthesized proteins were degraded into peptides to determine the specific radioactivity of the injected amino acid in the purified peptides. Soon afterward the answer became clear with the discovery of the double helix, mRNA, and its translation. Meanwhile, milk proteins were hard to purify: Casein sticks to everything—including [35 S] methionine—and goats do not like to be milked every few minutes. Lactoglobulin had to be crystallized from each very small milk sample, and although it forms the most beautiful crystals (particularly when you leave the tube on the bench while taking a week's holiday), this provided some difficulty. However, we managed to establish that the specific radioactivity of a given amino acid was the same along the whole peptide chain, indicating its very rapid synthesis in one piece (1).

With a background in pure biochemistry, I hardly knew the difference between the terms antigen and antibody, and immunology was not particularly in the news. Not really keen on remaining a milk maid, I wanted to develop a cell-free system for protein synthesis. In this effort, the multidisciplinary nature of NIMR was a great advantage. John Humphrey, who was then part of the Biological Standards division at the Institute, gave a lecture (then called a colloquium) in 1954 on antibodies against Pneumococcal polysaccharide(PnPS). During this talk he gave a demonstration: He mixed two clear solutions, one containing PnPS and the other purified rabbit antibodies. Within seconds there was a large white precipitate which settled. This was the inspiration to me: Why not use antibody formation as a model for protein synthesis? It is so much easier to purify antibodies by precipitation with antigen than to crystallize protein in numerous tiny samples of milk. We could be inducing antibodies in small animals rather than having to milk Ursula, the recalcitrant goat.

The next step was to consult John Humphrey as to which tissue would be best for in vitro and in vivo study of antibody formation. John was exceedingly enthusiastic, encouraging, and helpful, but he could not answer this question. It was known that removal of some lymphnodes or the spleen did not really affect the level of circulating antibody in the serum. The dynamics of lymphocyte circulation was not yet known. However, my decision was taken in 1955 to try to work on antibody formation, and I never looked back. John Humphrey became my mentor in immunology over many years. His knowledge of the immune system and his scientific generosity were unique. He attracted postdocs from all over the world to the laboratory. When Sir Charles Harington, our director, inquired in 1957 whether I would like to join the newly founded Immunology Division at Mill Hill, headed by John Humphrey, I did not hesitate for a second before agreeing. I became immersed more and more in the complexities of the immune system and its cells, yet I remained most grateful for my biochemical training. It has continued to influence my experimentation and critical analysis of results until the present day.

It was of course a stroke of great good fortune to be associated with John Humphrey. All who met John recognized his outstanding qualities. What made him so special to a postdoc? Generosity with his time and ideas, his enthusiasm for all aspects of immunology, his encyclopedic knowledge which he shared with everyone, and his encouragement to follow one's own ideas. In 1961 Sir Peter Medawar became Director of NIMR, and with him a second immunologically oriented division arrived under Av Mitchison. Peter Medawar was a great scientist who inspired us all, and his positive attitude and unfailing encouragement created a marvellous and exciting atmosphere. The spirit at the Institute reflected his personality, and he exerted a profound influence. It was thanks to Peter Medawar that I became a permanent member of the Institute staff. His illness was a great blow to British immunology and the NIMR.

THE SITES OF ANTIBODY FORMATION; IN VITRO EXPERIMENTS (1955–1959)

In 1948 Astrid Fagraeus (2) had completed histological analyses of lymphoid tissues after immunizing rats with horse serum. The results suggested that the large mature cells (plasma cells) arising in the lymphoid tissues correlated with increases in circulating antibodies. That these were the antibody-forming cells was then confirmed by the fluorescence staining technique developed by Al Coons and associates in the early fifties (3). This technique enabled the localization of antibodies within cells of lymphoid organs. R. G. White, just returned to the London Hospital after spending time with Coons, had found antibody-forming cells in the local granuloma after immunizing guinea pigs with ovalbumin, using the Lederer wax from tubercle as adjuvant. John Humphrey thought that granuloma might be a good tissue for in vitro pulsing. It was, of course, appallingly necrotic, but Bob White and I then decided to look at other guinea-pig tissues. After slicing them (mostly chopping), I pulsed them with radioactive amino acids in vitro and cleared supernatants with a heterologous antibody-antigen precipitate before precipitating the antibodies formed in vitro with ovalbumin or the appropriate antigen used for immunization. The radioactivity of each antibody/antigen sample had to be counted on a hand counter. This process created a lot of thinking time.

Our approach worked well, and we obtained a picture of where antibodies were formed. When footpads were injected, the most active tissue was the lymphnode contralateral to the footpad site of injection, because macrophages (Mo) had taken over the local lymphnode (4). John Humphrey and I looked in the same way at rabbits immunized by different routes—certainly easier than doing histology on all organs. What then seemed surprising was that, after intravenous immunization, bone marrow and lung turned out to be potent antibody-forming organs (5). In fact, perfusion of lung yielded milligram quantities of antibody to PnPS (6). In these experiments we also found that the immunization induced not only specific antibodies but also high levels of antigen-nonspecific Ig. This became a pet subject of John Humphrey who wanted to find out the reason for this "bystander effect." Now we know, of course, that macrophage and T cell–derived mediators, which are not antigen specific, can be mitogenic to B cells when the system is highly activated.

It is very difficult to think back and to believe how little was known about antibody molecules and the cells of the immune system in the mid-1950s when I started. The heterogeneity, structure, and subclasses of Ig were not understood. Rodney Porter described the antigenic and antibody properties of papain fragments of Ig in 1958 (7), and Edelman & Poulik in 1961 found that Ig could be reduced to light and heavy chains (8). I heard a lecture in London by Jim Gowans late in 1959 on the circulation of small lymphocytes and their immunocompetence (9), a revelation to me. By this time the clonal selection theory had been proposed by Sir Macfarlane Burnet (10). It was not until 1968–1969, after the Cold Spring Harbor meeting in 1967, that lymphocytes were functionally subdivided into B and T cells. Their collaboration in antibody formation was then elucidated elegantly by Henry Claman and Jaques Miller and associates (see review 11).

MOUSE PLASMA CELL TUMORS AS MODELS FOR ANTIBODY FORMATION (1959–1961)

The mixture of cells in lymphoid tissues and the low frequency of memory B cells specific for a particular antigen provided problems in my effort to study cell-free protein synthesis at the biochemical level. In fact it took many years of work by several groups to set up such a system. In the late 1950s Mike Potter, Thelma Dunn, and associates described plasma cell tumors transplantable in inbred mice (e.g. 12). Most generously they immediately sent us the X5563 tumor line forming an IgG2a myeloma protein, and later, additional plasma cell tumors forming different Ig subclasses as well as the inbred mouse strains. The 5563 tumor in C3H mice was highly differentiated, with a well-developed endoplasmic reticulum. This tumor tissue enabled us to study the kinetics of secretion of myeloma protein in vitro, although at that time myeloma proteins were not yet generally accepted as equivalent to antibody molecules. However, they were essential for many studies worldwide in subsequent years, and eventually they were fully accepted by the immunological world, once antigen reactivity was demonstrated.

When John Fahey came as a visiting worker we examined the micro-heterogeneity and papain fragments of the different tumor lines and their serological interrelationships (e.g. 13). We saw that excess light-chain secretion was equivalent to Bence Jones protein production in human myeloma patients.

MACROPHAGES AND THEIR ROLE IN ANTIGEN PRESENTATION (1962–1968)

More than a hundred years ago Metchnikoff discovered phagocytosis of foreign cells and microbes by macrophages (14). For almost a hundred years macrophages were thought to have a purely degradative role. About 25 years ago two American workers reported that macrophages transferred mRNA-encoding antibody to lymphoid cells. In 1961–1962 I spent a sabbatical year at Harvard Medical School, in Mahlon Hoagland's laboratory; this was for me a good refresher course on the basic biochemistry of protein synthesis and mRNA. This work made me realize that transfer of information between cells via intact mRNA was most unlikely. However, I supposed that the RNA extracts might contain antigen, and so on return to Mill Hill, I decided to study antigen handling by macrophages and their role in antibody formation. We iodinated our pet antigen, the large mol wt Maia squinada hemocyanin (HCY) and fed it to macrophages in vitro or in vivo. We then cultured the macrophages and looked at release of [131 I] from the cells as well as the state of the intracellular radiolabeled material. Most of the protein taken up by the cells was degraded within 1–2 hours, but 5–10% of [131 I] remained associated with the cells in TCA precipitable form, though of smaller molecular weight than the original HCY. Joan Rhodes and I then found that RNA extracts had protein fragments attached and were highly immunogenic (we called them super antigens) (15; 1964–1965).

Another stroke of luck—in 1966 Emil Unanue joined me as a postdoc on a Helen Hay Whitney fellowship on the macrophage project. We transferred macrophages 24–48 hours after antigen uptake and in vitro culture into mice. We found that these cells were highly effective in inducing antibody formation in the host (16). We could conclude that although most of the antigen taken up had been degraded and lost from the cells, the persisting material could still induce antibodies. After his return to the United States, Emil at the Scripps Institute, and Lydia Jaroskova and I at Mill Hill, showed that macrophages could retain immunogenicity and induce antibodies in vivo after removal of surface antigen. I greatly admire Emil for persisting with this fascinating problem and more recently for unearthing with Paul Allen and colleagues the importance of antigen fragmentation for T-helper-cell recognition of peptide epitopes (17). In the sixties most immunologists found it impossible to believe that a single cell could have the multiple functions of phagocytosis, antigen degradation, and antigen presentation. Many more macrophage activities have been defined since then. Truly it is an amazing cell even if we now know that dendritic cells are more effective in activating primary T-cell responses (e.g. 18).

THE FATE OF ANTIGEN IN RELATION TO ANTIBODY FORMATION (1963–1966)

In 1959 John Humphrey attended one of the first modern immunology meetings outside Prague, which was organized by J. Sterzl. This was the first of three important Czechoslovak conferences that allowed newcomers to meet the then-small community of distinguished immunologists. John spoke about the fate of antigen and posed clearly the question as to where the antigen relevant to induction of antibodies might localize while much of the administered antigen was being cleared by irrelevant cells (19). Is it necessary for antigen to be present in the antibody-forming plasma cell? In that era many people still believed that antigen determined the folding of antibody molecules.

John Humphrey was keen to study this question, and we initiated a collaboration with Michael Sela and Israel Schechter who were studying the immunogenicity of synthetic peptide copolymers, such as (T,G)-A-L, in rabbits. It was hoped that these could be synthesized with high specific radioactivity, by use of tritiated amino acids, and then localized in tissue sections by combined autoradiography and fluorescence staining for antibodies. Actually, quite unexpectedly, the hot tritiated TGAL fell to pieces, and the experiments had to be carried out with iodinated TGAL. We could not detect antigen in the plasma cells (20), but we also began to realize the limit of this method. With Gordon Ada's help at Mill Hill, and with Gus Nossal, we set sensitivity limits. Lack of detection did not mean total absence, but the results were compatible with the clonal selection theory. At the beginning of this project late in 1962, Hugh McDevitt arrived as a postdoctoral student. John Humphrey had found differences in immunogenicity of TGAL between sandy-lop and New Zealand black rabbits. This observation was followed up by Hugh and Michael, and it later formed the basis of the major discovery by Hugh McDevitt and collaborators of the H-2 linkage of antibody responsiveness (21). Slightly different versions of this collaboration are described already by John Humphrey in Volume 2 and by Michael Sela in Volume 5 of the *Annual Review of Immunology*. From this project we learned a great deal about the limitations of techniques.

BACK TO B CELLS (1965–1970)

In parallel to the macrophage experiments in the mid-sixties, I also worked together with Alan Williamson, who arrived in 1965 to join the Biochemistry Division for the first year on an Immunology appointment. This led to a very happy collaboration over many years as our experimental knowhow was totally complementary. The study of cell-free systems for protein synthesis had made considerable progress, but nothing had been achieved in Ig formation. We found that isolated polyribosomes forming light or heavy chains of myeloma protein 5563 could be precipitated with specific antisera (22); we examined the assembly of IgG, with free light chains and S-S-bonded heavy-chain dimers as intermediates. The 5563 tumor line showed a balanced synthesis of light and heavy chains, and the same was true for lymphoid tissue of immunized mice (23). A controversy with M. Scharff ensued about the balance of Ig-chain synthesis. He was studying other plasma cell tumors. Eventually it turned out that both groups were right but that some of the plasma cell tumors were far less differentiated than others.

Mike Parkhouse joined us in 1967, and we did parallel experiments on the biosynthesis of IgM (with 7S subunits as intermediates). During his PhD studies Mike Bevan tackled IgA assembly and cell-free systems (24); his independence and thoughtfulness were obvious from the beginning. Meanwhile Zuhayr Awdeh (from Lebanon) had joined us in 1965, and he put up with two PhD supervisors, Alan Williamson and myself. He was particularly interested in the microheterogeneity of myeloma protein formed by a single plasma cell tumor line. He and Alan developed isoelectric focusing (IEF) on thin layer acrylamide gels (25). This technique proved clearly that a single protein was formed by the X5563 plasmacytoma, while subsequent postsynthetic changes accounted for the family of protein bands characteristic of each myeloma (26). Characteristic gel patterns could then be defined for products of single antibody-forming clones, by the IEF gels overlayed with antigen, particularly haptens, followed by autoradiography (see below).

THE LIFE-STYLE OF B-CELL CLONES (1968–1971)

In spite of extensive trials in the late 1960s we could not clone B cells in vitro, and this still has not been achieved. With Brian Wright's help Alan Williamson and I attempted to clone B cells in vivo by transfer into syngeneic mice of limiting dilutions of spleen memory cells from donors primed with haptenated carrier antigens. The readout was by gel IEF of serum samples, and we chose DNP reactivity because we could detect the antibodies by overlaying the gel with I-131 hapten. Luck was with us, and the second transfer yielded host mice with a very strong single anti-DNP-forming clone that could be transferred into further host mice over many generations (27). This was clone E9, forming IgG2a anti-DNP. We subsequently selected many B-cell clones. This approach threw light on many aspects of B cells.

Kreth & Williamson (28) analyzed limiting dilution transfers from several CBA/H donors by IEF to estimate the diversity of the antibody repertoire for a single hapten within a mouse strain. The majority of anti-NIP molecules elicited in each donor were not produced by other donors of the same strain; there were rare overlaps in the IEF patterns of the antibodies. It was concluded that more than 3000 different antibody molecules reactive with NIP can be produced by CBA mice. Of course a small molecule can interact with many antibodies, but larger foreign proteins such as influenza HA also yield a very large number of different antibodies within a mouse strain (29). Somatic mutation will contribute to the diversity.

What else did we learn from these experiments? One of the findings was that B-cell clones do not have unlimited potential to proliferate; clone E9 expanded for about 70 cell generations and then stopped expanding at which point different DNP specific clones arose. These events showed that lack of T helper activity was not responsible for the senescence of clone E9. Our experiments also illustrated clonal dominance, which we analyzed further with A. McMichael during his PhD studies. The characteristic IEF pattern of the E9 antibody did not alter throughout the seven transfer generations in 300 mice. Although we could not discriminate mutations that would not alter the charge of the molecule, the stability of the IEF pattern suggested stable antibody production at the memory cell stage (30). Rajewsky and associates reached the same conclusions in their system (31). However, somatic mutation certainly occurs during maturation of the antibody response (32). A high affinity antibody response is likely to dominate. Earlier experiments by Krause and associates in rabbits also showed long-term clonal dominance (33).

We did not succeed, in hurried trials when E9 was senescing, in fusing our clonal B cells with fusion partners to eternalize the antibody-forming clone. This was most successfully achieved by Kohler & Milstein in 1975 (34).

BASEL INSTITUTE (1971–1972)

I had a wonderful year and a half at the Basel Institute in its infancy, and I am forever grateful to Niels Jerne for giving me this opportunity when he built up the Institute. He acted on his philosophy that everyone, whether they were young or old, should be encouraged to pursue their own ideas, and no one was permitted to build an empire. Georges Roelants came with me to Basel to continue the work with B-cell clones. He had first worked with me at Mill Hill on macrophages and T-cell suicide, using highly radiolabeled antigens. I interacted widely with many people to broaden my knowledge and experimental approaches. Particularly enjoyable (with lots of giggles) was a collaboration with Eberhard Wecker and Anneliese Schimpl, in Wurzburg, to follow up their experiments with T cell–replacing factor(s), which was antigen-nonspecific and which induced antibody and Ig biosynthesis in spleen cells of athymic mice (35).

TRANSITION PERIOD (1973–1976)

After Basel, on my return to Mill Hill, we continued our attempts to develop longer term cultures of antibody-forming B cells. We studied in detail the triggering of B cells into proliferation and maturation, and particularly the control mechanisms responsible for the switch from IgM to IgG formation. (This work was done with John North, Jackie Hunter, and John Kemshead.) IgG production was always very low, however, even when longer term cultures, initiated by Maurice Zauderer, showed clear waves of B-cell proliferation (36).

I could give you all sorts of theoretical reasons why I decided to change tack in 1976 when I took over the Division of Immunology at NIMR from John Humphrey—I moved away from antibodies and B-cell triggering after 21 years of research on this subject. If I am honest, I need to admit that the then trendy (rather fictitious) antigen-specific factors derived from T cells were never detected in our in-vitro antibody-forming cultures. All our data were compatible with the view that antigen nonspecific factors amplify B-cell responses. In any case it seemed an excellent moment to abandon this field and the pet antigens of the immunological world, i.e. haptenated carrier proteins, in favor of pathogens. We aimed to study antigens important in real life and to examine the defense of humans and other hosts against infections. I can highly recommend such a change, every so often in one's career, as refreshing and intellectually challenging. Molecular biology, gene cloning, and an understanding of the structure of protein components of pathogens of course made this move possible, as well as the tremendous progress in basic immunology that had by now occurred: i.e. T- and B-cell collaboration and tissue culture, MHC restriction, T-cell subpopulations, lymphoid cell markers, and cell separation techniques.

I took up two projects, uncertain whether either would succeed but with the hope that one of them might yield some interesting results. These projects were: (*a*) The cellular basis of immunosuppression in African Trypanosomiasis, and (*b*) cytotoxic T cells (CTL) in a mouse model of influenza infection. What are the viral recognition patterns of CTL; do virus-specific CTL alter the course of virus infection?

MACROPHAGES AS MEDIATORS OF IMMUNE SUPPRESSION IN TRYPANOSOMIASIS (1976–1983)

It is difficult to remember the considerations influencing the choice of a project. The breadth of immunology provides almost unlimited possibilities. Eventually one has to take the bull by the horns, to think about projects to which a small group such as ours might contribute, and unpredictable circumstances can lead to the final decision. Bridget Ogilvie (Parasitology Division) and I had been considering a collaboration on protozoal immunology. Why choose the induction of a general immunosuppression in African trypanosomiasis? I had a PhD vacancy and was interviewing new graduates. Along came Christine Clayton who had done very well in Cambridge biochemistry. She was highly motivated to work on African trypanosomes because of their continuous antigenic variation and resulting evasion from immune control. Christine's arrival was the moment to initiate a project. African trypanosomes are extracellular parasites, the cause of sleeping sickness with its waves of variant parasites. Bridget very rightly insisted that we must work with parasite clones in view of their variation in surface glycoprotein coat and virulence. Carlos Corsini (from Brazil) joined us and concentrated on the immunology side of the project. Bridget then left for the Wellcome Trust, and work continued for a few more years with the help of David Sacks, Josh Fierer, and two PhD students, Clem Grosskinsky and Greg Bancroft. Trypanosomes have profound effects on all aspects of B- and T-cell function; as long as IgM antibodies against the variant parasites are formed, each wave of parasitemia is controlled. With so many functional changes, cause and effect are difficult to disentangle. Contrary to other people's view, we found no direct B-cell mitogen associated with the parasite, but we obtained evidence that macrophages are the primary target cell for the parasite action. Their strong activation then mediates suppression of specific antibody and T-cell responses. The parasitic component causing havoc in the immune system was associated with the membrane fraction of the parasite, but we did not manage to purify it (reviewed in 37). The mechanisms involved in the multiple effects of the infection have not been resolved as yet in recent investigations by other groups.

CYTOTOXIC T CELLS IN INFLUENZA INFECTION (1976–present)

The inspiration to study antiviral cytotoxic T cells (CTL) came of course from the exciting discovery by Zinkernagel & Doherty in 1974 that CTL specific for lymphocytic choriomeningitis virus (LCMV) primed by infection of mice were class-I MHC restricted (38). I had been primed about the importance of the MHC complex particularly by Hugh McDevitt and Morten Simonsen at various meetings and then at the first IUIS symposium for young people in 1971. This was organized by Morten Simonsen, James Howard, Vlatko Silobrcic and myself. I was the first IUIS Symposium chairman. The symposium was called "The biological significance of transplantation antigens," and during the planning stage it was thought by some senior immunologists to be far too academic for the young! It brought together transplantation and basic immunologists at last and was most memorable.

In 1975 when Alex Matter visited we took the first step toward a CTL project by optimizing culture conditions for the generation of CTL, using alloreactivity as a model. Since we had and still have influenza experts at Mill Hill—John Skehel and associates—and since Mike Crumpton wanted to look at T-cell receptors of CTL, we decided to collaborate on influenza. This turned out to be the ideal choice—describing the three-dimensional structure of the hemagglutinin (HA) was underway, and the variable antigenic sites of HA seen by neutralizing antibodies were being defined (39, 40); the genes encoding the influenza proteins were being sequenced, and recombinant and variant viruses were available.

My original aim was to define whether CTL and B cells recognized the same regions on the HA molecule. In 1976 the one receptor or two receptor models for T-cell recognition were hotly debated. As is usual in research, the original plans failed to materialize, but other leads became important and could be followed up. It was fortunate that in autumn 1976 Hans Zweerink, a virologist from Duke University, visited to learn immunology; he involved himself in this project with Dianne Millikan's help. With our optimized culture conditions the influenza system was working within six weeks. Intranasal influenza infection of mice led to CD8+ memory CTL in the spleen that, on stimulation in vitro with influenza-infected APC, generated strong class-I MHC-restricted cytotoxic responses. The immediate surprise was that, unlike neutralizing antibodies, the majority of the CTL were type A–virus cross-reactive [i.e. they did not discriminate between serologically distinct type-A influenza viruses and failed to recognize targets infected with type-B influenza (41)]. Doherty and associates made the same observations simultaneously (42). In collaboration with

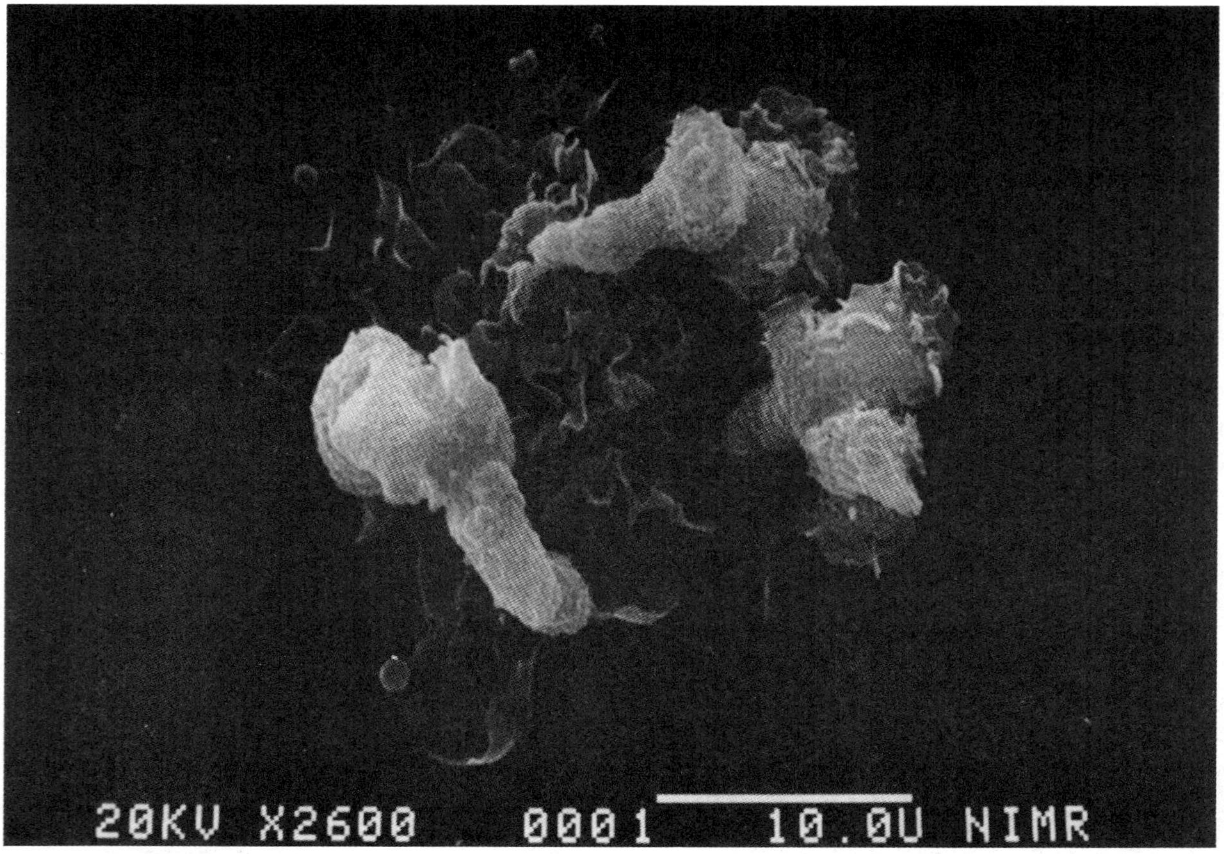

Figure 1   Scanning electron micrograph of an influenza nucleoprotein-specific CTL clone attacking an influenza-infected macrophage (Pat Taylor and Katie Sullivan).

Andrew McMichael, we found that human CTL were also type-A virus cross-reactive (43). Andrew has continued this project most successfully at Oxford, and the two labs have remained closely connected.

So rather than pursuing the original aims of the project, our attention was directed to the following questions: Which influenza proteins are recognized by type-A virus cross-reactive CTL? Are the CTL responsible for heterotypic immunity? Do they aid recovery; do they protect, or do they exacerbate pathology and disease? It soon became obvious that the inactivated influenza vaccines in present use do not induce CTL. Influenza infection in human and mouse, as well as other viral infections, prime for class-I MHC-restricted CTL and CD8 phenotype. Memory CTL are not cytolytic per se but become so after 5 days' stimulation with antigen. However, in certain infections, e.g. measles, CD4+ T cells are also primed and become cytolytic within a few days of antigen stimulation (44). This difference may relate to the site of infection: While systemic infections clearly prime CD8+ CTL, it is possible that local infections might also activate CD4+ cells to become cytolytic precursors. After longer term cultures most CD4+ cells are able to lyse appropriate target cells.

Very early in our work we cloned influenza-specific CTL, following the discovery of Gillis & Smith in 1977 (45) that T cells could be cloned and maintained in vitro in the presence of a source of TCGF (IL2). We were lucky; our first trial culture in 1979 yielded the influenza-specific CD8+ clone L4, restricted to L^d, which showed that cross-reactivity for all type-A viruses was a clonal property. Y.-L. Lin (from China) nursed the clone for many different experiments in vivo and in vitro (see below). After that it took us another year to select further clones to be used to define viral target antigens and the in vivo role of CTL. In later years many associates, not all mentioned in this narration, contributed greatly to our influenza studies while they were at Mill Hill: Chuck Hackett (USA), David Wraith, Alain Townsend, Pietro Pala (Italy), Pat Taylor, Roy Pemberton, Fernando Esquivel (Mexico), Sue Lightman, Hui-Fang Wu (China), and Helen Bodmer. Influenza T cells have become a big subject and only a few points can be discussed here. This also leaves out many contributions to this field by the groups of Gordon Ada, Tom Braciale and Jack Bennink, Jonathan Yewdell and Walter Gerhard.

CTL RECOGNITION OF INFLUENZA PROTEINS: DO CTL AND B CELLS RECOGNIZE THE SAME VIRAL DETERMINANTS?

Because hemagglutinin (HA) is so immunodominant for antibody responses after infection in mouse and human, everyone assumed this would be the case also for CTL. Do type A-virus cross-reactive CTL recognize well-conserved HA residues? Later it became obvious that CD8 + HA-specific CTL represent only a very small part of the repertoire induced by infection. They can be selected for but are influenza subtype specific.

In 1979 Rob Webster spent his sabbatical with us, and we tested his numerous polyclonal and monoclonal antibodies to influenza surface glycoproteins and internal matrix and nucleoprotein (NP). No well-defined antibody tested by us or Doherty et al blocked cytotoxic-T-cell recognition per se. The obvious conclusion was that viral epitopes seen by B cells and CD8 + CTL differed. So far in influenza this conclusion has held up, but we clearly need more data on other viruses before we can generalize. CD4 + T helper cells mount a strong anti-HA response on infection, and interestingly, Brian Thomas and associates at Mill Hill show a striking overlap between HA epitopes seen by B and CD4 + T cells in mice (46).

It took several more years to elucidate the target antigens for CTL cross-reactive for type-A influenza virus. In fact it was the rare CTL clones, not type-A influenza cross-reactive, that provided the first clues that internal viral proteins were being recognized: i.e. polymerase (47) and nucleo-protein (NP) (48). Since then, and with the help of recombinant vaccinia viruses encoding individual influenza protein genes, made by Geoff Smith and Bernie Moss, it has been shown that most internal and nonstructural influenza proteins could be seen by CTL in mouse and human with the usual Ir gene effects in different haplotypes (49, 50).

Why should internal virus proteins be so immunodominant for CTL responses? This question still remains unanswered. The discovery by Alain Townsend and associates in 1986 that CTL recognition can be defined by viral peptide epitopes in association with class-I MHC molecules (51) suggests the importance of intracellular degradative pathways to fragment endogenously produced viral proteins in infected cells. Of the four type-A influenza NP peptides so far defined, each class-I MHC molecule dictates recognition of a unique NP epitope (e.g. 52). In contrast to the internal proteins, HA is strongly expressed on the surface of infected cells, but appropriate fragments may not be produced at the site of class-I MHC interaction.

In other virus infections, recognition of internal viral proteins has also been established, but in some cases strong CTL responses specific for surface glycoproteins are induced (e.g. the envelope protein of HIV-1) (53). At present we cannot establish any rules for the immunodominance for CD8 + T cells of a given viral protein in a given haplotype. The striking differences in the requirements for induction of CD4 + T cells, compared to those of CD8 + cells, and in antigen presentation to these T cell subsets need further explorations (e.g. 54). We need to know much more about intracellular pathways and protein fragmentation and the site of interaction between class-I MHC and protein fragments. This is now the subject of intensive studies (e.g. 55–57).

CTL RECOGNITION OF RESPIRATORY SYNCYTIAL VIRUS

More recently we have carried out comparative studies on T cells in a mouse model of respiratory syncytial (RS) infection. RS virus is a paramyxovirus that causes serious illness in infants and the elderly. Antibodies and protection do not correlate well on the whole. The virus is difficult to grow in quantity and to purify. Charles Bangham started this study with us, with advice from David Karzon who spent a sabbatical here (58). Gail Wertz, Andy Ball, and colleagues constructed, for immunization and target-cell infection, recombinant vaccinia viruses encoding genes for individual RS virus proteins. Our collaboration showed that CTL in mouse and human recognized the fusion and nucleoprotein, but not the glycoprotein that is highly glycosylated (59, 60). The low level of target lysis suggested that additional target proteins still need to be defined.

THE ROLE OF T-CELL SUBPOPULATIONS IN INFLUENZA AND RS VIRUS INFECTION

The transfer of cloned CD8+ CTL into syngeneic influenza–infected mice, started by YL. Lin (e.g. 52, 61, 62), had beneficial effects on the infection. Host mice survived a lethal influenza challenge; the virus was cleared more rapidly from lung and trachea, and recovery from lung pathology was accelerated (63). CTL do not prevent infection but are likely to be responsible for partial protection, since we do not suffer regular infections with new influenza variants. It was reassuring that NP-specific CTL recognizing a defined peptide epitope also recognized cells replicating the virus (63), because all the in-vitro work had been done with target cells infected abortively.

We carried out parallel experiments on another respiratory infection. Mice were not thought to be a particularly useful model for this infection because RS virus does not induce obvious symptoms. However, they turned out to be useful to study immunopathology. Martin Cannon and Peter Openshaw transferred RS virus–specific CD8+ T-cell lines or clones, in numbers beneficial in influenza, into RS virus–infected mice. This resulted in their death or symptoms of severe illness and hemorrhagic pneumonitis, but in survivors the virus was cleared from the lung (64).

Why this difference between two respiratory virus infections? A likely explanation, not proven, may reflect the differences in the spread of the

RS virus infection to the finer airpassages of the lung, which does not occur with influenza. T-cell aggression will cause transient pathology wherever the site of virus infection, but at a sensitive site, this may prove to be detrimental, for example, in intracerebral LCMV infection (65).

Do class-II MHC-restricted T cells also have short-term effector functions similar to CD8+ cells? Much less work has been done in this respect. Cloned influenza-specific CD4+ T cells varied in function in vivo and in vitro on long-term culture. This was not related to antigen specificity or interferon γ production, but probably was caused by changes in interleukin production and cytolytic activity during the long-term maintenance of the clones in culture (52, P. Taylor and F. Esquivel, unpublished results).

IN CONCLUSION

I am now in another transition period as I have retired from Mill Hill and the Division of Immunology. I am very pleased to have the opportunity to continue my association with problems of viral immunology by spending time in Oxford with Andrew McMichael and Alain Townsend and their associates, at the Institute of Molecular Medicine, John Radcliffe Hospital. Immunology remains as challenging as ever and as fascinating. The present era is exciting because important molecules are being defined and new ones discovered. We do not have to worry about a shortage of problems; the biology of the complex immune system and its regulation, the challenge of individual pathogens and their host relationship, will keep us busy in the future.

Many virus infections are presently more serious than influenza, but the influenza model has become a guide for studies of other viral infections in humans and animals. Some pathogens are particularly crafty in escaping immune control and in causing persistent infections. Such organisms do not allow the use of attenuated organisms for vaccination, and hence great efforts are now directed toward development of simpler vaccines. If we need CD8+ cells without strong DTH induction by CD4+ cells or vice versa, how do we go about that? If Ab/Ag complexes cause problems, how can we avoid excess antibody formation? The success of the hepatitis B surface antigen vaccine (which is highly effective at least for several years) gives us hope that it will be possible to devise subunit vaccines. Each pathogen is a law unto itself in its tissue tropism and replication; we need to define protective immunity and learn to induce balanced T- and B-cell responses, while avoiding enhanced pathology.

I have greatly enjoyed the company of and stimulating discussions with my long-term colleagues and their associates in the Division of Immunology (Mike Parkhouse, David Dresser, Brian Thomas, and Gerry

Klaus). I have learned a lot from them about different areas of immunology and different experimental approaches. All along I have appreciated their cooperation and consistently professional attitude to science. In fact it is not possible to list all the colleagues at Mill Hill who at one point or another have given me a great deal of help. I have also much appreciated the influence of and stimulating discussions with so many friends and colleagues from abroad and in the United Kingdom, and it has always been a pleasure to meet them.

Literature Cited

1. Askonas, B. A., Campbell, P. N., Work, T. S. 1954. The biosynthesis of proteins. 2. Synthesis of milk proteins by the goat. *Biochem. J.* 58: 326
2. Fagraeus, A. 1961. The plasma cellular reaction and its relation to the formation of antibodies *in vitro. J. Immunol.* 58: 1
3. Coons, A. H., Leduc, E. H., Connolly, J. M. 1955. Studies on antibody production. I. A method for the histochemical demonstration of specific antibody and its application to a study of the hyperimmune rabbit. *J. Exp. Med.* 102: 49
4. Askonas, B. A., White, R. G. 1956. Sites of antibody production in the guinea pig. The relation between *in vitro* synthesis of anti-ovalbumin and γ-globulin and distribution of antibody-containing plasma cells. *Br. J. Exp. Pathol.* 37: 61
5. Askonas, B. A., Humphrey, J. H. 1958. Formation of specific antibodies and γ-globulin *in vitro.* A study of the synthetic ability of various tissues from rabbit immunized by different methods. *Biochem. J.* 68: 252
6. Askonas, B. A., Humphrey, J. H. 1958. Formation of antibody by isolated perfused lungs of immunized rabbits. The use of (^{14}C) amino acids to study the dynamics of antibody secretion. *Biochem. J.* 70: 212
7. Porter, R. R. 1958. Separation and isolation of fractions of rabbit gamma-globulin containing the antibody and antigenic combining sites. *Nature* 182: 670
8. Edelman, G. M., Poulik, M. D. 1961. Studies on structural units of the γ-globulins. *J. Exp. Med.* 113: 861
9. Gowans, J. L., McGregor, D. D., Cowen, D. M., Ford, C. E. 1962. Initiation of immune responses by small lymphocytes. *Nature* 196: 651
10. Burnet, F. M. 1959. *The Clonal Selection Theory of Acquired Immunity.* Nashville and Cambridge: Vanderbilt Univ. Press and Cambridge Univ. Press
11. Miller, J. F. A. P., Mitchell, G. F. 1969. Thymus and antigen reactive cells. *Transpl. Rev.* 1: 3
12. Potter, M., Fahey, J. L. 1960. Studies on eight transplantable plasma-cell neoplasms of mice. *J. Natl. Canc. Inst.* 24: 1153
13. Fahey, J. L., Askonas, B. A. 1962. Enzymatically produced subunits of proteins formed by plasma cells in mice. II. γ$_{2A}$-myeloma protein and Bence-Jones protein. *J. Exp. Med.* 115: 641
14. Metchnikoff, E. 1887. Sur la lutte des cellules de l'organisme contre l'invasion des microbes. *Ann. Inst. Pasteur* 7: 321
15. Askonas, B. A., Rhodes, J. M. 1965. Immunogenicity of antigen-containing ribonucleic acid preparations from macrophages. *Nature* 205: 470
16. Unanue, E. R., Askonas, B. A. 1968. Persistence of immunogenicity of antigen after uptake by macrophages. *J. Exp. Med.* 127: 915
17. Allen, P. M., Strydom, D. J., Unanue, E. R. 1984. Processing of lysozyme by macrophages; identification of the determinant recognized by two T cell hybridomas. *Proc. Natl. Acad. Sci. USA* 81: 2489
18. Inaba, K., Steinman, R. M. 1984. Resting and sensitized T-lymphocytes exhibit distinct stimulatory (antigen-presenting cell) requirements for growth and lymphokine release. *J. Exp. Med.* 160: 1717
19. Humphrey, J. H. 1960. The fate of antigen and the importance of different routes of administration. In *Mechanisms of Antibody Formation. Proc. Symposium, Prague*, p. 39
20. McDevitt, H. O., Askonas, B. A., Humphrey, J. H., Schechter, I., Sela, M. 1966. The localization of antigen in relation to specific antibody-producing cells. I. Use

of a synthetic polypeptide [(T,G)-A-L] labelled with iodine-125. *Immunology* 11: 337

21. McDevitt, H. O., Chinitz, A. 1969. Genetic control of the antibody response: Relationship between immune response and histocompatibility (H-2) type. *Science* 163: 1207

22. Williamson, A. R., Askonas, B. A. 1967. Biosynthesis of immunoglobulins: The separate classes of polyribosomes synthesizing heavy and light chains. *J. Mol. Biol.* 23: 201

23. Askonas, B. A., Williamson, A. R. 1967. Balanced heavy and light chain synthesis in immune tissue and disulphide bond formation in IgG assembly. In *Gamma Globulins, Structure and Control of Biosynthesis. Nobel Symposium 3*, ed. J. Killander, p. 369. Interscience Publ.

24. Bevan, M. J., Parkhouse, R. M. E., Williamson, A. R., Askonas, B. A. 1972. Biosynthesis of immunoglobulins. *Progr. Biophys. Mol. Biol.* 25: 131

25. Awdeh, Z. L., Williamson, A. R., Askonas, B. A. 1968. Isoelectric focusing in polyacrylamide gel and its application to immunoglobulins. *Nature* 219: 66

26. Awdeh, Z. L., Williamson, A. R., Askonas, B. A. 1970. One cell-one immunoglobulin: Origin of limited heterogeneity of myeloma proteins. *Biochem. J.* 116: 241

27. Askonas, B. A., Williamson, A. R., Wright, B. E. G. 1970. Selection of a single antibody-forming cell clone and its propagation in syngeneic mice. *Proc. Natl. Acad. Sci. USA* 67: 1398

28. Kreth, H. W., Williamson, A. R. 1973. The extent of diversity in inbred mice: anti NP (4-hydroxy-5-iodo-3 nitrophenacetyl) antibodies in CBA/H mice. *Eur. J. Immunol.* 3: 142

29. Staudt, L. M., Gerhard, W. 1983. Generation of antibody diversity in the immune response of BALB/c mice to influenza virus haemagglutinin. *J. Exp. Med.* 157: 687

30. Williamson, A. R., Askonas, B. A. 1972. Senescence of an antibody-forming cell clone. *Nature* 238: 337

31. Kocks, C., Rajewsky, K. 1989. Stable expression and somatic hypermutation of antibody V regions in B cell differentiation. *Annu. Rev. Immunol.* 7: 537

32. Berek, C., Milstein, C. 1987. Mutation, drift and repertoire shift in the maturation of the immune response. *Immunol. Rev.* 96: 23

33. Eichmann, K., Lackland, H., Hood, L., Krause, R. M. 1966. Induction of rabbit antibody with molecular uniformity after immunization with group C streptococci. *J. Exp. Med.* 131: 207

34. Kohler, G., Milstein, C. 1975. Continuous cultures of fused cells secreting antibody of predefined specificity. *Nature* 256: 495

35. Schimpl, A., Wecker, E. 1972. Replacement of T-cell function by a T-cell product. *Nature* 237: 15

36. Askonas, B. A., North, J. R. 1976. The life style of B cells—cellular proliferation and the invariancy of IgG. *Cold Spring Harbor Symp. Quant. Biol.* 41: 749

37. Askonas, B. A. 1985. Macrophages as mediators of immuno-suppression in murine African trypanosomiasis. *Curr. Top. Microbiol. Immunol.* 117: 119

38. Zinkernagel, R. M., Doherty, P. C. 1974. Restriction of *in vitro* T cell mediated cytotoxicity in lymphocytic choriomeningitis within a syngeneic or semiallogeneic system. *Nature* 248: 701

39. Gerhard, W., Yewdell, J., Frankel, M. E., Webster, R. G. 1981. Antigenic structure of influenza virus haemagglutinin defined by hybridoma antibodies. *Nature* 290: 713

40. Wiley, D. C., Wilson, I. A., Skehel, J. J. 1981. Structural identification of the antibody binding sites of Hong Kong influenza haemagglutinin and their involvement in antigenic variation. *Nature* 289: 373

41. Zweerink, H. J., Courtneidge, S. A., Skehel, J. J., Crumpton, M. J., Askonas, B. A. 1977. Cytotoxic T-cells kill influenza virus infected cells but do not distinguish between serologically distinct Type A viruses. *Nature* 267: 354

42. Effros, R. B., Doherty, P. C., Gerhard, W., Bennink, J. R. 1977. Generation of both cross-reactive and virus-specific T cell populations following immunization with serologically distinct influenza A viruses. *J. Exp. Med.* 145: 557

43. McMichael, A. J., Askonas, B. A. 1981. Influenza virus-specific cytotoxic T cells in man; induction and properties of the cytotoxic T-cell. *Eur. J. Immunol.* 81: 705

44. McFarland, H. F., Goodman, A., Jacobson, S. 1988. Virus specific cytotoxic T cells in multiple sclerosis. *Ann. N.Y. Acad. Sci.* 532: 273

45. Gillis, S., Smith, K. A. 1977. Long-term culture of tumour specific cytotoxic T-cells. *Nature* 268: 154

46. Thomas, D. B., Burt, D. S., Barnett, B. C., Graham, C. M., Skehel, J. J. 1989. B and T cell recognition of influenza haemagglutinin. *Cold Spring Harbor Symp. Quant. Biol.* In press

47. Bennink, J., Yewdell, J., Gerhard, W.

1982. A viral polymerase involved in recognition of influenza virus-infected cells by a cytotoxic T-cell clone. *Nature* 296: 75

48. Townsend, A. R. M., Skehel, J. J. 1982. Influenza A specific cytotoxic T-cell clones that do not recognize viral glycoproteins. *Nature* 300: 655

49. Bennink, J. R., Yewdell, J. W. 1988. Murine cytotoxic T-lymphocyte recognition of individual influenza virus proteins. *J. Exp. Med.* 168: 1935

50. Gotch, F., McMichael, A., Smith, G., Moss, B. 1987. Identification of viral molecules recognized by influenza specific human cytotoxic T lymphocytes. *J. Exp. Med.* 165: 408

51. Townsend, A. R. M., Rothbard, J., Gotch, F. M., Bahadur, G., Wraith, D., McMichael, A. J. 1986. The epitopes of influenza nucleoprotein recognized by cytotoxic T lymphocytes can be defined with short synthetic peptides. *Cell* 44: 959

52. Askonas, B. A., Taylor, P. M., Esquivel, F. 1988. Cytotoxic T cells in influenza infection. *Ann. N.Y. Acad. Sci.* 532: 230

53. Plata, F., Autran, B., Martins, L. P., Wain-Hobson, S., Raphael, M., et al. 1987. AIDS virus specific cytotoxic T-lymphocytes in lung disorders. *Nature* 338: 348

54. Wraith, D. C. 1987. The recognition of influenza A virus infected cells by cytotoxic T lymphocytes. *Immunol. Today* 8: 239

55. Townsend, A. R. M., Bodmer, H. 1989. Antigen recognition by class I-restricted T-lymphocytes. *Annu. Rev. Immunol.* 7: 601

56. Rothbard, J. B., Pemberton, R. M., Bodmer, H. C., Askonas, B. A., Taylor, W. R. 1989. Identification of residues necessary for clonally specific recognition of a cytotoxic T-cell determinant. *EMBO J.* 8: 2321

57. Bjorkman, P., Saper, M., Samradi, B., Bennett, W., Strominger, J., Wiley, D. 1987. The foreign antigen binding site and T cell recognition regions of class I histocompatibility antigens. *Nature* 329: 512

58. Bangham, C. R. M., Cannon, M. J., Karzon, D. T., Askonas, B. A. 1985. Cytotoxic T cell response to respiratory syncytial virus in mice. *J. Virol.* 68: 2177

59. Pemberton, R. M., Cannon, M. J., Openshaw, P. J. M., Ball, L. A., Wertz, G. W., Askonas, B. A. 1987. Cytotoxic T cell specificity for RS virus proteins: fusion protein is an important target antigen. *J. Gen. Virol.* 68: 2177

60. Bangham, C. R. M., Openshaw, P. J. M., Ball, L. A., King, A. M. Q., Wertz, G. W., Askonas, B. A. 1986. Human and murine cytotoxic T cells specific to RS virus recognize the viral nucleoprotein, but not the major glycoprotein expressed by vaccinia virus recombinants. *J. Immunol.* 137: 3973

61. Lin, Y.-L., Askonas, B. A. 1981. Biological properties of an influenza A virus specific killer T cell clone. 1986. *J. Exp. Med.* 154: 225

62. Luckacher, E., Braciale, V. L., Braciale, T. J. 1984. In vivo effector function of influenza virus specific cytotoxic lymphocyte clones is highly specific *J. Exp. Med.* 154: 225

63. Mackenzie, C. D., Taylor, P. M., Askonas, B. A. 1989. Rapid recovery of lung histology correlates with clearance of influenza virus by specific CD8+ cytotoxic T cells. *Immunology* 67: 375

64. Cannon, M. J., Openshaw, P. J. M., Askonas, B. A. 1988. Cytotoxic T cells clear virus but augment pathology in mice infected with respiratory syncytial virus. *J. Exp. Med.* 168: 1163

65. Leist, T. P., Cobbold, S. P., Waldmann, H., Aguet, M., Zinkernagel, R. M. 1987. Functional analysis of T lymphocyte subsets in antiviral host defense. *J. Immunol.* 138: 2278

Eva Klein
Georg Klein

Ann. Rev. Immunol. 1989. 7: 1–33

HOW ONE THING HAS LED TO ANOTHER

George Klein and Eva Klein

Department of Tumor Biology, Karolinska Institutet, S-104 01 Stockholm, Sweden, and Lautenberg Center for General and Tumor Immunology, Hadassah Medical School, Jerusalem, Israel

GEORGE KLEIN WRITES:

Dawn

This story starts on the 10th of January, 1945, when I emerged from a cellar on the outskirts of Budapest where I had been hiding, with false papers, during the last weeks of the German occupation. With a totally new feeling about the sunshine that was floating over the snow, the ruined houses, the dead and frozen soldiers, civilians, and horses, I suddenly realized, with a mixture of surprise, guilt, and delight, that I had survived in spite of an 80% chance that I would end my 19 years in the gas chambers or in a military slave labor camp. After a few quick walks in the newly liberated area of the still besieged capital, I decided that it was time to start my medical studies, already delayed by almost two years. During the first year after my graduation from middle school, it was impossible for a Jewish boy to enter medical school. After the German occupation nothing mattered except survival.

We were free at last, but it was a complicated freedom. After a few more days, the Eastern side of the city, Pest, was all in Russian hands. I moved around relatively freely but I was caught twice, like other young men who were automatically regarded as disguised soldiers. In comparison with my earlier escape from a Nazi labor camp, it was an easy matter to run away from the improvised, loosely organized Russian patrols. It was a wise move. Several friends of mine who went out to get a loaf of bread returned years later from Russia.

As soon as the streets were open, I walked to the University to see whether it would open its doors for me now. I found deserted buildings, broken windows, and dead soldiers. Together with a friend we therefore decided that we should try to reach Szeged.

The journey of less than 300 km took more than five days. We walked long stretches, hitched on horsedrawn carriages and every other vehicle that we could get on, including a Russian military truck. We arrived in Szeged on February 4. It was a cold and beautiful morning. The city was intact, and we were admitted to the University on the same day. It was a strange place. All the professors had fled to the West. An assistant professor of forensic medicine with a Christlike head and very sad eyes was teaching anatomy, pathology, and forensic medicine all by himself. Students kept arriving from all former theaters of war, labor camps, and illegal hiding. Cadavers were abundant. The large dissection hall of the Anatomy Department was crowded. The smell of formalin, the half dissected or fully prepared body parts, and even the continually tipsy attendant appeared to me as parts of a magic, enchanting landscape, a previously forbidden paradise that was now all mine.

Two years passed as a single wave of febrile activity. I finished three terms during three months in Szeged and returned to Budapest when the university reopened there. I wanted to start research work, but the departments were still paralyzed. They had no resources and the routine work consumed the energy of all staff. Still, I got a first decisive inspiration from the professor of histology, Tivadar Huzella, one of the few internationally known scientists in Hungary and also one of the few true liberals among the medical professors of his generation. In spite of his consistent anti-Fascist stance, and his strong opposition to any form of discrimination during the war, he became a suspected person in the eyes of the new rulers. His uncompromising individualism and his democratic value system invited the enmity of the political opportunists who wanted to see a more compromising person in his position. His arch-enemy, the professor of anatomy, a political opportunist and a scientific nonentity who had resented Huzella's international fame for many years, delivered a list of accusations against him to the "people's court." The sympathies of all the students were on Huzella's side. The crucial trial, where all the absurd accusations—exemplified by the charge that Huzella ate eggs ordered for tissue culture—were readily dismissed, ended in tragedy when the presiding lay judge asked whether Huzella still believed a sentence he wrote during the war. Huzella had stated (an act of great courage at the time) that Hitler, Stalin, and Salazar were equally abominable dictators. If he would have been willing to exempt Stalin and admit his "mistake," he would have been cleared. But he stuck to his words and was summarily dismissed. He died a few years later. Today he has been "rehabilitated."

His home and laboratory are kept as a public memorial. They also house the leading immunological laboratory of Hungary.

Huzella had an exceptional ability to convey his own deep interest in biology to his students. He was convinced that the time had come when biology could be converted from "metaphysical speculation" into a natural science with precision and dignity similar to those of chemistry and physics. He believed that the biology of the interstitial space would turn into detailed biochemistry in a few decades but that the cell interior would remain a black box during the rest of the century. Before blaming him for a lack of foresight, we must realize that most biologists of the time were unwilling to accept his "optimistic" view even about the connective tissue.

I learned some tissue culture, but my practical experience remained rudimentary, and I compensated only slightly by avid reading in the still quite deficient library. After Huzella's removal, I realized that I could not learn more in the now largely nonfunctional department, and so I moved to Pathology. After a few weeks I found myself totally immersed in autopsies. There was a great abundance of cadavers here and very few pathologists. The large postwar classes of medical students had to be taught quickly. I greatly enjoyed the double task of teaching the little I knew and trying to explain to the rushed and often very nervous clinicians what their patients had died of.

In the early spring of 1947, one of "my" students approached me after an autopsy. He said something appreciative about my demonstration and asked whether I would be interested to visit Sweden with a student group. I was amused by his naiveté. Who would not like to visit Sweden? But were we not all aware of the fact that foreign travel was the exclusive privilege of important functionaries and people with much money and many good connections?

He replied that he was currently organizing a trip for students and that he would include me. Hungary still had an elected coalition government at this time. It was possible to get a passport, but this was not sufficient to leave the country. A special exit permit had to be issued by the "Allied" forces, i.e. the Soviet Army. It was very difficult to get this permit, and it was nearly impossible to obtain foreign currency.

I mailed my papers to my student who was interested in Sweden and totally forgot about our conversation.

Decisive Summer

In June 1947 my boss, Professor Baló, told me that I would be responsible for the autopsies during the coming month, virtually alone. I was happy, proud and frightened. I was not yet 22, far from being an MD, but the night's sleep of a professor in surgery could depend on what I was going to find. The combined feeling of responsibility and awe turned every autopsy into an exciting detective story. During my minimal "spare hours" I also started my first attempts to do some experiments. I was sitting in a corner of the laboratory with a small water bath and a stalagmometer, trying to follow a lead that had been opened up by my chief.

The most important messengers of my future destiny appeared in the shape of two house painters in the middle of July. They had been ordered to repaint the laboratories. I was chased from room to room with my water bath, but I refused to give up. Finally, I was squeezed into a small corner in a tiny windowless alcove that I refused to leave. The painters complained to Professor Baló. With an irritated "you can take two weeks vacation for once" he ordered me to leave my paradise. A senior colleague was to take care of the autopsies. I was angry and disappointed. What was I to do during two whole weeks?

By coincidence I learned that some fellow students, two couples from the Pharmacology Department, were planning to spend the forthcoming week at the Lake Balaton. I was also told that they had invited some other friends and that I was welcome to join them. We were allowed to use the terrace of a bombed summer house and were going to sleep on mattresses, spread out on the terrace. It was quite warm during the first week in August, and we would have a roof over our head. After considerable hesitation, I decided to join them, but I felt ambivalent and uninterested.

The place was unexpectedly pleasant and my fellow students were much nicer in private life than at the University. On the second day, the two other boys went down to the train to meet another student from the Pharmacology Department, who was to join us. I did not know who it was, and since the Hungarian language does not distinguish between *he* and *she*, I did not even known whether we were expecting a boy or a girl. After a while I saw them walking up the hill with the new guest: a dark girl with a strange, breathtaking beauty. I perceived a most unusual combination of hilarity and sorrow, seriousness and play in her eyes. It was Eva, my future wife and colleague until this day.

I had seen her before at the university, but my obsessive preoccupation with work prevented me from giving her or any other girl much attention. Still, I could remember very well how I met her the first time. On the second day of my medical studies in Szeged, I was standing in the Dean's office, to get my papers. She entered, dressed in a skiing outfit, having arrived in the city after a long and adventurous trip from Budapest, like my own. She asked me how to get papers. I saw that she was very beautiful. Her direct way of talking to a strange boy—very unusual for a girl in Hungary at the time—struck me as original and sympathetic. During the forthcoming weeks I saw her at some lectures, but then she disappeared.

Later I saw her name on the posters of the city theater. She was playing small roles in Pirandello and Molière plays. Half a year later I saw her again in Budapest. She had returned to medical studies and came sometimes to my autopsy demonstrations. I knew that she belonged to the same group of students in the Pharmacology Department as my married friends and temporary hosts. Their "gang" treated me with friendly tolerance, and even with a trace of respect for my "knowledge"—in spite of their "objections" to the "dead morphology" that pathology represented in their eyes. I respected their intelligence and their dynamic experimentation and could therefore forgive their blatant ignorance of pathology and clinical medicine.

But this time everything was different. There was one table but only three intact chairs in the ruined villa, and we were six. We had to place a board on each chair to hold two. Eva and I were placed on the same board and had to coordinate our movements to prevent each other from falling down. This trivial problem initiated a contact that metamorphosed after only a few hours into a passion that conquered my entire consciousness with the force of an elementary power. All other interests and problems vanished as if they had never existed. I spent eight days at the lake, intoxicated, overwhelmed, cut-off from all earlier reality.

An unexpected telegram arrived on the seventh day. Everything was settled for the trip to Sweden! My former pathology student or, as we were soon to call him, Our Leader, had succeeded against all odds. He had pursued his plan with obstinate ingenuity and obtained all the exit permits for a group of seventeen students selected by himself with the arbitrariness of a sovereign. We came from different faculties and were to visit Stockholm and Gothenburg as the guests of the Jewish Student Club there, in order to see a country that was saved from the war.

Now I did not have the slightest wish to go. I felt very bitter about having to leave the person who had become more important than anything else in my life so far. The week at the Balaton appeared as an eternity; everything before was unreal. But vague feelings of responsibility and premonition commanded me to go. I left at dawn on a Sunday morning. Eva told me later that she heard the train whistle while half asleep and thought that a beautiful summer episode was now over. She did not believe that I would ever come back from Sweden or that she would see me again.

Cell Biology 1947

The first International Congress of Cell Biology had just terminated when I arrived in Stockholm. I was told that Torbjörn Caspersson was one of the most important figures at the Congress. His recent development of ultraviolet microspectrophotometry on fixed cells created much attention.

The method was based on his doctoral thesis, written in 1936 in German and largely unavailable to English speaking readers during the war years. It was the first major attempt to combine morphology and cytochemistry. Cells were photographed in monochromatic UV light under standardized conditions. A semiquantitative method was developed to map the localization of nucleic acids and proteins in different cell types. Jack Schultz, one of J. H. Morgan's last disciples, was the first American geneticist who saw the potentialities of the new approach. He traveled to Stockholm to work with Caspersson shortly before the outbreak of the war. He brought genetic thinking to the biophysically oriented group. His studies with Caspersson on the banding patterns of polytenic insect salivary gland chromosomes gave the first information about the distribution of nucleic acids and chromosomal proteins and set the conceptual basis for the development of the chromosome banding technique by Caspersson and Zech three decades later.

The chemistry of the genetic material was still unknown at the time of the Cell Biology Congress in Stockholm. Most biologists believed that only proteins could provide the necessary diversity. Nucleic acids were considered as repetitive, boring molecules. Levene and Bass pronounced the death sentence on the coding capacity of the nucleic acids already in the 1930s. The mistaken analogy between the "4-letter alphabet" of the nucleic acids and the phonetic alphabet served as a roadblock: how could one build a language from four letters? Caspersson's semiquantitative measurements of nucleic acids and proteins in different cell organelles led him to conclude that there was a definite relationship between nucleic acid and protein synthesis and that the former might actually govern the latter. This visionary insight was widely disbelieved, however. The idea that nucleic acids might carry genetic information that could be translated into proteins was totally foreign, even to Caspersson. The fundamental discovery of Avery, McLeod, and McCarthy on DNA-mediated transformation in Pneumococcus, published in 1944, was widely ignored or discarded as an artefact.

The Cell Research Department of Karolinska Institute had just moved to the newly built campus on the northern edge of the city; there I was to spend all my scientific years, up to the present day. I visited it first in the middle of August, 1947, the peak of the vacation season and soon after the Congress participants had left town. Members of the Department who happened to be in town were frantically trying to get settled in the new building. As I made my entry, tall, blond, 37-year-old Torbjorn Caspersson was lying under a large instrument in a blue overall, trying to fix the wires. I thought that he was an electrician or a technical assistant. His identity was not revealed to me and I was not introduced to him. After I had learned the difficult art of protecting him from uninvited visitors a few years later, I could understand the reasons. In 1947, I was desolate when I had learned the next day that he had left for the USA. Only after a long series of complications did I get in touch with him, several weeks later. But my first conversation with him was decisive. Thanks to the rudimentary and largely theoretical knowledge of tissue culture, acquired in the Huzella laboratory two years earlier, I got the best-paid job of my life (if the importance of the salary is considered). I was employed as a junior research assistant, on 500 Sw Crs (about US $100) per month.

I still remember the mixture of ecstatic happiness and enormous anxiety. My situation appeared totally hopeless. I knew virtually nothing. I was halfway through my medical studies, still far removed from an MD. I was desperately in love with a girl whom I had only known during a summer vacation of eight days and who was on the other side of an increasingly forbidding political barrier. I did not know a word of Swedish. Still, I was firmly decided to resist the more comfortable possibility of continuing my studies in Hungary.

My motivation was reinforced by a series of articles that kept appearing in the major Swedish daily, *Dagens Nyheter*, translated for me by my temporary host. The Prime Minister of Hungary, Ferenc Nagy (not to be confused with Imre Nagy) of the Smallholder's Party has just fled to the West, and he gave a series of interviews to the Swedish paper. In contrast to the rosy optimism that prevailed among my friends in Budapest who hoped that Hungary would become a democratic country, Nagy's statements had an ominous ring. He said that the influence of the Communist Party was increasing continuously behind the scenes. The Stalinist party leader, Rákosi, was acting under the protection of the Russian forces. The politicians of the other parties were frightened. Several of their leading representatives were arrested on false charges and deported to unknown destinations. Those who remained were increasingly inclined to give in. The police were infiltrated by party members. Nagy did not have the slightest doubt that a Communist takeover was imminent. Similar signals reached me indirectly from one of my teenage idols, Nobel Prize winning biochemist Albert Szent-Györgyi. He was still holding many high posts in Hungary at the time, but he had told his nephew, who was a friend of mine, that the days of freedom were numbered. If you were young and wanted to have a future in science, you should get your degree as soon as possible and leave the country.

Farewell, My Native Land

In mid-September, I decided to go back to Budapest and try to get out for good. My most important acquisition was safely tucked away in my breast pocket: a re-entry visa to Sweden and a labor permit for continued work in Caspersson's department. My passport was still valid for a few months.

The reunion with Eva confirmed what we both knew already: we wanted to live and work together. The day after my arrival, some of our friends gathered at my home to hear the latest news from the "great world." I told them about Nagy's report and the iron curtain that was about to descend over Hungary. The reaction was mixed. Those who were already preparing to leave believed me. Others wanted to stay and hoped that my report was exaggerated. One of them—still a good friend today—declared that I was probably right, and for that reason, he was going to break all further contact with me. This was his country, Hungarian was his language, his historical roots were here. I should leave, if I felt so inclined, but he had to stay and do the best he could. Today he is the foremost medical historian of Hungary.

I had none of his historical perspectives. I had only one goal, to get married and leave the country.

But how to get married? It had to be in secret, because nobody would understand why two 22-year-old students who had known each other for only a short time and had no income would want to get married. And how could my future wife join me? She had no passport and the difficulties in getting one were now increasing day by day. We agreed that I would go back to Stockholm before my own passport expired and try to obtain letters of invitation for Eva that could help her to get a passport.

The last weekday before my trip was a Friday. Eva and I met outside the pharmacological institute to go to the day's lecture. I suggested that we should go to the prefecture instead and ask how one gets married. We got a list of the many documents you needed. It looked hopeless. It would take months to get them. I suggested that we ask for the first document, a certificate to show that we had no police records. We went to the police station. "It takes at least three weeks." Suddenly I acted on impulse. I had always heard others tell of such things but I myself had neither seen nor done it. I pulled a fairly modest bill out of my pocket and put it in the policeman's hand. "Pardon me, how much time was it, you said?" "I'll go and get it at once," he answered.

It was now 11 AM. We continued from office to office.

Everywhere the same answer: one week, four weeks, six weeks. A little bill in the hand—the certificate was completed within a few minutes. I was amazed to find that the shyness I usually exhibited before persons of authority vanished completely. I learned a lesson about the importance of motivation and the unsuspected possibilities it may open to surpass one's limitations.

By 3 PM only one document was missing: a medical certificate that neither of us had venereal disease. The tests would take several weeks. What to do now?

We went to a slightly older colleague who had recently finished his medical studies. He had just started his first assignment in the Children's Hospital. We told him, in the strictest confidence, about our situation. He had a good laugh and wrote the certificate on the hospital stationary. By 4 PM we were at the prefecture again. We had all the papers and wanted to get married that second. Two other friends, sworn to the highest secrecy, came along as witnesses to the wedding. The official had just finished the day's work and had taken off the broad Hungarian tricolor from his corpulent chest when we rushed in. We heard him telling his wife on the phone that he was on his way home for dinner. Marry us at this time of day? Not a chance! Come back on Monday!

I started to appeal to his human feelings. I had to leave the country on Sunday. How could I leave my young bride alone if we didn't get married? He was noticeably irritated and doubted that we had all the papers. While leafing through the documents, he caught sight of the doctor's certificate that had been drawn up at the Children's Hospital. He laughed until tears ran down his cheeks. This was the funniest thing he had seen during his whole time in service. Now he was in splendid spirits. The flag resumed its place on the large body. We promised to love one another til death us did part.

Afterwards we ate our wedding dinner on the hall bench together with our witnesses. There was only one dish: my mother's carefully packed goose liver sandwiches. In the evening we went back to our parents' homes where no one suspected anything.

That Sunday I returned alone to Stockholm. Eva joined me, after many complications, in March 1948, after the Iron Curtain had already descended over the country.

GEORGE AND EVA WRITE:

The Genetics Congress

In August 1948, several months after we were happily settled in our rented room and Eva had also started to work in Caspersson's department, the International Congress of Genetics took place in Stockholm. The presidential address of J. H. Muller was a scathing denunciation of the abuse of genetics in the Soviet Union. The scientific world was still largely unaware of the fact that the "theories" of a charlatan, Lysenko, had been declared "official" by the Central Committee of the Communist Party, meaning that it became essentially illegal to do any scientific work in genetics. Muller himself had been the first to introduce Drosophila genetics into Russia, and he was still a member of the Soviet Academy of Sciences at this time. He called Lysenko "a paranoic and half educated young demagogue who had done some work in raising plants but who was in fact ignorant of scientific principles and incapable of understanding them." He added that many of the outstanding Russian geneticists had disappeared, and some had lost their lives in unexplained ways. His speech ended with his resignation from the Soviet Academy. The reaction from Moscow came the day after. They refused to accept his resignation and expelled him.

On the last day of the congress, the Bulgarian delegate asked to make a statement at the concluding plenary session. Speaking in the name of the delegates from Bulgaria, Roumania, Poland, and Czechoslovakia he delivered a strong protest against Muller's introductory speech that was "ill-suited to favor international understanding." His protest was taken to the protocol.

After the session was closed, the representative of Hungary came to us. He did not understand English well, and we had previously helped him. He wanted to know what the Bulgarian delegate said. When he heard our interpretation he became extremely upset. It was typical for the Slavic delegates to leave out the Hungarians! He had to join the protest, he had to think of his family! How could he return without having signed it! But he was out of luck, the congress was over, nothing could be done for him. His panic showed us how the fear imposed by Stalinism had descended on the country we had left only a few months earlier. It was also a reminder of the eternal strife among the nations that have risen from the ruins of the Hapsburg monarchy.

During the Congress we learned about the startling progress in microbial genetics. Bacteriology had been the last citadel of Lamarckism. At this time, when proteins were regarded as the vehicles of genetic information, when notions about a bacterial nucleus were regarded with great suspicion, induced enzyme adaptations and drug resistance were widely attributed to the inheritance of acquired characteristics. But the rapidly growing evidence of clonal variation and Darwinian selection was now definitely gaining ground, integrating microbiology with the rest of biology. Our knowledge about cancer was rudimentary, but we started nevertheless wondering whether the population dynamics of microorganisms and the phenomenon of cancer might share some common denominator(s). Cancer cells are resistant against growth control of the organism. Can they be compared to drug resistant microorganisms? Could cancers also arise by a series of mutations? These seemingly puerile notions were to play an important role for our work later on.

The Cell Research Department

Back at the laboratory, we found ourselves in an exciting environment but facing another impossible situation. We were still medical students in mid-course. We were struggling hard to get into a Swedish medical school and finish our studies. At first, this looked impossible but eventually we succeeded, one by one, taking turns between the school and the lab work. Worse than that, the project given us turned out to be quite unmanageable. Caspersson's methodology was based on the absorption of monochromatic ultraviolet light in fixed cells. Shortly before our arrival, it was heavily criticized by Barry Commoner and other biophysicists. They suggested that the loss of UV light registered by Caspersson's optical system was not due to absorption but to light scattering from the denatured proteins. Due to this artefact, part of the nonabsorbed light would never reach the objective, leading to false conclusions about the localization of nucleic acids and proteins. Their distribution in living cells could be totally different from the pattern suggested by Caspersson's measurements.

Our task was to measure light absorption in living cells. But this was more easily said than done. Tissue culturing of the times followed the dogmas laid down by Alexis Carrell. The plasma clot and the embryonic extract were regarded as essential substrates. Nobody in his right mind would have thought of culturing cells directly on glass, even less on quartz slides. The plasma clot was not transparent to UV. It turned out that my sudden and unexpected employment after my first conversation with Caspersson was due to the fact that I had some experience of growing cells on collagen, Huzella's favorite method. Collagen is poor in aromatic amino acids, and it was therefore expected to provide less of a problem for UV-microscopy.

We struggled frantically to obtain some results. We had no experience, no assistance, and virtually no apparatus. The large UV-equipment was not suitable. The cells were killed by UV long before we could take a picture. Washing and sterilization of the glassware, preparation of the embryonic extract, and most difficult of all, collecting plasma from the carotid of our single rooster with a primitive paraffin oil canule were a neverending struggle. At last, we managed to take a few pictures before the cells died, but we were still far from the number of monochromatic exposures needed for a spectrum. Our future looked dim. Salvation came in the form of two unexpected events. The first was a lecture given by Hans Lettré of Heidelberg on the Ehrlich ascites tumor, which he used for biochemical studies. Eva immediately pointed out that we might obtain homogenous populations of living cells from the peritoneal cavity of the mouse, without having to do any tissue culture at all! Our first attempt to propagate the tumor, kindly sent by Lettré in the form of a single mouse, ended in total failure, however. The first of our inoculated mice developed a nice round belly that turned out to carry a lovely litter of eight, instead of the expected tumor cells. But the second mouse developed a tumor, and we were in business. But as we were getting ready for the UV pictures, a paper was published by Brumberg & Larionov in the USSR. They used a new, reflecting optical system that avoided the killing of the cells during UV-exposure. They had done all the experiments that we had planned and showed that Caspersson was right and the critiques were wrong. UV-microscopy did measure nucleic acids, and they were localized exactly in the organelles where Caspersson had found them in fixed cells. Our project had become obsolete overnight. What were we going to do?

We expected the worst, but Caspersson suggested that we continue to work with ascites tumors and try to formulate our own project. The early experience of the Genetics Congress came to our rescue. Why was the Ehrlich ascites carcinoma unique? Why could other tumors not be propagated in this freely dissociated "fluid" form? Did most tumor cells require a solid substrate and/or the microenvironment of a solid tissue? We had some ideas about how to start looking at this, but our mouse and tumor facilities were very limited. Inbred mice were totally unknown in Sweden at this point.

Salvation came again unexpectedly. In the summer of 1950 we participated in the International Cancer Congress in Paris. The week was occupied by frantic and hopeless efforts to get acquainted with the entire cancer field, interspersed with meetings with old friends who had left Hungary after us. At the end of the week, we felt definitely reassured about two earlier, disparate but equally important, conclusions: (i) It was very fortunate that we had left Hungary in time, for the Stalinistic system now had a firm grip, and (ii) tumor cells could be definitely regarded as genetically heterogeneous populations with extensive subclonal variation. We also felt proud as contributors to the Congress. George lectured in broken and very slow English in the 32nd parallel section, the late afternoon of the last day, with six persons in the audience (1). It turned out later, however, that this was an extremely important event. One of Otto Warburg's assistants had been in the audience and brought home the great news that the mouse ascites tumor cell is an equally good tool for large-scale experimentation on such relatively homogeneous cell populations as the famous Chlorella algae of the Great Master. Warburg immediately requested the cells and was later very helpful in supporting us. In a letter written in 1956, he stated that we had made a very important contribution, because we had sent him the cells that made it possible for him to solve the cancer problem. All bureaucrats were deeply impressed!

During the Congress week, we also enjoyed frantic, colorful, decadent, exhilarating, and slightly putrescent Paris. Sitting at a cafe on Boulevard St. Michel the evening of Bastille Day we exclaimed: "How wonderful! how crazy! how can one possibly live in a sterile country like Sweden?" Rattling home a week later in a third-class wagon across strike-torn Belgium we said: "How marvelous that we can return to quiet, boring, aseptic, polite Sweden with its thousands of lakes, endless forests, and luminescent nights!"

We had hardly opened the door to our rented room in Stockholm when I saw the new miracle: an express letter in Caspersson's own handwriting. "Get in touch with me immediately on arrival."

One of the main private research foundations in Sweden, established in the memory of Knut and Alice Wallenberg, had asked Caspersson to choose two young men for an urgent mission. They were to go to the United States for several months and report about recent advances in cancer research. Caspersson chose one of my older colleagues and myself. But Eva had to stay home—there was not enough money. My colleague was to travel around from center to center. My task was to work with Jack Schultz at the Institute for Cancer Research in Fox Chase, Philadelphia, on my own project, and to make short visits to some of the major centers in the neighborhood.

We received the news with a mixture of joy and sorrow. It was a fantastic opportunity. But the sorrow and anxiety of being separated again from my young wife, and for quite some time, were further aggravated by the sudden outbreak of the Korean war with its forebodings of a possible world war. We shared our vision of an approaching Apocalypse with most other survivors of the Second World War and the Holocaust. Our officially stateless status added fuel to the nightmares. Still, I knew I had to go.

The Statue of Liberty

The Institute for Cancer Research has developed from a small private research group at Lankenau Hospital, due to the great foresight of Stanley Reimann. When I got there, they had just finished a major expansion from a small group of scientists to a large research center in a magnificent new building. Several prominent biologists had joined the laboratory. The leitmotiv was to look at the cancer problem from the biological point of view. My own boss was Jack Schultz, a lively little man in his sixties. Jack exuded boundless curiosity, joy of life, and great human warmth. He received me as if I were his long lost, finally recovered son. During my stay he often gave me a lift from my rented room to the laboratory. Most of what I know about genetics can be traced to those car rides. But the trip was not over when we arrived. Jack's office was at the far end of a long corridor. Walking down the hallway he would stick his head into every lab and stop and talk with people on the way. He asked them about everything, the health of their kids, mother's broken leg, the weekend excursion, but first and foremost about the latest experiment. The people brightened visibly when they saw him and were always ready to stop for a chat or to ask him to come in and look into the microscope, at a bacterial plate or at a Drosophila progeny. Jack looked, listened, discussed, interpreted, proposed new experiments. Under his arm he carried his brief-case with all the papers he planned to finish during the day. Sometimes half a day passed before we arrived at his office where his secretary waited in despair!

I visited Jack's office 25 years later, long after his death. It has been refurnished as a conference room. It bears Jack's name. A silver plate on the wall reminds us of the unselfish inspiration he provided to everybody in his environment.

Jack succeeded in communicating the notion that biology is the most exciting science. He told me about whole worlds I had never heard about. Barbara McClintock's discovery of transposons in maize was one of them. Jack was one among the dozen or even fewer geneticists who understood what McClintock was talking about. He already knew, 10–15 years before most others, that her findings were going to revolutionize biology.

Jack's corridor was a wonderland for me at the age of 25. Briggs and King experimented with nuclear transplantation to enucleated frog's eggs. The question was whether the nuclei remained totipotent during differentiation. This was also relevant for cancer research. Could cancer cells contain a totipotent nucleus? The question was answered several decades later, by Beatrice Mintz at the same institute—at least answered so far as diploid teratoma cells were concerned. A cartoon appeared on the wall of the same corridor where Jack and I so often walked in the morning. Two mice were talking to each other. One of them said: "My father was a cancer, what does your father do for a living?"

The Mintz experiment is still unique in showing that at least some cancers can develop by epigenetic changes. The majority are no doubt due to changes at the DNA level, however.

My other important master at the ICR in Philadelphia was the mouse geneticist Theodore Hauschka. Through him I became acquainted with the inbred mouse. He had also taken a direct interest in my experiments. He gave me my own room in the perfectly organized mouse colony where I was in full operation for days on end. I compared the ability of different solid tumors to grow in the fluid form in the abdominal cavity. When they were reluctant to behave according to my wishes, I tried to select variants that would. At the same time I began to wonder whether the "histocompatibility genes" that were shown to govern the transplantability of tissues might provide me with the right system to substantiate our speculations on variation and selection within populations of tumor cells?

Despite my loneliness and separation from Eva, alleviated only somewhat by the letters I mailed her daily, I enjoyed being in America. In addition to the positive attitude of Schultz and Hauschka, the environment of the whole laboratory was highly supportive for a young man. There was a wholesome difference compared to European laboratories, particuarly with regard to teacher-student relationships. It can best be summarized by a statement of the Danish biochemist, Lindeström-Lang: "The greatest accomplishment of the American revolution was to establish the right of young students to ask foolish questions."

During my stay in the United States I lost part of my emigrant complex. Hungarian emigrants in a comparable situation commented later that my initial shyness has turned into its opposite in the American setting. It may have seemed so. I was no longer afraid to ask questions, to inject myself into the conversation of learned professors, to speculate, and to risk making a fool of myself. I began to feel that this was not only my natural right but my responsibility.

Tumor Progression by Variation and Selection

The work on the conversion of solid into ascites tumors turned out to be quite interesting (2). Lymphomas and leukemias often converted immediately, while carcinomas and sarcomas refused to grow in the ascites form at first. Some of them could be converted gradually, however, by passaging the few desquamated, freely floating tumor cells in the peritoneal fluid. Using a simplified form of the Luria-Delbrück fluctuation analysis, I could show that this conversion was due to the selective enrichment of a small number of spontaneously occurring variants.

After four months in the United States, I returned to Sweden with 200 mice, anxiously guarded in my New York hotel room overnight and during the plane trip of more than 24 hours, to the great displeasure of my fellow passengers.

Back in Stockholm, the ascites tumor variants turned out to be stable. They retained the ability to grow in the peritoneal fluid immediately after inoculation, even after reconversion to the solid form and subcutaneous propagation over extended periods of time. The ascites adapted tumors were also more metastatic, less adhesive, and had a higher surface charge than their original nonadapted counterparts (3, 4). A comparison of our findings with Leslie Foulds' (5) work on tumor progression and Jacob Furth's studies (6) on the change of hormone dependent to autonomous tumors convinced us that we had hit an unusually well-defined case of progression (7). It appeared to have a certain clinical relevance, at least at the conceptual level, because it showed that tumor cell populations were heterogeneous, and subpopulations could differ in their metastatic properties. But where did we go from here?

Tumor Immunology

To study variation and selection in tumor cell populations, it was obviously necessary to study variation first. We were looking for cellular markers, determined by known genes that could be detected at the cellular level. We found them in the recently discovered H-2 antigens of the mouse. George Snell has just started to distribute his first H-2 congenic mouse strains. We had induced tumors in H-2 heterozygous but otherwise congenic F_1 hybrids and isolated haplotype-loss variants by transplantation to the parental strains (8). Single haplotype-loss variants could be readily obtained, but in frequencies that varied widely between different tumors, even if they had been induced by the same agent and in the same host genotype. This biological variability was no longer a surprise to us, after the variations in ascites convertibility that we had encountered previously. Double H-2 haplotype losses were extremely rare.

Around this time, in the mid-1950s, a former colleague from medical school started to make extravagant claims concerning the prospects for the immunological prevention and cure of human cancer. He had immunized a horse with pooled tumor tissue and was firmly convinced that his serum reacted with a universal tumor antigen. He advocated immediate vaccination against cancer. The newspapers made a big splash. He was supported by some of the most powerful professors of microbiology and virology who had no experience in cancer. He injected himself with a HeLa cell derived "cancer vaccine" on TV. The public regarded him as a hero, particularly since the newspapers started to accuse the "cancer establishment" of lacking any concern for preventing cancer, due to vested interests.

The few of us who actually worked with cancer cells were profoundly sceptical. This could just not be true. But what was the real situation? Would tumors elicit immunity in their own inbred strain of origin?

My previous work with Hauschka left me imbued with a healthy scepticism against most earlier research in tumor immunology. The field was dominated by misinterpreted artefacts of experimentation with noninbred mice. The confusion between transplantation immunology and tumor immunology prevailed during the entire first part of the century. Only several decades after the development of the inbred mice by Little, Strong, Tyzzer, McDowell and others, and after the formulation of the "transplantation laws" by George Snell, was it gradually realized that the so-called "transplantable tumors" violated histocompatibility barriers because serial homografting had selected them to outpace the rejection response. If the balance was tilted in favor of the host, e.g. by pre-immunization with attenuated tumor cells, it could reject the tumor. This easily won immunity could not be reproduced with tumors that had arisen in homozygous mice and were tested within their own strain. But what would happen at a more modest level of ambition? Could immunity protect the syngeneic host against near-threshold numbers of tumor cells? Clinicians and pathologists have always maintained that only a small proportion of disseminated tumor cells could grow into metastases in the human patient. Could an immune response that fell short of protecting the host against an established tumor still reject disseminated cells, in analogy with concomitant immunity in antiparasite responses?

Just as we started to think about these matters, Foley (9) and Prehn & Main (10) suggested that chemically induced mouse sarcomas, but not spontaneous mammary carcinomas, could elicit a state of immunity in syngeneic mice. The data were persuasive but still not fully convincing. Did chemically induced tumor cells really possess a distinct antigenicity of their own, or did these experiments merely reflect a residual heterozygosis in the inbred strains? It was obvious that the question could be decisively settled if it could be shown that the primary host could be immunized against its own tumor.

Using a combined scheme of tumor induction, operative removal, immunization with irradiated autologous tumor cells, and challenge with graded numbers of viable cells, we could show that methylcholanthrene-induced sarcoma cells were indeed capable of inducing true rejection reactions in the original host (11). Different tumors varied in their immunogenicity over a 5 log range of cell doses, required to break the state of immunity. Another and even more striking manifestation of biological individuality concerned the individual distinctness of the tumor antigens, also noted by Prehn, Baldwin, and Old (12–14). Each tumor could only immunize against itself. Cross-reactions were rare and irregular. The total number of possible specificities is still not known. We found no cross-reactions among more than 20 tumors. Hellström could not immunize mice against MC-carcinogenesis by using pools of a dozen tumors for immunization, while Old reported a certain preventive effect after the use of nonspecific immunomodulators that acted presumably by boosting the host's own responsiveness.

The nature of the carcinogen was not immaterial in determining the immunogenicity of the chemically induced tumors. Among the aromatic hydrocarbons, MC, BP, and DMBA induced sarcomas with decreasing immunogenicity, in that order. Sarcomas induced by the implantation of cellophane film were hardly immunogenic at all (15). In the rat, Baldwin found that most azo dye–induced tumors were highly immunogenic, whereas acetylaminofluorene-induced tumors and spontaneous fibrosarcomas were not immunogenic at all (16).

Several decades have passed since these findings, but the nature of the TSTA (tumor specific transplantation antigen) of the chemically induced tumors is still a mystery.

Antigenicity of Virus-Induced Tumors

In 1958 I went to the Canadian Cancer Conference, in Honey Harbor, Ontario. Stewart and Eddy's pioneering work on the polyoma virus was still very new. Most participants were flabbergasted by the number and variety of the tumors that arose after the inoculation of the virus into newborn mice. Burnet was one of them. "Sir Mac" had recently shifted from virology to immunology and had developed a very negative view of the role of viruses in cancer in the course of this transition; he considered all virus-induced tumors as laboratory artefacts. Viruses were essentially cytopathic, and he saw no place for any true tumor inducing effect. Confronted with the polyoma story, he formulated immediately a new hypothesis. It was based on the only observation of Stewart and Eddy that turned out to be incorrect. They claimed that polyoma tumors were not transplantable. This was due to the accidental use of heterozygous mice, however.

Burnet suggested that polyoma virus may destroy some unknown, systematic "growth-controlling center," a possible "hypothalamus-like" homeostatic regulator of cell renewal in many different tissues. This would explain the ability of the virus to cause tumors in many different tissues. These tumors would not be transplantable to mice that have not been similarly conditioned by polyoma virus.

Hans-Olof Sjögren had just started to work with us at this time. Stimulated by Burnet's idea, I asked him to test the transplantability of polyoma tumors in unmanipulated and polyoma-infected syngeneic mice. The result was the exact opposite of what was predicted by Burnet's hypothesis: the tumors were readily transplantable to untreated mice, while small, graded numbers of cells were rejected by virus-inoculated syngeneic mice (17). The resistance of the virus-infected mice could be transferred adoptively with lymphocytes but not with serum. Both Karl Habel and our group later showed that antiviral immunity was neither necessary nor sufficient to induce rejection. Polyoma-induced tumors or transformed cells induced rejection, whether they released virus or not. All polyoma-induced tumors were rejected by the immunized mice, irrespective of tissue origin, but they did not reject tumors induced by other viruses or by chemical agents. We have therefore developed the concept of a polyoma-specific transplantation antigen (TSTA) that was present in all tumors induced by polyoma, but not in tumors induced by other agents. We and others later found that similar group-specific rejection-inducing antigens were present on other virus-induced tumors (18). The retrovirus induced leukemias were particularly useful for the study of both humoral and cell-mediated reactions, as was shown by Old et al (19), and by our group.

Moloney virus-induced lymphomas were particularly useful for these studies, since they gave a brilliant membrane fluorescence reaction with the sera of preimmunized, syngeneic animals. Nevertheless, it was not possible to distinguish the rejection inducing antigen from the viral glycoprotein that accumulated on the surface. Different Moloney lymphomas induced in the same inbred strain differed in their rejection inducing potential. This system has permitted a distinction between immunogenicity and immunosensitivity, and we could show that they were independent variables. The former correlated with virus release, while the latter did not.

The Department of Tumor Biology

During the years of our transition from H-2 antigens to tumor immunology, our department developed rapidly. It was formally established in 1957, against all odds. Previously, both George and Eva had become assistant professors in Caspersson's Department of Cell Research (in 1951 and 1955, respectively), but our appointments were limited to a maximum of 6 years. Unless one acquired a tenured position, one was out of the research system. But no tenured positions were available in our field, which had not been previously represented at the Swedish universities. To circumvent the inflexibility of the university system, a number of "personal professorships" had been established for individual scientists, but some years before this time, the government decided to stop creating new positions of this type. Science was too expensive already for a country of eight million, they said; and it was also undesirable to continue the traditional recruitment of medical students into research. There was a shortage of doctors that made the authorities very sensitive about this, particularly since all higher education was financed by the taxpayers, and there was heavy competition for admission to the medical schools.

I had a good offer from the ICR in Philadelphia, and we seriously considered moving to the United States. Meanwhile, the Karolinska Institute, the Medical Research Council and the Swedish Cancer Society joined forces to initiate a parliamentry move, requesting the establishment of a Department of Tumor Biology with George Klein as its first head. This move was supported by representatives from the four major political parties, but it failed to convince the Government. Decisions about budgetary matters rest with the Parliament, however. A parliamentary committee dealt with the matter on April 30, 1957. The odds were against us. The committee had 13 members of the ruling Labour Party and 12 members from the three major opposition parties. It was expected that the move would fail with a majority of at least one vote. In fact, the opposite happened. One Labour Party member (unknown to us) decided to vote with the opposition parties and the Department was established, as of July 1, 1957. Numerous medical and PhD students interested in research joined our group. With the support of the National Institutes of Health of the United States and the Swedish Cancer Society, the department expanded rapidly. The accumulation of married couples who pursued research together was a peculiar feature of the lab that has remained with us ever since. In the early years, the Hellströms, the Möllers, the Sjögrens, the Nordenskjölds, the Nadkarnis, and the Ozers were some of the examples. At one point we had seven married couples working at the lab at the same time, surely a world record.

The problem of space became overwhelming in the late 1950s. Again, there was no provision or precedent for the type of support that was needed. We were facing the possibility of having to return the first major NIH grant we had received under the Virus Cancer Program. I turned to the Swedish Cancer Society, although with little hope since the statutes of this essentially private organization explicitly discouraged the support of building facilities. But the Chairman, Professor Hilding Bergstrand, drove through a positive resolution against all odds. A new laboratory building was constructed in 1961. It houses the Department even today.

Burkitt's Lymphoma

Sometime in the mid-1960s, Eva suggested that we should use our experience on virus-induced murine lymphomas to examine a human lymphoma with a presumptive viral etiology. Could we detect group specific antibody responses that might be helpful in tracing a virus? Burkitt's lymphoma (BL) was the obvious choice. The recent description of the highly endemic occurrence of the African form which is climate-dependent strongly supported the idea of a possible viral etiology.

I wrote letters to numerous hospitals in Africa and to international organizations, explaining our project and asking for tumor, blood, and serum. I received some polite letters in reply, promises of material, and lovely stamps which made my son happy. But the material was not forthcoming at all, apart from an occasional shipment that arrived broken or infected. Then somebody—I have forgotten who—advised me to write to Peter Clifford, ENT surgeon at the Kenyatta National Hospital in Nairobi.

I got no letter and no stamps in reply, but the material started coming in a continuous flow. It arrived with chronometric precision on the single direct flight from Nairobi, late Tuesday afternoon. Large dry ice boxes carried hundreds of sera, and a special wet ice package contained fresh biopsy material. There was always a long list in Clifford's own handwriting with all the essential details and a brief "good luck" message.

We worked together with Peter over a period of more than 10 years. We have published 45 joint papers, the first in 1966 (20), the last in 1974 (21). We worked and published together for several years before we had a chance to meet in person. This taught us a new lesson. For collaborative studies, we tried to find a colleague who was motivated to study the problem and to collaborate with us, no matter where he or she resided. But we hasten to add that we have never encountered another clinical collaborator like Peter Clifford. He had a profound interest in BL, ever since he introduced chemotherapy in the treatment of the disease and became fascinated by the remarkably good regression in most of the patients. Their long-term survival eventually turned out to be complete cure in 15–20% of patients, including those who had only received incomplete chemotherapy. This was quite different from the effect of chemotherapy on other types of B-cell lymphomas. Clifford was convinced that the immunological response of the patient was decisive. If it was effective, even incomplete chemotherapy could induce total and long-lasting remission. If it was not, even more effective forms of chemotherapy were ultimately unsuccessful. Peter hoped that we would find evidence for an antitumor response in his patients.

We changed our working habits. Every Tuesday night was "Burkitt night." We made living cell suspensions from the fresh tumors, reacted them with the patient's own serum and other sera, and tried to read the tests immediately to obtain clues for the continued work. It was not difficult to motivate our personnel to work through the night every Tuesday.

Eventually, numerous other laboratories requested material, in the United States, England, and Japan, and some of them became engaged in collaborative projects. We could identify a membrane antigen (MA) that was expressed in some Burkitt lymphoma–derived cultures, but not in others (20). When I presented these data at an ACS Conference in Rye, New York, in 1967 (22), Werner Henle gave a talk in the same session. He reported his results, obtained with an immunofluorescence test on fixed BL cells that he and Gertrud Henle had recently developed, later known as the VCA (viral capsid antigen) test (23). They already knew that the reaction was due to structural antigens of a newly discovered herpes virus, first seen by Epstein, Barr, and Achong in the electron microscope. Henle showed that it was antigenically distinct from previously known herpes viruses (24). We decided to call it EBV.

The Henles' VCA test and our MA test showed a certain concordance. The same lines appeared to react or failed to react in both tests. At the Rye meeting we agreed to collaborate. This initiated a highly productive association that has lasted for 20 years, terminated only by Werner Henle's death in 1987.

Already in the beginning of this work we obtained definite evidence that MA was encoded by EBV (25). It is now known as one of the viral envelope glycoproteins. It assembles within the membrane of the virus-producing cells, and after virus release, it can also attach to other cells in the same culture if they carry EBV-receptors. With Jondal, Yefenof, and Oldstone, we later identified the B-cell specific C3d (CR2) receptor as the attachment site of the viral glycoprotein (26, 27).

By 1970, it was clear that Epstein, the Henles, and ourselves had only seen the top of the iceberg when we looked at viral particles, VCA or MA. They only appear in virus-producing cell lines, and only in some of the cells. With Harald zur Hausen, we found in 1970, however, that more than 90% of the African BLs and all low differentiated or anaplastic nasopharyngeal carcinomas (NPC) contained multiple EBV-genomes per cell, no matter whether they produced virus or not (28). In 1973 I [GK] have found with Beverly Reedman that 100% of the cells in EBV-DNA positive BL biopsies and cell lines contained an EBV-encoded nuclear antigen, which we decided to call EBNA (29). Today we know that EBNA consists of a family of at least six different proteins (30).

Several important discoveries have been made by others in the meanwhile. The Henles, Pope et al, and Nilsson et al found that EBV could readily immortalize normal B cells in vitro (31–33). Departing from a serendipitous observation on a laboratory assistant, the Henles discovered (75) that EBV is the causative agent of infectious mononucleosis (IM). With Svedmyr, we could readily detect EBNA-positive cells in the peripheral blood of mononucleosis patients (34), and the Henles and George Miller found that the saliva of these patients contained transforming virus. Transformation was thus a natural property of the virus, not a laboratory artefact due to the accidental isolation of a defective strain, as our colleagues in the lytic herpes virus fields initially surmised. Miller & Epstein have also shown that EBV can cause lethal lymphoproliferative disease in immunologically naive marmoset and owl monkeys (35, 36).

Mononucleosis appeared as an acute rejection reaction of the "immunologically prepared" human host, selectively conditioned by a nearly symbiotic relationship with EBV over millions of years, against the virally transformed B cells. We found that the peripheral blood of the acute IM patient contains activated killer cells that can lyse EBV-carrying and other target cells (37, 38). Moreover, autologous mixed lymphocyte cultures between EBV-transformed B-cell lines and T cells of the same normal donor generated a proliferative and cytotoxic response equally as strong as that of MHC-incompatible allogeneic MLC (39). Later, Rickinson, Moss and Pope showed that the autologous mixed cultures generated specific MHC class I–restricted CTLs by repeated stimulation (40). Eva's group, Sigurbjörg Torsteinsdottir, and Maria Grazia Masucci in particular, showed that the CTL response was heterogenous, directed against different target epitopes (41). The nature and specificity of the relevant targets have not been clearly defined yet in terms of the known virally encoded proteins, although current evidence by Moss et al and by Thorley-Lawson, respectively, indicates that both EBNA-2 and LMP epitopes may serve in this capacity (42, 43).

Since the work of Townsend et al (44, 45) has shown that MHC class I–associated peptides of processed endogenous or viral proteins can serve as immunogens and CTL targets, it would not be surprising if even more among the seven known growth transformation–associated EBV proteins could serve as CTL targets. A similar reasoning can be applied to the polyoma virus–induced TSTA, discussed above. The recent work of Dalianis et al in our laboratory suggests that all three polyoma encoded T-antigens can elicit rejection responses of the TSTA-type.

The hypothesis that T cell–mediated responses inhibit the proliferation of EBV-carrying B cells in healthy seropositives and in IM patients was reaffirmed when we found with David Purtilo (46) that most and perhaps all lymphoproliferative diseases that appear in congenitally or iatrogenically immunodefective patients, like children with the X-linked lymphoproliferative syndrome or organ transplant recipients, carry EBV-genomes. Hanto, Ho, and others have later shown that these initially polyclonal immunoblastic proliferations may progress to monoclonal lymphoma (47, 48).

While the tumorigenic potential of EBV was clearly established by these and related findings, its lifelong, innocuous latent presence in more than 80% of all human populations has also suggested that disease occurs only as an accident. Even mononucleosis appears as an "accident" of civilization. Modern hygienic conditions have apparently interfered with the normal, disease-free ecology of the virus-host relationship, with its predominant symptom-free early childhood infection.

The "accident" of the EBV-associated tumors has now been largely clarified for Burkitt's lymphoma, as described below, while the pathogenesis of nasopharyngeal carcinoma, the most regularly EBV-carrying human tumor, is still not understood.

Oncogene Activation by Chromosomal Translocation

By 1970, it was clear that some important element was missing from the BL scenario. EBV has clearly contributed to the genesis of the high endemic form of the disease, since 97% of the African BLs carried the viral genome, whereas non-BL lymphomas did not (49). Moreover, the prospective study of Geser and de The (reviewed in 50) showed that children with a high EBV-load are at a greater risk to develop BL than are their brothers and sisters with a low EBV-load, as indicated by the antibody titers.

Since the number of EBV-infected B cells represents only a minor fraction of the total B-cell population even in persons with a high EBV-load, the presence of the virus in the majority of the African BLs can only be interpreted to mean that an EBV-carrying B cell runs a greater risk of turning into a BL cell under the conditions prevailing in the "high BL belt" of Africa than does its EBV-negative counterpart. This is to say that EBV contributes to the etiology of the tumor. But this is still not a satisfactory explanation; some essential element is obviously missing. BLs differ from the true EBV-induced lymphoproliferative diseases like fatal mononucleosis or the immunoblastic lymphoproliferative diseases in organ transplant recipients, with regard to their cellular phenotype (51). The latter resemble the EBV-transformed B-cell lines of nonneoplastic origin (LCLs). LCLs are permanently growing immunoblasts that express a set of activation markers but not CALLA or BLA. BL cells, on the other hand, carry surface antigen and glycoprotein markers that resemble resting B-cells, rather than immunoblasts (52, 53). They express CALLA and BLA but no activation markers (unless they drift to a more LCL-like phenotype during prolonged cultivation). Recently, Gregory et al found normal B cells with a corresponding phenotype in tonsil germinal centers (54).

For the understanding of BL pathogenesis, it is also important to remember that approximately 3% of the African BLs, and 80% of the sporadic BLs that occur all over the world are EBV-negative. Among the recent, AIDS-associated BLS, the incidence of the EBV-carrying form is currently estimated as 40–50%.

The discovery of the "missing factor" in the "Burkitt equation" started when Manolov and Manolova reported in 1972 (55) that a 14q + chromosomal marker was present in about 80% of the tumors. The Manolovs came from Sofia, Bulgaria, to work with us in 1970, at the time when the chromosome banding technique was discovered by Caspersson and Zech. I suggested that they apply the banding technique to the cytogenetically unexplored BL that kept coming in from Clifford every Tuesday in excellent condition. They agreed rather reluctantly since they had hoped to learn some immunology. But their cytogenetic work soon picked up momentum, particularly after Albert Levan agreed to consult and guide them. When George Manolov showed me the extra band that he found attached to the distal part of the long arm of one chromosome 14 in a BL biopsy, I first suspected some trivial reason, perhaps a constitutional variation (iso-chromosome), and suggested that the Manolovs should take a look at the fibroblasts of the patient. So they did, but they found that the anomaly was totally restricted to the clonal tumor.

After the Manolovs returned to Bulgaria, we continued the work with Lore Zech. She soon showed that the "extra piece" was derived from chromosome 8; the 14q+ marker was thus a product of a reciprocal 8; 14 translocation (56). Several groups found subsequently that approximately 20% of the BLs that had no 14q+ marker carried one of two variant translocations instead (for review, see 57). Chromosome 8 broke at the same site (8q24) and entered into a reciprocal translocation either with the short arm of chromosome 2 or with chromosome 22. All BLs were found to carry one of the three translocations, no matter whether they were high endemic or sporadic, EBV-positive or negative. The same translocations were only exceptionally found in non-BL-lymphomas, although 14q+ markers are quite common; they usually arise by reciprocal translocations between chromosome 14 and some other chromosome, with 11 and 18 as the most frequent participants. But BL-type translocations were also found in the form of B cell–derived ALL that resembles Burkitt lymphoma cells phenotypically and is often called Burkitt leukemia.

Meanwhile, another, quite independent cytogenetic study, entirely confined to mouse tumor cells, was progressing in our laboratory. It started when the Hungarian-Rumanian pathologist, Francis Wiener joined our group in 1970. He is still one of our closest coworkers. Wiener became interested in the role of chromosome 15 trisomy in mouse T-cell leukemia, and he was also the main cytogeneticist in the somatic hybrid studies, together with Henry Harris, mentioned below. In the late 1970s Wiener examined a series of pristane oil–induced mouse plasmacytomas (MPCs); he was working together with a Japanese guest worker, Shinsuke Ohno, and in collaboration with Michael Potter's group at the NIH. Our 1979 *Cell* paper described the MPC-associated typical (12; 15) and variant (6; 15) translocations (58).

Mouse plasmacytomas are very different from Burkitt lymphomas. The only common denominator is that both originate from cells of the B-lymphocyte series. We never expected to find anything in common between the two. Therefore, the fact that two apparently unrelated research projects, carried out by different cytogeneticists, led to the discovery of a common pathogenetic mechanism, based on almost exactly homologous chromosomal translocations, was one of the greatest and most pleasant surprises of my entire scientific career. It was even more surprising that the highly speculative working hypothesis, formulated to explain the mechanism whereby the translocations contribute to the tumorigenic process in such a decisive fashion, turned out to be essentially correct.

The hypothesis was built on the fact that the recipient murine chromosomes of the dislocated fragment from chromosome 15 were known to carry the IgH (chromosome 12) and the kappa (chromosome 6) gene, respectively. Likewise, human chromosome 14 was known to carry the IgH cluster. We have therefore speculated that a proto-oncogene and probably *the same* proto-oncogene could be localized at the breakpoint of the murine chr 15 and the human chr 8. Accidental translocation of the putative gene to one of the immunoglobulin loci might have led to the constitutional activation of the gene, in analogy with the retroviral activation of the c-*myc* gene by the insertion of an ALV-derived LTR in the chicken bursal lymphoma, as described by Hayward et al.

I started to expose the hypothesis to the test of peer criticism in 1979. An outstanding molecular biologist, a good friend of mine, called it the "most hair-raising extrapolation from the centimorgans to the kilobases." It was. Still, the hypothesis was published in *Nature* in 1981 (59), but I was not fully convinced of it myself, until the critical moment during the summer of 1981, when I was waiting for a plane at Washington airport to take me to Tokyo. The waiting hall was full of people, mostly Japanese. There were only two telephones on the other side of the security check. They were busy most of the time. The plane was called up. Finally, one of the telephones was free. I tried to get hold of Philip Leder at the NIH. I wanted to hear whether he knew anything about the chromosomal location of the immunoglobulin light chain genes in humans. Leder came to the telephone. No, he hadn't heard anything; it was still unknown. But one of his colleagues had just come back from the recently held Human Chromosome Mapping meeting in Oslo. If I waited, he would try to ask if the colleague had heard anything.

"Final call." The last Japanese walked aboard, and I had to leave. At the moment when I was about to hang up the phone, Leder's voice came back: Yes, there were two small reports in Oslo. An English group had found that kappa is on chromosome 2. An American group had proved that lambda is on chromosome 22.

I ran on board. It was an intoxicating feeling! I knew for certain that the hypothesis was correct.

The molecular confirmation and clarification came in a virtual avalanche during 1982. Taking off from quite different points, Jerry Adams with Susan Cory in Australia and Kenneth Marcu in New York showed for

MPC, and Carlo Croce and Phil Leder for BL, that the translocations resulted in the juxtaposition of donor chromosome derived sequences and immunoglobulin gene sequences. Michael Cole's group has identified the transposed gene as c-*myc* (for review, see 60).

The subsequent development has led to many new insights, but it has also created some puzzles and paradoxes with regard to *myc*-regulation, constitutive activation, and certain details of the timing and regulation of Ig-gene rearrangement (for review see 65). With Francis Wiener and Janos Sümegi, we have also found a third Ig/*myc* translocation system (61–63), the spontaneous immunocytoma of the Louvain rat (RIC), developed by Hervé Bazin. A comparison of the translocations in MPC, RIC, and BL at the molecular level reveals more similarities than differences. In fact, it would be hard to find a comparable situation in cancer biology where three pathogenetically different tumors that arise from the same cell lineage in three different species show a similarly close pathogenetic mechanism at the molecular level.

The causal, i.e. rate limiting, involvement of constitutive *myc* activation in the genesis of the three tumors was deduced from the regularity of the Ig/*myc* juxtaposition that extended to cryptic translocations and complex rearrangements, where two or three successive genetic events had occurred (61, 64). Further confirmation came from recent facsimile experiments. Michael Potter and Francis Wiener showed (66) that introduction of an activated *myc* gene within a retroviral (J3) construct into pristane oil–treated Balb/c mice induced plasmacytomas that did not carry any translocations, provided they expressed the inserted (v-*myc*) gene. Meanwhile, Adams & Cory's group generated transgenic mice that carried the *myc*-gene coupled to the IgH enhancer (67). The mice developed more than 90% pre-B- or B-cell-derived lymphomas. Using the Australian transgenic mice, Francis Wiener recently found that Abelson virus infection, already known to increase the incidence and shorten the latency period of pristane oil–induced mouse plasmacytoma, has led to the appearance of plasmacytomas in the Emu-*myc* transgenic mice. The virus has obviated the pristane requirement and lifted the genetic restrictions to MPC susceptibility. These plasmacytomas were also translocation free.

That introduction of an activated *myc* construct was tumorigenic for B cells and obviated the need for the translocations could be only interpreted to mean that the naturally occurring constitutive activation of *myc* by the Ig-translocations provided an essential, rate-limiting step within the carcinogenic process. But it is not the only step. All tumors were monoclonal, even in the transgenic mice where *myc* was activated in all B and pre-B cells. Sequential activation of several oncogenes or, alternatively, loss of suppressor genes may provide additional steps. Feedback inhibition by the clone that happens to get the upper hand first would be another alternative.

The Burkitt lymphoma story has also developed further in the meanwhile and has posed some new fascinating questions. We have suggested, for both conceptual and factual reasons, that the BL progenitor is a long-lived B-memory cell. In this scenario, antigenically stimulated B-cell clones that have previously expanded as immunoblasts, were in the process of switching their phenotype to CALLA- and BLA-positive, activation marker negative memory cells when, upon the waning of the antigenic stimulus, the translocation accident occurred. Due to the linking of *myc* to a constitutively active Ig-locus, the cells were unable to leave the cycling compartment, however. It could be shown that the translocation carrying "suspended resting cell" had several additional phenotypic properties that could facilitate its evasion from immunological control. Certain MHC class I polymorphic specificities were down-regulated in the BL cells, compared to EBV-transformed B-cell lines of normal origin. The BL cells also failed to express certain adhesion molecules present on the LCLs or expressed them at a low level. Even the EBV-encoded, growth-transformation associated nuclear and membrane antigens were down-regulated in the BL cells, with the exception of EBNA-1. This was paralleled by a relative resistance of the BL cell to CTL-mediated lysis (56).

It thus appears that the *myc*/Ig translocation promotes the malignant growth of the BL cell by several mechanisms. This may explain the extraordinary regularity of its presence in all typical BLs so far studied.

Tumor Suppressor Genes

I have recently reviewed this field in some detail (68) and concluded that we are probably approaching an era when the study of genes that can antagonize tumorigenic behavior will be equally as, if not more rewarding than, the study of the oncogenes. Our own commitment to this field started with a decade of another long distance collaboration, initiated by Henry Harris in 1969 (69). We have inoculated a large number of somatic cell hybrids, derived from the fusion of high malignant with normal or with low malignant cells, into genetically compatible and/or immunosuppressed mice. The hybrids were generated by Harris, and Wiener examined their chromosomes. These studies have firmly established the notion that tumorigenicity is suppressed by fusion with normal cells. It reappears after some critically important chromosomes, contributed by the normal cell, have been lost. Others have extended this work to human/human hybrids more recently and obtained similar results. Chromosomes that carry tumor suppressor genes have been identified by Stanbridge, Klinger, Sager and their associates (70–72). The field is now moving towards a more reductionistic analysis where microcell hybrids are taking the place of whole cell hybridization and c-DNA transfections are initiated to identify the suppressor genes and their products. Meanwhile, evidence for tumor antagonizing genes has also emerged from the study of revertants and particularly from the rapidly moving field of "recessive cancer genes" that contribute to tumorigenesis by their loss (73, 74). It is not clear if or to what extent there is a relationship between the genes identified by these three approaches.

Whither Tumor Immunology?

It is often asked if or to what extent the spectacular development of the oncogene field during the last decade may provide some new handles for targeting the antitumor response. The answer may differ in relation to oncogenes activated by regulatory or by structural changes, respectively. Up-regulation of a structurally normal oncoprotein is less likely to provide a rejection target than oncoproteins activated by structural changes, e.g. the products of the *ras*-mutations or the truncated growth factor receptors, exemplified by the tumorigenic variants of erbB or fms. Following Townsend's discovery that intracellular, endogenous proteins can be processed to peptides that combine with class I or class II molecules and can then serve as immunogens and/or as CTL targets, the structurally changed oncoproteins deserve serious consideration. Progress will depend on the expression of mutation-activated (compared to normal) oncogenes in non-immunogenic tumor cells—of which there are many—followed by the assessment of their immunogenicity and rejectability in syngeneic hosts.

Epilogue

As each of us is moving towards the approaching darkness, the sun is never setting over the vast oceans of science. It has been a rare privilege to live and work through the times when the genetic material turned from protein to DNA, when adaptive changes in cell populations—including antibody production—were unmasked as Darwinian variations and selection, when GOD became the rearrangement of immunoglobulin genes, violating the dogma that all somatic cells have the same DNA. Another central dogma was abolished when the RNA tumor viruses became DNA proviruses. Following closely in the wake of this discovery, the enthusiastic retrovirologists, searching for the universal cause of cancer, permitted the great cuckoo egg, the oncogenes, to hatch—almost imperceptibly at first, but with a rapidly increasing crescendo, towards the triumphant emphasis on the regulatory genes of the cells and their dysfunction as the key factor in the oncogenic process. Departing from even greater obscurity, the MHC system, once the esoteric pet of a few mouse geneticists, now occupies a central place in virtually every area of immunology. It was a great time, and it still is, but it is only the stumbling, stuttering, premature fore-shadowing of what lies ahead. We have barely scratched the surface.

Literature Cited

1. Klein, G., Klein, E. 1951. The nucleic acid content and the growth rate of mouse ascites tumor cells. *Acta Union Int. Contra Cancrum* 7: 376–85
2. Klein, G. 1951. Comparative studies of mouse tumors with respect to their capacity for growth as "ascites tumors" and their average nucleic acid content per cell. *Exp. Cell. Res.* 11: 518–73
3. Purdom, L., Ambrose, E. J., Klein, G. 1958. A correlation between electrical surface charge and some biological characteristics during the stepwise progression of a mouse sarcoma. *Nature* 181: 1586–87
4. Ringertz, N., Klein, E., Klein, G. 1957. Histopathological studies of peritoneal implantation and lung metastasis at different stages of the gradual transformation of the MCIM mouse sarcoma into ascites form. *J. Natl. Cancer Inst.* 18: 173–99
5. Foulds, L. 1954. The experimental study of tumor progression: A review. *Cancer Res.* 14: 327–39
6. Furth, J. 1953. Conditioned and autonomous neoplasms: A review *Cancer Res.* 13: 477–92
7. Klein, G., Klein, E. 1957. Evolution of independence from growth stimulation and inhibition . . . *Symp. Soc. Exp. Biol.* 11: 305
8. Klein, G., Klein, E. 1958. Histocompatibility changes in tumors. *J. Cell Comp. Physiol. Suppl.* 1, 52: 125–68
9. Foley, E. J. 1953. Antigenic properties of methylcholanthrene-induced tumors in mice of the strain of origin. *Cancer Res.* 13: 835–37
10. Prehn, R. T., Main, J. M. 1957. Immunity to methylcholanthrene-induced sarcomas. *J. Natl. Cancer Inst.* 18: 769–78
11. Klein, G., Sjögren, H. O., Klein, E., Hellström, K. E. 1960. Demonstration of resistance against methylcholanthrene-induced sarcomas in the primary autochthonous host. *Cancer Res.* 20: 1561–72
12. Baldwin, R. W. 1955. Immunity to methylcholanthrene-induced tumors in inbred rats following atrophy and regression of implanted tumors. *Br. J. Cancer* 9: 652–57
13. Old, L. J., Boyse, E. A., Clarke, D. A., Carswell, E. A. 1. 1962. Antigenic properties of chemically induced tumors. *Ann. NY Acad. Sci.* 101: 80–106
14. Prehn, R. T. 1962. Specific isoantigenicities among chemically induced tumors. *Ann. NY Acad. Sci.* 101: 107–13
15. Klein, G., Sjögren, H. O., Klein, E. 1963. Demonstration of host resistance against sarcomas induced by implantation of cellophane films in isologous (syngenic) recipients. *Cancer Res.* 23: 84–92
16. Baldwin, R. W. 1973. Immunological aspects of chemical carcinogenesis. *Adv. Cancer Res.* 18: 1–75
17. Sjögren, H. D., Hellström, I., Klein, G. 1961. Transplantation of polyoma virus-induced tumors in mice. *Cancer Res.* 21: 329–37
18. Klein, G. 1966. Tumor antigens. *Ann. Rev. Microbiol.* 20: 223–52
19. Old, L. J., Boyse, E. A., Stockert, E. J. 1963. *J. Natl. Cancer Inst.* 31: 977–86
20. Klein, G., Clifford, P., Klein, E., Stjernswärd, J. 1966. Search for tumor specific immune reactions in Burkitt lymphoma patients by the membrane immunofluorescence reaction. *Proc. Natl. Acad. Sci. USA* 55: 1628–35
21. Gunven, P., Klein, G., Clifford, P., Singh, S. 1974. Epstein-Barr virus-associated membrane-reactive antibodies during long term survival after Burkitt's lymphoma. *Proc. Natl. Acad. Sci. USA* 71: 1422–26
22. Klein, G., Klein, E., Clifford, P. 1967. Search for host defenses in Burkitt lymphoma: Membrane immunofluorescence tests on biopsies and tissue culture lines. *Cancer Res.* 27: 2510–20
23. Henle, G., Henle, W. 1966. Immunofluorescence in cells derived from Burkitt's lymphoma. *J. Bacteriol.* 91: 1248–56
24. Henle, G., Henle, W. 1967. Immunofluorescence, interference and complement fixation technics in the detection of herpes-type virus in Burkitt tumor cell lines. *Cancer Res.* 27: 2442–46
25. Klein, G., Pearson, G., Nadkarni, J. S., Nadkarni, J. J., Klein, E., Henle, G., Henle, W., Clifford, P. 1968. Relation between Epstein-Barr viral and cell

membrane immunofluorescence of Burkitt tumor cells. I. Dependence of cell membrane immunofluorescence on presence of EB virus. *J. Exp. Med.* 128: 1011–20

26. Jondal, M., Klein, G., Oldstone, M. B. A., Bokish, V., Yefenof, E. 1976. Surface markers on human B and T lymphocytes. VIII. Association between complement and Epstein-Barr virus receptors on human lymphoid cells. *Scand. J. Immunol.* 5: 401–10

27. Yefenof, E., Klein, G., Jondal, M., Oldstone, M. B. A. 1976. Surface markers on human B and T-lymphocytes. IX. Two-color immunofluorescence studies on the association between EBV receptors and complement receptors on the surface of lymphoid cell lines. *Int. J. Cancer* 17: 693–700

28. zur Hausen, H., Schulte-Holthausen, H., Klein, G., Henle, W., Henle, G., Clifford, P., Santesson, L. 1970. EBV-DNA in biopsies of Burkitt tumors and anaplastic carcinomas of the nasopharynx. *Nature* 228: 1056 58

29. Reedman, B. M., Klein, G. 1973. Cellular localization of an Epstein-Barr virus (EBV)-associated complement-fixing antigen in producer and non-producer lymphoblastoid cell lines. *Int. J. Cancer* 11: 499–520

30. Ricksten, A., Kallin, B., Alexander, H., Dillner, J., Fåhraeus, R., Klein, G., Lerner, R., Rymo, L. 1988. The BAMHI-E region of the Epstein-Barr virus genome encodes three transformation-associated nuclear proteins. *Proc. Natl. Acad. Sci. USA* 85: 995–99

31. Henle, W., Diehl, V., Kohn, G., zur Hausen, H., Henle, G. 1967. Herpestype virus and chromosome marker in normal leukocytes after growth with irradiated Burkitt cells. *Science* 157: 1064–65

32. Nilsson, K., Klein, G., Henle, W., Henle, G. 1971. The establishment of lymphoblastoid lines from adult and fetal human lymphoid tissue and its dependence on EBV. *Int. J. Cancer* 8: 443–50

33. Pope, J. H., Horne, M. K., Scott, W. 1968. Transformation of foetal human leukocytes in vitro by filtrates of a human leukaemia cell line containing herpes-like virus. *Int. J. Cancer* 3: 857–66

34. Klein, G., Svedmyr, E., Jondal, M., Persson, P. O. 1976. EBV determined nuclear antigen (EBNA)-positive cells in the peripheral blood of infectious mononucleosis patients. *Int. J. Cancer* 17: 21–26

35. Epstein, M. A., Hunt, R., Rabin, H. 1973. Pilot experiments with EB virus in owl monkeys (aortus-trivigatus). I. Reticulo proliferative disease in an inoculated animal. *Int. J. Cancer* 12: 309–18

36. Shope, T., De Chiaro, D., Miller, G. 1973. Malignant lymphoma in cotton-top marmosets after inoculation with Epstein-Barr virus. *Proc. Natl. Acad. Sci. USA* 70: 2487–91

37. Adams, J. M., Cory, S. 1985. Myc oncogene activation in B and T lymphoid tumours. *Proc. R. Soc. London Ser. B* 226: 59–72

38. Svedmyr, E., Jondal, M. 1975. Cytotoxic effector cells specific for B cell lines transformed by Epstein-Barr virus are present in patients with infectious mononucleosis. *Proc. Natl. Acad. Sci. USA* 72: 1622–26

39. Svedmyr, E. A., Deinhardt, F., Klein, G. 1974. Sensitivity of different target cells to the killing action of peripheral lymphocytes stimulated by autologous lymphoblastoid cell lines. *Int. J. Cancer* 13: 891–903

40. Rickinson, A. B. 1966. Cellular immunological responses. In *The Epstein-Barr Virus: Recent Advances*, pp. 77–125. London: Heinemann Medical

41. Torsteinsdottir, S., Masucci, M. G., Ehlin-Henriksson, B., Brautbar, C., Ben-Bassat, H., Klein, G., Klein, E. 1986. Differentiation-dependent sensitivity of human B-cell derived lines to major histocompatibility complex-restricted T-cell cytotoxicity. *Proc. Natl. Acad. Sci. USA* 83: 5620–24

42. Moss, D. J., Misko, I. S., Burrows, S. R., Burman, K., McCarthy, R., Sculley, T. B. 1988. Cytotoxic T-cell clones discriminate between A- and B-type Epstein-Barr virus transformants. *Nature* 331: 719–21

43. Thorley-Lawson, D. A., Israelsson, E. S. 1987. Generation of specific cytotoxic T cells with a fragment of the Epstein-Barr virus-encoded p63/latent membrane protein. *Proc. Natl. Acad. Sci. USA* 84: 5384–88

44. Townsend, A. R. M., McMichael, A. J., Carter, N. P., Huddeston, J. A., Brownlee, G. G. 1984. Cytotoxic T-cell recognition of the influenza nucleoprotein and hemagglutinin expressed in transfected mouse L-cells. *Cell* 39: 13–25

45. Townsend, A. R. M., Rothbard, J. M., Frances, M., Gotch, G., Bahadur, J., Wrath, D., McMichael, A. J. 1986. The epitopes of influenza nucleoproteins recognized by cytotoxic lymphocytes

can be defined with short synthetic peptides. *Cell* 44: 959–68

46. Purtilo, D. T., Klein, G. 1981. Introduction to Epstein-Barr virus and lymphoproliferative diseases in immunodeficient individuals. *Cancer Res.* 41: 4209

47. Hanto, D. W., Frizzera, G., Gajl-Peczalska, K. J., Sakamoto, K., Purtilo, D. T., Balfour, H. H., Simmons, R. L., Najarian, J. S. 1982. Epstein-Barr virus-induced B-cell lymphoma after renal transplantation. *N. Engl. J. Med.* 306: 913–18

48. Ho, M., Jaffe, R., Miller, G., Breining, M. K., Dummer, J. S., Makowka, L., Atchison, R. W., Karrer, F., Nalesnik, A., Starzl, T. E. 1988. The frequency of Epstein-Barr virus infection and associated lymphoproliferative syndrome after transplantation and its manifestation in children. *Transplantation* 45: 719–27

49. Klein, G. 1975. Studies on the Epstein-Barr virus genome and the EBV-determined nuclear antigen in human malignant disease. *Cold Spring Harbor Symp. Quant. Biol.* 39: 783–90

50. De The, G. 1980. Role of Epstein-Barr virus in human diseases: Infectious mononucleosis, Burkitt's lymphoma and nasopharyngeal carcinoma. In *Viral Oncology*, ed. G. Klein, pp. 769–97. New York: Raven

51. Nilsson, K., Klein, G. 1982. Phenotypic and cytogenetic characteristics of human B-lymphoid cell lines and their relevance for the etiology of Burkitt's lymphoma. *Adv. Cancer Res.* 37: 319–80

52. Ehlin-Henriksson, B., Klein, G. 1984. Distinction between Burkitt lymphoma subgroups by monoclonal antibodies: relationships between antigen expression and type of chromosomal translocation. *Int. J. Cancer* 33: 459–63

53. Rowe, M., Rooney, C. M., Edwards, C. F., Lenoir, G. M., Rickinson, A. B. 1966. Epstein-Barr virus status and tumour cell phenotype in sporadic Burkitt's lymphoma. *Int. J. Cancer* 37: 367–73

54. Gregory, C. D., Tursz, T., Edwards, C. F., Tetaud, C., Talbot, M., Caillou, B., Rickinson, A. B., Lipinski, M. 1987. Identification of a subset of normal B cells with a Burkitt's lymphoma (BL)-like phenotype. *J. Immunol.* 139: 313–18

55. Manolov, G., Manolova, Y. 1972. Marker band in one chromosome 14 from Burkitt lymphomas. *Nature* 237: 33–34

56. Zech, H., Haglund, U., Nilsson, K., Klein, G. 1976. Characteristic chromosomal abnormalities in biopsies and non-Burkitt lymphomas. *Int. J. Cancer* 17: 47–56

57. Bernheim, A., Berger, R., Lenoir, G. 1981. Cytogenetic studies on African Burkitt's lymphoma cell lines: t(8; 14), t(2; 8) and t(8; 22) translocations. *Cancer Genet. Cytogenet.* 3: 307–15

58. Ohno, S., Babonits, M., Wiener, F., Spira, J., Klein, G., Potter, M. 1979. Nonrandom chromosome changes, involving the Ig gene-carrying chromosomes 12 and 6 in pristane-induced mouse plasmacytomas. *Cell* 18: 1001–7

59. Klein, G. 1981. The role of gene dosage and genetic transpositions in carcinogenesis. *Nature* 294: 313–18

60. Klein, G. 1983. Specific chromosomal translocations and the genesis of B-cell-derived tumors in mice and men. Minireviews. *Cell* 32: 311–15

61. Pear, W. S., Wahlström, G., Nelson, S. F., Axelson, H., Szeles, A., Wiener, F., Bazin, H., Klein, G., Sumegi, J. 1988. 6-7 chromosomal translocation in spontaneously arising rat immunocytomas: evidence for c-myc breakpoint clustering and correlation between isotypic expression and the c-myc target. *Mol. Cell. Biol.* 8: 441–51

62. Sumegi, J., Spira, J., Bazin, H., Szpirer, J., Levan, G., Klein, G. 1983. Rat c-myc oncogene is located on chromosome-7 and rearranges in immunocytomas with t(6; 7) chromosomal translocation. *Nature* 306: 497 98

63. Wiener, F., Babonits, M., Spira, J., Klein, G., Bazin, H. 1982. Nonrandom chromosomal changes involving chromosomes-6 and 7 in spontaneous rat immunocytomas. *Int. J. Cancer* 29: 431–37

64. Fahrlander, P. D., Sumegi, J., Yang, J. Q., Wiener, F., Marcu, K. B., Klein, G. 1985. Activation of the c-*myc* oncogene by the immunoglobulin heavy-chain gene enhancer after multiple switch region-mediated chromosome rearrangements in a murine plasmacytoma. *Proc. Natl. Acad. Sci. USA* 82: 3746–50

65. Klein, G., Klein, E. 1985. Myc/Ig juxtaposition by chromosomal translocations. *Immunol. Today* 6: 208–15

66. Potter, M., Mushinski, J. F., Mushinski, E. B., Brust, S., Wax, J. S., Wiener, F., Babonits, M., Rapp, U. R., Morse, H. C. III. 1987. Avian v-myc replaces chromosomal translocation in murine plasmacytoma-genesis. *Science* 235: 787–89

67. Adams, J. M., Harris, A. W., Pinkert, C. A., Corcoran, L. M., Alexander, W. S., Cory, S., Palmiter, R. D., Brinster, R. L. 1985. The c-*myc* oncogene driven by immunoglobulin enhancers induces

lymphoid malignancy in transgenic mice. *Nature* 318: 533–38
68. Klein, G. 1987. The approaching era of the tumor suppressor genes. *Science* 238: 1539–45
69. Harris, H., Miller, O. J., Klein, G., Worst, P., Tachibana, T. 1969. Suppression of malignancy by cell fusion. *Nature* 223: 363–68
70. Klinger, H. P. 1982. Suppression of tumorigenicity. Sixth International workshop on human gene mapping. *Cytogenet. Cell Genet.* 32: 68–84
71. Sager, R. 1985. Genetic suppression of tumor formation. *Adv. Cancer Res.* 44: 43–68
72. Stanbridge, E. J. 1987. Genetic regulation of tumorigenic expression in somatic cell hybrids. *Adv. Viral Oncol.* 6: 83–87
73. Benedict, W. F. 1987. Recessive human cancer susceptibility genes (retinoblastoma and Wilms' loci). *Adv. Viral Oncol.* 7: 19–34
74. Knudson, A. G. 1987. A two-mutation model for human cancer. *Adv. Viral Oncol.* 7: 1–17
75. Henle, W., Henle, G. 1973. Epstein-Barr virus and infectious mononucleosis. *New Engl. J. Med.* 288: 263–64

Elvin A. Kabat

Ann. Rev. Immunol. 1988. 6 : 1–24

BEFORE AND AFTER[1]

Elvin A. Kabat

Departments of Microbiology, Genetics and Development and the Cancer Center/Institute for Cancer Research, Columbia University College of Physicians and Surgeons, New York, New York 10032, and the National Institute of Allergy and Infectious Diseases, National Institutes of Health, Bethesda, Maryland 20892

Introduction

The earlier portion of my autobiography (1) dealt with the period from the time I started to work with Michael Heidelberger as a laboratory helper on January 2, 1933 until my grants from the US Public Health Service were summarily cancelled in 1952. This chapter continues my story from that point. First though, I would like to add a few notes about my origins and earliest days.

Beginnings

My mother and father had married·in 1913 and I was born on September 1, 1914. Both of my parents had come to the United States as small children. My father, Harris Kabatchnick, was born in 1872 and emigrated to the United States from Lithuania as a small boy. His first recollection after landing in the United States by boat from Hamburg was that the flags were at half mast because President Garfield had just died. His family settled on the lower East Side. My mother, Doreen Otesky, came to the United States from Kiev in 1893 at the age of seven. My father completed elementary school, went to work, and with his two brothers, Samuel and Joseph, started manufacturing women's dresses. In 1908 they changed their name to Kabat and the firm was called Kabat Bros.

I know little of my maternal grandfather. He was killed in an accident before I was born. My grandmother and the children lived with his brother Morris. It seems that the generation that emigrates to a foreign country, and especially the first generation born in the United States, tends to ignore

---

[1] The US Government has the right to retain a nonexclusive, royalty-free license in and to any copyright covering this paper.

its history and background until a complete story becomes difficult or impossible to get.

My mother completed high school and played the piano very well. Her mother lived with us until her death, when I was 11 or 12 years old. We had a five-room apartment on the fourth floor at 24 West 111th Street.

Both my mother's and father's families were very close. Many lived in Harlem within four or five blocks of our home; others lived in the Bronx or in Brooklyn. Aunts, uncles, and their children visited one another frequently. My cousin Saul Meylackson, a physician who had been a captain in the Medical Corps, US Army, during World War I, became my role model. My interest in medicine grew largely from him, and when he visited, as he and his wife Pearl did frequently, I constantly plied him with questions about his patients and about medical research. I used to visit his office frequently on Saturdays, especially during my high school and college years, and looked at the smears he made of urethral exudates to diagnose gonorrhea. Pearl's brother was Dr. Murray Peshkin, a well-known allergist at Mount Sinai hospital, whom I also questioned incessantly about medicine.

One grew up in the 1910s and 1920s keenly aware of the role of infectious disease. I lost a brother who died of pneumonia at a few weeks of age in 1918; a cousin died of polio in the 1918 epidemic; my father was very sick in the influenza pandemic of 1917; a friend in our apartment house died of diphtheria, and many families lost a child or young relative. Epidemics of whooping cough, chicken pox, scarlet fever, measles, and diphtheria were frequent. When the Schick test and immunization with diphtheria toxin-antitoxin were first introduced in New York City Schools in 1924, I was Schick negative, an early indication of my potentiality as an antibody former.

My parents were very devoted to me and to my sister Harriet, born May 8, 1920. I had everything I wanted for the first 12 years of my life. My mother tended to be somewhat over-protective. At the age of 10 or 11 I went to a school on 117th Street, and had to cross Lenox and St. Nicholas Avenues on the way. She wanted to accompany me, but I absolutely refused. She then followed me at some discrete distance. When I turned around and saw her, I laid down in the middle of the road and motioned to her to go back before I would stand up.

When I was seven or eight I went through a religious phase and asked my mother to get me a Hebrew teacher; I studied with him until I became Bar Mitzvah at 13. Unfortunately, I was taught to read and memorize but never to know the structure of the language. This knowledge would have been of great value in my extensive contacts with the Weizmann Institute and Israel.

My interests in chemistry began early. One of my older friends who lived on the same floor was given a chemistry set and let me help him. I soon got my own set and was always experimenting. Many of the chemicals in the sets kids used in those days are now highly restricted even for laboratories. My father would stop off at Macy's on Saturday afternoon to buy me some science books. I was greatly influenced by Paul de Kruif's *Microbe Hunters.*

When I was thirteen life became very difficult for us. The family firm had prospered during the second and much of the third decade of the century, manufacturing dresses selling for $24.95 to $89.95 wholesale. Then cheap dresses began to flood the market, and the firm went bankrupt in 1927. They tried to get started again with money borrowed from relatives, but this enterprise also failed. From 1929 to 1933 my father did odd jobs, our income declined continuously, and we were dispossessed from one apartment after another because we could not pay the rent. When we moved into a small apartment, we had to put goods in storage and these were lost when we could not pay the monthly charges. They contained most of the papers of my early life. One landlord turned off our electricity so we were in the dark. We had no food but a small piece of butter which we kept from getting rancid by wrapping it and letting cold water run over it. My father and I went to court and the judge ordered the electricity turned on. One apartment hotel held on to our furniture, so we then had to find a furnished apartment. Two cousins who were dress manufacturers, Arthur and Dick Shill, gave my father a small job, helping to wait on customers in their showroom. He started in 1932 at five dollars per week, but as things improved during the New Deal he continued to work and ended up with a more respectable salary—perhaps $40 to $50 per week.

My survival and ability to continue my education were largely consequences of the educational concepts of the early 1920s. Lewis M. Terman at Stanford during World War I had emphasized the use of intelligence testing to select gifted children for special attention. The New York Public School System was allowing bright children to skip grades (half years) if they felt that they could do the work. When I entered elementary school in September 1920, I was already able to read and to do arithmetic. This was in part because the kids in our apartment house played school, with some of the older girls as teachers. By skipping four times in elementary school I saved two years. When I entered DeWitt Clinton High School at the age of 12, Terman's extensive study of 1000 gifted children who were to be followed for a good portion of their lives was well underway (2). The Terman wave had reached high school. By getting good grades and taking some summer classes, I completed high school in three years instead of four. Thus, in the fall of 1929, at the age of 15 I entered City College. I had applied for a Pulitzer Scholarship to Columbia but was not selected. At City College I also did well. Since I received extra credits for A and B grades, applied toward the 128 required for the bachelors degree, and since I took one or two summer courses, I was again able to complete the four year course in three. By June of 1932, I was graduated from City College with a B.S. degree. I applied to two medical schools, Cornell and Columbia, whose catalogs said that they had scholarships. Indeed, Columbia had a scholarship specifically for a City College graduate, but it was not being awarded in those days. (It is of interest that I was subsequently on the medical faculty of both these institutions.) Regardless of the extensive subsequent changes in psychologists' ideas about the limitations and validity of IQ tests, the Terman ideas had a very crucial influence on what happened to me.

From 1931 to the time I was graduated, the relatives who were helping to support us applied substantial pressure on me to quit college and get a job. I was picking up small amounts of money by working during college registration, and I worked as an usher in the old Loew's New York movie theatre during one summer. However, the pressure kept growing for me to quit college. I finally told one of my mother's brothers to "get the hell out and don't come back." It was many years before we made up.

Two good friends, Joseph Silagy and Joel Hartley, graduated in 1931 and went to New York University Medical School. I used to visit them and sit in on their classes on Saturday mornings. I heard lectures by Homer Smith and R. Keith Cannan. I later became very friendly with Cannan while he was at New York University as chairman of the Biochemistry Department and at the National Research Council. We also met during summers at Woods Hole.

While at City College, I had the opportunity to work with Professor Leo Lehrman in analytical chemistry. Later I started working with Michael Heidelberger, but I continued at City College on Tuesday and Thursday evenings, when Leo taught evening classes. My earliest papers were published with him. I also became very friendly with the professor of organic chemistry, W. L. Prager, and with a biology instructor, Alexander S. Chaikelis. I took his very popular course in physiology during one summer. As a way of trying to help me financially, we constructed a large flow chart of synthetic reactions of aromatic compounds and tried, unsuccessfully, to interest publishers in it. I did show it to Michael Heidelberger while I was looking for a job; he liked it and pointed out sections that represented his earlier work. It probably influenced his decision to take me. Charts of this type later became popular teaching tools. I was a good German student and became very friendly with my German teacher, Mark Waldman. Leo

Lehrman tried, also unsuccessfully, for us to get the rights to translate Fritz Feigl's book on spot tests into English. I visited the Feigls many years later in Brazil.

As a freshman at City College, I brashly decided to volunteer for the curriculum committee, a student group that made suggestions to the faculty. They were generally juniors and seniors; they had never had a freshman on the committee, but they took me. I enjoyed their meetings and did make some suggestions. David Rittenberg was also on the committee during my first year.

The Story Continues

I will now take up my story where the first chapter concluded.

Only in 1981 did I learn, under the Freedom of Information Act (Figures 1, 2), that the Criminal Investigation Division of the Department of Justice had placed me on a "list for the apprehension and detention of prominent individuals considered dangerous to the security of the United States." The FBI document implies that by 1954 they were at least looking for information that might lead to my removal from this Security Index. I include this since most Americans are probably unaware of the existence of such a list. Considering the admitted injustice of the forced resettlement of Japanese Americans during World War II, one wonders how useful such a list actually is. Needless to say, I was never apprehended or detained.

The last months of 1953 and 1954 were spent trying to continue work without support from the Public Health Service. The studies on cerebrospinal fluid gamma globulin had proven very useful as an aid in the diagnosis of multiple sclerosis, and Houston Merritt arranged for the Presbyterian Hospital to make it a routine clinical laboratory determination. They provided me with funds for a technician and some parttime help in washing glassware, drawn from the fees charged for the tests. My laboratory continued to do this until the late 1970s when automated methods were substituted. I received a small grant from United Cerebral Palsy, which helped. I activated my ONR contract, which allowed us to continue work on the blood group and dextran problems. The NSF grant in 1954 made it possible to keep my technicians, graduate and postdoctoral students. The University took responsibility for my salary (1). It was impossible, however, to support the monkey colony; the allergic encephalomyelitis work was discontinued, just when we were planning to isolate the encephalitogenic antigen. Fortunately, this problem was taken up by many other workers.

Blood Group Substances

We resumed work on the structure of the Blood Group A, B, H, Le^a, Le^b, I and i glycoproteins. My book *Blood Group Substances* (3) was written

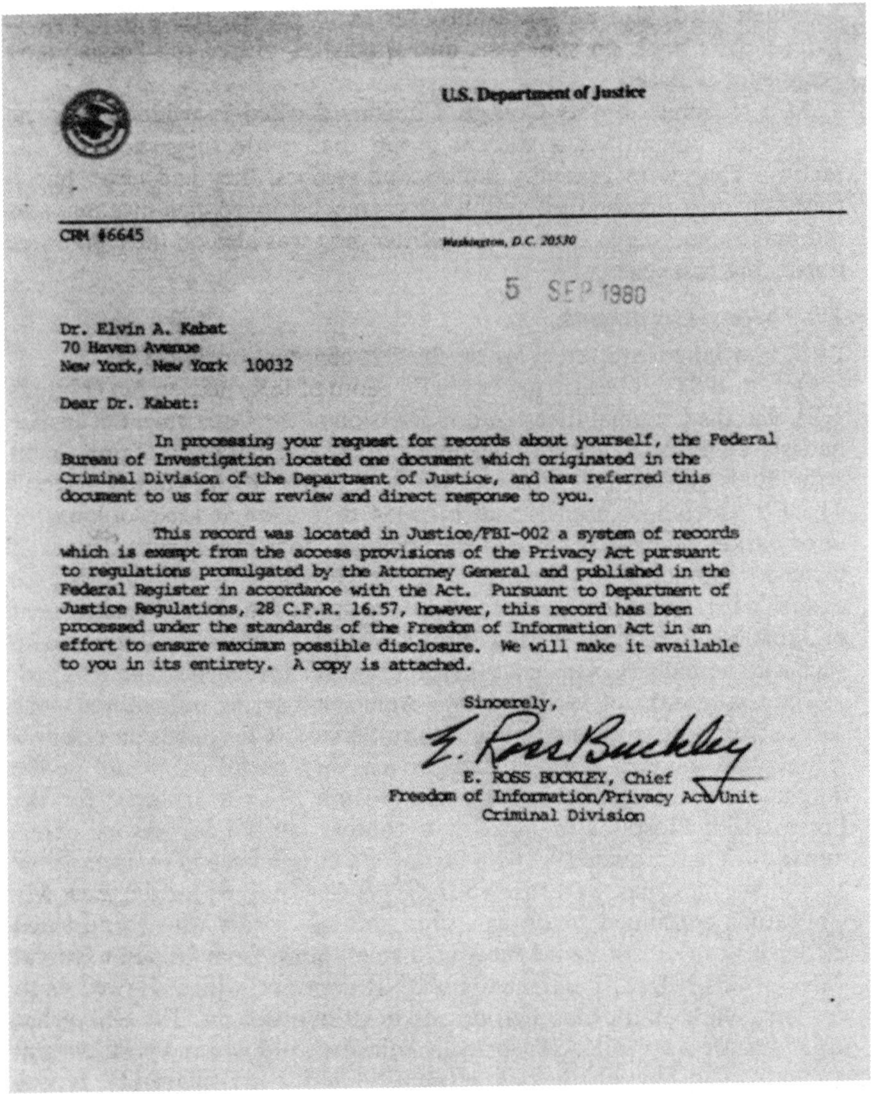

U.S. Department of Justice

CRM #6645

Washington, D.C. 20530

5 SEP 1980

Dr. Elvin A. Kabat
70 Haven Avenue
New York, New York 10032

Dear Dr. Kabat:

In processing your request for records about yourself, the Federal Bureau of Investigation located one document which originated in the Criminal Division of the Department of Justice, and has referred this document to us for our review and direct response to you.

This record was located in Justice/FBI-002 a system of records which is exempt from the access provisions of the Privacy Act pursuant to regulations promulgated by the Attorney General and published in the Federal Register in accordance with the Act. Pursuant to Department of Justice Regulations, 28 C.F.R. 16.57, however, this record has been processed under the standards of the Freedom of Information Act in an effort to ensure maximum possible disclosure. We will make it available to you in its entirety. A copy is attached.

Sincerely,

E. ROSS BUCKLEY, Chief
Freedom of Information/Privacy Act Unit
Criminal Division

Figure 1

during several summers at Woods Hole and was published in 1956. Oligosaccharides were isolated, after mild acid hydrolysis, and later by alkaline borohydride degradation. The use of periodate oxidation followed by Smith degradation made it possible to propose (with Kenneth O. Lloyd)

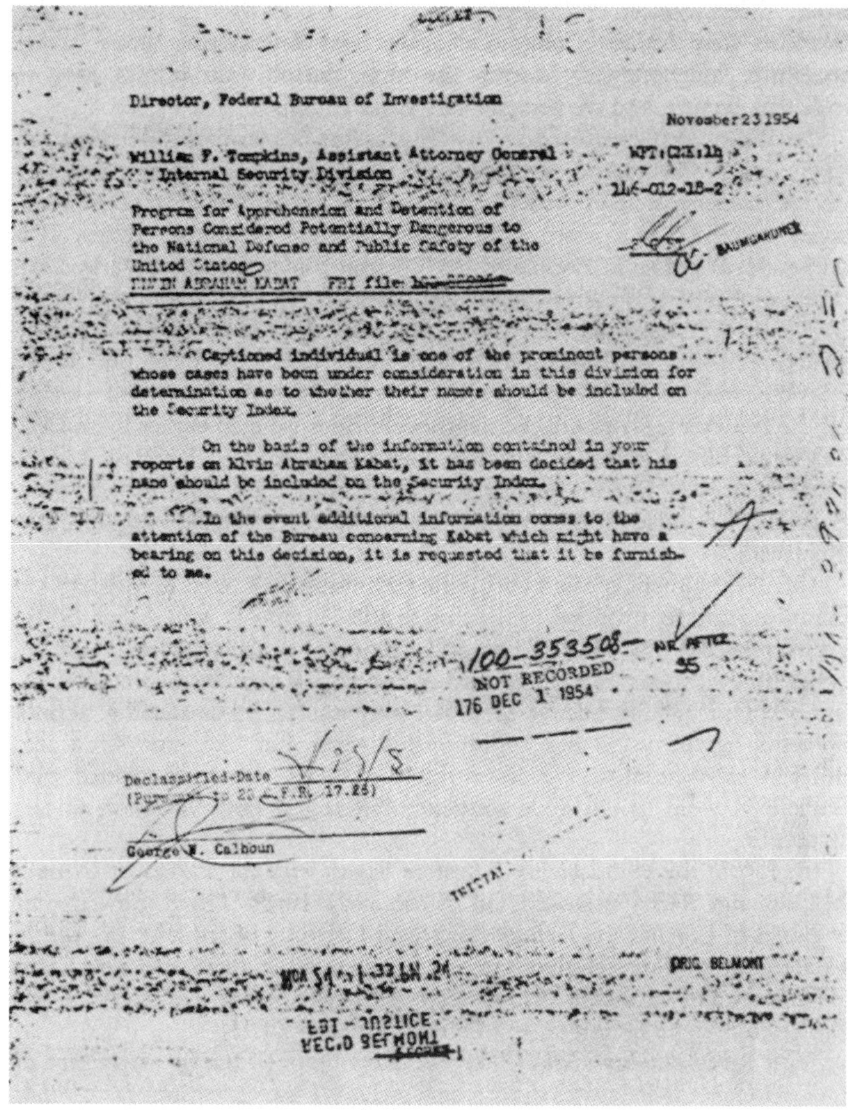

Figure 2

a composite structure for the carbohydrate moiety of the blood group substances. Later, Don Carlson developed improved conditions that prevented peeling after alkaline elimination and made possible isolation of oligosaccharides with reduced N-acetylgalactosaminitol, so that we could isolate intact reduced oligosaccharide chains. Walter Morgan and Winifred Watkins were isolating oligosaccharides and determining blood group structures independently during the same period. Our results were in excellent accord, and we became very close friends.

Our oligosaccharides isolated by ^{3}H-alkaline borotritiide differed extensively; many side chains resulted from incomplete biosynthesis, and this revealed extensive heterogeneity. The findings up to 1982 have been extensively described in a more personalized account (4). Later studies with Albert M. Wu, David Zopf, and Bo Nilsson, Flavio Gruezo & Jerry Liao on six grams of a human ovarian cyst blood group A substance subjected to Smith & Carlson degradations yielded many more oligosaccharide structures from the interior of the blood group substances and made possible extension of the composite structure. This also permitted placing on the composite structure the oligosaccharides isolated earlier from various intact blood group substances. These were done by Kenneth Lloyd, Luciana Rovis, Byron Anderson, Marilynn Etzler, and by Sherman Beychok, who also studied their optical rotatory dispersion and circular dichroism.

The development of the hybridoma technique by Kohler & Milstein led many groups to produce monoclonal anti-A, anti-B, and other blood group reagents. Dr. H. T. Chen and I mapped the fine structures of the combining sites of monoclonal anti-A and anti-B produced commercially in Canada, England, and Sweden. All were equivalent as blood grouping reagents by the usual hemagglutination tests, but their fine structures differed substantially. We have made cDNAs and have cloned and sequenced them to correlate sequence differences with variation in site structure.

Interest in the blood group Ii system began with the studies of Donald Marcus and Richard Rosenfield in the early 1960s. These showed that enzymes of *Clostridium tertium* destroyed I activity of red cells, liberating galactose and N-acetyl-glucosamine. Subsequent studies, with Ten Feizi, classified I and i antigens into groups. Anti-IMa (group 1) specificity was shown to involve $DGal\beta(1 \rightarrow 4)DGlcNAc\beta(1 \rightarrow OCH_2-$.

With Ray Lemieux's laboratory we have mapped the fine structure of monoclonal anti-IMa sites more extensively. We used a substantial number of synthetic oligosaccharides, each modifying a distinct portion of the antigenic determinant, to evaluate its contribution to binding. Most ingenious was Lemieux's replacement of one of the two H in the $-OCH_2-$ moiety by deuterium to give two optical isomers, one of which was highly active and the other inactive. After leaving our laboratory, Dr. Feizi continued studies on other I and i determinants, and Dr. Marcus clarified the nature of the determinants of the blood group P system.

Interlude

The Josiah Macy Foundation ran a series of conferences on Polysaccharides in Biology during 1955–1959. We all exchanged views, using the Frank Fremont-Smith principle, "Don't speak while I am interrupting."

One amusing incident in 1955 comes to mind. K. G. Stern and I differed on the question of whether the microheterogeneity of proteins was ascribable to mistakes in synthesis. I took the position that this could not be unequivocally established since small amounts of impurities might give the appearance of microheterogeneity.

> Stern replied, "I cannot help but feel that nature cannot work so precisely that it can reproduce a molecule of a million or twenty million molecular weight exactly in its own image as, for example, the imprint of a gramophone record. By way of example, if two typewriters of the same make and the same type are used, and the same letters are written with them, at first glance, it will be said that the imprint is identical, but it is a well-known fact that criminological investigation can distinguish between different individual typewriters by microscopic investigation, because there exist always minute irregularities; so I would say this is a microheterogeneity, something which at first approximation looks homogeneous but cannot stand up to the most searching criteria."
>
> Boyd: "You mean that nature does not always produce homogeneity, or can never produce homogeneity?"
>
> Stern: "I would say that at the level of the macromolecule, it is too much to expect of nature that such structures should be absolutely identical. This is a carry-over from the experience with small molecules."
>
> Kabat: "I think I was very careful to say that the evidence did not permit us to infer whether or not nature produced a homogeneous macromolecule. However, for the record I am perfectly willing to state that Dr. Stern has completely convinced me of the microheterogeneity of typewriters."

Georg Springer who was editing the conference, indexed this under: "Typewriters, microheterogeneity of, p 163."

Lectins

I had been interested in lectins since the studies on ricin during World War II carried out with Michael Heidelberger and Ada Bezer. While I was in Sweden in 1967, Sten Hammarström, from Peter Perlmann's laboratory, proposed to become a postdoctoral student. He mentioned that he had isolated a lectin from *Helix pomatia* which agglutinated A red cells, so I told him to bring some along. We studied its combining site and determined its association constant by equilibrium dialysis. Marilynn Etzler studied the site specificity of *Dolichus biflorus*. Arne Lundblad came from Upsala, bringing some of his blood group oligosaccharides from urine which were of great help in mapping A, B, H specific combining sites of lectins and antibodies.

When Miercio Pereira, with an M.D. degree, came to the laboratory from Brazil, he expanded the lectin studies to purify and characterize the combining sites of blood group specific lectins from Lotus, Ulex, soybean, and peanut. Hagen Bretting studied sponge lectins from *Axinella*; Jerzy Petryniak and Pereira studied *Evonymus europeus*; Shunji Sugii did *Wistaria floribunda*; and Paul Kaladas studied *Vicia villosa* with Kimura and Erson from Hans Wigzell's laboratory in Sweden. Hagen Bretting and Stephanie Phillips studied the mutagenic effects of *Axinella* lectin. Charles Wood with S. Ebisu, L. A. Murphy, and Irwin Goldstein investigated the lectins of *Bandeireae simplicifolia* A4, B4, and BSII. Santosh Sikder with C. J. Steer and Gil Ashwell studied the chicken hepatic lectin, and he did lima bean lectin with Goldstein and Dave Roberts. We were very fortunate to have collaborations with Nathan Sharon of the Weizmann Institute and Irwin Goldstein at Ann Arbor and their colleagues on many of these studies.

Many similarities exist between the fine structures of antibody combining sites to A, B, H glycoproteins and those of the blood group specific lectins as determined by immunochemical mapping. Of special interest is the finding that lima bean lectin (5) and a hybridoma antiblood group A (6) are very similar in reactivity with a monofucosyl hexasaccharide from human urine but behave entirely differently when a second fucose is substituted on the third sugar from the nonreducing end. The comparative mapping of the fine structures of lectin and antibody sites of similar specificities and correlating these differences with amino acid sequences and X-ray crystallographic structures should throw much light on the evolutionary emergence and functions of these two very important and diverse groups of molecules.

Studies on Antipolysaccharides

The immunochemical mapping of the combining sites and later recombinant DNA studies on antibodies to $\alpha(1 \rightarrow 6)$dextran became the predominant problem of the laboratory, with ancillary studies on anti-$\alpha(1 \rightarrow 3)\alpha(1 \rightarrow 6)$dextran (B1355S), antilevan, and anti-SIII. Rose Mage studied rabbit anti-$\alpha(1 \rightarrow 6)$dextran and could also show that the rabbit anti-SIII site was complementary to a hexasaccharide. Gerald Edelman studied the various antibodies I produced in myself—antidextran, antilevan, anti-A— as well as other human antidextrans, after reduction and alkylation, using starch gel electrophoresis. These were studied in acrylamide, by Bill Yount and Marianne Dorner, and their subgroup compositions and genetic factors were determined by James Allen, Bill Yount, Marianne Dorner, and Henry Kunkel. My antilevan turned out to be a monoclonal IgG; the others were pauciclonal. Amino acid sequencing of antibodies was just beginning. Hoping to be able to get enough of my antibodies to sequence, Dick Rosenfield kindly took 15 one-liter plasmaphereses from me, each a week apart, so that eventually I had about 7.5 liters of my serum to fractionate. Stuart Schlossman, Justus Gelzer, and Marianne Dorner tried to fractionate my antidextran to obtain monoclonal populations. While some fractionation into populations with smaller and larger antidextran sites was achieved, monoclonal populations were not obtained, and the amounts produced by fractionation were too small for the sequencing methods of the 1960s.

Michael Potter at the National Institutes of Health and Melvin Cohn and Martin Weigert at the Salk Institute had isolated mouse monoclonal anti-$\alpha(1 \rightarrow 6)$dextrans. The fine structures of the combining sites of three IgAs were studied with John Cisar, Jerry Liao, Marianne Dorner, and Michael Potter. In 1970 John Cisar had taken a summer course that I gave at Oregon State University at Corvallis, and he had come to the lab after completing his PhD degree. We showed that the combining sites of W3129 and QUPC52 differed. W3129 and W3434 were complementary to five $\alpha(1 \rightarrow 6)$ linked glucoses, whereas QUPC52 was complementary to six. Despite this, the binding constant of QUPC52 was only 1/30 that of W3129. By competition assays by equilibrium dialysis, methyl α-D-glucoside and isomaltose contributed 60% of the binding energy of the pentasaccharide with W3129, but only 5% of the binding energy of the hexasaccharide with QUPC52. This indicated that the specificity of W3129 was directed toward the terminal nonreducing end of the $\alpha(1 \rightarrow 6)$ linked oligosaccharide, whereas QUPC52 was specific for the internal chain of $\alpha(1 \rightarrow 6)$ linked glucoses, loosely termed cavity-type but also called end binders and groove-type sites, respectively. This distinction could be readily demonstrated with a synthetic linear dextran synthesized by Ruckel & Schuerch. Since the linear dextran had but one nonreducing end, it inhibited precipitation of W3129 by dextran as well as the pentasaccharide; toward QUPC52, however, it was multivalent and so formed a precipitate with hybridoma ascitic fluids. This provides a rapid screening method for classifying the two kinds of monoclonal anti-$\alpha(1 \rightarrow 6)$dextrans.

Human Monoclonal Antibodies

The demonstration by Waldenström that macroglobulins and gamma globulins could occur as monoclonal homogeneous proteins in the serum of individuals with various blood dyscrasias, and by Edelman and Gally that Bence Jones proteins were the light chains of immunoglobulins, led many investigators to search for antibody activities. A number were soon reported. Several laboratories acquired large collections of human monoclonal proteins, but unlike the mouse gammopathies, relatively few of the human monoclonals had identifiable antibody activities. My interest in screening such human monoclonals arose from two directions. Elliott Osserman had a large collection of sera from individuals with multiple myeloma and Waldenström macroglobulinemia. We decided to screen these serums with my collection of polysaccharides, blood group substances, etc. Edward Franklin at New York University also had a large collection of human sera that had been screened for various antibody activities, but without success. Marian Koshland had asked him for a human IgM; in an attempt to purify it over a Sepharose column, it did not come off. It seemed to her that some might be eluted with a number of sugars, so she thought I would be interested. In screening by gel diffusion we found that the IgM reacted strongly with a water extract of agar prepared at 20°C. Since agar contained 4,6-pyruvylated galactose, I asked Michael Heidelberger, who was studying rabbit antisera to Klebsiella polysaccharides, for a number of samples. The purified Franklin monoclonal IgM^{WEA} reacted much more strongly with Klebsiella K21 than it did with agar in quantitative precipitin assays; only about one twentieth as much of these Klebsiella K21 polysaccharides was needed. At the same time we had found another human monoclonal IgM^{MAY} in the Osserman collection that also reacted with agar and with the same Klebsiella polysaccharide. Unlike IgM^{WEA} which was studied as purified IgM, IgM^{MAY} was studied using whole serum. We tested other Klebsiella K polysaccharides, and K30 and K33 reacted in both sera in a way identical to K21. On the basis of these studies, which had been accepted by the *Journal of Experimental Medicine* (7), the American Cancer Society gave me a research development grant for a postdoctoral fellow, Dr. Arapalli S. Rao, who came from Calcutta. Since he was to learn the quantitative precipitin method to continue the work, I thought it would be logical for him to repeat the studies on IgM^{WEA} and IgM^{MAY}. We had run out of the first sample of K21 and made up a solution from another sample. He learned the technique rapidly and was able to confirm all of our data except that K21 gave no precipitate with IgM^{WEA}; he became convinced that our curve with K21 was wrong. By this time the proof of the *JEM* paper was due to arrive, and I had to decide whether to take out the curve of K21 with IgM^{WEA}. In worrying about this I remembered that Michael Heidelberger had first sent me six samples, of which only K21, which had the same 4,6-pyruvylated galactose as agar, reacted, thus giving us our first clue as to the specificity of IgM^{WEA}. Only later did we get K30 and K33 which had 3,4-pyruvylated galactose and precipitated equally well per unit weight. I told Rao that there must be some other explanation for the difference between the two K21 samples, since if the reaction with K21 had not provided the initial clue to the specificities of the two IgM, I would never have been able to apply for the grant and he would not have come to the laboratory. The explanation finally emerged. It turned out that the first K21 was the neutral sodium salt, but the second K21 preparation had been isolated as the free acid and was progressively lowering the pH of the unbuffered purified IgM^{WEA} but was not altering the pH of the serum of IgM^{MAY}. Rao carried out a study of the effect of pH on IgM^{WEA} and IgM^{MAY} and found that IgM^{WEA} gave identical precipitin curves with K21, K30, and K33 at pH 4, but K21 did not react at higher pH. With IgM^{MAY} all three polysaccharides reacted identically from pH 7.0 to 4.0, indicating differences in the structures of the two antibody combining sites.

Subsequent screening of the Osserman collection revealed a number of other human IgM monoclonals with various anticarbohydrate specificities including antibodies specific for the interior of the water soluble blood group A and B substances and for chondroitin sulfates. Most exciting was the discovery of IgM^{NOV} (8), an antibody specific for poly-$\alpha(2 \rightarrow 8)$N-acetylneuraminic acid, the specific capsular polysaccharide of the group B meningococcus and of *E. coli K1*, against which vaccination has not been successful in humans. This antibody is as protective per μg antibody in rats infected with *E. coli* as is a standard horse antigroup B meningococcal serum. IgM^{NOV} also reacts with poly A, poly I, and with denatured DNA probably ascribable to similar oligomeric patterns of the carboxyls in the group B polysaccharide and the phosphates in the polynucleotides and in DNA, permitting reactivity in the IgM^{NOV} antibody combining site. Michael Heidelberger in 1983 put forward a similar explanation of the cross-reaction of type 8 and 19 pneumococcal polysaccharides; he suggested it was due to the ability of negatively charged phosphoryl-β-D-N-acetylmannosamine to enter the type 8 site specific for cellobiuronic acid with its negatively charged COO^- and vice versa. IgM^{NOV} has potential as adjunct serotherapy together with standard antibiotic therapy for group B meningococcal meningitis, a possibility suggested by studies in the first three decades of this century of reductions in the death rate due to intra-thecal horse antimeningococcal serum. Giampaolo Merlini, Steve Birken, and Marian Gawinowicz are determining the amino acid sequence of the V_H and V_L regions of IgM^{NOV} so that the antibody can be synthesized by recombinant DNA techniques. Stanley Hoffman and Gerald Edelman have shown that IgM^{NOV} is antibody to N-CAM, and Karl Pfenninger has found it to show the histochemical distribution of N-CAM.

The Variability Plot

Careful amino acid composition analyses of antibodies of different specificities by Marian Koshland had revealed differences. In 1965 Hilschmann and Craig presented the first amino acid sequences of two human Bence

KABAT

Jones proteins and showed that they differed extensively in the first 107 residues from the amino terminus but were essentially identical throughout the C-terminal half of the molecule. These are now termed variable and constant regions.

At an antibody workshop (9) at the Weizmann Institute in 1965 Dreyer & Bennett suggested that these two domains were coded by separate genes; additional sequences were presented by Putnam, Milstein, Hood, Gray, Dreyer, and their colleagues. I became interested in these developments and in 1966–1967 began a Sabbatical in Pierre Grabar's laboratory at the Institut Pasteur, writing the first edition of "Structural Concepts in Immunology and Immunochemistry." I thought it would be important to include the sequence data. At that time two mouse κ and several human k Bence Jones sequences had been published. On aligning the data it was evident that in variable regions of human and mouse very few amino acid differences distinguish mouse from human, e.g. there were very few species-specific residues (now termed "species-associated residues," as later suggested by Capra). The constant region, however, contained 43 differences between human and mouse; this was comparable to findings with hemoglobins, cytochromes, etc, from different species. A sequence of a human λ Bence Jones protein by Titani, Wikler, and Putnam appeared, which further reduced the number of species-associated residues and was cited in a "note added in proof." Francois Jacob and Vernon Ingram both thought the finding important, so I sent it to the Proceedings of the National Academy of Sciences. Shortly thereafter, in comparing the V and C regions, I noted the occurrence of more invariant glycines in the V-region, including those at residues 99 and 101 which have now been found in homologous positions of V-regions in essentially all immunoglobulin light and heavy chains, in the α and β chains of the T-cell receptor for antigen, and in T cell surface antigens T4 and T8. I suggested that the role of these two variant glycines might be to provide flexibility and permit movement of the rest of the V-region to make maximum contact with an antigenic determinant.

Cesar Milstein, Niall & Edman, and the Hood group published sequences of several V_κ chains which indicated that there were subgroups largely based on the first 23 amino acids from the N-terminus. It then became evident from the frequency of identical repeats that this segment of the chain did not show the variability needed for it to be involved in generating antibody complementarity and diversity. This led to my suggesting that the first 23 amino acids were involved only in three-dimensional folding. Also, Milstein had noticed that beyond amino acid 94 he could no longer define subgroups and suggested that frequent crossing over might be occurring beyond residue 94, a finding prophetic of V-

J joining, which was largely ignored. The 1967 Cold Spring Harbor and Nobel Symposia provided forums for these exciting developments on amino acid sequences of antibodies.

In 1969 I received a letter from a Dr. Tai Te Wu, then at Cornell University Medical College and Sloan-Kettering in New York, asking if he could spend some time in the laboratory to learn quantitative immunochemistry. When he visited the laboratory I asked about his background. He said he was a mathematical biophysicist and had some computer experience. I proposed that we collaborate on the sequence variability. I had previously been entering and aligning the sequence data by hand using pencil and paper. We began to meet one day a week to evaluate and discuss results. By this time more complete sequences were beginning to pour into the literature from the laboratories of Capra, Edelman, Hilschmann, Hood, Metzger, Potter, Milstein, and their coworkers.

In June 1969 at a symposium in Prague, Edelman, Franek, and I discussed positions containing more substitutions than were seen in the subgroups. Franek had tabulated dissimilarity in the subgroups, whereas Wu and I had searched for maximum variability, and our data were in good agreement for three hypervariable regions. In June 1970 (10) we introduced an equation measuring variability at each position in a set of V_L sequences aligned for maximum homology as follows:

$$\text{Variability} = \frac{\text{Number of different amino acids at a given position}}{\text{Frequency of the most common amino acid at that position}}.$$

At this time some 77 complete and partial sequences of V_L regions of human V_κ and V_λ and of mouse V_κ chains were available. This led to the recognition of three linear stretches of hypervariability, which we predicted would fold to form the antibody combining site. When a sufficient number of V_H sequences were examined, they too had three hypervariable regions (11). X-ray crystallographic studies on Fab fragments, Bence Jones dimers, and more recently on the lysozyme-antilysozyme site by the Poljak group at the Institut Pasteur have all confirmed this prediction almost on a residue-by-residue basis. These hypervariable regions are now termed complementarity-determining regions (CDR). Wu and I proposed (10) in 1970 an insertional mechanism by which nucleotides coding for the CDRs would be recombined into the framework residues of the various subgroups.

The variability plot was based on the assumption that, since the antibody forming system was universal among vertebrates, combining data from different subgroups, species, etc, was justifiable, and the combining sites would be in one place rather than in different portions of the variable regions. When Martin Weigert and Melvin Cohn with Cesari and Yonkevich described the mouse V_λ chains, they found that most of their sequences were identical throughout the V_λ region. They postulated that these were the product of a single germ-line gene. Those chains that showed variation from the germ-line sequence were localized to the CDRs and could have been formed by point mutations. This provided the first evidence for somatic mutation in immunoglobulin V_λ regions. The assumption is still valid for the very large number of V_L and V_H chains now available.

Variability plots have since been used to define better allelisms in the MHC class I and II systems and also to locate sites of variability in AIDS viruses. Variability plots for the T cell receptor require further analysis. Variability plots for other proteins, such as cytochromes, did not show hypervariable regions. A statistical examination of each residue in the CDRs made it possible to designate certain positions as structural and others as probably in contact with the antigenic determinant.

Compilation of Sequences of Immunoglobulin Chains

In 1973 Harry Rose retired as chairman of the Microbiology Department, and Harold S. Ginsburg whom I knew very well from Woods Hole was appointed. Since he was just getting started, I delayed my Sabbatical one year so that I could participate in the teaching. I was chosen as a Fogarty Scholar and spent 1974–1975 at the National Institutes of Health, writing the second edition of *Structural Concepts in Immunology and Immunochemistry*. As it is customary for Fogarty Scholars to meet with the director of NIH, I was given an appointment with Dr. Robert S. Stone, who inquired as to my research interests. When I replied that I was interested in the structure of antibody combining sites and had been tabulating variable region sequences, he suggested that NIH had the PROPHET computer system and that this might help me. DeWitt Stettin, Jr., then Deputy Director for Science at NIH, introduced me to William F. Raub, then with the Division of Research Resources. Now Deputy Director of NIH, he took a great interest in my activity. To evaluate PROPHET, Howard Bilofsky came down from Bolt Beranek and Newman to help me in preparing the tables of sequences to be used in *Structural Concepts*. It soon became clear that PROPHET had greatly superior potentialities for tabulating and keeping track of variable region sequences. When my term as a Fogarty Scholar was up, I was asked to serve as an Expert, first to the National Cancer Institute and later to the National Institute of Allergy and Infectious Disease, spending two days a week at NIH to organize and maintain a data bank of variable region sequences. I rented an apartment in Building 20 at NIH. T. T. Wu came from Northwestern and Howard Bilofsky from Cambridge to spend two days a month keeping track of and checking the data that I collected from the literature. The first edition (12) appeared in 1976 and contained only amino acid sequences of variable regions; the second edition (13) was expanded to include constant regions; the third (14) also contained nucleotide sequences and derived amino acid sequences and included other members of the immunoglobulin superfamily. The fourth edition (15) appeared in April 1987, included T cell receptors, and had grown to 804 pages; Wu and I, along with Harold M. Perry and Kay Gottesman, are now working on a fifth edition. Keeping track of exponentially increasing data is extraordinarily difficult. The compendium has been distributed free to workers in the field and has been considered very useful.

While I am very happy that so many colleagues find it valuable, I would not have undertaken it had I not felt it was essential to my research interests. In addition to providing an enormous body of data supporting our prediction (10, 11) that the CDRs in the light and heavy chains would fold up to form the antibody combining site, we were able to demonstrate independent assortment of framework (FR) amino acids (16). This led us to formulate the minigene hypothesis, in which nucleotides coding for the three CDRs were assorted by recombination or insertion into the framework nucleotides. This hypothesis was put forward just as Tonegawa et al. (17) had shown that, in 12-day-old mouse embryo DNA, the V-region nucleotides only coded through amino acid 95. Since adult myeloma coded through amino acid 107, we postulated a mechanism of somatic joining of a minigene that coded for the remaining amino acid to give the intact V gene. The J minigene coding for amino acids 96 to 107 was found shortly thereafter by Tonegawa in 12-day-old embryo DNA. In the expressed assembled gene from adult myeloma DNA, it had been joined to the V-region (18). D and J minigenes coding for segments of the V_H gene were found shortly thereafter (19). In the T-cell receptor for antigen, J minigenes are also present in α, β, and γ chains and D minigenes in the β chains. These findings hold for all species examined.

Since genes coding for the rest of the V_L and V_H occurred in the germline as single stretches of nucleotides, considerable controversy existed as to whether the minigene hypothesis applied to the V-region. David Baltimore and Richard Egel proposed that gene conversions could explain our assortment data. The Rajewsky laboratory (20) described a double recombinant in which CDR1 of one germline gene was recombined with CDR2 and 3 from another. H. G. Zachau's laboratory (21) showed by nucleotide sequencing and comparisons of V-regions that our assortment data could be accounted for by gene conversions in the DNA coding for the FR segments. Most relevant are the studies of Reynaud et al (22) in J. C. Weill's laboratory at the Institut Pasteur on chicken V_λ chains. The chicken has only a single functional germline V_λ gene and a large number of pseudogenes; thus diversity and the antibody repertoire must be generated by rescue of functional CDRs of pseudogenes by gene conversions— essentially our original minigene hypothesis (16).

Work with WHO

In 1963, the World Health Organization became interested in research and training in immunology. They organized five meetings of scientific groups. I was invited to participate at one of these at Ibadan in December 1964. Niels Jerne was one of the prime movers. I met Professor Ian McIntyre, then Dean of the Faculty of Veterinary Science at University College, Nairobi, and Professor and later Dean in the Veterinary School at Glasgow. We become close friends. A high priority was assigned to setting up research and training centers in immunology in developing countries in various parts of the world. These were to be coordinated with centers in Geneva and elsewhere. Howard Goodman, Zdenek Trnka, and I went to Nigeria, Uganda, and Gambia. We selected Ibadan as the first center. Ada E. Bezer, who had been a technician with me for many years, and Dr. Ivan Riha from Prague staffed the center, which was in the Department of Chemical Pathology. A course in Immunology was given for six to eight months followed by opportunities for students to do research. The centers initially took students from all of West Africa and a number of students completed PhDs. In 1966 Howard Goodman and I went around the world and selected Singapore as another center. Centers were ultimately selected in Sao Paolo, Mexico City, Nairobi, Beirut, Teheran, and New Delhi, with support centers in Lausanne and Geneva. From 1968 to 1971, I spent two weeks each year (the first three years) with Dr. Otto G. Bier, and the last year with Ivam Mota, directors at the Sao Paolo center located at the Escola Paolista and later moved to the Institute Butantan. I conducted a review of the course work, gave the final exam, and tried to get a regular research seminar started. It was very difficult to maintain a critical mass and sustained enthusiasm for continuing the seminars when I was not there.

Dr. Vasek Houba and Ada Bezer staffed the Nairobi center for several years. Later, regional centers tended to draw students more exclusively from the country in which they were located, and Thailand set up its own immunological center. Dr. Georgio Torrigiani is currently in charge of the program, but WHO has difficulties supporting the centers (23). I continue as a member of the WHO Expert Advisory Panel on Immunology and have recently been involved with work on immunization with bacterial polysaccharides. A further development was the setting up of ILRAD (International Laboratory for Research in Animal Diseases) in Nairobi.

This began work in 1978 under the auspices of the World Bank, the Rockefeller Foundation, and the Agency for International Development, to develop vaccines against east coast fever and trypanosomiasis. I was involved in the early planning.

Interlude

On December 26, 1970, the front page of the *New York Times* contained an appeal from 14 Soviet scientists to President Nixon to see that Angela Davis received a fair trial. I thought: "We have a free judicial system, the trial will be open to the public; President Nixon ought to invite them to come and see for themselves." I reflected on how one might get him to do this. It became clear that one could not write to the President directly; he would never see the letter. I decided that one had to address the suggestion to someone on his staff, who at least could bring it to his attention if he thought it a good idea. I sent the following telegram:

Dec. 26, 1970

John Ehrlichman
The White House
    Suggest President Nixon reply to fourteen Soviet scientists' appeal about Angela Davis by inviting them or any Soviet lawyers they nominate to attend her trial. Stop. Also call attention to request of five New York City District Attorneys to attend trials of Jews accused by Soviet Union of plotting highjacking.

Elvin A. Kabat
Member, National Academy of Sciences

The idea evidently was favorably received. Figure 3 is the reply from John Ehrlichman. On Sunday, January 3, 1971, the six o'clock evening news announced that President Nixon had invited the 14 Soviet scientists to attend Angela Davis's trial. Figure 4 shows a portion of Monday's *New York Times* article and its editorial of January 8. It still seems remarkable that a suggestion from a private individual could reach the highest levels and be acted upon so rapidly.

Current Activities and Future Plans

Although I became Emeritus on December 31, 1984, I have been permitted to keep my laboratories and to maintain my group at about the same size. I decided a few years back that the laboratory had to go into cloning and sequencing if we were ever to be able to formulate a detailed concept of the interactions in the CDRs of antibody combining sites to explain antibody diversity. Accordingly I spent a short time in Mark Davis' laboratory, then at NIH, learning to make and clone cDNA. This was followed by about 10 days at the Weizmann Institut learning nucleotide sequencing in David Givol's laboratory. I worked closely with the carbohydrate chemists

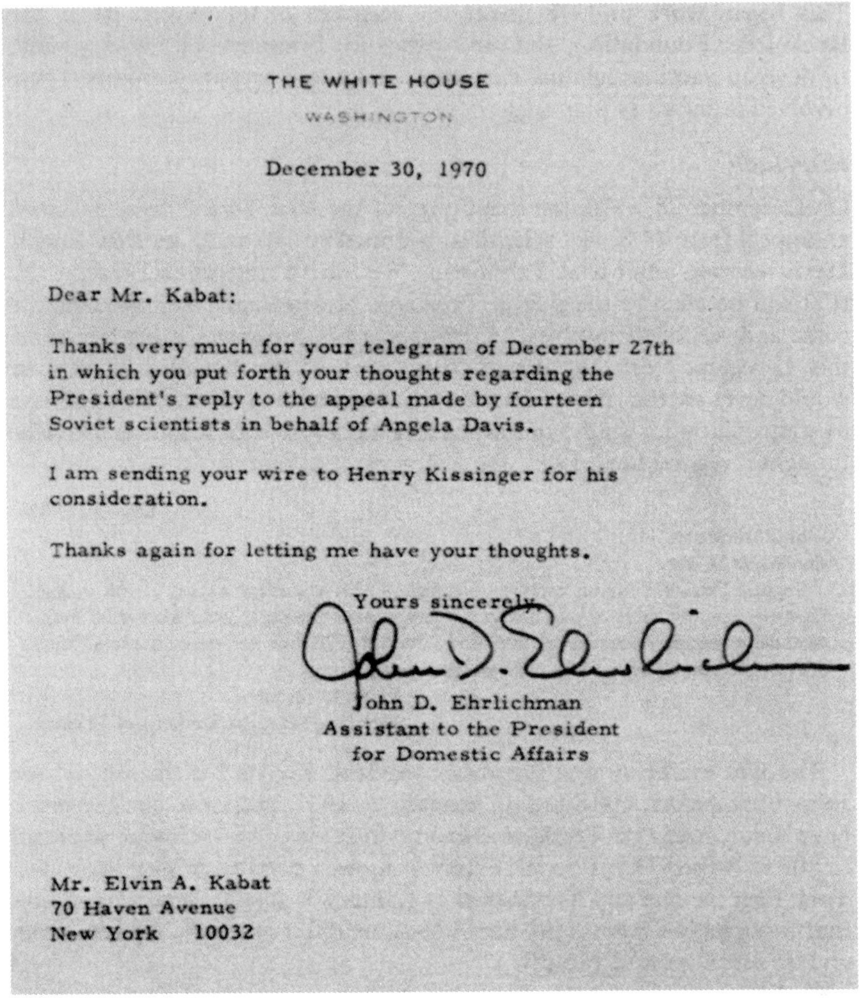

THE WHITE HOUSE

WASHINGTON

December 30, 1970

Dear Mr. Kabat:

Thanks very much for your telegram of December 27th
in which you put forth your thoughts regarding the
President's reply to the appeal made by fourteen
Soviet scientists in behalf of Angela Davis.

I am sending your wire to Henry Kissinger for his
consideration.

Thanks again for letting me have your thoughts.

Yours sincerely,

John D. Ehrlichman
Assistant to the President
for Domestic Affairs

Mr. Elvin A. Kabat
70 Haven Avenue
New York   10032

Figure 3

in the laboratory, Santosh Sikder and Pradip Akolkar, to establish the
making, cloning, and sequencing of cDNAs on a routine basis. They then
taught the other post-doctorals, graduate students, and technicians.

We set ourselves the goal of trying to probe the repertoire of antibody
sites formed to a single antigenic determinant, the epitope of $\alpha(1 \rightarrow 6)$-
linked glucoses from dextran B512. Graduate students Jacqueline Sharon
and Barbara Newman had been engaged in mapping the antibody com-
bining sites of hybridoma antibodies prepared by injecting $\alpha(1 \rightarrow 6)$dextran

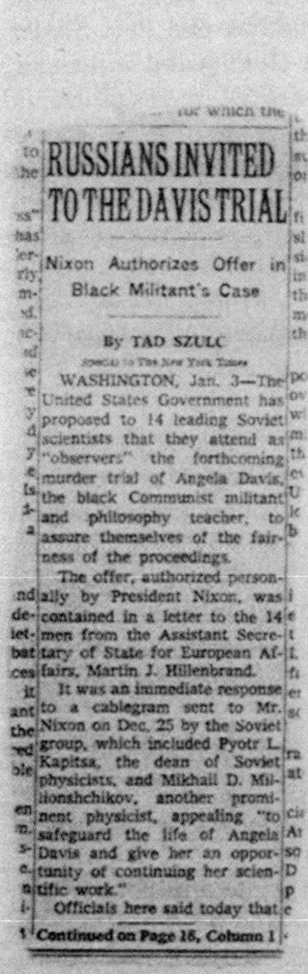

RUSSIANS INVITED TO THE DAVIS TRIAL

Nixon Authorizes Offer in Black Militant's Case

By TAD SZULC
Special to The New York Times

WASHINGTON, Jan. 3—The United States Government has proposed to 14 leading Soviet scientists that they attend as "observers" the forthcoming murder trial of Angela Davis, the black Communist militant and philosophy teacher, to assure themselves of the fairness of the proceedings.

The offer, authorized personally by President Nixon, was contained in a letter to the 14 men from the Assistant Secretary of State for European Affairs, Martin J. Hillenbrand.

It was an immediate response to a cablegram sent to Mr. Nixon on Dec. 25 by the Soviet group, which included Pyotr L. Kapitsa, the dean of Soviet physicists, and Mikhail D. Millionshchikov, another prominent physicist, appealing "to safeguard the life of Angela Davis and give her an opportunity of continuing her scientific work."

Officials here said today that

Continued on Page 16, Column 1

Invitation to a Trial

President Nixon's invitation to fourteen outstanding Soviet scientists to attend Angela Davis's trial, to see for themselves whether she is fairly judged, is a master propaganda stroke in the best sense of the word. Since the message the Soviet scientists sent to the President on Christmas Day implies that Miss Davis was in danger of being railroaded and punished for a crime she did not commit, it would be highly desirable for these observers to attend the trial. We hope the Soviet Government will accept Mr. Nixon's challenge, and permit them to come.

The appeal of the Soviet scientists may merely be a maneuver to assist the calculated campaign of American Communists to turn the Davis trial into a political circus and to establish the defendant as a martyr before any of the evidence is even heard. Despite shrill cries of "political repression," Miss Davis still has all of her constitutional rights including the ultimate right of appeal to the Supreme Court if due process is violated.

Appearances in this world are sometimes deceptive, however, so it is worth reflecting on several interesting characteristics of the appeal and those who signed it. Academicians Kapitsa, Millionshchikov, Tamm and the others who supported the statement are among the most brilliant and sophisticated people in the Soviet Union. Most, perhaps all, of them have spent enough time in the United States or Western Europe to know how to evaluate the crude Communist accusations of "capitalist frameups." Moreover, a large proportion of the signers of the message are outstanding Soviet liberals and several are Jews—and the appeal to President Nixon on Miss Davis's behalf was received just after a Leningrad court had sentenced two Jewish would-be hijackers to death.

These circumstances suggest that the message to President Nixon may have been a way of speaking up for all political prisoners, including the Jews tried in Leningrad, whose sentences were subsequently softened. Certainly the possibility must have occurred to Mr. Nixon. His invitation to the Soviet scientists to attend Miss Davis's trial implies that the Soviet Union should show reciprocity and invite American observers to Soviet trials, the fairness of which is strongly doubted by many millions of persons in this country who hardly think it worthwhile to cable their views to the Kremlin.

Figure 4

into BALB/c and C57BL/6 mice. Eric Lai had carried out similar studies of C57BL/6 hybridomas from mice immunized with stearylisomaltosyl oligosaccharides. H. T. Chen and D. Makover characterized hybridomas to the stearylisomaltosyl oligosaccharides in a nude mouse and in C58 mice. Thus, we had a sampling of hybridomas with groove-type sites complementary to six and seven $\alpha(1 \rightarrow 6)$-linked glucoses. These were about equally divided between IgM and IgA, but two IgG3 hybridomas were obtained. Both antigens used for immunization yielded T-independent hybridomas. Flavio Gruezo prepared mRNA and Drs. Sikder, Akolkar, Bhattacharya, and Jerry Liao have been cloning and sequencing them. Sherie Morrison is collaborating. She and her colleagues pioneered in preparing transfectomas substituting C-region genes of other species onto mouse V_H genes. With Amelia Black, a graduate student, Sherie and I are trying to express the transfection products in *E. coli* and in mammalian cells. Tsukasa Matsuda is studying the repertoire by the use of the isomaltose oligosaccharides coupled to BSA or to KLH to obtain a set of hybridomas to T-dependent antigens with an $\alpha(1 \rightarrow 6)$ epitope sufficiently large to fill the antibody combining site. Paula Borden has prepared antiidiotypic hybridomas and is studying the idiotypic determinants of the various anti-$\alpha(1 \rightarrow 6)$dextrans.

One important finding is that the T-independent hybridomas have V_H chains belonging to three major germline gene families of Brodeur and Riblet, J558, J606, and J36-60 and that the light chains also belong to several germline gene families although their combining sites are very similar. This would tend to indicate that the antibody-forming system is very well protected against germline gene loss. Since most other epitopes prepared by coupling small molecules to a protein create heterogeneous populations of antibodies, the finding that different germline gene families were used could not be interpreted unambiguously.

Paula Borden found that IgG3 antiidiotypes to an anti-$\alpha(1 \rightarrow 6)$dextran hybridoma precipitated with antidextran. The reaction differs from the usual precipitin reactions in that it is highly dissociated, and its role in idiotype antiidiotype reactions requires further study.

H.-T. Chen has mapped the combining sites of various antiblood group A and B hybridomas and, with Arne Lundblad and R. M. Ratcliffe, has cloned and sequenced several. These blood group epitopes are more complicated structurally and should also provide an intimate picture of those antibody combining sites.

We sorely need high resolution X-ray studies of one or more antibody combining sites to carbohydrate epitopes to be able to correlate immunochemical mapping data with the interactions of the dextran or blood group oligosaccharides in their respective sites. We also intend to study anti-antiidiotypes in this system in relation to the "internal image" antibody to a carbohydrate. I intend to stay in the laboratory for as long as I feel I can do creative work.

My life at Columbia has always been most satisfying. Over the years I have been able to collaborate with many colleagues at Columbia and throughout the world. The University provided me with the opportunity to do as I wished scientifically and always defended academic freedom in general and mine in particular. I am indebted to the Office of Naval Research for seventeen years of support; to the molecular biology section of the National Science Foundation for 36 years of continuous support; and since 1974 to the National Institutes of Health for providing the opportunity for me to pursue my interests there as well as at Columbia.

I thank Mr. Darryl J. Guinyard for typing the manuscript.

Literature Cited

1. Kabat, E. A. 1983. Getting started 50 years ago—Experiences, prospectives, and problems of the first 21 years. *Ann. Rev. Immunol.* 1: 1–32
2. Boring, E. G. 1956. Lewis Madison Terman. From *Biographical Memoirs of the National Academy of Sciences* 1959, Vol. 33: 414–61. New York: Columbia Univ. Press
3. Kabat, E. A. 1956. *Blood Group Substances—Their Chemistry and Immunochemistry.* New York: Academic
4. Kabat, E. A. 1982. Contributions of quantitative immunochemistry to knowledge of blood group A, B, H, Le, I and i antigens. *Am. J. Clin. Pathol.* 78: 281–92
5. Sikder, S. K., Kabat, E. A., Roberts, D. D., Goldstein, I. J. 1986. Immunochemical studies on the combining site of the blood group A-specific lima bean lectin. *Carbohyd. Res.* 151: 247–60
6. Chen, H.-T., Kabat, E. A. 1985. Immunochemical studies on blood groups. The combining site specificities of mouse monoclonal hybridoma anti-A and anti-B. *J. Biol. Chem.* 260: 13208–17
7. Kabat, E. A., Liao, J., Bretting, H., Franklin, E. C., Geltner, D., Frangione, B., Koshland, M. E., Shyong, J., Osserman, E. F. 1980. Human monoclonal macroglobulins with specificity for *Klebsiella* K polysaccharides that contain 3,4-pyruvylated-*D*-galactose and 4,6-pyruvylated-*D*-galactose. *J. Exp. Med.* 152: 979–95
8. Kabat, E. A., Nickerson, K. G., Liao, J., Grossbard, L., Osserman, E. F., Glickman, E., Chess, L., Robbins, J. B., Schneerson, R., Yang, Y. 1986. A human monoclonal macroglobulin with specificity for $\alpha(1 \rightarrow 8)$-linked poly-N-acetyl neuraminic acid, the capsular polysaccharide of group B meningococci and Escherichia coli K1, which crossreacts with polynucleotides and with denatured DNA. *J. Exp. Med.* 164: 642–54
9. Porter, R. R. 1986. Antibody structure and the antibody workshop 1958–1965. Symposium on The Role and Significance of International Cooperation in the Biomedical Sciences, eds. G. Salvatore, H. K. Schachman, in *Perspectives in Biology and Medicine* 29: Part 2, S161–S165
10. Wu, T. T., Kabat, E. A. 1970. An analysis of the sequences of the variable regions of Bence Jones proteins and myeloma light chains and their implications for antibody complementarity. *J. Exp. Med.* 132: 211–50
11. Kabat, E. A., Wu, T. T. 1972. Attempts to locate complementarity-determining residues in the variable positions of light and heavy chains. *Ann. N.Y. Acad. Sci.* 190: 382–93
12. Kabat, E. A., Wu, T. T., Bilofsky, H. 1976. Variable regions of immunoglobulin chains. *Medical Computer Systems.* Cambridge, MA: Bolt Beranek & Newman
13. Kabat, E. A., Wu, T. T., Bilofsky, H. 1979. *Sequences of Immunoglobulin Chains. NIH Publication 80-2008.* National Institutes of Health, Bethesda, Md.
14. Kabat, E. A., Wu, T. T., Bilofsky, H., Reid-Miller, M., Perry, H. 1983. *Sequences of Proteins of Immunological Interest.* National Institutes of Health, Bethesda, Md.
15. Kabat, E. A., Wu, T. T., Reid-Miller, M., Perry, H. M., Gottesman, K. S. 1987. *Sequences of Proteins of Immunological Interest.* U.S. Department of Health and Human Services, Public Health Service, National Institutes of Health, No. 165-462, pp. 1–804. Washington, DC: USGPO
16. Kabat, E. A., Wu, T. T., Bilofsky, H. 1978. Variable region genes for the immunoglobulin framework are assembled from small segments of DNA—A hypothesis. *Proc. Natl. Acad. Sci. USA* 75: 2429–33
17. Tonegawa, S., Maxam, A. M., Tizard, R., Bernard, O., Gilbert, W. 1978.

Sequence of a mouse germ-line gene for a variable region of an immunoglobulin light chain. *Proc. Natl. Acad. Sci. USA* 75: 1485–89

18. Bernard, O., Hozumi, N., Tonegawa, S. 1978. Sequences of mouse immunoglobulin light chain genes before and after somatic changes. *Cell* 15: 1133–40

19. Early, P., Huang, H., Davis, M., Calame, K., Hood, L. 1980. An immunoglobulin heavy chain variable region gene is generated from three segments of DNA: V_H, D, and J_H. *Cell* 19: 981–92

20. Krawinkel, U., Zoebelein, G., Bruggemann, M., Radbruch, A., Rajewsky, K. 1983. Recombination between antibody heavy-chain V and variable V region genes: Evidence for gene conversion. *Proc. Natl. Acad. Sci. USA* 80: 4997–5001

21. Jaenichen, H.-R., Pech, M., Lindenmaier, W., Wildgruber, N., Zachau, H. G. 1984. Composite human V_k genes and a model of their evolution. *Nucleic Acids Res.* 12: 5249–63

22. Reynaud, C.-A., Anquez, V., Grimal, H., Weill, J.-C. 1987. A hyperconversion mechanism generates the chicken light chain preimmune repertoire. *Cell* 48: 379–88

23. Kabat, E. A. 1986. A tradition of international cooperation in immunology. *Proc. Int. Symp. "The Role and Significance of International Cooperation in the Biomedical Sciences,"* ed. G. Salvatore, H. K. Schachman. *Perspect. Biol. Med.* 29: S159–S160

David Kingery

Annu. Rev. Mater. Sci. 1989. 19:1–20

CERAMIC MATERIALS SCIENCE IN SOCIETY

W. David Kingery

Department of Materials Science and Engineering, and Department of Anthropology, University of Arizona, Tucson, Arizona 85721

INTRODUCTION

It is easy for an academic or research-minded materials scientist to become so fascinated with the process of solving materials science problems that this activity is seen as an end in itself. Thomas S. Kuhn has suggested that, "Bringing a normal research problem to a conclusion is achieving the anticipated in a new way and it requires the solution of all sorts of complex instrumental, conceptual and mathematical puzzles. The man who succeeds proves himself an expert puzzle-solver and the challenge of the puzzle is an important part of what usually drives him on" (1). But materials science is much more than solving puzzles. It is an enabling science that makes possible new, improved, more reliable, and effective materials technology. The discoveries and scientific explanations of materials science are important primarily as they relate to this technology—the alteration and manipulation of the material world to obtain socially desired objectives. In sequence, materials technology is much more than manipulating materials; it is an enabling technology that makes possible new and improved devices, products, and systems. That is, the materials produced are important as they relate to devices and products that incorporate them and fit into larger systems that meet objectives and desires of the society in which they are embedded. This hierarchical nature of materials science—materials technology—devices and products—technological systems—societal needs and desires suggests that an appreciation of the relationships between these different components should be an essential part of the intellectual kit of every materials scientist.

For this purpose ceramic science and technology is a wonderful field. It extends from 25,000 BC up to the present day. Ceramic manufacture was one of man's earliest technical successes. European porcelain development and marketing were central to the scientific, industrial, economic, and consumer revolutions of the eighteenth century. New high-tech ceramics were important for the development of electrical technology a hundred years ago and have been central to the post-World War II electronic information and communications revolution. Ceramic superconductors may revolutionize technology of the next century. With opportunities to control form, texture, translucency, opalescence, and color, ceramics are a wonderful art medium. Pottery shards are the fundamental data of archaeology. In short, ceramics are a crossroads for art, archaeology, history, science, and technology and their relation to societal and human values. As such, they present a marvelous opportunity to investigate the interaction of materials science with the technical, economic, and human concerns of society.

CLASSIFICATION OF TECHNOLOGY

Technology and its history have come to be studied in categories more or less reflecting departmental structures of university schools of engineering—agriculture, transportation, building construction, electrical engineering, and so forth. Remembering that the encyclopedists of the eighteenth century changed the way that people thought about branches of knowledge (2), there may be value in considering alternate classification schemes. The usual industry-focused classifications of technology are shown as vertical columns in Figure 1. Peer groups based on university degree join professional societies, attend professional meetings, are elected to national academies, and otherwise reinforce the scheme. Another equally rational classification of technology is what we shall call task-focused and is illustrated by the horizontal cuts in the matrix of Figure 1.

	Generalized Technology	Textile Technology	Electronic Technology	Civil Engineer Technology
Materials Acquisition Technology	Starting Material	Silk Fiber	Silicon Powder	Iron Bar
Materials Engineering Technology	Shaped Object	Silk Thread	Silicon Crystal	Iron I-beam
Device Production Technology	Intermediate Devices	Silk Cloth	Transister Chip	Iron Truss
Product Manufacture Technology	Product	Wedding Gown	P.C. Computer	Railroad Bridge
System Design Technology	System	Marriage System	Publication System	Transportation System

Figure 1 The usual industry-focused classifications of technology are shown as vertical columns. Task-focused classifications are shown as horizontal rows.

In every industry there are a series of tasks of increasing structural complexity that begin with materials acquisition, preparation, and shaping. These materials are assembled to form products and devices that are the industry's output. Products are used in systems that may be based on one or several different industries. Materials science and engineering, systems design and, increasingly, manufacturing engineering are firmly in place or are becoming established as engineering departments at many universities. One of the advantages of a task-focused classification is that it recognizes commonalities in the tasks faced in manufacturing quite different product lines. It may also be appropriate for teaching technology to humanities students, management professionals, and other non-technologists. Task focusing achieves an industry-free generality and allows industry-craft-archaeology-ethnography continuities and comparisons to be seen more easily.

A different and equally rational classification is to focus on the technological systems associated with a specific individual product—a ceramic capacitor, a semiconductor chip, a temple, a television set. This product-focused technology is a particularly natural classification for archaeology and material culture studies; it implies the existence of a technological system consisting of materials acquisition, materials distribution, design, manufacturing, distribution, as well as product reception, perception, and use and various reuse and discard technologies (Table 1). A general illustration of such a system is shown as Figure 2.

In contrast to the two previous typologies, this product-focused classification scheme has little to offer in a direct way for the primary organization of technologist training or for professional societies. I believe it has much to offer in understanding how technology exists as a system within society—the first step in appreciating the interactions of materials science, materials technology, and society as components of culture. It also provides a way for seeing the system in which the work of materials scientists

Table 1 A series of technologies are required for the design, production and use of any product

Product-Centered Classification of Technologies
Materials acquisition technology
Design technology
Manufacturing technology
Distribution technology
Use technology
Re-use technology
Discard technology

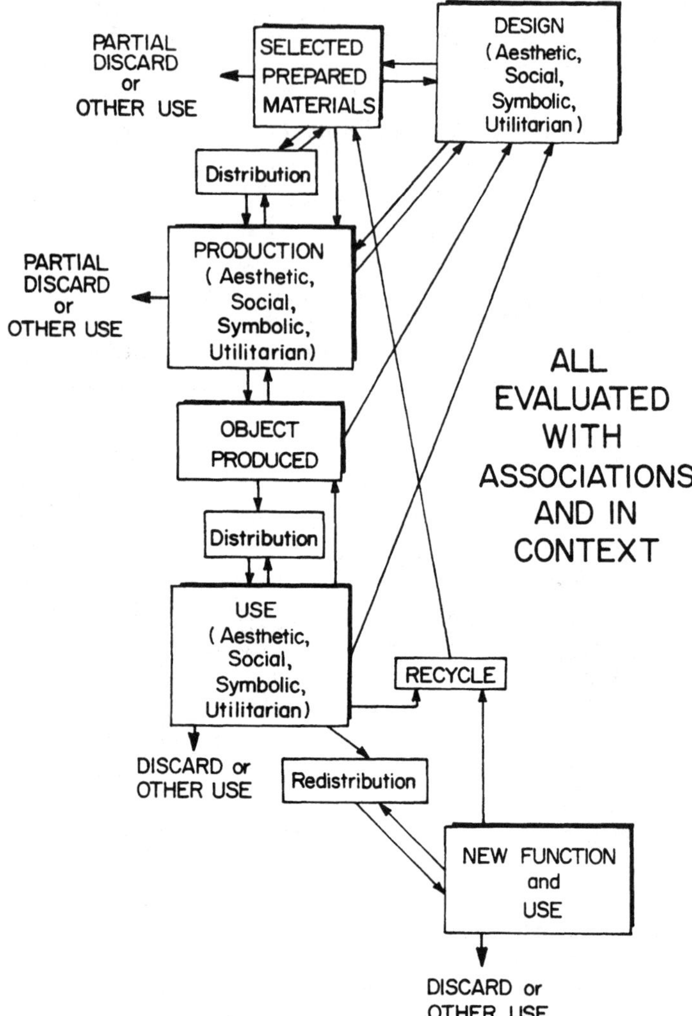

Figure 2   The components of a product-focused classification of technology are interrelated and interact to form a technological/social system.

and materials technologists must be embodied. In particular, it shows that the principal feedback loop affecting design technology and manufacturing technology is generated by the reception and perception of a product by its users. This is not surprising—sometime ago I discussed the tendency of technology to reach the level of its market (3), and von Hippel has recently shown that industrial innovation is mostly a user-inspired activity

(4). However, this understanding that not only the needs, but also the reactions and perceptions of users are an essential context of design and manufacturing technology is often a new thought to engineering students.

In the history of ceramics the influence of Augustus the Strong, Elector of Saxony and owner of the largest European collection of Oriental ceramics, in directing the eighteenth century invention of European porcelain is well known (5). Modern ceramics was born in response to the needs of the age of electricity (6). The recent dominant role of Japanese ceramic technology has been explained by Hiroaki Yanagida of Tokyo University (7): "The rapid growth of the electronics industry has created an intense requirement for better materials."

Thomas Alva Edison's first invention, patented in 1868, was an electrical vote recorder. He took the device to Washington and demonstrated it to a committee of the House of Representatives. It worked very well. Edison recounted that the chairman of the committee, after seeing how perfectly it worked, said "Young man, if there is any invention on earth that we don't want down here it is this. One of the greatest weapons in the hands of a minority to prevent bad legislation is filibustering on votes, and this instrument would prevent that" (8). The experience taught Edison to "turn his efforts towards inventing things that not only were needed, but were wanted as well." Good advice.

EUROPEAN PORCELAIN

Throughout the seventeenth century Chinese porcelain was extensively imported into Europe. It had a translucence, whiteness, and ring that could not be reproduced using European methods. Porcelain was beginning to be used for utilitarian purposes as coffee, chocolate, and tea were becoming more common, but mostly it was displayed as exotica from the East. This was a period of centralization of authority and establishment of autocratic rule after the model of Louis XIV in France. There were chemistry and physics experiments exploring new phenomena as well as exploitation of natural resources and development of new manufacturing methods as part of a mercantile economic policy. In Saxony, Augustus the Strong, Elector from 1694–1735, and also King of Poland, had a special interest in porcelain; his personal collection was the largest in Europe. He and his court supported one of the first truly modern research efforts; it was aimed at porcelain manufacture (9).

An important participant in this study was Count von Tschirnhaus, who had studied mathematics and physics at Leyden and carried out many experiments on materials behavior using high temperatures achieved by focusing sunlight in a solar furnace such as illustrated in Figure 3. He had found that calcium oxide was not melted in his furnace, nor was pure quartz. However, a mixture of the two, for which the lower melting temperature or eutectic is 1436°C, could be fused. Thus, his solar furnace reached at least this temperature, much higher than those used in ordinary practice. A budding alchemist, Johann Friedrich Böttger, whose demonstration transforming mercury into gold had got him into difficulties with the authorities, was enrolled as principal investigator. In 1705 he was given some assistants to work with him and entered into a research program aimed at porcelain manufacture. One of his helpers was Paul Wildenstein, a Freiburg miner, who reported to an investigation commission in 1736 (10). He recalled:

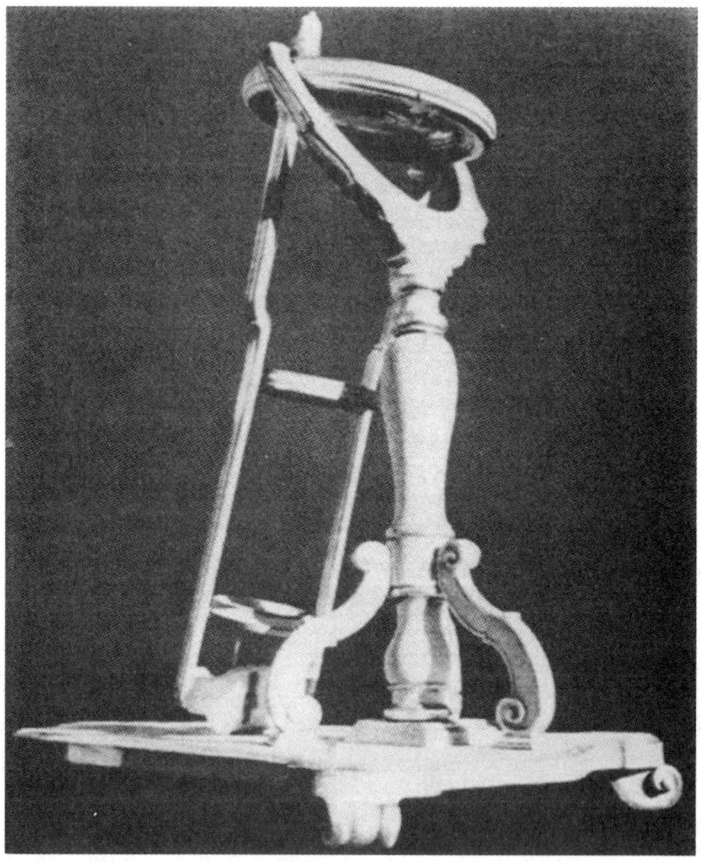

Figure 3 Solar furnace or "burning lens" such as used by Count von Tschirnhaus for porcelain experiments.

In 1706, I came to Meissen to the Baron Böttger, to the secret laboratory, and we were shut in there for 18 weeks. Even the windows had been walled up to half of their height, and Herr von Tzschirnhaussen [sic] from Dresden was often with us as well as the mining councillor Pabst from Freyberg. We had a laboratory with 24 kilns, and the baron and Tzschirnhaussen had already made specimens of red porcelain in the shape of small slabs and marbled slab stones.
Herr von Tzschirnhaussen, too, was giving instructions, and they began to research. Among other things, specimens of red porcelain were made, as well as white. Köhler and I had to stand nearly every day by the large burning-glass to test the minerals. There I ruined my eyes, so that I now can perceive very little at a distance.

After numerous experiments based on the idea of a flux to provide the lower temperature eutectic melting, there is little known about the details of experiments studying the response of different clays and clay mixtures under the high temperatures developed by von Tschirnhaus' lenses. The use of lime as the fluxing material required high temperatures and, in fact, this was the main sticking point for the experimental program. Wildenstein testified (10):

We couldn't manage to make a strong fire in the new kiln; all our toil was fruitless and the fire remained weak. While it was burning, we had to make the fire walls sometimes higher, sometimes lower, but it was no use until we finally discovered the fault in the casing. The coals wouldn't burn all the way down, so we had to pull them out every thirty minutes. . . . Our hair was scorched and the floor had grown so hot that our feet were covered with large blisters.

While most descriptions of the invention of European porcelain focus on the composition and clay used, there were really three principal requirements leading to success: first, the concept of partially fluxing a white clay with lime; second, the experimental testing of suitable compositions with von Tschirnhaus' burning lenses; third, achieving the high kiln temperature necessary for satisfactory firing. A temperature near 1400°C is required to fire the composition of Böttger porcelain so that a viscous liquid phase bonds the material together. This was an unheard of temperature for firing at the time and the secret of success was more related to the design of a kiln with multiple fireboxes of good design than to the particular clay used.

The resulting microstructure illustrated in Figure 4 is a mass of mullite crystals held together with a refractory glass, giving strength and resistance to thermal shock that remained a characteristic of the Meissen formula, one which is quite different from the Chinese compositions. One of the results of this is an extraordinary resistance to thermal shock. Wildenstein (10) described a visit of Augustus the Strong to Böttger's Jungfernbastei laboratory to view a firing.

His Majesty arrived with the Prince of Fürstenberg, but when they entered the laboratory and felt the terrible fire, they would rather have turned back. Since, however, the baron—looking like a sooty charcoal-burner—was so close to him, His Majesty entered

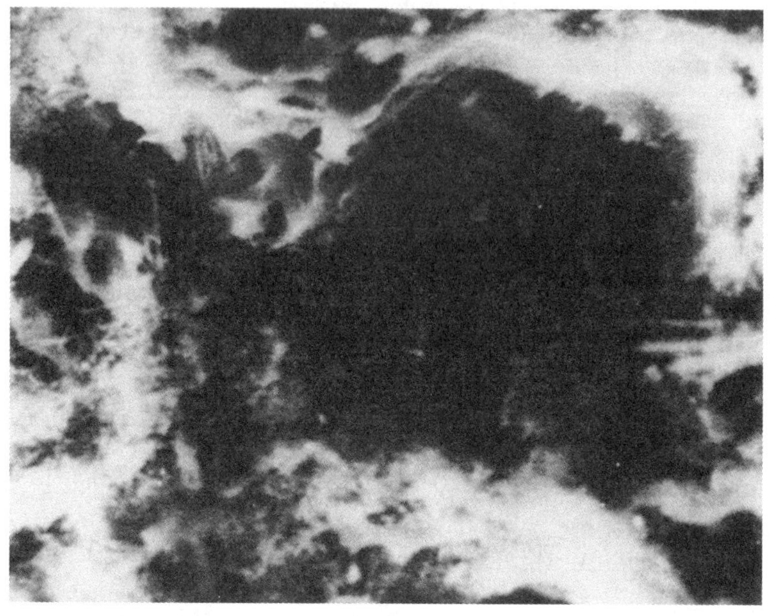

(a)

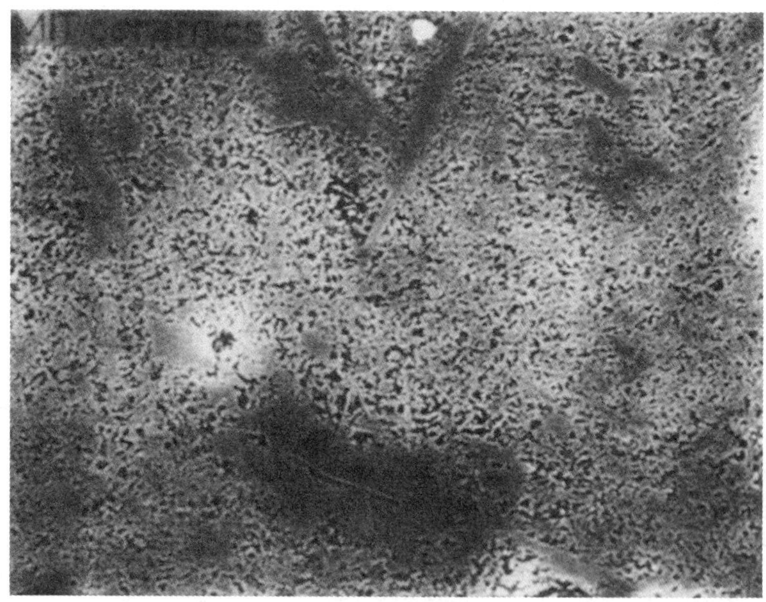

(b)

and urged the prince to come in, too. The baron told us to stop firing for a while and to open the kiln, and during this time the prince said several times, "Oh, Jesus." The king, however, laughed and said to him that it was in no way to be compared to Purgatory! The kiln was opened, and all was bathed in white heat so that nothing could be seen.[1] The king looked in and said to the prince, "Look Egon, they say that porcelain is in there!" The prince said he couldn't see anything either but finally the kiln grew red, since it was open, so they could see the porcelain. I had to draw out a specimen, which was a saggar containing a small teapot.

A tubful of water stood nearby, so that the flowing iron could be extinguished. The baron immediately seized the tongs, drew the teapot out, and threw it into the water. Suddenly a loud bang was heard, and the king said, "Oh, it's smashed." But the baron replied, "No, Your Majesty, it must stand this test." He then rolled up his sleeves and took it out of the tub. It indeed proved to be intact.

I was a bit doubtful about this story, but students in my laboratory repeated the experiment with the same result. The composition and microstructure are similar to some of the spark plug porcelains developed during the early years of high-compression automobile and aircraft engines.

The method of manufacturing hard porcelain developed at Meissen spread throughout Western Europe entirely by the movement of workers, first from Meissen to Vienna, who took with them the necessary compositional and kiln technology. Workmen went from Vienna to Venice and later other workmen to Höchst, Fürstenberg, Nymphenberg, Strasburg, Frankenthal, St. Petersburg and so forth. In 1762 at the end of the Seven Years War, Frederick the Great occupied Dresden and brought back with him models, molds, and workers to establish the royal porcelain manufactory in Berlin.

All in all, the development of European hard porcelain was an epochal event in the history of materials science and technology. To achieve the mercantile, social, and symbolic goals of Augustus the Strong, a dedicated research program led to new materials science, new materials technology, and new manufacturing methods.

THE AGE OF ELECTRICITY

Until the end of the eighteenth century the only source of electricity was generation by friction, a source known since ancient times. In 1600,

---

[1] Note: color temperatures are 1300°C, dull white; 1400°C, bright white.

Figure 4 Microstructures of a Böttger porcelain. (*a*) A sample from 1715 heavily etched (12 minutes in 2% hydrofluoric acid) shows mullite needles in a glass matrix that has been etched away (1500 ×). (Courtesy of W. Schulle & B. Ullrich.) (*b*) Replica prepared in author's laboratory that has been polished and lightly etched (ten seconds, 1% hydrofluoric acid) that also shows mullite crystals in a glass matrix (2000 ×).

electricity and magnetism were the topics of one of the first books on experimental science (11). A machine for providing electricity was first made by Otto von Guericke in 1660; many more elaborate and effective models were manufactured in the eighteenth century. Then, about 1750, E. G. von Kleist devised the Leyden jar for storing the elusive electrical fluid; but electricity remained a substance for scientific, scholarly, and public amusement. In a famous late century demonstration Abbé Jean Antoine Nollet charged a Leyden jar, formed a rank of 180 soldiers holding hands and had the ends of the rank grasp wires connected to the Leyden jar; the entire company jumped simultaneously in perfect unison. The experiment was so impressive that King Louis repeated it with 700 Carthusian monks (12).

This situation changed dramatically on March 20, 1800 when Alessandro Volta sent a communication to the Royal Society of London, which was read on June 26 of that year (13). He described a method of continuous current generation from a "pile" of dissimilar metals. The reaction was sensational. In England William Nicholson and Anthony Carlisle had seen part of the letter and, in less than a month (on April 30, 1800), observed the electrolytic decomposition of water. Volta himself was invited to Paris and demonstrated his discoveries at the Institute of France during November 1801. The emperor Napoleon attended and witnessed experiments that included the decomposition of water and heating an iron wire to incandescence. It had long been known that an electric current could be transmitted by wire long distances, and by 1805 Salva used electrolytic decomposition as an indicator in an elementary method of communicating signals over a long distance. Sir Humphrey Davy, lecturer at the Royal Institution in London, installed the largest voltaic apparatus of the time. He showed that impurities were needed for the electrolysis of water (1806), electrolysed fused salts to form elemental potassium (1807), discovered chlorine (1808), heated platinum to incandescence (1802), and displayed the first electric arc lamp (1808). Davy was an exciting lecturer who used lots of demonstrations and whose performances were major public attractions. Reminiscent of the recent discovery of high-temperature ceramic superconductors, there was intense scientific and public excitement and optimism about the nature and uses of electricity. In 1812 Davy hired Michael Faraday, who became his successor, and went on to discover the principle of the dynamo in 1831—that is, the possibility of transforming mechanical energy into electrical current.

The basic ideas for using electricity for chemical purposes, communication, and lighting were all well established in the first few years after Volta's cell was announced. However, it was a century or more before the possibilities of electricity were fully realized and the world had telegraphic, telephone and radio communication, electric light, recorded sound, electric motors, widespread electric power distribution, radios, t.v. and, more recently, transistors, computers, and a world full of electronic devices. The practical achievement of all these things was limited by, and required the discovery and development of new materials.

Edison's Carbon Microphone

After a series of inventions and successful manufacturing of various electrical devices, Thomas Alva Edison set himself up in a laboratory at Menlo Park devoted to the business of invention (14). That same year, 1876, Alexander Graham Bell announced the telephone at the Philadelphia Centennial Exhibition. His device was astounding, but crude and ineffective. A single instrument, both transmitter and receiver, had a flexible diaphragm that vibrated an armature facing an electromagnet, thus inducing a modulated current transformed back to sound by the reverse process. The current generated was weak. The problem of developing a better transmitter was immediately recognized by many inventors in the emerging field of electrical devices. Long distance telegraphy had grown to become a big business and there was no question in anyone's mind that a major market awaited a practical telephone.

In his new Menlo Park laboratory early in 1877, Edison characteristically tried a number of tacks for transforming sound vibrations into electrical oscillations. He tested moving coils, vibrating condensors, needles immersed in mercury and in electrolytic solutions. His success, the carbon microphone, was based on earlier experiments in which he had discovered the effect of pressure on contact resistance, when he constructed a carbon rheostat in 1873 (15). For that, he used layers of silk discs saturated with fine particles of graphite contained in an insulating cylinder. When pressure was applied with a screw drive to a plate that compressed the stack, its resistance changed by an order of magnitude. When experiments were begun to develop a better telephone transmitter, a pressed tablet of graphite mounted behind a vibrating platinum disc gave good initial results when included in the primary circuit of an induction coil (16). This touched off a whole series of electromechanical improvements. In addition, Edison instituted a materials research program on the influence of pressure on contact resistance; it was this that set him apart from other electromechanical inventors of the time.

Edison tested thousands of different materials placed between plates connected to a battery and galvanometer; the change in resistance was measured as weights were added. The results of these experiments were described by Prescott (1879):

The value of different substances to be used as buttons in the telephone is given below, the first mentioned being the best, and the others in the order given:

Lampblack
Hyperoxide of lead
Iodide of copper
Graphite
Gas carbon
Platinum black

Finely divided materials which do not oxidize in the air, such as osmium, ruthenium, silicon, boron, iridium and platinum, give results proportionate to this minute division, but many of them are such good conductors that it is necessary to mix some very fine nonconducting material with them before moulding (17).

Edison patented the use of lampblack in the telephone transmitter in February 1878 (18). Western Union manufactured the Edison telephone transmitter in New York until they retired from the telephone business in a compromise with the Bell companies in 1879. During this period the critical lampblack carbon discs continued to be produced at Menlo Park. A battery of smoking kerosene lamps produced soot that was scraped off, carefully weighed and pressed into 300 mg discs. According to Francis Jehl,

> The process of smoking chimneys was not so simple as it sounds. It was very essential that the soot should be deposited at the lowest possible temperature, and the flame could not be allowed to play upon the deposit, as otherwise it acquired a high resistance and was wholly unsuited to use in the transmitters.
>
> The first step after the chimney had been scraped was to take away the portions that had a brownish tinge. The remainder was then finely ground and placed in the press, or mould. Each button was supposed to weigh three hundred milligrams. Afterward, the buttons were packed in shallow wooden boxes padded with cotton, and shipped thus to New York City (19).

This was the first application of modern high-technology materials science, research, development, and production by an electrical company completely independent of the metal and ceramic industries (Figure 5).

Edison's Electric Light

One of the results of relatively successful arc lighting systems developed in the nineteenth century was an invigoration of the research for an incandescent lamp. Thomas Edison visited the factory of William Wallace on September 8, 1878 and saw a demonstration of the Wallace dynamo lighting, a system of eight arc lamps. Beginning the next day, he devoted the resources of his Menlo Park invention factory to producing a practical incandescent lamp (20). One major challenge of electric lighting was the invention of a workable filament. To accomplish this, Edison had available at Menlo Park not only the capability for electrical experimentation, but

Figure 5 A magazine illustration showing how lamp chimneys were smoked to produce the soot used in the carbon telephone transmitters. This was the beginning of modern materials technology. [Jehl (8), p. 131.]

also an in-house machine shop, vacuum equipment, chemical laboratory, and glass blower.

Edison first saw the problem as one of feedback controls to allow the use of platinum at a temperature near, but not exceeding, its melting point. Then he found that platinum worked better in vacuo, particularly if it were heated to remove occluded gases prior to having the vacuum sealed. Since a vacuum tube was indicated in any event, and Edison wanted a relatively high resistance (to use lamps connected in a parallel circuit), he turned back to carbon as the filament material. In order to achieve a long life, he specified a completely sealed glass envelope with platinum lead-through wires. This combination of carbon filament, platinum lead-through wires, and heating the filament while evacuated in a glass bulb that was then sealed to maintain the vacuum, was successful—a success that depended on the entire lamp system, all elements of which were essential (21).

The Nernst Lamp

Walther Hermann Nernst was only 29 years old in 1893 when the first edition of his book, *Theoretical Chemistry*, was helping establish the new field of physical chemistry; his work integrating electrochemistry and thermodynamics was well underway. A new laboratory was established for him at the University of Göttingen and the brilliant young professor had a team of research students working for the doctorate degree, most of whom were destined for careers in industry. By this time, the unification of Germany under Bismarck's Prussian leadership had led to an expansionist and nationalist fervor that affected all fields of endeavor. There was rapid industrial growth and university programs were promoted to produce the scientists and engineers needed for industry. The growth of chemical manufacturing and chemistry in the universities was particularly notable. In addition to providing for students, active faculty participation in new technological enterprises was both expected and encouraged. When an opportunity to develop the practical benefits of his new field of physical chemistry appeared, it is hardly surprising that Nernst rose to the occasion.

In the 1890s, Edison's basic patents had run out and there was a surge of research and invention aimed at a better light bulb (22). In 1897, Nernst filed a basic patent application for such a solid-state incandescent conductor "of the second class" (i.e. ionic, such as sulfuric acid rather than electronic, such as carbon or metal) (23). In this first patent, the conductor (Figure 6) was illustrated as heated with a Bunsen burner. Mendelssohn (24) recounts that Nernst amused the Kaiser by lighting his electric lamp with a match and then demonstrating that it could be blown out like a candle. For the solid electrolyte, Nernst indicated, "Such substances as lime, magnesia, zirconia, and other rare earths." His example was "burnt magnesia." [However, pure magnesia such as would be found in a physical chemistry laboratory is a good insulator and does not work, as was quickly discovered by those trying to imitate his art (25).] All in all, the first disclosure indicates an idea with good possibilities, but entirely impractical.

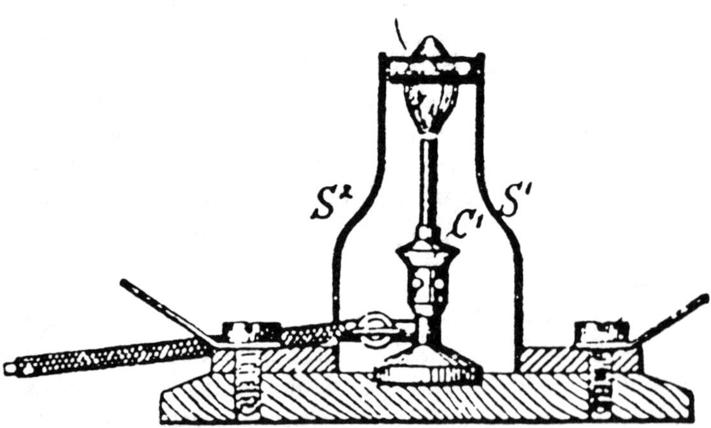

Figure 6 Drawing from Nernst's first patent application for the Nernst Glower Lamp. *US Patent No. 653,349.*

Emil Rathenau, who had bought the rights for Edison's lamp seventeen years before, also bought the German rights to Nernst's lamp. (He is reported to have paid a million marks which made Nernst a rich professor. Nernst indulged himself with sport cars and comfortable estates.) A Westinghouse employee at Göttingen, Henry Noel Potter, brought Nernst to East Pittsburgh early in 1898 where George Westinghouse bought the American rights to his invention (26).

When George Westinghouse bought the American rights, he set up a technical staff of chemists, physicists, and engineers working in co-operation with researchers at Göttingen (26, 27). Fortunately, it was soon discovered that a solid solution of zirconia with yttria is highly refractory, can be repeatedly heated and cooled, and has a good electrolytic electrical conductivity. [E. Ryskewitch reports that this discovery was made by A. Lukas in the Nernst Laboratory (28).] In a patent filed November 9, 1900, a mixture of 70–90% zirconia with 30–10% yttria is specified. In subsequent patents, 85% zirconia-15% yttria is indicated (29). Analysis of a 1901 lamp indicates the actual composition was 88% ZrO_2-12% Y_2O_3 with solid-state sintered microstructure shown in Figure 7 (30). In addition to developing the composition used for the glower, new manufacturing

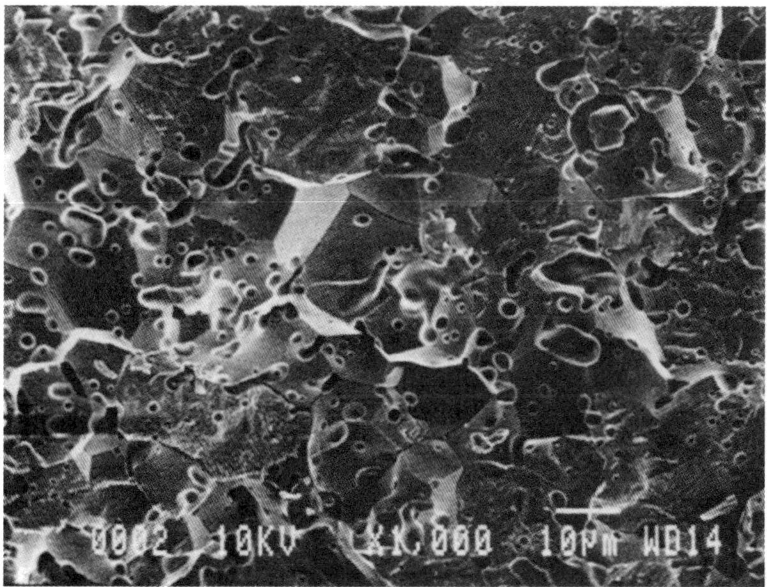

Figure 7 Microstructure of the zirconia-yttria solid solution Nernst electrolytic glower from a 1901 lamp (1000 ×).

methods were required. As part of the East Pittsburgh manufacturing plant, there was a chemical department in which oxides were prepared, mixed with a binding material and water; the plastic mass was forced through a small hole to extrude a filament using the same general method as was used in "squirting" incandescent carbon filaments of the time. After drying, the glowers were slowly heated to eliminate the binder and then sintered at high temperature in an electrical furnace or by passing a fiber through an electric arc for "fast firing."

New methods had to be developed for attaching platinum conductor wires to the glowers (31). Heaters were required to bring the glowers to a low-red heat where they would conduct; platinum wire wound on porous low-heat-capacity insulators in varying configurations was used (32). Special machines for winding the discs for extrusion and resistance on an extended rod were required, Figure 8 (33). During the years 1900–1902 more than 70 patents describing materials and design details of the Nernst lamp were assigned to George Westinghouse and the Nernst Lamp Company (34). They, and the lamps themselves, evidence careful materials selection and manufacture of the essential lamp components from chemically processed materials much in advance of ceramic industry practices. The Nernst Lamp Company chemical department prepared the carefully

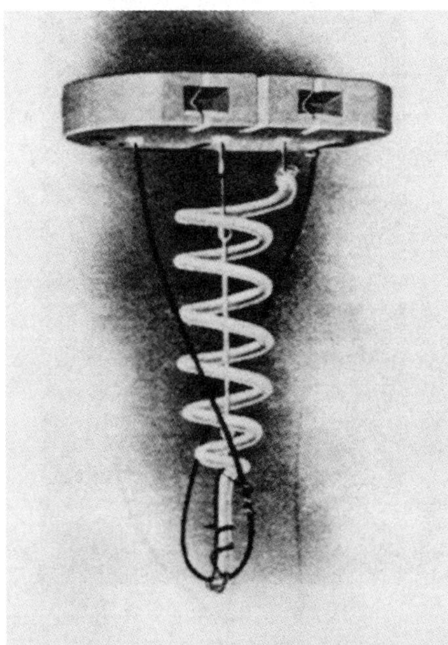

Figure 8 Vertical single glower lamp with platinum resistance heater wound around rod. *US Patent No. 652,638.*

controlled new materials. Advanced methods for shaping, firing, and testing the ceramic components were employed. In every way, the Nernst Lamp Company had a fully articulated high-tech fine ceramic processing unit as part of its manufacturing operation. For a few years it was the outstanding lamp on the market.

Improved Lamp Filaments

With the invention of Edison's (and Swan's) carbon filament lamps in 1878, the concerns of electric companies turned to other system components. Swan in 1883 developed a method of squirting a viscous nitro-cellulose solution through a die to form a "structureless" filament; Edison Electric continued to use bamboo filaments until 1892. Initially used by Sawyer and Man, "treated" carbon filaments were "flashed" by a chemical vapor deposition process. First evacuated, gasoline vapor was admitted to a filament chamber; when the filament was heated, a surface layer of graphite was deposited that adjusted the filament resistance and improved its emissivity. In the 1890s, particularly on the Continent where patent restrictions were less stringent, a growing number of efforts were aimed at obtaining filaments with long life at higher operating temperatures (35, 36).

In 1906, Carl Auer von Welsbach, inventor of the 99% ThO_2-1% CeO_2 gas mantle, patented a filament made of sintered osmium metal. Osmium has a high melting point, but is brittle and cannot be drawn into wire. Welsbach mixed osmium powder with an organic binder and extruded the paste through a die as had been done with cellulose for carbon filaments and with zirconia-yttria for Nernst glowers. After heating to remove the binder, the filament was sintered by passing current through the wire in a hydrogen atmosphere. The filament was weak and friable; in one configuration, Auer used a sintered thoria rod as a ceramic oxide refractory support (37). Auer later used a mixture of osmium (enormously expensive) and tungsten, gradually shifting his production completely over to tungsten. von Bolton, working at Siemens, found that filaments could be made of tantalum if sufficiently purified. The tantalum filament was patented in 1902. At the General Electric Research Laboratory W. R. Whitney found that if a carbon filament is first heated in vacuo, then flashed, and finally heated in a carbon furnace at an even higher temperature to consolidate the coating, it has a longer life at a higher operating temperature; this was the GEM filament.

All of these materials and processes were overwhelmed by the development of tungsten filaments. Various manufacturing methods were devised that included chemical vapor deposition, solid state sintering, and sintering with a liquid phase. Tungsten filament lamps were marketed in Europe during 1906 and in the United States the following year. About that same time, W. D. Coolidge at the General Electric Research Laboratories began experiments that led to a successful process for making drawn tungsten wire, which was much stronger than the sintered variety. These lamps were put on the market in 1911; unsold sintered filament lamps were withdrawn and scrapped.

DISCUSSION

In the early eighteenth century the performance goals of high-value-added ceramics in western Europe were the visual and mechanical properties associated with Oriental porcelain. Its manufacture was perceived as a social, symbolic, economic, nationalist goal. That goal was achieved by application of new materials science and the development of new manufacturing technology in a classic materials research program. The application of materials science led to a new materials technology in response to a perceived goal of western European society.

Toward the end of the nineteenth century the successes of telegraphy and arc lighting made it clear that there was a tremendous demand and an assured market for a feasible electrical technology. Among other parts of the overall system, this required new materials and new materials technology.

The use of manufactured lampblack for the carbon microphone, the use of extruded carbonized lamp filaments, the use of thorium oxide cerium oxide alloys in Auer's incandescent gas lamp mantle, the use of zirconia-yttria glowers in the Nernst lamp, and the use of thorium oxide supports in Auer's osmium lamp are all rational candidates as progenitors of modern high-tech ceramic materials technology. Perhaps we should also include the chemical vapor deposition flashing process for treating carbon filaments in a hydrocarbon atmosphere. One thing all these novel products and processes have in common is a role in bringing forth practical fruits of the new age of electricity. A second common element is that they were all done independent of the traditional ceramic clayworking industry; no one at the time thought of them as ceramics. Finally, each of these developments owes a great debt to the evolving chemical science.

This role of ceramics as critical components determining the effectiveness and value of larger devices and systems was a new one. Novel arrangements and accommodations were required. At the outset, in order to control the properties of critical new materials, ceramics were manufactured and shaped by the electrical companies. This was done by Westinghouse at the Nernst Electric Lamp Company; subsequently tungsten filaments were manufactured at General Electric. This integration of chemical, ceramic,

and metallurgical manufacturing within the developing electrical industry   593
began with Edison, for whom chemical studies were an integral part of his
invention factory. [He had a copy of Nernst's *Theoretical Chemistry* on
his bookshelf (38).] It continued and expanded with the development of
corporate research laboratories by the large electrical companies. These
research laboratories attended to all the needs of their systems, including
development of new materials and processes. The successful plan by C. P.
Steinmetz in 1900 that the General Electric Company set up a central
research laboratory "entirely separate from the factory" proposed four
targets for the initial work that included a substantial commitment to
materials science: mercury vapor lamps, Nernst-type lamps, new filaments,
and new materials for arc lamp electrodes (39). This combination of
integrated high-tech ceramic manufacturing and integrated research programs, including materials, was the innovation that assured the development of modern materials science, technology and engineering in its present
form.

The age of electricity gradually metamorphosed into the age of electronics with a continuing need for new materials and opportunities for
applying novel materials discoveries. The pattern continued of new
materials research being focused in user laboratories, which have included
those of national defense and atomic energy establishments, steel makers,
automobile manufacturers, the nuclear power industry, and others. But
the prime mover has been electronics.

In histories of materials science, that enterprise has most often been
pictured as a development of metallurgy and applied physics—microstructures of steel, X-ray diffraction, spectroscopy, quantum theory. In
contrast, high-tech ceramics (and modern materials technology) developed
in response to needs of the growing electrical industry. Its success is almost
entirely a story of chemists and applied chemistry, beginning with Edison
and continuing with Acheson, Moisson, Auer, von Welsbach, and Nernst.
The first complete research, development, and manufacturing program
meeting all the criteria of modern high-tech fine ceramics was the electrolytic conductors and high-temperature insulators developed for the
Nernst lamp. If Edison, famous as an inventor, is the founder of modern
materials technology, perhaps Nernst, famous as a physical chemist,
should be considered the father of modern ceramics. This would be particularly fitting if it reinforced the importance of chemistry for the future
of ceramic materials science and technology.

ACKNOWLEDGMENTS

Dr. Joseph E. Burke was the person who first suggested to me that the
source of modern materials science and technology lay in the needs of the

594 lighting and electrical industry at the turn of the century, and I am grateful for his stimulating ideas and suggestions. Susan Rosevear at M.I.T. and Mary Voss at Johns Hopkins University have been enormously helpful as research assistants and in manuscript preparation.

Literature Cited

1. Kuhn, T. S. 1970. *The Structure of Scientific Revolution*, p. 36. Chicago: Univ. Chicago. 2nd ed.
2. Lough, J. 1971. *The Encyclopedie*. New York: McKay
3. Kingery, W. D. 1967. *Transactions of the 10th International Ceramic Congress*, ed. C. Brosset, C. Helgesson, pp. 3–17. Gothenberg, Sweden
4. von Hippel, E. 1987. *Ceramics and Civilization*, Vol. 3. *High Technology Ceramics: Past, Present and Future*, ed. W. D. Kingery, pp. 325–34. Westerville, Ohio: The Am. Ceram. Soc.
5. Kingery, W. D. 1987. See Ref. 4, pp. 153–80
6. Kingery, W. D. 1988. *An Unseen Revolution: The Birth of High-Tech Ceramics*. Presented at Ann. Meet. Am. Ceram. Soc., Cincinnati
7. Yanagida, H. 1987. *Fine Ceramics*, ed. S. Saito, p. 239. New York: Elsevier
8. Jehl, F. 1937. *Menlo Park Reminiscences*, p. 41. Dearborn, Mich: Edison Inst.
9. Kingery, W. D. 1987. See Ref. 4, pp. 153–80
10. Wildenstein, P. 1736. Manufactory Commission Report (WA 1 A 24a/312 ff), cited in Walcha, O. 1980. *Meissen Porcelain*, p. 440. London: Studio Vista/Christies
11. Gilbert, W. 1600. *On the Magnet*. London: Chiswick (1900 reprint)
12. Dibner, B. 1964. *Alessandro Volta and the Electric Battery*, p. 18. New York: Franklin Watts
13. Dibner, B. See Ref. 12, pp. 111 ff
14. Josephson, M. 1959. *Edison*. New York: McGraw-Hill
15. Jehl, F. 1937. See Ref. 8, p. 41
16. Edison, T. A. 1878. *US Patent No. 203013*
17. Jehl, F. 1937. See Ref. 8, pp. 122–23
18. Edison, T. A. 1878. *US Patent No. 203015*
19. Jehl, F. 1937. See Ref. 8, p. 132
20. Friedel, R., Israel, P., Finn, B. S. 1987. *Edison's Electric Light*. New Brunswick, NJ: Rutgers Univ. Press
21. Edison, T. A. 1880. *US Patent No. 223,898*
22. Bright, A. A. Jr. 1949. *The Electric Lamp Industry*. New York: Macmillan
23. Nernst, W. 1899. *US Patent No. 653,349*
24. Mendelssohn, K. 1973. *The World of Walther Nernst*, p. 46. Pittsburgh: Univ. Pittsburgh Press
25. Davenport, C. B. 1899. *Science* NS 9(221): 456
26. Wurts, A. J. 1901. *AIEE Trans.* 18: 545–71
27. "The Nernst Lamp." 1904. anon. *Electrical World* 43: 981–85
28. Ryskewitch, E. 1960. *Oxide Ceramics*, p. 391. New York: Academic
29. Nernst, W. 1901. *US Patent No. 685,725*; *US Patent No. 685,729*; *US Patent No. 685,730*
30. Kingery, W. D. 1989. *Ceramics and Civilization*, Vol. 5, ed. W. D. Kingery. Westerville, Ohio: The Am. Ceram. Soc. In press
31. Hanks, M. W. 1900. *US Patent No. 652,607*
32. Potter, H. N. 1900. *US Patent No. 652,637*
33. Potter, H. N. 1900. *US Patent No. 652,638*
34. *The Nernst Lamp, More Light for Less Money*. 1902. anon. Pittsburgh: Nernst Lamp Co.
35. Bright, A. A. Jr. 1949. See Ref. 22
36. Howell, J. W., Schroeder, H. 1927. *History of the Incandescent Lamp*. Schenectady, New York: The Maqua Co.
37. von Welsbach, C. A. 1906. *US Patent No. 814,632*
38. Mendelssohn, K. 1973. See Ref. 24, p. 43
39. Wise, G. 1985. *Willis R. Whitney, General Electric and the Origins of U.S. Industrial Research*, p. 76. New York: Columbia Univ.

Ann. Rev. Mater. Sci. 1988. 18 : 1–24
Copyright © 1988 by Annual Reviews Inc. All rights reserved

DISLOCATIONS AND DISCLINATIONS PAST AND PRESENT:
Some Personal Views and Reminiscences

J. Friedel

Laboratoire de Physique des Solides de l'Université Paris Sud,
91405 Orsay, France, Associé au C.N.R.S.

INTRODUCTION

Dislocations and disclinations, the singularity lines in long-range translational or rotational order in materials, date back as concepts to the early years of this century. Their golden age was probably early after the Second World War, when many concepts were forged by a small group of physicists, chemists, and metallurgists. Dislocations and disclinations have recently regained impetus with the development of new observational techniques, such as high resolution electron microscopy, with a better understanding of their applications outside metallurgy, in fields such as semiconductors, surfaces, geophysics, liquid crystals, or biological materials, and with the realization that ideas and models developed for dislocations or disclinations have their counterparts for singularity points, lines, and walls in the order parameter of many materials systems. These include Bloch walls, Neel lines, and Bloch points in ferromagnets; vortex lines in Bénard instabilities of normal fluids, in superfluids, and superconductors, etc. Furthermore it is in this context that the concept of a hierarchy of successive orders was first imagined: the regular spacing of epitaxial dislocations builds a new lattice that can be incommensurate with the underlying crystal period; disorder in such a spacing can be analyzed in terms of dislocations in the lattice of epitaxial dislocations. Such concepts now emerge in all cases of incommensurate or weakly commensurate long-range order, whether crystalline and liquid crystalline, magnetic, superconductive, or hydrodynamic. They contribute to long-range order of increasingly large period and complexity that obey very definite scaling laws, such as the passage from the Frank and Kasper phases to quasicrystalline order.

These new developments would, I believe, justify a full review by themselves. They are, however, fairly well documented, by papers easily found in most good departmental libraries. Surprisingly this is not the case for the early literature of the field, which covers many of the essential concepts that developed just after the Second World War. The world of the scientists involved was small then; they had recently regained their freedom to work, travel, and communicate. As a result, most of the really important papers of the period can only be found in reports of small scale meetings, which only a few of the "old guard" still keep in their closets.

Finally, the early researchers in disclinations in liquid crystals were biologists and crystallographers; researchers in dislocations in crystals were mechanicians and metallurgists. This was in the prewar time when the concept of "solid state" was only slowly emerging and that of "materials" had not yet developed. I believe strongly that it is the meeting of such people with different backgrounds and, after the war, primarily the meeting of physicists and chemists, that has been the essential driving force in this field. As such, it is a very early example of what many people understand as materials science.

Today many reasons make it natural to produce a review of materials sciences centered on dislocations and disclinations. This review could be approached in a number of ways. People such as Sir Charles Frank, who produced many of the postwar concepts, would be best armed to discuss in depth some of the basic developments. My own reasons for writing are somewhat different. Having been in active contact with many aspects of the field, I can try to recapture the evolution in the way of thinking about this field of the average European engineer or scientist, from the late 1940s to today. Therefore this article centers on qualitative reminiscences and comments. It is historical and not didactic and indeed very personal and partial, as any good history must be!

FIRST STEPS IN METALLURGY: LOW ANGLE BOUNDARIES

Work in Paris

My first active contact with research was in 1948–1949, in the laboratory of physical metallurgy established less than ten years before at the School of Mines in Paris by my cousin Charles Crussard. Research was a family business—a chemist great-grandfather Charles and a crystallographer grandfather Georges both well known in their fields, and a father Edmond, who worked in Maurice de Broglie's famous laboratory before switching to geology and mining—that I entered with some trepidation at the age of 27.

Crussard had shown some years before that work-hardened crystals could recover a more ordered crystalline state by two different processes. Besides the well-known nucleation and growth process of nearly perfect crystals, one could also produce, by a more moderate heat treatment, a general reorganization of the work-hardened crystal into subgrains separated by low angle grain boundaries, This "recrystallization in situ," as he termed it, led to a state that Guinier, using X rays, and Cahn also characterized soon after and termed "polygonization." A study of these polygonized boundaries by Lacombe using surface etching had shown a distribution of pits whose separation (d) often varied inversely with the misfit angle (θ) between the corresponding subgrains. This was interpreted as pitting at emerging dislocations and was in agreement with the atomic structure observed by Bragg & Lomer on bubble crystal rafts. A computation by Shockley & Read soon showed that for simple edge dislocations one should expect the surface tension (γ) of the boundary to vary as $\theta \ln \theta$, with $\theta \cong b/d$ and b the interatomic distance, up to a maximum near $\theta \cong e^{-1}$. For larger angles, γ was expected to level off at a large angle boundary tension where the concept of dislocation loses its meaning because d becomes of the order of b.

Working on a single thin platelet of aluminum that had been work-hardened and recrystallized by Cullity, an American visitor to the laboratory, I measured the angles at the triple points of the grain boundaries and the corresponding misorientation between the grains. I worked with an old Baudouin X-ray camera that had been used by E. Friedel in the 1920s to test the molecular arrangement predicted by G. Friedel for smectic phases. I also developed a much quicker method by etching of the grains. I showed, I think for the first time, that γ varies with the general misorientation (Ω) in ways very similar to Shockley & Read's formula for single edge boundaries; I observed the flattening of $\gamma(\Omega)$ for $\Omega > e^{-1}$ and sharp decreases of γ (approaching zero) near simple twin orientations.

I particularly enjoyed this very simple piece of experimental work—my only one. I tried to learn everything known at the time on dislocations. This information was mostly contained in a series of review articles written by Seitz & Read (1) during the war, which in Paris could only be found in the recently opened library of the American Center. But my work also led me to worry about how to compute surface and twin tension in terms of "reasonable" interatomic interactions and how to analyze possible atomic relaxation from perfect crystal order. I tried to extract suitable ideas from Mott and Jones' *Theory of the Properties of Metals and Alloys* and from Seitz' *Modern Theory of Solids*. But I had little knowledge of basic quantum mechanics; university lectures were nonexistent or worse at that time in Paris except for Edmond Bauer's lectures in physical chemistry, and Dirac's book proved rather difficult to use practically. I then accepted Crussard's proposal in 1949 that I go to Bristol and learn the field with his friend Nevill Mott.

Later Works on Subboundaries

My first talk in Bristol was, quite naturally, on my recent and only piece of research. I think this was appreciated by Frank and might have been an impetus to develop his analysis of the geometry of grain boundaries in terms of minimal dislocation networks.

It was somewhat later that the systematic study of subboundaries developed, with the evidence of mobility under stress in hexagonal metals by Washburn and extension to twinning and martensitic transformations by Frank, Cahn, Cottrell, and others. It is noteworthy that dislocations were first seen in bulk somewhat later by Mitchell and Amelinckx as subboundaries decorated by precipitates in transparent ionic crystals.

A second wave of research explored the actual atomic structure and dislocation content of large misorientation boundaries after Bollmann's work on coexistence boundaries. This generalized the remark that two grains having a large density of atomic sites in common along the boundary should have a specially low energy; this situation is obtained if the two crystals have a common superlattice, a model already put forward for twins by G. Friedel in the 1920s.

Recent work in the field has centered more on the role of such polygonization processes in medium temperature creep and strain, which, in the components of rocks as well as in metals like Al or Cu, leads to a cell structure whose size is inversely proportional to stress and whose angle increases with strain. The exact mechanisms of cross-slip involved in the building and destruction of the polygonized walls have been studied in detail by electron microscopy, notably in Toulouse, Poitiers, and Lausanne.

POSTWAR BRISTOL: THE MECCA OF INDIVIDUAL DISLOCATIONS

Arrival in Bristol

Coming to Bristol from Paris in 1949 was extraordinarily exhilarating. One really felt in the middle of a whirlwind of research on high energy physics as well as on electrons in metals and defects in crystals, all of which had immediate international repercussions.

I arrived for the famous 1949 meeting on dislocations where experimental evidence immediately followed Frank, Burton, and Cabrera's prediction of crystal growth through spiraling of surface steps around points of emergence of dislocations. What made the impact of this theory remarkable was that the authors had considered in detail the consequences of interactions between neighboring dislocations of various signs, of interaction between successive arms of the spirals, and of possible anisotropy of step tension, with results that rendered independent experimental and theoretical figures nearly interchangeable. But there were many other works on dislocation motion presented at that conference, especially work on relativistic speeds and on interactions with impurities. What had the most profound impact on me was actually Frank's analysis of dislocation line energy in terms of a line tension that was only weakly dependent of its geometrical form. This new concept allowed one to consider dislocations as flexible lines and was to prove extraordinarily fruitful for further qualitative and semiquantitative estimates.

My first task on arrival in the theoretical group was to read the long paper that Burton, Cabrera, and Frank were writing, as a test to see if it was too complex for an average scientist. Because of my training at École Polytechnique to imbibe quickly a mass of rather esoteric matters, I was rather proud to be able to understand the paper well enough in a day to point out a number of small errors and misprints. This helped me to establish myself as a serious scientist, although it took me nearly two years to be able to publish any work of my own!

Frank's Lectures

As all postgraduate students, I was following lectures every day of the week, a system which I helped to develop in our 'third cycle teaching' after my return to France. I especially enjoyed Frank's lectures on dislocations, in which he mostly presented his thoughts from the last few weeks, if not from the night before! His oral presentation tended sometimes to be muddy, but the contents were so stimulating and fundamental in their simplicity as to be fascinating. It was with Frank that I really learned the essentials of what I know in the field, but I also learned how to try to reduce to simple terms an a priori complex question. Frank's essentially geometrical mind, coupled with simple but pertinent mathematics, was in a way surprising for somebody trained in chemistry. His insistence on trying to go back in time to the real initiator of a given concept also was great teaching, too often neglected nowadays!

It was, I think, in these lectures that Frank introduced his concept of dislocation associated with twinning or epitaxial boundaries and his extension of the Heidenreich & Shockley concept of splitting that were to influence so much research in the field. In Bristol there was considerable activity on such imperfect dislocations, which were also clarified by Thompson's concept of tetrahedron of stacking faults for face-centered cubic (fcc) metals.

But for us one of the most dramatic events was Frank's exposition, on returning from a trip to the States, of the process of dislocation multiplication by nucleation and growth that Read and he had thought of independently the day before the meeting they were attending. Without wishing to diminish Read's credit, it can be said in fairness that the Frank and Read "dislocation mill" concept follows as a straightforward extension of two of Frank's previous results: his idea that the most perfect macroscopic crystals do contain a minimum density of dislocations produced during growth or recrystallization, and arranged in a more or less isotropic network, and the Frank, Burton, and Cabrera nucleation and growth process for crystal. It was necessary only to replace the surface step by a piece of dislocation and the vapor supersaturation by the applied stress to translate one process into the other. At the same time, this process explained why low stresses were sufficient for dislocation multiplication and why many dislocation loops could be formed locally on the same or neighboring slip planes; these two aspects did not fit with Frank's earlier belief that dislocations could be multiplied by surface reflection if moving at high speed. Such Frank & Read sources were observed fairly soon by Dash using precipitation and infrared inspection in semiconductors; the concept also fits high elastic limits observed in whiskers and thin films in which the Frank network was absent or imperfect.

Work on Frank Network, Return to Paris

The Frank network was used by Mott to predict a lowering of the apparent elastic modulus of crystals caused by the elastic bowing of dislocation arcs under applied stress. At the end of my stay in Bristol, following a visit to the National Physical Laboratory at Teddington, I suggested that the change in apparent elastic constants of Al crystals was due to different conditions of precipitation on dislocations that more or less blocked their motion. I was to elaborate on this theme in the next few years. Polygonized samples should have a much larger anomaly in elastic constants (observed at IRSID laboratory in Saint Germain en Laye), and even randomly spaced impurities could pin down the dislocations effectively, thus raising the elastic limit to large values (at very low temperatures). This last work, started from a model initiated by Mott, distinguished between strong but short-range pinning that should produce a characteristic zigzagging of the dislocations and a very strongly temperature dependent hardness, and weak long-range interactions responsible for a small but temperature-independent hardness. I also extended this type of model to a study of amplitude-dependent internal friction, which completed Lücke's treatment by including thermal activation.

My contacts with dislocations also came from sharing the daily morning coffee with other theoreticians, especially Stroh and Eshelby, and with many visitors, including Seeger, Nabarro, and Liebfried.

This period ended for me with two events. Shortly after my return to the School of Mines in Paris in 1955, a second conference on dislocations was held in Bristol, where I met again my friend Seeger presenting his model of thermally activated cross-slip with Schoeck. Also, after my French thesis in the same year, I felt bound to present to a French audience what I had learned about dislocations. I gave each lecture twice, once at the School of Mines and once at IRSID, the steel research laboratory that was then developing under the impetus of Crussard. I wrote the text verbatim on the train back to Paris; this manuscript became a book published (5) at the suggestion of Y. Cauchois. The first edition was soon sold out. For me, the book was to stimulate my research in the field. In its planning as in much of its contents, it was strongly influenced by Cottrell's earlier book (3), more so than by Read's rather geometrical essay (4). Its success was probably due to a combination of the simplicity retained from Frank's lectures and the practical metallurgical knowledge gathered in Crussard's laboratory and through later contacts with Lacombe and especially Jaoul, who at the same time wrote the macroscopic version of the same theme. The second edition, written with more difficulty, was full of misprints that I tried to correct in later editions and translations. It was soon superseded by Nabarro's more mathematical book (6), and this brought to an end my participation in the classical field of dislocation research.

DISLOCATION ENSEMBLES, PLASTICITY OF METALS, RADIATION DAMAGE

In the ten years after my return to Paris (1955–1965), the dislocation specialists felt confident enough about the main properties of dislocations to try to return to the macroscopic plastic properties that in the 1920s had induced Taylor, Orowan, and Polanyi to invoke the presence of Volterra's dislocations in crystals.

Stress-Strain Curves of Metals, Cross-Slipping

The first effort of the dislocation specialists was to characterize in more detail the stress-strain curves of single (and poly) crystals of fcc metals at different orientations and temperatures. This extension of Taylor's initial work was mostly done by the German metallurgists in Stuttgart, but also by Dorn in Berkeley. For single crystals, various stages, their strain rates, their energies of activation, and the related slip systems were clearly characterized. These findings stimulated the theoretical models of Seeger and his collaborators that, I think, explained the essential thermal activation properties: logarithmic creep in the fast-hardening stage II and cross-slipping in stage III.

In the late 1950s, the German metallurgists asserted themselves in the continental scene where, just after the war, Dutch, Belgian, and French metallurgists had started periodical meetings. I remember vividly some of the later meetings I attended. In one, I observed our German colleagues systematically going up and down stairs as a short cut from the meeting room to the dining room, while the lazier French went a longer but level way. This perfectly illustrated our difference of approach to cross-slip: Schoeck and Seeger assumed a finite length of dislocation to recombine before splitting in another slip plane. I considered, along with Escaig, the more complex but less energetic possibility of mere points of constriction.

I later extended this model to the production of prismatic and pyramidal slip in some hexagonal metals; at the same time as Vitek in Prague, Escaig and I also applied this model to explain the strong temperature variation of the macroyield stress in body-centered cubic (bcc) metals by thermal recombination of split screw dislocations. This model explained the low-temperature fragility of steels, well known since the failures of Liberty ships in winter during the last war; it superseded, after some controversy, Cottrell's earlier explanation, namely the blocking of dislocations by "Cottrell's clouds" of carbon. Final macroscopic verification was given by detailed studies under a high voltage electron microscope by Kubin in Toulouse and others. The same studies also explained how point defects, such as dilute carbon or radiation damage, could lower the macroyield stress by helping to form preexisting constrictions. Together with French metallurgists from Toulouse, Poitiers, and Lausanne who used electron microscopy techniques, I had also been involved in a dispute with people like Weertman on the related nature of the creep of metals at medium temperatures. I believed, I think rightly, that cross-slipping had a decisive role, while other mechanisms, for instance using pipe diffusion along dislocations, were also proposed. A sort of tradition had thus established itself in France on the study of cross-slipping; one of the last remarkable works has been Legrand's a priori computation of the energetics of various possible splittings in hexagonal metals.

The easier cross-slip in Al than in Cu, predicted by Seeger from differences of plasticity in stage III, was amply confirmed by the first observations by Hirsch and his collaborators of moving dislocations by transmission electron microscopy. I was in Cambridge shortly after that discovery and remember vividly the excitement that it provoked. The first dislocation lines, observed mostly through the surface contrast due to the steps they left behind in slipping, were barely visible. But at last one could catch these objects in the act; it was a tremendous stimulation to note that the main characteristics were just as expected.

Stress-Strain Curves of Metals, Work-hardening

Transmission electron microscopy was to prove an increasingly powerful tool to study the detailed properties of dislocations. I am not so sure, however, that it did not bias the discussion about the structure and properties of dislocation ensembles. I am thinking especially of the development and structure of the dislocation networks responsible for the very characteristic fast-hardening stage II of fcc metals. I was involved from the start in that controversy, which later opposed Hirsch and Seeger, and which by its lack of clear-cut conclusion encouraged people to try macroscopic short cuts. However, there lies in this question a fundamental problem that must be solved from a microscopic point of view if metallurgists and mechanicians are to speak the same language. I held successive views that might seem opposite extremes but were in my mind rather complementary approaches. I find it useful to recall what happened.

My active interest in the field arose when Jacquet, a French specialist in electrolytic polishing, showed me systems of etch pits on strained α-CuZn alloys that clearly were due to piled-up dislocations; such piling-up was recently computed by Stroh for loops emitted by a Frank-Read source and blocked on an obstacle. Jacquet could not, at the time, be convinced to put this interpretation in his publication. But his photographs were beautiful and showed piled-up groups not only on grain boundaries but also inside grains: two piled-up groups were then clearly meeting at a "Cottrell Lomer lock," which was also predicted some time before. I brought these photographs to Mott and Frank during my last visit to Bristol before Mott went to Cambridge; I must say they did not please Frank, who was then trying to develop a theory of hardening by a simple three-dimensional "Frank network" whose size decreased with increasing strain.

The next summer, I spent some time in Cambridge and, when Mott was away, tried to develop a model of stage II with piled-up groups. I analyzed, I think rightly, the conditions in which Cottrell Lomer locks can be formed between intersecting slip systems; I also rightly predicted that, at least in metals such as Cu or Al, piled-up groups of any significant size should relax most of their stresses by activating neighboring Frank-Read sources of all slip systems and directions. However, I could not make a quantitative prediction of the number and length of the slip bands thus formed, nor therefore of the hardness and work-hardening rate. Seeger, who developed later models of similar kinds, stumbled on the same difficulty.

Hirsch, on the other hand, in his later observations of strained metals, saw little evidence of piled-up groups, except in specific cases of alloys with very low stacking fault energies (as indeed αCuZn was!). He concluded, a bit hastily, that one could forget about them. Indeed, relaxed piled-up groups would not be easy to distinguish and might even be disturbed by the thin film technique used. Hirsch was then a proponent of a model of hardening by a simple Frank network. With my first dislocation PhD student, Saada, I explored in detail the consequences of that model, as far as hardness, temperature-dependence, and creation of point defects were concerned. That model certainly gives the main characteristics of stage II. However, it had three major drawbacks: (*a*) One cannot deduce from the model the work-hardening rate, i.e. the way the size of the network reduces with increasing strain, (*b*) the model does not explain the production of definite slip lines, with length and density a function of strain, and (*c*) in a more macroscopic way, it neglects effects such as the Bauschinger effect which show hardening to contain a long-range part dependent on the history of previous strain (sign as well as direction).

Continuous Models

The inconclusive Hirsch-Seeger controversy encouraged many people to abandon hope of understanding in detail the mechanisms of work-hardening and to try to define it in simpler terms.

Indeed before the end of that period, another point of view used Volterra's initial concept of a continuous distribution of infinitesimal dislocations. This starting point was to prove essential to discuss extensions to linear singularities in more complex cases with at least a partly continuous order parameter, such as magnetoelastic effects and liquid crystals. The initial belief of promoters such as Bilby or Kröner was that work-hardening of crystals could be analyzed in a simple way with such techniques. One could consider the spatial variation of the density of dislocations averaged over distances large compared with the average distance between dislocations. This quantity is certainly what describes the macroscopic rotation in crystals. Hardness, on the other hand, is related more to the local fluctuations around this quantity; if this line of thought were pursued, one should average the second moment of the distribution of stresses, not the stress itself.

Many people then reverted to the classical macroscopic description of work-hardening used by mechanicians. However, the fact that the parameters involved have to be measured in each case, and that there is no hope of understanding and predicting effects such as Bauschinger's, show that the classical mechanical models are in fact not sufficient. To me, the problem is still open both for more experimental work and for a better modeling process.

Fracture

This disappointment with work-hardening led, at the same time, to a dramatic switch in the way of analyzing brittle cracks. My only contact with the field was at the fracture conference in 1959. This was the last conference for years to worry seriously about the microscopic aspects of plastic relaxation in front of a static or moving crack. With many others, Tetelman, then in Orsay doing postdoctoral work, was shortly going to switch, before his premature death, from microscopic to macroscopic points of view. It was perhaps a necessary stage in the development of the field, and the pressure of applications was strong indeed after the failure of the Comets; but it seems clear now that a serious understanding of plastic relaxation in terms of dislocation emission or multiplication is both still lacking and necessary to understand the values of the parameters of phenomenology.

Radiation Damage

During the same period 1955–1965, fundamental metallurgical research grew in nuclear energy centers and concentrated mainly on the numerous effects of radiation damage. As a consultant to the French Commissariat of Atomic Energy, I started thinking about the processes where point defects are absorbed or emitted by dislocations.

The concept of jogs, possibly created by the cutting of one dislocation by another, had been introduced by Frank, their constricted structure in split dislocations had been studied by Stroh, and their role as sources and sinks of point defects had been shown, especially by Herring, to be analogous to the role of kinks on surface steps in crystal growth.

I tried to define, as a function of the geometry of dislocation networks and the applied stresses, the slowest and therefore rate-determining process: transport between dislocations or absorption and emission by dislocations, the latter processes with and without pipe diffusion along the dislocations. I essentially showed that, in most practical cases, transport in perfect crystal was the rate-determining process, with an activation energy equal to that for volume diffusion. I applied this analysis to various situations, such as the size of the minimal Frank network, polygonization processes, high temperature Nabarro creep of polygonized crystals; with

Coulomb and then Grilhé, I also looked at precipitation of radiation damages on dislocations.

These are minor aspects of a wide field whose highlights were the competitive recovery after irradiation of interstitials and vacancies in the presence of dislocations, the possible creation of dislocations, loops or bubbles by precipitation of point defects, and the corresponding macroscopic effects (radiation hardening, radiation growth, radiation creep, and expansion by radiation). In all these aspects the whole network of dislocations, loops, or bubbles has to be treated simultaneously, i.e. the boundary conditions around each source or sink play an important role. The cooperative character of the processes considered leads to quasi-permanent kinematic regimes with spatial regularities. This is reminiscent of thermodynamical phases, with the possibility of bifurcation akin to second order phase changes, a point of view first developed by Martin.

HAVE DISLOCATIONS REVEALED THE SECRETS OF THEIR CORES?

The Early Peierls Model

The last topic on crystal dislocations I want to mention is the structure and properties of their cores (5). Volterra's infinitesimal dislocations had no proper cores but a mere mathematical short-range cutoff. It was Peierls who first seriously asked what the proper atomic structure of a core could be in a simple case. His simple soliton picture of an edge dislocation had to be made somewhat more complex by taking into account the discreetness of the crystal in order to obtain finite crystal friction.

The model was extended by Cottrell, Seeger, and Kochendörfer, to take into account the kinks associated with dislocations of arbitrary orientations and shapes. There developed in the 1950s and 1960s a way of thinking that used this kind of model with fairly high Peierls stresses to fit the phenomenological parameters and thus tried to explain the strong temperature variations of, for example, prismatic slip of some hexagonal metals or the macroyield of bcc metals.

Microscopic Approaches with Strong Splitting

I think that everybody now agrees on a somewhat more subtle approach in which a Peierls model can only be used, if at all, to simplify more complex microscopic details.

Following Seeger and Schoeck, with Escaig and Fontaine, I first developed a picture that took into account the core-splitting expected in metals and in ionic solids and explained the preference of dislocations for certain definite initial states of splitting (basal or prismatic depending on the metal in the hexagonal family, triply split sessile for screws in bcc metals, etc). We then described the kinks necessary for "difficult slip" in terms of thermally activated constrictions, as first discussed by Stroh for jogs.

In this kind of approach, for metals one can assume that the Peierls friction on the partial dislocations is negligible, except perhaps for some rare orientations possibly responsible for a specific Bordoni peak of internal friction.

Atomic Structure of Cores in Covalent Systems

When extending such a picture to covalent crystals, one first had to decide whether splitting occurred, as predicted by Shockley for Si and Ge. This was hotly disputed for a time, despite clear early evidence of splitting of dislocation nodes by Amelinckx' group. Splitting might be absent for dynamical reasons, even if, as seems now well established, the split configuration is the most stable. A second factor to take into account is a strong Peierls friction on the partials. This is expected qualitatively to be large at low temperatures because of the strong directionality of the bonds. However, it should be temperature-sensitive because during kink slipping bonds can be broken one after the other. This friction, which is not necessarily the same for the two partials, can induce moving dislocations to take core configurations out of equilibrium.

The German school, led by Haasen and Alexander, together with groups such as Amelinckx', George's, and Hirsch's, played the essential role in unravelling these points for Si and Ge. The effect of light or doping on plasticity, first established by Russian physicists, clearly shows that the detailed structure of partial dislocations must be taken into account: the core might be reconstructed to avoid broken bonds away from kinks. Indeed the strong electrical effects of dislocations on semiconductive properties was studied early by the American and Russian schools with no reference to such core splitting and reconstruction; it might be taken up again in better defined geometries and heat treatments. The similar effect of water vapor on the plasticity of quartz, discussed independently by Hirsch and myself, is still not understood in detail. Contrariwise, French metallurgists and geophysicists have clearly related the hardness induced by recovery in a number of ionocovalent oxides such as olivine to a characteristic splitting of dislocations out of their slip planes.

Conclusions

It is now clear that the detailed core properties of dislocations influence very markedly the plastic properties of the corresponding materials. In each case a specific microscopic study must be made. The only general statement that can be made is that lattice friction is expected to be less severe at low temperatures for crystals with close-packed structures and central interatomic forces than for those with looser packing and directional bonding. Thus pure metals, ionic solids, rare gases, and molecular solids should be more plastic than covalent or ionocovalent crystals. But systematic studies are still required in specific cases.

A last remark can be made about the core structure. Frank was the first to describe it as a cylinder of amorphous matter, and many of us were influenced by this point of view. Indeed my interest in amorphous structures came from my interest in the core of dislocations and in the structure of large angle grain boundaries. Modern detailed computer studies of the atomic structures of such cases have shown that the structure is usually more organized locally than what we then imagined amorphous or liquid structures to be. This fits with the more modern view of amorphous structures, which are thought to have a network of defects on a scale large compared with interatomic distances (see below). But this makes nonsense of any detailed comparison of the core structure of a dislocation with that of macroscopic amorphous matter.

VORTEX LINES AND DISCLINATIONS

After the middle of the 1960s, the second edition of my book on dislocations had been written and most of my PhD students working on dislocations had left Orsay. I progressively stopped thinking about crystal dislocations. I was, however, forced to come back to the field in a somewhat indirect way because of the activities of our Orsay laboratory in fields where similar problems arose. For me, this meant first switching from crystals to liquids and from dislocations to vortex lines or disclinations. But from then on, I remained somewhat on the sideline!

Superfluids, Superconductors; Hydrodynamic Instabilities in Normal Fluids

My first contact with singular lines in superfluids was, as I remember vividly, some fine experiments on the slowing down of a rotating cylinder in superfluid ^{4}He, shown to me in the Mond laboratory in Cambridge. But I came in closer contact with vortex lines in superconductors when de Gennes began systematic research in this field in our laboratory.

At the time, these quantized singular lines had been predicted by Abrikosov in preference to singular walls. They were to appear under a large enough magnetic field, in superconductors of the second kind, for which the coherence length is less than the penetration depth of the field. These vortices are quantized singularities of phase of the superconductive order parameter, but they could also be considered as rotation singularities of the velocity of the fluid concerned. From that point of view, they fell into the classification of disclinations or rotation dislocations introduced by Frank in the late 1940s. Their detailed electronic structure and interactions were studied in Orsay.

One of their most interesting properties was that, in pure enough crystals, they were predicted to build a regular close-packed lattice, with a lattice parameter decreasing in increasing magnetic field and in a direction parallel to that field. This lattice of vortices, in general completely incoherent with the crystal lattice, gave characteristic Bragg reflexions by neutron scattering, which were first observed by Cribier and Jacrot in Saclay. Later a surface decoration technique allowed Essmann in Stuttgart to study in detail the perfection (and imperfections!) of such a lattice. Characteristic defects, as in the Bragg bubble experiments, were missing vortices, analogous to lattice vacancies, and close pairs of opposite disclinations in the vortex lattice, forming edge dislocations. In imperfect lattices, the lattice of vortices broke down into a quasiamorphous structure, where only a short-range order of the vortices was respected.

These observations were strong incitements to build models to describe the hysteretic penetration or escape of a magnetic field in or out of such superconductors. As pointed out by de Gennes and myself, there is a close analogy of the action of the field on vortices with the Peach and Koehler force on dislocations. The models developed by Labusch were inspired by somewhat similar models for the propagation of dislocations through crystal imperfections.

I did not follow later developments on the more complex vortex lines imagined for superfluid ^{3}He, which were studied in various laboratories, including studies by Varoquaux in Orsay. But I was interested from the start in the Bénard instabilities in (normal) fluids, which Pieranski and Guyon had first studied for liquid crystals in Orsay. It seemed to me useful to think about nonlaminar flows in terms of the vortex lines they contained, their stability, and their interactions. For Bénard instabilities, the periodic rolls were vortex lines in interaction, quantized by the thickness of the sample. Indeed as shown by later works by Guyon's group, by Prost and by Ribotta, there is again the possibility of creating dislocations in the lattice of vortices. Depending on conditions, these are normal or split dislocations; they can play a role in the transition with increasing instability to more complex configurations.

Epitaxy; Lattice Modulation

The lattice of vortices and their defects in Bénard's instability of fluids are reminiscent of similar effects first studied by Frank and Van der Merwe in two-dimensional epitaxial crystal layers. These were later extended to three-dimensional crystalline or magnetic superstructures.

Epitaxial layers with an equilibrium lattice parameter different enough from that of the underlying lattice have a vernier effect which can be analyzed in terms of epitaxial dislocations. When moving, each such dislocation undergoes friction arising from either the lattice periodicity (Peierls friction) or lattice imperfections. For friction that is weak compared with the elastic interactions between dislocations, one expects a regular lattice of epitaxial dislocations that can move easily as a whole if incoherent with the crystal lattice. For strong friction one expects the epitaxial dislocations to show only short-range order; motion should occur with more difficulty and piece by piece. These general considerations were in fact already contained implicitly in Frank and Van der Merwe's paper on epitaxy and in Mott and Nabarro's study of dislocation motion in a randomly strained crystal. They were discussed more quantitatively in the 1970s, especially by Aubry.

Three-dimensional atomic or magnetic modulations can be similarly analyzed in terms of a lattice of walls that are not equidistant but distort under large enough friction. The corresponding incoherent motion can again be described in terms of nucleation and motion of dislocation loops in the lattice of walls. This point of view was especially developed by Rice and Maki, who stressed the mathematical equivalence of such dislocation lines with the vortex lines of a superconductor. In some cases, the lattice modulation can transform a metal into a insulator by introducing a gap at the Fermi level. It then corresponds to a charge density wave; the glide of such a wave produces a current in the sample. A dislocation loop can be nucleated under an applied field in such a charge density wave; its climb out of its glide plane corresponds to the emission or absorption of electronic carriers, just as the climb of crystal dislocation emits or absorbs point defects, an analogy which I stressed recently (13).

Liquid Crystals; Early Results

The main development on disclinations in recent years was in the field of liquid crystals in which our laboratory has played an active role since the end of the 1960s.

Liquid crystals offer a clear example of the way research proceeds in a cycle, with phases of great activity followed by periods when the field lies fallow. Liquid crystals were studied at the end of last century and during a brilliant period between 1910 and 1925 dominated by Lehmann, Grandjean, and G. Friedel. Interest in them then progressively died out until it was revived in the United States in the early 1960s, with an emphasis on industrial applications, especially in displays. The reasons for disaffection were that these compounds are rather complex and unstable, with a liquid crystal phase present over a very limited range of temperatures. Applications did not look numerous, and the fundamental simplicity of the long-range order developed did not strike physicists as being of any special interest. At the time similar ordering in magnetism and superconductivity was still poorly studied. Disinterest was not due to a lack of suitable instrumentation. The main tools still in use are the optical microscope, optical scattering, and X-rays, with only a limited use of nuclear magnetic resonance (NMR) or neutron scattering.

Singular lines and points were first observed by Lehmann and understood as such by G. Friedel around 1920. Thus the first dislocations and disclinations observed and understood in nature were in liquid crystals; they antedate by nearly thirty years the observation of crystal dislocations!

The simplest mesomorphic phase (to call liquid crystals by their proper name) is the nematic one. Elongated molecules have a long-range orientational order, but only a short-range repulsive positional order. G. Friedel described correctly the singular points and singular lines of this orientational order, using the optical contrast they produce; indeed he called that phase nematic because of the numerous "threads" (Greek nêmata) that were always observed in it. Frank then cited these lines in his fundamental study as examples of disclinations.

Grandjean had also observed what he took as singular walls in the structure of a sheet of "cholesteric" phase forced to lie with its molecules parallel to a given direction at the contact of two glass plates forming a dihedron of small angle. These walls appeared each time the thickness of the sample had increased by one period of the cholesteric phase, i.e. one turn of the helical twist characteristic of the structure. In such a vernier effect, reminiscent of that mentioned above for epitaxial layers, the singularities of the walls were already understood by G. Friedel to concentrate along lines running parallel to the glass plates. Thus Grandjean's walls are in fact the first observed example of a small angle boundary resolved in a series of parallel edge dislocations.

Liquid Crystals; Work in Orsay and Elsewhere

Frank was the first to give the general expression for the energy of distortion in liquid crystals. de Gennes was then the first to apply it to compute the energy of simple singularity lines in liquid crystals. He had been attracted to the field by the work Durand brought back to Orsay from Bloembergen's laboratory at Harvard. For him, singular lines were I think somewhat of a sideline, as they are not essential to understand most properties, but usually a mere nuisance in practical applications! The detailed work done in Orsay was mostly carried out around Kleman, with many ideas coming from Bouligand. A fundamental result on topological stability was expressed in the proper mathematical and general form by Toulouse. We were very stimulated by visitors such as Meyer or Frank and outside collaborators such as Michel.

From this and similar works carried out mostly in the US and in the USSR, there emerged a corpus of knowledge of singularities in liquid crystals that refer to disclinations primarily in nematics and cholesterics and less in smectics where the usual singularity lines are the Dupin cyclides, which have a different nature and origin. Only a few points are discussed here. Most of the work is analyzed in Kleman's *Points, Lines, Walls* (7).

Frank and de Gennes' definition of a disclination in the nematic phase properly stressed the orientational singularity. This was not completely suitable for the cholesteric phase, which preserves some translational periodicity. I found the following definition illuminating (8): "freeze" the substance, apply a Volterra cut process to produce a singular line in the solid, and finally allow some relaxation of the stresses and strains by "melting" the substance into the liquid crystal phase again. In such a process, one retains only the dislocations and disclinations that correspond to quantized symmetries of the perfect phase, a direct extension of the rule for crystals. The same extended Volterra process allows one to predict how disclinations move by emitting or absorbing continuous distributions of infinitesimal dislocations, which could be completely relaxed in nematics but not, in general, in cholesterics. This method clearly shows the relations between disclinations and dislocations as observed in actual cases.

A further step was the realization that some of the singular lines thus defined could relax their core into a continuous molecular arrangement of lower energy. This was first realized in the context of liquid crystals by Rault, through inspection of some examples of singular lines in cholesterics. Practically at the same time, Meyer's and Kleman's groups developed beautiful experiments on the core of disclinations of even parity in nematics enclosed in tubes so that the molecules were aligned normal to the surface of the tube. Contrary to Frank's earlier assumption, the molecules deviated from their radial distribution so as to approach the center of the cores in a continuous way by escaping in the third dimension. There were two possibilities of escape (upward and downward); planes separating two regions of opposite character correspond to points of singularity already observed by G. Friedel in 1920 at the surface of nematics. It was also remarked by Kleman and Bouligand that for topological reasons such relaxation of the core was not possible in nematic disclinations of odd parity, which should remain discontinuous.

The next step was then Toulouse & Kleman's paper on the topological stability of singularities of order parameters, which was the result of a common effort in teaching and systematized the previous results in a proper mathematical way. Curiously, at the same time Dyaloshynski in the USSR reached the same conclusions! This work allowed one to understand in what conditions a real discontinuity could be suppressed by a continuous distortion of the order parameter. The fundamental interest of this result came from the fact that topological instability is usually coupled with real energy instability. Thus it allows one to predict the really stable singularities in magnetism, superconductivity, lattice and spin modulations, as well as in liquid crystals. Indeed previous works on Bloch walls, Néel lines, and Bloch points in ferromagnetism fit nicely into this general scheme, as did the reduction of Grandjean walls into edge dislocations.

It was soon realized that the concepts involved in classical thermotropic liquid crystals could be extended to other similar cases: soaplike lyotropic liquid crystals combining water and hydrophobic long chain molecules with ionic hydrophilic heads, thermotropic mesophases with platelets or with polymeric chains, biological materials . . . . The plasticity of such materials remains largely to be understood, but the role of line defects is again obvious.

Curiously, Lehmann, the first serious worker on liquid crystals, thought them to be somehow a sign of life; he compared the varied and often beautiful droplets emerging in complex forms to flowers. To Grandjean and G. Friedel it became clear that these liquid crystal configurations originated from mesophases at equilibrium; they were not the results of kinetics of growth, even less the result of a living impetus. Now recent works have clearly shown that biological materials can, at least sometimes, take textures similar to those of liquid crystals. The cholesteric textures shown by some DNA molecules, either in chromosomes or in solutions, are quite clearly equilibrium forms, with exactly the same types of disclinations as their classical thermotropic counterparts. Cholesteric structures of elongated biological cells are also observed in many skins and wings of crabs and insects and are responsible for their iridescent colors. They also present typical linear defects connected with pores. However, it is less clear by what mechanism such structures develop, although the mechanical advantages of such "helical plywood" textures are obvious.

DISCLINATIONS AND CURVED SPACE

My last topic, slightly more qualitative and somewhat tentative in parts, concerns the relation between disclinations and curving of space. These concepts might possibly be used in describing clusters, amorphous and liquid structures, some crystal phases with large unit cells, and quasi-crystals.

Here again, my interest arose from the fact that people from Orsay, notably Sadoc and Mosseri, Kleman and Charvolin, took an active part in developing the fundamental concepts. These men, together with Frank, Rivier, Gaspard, and Nelson were some of the leading people in a growing field.

Short-Range Order with Fivefold Symmetries in Some Clusters and Amorphous Phases

It was realized before the war that a system of 13 slightly compressible balls attracting each other by long-range central forces are more stable in the icosahedral arrangement, with 6 fivefold symmetry axes, than in the nuclei of the fcc or hexagonal close-packed (hcp) phases. The increase in stability comes from the fact that each of the 12 surface balls has one additional bond, at the expense of a slight compression of radial distances compared with the distances between surface balls. Shortly after the war, Frank and Bernal proposed that the short-range order of monatomic liquids could show a preference for such icosahedral arrangements. X-rays and neutron data from studies by Sadoc and many others, together with computer studies, seemed to confirm such a point of view for metallic liquids and amorphous phases based on transitional metals, although more recent works have shown there is a local order of the minority elements that goes further than Frank's idea. Frank and Kasper also studied a number of crystal phases with large unit cells that showed local short-range order with some icosahedral symmetry.

It was in this context that Farges showed by electron scattering and computer simulations that clusters of rare gases such as Ar, between 15 and 40 atoms in size, have quasiamorphous structures that could be described in terms of incomplete and interpenetrating icosahedra, with scattering properties near to those of the liquid phase.

This prompted Sadoc to speculate again on Frank's idea of amorphous structures. With Kleman, he first showed that the icosahedron could be considered as a local projection, in real space, of part of a regular lattice in a positively curved space. This projection introduces elastic strains that would become insuperably large if a much larger cluster was taken. This situation is strictly equivalent to the fact that, if one tries to project on a plane the regular two-dimensional lattice inscribed on the outside sphere of the icosahedron, one can only do it locally if one wishes to avoid excessive elastic strains. Sadoc and Mosseri showed that this process of projection from a curved positive space was helpful in predicting many structures of clusters, whether close-packed as in rare gases and metals, or tetrahedrally bonded as for covalent elements.

Sadoc then proposed that an amorphous close-packed substance could be thought of as a close-packing of such clusters with local fivefold symmetry axes. His initial idea was that the packing introduced grain boundaries. However, it was soon clear to me that bonding of lower energy would be obtained by building continuous bonds at the center of each grain boundary. The singularities are then reduced to a network of lines following the triple points of Sadoc's grain boundaries. The situation is analogous to the reduction of a low angle grain boundary into a network of dislocations; the only difference is that the lines are now disclinations, as already described in amorphous substances by Rivier in London. It is again an application of the Toulouse theorem of topological stability. Kleman also pointed out that aggregates projected from regular lattices inscribed in a negatively curved space could fit together by a network of disclinations of opposite sign from the previous case. Gaspard from Liège generalized the concept with clusters of spatially varying curvatures.

Disclination Networks and Curving of Space

Projection from curved space and disclination networks are intimately connected. This is nothing new, as any infinitesimal disclination is equivalent to a distribution of transition dislocations, and projection from curved space was systematically used in the 1950s to represent such continuous distributions of translation dislocations.

In the present case, projecting a close-packed atomic network on a plane, with six neighbors, onto the close-packed atomic network on the outside surface of an icosahedron, with five neighbors, requires introducing six concentric disclinations of strength $2\pi/6$ each. Such disclination lines fit from cluster to cluster into a network that interpenetrates the previous one, which surrounds each cluster. These two networks are formed by disclinations of opposite signs; they compensate their curvatures at long range in such a way that one can talk in this context of a long-range orientational order.

If one generalizes these concepts, it is tempting to describe two types of situations (14).

In general, the two interpenetrating disclination networks show local deviations from perfect arrangements. This is due to positional entropy if one is in a liquid. If this disorder is frozen from the liquid phase (or directly from the vapor), one would be in an amorphous structure. The entropy due to this disorder in the disclination network might stabilize the liquid phase even if the local arrangement of bonding does not favor this arrangement with respect to a perfect crystal phase. This might be the case especially for amorphous Si or Ge, for which there is no evidence that the distorted tetrahedral bonding of the amorphous phase is in any way more stable than the regular crystal bonding.

However, one might imagine special cases where the networks of disclinations build regular lattices. To be a reasonably stable phase, this must show locally favorable arrangements of atoms in the cores of the disclinations. This might be the case in close-packed metallic phases, especially if a smaller minority component stabilizes the icosahedral arrangement by occupying its center. If then the disclination networks are commensurable with the cluster structure, i.e. if one of its multiple cells contains an integral number of atomic periods, one has a Frank and Kasper crystal phase, with a large cell and quasiamorphous local order. If the disclination networks are incommensurable with the cluster structure, one obviously has something akin to a quasicrystal, with orientational but no translational order and with quasiamorphous local order. This might apply to AlMn or AlLiCu quasicrystals.

A remark by Sadoc and Mosseri applies to this last case (12). Starting from a regular lattice in a curved space, one can decurve the space in a rough way by introducing a network of finite disclinations with a finite mesh. The major part of the remaining stresses can then be removed by introducing a second network of disclinations with a larger mesh, and so on. If a finite number of such operations removes all strains, one obtains a Frank and Kasper phase; if one goes to infinity in this scaling process, one obtains a quasicrystal.

Other Similar Frustration Effects Involving Networks of Disclinations

These concepts initially developed for amorphous metals and covalent structures seem helpful in other contexts as well. For instance, in cholesteric phases, the helical twist of the structure is fundamentally due to the fact that piling up screws, which possess a certain twist, necessarily leads to a twist of the screws of the opposite sign. However, there is no way locally for these screws to decide to twist around one normal axis or another; indeed the optimal local configuration is a double twist of the screws along two axes normal to the screw axis and normal to each other. But such a double twist introduces splay strains in the material which are only acceptable over short distances. The situation can be analyzed in terms analogous to those for icosahedral aggregates, but with a projection from a negatively curved space (11). A number of consequences follow, which were stressed notably by Kleman & Boulignand, and Pieranski & Williams.

Double twist can exist in cylindrical clusters of finite diameter. It leads to a characteristic toronlike arrangement that can be described as a helical arrangement of disclinations in the cholesteric structure, with a continuous distribution of translation dislocations responsible for the double twist. I first pointed out (10) that this could apply to the twisting of the DNA molecule in some chromosomes.

Double twist is usually suppressed in macroscopic samples of cholesteric matter, except possibly near surfaces or defects. However, it can be retained if toronlike pieces of matter, with their characteristic helical disclinations, fit together in a close-packed array. This is in essence the origin of the blue phases, which can be crystalline or amorphous depending on whether the ensuing disclination network is regular or not.

Another example of the effects that prevent macroscopic icosahedral order or double twist is the frustration that prevents the local curvature of lyotropic liquid crystals from taking the optimal value compatible with the arrangement of ionic heads and hydrophobic tails. This situation, well known experimentally from the work by Luzzati, leads to many possible regular arrangements of clusters, cylinders, or membranes. Recently it has been analyzed by Charvolin and Sadoc in terms very similar to those of the blue phases (15).

CONCLUSION

These patchy and qualitative impressions and remembrances were written purposely without looking at any reference. They are therefore necessarily very personal and incomplete; my memory might have picked up other aspects at a different time. The main conclusions I wish to draw are obviously on another plane.

The first is the extraordinary richness and variety contained in the concept of singular lines of the order parameter that constitute dislocations and disclinations. This comes partly no doubt from the non-euclidian character of these objects, which was difficult to apprehend by the first people working in the field, but also is rich in possible consequences and complex arrangements.

The other lesson I wish to stress is how much research, in this field as in others, proceeds in a cyclic manner, with phases of expansion then of slowing down before developing again. This means, as G. Friedel liked to stress, that a given concept gains by being considered from different points of view. This should also encourage young researchers to make the effort to open the old reviews and search for the old books.

Finally most of what was reported here was of fundamental interest. But applications were not far away. Indeed it is the essence of a good materials research activity to draw from knowledge of chemists, physicists, and mechanicians about a given material and then to transfer the knowledge and ideas from one type of material to another. It is clear that in the last twenty years much transfer of this type has occurred for dislocations from metallurgy to semiconductor technology or to geophysics, and application to plasticity of butter or to meiosis of some chromosomes are probably not far away. Finally, the concept of superstructures of defects, with their own defects or superstructures, is very rich in building large size regular order from small-scale objects and interactions; it will probably be at the root of other developments in all natural sciences.

Literature Cited

1. Seitz, F., Read, W. T. 1941. *J. Appl. Phys.* 12: 100, 170, 470, 538
2. Frank, F. C. 1952. *Adv. Phys.* 1: 91
3. Cottrell, A. H. 1953. *Dislocations and Plastic Flow*. London: Oxford Univ. Press
4. Read, W. T. 1953. *Dislocations in Crystals*. New York
5. Friedel, J. 1956. *Les Dislocations*. Paris: Gauthier Villars; Friedel J. 1964. *Dislocations*. London: Pergamon. 2nd ed.
6. Nabarro, F. R. N. 1967. *Theory of Crystal Dislocations*. London: Oxford Univ. Press
7. Kleman, M. 1983. *Points, Lines, Walls*. New York: Wiley
8. Friedel, J. 1983. Dislocations an introduction. In *Dislocations*, ed F. R. N. Nabarro, Amsterdam: North-Holland
9. Friedel, J. 1982. *Dislocations and Walls in Crystals*. Amsterdam: North-Holland
10. Friedel, J. 1984. Disclination lines in condensed matter. In *Trends in Physics*, ed. J. Santa, J. Pantoflicek, Prague: Union Czech. Math. Phys.
11. Kleman, M. 1985. *J. Phys. Lett.* 46: L723
12. Gaspard, J. P., Mosseri, R., Sadoc, J. F. 1984. *Philos. Mag. B*
13. Feinberg, D., Friedel, J. 1987. Imperfections of charge density waves in blue bronzes. In *Low Dimensional Electronic Properties of Molybdenum Bronzes and Oxides*, ed. C. Schlenker, New York: Reidel. In press
14. Friedel, J. 1987. *Atomic Bonding in Clusters*. Amsterdam: North-Holland. In press
15. Charvolin, J., Sadoc, J. F. 1987. *J. Phys.* In press

Ralph S. Wolfe

Annu. Rev. Microbiol. 1991. 45:1–35

621

MY KIND OF BIOLOGY

Ralph S. Wolfe

Department of Microbiology, University of Illinois, Urbana, Illinois 61801

KEY WORDS: methanogenic bacteria, ferredoxin, Archaebacteria, coenzymes of
methanogenesis, interspecies hydrogen transfer

CONTENTS

Because the rigid format of scientific publishing renders the published product
nearly sterile of the fun of doing science, I have attempted to provide here
some background and perspective on the contributions to science in which my
laboratory played a part.

Toward Microbiology

I was 25 years old before I saw a living bacterial cell. As an undergraduate, I
had majored in biology at Bridgewater College in Virginia, a small liberal arts
college. Undergraduate biology at that time was mostly descriptive, but
"Doc" Jopson in the biology department insisted that I take organic chemistry

0066-4227/91/1001-0001$02.00

and minor in chemistry; this advice was sound. I had been raised on the campus of a small college where my father taught religion and philosophy. He would go to the campus, present a lecture, and return home to work in his shop or garden. His summers were free of college obligations. To me in my late teens, this seemed an ideal way to spend one's life. The pay was poor, but the freedom was fantastic—especially with a three-month summer. After I had taught high school for a brief period, a friend who had just received his Master's degree in history at the University of Pennsylvania recommended Penn as a "good university"; so I hitchhiked to Philadelphia intent on getting a MS degree and then teaching in a small college. (I never reached the small college.)

In my youth, I had been fascinated by petrified bones and had thought that working in a museum or digging up petrified skeletons would be exciting. In Philadelphia, I went to the Academy of Natural Sciences, visited the paleontology research area of the museum, and talked to a curator. The experience was rather sobering. Here was an investigator in a dimly lit area working under a light bulb, surrounded by what seemed to be acres of petrified bones. I remember thinking on my way out that I would have to find something more alive than this.

A fixture at many universities are frustrated individuals who enjoy keeping graduate students in line. When I inquired at the graduate college office of the university about the possibility of doing graduate work in biology, I was told, "We don't have a department of biology. We have a department of botany and a department of zoology; now which will it be?" Off the top of my head I blurted out, "botany." "Well in that case, you must go to McFarlane Hall and talk with Dr. Schramm." Professor Schramm was a white-haired kindly man who seemed interested in me. Instead of asking what courses I had taken, he inquired as to what I knew about various subjects. One question in particular stands out in my memory: "Do you know anything about bacteriology?" I replied that I knew nothing, but that I had often thought that might be an interesting subject. So, he signed me up for general bacteriology. This was the only course I could take because I needed to find employment to support myself, and I asked him if any jobs were available in the department. A job in the herbarium involved mounting pressed flowers on sheets of paper; would I be interested? A job was a job, and since I was to be a student in botany this seemed to be an ideal way to learn the names of flowers. Besides, I could set my own hours. So my career as a graduate student in botany was launched.

I was enthusiastic, and on my way to the first class in bacteriology I wondered if I would get to see a living bacterial cell. I did; I was fascinated. As the laboratory course unfolded, the fact that one could start to grow these living cells in the afternoon and the next morning could read the results seemed incredible to me. This was my kind of biology. The course was taught by W. G. Hutchinson, and he turned me on to bacteriology. I did well—after all I was only taking one course—and when an opening occurred for a teaching assistant in bacteriology, Professor Hutchinson invited me to assist him. I was elated and did not hesitate to leave the job of mounting flowers. It had become painfully obvious to me that something was wrong; I could not remember the names of the plants! In contrast, I had no trouble remembering the names of bacteria.

So as my second year in graduate school began, I moved to the old public health laboratory on 34th Street where Professor Hutchinson introduced me to a young assistant professor, D. J. O'Kane, who had just arrived from Cornell and was setting up his laboratory. I was to be a teaching assistant in general bacteriology and would take a new course being developed by O'Kane to give students some exposure to biochemical activities of microbes. I never had a formal course in biochemistry, and sometimes it shows.

As a teaching assistant, I soon found that a successful laboratory course was based on thorough preparation. The degree of success depended on attention to details; everything must be checked: the cultures, the media, the glassware, the incubator. Assume nothing! It was great when everything worked. I especially enjoyed being a teaching assistant with W. G. Hutchinson. However, his interests were not in the area of bacterial physiology and metabolism, an area that was on the forefront at the time and that seemed attractive to me. So O'Kane accepted me as a graduate student, one intent on getting a terminal Master of Science degree. I worked on the enzyme hippicurase, from *Streptococcus*, a hydrolytic enzyme that cleaved a peptide-like bond and was considered of interest because the mechanism of peptide bond synthesis was unknown at the time. My initial attempts at research were rather painful for my professor, I'm afraid, especially when I didn't even know how to plot the data I generated. I needed to be spoonfed. Eventually, I found that doing the experiment on my own without letting my professor know was fun. When I thought I had established a scientific fact, we would talk. I liked this system; the thrill of discovery on one's own is the best motivating force.

After receiving a MS degree, I decided to take a year off from graduate school. Graduate students were paid $950 for nine months, so it was necessary to save enough money during that period to live during the summer. I wanted to see what nonacademic life was like and try to shore up my financial condition. I accepted a position as a technician in Ruth Patrick's laboratory at the Academy of Natural Sciences. She had become interested in devising ways for assaying the toxicity of stream pollutants. This work broadened my appreciation of biology. My professors at Penn encouraged me to continue graduate studies toward a PhD degree, but before returning to the university, my life became complete; Gretka Young, who worked for the American Friends Service Committee, and I were married in September, 1950. My goal of teaching in a small college would become untenable in the next three years as my professors made me aware that the PhD degree was a research degree, that to do research one needed proper equipment, and that adequate facilities were rarely found in a small college. I eventually would realize that I had been scientifically seduced.

Professor O'Kane offered me a choice of two thesis topics involving a new factor, the pyruvate oxidation factor, which he had worked on in the laboratory of I. C. Gunsalus at Cornell. This factor was later identified as lipoic acid. He wanted to know whether lipoic acid was involved in the oxidation of pyruvate by *Escherichia coli* and *Clostridium butyricum*. I instantly chose the clostridial system because it seemed challenging and unknown. I was able to show with treated extracts that diphosphothiamin, coenzyme A, and ferrous ions were required for the oxidation of pyruvate, but I could find no role for lipoic acid (166). Various dyes could be used as electron acceptors to bypass hydrogenase; however, I could not make progress on the natural electron acceptor. Exchange of $^{14}CO_2$ with the carboxyl group of pyruvate occurred readily (167). These experiments introduced me to the use of radioisotopes. I completed my thesis work in June 1953, a few months after our first child, Danny, was born.

The event that encouraged me to consider the possibility of an academic position in a research environment involved the preparation of my first manuscript from part of my thesis. I was apprehensive about submitting a manuscript to the *Journal of Biological Chemistry* because I lacked confidence in writing. However, I carefully patterned the manuscript in the style of the journal and gave it to my professor. He returned it and announced, "I think this is fine, Ralph; let's send it in." We did, and it was accepted! This was the first time anyone had ever expressed approval of my writing and gave me confidence that perhaps I could become successful at scientific writing. I owe much to D. J. O'Kane, who made me appreciate the importance of hard data in nailing down a concept as well as the importance of freedom in exploring and making discoveries. Later, I would use this same philosophy in running my own laboratory.

Very few academic jobs were available in the summer of 1953. I had two interviews, one at a small college, which had only a case-full of student microscopes as equipment, and one at the University of Illinois. I was hired as an instructor at Urbana by H. O. Halvorson. My case had been presented in a sufficiently positive way through the efforts of my scientific grandfather, I. C. Gunsalus (known to everyone as Gunny). The Department of Bacteriology was an exciting place; with recent appointments of Halvorson, Spiegleman, Luria, Gunsalus, and Juni, the department was considered one of the best in the country, and I was fortunate to join it. A heavy teaching load didn't leave much time for research, and my program was rather slow in evolving. Before

I started anything, I was determined to prepare a manuscript for publication from the last part of my thesis. So I closed my office door whenever I had a chance and worked on the manuscript.

This action was interpreted as inaction but led to a revelation of one of the truisms of academic life. One day there was a knock on my office door. Professor Halvorson entered, sat down, and in a very concerned manner said, "I just want to tell you one thing—you are paid to teach; you get promotions for doing research." He departed immediately, and I pondered these words of wisdom. They are as true today as they were 38 years ago.

The faculty wanted to augment their specialties with someone who had a real interest in diverse organisms and who would want to teach a van Niel–type course. Largely through persuasion by Gunny, van Niel accepted me as an observer in his course at Pacific Grove for the summer of 1954. The class was a fantastic experience that opened my eyes to a microbial world of unfamiliar organisms and made it possible for me to attempt to fill the niche for which I had been hired at Urbana. I returned to Illinois with many ideas from van Niel that, together with some from Gunny, Luria, Sherman, and myself, became an organisms course that would be taught for nearly three decades.

When I left Penn, O'Kane generously allowed me to take the pyruvate clastic system with me to serve as a basis for my research program. In response to an inquiry about equipment that I would need to get started, I had suggested to Professor Halvorson a colorimeter, a vacuum pump (with which to freeze dry cell extracts) and a Warburg apparatus. He seemed a little dismayed by this, and I thought his response was a bit curious, for the request seemed modest to me. A year later, I was told, "We have enough biochemists around here; I hired you because I thought you weren't one." The message was clear—I had better begin visibly studying unusual organisms or I did not have a future at Illinois. Having a mandate to study such organisms was great, but I also instinctively knew that to gain the respect of the scientific community I must be involved in an in-depth study of a biochemical phenomenon. Isolation and cultivation of "funny bugs" (although challenging and rewarding) alone was not enough. I knew my abilities would limit how far I could go, but I was determined to become a respectable microbiologist.

Ferredoxin

As a graduate student I poured ferredoxin down the sink for three years. Fortunately, much later one of my own graduate students was involved in its discovery. At the time of my studies, the research community knew that clostridia did not possess cytochromes. Because no other protein electron carriers were known, we assumed that the unknown electron acceptor for pyruvate oxidation would be a soluble cofactor. Formation of carbon dioxide and acetyl phosphate from pyruvate could readily be followed in treated cell extracts upon addition of dyes, but I could find no evidence for a coenzyme that could play a role similar to nicotinamide adenine dinucleotide (NAD) or flavin adenine dinucleotide (FAD). When I accepted my first graduate student, Robert Mortlock, at Urbana, I suggested that he identify the electron acceptor and study the reversal of the reaction. He obtained convincing evidence for synthesis of pyruvate from acetate and carbon dioxide (99, 100). The nature of the electron acceptor remained obscure, however.

In the summer of 1957, an undergraduate student, Raymond Valentine, expressed an interest in learning about research with bacteria, and I suggested that he begin by helping Mortlock with studies on the pyruvate clastic reaction. Valentine was highly motivated, and the two developed an isopropanol precipitate of crude cell extract that actively decarboxylated pyruvate with methyl-viologen as electron acceptor. He realized the significance of this resolved extract and wanted to pursue study of the natural electron acceptor for a thesis.

One of the most difficult decisions in science is deciding when to quit one line of research and start a new one. There is always the possibility that one more cast might do it. I had been casting for a strike on the electron acceptor for about eight years and was eager to try something new. So, much to Valentine's disappointment, we dropped the clastic reaction and began to study anaerobic allantoin degradation. To augment Valentine's training, arrangements were made for him to spend a summer working on organic synthesis with Roe Bloom, an organic chemist with whom I was acquainted at the DuPont Company in Wilmington, Delaware. In the meantime, a paper by the DuPont nitrogen fixation group headed by Carnahan appeared. Mortenson had found that the addition of pyruvate to extracts of *Clostridium pasteurianum* greatly stimulated nitrogen fixation. We realized that the electron acceptor from pyruvate oxidation for which we had an assay might be the electron donor for nitrogenase, and Valentine communicated this to Gunny, who was a consultant to DuPont at the time. Gunny picked up the phone and arranged for Valentine to join the nitrogen fixation group for the summer. He applied the isopropanol precipitation procedure developed in Urbana to extracts of *C. pasteurianum* and produced a methyl-viologen-dependent pyruvate clastic reaction. With this assay, he began to test various nitrogenase preparations and fractions. Mortenson suggested that he should test a brown preparation stored in a certain refrigerator, a fraction prepared by Ralph Hardy before he left the company. When he opened the refrigerator, he found two bottles, each of which contained a brown fluid. The first one tested instantly denatured the proteins of the assay mixture; it turned out to be chromate cleaning solution. A sample from the second complemented the resolved enzyme preparation; the electron carrier had been located and could donate electrons to hydrogenase. (The first name considered for the carrier was "co-hydrogenase.") Under Mortenson's leadership, the protein was fractionated, purified, and shown to be an iron-sulfur protein, a new type of electron carrier. The naming group at the DuPont Company named this new product ferredoxin.

Upon returning to Urbana, Valentine wanted to pursue ferredoxin as a thesis, but I felt that the discovery had been made in DuPont laboratories, and to continue to study the ferredoxin of a saccharolytic *Clostridium* in my laboratory would be inappropriate. However, looking for ferredoxin in other anaerobes would not be a problem. We used the finding of ferredoxin in *Micrococcus lactilyticus* as a lever with which to pry the original proprietary data on *C. pasteurianum* into publication, and the papers were published back-to-back (98, 139). In the next 10 years, hundreds of papers on this new type of iron sulfur protein electron carrier, ferredoxin, appeared. We explored several ferredoxin-dependent reactions (18, 19, 136, 137, 140, 145).

I wished to know if the pyruvate reactions in *Bacillus macerans*, one of the sugar-fermenting members of the genus *Bacillus*, were similar to those of *C. butyricum*. Raymond Hamilton (65) studied the exchange reactions of CO_2 and of formate with the carboxyl group of pyruvate and found that the CO_2 exchange reaction was similar to that in *C. butyricum*, but that a separate formate-pyruvate exchange reaction also occurred.

Methanobacillus omelianskii

In 1960, I arranged to spend my first sabbatical leave in the laboratory of Sidney Elsden at Sheffield, England. He had been a student of Marjorie Stephenson and C. B. van Niel and was interested in microbial physiology. I thought he would be an ideal sounding board for the major purpose of my sabbatical, which was to write a guide to the isolation of unusual organisms from nature, for which I had received a Guggenheim Fellowship. After about a month in the library, I realized that this project wasn't much fun, so I threw down the pencil and moved to a lab bench. We decided that methane bacteria would be a good project. H. A. Barker kindly sent us a culture of *Methanobacillus omelianskii*, which I was able to culture. Our purpose was to study the synthesis of amino acids from [^{14}C]-labeled CO_2 and acetate by this unusual anaerobe. We got off to a good start; the labeling patterns were definitive on several amino acids, and Martin Knight continued the project for his thesis (85).

I returned to Urbana convinced that *M. omelianskii* could be mass cultured, and that the time was ripe to go after the biochemistry of methanogenesis. My colleague, M. J. Wolin, in the Department of Dairy Science, was interested in this subject, so I invited him to collaborate. We hired Eileen Wolin to initiate cultures and scale them to the 3-liter–florence flask stage. They were then brought to my laboratory where carboys were inoculated. We developed a production line so that we could harvest a carboy of cells at least three times a week, each carboy yielding enough cells for one experiment. Cells were immediately broken in a Hughes press, and a dark brown cell extract was prepared. Norman Ryckman played an essential role in growing and harvesting cells and in preparing cell extracts. Old Warburg flasks fitted with serum stoppers served as reaction vessels, and samples of the flask's atmosphere were transferred by syringe to a gas chromatograph to test for methane formation. The first extract was tested in October 1961, and for the next five months the recorder pen never moved from the baseline except to respond to a standard injection of methane. Extracts refused to oxidize ethanol or acetaldehyde and reduce CO_2 to methane under any condition. In March, out of desperation, I tipped my old friend, pyruvate, into the extract. I shall never forget the zing of the recorder as the pen soared to the top of the chart and back precisely at the time methane should elute from the column. The first formation of methane by a cell-free extract had occurred (171).

This assay allowed us to optimize the system and to show that the role of pyruvate was to provide electrons, carbon dioxide, and ATP. The discovery by Blaylock & T. Stadtman (11) that the methyl group of methylcobalamin could be reduced to methane by extracts of *Methanosarcina* in the presence of pyruvate was a major breakthrough. M. J. Wolin synthesized methylcobalamin, and we showed the ATP-dependent reduction of the methyl group to methane by extracts of *M. omelianskii* (172) and that B_{12r} was the product (173). John Wood (178) studied the reaction and showed that the cobamide derivatives, methyl-Factor B, and methyl-Factor III also were effective methyl donors for methanogenesis (180). The ferredoxin-dependent conversion of formate or acetaldehyde was worked out by Winston Brill (18, 19), and Wood (176) showed that the methyl group of methyl-tetrahydrofolate was reduced to methane. Additional studies suggested that carbon-3 of serine was a precursor of methane via conventional C_1-tetrahydrofolate intermediates (175). Wood also obtained evidence for alkylation (177) and propylation (179) of a cobamide enzyme involved in methanogenesis. Wolin (170) found that viologen dyes were potent inhibitors of methanogenesis. These experiments were exciting as we groped to figure out how extracts of "*M. omelianskii*" made methane from carbon dioxide. The laboratory of T. Stadtman was the only other group actively working on this project.

Interspecies Hydrogen Transfer

In the summer of 1965, I asked Marvin Bryant, who had recently joined the Department of Dairy Science, if he would teach me the Hungate technique. For the roll tube experiment, he suggested that we carry out an agar dilution series using *M. omelianskii* because one of his colleagues was not satisfied that the culture met the criteria of a pure culture. We decided to use a rich medium that contained rumen fluid to encourage the growth of any contaminants. In one series of roll tubes, we added H_2 and CO_2. We picked isolated colonies back into the ethanol carbonate medium of Barker, but nothing grew—not an unusual observation when working with methanogenic bacteria at that time. Soon the summer was over, and we returned to other duties. About seven months later, Bryant found a rack of roll tubes from our experiments in the incubator. One tube contained a large isolated colony, and, when he removed the rubber stopper, he found a strong negative pressure. The organism oxidized hydrogen, and a serial dilution in agar roll tubes established that the culture was pure and that it made methane. The organism was labeled strain M.o.H. and would not grow in the ethanol medium of Barker, so Hungate suggested to Bryant that the original culture must contain a companion organism that used ethanol as a substrate, and he should go after it. After many difficulties, Bryant succeeded in isolating the S organism, which was inhibited by the hydrogen that it produced.

M. J. Wolin realized what was going on. He blew the dust from his physical chemistry book and calculated the free energy change for the oxidation of ethanol to acetate and hydrogen vs the partial pressure of hydrogen. *M. omelianskii* was a symbiotic association of two organisms. The S organism oxidized ethanol to acetate and hydrogen; the methanogen, *Methanobacterium* strain M.o.H., lowered the partial pressure of hydrogen by oxidizing it to reduce CO_2 to CH_4; this allowed the anaerobic oxidation of ethanol by the S organism to become thermodynamically favorable. Thus, interspecies hydrogen transfer (the importance of the partial pressure of hydrogen in anaerobic biodegradation), one of the first principles of anaerobic microbial ecology, had been discovered. We thought this paper would be a suitable way to honor C. B. van Niel (22).

With the discovery that *M. omelianskii* was a mixed culture, the roof of my research program more or less collapsed. Much of Winston Brill's thesis could not be published because we did not know which enzymes came from which organism. During this work, I could not figure out why such a bright, dedicated student was having so much difficulty with variability of cell extracts. I have always regretted this and feel that I should have been more astute about the culture. Much of Abdel Allam's thesis could not be published, and Richard Jackson's work with Lovenberg on the amino acid sequence of ferredoxin was in doubt concerning which organism produced the iron sulfur protein. Five years of work on methanogenesis by extracts of *M. omelianskii* needed to be reinterpreted.

The real challenge now was to develop a mass culture technique and a method for growing cells on H_2 and CO_2. I accepted this challenge, but all attempts to culture strain M.o.H. by sparging gas through a liquid medium failed. In desperation, I developed a closed system in which a diaphram pump recirculated the gas atmosphere over heated copper and back into the culture vessel. I named this gadget the gaspirator. (Spiegleman referred to it as Wolfe's last gaspirator.) It provided a few good runs, but its performance was erratic, and it was abandoned. However, I developed a system for slowly shaking 200 ml of medium under a gas atmosphere of H_2 and CO_2 in a flask with a small continuous addition of gas (21). I was delighted that the inoculum grew when transferred to a 12-liter fermentor and could then be harvested or used to inoculate a 200-liter fermentor. Soon we had kilogram quantities of cells. This work could not have been done without Marvin Bryant, who patiently provided us with inocula of strain M.o.H. Langenberg (87) documented the electron microscopy of strain M.o.H. Anthony Roberton (109) studied the ATP requirement for methanogenesis from methylcobalamin by cell extracts of strain M.o.H. He showed that intracellular pools of ATP increased when cells were oxidizing hydrogen and carrying out methanogenesis, but decreased when hydrogen was removed (110).

Coenzyme M, the Terminal Methyl Carrier

Barry McBride arrived from the University of British Columbia, Vancouver, at a propitious time. A technology for growing cells on H_2 and CO_2 had been developed; he was able to contribute to the mass culture of cells at the 12-liter stage and was the first one to have the courage to grow a 200-liter batch of strain M.o.H on H_2 and CO_2. He discovered a new cofactor that was required for the formation of methane from the methyl group of methylcobalamin by cell extracts. Evidence suggested that this cofactor was involved in methyl transfer, so we named it coenzyme M (CoM) (94). He found a curious inhibition of methanogenesis by DDT (95). In testing various buffers to optimize the assay for CoM, he noticed that a strong garlic-like odor was produced in arsenate buffer, and we (96) documented the synthesis of dimethylarsine by cell extracts from methylcobalamin and arsenate. His work opened up a new era in the biochemistry of methanogenesis, one that we would follow for 20 years. We owe much to Barry McBride, who later pioneered use of the Frêter chamber for handling methanogens, the growth of methanogens on Petri dishes, the fluorescence of methanogen colonies, the use of the epifluorescence microscope to detect individual cells of methanogens, and the bright fluorescence of protozoa. Unfortunately, he did not receive proper acknowledgment for the discovery that individual cells of methanogens fluoresce.

The study of CoM was taken up by Craig Taylor, who over a five-year period purified the factor to homogeneity and determined its structure as 2-mercaptoethanesulfonic acid (128). He showed that the coenzyme was methylated on the reduced sulfur atom to form 2-(methylthio)ethanesulfonic acid, CH_3-S-CoM. Both forms of the coenzyme were chemically synthesized. CoM was identified as the unknown growth factor required by *Methanobrevibacter ruminantium* (126). Later, William Balch (8) studied its transport. A simplified enzymatic assay was developed for the synthesis of CH_3-S-CoM with methylcobalamin as the methyl donor, and the transmethylase that catalyzed this reaction was purified (127).

Because CH_3-S-CoM served as the substrate for methane formation in the absence of methylcobalamin, we considered both methylcobalamin and the transmethylase to be outside the natural pathway of methanogenesis. Attention now became focused on the CH_3-S-CoM methylreductase reaction, and Robert Gunsalus pursued this reaction as a thesis topic (64). He collaborated with James Romesser in synthesizing a variety of CoM analogues (59). We wanted to know just how specific the coenzyme was, and of all the derivatives tested only ethyl-CoM showed a positive effect, producing ethane at 20% of the rate of methane. We had noted that certain preparations of CH_3-S-CoM would not serve as a substrate for the methylreductase reaction, and we did not understand why until bromethanesulfonate (BES) was found to be a potent inhibitor of methanogenesis (59). We then made sure that any excess BES, the starting compound for the synthesis, was removed from preparations of CH_3-S-CoM.

Archaebacteria

As a result of graduate training and exposure to ideas of the Delft school, I had become a firm believer in the unity of biochemistry for the biological world. In my own laboratory, the 1926 paper by Kluyver & Donker (84) had a special impact on students during our studies on electron transport in clostridia, especially since we felt we were a part of the family tree to Kluyver through contact with van Niel. Perhaps students today should be made more aware that they are part of a continuum—that their experiments have scientific roots! For vitamins and coenzymes, the case for biochemical unity was particularly strong.

The discovery of coenzyme M and the elucidation of its structure revealed a chance to document the distribution of a new vitamin-coenzyme relationship with a classic microbial growth–dependent assay. Not since lipoic acid had there been such an opportunity. Perhaps CoM would have a similar distribution as well as an important role in methylation reactions. However, growing a hydrogen-oxidizing methanogen using the Hungate technique was difficult. Each tube had to be opened and regassed with H_2 and CO_2 more than twice a day for 4 or 5 days. The negative pressure developed inside each tube made it especially difficult to prevent contamination by O_2 and bacteria when the stopper was removed. The single figure (126) showing that CoM was the growth factor for *M. ruminantium* represented efforts made over a large portion of a year. Assay after assay failed in Bryant's laboratory, and finally the master himself took his spring break to generate the data. I thought there must be a better way. I suggested to a new student, William Balch, who was tooling up to document the distribution of CoM in the biological world for part of his thesis work, that to avoid the pitfalls of this assay (which only an expert like Bryant could handle) we should try to develop a system in which cells could be grown in a pressurized atmosphere.

Macy, Snellen, & Hungate (93) had pioneered the use of syringes for the transfer of oxygen-sensitive bacteria in the Hungate technique. Miller & Wolin (97) extended the use of syringes to the inoculation of media contained in a standard serum bottle with an aluminum-crimped seal. In addition, they had the manufacturer put the serum vial top on standard culture tubes. This procedure worked well for fermentative microbes, which produced gas, but the seal was not designed to hold for organisms such as methanogens that created a negative pressure during growth. Balch replaced the standard serum rubber seal with a solid rubber stopper, and pressurized the H_2 and CO_2 atmosphere above the medium to 2 atm. More carbonate was added to the medium to increase the buffering capacity. The medium was inoculated by use of a syringe, and the atmosphere could be repressurized aseptically so that a standard growth curve was produced. Contamination was no longer a problem. A special stopper with a lip was designed, and Bellco Glass agreed to market it (6). The Balch modification of the procedures of Macy & Hungate and Miller & Wolin became standard procedure for the field (3), and in the hands of Karl Stetter proved to be equally valuable for isolation of extremely thermoacidiophilic archaebacteria (163).

In Balch's hands, the CoM growth-dependent assay became routine and could easily detect 10 pmol. With this sensitive assay, over the next two years he tested animal tissues of all types as well as a wide range of plants and microbes (7). The answer was clear-cut: the new vitamin-coenzyme was present only in methanogens! I was disappointed; not only had the unity of biochemistry thesis let me down, but it appeared as if we had spent two years on a fruitless endeavor. However, Balch had thoroughly documented a fact that later would become important and had perfected the new system for growing methanogens in a pressurized atmosphere. His expertise would soon reap unexpected collaborative discoveries that otherwise might not have been possible. Carl Woese and I had been discussing a proposed analysis of the 16S rRNA of methanogens. Because a sealed reliable growth procedure had been developed, we entered into a collaboration in which the ability to label the nucleic acids of methanogens with high ^{32}P-specific activity was the limiting parameter.

Woese had found that growth of slow-growing organisms could easily cease because of radiation damage before sufficient label could be incorporated. Balch, with his pressurized atmosphere technique, had the key to success in coaxing each methanogen to take up sufficient label before dying. He developed a close working relationship with George Fox in Woese's laboratory, and each methanogen was a special challenge. When I asked Woese about the results of the first attempt to label the 16S rRNA of a methanogen, he replied that something had gone wrong with the extraction—perhaps they had isolated the wrong RNA. The experiment was repeated with special care, and this time Carl's voice was full of disbelief when he said, "Wolfe, these things aren't even bacteria."

When we started these experiments in 1976, Woese's laboratory had previously analyzed the T_1 endonuclease–generated 16S rRNA oligonucleotides of 60 species of microbes representing a wide variety. Against this wealth of background information, the data from the methanogens were perceived to be clearly different. Analysis of other methanogenic species supported the conclusion that methanogens were only distantly related to typical bacteria with regard to their 16S rRNA (54). And now the curious coenzymes assumed special importance. CoM had been shown to be unique to methanogens, and Dudley Eirich had just finished working out the structure of the first natural deazaflavin, the unique coenzyme F_{420}.

So the concept that methanogens represented an ancient divergence in evolution was initially a two legged stool—one leg supporting the concept was the 16S rRNA oligonucleotide data and the other was the "crazy coenzymes." We needed more legs on the stool, and I thought that if methanogens were this different in these areas, they also should exhibit other unusual properties. I wrote to Otto Kandler in Munich asking if he would be interested in examining cell walls of methanogens that we could send to him. He was enthusiastic for the project, and his laboratory soon showed that, indeed, *Methanobacterium* species not only lacked the typical peptiodglycan cell wall structure, which was known to be a characteristic of all bacteria except the mycoplasma, but possessed instead a pseudomurein; other methanogens had no peptidogylycan at all. We now had a third leg on the stool supporting the concept that methanogens were only distantly related to typical bacteria (4). Soon, many legs would support the concept, but all of these data only made sense when Carl Woese proposed that methanogens, extreme halophiles, and certain thermoacidophiles belonged to a distinct phylogeny, the archaebacteria (161), now known as the Archaea (162).

Many researchers were sceptical about this concept and about the the use of 16S rRNA to establish a phylogeny for the microbial world. We were fortunate at Illinois to have Marvin Bryant, who had written the section on methanogenic bacteria for *Bergey's Manual of Determinative Bacteriology*, as a colleague. The acceptance of the methanogens as a distinct group was controversial; the popular press did not help by calling them a "third form of life."

WOLFE

We decided to prepare an article for *Microbiological Reviews* in which all evidence that supported the uniqueness of the methanogens could be evaluated readily by the scientific community (3). The backbone of the paper was the oligonucleotide analysis prepared in Woese's laboratory. This article gained the the respect of the scientific community. Working with Carl Woese was a unique experience, one of the high points in my scientific life. I gained respect and admiration for him, one of the most dedicated scientists I have known.

Coenzyme F_{420}, the Deazaflavin

While fractionating cell extracts of strain M.o.H., Paul Cheeseman observed an abundant yellow compound that exhibited a bright blue-green fluorescence under UV light. He purified this compound to homogeniety and Ann Toms Wood continued these studies (27). The compound was named factor-420 (F_{420}) because of the strong absorption maximum at 420 nm. Godfried Vogels chose to study the fluorescence spectra and other properties of F_{420} during a sabbatical leave at Urbana. In Bryant's laboratory, extracts of *M. ruminantium* grown on formate as substrate exhibited both formic dehydrogenase and hydrogenase activity. During fractionation, NADP reduction by hydrogenase was found to require an unknown factor. A sample of our preparation of F_{420} substituted; the K_m for F_{420} was 5×10^{-6} M (133). A similar story developed for the F_{420}-dependent reduction of NADP by formic dehydrogenase in which the enzyme also showed a high specificity for F_{420} (132). These papers documented the role of F_{420} as an electron carrier in methanogens.

I suggested to Dudley Eirich that the structure of F_{420} would make a good thesis topic. After a course in organic qualitative analysis, he began to analyze hydrolytic fragments of F_{420}, and after a five-year dedicated effort, he had a structure in hand (36). The kind interest of K. L. Rinehart, Jr. as well as his willingness to be a sounding board were essential to the success of the project. The coenzyme is the N-(N-L-lactyl-γ-L-glutamyl)-L-glutamic acid phosphodiester of 7,8-didemethyl-8-hydroxy-5-deazariboflavin 5' phosphate. The distribution of F_{420} and properties of its hydrolytic fragments were published (37). F_{420} was the first naturally occurring deazaflavin to be described. In contrast to CoM, it is found in certain organisms other than methanogens, although the quantities in methanogens may be vastly greater.

Factor 342, Methanopterin, the Coenzyme of Methyl Transfer

When fractionating cell extracts by column chromatography, Robert Gunsalus noted a compound that fluoresced bright blue under UV light (63). This compound was named factor 342 (F_{342}) because of its absorption at 342 nm. No role was found for the factor, but Vogel's laboratory showed that it was a pterin derivative. James Romesser started another line of investigation when he began to study methanogenesis from formaldehyde. Hydroxymethyl-CoM, which was readily synthesized from formaldehyde and HS-CoM, was proposed as an intermediate in methanogenesis (112). The compound was found to hydrolyze in water to yield HS-CoM and formaldehyde, but the rate was much slower than the rate of methanogenesis from hydroxymethyl-CoM catalyzed by cell extracts, so we believed at the time that it was an intermediate. Later, Jorge Escalante showed that, because formaldehyde was removed from solution by methanogenesis, the equilibrium was drastically displaced toward hydrolysis; hydroxymethyl-CoM was not an intermediate in methanogenesis (48).

However, Escalante discovered a factor that was required for methanogenesis from formaldehyde that was named the formaldehyde activation factor (FAF). The factor was purified anaerobically to homogeniety; in the reduced form it had a molecular weight of 776, whereas that of the oxidized form was 772 (45, 46). The factor had properties very similar to methanopterin described by Vogel's laboratory (83), and Escalante showed that FAF was tetrahydromethanopterin (45, 46, 49). Escalante documented the role of the methenyl, methylene, and methyl derivatives of tetrahydromethanopterin in methanogenesis (47). In Neijmegen, the structure of the yellow fluorescent compound of Daniels' was firmly established to be methenyl-tetrahydromethanopterin (146). Mark Donnelly (33, 34) documented the role of 5-formyl-tetrahydromethanopterin as a C_1 intermediate in methanogenesis. So the blue fluorescent compound F_{342} led to the discovery of the central C_1 carrier of methanogenesis.

Coenzyme F_{430}, the First Nickel Tetrapyrrole

In 1977, we sent some frozen cells of *Methanobacterium* to Gregory Ferry, who was pursuing postdoctoral studies with Harry Peck at Athens, Georgia. Somehow an extract of these cells ended up in the hands of Jean LeGall. He communicated to us the presence of a yellow nonfluorescent compound in these extracts and provided us with a sample, stating that he had a hunch the compound might be a cobamide derivative. Robert Gunsalus repeated the isolation, but the compound was uninteresting to me: when added to extracts, it neither inhibited nor stimulated methanogenesis. I thought it might be a carotenoid. We named it factor 430 (F_{430}) because of its dramatic absorption at 430 nm (63). Later, William Whitman submitted F_{430} for neutron activation analysis, and the results suggested that the compound contained one nickel atom per molecule (152). This finding was of considerable interest, and we were pondering experiments when we received a note from Rolf Thauer saying that F_{430} contained nickel and that the incorporation of ^{63}Ni into F_{430} during growth was documented (29). In addition, his laboratory produced convincing evidence for the incorporation of δ-aminolevulinic acid into F_{430}

(28). This work stimulated Eschenmoser's group to document the structure of F_{430} as a pentaacid tetrahydrocorphin (92, 107). But we still had no role for F_{430} until William Ellefson observed that homogeneous CH_3-S-CoM methylreductase was yellow and had an absorption maximum at 425 nm. Ellefson & Whitman grew cells with ^{63}Ni in the medium, and when the homogeneous radioactive methylreductase was allowed to react with specific antibodies in an immunodiffusion plate, the precipitated protein-antibody band was labeled with ^{63}Ni. The methylreductase contained two molecules of F_{430} per M_r 300,000 protein (38). So F_{430}, the compound that wouldn't do anything, now is considered to be at the very heart of methanogenesis. Recently, Karl Olson collaborated with Michael Summers in Baltimore on the 2D and 3D NMR analysis of native F_{430}, the 12-13 dieprimer, and the reduced derivative F_{560} (106, 174).

Methanofuran, the Coenzyme of Formylation

Although progress had been made in other areas of C_1 reduction, nothing was known in 1980 about the activation and reduction of CO_2 to the formyl level. James Romesser observed that cell extracts could be resolved for a factor that was required for the reduction of carbon dioxide to methane after passage through a short Sephadex G-25 column. This factor was partially purified and named the carbon dioxide reduction factor, CDR (113). John Leigh took up the study of this factor for his thesis project, a study that would take five years. He resolved the fraction from methanopterin and showed that the CDR factor had properties that were quite distinct (91), including the lack of absorption in the visible or long UV spectrum. This finding meant that each fraction from column chromatography had to be tested in a methanogenic assay to locate the active fraction, a laborious and time consuming process. After analysis of its hydrolytic fragments, a structure for the coenzyme was proposed (89) as 4[N-(4,5,7-tricarboxyheptanoyl-γ-L-glutamyl-γ-L-glutamyl)-p-(β-aminoethyl)phenoxymethyl]-2-(aminomethyl)furan.

The counsel of K. L. Rinehart, Jr. was essential to the success of the project. The compound was essentially a long linear molecule with a hydrophilic tetracarboxylic acid moiety on one end and a furan ring with a primary amine on the other end. Aharon Oren suggested the name methanofuran, and we liked it. Later, the methanofuran of *Methanosarcina barkeri* was found to contain glutamyl units instead of the tetracarboxylic acid structure (12). But how did the compound function? John Leigh thought it might be involved in electron transport. But now and then I would ask, "John, ol' boy, does CDR carry a C_1 group?" It was great fun when he showed that formyl-methanofuran was the product of CO_2 activation and reduction (90) and that the formyl group was carried on the primary amine of the furan moiety. A new method of CO_2 fixation had been discovered.

7-Mercaptoheptanoylthreonine Phosphate, the Coenzyme That Donates Electrons to the Methylreductase

Robert Gunsalus & S. Tandon developed a successful procedure for anaerobic column chromatography of cell extract at room temperature (60). Gunsalus separated the methylreductase system into three fractions, each of which was required for reconstruction of the methylreductase reaction from CH_3-S-CoM and molecular hydrogen (64). Component A consisted of large oxygen-sensitive, heat-labile protein complexes with hydrogenase activity. Component B was an oxygen-labile, heat resistant, dialyzable cofactor. Component C was an oxygen-stable, heat-labile protein. Over 10 years passed between the discovery of component B and elucidation of its structure, the most frustrating experience of my academic life. Component B, the small, heat-stable molecule that was absolutely required for reconstitution of the methylreductase reaction was rather stable in crude preparations. However, the factor proved to be highly unstable on fractionation. Ralph Tanner invested several years in attempts to purify the factor to homogeneity but finally had to settle for a partially characterized compound. Kenneth Noll then continued studies on component B with four years of discouraging results. As with methanofuran, no spectral properties were available that could be used to follow the compound during fractionation. The only assay available was methanogenesis by the reconstituted methylreductase system. Years of frustration centered around loss of activity during fractionation. A complex molecule appeared to be decomposing during fractionation, but no conditions could be found to prevent the decomposition. One day, Kenneth Noll showed me a recording from HPLC that exhibited an isolated peak about 2 mm high, and he said with an air of resignation in his voice, "That's B." This was what I had been waiting for. We increased the manpower and ran the column repeatedly until we had enough material to make some measurements.

The breakthrough had been made. We would eventually find that the compound was not decomposing; instead it was forming heterodisulfides during fractionation with any HS-compound that was available. Elucidation of the structure by Ken Noll as 7-mercaptoheptanoylthreonine phosphate, HS-HTP (103), with the counsel of K. L. Rinehart and its confirmation through chemical synthesis by Noll & Mark Donnelly (102) were rewarding experiences after 10 years of frustration. Tanner studied the growth factor required by *Methanomicrobium mobile* (124), and my last graduate student, Carla Kuhner, extended these studies to show that the unknown growth factor is HS-HTP.

Enzymology of Methanogenesis from CO_2 and H_2

Weaving the role of these six new coenzymes into a metabolic pathway for the reduction of carbon dioxide to methane was an intriguing experience that spanned about 15 years. Some of the coenzymes were C_1 carriers; others were electron carriers. It became evident that nature had chosen the strategy of keeping the C_1 group bound to a coenzyme as it was sequentially reduced. The first reaction of the methanogenic pathway to be studied was the CH_3-S-CoM methylreductase reaction (62, 64, 94, 127). Component C was shown to be a protein of M_r 300,000 with an $\alpha_2\beta_2\gamma_2$ subunit configuration with 2 mol of F_{430} (38, 39). The enzyme contained CoM, and later Patricia Hartzell determined by the *M. brevibacter* assay that a preparation of component C prepared in Walsh's laboratory contained 2 mol of bound CoM per mol of protein (71). Hartzell et al (66) could find no stoichiometry for CoM incorporation into component C. Kenneth Noll showed that component C also contained bound component B (HS-HTP) (104). Although we could find no evidence for cobamides or corrins in component C, Whitman (155) found that the addition of corrins to the reaction mixture activated the methylreductase reaction. He studied the activation of the methylreductase by ATP (153) as well as its inhibition by corrins (154). Hartzell performed a comparative study of component C from various methanogens and revealed minor differences (68). Later, Pierre Rouvière used the variation in the subunits of component C to construct a taxonomy for the methanogens (118) that is in remarkable agreement with that derived from analysis of 16S rRNA.

Because component C from *Methanobacterium thermoautotrophicum* strain ΔH required component A proteins for conversion to the active state, Hartzell was able to dissociate and purify separately the α, β, and γ subunits as well as F_{430}. She then showed that these constitutents could be reassembled into component C, which in the presence the A proteins showed a 70% recovery of the methylreductase activity (69). F_{430} was absolutely required, showing for the first time that it really was a coenzyme in the methylreductase reaction. Homogeneous component C from strain ΔH required protein fraction component A and coenzyme component B for activity (64). In a series of very demanding experiments, David Nagle (101) separated component A into three protein fractions, A1, A2, and A3. This was an especially difficult task because the only assay available was the reconstituted methylreductase system. Each fraction that was purified required three other active fractions for its assay. Fraction A3 was found to be specifically inhibited by 2',3'dialdehyde of ATP (119). Later, fraction A3 was separated into two protein fractions, and A2 was purified to homogeneity by Rouvière (117, 121). So far, no specific enzymatic reaction has been assigned to these fractions, but A2 and A3 appear to be involved in converting inactive component C (C_i) to its active form (C_a) (120). Attempts to simplify the assay (67) or study the reductive activation (116) did not produce a breakthrough. A breakthrough had occurred when Ankel-Fuchs in Thauer's laboratory prepared an active form of component C from the Marburg strain. These experiments proved what everyone had suspected: that component C was the methylreductase (2). We sent Ankel-Fuchs a sample of HS-HTP to see if it substituted for component B in a reaction mixture with C_a; it did (1). Because HS-HTP worked in this ATP-free system with C_a, it obviously was not adenylated. When 7-[methylthio]heptanoylthreonine phosphate was tested, the methyl group was not reduced to methane (105). Noll (105) showed that HS-HTP was the electron donor for CH_3-S-CoM reduction, and Bobik & Olson (13) found that the products of the reaction were methane and the heterodisuflide, CoM-S-S-HTP. Similar results were found in Thauer's laboratory (40, 41). Finally, after nearly 20 years of work on CoM, we could write a reaction for the formation of methane:

$$CH_3\text{-S-CoM} + HS\text{-HTP} \rightarrow CH_4 + CoM\text{-S-S-HTP}.$$

In 1977, Robert Gunsalus made a curious discovery; the conversion of carbon dioxide to methane was greatly stimulated by the addition of CH_3-S-CoM to the reaction mixture (61). We believed this observation meant that in some manner the terminal methylreductase reaction was coupled to the activation of carbon dioxide and named this phenomenon the RPG effect after his initials. Romesser showed that the RPG effect in three species of methanogens could be initiated by methanogenic precursors (114). With the discovery that CoM-S-S-HTP was a product of the methylreductase reaction, Bobik found that this heterodisulfide specifically initiated the RPG effect in the absence of the methylreductase reaction (14). At present, the mechanisms of the activation of CO_2 reduction by the heterodisulfide is not understood, but Bobik has proposed that CoM-S-S-HTP activates an electron carrier that donates electrons to formylmethanofuran dehydrogenase (15). Another curious observation, made by Romesser (111), was that addition of certain intermediates of the citric acid cycle such as malate initiated the RPG effect. At that time, we believed that certain carbon atoms of these compounds, because of metabolic reactions, eventually ended up as a methyl group on CoM. However, Bobik discovered a very unusual fumarate reductase, a thiol-driven reaction in extracts of strain ΔH that required both HS-CoM and HS-HTP as electron donors in which succinate and CoM-S-S-HTP are formed as products (16). Thus, CoM-S-S-HTP may be formed by more than one reaction. Bobik also showed that the activation of CO_2 by the heterodisulfide resulted in the production of formylmethanofuran (15).

When Mark Donnelly joined the laboratory, he was attracted to the fate of the formyl group on formylmethanofuran. Jorge Escalante had just shown that tetrahydromethanopterin (H_4MPT) reacted with formaldehyde to form methylene-H_4MPT, which could be oxidized by extracts to form methenyl-H_4MPT (45). Transfer of the formyl group from formylmethanofuran to H_4MPT to produce formyl-H_4MPT seemed to be a logical reaction to look for, and Donnelly found it. He developed an assay to follow the formyltransferase reaction and purified the enzyme to homogeneity (34). Formation of 5-formyl-pterin was a novel enzymatic reaction, one not found in folate biochemistry. Subsequently, the formyltransferase was cloned and sequenced by Anthony DiMarco with Karen Sment from Jordan Konisky's laboratory (32). The enzyme was expressed in *E. coli* and was found to be active at 60°C. A search of gene banks revealed that the enzyme is unique and unrelated to any previously described enzyme. Donnelly & Escalante (33) found the 5-10 methenyl-H_4MPT cyclohydrolase and showed that 5-formyl-H_4MPT was formed as a product; DiMarco & Donnelly (31) purified the enzyme to homogeneity and described its properties. By use of both homogeneous enzymes, Donnelly then showed that formylmethanofuran could be converted to methenyl-H_4MPT (34), indicating that these two enzymes were components of the methanogenic pathway.

Escalante (45) had shown that methylene-H_4MPT could be oxidized to methenyl-H_4MPT, and Hartzell et al (70) found a methylene-H_4MPT dehydrogenase that required coenzyme F_{420} to carry out the oxidoreductase reaction. Aharon Oren & Jeffrey Hoyt (75) purified and described the H_4MPT-serine transhydroxymethylase that formed methylene-H_4MPT. Escalante (47) had shown that the methylene group of methylene-H_4MPT could be reduced chemically to form 5-methyl-H_4MPT, but had difficulty showing the enzymatically catalyzed reduction; however, the methyl group of 5-methyl H_4MPT was converted to methane by cell extracts (45, 47). Current concepts are that the methyltransferase contains a cobamide group in which the cobalt atom accepts the methyl group from methyl-H_4MPT (108); this enzyme was named methyltransferase-1. Earlier, Taylor reported the purification of an enzyme that transferred the methyl group from methylcobalamin to HS-CoM (127). Taylor's enzyme is now believed to transfer the methyl group from the cobamide of methyltransferase-1 to HS-CoM. Evidence for the methylene-H_4MPT reductase and its requirement for reduced F_{420} was obtained by TeBrömmelstroet et al (129) in Vogel's laboratory. Shapiro (122) showed that CH_3-S-CoM was an intermediate in methanogenesis from methanol by *Methanosarcina*, and Baresi (9) documented levels of coenzyme F_{420}, CoM, hydrogenase, and CH_3-S-CoM reductase in acetate-grown cells. So, it appears that the pathway we had proposed (30, 79, 120) for methanogenesis from carbon dioxide and hydrogen is essentially correct.

Microbial Diversity

Gallionella had interested me as a graduate student, so in 1954, I looked forward to studying this iron bacterium in Pacific Grove. van Niel was enthusiastic, and upon my arrival handed me a file of reprints, saying as he pointed to one, "This is an important paper. Unfortunately it is in Russian, but it has a good German summary." He loved to remind American students of their language limitations. *Gallionella* secretes a twisted ribbon-like sessile stalk of ferric hydroxide, and van Niel had suggestions about setting up enrichments in Carrell flasks. I followed his suggestions, and, in a few weeks, had my first successful enrichments; van Niel seemed impressed. These cultures did not survive the return trip to Illinois, so I was anxious to find a source of organisms.

During a weekend at Turkey Run State Park in Indiana, Gret saved the day by finding rusty patches along the Rocky Hollow trail; these proved to be an excellent source of *Gallionella*. Al Vatter took some electron micrographs that showed that the stalks were not solid ribbons but were composed of many strands secreted from one side of the cell (149). We developed a defined medium, and Sonia Kucera worked on *Gallionella* for her MS thesis (86). I prepared an article on iron bacteria for the *Journal of the American Water Works Association* (164); over the years I would take some kidding from scientific colleagues for publishing in this "prestigious" journal. I used this article to encourage water plant operators to send me samples of problem organisms. A beautiful example of the chlamydobacterium, *Crenothrix*, arrived from Sweden (165), but I could not cultivate it. Another sample contained a large *Leptothrix*. Because of problems with obtaining pure cultures and with mass culture, these organisms had to be abandoned as serious research subjects, but iron bacteria continue to be a scientific hobby. Our children continue to send me samples of rusty deposits they find in nature.

Phototrophs, my favorite enrichments in the organism course, were always fun to isolate, and, when Al Vatter became interested in the ultrastructure of *Rhodospirillum rubrum*, I suggested that we collaborate on a study of representative photosynthetic bacteria to discern their cellular anatomy. We published a pioneering article on the structure of photosynthetic bacteria, showing that *R. rubrum*, *Rhodopseudomonas sphaeroides*, and *Chromatium* all contained differentiated membranes, chromatophores (150). We also showed that *Chlorobium* was different, but we missed the chlorosome. While on a summer fellowship at the University of Washington, I set up enrichments for *Rhodomicrobium vannielii*, a most unusual phototroph discovered by Howard Douglas. We were surprised to find that the organism was motile (35) and that it contained lamellae rather than chromatophores (148). When Robert Uffen arrived for postdoctoral study, I suggested that he set up enrichments for methanogens that could grow on carbon monoxide. A technique was developed for streaking enrichment cultures on agar in bottle plates to which we could add an atmosphere of carbon monoxoide. Much later an improved bottle plate was developed and Bellco Glass produced it (74). Uffen could detect no methane in the bottle plates, but one plate contained a large red colony. I knew that Howard Gest and other investigators had been trying to grow phototrophs anaerobically in the dark without success. The organism proved to be *Rhodopseudomonas palustris*, a purple nonsulfur phototroph. A very careful study to insure that utilizable radiation was not available to the cells revealed that by use of pyruvate and H_2, representative nonsulfur phototrophs also could be cultured anaerobically in the dark indefinitely and that their photosynthetic apparatus could not be differentiated from that of anaerobic light-grown cells (134). Mutants also were studied (134a).

Beggiatoa, with its gliding trichomes stuffed with sulfur, was a fascination for me. I thought it would be fun to isolate, mass culture, and study the metabolism of this organism. We followed the method of Cataldi for isolation by allowing a trichome to glide over an agar surface; a cut-out agar block with the trichome on it was then transferred to sterile medium. Lois Faust developed a defined medium for the organism. By 1959, I realized that it was unrealistic to ever expect to mass culture this organism in quantities required for enzyme studies, so I documented its characteristics with photographs and motion pictures (50) and used these in teaching for many years.

Sarcina ventriculi, which carries out an alcoholic fermentation of sugar at pH 2.5, was studied by Ercolé Canale-Parola, my second PhD student. The organism was isolated from sediments of Boneyard Creek, a somewhat polluted stream that runs through our campus. The real challenge was to maintain the organism, for it could die quickly if the culture was not transferred properly. He developed a stock culture technique (24), evolved a synthetic medium (25), and studied the localization of cellulose (23) as well as its synthesis (26).

Streptococcus allantoicus was isolated by Ray Valentine from a duck pond at Monticello, Ill., to study allantoin degradation. H. A. Barker had discovered the organism and suggested that possibly one of the ureido groups of allantoin could be cleaved in an energy-yielding reaction. Valentine discovered the enzyme, oxamic transcarbamylase (141), and showed that the phosphorolysis of oxalurate yielded oxamate and carbamyl phosphate, the latter compound donating a phosphate group to ADP in a kinase reaction to yield ATP (142). In studying the phosphate-dependent degradation of urea, ureidoglycolate was found to be an intermediate (143, 144), and its NAD-dependent oxidation to oxalurate was proposed as the only oxidative step carried out by the organism when it grew anaerobically on allantoin (135). Elizabeth Gaudy (55, 58) characterized the enzyme ureidoglycolate synthetase for her thesis and implicated the enzyme in the allantoin degradation pathway. Robert Bojanowski & Gaudy (17) characterized the oxamic transcarbamylase, and Valentine & Harvey Drucker (138) documented the conversion of glyoxylate to tartronic semialdehyde and CO_2.

Sphaerotilus natans was studied by Elizabeth Gaudy for her MS thesis. In rich media, sheath formation by this chlamydobacterium ceased and the organism produced copious amounts of slime (56). The slime was purified and found to be composed of equal amounts of fucose, galactose, glucose, and glucuronic acid; we suggested that the latter two components occurred as an aldobiuronic acid (57). We then received a letter from Michael Heidelberger, who suggested that the aldobiuronic acid could be identified using antibodies that he had. The aldobiuronic acid precipitated with Type III and Type VII antipneumococcal horse sera and was identified as cellobiuronic acid (73).

Arthrobacter crystallopoietes was brought to my laboratory by a postdoctoral student, Jerald Ensign. *Arthrobacter* goes through a rod to coccus differentiation during growth, and he wished to study this phenomenon. He developed a chemically defined medium in which the organism grew only in the coccoid form. He discovered that addition of certain amino acids or organic acids to the defined medium resulted in the formation of the rod-shaped stage (42).

Myxobacter strain AL-1 was isolated by Ensign (43) by placing soil inocula on a lawn of *Arthrobacter*, where the myxobacter produced large clear areas; the lytic enzyme was found to readily lyse a number of gram-positive species (43). The enzyme was found to be a small protease that hydrolyzed proteins as well as peptide bonds in the glycosaminopeptide of cell walls (44). Richard Jackson studied the amino acid composition and characterized the enzyme (78). Strominger found the enzyme, Myxobacter AL-1 Protease I, to be valuable in determining the structure of the cell walls of gram-positive bacteria. A second protease, Myxobacter AL-1 Protease II, was discovered when the enzyme crystallized in certain fractions as it eluted from a column. We were somewhat stunned; usually an enzyme yields to crystallization only after a battle, but here we had crystals and no assay. Marilyn Wingard (158) determined that the enzyme was a protease that had the unusual specificity to cleave the peptide bond on the amino side of lysine. A third enzyme was discovered when in the course of dialyzing myxobacter extracts the dialysis tubing disintegrated. Allan Hedges (72) showed that this enzyme had both β-1,4 gluconase as well as chitosanase activity.

Clostridium tetanomorphum was studied by Robert Twarog, who showed that the organism makes only 1 mol of ATP per mol of glutamate and that ATP is generted via phosphostransbutyrylase and butyrate kinase reactions (130, 131).

Geodermatophilus was brought to the laboratory by Edward Ishiguro; he had found the organism in soil obtained from the lower reaches of Mt. Everest. From a clump of large coccoid cells, certain cells differentiated small motile rods, which multiplied as rods but later differentiated to cocci, forming large clumps. He studied this phenomenon and found that the life cycle could be controlled; a factor in certain batches of Tryptose (Difco) was required for differentiation to the motile-rod form (76). This factor was difficult to purify because the growth-differentiation assay was not easy to quantify. After a few years, the factor was in hand and the assay was reproducible (77), but Ishiguro was rather depressed; the factor was an inorganic cation (NH_4^+) and other inorganic as well as organic cations induced morphogenesis.

Methanobacterium thermoautotrophicum was isolated by Gregory Zeikus. In 1970, Leon Campbell encouraged me to set up some enrichments for thermophilic methanogens. These enrichments readily produce methane, but I could not pursue a successful isolation at the time. When Zeikus came to the laboratory, I suggested that he continue this project of isolating a thermophilic methanogen. He proceeded to enrich and isolate an organism from sewage sludge that grew well at 60–65°C. In spite of our poor latin scholarship (181), isolation of this organism, *Methanobacterium thermoautotrophicum* strain ΔH, was a major breakthrough. The organism had an interesting ultrastructure (182). Strain ΔH became the backbone of my research program. It could be readily scaled up to the 200-liter fermentor stage, it grew more rapidly than mesophilic strains, and, above all, enzymes were stable at room temperatures, 40° below the optimal growth temperature.

Methanogenium cariacii and *Methanogenium marisnigri* were isolated by James Romesser (115). Since I was a child, I have had an unfortunate propensity for motion sickness, so when Holger Jannasch invited me to join an expedition to the Cariaco Trench and later to the Black Sea, I was forced to decline. However, Romesser was pleased at the possibility for adventure and returned with samples from water columns of these anaerobic habitats.

Methanococcus voltae was sent to us by J. M. Ward, who had isolated it for his MS thesis, and we decided to name it after Alessandro Volta, who first described the formation of "combustible air" in sediments (3). Whitman and colleagues defined its nutrition and carbon metabolism (151) as well as its plating efficiency (82). We suggested that this organism might be the organism of choice for initiating genetic studies with methanogens.

Methanococcus jannaschii was isolated by John Leigh from samples obtained by Holger Jannasch at a geothermal vent on the east Pacific rise by use of the submersible, Alvin. This unusual organism grew optimally at 85°C with a generation time of 25 min. Its properties were reported by Jones et al (81).

Methanogenium thermophilum was isolated by Friedrich Widdel. He had stunned the field by showing that methanogens could oxidize alcohols and reduce CO_2 to CH_4. In Urbana, he isolated mesophilic and thermophilic alcohol-oxidizing methanogens and with Rouvière studied their taxonomy (156). He showed that F_{420} served as an electron acceptor for alcohol dehydrognease (157).

Acetobacterium woodii was discovered in enrichments set up for methanogens in which a gas mixture of 80% H_2:20% CO_2 was bubbled through a series of tubes that had been inoculated with sediment from Crystal Lake, Urbana, Ill. To save gas, the effluent of one tube was connected to the inlet of the next tube. The tubes were placed in a fume hood. About a month later, I happened to notice these tubes with their slowly bubbling contents; I had forgotten them. I could find no evidence of methane formation, but the black rubber stoppers were swollen and gnarled, and the amount of acetic acid coming out of the tubes was fantastic. This was obviously not the way to enrich for methanogens! At that time, no H_2-oxidizing, CO_2-reducing acetogen was in culture. One summer at Woods Hole, we set up enrichments from sediments at Oyster Pond Inlet, and Balch succeeded in isolating a cocco-bacillus-like organism. We named the organism after Harland Wood (5), who had spent 20 years trying to figure out how acetic acid was synthesized from CO_2, and presented him with a picture of the organism on his 70th birthday. Siegfried Schoberth thoroughly documented the substrate range and properties of the organism, and Ralph Tanner determined the presence of tetradhydrofolate enzymes in the organism (125). *A. woodii* opened the modern era of hydrogen-oxiding acetogens, the ecological importance of which is becoming increasingly appreciated. Later, John Leigh isolated a thermophilic acetogen on H_2 and CO_2 from sediments of Lake Kivu, Africa. We named this organism *Acetogenium kivui* (88).

Methanospirillum hungatei, strain JF, was isolated by Gregory Ferry from a stable consortium of cells that degraded benzoate anaerobically to CH_4 and CO_2 (52). He resolved the culture into three organisms. One was a pseudomonad that used benzoate as its substrate. Another organism was a spiral-shaped methanogen that grew on H_2 and CO_2 or formate. The other organism could not be isolated but was an acetophilic methanogen. The spirillum was similar to one described by Paul Smith in an abstract, but a publication never appeared, so we collaborated on its description and named the organism *Methanospirillum hungatei*, in honor of Robert Hungate (51). Later, the nutritional and biochemical characteristics of *M. hungatei* were described (53).

Josef Winter joined the laboratory and achieved complete degradation of carbohydrate to methane by a syntrophic culture of *A. woodii* and *M. barkeri* (160). In another study, he followed methane formation from fructose by *A. woodii* and other methanogens in a chemostat (159). I believe he was the first person to use a chemostat to study methanogens. He returned to Otto Kandler's laboratory and trained his colleagues to use the techniques developed by Balch. One brilliant pupil was Karl Stetter. Other consortia were studied by Jack Jones and Jean Pierre Guyot (80).

While on sabbatical leave in 1975, Norbert Pfenning suggested that I

WOLFE

should isolate and study a spirillum-like organism that was persistent in enrichment cultures of *Desulfuromonas*. We named the organism spirillum 5175 and found that it could grow anaerobically by the oxidation of formate with the reduction of fumarate to succinate. It could also use H_2S as electron donor for this reaction. In addition, it could anaerobically oxidize hydrogen or formate and reduce elemental sulfur, thiosulfate, or sulfite, but not sulfate. Since *Chlorobium* does not use formate, we set up a syntrophic culture in which the spirillum oxidized formate and reduced elemental sulfur to H_2S; *Chlorobium* used the H_2S as its electron donor, producing elemental sulfur. Sulfur was limiting in the medium and was recycled through H_2S and elemental sulfur, so neither organism could grow alone, but together they grew beautifully (168).

Richard Blakemore discovered magnetotactic bacteria in sediments that had been collected at Woods Hole, but he was unable to pursue their isolation. Before his arrival in Urbana as a postdoctoral student, I spent the summer studying the survival of these organisms from enrichments in various media and conditions. Their survival was a matter of minutes or hours not days. When he arrived in September, 1976, I suggested that we set Christmas as a goal for achieving the first pure culture; by February he had a magnetic spirillum in hand (10). It would be 10 years before a second species would be captured by others. While on a von Humboldt Fellowship in Germany, I continued the study of magnetic bacteria; why were they so difficult to isolate? An undergraduate student in Rolf Thauer's laboratory, Alfred Spormann, was interested, and we found an excellent natural enrichment. Using reducing agents, we showed that chemotaxis could maintain a magnetically oriented cell at a certain oxygen tension, explaining why certain magnetotactic cells may not follow the magnetic lines of force deep into anaerobic sediments (123). I was able to perfect a capillary racetrack method for separating magnetic bacteria from typical bacteria (169), but an appropriate medium for their culture could not be formulated.

Reflections

Some believe it takes as much work to study an unimportant problem as an important one, so one should only work on important problems. Perceived important problems are usually in good hands, which means that most of us, fortunately, have the freedom to work on what interests us, and sometimes this turns out to be important. My goal has been to get as many persons as possible interested in working on methanogens. After a sabbatical leave in Urbana, Godfried Vogels returned to start his own program on methanogens at Nijmegen. Rolf Thauer became interested in methanogens after a two-week visit to Urbana. Josef Winter returned to Kandler's laboratory in Munich. Ziegfried Schoberth returned to Gottschalk's laboratory at Göttingen. Many individuals were trained by Balch, including visitors from Ottawa. Chris Walsh, Orme-Johnson, and students visited from MIT. John Reeve and students visited from Columbus. Jordan Konisky was encouraged to start his own program at Urbana. By 1982, there was a critical mass of investigators, and we began to organize our first Gordon Conference on methanogenesis.

During the years of teaching the organisms course at Urbana, I was frequently reminded of the limitations of formal laboratory scheduling where a two- to three-hour laboratory is wedged between classes, and where students have limited opportunities to make discoveries. I had always wanted to try a van Niel–type summer course, so when the opportunity came to be Director of the Microbiology Course at the Marine Biological Laboratory at Woods Hole, I asked Peter Greenberg, who would bring genetic and molecular components to the course, to join me as codirector with the purpose of starting a 1985 version of a van Niel–type summer organisms course. He agreed. We asked Bernhard Schink from Konstanz to join us for the first summer, and Carrie Harwood was course coordinator. Holger Jannasch was instrumental in urging students on the national and international scene to take the course. In subsequent years, we were joined by Norbert Pfenning, Friedrich Widdel, Bernhard, and Andrew Kropinski as course coordinator. We found that the Volta experiment on the third evening of the course not only focused interest on anaerobic physiology but was a great social mixer for students and staff. Student response and faculty dedication made these five summers the most satisfying teaching experiences of my academic career.

Over the years, I have formulated a few truisms that I refer to as Wolfe's Laws of Thermodynamics: 1st Law: Unpublished data do not improve with age. 2nd Law: If you are first on the scene, it is easy to make discoveries. 3rd Law: The emotion generated in scientific discussion increases proportionally with the softness of the data being discussed. 4th Law: If you join a parade, you become one of the marchers. Graduate students like to join highly visible parades, to be part of the current scene, not realizing that their visibility is reduced by the number of marchers. As an independent investigator, each should realize the importance of choosing a problem that isn't moving and move it.

Throughout my academic life, I have more or less been driven in some manner "to go back to the lab." This common syndrome of persons in science needing to be in the lab could be simply a security blanket or a necessary component of scientific survival. I have felt that, for me, the latter was the case, coupled with a feeling that possibly my presence somehow might encourage students. Gret has spent many lonely evenings sharing her life with the lab, and her support was pivotal. Our children, Danny, Jon, and Sue, shared a portion of their lives with the lab. However, one of the enjoyable aspects of academic life is the sabbatical; we have had interesting sabbaticals in England, Hawaii, France, and Germany. My last sabbatical, a short one, was spent with John Breznak, where I was introduced to the fascinating world of anaerobic protozoa from termites.

As these scientific memoirs have shown, I was fortunate to have a series of talented graduate and postdoctoral students to work with me. I thank them, my colleagues, Gret, and our children for making this experience a rewarding one. In operating my laboratory, my purpose has not been to play the role of the brilliant intellectual leader, but rather to stay in the background and try to create an atmosphere in which students could develop into independent investigators; this, for me, is what it's all about.

Literature Cited

1. Ankel-Fuchs, D., Böcher, R., Thauer, R. K., Noll, K. M., Wolfe, R. S. 1986. 7-Mercaptoheptanoylthreonine phosphate functions as component B in ATP-independent methane formation from methyl-CoM with reduced cobalamin as electron donor. *FEBS Lett.* 213:123–27

2. Ankel-Fuchs, D., Huster, R., Mörschel, E., Albracht, S. P. J., Thauer, R. K. 1986. Structure and function of methyl-coenzyme M reductase and of factor F_{430} in methanogenic bacteria. *Syst. Appl. Microbiol.* 7:383–87

3. Balch, W. E., Fox, G. E., Magrum, L. J., Woese, C. R., Wolfe, R. S. 1979. Methanogens: reevaluation of a unique biological group. *Microbiol. Rev.* 43:260–96

4. Balch, W. E., Magrum, L. J., Fox, G. E., Wolfe, R. S., Woese, C. R. 1977. An ancient divergence among the bacteria. *J. Mol. Evol.* 9:305–11

5. Balch, W. E., Schoberth, S., Tanner, R. S., Wolfe, R. S. 1977. *Acetobacterium*, a new genus of hydrogen-oxidizing CO_2-reducing anaerobic bacteria. *Int. J. Syst. Bacteriol.* 27:355–61

6. Balch, W. E., Wolfe, R. S. 1976. New approach to the cultivation of methanogenic bacteria: 2-mercapto-ethanesulfonic acid (HS-CoM)-dependent growth of *Methanobacterium ruminantium*. *Appl. Environ. Microbiol.* 32:781–91

7. Balch, W. E., Wolfe, R. S. 1979. Specificity and biological distribution of coenzyme M (2-mercaptoethanesulfonic acid). *J. Bacteriol.* 137:256–63

8. Balch, W. E., Wolfe, R. S. 1979. Transport of coenzyme M (2-mercapto-ethanesulfonic acid) in *Methanobacterium ruminantium*. *J. Bacteriol.* 137:265–73

9. Baresi, L., Wolfe, R. S. 1980. Levels of coenzyme F_{420}, coenzyme M, hydrogenase and methylcoenzyme M methylreductase in acetate-grown *Methanosarcina*. *Appl. Environ. Microbiol.* 41:388–91

10. Blakemore, R. P., Maratea, D., Wolfe, R. S. 1979. Isolation and pure culture of a freshwater magnetic spirillum in chemically defined medium. *J. Bacteriol.* 140:720–29

11. Blaylock, A. B., Stadtman, T. C. 1963. Biosynthesis of methane from the methyl moiety of methylcobalamin. *Biochem. Biophys. Res. Commun.* 11:34–38

12. Bobik, T. A., Donnelly, M. I., Rinehart, K. L. Jr., Wolfe, R. S. 1987. Structure of a methanofuran derivative found in cell extracts of *Methanosarcina barkeri. Arch. Biochem. Biophys.* 254:430–36

13. Bobik, T. A., Olson, K. D., Noll, K. M., Wolfe, R. S. 1987. Evidence that the heterodisulfide of coenzyme M and 7-mercaptoheptanoylthreonine phosphate is a product of the methyl-reductase reaction in *Methanobacterium. Biochem. Biophys. Res. Commun.* 149:455–60

14. Bobik, T. A., Wolfe, R. S. 1988. Physiological importance of the heterdisulfide of coenzyme M and 7-mercaptoheptanoylthreonine phosphate in the reduction of carbon dioxide to methane of *Methanobacterium. Proc. Natl. Acad. Sci. USA* 85:60–63

15. Bobik, T. A., Wolfe, R. S. 1989. Activation of formylmethanofuran synthesis in cell extracts of *Methanobacterium thermoautotrophicum. J. Bacteriol.* 171:1423–27

16. Bobik, T. A., Wolfe, R. S. 1989. An unusual thiol-driven fumarate reductase in *Methanobacterium* with the production of the heterodisulfide of coenzyme

M and N-(7-Mercaptoheptanoyl)threonine-O^3-phosphate. *J. Biol. Chem.* 264: 18714–18

17. Bojanowski, R., Gaudy, E., Valentine, R. C., Wolfe, R. S. 1964. Oxamic transcarbamylase of *Streptococcus allantoicus. J. Bacteriol.* 87:75–80

18. Brill, W. J., Wolfe, R. S. 1966. Acetaldehyde oxidation by *Methanobacillus*— a new ferredoxin-dependent reaction. *Nature* 212:253–55

19. Brill, W. J., Wolin, E. A., Wolfe, R. S. 1964. Anaerobic formate oxidation a ferredoxin-dependent reaction. *Science* 144:297–98

20. Deleted in proof

21. Bryant, M. P., McBride, B. C., Wolfe, R. S. 1968. Hydrogen-oxidizing methane bacteria. I. Cultivation and methanogenesis. *J. Bacteriol.* 95:1118–23

22. Bryant, M. P., Wolin, E. A., Wolin, M. J., Wolfe, R. S. 1967. *Methanobacillus omelianskii*, a symbiotic association of two species of bacteria. *Arch. Mikrobiol.* 59.20–31

23. Canale-Parola, E., Borasky, R., Wolfe, R. S. 1961. Studies on *Sarcina ventriculi.* III. Localization of cellulose. *J. Bacteriol.* 81:311–18

24. Canale-Parola, E., Wolfe, R. S. 1960. Studies on *Sarcina ventriculi.* I. Stock culture method. *J. Bacteriol.* 79:857–59

25. Canale-Parola, E., Wolfe, R. S. 1960. Studies on *Sarcina ventriculi* II. Nutrition. *J. Bacteriol.* 79:860–62

26. Canale-Parola, E., Wolfe, R. S. 1964. Synthesis of cellulose by *Sarcina ventriculi. Biochim. Biophys. Acta* 82:403–5

27. Cheeseman, P., Toms-Wood, A., Wolfe, R. S. 1972. Isolation and properties of a fluorescent compound, Factor 420, from *Methanobacterium* strain M. o.H. *J. Bacteriol.* 112:527–31

28. Diekert, G., Jaenchen, R., Thauer, R. K. 1980. Biosynthetic evidence for a nickel tetrapyrrole structure of factor F_{430} from *Methanobacterium thermoautotrophicum. FEBS Lett.* 119:118–20

29. Diekert, G., Klee, B., Thauer, R. K. 1980. Nickel, a component of factor F_{430} from *Methanobacterium thermoautotrophicum. Arch. Microbiol.* 124:103–6

30. DiMarco, A. A., Bobik, T. A., Wolfe, R. S. 1990. Unusual coenzymes of methanogenesis. *Annu. Rev. Biochem.* 59:355–94

31. DiMarco, A. A., Donnelly, M. I., Wolfe, R. S. 1986. Purification and properties of the 5,10-methenyltetrahydromethanopterin cyclohydrolase from *Methanobacterium thermoautotrophicum. J. Bacteriol.* 168:1372–77

32. DiMarco, A. A., Sment, K. A., Konisky, J., Wolfe, R. S. 1990. The formylmethanofuran: tetrahydromethanopterin formyltransferase from *Methanobacterium thermoautotrophicum* ΔH. *J. Biol. Chem.* 265:472–76

33. Donnelly, M. I., Escalante-Semerena, J. C., Rinehart, K. L. Jr., Wolfe, R. S. 1985. Methenyl-tetrahydromethanopterin cyclohydrolase in cell extracts of *Methanobacterium. Arch. Biochem. Biophys.* 242:430–39

34. Donnelly, M. I., Wolfe, R. S. 1986. The role of formylmethanofuran: tetrahydromethanopterin formyltransferase in methanogenesis from carbon dioxide. *J. Biol. Chem.* 261:16653–59

35. Douglas, H. C., Wolfe, R. S. 1959. Motility of *Rhodomicrobium vannielii. J. Bacteriol.* 78:597–98

36. Eirich, L. D., Vogels, G. D., Wolfe, R. S. 1978. Proposed structure for coenzyme F_{420} from *Methanobacterium. Biochemistry* 17:4583–93

37. Eirich, L. D., Vogels, G. D., Wolfe, R. S. 1979. Distribution of coenzyme F_{420} and properties of its hydrolytic fragments. *J. Bacteriol.* 140:20–27

38. Ellefson, W. L., Whitman, W. B., Wolfe, R. S. 1982. Nickel-containing factor F_{430}: Chromophore of the methylreductase of *Methanobacterium. Proc. Natl. Acad. Sci. USA* 79:3707–10

39. Ellefson, W. L., Wolfe, R. S. 1981. Component C of the methylreductase system of *Methanobacterium. J. Biol. Chem.* 256:4259–62

40. Ellerman, J., Hederich, R., Böcher, R., Thauer, R. K. 1988. The final step in methane formation: investigations with highly purified methyl-CoM reductase (component C) from *Methanobacterium thermoautotrophicum* (strain Marburg). *Eur. J. Biochem.* 172:669–77

41. Ellerman, J., Kobelt, A., Pfaltz, A., Thauer, R. K. 1987. On the role of N-7-mercaptoheptanoyl-O-phospho-L-threonine (component B) in the enzymatic reduction of methyl-coenzyme M to methane. *FEBS Lett.* 220:358–62

42. Ensign, J. C., Wolfe, R. S. 1964. Nutritional control of morphogenesis in *Arthrobacter crystallopoietes. J. Bacteriol.* 87:924–32.

43. Ensign, J. C., Wolfe, R. S. 1965. Lysis of bacterial cell walls by an enzyme isolated from a myxobacter. *J. Bacteriol.* 90:395–402

44. Ensign, J. C., Wolfe, R. S. 1966. Characterization of a small proteolytic enzyme which lyses bacterial cell walls. *J. Bacteriol.* 91:524–34

45. Escalante-Semerena, J. C. 1983. *In vitro methanogenesis from formaldehyde. Identification of three C_1 intermediates of the methanogenic pathway.* PhD thesis. Urbana: Univ. Ill. 229 pp.

46. Escalante-Semerena, J. C., Leigh, J. A., Rinehart, K. L. Jr., Wolfe, R. S. 1984. Formaldehyde activation factor, tetrahydromethanopterin, a coenzyme of methanogenesis. *Proc. Natl. Acad. Sci. USA* 81:1976–80

47. Escalante-Semerena, J. C., Rinehart, K. L. Jr., Wolfe, R. S. 1984. Tetraphydromethanopterin, a carbon carrier in methanogenesis. *J. Biol. Chem.* 259: 9447–55

48. Escalante-Semerena, J. C., Wolfe, R. S. 1984. Formaldehyde oxidation and methanogenesis. *J. Bacteriol.* 158:721–26

49. Escalante-Semerena, J. C., Wolfe R. S. 1985. Tetrahydromethanopterin-dependent methanogenesis from non-physiological C_1 donors in *Methanobacterium*. *J. Bacteriol.* 161:696–701

50. Faust, L., Wolfe, R. S. 1961. Enrichment and cultivation of *Beggiatoa alba*. *J. Bacteriol.* 81:99–106

51. Ferry, J. G., Smith, P. H., Wolfe, R. S. 1974. *Methanospirillum*, a new genus of methanogenic bacteria, and characterization of *Methanosprillum hungatii* sp. nov. *Int. J. Syst. Bacteriol.* 24:465–69

52. Ferry, J. G., Wolfe, R. S. 1976. Anaerobic degradation of benzoate to methane by a microbial consortium. *Arch. Microbiol.* 107:33–40

53. Ferry, J. G., Wolfe, R. S. 1977. Nutritional and biochemical characterization of *Methanospirillum hungatii*. *Appl. Environ. Microbiol.* 34:371–76

54. Fox, G. E., Magrum, L. J., Balch, W. E., Wolfe, R. S., Woese, C. R. 1977. Classification of methanogenic bacteria by 16S ribosomal RNA characterization. *Proc. Natl. Acad. Sci. USA* 74:4537–41

55. Gaudy, E. T., Bojanowski, R., Valentine, R. C., Wolfe, R. S. 1965. Ureidoglycolate synthetase of *Streptococcus allantoicus*. I. Measurement of glyoxylate and enzyme purification. *J. Bacteriol.* 90:1525–30

56. Gaudy, E. T., Wolfe, R. S. 1962. Factors affecting filamentous growth of *Sphaerotilus natans*. *Appl. Microbiol.* 9:580–84

57. Gaudy, E., Wolfe, R. S. 1962. Composition of an extracellular polysaccharide produced by *Sphaerotilus natans*. *Appl. Microbiol.* 10:200–5

58. Gaudy, E. T., Wolfe, R. S. 1965. Ureidoglycolate synthetase of *Streptococcus allantoicus*. II. Properties of the enzyme and reaction equilibrium. *J. Bacteriol.* 90:1531–36

59. Gunsalus, R. P., Romesser, J. A., Wolfe, R. S. 1978. Preparation of coenzyme M analogues and their activity in the methylcoenzyme M reductase system of *Methanobacterium thermoautotrophicum*. *Biochemistry* 17:2374–77

60. Gunsalus, R. P., Tandon, S. M., Wolfe, R. S. 1980. A procedure for anaerobic column chromatography employing an anaerobic Freter-type chamber. *Anal. Biochem.* 101:327–31

61. Gunsalus, R. P., Wolfe, R. S. 1977. Stimulation of CO_2 reduction to methane by methylcoenzyme M in extracts of *Methanobacterium*. *Biochem. Biophys. Res. Commun.* 76:790–95.

62. Gunsalus, R. P., Wolfe, R. S. 1978. ATP activation and properties of the methyl-coenzyme M reductase system in *Methanobacterium thermoautotrophicum*. *J. Bacteriol.* 135:851–57

63. Gunsalus, R. P., Wolfe, R. S. 1978. Chromophoric factors F342 and F430 of *Methanobacterium thermoautotrophicum*. *FEMS Microbiol. Lett.* 3:191–93

64. Gunsalus, R. P., Wolfe, R. S. 1980. Methyl coenzyme M reductase from *Methanobacterium thermoautotrophicum:* resolution and properties of the components. *J. Biol. Chem.* 255:1891–95

65. Hamilton, R. D., Wolfe, R. S. 1959. Pyruvate exchange reactions in *Bacillus macerans*. *J. Bacteriol.* 78:253–58

66. Hartzell, P. L., Donnelly, M. I., Wolfe, R. S. 1987. Incorporation of coenzyme M into component C of methylcoenzyme M methylreductase during in vitro methanogenesis. *J. Biol. Chem.* 262: 5581–86

67. Hartzell, P. L., Escalante-Semerena, J. C., Bobik, T. A., Wolfe, R. S. 1988. A simplified methylcoenzyme M methylreductase assay with artifical electron donors and different preparations of component C from *Methanobacterium thermoautotrophicum*. *J. Bacteriol.* 170: 2711–15

68. Hartzell, P. L., Wolfe, R. S. 1986. Comparative studies of component C from the methylreductase system of different methanogens. *Syst. Appl. Microbiol.* 7:376–82

69. Hartzell, P. L., Wolfe, R. S. 1986. Requirement of the nickel tetrapyrrole F_{430} for in vitro methanogenesis: Reconstitution of methylreductase component C from its dissociated subunits. *Proc. Natl. Acad. Sci. USA* 83:6726–30

70. Hartzell, P. L., Zvilius, G., Escalante-

Semerena, J. C., and Donnelly, M. I. 1985. Coenzyme F_{420} dependence of the methylene H_4MPT dehydrogenase. *Biochem. Biophys. Res. Commun.* 133:884–90

71. Hausinger, R. P., Orme-Johnson, W. H., Walsh, C. 1984. The nickel tetrapyrrole cofactor F_{430}: comparison of the forms bound to methyl-S coenzyme M methylreductase and protein-free in cells of *Methanobacterium thermoautotrophicum. Biochemistry* 23:801–4

72. Hedges, A., Wolfe, R. S. 1974. Extracellular enzyme from myxobacter AL-1 that exhibits both β-1, 4-gluconase and chitosanase activities. *J. Bacteriol.* 120:844–53

73. Heidelberger, M., Gaudy, E., Wolfe, R. S. 1964. Immunochemical identification of the aldobiuronic acid of the slime of *Sphaerotilus natans. Proc. Natl. Acad. Sci. USA* 51:568–69

74. Hermman, M., Noll, K. M., Wolfe, R. S. 1986. Improved agar bottle plate for isolation of methanogens or other anaerobes in a defined gas atmosphere. *Appl. Environ. Microbiol.* 51:1124–26

75. Hoyt, J. C., Oren, A., Escalante-Semerena, J. C., Wolfe, R. S. 1986. Tetrahydromethanopterin-dependent serine transhydroxymethylase from *Methanobacterium thermoautotrophicum. Arch. Microbiol.* 145:153–58

76. Ishiguro, E. E., Wolfe, R. S. 1970. Control of morphogenesis in *Geodermatophilus:* ultrastructural studies. *J. Bacteriol.* 104:566–80

77. Ishiguro, E. E., Wolfe, R. S. 1974. Induction of morphogenesis in *Geodermatophilus* by inorganic cations and by organic nitrogenous cations. *J. Bacteriol.* 117:189–95

78. Jackson, R. L., Wolfe, R. S. 1968. Composition, properties, and substrate specificities of Myxobacter AL-1 protease. *J. Biol. Chem.* 243:879–88

79. Jones, W. J., Donnelly, M. I., Wolfe, R. S. 1985. Evidence of a common pathway of carbon dioxide reduction to methane in methanogens. *J. Bacteriol.* 163:126–31

80. Jones, W. J., Guyot, J. -P., Wolfe, R. S. 1984. Methanogenesis from sucrose by defined immobilized consortia. *Appl. Environ. Microbiol.* 47:1–6

81. Jones, W. J., Leigh, J. A., Mayer, F., Woese, C. R., Wolfe, R. S. 1983. *Methanococcus jannaschii* sp. nov., an extremely thermophilic methanogen from a submarine hydrothermal vent. *Arch. Microbiol.* 136:254–61

82. Jones, W. J., Whitman, W. B., Fields, R. D., Wolfe, R. S. 1983. Growth and plating efficiency of methanococci on agar media. *Appl. Environ. Microbiol.* 46:220–26

83. Keltjens, J. T., Huberts, M. J., Laarhoven, W. H., Vogels, G. D. 1983. Structural elements of methanopterin, a novel pterin present in *Methanobacterium thermoautotrophicum. Eur. J. Biochem.* 130:537–44

84. Kluyver, A. J., Donker, H. J. L. 1926. Die Einheit in der Biochemie. *Chem. Zelle Gewebe* 13:134–90

85. Knight, M., Wolfe, R. S., Elsden, S. R. 1966. Synthesis of amino acids by *Methanobacterium omelianskii. Biochem. J.* 99:76–86

86. Kucera, S., Wolfe, R. S. 1957. A selective enrichment method for *Gallionella ferruginea. J. Bacteriol.* 74:344–49

87. Langenberg, K. F., Bryant, M. P., Wolfe, R. S. 1968. Hydrogen-oxidizing methane bacteria. II. Electron microscopy. *J. Bacteriol.* 95:1124–29

88. Leigh, J. A., Mayer, F., Wolfe, R. S. 1981. *Acetogenium kivui,* a new thermophilic hydrogen-oxidizing acetogenic bacterium. *Arch. Microbiol.* 129: 275–80

89. Leigh, J. A., Rinehart, K. L. Jr., Wolfe, R. S. 1983. Structure of methanofuran (carbon dioxide reduction factor) of *Methanobacterium. J. Am. Chem. Soc.* 106:3636–40

90. Leigh, J. A., Rinehart, K. L. Jr., Wolfe R. S. 1985. Methanofuran (carbon dioxide reduction factor), a formyl carrier in methane production from carbon dioxide in *Methanobacterium. Biochemistry* 24:995–99

91. Leigh, J. A., Wolfe, R. S. 1983. Carbon dioxide reduction factor and methanopterin, two coenzymes required for CO_2 reduction to methane by extracts of *Methanobacterium. J. Biol. Chem.* 258: 7435–40

92. Livingston, D. A., Pfaltz, A., Schreiber, J., Eschenmoser, A., Ankel-Fuchs, D., et al. 1984. Zur kenntnis des faktors F_{430} aus methanogenic bakterien: Struktur des proteinfreien faktors. *Helv. Chim. Acta* 67:334–51

93. Macy, J. M., Snellen, J. E., Hungate, R. E. 1972. Use of syringe methods for anaerobiosis. *Am. J. Clin. Nutr.* 25: 1318–23

94. McBride, B. C., Wolfe, R. S. 1971. A new coenzyme of methyl transfer, coenzyme M. *Biochemistry* 10:2317–24

95. McBride, B. C., Wolfe, R. S. 1971. Inhibition of methanogenesis by DDT. *Nature* 234:551–52

96. McBride, B. C., Wolfe, R. S. 1971. Biosynthesis of dimethylarsine by

Methanobacterium. Biochemistry 10: 4312–17

97. Miller, T. L., Wolin, M. J. 1974. A serum bottle modification of the Hungate technique for cultivating obligate anaerobes. *Appl. Microbiol.* 27:985–87

98. Mortenson, L. E., Valentine, R. C., Carnahan, J. E. 1962. An electron transport factor from *Clostridium pasteurianum. Biochem. Biophys. Res. Commum.* 7:448–52

99. Mortlock, R. P., Valentine, R. C., Wolfe, R. S. 1959. Carbon dioxide activation in the pyruvate clastic system of *Clostridium butyricum. J. Biol. Chem.* 234: 1653–56

100. Mortlock, R. P., Wolfe, R. S. 1959. Reversal of pyruvate oxidation in *Clostridium butyricum. J. Biol. Chem.* 234: 1657–58

101. Nagle, D. P. Jr., Wolfe, R. S. 1983. Component A of the methyl coenzyme M methylreductase system of *Methanobacterium:* Resolution into four components. *Proc. Natl. Acad. Sci. USA* 80:2151–55

102. Noll, K. M., Donnelly, M. I., Wolfe, R. S. 1987. Synthesis of 7-mercaptoheptanoylthreonine phosphate and its activity in the methylcoenzyme M methylreductase system. *J. Biol. Chem.* 262:513–15

103. Noll, K. M., Rinehart, K. L. Jr., Tanner, R. S., Wolfe, R. S. 1986. Structure of component B (7-mercaptoheptanoylthreonine phosphate) of the methylcoenzyme M methylreductase system of *Methanobacterium thermoautotrophicum. Proc. Natl. Acad. Sci. USA* 83: 4238–42

104. Noll, K. M., Wolfe, R. S. 1986. Component C of the methylcoenzyme M methylreductase system contains bound 7-mercaptoheptanoylthreonine phosphate (HS-HTP). *Biochem. Biophys. Res. Commun.* 139:889–95

105. Noll, K. M., Wolfe, R. S. 1987. The role of 7-mercaptoheptanoylthreonine phosphate in the methylcoenzyme M methylreductase system from *Methanobacterium thermoautotrophicum. Biochem. Biophys. Res. Commun.* 145: 204–10

106. Olson, K. D., Won, H., Wolfe, R. S., Hare, D. R., Summers, M. F. 1990. Stereochemical studies of coenzyme F_{430} based on 2D NOESY back calculations. *J. Am. Chem. Soc.* 112:5884–86

107. Pfaltz, A., Jaun, B., Fassler, A., Eschenmoser, A., Jaenchen, R., et al. 1982. Zur Kenntnis des Faktors F_{430} aus methanogenen Bakterien: Struktur des

porphinoiden Ligandsystems. *Helv. Chim. Acta* 64:828–65

108. Poirot, C. M., Kengen, S. W. M., Valk, E., Keltjens, J. T., van der Drift, C., Vogels, G. D. 1987. Formation of methylcoenzyme M from formaldehyde by cell-free extracts of *Methanobacterium thermoautotrophicum.* Evidence for the involement of a corinoid-containing methyltransferase. *FEMS Microbiol. Lett.* 40:7–13

109. Roberton, A. M., Wolfe, R. S. 1969. ATP requirement in methanogenesis in cell extracts of *Methanobacterium. Biochem. Biophys. Acta.* 192:420–29

110. Roberton, A. M., Wolfe, R. S. 1970. Adenosine triphosphate pools in *Methanobacterium. J. Bacteriol.* 102:43–51

111. Romesser, J. A. 1978. *Activation and reduction of carbon dioxide to methane in* Methanobacterium thermoautotrophicum. PhD thesis. Urbana: Univ. Ill. 183 pp.

112. Romesser, J. A., Wolfe, R. S. 1981. Interaction of coenzyme M and formaldehyde in methanogenesis. *J. Biochem.* 197:565–71

113. Romesser, J. A., Wolfe, R. S. 1982. CDR factor, a new coenzyme required for CO_2 reduction to methane. *Zentralbl. Bakteriol. Parasitenkd. Infektionskr. Hyg. Abt. 1: Orig.* C3:271–76

114. Romesser, J. A., Wolfe, R. S. 1982. Coupling of methylcoenzyme M reduction with carbon dioxide activation in extracts of *Methanobacterium. J. Bacteriol.* 152:840–47

115. Romesser, J. A., Wolfe, R. S., Mayer, F., Spiess, E., Walther-Mauruschat, A. 1979. *Methanogenium,* a new genus of marine methanogenic bacteria, and characterization of *Methanogenium cariaci* sp. nov. and *Methanogenium marisnigri* sp. nov. *Arch. Microbiol.* 121:147–53

116. Rouvière, P. E., Bobik, T. A., Wolfe, R. S. 1988. Reductive activation of the methylcoenzyme M methylreductase system of *Methanobacterium thermoautotrophicum. J. Bacteriol.* 170:3946–52

117. Rouvière, P. E., Escalante-Semerena, J. C., Wolfe, R. S. 1985. Component A2 of the methylcoenzyme M methylreductase system from *Methanobacterium thermoautotrophicum. J. Bacteriol.* 162: 61–66

118. Rouvière, P. E., Wolfe, R. S. 1987. Use of subunits of the methylreductase protein for taxonomy of methanogenic bacteria. *Arch. Microbiol.* 148:253–59

119. Rouvière, P. E., Wolfe, R. S. 1987.

2',3' dialdehyde of ATP: a specific, irreversible inhibitor of component A3 of the methylreductase system of *Methanobacterium thermoautotrophicum*. *J. Bacteriol.* 169: 1737–39

120. Rouvière, P. E., Wolfe, R. S. 1988. Novel biochemistry of methanogenesis. *J. Biol. Chem.* 263:7913–16

121. Rouvière, P. E., Wolfe, R. S. 1989. Component A3 of the methylcoenzyme M methylreductase system of *Methanobacterium thermoautotrophicum* ΔH: Resolution into two components. *J. Bacteriol.* 171:4556–62

122. Shapiro, S., Wolfe, R. S. 1980. Methylcoenzyme M, an intermediate in methanogenic dissimilation of C_1 compounds by *Methanosarcina barkeri*. *J. Bacteriol.* 141:728–34

123. Spormann, A. M., Wolfe, R. S. 1984. Chemotactic, magnetotactic and tactile behavior in a magnetic spirillum. *FEMS Microbiol. Lett.* 22:171–77

124. Tanner, R. S., Wolfe, R. S. 1988. Nutritional requirements of *Methanomicrobium mobile*. *Appl. Environ. Microbiol.* 54:625–28

125. Tanner, R. S., Wolfe, R. S., Ljungdahl, R. S. 1978. Tetrahydrofolate enzyme levels in *Acetobacterium woodii* and their implication in the synthesis of acetate from CO_2. *J. Bacteriol.* 134:668–70

126. Taylor, C. D., McBride, B. C., Wolfe, R. S., Bryant, M. P. 1974. Coenzyme M, essential for growth of a rumen strain of *Methanobacterium ruminantium*. *J. Bacteriol.* 120:974–75

127. Taylor, C. D., Wolfe, R. S. 1974. A simplified assay for coenzyme M ($HSCH_2CH_2SO_3$). *J. Biol. Chem.* 249: 4886–90

128. Taylor, C. D., Wolfe, R. S. 1974. Structure and methylation of coenzyme M ($HSCH_2CH_2SO_3$). *J. Biol. Chem.* 249:4879–85

129. TeBrömmelstroet, B. W., Hensgens, C. M. H., Keltjens, J. T., van der Drift, C., Vogels, G. D. 1990. Purification and properties of 5,10-methylenetetrahydromethanopterin reductase, a coenzyme F_{420}-dependent enzyme, from *Methanobacterium thermoautotrophicum* strain ΔH. *J. Biol. Chem.* 265: 1852–57

130. Twarog, R., Wolfe, R. S. 1962. Enzymatic phosphorylation of butyrate. *J. Biol. Chem.* 237:2474–77

131. Twarog, R., Wolfe, R. S. 1963. Role of butyryl phosphate in the energy metabolism of *Clostridium tetanomorphum*. *J. Bacteriol.* 86:112–17

132. Tzeng, S. F., Bryant, M. P., Wolfe, R. S. 1975. Factor 420-dependent pyridine nucleotide linked formate metabolism of *Methanobacterium ruminantium*. *J. Bacteriol.* 121:192–96

133. Tzeng, S. F., Wolfe, R. S., Bryant, M. P. 1975. Factor 420-dependent pyridine nucleotide linked hydrogenase system of *Methanobacterium ruminantium*. *J. Bacteriol.* 121:184–90

134. Uffen, R. L., Wolfe, R. S. 1970. Anaerobic growth of purple nonsulfur bacteria under dark conditions. *J. Bacteriol.* 104:462–72.

134a. Uffen, R. L., Wolfe, R. S. 1971. Mutants of *Rhodospirillum rubrum* obtained after long-term anaerobic dark growth. *J. Bacteriol.* 108:1348–56

135. Valentine, R. C., Bojanowski, R., Gaudy, E., Wolfe, R. S. 1962. Mechanism of the allantoin fermentation. *J. Biol. Chem.* 237:2271–77

136. Valentine, R. C., Brill, W. J., Wolfe, R. S. 1962. Role of ferredoxin in pyridine nucleotide reduction. *Proc. Natl. Acad. Sci. USA* 48:1856–60

137. Valentine, R. C., Brill, W. J., Wolfe, R. S., San Pietro, A. 1963. Activity of PPNR in ferredoxin-dependent reactions of *Clostridium pasteurianum*. *Biochem. Biophys. Res. Commun.* 10:73–78

138. Valentine, R. C., Drucker, H., Wolfe, R. S. 1964. Glyoxylate fermentation by *Streptococcus allantoicus*. *J. Bacteriol.* 87:241–46

139. Valentine, R. C., Jackson, R. L., Wolfe, R. S. 1962. Role of ferredoxin in hydrogen metabolism of *Micrococcus lactilyticus*. *Biochem. Biophys. Res. Commun.* 7:453–56

140. Valentine, R. C., Mortenson, L. E., Mower, H. F., Jackson, R. L., Wolfe, R. S. 1963. Ferredoxin requirement for reduction of hydroxylamine by *Clostridium pasteurianum*. *J. Biol. Chem.* 238:857–58

141. Valentine, R. C., Wolfe, R. S. 1960. Biosynthesis of carbamyl oxamic acid. *Biochem. Biophys. Res. Commun.* 2: 384–87

142. Valentine, R. C., Wolfe, R. S. 1960. Phosphorolysis of carbamyl oxamic acid. *Biochim. Biophys. Acta.* 45:389–91

143. Valentine, R. C., Wolfe, R. S. 1961. Glyoxylurea. *Biochem. Biophys. Res. Commun.* 5:305–08

144. Valentine, R. C., Wolfe, R. S. 1961. Phosphate-dependent degradation of urea. *Nature* 191:925–26

145. Valentine, R. C., Wolfe, R. S. 1963. Role of ferredoxin in the metabolism of

molecular hydrogen. *J. Bacteriol.* 85:1114–20

146. van Beelen, P., Stassen, A. P. M., Bosch, J. W. G., Vogels, G. D., Guijt, W., Haasnoot, C. A. G. 1984. Elucidation of the structure of methanopterin, a coenzyme from *Methanobacterium thermoautotrophicum,* using two-dimensional-nuclear-magnetic reasonance techniques. *Eur. J. Biochem.* 138:563–71

147. Deleted in proof

148. Vatter, A. E., Douglas, H. C., Wolfe, R. S. 1959. Structure of *Rhodomicrobium vannielii. J. Bacteriol.* 77:812–13

149. Vatter, A. E., Wolfe, R. S. 1956. Electron microscopy of *Gallionella ferruginea. J. Bacteriol.* 72:248–52

150. Vatter, A. E., Wolfe, R. S. 1958. The structure of photosynthetic bacteria. *J. Bacteriol.* 75:480–88

151. Whitman W. B., Ankwanda, E., Wolfe, R. S. 1982. Nutrition and carbon metabolism of *Methanococcus voltae. J. Bacteriol.* 149:852–63

152. Whitman, W. B., Wolfe, R. S. 1980. Presence of nickel in Factor F_{430} from *Methanobacterium bryantii. Biochem. Biophys. Res. Commun.* 92:1196–201

153. Whitman, W. B., Wolfe, R. S. 1983. Activation of the methylreductase system from *Methanobacterium bryantii* by ATP. *J. Bact.* 154:640–49

154. Whitman, W. B., Wolfe, R. S. 1987. Inhibition by corrins of the ATP-dependent activation and CO_2 reduction by the methylreductase system in *Methanobacterium bryantii. J. Bacteriol.* 169:87–92

155. Whitman, W. B., Wolfe, R. S. 1985. Activation of the methylreductase system from *Methanobacterium bryantii* by corrins. *J. Bacteriol.* 164:165–72

156. Widdel, F., Rouvière, P. E., Wolfe, R. S. 1988. Classification of secondary alcohol-utilizing methanogens including a new thermophilic isolate. *Arch. Microbiol.* 150:477–81

157. Widdel, F., Wolfe, R. S. 1989. Expression of secondary alcohol specific enzyme from *Methanogenium thermophilum* strain TCI. *Arch. Microbiol.* 152:322–28

158. Wingard, M., Matsueda, G., Wolfe, R. S. 1972. Myxobacter AL 1 protease II: specific peptide bond cleavage on the amino side of lysine. *J. Bacteriol.* 112:940–49

159. Winter, J., Wolfe, R. S. 1980. Methane formation from fructose by syntrophic associations of *Acetobacterium woodii* and different strains of methanogens. *Arch. Microbiol.* 124:73–79

160. Winter, J., Wolfe, R. S. 1979. Complete degradation of carbohydrate to carbon dioxide and methane by syntrophic cultures of *Acetobacterium woodii* and *Methanosarcina barkeri. Arch. Microbiol.* 121:97–102

161. Woese, C. R., Fox, G. E. 1977. Phylogenetic structure of the prokaryotic domain: the primary kingdoms. *Proc. Natl. Acad. Sci. USA* 74:5088–90

162. Woese, C. R., Kandler, O., Wheelis, M. L. 1990. Towards a natural system of organisms: proposal for the domains Archaea, Bacteria, and Eucarya. *Proc. Natl. Acad. Sci. USA* 87:4576–79

163. Woese, C. R., Wolfe, R. S., eds. 1985. *The Archaebacteria*, Vol. 8, *The Bacteria*, ed. I. C. Gunsalus, J. R. Sokatch, L. N. Ornston. Orlando: Academic. 582 pp.

164. Wolfe, R. S. 1958. Cultivation, morphology and classification of the iron bacteria. *J. Am. Water Works Assoc.* 50:1241–49

165. Wolfe, R. S. 1960. Observations and studies of *Crenothrix polyspora. J. Am. Water Works Assoc.* 52:915-18

166. Wolfe, R. S., O'Kane, D. J. 1953. Cofactors of the phosphoroclastic reaction of *Clostridium butyricum. J. Biol. Chem.* 205:755–65

167. Wolfe, R. S., O'Kane, D. J. 1955. Cofactors of the carbon dioxide exchange reaction of *Clostridium butyricum. J. Biol. Chem.* 215:637–43

168. Wolfe, R. S., Pfennig, N. 1977. Reduction of sulfur by spirillum 5175 and syntrophism with *Chlorobium. Appl. Environ. Microbiol.* 33:427–33

169. Wolfe, R. S., Thauer, R. K., Pfenning, N. 1987. A "capillary racetrack" method for isolation of magnetotactic bacteria. *FEMS Microbiol. Ecol.* 45:31–35

170. Wolin, E. A., Wolfe, R. S., Wolin, M. J. 1964. Viologen dye inhibition of methane formation by *Methanobacillus omelianskii. J. Bacteriol.* 87:993–98

171. Wolin, E. A., Wolin, M. J., Wolfe, R. S. 1963. Formation of methane by bacterial extracts. *J. Biol. Chem.* 238:2882–86

172. Wolin, M. J., Wolin, E. A., Wolfe, R. S. 1963. ATP-dependent formation of methane from methylcobalamin by extracts of *Methanobacillus omelianskii. Biochem. Biophys. Res. Commun.* 12:464–68

173. Wolin, M. J., Wolin, E. A., Wolfe, R. S. 1964. The cobalamin product of the conversion of methylcobalamin to CH_4 by extracts of *Methanobacillus omelianskii. Biochem. Biophys. Res. Commun.* 15:420–23

174. Won, H., Olson, K. D., Wolfe, R. S., Summer, M. F. 1990. Two-dimensional NMR studies of native coenzyme F_{430}. *J. Am. Chem. Soc.* 112:2178–84

175. Wood, J. M., Allam, A. M., Brill, W. J., Wolfe, R. S. 1965. Formation of methane from serine by extracts of *Methanobacillus omelianskii*. *J. Biol. Chem.* 240:4564–69

176. Wood, J. M., Wolfe, R. S. 1965. Formation of CH_4 from N^5-methyltetrahydrofolate monoglutamate by cell extracts of *Methanobacillus omelianskii*. *Biochem. Biophys. Res. Commun.* 19:306–11

177. Wood, J. M., Wolfe, R. S. 1966. Alkylatin of an enzyme in the methane-forming system of *Methanobacillus omelianskii*. *Biochem. Biophys. Res. Commun.* 22:119–23

178. Wood, J. M., Wolfe, R. S. 1966. Components required for the formation of CH_4 from methylcobalamin by extracts of *Methanobacillus omelianskii*. *J. Bacteriol.* 92:696–700

179. Wood, J. M., Wolfe, R. S. 1966. Propylation and purification of a B_{12} enzyme involved in methane formation. *Biochemistry* 5:3598–603

180. Wood, J. M., Wolin, M. J., Wolfe, R. S. 1966. Formation of methane from methyl Factor B and methyl Factor III by extracts of *Methanobacillus omelianskii*. *Biochemistry* 5:2381–84

181. Zeikus, J. G., Wolfe, R. S. 1972. *Methanobacterium thermoautotrophicus* sp.n., an anaerobic, autotrophic, extreme thermophile. *J. Bacteriol.* 109:707–13

182. Zeikus, J. G., Wolfe, R. S. 1973. Fine structure of *Methanobacterium thermoautotrophicus*: effect of growth temperature on morphology and ultrastructure. *J. Bacteriol.* 113:461–67

Annu. Rev. Microbiol. 1990. 44:1–25

A LIFE WITH BIOLOGICAL PRODUCTS[1]

Margaret Pittman

Center for Biologics Evaluation and Research, HFB-210, Food and Drug Administration, 8800 Rockville Pike, Bethesda, Maryland 20892

KEY WORDS: development of potency and safety tests for bacterial vaccines and antisera, standards for biological products, sterility test for biologics

CONTENTS

In 1925, I took my first course in bacteriology. The time was propitious. In the later part of the nineteenth century, bacteriology became a distinct biologic science. The germ theory of disease had been proven and the causative organisms of many diseases had been isolated. During the first quarter of this century, many investigations focused on the role of bacteria and their antigens in infectious diseases and their immunological prevention and treatment.

In my 64 years of association with bacteria, bacteriology came of age and

[1]Dedicated to Dr. Roderick Murray (1909–1980).

0066-4227/90/1001-0001$02.00

recently entered the molecular biological phase. It has been my high privilege to participate in the development and standardization of biological products both nationally and internationally to insure their safety, purity, and potency. My emphasis has been on bacterial vaccines, and antisera and general standards.

During these years, it was my good fortune to become acquainted with outstanding pioneer scientists worldwide who were devoted to problem-solving of the mysteries of how bacteria effect infectious diseases and their control, and the improvement of public health in general. I am deeply indebted to them for their stimulus and help, as well as to associates, and my staff scientists and technicians.

EARLY DAYS IN ARKANSAS

I was born January 20, 1901 in a Christian home where discipline, truth, love, and spiritual growth shaped my adult life. Both parents descended from early British emigrants. The name Pittman appeared in New England as early as 1653. In 1834, Samuel settled on a farm in the Prairie Grove valley of the Ozark mountains in northwest Arkansas. My father, James Pittman, grandson of Samuel, was born in 1871 in a large log cabin that was the first post office in the valley. He was the first of seven children. His higher education was at the University of Arkansas, in nearby Fayetteville, and St. Louis Medical College. Work as a street car conductor helped defray expenses.

My mother, Virginia Alice McCormick, the youngest of seven children, was born in 1871 in Churchville in the Shenandoah Valley of Virginia. She was a cousin of Cyprus Hall McCormick, inventor of the harvester reaper. Her parents migrated via the Ohio, Mississippi, and Arkansas rivers to Evansville, Arkansas in the Ozarks. There her father operated a sawmill. Attracted to the sawmill by the machinery, Mother learned to identify trees by bark on the logs. Later, the family moved to Prairie Grove where her parents contracted typhoid fever and died. Mother had many talents: mechanical ability was used to repair her sewing machine and the brass front 1914 Ford car (there was no garage in the township). She had a "green thumb," was very observant and made logical interpretations, e.g. weather forcasts or the character of a person; was an excellent seamstress; did fancy needle work and oil painting; was hospitable, a quiet leader, and promoter of education; kept a small lending library; and taught rudiments of piano playing.

After marriage in 1899, Father began medical practice in Allen, Cherokee Nation (Oklahoma). Malaria infection influenced their return to Prairie Grove within the year. They had three children. About 1909, the family moved to Cincinnati (114), a country village that had been a large Indian trading center before the Frisco Railroad was constructed nearby. Father was the only doctor for miles around. He responded to calls, day or night, sunshine or rain, a true horse and buggy doctor (34) who passionately served all people. One time a cholecytectomy was performed on a patient without funds in a room of our house.

My first medical experience was administration of anesthesia during the setting of a fractured bone. The ether was poured on cotton placed over an openly woven egg beater and held over the patient's nose. Then, when a federal law required that all children be vaccinated against smallpox, I served as assistant in the vaccination of school children in about 10 rural schools.

My early education was in a two room school and at home. We were taught the fundamental subjects; most importantly, the motivation to succeed and discipline.

At home there was oral reading of current and classic novels and the Bible at night by kerosene lamp light. College education was considered essential.

Three years of high school was in Prairie Grove. I lived with grandparents on their farm. Then I spent two years at a newly organized college in Siloam Springs, 12 miles north of Cincinnati with a focus on piano music. Long hours of practice precipitated a spinal complication that has been a life companion. Music study was discontinued. This was good; I was not truly talented.

In December 1919, my father died after an appendectomy from peritonitis, said to be caused by infected catgut. On his death bed, he requested Mother to take the three children to Hendrix College, a small accredited Methodist College with a ratio of ten boys to one girl. Available money was supplemented with canned vegetables and fruit that were prepared during the summer and with Mother's sewing. Each child graduated with honors and, after graduate study, entered into medical work. My sister Helen became a public health nurse and teacher; my brother James, a general surgeon; and myself, a microbiologist engaged in the federal control of biologic products.

At Hendrix, I majored in mathematics and was awarded the college Mathematic Medal. One year, I taught a class in algebra. I also majored in biology. Graduation, *magna cum laude,* was in 1923. The dedicated professors stimulated and motivated students to excel at college and throughout life. After serving for two years as teacher and principal of the Academy of Galloway Female College, I went to the University of Chicago for graduate study to prepare for some form of medical work. Funds for medical school were not available.

THE UNIVERSITY OF CHICAGO, 1925–1928

I arrived at the University of Chicago in late August 1925 with $600 saved from the $2100 received for the two years of teaching. Without prior arrangements, I was accepted as a candidate for a MS in the Department of Bacteriology and Hygiene. Payment of the tuition of $75 per quarter was supplemented with funds received for babysitting at night at 35¢ per hour.

Bacteriology was a relatively new and exciting biologic science. Status of development in the field was well described by 83 contributors to *Newer Knowledge of Bacteriology and Immunology* edited by Jordan & Falk (38).

Dr. Edwin Oakes Jordan (1866–1930) (42) was selected by President Harper as one of the original faculty members of the University, in the Department of Zoology. He was a student of William Thompson Sedgwith, a leader of his day in applied biology and sanitation (100). He taught first a course in Sanitary Biology and, in 1894–95, courses in General Bacteriology. In 1900, bacteriology became a part of the Department of Pathology and Bacteriology. A number of students became outstanding leaders. In 1913, the Department of Bacteriology and Hygiene was created with Dr. Jordan as Chairman (42).

The spirit of research permeated the Department. As a student it was my high privilege to be associated with the scholarly faculty and the graduate students that were making outstanding contributions. The early days are well described in a history of bacteriology at the University by Koser (42).

My courses and research in this Department and in other Departments of the University provided a fitting foundation for post-graduate work. I was impressed with a course in vital statistics. The reference book was the PhD thesis of I. S. Falk (25) who received this degree at the age of 23. Research on the pneumococcus was done under his supervision. My MS thesis project was on the S and R colonial forms of the pneumococcus and their pathogenicity (89).

A few months before I was to receive the MS degree, the department offered me a fellowship to continue my studies for a PhD: an opportunity undreamed of. After the 1918–19 influenza pandemic, the Metropolitan Life Insurance Company had provided funds for two fellowships for investigations on respiratory diseases. The $75 received per month paid for tuition and one good meal per day.

Courses in the Department of Bacteriology and Hygiene were supplemented with very valuable courses in the Medical School. They included biochemistry, human anatomy dissection (thorax and abdomen), histology, and pathology. My thesis research was directed by Professor Mercy A. Southwith in the Department of Pathology. The title was Pathogenesis of Experimental Pneumococcus Lobar Pneumonia (97).

In the fall of 1928, I left the University to work with Dr. Rufus Cole at the Hospital of the Rockefeller Institute for Medical Research (RIH) (predecessor of the Rockefeller University). Dr. C. P. Miller, Department of Medicine, recommended my appointment after a consultation with the bacteriology department. I returned to Chicago during my 1929 vacation, defended my thesis—written in New York after work hours. The PhD was conferred at the August Convocation.

THE ROCKEFELLER INSTITUTE FOR MEDICAL RESEARCH, 1928–1934

It was a privilege and great opportunity to spend six years of post graduate work at the hospital, a department of the Rockefeller Institute (RI). My life with biological products started here. The hospital was established so that disease could be studied as it actually occurs (102). It opened in 1910 with Dr. Rufus Cole as Director. Pioneers had made highly significant contributions, especially on the characteristics of the pneumococcus and treatment of pneumococcus lobar pneumonia.

In the Acute Respiratory Group, prudently led by O. T. Avery (from 1913 until retirement in 1948), there was unity of purpose and close cooperation of the staff members. How fortuitous that I was associated with this group when I discovered the capsule of *Haemophilus influenzae* and the "S" and "R" colonies of *H. influenzae* that were similar to the colonies of the pneumococcus that I had studied in Chicago and that had been extensively studied by Avery et al (4).

When I went to the RIH, the role of *H. influenzae* as the cause of influenza had not been definitely settled. My project was to determine if *H. influenzae* is the cause of influenza. Numerous studies on this organism had been made during the 1918–19 catastrophic influenza epidemic. In his 1926 comprehensive review of about 1100 articles, Jordan (37) "concluded that it could not be deducted that *H. influenzae* was the cause or not the cause of influenza." The most encouraging results had been reported by Blake & Cecil (10, 19). Blake had been on the staff of RIH and was closely associated with RI. A strain of *H. influenzae* isolated from empyema fluid of a case of influenza, passed in mice and monkeys, induced a severe respiratory infection in monkeys that clinically resembled influenza. The authors, however, cautiously declined to infer that the *H. influenzae* strain used was the cause of influenza.

H. influenzae was not a stranger at the Hospital. A number of studies, published by Stillman (110) and by Rivers (101), did not promise supporting evidence.

With this background, it was not clear where I should start. I began by getting acquainted with the characteristics of the 28 cultures that had been collected. Out of floundering experiments emerged the discovery that some strains of *H. influenzae* were capsulated (62, 63).

The Capsule of H. influenzae

I had observed that the hemophilic bacteria were lysed by serum from laboratory animals, that the bacterial strains differed in susceptibility, and that the serum of animal species differ in activity. Guinea pig serum was especially active.

In order to compare the susceptibility of different strains, the transparent medium of Levinthal & Fernbach (43), which they used to compare the colonies of the Pfeiffer bacillus and the Koch-Weeks bacillus, was selected to determine if all bacteria were killed. My able technician Salvatorus Spatolli and I simplified preparation of the medium by combining equal parts of filtered boiled 10% blood broth and 4% melted nutrient agar instead of filtering chocolate agar.

In the very first test with the medium, encapsulated *H. influenzae* was discovered. Two types of colonies were demonstratable on a plate culture from the lytic test of culture no. 35. Both were much larger than on blood or chocolate agar. One colony, "S," was opaque, mucoid, and iridescent in obliquely transmitted strong light. The other colony, "R," was smaller in size, faintly rough, more translucent, and had a bluish sheen in transmitted light.

Two more cultures in the collection had S and R forms. These three cultures had been carried for 5 to 9 months since isolation. The first culture studied, no. 35, had been isolated from the heart blood of a mouse inoculated with sputum from a case of bronchopneumonia by the routine procedure used at RIH for the isolation of the pneumococcus (5). The second culture was from a case of meningitis and the third from the blood of a case of bronchopneumonia. The latter two cultures differed serologically from no. 35. No. 35 was designated type a and the other two type b (63).

The project on the causal relation of *H. influenzae* to influenza was discontinued. Emphasis was placed on the determination of the incidence of encapsulated *H. influenzae*, their serotype specificity and relation to pathogenesis, and on the production of type b antiserum and its efficacy.

During a period of less than four years, 521 hemophilic cultures were isolated largely from the sputum of cases of pneumonia admitted to RIH for research on cause and prevention of pneumonia. Some cultures were from outside sources. Only a few were encapsulated. I identified six serotypes. Type b was dominant and the most pathogenic. I found only one strain each of type d and type e. Peculiarly, no additional types have been found during more than 50 years. My study was intensive. Three media and the mouse were used for the isolation of the bacteria in each sputum specimen from many patients. This study showed that non-type-specific (nontypable) *H. influenzae* was pathogenic. Avery et al (5) had pointed out the association of *H. influenzae* with the pneumococcus microscopically and in the heart blood of the mouse inoculated with sputum. Pathogenicity of *H. influenzae* was shown during the 1918–19 influenza pandemic. The influenza bacillus was isolated in pure culture from lungs of patients at autopsy but not from the blood (115). This lack of blood invasion and the difference in rabbit pathogenicity of these bacteria from cases of influenza and cases of meningitis (116) confirm that *H. influenzae* strains associated with influenza were not encapsuated (M. Pittman, review in preparation). Nevertheless, there has been some confusion that nontypable *H. influenzae* is an "R" form that has lost most of its original capsule (105). Recently, Hoiseth & Gilsdorf (36) provided evidence that nontypable organisms lack genetic ability to be encapsulated. Murphy & Apicello have pointed out the role of this nonencapsulated pathogen (52).

Production of Type b Antiserum

The discovery that the strains of *H. influenzae* from meningitis patients were usually type-specific and of one type (type b), suggested that a highly immune serum might have therapeutic value. Earlier attempts by other investigators did not find *H. influenzae* antiserum effective (cf. 64). Notwithstanding, in the light of newer knowledge, a horse was immunized. Type-specific anti-pneumococcus serum (horse) was in use in the hospital for treatment of pneumococcus pneumonia. To produce the *Haemophilus influenzae* type b antiserum, I followed in principle the method used to produce the pneumococcus antiserum at RI (6).

During immunization of the horse, type-specific antibody progressively rose, as shown by the precipitation of type b carbohydrate purified by W. F. Goebel, anti-infectious action in mice and rabbits, and prevention of skin lesions in rabbits induced by type b bacteria (64).

The therapeutic action of the antiserum was examined by cooperative physicians outside of RIH. In spite of widely varying conditions, their results indicated, as did the laboratory results, that the serum had definite anti-infectious activity. One patient recovered. Septicemia of the treated patients was promptly cleared and spinal fluid clearance might last up to 14 days. Then the bacteria returned and the patients died.

Horse antiserum has a deficiency in complement activity. Pittman & Goodner (94) showed that the type b antiserum did not fix complement in the presence of type b, capsular carbohydrate. With the bacterial protein, complement was fixed. In contrast, rabbit antiserum in the presence of both the carbohydrate and protein fixed complement. The agglutinin titer with type b bacteria did not differ between the two immune sera: neither did the precipitin titer with capsular carbohydrate.

Rabbit antiserum treatment was introduced by Alexander (2). Subsequently, therapy with antiserum was replaced by sulfonamide compounds (67) and antibiotics (35). We (35, 67) showed experimentally that both agents had antibacterial action on *H. influenzae*. Currently it appears that type b carbohydrate conjugated with a protein antigen will contribute to the prevention of type b infection (104).

TAXONOMY OF GENUS *HAEMOPHILUS* AND GENUS *BORDETELLA*

The study of the characteristics of more than 500 hemophilic cultures relative to incidence of encapsulated strains and related publications led to an invitation to prepare "Genus *Hemophilus* Winslow et al 1917" for *Bergey's Manual of Determinative Bacteriology,* editions 5 (1939), 6 (1948), and 7 (1957). In the 7th edition, *H. aegyptius* (Trevisan, 1899) Pittman & Davis, 1950 was listed as a separate species, and *H. pertussis* with related species was transferred to the new genus *Bordetella.* I prepared the chapter, "Genus *Bordetella Moreno-Lópes* 1952, 178[AL]," for *Bergey's Manual of Systematic Bacteriology* (84). As a member of the International Committee on Genus *Haemophilus,* I was closely associated with national and international culture collections, in which I deposited cultures. These worldwide contacts broadened my information on infectious diseases. Cultures were received from local, national, and international sources, often with a request for an opinion on classification. The first cultures were the hemophilic bacteria isolated from pigs with swine influenza by Dr. Richard Shope (108). The bacteria and virus acted symbiotically. Alone they were not pathogenic. Later, cultures from swine were received from Europe. According to my records, the first and latter cultures were different. There were also cultures from fish and turkeys as well as from humans. The cultures isolated from cases of epidemic acute conjunctivitis in the Rio Grande Valley of Texas (22, 88) were of special interest from the standpoint of the clinical disease, epidemiology, and taxonomic classification.

Classification of the Koch-Weeks Bacillus

Since the primary isolations of the Koch-Weeks and the Pfeiffer bacilli, opinions on their taxonomic relationship have varied (cf. 88). One confounding factor was the custom of designating all *Haemophilus* bacteria from conjunctivitis the Koch-Weeks bacillus, at least in New York City. Cultures submitted to me at the RIH did not differ from the non-type-specific *H. influenzae* that I was studying. Later, Dr. Dorland J. Davis and I studied cultures from epidemic acute conjunctivitis that differed from *H. influenzae.*

In 1947, Dr. Davis was requested by the US Public Health Service to investigate conjunctivitis in the Rio Grande Valley in Texas (22). Cultures sent to me were hemophilic but within two weeks I had a vague feeling that they differed from any *H. influenzae* that I had ever examined. The following year we isolated both the "Koch-Weeks bacillus" and the influenza bacillus from children with conjunctivitis (88). One helpful criteria in the field was that the colony of the Koch-Weeks bacillus in semi-solid media was small and fluffy with a comet-like tail. In contrast, the colony of nontypable *H. influenzae* is small and granular, while the colony of encapsulated bacteria is large and fluffy (88).

Before publication of the identifying characters of the bacillus as *Haemophilus aegyptius* we consulted authorities on systematic bacteriology (M. Pittman, personal letters). *Haemophilus aegyptius* is on the Approved List of Bacterial Names (109) and Kilian & Biberstein (41) used this name in *Bergey's Manual of Systematic Bacteriology*. In 1982, Mazloum et al (47) noted differential criteria.

Phylogenetic studies, however, have failed to show a difference in the DNA of the two organisms (29) and the name *Haemophilus influenzae* biotype aegyptius is used by some scientists (33). The gene that effects the marked differences in pathogenicity has not been identified. *H. aegyptius* infection is limited to the conjunctiva and occurs only during the hot climate breeding season of the "sore eye" gnat or fly (9). Whereas *H. influenzae* infections occur in many sites, eye infection is an extension from the nasopharynx (88). Between epidemics no carrier has been found (3). Is there a nonculturable stage? Colwell et al (20) reported that from Bangladesh mud no *Vibrio cholerae* grew on culture media but they were recovered from inoculated animal and humans. *Campylobacter jejuni* also has a nonculturable stage (103). It appears that *H. aegyptius* and *H. influenzae* are engaged but marriage has not, and probably will not be consummated.

THE NATIONAL INSTITUTES OF HEALTH, 1936–1972

Regulations of Biological Products

The regulations of biologicals in the United States began with the Congressional Biologics Control Act in 1902. The purpose of the Act was to insure the continued safety, purity, and potency of all vaccines, toxins, antitoxins, therapeutic serums, or analogous products applicable to the prevention, treatment, or cure of diseases or injuries (83). The year before, 10 children had died from tetanus after treatment with diptheria antitoxin.

The National Institute of Health (successor of the Hygienic Laboratory) was designated to regulate the Control Act. This responsibility was transferred to the Food and Drug Administration in 1972. I prepared a short history of the Regulation of Biological Products, 1902–1972 for the NIH Centennial Activities (1987) (83). Developments in regulations were made possible by advancements in knowledge. I was Chief, Laboratory of Bacterial Products, Division of Biologics Standards from 1957 until official retirement in 1971. By invitation, I have remained as a Guest Worker without official responsibilities or pay. Apparently, I was the first woman to be named chief of a NIH laboratory and among the first to be promoted to super grade GS-16.

The story of the Biologic Control Laboratory is a story of an amazing safety record, as told in 1950, by R. D. Goldthrope, Scientific Branch, NIH, (unpublished). "The painstaking work and perpetual vigilance of its small group of scientists and technicians have protected the lives of millions of sick and well Americans." I was privileged to be a member of this group. The group remained small, not above 50, until the introduction of tissue culture of viruses for viral vaccines made a larger staff necessary. The Division of Biologics Standards was established at NIH in 1955 to meet the growing needs. Dr. Roderick Murry ably served as Director until retirement in 1972. He published an excellent and comprehensive view of the activity of DBS in 1968 (53).

My experiences in the development of requirements and standards for bacterial products and general requirements include the collateral research that supported the development and improvement of standards for bacterial vaccines.

REQUIREMENTS FOR BACTERIAL ANTISERA *Meningococcus Antiserum* In 1936, when the US was coming out of the Great Depression, funds from the Social Security Act were made available to expand medical research. I was among the microbiologists employed. My assignment was to work with Sara E. Branham (14). She had been one of my instructors at the University of Chicago. The project was to develop a potency test for meningococcus antiserum (horse) that was less efficacious in the current meningitis epidemic than in earlier epidemics (13).

A mouse potency assay was developed (15, 16) but never promulgated. The newly introduced sulfonamide derivatives were more efficacious than this antiserum (13). However, much was learned about mouse potency assay techniques (66). Most important was the application of the Petri technique to quantitate the level of specific meningococcus antibodies (86). With 100 lots of antiserum, there was a 94% correlation between the mouse protection assay value and the precipitation estimate (68). The discrepancy was obtained with a serotype of the meningococcus, which at that time had not been clearly defined. Some 30 years later, Mancini (45), unaware of our study (68), referred to a practical method for quantitation of antigens by the Ouchterlony technique. Our precipitin test has been adopted for the isolation of *H. influenzae* (105) and *Salmonella typhi* (59).

Study of *Neisseria meningitis* was facilitated by the procurement in 1937 of the first freeze-dry apparatus at NIH. Dr. Branham had >100 cultures that had to be transferred at least twice a week. Cultures of *H. influenzae* that I dried at that time are still viable.

Haemophilus influenzae **antiserum** Requirements were also developed for type b *H. influenzae* antiserum (rabbit): 1000 units/mg of capsular carbohydrate was specified. Then production of the therapeutic rabbit antiserum was discontinued. Sulfonamide derivatives were antibacterial against both type b (67) and non-type-specific bacteria (65). Later, antibiotics were used (35). To assure that the capsular serotype of diagnostic sera were properly labeled, requirements were issued. The quellung or precipitin test was specified and type-specific strains were supplied.

Bordetella pertussis **antiserum** Beginning with the first trial use of pertussis antiserum therapy (12), the results of treatment and prophylaxis with various antipertussis products were enigmatic (93). Nevertheless, an intracerebral challenge potency assay was developed and a Standard Antipertussis Serum (rabbit) was designated. The assigned unit value was based on the Netherlands unit of Antipertussis Reference. Manufacturers of antipertussis serum products were notified of the US Reference Standard Lot 2 and of the proposed potency assay. The realization that the clinical symptoms of pertussis were effected by pertussis toxin (80) cast light upon the variable results obtained with pertussis antiserum and the nontherapeutic effect obtained with pertussis immune globulin (93). Like with other bacterial-toxin diseases, pertussis antitoxin is not therapeutic after the toxin has attached to cells and altered their regulatory function (113).

Serotype antibodies of *B. pertussis* were considered essential for immunity against whooping cough (98). Their role in prolonged immunity in humans, however, remains uncertain. To insure that the types were represented in pertussis vaccine, Dr. Grace Eldering, an authority on typing (23), prepared, on contract, type specific antiserum for each of the six types. She also typed each of the production *B. pertussis* strains used by each manufacturer (G. Eldering, unpublished data). The role of agglutinogens in prevention of pertussis in humans remains controversial.

REQUIREMENT FOR BACTERIAL VACCINES Following the early successful treatment with diphtheria antitoxin, attention turned to use of antisera and vaccines prepared from nontoxin producing bacteria. Licenses were issued for bacterial antisera, followed by vaccines against practially all known pathogenic bacteria. Today, the Code of Federal Regulation (27) specifies specific requirements (Additional Standards) for only five bacterial vaccines—pertussis, typhoid, cholera, anthrax, and BCG.

Although concerned with regulations of all bacterial vaccines at one time or another, my research on both the development of potency assays and participation in the correlation of potency with human efficacy was largely limited to pertussis, typhoid, and cholera vaccines.

Pertussis vaccine   In 1943, Dr. Milton V. Veldee, Director, Biologics Control Laboratory, requested that I develop a potency assay for pertussis vaccine. Others had failed, including Dr. J. W. Hornibrook, formerly assigned to this laboratory, and Dr. Pearl Kendrick, Michigan State Bureau of Laboratories. Both had tried the intranasal route of challenge and Kendrick had also tried the intraperitoneal route (40).

My acquaintance with *Bordetella pertussis* was largely limited to whooping cough at age 4 years and the confusing literature I had reviewed in 1926 for a term paper at the University of Chicago. Where should I start? The answer came from Dr. John Foote Norton, one of my teachers at the University of Chicago. He had observed encouraging pertussis-vaccine protection of mice against the intracerebral (ic) route of challenge. He and Dr. John Dingle (59a), both then at Upjohn Laboratories, were using this route in experimental potency tests of typhoid vaccine. In January, 1944, Dr. Norton suggested to Dr. Kendrick and to me that we try this route of challenge in potency testing of pertussis vaccine. Results were good. We worked independently but with free exchange of developments. Dr. Kendrick used two doses of vaccine to immunize mice (40) and I used one dose to avoid an inflated estimate of potency (57).

Manufacturers were informed of progress in 1945 (55), and in January, 1946, a tentative mouse protection test was sent to them (56). This test became official when Minimum Requirements were issued in 1948 and became effective on January 1, 1949 (57). Manufacturers quickly made adjustments in the preparation of their product. Samples that I tested in 1945–46 showed that three products had no potency and others varied as much as 10 fold within and between products.

Standardization of pertussis vaccine   1. Opacity. The first step in the development of the potency test was to prepare an opacity standard to be used in estimating the bacterial content of the vaccine and of the challenge culture for the potency test. Together, Dr. Kendrick and I adjusted the turbidity of a suspension of Pyrex glass particles (18) to be equivalent to that of a specified number of bacteria in an aged vaccine determined by direct count (78). The preparation was designated as the US Opacity Standard, and subsequently it was used as the International Opacity Reference Preparation (120), with an assigned value of 10 units without bacterial equivalence. The use of this Standard reduced significantly the variation in the turbidity of vaccines claimed to have equal numbers of bacteria (78).

2. Potency. From the beginning of the use of bacterial vaccine, the human dose was expressed in numbers of bacteria. Hence, the potency of the bacteria was estimated, which was not good. The number of bacteria per total human immunizing dose (THD) varied between products. Adjustment was made by

specifying that the THD be 12 units (74, 75). The value of 12 units was estimated to be equivalent to the potency of the vaccine used by Kendrick & Eldering (39) that was protective in a field study. *Minimum Requirements: Pertussis Vaccine* were revised effective May 1, 1953 (74).

To determine if the mouse potency estimate reflected human protection and to assess the efficacy of 12 U per THD, a field trial was designed. No money was available for a trial. Felton & Verwey (26) were the first to show with a noncellular, nontoxic vaccine (15 U/THD), the relationship between unitage and human protection. The extensive field trials of the Medical Research Counsel of England showed that there was a direct correlation between mouse potency and protection against home exposure (48). My estimate of the unitage of some of the trial vaccines, determined relative to the US Standard Pertussis Vaccine Lot 4, showed that unitage reflected the level of protection against home exposure (76). In the United States and other countries, efficacy has been $\geq 90\%$.

The International Standard Pertussis Vaccine was calibrated against the US Standard Vaccine and the requirements followed largely the US requirements except not less than 4 PU per single dose was specified (123). The US requirement has an upper limit on potency.

3. Statistics. At the time the potency test was being developed, statistical methods were not in general use by manufacturers to estimate the ED_{50} from the results of the test. The manufacturers were provided with the Wilson-Worcester Method (58).

Pittman & Lieberman (96) had analyzed the results of a large number of protection tests using the methods of Wilson & Worcester and of Reed & Muench in comparison with the Probit estimates. The Wilson-Worcester estimates of the ED_{50} were as exact as and consistent with the Probit estimates. The formulae of the former method are simpler, calculation less complicated and less time consuming. The Reed-Muench estimates were not entirely consistent with the Probit estimates.

The *Memorandum* revised in 1956 (58), with two tables from which the $ED_{50} \pm SD$ may be obtained directly when 16 or 32 mice are used per immunizing dose, were widely used until computers were readily available. They were reproduced by the World Health Organization (WHO) and distributed worldwide. They are still being used in some countries (30).

4. Toxicity and Antigenicity. Before isolation, *B. pertussis* showed one of its abnormalities, nongrowth on medium containing peptone. The next anomaly was that rupture by *B. pertussis* caused high parental fatality and local dermonecrosis in animals (11), whereas intact bacteria were not dermonecrotic (12). In 1951, Gordon & Hood (28) pointed out many epidemiological abnormalities.

Relative to NIH responsibility for the requirements for pertussis vaccine, I

PITTMAN

participated in a number of studies of reactivity. These and studies of others (cf. 81) pointed out the nature of the two-stage disease and of the reactions in animals. Reactions included increased susceptibility to a number of agents (histamine, serotonin, endotoxin, and bacterial infections); metabolic alterations (hypoglycemia, refractoriness to epinephrine hyperglycemia, and hyperinsulinemia); and potentiation of immune response to protein antigens (cf. 80). Four or more toxins had been described. Did one or more toxins effect the multiple responses?

I was a Guest Scientist at the University of Glasgow and was invited to present a paper entitled "Antigenic Specificity of Bacterial Infections" at the seminar on Epidemiology of Communicable Diseases, May 10, 1977. Some bacteria had a capsule, others secreted an exotoxin. What was the specific antigen of *B. pertussis?* It suddenly dawned on me that it was an exotoxin. I was chagrined. Of course, there was a thread of evidence running through the literature. However, not until my fourth presentation of this hypothesis was its significance recognized (79).

Shortly after Parfentjev & Goodline (60) reported histamine shock in mice injected with *Hemophilus pertussis* vaccine, I published several articles on histamine sensitivity (HS) of the mouse. Mice intranasally infected with *H. pertussis* showed that the degree of HS was directly related to inoculum and severity of infection, that duration paralleled paroxysmal coughing of the child, and that after recovery from infection, mice were immune to ic challenge (70). With plain pertussis vaccine, the HS paralleled protective activity (72), and female mice were more susceptible to HS than male mice (71).

Another study (111) demonstrated, in contrast to the marked increase in HS in the mouse induced by pertussis vaccine, that both the rabbit and the guinea pig showed a slight but significant decrease in normal HS; the guinea pig more than the rabbit. The mouse showed a decrease in blood glucose, while the rabbit showed no change. Why?

Numerous studies support the essential role of pertussis toxin (PT) in whooping cough. Foremost is the molecular structure of PT and its activity that is effected by catalytic action of ADP-ribosyltransferase on a specific membrane protein that controls the function of cells (107), leading to multiple types of clinical reactions.

The mouse strain selected for potency (77) and toxicity (80a) evaluation is very important. With the N:NIH (SW) strain I routinely used, there was a direct correlation between the histamine sensitizing dose (HS_{50}) and the ED_{50} responses. By selective breeding, high and low HS strains (HSF-S and HSF-R) were developed (46). The ratio, R/S, of the SD_{50} and the ED_{50} of the test vaccine remained constant (cf. 81). Mouse strains with highest sensitizability have the highest immunizability for pertussis vaccine, as well as for diphtheria and tetanus toxoids, which was shown by Csizmas (see 77).

An analysis of the results of 9.5 years of toxicity testing (87) and other studies point out the relationship between mouse weight gain and human toxicity, both of which are related to the HS unitage (cf. 81), and that stability of potency of the vaccine is effected by the preservative (87). Merthiolate is the preservative of choice. Whooping cough is a neurotoxic disease (82). The pharmacologic mode of action of the metabolic changes is only beginning to be revealed.

Typhoid vaccine   The participation of Pittman & Bohner (85) in the laboratory assessment of the potency of typhoid vaccines that were being tested in WHO field trials contributed to promulgation of US Standards for Typhoid Vaccine. The mouse potency test in routine use differentiated two trial vaccines (K, acetone killed and dried, and L, heat-killed, phenolized and dried) in the same order as their efficacy for humans. The ratios, K/L, were the same, 3.69. Melikova et al (49) confirmed our findings that mouse potency and human protection reflected the same difference as the mouse in protective activity of K and L vaccines. They used the same mouse test we used.

The US typhoid vaccine reference was tested in the laboratory concurrently with vaccines K and L (85). It was much lower in potency and was replaced with a K-type vaccine that was comparable in potency to the K vaccine that was efficacious in protection of humans. Publication of "Additional Standards for Typhoid Vaccine" in the Federal Register followed.

Studies on typhoid vaccine continued with colleagues. One study showed the importance of the mouse strain used for the potency test (24). Two studies indicated that Vi antigen was the most important protective antigen for the mouse (117, 118). Purified Vi was prepared shortly after I retired (119).

Dr. John B. Robbins and associates have been promoting the assessment of the Vi capsular polysaccharide for protection of humans against typhoid. In a recent study in Nepal, the Vi antigen, tested in comparison with pneumococcus capsular polysaccharide, provided about 75% protection (1). Both antigens were nonreactive. It is anticipated that Vi convalently bound to a protein (112) will increase immunogenicity and the use of this antigen will contribute to the control of typhoid fever in endemic areas. It is gratifying that our earlier studies on the protective activity of the Vi antigen for the mouse supported the studies that have shown the protective activity of Vi vaccine for humans.

Cholera vaccine   My interest in cholera vaccine started in 1958 when I participated in the development of the first international requirements for biological substances—yellow fever vaccine and cholera vaccine. After further considerations, the requirements for cholera vaccine were published in 1959 (121). Other meetings with World Health Organization and cooperative tests followed. Committee members were concerned with the relation of the laboratory assessed potency of the vaccine and its prevention of cholera. The requirements were revised in 1968 (121).

The years from 1960 to 1970 were of special interest, stimulated by the Southeast Asia Treaty Organization (SEATO)-Pakistan Cholera Research Laboratory (CRL) in Dacca, East Pakistan (now Dhaka, Bangladesh). SEATO, after effecting an improvement of small pox vaccine in Southeast Asia, decided to focus on cholera. Dr. Joseph Smadel, an eminent research scientist, especially on Southeast Asian infectious diseases, was the power behind the project. I had known him at RIH. Dr. John C. Feeley, a new staff member, and I were brought in to help design the laboratories and equipment for CRL that would be located in a building erected by the US Public Health Service.

Through the activities of two committees, I was closely connected with the cholera investigations. The NIH Cholera Advisory Committee members were actively engaged in cholera research. Their evaluation and coordination of the projects were extraordinary and very beneficial.

As a member of the Panel of Expert Consultants to Technical Committee for Pakistan-SEATO Cholera Research Laboratory, I followed closely the work at this laboratory. All aspects of the disease were under investigation. Twice I visited CRL and the CRL hospital at Matlab.

In 1965, after Dr. Smadel's death from cancer, I served as Project Officer for CRL until 1970. Responsibility was largely related to NIH financial agreement.

With Dr. Feeley, I participated in several studies on *Vibrio cholerae* and on the El Tor strains (91). We evaluated the potency of the cholera vaccines used in the first field trial of cholera vaccine by CRL (25a). Mosely et al (50) pointed out the interrelations of serological responses in humans and the active mouse protection test to cholera vaccine effectiveness. For several years, Dr. J. C. Feeley participated with the U.S. Cholera Panel of the United States–Japan Cooperative Medical Science Program.

CRL has been superceded by the International Centre for Diarrhoeal Disease Research, Dhaka, Bangladesh.

Other bacterial products Among other product studies, I developed the guinea pig skin potency assay that was included in the first requirements for tuberculins, and participated with a Public Health Service Committee on the required strength for the skin test dose and in the standardization of the Purified Protein Derivative (PPD-S) prepared by Dr. Florence B. Seibert (106) for the United States and the International Standard.

A diphtheria toxin was selected and standardized for the Shick Test, and the erythema potency assay for the Shick Test dose was developed by Barile et al (8).

Then there were studies on tetanus toxoid with Dr. M. C. Hardegree and associates and on requirements for anthrax vaccine and *Clostridium* gas-gangrene vaccines that have limited use.

General Standards

During the war years, the work load increased in order to meet the needs of the Armed Forces for immunization and treatment products. My participation largely concerned pyrogenicity of plasma and its contamination, and the formula for the sterility test.

PYROGENICITY   Intravenous therapy, especially plasma, greatly increased. Occasionally febrile reactions occurred. It was known that microorganisms present in distilled water effected pyrogenicity (124). But there was little information about the pyrexial characteristic of contaminates from biologics: I studied 59 cultures received from eight processing laboratories. Twenty-eight representative strains were selected for evaluation of their pyrexial property in collaboration with Thomas F. Proby (99). Viable and killed bacteria and culture filtrates, all in serial dilutions, were examined using the rabbit pyrogen test as described by the US Pharmacopeia XII, with modifications. All microorganisms were capable of inducing fever but they varied widely both qualitatively and quantitatively. Counts of bacteria did not furnish an index. Gram-negative bacteria were the most pyrogenic: 6000 cfu/ml were pyrogenic. A rabbit pyrogenic test that was applicable to all intravenous fluid products was promulgated.

THE STERILITY TEST   The early test to determine the sterility of biologics used infusion broth in the Smith container with a side arm to provide both aerobic and anaerobic conditions. In 1941, fluid thioglycolate medium (FTM) developed by Brewer (17) was adopted. It had two advantages: neutralization of mercurial preservatives and support of the growth of both aerobic and anaerobic bacteria. Shortly thereafter, investigators in two establishments reported the recovery of more contaminants in the infusion broth than in the new FTM. Inhibition was limited by methylene blue, the Eh indicator.

An evaluation of the growth-promoting property of each ingredient in FTM was made. Many consulted persons provided valuable suggestions and assistance. The suggestion to use resazurin for the Eh indicator came from R. H. Breed through H. C. Dunham of Difco. The dairy industry used resazurin to evaluate the degree of contamination in milk. The 26 cultures selected for the study were either from contaminated products or were considered possible contaminants. Each culture was tested to determine the minimum number of bacteria that would grow in each test medium, at different pH and at incubation 36°C and 22–25°C. The selected formula (69) has remained unchanged except for a reduction of L-cystine from 0.75 to 0.5 mg per liter. This medium is listed in the WHO General Requirements for Sterility of Biological Substances for detection of both aerobic and anaerobic contaminants (122). It has been given in the US Pharmacopeia since 1970.

The use of two temperatures of incubation for the sterility test was not adopted immediately. The use of 37°C to detect the presence of pathogenic contaminants was so ingrained that it took a dramatic event to specify the use of a lower temperature. A patient being treated with plasma developed a febrile shock. This plasma was tested for sterility at 37°C; no growth. At room temperature, growth was obtained from 1.0 ml of a 10^{-5} dilution.

The final study on contaminants was prompted by recurring reports of severe or fatal reactions following administration of bacteriologically contaminated blood. This study provided a better understanding of the problem of bacterial contamination in the handling of blood and blood products. The results were presented in 1952 before the Fifth Annual Meeting of the American Association of Blood Banks (73). Emphasis was placed on the importance of contaminants from the environment, not the donor, and the need to identify the contaminant in order to trace its source. The report also emphasized the need to use two temperatures of incubation for the sterility test. All of the recovered gram-negative bacilli and some gram-positive cocci grew at 10°C, the upper limit of storage of blood. Earlier we had reported (99) that all contaminants tested had the ability to induce a febrile reaction. Furthermore, that of the 13 blood contaminants associated with severe or fatal reactions, 11 grew at 2.5°C. The range of temperatures at which all cultures grew readily was limited to above 25°C and below 30°C. Hence, adjustment of the 37°C temperature of incubation of the sterility test was necessary in order to detect the presence of psychrophilic bacterial contaminants.

The inclusion of the medium to support the recovery of fungal contaminants was based on studies by Pittman & Feeley (90) and Cox et al (21).

Cooperative Studies

It is imperative that biological standards be uniform within a nation and throughout the world, and best accomplished through cooperation.

After World War I, the international cooperation of NIH began. Dr. George W. McCoy, Director of NIH served as an invited unofficial member of the Health Organization, League of Nations Committee at Copenhagen, and NIH furnished gas-gangrene antitoxins for International Standards (cf. 83).

Dr. Roderick Murray, Director, DBS, NIH, was one of the first to serve on the WHO Expert Committee on Biological Standardization of the United Nations, which succeeded the League of Nations Committee. I was involved as Consultant of the WHO Secretariat in the preparation of the requirements for the bacterial vaccines and the general requirements for the Sterility of Biological Substances and their revision in 1973 (122).

A number of WHO collaborative studies related to the International Pertussis Vaccine Standard, the Cholera Vaccine, the Opacity Standard, and the variables that influence the mouse potency test of pertussis vaccine (51), as well as the relation of mouse toxicity to human toxicity (61, cf. 81).

The studies relating to typhoid vaccine, cholera vaccine, and epidemic acute conjunctivitis were discussed earlier.

The US Pharmacopeia Panel on Sterility Tests and the Panel on Biological Indicators (to control effectiveness of steam and dry heat) contributed to ensuring the sterility of biologics and other products. I was a consecutive member of the Panels during 1967–73.

As an associate member of the Commission on Immunization of the Armed Forces Epidemiological Board, I kept abreast of the immunological program in the Armed Forces (1969–1970).

Participation at workshops, symposiums, national and international scientific society meetings, especially the meetings of the International Association of Biological Standardization and its journal, were also helpful.

The most extensive study was on "Immunization against neonatal tetanus in New Guinea" with Dr. F. D. Schofield and Dr. R. MacLennon (7, 31, 32, 44, 95). Besides the success in preventing neonatal tetanus with tetanus toxoid, there was some untoward reactivity with the use of tetanus toxoid containing mineral oil adjuvant. At that time, there was much interest in the use of mineral oil in allergens and other products. DBS initiated a series of studies under contract to evaluate safety. The final DBS report indicated that the use of mineral oil adjuvants in the human population may be hazardous and should not be recommended for general use in humans (54).

AFTER RETIREMENT, 1971–1990

To paraphrase the title of the book *Life Begins at Forty*, I might use the title *Life Continues at Seventy*. The years have been full, challenging, and gratifying.

It has been an unusual privilege to be a Guest Worker (without payment) with an office and a parking space on the NIH campus. Without official responsibility, I was able to clear my desk of pending manuscripts. Now two reviews are pending—early work on *H. influenzae* and on the Koch-Weeks bacillus. In addition, I have been invited to give lectures, participate in workshops, and write articles for journals and books.

Serving as consultant or Guest Scientist in nine countries has been stimulating and pleasant. With WHO, I completed the revision of requirements for the sterility of biological substances, and consulted on cholera vaccine in Cairo, Egypt and Madrid, Spain for three months each. Three months at the Razi Institute, Teheran, Iran were very interesting and informative, historically and politically. The most productive work was at the University of Glasgow.

There were four visits. A study started in the Netherlands was completed on the pathophysiological response of mice to *B. pertussis* respiratory infections (92). The participating pharmacologist, Dr. Furman, has continued the work on insulin and glucose metabolism. There is some evidence that hypoglycemia may play an important role in pertussis encephalopathy. I was at the University of Glasgow when the hypothesis was conceived that pertussis toxin was the cause of the harmful effects and prolonged immunity of whooping cough (80). More pieces are yet to be fit into the jigsaw puzzle of pertussis. Pieces are still missing in the jigsaw puzzle of epidemic acute conjunctivitis as well as other infectious diseases.

A statement made by Sir J. Willard in a letter to the Academy of Arts and Sciences in January 1986 has been a helpful guide in doing and directing research. It reads, "one of the principal objects of theoretical research is to find the point of view from which the subject appears in its greatest simplicity." There is a truism to be remembered. "That whatsoever is known has always seemed indeductable to the known, and equally whatever is alien has seemed fanciful and arbitrary." We may see but not perceive if the mind is not prepared.

With home training and inspiratory education, my life's journey led to biological products. Many scientific and social friends and church activities provided inspiring vistas. Life has brought satisfaction and gratitude for many honors. Lacking has been the enhancement of the inborn spirit of youth. Hence, I have helped to initiate and am promoting the Urban Ministry Program at Wesley Theological Seminary.

Literature Cited

1. Acharya, I., Lowe, C. U., Thapa, R., Gurubacharya, V. I., Shrestha, M. B., et al. 1987. Prevention of typhoid fever in Nepal with Vi capsular polysaccharide of *Salmonella typhi*. *New Engl. J. Med.* 317:1101–4
2. Alexander, H. E. 1939. Type "B" anti-influenzal rabbit serum for therapeutic purposes. *Proc. Soc. Exp. Biol. Med.* 40:313–14
3. Attiah, M. A. H. 1935. Seasonal epidemics of acute Koch-Weeks and gonococcus ophthalmlas. *Bull. Ophthalmol. Soc. Egypt* 38:66–72
4. Avery, O. T., Chickering, H. T., Cole, R., Dochez, A. R. 1917. *Acute Lobar Pneumonia: Prevention and Serum Treatment*, Monogr. No. 7. New York: The Rockefeller Inst. Med. Res. 110 pp.
5. Avery, O. T., et al. 1917. See Ref. 4, pp. 22–25
6. Avery, O. T., et al. 1917. See Ref. 4, pp. 42–50
7. Barile, M. F., Hardegree, M. C., Pittman, M. 1970. Immunization against neonatal tetanus in New Guinea. 3. The toxin-neutralization test and the response of guineapigs to the toxoids as used in the immunization schedules in New Guinea. *Bull. WHO* 43:453–59
8. Barile, M. F., Kolb, R. W., Pittman, M. 1971. United States Standard Diphtheria Toxin for the Schick Test and the erythema potency assay for the Schick Test dose. *Infect. Immun.* 4:295–306
9. Bengtson, I. A. 1933. Seasonal acute conjunctivitis occurring in the southern states. *Public Health Rep.* 48:917–26
10. Blake, F. G., Cecil, R. L. 1920. Studies on experimenal pneumonia. IX. Production in monkeys of an acute respiratory

disease resembling influenza by inoculation with *Bacillus influenzae*. *J. Exp. Med.* 32:691–717

11. Bordet, J., Gengou, O. 1906. Le microbe de la coqueluche. *Ann. Inst. Pasteur* 20:731–41

12. Bordet, J., Gengou, O. 1909. L'endotoxine coquelucheuse. *Ann. Inst. Pasteur (Paris)* 23:415–19

13. Branham, S. E. 1938. Serums, antitoxin, and drugs in the treatment of meningococcus meningitis. *Public Health Rep.* 53:645–51

14. Branham, S. E. 1986. See Ref. 59b, pp. 141–48

15. Branham, S. E., Pittman, M., 1940. A recommended procedure for the mouse protection test in evaluation of antimeningococcus serum. *Public Health Rep.* 55:2340–46

16. Branham, S. E., Pittman, M., Rake, G., Sherp, H. W. 1938. A proposed mouse protection unit for anti-meningococcus serum. *Proc. Soc. Exp. Biol. Med.* 39:348–50

17. Brewer, J. H. 1940. A clear liquid medium for "aerobic" cultivation of anaerobes. *J. Am. Med. Assoc.* 115:598–600

18. Brewer, J. H., Cook, E. B. 1939. A permanent nephelometer from pyrex glass. *Am. J. Public Health* 29:1147–48

19. Cecil, R. L., Blake, F. G. 1920. Studies on experimental pneumonia. X. Pathology of experimental influenza and of *Bacillus influenzae* in monkeys. *J. Exp. Med.* 32:719–44

20. Colwell, R. R., Brayton, P. R., Grimes, D. J., Roszak, D. B., Huq, S. A., et al. 1985. Viable but nonculturable *Vibrio cholerae* and related pathogens in the environment: Implications for release of genetically engineered microorganisms. *Bio/Technol.* 3:817–20

21. Cox, C. B., Feeley, J. C., Pittman, M. 1973. Sterility testing: detection of fungi and yeasts in the presence of preservatives. *J. Biol. Stand.* 1:11–19

22. Davis, J. D., Pittman, M. 1950. Acute conjunctivitis caused by *Hemophilus*. *Am. J. Dis. Child.* 79:211–19

23. Eldering, G., Hornbeck, C., Baker, J. 1957. Serological study of *Bordetella pertussis* and related species. *J. Bacteriol.* 74:133–36

24. Esposito, V. N., Feeley, J. C., Leeder, W. D., Pittman, M. 1969. Immunological responses of three mouse strains to typhoid vaccine and Vi antigen. *J. Bacteriol.* 99:8–12

25. Falk, I. S. 1923. *The Principles of Vital Statistics*. Philadelphia: Saunders. 258 pp.

25a. Feeley, J. C., Pittman, M. 1965. Laboratory assays of cholera vaccine used in field trial in East Pakistan. *Lancet* 1:449–50

26. Felton, H. M., Verwey, W. F. 1955. The epidemiological evaluation of a non-cellular pertussis vaccine. *Pediatrics* 16:637–51

27. Food and Drug Administration, Code of Federal Regulations. 1989. Title 21. Parts 600 to 799.

28. Gordon, J. E., Hood, R. I. 1951. Whooping cough and its epidemiological anomalies. *Am. J. Med. Sci.* 222:333–61

29. Grimont, F., Grimont, P. A. D. 1986. Ribosomal ribonucleic acid gene restriction patterns as potential taxonomic tools. *Ann. Inst. Pasteur Microbiol.* 137B:165–75

30. Gupta, R. K., Sharma, S., Ahuja, S. S., Saxera, S. N. 1987. Evaluation of repeated immunoassays (mouse intracerebral potency tests) of the second international standard of pertussis vaccine. *J. Immunoassay* 8:309–18

31. Hardegree, M. C., Barile, M. F., Pittman, M., Maloney, C. J., Schofield, F., et al 1970. Immunization against neonatal tetatnus in New Guinea. 4. Comparison of tetanus antitoxin titres obtained by hemagglutination and toxin neutralization in mice. *Bull WHO* 43:461–68

32. Hardegree, M. C., Barile, M. F., Pittman, M., Schofield, F. D., MacLennan, R., et al. 1965. Immunization against neonatal tetanus in New Guinea. 2. Duration of primary antitoxin responses to adjuvant tetanus toxoids and comparison of booster responses to adjuvant and plain toxoids. *Bull. WHO* 43:439–51

33. Harrison, L. H., Da Silva, G. A., Pittman, M., Fleming, D. W., Vranjac, A., et al. 1989. Historical views of *H. aegypticus*, Epidemiology and clinical spectrum of brazilian purpuric fever. *J. Clin. Microbiol.* 27:599–604

34. Hertzler, A. E. 1938. *The Horse and Buggy Doctor*. New York: Harper & Brothers. 322 pp.

35. Hewitt, W. L., Pittman, M. 1946. Antibacterial action of penicillin, penicillin X, and streptomycin on *Hemophilus influenzae*. *Public Health Rep.* 61:768–78

36. Hoiseth, S. K., Gilsdorf, J. R. 1988. The relationship between type b and nontypable *Haemophilus influenzae* isolated from the same patient. *J. Infect. Dis.* 158:643–45

37. Jordan, E. O. 1927. In *Epidemic influenza: A Survey*, p. 411. Chicago: Am. Med. Assoc. 599 pp.

38. Jordan, E. O., Falk, I. S., eds. 1927. *The Newer Knowledge of Bacteriology and Immunology.* Chicago: Univ. Chicago Press. 1196 pp.

39. Kendrick, P. K., Eldering, G., Borowski, A. 1939. A study in active immunization against pertussis. *Am. J. Hyg.* 29(Sect. B):133–53

40. Kendrick, P. L., Eldering, G., Dixon, M. K., Misner, J. 1947. Mouse protection tests in the study of pertussis vaccine: A comparative series using the intracerebral route for challenge. *J. Public Health* 37:803–10

41. Kilian, M., Biberstein, E. L. 1984. 2. *Haemophilus aegypticus* (Trevisan 1889) Pittman and Davis 1950, 413[AL]. In *Bergey's Manual of Systematic Bacteriology,* ed. N. R. Krieg, J. H. Holt, 1:563–64. Baltimore/London: Williams & Wilkins. 964 pp.

42. Koser, S. A. 1952. Bacteriology at the University of Chicago. *Bios* 23:175–91

43. Levanthal, W., Fernbach, H. 1922. Morphologische studien an influenzabacillen und das aetiologische grippeproblem. *Z. Hyg. Infectionskr.* 96:456–519

44. MacLennan, R., Schofield, F. D., Pittman, M., Hardegree, M. C., Barile, M. F. 1965. Immunization against neonatal tetanus in New Guinea. Antitoxin response of pregnant women to adjuvant and plain toxoids. *Bull. WHO* 32:683–97

45. Mancini, G., Nash, D. R., Herman, J. E. 1970. Further studies on single radial immunodiffusion. III. *Immunochemistry* 7:261–64

46. Manclark, C. R., Hansen, C. T., Treadwell, P. E., Pittman, M. 1975. Selective breeding to establish a standard mouse for pertussis vaccine bioassay. II. Bioresponses of mice susceptible and resistant to sensitization by pertussis vaccine HSF. *J. Biol. Stand.* 3:353–63

47. Mazloum, H. A., Kilian, M., Mohamed, Z. M., Said, M. D. 1982. Differentiation of *Haemophilus aegypticus* and *Haemophilus influenzae. Acta Pathol. Microbiol. Immunol. Scand. Sect. B* 90:109–12

48. Medical Research Council. 1956. Vaccination against whooping cough: Relation between protection in children and results of the laboratory tests. *Br. Med. J.* 2:454–62

49. Melikova, E. N., Vasil'eva, I. G., Lesniak, S. V. 1965. Comparative characterization of the methods employed in a laboratory evaluation of the effectiveness of WHO dry typhoid vaccines. *Zh. Mikrobiol. Epidemiol. Immunbiol.* 5:58–65

50. Mosley, W. H., Feeley, J. C., Pittman, M. 1971. The interrelationships of serological responses in humans, and the active mouse protection test to cholera vaccine effectiveness. In *International Symposium on Enterobacterial Vaccine, Berne, 1968,* pp. 185–96. Basel/New York: Karger

51. Murata, R., Perkins, F. T., Pittman, M., Scheibel, I., Sladky, K. 1971. International collaborative studies on the pertussis vaccine potency test. *Bull. WHO* 44:673–87

52. Murphy, T. F., Apicella, M. A. 1987. Nontypable *Haemophilus influenzae:* A review of clinical aspects, surface antigens, and human immune response to infection. *Rev. Infect. Dis.* 9:1–15

53. Murray, R. 1968. The division of biologics standards. *Public Health Serv. Publ. No. 1744*

54. Murray, R., Cohen, P., Hardegree, M. C. 1972. Mineral oil adjuvant: Biological and chemical studies. *Ann. Allergy* 30:146–51

55. National Institutes of Health. 1945. Memorandum to manufacturers of pertussis vaccine

56. National Institutes of Health. 1946. *Minimum Requirements: A tentative mouse protection test for determining the antigenicity of pertussis vaccine*

57. National Institutes of Health. 1948. *Minimum Requirements: Pertussis vaccine*

58. National Institutes of Health. 1948. Memorandum: Application of the method proposed by Wilson and Worcester for determining the ED_{50} to the evaluation of the potency of pertussis vaccine. 1st Revis. 1956. A table for use in calculating the 50-percent effective dose (16 mice per dose). 1953. A table for use in calculating the 50-percent effective dose (32 mice per dose). 1954

59. Nolan, C. M., LaBorde, E. A., Howell, R. T., Robbins, J. B. 1980. Identification of *Salmonella typhi* in faecal specimens by an antiserum-agar method. *J. Med. Microbiol.* 13:373–77

59a. Norton, J. F., Dingle, J. H. 1935. Virulence tests for typhoid bacilli and antibody relationships in antityphoid sera. *Am. J. Public Health* 25:609–17

59b. O'Hern, E. M., ed. 1986. *Profiles of Pioneer Women Scientists,* Washington, DC: Acropolis. 264 pp.

60. Parfentjev, I. A., Goodline, M. A. 1948. Histamine shock in mice sensitized with *Hemophilus pertussis* vaccine. *Pharmacol. Exp. Ther.* 92:411–13

61. Perkins, F. T., Sheffield, F., Miller, C. L., Skegg, J. L.. 1970. The comparison of toxicity of pertussis vaccine in children and mice. In *International Symposium on Pertussis, Symp. Ser. Immunobiol. Stand.* ed. P. A. van Hemert, J. D. van Ramshorst, R. H. Regamy. 13:141–49. Basel: Karger

62. Pittman, M. 1930. The "S" and "R" forms of *Hemophilus influenzae. Proc. Soc. Exp. Biol. Med.* 27:299–301

63. Pittman, M. 1931. Variation and type specificity in the bacterial species *Hemophilus influenzae. J. Exp. Med.* 53:471–92

64. Pittman, M. 1933. The action of type-specific *Hemophilus influenzae* antiserum. *J. Exp. Med.* 58:683–706

65. Pittman, M. 1939. The protection of mice against *Hemophilus influenzae* (non-type-specific) with sulfapyridine. *Public Health Rep.* 54:1769–75

66. Pittman, M. 1941. A study of certain factors which influence the determination of the mouse protective action of meningococcus antiserum. *Public Health Rep.* 56:92–110

67. Pittman, M. 1942. Antibacterial action of several sulfonamide compounds on *Hemophilus influenzae* type b. *Public Health Rep.* 57:1899–1910

68. Pittman, M. 1943. Mouse protective values of antimeningococcus serum in comparison with precipitation in immune serum agar plates. *Public Health Rep.* 58:139–42

69. Pittman, M. 1946. A study of fluid thioglycoliate medium for the sterility test. *J. Bacteriol.* 51:19–32

70. Pittman, M. 1951. Sensitivity of mice to histamine during respiratory infection by *Hemophilus influenzae. Proc. Soc. Exp. Biol. Med.* 77:70–74

71. Pittman, M. 1951. Influence of sex of mice on histamine sensitivity and protection against *Hemophilus pertussis. J. Infect. Dis.* 89:296–99

72. Pittman, M. 1951. Comparison of the histamine-sensitizing property with protective activity of pertussis vaccines for mice. *J. Infect. Dis.* 89:300–4

73. Pittman, M. 1953. A study of bacteria implicated in transfusion reactions and of bacteria isolated from blood products. *J. Lab. Clin. Med.* 42:273–88

74. Pittman, M. 1954. Variability of the potency of pertussis vaccine in relation to number of bacteria. *J. Pediatr.* 45:57–69

75. Pittman, M. 1956. Pertussis and pertussis vaccine control. *J. Wash. Acad. Sci.* 46:234–43

76. Pittman, M. 1958. Variations du pouvoir protecteur des differente vaccins anticoquelucheux: leur rapport avec la protection de l'être humain. *Rev. Immunol.* 22:308–22

77. Pittman, M., 1967. Mouse strain variation in response to pertussis vaccine and tetanus toxoid. In *International Symposium on Laboratory Animals, London 1966. Symp. Ser. Immunobiol. Stand.* 5:161–66. Basel/New York: Karger. 204 pp.

78. Pittman, M. 1976. History, benefits and limitations of pyrex glass particle opacity references. *J. Biol. Stand.* 4:115–25

79. Pittman, M. 1978. Discussion: Control testing of pertussis vaccine, pertussis toxin. In *International Symposium on Pertussis,* ed. C. R. Manclark, J. C. Hill, p. 240. Washington, DC: DHEW Publ. No. (NIH) 79-1830

80. Pittman, M. 1979. Pertussis Toxin: The cause of harmful effects and prolonged immunity of whooping cough. A hypothesis. *Rev. Infect. Dis.* 1:401–12

80a. Pittman, M. 1980. Mouse breeds and the toxicity test for pertussis vaccine. In *Standardization of Animals to Improve Biomedical Research, Production and Control,* 16th IABS Congress, San Antonio, 1979. *Dev. Biol. Stand.* 45:129–35. Basel: Karger

81. Pittman, M. 1984. The concept of pertussis as a toxin-mediated disease. *Pediatr. Infect. Dis.* 3:467–86

82. Pittman, M. 1986. Neurotoxicity of *Bordetella pertussis.* In *Neurotoxicity in the Fetus and Child,* ed. J. M. Cranmer, pp. 53–67. Roland, AR: Intox. 681 pp.

83. Pittman, M. 1987. The regulations of biologics standards. 1902–1972. In *National Institute of Allergy and Infectious Diseases: Intramural Contributions, 1887–1987,* ed. H. R. Greenwald, V. A. Harden, pp. 61–70. Washington, DC: US Dept Health Hum. Serv. 133 pp.

84. Pittman, M. 1988. Genus *Bordetella* Moreno-López 1952, 178[AL] In *Bergey's Manual of Systematic Bacteriology,* ed. N. R. Krieg, J. H. Holt, 1:388–93. Baltimore/London: Williams & Wilkins. 964 pp.

85. Pittman, M., Bohner, H. J. 1966. Laboratory assays of different types of field trial typhoid vaccines and relationship to efficacy in man. *J. Bacteriol.* 91:1713–23

86. Pittman, M., Branham, S. E., Sockrider, E. M. 1938. A comparison of the precipitation reaction in immune serum agar plates with the protection of mice by antimeningococcus serum. *Public Health Rep.* 53:1400–8

87. Pittman, M., Cox, C. B. 1965. Pertussis vaccine testing for freedom-from-toxicity. *Appl. Microbiol.* 13:447–56

88. Pittman, M., Davis, J. D. 1950. Identification of the Koch-Weeks bacillus *(Hemophilus aegypticus)*. *J. Bacteriol.* 59:413–26

89. Pittman, M., Falk, I. S. 1930. Studies on respiratory diseases. XXXIV. Some relations between extracts, filtrates and virulence of pneumococci. *J. Bacteriol.* 19:327–61

90. Pittman, M., Feeley, J. C. 1962. Sterility testing: Detection of fungal and yeast contamination in biological preparations. In *Proc. 7th International Congress for Microbiological Standardization, London, 1961*, ed. A. F. B. Standfast, D. G. Evans, B. G. F. Weitz, pp. 207–14. Edinburgh/London: E. & S. Livingstone. 556 pp.

91. Pittman, M., Feeley, J. C. 1963. Protective activity of cholera vaccines against El Tor cholera vibrio. *Bull WHO* 28:379–83

92. Pittman, M., Furman, B. L., Wardlaw, A. C. 1980. *Bordetella pertussis* respiratory tract infection in the mouse: Pathophysiological responses. *J. Infect. Dis.* 142:56–66

93. Pittman, M., Gardner, R. A., Marshall, J. E. 1979. Evaluation of the potency of antipertussis serum products. *J. Biol. Stand.* 7:261–71

94. Pittman, M., Goodner, K. 1935. Complement-fixation with the type-specific carbohydrate of *Hemophilus influenzae* type b. *J. Immunol.* 29:239–47

95. Pittman, M., Kolb, R. W., Barile, M. F., Hardegree, M. C., Seligmann, E. B. Jr., et al. 1970. Immunization against neonatal tetanus in New Guinea. 5. Laboratory assayed potency of tetanus toxoids and relationship to human antitoxin response. *Bull. WHO* 43:469–78

96. Pittman, M., Lieberman, J. E. 1948. An analysis of the Wilson-Worcester Method for determining the median effective dose of pertussis vaccine. *Am. J. Public Health* 38:15–21

97. Pittman, M., Southwick, M. A. 1930. Studies on respiratory diseases. XXXV. The pathology of pneumococcus infections in mice. *J. Bacteriol.* 19:363–74

98. Preston, N. W. 1965. Effectiveness of pertussis vaccine. *Br. Med. J.* 1:11–13

99. Probey, T. F., Pittman, M. 1945. The pyrogenicity of bacterial contaminants found in biologic products. *J. Bacteriol.* 50:397–411

100. Pupils of William Thompson Sedgwick, 1906. *Biological Studies.* Chicago: Univ. Chicago Press. 329 pp.

101. Rivers, T. M. 1922. Influenzal meningitis. *Am. J. Dis. Child.* 24:102–15

102. Rockefeller, J. D. 1938. *Address—Dinner in Honour of Dr. Rufus Cole*, p. 5, New York: The Rockefeller Inst. Med. Res. 75 pp.

103. Rollins, D. M., Colwell, R. R. 1986. Viable but nonculturable state of *Campylobacter jejuni* and its role in survival in the natural aquatic environment. *Appl. Environ. Microbiol.* 52:531–38

104. Schneerson, R., Barrera, A., Sutton, A., Robbins, J. B. 1980. Preparation, characterization, and immunogenicity of *Haemophilus influenzae* type b polysaccharide-protein conjugates. *J. Exp. Med.* 152:361–76

105. Schneerson, R., Robbins, J. B., Parke, J. C. Jr. 1973. Studies on *Haemophilus influenzae* isolated from otitis media. In *Hemophilus influenzae*, ed. S. W. Sell, D. T. Karzon, pp. 13–20. Nashville: Vanderbilt Univ. Press. 325 pp.

106. Seibert, F. B. 1968. Preparation of the government and international standards for tuberculins (PPDS). In *Pebbles on the Hill of a Scientist*, pp. 73–87. St. Petersburg Florida: St. Petersburg Print. Co. 162 pp.

107. Sekura, R. D., Moss, J., Vaughn, M., eds. 1985. *Pertussis Toxin.* New York/London: Academic. 225 pp.

108. Shope, R. 1931. Swine influenza. I. Experimental transmission and pathology. *J. Exp. Med.* 54:372–85

109. Skerman, V. B. D., McGowan, V., Sneath, P. H. A., eds. 1980. Approved lists of bacterial names. *Int. J. Syst. Bacteriol.* 30:225–40

110. Stillman, E. G. 1922. The frequency of *Bacillus influenzae* in the nose and throat in acute lobar pneumonia. *J. Exp. Med.* 35:7–15

111. Stronk, M. G., Pittman, M. 1955. The influence of pertussis vaccine on histamine sensitivity of rabbits and guinea pigs and on the blood sugar in rabbits and mice. *J. Infect. Dis.* 96:152–61

112. Szu, S. C., Stone, A. L., Robbins, J. D., Schneerson, R., Robbins, J. B. 1989. Vi capsular polysaccharide-protein conjugates for prevention of typhoid fever. *J. Exp. Med.* 166:1510–24

113. van Heyningen, W. E. 1970. General characteristics. In *Microbial Toxins. Bacterial Protein Toxins*, ed. S. J. Ajl, S. Kadis, T. C. Montie, 1:1–28. New York: Academic. 517 pp.

114. Wilson, J., Compiler. 1986. *History of Cincinnati, Arkansas, 1836–1986, and*

Illinois Township. Siloam Springs, AK: Siloam Springs Print. 304 pp.

115. Winternitz, M. C., Wason, I. M., Mac-Namara, F. P. 1920. *The Pathology of Influenzae*, p. 51. New Haven: Yale Univ. Press. 61 pp.

116. Wollstein, M. 1919. Epidemic influenza in infants. *Am. J. Dis. Child.* 17:166–73

117. Wong, K. H., Feeley, J. C., Pittman, M. 1972. Effect of a Vi-degrading enzyme on potency of typhoid vaccines in mice. *J. Infect. Dis.* 125:360–66

118. Wong, K. H., Feeley, J. C., Pittman, M., Forlines, M. E. 1974. Typhoid vaccines, acetone-inactivated versus heat-phenol inactivated: Vi antigen and toxicity. *J. Infect. Dis.* 129:501–6

119. Wong, K. H., Feeley, J. C., Northrup, R. S., Forlines, M. E. 1974. Vi antigen from *Salmonella typhosa* and immunity against typhoid fever. I. Isolation and immunologic properties in animals. *Infect. Immun.* 9:348–53

120. World Health Organization. 1958. The International Opacity Reference Preparation. *WHO Tech. Rep. Ser.* 148: (Third Reference preparation. 1966. 329:21)

121. World Health Organization. 1959. Requirements for Biological Substances No. 6. Cholera Vaccine. *WHO Tech. Rep. Ser.* 179:41–46. (Revised 1968. 413:28–44)

122. World Health Organization. 1960. Requirements for Biological Substances No. 6. General requirements for the Sterility of Biological Substances. *WHO Tech. Rep. Ser.* 200:1–31. (Revised 1973. 530:41–56)

123. World Health Organization. 1964. Requirements for Biological Substances No. 8. Pertussis Vaccine. *WHO Rep. Ser.* Annex 1, 274:25–40. (Revised 1979, 683:60–80)

124. Seibert, F. B. 1968. See Ref. 106, pp. 14–20

Annu. Rev. Microbiol. 1989. 43:1–22
Copyright © 1989 by Annual Reviews Inc. All rights reserved

THE MOUNTING INTEREST IN BACTERIAL AND VIRAL PATHOGENICITY

H. Smith

The Medical School, The University of Birmingham, Birmingham B15 2TJ, England

CONTENTS

INTRODUCTION

Today pathogenicity is a popular and exciting area of microbiology. Forty years ago, when I first became interested, studies on bacterial pathogenicity were moribund after their brilliant initial phase on toxins before the first world war. Those on viral pathogenicity had hardly started.

My involvement with microbiology in general and pathogenicity in particular was by chance. In 1947, after completing a PhD on nucleotide synthesis under the late J. M. Gulland at University College, Nottingham, I was set for a career in organic chemistry as a Lecturer in Gulland's department. I then received an invitation to join the Microbiological Research Establishment (MRE) at Porton Down, United Kingdom. I knew nothing about MRE and had not applied for a job there. I assume that the invitation was instigated by one of my PhD examiners, either C. K. Ingold or A. R. Todd (later Lord Todd, and President of the Royal Society). Having just married, I took the job at MRE because the salary was higher than that at Nottingham (£100 a year) and a house was promised. (After the war housing was scarce.) My only formal training in microbiology was a short course in the fall of 1947 at the Royal Postgraduate Medical School, Hammersmith, under the late Lord Stamp (an adviser to MRE) and Mary Barber.

The function of MRE introduced me to microbial pathogenicity. Unlike departments of medical microbiology that were interested in protection against disease, MRE was concerned with the initiation and progress of infection. This different attitude to problems of infectious disease bore much fruit, especially insights into mechanisms of respiratory infection and pathogenesis. This point was made in a lecture to the Royal Society in 1955 (26) by the then director of MRE, Dr D. W. Henderson. There were two other advantages to working at MRE, unique facilities for dealing with dangerous microorganisms and advice from some of the most eminent microbiologists and pathologists in the United Kingdom. Sir Roy Cameron, Sir Paul Fildes, Sir Howard Florey, Sir Ashley Miles, Sir Graham Wilson, and Professor D. D. Woods were all advisers to the department.

Soon after joining MRE, I suggested to Henderson that we should examine organisms grown in vivo to learn more about microbial pathogenicity. He backed the idea but, thinking no doubt that an organic chemist might be unsafe on such a project, coupled me to a veterinary bacteriologist, the late James Keppie. It was a synergistic relationship. We began working on anthrax, trying to discover the cause of death, which had remained an enigma since the time of Koch. The possibility of explaining in chemical terms the intricacies of host-pathogen interactions in vivo fascinated me. This fascination explains why I persevered with what was then a backwater of microbiology rather than changing to a more fashionable area of the subject.

Pathogenicity was studied in some government institutes but in hardly any universities. The people engaged were few. Gradually during the 1960s and then rapidly in the 1970s the enthusiasm spread to others. The flame of interest was fanned by the failure of antibiotics to eliminate bacterial diseases. Mostly, however, it was due to the realization that the techniques of cell biology, genetics, and molecular biology, which had developed elsewhere, could be applied effectively to a practically important field. For me, the watershed of the increased interest was an invitation in the fall of 1964 to give a course on pathogenicity at the University of California at Berkeley and my appointment in 1965 to the Chair of Microbiology at the University of Birmingham, there to establish a department involved with pathogenicity. Nothing has given me more pleasure than to see in my lifetime the re-awakening and blossoming of the subject that hooked me so many years ago.

Here I describe the development of the subject to its present state and then look toward the future. No attempt is made to cover the subject comprehensively, and readers will need some knowledge of pathogenicity to appreciate all the points. Epidemiology is not mentioned despite its close connection with pathogenicity and the present erudition in that field. Sometimes, I refer to reviews and key, multiauthor books because of shortage of reference space. The excuse is that I am dealing with principles rather than detail. I ask the forgiveness of both original authors and readers finding difficulty in identifying specific references.

In the first part I enquire how far we have accomplished the logical steps toward the goal of studies of pathogenicity, which is to explain in molecular terms the various biological aspects of this multifactorial property. These steps are as follows. (a) Methods must be established for comparing the virulence of strains either in the natural host or in a relevant animal model. (b) Strains of high and low virulence must be identified. (c) The strains must be compared in biological tests related to the cardinal requirements for pathogenicity: infection and penetration of mucous surfaces (sometimes bypassed by vectorial or traumatic puncture of the skin), multiplication in vivo, interference with host defenses, and causation of damage to the host. (d) The determinants that cause these biological properties must be identified. (e) The relevance of the property and its determinant to infection in vivo must be proved. (f) The chemical structure of the virulence determinant must be related to the mechanism of its biological action. Only for a few bacterial toxins and for certain aspects of viral replication have all these steps been achieved. The difficulties involved at each stage and the extent of fulfillment are discussed. Then follows a section on genetic manipulation because of its special contribution to the understanding of pathogenicity. Finally, in considering the future I indicate facets of pathogenicity that should be moved up the research ladder and then designate neglected areas that need attention.

Throughout, bacteria and nononcogenic viruses are dealt with side by side. Why, when they differ so much in size and mode of replication? To be pathogenic, bacteria and viruses have to satisfy the same biological requirements (see above). These requirements might be fulfilled in different ways, but there is enough common ground for bacteriologists and virologists to learn from one another. This is why I have worked with both types of microbes. I am, however, in the minority. Most people work on either viruses or bacteria, and there is little cross-communication. This article may show that there is some profit in the exercise.

STEPS IN STUDIES OF BACTERIAL AND VIRAL PATHOGENICITY

Establishing Methods for Comparing Virulence

The principle is simple. Animals are inoculated with graded doses of different strains, and the doses that produce a defined disease effect are compared among strains by statistical methods. The requirements are reliable methods for propagating and quantifying the pathogen, the availability of either the natural host or a valid animal model, and an easily measured disease effect.

Assessing bacterial virulence has one advantage over similar studies with viruses. Viable counts on test cultures, being usually similar to total counts, detect all the potentially infective organisms. Also, assessment of disease production is usually satisfactory. The natural host and route of inoculation can be used for veterinary diseases such as anthrax and the bacterial diarrheas, from which much knowledge has been gained. Trouble arises when animal models for human diseases are either nonexistent or of questionable validity because of differences in syndromes among species and unnatural routes of infection. Some models, such as aerosol infection of guinea pigs with *Legionella pneumophilia*, are better than others, e.g. intracerebral infection of mice with *Bordetella pertussis*. Sometimes a human pathogen that produces an unnatural syndrome in animals, e.g. *Salmonella typhi* in mice, can be replaced by a related species whose behavior in animals mimics the human disease, e.g. *Salmonella typhimurium*. No model can, however, substitute entirely for the natural disease. Some bad models persist because of convenience, habit, and sometimes unexplained parallels with field results.

Comparisons of viral virulence are not as accurate as those for bacteria. Focal or quantal assays of infectious virions detect only a few (from 1 in 10 to 1 in more than 1000) of the total particles present and therefore may not measure all virions capable of multiplying in animals. Assessments of virulence based on such counts can therefore be misleading. Unfortunately, not only are the numbers of particles needed to infect animals unknown, but the differences between these doses and plaque counts may vary from strain to strain. The only recourse is to make sure that strains show very large differences in virulence based on the best available counting methods. Another disadvantage is that nonfatal viral diseases must be assessed by inaccurate means such as clinical score. Finally, despite exceptions such as the ferret for influenza virus (56), the absence or questionable validity of animal models for human diseases (e.g. measles) is as troublesome for viruses as it is for bacteria, if not more so. Hence, veterinary diseases have been used for fundamental studies on virus pathogenicity (5) and especially infections of mice with reovirus (19).

To sum up, methods of assessing bacterial and particularly viral virulence are fraught with difficulties. Animal diseases offer advantages for fundamental studies. Good animal models for human diseases are not readily available. This oft-seen Achilles heel of studies of pathogenicity will persist. It is epitomized by the absence of suitable animal models for human immunodeficiency virus (HIV).

Obtaining Strains of High and Low Virulence

The multifactorial nature of pathogenicity can be analyzed better by comparing strains of high and low virulence than by observations on a single strain. Screening of isolates for virulence differences was soon replaced by random, then specific, manipulation of strains.

The virulence of bacterial isolates has been lowered by repeated subculture and raised by animal passage since the time of Pasteur (54). Originally, the basis of these methods, selection of phenotypic and genotypic variants by conditions in vitro and in vivo, was not realized. Colony morphology, e.g. the smooth form, was sometimes correlated with virulence, but the presence or absence of particular virulence attributes was a matter of chance. Manipulation of pneumococcal virulence by Griffith in 1928 (cited in 54) heralded the identification of DNA as the genetic material in the 1940s. Paradoxically, it was some time before genetics was used widely for producing strains of differing virulence. The impetus came with plasmid-mediated transfer of virulence attributes in the late 1960s (58), which was soon followed by recombinant DNA technology and sophisticated mutagenesis.

Virus strains of differing virulence were obtained by methods similar to those used for bacteria. Attenuation of human isolates was achieved randomly by repeated growth in cell culture (yellow fever virus), eggs (mumps virus), or mice (poliovirus) (1, 19). Reassortment, i.e. genetic exchange between the segmented genomes of parent viruses, was shown originally for influenza virus (9). This method has been used as frequently as plasmids have been used in bacteriology for producing strains of differing virulence (19). Blunderbuss mutagenesis (17) has been superceded by selective methods and by recombinant nucleic acid technology (1, 19).

There are now fewer problems in obtaining bacterial and viral strains of varying virulence; they can often be tailored to specific needs.

Comparing Strains in Biological Tests Related to Pathogenicity

Chemical comparison of strains showing high and low virulence in animals relates genome components or gene products to overall virulence. It does not identify the role of these components in the complex, multifactorial mechanisms of virulence. Studies must thrust deeper to attain this goal. Strains must be compared in biological tests pertinent to mucosal infection, mucosal invasion, multiplication in vivo, interference with host defense, and causation of host damage. Such in-depth analysis of the multifactorial nature of pathogenicity has progressed more for bacteria than for viruses.

During the Golden Age of medical bacteriology (1876–1918), resistance to serum killing, interference with phagocytosis, and toxicity in animals received brilliant attention, but from 1920 to 1960 there were hardly any additions to the repertoire (54). Now, biological tests span all facets of pathogenicity. Tests related to mucosal adherence and invasion abound (21, 44). Factors affecting multiplication in vivo have been revealed by bacterial growth in body fluids with and without appropriate nutrients (52). As regards counteracting host defense, inhibition of serum killing and prevention of phagocytosis have been supplemented by prevention of complement activation or action, inhibition of chemotaxis and intracellular killing by phagocytes, and interference with or suppression of the immune responses (15, 38, 44). Toxins are still detected by animal experiments, but increasingly also by cytotoxicity to tissue culture cells (44, 57, 61). Finally, there is a panoply of immunological tests for cytotoxic, Arthus-type, and delayed-hypersensitivity reactions.

In studies of viral pathogenicity the spectrum of tests is much narrower, and comparisons between virulent and attenuated strains are less frequent. Whole-animal inoculation and replication have had the forum (45). The former gives little guidance on particular aspects of virulence, although unnatural routes of inoculation (e.g. intraspinal injection of poliovirus) can bypass some aspects (mucosal invasion and withstanding of defense mechanisms) for concentration on others (capacity to grow in and damage neural tissues) (1). Studies on replication have achieved spectacular results. For example, they have revealed the association of replication and virulence with ease of proteolytic cleavage of influenza virus hemagglutinin (29) and the role of the σ1 protein of reovirus in tissue tropisms of virulent and less virulent strains (19). Unfortunately, there has been little probing of other facets of viral pathogenicity. Access and adherence to mucous membranes, survival, and subsequent penetration have hardly been studied (15, 65). Nonspecific and immunospecific defenses against viruses and virus-infected cells are well understood (5, 19, 53). But apart from interaction with macrophages (15) and interferon (53), the relative abilities of virulent and attenuated virus strains to overcome these host defenses have received little attention. The types of tests that have yielded so much information on bacterial interference with host defense have not been used. For instance, whether or not virulent viruses resist ingestion and killing by polymorphonuclear (PMN) phagocytes more than less virulent strains has not attracted interest. Also, no one has enquired whether cells infected with virulent viruses are less easily killed by natural killer cells or cytotoxic T cells than those infected with attenuated viruses.

Cytotoxicity to tissue culture cells can be assessed by morphological effects or inhibition of macromolecular synthesis (2), but strains of differing virulence have not been compared in such tests. Virus-induced immunopathology has been well studied (5, 19), but again comparisons of virulent and less virulent strains have been rare. Against this negative background, studies on reovirus pathogenicity for mice (19) stand out. Resistance to proteolytic enzymes on mucous surfaces, penetration of mucous surfaces via M cells, replication in different tissues in vivo, virus spread from the gut (which implies interference with host defense), and shutoff of host cell macromolecular synthesis have all received attention. Then, in studies on influenza virus in ferrets, C. Sweet and I compared the abilities of virulent and attenuated strains to attack bronchial mucosa, resist fever temperatures, induce interferon, adsorb to and survive within phagocytes, and liberate endogenous pyrogen (reviewed in 56).

Identifying the Determinant Causing the Biological Property

Two methods have been used for identifying virulence determinants, chemical purification and genetic manipulation. The second now takes precedence in bacteriology as it has done in virology.

The chemical approach was first used for the classical bacterial toxins. Culture filtrates were fractionated using toxicity as an assay until purity was achieved. Neutralization of toxicity by specific antibody sealed the proof of causation. Identification of toxins and proof of causation is now commonplace (44, 61). Identifying a surface virulence determinant and proving that it causes the biological property are more difficult. The first step is demonstrating an association. More and less active strains are examined for differences in capsulation, pilation, and outer-membrane components. Also, the biological consequences of removing or destroying selected components enzymically or chemically are noted. Originally used to associate surface components with resistance to phagocytosis (54), such experiments have been popularized for all aspects of pathogenicity by the advent of sodium dodecyl sulfate polyacrylamide gel electrophoresis (SDS-PAGE) and Western blotting. But they demonstrate association: they do not prove causation (51, 60).

The putative determinants must now be removed from the bacteria, purified, and examined chemically. This is not usually difficult with modern methods. Proving causation is the real problem. How does one establish that the isolated, putative determinant causes the particular biological effect when present in situ on the bacterial surface? A common practice has been to demonstrate that the putative determinant confers biological activity (e.g. resistance to phagocytosis) on an otherwise inactive strain in tests where a solution of the determinant is present throughout (54). The determinant, however, may act extracellularly, as in opsonin neutralization by pneumococcal polysaccharides (38). The avirulent strain must be removed from the solution of the isolated determinant before being examined. Biological activity will then be due to the surface-attached determinant, which can be checked by fluorescent or gold-labeled antibody. Reattachment does not usually occur, but sometimes it happens when the test bacteria are closely related to the source of the putative determinant (51). The main method of proving causation is to neutralize the biological activity of intact virulent bacteria by antibody specific to the purified, putative determinant (60). Antibody specificity is vital. Monoclonal antibodies have an obvious role, e.g. in proving that lipopolysaccharide (LPS) is involved in the resistance to phagocytosis of the K1 strains of *Escherichia coli* (28).

My laboratory's work on gonococcal pathogenicity (39) exemplifies the chemical approach. Gonococci selected in vivo are more resistant to intracellular killing by human phagocytes than subcultured laboratory strains. Surface washes of the former but not the latter strains neutralized the ability of antigonococcal serum to abolish the intracellular resistance of the gonococci selected in vivo. SDS-PAGE of the surface washes of the strain selected in vivo associated a 20-kd component with intracellular resistance. This material, purified to form a single band on SDS-PAGE, was analyzed as a lipoprotein with a high glutamic acid content. When the susceptible laboratory strain was pretreated with the lipoprotein, intracellular resistance was conferred. Also, monospecific mouse antiserum to the lipoprotein abolished the intracellular resistance of the strain selected in vivo.

The relatively few chemical investigations related to pathogenicity of viruses show parallels with those for bacteria. For example, vesicular stomatitis virus lost its infectivity and glycoprotein spikes when treated with bromelain or pronase; infectivity was restored by treating the spikeless virus with purified glycoprotein and was neutralized by antiserum to it (6).

Proving Relevance to Infection In Vivo

We should be mindful here of the difference between relevance to infection in an animal model and in the natural host. For serious human diseases we often cannot prove relevance in the latter. Also, this section applies more to studies of bacterial than viral pathogenicity. The former are conducted largely with organisms grown in vitro, whereas the latter involve either virulence tests in animals or replication in live animal cells. These cells have some characteristics of cells in vivo, but they may have different virus susceptibilities (17). Also, virion components produced in subcultured cell lines may be different from those formed in vivo, e.g. the glycoproteins of influenza virus (45). Experiments with primary cells or organ cultures are better than those with cell lines for studies on pathogenicity.

The scene for proving relevance should be set before the quest for the determinant begins, by choosing a biological test related to a well-observed aspect of disease produced by the pathogen being considered. Reliable experimental pathology is paramount. Light and electron microscopy of early interactions of pathogens with mucous surfaces and with host defenses in subcutaneous lesions, the peritoneal cavity, the liver, or the spleen can form the basis of relevant tests. Quantitative surveys of tissues for bacteria can reveal localized growth that may be due to preferred nutrients such as erythritol for brucellae (52). The histology of subcutaneous lesions and infections of rabbit intestinal loops as well as clinical effects in humans can indicate tests for toxins. The logic of making such preliminary observations to choose relevant biological tests is inescapable. Sometimes, however, the tests are chosen for ease of operation rather than relevance to the disease. Testing lethality for mice by unnatural routes of inoculation and resistance to serum killing are examples. Also, tissue culture tests are increasing because of convenience and repeatability. Sometimes the cells are related to the situation in vivo, e.g. endometrial cell lines (47), but often unrelated HeLa or Vero cells are used (24, 57). Tissue culture tests vary in their reflection of the disease process. Bacterial adherence to cultured cells has little meaning if the pathogen infects a mucous surface with an overlying glycocalix (21, 44). Penetration of HeLa cells showed the relevance of a 140-Md plasmid to mucosal invasion by shigellae, but experiments with intestinal loops or whole animals demonstrated that the plasmid was not the sole requirement (24, 33). Vero toxins, discovered by their cytotoxicity to Vero cells but not some other cell lines, were later shown to be associated with human disease, enterotoxic in ligated gut loops, and lethal for mice and pigs (57).

The presence in vivo of putative virulence determinants should be demonstrated. This once-difficult task has been resolved by modern methods for examining small numbers of bacteria from animals (see below). Demonstrating biological activity of bacterial toxins in vivo has been relatively easy; toxin effects can be compared with those of the naturally occurring disease, and passive and active immunoprotection based on antibodies to toxins can be investigated (44, 61). Direct demonstration of the biological effect of surface virulence determinants in vivo is virtually impossible. Their interactions with mucous surfaces and/or host defenses are too complex and may be concurrent with interactions of other determinants, as with the antiphagocytic and anti-complementary effects of the LPS and capsular polysaccharides of *E. coli* K1 (28). The aggressin test (promotion by the putative determinant of lethal infection with an otherwise sublethal dose of organisms), once much used (54), proves overall interference with defense, not interference with a particular aspect. Biological relevance in vivo is proved by protecting animals against infection with specific antibody that neutralizes, in a separate test, the determinant's particular biological effect. Monoclonal antibodies have been increasingly used (28, 44).

SMITH

Now I come to my research theme for 40 years, the need to search in vivo for hitherto unknown virulence determinants, which I expressed in this review series in 1958 (49). Bacteria grown in vitro may be deficient in some of the virulence determinants present in vivo because of phenotypic change and selection of types. When in vitro experiments fail to provide answers to important aspects of pathogenicity, direct examination of organisms grown in vivo can provide further clues. This philosophy solved one of the enigmas of pathogenicity, the cause of death from anthrax. A relevant toxin was found in the blood of guinea pigs dying of the disease (55). It was later produced in vitro and shown to have three components. Also, the cholera toxin was indicated by fluid accumulation in ligated rabbit gut loops infected with *Vibrio cholerae* (54). The philosophy was slow to catch on (54), but it is now well recognized owing to techniques such as SDS-PAGE and Western blotting, whereby unfractionated surface components can be analyzed in the relatively few bacteria obtained from in vivo sources (8). There have been many such examinations (44, 54). For example, the pilin of gonococci obtained from urethral infection of volunteers was different from that of the inoculated gonococci (63). Most experiments, however, have dealt with the unspecified effect of the in vivo environment on overall virulence, a related biological property, or a putative determinant. Specific host factors that induce particular virulence determinants have only been identified in two cases. Iron restriction in patients and experimental animals was found to cause *E. coli*, *Klebsiella pneumoniae*, *Proteus mirabilis*, *Pseudomonas aeruginosa*, and *V. cholerae* to produce one or more outer membrane proteins, some of which help the pathogen to derive iron for growth in vivo (23). The second case was my laboratory's recent work on host-induced resistance of gonococci to complement-mediated killing by human serum (36, 40). This resistance, shown by gonococci isolated directly from urethral exudates, was lost phenotypically on subculture in vitro but was restored by incubation with human urogenital secretions, serum, and red and white blood cell lysates. The resistance-inducing activity is due to nanogram quantities of a sialylating agent, cytidine 5'-monophospho-*N*-acetyl neuraminic acid (CMP-NANA) or a closely related compound. The silver staining characteristics of gonococcal LPS were profoundly altered during induction of resistance, and sialylation of LPS was demonstrated by using CMP-NANA radiolabeled on the NANA moiety. Serum resistance almost certainly results from the change in LPS, which prevents it from acting in its normal role as the target antigen for bactericidal immunoglobulin M (IgM) in human serum. These results, important in relation to gonorrhoea, may have wider implications because CMP-NANA is widespread in mammalian tissues and LPS is ubiquitous among gram-negative pathogens.

Relating Chemical Structure to Biological Function

The three-dimensional chemical structures of some bacterial toxins have been related to their biological action by the most sophisticated molecular biology. Diphtheria and cholera toxins were in the vanguard, with *E. coli*, tetanus, and botulinum toxins close behind. Intricate studies of protein, enzyme, and membrane chemistry have been followed by DNA and peptide sequencing, X-ray analysis, site-directed mutagenesis, and mapping of epitopes by monoclonal antibodies (4, 61). The objective, pinpointing sites that determine toxicity, has been achieved in some cases (13, 64). Shiga, pasteurella, Vero, and other toxins are now being investigated in the same way (44, 57). The structures of adhesins from gonococci, *E. coli,* and streptococci are known, and information on the structures of their receptors is emerging (3, 42, 44, 59, 62). Nevertheless, examination of adhesin-receptor interaction is not yet at the same level as work on toxins. The determinants of mucosal invasion are not sufficiently defined for such investigations (24, 33), and molecular studies of multiplication in vivo have been confined to iron limitation of *E. coli* growth (23). Regretably, interference with host defense is not yet being studied at the structural level even though the chemistry of some relevant bacterial surface components is clear and a few host defense mechanisms are fairly well defined, e.g. the complement cascade and intracellular killing by phagocytes (38, 44, 52). Even the manner by which some defenses are counteracted is known, e.g. prevention of oxygen burst and phagosome lysosome fusion (44, 52), but the underlying chemical reactions have not been identified.

The molecular biology relating hemagglutinin structure to the replication and pathogenicity of influenza virus is as renowned and brilliant as that on bacterial toxins (19, 29, 45, 68). The hemagglutinin determines virus attachment and entry into the cell. The attachment site for the sialic acid of the cell receptor is in a surface pocket of conserved amino acids at the distal end of the protein. This site was identified after sequencing of the hemagglutinin via DNA copies of the RNA encoding segment and X-ray analysis of both the hemagglutinin alone and its complex with sialic acid. Cell entry relies on a conformational change in the hemagglutinin induced by acidity of endocytotic vesicles which promotes fusion of their membranes with the virus envelope. Proteolytic cleavage of the hemagglutinin at maturation and budding is necessary for virus infectivity. Cleavage occurs more easily and in a greater variety of animal cells in the virulent avian viruses than in the less virulent human viruses (29). Ease of cleavage depends on the number of basic amino acids at the cleavage site and stereochemical interactions with neighboring amino acids or oligosaccharides (29). The surface F glycoproteins of the paramyxoviruses, which also determine cell entry, have been subjected to similar studies (45). Facets of viral pathogenicity other than replication have not been subjected to molecular biological study, but the procedure of relating virus genome sequences and polypeptide primary structures to overall virulence is popular (19, 45). For example, comparisons of nucleotide sequences of cDNA from the genomes of neurovirulent and vaccine poliovirus strains, coupled with similar observations on recombinant viruses prepared with infectious cDNA, pinpointed the attenuating mutations (1). Only two were needed for poliovirus type 3. A change from cytidine to uridine at position 472 in the noncoding region of the genome caused a 70–80% reduction in neurovirulence because of lack of replication in the gut and spinal cord. Completion of the attenuation occurred when cytidine was changed to uridine at position 2034, which resulted in a serine-to-phenylalanine substitution in virus protein 3.

IMPACT OF GENETIC MANIPULATION

Genetic manipulation, particularly mutagenesis, has been used in studies of pathogenicity since the 1950s, but recently its versatility and precision have mushroomed. It can be used both to identify virulence determinants and to pinpoint their active sites. The essence of the first procedure is to remove a virulence-related biological property from a positive strain or to confer it on a deficient strain by deletion or transfer of a single gene or gene cluster whose product (the virulence determinant) is known or can be identified (16). It is paramount that only one genetic locus be involved. The second procedure involves detecting, deleting, or changing codons within genes that affect the incorporation into peptides of a few or even a single amino acid. The methods of genetic manipulation in studies of pathogenicity have been described elsewhere (1, 19, 60, 69). I concentrate on the successes and difficulties in their application.

Bacterial Pathogenicity

Genetic manipulation has been used mainly for proving causation and relevance in vivo of putative virulence determinants suggested by other studies, rather than for finding new determinants. There are, however, exceptions, such as the cell-invasion determinants of shigellae (33) and *Listeria monocytogenes* (M. Khun & W. Goebel, paper submitted), which foreshadow further discoveries using this technique. The method has a particular advantage over the chemical approach for dealing with surface determinants. Whether determinants form on the intact organism depends on the presence or absence of the gene. Reattachment of an isolated determinant to the surface is not needed. Originally, methods that involved gene transfer were largely confined to enteric bacteria (69), but now the range of bacterial species has been extended. Techniques such as electroporation will continue to extend it (18).

Chemical mutagenesis produces random multiple mutations and, used unguardedly, can complicate attempts to identify virulence determinants (60). The mutations should be introduced into nonmutagenized isogenic strains to reduce silent mutations and then critically analyzed to demonstrate only a single mutation. The more specific transposon-induced mutagenesis has now largely superceded chemical methods. Although Tn5 or Tn7 insertions, for example, are random, the particular genes inactivated can be identified either by mapping the transposon marker or by DNA hybridization (11, 70). The technique has wide application, e.g. in proving causation of biological effect and relevance in vivo for the cytolysins of *E. coli, Aeromonas hydrophilia,* and *L. monocytogenes* (10) and in producing salmonella mutants less able to grow in vivo because of impaired metabolic pathways (11, 37). There are snags. Sometimes transposon insertion does not occur or is followed by polar effects; either event leads to uncertainty as to whether the phenotype results from inactivation of the appropriate gene. Site-directed mutagenesis is even more specific. It has many variations, but all rely on knowing where the target DNA area is located. Two examples among many are single amino acid substitutions in diphtheria toxin fragment A and in the exotoxin of *P. aeruginosa,* which showed that glutamic acid residues at positions 148 and 553, respectively, are crucial for ADP ribosylation and toxicity (13, 64).

Plasmid transfer of virulence attributes was discovered and became popular in contributing to proving that the adherence antigens, toxins, and hemolysins of *E. coli* are virulence determinants active in vivo (14, 58). Now it is important in studies of all aspects of bacterial pathogenicity (44). Finding that a plasmid carries a virulence attribute is, however, only the first stage, since a plasmid can contain many genes coding for numerous peptides. The plasmid must then be analyzed to identify the single gene or gene cluster responsible for biological activity. This identification is relatively easy for determinants located on smaller plasmids such as those containing the genes coding for the adherence antigens and hemolysins of *E. coli* (10, 12, 44). The large 140-Md plasmid associated with epithelial cell invasion of shigellae is proving more difficult to analyze, although some progress has been made (24, 33). It now appears that proteins b (57 kd) and c (43 kd), products of a 22-Md fragment of the large plasmid, may be the invasion determinants (24).

Gene cloning has been a great boon (4, 16, 69). Cell components of fastidious pathogens that previously were not readily available, e.g. those of *Treponema pallidum* (66), can be produced in quantity for meaningful demonstration of biological activity and influence on virulence. More importantly, the cloned gene for the putative virulence determinant can be reintroduced into deficient strains of the relevant pathogen to see if biological activity and virulence are conferred. Then, before reintroduction the gene can be deleted by site-directed mutagenesis with consequent loss of biological effect and virulence. The numerous successes of gene cloning (44) include the characterization of not only genes coding for protein determinants, but also those responsible for capsular polysaccharide production by, for example, *E. coli* (43, 48) and *Neisseria meningitidis* (G. J. Boulnois, unpublished observation). Successful use of gene cloning for pathogenicity studies depends on (*a*) paring DNA fragments down to the smallest unit, optimally one gene; (*b*) accurate transcription and translation of the cloned gene in the host cell; (*c*) adequate expression of the gene; and (*d*) effective retransfer of the cloned gene from a convenient host cell back to the appropriate pathogen. Sometimes these goals are difficult to achieve (10, 69, 70).

Viral Pathogenicity

Viruses have smaller genomes than bacteria, so genetic manipulation, where possible, is less complex.

Chemical mutagenesis, with selection of mutants by temperature sensitivity, host dependence, plaque size, and other methods (17, 19), is still used in studies of pathogenicity (45) despite the availability of more specific methods. Such mutants provided a strong indication that early gene function of adenovirus causes pneumonia, at least in the cotton rat model (22). Now, however, deletion, point, and insertion mutagenesis techniques dominate the viral scene (19, 45). These methods have been used to probe the functional domain of the attachment $\delta 1$ protein of reovirus (35) and to identify a region of the envelope gene of HIV that is critical for infectivity (71).

Reassortants are now used in pathogenicity studies on many viruses (19, 45). The original work with reovirus and influenza virus shows the potential and limitations, respectively. Comparison in mouse tests of parent reoviruses (types 1 and 3) with reassortants having genome sections of known parental origin indicated several virulence determinants (19). The $\delta 1$ polypeptide determines spread from the gut to the spleen of 10-day-old mice and neuronal cell tropism leading to fatal encephalitis. The shutoff of host-cell RNA and protein synthesis appears to be due to the $\delta 3$ protein. Studies with influenza virus have been less rewarding. Occasionally particular genome segments from virulent parents appear to influence a facet of pathogenicity. For example, the P1 and nucleoprotein segments appear to influence ability to replicate at fever temperatures (56). However, viewed broadly over numerous experiments with avian and human viruses, no single gene appears to determine either full virulence or particular facets of it (29, 56). An optimal combination of genes is needed. Nonpathogenic reassortants can be derived from two pathogenic parents and vice versa. In using reassortants, SDS-PAGE can reveal subtle differences in the same gene segments between the parent viruses. Also, owing to variants in the parent populations or to point mutations during reassortment, reassortants with genome segments of the same parental origin can differ in some facets of virulence. For example, a pair of reassortants from an A/Finland/4/74–A/Okuda/57 system differed in ability to produce fever in ferrets (56).

Strains with instructive differences in pathogenicity can be produced either by recombination between single genome viruses or by cotransfection with nucleic acid fragments. Genome differences are identified by sequencing and restriction endonuclease mapping. As an example, an intertypic recombinant of virulent herpes simplex virus 1 and 2 was nonvirulent for mice (32). Cotransfection with DNA fragments showed that the virulence gene was between 0.71 and 0.83 genome map units. The ability of a fragment containing these units to restore the virulence of the recombinant was destroyed by cutting with restriction enzyme *Eco*RI at 0.72 units. Since virulence was also restored by another fragment of 0.71–0.75 units, the virulence gene appeared to be near the *Eco*RI site.

To sum up, genetic manipulation has transformed the study of pathogenicity. Difficulty may arise in using a particular method, but now there is such a variety of powerful techniques that if one does not work, another will probably substitute.

FUTURE OF STUDIES ON PATHOGENICITY

Two measures should be taken to promote progress. First, ongoing studies and additional examples should be taken up the research ladder, i.e. from detection of a relevant biological property through identification of the putative determinant and proof of causation and relevance in vivo to investigations of structure and function. Second, neglected areas should receive attention.

Mounting the Research Ladder

Bacterial adherence is ready for structure and function studies. The chemistry of some adhesins and their receptors is clearer (see above), and investigation of their interaction is the next step. Invasion of mucosal epithelium by shigellae, gonococci, salmonellae, and other bacteria is receiving much attention at present using cell models of varying relevance (20, 24, 33, 44, 47). The determinants must be identified before research can progress further. Bacterial multiplication in vivo has been neglected. Structure and function studies on interference with host defense using knowledge of the bacterial and host factors involved (see above) are long overdue. The progress of studies on toxins toward the top of the ladder will not slacken.

The present spate of work relating virus genome and protein structure to overall virulence will continue; there are many different viruses to examine, and analysis of primary and tertiary chemical structures is well under way.

Also, more virion components and their structures will be related to replication and pathogenicity. In addition, replication processes that follow adsorption and penetration may receive more attention (53).

One ever-present problem in mounting the research ladder, that of obtaining relevant animal models, may be mitigated in the future by use of transgenic animals (G. F. Boulnois, private communication). For example, mice might be constructed that express human glycoproteins on their mucosal cells or various components of the human defense system, thus allowing some aspects of pathogenicity to be studied in a more natural host.

Neglected Areas

BACTERIA   Enteric, respiratory, and urogenital infections cause worldwide distress, yet a crucial first stage in these debilitating infections remains a mystery. How do small numbers of shigellae, streptococci, *E. coli,* and other bacteria overcome the protective effects of the normal mucosal flora (44, 50)? This phenomenon should be investigated at the biological and determinant level. The work of Freter (reviewed in 44) provides a starting point.

Recent meningococcal outbreaks raise the old problem of commensal-pathogen conversion. What microbial and host factors determine the change? Does it occur in individuals, or must the microbe move to another host? The absence of animal models and the rarity of natural occurrences handicap research. Studies should concentrate on natural outbreaks, with potential virulence attributes of the isolates and the nutritional environment and defense mechanisms of the patients coming under scrutiny.

Close to commensal-pathogen conversions are opportunistic infections, e.g. by coagulase negative staphylococci. What are the reasons for sickness? Do the microbes produce toxins or liberate endogenous pyrogen? If the latter, what is the microbial product concerned? The major handicap to research is low overall pathogenicity manifest only in compromised patients. Relevant animal models are hard to find, and so are, strains of sufficiently different virulence for meaningful comparisons in biological tests. Nevertheless, answers must be obtained, because these opportunist infections in compromised patients are the main problems in hospitals today.

Rate of growth is important in pathogenicity. If growth is slow, host defenses are more able to cope than if it is fast. What nutrients are present in vivo? Which determine growth rate at different stages of infection? How do they affect metabolism? The lack of research is in stark contrast to the big effort on viral replication. It is also surprising because of the success of studies on iron availability and pathogenicity (23). Other limiting nutrients should be recognized, and their effects should be investigated. Nutrients responsible for rapid growth later in infection should also be identified, and their effects should be examined. Preferred nutrients localized in certain tissues can determine tissue specificity. My colleagues and I showed erythritol to be the cause of localization of brucellae in fetal tissues of domestic animals, which leads to abortion (52). This work, like that on iron limitation, pointed to further studies on nutrition and pathogenicity, but they have not occurred. To remedy matters, experts on microbial metabolism are needed. Some approaches are indicated above. Another comes from an unexpected source. To provide *S. typhimurium* of reduced virulence as a vehicle for genetically engineered vaccines, mutants unable to synthesize aromatic amino acids and purines were prepared (11, 37). These mutants have impaired growth in vivo. By comparing rates of growth in vivo of mutants blocked at differing points along a metabolic chain, clues should be gained about the nutrients available and their influence on growth rate. Such an approach need not be limited to aromatic amino acid and purine metabolism.

The increased attention paid to the influence of the in vivo environment on pathogenicity is gratifying, but the time has come to be more specific. At present, there have been only two identifications of host factors responsible for producing particular virulence determinants. Iron limitation of gram-negative pathogens induces new outer-membrane, iron-processing proteins (8, 23). CMP-NANA or a closely related compound in human blood cells induces LPS alteration in gonococci while converting them to serum resistance (36, 40). This work should be extended to more pathogens and other aspects of pathogenicity. The effects of nutrients or nutrient deficiencies in vivo (see above) on the production of virulence determinants should be examined. Chemostat cultures could help; when iron was limited in such cultures, in vivo conditions were simulated and new outer-membrane proteins were formed that were identical to those in bacteria from patients (8, 23). Another recent development is exciting. The observation that the *toxR* gene of *V. cholerae* regulates production of both toxin and pilus-colonization factor and similar observations on other pathogens suggest that production of different virulence determinants may be under a common control system influenced by environmental factors (4, 34). The environmental factors in vivo could therefore affect the control systems, but at present neither these factors nor their influence has been defined. The effect on such control systems of environmetal conditions known to occur in vivo should be examined.

Carrier and latent states, i.e. long-term survival of pathogens in the host in the presence of immunospecific defenses, are important in human and veterinary medicine because they spawn future epidemics. In contrast to great interest in viral persistence (41), bacterial carrier states are often neglected. I have suggested that immunosuppression, poor antigenicity of virulence determinants, antigenic shift, and intracellular habitat either in epithelial cells or in impaired macrophages might have roles (15). However, there is neither proof of their involvement nor any idea of the determinants responsible. The carrier state remains a mystery. Research is necessarily long term, and this deters university staff, who rely on 3–5-year grants. The problem is, however, too important to neglect.

Finally, I describe three underworked areas of bacterial pathogenicity. (*a*) Cytotoxic, Arthus-type, delayed hypersensitivity, and autoimmune reactions are clinically important in bacterial infections. The determinants of immunopathological damage are, however, largely neglected except for streptococcal cell-wall antigens (52). (*b*) Tissue and host specificities of bacterial infection have been insufficiently studied despite experimental success in some instances. Possible explanations are the presence or absence in the tissue or host of receptors for adherence antigens (44), mucosal enzymes that can destroy invasion determinants (24), preferred nutrients (52), and competent defense mechanisms (31). Here, the use of transgenic animals (see above) is obvious. (*c*) Clinically important interactions among bacteria have received little attention compared with those between viruses and bacteria. Whenever bacterial interactions have been investigated, e.g. in heel abscess in sheep, periodontal disease, and abdominal abscess, determinants have been identified from experiments with natural hosts or animal models (7, 50). These successes should encourage further work.

VIRUSES   Knowledge of viral replication is outstanding, and the attention given to some subjects, e.g. virus persistence (41), is commendable. Overall, however, viral virulence has yet to be analyzed in the depth achieved for bacteria. Four of the cardinal requirements for pathogenicity need attention.

Respiratory and enteric virus diseases are rife, yet the mechanisms whereby viruses first establish themselves on mucous surfaces are largely unknown (15, 53, 65). What viral and host factors are responsible for viruses gaining access to epithelial surfaces through the overlying mucus, adhering to the surfaces, and overcoming there the inhibitory mechanisms provided by either the host or possibly commensal microorganisms? Virulent and attenuated strains have not been compared in any of these events, and there have only been speculations on what might happen. Experiments with reovirus in mice on resistance to intestinal enzymes (see above) and binding to epithelial cells (67) show the way.

Penetration of mucous surfaces by viruses such as rubella and varicella, which are not known to infect epithelial cells, is a mystery (15, 53, 65). Comparisons between virulent and less virulent viruses have not been made, nor has radiolabeled virus been used to track penetration.

Little is known, either biologically or molecularly, about the factors that enable virulent viruses but not attenuated ones to interfere with host defense. A few pertinent results show what can be done by applying the methods already described. A greater content of sialic acid in Sindbis virus envelope glycoproteins reduced complement activation and hence bloodstream clearance (19). Sialic acid on the surface of caprine arthritis-encephalitis virus probably protects against host proteases (27). The p15E protein of feline leukemia virus suppresses neutrophil and natural killer cell functions (25, 30).

Host-cell damage results from either viral cytotoxicity or immunopathology. Only a few determinants of cytotoxicity, either morphological damage or impaired macromolecular synthesis, have been identified (2, 46, 53). Most are virion components, e.g. the tubular protein of vaccinia virus, the structural proteins of frog virus 3, and the $\delta 3$ protein of reovirus. One virus-induced cytotoxin has been purified from infected cells, namely short RNA transcripts from vaccinia virus that shut off host-cell protein synthesis. Whenever host-cell damage occurs without virus replication, cytotoxins should be suspected. Recent examples are lung damage resulting from early gene products of adenovirus (22) and liberation of endogenous pyrogen (the cause of sickness in influenza) from phagocytes by inactivated influenza viruses (56). Although viral immunopathology has attracted much interest (5, 19), attempts to identify the viral determinants have yet to begin.

CONCLUSIONS

In the past 40 years there has been a gigantic increase of interest and progress in bacterial and viral pathogenicity. Obtaining valid animal models remains a problem, and use of relevant biological tests must be constantly guarded. Identification of virulence determinants and proof of causation and relevance in vivo have been transformed by genetic manipulation. More attention is being given to pathogens in vivo, and structure and function studies have been increasing. The major gap in our understanding of bacterial pathogenicity is lack of knowledge of the nutrients and metabolism that underlie growth in vivo. In viral pathogenicity, overwhelming attention has been given to replication, but mucosal invasion, mucosal penetration, and interference with host defense mechanisms have been neglected, and damage to host tissue has been understudied. These omissions should be remedied. In all aspects, the high level of biochemistry, genetics, and molecular biology must be matched by sound and innovative experimental pathology, which lays the basis for relevant biological tests. I am well satisfied with the progress so far and the future prospects of the subject I love.

ACKNOWLEDGMENTS

I am grateful to T. W. Almond, D. H. L. Bishop, C. F. Boulnois, D. Coates, A. Cockayne, J. A. Cole, G. Dougan, C. W. Penn, D. Strugnell, and C. Sweet for their critical reading of the manuscript.

Literature Cited

1. Almond, J. W. 1987. The attenuation of poliovirus neurovirulence. *Annu. Rev. Microbiol.* 41:153–80
2. Bablabian, R. 1975. Structural and functional alterations in cultured cells infected with cytocidal viruses. *Prog. Med. Virol.* 19:40–83
3. Beachey, E. H., Courtney, H. S. 1987. Bacterial adherence: the attachment of group A streptococci to mucosal surfaces. *Rev. Infect. Dis.* 9:S475–81
4. Betley, M. J., Miller, V. L., Mekalanos, J. J. 1986. Genetics of bacterial enterotoxins. *Annu. Rev. Microbiol.* 40:577–605
5. Bielefeldt-Ohmann, H., Babuik, L. A. 1986. Viral infections in domestic animals as models for studies of viral immunology and pathogenesis. *J. Gen. Virol.* 66:1–25
6. Bishop, D. H. L., Repik, P., Obijeski, J. F., Moore, N. F., Wagner, R. R. 1975. Restitution of infectivity to spikeless vesicular stomatitis virus by solubilized viral components. *J. Virol.* 16:75–84
7. Brook, I. 1986. Encapsulated anaerobic bacteria in synergistic infections. *Microbiol. Rev.* 50:452–57
8. Brown, M. R. W., Williams, P. 1985. The influence of environment on envelope properties affecting survival of bacteria in infection. *Annu. Rev. Microbiol.* 39:527–56
9. Burnet, F. M., Lind, P. E. 1951. A genetic approach to variation in influenza viruses. 4. Recombination of characters between the influenza virus A strain NWS and strains of different serological subtypes. *J. Gen. Microbiol.* 5:67–82
10. Chakraborty, T., Kathariou, S., Hacker, J., Hof, H., Hickle, B., et al. 1987. Molecular analysis of bacterial cytolysins. *Rev. Infect. Dis.* 9:S456–66
11. Curtiss, R. III, Kellig, S. M., Gulip, P. A., Gentry-Weeks, C. R., Galan, J. E. 1988. Avirulent salmonellae expressing virulence antigens from other pathogens for use as orally administered vaccines. See Ref. 44, pp. 311–28
12. Dougan, G., Sellwood, R., Maskill, D., Sweeney, K., Liew, F. Y., et al. 1986. In vivo properties of a cloned K88 adherence antigen determinant. *Infect. Immun.* 52:344–47
13. Douglas, C. M., Collier, R. J. 1987. Exotoxin A of *Pseudomonas aeruginosa:* Substitution of glutamic acid 553 with aspartic acid drastically reduces toxicity and enzymic activity. *J. Bacteriol.* 169:4967–71
14. Elwell, L. P., Shipley, P. L. 1980. Plasmid-mediated factors associated with virulence of bacteria to animals. *Annu. Rev. Microbiol.* 34:465–96
15. Falconi, G., Campa, M., Smith, H., Scott, G. M., eds. 1984. *Bacterial and Viral Inhibition and Modulation of Host Defences.* London: Academic
16. Falkow, S. 1988. Molecular Koch's postulates applied to microbial pathogenicity. *Rev. Infect. Dis.* 10:S274–76
17. Fenner, F., McAuslan, B. R., Mims, C. A., Sambrook, J., White, D. O. 1974. *The Biology of Animal Viruses.* London: Academic
18. Fielder, S., Wirth, R. 1988. Transformation of bacteria with plasmid DNA by electroporation. *Anal. Biochem.* 170:38–44
19. Fields, B. N., Knipe, D. M., Chanock, M., Melnick, J. L., Roizman, B., et al. 1985. *Virology.* New York: Raven
20. Finlay, B. B., Gumbiner, B., Falkow, S. 1988. Penetration of *Salmonella* through a polarized Madin-Darby canine kidney epithelial cell layer. *J. Cell Biol.* 107:221–30
21. Freter, R., Jones, G. W. 1983. Models for studying the role of bacterial attachment in virulence and pathogenesis. *Rev. Infect. Dis.* 5:S647–58
22. Ginsburg, H. S., Lundholm-Beauchamp, U., Prince, G. 1987. Adenovirus as a model of disease. *Symp. Soc. Gen. Microbiol.* 40:245–58
23. Griffiths, E., Chart, H., Stevenson, P. 1988. High-affinity iron uptake systems and bacterial virulence. See Ref. 44, pp. 121–37
24. Hale, T. L., Formal, S. B. 1988. Virulence mechanisms of enteroinvasive pathogens. See Ref. 44, pp. 61–69
25. Harris, D. T., Cianciolo, G. J., Snyderman, R., Argov, S., Koren, H. S. 1987. Inhibition of human natural killer cell activity by a synthetic peptide homologous to the conserved region in the retroviral protein p15E. *J. Immunol.* 138:889–94
26. Henderson, D. W. 1955. The Microbiological Research Department, Ministry of Supply, Porton, Wilts. *Proc. R. Soc. London Ser. B* 143:192–202
27. Huso, D. L., Narayan, O., Hart, G. W. 1988. Sialic acid on the surface of caprine arthritis-encephalitis virus defines the biological properties of the virus. *J. Virol.* 62:1974–80

28. Kaufmann, B. M., Cross, A. S., Futrovsky, S. L., Sidberry, H. F., Sadoff, J. C. 1986. Monoclonal antibodies reactive with K1-encapsulated *Escherichia coli* lipopolysaccharide are opsonic and protect mice against lethal challenge. *Infect. Immun.* 52:617–19

29. Klenk, H. D., Rott, R. 1988. The molecular biology of influenza virus pathogenicity. *Adv. Virus Res.* 34:247–81

30. Lefrado, L. J., Lewes, M. G., Mathes, L., Olsen, R. G. 1987. Suppression of *in vitro* neutrophil function by feline leukemia virus (FeLV) and purified FeLV-p15E. *J. Gen. Virol.* 68:507–13

31. Lissner, C. R., Swanson, R. N., O'Brien, A. D. 1983. Genetic control of the innate resistance of mice to *Salmonella typhimurium:* expression of the *ity* gene in peritoneal and splenic macrophages isolated *in vitro. J. Immunol.* 131:3006–13

32. Marsden, H. S. 1987. Herpes simplex virus glycoproteins and pathogenesis. *Symp. Soc. Gen. Microbiol.* 40:259–88

33. Maurelli, A. T., Sansonetti, P. J. 1988. Genetic determinants of *Shigella* pathogenicity. *Annu. Rev. Microbiol.* 42:127–50

34. Miller, J. F., Mekalanos, J. J., Falkow, S. 1988. Coordinate regulation and sensory transduction in the control of bacterial virulence. *Science* In press

35. Nagata, L., Masni, S. A., Pon, R. T., Lee, P. W. K. 1987. Analysis of functional domains of reovirus cell attachment protein δ1 using cloned S1 gene deletion mutants. *Virology* 160:162–68

36. Nairn, C. A., Cole, J. A., Patel, P. V., Parsons, N. J., Fox, J. E., et al. 1988. Cytidine 5'-monophospho-*N*-acetyl neuraminic acid or a related compound is the low M_r factor from human red blood cells which induces gonococcal resistance to killing by human serum. *J. Gen. Microbiol.* 134:3295–3306

37. O'Callaghan, D., Maskill, D., Liew, F. Y., Easmon, C. S. F., Dougan, G. 1988. Characterization of aromatic- and purine-dependent *Salmonella typhimurium:* attenuation, persistence and ability to induce protective immunity in BALB/c mice. *Infect. Immun.* 56:419–23

38. O'Grady, F., Smith, H., eds. 1981. *Microbial Perturbation of Host Defences.* London: Academic

39. Parsons, N. J., Kwaasi, A. A., Patel, P. V., Nairn, C. A., Smith, H. 1986. A determinant of resistance of *Neisseria gonorrhoeae* to killing by human phagocytes: an outer membrane lipoprotein of about 20 kDa with a high content of glutamic acid. *J. Gen. Microbiol.* 132:3277–87

40. Parsons, N. J., Patel, P. V., Tan, E. L., Andrade, J. R. C., Nairn, C. A., et al. 1988. Cytidine 5'-monophospho-*N*-acetyl neuraminic acid and a low molecular weight factor from human blood cells induce lipopolysaccharide alteration in gonococci when conferring resistance to killing by human serum. *Microb. Pathog.* 5:303–9

41. Rapp, F., Cory, J. M. 1988. Mechanisms of persistence in human virus infections. *Microb. Pathog.* 4:85–92

42. Reid, G., Sobel, J. D. 1987. Bacterial adherence in the pathogenesis of urinary tract infection: a review. *Rev. Infect. Dis.* 9:470–87

43. Roberts, I. S., Mountford, R., Hodge, R., Jann, K. B., Boulnois, G. J. 1988. Common organization of gene clusters for production of different capsular polysaccharides (K antigens) in *Escherichia coli. J. Bacteriol.* 170:1305–10

44. Roth, J. A., ed. 1988. *Virulence Mechanisms of Bacterial Pathogens.* Washington, DC: Am. Soc. Microbiol.

45. Russell, W. C., Almond, J. W., eds. 1987. *Molecular Basis of Virus Disease. Symp. Soc. Gen. Microbiol.* Vol. 40

46. Shatkin, A. J. 1983. Molecular mechanisms of virus-mediated cytopathology. *Philos. Trans. R. Soc. London Ser. B* 303:167–76

47. Shaw, J. H., Falkow, S. 1988. Model for invasion of human tissue culture cells by *Neisseria gonorrhoeae. Infect. Immun.* 56:1625–32

48. Silver, R. P., Vann, W. F., Aaronson, W. 1984. Genetic and molecular analyses of *Escherichia coli* K1 antigen genes. *J. Bacteriol.* 157:568–75

49. Smith, H. 1958. The use of bacteria grown in vivo for studies on the basis of their pathogenicity. *Annu. Rev. Microbiol.* 12:77–102

50. Smith, H. 1982. The role of microbial interactions in infectious disease. *Philos. Trans. R. Soc. London Ser B* 297:551–61

51. Smith, H. 1983. The elusive determinants of bacterial interference with nonspecific host defences. *Philos. Trans. R. Soc. London Ser. B* 303:99–113

52. Smith, H. 1984. The biochemical challenge of microbial pathogenicity. *J. Appl. Bacteriol.* 57:395–404

53. Smith, H. 1986. The little known determinants of viral pathogenicity. In *Clinical Investigation of Infectious Diseases,* ed. D. A. J. Tyrrell, pp. 29–46. London: R. Soc. Med.

54. Smith, H. 1988. The development of studies on the determinants of bacterial pathogenicity. *J. Comp. Pathol.* 98: 253–73

55. Smith, H., Keppie, J. 1954. Observations on experimental anthrax: demonstration of a specific lethal factor produced in vivo by *Bacillus anthracis. Nature* 173:869–70

56. Smith, H., Sweet, C. 1988. Lessons for human influenza from pathogenicity studies with ferrets. *Rev. Infect. Dis.* 10:56–75

57. Smith, H. R., Scotland, S. M. 1988. Vero cytotoxin-producing strains of *Escherichia coli. J. Med. Microbiol.* 26:77–85

58. Smith, H. W., Linggood, M. A. 1971. Observations on the pathogenic properties of the K88, HLY and ENT plasmids of *Escherichia coli*, with particular reference to porcine diarrhoea. *J. Med. Microbiol.* 4:467–85

59. So, M., Billyard, E., Deal, C., Getzoff, E., Hagblom, P., et al. 1985. Gonococcal pilus: genetics and structure. *Curr. Top. Microbiol. Immunol.* 118:13–28

60. Sparling, P. F. 1983. Application of genetics to studies of bacterial virulence. *Philos. Trans. R. Soc. London Ser. B* 303:199–207

61. Stephen, J., Pietrowski, R. A. 1986. *Bacterial Toxins.* Walton-on-Thames, UK: Nelson. 2nd ed.

62. Stromberg, N., Deal, C., Nyberg, G., Normark, S., So., M., et al. 1988. Identification of carbohydrate structures that are possible receptors for *Neisseria gonorrhoeae. Proc. Natl. Acad. Sci. USA* 85:4902–6

63. Swanson, J., Robbins, K., Barrera, O., Corwin, D., Boslego, J., et al. 1987. Gonococcal pilin variants in experimental gonorrhoea. *J. Exp. Med.* 165:1344–57

64. Tweten, R. K., Barbieri, J. T., Collier, R. J. 1985. Diphtheria toxin. Effect of substituting aspartic acid for glutamic acid 148 on ADP-ribosyltransferase activity. *J. Biol. Chem.* 260:10392–94

65. Tyrrell, D. A. J. 1983. How do viruses invade mucous surfaces? *Philos. Trans. R. Soc. London Ser. B* 303:75–83

66. van Embden, J. D., van der Donk, H. J., van Eijk, R. V., van der Heide, H. G., de Jong, J. A., et al. 1983. Molecular cloning and expression of *Treponema pallidum* DNA in *Escherichia coli* K-12. *Infect. Immun.* 42:187–96

67. Weiner, D. B., Girard, K., Williams, W. V., McPhillips, T., Rubin, D. H. 1988. Reovirus type 1 and type 3 differ in their binding to isolated epithelial cells. *Microb. Pathog.* 5:29–40

68. Weis, W., Brown, J. H., Cusack, S., Paulson, J. C., Skehel, J. J., et al. 1988. Structure of influenza virus haemagglutinin complexed with its receptor, sialic acid. *Nature* 333:426–31

69. Weiss, A. A., Falkow, S. 1983. The use of molecular techniques to study microbial determinants of pathogenicity. *Philos. Trans. R. Soc. London Ser. B* 303:219–25

70. Weiss, A. A., Hewlett, E. L. 1986. Virulence factors of *Bordetella pertussis. Annu. Rev. Microbiol.* 40:661–86

71. Willey, R. L., Smith, D. H., Lasky, L. A., Thiodore, T. S., Earl, P. L., et al. 1988. In vitro mutagenesis identifies a region within the envelope gene of the human immunodeficiency virus that is critical for infectivity. *J. Virol.* 62:139–47

NEUROSCIENCE

Ann. Rev. Microbiol. 1988. 42:1–34

A STRUCTURED LIFE

Robert G. E. Murray

Department of Microbiology and Immunology, University of Western Ontario, London, Ontario, Canada N6A 5C1

CONTENTS

> One of the profound lessons to be learned from science is not what it has done, but the way it goes about doing it.
>
> Hanbury Brown: *The Wisdom of Science*

I may not have been introduced to thinking about microbes at birth, but this was a likely event soon afterwards. I count myself lucky that bacteriology, its history, and science in general lived for me because my father, E. G. D. Murray (1890–1964) (88), was a bacteriologist whose teachers were young contemporaries of the giants of the early days. A few of that earlier generation were still teaching at Cambridge when I was there as a student, which emphasized the bridge between scientific generations. They were long past

0066-4227/88/1001-0001$02.00

retirement but had valuable things to say about science and the early days. This gives me a good feeling about being on postretirement appointment, as I am now, and still having contact with dedicated students. Between us, my father and I have enjoyed devotion to bacteriology for almost three quarters of this century.

It is hard to decide from biographies whether or not a career in science is assisted by having a parent a scientist. At the least, it gives a more defined set of prejudices as a basis for deciding what to do in life, or what to avoid. At best, it provides a strong set of interests, standards, stimuli and concerns which shape the inevitable explorations of the world around us and direct the neophyte's progress insensibly and in subtle ways from tentative to defined courses of action. A wealth of understanding gained informally provides the "gilt on the gingerbread" that amplifies capabilities.

BEGINNINGS

The Murrays who were my forebears lived in South Africa from 1835 and for the most part were dedicated to the land; my mother's family was English and of mercantile bent. My grandfather, G. A. E. Murray, departed from those norms by qualifying in medicine, and he set up for surgical practice and family life in the infant city of Johannesburg in 1888. My father followed in his footsteps and studied medicine at St. Bartholomew's Hospital, London. He did a degree at Cambridge (Christ's College, 1912), with a special interest in zoology, which I was to profit from in years ahead. While still a student at St. Bartholomew's he was much influenced by Dr. Mervyn Gordon, and the two worked together on studies of meningococcal meningitis. This interest continued during World War I after my father's qualification, marriage, and army service. He was in the Royal Army Medical College Vaccine Laboratories and was concerned with the making of vaccines when I was born in 1919 in Ruislip, a London suburb. The war was over and the bacteriological and epidemiological emergencies of wartime were displaced by the influenza pandemic with all its alarming complications. Soon after I arrived on the scene the family moved to Cambridge, where my father was appointed a research bacteriologist for the Medical Research Council (1920–1926). He revived his wartime interests in the meningococcus while supervising the production of therapeutic antisera. My very early memories include attending the bleeding of large and elegant horses at the Field Laboratories outside Cambridge.

My father was soon back to Cambridge academic life in Christ's College and in the Department of Pathology, where he became a lecturer. His chief, and later my teacher, was the redoubtable Professor H. R. Dean. We shared energetic summer vacations with his family for several years. Professor Dean was a pathologist with interests in bacteriology and immunology. (He introduced the concept of optimal proportions into serology, and he also studied complement.) It was my father's task to help him develop an honors course in pathology, which included medical microbiology in those days and still does. The first class graduated in 1925, and the nine students, now very distinguished British medical microbiologists, included A. A. Miles, C. H. Andrewes, and two who taught me later on, E. T. C. Spooner, who later returned to Cambridge, and Frederick Smith, who was to join my father at McGill. They were forerunners of a considerable company of medical scientists, and I was to join them myself.

My childhood was that of a homebody, but life changed considerably in 1927 when I was sent to Summer Fields, a boarding school at Oxford devoted to helping as many pupils as possible gain scholarships to enter one or another of the great "public" schools. It was a tough time because I had remarkably little experience of children my own age. It was a good school and there were miseries; but we worked hard and played hard as required by the school's motto, *"Mens sana in corpore sano."* I have every reason to bless the school for the basics (in everything but music, alas) that have served me well.

In 1930 my father became professor and head of the Department of Bacteriology and Immunology in the Faculty of Medicine of McGill University, Montreal. It was an exciting transition for all of us and one we embraced enthusiastically, even if I never managed more than minor modifications to an English accent. I was old enough to be more aware, and being at home for much of the year, I could now hear what was going on in academic life. So Montreal became my second home. As a great metropolis encompassing two cultures, the city provided an ambience of great importance to my development. The change in school systems was a traumatic experience, but happily I went to Lower Canada College, an English-type boy's school that offered a suitable program and games familiar to me, such as soccer and cricket. What helped me most in becoming Canadian without too many tears was the enthusiasm with which my parents took to the country and its pace, space, and natural beauties.

By 1936 I was ready for McGill University and completed the matriculation examinations, which then still required Latin. There was no science requirement, and indeed I did only a minimal combined chemistry and physics course at school. So McGill provided the grand and eagerly awaited start in the nuts and bolts of science and the opportunity to learn under senior and very capable teachers whom, in many instances, I already knew and admired. Largely because of my father's early interest in and retentive memory for zoological information, I set out with an intention to do zoology. However, all courses were interesting, and I performed well in the examinations for the two years 1936–1938.

There were omens of war in Europe, and evidently it would be now or never if I were to go to Cambridge University, which meant so much to my father and which I wanted with all my heart. Arrangements were made for me to enter Christ's College in October 1938. However, there was a summer to fill in, and I had the opportunity to go to a marine biological station for a course in invertebrate zoology at Salisbury Cove, near Bar Harbor in Maine, under Professor Ulric Dahlgren of Princeton University. The course was fun and influential; I was later to marry (in 1944) one of my classmates (Doris Marchand, also from a scientific family). Also, the scientific community provided much informal discourse with first class people such as Homer Smith and Gairdner Moment. After this very educational interlude I set off for England by ship during the Munich Crisis in September 1938.

Cambridge was a revelation to me, not just because of the quality of teaching but because of what was expected of the students. It was a matter of attitude and preparation. The attitude was expressed in a way by my tutor, C. P. Snow, who said (and he didn't say much more), " . . . you realize you have to present yourself for examination in May, two years hence." We had to pace ourselves both in and out of term, keep up, and do a wide range of independent reading, quite apart from lectures and laboratories. In this we were assisted by a supervisor of studies (I was assigned to John Yudkin, later a distinguished nutritionist, who then had interests in *Escherichia coli* physiology). We were expected to have the basics or quickly acquire them; I thought myself inordinately well prepared, but to my chagrin, my scholarly classmates just out of school had a remarkably well-honed understanding of basic physics and chemistry. The teaching also reflected attitude. For instance, the lectures and laboratories in introductory biochemistry were given by Ernest Baldwin. The first-term course was on comparative biochemistry, given in a fascinating fashion. Baldwin later went on to systematic biochemistry and the outlines of what was to become a distinguished textbook. It was good stuff and we learned a lot.

I was enrolled in medical studies, which meant the inclusion of anatomy, biochemistry, physiology, pathology (including bacteriology), and pharmacology in the first two years. Despite a rather miserable performance in anatomy, I was able to go on to a third year (Part II of the Natural Sciences Tripos) spent entirely in one disciplinary area, which was in my case pathology and bacteriology, still under Professor Dean. The Part II course in pathology was a combined effort coordinated for the most part by G. Williamson (pathology), E. T. C. Spooner, and A. W. Downie (medical microbiology). Downie was then running the Emergency Medical Service Laboratory in Cambridge. He was dealing with problems ranging from streptococcal diseases to leptospirosis, which he and his colleagues shared with the five of us students in exemplary fashion. Likewise, A. M. Barrett saw that we got all the tissue and bacteriological specimens that we needed from the autopsies he did for Addenbrooks Hospital. It was an extraordinarily thorough experience in the details and practice of medical bacteriology and pathology, full of interests to suit each of us. There were special additions: parasitology at the Molteno Institute and special series of lectures by senior (some very senior) colleagues such as G. S. Graham-Smith (toxins and venomous animals and insects), Louis Cobbett (tuberculosis), R. I. N. Greaves (immunology and the preservation of blood and sera), R. R. Race (blood grouping), and L. Faulds (cancer research). To me it was sad but not distressing that they excluded much general microbiology, which I heard a little about from a few fellow students in botany. The biochemistry department included some distinguished contributors to microbial biochemistry, and the group of us arranged an informal session or two with Marjory Stephenson and W. E. van Heynigen. As it turned out, I did not cover myself with glory and received a Class II overall, which was a bit humbling since my classmates John Stowers, Leo Wolman, and Alec Comfort all were Class I, as they have shown by their distinctions in later life. But after this year I understood what kind of advanced bacteriology course my father had developed at Cambridge (when Part II was initiated) and later introduced in refined form at McGill University.

Graduation in wartime brought the necessity of a clear track in medicine, in research, or in military service. My decision, encouraged by Alan Downie, came with the news of my acceptance at the McGill Medical School. The issuance of an exit permit and my assignment to a ship for the Atlantic crossing was slow, and I did not sail until late October 1941. I effectively spent the four-month interval on the wards of Addenbrooks Hospital learning physical diagnosis, which helped me considerably. I arrived back in Canada just before the invasion of Pearl Harbor changed the aspect of the war, after an exciting three-week voyage in convoy. We had few losses thanks to Canadian corvettes and the newly acquired lend-lease destroyer escorts supplied by the United States.

McGill had introduced an accelerated medical course with no vacations, and I settled down to a rather exhausting routine. Graduation came late in 1943, followed by internship at Royal Victoria Hospital. My internship was unconventional because it was based in the clinical laboratory of the Department of Bacteriology, which provided services for the hospital. This internship gave me clinical experience in most departments and laboratory experience in the fullest possible range of microbial diseases. The experience was all the better for the teaching of Fred Smith (who was holding the fort for my father while he was on war service) and for the practical clinical laboratory instruction of Gertrude Kalz, who really put the polish on my bacteriological training, for which I am ever grateful. Internship also brought me marriage and the start of family life.

The army called in 1944 with suggestions of using my laboratory competence, but predictably, after the usual training courses I ended up as a medical officer in tank training regiments. The war was all but over when an offer came from the dean of medicine at the University of Western Ontario.

INFLUENCES

Small things, thoughtful people, and lucky happenstances enrich life, teach us in subtle ways, establish values and interests, prepare the mind, and without intention guide the footsteps. Some of these influences I identify in what follows, with the certainty that others of equal importance are left out or were never recognized.

Nature was revealed to me in my family's series of well-tended gardens and country walks. My mother's keen eye for birds, beasts, and natural wonders, and my father's habit of collecting interesting specimens and sharing his knowledge about them were a part of many days' outings. Fishing was important to my father and it became so to me; his interest in the sport and his competence were infectious, and we fished together for many years. I acquired a great deal from the opportunity for gentle but serious conversation and for airing what we knew or thought we knew about life around us.

Microscopes had a pervasive, persuasive, and recurring role in my life because of my father's interest in microscopes and microscopy. Certainly I must have seen them early on just as I admired them on later occasions when visiting laboratories. However, I have a clear memory (identifiable by circumstance to the time when I was three or four years old) of seeing Leeuwenhoekian animalcules down an old brass microscope; their source was an inverted bell-jar aquarium full of weeds, water beetles, copepods, and other small forms. That same microscope was in more regular summertime use when I was seven to ten years old, and I still have it.

I did no formal biology until I went to McGill. Two kindnesses were of particular importance: I was attending N. J. Berrill's invertebrate zoology course and he asked me, "Would you like to have six feet of bench?" For sure I would. He gave me space in a corner, the use of a microscope, and facilities for fixing, paraffin-embedding, sectioning (an old "Cambridge rocker"), and staining preparations for microscopy. I enjoyed myself, learned by doing, and produced good sections of a frog kidney adenoma among other things. My father's kindness was no less formative; it was the present of a Leitz binocular microscope with appropriate optics and slide boxes of the preparations of microscopic invertebrates that he had made for himself as a student in 1909–1912. Thus I became familiar with preparative techniques and with an instrument that I have and use to this day, and which has provided all of my light micrographs to date.

I cannot remember life without books and discourse about books. My father was an inveterate book-buyer and reader of all sorts of subjects from Middle English to evolution. Books were treasured. Even the bad books were kept because the argument and the horrible example might be needed; they were annotated with marginalia, incisive comments on the title page if they were bad enough, and a personal index to interesting titbits noted on a back flyleaf. The home and the laboratory contained eclectic collections of books, journals, and reprints; as a result I became a card-carrying, second-generation pack rat and omnivorous reader. Fortunately few near disasters resulted, but a few there were, from such dilettantism. As long as I can remember the dinner table was a place to talk, to try out outrageous arguments, to remember interesting facts, and to embroider a good story. My mother had to contend with an impossibly elastic timetable for the course of a dinner.

I was lucky indeed that at home, at secondary school, and at the universities I was blessed with mentors who cared enormously about how we expressed ourselves in speech and in writing. At Summer Fields I took classes from L. A. G. Strong (a substantial novelist) and Cecil Day-Lewis (a major poet). At school in Montreal I was helped enormously by Hugh MacLennan (a major Canadian novelist and academic), D. S. Penton (a fine teacher), and V. C. Wansbrough (a scholarly headmaster), who were as much friends as teachers. Behind-the-scenes work on the school magazine, which was a rather elaborate annual, introduced me to the life of an editor. The interest must have been latent because I was to spend much time on the *McGill Medical Journal*, then a substantial quarterly, and one year as its editor. My involvement with the journal provided experience every step of the way from soliciting of contributions, to copy editing, proofreading, and production, to review writing. Only seven years later I was to be on the Editorial Board of the *Journal of Bacteriology*, enjoying the first of a number of appointments in scientific editing.

I was most fortunate to have done half of my schooling and half of my university training at first-class institutions on each side of the Atlantic. These opportunities were given at no little expense to my kind, generous parents, who gave me the best that they could of both worlds and all their support. I can only imagine a certain anguished concern at sending an only son as a student, not a serviceman, in the direction of crisis and war in 1938.

The final good fortune was to get started in teaching and research at a small medical school in a small university that had only one way to go and that was up, fast. There was infectious energy and ability at the top in the person of G. E. Hall, a fine scientist turned administrator, who knew what he wanted, usually got it, and made good decisions about people more of the time than do most administrators. He brought in able young professors to guide into the bright future the small (two–faculty member) departments, such as R. J. Rossiter (biochemistry) and A. C. Burton (biophysics), who became my very good friends and colleagues, as well as even younger faculty (of whom I was one) to back them up. His successor as dean of Medicine, J. B. Collip, a most distinguished endocrinologist, was even more influential and also pushed hard for productive involvement in basic research. Collip strengthened the school, with the help of Hall, then the university president, to a level to be proud of. There was no temptation to look for another job; promotion came fast, and the school was, and still is, an exciting place to be. As a result I lost no time along the way due to relocation and I received every possible encouragement. We prospered intellectually; we had to cooperate in all things because there were so few of us (in 1950 there were 18 full-time faculty in the Medical School), and we enjoyed each other's company.

A FLYING START IN ACADEME

The invitation to the University of Western Ontario Medical School was urgent because the professor of bacteriology and immunology, I. N. Asheshov, was in the hospital and was likely to be there for a few weeks. It was September 1945, and the course for 44 medical students had just started. Dean Hall, himself recently appointed after a senior posting in the Royal Canadian Air Force, knew how to pull strings. A brief weekend visit was enough for me to say yes to the post of lecturer in the department with duties to include teaching medical students, assisting in clinical bacteriology for the hospital, and involvement in research. Two weeks later I was seconded to the Wolsey Barracks, London, Ontario, for "special duties." My wife and infant daughter, with the help of friends, found us a temporary living place. Within a week of our being together again I was embarked on a 12-week stint of four lectures and four labs a week, in the midst of which I was demobilized. Many of this first class were ex-servicemen, and not a few were older than I was. We all survived the experience and I learned more from and with them in a shorter time than ever before or since then. Three of that class are senior colleagues still in this faculty.

Asheshov was out of the hospital before the end of the term, in time for a few lectures in his remarkable style, and he was marvelously helpful and supportive to me. He was a lively minded scientist, then engaged in seeking antibiotics, but at heart he was interested in bacteriophages, an area in which he had worked productively since 1922. He and I taught hard for five months of the year. It was a good old-fashioned bacteriology course topped off with minor dollops of virology, parasitology, and mycology. In many ways it was the oral examination rather than written papers that taught me the most about my subject, our teaching, and students. Whatever may be said about sub-jectivity in examinations, the hard work involved in these oral examinations was repaid in knowledge of the student and understanding of the strengths and weaknesses in teaching. Even more importantly, after the war we had mature veteran students who were remarkable people: dedicated, serious, accomplished, and worth every effort. They taught me a lot.

I had to get started on research in 1946 because there was time and I wanted to do it. Dr. Asheshov gave me every encouragement, without demanding any service or direct assistance in his own work, which was mighty kind of him. I was equally encouraged by Dean Hall and by Dean Collip (who succeeded Hall in 1948), who provided me with an initial grant for a phase microscope. Funds were very hard to get, so I had to make use of the sorts of things we worked with every day in the clinical laboratory. Each of my first four independent forays into research were important to my life and work in the laboratory. I write about them with that in mind.

The first project was prompted by the isolation of a mucoid Group A *Streptococcus pyogenes* from a chronic lung infection. This reminded me that when I was an intern I had seen similar mucoid colonies collapse flat in the neighborhood of some colonies of staphylococci and pneumococci. Knowing then that the capsular polysaccharide was hyaluronic acid provided a method for ready detection of hyaluronidase ("spreading factor") and also a diffusion assay for the enzyme, which was of great interest at that time as a component of virulence targeted on the intercellular matrix of connective tissue. Fortunately, R. H. Pearce, an equally young biochemist in the Department of Pathological Chemistry, was interested in acid mucopolysaccharides. The resulting joint paper was good experience and was accepted without demur by the *Canadian Journal of Research* (64). So I started out in collaborative research, and I am grateful to Pearce because I learned from him some of the many things I should have known had I had doctoral training in science. Entering PhD studies was an option, but I decided to get on with research because there was time and opportunity.

The next project arose from observing the remarkable motile colonies shown by a fortuitous isolation of *Bacillus circulans*, and it initiated a lifelong interest in swarming and motility. Both the rotating colonies and the curving paths of the bullet-shaped colonies showed an exact 2:1 ratio of counterclockwise to clockwise motion, which is still unexplained. For the first time I had the assistance, in the summers of 1947 and 1948, of a veteran medical student with laboratory experience, R. H. Elder, now a senior and respected clinical microbiologist in Ottawa. We wrote a neat paper which was accepted by the *Journal of Bacteriology* (60), and we made a short 16-mm movie, using a haywire rig, to illustrate a paper given at the 1948 ASM Meeting in Minneapolis. Most importantly, Carl Robinow introduced himself after that presentation, and we discussed swarming but also his observations on bacterial nuclei and endospores. So this project introduced me to a fascinating phenomenon which is still in mind, and to a dear colleague whose association with me and our students has been nothing but rewarding and a pleasure for 40 years.

Igor Asheshov taught me about phages and how to handle them just at the time the papers on the T-phages were coming out. The excitement of following the fate of the phage and the host components in these papers induced thoughts about research projects. Asheshov was advising Fred Heagy (an MD interested in science), a PhD candidate studying T2 phage infection, and suggested a parallel study of changes in the host cell in advance of lysis. Indeed, there were indications that these changes should be observable by microscopy (46). We tried Robinow's cytological techniques because I had been reading his papers and his "Addendum," which was a remarkable seminal chapter on bacterial cytology appended to *The Bacterial Cell* by René Dubos (75). We had hardly started on this in 1948 when Asheshov and his technical forces left to work on a March of Dimes project seeking antiviral agents at the Bronx Botanical Garden. So I was left willy-nilly to elaborate my own expertise and to supervise a doctoral candidate my own age. It was another learning experience, to use the language of educators, to become acting head of the department.

That summer and the next, Heagy and I were helped by D. H. Gillen, a medical student, and we made lots of preparations at timed intervals following infection of *Escherichia coli* B with phage T2. Our preparations were terrible but encouraging. The situation improved dramatically after Robinow paid us a visit and showed us in his inimitable fashion delightfully simple technical stratagems, which paid off in the quality of the preparations and the photomicrographs. There was a progression of cytological events involving a shift in the distribution of chromatin with eventual dissolution (host-cell DNA), gradual depletion of the cytoplasmic basophilia (host-cell RNA), and developing granularity in the last two thirds of the cycle due to synthesis of phage DNA. We published a good descriptive paper (62), and we were only slightly sorry to have been scooped by Luria & Human (45), who compared cytological events due to several of the T-phages. It was evident that the effect of phages on host cells was visible and was determined by the virus rather than the host. Our paper was appreciated and brought me into touch with S. E. Luria and his exciting group, paved the way for a collaboration with G. Bertani, and gained me an introduction to their colleagues at the University of Illinois, where laboratory visits were exhaustingly educational, night and day. The project was also the beginning of real cytological studies in our department because Dean Collip's unstinting support allowed me to invite Carl Robinow to join me in the department, which he did in 1949.

The fourth seminal research project followed the invitation by Charles Evans for me to spend the summer of 1949 in the Department of Microbiology, University of Washington, Seattle, as a sessional lecturer. There were fascinating people to learn from such as Erling Ordal and Evans. More to the point, I was able to put my newfound cytological competence to good use in helping Howard Douglas study the growth and division of an interesting budding and phototrophic bacterium, *Rhodomicrobium vanielii* (59). It was fun to find out that nuclear division took place in the mother cell and that one of the nuclei so formed was exported to the new bud through the long, 0.3-μm–diameter hyphal tube, which had the newly budded cell at its end. We spent happy hours observing fixed and stained preparations and taking time-lapse photomicrographs of living cells under phase contrast. I learned much that was useful from Douglas and his colleagues as well as from my teaching assistants, Wesley Volk and Quentin Myrvik. While I was there a telegram came from Dean Collip telling me I had been appointed professor and head of the department. This meant, I suppose, that the invitation from Seattle was a stimulus to decision. There was considerable rejoicing; I was just 30 years old and an expression of trust had been given by the university. As I see it, I was still drying behind the ears, as the saying goes.

RESPONSIBILITIES

I was lucky to be given from the outset as much responsibility as I could manage, such as the provision of a clinical microbiology service to the Children's Hospital and Victoria Hospital, our main teaching hospital, in which the department was then situated. I was responsible for this service from 1948 until 1965, when the department moved to the university campus and broadened its horizons. The clinical work kept us busy, and the major clinical problems came to me, day and night, in those days before we had physicians who specialized in infectious diseases. Antibiotics were with us and so were the developing problems concerning policy in treatment, spread of resistant strains within the hospitals, and infection control. This meant that research was a source of enjoyment and recreation in the midst of a busy life. Rightly or wrongly, the research projects grew more and more general and drifted further than might have seemed proper from the medical applications which were our clinical and teaching preoccupations.

Relief came in the late 1950s and early 1960s, when increased professional and technical forces were appointed to cope with the burgeoning use of clinical microbiology due to the introduction of a hospital insurance program in Ontario.

In those days there was much greater interdepartmental cooperation and we often had to teach what we could with any help we could get. I note with some astonishment that I gave courses in epidemiology (1948–1950), human genetics (1955–1958), and history and methods of medical science (for graduate students, 1958–1964). Luckily we were mostly young and energetic as well as communicative with each other owing to a lively lunchroom. This spirit and the efficiency and some of the enjoyment that went with it disappeared in a very few years with the increasing size and complexity of departments.

The move to the university campus in 1965 brought with it new faculty members and the integration of an honors course in microbiology and immunology into the biology programs. We were interlopers, and the introduction of six new programs by the basic medical sciences group was seen as a threat to the biologists and was not happily received. Time has resolved most of the problems, but increasing devotion of energy to administrative matters was required, which brought with it devotion to university affairs. I sat on the senate and many committees and spent a year as acting dean of science. So it was a degree of relief to give up being head of the department in 1974, after 25 years at it, and to administer thereafter nothing bigger than my own research group.

Another change came because of vastly increased faculty forces in medical microbiology, which drifted my teaching from medical microbiology to general bacteriology and systematics. It is not a bad thing to have such a shift in the midstream of a career and to be able to step away from administration before it becomes tedious or all-consuming. Too many good people regret losing their foothold in their academic discipline, and I have seen them get tired before their time. Life has to be lived with some care for what is best for the individual. Fortunately some good administrators and some good researchers and teachers find their appropriate niches and avoid the pitfalls of academic life; others are not as lucky as I was.

EXPLORING BACTERIAL CYTOLOGY

My feet were well set on cytological trails, but the diversity of my studies through the 1950s was needed to fix the major direction of my work. Observations on the cytology of phage infections continued with J. F. Whitfield and were extended by collaboration with G. Bertani to follow the cytological events during the establishment of a lysogenic state for P1 and P2 phages (94). Interesting as these phenomena were, perhaps the most fruitful derivative of questions about why changes in nuclear form accompanied phage infections turned out to be a study of the effect of the influx or efflux of cations on the conformation of the nucleoids of bacteria (95). As we and Kellenberger's group were to remember years later (35), an understanding of ionic effects on charged polymers is crucial to assembly and conformation of working structures and to lifelike preservation during cytological fixation of specimens. In fact, the crucial events involved in cellular structures and processes are nature's physicochemical experiments. So it was natural that my interest turned to trying to learn more about the structure of the host cell and that my studies broadened to the general cytology of bacteria. In this effort the daily discussions, suggestions, and joint efforts with Carl Robinow were determinative.

We started out in two directions, based much more on Robinow's experience than on mine. One was to do the best possible light microscopy of the components of what would now be called the cell envelope, i.e. the cytoplasmic membrane/cell wall complex (76). This was a necessary step because electron-microscope preparations of the time did not yet provide the contrast and differential staining in sections needed to demonstrate such membranes, although they soon would do so. The other approach, which was undertaken by P. C. Fitz-James (29), was to initiate studies of the biochemistry of a distinct cytological entity, the endospore of a *Bacillus* species. These studies of spores were influential then because they broke new ground and because they contributed to bacterial cytology (29). They indicated to us how powerful combined structural/biochemical research was in providing seminal data even with the restriction of resolution due to the wavelength of light. At this point it became obvious, despite our considerable pride at being able to attain excellent photomicrographs, that the potential of electron microscopy was developing rapidly and that high resolution was essential.

It is hard to realize now what we did not know about the structure of bacterial cells in 1950–1953. The microscopy of the day was giving a new dimension to understanding of cellular components and their functions by correlation with biochemical studies of many kinds of higher cells, and this was a stimulus to finding out what was hidden behind the somewhat undramatic facade of bacteria. The technical and intellectual essentials for effective electron microscopy were coming into focus. Metal shadowing (96) had been available for a few years and showed the topography of bacterial surfaces (36) and external structures such as flagella. Fractionation and chemical analysis of walls following disintegration of the cell was giving exciting data on the nature of bacterial walls in the capable hands of Salton & Horne (79) and could be monitored by electron microscopy even if the techniques were poorly developed. Sections thin enough for effective electron microscopy were obtainable from embeddings in methacrylate (70), and cytological fixatives such as osmium tetroxide (73) were being recognized as suitable preliminaries to embedding and sectioning tissues and bacteria for electron microscopy. I remember still my pleasure and excitement on viewing the first good micrographs of sections of a bacterial cell in the paper by Chapman & Hillier (16) submitted to me as an editorial board member for the *Journal of Bacteriology*. Importantly, the paper demonstrated that the images were recognizable in terms of the light-microscope cytological preparations that we and others were studying and relating to the biochemistry of fractions. Establishing the technical basis and accumulating the experience necessary for adequate preparation of bacterial cells and cell fractions for electron microscopy took more than a decade, and the techniques are still being refined.

Although we recognized the electron microscope as the coming instrument of cytology, Carl Robinow and I firmly believed then, as we do now, that the light microscope is here to stay. I remember brashly trying to persuade G. E. Palade that light microscopy could still do good things for understanding the insides of bacteria. Certainly, light microscopy gave us useful first approximations to the arrangement of the elements of the cell surface in studies of *Bacillus cereus* and *Bacillus megaterium*. A chance but very useful observation (65) showed us that when fixed cells were crushed on the cover slip and then stained, the walls were a rigid sleeve containing septa and the poles were hemispherical caps. The next step was based on Robinow's observation that these cells could be fixed in the plasmolyzed state to retain separation of the surface of the protoplast and the enveloping wall structures. Interpretation of the nature of the protoplasmic interface was greatly assisted by the partition of polar dyes (notably the Victoria blues) into an infinitely thin layer at the protoplasmic interface. This convinced us that this interface was an osmotic barrier, as had been known since 1895, and also was a differentiated structure equivalent to the cytoplasmic membrane of higher cells; the membrane profile was detected rather than resolved by light microscopy (76). We thus had an anatomical concept but were disappointed that the Chapman & Hillier (16) electron microscopic image and others of the time did not show a distinct cytoplasmic membrane at the surface of the protoplast. But most of the elements were there even if their topological features might be dim or distorted owing to inadequacies of fixation and processing; we could see a wall, a remarkably dense cytoplasm with some possible inclusions or spaces for them, cell wall septa accomplishing division, and something that could represent the Feulgen-positive nucleoids.

Salvation was around the corner. Robinow took his spore structure questions to K. R. Porter's laboratory at the Rockefeller Institute and found that sectioning was rewarding because it gave access to the hitherto inaccessible structure of the spore and revealed the cortex beneath the coats (75a). I took my interest in making preparations of infected host cells to R. W. G. Wyckoff's laboratory at the National Institutes of Health in Bethesda, Maryland, and found that whole-cell preparations were not going to take us far in phage or host-cell studies. These first experiences in electron microscopy and visits to James Hillier at the RCA Laboratories at Princeton, New Jersey, convinced us to seek funds for an electron microscope as an essential tool. A Philips EM-100 arrived early in 1954, which we shared for some years with our histologist friend and colleague, R. C. Buck. Learning was fun, but getting the very best out of that instrument (our only electron microscope until 1965) would have been slow indeed without the skill and experience of Aksel

Birch-Andersen (State Serum Institute, Copenhagen), who spent six months with us in 1955 and taught us more than the basics. Many tricks were needed then, including knowledge of when and where to kick an EM!

Carl Robinow's research interests turned to fungal cytology after 1953, and P. C. Fitz-James, now a department member, continued to utilize light and electron microscopy as a monitor and guide to recognizing the steps in spore morphogenesis and fractionation for biochemical studies. From then on each of us developed our own lines of inquiry, but they were close enough for mutual stimulus.

For my part, my graduate students and I explored the anatomy of a variety of bacteria including "blue-green algae." There was thought of taxonomic returns, then, as well as general insights into bacterial structure because J. P. Truant, J. W. Costerton, and I looked at a far wider range of physiological groups than was strictly necessary for the purposes of their theses. We worked toward better cellular descriptions for some bacterial genera such as *Moraxella* (67) and *Vitreoscilla* (19), with comments that applied to taxonomy. Obviously these experiences induced some broad intentions, because I wrote in 1960 (51): " . . . we can now glimpse the bare outline of what can be done to bring our knowledge of bacterial structure to a point where it can be incorporated into the whole picture we should like to have of the kingdoms of living things. . . . All the approaches to the unravelling of structure are bringing to light unique properties of bacteria that will stimulate more effective research." These forays were to be continued thereafter by increasingly focused inspections of *Lampropedia* (17), *Thioploca* (48), nitrifying bacteria (68), *Listeria* (31), and *E. coli* (66), each of which turned our thoughts to details of the cell wall, membranes, and protoplasmic structure.

At the end of all this work I had an unparalleled chance to assess the state of a large part of bacterial cytology by writing a chapter (51) for *The Bacteria*, a benchmark series edited by I. C. Gunsalus and R. Y. Stanier and published in 1960. I believe that volume, devoted entirely to structure and part of a still-significant series, put structural studies into the thinking of many microbiologists.

The major factors impeding the revelation and interpretation of cellular ultrastructure involved preparative techniques. Bacterial cells, particularly the nuclei, were sensitive to the ionic environment during fixation, and most fixation schedules gave unpredictable results. A suitable routine was not available until Ryter & Kellenberger (78) published in 1958 their still-useful method for using osmium tetroxide followed by uranyl acetate in a defined veronal buffer system. This method was not improved upon until dialdehydes, e.g. glutaraldehyde, were introduced in the mid 1960s and used in a prefixation step before osmium tetroxide postfixation. Embedding in a synthetic resin evaded the disturbing artefacts generated by differential polymerization rates in the methacrylates used up to that time.

The sections that we cut were good enough, but the structural contrast due to scattering of electrons by the embedded cell substance was inadequate and could be even less than that of the embedding plastic. Many microscopists realized gradually that scattering could be increased by treating tne sections with heavy metal salts to be taken up by ligands on structural polymers; for some time we favored lanthanum or uranyl salts rather than lead, which tended to precipitate. What was so surprising at the time was the discrimination of layers and substructure that became possible with these staining methods. An early outcome due to applying metal salts was our recognition in 1957 that a double-track unit membrane did indeed enclose the protoplast of the bacterial cell (50); not less important as a lesson was our then finding that earlier papers had shown such a membrane but it was not recognized. Generally we saw what we were prepared to see at each stage of our understanding of ultrastructure. We applied the technique usefully, for example in my study with Francombe and Mayall (61) of the remarkably direct effect of penicillin poisoning on the structure and integrity of staphylococcal cell walls, which helped to focus attention on the cell wall and peptidoglycan as the target of that remarkable antibiotic. Eventually in 1963 Reynolds (74) put into general use a method of combining lead citrate and uranyl acetate for staining sections, which became a standard treatment to give contrast.

We needed an effective method for following the fractionation of cells and cell walls and assessing the nature of the pieces of cells displayed after ballistic or sonic disruption. We used light microscopy to monitor cells and fractions negatively stained with nigrosin, a most useful technique then and now, but limited as to resolution; and we sectioned pellets to get profiles. Despite the amount of nigrosin that we used it never occurred to us that negative staining for electron microscopy was possible; the demonstration of this technique and practical applications by Brenner & Horne (7) was a scientific bombshell with repercussions continuing today. Brian Mayall and I tried it and got immediate results looking at a moiré pattern of an array on a wall fragment of a strange coccus, later identified as *Deinococcus (Micrococcus) radiodurans*. We described this in a review (52) and demonstrated it in a paper at the 1958 Microbiology Congress in Stockholm. It was exciting enough even if no formal publication resulted, but it did lead us into other applications and particularly studies of the nature of wall arrays in *Aquaspirillum serpens* (53), which have engaged us ever since.

These general and technical developments, which I hasten to say we participated in but did not originate, allowed us to enter into more detailed and complex studies. A worthwhile preliminary to our relation of structural studies and biochemical fractionation was determination of the location of the peptidoglycan component in the sectioned wall profile of *E. coli* and some other gram-negative bacteria (66). The location correlated exactly with the order established in the elegant biochemical dissection of the *E. coli* wall initiated by Weidel and colleagues (93). Then to carry this further I studied what happened to these layers during cell wall septum formation and cell division with Pamela Steed-Glaister, then doing her PhD studies, and we established an effective baseline for the wall studies to come (86). This work allowed us to understand the various forms of septum that we see. Septa incorporate a peptidoglycan layer as the primary component, which intrudes either as a double loop of peptidoglycan or as a single layer that thickens and differentiates into a doublet (laughingly attributed to "zipperase") to allow intrusion of outer layers and the ultimate separation of the cells. However, the constrictive divisions of *E. coli* and all but a few mutant enteric bacteria were a puzzle until we were able to show (first serendipitously, by my having let a water bath get too hot) some 10 years later that an appropriate fixation regime revealed septum formation of the basic type involving the peptidoglycan layer. We had to be confident and persuasive, thanks to the clear experiments of Ian Burdett (14), that this was an image that could not possibly be an artefact; yet the way some of our colleagues refer to septation in *E. coli* makes one wonder whether or not they are convinced. However, the structural details of these events are now being resolved in great detail for both gram-negative and gram-positive models (34, 38, 47).

Electron microscopy, although limited for cytochemical analysis, was adept at revealing structures in search of a function. Among these were the membranous intrusions often found near the site of septum formation and at the poles of the dividing cell. These lamellated structures were named mesosomes by Fitz-James (29) and were recognized as having continuity with the cytoplasmic membrane. These may have been among the sites identified earlier as bacterial mitochondria, because the granules exhibited strong oxidation-reduction reactions (49). However, cell fractionation indicates that respiratory activity is a general property of the cytoplasmic membrane. The polemical arguments died and left behind the interesting problems raised by the variety of membranous structures becoming visible in the cytoplasm. Some of these structures were obviously functional, as was supported by biochemical data, providing for photosynthesis or for complex metabolic processes requiring coordinated energy transfers such as nitrification.

Stanley Watson and I (68) were excited by the elaborate membranes of nitrifying species of *Nitrosocystis*, *Nitrosomonas*, and *Nitrobacter*. It was clear from these and other examples that there were membranous organelles, but in most cases they arose from and were still continuous with the cytoplasmic membrane. The exceptions, where the internal membranes separated from the peripheral membrane, still originated from the cytoplasmic membrane, however specialized the final function. Thylakoid membranes in cyanobacteria may yet prove to be the only bacterial membranes independent of the cytoplasmic membrane. In general, a major demand for a membrane-based function results in more membrane, and space for it has to be gained by internal projection, which may be structurally differentiated for whatever energy-linked process the cell may require. We now realize that not all membrane functions can be everywhere in the periphery and that a degree of area specialization is a necessity.

What is really involved in the formation of mesosomes is still far from certain; they may be artefacts of fixation because they are absent in freeze-cleaved preparations not exposed to a fixative (99). Yet they occur in persuasive sites, and the bits of membrane that give rise to them, even as artefacts, may be associated with some specific function. One concept that may not be too farfetched is that their function derives from their association with nucleoids (77) and the probability that DNA replication and nucleoid segregation requires an association with membrane sites.

VENTURING INTO MACROMOLECULAR ASSEMBLIES

An important outcome of the introduction of electron microscopy has been our ability to recognize specific cellular components and, by their structural characteristics, to recognize them also in the fractions generated after cell disintegration. This discriminatory function has been as essential to many cell biology structure/function studies as it has been to virology. Discrimination was made immensely more effective when negative staining with appropriate heavy metals (notably phosphotungstates, molybdates, and uranyl salts) was added to the techniques of metal shadowing of specimens and freeze-cleaved and etched preparations (7). What then became possible was the outlining of the shape and form of some kinds of macromolecules, mainly proteinaceous, to a level of detail limited by the resolution attainable on that type of specimen and the potential of that electron microscope. So an effective resolution of about 2.5 nm became possible on biological specimens, which cannot be infinitely thin. Negative staining was applied in many fields, but in microbiology there was a rapid evolution of structural studies on viruses, membranes, ribosomes, and in our case bacterial surface arrays.

I was very impressed by the images we obtained early in the 1960s from negatively stained specimens of cell wall fragments of *Deinococcus (Micrococcus) radiodurans* (52), *Lampropedia hyalina* (17), and *Aquaspirillum serpens* (53). The fragments showed as two-dimensional, hexagonal arrays of linked subunits with paracrystalline regularity that enveloped the external surface of the cells. The arrays had every possibility of being amenable to fractionation and analysis after the fashion of the techniques applied to *E. coli* by Weidel's group (93). Other possible arrays were assessed; M. V. Nermut, who was visiting in 1966, and I found that *B. polymyxa* possessed a tetragonal array (69). We explored the stability of this array on exposure to various chaotropes (e.g. types of detergents), which we thought might be of use in attempting isolation of the components. We found, as did Baddiley's group (32), that the fraction containing the array was mostly protein. We also took a first step toward image processing by enlisting the help of Klug's group at Cambridge. They found that each unit forming the tetragonal array was formed by four centers of mass arranged in p4 symmetry with the whole array (28), in which the lattice frequency was 10 nm. The units are now better resolved (15).

It was obvious that these arrays, or S-layers as they are now termed, were a common feature of bacteria in nature and, as a major protein component of the cell, would be retained on these cells for reasons of selective advantage. They needed serious study, and we chose to study the S-layer of *A. serpens* because it was a single layer and seemed to have properties useful for the manipulations required for isolation, as Pamela Steed-Glaister's studies of stability during growth in fluid media had shown (87). Francis Buckmire and I (11–13) isolated the *A. serpens* VHA S-layer protein and explored its properties, and these studies formed the basis of a study that I still continue in association with Susan Koval (42). The protein proved to be a large, 140-kd, acidic protein that can self-assemble into a lifelike array with the help of Ca^{2+} and the template provided by the outer membrane of the organism. This protein has been subjected recently to further biochemical and assembly studies (39–41), and its properties help to provide a basic description of many of the S-layer proteins (81, 82). I explored the variety of structures possible on the walls of a number of species of *Aquaspirillum* with T. J. Beveridge. Among them *A. putridiconchylium* showed p2 symmetry (4, 89), and several had two or more layers. *A. serpens* MW5 had a double layer and, in similar fashion, was subjected to image analysis in a collaboration with Murray Stewart (90). The MW5 array showed in freeze-etched preparations as a linear structure and in negative stains as a linear moiré. The image analysis resolved this moiré as two hexagonal arrays with similarly sized units superimposed but slipped a half interval along one three-fold axis. I studied the proteins with Marion Kist (37), and we found that the inner layer protein (150 kd) would self-assemble in vitro but would also form a directly linked array on the outer membrane. But the outer layer protein (125 kd) would assemble only on the formed inner layer, and these assemblies (like that of strain VHA) required either Ca^{2+} or Sr^{2+} even if the inner layer hexamer unit did not. The proteins have few or no sulfur-containing amino acids, and it is apparent in our model systems that divalent cations are required for both assembly and the conformation of the monomers in most cases (41). The structural studies of the

VHA S-layer were only two-dimensional or planar until recently; in 1986 some fortunate preparations allowed Robert Glaeser's group (24) to undertake filtered Fourier transform reconstructions from a tilt series of micrographs to give a three-dimensional model of the assembled units, each consisting of six monomers. However, other aspects of understanding are primitive or nonexistent, and these include molecular structure, transport, and regulation.

The other early examples of arrays that started us off have not been neglected: *D. radiodurans* turned out to have some considerable biochemical and taxonomic peculiarities, as I shall recount, and less attention was paid to the S-layers (43, 91, 100). However, Baumeister's group (2) has undertaken detailed structural analyses. *B. polymyxa* engaged me again in a comparative study with Stephen Burley (15), which served as an exercise in establishing our image-processing facility. Eventually, I have returned to the double S-layer that surrounds the cells of *Lampropedia hyalina* (1); this was much too complex a structure to consider before, and even now it is a challenge to work out the order of assembly and the identity of the multiple components of the outermost layer.

The considerable amount of information now available about S-layers is not susceptible to summation in this essay. They have been described as a component of the walls of about 200 species of bacteria from most major phylogenetic groups, and the phenomenon must be considered as a general attribute with varied functions. Fortunately there are several recent reviews (42, 81, 82) to provide the details. To a considerable extent my early work sparked an interest in these models of structural assembly of macromolecules, and the 25 years since then have involved me with 25 coworkers; sadly, it is not possible to give appropriate credit to all of them in this essay.

I had hoped at one time to be able to contribute to a structural and macromolecular description of the basal complex, the motor, of flagella in studies undertaken with James Coulton (20). It is more than a challenge to resolve components of this puzzling rotor-stator mechanism, which is an organelle not much more than 25 nm in total depth and diameter, effectively described in principle by DePamphilis & Adler (23). The closest we got to any macromolecular description of the operative part in the plasma membrane was to see that the central rod was made of triads forming a hole down the center, and to find a circlet of studs around the M-ring in the freeze-cleaved membrane. I remain sceptical, as are others (27), about the exact relationships of the structure to the plasma membrane, and I believe that we are missing the boundaries of a compartment and some part of the complex penetrating to the cytoplasm.

The bacterial wall is a remarkably complex and dynamic structure interposed between the cell and an often hostile environment. The export-import services and machinery such as flagella that traverse the wall put special demands on the maintenance of integrity and the assembly-disassembly capabilities of the entire complex. Enormous effort has been given to understanding the governance of the murein covalently cross-linked network because it is a target for antibiotics. The entropy-driven assemblies of membranes and porins and the integral transport mechanisms have had well-deserved attention in recent years. The S-layers are equal participants in the dynamics of walls in growth and division, and go by a simpler but no less sophisticated set of rules. Furthermore, they draw attention to functions other than strength (42), and the more complex of them may act as generalized models for the assembly of cell structures.

TAXONOMY IN TRANSITION

There were compelling reasons for my getting involved in taxonomy: My father was a trustee of *Bergey's Manual of Determinative Bacteriology* (1936–1964) and we talked about the problems; my work involved diagnostic bacteriology and wrestling with identification, and the study of the structure of bacteria inevitably drew attention to the inadequacies of descriptions. An early direct contact with the *Bergey's Manual* trust was an invitation from R. S. Breed in 1955 to join in discussing the description of the "Schizomycetes" and how it might be revised; my advice was not taken.

Work on structural aspects of bacteria had brought both Carl Robinow and me into contact with Roger Y. Stanier, whose thoughts about the nature of bacteria and approaches to taxonomy, sharpened by his association with C. B. van Niel, were particularly penetrating. His papers shaped my views and my intentions toward the definition of unique features of bacteria. The opportunity to do something about them came hot on the heels of my completing the review of structure for *The Bacteria* (51) with the invitation to contribute to the 1962 *Symposium of the Society for General Microbiology*. This essay, "Fine structure and taxonomy of bacteria" (52), had some good ideas, but the conclusions were not as strong and definitive as those of the almost coincident essay, "The concept of a bacterium," by Stanier & van Niel (85). Neither essay developed a formal taxonomic proposal, but both were disposed to accept any nomenclatural arrangement that recognized the relationship of blue-green algae and bacteria, that incorporated them both into a grouping distinct from all other microbes and macrobes, and that recognized their unique features of organization. Stanier & van Niel referred to them consistently as "prokaryotic organisms," which reflected Stanier's decision the year before (84) to describe bacteria as cells and to use Chatton's vernacular terminology, introduced in 1937 (18), for the major divisions of cellular organization, eukaryotes and prokaryotes. I feel that I rode on the shoulders of worthier colleagues because, almost inadvertently, I was the author of the formal name, the kingdom Procaryotae (54). I had something to do with establishing the generalizations that supported a position that became agreeable to most microbiologists.

It was clear to me that taxonomy was a worthwhile endeavor, even if it was still unpopular, because it required the use of all that one knew (or thought one knew) about organisms and because it drew attention to great gaps in basic knowledge. At that time objective means of checking the relationships in an arrangement of taxa or within any taxon were only just developing. Both the selection and the reliability of the phenotypic characters were largely a matter of faith. Mechanisms evolved over the past 30 years have given a scientific aspect to the assessment of characters and, far beyond the powers of serology, the definition of taxa. These mechanisms are now well known and came to include, successively, numerical (computer-assisted) taxonomy, murein (peptidoglycan) types, the G+C ratio in DNA, DNA/DNA homology, DNA/RNA hybridization, and the exploitation of the highly conserved RNA cistrons and the ribosomal RNA sequences. The power of these approaches to taxonomy has only been fully realized in the past decade, and the data supporting a phylogenetic assessment are still only partially complete.

I joined the *Bergey's Manual* Board of Trustees in 1964. The chairman, R. E. Buchanan, was strongly oriented to nomenclature and classical approaches. Most of the trustees were practical bacteriologists, but there was one grand heretic among them in the person of S. T. Cowan (22). A consensus in views was hard to accomplish, and arguments prolonged the gestation of the 8th edition, which did not appear until 1974. The turning point came in 1969, by which time R. Y. Stanier was also a member, with the decision to use vernacular names for the chapter headings of the 8th edition because the higher taxa were of dubious validity. However, our expert authors were by no means convinced, or if they were in principle, many traditional arrangements were nonetheless maintained.

We were by no means satisfied, scientifically, with what we had done, despite all the good intentions of N. E. Gibbons (the editor, following Buchanan's death) and the board of editor-trustees. I wrote at the time (55): "The future will have to bring a regrouping of higher taxa to express a more coordinate view. . . . Haste is unwise; all previous classifications seem to have suffered infinite rearrangement due to insufficient information. . . . The new insights are likely to come from a clear understanding, on a comparative level, of the components of the genome of procaryotic cells." That goal is closer now.

The ten years after 1974 saw some remarkable changes in the rules that govern the nomenclature that expresses the taxonomic conclusions of bacteriologists. The 1975 *International Code of Nomenclature of Bacteria* (44) allowed for clearing the records of the enormous list of synonyms and useless names by declaring January 1, 1980 to be the new starting date for bacterial nomenclature and arranging for the *Approved Lists of Bacterial Names* (80) to be available from that date. We owe this most unusual form of taxonomic housekeeping (which generated envy in taxonomists devoted to plants and animals) to the international efforts of V. B. D. Skerman and P. H. A. Sneath, in particular, through the Judicial Commission of the International Committee for Systematic Bacteriology. It was time for an assessment of taxonomic problems in the groupings of bacteria, and the *Bergey's Manual* trustees resolved to produce a systematic manual.

A new view of bacterial taxonomy has arisen since Carl Woese and his group proposed (30) that the computer-assisted comparison of T_1 ribonuclease–resistant oligonucleotide sequence catalogs of the 16S ribosomal RNA of representative bacteria gave clear evidence for the phylogenetic lineages emerging from the main stem of the clonal evolution of bacteria and, in fact, all living things (97). They all share the ribosomal mechanism for constructing proteins. The ribosomal RNA molecules are up to now the most powerful, universal, and practical biological semantides or molecular sequences documenting evolutionary history in the sense of Zuckerkandl & Pauling (101). So the past decade has seen the accumulation of data on some 500 representative strains from many but not all taxa, and many old taxonomic assumptions can be tested. There are several distinct high-level phylogenetic taxa in both of the major phylogenetic divisions, the Archaebacteria (at least three) and the Eubacteria (at least 10). Some of the rRNA groupings of Eubacteria (98) appear to pose few problems of phylogenetic and phenotypic interpretation (e.g. the spirochetes and the cyanobacteria); some provide associations that are hard for traditional bacteriologists to assimilate (e.g. among gram-positive bacteria, *Micrococcus* closely associated with *Arthrobacter*); and others show such a remarkable diversity of morphology and physiology that there will be difficulty in developing phenotypic consistency [e.g., notably, the purple photosynthetic bacteria and their relatives (92)]. One can be optimistic, as Woese is (97), that the bacteria will eventually fall into naturally (i.e. phylogenetically) defined taxa and that an appropriate search will reveal unifying phenotypic characters. I hope that we will soon discover other cistrons determining nearly universal, complex, and essential cellular functions that are highly conserved. We must try to cross-check the phylogenetic conclusions, now based almost entirely on the RNA cistrons, especially in the diverse and rapidly evolved groups to detect anomalies or lateral transfer. This is the more important because transfer of heterologous ribosomal RNA genes (*Proteus vulgaris* to *E. coli*) has been attained with a plasmid vector (71) and as a result 25% of ribosomes contained the heterologous rRNA. Genomic integration has also been claimed, and if this is true, a phylogenetic chimera is possible. Who knows what strange experiments have succeeded in nature's laboratory?

With all this ferment, the development of *Bergey's Manual of Systematic Bacteriology* in four volumes appearing serially 1984–1988 presented unusually difficult problems, which came during my time as chairman of the trustees, 1976–1988. The data were compelling but patchy. Of course, it is sad that one cannot instantly and dramatically reform the whole of bacterial taxonomy, as was the fervent hope of the main laborers in the field of molecular phylogeny (98). It was possible to institute no more than cosmetic changes in many groups, major revisions in only a few, and phylogenetic consistency only in the *Archaebacteria*. Now it can be recognized that it is important to express the principle of *festina lente* (make haste slowly), as exemplified by the approach to the problems in the *Bacillaceae* (83), and keep the complexities in reserve until a more complete survey and decisions are accomplished. This attitude is helpful for the retention of a useful framework of classification for groups, including several genera threatened with major surgery.

I have been interested in applying modern forms of taxonomy to the *Deinococcus* group (9), seemingly gram-positive cocci with gram-negative characteristics, which proved (10) to be a distinctive and ancient lineage but phenotypically deceptive. If it is an ancient set of clones there should be relatives which might be hard to recognize. The superficially distinctive feature of radiation resistance is a mutable character, so the usual selective mechanism of using radiation is likely to reveal only clones similar to the extant species. The distinctive polar lipid profile (21) may aid recognition, in addition to the ribosomal RNA sequences and select signatures, which have already identified a gram-negative relative in *Deinobacter grandis* (72). I am prepared to follow with great interest the next steps in developing a new natural taxonomy of bacteria.

Fortunately, what appears to be accepted in classificatory schemes is not immutable. Taxonomy has to represent the best of science and for that reason is bound to change with new knowledge. I have speculated on the need for an academic taxonomy as opposed to a practical classification to serve the bench worker who identifies bacteria (57). However, there is need for only one taxonomy (92), soundly based in phylogenetic terms and phenotypically recognizable at all ranks, expressing a best view of a natural order. The bench worker does not need a complete hierarchy but operates within a framework of experience and needs only a number of identifiable vernacular groupings. No doubt, we must eventually develop new molecular and phenotypic markers for the recognition of all taxa. The tests used should be available and practical.

There will be changes in approaches to identification, but it is to be hoped that whatever systems are used, they will not inhibit the recognition of new taxa. Almost every ecological niche includes species that have not been cultivated and recognized, which should dispel any feeling of diminishing returns in the study of natural populations or in taxonomy.

ODDS WITHOUT END

Research may owe its continuing support to high-profile projects, but the next generation of research and so-called innovation will owe a lot to the odds and ends undertaken out of sheer curiosity, to satisfy a hunch, to provide experience for a summer student, and for many other ostensibly trivial but important reasons. These items enliven what otherwise might be periods of diminishing returns and spawn unexpected new research; so they become "odds without end." All research programs should devote a proportion of their funds to free-wheeling exploration (What the X Foundation doesn't know isn't good for it), much as buildings need a proportion of cost devoted to art. Some experiments should be done with "controlled sloppiness," as S. E. Luria told me years ago in order to encourage the unexpected. (If you do the same old thing you will get the same old answer.)

When T. J. Beveridge was working with me on his doctorate and studying the wall structure of spirilla we had many an occasion to discuss the problem of revealing substructure by staining with metal salts. So we experimented with isolated *Bacillus subtilis* walls to see what sort of capabilities the durable cell-wall polymers and heteropolymers might have in capturing and sequestering metal ions from solutions of their salts. The results (5) showed that substantial amounts of many metals were taken out of solution (including those important to metal enzymes such as Fe^{3+}, Cu^{2+}, Mn^{2+}, and Zn^{2+}) but some were not absorbed (such as Li^+, Ba^{2+}, Co^{2+}, and Al^{3+}). Furthermore, if the walls were linked to a column and a series of metal-salt solutions was run through it, some metals were strongly bound (including Mg^{2+}, Ca^{2+}, Fe^{3+}, and Ni^{2+}) and others were displaced or replaced. We were interested in the effects of modifying ligands (6) or, for instance, of saturation with Mg^{2+} on the subsequent staining of structures with ruthenium (severe) or lead and uranyl acetate (minimal). But our geologist colleague, Professor W. S. Fyfe, was more interested in why many ore bodies have a high percentage of organic residues and particular selections of metals. The mediation of biopolymers provided a stimulating hypothetical mechanism. The observations were extended by collaboration on geological diagenesis (3) and have been explored further in recent years (26), as have the consequences for the physicochemical well-being of wall components (25). Of course, parallel work has been and is going on in many laboratories concerned with ore leaching and metal transport in waters, to which our observations have contributed.

A discussion with Carl Robinow, C. L. Hannay, and Philip Fitz-James about why *Bacillus laterosporus* spores were eccentric in the cell led to the recognition of parasporal bodies and to the rediscovery of the spore-associated crystals in *Bacillus thuringiensis* (33). This is certainly a fertile field of research and practical application today.

On a lesser scale, an early study with Aksel Birch-Andersen (58) on the nature of flagella basal structures was diverted somewhat by the finding of a decoration on the inside of the plasma membrane of *Aquaspirillum serpens,* which we called polar membrane because of its position surrounding the polar tuft of flagella. An exactly similar structure is associated with many lophotrichous bacteria. We have now tried to understand this "structure in search of a function" and have found that the polar membrane of *Campylobacter jejuni* is a close-packed array of ATPase (8). What is consequential is that this particular ATPase turned out to be unusual by being specifically directed to ATP, activated by Mg^{2+} but inhibited by Ca^{2+}, made of a single subunit, and serologically distinct from *E. coli* ATPase; it is also likely to be a phylogenetically distinct lineage of that highly conserved enzyme.

As at the beginning of my time in science, the odds and ends still prove to be a stimulus. The current example is the lucky finding that high growth temperature stimulates cyst formation by *Azospirillum brasilense* (63). Now we know that a heat shock is equally effective for cyst formation at a growth temperature that is usually ineffective.

SOME SERVICES TO SCIENCE

I was brought up to believe that science knew no boundaries and that it was a duty of scientists to communicate freely. So from the outset I joined with the major societies in my orbit and took part in the meetings of the Laboratory Section of the Canadian Public Health Association and the Society of American Bacteriologists, as they were known then. The friendly winter meetings of the former, with extensive discussions of medical microbiology, and the comprehensive coverage of microbiology each springtime at the meetings of the latter were just what I needed as a stimulus to get started. But I was soon (1950–1951) in the midst of organizing a Canadian Society of Microbiologists, a much-needed catalyst and unifier for the diverse applied and basic microbiologists of the country. It took off owing in large part to the efforts of N. E. Gibbons, the founding secretary. I was put by my senior colleagues in the position of chairman of the organizing committee and the inaugural meeting in Ottawa in 1951. This was unforgettable for me, not just because it gave me a fine experience and the confidence of my colleagues, but because my headache on the last day of that meeting was due to encephalitis, which bedded me in that city for three weeks and took seven months out of my working life. The charter members elected me the founding president, perhaps as an act of kindness. Societies and their journals have been important to me and my work ever since, particularly the Canadian and the American societies.

I had an interest in editing, but it is not clear to me why I was asked to join the editorial board of the *Journal of Bacteriology* by J. Roger Porter in 1951; the invitation was unexpected because I had published less than a handful of papers and only three in that journal. Membership on the editorial board was good experience for my appointment as founding editor of the *Canadian Journal of Microbiology* (1954–1960); this appointment was less mysterious because I had a role in persuading the National Research Council of Canada that the journal was justifiable. In these years I learned the basics of editing from experienced colleagues and by doing it. My most challenging assignment was as editor of *Bacteriological* (later *Microbiological*) *Reviews, 1969–1979*, when I learned properly that a scientific editor's crucial job was not just to adjudicate the reports of referees but was to help authors to do their best (56). It was no mean exercise in diplomacy to deal with the sensitivities of undoubted authorities in their fields, both authors and referees, who could produce not only an appalling text but also a distressing narrowness of view. This was one aspect of my education; the other was association with people of marvelous ability and judgment both on the editorial boards and on the Publications Board of the American Society for Microbiology, notably L. Leon Campbell and Robert A. Day, who really taught me about the management and production of scientific publications. My association with the International Committee for Systematic Bacteriology and the *International Journal of Systematic Bacteriology* has brought experience in the additional stringencies of monitoring the description of bacteria and the application of rules of bacterial nomenclature. There is a great need to help authors in a field unfamiliar to most of them. I feel strongly that an editor is not the savior of science; if the editor is any good, the role is more of a "friend in court" for the author, who is the only person really responsible for what is written. The editor is the author's most concerned critic and has the advantage of taking or refusing the comments of referees to formulate rational advice.

My involvement in *Microbiological Reviews* increased my interest in the affairs of the American Society for Microbiology and my interest in what societies do for science and society.

Being an editor of a first-rate journal has an enhancing effect on one's scientific profile. I was most honored to be a candidate and to be elected president of the American Society for Microbiology, 1972–1973, which was in all respects a remarkable experience. I am sure I owe this signal recognition by the membership as much to my activities as an editor as to my scientific contributions.

I regret no part of these varied experiences and recommend that all aspiring young scientists owe a duty to scientific societies and their journals. We have to be prepared to communicate and to promote the highest standards in both the meetings of scientists, in the broadest arenas possible, and the journals of science. It is not good practice to speak only to a group of connoisseurs. We must play the roles of critic and of supporting actor in the interplay of science and society.

ENVOI

Science has shifted in the past 50 years from being the province of scientists to being a crucial part of our culture, permeating all conditions of life and living. Microbiology may have been ahead of most disciplines in this matter because of the sensitivity to matters affecting human health. The shift should be encouraging because one might hope that wisdom should accompany or be encouraged by understanding and technical competence. But there has been no more than a minimal attainment of that goal, and man's place in nature's world (and vice versa) seems ever more insecure, much as we would prefer to interpret it otherwise. My own contributions to the fostering of wisdom have been minimal, especially on the public and political fronts, which are so critical. At one time I scorned taking on public enlightenment, but I am convinced now that it is the only route to the heart of the political animal, which is effectively indifferent to anything beyond votes and short-term gains. I now believe that scientific societies and institutions should balance their concerns for communication within the house of science with an equal concern for public understanding as a stimulus to political action. I am not proud of my inaction in the public arena, nor am I impressed with what we have managed to do in our universities and colleges to inculcate a high level of understanding in the 15–25% of the new generations that transit these institutions seeking enlightenment.

A major factor in fostering my scientific activity and productivity has been the continuing support by research grants thoughtfully administered, and I owe thanks to the old Medical Committee of the National Research Council of Canada and its successor, the Medical Research Council of Canada. The former did me a great service for ten of my early years by awarding a block grant, which had no strings attached and only required a letter at year's end asking that it be continued, with a minimal accounting. It was a fixed sum and I had to abandon the award due to inflation of costs, but it did a great deal of good. Life is not so easy for even the most fortunate of young scientists today, who must spend 10% or more of their year in supplicating for funds. From the way such applications are treated one wonders whether or not anything should be done for the first time.

An unwitting scientific reward resulted from my study of bacterial structure because understanding of the structure of cells was an essential accompaniment to the development of cell biology and modern microbiology. As I have pointed out, a part of the transformation of bacterial taxonomy involved the realization that the unique features of bacteria as cells are crucial to description, and this was equally true of the understanding of function. The microscopist was, therefore, in a fortunate position on the sidelines of the great game between the cell biologists and the molecular biologists, and had entry to the game in some of the interesting plays. I may not have contributed more than peripheral structure to the game, but some friends and colleagues in cytology were major players, such as Keith Porter and Edward Kellenberger, to each of whom we owe a lot of our understanding. The great importance of correlating structure and function will be no less tomorrow than it is today.

Many of my contemporaries joined in the great exploitation of *E. coli* and a few other "handmaidens of science," doing a great service to science and mankind in the doing. However, their lives had a hectic and unenviable pace compared to the life I led without the hot breath of competitors on my neck. My colleagues and I protected ourselves from that fate because, it may be noted, we usually worked on organisms that were not commonly exploited and we could work at our own rate. This attitude also allowed a diversity of research topics, from biochemical cytology to taxonomy, which could be pursued as and when competence allowed. There was opportunity for me to be helpful to colleagues and for them to be helpful to me, which provided scientific pleasures in good company.

Where is the bacteriology I once knew? It is changing in a technological fashion in the medical diagnostic laboratories where I worked in the past. I fancy I would be more comfortable now with the microbial ecologists and those interested in organisms and their associations; their studies give rise to vistas broader than the most direct route to diagnosis. Times change and attitudes must change with them, but our old bacteriologists have done their duty and assisted the birth and development of microbiology and a remarkable range of disciplines and subdisciplines of biology. Whatever happens to current developments, and despite all diminishing interdisciplinary distinctions, we must see that there is a strong disciplinary base of teaching about the life and interrelationships of microbes. Microbiology departments should not be indistinguishable from departments of biochemistry, however well we get on with each other. The biosphere has been explored only in part, and there is much that we need to learn about the life and nature of the microbes that are no small part of it.

It is obvious that I owe much to the people in my life, and I can mention only a few of them in the context of even fewer activities. The roles of some, such as my parents, have been made clear and must not be underestimated. However, my life would not have been half as productive without the understanding of my first wife, Doris, who was my helpmate for just short of 40 years, or, in the past few years, the supportiveness of my second wife, Marion. My life has been enlivened and the work in the laboratory has prospered because of the talent, energy, and adventurousness of the graduate students and postdoctoral fellows who have spent their time with us. I am ever grateful to a passing parade of friends in and around the laboratory; but a few, who are all colleagues whatever their calling, have contributed more to me over the years than I can possibly acknowledge: Igor N. Asheshov[†], Terrance J. Beveridge, Aksel Birch-Andersen, Myrtle Hall, Phyllis Hobson, Gertrude G. Kalz, Susan F. Koval, Marion I. Luney, John F. Marak, Dianne Moyles, Carl F. Robinow, Roger J. Rossiter[†], and Gertrude Vaughan-Dragon.

Microbiology, in its own ways, tells us about life as it is reflected in the microbial cells, which have been in existence for longer than the imagination can appreciate. It tells us about how and where life can be lived, about the extremes of survival, and something about how life has evolved; but how it came about and where it came from is among the mysteries. Mankind must listen to the messages.

> My dream is faded now, and I am through
> With dreaming . . . yet I know
> The iris still will keep its gorgeous hue.
>
> Shushiki (Transl. C. H. Page)

Literature Cited

1. Austin, J. W., Murray, R. G. E. 1987. The perforate component of the regularly structured (RS) layer of *Lampropedia hyalina. Can. J. Microbiol.* 33(12): 1039–45
2. Baumeister, W., Karrenberg, F., Engel, R. R., Tentteggeler, B., Saxton, W. D. 1982. The major cell envelope protein of *Micrococcus radiodurans* (R_1): Structural and chemical characterization. *Eur. J. Biochem.* 125:535–44
3. Beveridge, T. J., Meloche, J. D., Fyfe, W. S., Murray, R. G. E. 1983. Diagenesis of metals chemically complexed to bacteria. *Appl. Environ. Microbiol.* 45:1094–108
4. Beveridge, T. J., Murray, R. G. E. 1974. Superficial macromolecular arrays on the cell wall of *Spirillum putridiconchylium. J. Bacteriol.* 119:1019–38
5. Beveridge, T. J., Murray, R. G. E. 1976. Uptake and retention of metals by cell walls of *Bacillus subtilis. J. Bacteriol.* 127:1502–18
6. Beveridge, T. J., Murray, R. G. E. 1980. Sites of metal deposition in the cell wall of *Bacillus subtilis. J. Bacteriol.* 141:876–87
7. Brenner, S., Horne, R. W. 1959. A negative staining method for high resolution electron microscopy of viruses. *Biochim. Biophys. Acta* 34:103–10
8. Brock, F. M., Murray, R. G. E. 1988. The ultrastructure and ATPase nature of polar membrane in *Campylobacter jejuni. Can. J. Microbiol.* 34(5):In press
9. Brooks, B. W., Murray, R. G. E. 1981. Nomenclature for "*Micrococcus radiodurans*" and other radiation-resistant cocci: Deinococcaceae fam. nov. and *Deinococcus* gen. nov. including five species. *Int. J. Syst. Bacteriol.* 31:353–60

[†]Deceased.

10. Brooks, B. W., Murray, R. G. E., Johnson, J. L., Stackebrandt, E., Woese, C. R., et al. 1980. Red-pigmented micrococci: A basis for taxonomy. *Int. J. Syst. Bacteriol.* 30:627–46

11. Buckmire, F. L. A., Murray, R. G. E. 1970. Studies on the cell wall of *Spirillum serpens*. I. Isolation and partial purification of the outermost cell wall layer. *Can. J. Microbiol.* 16:883–87

12. Buckmire, F. L. A., Murray, R. G. E. 1973. Studies of the cell wall of *Spirillum serpens*. II. Chemical characterization of the outer structured layer. *Can. J. Microbiol.* 10:59–66

13. Buckmire, F. L. A., Murray, R. G. E. 1976. Substructure and *in vitro* assembly of the outer structural layer of *Spirillum serpens*. *J. Bacteriol.* 125:290–99

14. Burdett, I. D. J., Murray, R. G. E. 1974. Electron microscope study of septum formation in *Escherichia coli* strains B and B/r during synchronous growth. *J. Bacteriol.* 119:1039–56

15. Burley, S. K., Murray, R. G. E. 1983. Structure of the regular surface layer of *Bacillus polymyxa*. *Can. J. Microbiol.* 29:775–80

16. Chapman, G. B., Hillier, J. 1953. Electron microscopy of ultrathin sections of bacteria. I. Cellular division in *Bacillus cereus*. *J. Bacteriol.* 66:362–73

17. Chapman, J. A., Murray, R. G. E., Salton, M. R. J. 1963. The surface anatomy of *Lampropedia hyalina*. *Proc. R. Soc. London Ser. B.* 158:498–513

18. Chatton, E. 1937. *Titres et Travaux Scientifiques*. Sète, France: Sottano

19. Costerton, J. W. F., Murray, R. G. E., Robinow, C. F. 1961. Observations on the motility and the structure of *Vitreoscilla*. *Can. J. Microbiol.* 7:329–39

20. Coulton, J. W., Murray, R. G. E. 1978. Cell envelope associations of *Aquaspirillum serpens* flagella. *J. Bacteriol.* 136:1037–49

21. Counsell, T. J., Murray, R. G. E. 1986. Polar lipid profiles of the genus *Deinococcus*. *Int. J. Syst. Bacteriol.* 36:202–6

22. Cowan, S. T. 1971. Sense and nonsense in bacterial taxonomy. *J. Gen. Microbiol.* 67:1–8

23. DePamphilis, M. L., Adler, J. 1971. Attachment of flagellar bodies to the cell envelope: specific attachment to the outer, lipopolysaccharide membrane and the cytoplasmic membrane. *J. Bacteriol.* 105:396–407

24. Dickson, M. R., Downing, K. H., Wu, W. H., Glaeser, R. M. 1986. Three-dimensional structure of the surface layer proteins of *Aquaspirillum serpens* VHA determined by electron crystallography. *J. Bacteriol.* 167:1025–34

25. Ferris, F. G., Beveridge, T. J. 1986. Physichochemical roles of soluble metal cations in the outer membrane of *Escherichia coli* K-12. *Can. J. Microbiol.* 32:594–601

26. Ferris, F. G., Beveridge, T. J., Fyfe, W. S. 1986. Iron-silica crystallite nucleation by bacteria in a geothermal sediment. *Nature* 320:609–11

27. Ferris, F. G., Beveridge, T. J., Marceau-Day, M. L., Larson, A. D. 1984. Structure and cell envelope associations of flagellar basal complexes of *Vibrio cholerae* and *Campylobacter fetus*. *Can. J. Microbiol.* 30:322–33

28. Finch, J. T., Klug, A., Nermut, M. V. 1967. Analysis of the fine surface structure of the macromolecular units on the cell wall of *Bacillus polymyxa*. *J. Cell Sci.* 2:587–90

29. Fitz-James, P. C. 1960. Participation of the cytoplasmic membrane in the growth and spore formation of bacilli. *J. Biophys. Biochem. Cytol.* 8:507–28

30. Fox, G. E., Pechman, K. R., Woese, C. R. 1977. Comparative cataloging of 16S ribosomal ribonucleic acid: molecular approach to procaryotic systematics. *Int. J. Syst. Bacteriol.* 27:44–57

31. Ghosh, B. K., Murray, R. G. E. 1967. Fine structure of *Listeria monocytogenes* in relation to protoplast formation. *J. Bacteriol.* 93:411–26

32. Goundry, J. A. L., Davison, A. R. A., Baddiley, J. 1967. The structure of the cell wall of *Bacillus polymyxa* (NCIB 4747). *Biochem. J.* 104:16

33. Hannay, C. L., Fitz-James, P. C. 1955. The protein crystals of *Bacillus thuringiensis* Berliner. *Can. J. Microbiol.* 1: 694–710

34. Higgins, M. L., Shockman, G. D. 1976. Study of a cycle of cell wall assembly in *Streptococcus faecalis* by three-dimensional reconstructions of thin sections of cells. *J. Bacteriol.* 127:1346–58

35. Hobot, J. A., Villiger, W., Escaig, J., Maeder, M., Ryter, A., Kellenberger, E. 1985. Shape and fine structure of nucleoids observed on sections of ultrarapidly frozen and cryosubstituted bacteria. *J. Bacteriol.* 162:960–71

36. Houwink, A. L. 1953. A macromolecular monolayer in the cell wall of *Spirillum* spec. *Biochim. Biophys. Acta* 10: 360–66

37. Kist, M. L., Murray R. G. E. 1984. Components of the regular surface array of *Aquaspirillum serpens* MW5 and their

assembly *in vitro*. *J. Bacteriol.* 157: 599–606

38. Koch, A. L., Doyle, R. J. 1986. The growth strategy of the gram-positive rod. *FEMS Microbiol. Rev.* 32:247–54

39. Koval, S. F., Murray, R. G. E. 1981. Cell wall proteins of *Aquaspirillum serpens*. *J. Bacteriol.* 146:1083–90

40. Koval, S. F., Murray, R. G. E. 1983. Solubilization of the surface protein of *Aquaspirillum serpens* by chaotropic agents. *Can. J. Microbiol.* 29:146–50

41. Koval, S. F., Murray, R. G. E. 1985. Effect of calcium on the *in vivo* assembly of the surface protein of *Aquaspirillum serpens* VHA. *Can. J. Microbiol.* 31:261–67

42. Koval, S. F., Murray, R. G. E. 1986. The superficial protein arrays on bacteria. *Microbiol. Sci.* 3:357–61

43. Lancy, P. Jr., Murray, R. G. E. 1978. The envelope of *Micrococcus radiodurans:* Isolation, purification, and preliminary analysis of the wall layers. *Can. J. Microbiol.* 24:162–76

44. Lapage, S. P., Sneath, P. H. A., Lessel, E. F., Skerman, V. B. D., Seeliger, H. P. R., Clarke, W. A., eds. 1975. *International Code of Nomenclature of Bacteria.* Washington, DC: Am. Soc. Microbiol.

45. Luria, S. E., Human, M. L. 1950. Chromatin staining of bacteria during bacteriophage infection. *J. Bacteriol.* 59:551–60

46. Luria, S. E., Palmer, J. L. 1946. Cytological studies of bacteria and bacteriophage growth. *Carnegie Inst. Washington Yearb.* 45:153–56

47. MacAlister, T. J., Cook, W. R., Wiegand, R., Rothfield, L. I. 1987. Membrane-murein attachment at the leading edge of the division septum: A second membrane-murein structure associated with morphogenesis of the gram-negative bacterial division septum. *J. Bacteriol.* 169:3945–51

48. Maier, S., Murray, R. G. E. 1965. The fine structure of *Thioploca ingrica* and a comparison with *Beggiatoa*. *Can. J. Microbiol.* 11:645–55

49. Mudd, S. 1953. The mitochondria of bacteria. *Bacterial Cytology, Symp. 6th Int. Congress Microbiol.* pp. 67–81, Suppl. Rend. Ist. Super. Sanità, Rome.

50. Murray, R. G. E. 1957. Direct evidence for a cytoplasmic membrane in sectioned bacteria. *Can. J. Microbiol.* 3:531–32

51. Murray, R. G. E. 1960. The internal structure of the cell. In *The Bacteria*, ed. I. C. Gunsalus, R. Y. Stanier, 1:35–96. New York: Academic

52. Murray, R. G. E. 1962. Fine structure and taxonomy of bacteria. *Symp. Soc. Gen. Microbiol.* 12:119–44

53. Murray, R. G. E. 1963. On the cell wall structure of *Spirillum serpens*. *Can. J. Microbiol.* 9:393–401

54. Murray, R. G. E. 1968. Microbial structure as an aid to microbial classification and taxonomy. *Spisy Prirodoved. Fak. Univ. J. E. Purkyne Brne* 43:249–52

55. Murray, R. G. E. 1974. A place for bacteria in the living world. In *Bergey Manual of Determinative Bacteriology*, ed. R. E. Buchanan, N. E. Gibbons, pp. 4–9. Baltimore, Md: Williams & Wilkins. 8th ed.

56. Murray, R. G. E. 1983. What *is* an Editor for? *CBE Views* 6:14–19

57. Murray, R. G. E. 1984. The higher taxa, or, a place for everything . . .? In *Bergey's Manual of Systematic Bacteriology*, ed. N. R. Krieg, J. G. Holt, pp. 31–34. Baltimore, Md: Williams & Wilkins.

58. Murray, R. G. E., Birch-Anderson, A. 1963. Specialized structure in the region of the flagella tuft in *Spirillum serpens*. *Can. J. Microbiol.* 9:393–401

59. Murray, R. G. E., Douglas, H. C. 1950. The reproductive mechanism of *Rhodomicrobium vanniellii* and the accompanying nuclear changes. *J. Bacteriol.* 59:157–67

60. Murray, R. G. E., Elder, R. H. 1949. The predominance of counterclockwise rotation during swarming of *Bacillus* species. *J. Bacteriol.* 58:351–59

61. Murray, R. G. E., Francombe, W. H., Mayall, B. H. 1959. The effect of penicillin on the cell structure of staphylococcal cell walls. *Can. J. Microbiol.* 5:641–48

62. Murray, R. G. E., Gillen, D. H., Heagy, F. C. 1950. Cytological changes in *Escherichia coli* produced by infection with phage T2. *J. Bacteriol.* 59:603–15

63. Murray, R. G. E., Moyles, D. 1987. Differentiation of the cell wall of *Azospirillum brasilense*. *Can. J. Microbiol.* 33:132–37

64. Murray, R. G. E., Pearce, R. H. 1949. The detection and assay of hyaluronidase by means of mucoid streptococci. *Can. J. Res. Sect. E* 27:254–64

65. Murray, R. G. E., Robinow, C. F. 1952. A demonstration of the disposition of the cell wall of *Bacillus cereus*. *J. Bacteriol.* 63:298–300

66. Murray, R. G. E., Steed, P., Elson, H. E. 1965. The location of the mucopeptide in sections of the cell wall of *Escherichia coli* and other gram-negative bacteria. *Can. J. Microbiol.* 11:547–60

67. Murray, R. G. E., Truant, J. P. 1954. The morphology, cell structure and taxonomic affinities of the *Moraxella. J. Bacteriol.* 67:13–22

68. Murray, R. G. E., Watson, S. W. 1965. Structure of *Nitrosocystis oceanus* and comparison with *Nitrosomonas* and *Nitrobacter. J. Bacteriol.* 80:1594–609

69. Nermut, M. V., Murray, R. G. E. 1967. The ultrastructure of the cell wall of *Bacillus polymyxa. J. Bacteriol.* 93: 1949–65

70. Newman, S. B., Borysko, E., Swerdlow, M. 1949. New sectioning techniques for light and electron microscopy. *Science* 110:66–68

71. Niebel, H., Dorsch, M., Stackebrandt, E. 1987. Cloning and expression in *Escherichia coli* of *Proteus vulgaris* genes for 16S rRNA. *J. Gen. Microbiol.* 133: 2401–9

72. Oyaizu, H., Stackebrandt, E., Schleifer, K. H., Ludwig, W., Pohla, H., et al. 1987. A radiation-resistant rod-shaped bacterium, *Deinobacter grandis* gen. nov., sp. nov., with peptidoglycan containing ornithine. *Int. J. Syst. Bacteriol.* 37:62–67

73. Palade, G. E. 1952. A study of fixation for electron microscopy. *J. Exp. Med.* 95:285–99

74. Reynolds, E. S. 1963. The use of lead citrate at high pH as an electron-opaque stain in electron microscopy. *J. Cell Biol.* 17:202–12

75. Robinow, C. F. 1945. Addendum: Nuclear apparatus and cell structure of rod-shaped bacteria. In *The Bacterial Cell*, by R. J. Dubos, pp. 355–77. Cambridge, Mass: Harvard Univ. Press

75a. Robinow, C. F. 1953. Spore structure as revealed by thin sections. *J. Bacteriol.* 66:300–11

76. Robinow, C. F., Murray, R. G. E. 1953. The differentiation of cell wall, cytoplasmic membrane and cytoplasm of gram positive bacteria by selective staining. *Exp. Cell. Res.* 4:390–407

77. Ryter, A. 1968. Association of the nucleus and the membrane of bacteria: A morphological study. *Bacteriol. Rev.* 32:39–54

78. Ryter, A., Kellenberger, E. 1958. Etude au microscope électronique de plasma contenant de l'acide désoxyribonucléique. I. Les nucléoides des bactéries en croissance active. *Z. Naturforsch.* 133:597–605

79. Salton, M. R. J., Horne, R. W. 1951. Studies of the bacterial cell wall. II. Methods of preparation and some properties of cell walls. *Biochim. Biophys. Acta* 7:177–97

80. Skerman, V. B. D., McGowan, V., Sneath, P. H. A., eds. 1980. *Approved Lists of Bacterial Names.* Washington, DC: Am. Soc. Microbiol.

81. Sleytr, U. B., Messner, P. 1983. Crystalline surface layers on bacteria. *Ann. Rev. Microbiol.* 37:311–39

82. Smit, J. 1986. Protein surface layers of bacteria. In *Bacterial Outer Membranes as Model Systems*, ed. M. Inouye, pp. 343–76. Chichester, UK: Wiley

83. Stackebrandt, E., Ludwig, W., Weizenegger, M., Dorn, S., McGill, T. J., et al. 1987. Comparative 16S rRNA oligonucleotide analyses and murein types of round-spore-forming bacilli and non-spore-forming relatives. *J. Gen. Microbiol.* 133:2523–29

84. Stanier, R. Y. 1961. La place des bactéries dans le monde vivant. *Ann. Inst. Pasteur Paris* 101:297–312

85. Stanier, R. Y., van Niel, C. B. 1962. The concept of a bacterium. *Arch. Microbiol.* 42:17–35

86. Steed, P., Murray, R. G. E. 1966. The cell wall and cell division of gram-negative bacteria. *Can. J. Microbiol.* 12:263–70

87. Steed-Glaister, P. D. 1967. *A study of cell wall and division of gram-negative bacteria.* PhD thesis. Univ. Western Ontario, London, Ontario, Canada. 209 pp.

88. Stevenson, J. W., Cowan, S. T. 1967. Obituary notice: E. G. D. Murray. *J. Gen. Microbiol.* 46:1–21

89. Stewart, M., Beveridge, T. J., Murray, R. G. E. 1980. Structure of the regular surface layer of *Spirillum putridiconchylium. J. Mol. Biol.* 137:1–8

90. Stewart, M., Murray, R. G. E. 1982. Structure of the regular surface layer of *Aquaspirillum serpens* MW5. *J. Bacteriol.* 150:348–57

91. Thompson, B. G., Murray, R. G. E., Boyce, J. F. 1982. The association of the surface array and the outer membrane of *Deinococcus radiodurans. Can. J. Microbiol.* 28:1081–88

92. Wayne, L. G., Brenner, D. J., Colwell, R. R., Grimont, P. A. D., Kandler, O., et al. 1987. Report of the *ad hoc* committee on reconciliation of approaches to bacterial systematics. *Int. J. Syst. Bacteriol.* 37:463–64

93. Weidel, W., Frank, H., Martin, H. H. 1960. The rigid layer of the cell wall of *Escherichia coli* strain B. *J. Gen. Microbiol.* 22:158–66
94. Whitfield, J. F., Murray, R. G. E. 1954. A cytological study of the lysogenization of *Shigella dysenteriae* with P₁ and P₂ bacteriophages. *Can. J. Microbiol.* 1: 216–26
95. Whitfield, J. F., Murray, R. G. E. 1956. The effects of the ionic environment on the chromatin structures of bacteria. *Can. J. Microbiol.* 2:245–60
96. Williams, R. C., Wyckoff, R. W. G. 1946. Applications of metallic shadow-casting to microscopy. *J. Appl. Phys.* 17:23
97. Woese, C. R. 1987. Bacterial evolution. *Microbiol. Rev.* 51:221–71
98. Woese, C. R., Blanz, P., Hahn, C. M. 1984. What isn't a pseudomonad: The importance of nomenclature in bacterial classification. *Syst. Appl. Microbiol.* 5: 179–95
99. Woldringh, C. L., Nanninga, N. 1976. Organization of the nucleoplasm in *Escherichia coli* visualized by phase-contrast light microscopy, freeze fracturing, and thin sectioning. *J. Bacteriol.* 127:1455–64
100. Work, E., Griffiths, H. 1968. Morphology and chemistry of cell walls of *Micrococcus radiodurans*. *J. Bacteriol.* 95: 641–57
101. Zuckerkandl, E., Pauling, L. 1965. Molecules as documents of evolutionary history. *J. Theor. Biol.* 8:357–66

Annu. Rev. Neurosci. 1990. 13:1–13

CHEMOSENSORY PHYSIOLOGY IN AN AGE OF TRANSITION

Vincent G. Dethier

Department of Zoology, University of Massachusetts, Amherst, Massachusetts 01003

In a recent published book entitled *Masks of the Universe*, the cosmologist Edward Harrison reminds us that we as a society create the universe in which we live and that each age fashions its own unique universe. Each of these universes is a mask of The Universe. Similarly, scientific verities are the creations of the society of contemporary scientists who believe as firmly as do cosmologists that their tenets represent reality. This reality changes with time. Each age tends to be but marginally charitable toward the "ignorance" of its predecessors and not especially sensitive to the possibility that, as before, some of today's truths may become tomorrow's heresies. This is not to deny that shards of truth are salvagable from each age to the next; it is to remind us that comprehension at any time is profoundly influenced by contemporary intellectual ambience.

Although historians of science can view from the mountain top of time the totality and comprehensiveness of the scientific endeavor and can discern the slow grandeur of its progress from one period to the next, the proximate eddies and currents that perturb the main stream are perhaps most acutely perceived by those who have lived through transitional periods. During a time of transition one is more deeply aware of the tenuous hold that science has on truth at any one moment, of the fragility as well as power of hypothesis and theory, of the influential role of improbabilities, and of the paradoxical relationship between concept and technology. The last point is particularly relevant in this time of high technology. Although new inventions and developments indubitably stimulate new ideas as well as provide means for solving old questions, the generation of ideas is not inextricably constrained by new technology. At most, realizations may be delayed. As Beidler (1987) has pointed out, technology cannot be substituted for keen insight, philosophical undcr-standing, and serendipity.

In some small measure the course of biology exemplifies the flux of conceptual cosmologies in general. Biology experienced a rapid transition from one age to another during the first three quarters of the present century. Classical biology was on the wane, molecular biology in the ascendancy, especially in the 1950s. In the 1960s, a new cohesive discipline began to emerge, neuroscience. I consider myself fortunate to have studied and practiced science during this period. It was a time of excitement, expectation, questioning, and discovery. It provided challenges, frus-trations, and expanding vistas of terra incognita.

In presenting this prefatory chapter I offer an account of my area of special interest, the chemical senses, as a vehicle to illustrate some general aspects of a period of transition, to record the importance of time, place, and informal eclectic personal interaction, to describe how ideas may develop, and to note their independence as well as their dependence on technology. Most especially I wish to present the thesis that cherishing a clear goal is essential but that at the same time one profits from peripheral vision. The "surround," to borrow a term from visual psychophysics, modulates and enriches the pursuit of a goal and ultimately places it in its clearest perspective. This essay, then, is not a review of the field. That has been done many times, the most recent being that of Finger & Silver (1987). It is a journey of inquiry through a period when electrophysiology was just coming of age and neuroscience had not yet emerged as a recog-nized discipline.

My early education in science began toward the end of the era of classical biology before any adumbrations of transition were apparent to the bio-logical community as a whole. The starting point for me was a fascination with nature awakened by the beauty and vibrancy of living things, by sensual delight in colors, scents, sounds, and the cycling of the seasons, all enhanced by the inability of adults to answer satisfactorily the childhood "whys." Later, under the stern tutelage of formal science, the "whys" would become "whats" and "hows."

The wonderment that sustained my interest in science throughout child-hood and adolescence was almost obliterated during undergraduate days at Harvard. There, young aspiring biologists absorbed heavy doses of anatomy and taxonomy, preponderantly, of course, by examining pickled, mummified, and skeletonized specimens—or fossils thereof. At the very least, however, one learned the names of what was out there in the plant and animal world and approximately what each looked like externally and internally. Function, on the other hand, remained a mystery and was not related to behavior in any but the most obvious ways. Physiology, particularly the machinery of the human body, flourished in the schools of medicine. Experimental zoology in this country and abroad was not held in high esteem in the first third of the century.

A nodding acquaintance with physiology was acquired en route through two physiology courses. One course offered to our appetites was served up by Cannon, from whom I learned the "wisdom of the body" (1939). He also kept us informed about the controversy between the "electrical" people and the "purveyors of soup" over the nature of transmission at the neuromuscular junction. New discoveries about the role of acetylcholine, esserine, and ATP were only beginning to trickle into the curriculum. The pioneering work of Adrian on electrophysiology in the 1920s had not yet attracted wide attention (Adrian & Forbes 1922, Adrian 1926).

None of this collegial enlightenment satisfied an eccentric interest that I had in the behavior of insects. The closest approach to animal behavior was provided by Welsh's course in invertebrate (mostly marine) physiology. It did not offer any help toward solving questions in the area of my special interest, the obsessive gourmet habits of herbivorous insects. The only explanation of this gastronomic phenomenon extant postulated the existence of a "botanical" sense, a mysterious sixth sense. By analogy with my own eating habits, I suspected that chemical senses were involved. Although we students were constantly warned against the evil of anthropomorphism, I learned over the years that there was considerable heuristic value in posing questions from this point of view.

At Harvard in the 1930s there were two specialists in the field of chemoreception, Parker and Crozier. Neither was currently active in that field, having turned to other matters, but each had written reviews (Parker 1922, Crozier 1934) that summed up knowledge about these least understood of all sensory systems. At this time, knowledge of the chemical senses was restricted to anatomy and histology as constrained by limits of the light microscope and gross anatomy. Only gold and silver impregnation and methylene blue staining were available for revealing the tracery of the nervous system. Generally speaking, investigators were focusing their attention on gross anatomy, histology, thresholds, classification of tastes and odors, relations of sensation to chemical structure, theories of action, psychophysics, flavor, and perfumery. The emphasis throughout lay on vertebrates.

Insofar as insects were concerned, zoologists occupied themselves in free speculation. Despite a great legacy of elegant nineteenth century histology, there was no unanimity of opinion regarding the loci of chemosensory organs, no consensus as to whether there were separate and distinct olfactory and gustatory senses, and no knowledge concerning the identity of the end-organs themselves.

If the problem of insect monophagy was to be solved, the identity and response characteristics of the chemoreceptors had to be revealed. The first and only conceivable approach was to combine ablation, anesthesia, and occlusion of putative end-organs with behavioral observations. By means of these techniques tentative identification of some olfactory organs, among them those of caterpillars, was made during the period 1921–1941 (von Frisch 1921, Dethier 1937, 1941). A new approach was needed.

Although I had acquired a vicarious acquaintance with the pioneering technique of electrophysiology of the 1920s by reading Adrian's *Basis of Sensation* (1928), I had neither the knowledge nor the equipment to apply this new approach to my problem. The instruments that existed were the Einthoven string galvanometer, the Lippmann capillary electrometer, and the Matthews oscillograph. They were available in a few laboratories only, notably Adrian's in Cambridge, Forbes' at Harvard, and Erlanger's and Gasser's at Washington University. Most investigators using this equipment were concerned with trying to understand the nature of the action potential (Gasser & Newcomer 1921, Gasser & Erlanger 1922). Motor nerves and the sensory nerves of mechanoreceptors were the preparations best suited for this research. Only a few attempts to record from chemosensory nerves were made in the early 1930s. Hoagland (1933) detected but could not resolve electrical activity in nerves of the lips and barbels of catfish in response to salt and acid; Zotterman (1935) recorded impulses in the chorda tympani and glossopharyngeal of the rat; Pumphrey (1935) recorded responses to salt and acid placed on the tongue of the frog; Adrian & Ludwig (1938) recorded activity in the olfactory nerves of catfish and carp.

At this time, through chance encounters, I made the acquaintance of Roeder at Tufts College and Prosser at Clark University. Roeder had arrived at Tufts in 1932 and Prosser at Clark in 1934. Roeder's entrée to electrophysiology came as a consequence of his taking a course at Woods Hole, where Prosser was instructing. Prosser was investigating electrical activity in the nerve cord of earthworms and marine invertebrates by means of a Matthews oscillograph and amplifier. Roeder applied this technique together with ablation to studies of copulatory behavior in praying mantids.

At Prosser's invitation I spent part of one summer at Clark, where we attempted to record from chemosensory nerves of caterpillars. Activity from mechanoreceptors was detected, but only physiological silence and instrumental noise followed chemical stimulation. Attempts continued intermittently and unsuccessfully until the advent of World War II.

In the meantime Roeder, a skilled and ingenious tinkerer, was perfecting his instrumentation and techniques for recording from the central nerve cord of the cockroach. Shortly after the war, he and I made another attempt to record from chemoreceptors. Having observed that the very long ovipositors of some parasitic wasps were sensitive to chemical stimulation, I thought that the correspondingly long sensory nerve might lend itself to recording. Again we failed to detect any action potentials.

At this point behavior seemed to offer a more promising approach to the problems of chemoreception. The turning point came as the result of two lucky decisions, namely, that the caterpillar was not the most cooperative animal and that gustation was more tractable than olfaction. The choice of the blowfly was serendipitous. In 1922 Minnich had observed that flies and butterflies extended their proboscises when the legs were touched with sugar. Four years later he reported in one short sentence that touching a single long curved hair on the proboscis of the blowfly with sugar elicited extension (Minnich 1926). Here was a preparation where single chemoreceptors could be isolated and gustation studied behaviorally without the usual confounding postingestional effects.

Early in 1953, a student, Grabowski, and I succeeded by microtopical application of sucrose to single hairs in demonstrating unequivocally that the gustatory end-organs of the tarsi were hairs (setae) similar in appearance to those on the labellum, which Minnich had identified as gustatory (Grabowski & Dethier 1954). We were able to prove that the tips were not covered with cuticle, that the hairs were innervated by three bipolar sensory neurons (later examination with an electron microscope revealed two additional neurons), and that each of the cells responded to a different sensory modality. It was immediately apparent that this preparation could serve as an excellent model of a gustatory apparatus because, unlike the vertebrate taste papillae, the receptors were primary neurons, the axons of which led directly without synapsing into the head ganglia.

During the next ten years it was possible with this preparation to examine the response spectra of the chemosensory cells, detail, separate, and measure peripheral and central adaptation, measure temporal and spatial summation, evaluate differential thresholds ($\Delta I/I$), relate responsiveness to the structural configuration of stimulating molecules, propose a provisional functional map of gustatory projections in the central nervous system, demonstrate and evaluate in an intact organism central excitatory and inhibitory states as Sherrington had done with the flexion reflex in spinal cats, advance promising hypotheses regarding the nature of molecular receptor sites and transduction, and lay a groundwork for elucidating mechanisms underlying hunger and satiation. All of this work was eventually summarized in the book, *The Hungry Fly* (Dethier 1976).

The goal throughout this period was to understand the neural mechanisms mediating chemosensory responses and related behavior. At the same time, it was the behavior that was providing insight to the mechanisms. The behavioral approach was an exciting game of wits. It proved to be a powerful tactic, and many of its findings were subsequently shown by electrophysiology to be gratifyingly accurate.

Through the period of the late 1940s to 1958 there had been no breakthrough in the impasse to successful electrophysiological recording from insect chemoreceptors. Investigation of vertebrate systems had been making considerable progress, as exemplified by Pfaffmann's beautiful recordings of single chorda tympani fiber responses in the cat. At the Johns Hopkins University, to which I had moved after the war, Bronk, Hartline and others of the group were fully engaged in electrophysiological work; however, the only experimentation in chemoreception was being conducted by Beidler (then a graduate student). He was studying the integrated response of multiple chorda tympani fibers of the rat. Most of the studies with vertebrates were concerned with the neural basis of the four classical taste modalities.

In Europe there was progress in the field of olfaction, beginning with Adrian & Ludwig's (1938) recording of electrical activity in the olfactory nerves of catfish and carp and Adrian's (1942) recording of activity in the olfactory projections in the brain of the hedgehog. In 1953, Boistel and Coraboeuf succeeded in recording mixed neural responses in insect antennae. It was Schneider in Germany, however, who finally made a breakthrough (1955). Realizing in 1952–1953 that in the silkworm moth's response to specific pheromones he had the perfect experimental animal and the perfect olfactory stimulus, he exploited that system fully and in 1955 recorded the first single unit olfactory responses to pheromones. This accomplishment paved the way for his own elegant work and that of Kaissling, Boeckh, and others (Boeckh, Kaissling & Schneider 1965). The insect gustatory system remained intransigent, but behavioral analysis continued.

In the course of measuring behavioral thresholds, a number of workers observed that sensitivity decreased as the duration of deprivation increased. The question of whether the sensitivity of the receptors themselves changed or some postingestional factors influenced response steered our research in a direction that was also being investigated in vertebrates. Physiologists and physiological psychologists were probing the central nervous system for answers. Hetherington & Ranson (1942) had discovered that hypothalamic lesions in the brain of the rat affected feeding. Anand & Brobeck (1951a,b) at Yale and Teitlebaum & Stellar (1954) at Hopkins were also lesioning the brain. At that time it was believed that the lateral hypothalamus was a hunger or feeding center and the ventromedial hypothalamus a satiety center. In time the mechanisms were discovered to be more complex. Eventually, in the years to follow, investigators of the vertebrate system gradually worked their way toward the periphery, while those of us studying the blowfly were working from the periphery inwards.

Still employing threshold as a monitor, my associates and I began to isolate by surgery, ligation, and intubation various parts of the digestive system as possible origins of signals modulating the response to chemosensory stimulation. Complementary tests involved injection of the haemocole and also parabiosis. One tremendous advantage of the fly over the rat was the possibility of removing the entire digestive tract together with the oral gustatory receptors and a bit of brain and studying ingestion in vitro. Thus, by a process of elimination we determined that one mechanism regulating behavioral threshold and therefore ingestion was resident in the stomatogastric system, the analogue of the vertebrate autonomic system.

At this point in the investigation I discussed the problem with Dietrich Bodenstein, who was then a civilian employee at the Army Chemical Center in Edgewood, Maryland. Bodenstein was a developmental biologist who worked both on amphibians and cockroaches. He was one of the pioneers in insect developmental biology and a superb microsurgeon. When I mentioned to him my plan to section the recurrent nerve, he assured me that the operation was simplicity itself and proceeded to demonstrate by operating successfully on some *Drosophila*. Following his technique I sectioned the nerve in blowflies; the result was spectacular, a rapid and extreme hyperphagia. The whole sequence of normal feeding then became explicable in terms of interaction between chemosensory excitation and central inhibition triggered by internal mechanoreceptors (Dethier & Bodenstein 1958). This work, together with elaboration in the 1960s by Gelperin (1966a,b, 1967) and more recent refinements in our laboratory, gave us one of the most complete pictures of neural mechanisms of feeding up to that time.

In 1957 I was able to return to the gustatory receptor itself because of a remarkable innovation in electrophysiological recording. Hodgson, who had left our laboratory to continue postdoctoral studies with Roeder, developed, together with Lettvin and Roeder (1955), a technique that solved the problem of recording from single chemosensory sensilla. By one of those odd coincidences of science, Morita and his associates at Kyushu University made the same discovery in 1957. A glass micropipette containing weak saline plus the stimulating compound to be tested was placed over the tip of the hair and served as a salt bridge to a Ag/AgCl wire. Shortly thereafter Morita (1959) developed an elegant refinement whereby he inserted an electrode through a hole drilled in the shaft of the hair.

At Hopkins, Wolbarsht, Evans, and I immediately applied these new techniques to studies of the sensitivity and action spectra of individual chemosensory cells. The early behavioral conclusion that there was a sugar-sensitive cell, a salt-sensitive cell, and a mechanoreceptor was confirmed. In addition, the existence of two other cells, a water receptor (Wolbarsht 1957) and a second but distinctive salt receptor, was revealed. The first electron microscopic studies by Larsen (1962) in our laboratory and Adams et al (1965) at Rutgers University confirmed the existence of these cells, thus correcting the early methylene blue evidence.

These technical developments now made possible pursuit of two problems that had constituted the crux of chemosensory physiology from the beginning, namely transduction and coding. A chance observation by Barnhard & Chadwick at the Army Chemical Center in 1953 had put us on the track. They had observed that bait frequented by flies was more attractive than bait that had been protected from visiting flies. The obvious reason seemed to be that regurgitation and defecation altered the material; however, flies with both proboscis and anus plugged also enhanced the attractiveness. On a hunch I eluted the legs of flies with water, added sucrose to the eluate, and tested for enzyme activity. The eluate contained an alpha-glucosidase. The idea that this enzyme might initiate the process of transduction was tempting; however, since it was already known that some sugars without alpha-glucosidase linkages and some pentoses and hexoses as well were adequate stimuli, we did not pursue the matter further. Hansen in Germany and Morita, Kijima, Koizumi, and their associates in Japan picked up the trial (Hansen & Kuhner 1972, Kijima et al 1973). They proved that the enzymes were intimately and exclusively associated with the receptors. Exactly what part the glucosidases play in the process of transduction is still a mystery.

The relation between the sensitivity of the sugar receptor and the configuration of carbohydrate molecules was also unknown. Neither von Frisch (1935) nor I (1955) had been able to make any sense of the comparative stimulatory effectiveness of the various sugars. The first clue appeared in 1955, when behavioral studies of mixtures revealed that some sugars synergized and others inhibited each other. Evans (1963) suggested that the sugar receptor cell contained multiple molecular sites, specifically one for pyranose and one for furanose sugars [the idea of different multiple sites had been proposed earlier by Biedler for the rat salt receptor (1957, 1962)]. Six years later electrophysiological support for this hypothesis was provided (Omand & Dethier 1969). The matter was finally settled by Morita & Shiraishi (1968) and Shimada et al (1974) by further electrophysiological and pharmacological studies. As the matter now stands, the sugar receptor of the blowfly is presumed to have four specific molecular sites.

Comparable studies of the nature of the salt receptor were stimulated by Beidler's work at Hopkins with the rat chorda tympani. His equation based on the Law of Mass Action was found by Evans & Mellon (1962) to apply to the salt receptor of the fly.

One remaining series of studies attempting to relate stimulating effectiveness and chemical structure is worth mentioning. Over the years, beginning in 1947, more than 200 aliphatic compounds were tested and found to cause flies to reject sugar solutions. It was believed that these compounds stimulated a receptor mediating rejection. The "rejection" threshold could be predicted with great accuracy from the structural formulae. Discussions with Brink at Hopkins prompted me to apply to these data the same thermodynamic analyses that he and Posternak had made of narcotics (Brink & Posternak 1948). It was not until later (1965) in our laboratory at the University of Pennsylvania that Hanson showed electrophysiologically that these aliphatic compounds, rather than stimulating a "rejection" receptor, were narcotizing the receptors mediating acceptance. Steinhardt et al (1966) came to a similar conclusion.

Despite these forays into the nature of transduction, the bearing of gustation on feeding continued to be a central theme in our work. Richter, at the peak of his studies of food preference by the rat and the relation between preference and nutritional need, often visited our laboratory and observed a "two-bottle" preference apparatus that a student (Rhoades) and I had designed in emulation of his apparatus for rats (Dethier & Rhoades 1954). He encouraged us to investigate long-term ingestion and preference as they related to nutrition. Seven years earlier at the Army Chemical Center, a comprehensive survey of the nutritional adequacy of various carbohydrates had been undertaken (Hassett, Dethier & Gans 1950), so the stage was set for a study of preference. The final results indicated that insofar as sugars were concerned, taste preference was not an infallible guide to nutritional value.

After eleven stimulating and productive years at Hopkins, where I learned electrophysiology, electron microscopy, and physiological psychology, I moved to the University of Pennsylvania, where there was a strong multidisciplinary group studying many and varied aspects of feeding behavior at the Institute of Neurological Science. There the lines of work already described were continued and expanded. Another set of gustatory receptors was discovered in the oral cavity of flies; the sensory basis of water, alcohol, and protein ingestion was investigated, and the existence and characteristics of central excitatory and inhibitory states initiated by chemosensory input were established.

Advances in our knowledge of the chemoreceptors of the fly and the evolution of modern techniques prompted a return to the investigations on caterpillars begun 30 years earlier. Caterpillar gustatory receptors were discovered to be more complicated and versatile than those of the blowfly. They also appeared to be less specific, more broadly tuned. Schneider had classified the olfactory receptors of the silkworm moth as "specialists" and "generalists." The receptors of caterpillars fell somewhere between these two extremes of a response spectrum. Extensive recording of responses to plant saps led to the hypothesis that differential preferences were mediated by patterns of impulses from multiple receptors. Explanations previous to this, molded by Verschaffelt's (1910) discovery that special compounds triggered feeding and by the prevailing ethological concept of "sign" stimuli, emphasized the specificity of receptors. A theory stressing the importance of multineuronal afferent patterns (across-fiber patterns) proposed by Pfaffmann (1941) and eloborated by him and by Erickson (1963) seemed to fit the case of caterpillars. This work continued for many years at Pennsylvania and then at Princeton. Eventually some progress was made in decoding chemosensory messages with the help of information theory and analysis of spike-interval distributions by autocorrelograms (Dethier & Crnjar 1982).

The investigative journey in this selected field of inquiry has come full circle in the period spanning the very early development of neurophysiology in the first third of the century to the sophisticated armamentarium of neuroscience of the present. The character of the pursuit of knowledge in the field of chemoreception and related behavior reflects the general nature and evolution of sensory physiological investigation in this period of transition. The enlistment of ingenuity and indirect approaches that characterized general physiology in the 1920s when direct approaches were technically impossible carried sensory physiology forward. Awareness of progress in apparently unrelated fields facilitated advance toward focused goals. A contrapuntal relation gradually developed between ideation and technology.

In the broad search for and understanding of behavior, the investigation had advanced contripetally from sense organs to central phenomena, from a proximate goal of understanding the mechanics of stimulation at the receptor level to decoding sensory spike trains. Furthermore, it moved from information transmission to the meaning of all this for behavior. I do not presume to imply that all the questions addressed had been answered. It is clear, however, that what began as a specialized, one might even say parochial, interest, expanded along the way to contribute some measure of insight to broader issues relating neural machinery to behavior.

At the beginning of this prefatory chapter I referred to Harrison's *Masks of the Universe*. In introducing universes that are impermanent cosmic belief-systems of societies, he alluded also to private world pictures. Among the data from which these private worlds are constructed are sensory data.

Though our perception of the world changes with our intellectual and introspective knowledge, it still is limited by our senses. Considering that advanced technologies enormously extend our senses, one might well wonder whether biologically evolved sensory systems any longer faithfully serve our needs. To what extent does the world as perceived directly by sense organs have reality and validity? The molecular and time/space world, which we are convinced exists, is obviously different from the biological world as perceived.

The perceived world of infrahuman animals is certainly a reality in that, insofar as they survive and evolve successfully in the physical world, they perceive it correctly and so attest its reality. We, on the other hand, can generate our own perceptions independently of what our sense organs tell us. When we know that there is a Chernobyl, we act as though we can indeed sense it (even without the benefit of aversion-learning).

Paramount though the mind/brain problem may be for understanding behavior, the brain is limited in its perfection by sensory input, even though it can stimulate itself by introspection. Studies on sensory deprivation prove the need for sensory input (Zubek 1969). Thus, sensory physiology contributes in a major way to human understanding. Sense organs provide our only *direct* contact with the universe. Granit (1955) touched the heart of the matter when he wrote that sensory physiology is "a branch of natural science which is actually capable of giving some meaning to 'meaning'."

Literature Cited

Adams, J. R., Holbert, P. E., Forgash, A. J. 1965. Electronmicroscopy of the contact chemoreceptors of the stable fly, *Stomoxys, calcitrans* (Diptera: Muscidae). *Ann. Ent. Soc. Amer.* 58: 909–17

Adrian, E. D. 1926. The impulses produced by sensory nerve endings. Part I. *J. Physiol.* 61: 49–72

Adrian, E. D. 1928. *The Basis of Sensation: The Action of the Sense Organs.* London: Christophers

Adrian, E. D. 1942. Olfactory reactions in the brain of the hedgehog. *J. Physiol.* 100: 459–73

Adrian, E. D., Forbes, A. 1922. The all-or-nothing response of sensory nerve fibers. *J. Physiol.* 56: 301–30

Adrian, E. D., Ludwig, C. 1938. Nervous discharges from the olfactory organs of fish. *J. Physiol.* 94: 441–60

Adrian, E. D., Zotterman, Y. 1926. The impulses produced by sensory nerve endings. Part 2. The response of a single end-organ. *J. Physiol.* 61: 151–71

Anand, B. K., Brobeck, J. R. 1951a. Hypothalamic control of food intake in rats and cats. *Yale J. Biol. Med.* 24: 123–40

Anand, B. K., Brobeck, J. R. 1951b. Localization of a "feeding center" in the hypothalamus of the rat. *Proc. Soc. Exp. Biol. NY* 1951: 323.

Barnhard, C. S., Chadwick, L. E. 1953. A "fly factor" in intractant studies. *Science* 117: 104–5

Beidler, L. M. 1954. A theory of taste stimulation. *J. Gen. Physiol.* 38: 133–39

Biedler, L. M. 1957. Physiological basis of taste psychophysics. *Fed. Proc.* 16: 9

Beidler, L. M. 1962. Taste receptor stimulation. In *Progress in Biophysics and Biophysical Chemistry*, ed. J. A. V. Butler, H. E. Huxley, R. E. Zirkle, 12: 107–51. Oxford: Pergamon

Beidler, L. M. 1987. Research directions in the chemical senses. In *Neurobiology of Taste and Smell*, ed. T. E. Finger, W. L. Silver, pp. 423–37. New York: Wiley

Boeckh, J., Kaissling, K. E., Schneider, D.

1965. Insect olfactory receptors. *Cold Spring Harbor Symp. Quant. Biol.* 30: 263–80

Boistel, J., Coraboeuf, E. 1953. L'activité électrique dans l'antenne isolée de Lepidoptère au cours de l'étude de l'olfaction. *CR Soc. Biol. Paris* 147: 1172–75

Brink, F., Posternak, J. M. 1948. Thermodynamic analyses of the relative effectiveness of narcotics. *J. Cell Comp. Physiol.* 32: 211–33

Cannon, W. B. 1939. *The Wisdom of the Body.* New York: Norton

Crozier, W. J. 1934. Chemoreception. In *A Handbook of Experimental Psychology,* ed. C. Murchison, pp. 987–1036. Worcester, Mass.: Clark Univ. Press

Dethier, V. G. 1937. Gustation and olfaction in lepidopterous larvae. *Biol. Bull.* 72: 7–23

Dethier, V. G. 1941. The function of the antennal receptors in lepidopterous larvae. *Biol. Bull.* 80: 403–14

Dethier, V. G. 1955. The physiology and histology of the contact chemoreceptors of the blowfly. *Q. Rev. biol.* 30: 348–71

Dethier, V. G. 1976. *The Hungry Fly.* Cambridge: Harvard Univ. Press

Dethier, V. G., Bodenstein, D. 1958. Hunger in the blowfly. *Z. Tierpsychol.* 15: 129–40

Dethier, V. G., Crnjar, R. M. 1982. Candidate codes in the gustatory system of caterpillars. *J. Gen. Physiol.* 79: 549–69

Dethier, V. G., Rhoades, M. V. 1954. Sugar preference—aversion functions for the blowfly. *J. Exp. Zool.* 126: 177–204

Erickson, R. P. 1963. Sensory neural patterns and gustation. In *Olfaction and Taste,* ed. Y. Zotterman, 1: 205–13. Oxford: Pergamon

Evans, D. R. 1963. Chemical structure and stimulation by carbohydrates. In *Olfaction and Taste,* ed. Y. Zotterman, 1: 165–92. Oxford: Pergamon

Evans, D. R., Mellon, DeF. 1962. Stimulation of a primary taste receptor by salts. *J. Gen. Physiol.* 4: 651–61

Finger, T. E., Silver, W. L. 1987. *Neurobiology of Taste and Small.* New York: Wiley

Gasser, H. A., Erlanger, J. 1922. A study of the action currents of nerve with the cathode ray oscillograph. *Am. J. Physiol.* 62: 496–524

Gasser, H. S., Newcomer, H. S. 1921. Physiological action currents in the phrenic nerve. An application of the thermionic vacuum tube to nerve physiology. *Am. J. Physiol.* 57: 1–26

Gelperin, A. 1966a. Control of crop emptying in the blowfly. *J. Insect Physiol.* 12: 331–45

Gelperin, A. 1966b. Investigation of a foregut receptor essential to taste threshold regulation in the blowfly. *J. Insect Physiol.* 12: 829–41

Gelperin, A. 1967. Stretch receptors in the foregut of the blowfly. *Science* 157: 208–10

Grabowski, C. T., Dethier, V. G. 1954. The structure of the tarsal chemoreceptors of the blowfly, *Phormia regina* Meigen. *J. Morph.* 94: 1–17

Granit, R. 1955. *Receptors and Sensory Perception.* New Haven: Yale Univ. Press

Hansen, K. 1968. *Untersuchungen über den Mechanismus der Zucker-Perzeption bei Fliegen.* Habilitationschrift der Universität Heidelberg

Hansen, K., Kuhner, J. 1972. Properties of a possible receptor protein of the fly's sugar receptor. In *Olfaction and Taste IV,* ed. D. Schneider, pp. 350–56. Stuttgart: Wissenshaftliche Verlagsgesellschaft MBM

Hanson, F. E. 1965. *Electrophysiological studies on chemoreceptors of the blowfly,* Phormia regina *Meigen.* Phd dissertation, Univ. Penna., Philadelphia

Harrison, E. 1985. *Masks of the Universe.* New York: Macmillan

Hassett, C. C., Dethier, V. G., Gans, J. 1950. A comparison of nutritive value and taste thresholds of carbohydrate for the blowfly. *Biol. Bull.* 99: 446–53

Hetherington, A. W., Ranson, S. W. 1942. The spontaneous activity and food intake of rats with hypothalamic lesions. *Am. J. Physiol.* 136: 609–17

Hoagland, H. 1933. Specific nerve impulses from gustatory and tactile receptors in catfish. *J. Gen. Physiol.* 16: 685–714

Hodgson, E. S., Lettvin, J. Y., Roeder, K. D. 1955. Physiology of a primary chemoreceptor unit. *Science* 122: 417–18

Kijima, H., Koizumi, O., Morita, H. 1973. α-Glucosidase at the tip of the contact chemosensory seta of the blowfly, *Phormia regina. J. Insect Physiol.* 19: 1351–62

Larsen, J. R. 1962. The fine structure of the labellar chemosensory hairs of the blowfly, *Phormia regina* Meigen. *J. Insect Physiol.* 8: 683–91

Minnich, D. E. 1922. The chemical sensitivity of the tarsi of the red admiral butterfly, *Pyrameis atalanta* Linn. *J. Exp. Zool.* 35: 57–81

Minnich, D. E. 1926. The organs of taste on the proboscis of the blowfly, *Phormia regina* Meigen. *Anat. Rec.* 34: 126

Morita, H. 1959. Initiation of spike potentials in contact chemosensory hairs of insects. III. D.C. stimulation and generator potential of labellar chemoreceptor of Calliphora. *J. Cell. Comp. Physiol.* 54: 182–204

Morita, H., Doira, S., Takeda, K., Kuwabara, M. 1957. Electrical responses of contact chemoreceptors on tarsus of the butterfly, *Vanessa indica*. *Mem. Fac. Sci. Kyushu Univ.* E2: 119–39

Morita, H., Shiraishi, A. 1968. Stimulation of the labellar sugar receptor of the flesh-fly by mono and disaccharides. *J. Gen. Physiol.* 52: 559–83

Omand, E., Dethier, V. G. 1969. An electrophysiological analysis of the action of carbohydrates on the sugar receptor of the blowfly. *Proc. Natl. Acad. Sci. USA* 62: 136–43

Parker, G. H. 1922. *Smell, Taste, and Allied Senses in the Vertebrates*. Philadelphia: Lippincott

Pfaffmann, C. 1941. Gustatory afferent impulses. *J. Cell. Comp. Physiol.* 17: 243–58

Pumphrey, J. 1935. Nerve impulses from receptors in the mouth of the frog. *J. Cell. Comp. Physiol.* 6: 457–67

Richter, C. P. 1942. Total self regulatory functions in animals and human beings. *Harvey Lect. Ser.* 38: 63–103

Schneider, D. 1955. Mikro-Electroden registrieren die electrischen Impulse einzelner Sinnesnervenzellen der Schmetterlingsantenne. *Ind. Electron. Forsch. Ferfigung* 3(3/4): 3–7

Shimada, I., Shiraishi, A., Kijima, H.,

Morita, H. 1974. Separation of two receptor sites in a single labellar sugar receptor of the flesh-fly by treatment with *p*-chloromercuribenzoate. *J. Insect Physiol.* 20: 605–21

Steinhardt, R. A., Morita, H., Hodgson, E. S. 1966. Mode of action of straight chain hydrocarbons on primary chemoreceptors of the blowfly. *Phormia regina*. *J. Cell. Physiol.* 67: 53–62

Teitlebaum, P., Stellar, E. 1954. Recovery from failure to eat produced by hypothalamic lesions. *Science* 120: 894–95

Verschaffelt, E. 1910. The cause determining the selection of food in some herbivorous insects. *Proc. R. Acad. Amsterdam* 13: 536–42

von Frisch, K. 1921. Über den Sitz des Geruchsinnes bei Insekten. *Zool. Jahrb. Abt. Zool. Physiol.* 38: 449–516

von Frisch, K. 1935. Über den Geschmackssinn der Biene. *Z. Physiol.* 21: 1–156

Wolbarsht, M. L. 1957. Water taste in *Phormia*. *Science* 125: 1248

Zotterman, Y. 1935. Action potentials from the chorda tympani. *Skand. Arch. Physiol.* 72: 73–77

Zubek, J. P. 1969. *Sensory Deprivation: Fifteen Years of Research*. New York: Appelton-Century-Crofts

Viktor Hamburger

Ann. Rev. Neurosci. 1989. 12 : 1–12

THE JOURNEY OF A NEUROEMBRYOLOGIST

Viktor Hamburger

Department of Biology, Washington University, St. Louis, Missouri 63130

How did one become an experimental neuroembryologist in the 1920s? As an American, one would go to Ross G. Harrison at Yale, or to one of his students. Harrison had pioneered the field of neuroembryology at the beginning of the century and had made his mark by designing the first tissue culture experiment, in order to provide indisputable evidence for the axon outgrowth theory. By the adoption of the limb transplantation experiment in amphibians he had provided us with a paradigmatic model for the analysis of the formation of nerve patterns. The handful of neuroembryologists who were active at that time were mostly his students. He had turned to other problems, but retained a strong interest in the field (Harrison 1935). As a student of Hans Spemann in Germany, whose domain was the analysis of determination and inductions in early amphibian development, I entered the field through the back door, so to speak. When Spemann accepted me as a PhD candidate, he assigned me a topic unrelated to his turf, because he wanted to assure me of an independent career from the start. But I have never found out what prompted him to have me repeat an experiment that had given rather questionable results. B. Duerken (1913) had done eye extirpations on young frog tadpoles and had found leg abnormalities in a fairly high percentage of cases. He had interpreted them as neurogenic, in the sense that the operation would cause a primary defect in the midbrain that would cause secondary deficiencies in the motor centers of the spinal cord, resulting in deficient leg innervation. Spemann was skeptical of these findings, but I surprised him by actually obtaining some slightly deformed legs, though at a low percentage (1925).[1] However, when I repeated the experiments the following year on another 1000 tadpoles, I obtained 2000 entirely normal legs (1927). Spemann was

---

[1] References that contain only the year of publication refer to my own publications.

convinced that Duerken's results were spurious, but I decided to do the crucial experiment of creating nerveless legs by extirpating the lumbar part of the spinal cord in early stages preceding nerve outgrowth. Such legs developed normally in every respect, though the musculature atrophied and eventually degenerated (1928). This, my first original contribution, seems to have settled the issue, since to my knowledge the experiment has never been challenged.

Hidden behind my clear-cut result was an uphill fight against the resolute efforts of the embryo to regenerate the defect I had inflicted on its spinal cord. In fact, I found myself in the possession of only a few nerveless legs but an abundance of cases with partial or poor innervation. I exploited them for information on nerve pattern formation. Among other observations, I found that even tiny nerves would manage to form a typical pattern, though they would not reach their targets. I concluded that the signals that guide nerves to their destinations are distinct from those responsible for the appropriate terminal connections (1929).

By that time I became aware that the developing nervous system provides the analytically minded embryologist with almost unlimited challenges to keep him busy for a life-time. In my first review article, entitled "Developmental-physiological Correlations Between the Amphibian Limbs and Their Innervation" (1927), I identified three major areas: the influence of innervation on limb development (which I considered as settled); the guidance of nerves to their targets; and a possible effect of the growing limb on the nerve centers that innervate it. This section was the shortest—less than half a page. I mentioned in passing a largely-forgotten study by one M. Shorey without even indicating that it had been done on chick embryos. She had removed the wing buds of chick embryos by electrocautery and found that the lateral motor columns and lumbar spinal ganglia (DRG) were severely defective (Shorey 1909). The only other experiments dealing with this topic had been done by Harrison's student, S. Detwiler, on salamander embryos. Following forelimb extirpation he had found a hypoplasia of the brachial DRG and, following the transplantation of a forelimb primordium to the flank, a hyperplasia of the "overloaded" thoracic DRG (Detwiler 1920). Strangely enough, he had not observed any changes in the motor region of the spinal cord. These experiments are of great historical significance. They initiated a trend of thought that led eventually to the discovery of the nerve growth factor.

While I realized that neurogenesis offered many opportunities to a budding experimental embryologist, my own role was not clear. The problem of the influence of the target on its nerve centers was in the hands of Detwiler, who worked assiduously in this field (Detwiler 1936), and there was then no obvious way of advancing the problem of nerve guidance. I turned to another, very different field that had fascinated me: developmental genetics (which at that time did not even have a name). It was clear to me—though not to many geneticists—that genes "produce" a particular phenotype by controlling the developmental processes that lead to it. It occurred to me that one might gain insight into the way genes do this by combining hybridization with the method of transplantation. Since there were then no known mutants in salamanders, I undertook hybridization of two local species and studied in the reciprocal hybrids the development of some characteristics in which the species differed markedly, such as the development of toes. The project was terminated before I got around to transplantations, when I moved to the United States. The results were published in English (1936). As fate would have it, one set of species hybrids showed leg abnormalities which again eluded an explanation (1935).

My penchant for stage series dates back to the earliest days; a limited one was created for the frog species that gave the material for my PhD thesis. I assigned the tedious and not particularly inspiring task of designing a stage series of the two *Triturus* species used for hybridization to the only PhD candidate I had in Germany, Salome Gluecksohn (1931). The only lasting result of her effort was that it started her career as an eminent developmental geneticist. My friend Salome Gluecksohn Waelsch has held the chair of Genetics at Albert Einstein Medical School for many years. Incidentally, I returned briefly to developmental genetics later, using the creeper mutant of the chick embryo. Reciprocal transplantations of limb and eye primordia between normal and mutant embryos gave some interesting results (1942).

My life took a fateful turn in 1932. I received a fellowship from the Rockefeller Foundation to spend a year at the Zoological Laboratory of the University of Chicago; its Director, Dr. Frank Lillie, was a friend of Spemann. The change of continents became permanent; it saved me from Nazi persecution, and at the same time I changed my allegiance from the amphibian to the chick embryo, and, fortuitously, all these changes brought about the return to my destiny: experimental neurogenesis. How did this happen? Lillie's classic book, *The Development of the Chick* (1908), had introduced the chick embryo to the classroom and to experimental research. When I had my first meeting with Dr. Lillie, in which my plans were discussed, he reminded me that M. Shorey had done her work in his laboratory, at his suggestion. Would it not be expedient for me to try out Spemann's glass neddle technique on the chick embryo by repeating her experiment? I would then have the chance to resolve the discrepancy between her and Detwiler's results, with respect to the effect of the ablation on the motor system. Dr. Lillie was at that time the Dean of the Bio-Medical Sciences. I was entrusted to his former student, B. H. Willier, then Professor of Embryology in the Zoology Department. He and his exceptionally skillful research associate, Dr. Mary Rawles, initiated me in the art of handling chick embryos. During the winter of 1932–1933, using Spemann's glass needle technique, I learned how to extirpate limb buds and to transplant them to the flank. I was thrilled to see them develop and grow to almost normal size and to quite advanced stages (1938). What a difference compared to the amphibian experimentation, to have results in a few days and not have to fight any longer the high mortality rates. Lillie and Willier and Mary Rawles followed my adventure with great interest. We all realized that since the Spemann technique worked beautifully on the chick embryo, it would now become available for a wide range of experimentation; but I did not anticipate that in neuroembryology the chick embryo would eventually play a more important role than the amphibian embryo. It should be mentioned that up to then the manipulation of the embryo itself had not been feasible. The embryos had been used, however, for growing embryonic tissues on their chorio-allantoic membrane, but this substrate did not permit the normal morphogenesis of structures such as limb or eye primordia.

By the end of the 1940s, the chick embryo had conquered wide areas of experimental research, far beyond the boundaries of experimental embryology. Now there arose the need for a precise characterization of the different stages of development. Up to then, embryos had been identified by their age in terms of hours or days of incubation. But the embryological staging was inadequate, partly because conditions of incubation varied widely, and partly due to individual differences in rate of development. A three-day embryo can show a wide range of morphological differences. In 1948, I saw my colleague Howard Hamilton of the University of Iowa at a meeting in Chapel Hill, North Carolina. We discussed a project he was engaged in: a revised edition of Lillie's *Development of the Chick*. We commented on the inadequate description of stages in the earlier editions, and I suggested that we produce a stage series that would meet the highest standards of precision. The series would have to fulfill two critical requirements: each stage had to be characterized unequivocally by a set of easily identifiable morphological features; and the continuum of development had to be transected into an arbitrary sequence of stages that would be so close to each other that they would indicate the smallest detectable changes. We ended up with 45 stages, presented by photographs and descriptions (Hamburger & Hamilton 1951). Obviously, we were very successful. The stage series soon became indispensable; it still is one of the most frequently quoted publications. Without a doubt it contributed materially to the precision whereby data obtained with chick embryos could be processed.

Returning to the wing extirpations, my results confirmed those of M.

Shorey. Both the dorsal root ganglia (DRG) and the lateral motor column (lmc) were distinctly hypoplastic. I became aware that I was dealing with a fundamental issue and that its analysis could be carried much further than Shorey and Detwiler had done previously. The clear demarcation of the lmc in the chick embryo, which one does not find in the salamander embryo, played into my hands, as did my imperfection as a beginner in my dealings with the chick embryo. It was easy enough to cut off the wing bud neatly; but I did not realize that the primordia of the massive pectoral muscles that were also supplied by the brachial nerves were situated beneath the surface. So I obtained a wide range of muscle deficiencies, which was reflected in the range of motor hypoplasia. When I quantified the data, I found a close correlation between the two sets of deficiencies. On the other hand, the hypoplasia of the DRG amounted to 50% with rather little variation. This I attributed to a fairly uniform loss of skin. I drew the conclusion that "every structure within the growing limb, muscles, as well as sense organs, send stimuli to the nervous system. Each part of the peripheral field controls directly its own nervous center, i.e. the limb muscles affect the lateral motor centers and the sensory fields control the ganglia" (1934, p. 470). From this conceptualization it follows logically that targets and their centers must be connected directly, and I suggested "that the nerve fibers themselves serve as mediators between the two links of the correlation" (1934, p. 475). To remove any lingering doubt that the neuronal deficiencies might have been caused by a nonspecific trauma inflicted by the operation, I made use of my extensive material of limb transplantations. I found that the most strongly innervated transplants were wings placed immediately behind the normal wing and supplied by the brachial plexus, and leg buds placed immediately in front of the normal leg and supplied by the lumbar plexus. A quantitative analysis of individual segments showed clearly that only those lmc segments and DRG which contributed to the innervation of the transplants were hyperplastic (1939). The model of growth-controlling agents produced by the targets and transported retrogradely to their respective nerve centers has stood the test of time well. It is often forgotten that concepts that are now taken for granted, at one time had to pass the test of authenticity.

The ground was now prepared for the pursuit of two specific problems: the nature of the growth-regulating agents, and their specific mode of action on the nerve centers. The first question did not seem to be manageable at that time. Several answers to the second question were proposed. Detwiler had suggested that the targets might regulate the size of centers by controlling proliferation. This option was ruled out for the lmc in the chick by mitotic counts in the spinal cord in cases of wing extirpation and transplantation that showed no difference between the left and right sides

(Hamburger & Keefe 1944). Then I formulated a "recruitment hypothesis," in which I postulated that the first pathfinder nerves that reached the target would explore its size and recruit from a pool of (hypothetical) undifferentiated neuroblasts the appropriate number that would saturate the target. The recruitment I imagined would take the form of an induction of undifferentiated neuroblasts by the mature pathfinder neurons (1934; Hamburger & Keefe 1944). The hypothesis had the merit that it could account for both hypo- and hyperplasia.

Soon after the end of the war, I became acquainted with the publications of two Italian investigators, Rita Levi-Montalcini and Guiseppe Levi (1942, 1944), who had proposed an entirely different explanation of hypoplasia in DRGs. They had repeated the limb bud extirpation and counted neurons in successive stages from 6 to 19 days of incubation. The counts provided evidence that the hypoplasia resulted from the degeneration of fully differentiated neurons rather than from some form of interference with progressive differentiation. This spelled the doom of my brain child, but it did not alter the underlying basic conception of the control of the nerve centers by their targets via the nerve fiber route. After I invited Dr. Levi-Montalcini to join me in St. Louis in 1947, we repeated both limb ablation and transplantation experiments. The former fully substantiated her claim, since large numbers of degenerating neurons were found in the brachial DRG. In the same material she made another observation: cervical and thoracic DRG that were not affected by the operation also displayed numerous degenerating cells. This was the seminal discovery of naturally occurring neuronal death. As far as the hyperplasia in the transplantation experiments was concerned, mitotic and cell counts seemed to indicate that we were dealing with the enhancement of proliferation; however, in this matter we were on the wrong track (Hamburger & Levi-Montalcini 1949).

Conceptually, our new data gave my model of the target-nerve center relationship a new slant, and we suggested that "substances necessary for . . . neuroblast growth and maintenance would not be provided in adequate quantities, when the limb bud is removed" (Hamburger & Levi-Montalcini 1949, p. 493). The hypothetical agent was thus identified as a *maintenance* factor. When we reflected on the cause of naturally occurring death, we were struck by the fact that the presence of degenerating cells was limited to a short period of a few days and that it occurred at exactly the same stages in the experimentally affected brachial DRG and the adjacent normal cervical and thoracic DRG. This strongly suggested a common denominator for both types of neuronal death. "In both instances, the early differentiating VL cells [a special subpopulation in DRG] are affected after their neurites have reached the periphery, and in both instances the breakdown occurs between 5 and 7 days. . . . In the experimental situation the reduction of the peripheral area is definitely responsible for degeneration. It is possible that the same mechanism operates in the case of the cervical and thoracic ganglia. This would imply that in early stages cervical and thoracic VL cells send out more neurites than the periphery can support. The excess of neurons would break down at the stage at which the VL cells are highly susceptible to environmental conditions" (Hamburger & Levi-Montalcini 1949, pp. 495–96). This was clearly an anticipation of the competition hypothesis.

In retrospect, we made the right choice not to pursue this topic further but to concentrate on the analysis of hyperplasia. The exuberant vitality of the DRG displayed in their overgrowth, far beyond their normal performance, seemed more worthy of our attention than the morbid story of neuronal death. But how to proceed? We realized that limb transplantation would yield no further insights. We needed a more homogeneous tissue with growth-stimulating activity. I then remembered that my former student, Elmer Bueker, with a similar idea in mind, had already provided us with an ideal solution to our quest, namely to use tumor tissue. Applying an experimental design that he had devised for his PhD thesis Bueker (1943) implanted fragments of mouse sarcoma 180 in the place of the leg bud, which amounted practically to a coelomic graft. The tumor had grown well and had been invaded by sensory axons but was avoided by motor fibers. The corresponding DRG showed a considerable hyperplasia. Bueker concluded "that sarcoma 180, because of intrinsic physicochemical properties and mechanics of growth, selectively causes the enlargement of spinal ganglia" (Bueker 1948, p. 382). This ingenious experiment was a crucial link to the discovery of NGF. It provided us with a growth-promoting, fast-growing tissue that was specific for a particular type of neuron and available in large quantities for future biochemical analysis. With Bueker's consent we repeated the experiment. The results far surpassed our expectations. The tumors were profusely invaded by fibers that could be traced not only to DRG but also to sympathetic chain ganglia. The former were enlarged up to $2\frac{1}{2}$ times and the latter up to 6 times their normal size. The enlargement did not begin until the fibers had spread in the tumor. Now for the first time the prospect of giving concrete reality to the paradigm that had guided me since the 1930s, namely that specific nerve-growth regulating agents were produced by the targets and transported retrogradely to the centers, seemed to be within reach. Indeed, a series of brilliant experimental masterstrokes by Dr. Levi-Montalcini, in collaboration with a superb biochemist, Dr. Stanley Cohen, whom I invited to join my laboratory in 1953, led within a few years to the discovery of the nerve growth factor and its identification as a protein. I actively participated in the early phases of this work and in the preparation of the first two publications (Levi-Montalcini & Hamburger 1951, 1953) but withdrew from the project in 1953 to pursue other interests. I later returned to the exploration of the role of NGF in naturally occurring cell death (see below).

Ever since I became acquainted with the chick embryo, I had noticed its stirrings, but had paid little attention to them. However, I was familiar with the general topic of embryonic motility and with the sharp controversy over this issue that had been fought in the 1920s and 1930s. The behaviorists who then dominated the scene contended that local reflexes are the primary units of embryonic behavior and that they become gradually integrated into coordinated performances. Among their opponents, the neuroembryologist G. E. Coghill stood out as an astute observer and an eminent theorist. He had done pioneer work on the correlation of neurogenesis and genesis of behavior. On the basis of these investigations he had developed the theory that motility is an integrated performance from the first bending of the head in the early embryo to the complex behavior patterns of locomotion, feeding, etc. Local reflexes would originate secondarily by "emancipation" from the total pattern (Coghill 1929). The controversy had ended in an impasse and not much had been done about it.

It occurred to me that since this was a fundamental issue, it could perhaps be revived by approaching it with the experimental method. So far observation and stimulation had been the standard implements. The project that I started in the late 1950s kept me and a group of very able co-workers busy for over a decade. My associates were M. Balaban, A. Bekoff, J. Decker, C. H. Narayanan, R. Oppenheim, R. Provine, S. Sharma, and E. Wenger. We realized immediately that the chick embryo does not comply with either one of the old paradigms. The main characteristic of its performance is that it is spontaneous, that is, nonreflexogenic. This is apparent from its two main features: the movements are entirely uncoordinated and they are performed intermittently, activity phases alternating with inactivity phases. They are performed in the absence of any apparent stimulation. By uncoordinated I mean that at any moment the head, trunk, wings, legs, and later beak and eyelids, can be active in any combination. The movements are jerky, convulsive-like, and they appear completely aimless. The activity phases get gradually longer and the inactivity phases shorter, until from day 13 of incubation on the embryo is almost continuously active.

We confirmed an important discovery that an eminent physiological psychologist of the last century, W. Preyer, had reported in his classic book *Spezielle Physiologie des Embryo* (1885). He had observed that in the chick embryo, motility begins at about 4 days of incubation, but responses to stimulation cannot be elicited until about day 8. Hence, there exists a prereflexogenic period of spontaneous motility. To determine whether sensory input plays a role at later stages we performed complete deafferentations of the leg level. We did this by removing the dorsal part of the lumbar spinal cord including the neural crest, which gives rise to the DRG, and by producing a gap in the thoracic cord to exclude descending inputs from more rostral levels. Overall activity and the periodicity pattern were the same in experimental embryos of different stages and in controls with thoracic gaps only. Thus we had established that spontaneous motility is the basic form of behavior in the chick embryo. This implies that any form of stimulation, including self-stimulation by the brushing of the legs against the head, which had been postulated by behaviorists as an essential ingredient, plays a very minor role, if any. Later on, we found that rat fetuses display the same uncoordinated, intermittent, jerky motility as the chick embryos, and we believe that this type of embryonic and fetal motility is paradigmatic for all warm-blooded forms.

If the motility is not elicited by stimulation, it must be generated by endogenous bioelectrical activity. This aspect was investigated in the second phase of our project. The main result of extensive extracellular recordings from the ventral regions of the spinal cord can be summarized briefly: The periods of motility and the duration of activity phases were reflected precisely in polyneuronal burst patterns. We could not have wished for a better correlation between overt motility and electrophysiological activity. This holds for all stages from the beginning of motility to near-hatching stages. One can consider the spontaneous, intermittent, self-generated motility of warm-blooded embryos and fetuses as the earliest manifestation of innate behavior patterns. This is a novel paradigm; it has created a solid foundation for all modern studies of fetal activity, including the human fetus (reviews in Hamburger 1963, 1970, 1973).

How does the embryo manage to extricate itself from the shell? Certainly not by the uncoordinated, jerky movements that it has displayed so far. When I searched the literature, I found to my surprise that my question pointed to unexplored territory. In the mid-1960s Ron Oppenheim and I embarked on the project to detect the embryo's way of solving a major problem in its life-history. We spent many months observing the embryos through enlarged windows in the shell and became witness to most remarkable events. They begin 3–4 days before the actual hatching occurs. While the uncoordinated random movements are temporarily suspended, the embryo performs a series of well-coordinated movements by which it maneuvers itself into what we have called the "hatching position." The embryo is oriented lengthwise in the shell, with the head near the blunt end. The head is tucked under the right wing and the beak points obliquely toward the shell, almost touching it. This position is attained by coordinated head, trunk, and leg movements with a rotatory component; they are synchronized with wing lifting. This positioning takes many hours and much trial and error and it is interrupted by long rest periods. Next, a hole is pierced in the shell by vigorous back thrusts of head and beak; this is referred to as "pipping." About a day later (on the twenty-first day of incubation), the hatching act begins rather suddenly with repetitions of the powerful back thrusts of head and beak, but this time they are combined with the slow rotation of the whole body. In this way, the shell is cracked along a circle at some distance from the blunt end. When pieces of shell are chipped off along about 2/3 of the circumference, the cap of the shell is loosened, and it soon breaks open and the chicken escapes. All this takes between 1/2 to 1 hour. Having watched the aimless fidgeting of embryos for several years, viewing this smooth, goal-directed performance held its fascination. I may mention that this was the only investigation of mine that did not involve experimentation (apart from the designing of the stage series, which was not a problem-solving project).

In the meantime, the nerve growth factor had become a cause célèbre, but interest in neuronal death had faded. I had brought the lateral motor column in line with the DRG by demonstrating that limb bud extirpation likewise causes massive degeneration of differentiated motor neurons, which had sent their axons to the target area (1958). But it was not until after my excursion to the field of embryonic behavior that I returned once more to my old favorite theme, the dialogue between nerve centers and their partners. While natural neuronal death had been established as an integral part of neurogenesis, a rationale for this strange phenomenon was not yet in evidence. When it was first discovered in 1949, we had suggested a possible explanation: that the target might produce a maintenance factor in insufficient amounts and that the neurons that lose in the competition for it might be the ones that die. Now, in the mid-1970s, I had the idea that the old stand-by, the limb transplantation experiment, could be used to test the competition hypothesis. An enlargement of the target area should result in the rescue of neurons that would otherwise die. The leg buds that M. Hollyday transplanted skillfully close to the normal leg buds developed to normal size and became innervated by nerves from the lumbar plexus. We succeeded in salvaging an average of 30% motor neurons, and in one case 53% neurons (Hollyday & Hamburger 1976).[2] Later on, others,

---

[2] The figures reported in our publication are considerably lower than those mentioned here. We had underestimated the gain by comparing the percentage difference of the surviving cells between the left (control) side and the right side, at the end of the degeneration period (day 12). The figures stated above represent a comparison between the total numbers of neurons before the onset of the degeneration period at day 6 and after its termination at day 12. I owe this rescue of our success to my friend J. Sanes.

using different experimental designs, managed to rescue up to 100% neurons that would have died. These results provide conclusive evidence that neuronal death in vertebrates is probabilistic and not programmed, as in some invertebrates.

Finally, we tried to link the two seemingly disparate events in neurogenesis: neuronal death and neuron-growth promotion by NGF. Is NGF identical with the hypothetical maintenance factor for DRG, produced by limb tissue? The best we could do with the techniques at our disposal was to provide indirect evidence. First, we showed that if labeled NGF is injected into the leg during the critical period of neuron degeneration, it is transported retrogradely and selectively to the lumbar DRG (Brunso-Bechtold & Hamburger 1979).

Next, we were able to reduce normally occurring neuronal death in both brachial and thoracic DRG by daily injections of NGF into the yolk sac from early stages on (Hamburger et al 1981). Finally, we subjected the competition hypothesis to the most severe test: we combined wing bud extirpations, which normally cause a near-total breakdown of the brachial DRG, with daily injections of NGF. With the dosage we used, we reduced the death rate in the subpopulation of large, early differentiating neurons (VL) by 50% and rescued nearly all neurons in the subpopulation of small, late-differentiating cells (DM) (Hamburger & Yip 1984). The fact that NGF is an effective substitute for the hypothetical trophic maintenance factor for DRG strengthened our belief that NGF is identical with this factor.

Some investigators are adventurous, and enjoy exploring uncharted territory. They make bold forays and get their deepest satisfaction from the unforeseen surprises that await them when they reach a clearing in the forest or a mountain top. The best are guided by an unfailing magnet or endowed with a green thumb. They are often rewarded by momentous discoveries. Others, with a different temperament, start with some general idea that may be nothing more than a dim hunch. They conquer their territory by patient step-by-step analysis, guided by the inner logic of their pursuit. They get their rewards when the story unfolds and has a happy ending, and when an unexpected discovery comes their way. Obviously, I belong to the second type; undoubtedly, inherent tendencies were reinforced by my mentor Spemann. But I had the good fortune to count among my close friends some very successful personifications of the first type, who have enriched my life.

At the end of my experimental exploits I realize that I have come full circle to my original ideas of 1934 about the influence of targets on the nerve centers that innervate them. With the book, *The Heritage of Experimental Embryology* (1988), in which the concept of embryonic induction plays a major role, I have closed another circle. It took me back to my student days in Spemann's laboratory where I was exposed to the major issues in experimental embryology, before I started my journey as an experimental neuroembryologist.

Literature Cited

Brunso-Bechtold, J., Hamburger, V. 1979. *Proc. Natl. Acad. Sci. USA* 76: 1494–96

Bueker, E. 1943. *J. Exp. Zool.* 93: 99–129

Bueker, E. 1948. *Anat. Rec.* 102: 369–90

Coghill, G. E. 1929. *Anatomy and the Problem of Behaviour.* London: Cambridge Univ. Press

Detwiler, S. R. 1920. *Proc. Natl. Acad. Sci. USA* 6: 96–101

Detwiler, S. R. 1936. *Neuroembryology.* New York: Macmillan

Duerken, B. 1913. *Z. Wiss. Zool.* 105: 191–242

Gluecksohn, S. 1931. *Roux' Arch. Entw. Mech.* 125: 341–405

Hamburger, V. 1925. *Roux' Arch. Entw. Mech.* 105: 149–201

Hamburger, V. 1927. *Naturwissenschaften* 15: 657–81

Hamburger, V. 1928. *Roux' Arch. Entw. Mech.* 114: 272–362

Hamburger, V. 1929. *Roux' Arch. Entw. Mech.* 119: 47–99

Hamburger, V. 1934. *J. Exp. Zool.* 68: 449–94

Hamburger, V. 1935. *J. Exp. Zool.* 70: 43–54

Hamburger, V. 1936. *J. Exp. Zool.* 73: 319–73

Hamburger, V. 1938. *J. Exp. Zool.* 77: 379–97

Hamburger, V. 1939. *J. Exp. Zool.* 80: 347–89

Hamburger, V. 1942. *Biol. Symp.* 6: 311–34

Hamburger, V. 1958. *Am. J. Anat.* 102: 365–410

Hamburger, V. 1963. *Q. Rev. Biol.* 38: 342–65

Hamburger, V. 1970. *The Neurosciences Second Study Program,* ed. F. O. Schmitt, pp. 141–51

Hamburger, V. 1973. In *Studies on the Development of Behavior and the Nervous System,* ed. G. Gottlieb, 1: 51–76. New York/London: Academic

Hamburger, V. 1988. *The Heritage of Experimental Embryology. Hans Spemann and the Organizer.* New York/Oxford: Oxford Univ. Press

Hamburger, V., Hamilton, H. 1951. *J. Morphol.* 88: 49–92

Hamburger, V., Keefe, E. L. 1944. *J. Exp. Zool.* 96: 223–42

Hamburger, V., Levi-Montalcini, R. 1949. *J. Exp. Zool.* 111: 457–502

Hamburger, V., Brunso-Bechtold, J. K., Yip, J. 1981. *J. Neurosci.* 1: 60–71

Hamburger, V., Yip, J. 1984. *J. Neurosci.* 4: 767–74

Harrison, R. G. 1935. *Proc. R. Soc. London Ser. B* 118: 155–96

Hollyday, M., Hamburger, V. 1976. *J. Comp. Neurol.* 170: 311–20

Levi-Montalcini, R., Hamburger, V. 1951. *J. Exp. Zool.* 116: 321–62

Levi-Montalcini, R., Hamburger, V. 1953. *J. Exp. Zool.* 123: 233–88

Levi-Montalcini, R., Levi, G. 1942. *Arch. Biol.* 53: 537–45

Levi-Montalcini, R., Levi, G. 1944. *Comment. Pontific. Acad. Sci.* 8: 627–68

Lillie, F. R. 1908. *The Development of the Chick.* New York: Holt

Preyer, W. 1885. *Specielle Physiologie des Embryo.* Leipzig: Grieben

Shorey, M. L. 1909. *J. Exp. Zool.* 7: 25–63

Annu. Rev. Nucl. Part. Sci. 1989. 39: xiii–xxviii

FISSION IN 1939: The Puzzle and the Promise

John Archibald Wheeler

Physics Department, Princeton University, Princeton, New Jersey 08544;
and Physics Department, University of Texas, Austin, Texas 78712

KEY WORDS: compound nucleus, channel count, fissility parameter, saddle
point, barrier penetration.

Never more actively, responsibly, and productively than today do historians of science study the evolution of modern physics. Their enterprise founds itself on written records, historical training, and scholarship. As they build, however, the living lamps of memory one by one go out. Therefore, the present account—by one early participant in fission physics—will perhaps be more useful if it is conceived, not as history, but as memories, impressions, atmosphere.

I stood in the winter cold at Pier 97, North River, New York on Monday, January 16, 1939 to welcome Niels Bohr (Figure 1), about to debark from the Swedish-American liner *Drottningholm*. In my wait Enrico and Laura Fermi (Figure 2) joined me. Fermi had been at Columbia University for less than a month since his Rome-to-Stockholm trip. As Bohr cleared customs we greeted him, his son Erik, and his long-time colleague Léon Rosenfeld. Their upcoming three-month stay at the Institute for Advanced Study in Princeton had for Bohr an overriding purpose: Clarify the quantum. To that end pursue the long-continued dialogue (1) with Einstein (Figure 3). Do everything possible, man to man, to reach agreement with him.

January 3rd, however, had opened to Bohr a second vista. That day, just before the *Drottningholm* sailed, Otto Robert Frisch—a friend of mine since my 1934–1935 year in Copenhagen—fresh back from Sweden, reported to him the conclusions to which he and his aunt, Lise Meitner, had been forced (2) by the not yet published findings of Otto Hahn and Fritz Strassmann (3). "I had hardly begun to tell him," Frisch writes (4), "when he struck his forehead with his hand and exclaimed, 'Oh what

0163–8998/89/1201–0xiii$02.00

idiots we have all been! Oh but this is wonderful! This is just as it must be.' "

Show that fission as then known really does proceed just as it must: This goal held itself out ever more invitingly to Bohr with each new day of pacing up and down the deck with the one shipboard companion or the other. Nothing of this new goal or of fission itself did I or anyone in America know when I greeted Bohr at the pier. How could I have anticipated that he would invite me to join him in this enterprise? Or that we would extend the work (5): **predict** as yet undiscovered features of fission?

Bohr felt obligated not to let out word of fission until Frisch with an ionization chamber or otherwise could demonstrate splitting and send in his findings for publication (6). Niels and Erik went into Manhattan with Enrico and Laura to visit old friends, father and son spending a night or two there before coming down to Princeton. Rosenfeld and I, however, took the next train. He was unaware of Bohr's self-imposed commitment to silence on fission. He revealed the exciting news on the one-hour trip. I got him to report the great discovery that very evening at the regular Monday 7–9 p.m. Journal Club (Figure 4).

Bohr arrived a day or two later, discovered that the cat was out of the bag, told me more and we got to work. The aim was straightforward. The burgeoning world of experimental findings: bring them into order within the framework of Bohr's compound-nucleus model of nuclear reactions. This model he had first enunciated in 1935, during my time in Copenhagen.

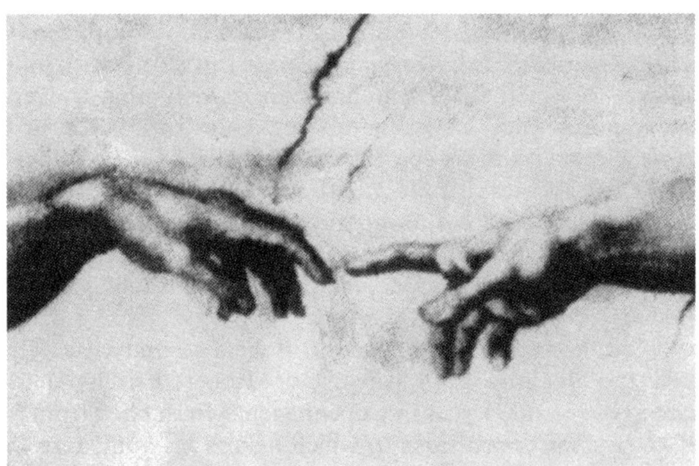

Figure 1   The Old World reached across to the New World—at Pier 97 North River, New York, Monday January 16, 1939—with the word of fission [Detail from Michelangelo's *The Creation of Adam*, Sistine Chapel.]

Figure 2 The Fermis shortly after their arrival in the United States. Courtesy of Wide World Photos, Inc.

Figure 3 Niels Bohr and Albert Einstein in dialogue in the Leyden home of Paul Ehrenfest in 1933. The restoration of the negative of Ehrenfest's photograph and the production of the print were done by William R. Whipple. Courtesy of the American Institute of Physics Niels Bohr Library.

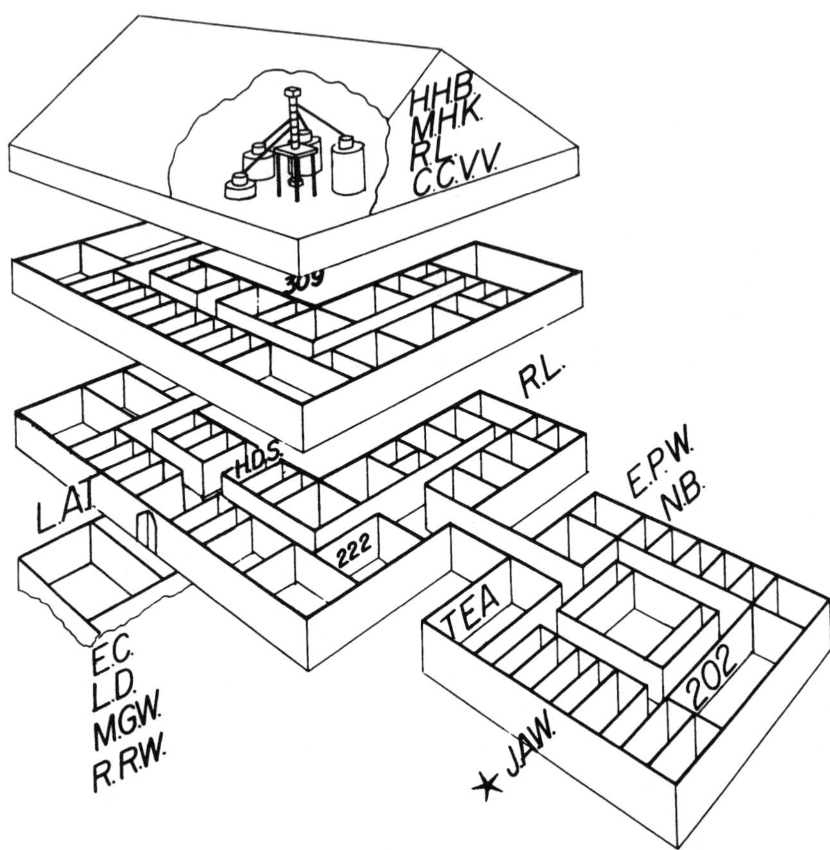

Figure 4 *Star*: location of the Fine Hall (today Jones Hall) ground floor seminar room where Rosenfeld made known the discovery of fission. Subsequent Princeton work on fission was divided between the main floor of Fine Hall (Eugene P. Wigner, Bohr, and the writer) and the adjacent Palmer Physical Laboratory. Cyclotron: Edward Creutz, Luis A. Delsasso, Milton G. White, and Robert R. Wilson. Main floor: Louis A. Turner, Henry D. Smyth, and Rudolf Ladenburg. Attic: Van de Graaff generator of neutrons: Henry H. Barschall, Morton H. Kanner, Ladenburg, and Cletus C. Van Voorhis. TEA: "where we explain to each other what we don't understand." Room 222: graduate courses in physics. By January 1939 Einstein had vacated his former Fine Hall office (E.P.W.) for new quarters at the Institute for Advanced Study, then abuilding a mile away, but nearby, over the fireplace in the Professors' Room (# 202), are engraved his famous words, "Raffiniert ist der Herr Gott, aber boshaft ist Er nicht." Of physics colloquia in Room 309 he attended occasional ones, including the one on the mechanism of fission; and there he gave his last talk (7) before his death.

Since then he and I had been working on the development and application of this model (Figure 5).

As we proceeded with our work, we found we had to introduce concepts new to nuclear theory: **fissility parameter**, nuclear **potential-energy** surface, **saddle-point energy** as threshold for fission, **channel** open for over-the-barrier fission, **channel count** as determiner of the contribution of fission to level width, and **spontaneous fission** via barrier penetration. The analysis culminated in a 25-page paper, "The Mechanism of Nuclear Fission." As if omen of a new world of weapons, it appeared in the issue of the *Physical Review* dated the first of September, 1939, the very day World War II began. What was the background for the collaboration of the American junior partner in this work?

No better symbol do I know than Michelangelo's great Sistine Chapel painting for the lightning stroke of January 16th, 1939 that brought the word of fission from Europe to America. No fitter image, either, can I offer for the electricity of learning that had flowed from the outstretched

Figure 5 A few of the participants in the Third Washington Conference on Theoretical Physics of February 18, 1937, where the compound-nucleus model of nuclear reactions received considerable attention. Niels Bohr, front; I. I. Rabi and George Gamow, second row; Fritz Kalckar and John Wheeler, third row; and Gregory Breit, directly behind Wheeler.

finger of Europe to the outheld finger of America for many decades before 1939.

In 1876 the new Johns Hopkins University in Baltimore, under the leadership of Daniel Coit Gilman, became the first great institution of higher learning in America explicitly to dedicate itself to the Europe-inspired research ideal. Henry A. Rowland brought to it preeminence in physics as other men brought to it a like spirit of exploration and discovery in other fields. What we would today call graduate-level training for research took first place in those days. Relative to it, any undergraduate education provided was regarded as preparatory, secondary, gap filling. Education was not a teacher facing a class. Education was colleagues, young and old, facing an issue.

Seminar courses provided the most stimulating source of education (8) even for one who had entered Hopkins so late as 1927, especially seminars focused on such books as Born and Jordan's *Quantenmechanik*, conducted by Gerhard Dieke, Maria Mayer, and Karl Herzfeld—relatively new arrivals from The Netherlands, Germany, and Austria; Compton and Allison's *X-Rays in Theory and Experiment*, guided by Joyce Bearden; Wintner's *Spektraltheorie der unendlichen Matrizen—Einführung in den analytischen Apparat der Quantenmechanik*, inspired by Wintner himself—from Leipzig; Ames and Murnaghan's *Theoretical Mechanics*, converted into tools for the future by Murnaghan; assorted Rayleigh works on optics, seminars given coherence by A. H. Pfund; and the then hot Rutherford, Chadwick, and Ellis *Radiations from Radioactive Substances*, illuminated by Norman Feather, in his early twenties, whom R. W. Wood had just recruited from Rutherford's own Cavendish Laboratory group. Nuclear physics was clearly the wave of the future. So, in the traditional Hopkins spirit of hands-on involvement in experimental research, after some small work in x rays with Bearden and in spectroscopy with Dieke, I learned from Feather how to determine for myself the 3.5-day half-life of radon. The most important requirement was simple. Sit in the dark for half an hour. Then the fully dilated pupil easily picks up the flash that the alpha particle makes when it hits the zinc sulfide screen. Come back every few hours and repeat the measurement of counting rate. Voilà—the decay curve.

How prophetic it was of the future that I should be asked to give a seminar report on the 1930 paper (9) of W. Bothe and H. Becker on the artificial excitation of nuclear "gamma radiation." How did the excitation get from A to B? A mystery, a puzzle, an enigma! This paper was a doorway to the discovery of the neutron as the neutron was a doorway to the discovery of fission.

A penetrating radiation with mysterious properties! What a challenge to try to understand it. We interested students, especially Robert T. K. Murray and I, followed the subsequent efforts to unravel the mystery, not least among them the attempt (10), the failed 1932 attempt, of Irene Curie and Frederic Joliot, "Émission de protons de grande vitesse par les substances hydrogénées sous l'influence des rayons γ très pénétrants." How brief the time was from the finding of Bothe and Becker to Chadwick's discovery (11) of the neutron! Minds were better prepared for the new particle at the Cavendish Laboratory than anywhere in the world because Rutherford—"Ce jeune homme dévine tout" in the words of Becquerel—had been arguing as early as 1920 that such a particle should exist.[1]

We graduate students raised with each other question after question about the neutron, and followed with excitement each week's new findings. Can neutrons be bottled? Are free neutrons present in the atmosphere? What will a neutron do to a nucleus?

Clearly nuclear physics was becoming wide open territory. So my first year of postdoctoral research found me in New York in the fall of 1933 with Gregory Breit, working on questions of nuclear barrier penetration, resonant and nonresonant nuclear reactions, and the production of electron-positron pairs out of the emptiness of the vacuum.

By the spring of 1935, with Breit's support, I was applying to go to Copenhagen for a second year of a National Research Council fellowship, to work with Niels Bohr, because—I wrote—"he sees further ahead than any man alive."

No one starting a year with Bohr could have had a more marvelous introduction to the great men and the great open issues of physics than the International Conference on Physics held at London and Cambridge in the fall of 1934: J. J. Thomson, seventy-seven, frail, and white haired, host at a reception at Trinity College. Ernest Rutherford, head thrown back, in impressive discourse, with a circle of delegates, a lordly presence at an evening reception in the rooms of the Royal Society. Max Born, deprived of his position in Göttingen and newly arrived in the United Kingdom, writing in huge letters on the blackboard, "NUCLEAR PHYSICS," then with eraser and chalk—to laughter—altering the title to read "UNCLEAR PHYSICS." Thirty-three-year-old Enrico Fermi reporting radioactivities produced in many an element by neutron irradiation. Less than a month later (11 a.m., October 27, 1934) came the Rome group's great discovery that hydrogenous substances moderate neutrons and this moderation of the neutrons "increases the activation intensity by a factor which, depending on the geometry used, ranges from a few tens to a few hundreds."

---

[1] Rutherford in his 1920 Bakerian Lecture to the Royal Society of London had predicted the neutron's likely properties.

There was not one of the many papers (8) in solid-state physics or nuclear physics which I did not find truly significant. One, by Gray and Tarrant (13)—on the anomalous back-scattering of gamma rays—incited me to prove the effect to be, not back-scattering at all, but a minishower. This work later in the Copenhagen year brought me into meeting with Lise Meitner (Figure 6), initiator of experiments of this kind (14) and a guiding spirit of the Hahn-Strassmann uranium work.

Bohr's institute, smaller than many a house, his small group, and the man (Figure 7) and his way of work I have described elsewhere (8, 15). Formulator of complementarity, he was also the personal embodiment of complementarity. Who could switch so totally—as occasion warranted—from one mode of operation to another? From boldness to caution? From breadth to concentration? From one who never makes an advance except when in solitary thought to one who never makes an advance except in discussion with another?

KAISER WILHELM-INSTITUT FÜR CHEMIE
Professor Dr. LISE MEITNER

BERLIN-DAHLEM, den 23.März 1935.
THIEL-ALLEE 63

Herrn I.A. W H E E L E R,

Institut für Theoretische Physik,

K O P E N H A G E N.

Lieber Herr Wheeler,

Leider ist Dr.v.Droste, der die Streuungsmessungen an der γ-Strahlung bei 60° gemacht hat, derzeit krank, so dass ich Ihnen jetzt nichts über die Einzelheiten der Kurven berichten kann. Ich schicke Ihnen gleichzeitig die Arbeit von — Dr.Kösters und hoffe, Ihnen nächste Woche, wenn Dr.v.Droste wieder im Institut ist, auch etwas näheres über dessen Messungen schreiben zu können.

Mit besten Grüssen

Lise Meitner

Figure 6 Meitner letter about gamma radiation scattered nearly backward by heavy nuclei.

Figure 7 Niels and Margrethe Bohr on motorcycle (Courtesy of Aage Bohr).

Eastertime 1935, Christian Møller returned from a visit to Fermi's group in Rome. Bohr called a seminar to hear and discuss the new findings, most impressive among them being the high cross section of many nuclei for the interception of a neutron. Møller had not gotten half an hour down the road when Bohr interrupted him and took his place. Head lowered, pacing back and forth, he murmured over and over, "Now it comes. Now it comes. Now it comes...". Suddenly it did come. Then and there Bohr sketched out the compound-nucleus model of nuclear reactions.[2] It stood totally at variance with the earlier conception of the nucleus as an open planetary system. The incoming particle, in the new view, has in nuclear matter a mean free path that, compared to nuclear dimensions, is not long but short. The new nucleus, the compound nucleus, retains no memory of how it was formed. How—and how quickly—it breaks up or deexcites depends only on its energy and its angular momentum, not on its history.

[2] Reference (16) contains a carefully documented chronology of the birth and elaboration of the compound-nucleus model. I was not present in Copenhagen for the developments that Bohr, and Bohr and Kalckar, made in this model subsequent to June 1, 1935, but I remember vividly the Eastertime seminar at which the idea arose, as well as the Bohr-Kalckar manuscript of some months later and their report on this work in Washington in February of 1937 (Figure 5).

By the time of the Third Washington Conference on Theoretical Physics of February 18, 1937, Bohr with Fritz Kalckar was well on the way to developing a comprehensive account of nuclear energy levels and nuclear reaction probabilities on the foundation of this compound-nucleus model. To think of the mean free path of nucleons inside the nucleus as short compared to the nuclear diameter was to have justification, they pointed out (17), for adopting a liquid-drop model for the nucleus. This droplet model had been advanced by George Gamow some years earlier [(18); after Bohr's compound-nucleus model (18a)]. However, no one had ever really pushed it until the compound-nucleus model converted it from vision to tool.

The liquid-drop model gave an approximate way to estimate the quantities that totally define the compound-nucleus model of nuclear reactions: (*a*) the energy levels of the nucleus, and (*b*) the probabilities—per second—that the state in question will send out this, that, or the other particle, or a photon, with this, that, or the other energy.

A less global approach to nuclear reaction processes I had imbibed from my 1933–1934 year with Breit: Let scattering be the key to knowledge! Deduce law of force from phase shifts in scattering. Deduce these phase shifts from the variation of scattering cross section with angle (19). Already while I was with him I had started working on the scattering of alpha particles in helium. I was inspired to go on in my Copenhagen time and at Chapel Hill (1935–1938) by new results on alpha-alpha scattering reported at the October 1934 London-Cambridge conference. It was obvious that neither helium nucleus is a simple particle. Therefore new methods had to be developed to treat the interaction between nuclei (20): the scattering matrix to give a precise description of the phenomenology of scattering, and the method of resonating group structure to define an effective interaction potential.

Despite the rich subsequent development of this clustering model by many workers, I—like others—soon found it quicker to make progress in understanding the broad features of nuclear reactions by applying the compound-nucleus model. That model I described and extended in work with graduate students at Chapel Hill and in lectures I gave at Princeton over a period of some weeks in 1937, as well as in the fall of 1938 after I moved to Princeton. Thus somehow fate had put me with the right ideas in the right place at the right time and with the right man when Bohr arrived on January 16, 1939 with the news of fission.

Once at work together, Bohr and I undertook a detailed analysis of fission regarded as an exciting new application of a compound-nucleus-plus-liquid-drop model. This work took not only the three months of Bohr's stay in Princeton but two additional months of finishing up until I could send it in for publication (June 28, 1939). The topics that had to be taken up are seen in this quotation from the paper (21):

> [We] estimate quantitatively in Section I by means of the available evidence the energy which can be released by the division of a heavy nucleus in various ways, and in particular examine not only the energy released in the fission process itself, but also the energy required for subsequent neutron escape from the fragments and the energy available for beta-ray emission from these fragments.
>
> In Section II the problem of the nuclear deformation is studied more closely from the point of view of the comparison between the nucleus and a liquid droplet in order to make an estimate of the energy required for different nuclei to realize the critical deformation necessary for fission.
>
> In Section III the statistical mechanics of the fission process is considered in more detail, and an approximate estimate made of the fission probability. This is compared with the probability of radiation and of neutron escape. A discussion is then given on the basis of the theory for the variation with energy of the fission cross section.
>
> In Section IV the preceding considerations are applied to an analysis of the observations of the cross sections for the fission of uranium and thorium by neutrons of various velocities. In particular it is shown how the comparison with the theory developed in Section III leads to values for the critical energies of fission for thorium and the various isotopes of uranium which are in good accord with the considerations of Section II.
>
> In Section V the problem of the statistical distribution in size of the nuclear fragments arising from fission is considered, and also the questions of the excitation of these fragments and the origin of the secondary neutrons.
>
> Finally, we consider in Section VI the fission effects to be expected for other elements than thorium and uranium at sufficiently high neutron velocities as well as the effect to be anticipated in thorium and uranium under deuteron and proton impact and radiative excitation.

A new feature of capillarity entered in the case of fission, the concept of fission barrier. The very idea was new and strange. More than one distinguished colleague objected that no such quantity could even make sense, let alone be defined. According to the liquid-drop picture, is not an ideal fluid infinitely subdivisible? And therefore cannot the activation energy required to go from the original configuration to a pair of fragments be made as small as one pleases? We obtained guidance on this question from the theory of the calculus of variations in the large, maxima and minima, and critical points. This subject I had absorbed over the years by osmosis from the Princeton environment, so thoroughly charged by the ideas and results of Marston Morse (22). It became clear that we could find a configuration space to describe the deformation of the nucleus. In this deformation space we could find a variety of paths leading from the normal, nearly spherical configuration over a barrier to a separated configuration. On each path the energy of deformation reaches a highest value. This peak value differs from one path to another. Among all these maxima, the minimum measures the height of the saddle point or fission threshold or the activation energy for fission. The fission barrier *was* a well-defined quantity!

Bohr knew from his own student research on water jets that a work of Lord Rayleigh would have something to say about the capillary oscillations of a liquid drop. We rushed up to the library on the next floor of Fine Hall and looked it up in the *Scientific Papers* of Rayleigh. This work furnished a starting point for our analysis. However, we had to go to terms of higher order than Rayleigh's favorite second-order calculations to pass beyond the purely parabolic part of the nuclear potential, that is, the part of the potential that increases quadratically with deformation. We determined—as soon also did Feenberg, von Weizsäcker, Frenkel, and others—the third-order terms to see the turning down of the potential. These terms enabled us to evaluate the height of the barrier, or at least the height of the barrier for a nucleus whose charge was sufficiently close to the critical limit for immediate breakup.

We found that we could reduce the whole problem to finding a function f of a single dimensionless variable x. This "fissility parameter" measures the ratio of the square of the charge to the nuclear mass. This parameter has the value 1 for a nucleus that is already unstable against fission in its spherical form. For values of x close to 1, by a power series development we could estimate the height of the barrier and actually give quite a detailed calculation of the first two terms in the power series for barrier height, or f, in powers of $(1 - x)$. The opposite limiting case also lent itself to analysis. In this limit the nucleus has such a small charge that the barrier is governed almost entirely by surface tension. The Coulomb forces give almost negligible assistance in pushing the material apart.

Between this case (the power series about $x = 0$) and the other case (the power series about $x = 1$) there was an enormous gap. We saw that it would take a great amount of work to calculate the properties of the fission barrier at points in between. Consequently we limited ourselves to interpolation between these points. In the decades since that time many workers (among them Wladyslaw J. Swiatecki at Berkeley, Vladimir Strutinski and his colleagues in the USSR, and Ray Nix and his colleagues at Los Alamos) have revealed many previously unsuspected features of the fission barrier. This is not the place to go into the deeper theoretical considerations on prompt neutrons, delayed neutrons, the physics of fission product decay, and many another topic that came up, nor to detail the many impressive experimental results that were obtained on these and other topics week by week.

Two other issues of comparable or greater challenge came up in doing our paper: (*a*) figuring the rate of decay associated with spontaneous fission, and (*b*) determining the probability for fission of a nucleus with excitation in excess of the barrier summit. The first of these forced us to introduce a measure for the inertia associated with a deformation. The second led us to the concept of channels for fission associated with the transition state. Both introduced lines of thought still under active exploration and development today.

For our immediate needs, however, our simple "poor man's" interpolation was adequate. With it, knowing—or estimating from observation—the fission barrier for one nucleus, we could estimate the fission barrier for all the other heavy nuclei, among them plutonium 239. Thanks to the questioning of Louis A. Turner (Figure 8), soon to write his great and timely review of nuclear fission (23), we came to recognize that this substance, which up to then one had never seen except through its radioactivity [McMillan & Abelson (24), June 15, 1940[3]] would be fissile. This conclusion was soon to lead to a preposterous dream: by means of a neutron reactor such as never before existed, one could manufacture kilograms of an element never before seen on Earth. By the fall of 1944 the E. I. duPont de Nemours Company had converted this dream into the reality of a double alchemy. In total silence neutrons flow from place to place, splitting some uranium nuclei, converting others (the U^{238}) to U^{239}. Then nature itself took hold to complete the alchemy. Over the ensuing hours and days it let the U^{239} spontaneously transform—through two beta transformations—to Pu^{239}. By this double miracle, duPont at Hanford, Washington, was able, week by week, to supply ponderable masses of a strange and totally new element, plutonium, to its Los Alamos, New Mexico customer.

The barrier height of a compound nucleus against fission was not the only factor relevant for fission. Equally important in governing the probability of this process was the excitation, or "heat of condensation," delivered by the uptake of a neutron to form the compound nucleus in the first place. On this point an important development occurred on a snowy morning when I was occupied with classes and not with Bohr. He, having breakfast at the Nassau Club with Rosenfeld and with an arrival of the night before, George Placzek, faced Placzek's continuing skepticism about the very existence of fission. Placzek asked, how can it possibly make sense that slow neutrons and fast neutrons cause uranium to split but not neutrons of intermediate energy? Bohr stopped but said not a word, left with Rosenfeld, crossed the campus to Fine Hall still without a word and there, when Placzek and I joined them, explained the great idea (25) that had just come to him: that the slow neutron fission takes place in the rare

---

[3] According to Irving (24a), J. Schintlmeister and F. Hernegger identified neptunium and plutonium in June 1940 but reported their findings only at the end of 1940.

Figure 8   Louis A. Turner, who pointed the way to the manufacture of plutonium. Courtesy of Argonne National Laboratory Photographic Archives.

isotope U^{235} and the fast neutron fission in the abundant isotope U^{238}. Thus an incoming neutron delivers up a high heat of condensation when it enters into a nucleus with 143 neutrons, because it can form a new neutron pair. This excitation puts the compound nucleus U^{236} over the barrier summit. Therefore, the U^{236} must split when it is formed by slow neutron capture. Moreover, the cross section for fission of the rare U^{235}, like the cross section for the "fission" of boron,

$$n^1 + B^{10} \rightarrow He^4 + Li^7,$$

must exceed by far the geometrical cross section of the nucleus for sufficiently slow neutrons. That circumstance makes it understandable why an isotope present to only one part in 139 imparts to natural uranium the observed substantial fission cross section. However, the cross section must fall off inversely as the velocity of the neutron for U^{235} as for B^{10}; hence the negligible fission cross section of natural uranium for neutrons of intermediate energy. In contrast, when a slow neutron enters U^{238} to form the compound nucleus U^{239}, no new neutron pair is formed. The heat of condensation delivered up is not enough to exceed the fission barrier. Only neutrons of a substantial kinetic energy striking U^{238} can produce U^{239} with enough energy to surmount the barrier. Hence, the existence of fast neutron fission in natural uranium.

Was it reasonable to expect so great a difference between U^{235} and U^{238} from the estimated odd-even difference in neutron binding? Could not the fission barrier differ equally drastically from the one nucleus to the other? Might not this difference be the dominant factor? How could one be sure that the proposed attribution of slow fission to U^{235} and fast fission to U^{238} really made sense until one was clear about these energies? Fortunately Bohr and I had just been through the systematics of nuclear energies in the course of calculating the release of energy in various actual and potential fission processes. Therefore, we could estimate the difference between the excitation developed by neutron capture in the two uranium isotopes as almost a million volts, in favor of fission of U^{235}. From our interpolation for fission barriers we estimated on the other hand a barrier almost 1 MeV lower for U^{235} than for U^{238}. Thus we concluded there was about a 2-MeV margin in favor of the fission of the rare isotope. In later years, after the development of the collective model it became clear that individual particle effects can modify significantly barrier heights and barrier shapes from the predictions of the simple liquid-drop model. However, the qualitative conclusions are not affected; U^{235} is the fissile nucleus.

Placzek, wonderful person that he was, a man of the highest integrity, often a thoroughgoing skeptic about new ideas, said to me over and over in those early spring days of 1939 that he could not believe that the small amount of U^{235} could be the cause of the slow neutron effects in natural uranium. I therefore bet him a proton to an electron, \$18.36 to a penny, that Bohr's diagnosis was correct. A year later Alfred Nier at Minnesota had separated enough U^{238} to make possible a test and sent it to John Dunning at Columbia to measure its fission cross section (26). On April 16, 1940, I received a Western Union money order telegram for one cent with the one-word message "Congratulations!" signed Placzek (27).

Literature Cited

1. Wheeler, J. A., Zurek, W. H., eds., in *Quantum Theory and Measurement.* Princeton Univ. Press, NJ (1983), pp. 3–49
2. Meitner, L., Frisch, O. R., *Nature* 143: 239–40 (1939)
3. Hahn, O., Strassmann, F., *Naturwissenschaften* 27: 11–15 (1939)
4. Frisch, O. R., *What Little I Remember*, Cambridge Univ. Press (1979), p. 116
5. Bohr, N., Wheeler, J. A., *Phys. Rev.* 56: 426–50 (1939)
6. Frisch, O. R., *Nature* 143: 276 (1939)
7. Wheeler, J. A., et al., "Mercer Street and Other Memories," in *Albert Einstein: His Influence on Physics, Philosophy and Politics*, ed. P. C. Aichelburg, U. R. Sexl. Braunschweig: Vieweg & Sohn (1979), pp. 201–11
8. Wheeler, J. A., "Some men and moments in the history of nuclear physics: The interplay of colleagues and motivations," in *Symp. on the History of Nuclear Physics, Univ. Minn., 1977*, Minneapolis: Univ. Minn. (1979), pp. 217–322
9. Bothe, W., Becker, H., *Z. Phys.* 66: 289–306 (1930)
10. Curie, I., Joliot, F., *Comptes Rendus* 194: 273–75 (1932)
11. Chadwick, J., *Nature* 129: 312 (1932); *Proc. Roy. Soc. London Ser. A* 136: 692–708 (1932)
12. Deleted in proof
13. Gray, L. H., Tarrant, G. T. P., *Proc. Roy. Soc. London Ser. A* 136: 662–91 (1932); 143: 681–706 (1934); 143: 706–24 (1934)
14. Brown, L. M., Moyer, D. F., *Am. J. Phys.* 52: 130–36 (1984)
15. Wheeler, J. A., "Niels Bohr: The Man and His Legacy," in *The Lesson of Quantum Theory: Niels Bohr Centenary Symposium October 3–7, 1985*, ed. J. de Boer, E. Dal, O. Ulfbeck. Amsterdam: North-Holland (1986), pp. 355–67
16. Rüdinger, E., gen. ed., Peierls, R., vol. ed., *Niels Bohr Collected Works: Volume 9 Nuclear Physics (1929–1952)*. Amsterdam: North-Holland (1986), pp. 14–27
17. Bohr, N., Kalckar, F., *Mat.-Fys. Medd. Dan. Vidensk. Selsk.* Vol. 14, No. 10 (1937)
18. Gamow, G., *Constitution of Atomic Nuclei and Radioactivity*, Oxford: Clarendon (1931), pp. 18–19
18a. Gamow, G., *Structure of Atomic Nuclei and Nuclear Transformations*, Oxford: Clarendon, 2nd ed. (1937), pp. 4, 26–38
19. Wheeler, J. A., *Phys. Rev.* 45: 746 (1934); *Phys. Rev.* 59: 16–26 (1941); *Phys. Rev.* 59: 27–36 (1934)
20. Wheeler, J. A., *Phys. Rev.* 52: 1107–22 (1937)
21. Bohr, N., Wheeler, J. A., *Phys. Rev.* 56: 427–28 (1939)
22. Morse, M., *The Calculus of Variations in the Large.* New York: Am. Math. Soc. (1934); *Functional Topology and Abstract Variational Theory.* Paris: Gauthier-Villars (1938); Morse, M., Cairns, S., *Critical Point Theory in Global Analysis and Differential Topology: an Introduction.* New York: Academic (1969)
23. Turner, L., *Rev. Mod. Phys.* 12: 1–29 (1940)
24. McMillan, E., Abelson, P. H., *Phys. Rev.* 57: 1185–86 (letter) (1940)
24a. Irving, D., *The German Atomic Bomb: The History of Research in Nazi Germany.* New York: Simon & Schuster (1967)
25. Bohr, N., *Phys. Rev.* 55: 418–19 (1939) (letter)
26. Nier, A., Booth, E., Dunning, J., Grosse, A., *Phys. Rev.* 57: 546 (1940) (letter)
27. Placzek, G., April 16, 1940, telegram, on deposit at the National Museum of Science and Technology, Washington, DC

Fred Stare

Annu. Rev. Nutr. 1991. 11:1–20

NUTRITION RESEARCH FROM RESPIRATION AND VITAMINS TO CHOLESTEROL AND ATHEROSCLEROSIS

Fredrick J. Stare

Department of Nutrition, Harvard School of Public Health,
Boston, Massachusetts 02115

KEY WORDS: Harvard, nutrition, School of Public Health, cholesterol, atherosclerosis

---

CONTENTS

The invitation to prepare this review mentioned that "prefatory chapters are more personal and philosophical than the usual critical reviews in ANNUAL REVIEWS." As the title of this review suggests, I shall recount some of the research with which I have been associated during the past 60 years.

0199-9885/91/0715-0001$02.00

THE EARLY YEARS

My research activities began in my senior year at the University of Wisconsin in September 1930. My advisor, C. A. Elvehjem, assigned me to study the phosphorus partition in the blood of rachitic and nonrachitic calves as a part of an ongoing cooperative study between the Departments of Agricultural Chemistry and Animal Husbandry. This research became the basis for my senior thesis and resulted in my first publication (36).

In the fall of 1930, Dr. Elvehjem returned from a year as a postdoctoral fellow in the laboratory of Sir Frederick Gowland Hopkins, chairman of the Department of Biochemistry, University of Cambridge, England, and brought with him 12 Warburg manometers which were not then available in this country. In addition to studying the partition of phosphorus in blood, I learned how to use these manometers to study tissue respiration. These studies resulted in a publication (37) in which we reported the oxygen uptake for various tissues of chicks and rats under different nutritional conditions. In normal chicks, the rates of oxygen uptake for liver, kidney, and brain tissue are all in the range of 1000 to 1500 μL per gram of fresh material per hour. Muscle, both cardiac and striated, has a considerably lower oxygen requirement with values ranging from 230 to 380 μL. With the exception of the cerebellum, the oxygen uptake of tissues from chicks deficient in vitamins A and B_1 showed little difference from tissues of normal chicks. We found no significant difference in the uptake of liver tissue from normal and anemic rats.

Other studies of tissue respiration in which I was involved as a graduate student concerned the relation of copper to tissue respiration and the effect of certain reducing substances in chronic fluorosis and scurvy in the guinea pig (29). The research on fluorosis and the use of sodium fluoride as an enzyme inhibitor in many of these respiration studies interested me because, at about the same time, others had found that fluoride in small concentrations in drinking water was markedly effective in reducing dental caries in children. Thus began my long interest in the fluoridation of water.

In 1934, I was awarded a PhD degree at the University of Wisconsin. My thesis, only 35 double-spaced, typewritten pages, was titled "Studies on the Respiration of Animal Tissues." A few months before obtaining my PhD, I received a postdoctoral fellowship from the Rockefeller Foundation and spent the following year in the laboratory of Philip Shaffer, chairman of the Department of Biochemistry, Washington University, St. Louis. With Shaffer, I studied hepatoflavin, thought to be a portion of what was then called the vitamin B_2 complex. We developed a procedure for the preparation of flavin from liver that was shorter and less laborious than previously used methods, and we also obtained crystals of the flavin. In experiments with both rats and chicks, we found that our flavin preparation did not prevent the development of dermatitis on a B_2-deficient diet, but another as yet unidentified substance did. Both this unknown substance and the flavin we had isolated were necessary for growth (33).

After a year with Philip Shaffer, my fellowship was extended a year by the International Health Division of the Rockefeller Foundation, and in 1936 I had the opportunity to work in the laboratory of David Keilin, Institute of Parasitology, University of Cambridge, England. Professor Keilin was not a parasitologist, but had his laboratories in the Institute of Parasitology. He was at that time a leading investigator in the study of the cytochrome enzymes that play a major role in tissue respiration. I shared a laboratory with another American, Carl Baumann, who had also received his graduate training in the Department of Agricultural Chemistry at Wisconsin. Professor Keilin assigned us to work on the role of fumaric acid in tissue respiration; again, this included extensive use of Warburg manometers.

At Cambridge we studied liver and kidney tissues of rabbits and heart muscle from pigeons and pigs in order to test the fumarate-oxaloacetate catalytic hypothesis of tissue respiration. We found that the oxygen uptake of these tissues was increased by the addition of very small amounts of fumarate and that this increase was much more than could be accounted for by the oxidation of the fumarate itself. Substances that yielded fumarate on contact with tissues, such as succinate, malate, oxaloacetate, had a similar action (35).

During the Christmas holiday of 1936, my late wife and I spent about 10 days in Germany living with a German family, primarily so we could improve our knowledge of German. Also, because I had used Warburg manometers so often in studies of tissue respiration, I wanted to meet the man who had devised them, Otto Warburg of Berlin.

Our studies on fumarate and succinate were forerunners of the now well-known Krebs citric acid cycle. Indeed Baumann and I often discussed hypotheses of tissue respiration with Hans Krebs, because during much of 1936 he worked at the Department of Biochemistry at Cambridge, just across the courtyard from the Institute of Parasitology. Another frequent visitor to Cambridge that year was Albert Szent-Gyorgyi of Szeged, Hungary. One day after I had given a seminar, Szent-Gyorgyi asked if I would like to spend some time at his laboratory in Szeged. I replied that I would, but I doubted my fellowship would be extended for a third year. He said he thought he could arrange that, and he did.

This third postdoctoral year in Europe was divided half in Szeged and half with Paul Karrer at the University of Zurich so that I might learn some organic chemistry. In Szeged, I continued manometric and chemical experiments on the roles of succinate, fumarate, and oxaloacetate in the respiration of liver and kidney tissues of rabbits (34). At the end of three years of study in Europe, I returned to the University of Wisconsin as a research associate in biochemistry at the Bowman Cancer Research Institute, later to be renamed the McArdle Cancer Institute. Here, I continued research on tissue respiration with Carl Baumann who had also returned to Wisconsin (35).

MEDICAL SCHOOL

After two years of additional work at the University of Wisconsin, I left to enter the University of Chicago School of Medicine as a third year student in the fall of 1939. Except for anatomy, the course work I had taken as a graduate student met the requirements for the first two years of medicine. I made up the requirement for anatomy by special instruction during the summer of 1938. As a medical student, I was able to find enough time to continue some tissue respiration studies (38) in the laboratory of Alberto Guzman Barron, who worked at the Albert Lasker Research Laboratories at the University of Chicago School of Medicine. These studies led to a review with Dr. Barron in the *Annual Review of Biochemistry* (1).

Perhaps of more importance for my future, by working in the Lasker Laboratories, I had the opportunity to meet Albert Lasker and his charming wife, Mary. Mary Lasker and I became good friends, particularly when we discovered that we both came from nearby small towns in Wisconsin, she from Watertown and I from Columbus. For years our two high schools had been rivals in basketball.

After receiving my MD from the University of Chicago in 1941, I spent a year as an intern in general medicine at Barnes Hospital in St. Louis. That year, I could find no spare time for any research, even though Philip Shaffer's laboratories where I had worked a few years before were just across the street. Near the end of my internship I took and passed the Missouri State Medical Examinations. If I had difficulty getting a job, I could always move to the Ozarks and practice medicine to support my family!

APPOINTMENT AT HARVARD

In February of 1942, after three visits to the Harvard Medical School during which I presented two seminars in the Department of Biochemistry, whose chairman was A. Baird Hastings, I was told that I would be considered for an appointment at Harvard.

At that time, there was no department named "Nutrition" in any health or medical center in the world. Most of the pioneering research in nutrition had been carried out in schools of agriculture, medicine, public health, and in universities by men and women who were in departments of chemistry, physiology, pathology, and bacteriology. In 1940, C. K. Drinker, dean of the Harvard School of Public Health, and A. Baird Hastings prevailed upon the Rockefeller Foundation to provide funds to establish a Department of Nutrition to be jointly in the schools of medicine and public health. It was specified that the chairman of this new department should have some training in medicine as well as in nutrition and biochemistry, should have had some experience in foreign laboratories, and also have an interest in public health. The department was to organize and supervise the teaching of nutrition in both schools. To my delight, I was invited by Harvard to take on that task and on July 1, 1942, became an assistant professor of nutrition and chairman of this new department. I selected two young scientists, D. Mark Hegsted and John M. McKibbin, as my first academic appointments in the Department of Nutrition. Both had received PhD degrees from the University of Wisconsin in nutritional biochemistry.

In 1942, our country was at war and much of the research in universities related to problems associated with the war. Our initial war-related research dealt with malaria and with the improvement of total parenteral nutrition. Because malaria was a disease of worldwide importance to public health, and was prevalent in many areas where the war was underway, it was important to determine whether atabrine, a relatively new and commonly used antimalarial drug at that time, in any way affected nutritional status and whether nutritional status, particularly certain vitamin deficiencies, affected the efficacy of atabrine. From 1942 to 1946, using ducks and chicks, we did several studies in this area but all essentially with negative results (10). The only convenient experimental animal susceptible to malaria is the duck. The chick, on the other hand, is relatively resistant to malaria. Thus, we compared the response to the same malarial parasite *(Plasmodium lophurae)* of a resistant species and a susceptible species fed identical diets including diets deficient in nicotinic acid, thiamin, choline, and vitamin A. In ducks or chicks, none of these deficiencies influenced the course of the malaria to a significant degree (31).

Our other war-related effort was the improvement of total parenteral nutrition (TPN). I became interested in this problem when I interned at Barnes Hospital in St. Louis. When I arrived in Boston in 1942, the Brigham Hospital had a shortage of physicians. The hospital had just appointed George Thorn as chief of the Department of Medicine, and he asked me to attend on the wards until the war was over and some of the Brigham physicians returned.

Another important development that was to influence my career was the formation of the Nutrition Foundation by a number of executives in the food industry and Karl Compton, president of M.I.T. in 1941. They persuaded Glen King, a co-discoverer of vitamin C, who was then professor of chemistry at the University of Pittsburgh, to accept the position of scientific director and president of the Nutrition Foundation. The goals of this new foundation, with principal support from the food industry, were to conduct an extensive program in education and research in experimental and clinical nutrition. As part of the education program for scientists, the Nutrition Foundation created a new journal, *Nutrition Reviews,* to provide an authoritative and unbiased review of worldwide progress in the science of nutrition. Dr. King invited me to be the editor of this new journal. I was pleased to accept, and the first issue was published in November 1942. My first editorial board consisted of ten working scientists in the field of nutrition and included Elmer Stotz at Harvard, Carl Moore at Washington University, St. Louis, and Max Wintrobe at the University of Utah, Salt Lake City.

At the time, the major difficulty in total parenteral nutrition was an inability to provide adequate amounts of calories. We tried to increase the calories available to persons on TPN by developing emulsions of fat that could be given intravenously.

RESEARCH DURING THE WAR

From 1942 to 1946 research on parenteral nutrition was generously supported by the US Army and for the next several years by the Upjohn Company. Our very active program to improve fat emulsions for TPN was initially directed by John McKibbin, who left us in 1945 to accept a faculty position at the University of Alabama School of Medicine where he later became professor and chairman of the Department of Biochemistry. I promptly recruited Robert P. Geyer, who had recently obtained his PhD in nutritional biochemistry from the University of Wisconsin, to carry on this research.

Our first published report on fat emulsions for intravenous use appeared in 1945 (19). In dogs, refined coconut oil emulsions stabilized with purified soybean phosphatides were the most successful in our infusion studies. Over the course of the next several years a total of 38 research and review papers in this area came from our laboratories. Our last paper on this subject, published in 1960, was by Dr. Geyer (6). In it, Dr. Geyer concluded, "Although oral feeding is still to be preferred wherever possible, parenteral nutrition has achieved a permanent place in both therapeutic and experimental applications."

From 1942 to 1945, our war-related research was generously supported by various branches of our government. With these funds our laboratories were greatly expanded, not only physically but also in staff. The physical expansion was made possible because Harvard University had bought the Huntington Hospital at 695 Huntington Avenue, near the medical school and school of public health, in Boston, and the Department of Nutrition moved into this new space. But with the reduction of funds for war-related research in 1946 and with an expanded department, it was necessary either to drastically reduce our activities and staff or find other sources of research funds.

RAISING FUNDS FOR NUTRITION RESEARCH

I opted to raise additional funds and decided to concentrate on various aspects of atherosclerosis and cardiovascular disease. There were several reasons for this decision: Atherosclerosis is an important part of the etiology of coronary heart disease, hypertension, and cerebral hemorrhage, the principal causes of death in our country; obesity is also known to be a hazard in these conditions, as well as in many other illnesses. Since nutrition played a role in these diseases, I thought there might be a good chance of attracting private support from the food industry. Financial support for nutrition research from the latter was already being obtained through the Nutrition Foundation.

As Editor of *Nutrition Reviews,* I was invited to the twice yearly meetings of their Board of Trustees and the annual meeting of their Food Industry Advisory Committee. The former consisted of the presidents of several of our largest food companies and the latter of their directors of research. This obviously provided me with opportunities to seek funds at the highest level in those companies that formed the Nutrition Foundation. In seeking support from the food industry, I had to emphasize that we were not a department of food technology but of nutrition and that any support they might provide had to be unrestricted for the support of both our teaching and research. Eventually, I succeeded in convincing some of the leaders of the food industry that they should consider it an obligation as well as an opportunity to support research in nutrition not only at Harvard but at other academic nutrition centers.

As mentioned earlier, when I was a medical student at the University of Chicago, I spent some time in the Lasker Laboratories and had met Mrs. Albert Lasker. After moving to Boston in 1942, occasionally I would see Mrs. Lasker when I was in New York. When support for war-related research began to lessen, it was natural for me to ask the Lasker Foundation for help. Beginning in the late 1940s, and for many years thereafter, the Lasker Foundation was very generous in providing unrestricted funds for our research.

TISSUE RESPIRATION AND VITAMINS

After the war I decided to extend my earlier work on "respiration and vitamins." Now that we were receiving generous, unrestricted private support (this was before the days of NIH grants), it was possible to continue some activities in tissue respiration; but because the long-range goal was "heart disease," I wanted to conduct research on basic factors affecting the respiration of heart muscle. In the mid 1940s, I recruited a bright, young postdoctoral fellow, Robert E. Olson, who was interested in metabolism and nutrition. I encouraged him to undertake studies of respiration of heart tissue available from ducks and chicks that had been made deficient in various vitamins. Olson et al found that the pyruvate utilization by cardiac muscle from rats and ducks is depressed in thiamin deficiency and that the addition of thiamin in vitro tends to restore pyruvate utilization to normal (26). Dr. Olson continued these respiration studies with heart muscle tissue from rats and ducks deficient in biotin, pantothenic acid, and pyridoxine (2, 22–28). Dr. Olson went on to chair departments of biochemistry and nutrition at the University of Pittsburgh and St. Louis University and eventually became the editor of *Nutrition Reviews*.

More on vitamins! By the mid 1940s, Dr. Hegsted had formed a friendly and cooperative arrangement with S. B. Wolbach, who was then professor and chairman, department of pathology at Harvard Medical School. In 1952, they reported that the skeletal responses to hypervitaminosis A in young chicks are precisely of the same nature as those in mammals (44). As in mammals, hypervitaminosis A in growing chicks accelerates all histologic sequences in bone growth in conformity with the normal growth patterns.

Several papers dealing with vitamins A, D, E, ascorbic acid, folic acid, and vitamin B_{12} were published by members of our department; the last one was published in 1985 (5).

CHOLESTEROL AND ATHEROSCLEROSIS

The first paper on cholesterol and atherosclerosis from our department was published in January 1953 by George V. Mann (17), a research associate. In the first sentence of that paper the author noted, "The factors that control cholesterol metabolism are poorly understood." Dr. Mann had also been involved with some of our studies with fat emulsions given intravenously and he wrote: "Chance observations of the effects of infusions of fat emulsions in dogs suggested that the cholesterol content of the blood was sensitive to stress [generated by intravenous fat emulsions] and that the esterified cholesterol component was particularly influenced by such stress."

Dr. Mann was also an author of our second paper in this area, published in June 1953 (41). To quote the first and last sentences of that paper published almost 40 years ago: "Determination of the role of weight reduction as a preventive and therapeutic measure in the control of atherosclerotic vascular disease is of considerable practical importance. . . . If elevated serum lipid levels contribute to the causation of atherosclerosis, weight reduction is a proper treatment for this disease."

Productive study of atherosclerosis or any disease is greatly facilitated if an animal model can be found that resembles the human and can be managed under laboratory conditions. For atherosclerosis such a species should have a susceptibility to the development of arterial disease resembling that seen in humans. The necessary and crucial transfer of information to humans would also be facilitated if the dietary habits of the experimental species resembled those of man. Certain species of subhuman primates would appear to be a likely choice, yet it was widely believed in the late 1940s that atherosclerosis is rarely seen in subhuman primates as a naturally occurring disease, and further that these animals are resistant to the experimental induction of atherosclerosis. Our initial studies of atherosclerosis in monkeys were done primarily by George V. Mann and by our pathologist, Steven B. Andrus.

We had intended to use rhesus monkeys, but soon found that they were nearly impossible to obtain and very expensive, because in the late 1940s the pharmaceutical companies were purchasing most available rhesus monkeys for use in making Salk polio vaccine. Hence, we decided to use the ordinary "organ grinder" monkcy obtainable from a variety of pet shops. They were imported from Latin America, mostly from Colombia, and were Cebus monkeys *(Cebus albifrons)*. The results of our first studies using monkeys were published in 1952 and 1953 (14, 16). To our knowledge, this was the first report of the experimental production of atherosclerosis by dietary measures in a subhuman primate.

We had produced the disease by feeding a purified diet high in cholesterol and low in sulfur amino acids over periods of 18 to 30 weeks. Within 2 to 8 weeks, this regimen caused the total serum cholesterol to rise to 300 to 800 mg/dl from the usual value of 140 mg/dl. The hypercholesterolemia could be largely prevented by feeding 1 g per day of *dl*-methionine or *l*-cystine as supplements to the diet. The vascular lesions were in the ascending aorta but extended from the valves of the left ventrical to the proximal portions of the carotid and femoral arteries. Minimal lesions were observed in the coronary arteries. Visceral cholesterolosis was not associated with this disease.

Having found that the hypercholesteremia we produced in monkeys could be largely prevented by adding supplements of methionine or cystine to the diet, we hastened to conduct human studies, but our results were negative (15). In 24 American males with moderate to marked elevations of serum cholesterol, 3 g of methionine added daily to their usual diet at mealtime for six weeks did not alter their serum cholesterol and lipoprotein levels.

Cooperative Study on Lipoproteins and Atherosclerosis

A major paper on cholesterol and beta-lipoproteins in the serum of healthy Americans and of those with coronary heart disease was published by members of our department in 1957 (11). During the period May 1951 through June 1954, serum lipids of 2,045 human subjects were examined as part of the Cooperative Study of Lipoproteins and Atherosclerosis (20). Our part of this study was concerned with 1,968 adults of various ages who were active and healthy as compared with 273 men who had experienced a myocardial infarction, 141 men with angina pectoris but without myocardial infarction, and 23 women with myocardial infarction—a sizable sample for a nonepidemiological study.

Our measurements of various lipoproteins and total cholesterol were related to sex, age, and body weight. We reported that the most characteristic attribute of serum lipid measurements in adults of similar age, sex, and clinical status is their large variability. Among healthy people under 50 years of age, men show higher levels of all these serum lipids than do age-matched women. The age trend of serum lipid levels is different for the two sexes. Women show a steady increase in levels with age throughout the age span studied. After age sixty, serum cholesterol levels of women frequently exceed those of men. The trends of serum lipid levels with age are partly attributable to fattening with age.

The 273 men with established myocardial infarction were found to have both serum cholesterol and lipoprotein levels that were higher on the average than those found in age-matched men without obvious disease. This finding supports the belief that clinical manifestations of atherosclerosis are associated with a disorder of lipid metabolism. The serum lipid levels of the 23 women with myocardial infarction were similar to those of men with the same disease and were higher than age-matched women without obvious disease. The serum lipid levels of the 141 men with angina pectoris only were intermediate between those of healthy men and men with myocardial infarction. The small size and great variability of these differences of serum lipid levels between well men and women and those with angina pectoris or myocardial infarction prevent efficient application of serum cholesterol and lipoprotein levels *by themselves* to the clinical prediction of coronary heart disease among individuals.

Clinical Studies

In the early 1960s, we made arrangements with the Danvers State Hospital, a mental institution located about 25 miles north of Boston, for a long-term study on the quantitative effects of dietary fat on serum cholesterol. Dr. Hegsted and others of our staff implemented this study. We provided a separate kitchen and dining room for our subjects who were all males. Selected from a large group, most of our subjects were chronic schizophrenic patients without evidence of physical disease and were between 34 to 57 years of age. Those with serum cholesterol values above 300 or lower than 200 mg/dl were excluded. The final selection was based primarily upon the probability that they would be available and cooperative over a long period of time. All the men were housed and fed in an isolated ward that included a recreation room in addition to the kitchen and dining facilities. They continued their usual supervised activities. The purpose of the study and the need for adherence to the diet were continually stressed to attendants and subjects. A dietitian from our staff supervised preparation of all meals.

The men were studied in groups of ten. They were fed a low-fat diet to which various fats were added. The data were analyzed to determine the serum cholesterol response to various fatty acids. An unusually strong feature of these studies was that the comparisons were made with the same men throughout the study. The test oils were used primarily by incorporating them into recipes for many products such as waffles, muffins, cakes, pie crust, biscuits, salad dressings, and spreads for bread. Considerable experimental work was necessary to obtain satisfactory products with certain oils. Filled milk and ice creams prepared from nonfat milk and the appropriate oil were used. Including the control diet, a total of 36 different fats and oils were used, each for a period of four weeks and each experimental diet was preceded by a four-week period on the control diet. The range of total fat in the diets varied from low, at 22% of total calories, to high, at 38 to 40%.

Among the more interesting results (9) are the following:

1. Approximately 67% of the total variance in the level of serum cholesterol was explained by changes in the dietary content of myristic acid alone. This appears to be the most important of the fatty acid components affecting total serum cholesterol levels.
2. Palmitic acid has significant but much lesser effects upon the level of serum cholesterol than does myristic acid.
3. No specific effects on serum cholesterol could be detected for stearic, lauric, or shorter chain saturated acids or for monounsaturated acids except that their presence in fats lowers the proportions of myristic, palmitic, and polyunsaturated fatty acids.
4. The amount of dietary fat tested between 22 and 40% of the total calories appeared to be without influence upon the level of serum cholesterol.
5. Dietary cholesterol appeared to be linearly related to the serum cholesterol. An increase in 100 mg in dietary cholesterol provoked a rise of serum cholesterol of approximately 5 mg/dl. This response was independent of the effects induced by dietary fat.
6. In view of these results we proposed that the most effective practical diets for lowering serum cholesterol should be those relatively high in total fat with a small proportion of myristic and palmitic acids, particularly myristic acid, a high proportion of polyunsaturated acids, and a small amount of cholesterol.

Experimental Studies

In the 1960s our experimental studies continued in parallel with these clinical studies. While our initial studies of diet and atherosclerosis in nonhuman primates were with *Cebus albifrons,* we soon began to use other species of New World monkeys. In fact, we sponsored an expedition to the jungles of Colombia to see how many different species of monkeys could be obtained, and then studied at autopsy 73 vascular specimens from Cebus and 4 other species of New World monkeys. We found substantial differences in the frequency and extent of early arterial lesions. When different species were studied in the laboratory, we found different responses to arterial lesions and levels of serum cholesterol with the same diets (30).

One of the puzzling aspects of atherosclerosis has been the occasional discrepancy between hypercholesterolemia and the incidence of atherosclerosis or ischemic heart disease. Whereas hypercholesterolemia is considered a primary risk factor in atherosclerosis, some individuals with elevated levels of cholesterol do not develop significant atherosclerotic vascular disease while other normocholesterolemic persons do.

In 1977 we reported that squirrel and Cebus monkeys fed a coconut oil diet develop comparable hypercholesterolemias, but the squirrel monkey primarily increases its high-density lipoprotein (HDL) pool of cholesterol (21). These results, coupled with the greater accumulation of aortic lipid, particularly cholesteryl ester, in the atherosclerotic-susceptible squirrel monkey, support the concept of the protective nature of high-density lipoproteins and the atherogenic potential of low-density lipoprotein (LDL). They also suggest that a species' genetic control of the lipoprotein response to diet is variable, which has important biological implications.

Our use of several species of New World monkeys greatly expanded, not only for research related to atherosclerosis but also for research on other problems concerning nutrition. Therefore we started breeding our own monkeys, first in Colombia, then in Guatemala, and finally in our own animal facilities in Boston. Here, for several years, an average of about 100 baby monkeys were born each year. Having a good supply of infant monkeys enabled us to study a number of psychological and behavioral problems during growth and development. Stephan & Hayes published one of our last papers related to diet and cholesterol in monkeys in 1985 (40).

Boston-Ireland Brothers Study

Any review of the contributions of our laboratory to studies of diet, atherosclerosis, and coronary heart disease should mention our Boston-Ireland Brothers Study. This was a large cooperative study with Trinity College School of Medicine, Dublin, Ireland, carried out in the 1960s and published in 1970 (3). It was a nutritional and epidemiological study involving 1,994 middle-aged men including over 500 pairs of brothers, one of whom still lived in Ireland whereas the other had lived in Boston for at least 10 years. All brothers were within 10 years of age of each other. The intake of calories, complex carbohydrates, magnesium, and fluoride (from tea) was higher in Ireland. The proportion of calories derived from fat and saturated fat, the serum cholesterol, blood pressure levels, and the amount of cigarette smoking did not differ markedly. The weight, skinfold thickness, and number of abnormal electrocardiograms were higher in the Boston subjects. A study of the pathology of coronary arteries and aortas from autopsies (of other individuals) revealed much earlier serious atheromatous involvement in Boston Irish than in Irish specimens. Increased physical activity appeared to be most important in reducing the risk of coronary heart disease in Ireland.

Although the major interest in atherosclerotic vascular disease has been directed toward males in the fifth decade of life and older, the period in which clinical complications become increasingly common, a consideration of the prolonged course of the underlying atherosclerosis suggests that attention aimed at prevention must ultimately be turned to the second decade. We were among the first to study the role of diet on total serum cholesterol levels in adolescent boys. Our aim was to demonstrate whether practical, realistic, acceptable modification in the usual diets of adolescent males could effectively prevent or lessen the usual rise in plasma cholesterol levels that characterizes American adolescent males. We began with no preset criteria or dietary goals, but instead aimed at making changes within limits of acceptability. Furthermore, we were not working in a metabolic ward situation and therefore interpreted the outcome against a background of considerable and variable nonadherence.

Study of Adolescent Boys

In the fall of 1969, we obtained cooperation of the administration, faculty, students, and parents of a boys' boarding school located an hour's drive from Boston. Much of this cooperation was due to the assistance of an internationally recognized cardiologist and Harvard professor, Dr. Paul Dudley White. The school had a student body of approximately 220 boys aged 13 to 18 years. The school was located about 3 miles from the nearest source of off-campus food, thus at least minimizing a major source of nonadherence. Meals were prepared in the school kitchen under our supervision. The first few months were spent in trying out various diet modifications at a "test table," where only nine boys ate, and gaining the cooperation of a few food companies to prepare special foods. In January 1970, before introduction of diet modifications to the entire school population, serum cholesterol determinations were made. They were repeated before and after the Easter vacation in the spring of 1970, again when the boys left for summer vacation, and also when they returned in the fall of 1970. We showed clearly that blood cholesterol, a risk factor for atherosclerosis, is readily lowered by realistic and acceptable modifications of dietary fats and cholesterol. Among boys whose baseline levels of total cholesterol were 200 mg/100 ml or higher, the average reduction was 15%; in those whose baseline level was 199 mg or lower, the mean decrease was 8%. As far as we know, this study represents the first attempt at lessening one of the risk factors associated with atherosclerotic vascular disease in later life at the time of adolescence when it may begin to develop (18). This study was repeated in 1970–1971 in a larger boarding school (448 students) with similar results (4).

From these two studies, it appears that the most significant dietary changes to initiate in feeding adolescent boys in order to lower serum cholesterol levels are, in order of importance:

1. Use of a low-fat milk with extra skim milk solids
2. Replacement of butter with a highly polyunsaturated margarine
3. Use of polyunsaturated oils and shortenings in baked goods and for frying
4. Use of low cholesterol or cholesterol-free egg products whenever possible, particularly for baking
5. Fewer eggs and more cereals
6. Use of a low-fat ice cream or an ice cream made with a polyunsaturated fat
7. Use of fish, veal, poultry, and carefully trimmed beef with some frequency

These suggestions should be given serious consideration by institutions, whether private or public, when planning menus for feeding adolescent boys and for any modification of existing legislation for school-feeding programs.

Abundant evidence is available that the risk of developing coronary heart disease is positively correlated with the plasma cholesterol level in men from 30–35 up to 60–65 years of age when the total cholesterol is 240–250 mg/dl or higher *and other coronary risk factors are present*. Evidence for similar findings beyond 65 years of age is skimpy.

The efficacy of diets altered in total calories, fat, and cholesterol in lowering plasma cholesterol has been demonstrated for more than three decades. The American Heart Association and the National Heart, Lung, and Blood Institute have urged Americans to modify their dietary habits. A logical application of these suggestions infers a change in family food choices. The cholesterol response of family groups to altered dietary regimens has seldom been studied. We were one of the first groups to report a study in which families participated in a dietary study for the purpose of lowering total serum cholesterol (43).

Boston Family Study

We recruited 46 families in a short-term (7 week) study. A family was defined as including an adult man, an adult woman, and one or more adolescents, all living at home and eating most of their meals there. The study consisted of two parts: a four-week baseline period and a three-week diet-change period. Each family was assigned a nutritionist who assisted with dietary problems and maintained contact throughout the study.

In this short-term, professionally supervised study, in which dietary cholesterol and saturated fat sources were decreased and sunflower oil and margarine were added as the major sources of polyunsaturated fat, an average reduction in serum cholesterol approximating 10% was achieved within 10 days and maintained throughout a three-week period. Adult and adolescent females had slightly greater cholesterol-lowering responses than did adolescent and adult males.

The high degree of cooperation among the free-living families in this study has positive implications for enlisting complete family involvement in preventive and therapeutic dietary programs. It should be remembered, however, that this was a short-term study and it was carefully and frequently monitored by a trained nutritionist.

A final blood sample was obtained on all but one of the study participants approximately 14 weeks after the test-diet ended. Serum cholesterol had returned to, or surpassed, baseline values. These results indicate the rapidity with which achieved goals may be negated and the need for continued adherence and monitoring if reductions in serum cholesterol are to be maintained.

NUTRITION EDUCATION

So far, I have commented briefly on a few of the research papers dealing with "respiration, vitamins, cholesterol, and atherosclerosis" from Harvard's Department of Nutrition, but the department has also been active in a variety of nutrition education pursuits. I would like to mention a few of them.

From the early 1940s, various members of the department prepared the sections on vitamins and nutrition in several encyclopedias, including the *Encyclopedia Britannica* and the *Britannica Book of the Year*. In 1952, at the request of the American Heart Association, we prepared a booklet, "Food for Your Heart, A Manual for Patient and Physician" (39). It was reviewed and endorsed by the Council on Foods and Nutrition of the American Medical Association. It has long been out of print and as of a few years ago I was interested to learn that neither the offices of the American Heart Association nor the National Heart, Lung, Blood Institute knew of its existence. It might be of interest to quote a few sentences: "Nutrition and food have a two-way relationship to normal heart function and to high blood pressure. First is the prevention or treatment of overweight or obesity—the result of eating more food energy (calories) than the body needs. Second is the treatment of hypertension. . . . There is little point in comparing the nutritive value of single foods because most foods supplement the nutritive value of others. The fundamental rules of good nutrition are these: First, eat a variety of foods; second, maintain your desirable weight."

The concept of the Basic Four Food Groups came from our department and was presented at the 38th Annual Meeting of the American Dietetic Association in St. Louis in October, 1955 (8). Two years later, this concept was adopted by the US Department of Agriculture.

Our most recent "education" paper (42) was directed to physicians and questions the efficacy and cost efficiency of the National Cholesterol Education Program, which recommends that the entire US population reduce dietary intake of saturated fats and cholesterol to reduce the risk of coronary heart disease (CHD). The role of diet in the causation of CHD remains unresolved, and overemphasis on diet may shift attention away from more promising regimens, such as quitting smoking, controlling hypertension, diabetes, and stress, and maintaining a reasonable body weight and a moderate level of physical activity.

THE LOWN RESEARCH GROUP

No comments about the activities of Harvard's Department of Nutrition related to heart disease would be complete without mentioning the outstanding activities of the "Lown Group." They have not been discussed thus far in this prefatory chapter because they do not deal primarily with nutritional aspects of atherosclerosis. Bernard Lown joined our group in 1958 as a research associate in medicine, in part to give us a more balanced approach to interpretation of our animal work relative to cholesterol and atherosclerosis and its application to heart disease in man. He and his group helped us gain the general acceptance of our department by the medical profession. Some years ago, Dr. Lown was promoted to professor of cardiology in public health. He gathered around him a group of brilliant young postdoctoral fellows whose productivity over the years has been tremendous, approximately 400 research and review papers. I shall mention only two of his papers. The first deals with the development of a method for terminating cardiac arrhythmias (13) that utilizes an electronic device employing direct current discharge. Transthoracic DC countershock was found to be 100% effective in controlling 550 episodes of ventricular fibrillation in 20 dogs, all of which survived. It was then used to treat 25 episodes of arrhythmias in 19 patients, and 23 of the 25 episodes were successfully reverted. The Lown Cardioverter is now standard equipment in most hospitals and in many ambulances. It is truly a life-saving device.

The second paper from the "Lown Group" that I mention is very simple, yet the recommendations made in it are commonly accepted today as part of the treatment of myocardial infarction (32). Myocardial infarction and its complications evoke the highest toll of human life of any single disease in the United States. Over the years, many different forms of therapy have been introduced. A vast array of drugs and sophisticated electronic devices have been developed, and these have important uses. However, one of the simplest, least expensive, and possibly most effective modes of therapy is the so-called "arm chair" treatment of myocardial infarction.

Prior to this simple procedure, complete bed rest had been the rule in treating patients with acute myocardial infarction, despite the lack of convincing evidence that this is beneficial to the heart. The "arm chair" treatment is based on the belief that strictly enforced bed rest is more taxing to the patient and to the damaged heart than a sitting position in a comfortable chair.

This report dealt with 184 patients with acute myocardial infarction who were admitted to a coronary care unit over a two and one-half year period and who were helped into an arm chair, most within 1–2 days of admission, all within one week, and daily thereafter. Patients tolerated this treatment well. Clinically, important changes in blood pressure and pulse were seldom observed. These results show that arm chair treatment can be utilized in patients with acute myocardial infarction without adverse effects upon the circulation and with great benefits to the patient's morale.

An accomplishment of Dr. Lown that has nothing to do with "respiration, vitamins, cholesterol, or atherosclerosis," but that, in part, involved our department and may be more important than all research from this department, was his founding of the International Physicians for the Prevention of Nuclear War (IPPNW) in 1979 in cooperation with his cardiology colleague in Moscow, Evugene Chazov. Drs. Lown and Chazov had collaborated in cardiologic work for more than 20 years. Now, they collaborated in developing a society for the prevention of nuclear war. IPPNW was formally organized in the seminar room of Harvard's Department of Nutrition in 1979. It now has groups in many cities and in 70 countries and concerns itself with only one issue: prevention of nuclear war. In 1985 IPPNW was awarded the Nobel Peace Prize, and it was accepted by Drs. Lown and Chazov (12).

CONCLUSION

My comments have been mainly confined to "respiration, vitamins, cholesterol, and atherosclerosis" and are autobiographical in part as requested in the invitation to prepare this prefatory chapter. I have confined this chapter to a review of a few of my papers published as a graduate and postgraduate student on these subjects and also to a few of the papers on these four subjects from scientists who were members of Harvard's Department of Nutrition from 1942 to 1986. In 1987, I prepared a monograph titled, "Harvard's Department of Nutrition 1942–86." It lists the complete references of all papers published from this department during that 44-year period, a total of 2,250, divided into various categories, and is preceded by a brief history of the department, followed by a listing of all sources of support divided into private, government, and industry.

Professor A. Baird Hastings was probably the key figure in motivating the International Health Division of the Rockefeller Foundation to provide a five-year grant that made possible the founding of Harvard's Department of Nutrition. He also helped nurture the department for its first few years. In his autobiography published in 1989 he states, "It is still my conviction that the training of able scientists is the way to secure progress in science, and I must confess that I am personally much prouder of the M.D.'s and Ph.D.'s whose scientific education and research activities I have had the privilege of helping than I am of the papers I have published myself" (7).

This chapter closes in the same spirit. I am more proud of the men and women who as students, postdoctoral fellows, staff members, and technicians were trained in Harvard's Department of Nutrition, in part by our senior staff but largely by themselves, in facilities and with fellowships and salaries provided by the department from funds raised largely by me, than I am of any studies suggested or done primarily by me. And they were trained in many aspects of nutrition other than respiration, vitamins, cholesterol, and atherosclerosis.

Literature Cited

1. Barron, A. G., Stare, F. J. 1941. Biological oxidations and reductions. *Annu. Rev. Biochem.* 10:1–30
2. Brin, M., Olson, R. E. 1952. Metabolism of cardiac muscle. VI. Competition between L (+)-lactate uniformly labeled with C^{14} and C^{14}-carbonyl-labeled pyruvate. *J. Biol. Chem.* 199:475–83
3. Brown, J., Bourke, G. J., Gearty, G. F., Finnegan, A., Hill, M., et al. 1970. Nutritional and epidemiologic factors related to heart disease. *World Rev. Nutr. Diet.* 12:1–42
4. Ford, C. H., McGandy, R. B., Stare, F. J. 1972. An institutional approach to the dietary regulation of blood cholesterol in adolescent males. *Prevent. Med.* 1:426–45
5. Franceschi, R. T., James, W. M., Zerlauth, G. 1985. 1α, 25-dihydroxy-vitamin D_3 specific regulation of growth, morphology, and fibronectin in a human osteosarcoma cell line. *J. Cell. Physiol.* 123:401–9
6. Geyer, R. P. 1960. Parenteral nutrition. *Physiol. Rev.* 40:150–86
7. Hastings, A. B. 1989. *Crossing Boundaries: Biological, Disciplinary, Human,* ed. H. N. Christensen. Grand Rapids, Mich.: Four Corners Press. 373 pp.
8. Hayes, O., Trulson, M. F., Stare, F. J. 1955. Suggested revisions of the basic 7. *J. Am. Diet. Assoc.* 31:1103–7
9. Hegsted, D. M., McGandy, R. B., Myers, M. L., Stare, F. J. 1965. Quantitative effects of dietary fat on serum cholesterol in man. *Am. J. Clin. Nutr.* 17:281–95
10. Hegsted, D. M., McKibbin, J. M., Stare, F. J. 1944. Nutrition and tolerance to atabrine. *J. Nutr.* 27:141–48

11. Lawry, E. Y., Mann, G. V., Peterson, A., Wysocki, A. P., O'Connell, R., et al. 1957. Cholesterol and beta lipoproteins in the serum of Americans: Well persons and those with coronary heart disease. *Am. J. Med.* 22:605–23

12. Lown, B. 1986. Nobel Peace Prize lecture: A prescription for hope. *New Engl. J. Med.* 314:985–87

13. Lown, B., Amarasingham, R., Newman, J. 1962. New method for terminating cardiac arrhythmias: Use of synchronized capacitor discharge. *J. Am. Med. Assoc.* 182:548–55

14. Mann, G. V., Andrus, S. B., McNally, A., Stare, F. J. 1953. Experimental atherosclerosis in cebus monkeys. *J. Exp. Med.* 98:195–218

15. Mann, G. V., Farnsworth, D. L., Stare, F. J. 1953. An evaluation of the influence of DL-methionine treatment on the serum lipids of adult American males. *New Engl. J. Med.* 249:1018–19

16. Mann, G. V., Watson, P. L., Adams, L. 1952. Primate nutrition. I. The cebus monkey—normal values. *J. Nutr.* 47:213–24

17. Mann, G. V., White, H. S. 1953. The influence of stress upon plasma cholesterol levels. *Metabolism* 2:47–58

18. McGandy, R. B., Hall, B., Ford, C. H., Stare, F. J. 1972. Dietary regulation of blood cholesterol in adolescent males: a pilot study. *Am. J. Clin. Nutr.* 25:61–66

19. McKibbin, J. M., Pope, A., Thayer, S., Ferry, R. J. Jr., Stare, F. J. 1945. Parenteral nutrition. I. Studies on fat emulsions for intravenous alimentation. *J. Lab. Clin. Med.* 30:488–97

20. Natl. Advisory Heart Counc. Tech. Group and Comm. on Lipoproteins and Atherosclerosis. 1956. An evaluation of serum lipoprotein and cholesterol measurements as predictors of clinical complications of atherosclerosis: a report of a cooperative study of lipoproteins and atherosclerosis. *Circulation* 14:691

21. Nicolosi, R. J., Hojnacki, J. S., Llansa, N., Hayes, K. C. 1977. Diet and lipoprotein influence on primate atherosclerosis. *Proc. Soc. Exp. Biol. Med.* 156:1–7

22. Olson, R. E., Deane, H. W. 1949. A physiological and cytochemical study of the kidney and the adrenal cortex during acute choline deficiency in weanling rats. *J. Nutr.* 39:31–55

23. Olson, R. E., Hirsch, E. G., Richards, H., Stare, F. J. 1949. Coenzyme A and citrate formation in homogenates of heart ventricle from normal and pantothenic acid-deficient ducklings. *Arch. Biochem.* 22:480–82

24. Olson, R. E., Kaplan, N. O. 1948. The effect of pantothenic acid deficiency upon the coenzyme A content and the pyruvate utilization of rat and duck tissues. *J. Biol. Chem.* 175:515–29

25. Olson, R. E., Miller, O. N., Topper, Y. J., Stare, F. J. 1948. The effect of vitamin deficiencies upon the metabolism of cardiac muscle *in vitro*. II. The effect of biotin deficiency in ducks with observations on the metabolism of radioactive carbon-labeled succinate. *J. Biol. Chem.* 175:503–14

26. Olson, R. E., Pearson, O. H., Miller, O. N., Stare, F. J. 1948. The effect of vitamin deficiencies upon the metabolism of cardiac muscle in vitro. I. The effect of thiamine deficiency in rats and ducks. *J. Biol. Chem.* 175:489–501

27. Olson, R. E., Schwartz, W. B. 1951. Myocardial metabolism in congestive heart failure. *Medicine* 30:21–41

28. Olson, R. E., Stare, F. J. 1951. The metabolism *in vitro* of cardiac muscle in pantothenic acid deficiency. *J. Biol. Chem.* 190:149–64

29. Phillips, P. H., Stare, F. J., Elvehjem, C. A. 1934. Study of tissue respiration and certain reducing substances in chronic fluoroses and scurvy in the guinea pig. *J. Biol. Chem.* 106:41–61

30. Portman, O. W., Andrus, S. B. 1965. Comparative evaluation of three species of New World monkeys for studies of dietary factors, tissue lipids, and atherogenesis. *J. Nutr.* 87:429–38

31. Roos, A., Hegsted, D. M., Stare, F. J. 1946. Nutritional studies with the duck. IV. The effect of vitamin deficiencies on the course of *P. lophurae* infection in the duck and the chick. *J. Nutr.* 32:473–84

32. Schmitt, Y., Hood, W. B. Jr., Lown, B. 1969. Armchair treatment in the coronary care unit: effect on blood pressure and pulse. *Nurs. Res.* 18:114–18

33. Stare, F. J. 1935. The preparation and nutritional value of hepatoflavin. *J. Biol. Chem.* 111:567–75

34. Stare, F. J. 1936. The effect of fumarate on the respiration of liver and kidney tissue. *Biochem. J.* 30:2257–61

35. Stare, F. J., Baumann, C. A. 1936–37. The effect of fumarate on respiration. *Proc. R. Soc. London Ser. B* 121:338–57

36. Stare, F. J., Elvehjem, C. A. 1932. The phosphorus partition in the blood of rachitic and nonrachitic calves. *J. Biol. Chem.* 97:511–24

37. Stare, F. J., Elvehjem, C. A. 1933. Studies on the respiration of animal tissues. *Am. J. Physiol.* 105:655–64

38. Stare, F. J., Lipton, M. A., Goldinger, J. M. 1941. Biological oxidations, XVIII. Citric acid cycle in pigeon-muscle respiration. *J. Biol. Chem.* 141:981–87

39. Stare, F. J., Trulson, M. F. 1952. *Food for Your Heart: A Manual for Patient and Physician.* New York: Am. Heart Assoc. 51 pp.

40. Stephan, Z. F., Hayes, K. C. 1985. Evidence for distinct precursor pools for biliary cholesterol and primary bile acids in cebus and cynomolgus monkeys. *Lipids* 20:343–49

41. Walker, W. J., Lawry, E. Y., Love, D. E., Mann, G. V., Levine, S. A., et al. 1953. Effect of weight reduction and caloric balance on serum lipoprotein and cholesterol levels. *Am. J. Med.* 14:654–64

42. Whelan, E., Stare, F. J. 1990. Nutrition. *J. Am. Med. Assoc.* 263:2661–63

43. Witschi, J. C., Singer, M., Wu-Lee, M., Stare, F. J. 1978. Family cooperation and effectiveness in a cholesterol-lowering diet. *J. Am. Diet. Assoc.* 72:384–89

44. Wolbach, S. B., Hegsted, D. M. 1952. Hypervitaminosis A and the skeleton of growing chicks. *Am. Med. Assoc. Arch. Pathol.* 54:30–38

Annu. Rev. Nutr. 1990. 10:1–20

NUTRITION SCIENCE FROM VITAMINS TO MOLECULAR BIOLOGY[1]

Thomas H. Jukes

Division of Biophysics and Cell Physiology, Department of Molecular and Cell Biology, and Department of Nutritional Sciences, University of California, Berkeley, California 94720

KEY WORDS: nutrition, vitamins, evolution, molecular biology, environmentalism

---

CONTENTS

Much have I travell'd in the realms of gold
And many goodly states and kingdoms seen

John Keats, "On First Looking Into Chapman's *Homer*"

[1] I dedicate this essay to my wife, Marguerite, our three children, and our seven grandchildren.

0199-9885/90/0715-0001$02.00

INTRODUCTION—EARLY YEARS

Nutrition is an unusual science. Most Americans, and probably most other people, believe they know something about it. They like talking about food, food preferences, and how food affects them. Food fads are common, and they vary from year to year. At present, oat bran, broccoli, and low dietary cholesterol are in the news, supported by segments of the "nutrition establishment." Nutrition, more than any other science, is exploited and used for money-making schemes based on superstition. A walk through a health food store is something like a visit to an intellectual Dante's inferno. Many newspapers have full sections on food but would never think of publishing a section on enzymes. Yet professors who work in departments of nutrition quite often list themselves as biochemists rather than as nutritionists.

The editors invited me to write a review on "Nutrition Science from Vitamins to Molecular Biology." I worked in the vitamin field from 1934 to 1957, during the "discovery period." In 1962, the emergence of the genetic code led me to transfer my main scientific interest to molecular evolution and molecular biology. Also that year, I devoted much time to defending science against environmental activists and later against creationists. During the same period, I was also active in exposing various forms of nutritional quackery.

In such an essay, I could describe how much harder life was in the old days, how science was just great 50 years ago, tell my readers what they should do, or describe what went on behind the scenes in research. I shall touch on all these matters, and I shall try to be entertaining.

I divide my scientific experiences into two periods. The first was the expansive era of 1930 to 1962, when science was the endless frontier and we strove to prevent hunger and disease. The second was the contentious years following 1962, during the rise of the ideas that human beings are intruders in the biosphere, that their growth must be held in check, and that science is harmful because, even at its best, it enables more people to live.

My memories and experiences of the past 60 years started in the golden age of nutrition: the period when the vitamins, the essential trace minerals, and the essential amino acids were identified and the vitamins were synthesized. These discoveries ended much suffering from nutritional deficiency diseases and led to further improvements in the health of people all over the world.

I came to Canada as an immigrant in 1924, at the age of 17, by myself. I hoped to make my way, on my own, in the New World. I started by working on a farm in southwestern Ontario, where the farmer and his family "lived off the land" in the traditional manner. Fuel was obtained by cutting trees in a wood lot; the crops, livestock, and home garden supplied most of the food; there was no electricity, but there was a party-line telephone and a crystal radio set with earphones. Relaxation was rare. Most of the time was spent in hard work: hoeing weeds, pitching hay, sawing wood, husking corn, harvesting grain, and picking tomatoes.

Across the border from Canada beckoned the glamorous city of Detroit, where wages were high for unskilled labor. Night shifts ran for 12 hours but for only five nights a week. It was exhausting work, but easier than farming, and I soon saved enough for my first year in college.

In the fall of 1926, I entered the Ontario Agricultural College, then part of the University of Toronto. After toiling on the farm and in the factories, life as a student was like a rest-cure, punctuated by my return to Detroit each summer to earn money for the next college year. At the end of my junior year, I took a job in a biochemical laboratory on campus, to study the effect of nutrition on the hatchability of hens' eggs. Various crude protein supplements gave different effects on hatchability, and Pollard & Carr (cited in Ref. 43) proposed that this was related to an effect of diet on the amino acid content of eggs. If we had confirmed this hypothesis, we would have contradicted the basis of the then-unknown science of biochemical genetics. Fortunately for science (and us), my boss, W. D. McFarlane, and I found no such effect, but we did record our observation that cod liver meal in the diet increased a yellowish tinge in egg whites, which, of course, was due to the then-unknown riboflavin. These eggs hatched well (43).

I spent the next three years as a graduate student in the Department of Biochemistry at the University of Toronto, studying proteins. With my PhD in hand, and with National Research Council support, I then headed for Berkeley as a postdoctoral fellow in biochemistry. The Depression caught up with me in 1934 in California: my postdoctoral fellowship was suddenly terminated because of lack of funds. I went back into nutrition as an instructor in poultry husbandry at the University of California, Davis, then known as the "University Farm."

I felt that I had been "sent to Siberia," except that Davis was noted for being too hot rather than too cold. I had no budget and no laboratory. I was told that my assignment was to study the calcium and phosphorus requirements of turkeys. This was at a time when most of the nutritional requirements of turkeys, or anything else, were unknown. Of course, turkeys had to have calcium and phosphorus to make bones; in retrospect, I should have written a memorandum saying that 2% of bone meal and 2% of ground limestone should be included in turkey diets, because this was what was actually being used—project completed and now let's get on with something else. Probably I would have been fired. Instead, I used patience and cunning. I put the turkey problem "on the back burner" and wrote a proposal for constructing a purified diet for chicks that would supply all the requirements for growth and reproduction. No one had been able to keep baby chicks alive for more than about 6 weeks on a purified diet, even one that contained up to

40% yeast. I was told that my proposal was unpromising, but somehow I was allowed to work on it, probably because I wasn't given any funds. A long list of vitamins and trace minerals had to be discovered before the objective was finally reached in 1948.

SORTING OUT THE B-COMPLEX VITAMINS: 1935 TO 1948

My first purified diet consisted mainly of starch, casein, and rice-bran extract. The chicks did fairly well for a while, but as a result of an article by Lee Kline and his coworkers at the University of Wisconsin (37) I decided to go in a different direction (41). Kline et al had produced in chicks a distinctive nutritional deficiency, characterized by dermatitis, by feeding a diet of corn, wheat middlings, and casein that had been subjected to prolonged dry heat. Their chicks did not develop polyneuritis, and this showed that the diet was not deficient in vitamin B_1. Instead they grew slowly and had a severe dermatitis. Thiamine had not yet been identified or named, and "vitamin B" was considered to consist of B_1, destroyed by autoclaving, and B_2 or G, a thermostable growth-promoting component, which was thought to be riboflavin. By good luck, I used a supply of acid-washed casein in the diet that was practically free from riboflavin. My chicks on the heated diet developed a double deficiency of riboflavin and of the antidermatitis factor that had been destroyed by dry heat treatment. In 1935, I started to collaborate with Sam Lepkovsky at Berkeley. We found that riboflavin could be removed from a water extract of beef liver by treating it with fuller's earth, and the filtrate contained the chick antidermatitis factor (41). I named the antidermatitis component "filtrate factor." Elvehjem and Koehn, in similar experiments, used the name "chick pellagra" for the dermatitis (7). We announced (42) that so-called "vitamin G" contained two vitamins, and that our experiments showed that a third one existed that was needed for prevention of dermatitis in rats. We called this "factor 1". Paul György named it "vitamin B_6" (12).

By 1936 there were four known members in the "vitamin B complex." A big prize loomed on the horizon—the anti-pellagra vitamin, or "PP-factor." After years of work, Joseph Goldberger and his colleagues had shown that pellagra was indeed a nutritional-deficiency disease (11). Their studies also showed that dogs in the households of families with pellagra developed an analogous nutritional deficiency disease, "black tongue." So dogs could be used in the hunt for the PP factor.

We were not allowed to keep dogs for nutritional research in a poultry department. The nearest approach was watchdogs to guard turkeys. However, Sam Lepkovsky at Berkeley had working relationships with the Lilly Research Laboratories at Indianapolis City Hospital, where pellagra was being studied in human patients.

The end of the search came at Madison, Wisconsin, where, as I have told elsewhere (23), a group of young researchers decided to test nicotinic acid on dogs with black tongue. This took place in the biochemistry department at the University of Wisconsin, where the use of dogs was permitted. Doug Frost has related how in 1936 he bought a small bottle of nicotinic acid from Eastman Kodak Company, with a skull and crossbones and "POISON" in capitals on the label. I purchased some nicotinic acid from Eastman Kodak at about the same time, but my bottle did not have a poison label. Interest in nicotinic acid had come from the discovery by B. C. J. G. Knight in England (38) that it was a growth factor for *Staphylococcus*. Now the Wisconsin group showed that nicotinic acid and nicotinic acid amide cured black tongue in dogs. My bottle of nicotinic acid went to Indianapolis, where Paul Fouts and Oscar Helmer showed that this new vitamin promptly cured pellagra in human patients, 1937 (10). Thus in 1937 we could list five vitamins in the B-complex.

In 1938, crystallization of vitamin B_6 was reported by Lepkovsky (40) at Berkeley and by four other groups elsewhere (12, 14, 33, 39). Vitamin B_6 had been previously crystallized, isolated, and its empirical formula determined by Ohdake in 1932 (45), but he did not recognize it as a vitamin. This suggests the precept that a vitamin is not a vitamin until it has been identified as such. Several substances, including carotene, lactoflavin or lactochrome (riboflavin), nicotinic acid, pyridoxine, hexuronic acid (ascorbic acid), and phthiocol (vitamin K) were known chemically before their nutritional role was discovered. Lepkovsky's isolation of B_6 (factor 1) was aided by using a "filtrate factor" supplement, tested for potency with chicks (41), from which factor 1 had been removed by fullers earth. The filtrate factor preparation probably also contained nicotinic acid amide. I was unable to get any crystalline vitamin B_6 from Lepkovsky, but soon it was synthesized at the Merck Laboratories, and John Keresztesy sent me a small sample. I put together a simplified diet that enabled me to produce uncomplicated vitamin B_6 deficiency in chicks. The deficient birds developed convulsions (16). Later, it was shown that infants on "formula diets" had convulsions as a result of vitamin B_6 deficiency.

In 1939, I found that a substance in yeast prevented turkeys from developing a bone deformity termed "perosis." In 1940, in two short experiments, I identified the substance as choline (17).

By 1939, the race was on for identification of the filtrate factor, or "chick antidermatitis factor." Early in 1939, R. J. Williams, H. Weinstock, and their collaborators at Oregon State College described the concentration of an unidentified substance present in liver and elsewhere that yeast cells needed for growth. They named the substance "pantothenic acid" in reference to its allegedly universal occurrence (65). The properties they recorded seemed like those of the filtrate factor. I wrote to Dr. Williams, and asked him for a sample of pantothenic acid concentrate. It did not have to be pure, as long as its potency was expressed in terms of equivalence to a known weight of whole liver. He kindly sent me a sample of a concentrate prepared from liver. By feeding it to chicks, I found within 43 hours that the filtrate factor was indeed pantothenic acid (15). Simultaneously the same conclusion was reached at the University of Wisconsin (67), and the two announcements appeared in the same issue of the *Journal of the American Chemical Society*. William's group showed that beta-alanine was a component of pantothenic acid. The other half of the molecule was identified by Jake Finkelstein at the Merck Laboratories, who found a description of it in Beilstein's compilation of organic chemicals. It was a lactone, 2,4-dihydroxy-3,3-dimethylbutyric acid-gamma-lactone (60). The Merck group synthesized pantothenic acid in 1940. I collaborated with Sidney Babcock, an organic chemist at the University of California, Davis, who synthesized the lactone and coupled it with beta-alanine in alkaline solution. The chick test showed that the product was the filtrate factor, synthetic pantothenic acid (2).

In the meantime, Esmond Snell and Frank Strong, in a ground-breaking paper, showed that *Lactobacillus casei* could be used as an assay organism for riboflavin (59). Animal tests were much slower and more cumbersome than overnight assays with bacterial cultures, and their procedure revolutionized the hunt for new B-complex vitamins. Esmond Snell described this at length in the 1989 *Annual Review of Nutrition* (58). Snell's work with lactic acid bacteria as assay organisms for growth factors pointed to a new member of the B-complex group—the "yeast norite eluate factor," later named folic acid. Its existence was first shown, in 1931, by Lucy Wills (66) in studies with pregnant women in India. Bob Stokstad pioneered in studies with chicks and called it "factor U" (62). He then switched from the slow chick tests to the use of *Lactobacillus casei* as an assay procedure, and eventually he isolated folic acid from liver (61). Folic acid was identified and synthesized by a large group of organic chemists and biochemists at Lederle Laboratories in 1945 (1). The same feat was performed simultaneously by scientists at Parke Davis & Co. (51).

World War II intervened during these years, but when it ended, I was put in charge of nutrition and physiology research at Lederle Laboratories, Pearl River, NY, a division of American Cyanamid Company. Robert Stokstad and Alfred Franklin joined my group in 1945, and Harry Broquist in 1949. The search started in earnest for the most elusive of the B-vitamins, the anti-pernicious anemia (APA) factor of liver. By this time, it was obvious that vitamins extended their domain over many living species, including bacteria that had become too finicky to live on simple media, rats and chicks on purified diets, and human beings who did not eat the right food. Rats needed a substance called "factor X," or "zoopherin." Chicks needed the "animal protein factor," sometimes called the "cow manure factor." Cattle and sheep developed a wasting disease caused by lack of cobalt in their diet or by living on crops grown on soils deficient in cobalt. An unknown dietary factor was needed by humans; its lack caused pernicious anemia. Patients with this disease needed both "extrinsic factor" and "intrinsic factor." The intrinsic factor was present in normal human gastric juice, but absent from the gastric juice of patients with pernicious anemia. Intrinsic factor enabled extrinsic factor to be absorbed from the intestine. Injection of purified liver extract bypassed the gastric juice-intestinal absorption route.

The big prize to be won was discovering the extrinsic (APA) factor. Patients with pernicious anemia died from the disease unless regularly treated. For hereditary reasons the disease was most common among people in the north temperate countries. It had no cure, and the only reliable treatment was injection with concentrated liver extract, which was not well tolerated by some patients. We produced a bacterial extract that was tolerated.

Eventually, a lactic acid bacterium was found that responded to the factor, but only one or two research groups knew how to do the test. Finally, APA factor was found to be bright red, and chromatography could be used visually to concentrate it.

In 1948, the APA factor was isolated by the Merck Laboratories in New Jersey (52) and the Glaxo Laboratories in England (57). It was named vitamin B_{12}. Everything fell into place; the red color was caused by cobalt. In our lab we found that we could pour commercial liver extract down a silicic acid column, and the red band of the active factor came immediately into view. The disease of ruminants, which resulted from cobalt deficiency, occurred because bacteria in their digestive tracts could not make the vitamin unless cobalt was supplied. Vitamin B_{12} was shown to be the only one of the B-complex vitamins that was not made by green plants. Except for pernicious anemia patients, vitamin B_{12} deficiency in humans is comparatively rare, occurring mainly in vegetarians who do not consume milk or eggs.

Thus ends the first part of my narrative. During the 1930s and 1940s, the fat-soluble vitamins A, D, E, and K, and all the nine water-soluble vitamins were chemically identified, and most of the vitamins were synthesized (24, 53). The golden age of discovery of vitamins had come to an end.

OTHER GROWTH FACTORS

Our group at Lederle continued the search for unidentified growth factors. Kidder & Dewey (34) found that the protozoan *Tetrahymena geleii* needed a substance in liver for growth. Later, members of our group, including John Brockman, Harry Broquist, Milon Bullock, Jack Pierce and Robert Stokstad, among others, showed that the substance was identical with the "acetate factor" and "pyruvate oxidation factor" needed by *Streptococcus faecalis* bacteria, and that its chemical structure, obtained by synthesis, was 6,8-thioctic acid (1,2-dithiolane-3-valeric acid) (4, 48). Seymour Hutner urged us to go after other growth factors needed by protozoa. Ernie Patterson et al identified and synthesized biopterin: 2-amino-4-hydroxy-6(1,2-hydroxy propyl)pteridine, needed for growth of *Crithidia fasciculata* (46).

We thought that a new vitamin might be needed for prevention of alimentary exudative diathesis in chicks fed a diet high in Torula yeast. The factor was present in pork liver. However, the ash of the pork liver was biologically active, and Patterson et al identified it as selenium (47). Almost immediately, selenium deficiency in livestock was then reported in many parts of the world. Prior to this, selenium was universally regarded as a toxic element without nutritional value.

ANTIBIOTICS AS FEED ADDITIVES

Introducing the antibiotic aureomycin (chlortetracycline) into animal feeds was very exciting and a great deal of fun. An article by Moore, Luckey, Elvehjem & Hart of University of Wisconsin in 1946 (44) described growth promotion in chicks by streptomycin and growth depression by streptothricin. This article was largely ignored; indeed, Elvehjem wrote a review a year later (6) on effects of intestinal bacteria on growth without mentioning the 1946 publication, of which he was a coauthor! Our finding, in 1949, was that addition of a crude aureomycin culture to the diet promoted growth of chicks in excess of that obtained with a complete diet. We distributed samples of aureomycin fermentation residues to many research groups that were working with farm animals; large growth responses were obtained, especially in animals with diarrhea. A strong commercial demand developed for the product. The year 1949 was a truly remarkable adventure for us, because of the severity of bloody diarrhea in pigs in the middle west of the United States, including the herds of the agricultural experiment stations of Illinois, Ohio, Florida, and Hormel Institute, Minnesota. The disease was rapidly cured by adding small amounts of aureomycin fermentation residues in the feed. An entrepreneur bought runt pigs in Iowa and brought them to normal size. A druggist in Minnesota prospered greatly when he purchased aureomycin residues from Lederle and repackaged them for sale to pig growers. Senator Wherry of Nebraska complained that an undue proportion of the available supply of the aureomycin supplement ("Aurofac") was going to Iowa (25). The next step was the finding by Bob White-Stevens that higher levels of aureomycin, up to 200 grams per ton of feed, prevented the common ailment air-sac disease, which was usually attributed to *Mycoplasma* infections in broiler chickens (64). Soon antibiotics, especially aureomycin (chlortetracycline), oxytetracycline, and penicillin became standard additives in animal feeds. These antibiotics are still in use for this purpose, despite numerous objections based on claims that low-level feeding of antibiotics to animals produces a spread of transferable resistance, passing from nonpathogenic to pathogenic bacteria. This issue has been debated for about 20 years without resolution. The most recent report on it was in 1989 by a committee appointed by the National Institute of Medicine (5). This voluminous report stated that "the committee was unable to find a substantial body of direct evidence that established the existence of a definite health hazard in the use of subtherapeutic concentrations of penicillin and the tetracyclines in animal feeds."

The antibiotic growth effect is produced by action on intestinal bacteria. This fact has been established in various ways, the most definite of which is that "germ-free animals" grow faster than "contaminated" (i.e. normal) controls, and that these controls show a growth response to antibiotics (9) but the germ-free animals do not. The growth effect occurs simultaneously with an increase in the count of both total and resistant intestinal bacteria. A sparing effect on the requirement for vitamins occurs on diets that are deficient in the vitamin being studied. There is also a sparing effect on the mineral requirement for some minerals (25). Therefore, antibiotic feeding has a definite relationship to nutrition. These two effects are not seen under practical conditions, for under these circumstances, diets deficient in vitamins and minerals are not fed. Instead, the level of "disease" is lowered by antibiotics, and animals can be reared more successfully in confinement. The proportion of total antibiotic production used in animal feeds was about 31% (9.9 million pounds) in 1983 (5).

METHOTREXATE, LEUCOVORIN AND CANCER

A biochemical adventure started for us in 1946, because of the synthesis of folic acid. The first truly successful chemotherapeutic drug was sulfanilamide, and its effect was due to blocking the synthesis of folic acid in bacteria. Unlike bacteria, animals need preformed, dietary folic acid, so sulfanilamide is without effect on animals. The synthesis of folic acid does not take place in animal cells and therefore cannot be blocked. The possibility arose that folic acid antagonists would block the utilization of folic acid rather than its synthesis. Perhaps, therefore, a chemical antagonist of folic acid could be used to stop the formation of DNA and hence the unwanted proliferation of cells, such as occurs in neoplastic diseases. Our colleagues in organic chemistry made a crude folic acid antagonist by adding a methyl group to one of the compounds used in its chemical preparation. We fed this crude compound, "X-methyl folic acid", to rats, and we found that it reduced their white blood cell count to very low levels; indeed, the granulocyte count went to zero.

Addition of large amounts of folic acid to the diet prevented all the changes. We thought we might have opened a door to the chemotherapy of leukemia, by the administration of a compound whose inhibitory effects on the proliferation of white blood cells could be reversed, if necessary.

Our hopes were short-lived. X-methyl folic acid was ineffective in patients with leukemia. This disappointment soon vanished when another folic acid antagonist, 4-amino folic acid, termed *aminopterin,* was synthesized in 1947 by Seeger, Smith & Hultquist (56). We found that this compound was intensely toxic to mice and killed them within a week or two when fed at only one part per million of the diet. In 1948, Dr. Sidney Farber at Children's Hospital in Boston found that aminopterin produced temporary remission in leukemia in young children. However, its high toxicity, which results from blocking dihydrofolic reductase, made it difficult to use in chemotherapy on a sustained basis. To make aminopterin safe for clinical use, we needed a compound that would block its action, just as folic acid blocked the action of X-methyl folic acid. A lead came from publications by Sauberlich & Baumann (54, 55), who reported that liver contained a substance, needed by the microorganism *Leuconostoc citrovorum.* The substance was evidently a chemical relative of folic acid because slow growth could be obtained by adding large amounts of folic acid to the culture. Sauberlich showed that the "citrovorum factor," but not folic acid, reversed the toxic effects of aminopterin on the test microorganism. Our research group, led by John Brockman and Harry Broquist, synthesized citrovorum factor as 5-formyltetrahydrofolic acid (3). We named it *leucovorin,* and it is used clinically in the procedure termed *leucovorin rescue,* in which methotrexate (a close relative of aminopterin, but somewhat less toxic) is used to inhibit the growth of cancer cells, and the danger to the patient caused by the toxicity of methotrexate is avoided by administering leucovorin at appropriate times.

Just as sulfanilamide opened the door to chemotherapy against pathogenic bacteria, so did the folic acid antagonists make a limited beginning in the chemotherapy of certain forms of cancer; notably childhood leukemia and choriocarcinoma in women. I have described the participation of J. M. Smith, D. Seeger, D. Cosulich, and M. E. Hultquist, organic chemists, and A. L. Franklin, M. Belt, and E. L. R. Stokstad in these studies (30). Methotrexate is now used also in the treatment of noncancerous diseases such as psoriasis and rheumatoid arthritis.

MOLECULAR EVOLUTION: THE GENETIC CODE: THE NEUTRAL THEORY

In 1962, I heard two lectures that described the beginnings of solving the genetic code. The science of biology was about to be transformed by molecular biology, and, for the same reason, evolution would be transformed into molecular evolution. I decided to enter a new field, so I resigned my job and returned in 1963 to the University of California, at Berkeley, to become Professor of Biophysics and principal investigator on a grant from NASA, "The Chemistry of Living Systems." NASA was interested in the origin of life and hence was willing to support fundamental studies of molecular biology and molecular evolution. This was in the early years after Sputnik, when NASA was well funded to support research. I hired several young molecular biologists, I put laboratories in an old building that the university owned, and I proceeded to write a book called *Molecules and Evolution* (21). In this book, I discussed the fact that cytochrome *c* in different species has different amino acid sequences, although its function is always the same. I asked if the differences in the amino acid sequences were needed by the individual species, or whether the molecule had diverged because the species, such as dogs and horses, had separated during evolution, and the changes had been "carried along" during the separation because they were the result of random mutations. The changes produced in proteins by mutations will in some cases destroy their essential functions, but in other cases the change allows the protein molecule to continue to serve its purpose. This was my first statement of the "neutral theory" of molecular evolution, 1966.

Earlier, N. Sueoka (63) in 1961 had shown, before the genetic code was known, that the nucleotide composition of DNA in bacteria varied over a wide range in different species. High A+T content of DNA in bacteria was correlated with increases in isoleucine, lysine, phenylalanine, and tyrosine in total proteins, and high G+T with increases in alanine, arginine, and glycine. When the code was solved, it showed that these increases corresponded to the composition of the codons; for example, UAU and UAC both are codons for tyrosine. A codon is the sequence of three nucleotides in RNA (where U replaces T in DNA) that specifies an amino acid in a protein molecule. Evidently, the amino acid composition of bacterial proteins could change without loss of viability. In 1965 (20), I pointed out that changing all the silent positions in codons, from A and U to G and C, for example, changing UAU to UAC, would change the base composition of the genes in DNA from 40% (high AT) to 73% (high GC) without any change in the amino acid content of the organism. This was even more striking evidence for neutral changes than Sueoka (63) had been able to obtain before the code was known.

In 1968, I collaborated with Jack King, a population geneticist, in further exploration of this possibility, and we published a long article in *Science,* "Non-Darwinian Evolution" (36). The idea behind the title was that Darwinian evolution was based on natural selection, implying that all evolutionary changes were adaptive. We pointed out that this was not necessarily always the case: The same proteins, such as cytochrome or hemoglobin, in different species were different according to the length of time the species had been separated, because molecular changes, without changing function, had become fixed by random drift. Our manuscript was rejected by *Science;* one referee said that the idea was so obvious that it did not need to be published, and the other referee said that the idea was completely wrong. We appealed this decision, a step that is not possible to take with *Science* today, and we won the appeal so that our article was published in 1969. In 1977, we received a "citation award" because so many references had appeared to our article. (A few months later Jack King tragically and suddenly died of leukemia.) The same idea of neutral changes was published by Motoo Kimura of Mishima, Japan in a short note in *Nature* in 1968 (35). Our 1969 publication was vigorously attacked by classical evolutionists, many of whom insisted that every change in proteins had an adaptive value. Kimura proceeded to support the neutral theory by means of mathematical studies based on population genetics, and in 1983 he wrote a textbook, *The Neutral Theory of Molecular Evolution.*

Determination of the sequences of codons in DNA molecules provided additional support for the neutral theory as I had predicted in 1965. From 1966 to 1989 I published various articles on evolution of the genetic code. It is now evident that the code itself is evolving, and this is shown by differences between the code in cells and the code in their mitochondria, also by departures from the usual code in certain bacteria and ciliated protozoa.

EVOLUTION AND VITAMIN REQUIREMENT

Jack King and I used the neutral theory in explaining a nutritional phenomenon. The ability to synthesize ascorbic acid exists in most terrestrial vertebrates. However, scurvy occurs because human beings have lost this ability. So have other members of Anthropoidea, also guinea-pigs, fruit-eating bats, and certain passerine birds. Obviously the loss of ASA (ascorbic acid synthesizing ability) could not be tolerated in a species unless its diet was adequately supplied with ascorbic acid. The guinea-pig, for example, eats green leaves and does not need to synthesize ascorbic acid. But goats also eat green leaves, and they are able to synthesize ascorbic acid. Among the birds, why did the red-vented bulbul, but not the omnivorous mynah bird, stop making its own vitamin C?

Our proposal was that species without ASA lost this ability by a neutral evolutionary change that occurred sporadically by mutation. The change was adopted by genetic drift in the DNA of a few groups of birds and mammals that are widely scattered in phylogeny (31).

In addition to the fairly recent loss of ascorbic-acid synthesizing ability by some species, there is the fact that many animals, including mammals, depend on food or microbial sources for other vitamins (except vitamin D) and for eight to ten amino acids. This dietary dependence probably means that during evolution, animals lost the metabolic pathways for synthesizing these substances. Here is a field for consideration by nutritionists. Were these losses neutral evolutionary changes? Why are some amino acids made by animals while others must be supplied in the diet? Are there some cases in which animals never had the necessary anabolic pathway, such as that of vitamin B_{12} which is not made by green plants? Vitamin A, retinal, is derived from carotene, which probably originated in cyanobacteria as an antioxidant to protect against oxygen produced in photosynthesis. Carotene is now recommended by some in nutrition for the same purpose: to detoxify free radicals produced by oxidation. We seem to be mimicking our remote ancestors who appeared 3.5 billion years ago.

DEFENSE OF PESTICIDES

In June 1962, *The New Yorker* published "Silent Spring" by Rachel Carson. This article was mainly an attack on DDT, which had saved more lives and prevented more illnesses than any single chemical in history. This fact was not mentioned by Rachel Carson. I saw that I had to take the side of the people of the Third World against American environmentalists and bird-watchers. I promptly wrote a short manuscript: "A Town in Harmony," which described the privations and dangers of life in the days before the control of pests. *The New Yorker* didn't even acknowledge receipt of it, but it was published in *Chemical Week* (18). I then wrote one of my favorite articles "People and Pesticides," an account of the effects of DDT, for *American Scientist,* the Sigma Xi journal (19). One member of Sigma Xi cancelled her subscription immediately in protest, and officials of the Audubon Society and the American Museum of Natural History retorted with a letter containing long, blistering attacks. These attacks were based on allegations of the effects of insecticides on birds and ignored the information I had presented on control of malaria and other vector-borne diseases. I also received many letters praising the article and asking for bulk supplies of reprints. My defense of pesticides and the attacks on me because of it were to be repeated for many years, but "People and Pesticides" pretty much summed up the whole story. In it, I drew on some of the Audubon bird counts that showed birds counted per observer had increased after DDT was introduced. The defense of DDT occupied me for 13 years, and I was joined in it by Bob White-Stevens, Gordon Edwards, and Norman Borlaug.

In 1972, *The New York Times* (NYT) published an article saying that five people—Gordon Edwards, Bob White-Stevens, myself, Norman Borlaug, and Donald Spencer—were paid liars acting on behalf of the pesticide industry. The first three of us filed a libel suit against Audubon and NYT. We won a jury verdict and an award of damages in a New York District Court. NYT appealed, and Judge Irving Kaufmann, who was a long-standing friend of *The New York Times,* presided over the appeal at his own request. He wrote an opinion reversing the lower court, saying, in effect, that although we had been unjustifiably slandered, we would have to put up with it so as to protect the freedom of the press. Ironically, NYT had refused to print my immediate rebuttal of the original "paid liars" article. Evidently "freedom of the press" is intended for publishers rather than those who are slandered. The US Supreme Court would not hear our appeal. The episode was a fascinating insight into the workings of the judiciary.

Despite repeated pleas of the World Health Organization (WHO), DDT was banned in 1972 by William Ruckelshaus, head of the Environmental Protection Agency, following long hearings. He gave no indication that he read the proceedings of the hearings, and his ban overturned the recommendations of his own Hearing Examiner. At the request of the editor of *Die Naturwissenschaften* (22) I wrote a long article describing the campaigns against pesticides and rebutting the anecdotal claims that pesticides were injurious to wildlife. Pesticides are essential to the production of enough food.

THE RISE OF ENVIRONMENTALISM

Starting in the 1960s, the public perception of science largely changed from acceptance to rejection. According to H. Fairlie, "the origins of the widespread refusal to accept . . . risk as a normal and necessary hazard of life began in the early 1970's . . . beyond this has been the growth of the larger belief that science . . . promises evil and not beneficence" (8). This change, influenced by the environmental movement, was so great that most younger people take it for granted, and only a dwindling group of their elders can recall a day when science was the "endless frontier," pushing back what we regarded as the bad old days. The phrase "contains no additives" is used on food labels as a sales aid. The media treat us to regular scares, and quite often pesticide residues are the target, even though their effects are trivial (22).

Nevertheless, in the United States, life expectancy steadily increases, year after year.

DIETHYLSTILBESTROL IN BEEF PRODUCTION

I had a long involvement with the ban on diethylstibestrol (DES) in beef production. Low levels of supplementary estrogens (rather unexpectedly) increase the production of lean meat by beef cattle. DES is a cheap synthetic estrogen that has this effect on an animal when a small pellet is implanted in its ear, which is discarded at slaughter, or when low levels are fed and then withdrawn.

This use of DES was approved on the basis of absence of residues in meat, but the FDA soon found itself "chasing a receding zero" (28) as methods of assay became more sensitive. DES became strongly politicized in 1971 when A. L. Herbst and co-workers in Boston reported (13) that a small number of young women developed cancer attributed to large doses of DES given to their mothers during pregnancy as an attempted but ineffective medication to prevent miscarriages. The wrath of Senator Edward Kennedy fell, not upon the clinical use of DES, but upon the luckless cattle growers, and he was helped by several prominent scientists including Roy Hertz, who said that one molecule of DES consumed in a portion of beef liver might be enough to start the cancerous process. Hertz did not point out that the daily endogenous production of estrogens in women is about 6 times 10^{17} molecules. The average amount of DES given to women during pregnancy was the same as the amount present in 122,500 tons of beef liver containing 0.1 ppb of DES, more than the total annual production of beef liver in the United States. I estimated the average daily intake of DES, resulting from its use in beef production, as between 4 and 5 nanograms. The total for a lifetime of 90 years would be about 0.16 mg. DES is metabolized, is not cumulative, and the clinical dose is several milligrams *daily*. These facts and figures led me to participate in the defense of the use of DES in beef production. The US Food and Drug Administration (FDA) Commissioner, A. M. Schmidt, MD, was against a ban, but he was succeeded by Donald Kennedy, PhD, who placed high priority on the ban. While the hearings were still in progress, Kennedy announced that the ban would take place (sentence first, verdict afterwards), and it went into effect in November 1979. In 1980, Mr. and Mrs. Jud Lackey were brought into court in Wichita, Kansas, accused by the FDA of implanting their cattle with DES a few days after the ban. During the trial, in which I was a witness, FDA witnesses alleged that 1 part per trillion of DES in beef was a finite carcinogenic hazard, but Judge Patrick Kelly was deaf to their blandishments (29). The FDA withdrew its appeal; the disputed beef was liberated from custody, and a barbecue party featuring the said beef was given by the Lackeys in Kansas for its defenders. Of course, the FDA did not rescind the ban.

This experience was my only court appearance against the FDA. I was witness for the FDA and other governmental prosecutors in several antiquackery trials, starting with laetrile.

QUACK REMEDIES FOR CANCER

My involvement with laetrile, a name under which the cyanogenic glycoside, amygdalin, was illicitly peddled for treating cancer, started when I wrote a short note for the *Journal of the American Medical Association* (26), rebutting the therapeutic claims for laetrile. Its use was promoted by Ernst Krebs, Jr., on the mythical basis that amygdalin would be broken down by an enzyme in cancerous tissue to liberate cyanide which would "kill" the cancer. This nonsense was exposed by David Greenberg, who showed there were only traces of beta-glucosidase in animal tissues and even less in experimental tumors. Even if cyanide were liberated from amygdalin, it would diffuse rapidly and poison surrounding normal tissues.

Krebs then alleged that laetrile was "vitamin B_{17}," although it had not the slightest resemblance to a vitamin; the crucial property of a vitamin is that its absence from the diet produces a specific deficiency disease in vertebrate animals. I testified to this effect in several court trials, all of which ended in victories for the prosecution. One was in San Francisco, where Krebs, the originator of the name laetrile, received a 6-months jail sentence, which he eventually served. Preparation for these trials involved much work in which the lawyers prepared me for the courtroom, and in turn, I lectured them about nutrition. I found that I was "on the list" as a witness for topics other than laetrile. Another fable concocted by Ernst Krebs with the support of Dean Burk was the claim that a hypothetical "pangamic acid" was "vitamin B_{15}". This claim was gladly accepted by many "health food" purveyors. Even the name "pangamic acid" was spurious; the actual product was usually a mixture of betaine and glucuronic acid.

In general, legal action against quack remedies is a protracted process. By the time a court decision is reached, the defendants have moved on to other, greener pastures.

Vitamin C came into prominence as a universal remedy. Bogus claims for vitamin C were made as a result of its hyperenthusiastic promotion by Dr. Linus Pauling (49), and I appeared as a rebuttal witness against him in San Francisco and Santa Rosa, California.

Linus Pauling published the following account of a discussion with me (50):

> I am reminded of an experience I had in 1984 on a radio medical program (on station KQED) in San Francisco. There was another guest on the program, a retired professor of nutrition from the University of California in Berkeley. I made a statement about the value of a high intake of vitamin C . . . The retired professor of nutrition said simply, "No one needs more than 60 mg of vitamin C per day," without giving any evidence to support his flat statement. I then presented some more evidence for my large intake, and he responded by saying, "Sixty mg of vitamin C per day is adequate for any person." After I had presented some more evidence, this retired professor said, "For fifty years I and other leading authorities in nutrition have been saying that 60 mg of vitamin C per day is all that any person needs!" There was just time enough left on the live radio program for me to say. "Yes—that's just the trouble: you are fifty years behind the times."

What actually happened is described on the official audiotape of the broadcast, May 7, 1984. (LP = Linus Pauling, DW = moderator, TJ = T. Jukes).

LP: Well, I failed to mention that if I am traveling and people are sneezing in my face and [I] get pretty tired and think that I may be coming down with a cold I go up to as much as 50 grams of Vitamin C in a day and that stops it.
DW: Do you think that is recommended for all people, Dr. Jukes, to take that much Vitamin C?
TJ: Well, I certainly don't.
DW: Based upon your previous observations.
TJ: Based upon my previous observations and my studies of literature and my conversations with vitamin experts for about the last 50 years.
DW: All right.
LP: You know it's too bad that these vitamin experts have been making the same statements for the last 50 years and haven't caught up with the times yet.
TJ: Yes, but their statements are based on controlled observations.

This shows a lack of agreement between what Pauling reported and what actually happened.

MEGAVITAMINS

The Food and Drug Administration proposed that an upper limit should be set on over-the-counter vitamins of 150% of Recommended Daily Allowances (RDAs) per tablet. This modest proposal was seen as a challenge by the megavitamin industry. The National Health Federation organized a letter-writing campaign to Congress for the support of the "Proxmire Bill" that specifically prevented FDA from making such a regulation. The Proxmire Bill was opposed by the American Institute of Nutrition and other scientific groups. It was supported by *Prevention* magazine, the National Health Federation, Linus Pauling, and Roger Williams. The Proxmire Bill passed the Senate by a vote of 81 to 10 in September 1974.

Another successful nutritional deception is the "organic food" industry. The term *organic* to describe food and a particular farming style was originated by the late Jerome Rodale, an electrical contractor who moved from New York City to Pennsylvania and decided to become an expert on farming and health. His ideas feature the promotion of garlic, dolomite, fertilized eggs, sunflower seeds, honey, and dried seaweed. His son, Robert Rodale, offered the following definition at a public hearing on "organic foods" in New York City in 1972: "Organically grown food is food grown without pesticides, grown without artificial fertilizers, grown in soil whose humus content is increased by the additions of organic matter, grown in soil whose mineral content is increased with applications of natural mineral fertilizers, has not been treated with preservatives, hormones, antibiotics, etc."

By incessant repetition, and conspicuous publicity in the media, the "organic" concept of farming has gained headway. It has been helped by unfounded fears of pesticide residues in food.

CREATIONISM AS AN ENEMY OF SCIENTIFIC EDUCATION

The creationists are a religious sect of biblical fundamentalists. They assert that any scientific conclusions that differ from the literal reading of the Book of Genesis are in error. Creationists distort and pervert science by fabricating "creation science," in which the age of the Earth (and the Universe) is placed at 10,000 years, all species of life were created instantaneously, the Great Flood of Noah was an actual event, fossils of animals are the remains of creatures drowned in the Great Flood, and nations and languages originated at the Tower of Babel. They lobby incessantly for the teaching of "creation science" on an equal-time basis with science in schools. Creationists harass schoolteachers who present subject matter on evolution in public school science classes. They intimidate publishers into deleting mention of evolution from school science textbooks.

Their main objective is to deny and discredit evolution as evil, but they extend this objective to include condemnation of geology, physics, and astronomy. By incessant Bible-quoting and appeals to "fair play," and aided by the mass media, creationists have succeeded in getting support from a substantial proportion of the US population. I have spent a lot of time during the past 20 years exposing the creationist attacks on science. Currently, creationists have been successful in obtaining revisions in the new Science Framework for California Schools. Creationism is an intolerable burden on the scientific community.

In 1982, I organized and chaired a symposium, "The Creationist Attack on Science" for the American Society of Biological Chemists at the Federation Meetings in New Orleans. One of the speakers was Julian Bartlett, D. D., Dean of Grace Cathedral, San Francisco. It was the first and only time that a clergyman addressed the Society on a religious topic (27).

SUMMARY

Nutrition is a science of great importance. Indeed, unless one studies it, one might be afraid to eat anything, in view of all the scare stories about food. Today is the age of molecular biology and above all of DNA. The human genome project will bring new understanding of genetic diseases, and many of these will be "inborn errors of metabolism," in which nutrition has an important role. The challenge of nutrition is to help provide a healthy diet for all the world.

ACKNOWLEDGMENT

I thank Carol Fegte for helping me prepare the manuscript.

Literature Cited

1. Angier, R. B., Boothe, J. H., Hutchings, B. L., Mowat, J. H., Semb. J., et al. 1946. The structure and synthesis of the liver L. casei factor. *Science* 103:667–69
2. Babcock, S. H. Jr., Jukes, T. H. 1940. The biological activity of synthetic pantothenic acid. *J. Am. Chem. Soc.* 62:1628
3. Brockman, J. A., Jr., Roth, B., Broquist, H. P., Hultquist, M. E., Smith, J. M., Jr. 1950. *J. Am. Chem. Soc.* 72:4325
4. Brockman, J. A., Stokstad, E. L. R., Patterson, E. L., Pierce, J. V., Macchi, M. E. 1954. Proposed structures for protogen-A and protogen-B. *J. Am. Chem. Soc.* 76:1827
5. Committee on Human Health: Risk Assessment of Using Subtherapeutic Antibiotics in Animal Feeds. 1989. *Human Health Risks with the Subtherapeutic Use of Penicillin or Tetracyclines in Animal Feeds.* Washington, DC: Natl. Acad. Press. 216 pp.
6. Elvehjem, C. A. 1948. *Fed. Proc.* 7:410
7. Elvehjem, C. A., Koehn, C. J., Jr. 1935. *J. Biol. Chem.* 108:709
8. Fairlie, H. 1989. Fear of living: America's morbid aversion to risk. (Forum) *Sacramento Bee*, April 23, 1989
9. Forbes, M., Supplee, W. C., Combs, G. F. 1958. Response of germ free and conventionally reared turkey poults to dietary supplementation with penicillin and oleandomycin. *Proc. Soc. Exp. Biol. Med.* 99:110
10. Fouts, P. J., Helmer, O. M., Lepkovsky, S., Jukes, T. H. 1937. Treatment of human pellagra with nicotinic acid. *Proc. Soc. Exp. Biol. Med.* 37:405–7
11. J. Goldberger, Wheeler, G. A., Lillie, R. D., Rogers, L. M. 1928. *US Pub. Health Serv. Pub. Health Reports* 43:1385
12. Gyorgy, P. 1938. *J. Am. Chem. Soc.* 60:983
13. Herbst, A. L., Ulfelder, H., Poskanzer, D. C. 1971. Adenocarcinoma of the vagina: Association of maternal stilbestrol therapy with tumor appearance in young women. *N. Engl. J. Med.* 248:878
14. Itiba, A., Miti, K. 1938. *Sci. Papers Inst. Phys. Chem. Res.* (Tokyo) 34:623
15. Jukes, T. H. 1939. *J. Am. Chem. Soc.* 61:975
16. Jukes, T. H. 1939. Vitamin B_6 deficiency in chicks. *Proc. Soc. Exp. Biol. Med.* 42:180–82
17. Jukes, T. H. 1940. Prevention of perosis by choline. *J. Biol. Chem.* 134:789–90
18. Jukes, T. H. 1962. A town in harmony. *Chem. Wk*, p. 5. August 18, 1962
19. Jukes, T. H. 1963. People and pesticides. *Am. Sci.* 51:355–61
20. Jukes, T. H. 1965. The genetic code, II. *Am. Sci.* 53:477–487
21. Jukes, T. H. 1966. *Molecules and Evolution.* New York: Columbia Univ. Press. 285 pp.
22. Jukes, T. H. 1974. Insecticides in health, agriculture and the environment. *Die Naturwissenschaften* 61:6–16
23. Jukes, T. H. 1974. Dilworth Wayne Woolley (Biographical Sketch). *J. Nutrition* 104:509–11
24. Jukes, T. H. 1977. Adventures with vitamins. *In Discovery Processes in Modern Biology*, ed. W. R. Klemm, pp. 152–70. New York: Krieger
25. Jukes, T. H. 1977. The history of the "antibiotic growth effect." *Fed. Proc.* 37:11, 2514
26. Jukes, T. H. 1979. Laetrile on trial. *J. Am. Med. Assoc.* 242:719–20
27. Jukes, T. H. 1983. The creationist attack on science. Introduction: The magnitude of the threat. *Fed. Proc.* 42:3022–4
28. Jukes, T. H. 1983. Chasing a receding zero: Impact of the zero threshold concept on actions of regulatory officials. *J. Am. Coll. Toxicol.* 2:147–160
29. Jukes, T. H. 1987. Carcinogenicity of female sex hormones, especially as related to use in pregnancy and in meat production. *In CRC Handbook of Endocrinology*, vol. 2 Part B, ed. G. H. Gass, H. M. Kaplan, pp. 143–58. Boca Raton: CRC Press
30. Jukes, T. H. 1987. Searching for magic bullets: Early approaches to chemotherapy—Antifolates, methotrexate—The Bruce F. Cain Memorial Award Lecture. *Cancer Res.* 47:5528–36
31. Jukes, T. H., King, J. L. 1975. Evolutionary loss of ascorbic acid synthesizing ability. *J. Human Evol.* 4:85–8
32. Deleted in proof
33. Keresztesy, J. C., Stevens, J. R. 1980. *Proc. Soc. Exptl. Biol. Med.* 38:64
34. Kidder, G. W., Dewey, V. 1944. *Biol. Bull.* 87:121
35. Kimura, M. 1968. *Nature* 217:624
36. King, J. L., Jukes, T. H. 1969. Non-Darwinian evolution. *Science* 164:788–98
37. Kline, O. L., Keenan, J. A., Elvehjem, C. A., Hart, E. B. 1932–1933. *J. Biol. Chem.* 99:295

38. Knight, B. C. J. G. 1937. *Biochem. J.* 31:731
39. Kuhn, R., G. Wendt. 1938. *Ber.* 71: 780, 1118
40. Lepkovsky, S. 1938. *Science* 87:169
41. Lepkovsky, S., Jukes, T. H. 1936. The effect of some reagents on the "filtrate factor" (a water-soluble vitamin belonging to the vitamin B complex and preventing a dietary dermatitis in chicks). *J. Biol. Chem.* 114:109–16
42. Lepkovsky, S., Jukes, T. H., Krause, M. E. 1936. The multiple nature of the third factor of the vitamin B complex. *J. Biol. Chem.* 115:557–66
43. McFarlane, W. D., Fulmer, H. L., Jukes, T. H. 1930. Studies in embryonic mortality in the chick. 1. The effect of diet upon the nitrogen, amino-nitrogen, tyrosine, tryptophan, cystine, and iron content of the proteins and on the total copper of the hen's egg. *Biochem. J.* 24:1611–31
44. Moore, P. R., A. Evenson, T. D. Luckey, E. McCoy, et al. 1946. *J. Biol. Chem.* 165:437
45. Ohdake, S. *Bull. Agr. Chem. Soc. Japan* 1932. 8:111
46. Patterson, E. L., Milstrey, R., Stokstad, E. L. R. 1956. The synthesis of a pteridine required for growth of *Crithidia fasciculata*. *J. Am. Chem. Soc.* 78: 5868–5871
47. Patterson, E. L., Milstrey, R., Stokstad, E. L. R. 1957. Effect of selenium in preventing exudative diathesis in chicks. *Proc. Soc. Exp. Biol. Med.* 95:617–20
48. Patterson, E. L., Pierce, J. V., Stokstad, E. L. R., Hoffman, C. E. et al. 1954. The isolation of protogen. *J. Am. Chem. Soc.* 76:1823
49. Pauling, L. 1970. *Vitamin C and the Common Cold*. San Francisco: W. H. Freeman
50. Pauling, L. *How to Live Longer and Feel Better*. 1986. New York: W. H. Freeman
51. Pfiffner, J. J., Calkins, D. G., Bloom, E. S., O'Dell, B. L. 1946. On the peptide nature of vitamin Bc conjugate from yeast. *J. Am. Chem. Soc.* 68:1392
52. Rickes, E. L., Brink, N. G., Koniuszy, F. R., Wood, T. R., Folkers, K. 1948. *Science* 107:396–97
53. Rosenberg, H. R. 1942. *Chemistry and Physiology of the Vitamins*. New York: Intersci. 674 pp.
54. Sauberlich, H. E., Baumann, C. A. 1948. A factor required for the growth of Leuconostoc citrovorum. *J. Biol. Chem.* 176:165–73
55. Sauberlich, H. E. 1949. Relationship of purine, folic acid, and other principles in the nutrition of Leuconostoc citrovorum 8081. *Fed. Proc.* 8:247
56. Seeger, D. R., Smith, J. M., Jr., Hultquist, M. E. 1947. Antagonist for pteroylglutamic acid. *J. Am. Chem. Soc.* 69:2567
57. Smith, E. L. 1948. *Nature* 161:638–39
58. Snell, E. E. 1989. Nutrition research with lactic acid bacteria: A retrospective view. *Annu. Rev. Nutr.* 9:1–19
59. Snell, E. E., Strong, F. M. 1939. A microbiological assay for riboflavin. *Ind. Eng. Chem. Anal. Ed.* 11:346–50
60. Stiller, E. T., Harris, S. A., Finkelstein, J., Keresztesy, J. C., Folkers, K. 1940. *J. Am. Chem. Soc.* 62:1785
61. Stokstad, E. L. R. 1979. Early work with folic acid. *Fed. Proc.* 38:2696–2698
62. Stokstad, E. L. R., Manning, P. D. V. 1938. *J. Biol. Chem.* 125:687
63. Sueoka, N. 1961. Correlation between base composition of deoxyribonucleic acid and amino acid composition of protein. *Proc. Natl. Acad. Sci. USA* 47: 1141–49
64. White-Stevens, R. H., Zeibel, H. G., Walker, N. E. 1956. *Cereal Sci.* 1:101
65. Williams, R. J., Weinstock, H. H., Rohrmann, E., Truesdail, J. H., Mitchell, H. K., Meyer, C. E. 1939. *J. Am. Chem. Soc.* 61:454
66. Wills, L. 1931. Treatment of "pernicious anaemia of pregnancy" and "tropical anaemia," with special reference to yeast extract as curative agent. *Br. Med. J.* 1:1059–1064
67. Woolley, D. W., Waisman, H. A., Elvehjem, C. A. 1939. *J. Am. Chem. Soc.* 61:977

Esmond E. Snell

Annu. Rev. Nutr. 1989. 9:1–19

NUTRITION RESEARCH WITH LACTIC ACID BACTERIA: A Retrospective View[1]

Esmond E. Snell

Departments of Microbiology and Chemistry, The University of Texas, Austin, Texas 78712

CONTENTS

INTRODUCTION

From 1935 to 1955 an almost explosive growth occurred in knowledge of the nutritional requirements of animals and bacteria. Results in each of these

---

[1]This chapter cites only a few original articles from the relevant literature. It emphasizes work on the lactic acid bacteria during the period between about 1936 and 1952 and is an extension of previous accounts of this subject (75, 79). Reviews that cover nutritional requirements of both these and other microorganisms in greater detail are available (16, 18, 55, 77, 34). Reviews dealing extensively with development and use of microbiologic assays are also available (1, 3, 77). A useful anthology of papers in bacterial nutrition has appeared (35).

0199-9885/89/0715-0001$02.00

fields contributed greatly to understanding the other. However, the extent to which studies of microbial nutrition contributed to recognition of the identity, distribution, and function of vitamins and related nutrients is not widely known today, and these are the contributions I emphasize here.

Today's biochemists have difficulty appreciating the many hurdles associated with the initial definition of the trace nutrients required by either animals or microorganisms. Imagine a time when the basis for nutritional requirements was unknown and their complexity unsuspected; when differences in the nutritional requirements of related species or even (in bacteria) different strains of one species also were unsuspected; when nutritionally important substances such as amino acids were mostly unavailable from supply houses; when present criteria of purity were lacking, and even supposedly pure compounds frequently contained sufficient impurities to obscure a given nutritional response; and when purification procedures—chromatography was barely beginning, ion exchange resins and molecular sieves were unknown—were primitive by today's standards. Small wonder that 27 years elapsed between the discovery in 1901 of "bios," a trace nutrient required for yeast growth (106), and identification of inositol as one of its components in 1928 (13); and that four years elapsed following the first isolation of thiamine by Jansen and Donath in 1926 and recognition of its identity with an unidentified growth factor for yeast (110).

One hotbed of research in nutritional biochemistry during this period was the University of Wisconsin. I was fortunate to receive a $400 Wisconsin Alumni Research Foundation Fellowship (the great Depression was still on) that permitted me to enroll there as a graduate student in the fall of 1935. There, to my own surprise, since I was an undergraduate chemistry major with no exposure to biochemistry, I chose biochemistry as a major field and W. H. Peterson, a fermentation biochemist, as my major professor. For a thesis problem, he suggested we explore the nutritional requirements of lactic acid bacteria. Neither he nor I suspected at that time the complexity of these requirements, which turned out to parallel to a remarkable extent those of animals and that required for their elucidation the efforts of many individuals and many more than the four years I spent under Peterson's guidance in Madison. I emphasize these findings in the following narrative, not because they were always the first or most important in characterizing a given nutrient, but rather because I am most familiar with them and because they illustrate the manifold contributions of such studies to animal nutrition. These contributions include (*a*) the initial or independent discovery and identification of several vitamins and related substances, (*b*) the provision of convenient, sensitive, and accurate methods for following purification of these substances and for determining their distribution in nature, and (*c*) the provision of clues to the metabolic origin, roles, and fate of several vitamins.

ACETATE, LIPOATE, RIBOFLAVIN, PANTOTHENATE, AND PANTETHEINE

Where were we to begin investigation of the lactic acid bacteria? Of the presently known B vitamins, only thiamine was available in pure form. Riboflavin had just been isolated in Europe, and Orla-Jensen et al reported in 1936 (51) that it and additional factors of unknown nature were required for growth of certain lactic acid bacteria in a charcoal-treated whey medium.

The experimental approach we chose was similar to that used in studies of animal nutrition, i.e. to use a crude medium (Medium A: peptone, 0.5%, glucose, 1%: inorganic salts) that did not permit good growth of our test organism (initially *Lactobacillus delbrueckii)* and to determine what supplements to the medium were necessary for growth. A potato extract was an excellent source of unidentified growth-promoting factors (Figure 1), one or more of which was not precipitated by mercuric acetate in sodium carbonate (Neuberg's reagent), could be extracted in part with ether from the acidified, dried extract, and was partially replaced by sodium acetate. These results demonstrated for the first time the peculiar efficacy of acetate in promoting growth of lactic acid bacteria and led to inclusion of acetate in essentially all later media designed for nutritional studies with these organisms. Clearly, however, acetate was present in insufficient amounts to account for the growth-promoting effects of the potato extract. Much later, in following up these observations, we partially purified a water-soluble, acetate-replacing factor that was far more active than acetate in promoting growth of organisms such as *Lactobacillus casei* (19) (Figure 2). By use of the bacterial assay thus provided, Reed et al (63, 64) isolated a crystalline compound that they named *lipoic acid* and that proved to be an essential cofactor in oxidation of pyruvate to acetate by the pyruvate dehydrogenase (49, 62). The same substance was

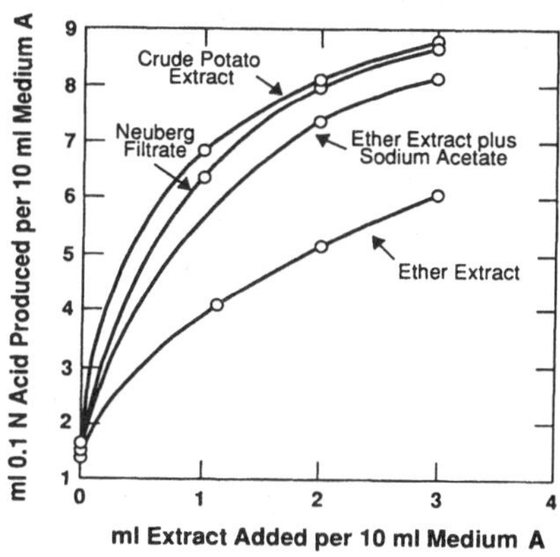

Figure 1 Effect of fractions of a potato extract and acetate on growth of *Lactobacillus delbrueckii* in Medium A. [From (93) with slight modifications.]

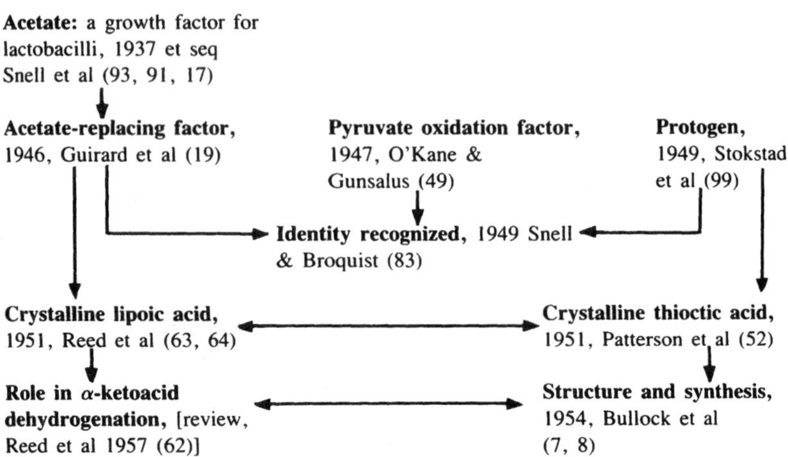

Figure 2 A partial chronology of findings related to the discovery, isolation, structure, and function of lipoic acid.

isolated independently at Lederle Laboratories as a growth factor for the protozoan *Tetrahymena geleii*. These studies eventually led to the synthesis of lipoic acid (7, 8).

When sodium acetate was added to Medium A, relatively good growth of *L. casei* was obtained. When the peptone was treated with alkali (1N NaOH) for 24 h at room temperature, however, no growth occurred unless small amounts of yeast or liver extracts were added. One of the substances necessary under these conditions was identified as riboflavin, which was known to be alkali labile and to be required by certain lactic acid bacteria (51). Addition of both riboflavin and acetate to the NaOH-treated peptone medium (Medium B), however, permitted growth only if an additonal, unidentified factor supplied by yeast or liver extracts was added. I spent more than two years as a graduate student purifying this substance. Toward the end of this period, apparent similarities in properties led us to exchange purified preparations with Professor R. J. Williams. We found his concentrates of an unidentified growth factor for yeast (named *pantothenic acid* although they, too, were not pure) were highly active for our test organisms and vice versa. The chronology of microbiologic studies in this area is shown in Figure 3. Williams's group had discovered pantothenic acid before we did and was substantially ahead of us in activity of their concentrates. They also found that β-alanine substituted for pantothenate in promoting yeast growth under some conditions. This observation led them to the discovery that pantothenic acid was an amide of an unidentified hydroxy acid with β-alanine. The Merck group later isolated and characterized this hydroxy acid as its lactone (97) from hydrolysates of pantothenic acid concentrates in work that led to the initial synthesis of pure pantothenic acid (96).

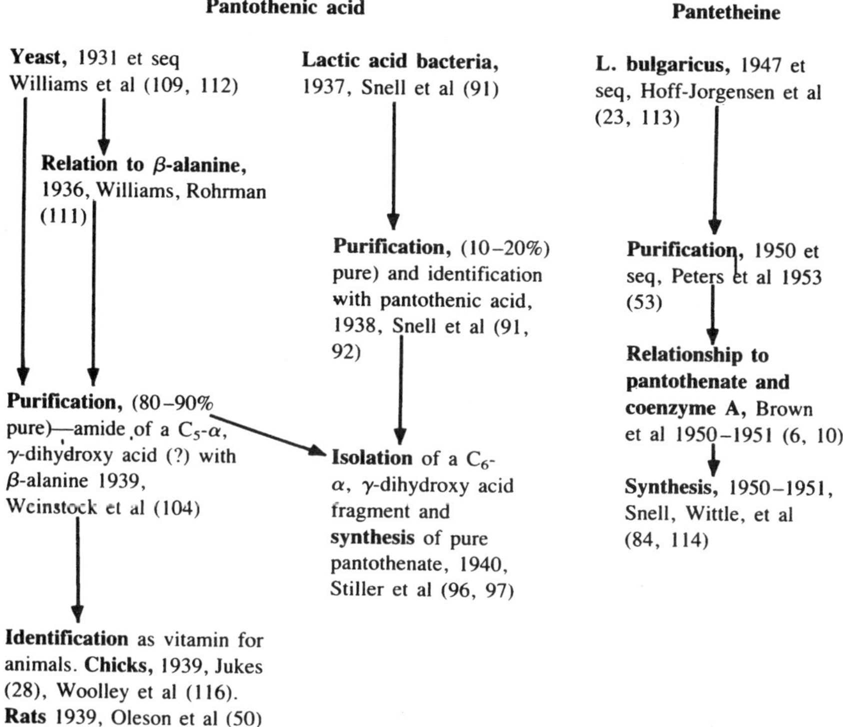

Figure 3 A partial chronology of studies leading from the discovery of pantothenic acid and pantetheine as microbial growth factors to their synthesis.

Some nine years later, we discovered another unidentified factor necessary for growth of *Lactobacillus bulgaricus* in a medium that contained pantothenate (Figure 3). Large amounts of pantothenate replaced this factor, however, which indicated a close relationship of the two substances. We eventually showed that the growth factor was an amide of pantothenic acid with the then-unidentified fragment of coenzyme A, β-mercaptoethylamine (cysteamine), a substance not previously known to occur in nature (6). We named the -SH form of the synthetic compound *pantetheine* and the disulfide form *pantethine* by euphoniously combining segments of the names *pan*-*t*othenic acid, β-mercapto*eth*ylamine, and cyst*eine*-cyst*ine* (84).

MICROBIOLOGIC ASSAYS

I have jumped ahead of events, however. To return to 1938, we had a medium in which excellent growth responses of *L. casei* to either pantothenate (if riboflavin were added) or to riboflavin (if pantothenate were added) could be obtained. The nutritional significance of riboflavin and its distribution in foods were topics of considerable interest at that time; we believed that our procedure for its determination with *L. casei* would be far superior to other methods then available. Since pantothenic acid was not yet commercially available, we supplied it (and perhaps other stimulatory, though nonessential, growth factors) by adding to Medium B a yeast extract (0.2%) from which riboflavin had been removed by adsorption on lead sulfide and photolysis (Medium C). The growth response of *L. casei* to riboflavin in this medium was excellent (Figure 4), whether measured acidimetrically or turbidimetrically, and values obtained for the riboflavin content of a series of standard samples (Table 1) were in excellent agreement with those obtained by the much more lengthy, cumbersome, and expensive rat assay. Recoveries of added riboflavin were excellent, and the organism responded to a series of riboflavin analogs in much the same way as rats. As a result, this assay was rapidly accepted, was widely used, and served as a prototype for similar methods developed later for each of the other B vitamins [for reviews see (1, 3, 77)].

We found later that the idea of using microorganisms for vitamin assay was not new. For example, R. J. Williams (107) had suggested use of yeast for assay of "vitamine" as early as 1919, before "vitamine" was known to be a complex and before any components of that complex had been identified. West & Wilson (105) had described use of *Staphylococcus aureus* for thiamine assay in 1938. That the latter method failed to gain wide recognition I believe stems from several factors: (*a*) other reasonably satisfactory traditional methods (e.g. the thiochrome method for thiamine) were available, (*b*) *S. aureus* is a potential pathogen, and (*c*) the method was not validated by careful recovery experiments or cross-checking against other available

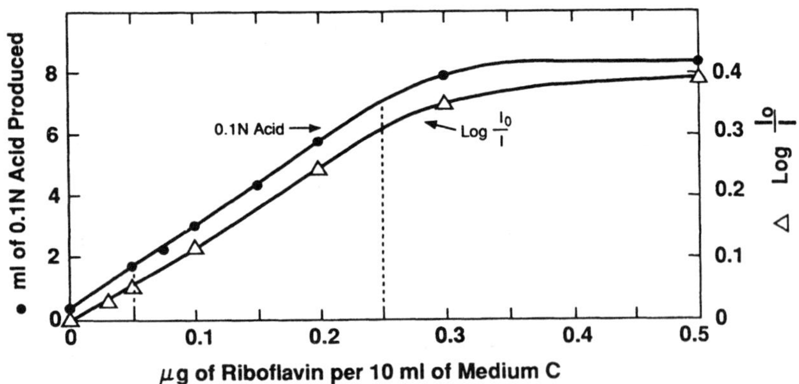

Figure 4   Response of *Lactobacillus casei* to pure riboflavin in the riboflavin-free assay Medium C. [Redrawn from (90).]

Table 1 Riboflavin content of crude reference samples as determined with *Lactobacillus casei* or with rats. From (90)

| | Microbiologic assay | | Rat assay | |
Material	Direct (μg/g)	Extract (μg/g)	Direct (μg/g)	Bourquin-Sherman units[a]
Dried grasses				
Sample I	—	24.4	20	—
Sample II	—	24.1	25	—
Sample III	—	22.6	22	—
Sample IV	—	23.9	20	—
Dried oat plant	30.3	31.2	35	—
Skim milk powder	17.1	—	17	—
Debittered yeast *a*	39.2	38.5	37	—
Debittered yeast *b*	—	36.5	(38.8)	17.7
Debittered yeast *c*	34.7	33.6	(34.4)	15.7
Yeast extract	149.6	—	(145.6)	66.5
Nondebittered yeast	31.8	32.8	(32.0)	14.6

[a] A value of 2.19 μg riboflavin/unit was used for conversion of these values to those in parentheses.

methods. In contrast, the lactic acid bacteria used for assay of riboflavin and many other vitamins are nonpathogens, and their growth can be followed either turbidimetrically or by titration of the lactic acid produced. The latter feature was important, since some samples were highly colored or even turbid; furthermore, appropriate photoelectric colorimeters or nephelometers for turbidimetric measurements of growth were expensive and not widely available at that time.

The complexity of the medium required for growth of lactic acid bacteria, sometimes cited as a disadvantage of these methods, actually is an advantage. Because such media contain many different physiologically important materials that could stimulate (or inhibit) response to an essential vitamin, their use minimizes aberrant responses that might result from addition of these same materials with the sample being assayed. Such aberrant responses were often observed in minimal media such as those used for the initial detection of the vitamins or in simple "synthetic" media such as those later used for demonstrating the response of auxotrophic mutants to a limiting nutrient. Assays with such mutants or with wild-type bacteria in minimal media have been very useful in limited areas, such as in purification of a growth factor or in studies of its biosynthesis or degradation. For quantitative assay of crude samples, however, complex assay media are preferable. The problems in use of minimal media with crude samples are akin to those in determining the pH of unbuffered samples; one gets a number, but its quantitative significance is doubtful.

NICOTINIC ACID, PYRIDOXINE, BIOTIN, AND AVIDIN

At this point, we could begin to analyze the unknown substances supplied by the peptone component of the medium. For this purpose, the NaOH-treated peptone of Medium B was replaced by an acid hydrolysate of purified casein plus tryptophan, and riboflavin and a pantothenic acid concentrate were added. Under these conditions a distinct growth response of several lactic acid bacteria to nicotinic acid was obtained (92). We tried nicotinic acid because it had already been established as an essential nutrient for some bacteria in reports that appeared during the spring of 1937 [for *Staphylococcus aureus* by Knight (32) and for *Corynebacterium diphtheriae* by Mueller (45)]. These reports followed the discovery in 1936 (38) that NAD^+ was essential for growth of *Hemophilus parainfluenzae*. I well remember Bob Madden (one of Professor Elvehjem's graduate students), Wayne Woolley, and myself discussing these discoveries in a late evening laboratory "bull session." Very shortly thereafter Madden gave a nicotinic acid supplement to his black-tongue-afflicted dogs. Forty-eight hours later the answer was in: it worked! Their important report (14) and subsequent reports of the effectiveness of nicotinic acid in human pellagra [see Bean (4)] appeared that same fall.

Although nicotinic acid stimulated growth and acid production on our hydrolyzed casein medium, this stimulation was not evident upon subculture, apparently because of exhaustion of additional, unidentified essential nutrients carried over with the inoculum. We therefore added nicotinic acid to our basal medium and examined the additional crude supplements necessary for growth. Unidentified factors present in both the Norite[2] eluate and the Norite filtrate of a liver extract proved essential for growth (87, 88). We first added a preparation of the Norite eluate factor to our basal medium and examined the nature of the Norite filtrate.

Immediately following the isolation of pyridoxine in 1938 as a vitamin for animals, Moeller (43) showed that it acted as an essential growth factor for some lactic acid bacteria. Following his lead, we found that pyridoxine partially replaced the Norite filtrate for *L. casei* but that another substance(s) also was required (88). Moeller & Schwarz (44), in a report we learned of only later because of war in Europe, had also shown that biotin, a substance originally discovered and isolated as a yeast growth factor (33), was required by lactic acid bacteria. We later confirmed this report (95) and found that biotin and pyridoxine together completely replaced the Norite filtrate for *L.*

[2]"Norite" is a trade name for a widely used activated charcoal.

casei. The resulting medium, with nicotinic acid omitted, was used in a successful assay for nicotinic acid with *L. arabinosus* (95).

I left Madison in 1939 for my first job as a postdoctoral research associate with R. J. Williams at the University of Texas. In Austin, I was to purify an unidentified growth factor for yeast that Williams called "biotic acid." These plans changed when we received a few micrograms of pure biotin from F. Kögl. It completely replaced "biotic acid" for yeast; it also replaced an unidentified growth factor required by *Clostridium butylicum* (94). The work with yeast was not a loss, however, for it provided a sensitive microbiologic assay for biotin (85) that we used in the first purification and characterization of avidin (11), the egg white injury factor studied earlier by Parsons and others. The assay was also used by du Vigneaud and co-workers (10a) in their reisolation of biotin from liver prefatory to their initial proof of its structure and synthesis.

THE NORITE ELUATE FACTOR, FOLIC ACID

Since biotin was the last of the new growth factors for yeast, Williams suggested that I return to investigation of unidentified growth factors for lactic acid bacteria. This change presented difficulties, since Professor Peterson planned to continue work on the Norite eluate factor in Madison, and we had agreed that I would not use *L. casei* as a test organism for this purpose in Austin. Little was then known about nutrition of the lactic acid–producing *Streptococci*, however, so I turned my efforts to this field with *Streptococcus faecalis (S. lactis R)* as the test organism. Within the year, however, we realized that an unidentified factor we were fractionating for this organism, which we extensively purified (41) and called "folic acid" (because spinach leaves were our source material), was the same as the Norite eluate factor required by *L. casei*. The two organisms thus introduced have continued to be those most widely used in study of this vitamin and its derivatives.

The chronology in elucidation of the nature of folic acid is shown (in part) in Table 2. Work on the Norite eluate factor for *L. casei* was continued in Peterson's laboratory by Hutchings & Bohonos (26), who later participated at Lederle Laboratories in isolation of two crystalline forms of this substance, pteroylglutamic acid and pteroyltriglutamate (27, 100). An interesting review that describes early work and early confusion in this field is that of Pfiffner & Hogan (58).

About the time most investigators thought that enough forms of folic acid were already known, Sauberlich & Baumann (69) discovered a new essential growth factor for *Leuconostoc citrovorum* related to folic acid. A closely related substance(s) was discovered by Shive and colleagues by studying growth inhibition of *S. faecalis* by folate analogs and its reversal by natural

Table 2   Assay organisms used in discovery and for monitoring purification of folic acid and some related compounds

---

Discovery and partial purification[a]
  1. *Lactobacillus casei* (Norite eluate factor, etc.) 1939 et seq (87, 88, 26, 98)
  2. Chicks (vitamin B_c) 1939 et seq (24)
     Chicks, *L. casei* (vitamin B_c conjugate) 1947 (57)
  3. *Streptococcus faecalis* (folic acid, etc.) 1941 et seq (41, 42)
  4. *S. faecalis* (Rhizopterin[b]) (29)
  5. *Leuconostoc citrovorum, S. faecalis* (Citrovorum factor, folinic acid, etc.) 1948 et seq
     (69, 5)
Crystallization of active compounds
  1. From Liver: 1943–1947[c] (2, 100, 56, 57)
  2. From *Corynebacterium* fermentation: 1946[d] (27)
Structure and synthesis: 1946 et seq
  1. Folic acid (pteroylglutamic acid)[c] (2, 103)
  2. Vitamin Bc conjugated[e] (57)
  3. Formyltetrahydrofolate[f] (5)

---

[a] A more complete account of the involved early history of these substances is given in several reviews (e.g. 65, 102, 108).
[b] N^{10}formyl pteroylglutamic hydroxide.
[c] Pteroylglutamic acid, PteGlu.
[d] $Pte(Glu)_3$.
[e] $Pte(Glu)_7$.
[f] $N^{10}-H_4PteGlu$ and related compounds.

extracts (5, 15). One of the effective compounds was identified as a formylated reduced folic acid, N^{10}-formyltetrahydrofolate. This work provided one key to elucidation of the role played by folic acid coenzymes in one-carbon metabolism (15, 25).

PYRIDOXAL AND PYRIDOXAMINE

By omitting pyridoxine and supplementing our basal medium for *L. casei* or *S. faecalis* with folic acid and biotin, we hoped to have a medium suitable for quantitative assay of pyridoxine. This medium did indeed permit excellent growth of these organisms when supplemented with pyridoxine. Much to our surprise, however, the values we obtained upon assay of tissues for pyridoxine were far higher than those obtained by assay of the same samples with a vitamin B_6–requiring yeast; evidently, something other than pyridoxine was contributing to growth of *S. faecalis* and *L. casei* under these conditions. We were able to prove (86) that the growth response of these organisms to pyridoxine was due not to pyridoxine per se but rather to one or more substances formed from it in minute yield during heat sterilization of the growth medium. Similar products were formed and partially excreted when pyridoxine was fed to rats or other animals. We provisionally termed this substance *pseudopyridoxine* (86) and showed (73, 86) that its activity resulted from an aldehyde and an amine formed by oxidation or amination, respectively, of pyridoxine. By appropriate tests we reduced the number of possible structures to two for the aldehyde and two for the amine (73, 21). With this information we enlisted the expertise of Folkers and his group at Merck, who had previously synthesized pyridoxine, and now synthesized our candidate compounds (21). Two of them, which we named *pyridoxal* and *pyridoxamine*, were in fact thousands of times more active than pyridoxine in supporting growth of *S. faecalis;* only pyridoxal showed such activity for *L. casei* (74).

The chronology of events in this field (Table 3) illustrates a point of general validity with respect to priority in scientific investigation. If methods, however primitive, become available, for studying a topic of interest, that topic will attract multiple investigators and its clarification becomes only a matter of time. One should note, for example, that following development of an appropriate animal assay for vitamin B_6, five different groups reported isolation of pyridoxine within a few weeks of each other (see 81, 102). Similarly, if we had not identified pyridoxal in 1944, Carpenter & Strong (9), who independently found that partial oxidation of pyridoxine increased its activity for *L. casei*, would probably have done so. Had they by chance failed, Gunsalus et al (20), who knew that the coenzyme for tyrosine decarboxylase was related to vitamin B_6, were in an excellent position to do so. Subsequent related work showed (*a*) that the two new forms of vitamin B_6 were as active as pyridoxine for animals, (*b*) that pyridoxal-5'-phosphate (PLP) was the coenzyme of the amino acid decarboxylases (20, 22), (*c*) that pyridoxamine-5'-phosphate (PMP) also occurs naturally (60), and (*d*) that PMP (PLP is less active) is itself an essential growth factor for some bacteria (40) that apparently cannot synthesize it even when supplied with an external source of pyridoxamine.

Table 3 A partial chronology of discovery of vitamin B_6 (pyridoxine, pyridoxal, and pyridoxamine)

1934:	Rat assay for *vitamin B_6* reported by György [reviews (81, 102)]
1938:	*Pyridoxine* (PN) isolated in 5 different laboratories [reviews (81, 82, 102)]
1938–1939:	PN reported as an essential growth factor for lactic acid bacteria (43) and yeasts (12, 70)
1939:	Synthesis of PN (see 102)
1942–1944:	Discovery of *pyridoxal* (PL) and *pyridoxamine* (PM) ("pseudopyridoxine"), formed from pyridoxine by oxidation or amination, as actual growth factors for lactic acid bacteria; pyridoxine inactive (86, 73)
1944:	Synthesis (21) and activity (74) of PL and PM as vitamins for bacteria and yeasts
1945–1951:	PL, PM occur naturally (76) and are about equally as active as PN for rats (89, 36, 67), chicks (37), and dogs (68)

Vitamin B_6 is thus a complex of three compounds (and their 5'-phosphates) that vary in their chemical properties and in their ability to support growth of different vitamin B_6–dependent organisms. *Lactobacillus casei* responds only to pyridoxal; *S. faecalis* responds about equally to pyridoxal and pyridoxamine but not to pyridoxine, whereas yeast (like animals) responds about equally to all three. For many years differential microbiologic assays with these three organisms (cf 60a) supplied the only quantitative estimates of the individual forms of the vitamin in natural products (81).

Although pyridoxal and pyridoxamine have now been known for more than 40 years, the multiple nature of vitamin B_6 frequently is not reflected in articles dealing with this vitamin, despite the recommendations of nomenclature committees (see 82). *Pyridoxine* is commonly (and incorrectly) used both as a synonym for vitamin B_6, and (correctly) to designate the single compound first characterized and still sold under that name. Imprecise nomenclature promotes imprecise thought. I've always believed that the unfortunate incident, in which a commercial milk formula produced vitamin B_6 insufficiency with attendant convulsions in infants, resulted in part from the belief that vitamin B_6 in milk should show stability characteristics of pyridoxine, rather than those of the much less stable pyridoxal and pyridoxamine, which are the predominant forms of vitamin B_6 in milk and many other foods. Many examples of similar confusion could be cited. A striking recent example is provided by an article (48) that describes the "pyridoxine" content of some archaebacteria, determined by assay with *L. casei* against a pyridoxine standard. Since both pyridoxine and pyridoxamine are inactive in promoting growth of this organism (their apparent activity depends upon the time and conditions of heating during sterilization of the medium), the assay values obtained suffice to show that vitamin B_6 is present in these organisms, but they are quantitatively meaningless.

The changes in activity for various bacteria that occur upon sterilization of pyridoxal, pyridoxamine, or pyridoxine with the growth medium result from a variety of reactions. Most interesting among these is the reversible reaction of pyridoxal with amino acids to form pyridoxamine and keto acids (76, 76a). Observation of this reaction led to the initial suggestion of the role of vitamin B_6 in enzymatic transamination (74), and eventually to formulation (40a) of a general mechanism that explains catalysis of vitamin B_6-dependent reactions of amino acids in terms of the chemical properties of pyridoxal.

THIAMINE AND VITAMIN B_{12}

Thiamine, although not essential for the lactobacilli used in the foregoing work, is required for many of the lactic acid–producing enterococci (47) and some lactobacilli, e.g. *Lactobacillus fermenti* (66). The same is true for vitamin B_{12}; in this case, two observations of Shorb (72), first, that injectable liver concentrates prepared for control of pernicious anemia in man contained an essential growth factor for *Lactobacillus lactis* Dorner, and second, that their microbial activity paralleled their activity in man, provided an assay procedure that greatly facilitated isolation of vitamin B_{12} in the Merck laboratories. Isolation of this substance in England, in contrast, was guided by assay with pernicious anemia patients alone.

AMINO ACID AND PEPTIDE REQUIREMENTS OF LACTIC ACID BACTERIA

Once all of the vitamins required by the lactic acid bacteria had been identified and became commercially available, they could be added to the basal medium, and the hydrolyzed casein could be replaced with a complete assortment of amino acids. By single omissions, individual amino acids required for growth could be identified, and organisms that required them could then be used for their quantitative determination. Such microbiologic methods for amino acids were widely used (review 71) until the automated ion exchange procedures, first developed by Spackman, Moore, & Stein (95a), were introduced. Even now, for repetitive assays of amino acids of special interest, e.g. of essential amino acids in foodstuffs during nutritional survey work, I believe the microbiologic methods would be simpler, more convenient, sufficiently accurate, and much cheaper, especially in undeveloped areas, than the chromatographic methods.

Quantitative comparison showed that peptides of a given amino acid were sometimes more active than the free amino acid in promoting growth of these bacteria. We identified three circumstances under which this phenomenon occurred (see 80): (*a*) a free amino acid, but not its peptides, may be partially degraded by cellular enzymes [e.g. by decarboxylases (30)]; (*b*) uptake of an amino acid may be blocked by antagonistic amino acids, while uptake of its peptides is not (59); and (*c*) a simple deficiency in ability to transport a given amino acid into the cell may be present (54). Once inside the cell, the peptides seem to release the limiting amino acid by simple hydrolysis. Whether similar phenomena occur in animals is unknown. An untested possibility is that the requirement of animal cell cultures for some of the numerous peptide growth factors now being reported may find its explanation in similar terms. A widely studied peptide growth factor for bacteria, strepogenin, was identified and its activity explained in these terms (31).

THE STATUS QUO

By 1955, the vitamin, amino acid, and fatty acid requirements of most lactic acid bacteria were known. Similarly, the basis for the highly varied peptide requirements of some of these bacteria had been mostly established, together with some of their requirements for inorganic ions (39). Consequently, most of them can now be grown in media of known composition, and dehydrated media deficient in single vitamins suitable for vitamin assay can be purchased commercially. As a result, my own interests turned to more biochemically oriented investigations, e.g. of the mechanism of catalysis by pyridoxal (78), by various pyridoxal phosphate-dependent enzymes, and by the functionally related pyruvoyl enzymes (61), and to associated problems of vitamin function, metabolism, and degradation (46).

Many important problems in bacterial nutrition remain, however. Some pathogens cannot yet be grown in media of known composition. The problems of commensal and symbiotic growth and of parasitic relationships have scarcely been touched. While the pace of discovery of new substances of nutritional importance has slackened markedly, recent discoveries of new and distinctive coenzymes in the archaebacteria (115) suggest that this rewarding chapter in the history of nutrition may not yet be closed.

SUMMARY

From the above discussion, it is apparent that one or another of the lactic acid bacteria requires each of the B vitamins required by animals and that assay methods developed during study of nutrition of bacteria and yeasts have played a large role in the initial or independent discovery, isolation, and characterization of vitamins, vitamin derivatives, and functionally similar substances. Clear examples include biotin, biocytin, lipoic acid, nicotinic acid, pantothenic acid, pantetheine, folic acid and tetrahydrofolic acid (and their derivatives), pyridoxal, pyridoxamine, and pyridoxamine phosphate. Improved assay methods that use these organisms also have provided much of the currently available information concerning distribution and stability of the vitamins in natural products, while quantitative inconsistencies between assays, when traced to their origin, have frequently revealed previously unknown metabolic precursors, products, or functions of the vitamins and have provided explanations of the mechanisms by which certain peptide growth factors act. Extension of such studies to organisms that cannot yet be grown in media of known composition should provide additional insights into currently obscure areas of nutrition.

Literature Cited

1. Adrian, J. 1959. *Le Dosage Microbiologique des Vitamins du Group B*. Paris: CNRS. 183 pp.
2. Angier, R. B., Boothe, J. H., Hutchings, B. L., Mervat, J. M., Semb, J., et al. 1946. The structure and synthesis of the liver *L. casei* factor. *Science* 103:667–69
3. Barton-Wright, E. C. 1952. *The Microbiological Assay of the Vitamin B-Complex*. London: Pitman & Sons. 179 pp.
4. Bean, W. B. 1982. Personal reflections on clinical investigations. *Annu. Rev. Nutr.* 2:1–20
5. Bond, T. J., Bardos, T. J., Sibley, M, and Shive, W. 1949. The folinic acid group, a series of new vitamins related to folic acid. *J. Am. Chem. Soc.* 71:3852–53
6. Brown, G. M., Craig, J. A., Snell, E. E. 1950. Relation of the *Lactobacillus bulgaricus* factor to pantothenic acid and coenzyme A. *Arch. Biochem.* 27:473–75
7. Bullock, M. W., Brockman, J. A. Jr., Patterson, E. L., Pierce, J. V., Stockstad, E. L. R. 1952. Synthesis of compounds in the thioctic acid series (Letter). *J. Am. Chem. Soc.* 74:3455
8. Bullock, M. W., Brockman, J. A. Jr., Patterson, E. L., Pierce, J. V., von Salza, M. H., et al. 1954. Synthesis in the thioctic acid series. *J. Am. Chem. Soc.* 76:1828–32
9. Carpenter, L. E., Strong, F. M. 1944. Determination of pyridoxine and pseudopyridoxine. *Arch. Biochem.* 3:375–88
10. Craig, J. A., Snell, E. E. 1951. The comparative activities of pantethine, pantothenic acid and coenzyme A for various microorganisms. *J. Bacteriol.* 61:283–91
10a. du Vigneaud, V., Hofmann, K., Melville, D. B., György, P. 1941. Isolation of biotin (vitamin H) from liver. *J. Biol. Chem.* 140:643–51
11. Eakin, R. E., Snell, E. E., Williams, R. J. 1940. A constituent of raw egg white capable of inactivating biotin in vitro. *J. Biol. Chem.* 136:801–2
12. Eakin, R. E., Williams, R. J. 1939. Vitamin B_6 as a yeast nutrilite (Letter). *J. Am. Chem. Soc.* 61:1932
13. Eastcott, E. V. 1928. Wildiers' bios: The isolation and identification of "Bios I." *J. Phys. Chem.* 32:1094–1111
14. Elvehjem, C. A., Madden, R. J., Strong, F. M., Woolley, D. W. 1937. Relation of nicotinic acid and nicotinic acid amide to canine black tongue. *J. Am. Chem. Soc.* 59:1767–68
15. Flynn, E. H., Bond, T. J., Bardos, T. J., Shive, W. 1951. A synthetic compound with folinic acid activity. *J. Am. Chem. Soc.* 73:1979–82
16. Guirard, B. M., Snell, E. E. 1962. Nutritional requirements of microorganisms. In *The Bacteria,* ed. I. C. Gunsalus, R. Y. Stanier, 4:33–93. London/New York: Academic. 459 pp.
17. Guirard, B. M., Snell, E. E. 1964. Nutritional requirements of *Lactobacillus* 30a for growth and histidine decarboxylase production. *J. Bacteriol.* 87:370–76
18. Guirard, B. M., Snell, E. E. 1981. Biochemical factors in growth. In *Manual of Methods for General Bacteriology,* ed. P. Gerhardt, R. E. Murray, T. N. Costilow, E. W. Nester, W. A. Wood, et al., pp. 79–111. Washington, DC: American Society of Microbiology. 529 pp.
19. Guirard, B. M., Snell, E. E., Williams, R. J. 1946. The nutritional role of acetate for lactic acid bacteria. II. Fractionation of extracts of natural materials. *Arch. Biochem.* 9:381–86
20. Gunsalus, I. C., Bellamy, W. D., Umbreit, W. W. 1944. A phosphorylated derivative of pyridoxal as the coenzyme of tyrosine decarboxylase. *J. Biol. Chem.* 155:685–86
21. Harris, S. A., Heyl, D., Folkers, K. 1944. The structure and synthesis of pyridoxamine and pyridoxal. *J. Biol. Chem.* 154:315–316
22. Heyl, D., Luz, E., Harris, S. A., Folkers, K. 1951. Phosphates of the vitamin B_6 group. I. The structure of codecarboxylase. *J. Am. Chem. Soc.* 73:3430–33
23. Hoff-Jorgensen, E., Williams, W. L., Snell, E. E. 1947. Preferential utilization of lactose by a strain of *Lactobacillus bulgaricus. J. Biol. Chem.* 168:773–74
24. Hogan, A. G., Parrott, E. M. 1940. Anemia in chicks caused by a vitamin deficiency. *J. Biol. Chem.* 132:507–517
25. Huennekins, F. M., Osborne, M. J. 1959. Folic acid coenzymes and one-carbon metabolism. *Adv. Enzymol.* 21:369–478
26. Hutchings, B. L., Bohonos, N., Peterson, W. H. 1941. Growth factors for

bacteria. XIII. Purification and properties of an eluate factor required by certain lactic acid bacteria. *J. Biol. Chem.* 141:521–28

27. Hutchings, B. L., Stokstad, E. L. R., Bohonos, N., Sloane, N. H., Subbarow, Y. 1948. The isolation of the fermentation *Lactobacillus casei* factor. *J. Am. Chem. Soc.* 70:1–3

28. Jukes, T. H. 1939. Pantothenic acid and the filtrate (chick antidermatitis) factor. *J. Am. Chem. Soc.* 61:975–76

29. Keresztesy, J. C., Rickes, E. L., Stokes, J. L. 1943. A new growth factor for *Streptococcus lactis* (Letter). *Science* 97:465

30. Kihara, H., Klatt, O. A., Snell, E. E. 1952. Peptides and bacterial growth. III. Utilization of tyrosine and tyrosine peptides by *Streptococcus faecalis*. *J. Biol. Chem.* 197:801–7

31. Kihara, H., Snell, E. E. 1960. Peptides and bacterial growth. VIII. The nature of strepogenin. *J. Biol. Chem.* 235:1409–14

32. Knight, B. C. J. G. 1937. Nicotinic acid and growth of *Staphylococcus aureus* (Letter). *Nature* 139:628

33. Kögl, F., Tönnis, B. 1936. Uber das Bios-Problem. Darstellung von Krystallisierten Biotin aus Eigelb. *Z. Physiol. Chem.* 242:43–73

34. Koser, S. A. 1968. *Vitamin Requirements of Bacteria and Yeasts*. Springfield, Ill: C. C. Thomas. 663 pp.

35. Lichstein, H. C. 1983. *Bacterial Nutrition*. Stroudsburg, Pa: Hutchinson Ross. 377 pp.

36. Linksweiller, H., Baumann, C. A., Snell, E. E. 1951. Effect of aureomycin on the response of rats to various forms of vitamin B_6. *J. Nutr.* 43:565–73

37. Luckey, T. D., Briggs, G. N., Elvehjem, C. A., Hart, E. B. 1945. Activity of pyridoxine derivatives in chick nutrition. *Proc. Soc. Exp. Biol. Med.* 58:340–44

38. Lwoff, A., Lwoff, M. 1936. Sur le role physiologique des codéhydrogénases pour *Hemophilus parainfluenzae*. *Compt. Rend. Acad. Sci., Paris* 203:520–22

39. MacLeod, R. A., Snell, E. E. 1950. The relation of ion antagonism to the inorganic nutrition of lactic acid bacteria. *J. Bacteriol.* 59:783–92

40. McNutt, W. S., Snell, E. E. 1948. Phosphates of pyridoxal and pyridoxamine as growth factors for lactic acid bacteria. *J. Biol. Chem.* 173:801–2

40a. Metzler, D. E., Ikawa, M., Snell, E. E. 1954. A general mechanism for vitamin B_6-catalyzed reactions. *J. Am. Chem. Soc.* 76:648–52

41. Mitchell, H. K., Snell, E. E., Williams, R. J. 1941. The concentration of "folic acid" (Letter). *J. Am. Chem. Soc.* 63:2284

42. Mitchell, H. K., Snell, E. E., Williams, R. J. 1944. Folic acid. I. Concentration from spinach. *J. Am. Chem. Soc.* 66:267–68

43. Moeller, E. F. 1938. Vitamin B_6 (Adermin) als Wuchstoff für Milchsäurebakterien. *Z. Physiol. Chem.* 245:285–86

44. Moeller, E. F., Schwarz, K. 1941. Growth of *Streptobacterium plantarum* in nutrient solutions of chemically exactly defined compounds. *Ber. Dtsch. Chem. Ges.* 74:1612–16. From 1943 *Chem. Abstr.* 37:356

45. Mueller, J. H. 1937. Substitution of β-alanine, nicotinic acid, and pimelic acid for meat extract in growth of *Diphtheria bacillus*. *Proc. Soc. Exp. Biol. Med.* 36:706–8

46. Nelson, M. J. K., Snell, E. E. 1986. Enzymes of vitamin B_6 degradation. Purification and properties of 5-pyridoxic acid oxygenase from *Arthrobacter sp.* *J. Biol. Chem.* 261:15115–20

47. Niven, C. F., Sherman, V. M. 1944. Nutrition of the enterococci. *J. Bacteriol.* 47:335–42

48. Noll, K. M., Barber, T. S. 1988. Vitamin contents of archaebacteria. *J. Bacteriol.* 170:4315–21

49. O'Kane, D. J., Gunsalus, I. C. 1947. Accessory factor requirement for pyruvate oxidation. *J. Bact.* 54:20–21 (Abstr.). Republished 1948 as Pyruvic acid metabolism: A factor for oxidation by *Streptococcus faecalis* in *J. Bacteriol.* 56:499–506

50. Oleson, J. J., Woolley, D. W., Elvehjem, C. A. 1939. Is pantothenic acid essential for growth of rats? *Proc. Soc. Exp. Biol. Med.* 42:151–53

51. Orla-Jensen, S., Otte, N. C., Snog-Kjaer, A. 1936. Der Vitaminbedarf der Milchsäurebakterien. *Zentralbl. Bakteriol. Parasitenkd. Infectionskr.* II, 94:434–47

52. Patterson, E. L., Brockman, J. A. Jr., Day, F. P., Pierce, J. V., Macchi, M. E., et al. 1951. Crystallization of a derivative of protogen B. *J. Am. Chem. Soc.* 73:5919–20

53. Peters, V. J., Brown, G. M., Williams, W. L., Snell, E. E. 1953. Isolation of the *Lactobacillus bulgaricus* factor from natural sources. *J. Am. Chem. Soc.* 75:1688–91

54. Peters, V. J., Prescott, J. M., Snell, E. E. 1953. Peptides and bacterial growth. IV. Histidine peptides as growth factors

for *Lactobacillus delbrueckii* 9469. *J. Biol. Chem.* 202:521–32

55. Peterson, W. H., Peterson, M. S. 1945. Relation of bacteria to vitamins and other growth factors. *Bacteriol. Rev.* 9:49–109

56. Pfiffner, J. J., Binkley, S. B., Bloom, E. S., Brown, R. A., Bird, O. D., et al. 1943. Isolation of the antianemia factor (vitamin B$_c$) in crystalline form from liver. *Science* 97:404–5

57. Pfiffner, J. J., Binkley, S. B., Bloom, E. S., O'Dell, B. L. 1947. Isolation and characterization of vitamin B$_c$ from liver and yeast. Occurrence of an acid labile chick antianemia factor in liver. *J. Am. Chem. Soc.* 69:1476–87

58. Pfiffner, J. J., Hogan, A. G. 1946. The newer hematopoietic factors of the vitamin B-complex. *Vitam. Horm.* 4:1–34

59. Prescott, J. M., Peters, V. J., Snell, E. E. 1953. Peptides and bacterial growth. V. Serine peptides and growth of *Lactobacillus delbrueckii* 9649. *J. Biol. Chem.* 202:533–540

60. Rabinowitz, J. C., Snell, E. E. 1947. The vitamin B$_6$ group. XII. Microbiological activity and natural occurrence of pyridoxamine phosphate. *J. Biol. Chem.* 169:643–50

60a. Rabinowitz, J. C., Snell, E. E. 1948. The vitamin B$_6$ group, XIV. Distribution of pyridoxal, pyridoxamine and pyridoxine in some natural products. *J. Biol. Chem.* 176:1157–67

61. Recsei, P. A., Snell, E. E. 1984. Pyruvoyl enzymes. *Annu. Rev. Biochem.* 53:357–87

62. Reed, L. J. 1957. The chemistry and function of lipoic acid. *Adv. Enzymol.* 18:319–47

63. Reed, L. J., DeBusk, B. G., Gunsalus, I. C., Hornberger, C. S. Jr. 1951. Crystalline α-lipoic acid: A catalytic agent associated with pyruvate dehydrogenase. *Science* 114:93–94

64. Reed, L. J., Getzendaner, M. E., DeBusk, B. G., Johnston, P. M. 1951. Acetate-replacing factors for lactic acid bacteria. II. Purification and properties. *J. Biol. Chem.* 192:859–65

65. Robinson, F. A. 1951. *The Vitamin B Complex.* New York: Wiley. 688 pp.

66. Sarett, H. P., Cheldelin, V. H. 1944. The use of *Lactobacillus fermentum* 36 for thiamine assay. *J. Biol. Chem.* 155:153–60

67. Sarma, P. S., Snell, E. E., Elvehjem, C. A. 1946. The vitamin B$_6$ group. VIII. Biological assay of pyridoxal, pyridoxamine, and pyridoxine. *J. Biol. Chem.* 165:55–63

68. Sarma, P. S., Snell, E. E., Elvehjem, C. A. 1946. The vitamin B$_6$ group. IX. Comparative growth and antianemic potencies of pyridoxal, pyridoxamine, and pyridoxine for dogs. *Proc. Soc. Exp. Biol. Med.* 63:284–86

69. Sauberlich, H. E., Baumann, C. A. 1948. A factor required for the growth of *Leuconostoc citrovorum*. *J. Biol. Chem.* 176:165–73

70. Schultz, A., Atkin, L., Frey, C. N. 1939. Vitamin B$_6$, a growth-promoting factor for yeast. *J. Am. Chem. Soc.* 61:1931

71. Schweigert, B. S., Snell, E. E. 1947. Microbiological methods for estimation of amino acids. *Nutr. Abstr. Rev.* 16:497–510

72. Shorb, M. S. 1947. Unidentified growth factors for *Lactobacillus lactis* in refined liver extracts. *J. Biol. Chem.* 169:455–56

73. Snell, E. E. 1944. The vitamin B$_6$ group. I. Formation of additional members from pyridoxine and evidence concerning their structure. *J. Am. Chem. Soc.* 66:2082–88

74. Snell, E. E. 1944. The vitamin activities of "pyridoxal" and "pyridoxamine." *J. Biol. Chem.* 154:313–14

75. Snell, E. E. 1945. The nutritional requirements of the lactic acid bacteria and their application to biochemical research. *J. Bacteriol.* 50:373–82

76. Snell, E. E. 1945. The vitamin B$_6$ group, IV. Evidence for the occurrence of pyridoxamine and pyridoxal in natural products. *J. Biol. Chem.* 157:491–505

76a. Snell, E. E. 1945b. The vitamin B$_6$ group. V. The reversible interconversion of pyridoxal and pyridoxamine by transamination reactions. *J. Am. Chem. Soc.* 67:194–97

77. Snell, E. E. 1950. Microbiological methods in vitamin research. In *Vitamin Methods*, ed. P. Gyorgy, 1:327–505. New York: Academic. 571 pp.

78. Snell, E. E. 1958. Chemical structure in relation to biological activities of vitamin B$_6$. *Vitam. Horm.* 16:77–125

79. Snell, E. E. 1979. Lactic acid bacteria and identification of B-vitamins: Some historical notes, 1937–40. *Fed. Proc.* 38:2690–93

80. Snell, E. E. 1980. Introduction. In *Microorganisms and Nitrogen Sources*, ed. J. W. Payne, pp. 1–6. New York: Wiley. 870 pp.

81. Snell, E. E. 1981. Vitamin B$_6$ analysis: Some historical aspects. In *Methods in Vitamin B$_6$ Nutrition*, ed. J. Leklem, R. Reynolds, pp. 1–19. New York: Plenum. 401 pp.

82. Snell, E. E. 1986. Pyridoxal phosphate:

History and nomenclature. In *Coenzymes and Cofactors*, ed. D. Dolphin, R. Poulsen, O. Avramovic, 1A:1–12. New York: Wiley. 725 pp.

83. Snell, E. E., Broquist, H. P. 1949. On the probable identity of several unidentified growth factors. *Arch. Biochem.* 23:326–28

84. Snell, E. E., Brown, G. M., Peters, V. J., Craig, J. A., Wittle, E. L., et al. 1950. Chemical nature and synthesis of the *Lactobacillus bulgaricus* factor. *J. Am. Chem. Soc.* 72:5349–50

85. Snell, E. E., Eakin, R. E., Williams, R. J. 1940. A quantitative test for biotin and observations regarding its occurrence and properties. *J. Am. Chem. Soc.* 62:175–78

86. Snell, E. E., Guirard, B. M., Williams, R. J. 1942. Occurrence in natural products of a physiologically active metabolite of pyridoxine. *J. Biol. Chem.* 143:519–30

87. Snell, E. E., Peterson, W. H. 1939. Properties of a new growth factor for lactic acid bacteria. *J. Biol. Chem.* 128:xciv–xcv

88. Snell, E. E., Peterson, W. H. 1940. Growth factors for bacteria. X. Additional factors required by certain lactic acid bacteria. *J. Bacteriol.* 39:273–85

89. Snell, E. E., Rannefeld, A. N. 1945. The vitamin B_6 group. III. The vitamin activity of pyridoxal and pyridoxamine for various organisms. *J. Biol. Chem.* 157:475–89

90. Snell, E. E., Strong, F. M. 1939. A microbiological assay for riboflavin. *Ind. Eng. Chem. Anal. Ed.* 11:346–50

91. Snell, E. E., Strong, F. M., Peterson, W. H. 1937. Growth factors for bacteria. VI. Fractionation and properties of an accessory factor for lactic acid bacteria. *Biochem. J.* 31:1789–99

92. Snell, E. E., Strong, F. M., Peterson, W. H. 1938. Pantothenic and nicotinic acids as growth factors for lactic acid bacteria (Letter). *J. Am. Chem. Soc.* 60:2825

93. Snell, E. E., Tatum, E. L., Peterson, W. H. 1937. Growth factors of bacteria. III. Some nutritive requirements for *Lactobacillus delbrueckii*. *J. Bacteriol.* 33:207–25

94. Snell, E. E., Williams, R. J. 1939. Biotin as a growth factor for the butyl alcohol producing anaerobes (Letter). *J. Am. Chem. Soc.* 61:3594

95. Snell, E. E., Wright, L. D. 1941. A microbiological method for the determination of nicotinic acid. *J. Biol. Chem.* 139:675–86

95a. Spackman, D. H., Stein, W. H., Moore, S. 1958. Automatic recording apparatus for use in the chromatography of amino acids. *Analyt. Chem.* 30:1190–1206

96. Stiller, E. T., Harris, S. A., Finkelstein, J., Keresztesy, J. C., Folkers, K. 1940. Pantothenic acid. VIII. The total synthesis of pure pantothenic acid. *J. Am. Chem. Soc.* 62:1785–90

97. Stiller, E. T., Keresztesy, J. C., Finkelstein, J. 1940. Pantothenic acid. VI. The isolation and structure of the lactone moeity. *J. Am. Chem. Soc.* 62:1779–84

98. Stokstad, E. L. R. 1943. Some properties of a growth factor for *Lactobacillus casei*. *J. Biol. Chem.* 149:573–74

99. Stokstad, E. L. R., Hoffman, C. E., Regan, M. A., Fordham, D., Jukes, T. H. 1949. Observations on an unknown growth factor essential for *Tetrahymena geleii*. *Arch. Biochem.* 20:75–82

100. Stokstad, E. L. R., Hutchings, B. L., Subbarow, Y. 1948. The isolation of the *Lactobacillus casei* factor from liver. *J. Am. Chem. Soc.* 70:3–5

101. Deleted in proof

102. Wagner, A. F., Folkers, K. 1964. *Vitamins and Coenzymes*. New York: Wiley Interscience. 532 pp.

103. Waller, C. W., Hutchings, B. L., Mowat, J. H., Stokstad, E. L. R., Boothe, J. H., et al. 1948. Synthesis of pteroylglutamic acid (liver *L. casei* factor) and pteroic acid. *J. Am. Chem. Soc.* 70:19–22

104. Weinstock, H. H. Jr., Mitchell, H. K., Pratt, E. F., Williams, R. J. 1939. Pantothenic acid. IV. Formation of β-alanine by cleavage of pantothenic acid. *J. Am. Chem. Soc.* 61:1421–25

105. West, P. M., Wilson, P. W. 1938. Biological determination of vitamin B_1 (thiamine) in *Rhizobium trifolii*. *Science* 88:334–35

106. Wildiers, E. 1901. Nouvelle substance indispensable au développemment de la levure. *Cellule* 18:313–31

107. Williams, R. J. 1919. The vitamine requirement of yeast. A simple biological test for vitamine. *J. Biol. Chem.* 38:465–86

108. Williams, R. J., Eakin, R. E., Beerstecher, E. Jr., Shive, W. 1950. *Biochemistry of B Vitamins*. New York: Reinhold. 741 pp.

109. Williams, R. J., Lyman, C. M., Goodyear, G. H., Truesdail, J. H., Holaday, D. 1933. "Pantothenic acid," a growth determinant of universal biological occurrence. *J. Am. Chem. Soc.* 55:2912–27

110. Williams, R. J., Roehm, R. R. 1930. The effect of antineuritic vitamin pre-

parations on the growth of yeasts. *J. Biol. Chem.* 87:581–90

111. Williams, R. J., Rohrmann, E. 1936. β-alanine and "bios" (Letter). *J. Am. Chem. Soc.* 58:695

112. Williams, R. J., Truesdail, J. H., Weinstock, H. H. Jr., Rohrmann, E., Lyman, C. M., McBurney, C. H. 1938. Pantothenic acid. II. Its concentration and purification from liver. *J. Am. Chem. Soc.* 60:2719–23

113. Williams, W. L., Hoff-Jorgensen, E., Snell, E. E. 1949. Determination and properties of an unidentified growth fac-tor required by *Lactobacillus bulgaricus*. *J. Biol. Chem.* 177:933–40

114. Wittle, E. L., Moore, J. A., Stipek, R. W., Peterson, F. E., McGlohon, V. M., et al. 1953. The synthesis of pan-tetheine-pantethine. *J. Am. Chem. Soc.* 75:1694–1700

115. Wolfe, R. S. 1985. Unusual coenzymes of methanogenesis. *Trends Biochem. Sci.* 10:396–99

116. Wooley, D. W., Waisman, H. A., Elvehjem, C. A. 1939. Nature and par-tial synthesis of the chick antidermatitis factor. *J. Am. Chem. Soc.* 61:977–78

Bernard B. Brodie

Annu. Rev. Pharmacol. Toxico. 1989. 29:1–21

BERNARD B. BRODIE AND THE RISE OF CHEMICAL PHARMACOLOGY

Erminio Costa

FIDIA-Georgetown Institute of Neurosciences, Georgetown University Medical School, 3900 Reservoir Road, Washington, D.C. 20007

Alexander G. Karczmar

Department of Pharmacology, Loyola University Medical Center, Maywood, IL 60153

Elliot S. Vesell

Department of Pharmacology, Pennsylvania State University, College of Medicine, Hershey, PA 17033

INTRODUCTION

It is difficult to describe the emotionally charged atmosphere of intellectual excitement and scientific commitment that pervaded the Laboratory of Chemical Pharmacology (LCP) in the National Heart Institute during the years between 1950 and 1970. It was during this period that the leadership of Bernard B. Brodie, together with his ingenious methodological innovations and dedication to scientific pursuit, inspired the intensely motivated and creative environment of LCP. In the following sections we attempt to characterize this period because we believe that historical documentation of science is not only interesting, but also necessary to understand the genesis of new scientific disciplines prior to our dependence on technology and computers.

0362-1642/89/0415-0001$02.00

BERNARD B. BRODIE

Any attempt to describe the atmosphere and guiding spirit permeating the LCP would be futile without first understanding the character and unusual intellect of its director. Born in Liverpool, England, on August 7, 1907, Bernard Beryl Brodie was the third of five children. In 1911, his family emigrated to Ottawa, Canada, where his father, Samuel, owned a small clothing store. Samuel Brodie worked hard and provided his family with all the basic necessities, but very few luxuries. Brodie's mother devoted herself to her children and was insistent in her determination that each child receive a sound education.

In school, Brodie did not distinguish himself in either general studies or science. Indeed, he fondly recalls his slow childhood development with a remarkable ability to laugh at himself. His language development was far from precocious; he did not begin to talk until the age of four. Even during adolescence, his record was far from exceptional; a high school chemistry teacher once refused to recommend him for a summer job.

When recounting stories about himself, Brodie related them with tremendous personal charm, magnetism, and a unique sense of humor. His eyes would twinkle in delight with a humorous remark, situation, or paradox that he enjoyed telling, often at his own expense. He enticed his colleagues to follow him and his scientific ideas; few could resist his offer to work in his laboratory. All his collaborators have fond memories of the time they spent with him, despite his exacting demands of time and dedication to science.

In 1926, after a disagreement with the principal, Brodie dropped out of high school and enlisted in the Royal Canadian Signal Corps. The army converted a shy, introspective teenager into a confident young man who learned to be aggressive, to box well, and defend himself. Indeed, he became an expert boxer and won the Canadian Army championship in his weight division. This ambition in boxing was not to win or to become a champion, he claims, but simply to avoid getting hurt. The discipline of achieving excellence at something he did not find truly enjoyable stood him in good stead in later years when he needed to master difficult subjects in college and graduate school.

With money won playing poker in the army, Brodie enrolled at McGill University. During his early college years, he found little incentive to work hard or make high grades, even in science courses. His interests had not yet been stimulated and he recounts falling asleep in chemistry class. However, a chance experience during his fourth year stimulated his interests and markedly changed his motivation.

His chemistry professor, W. H. Hatcher, needed assistance with an experiment requiring long periods of careful monitoring. Brodie agreed to help, and stayed up night after night to obtain data for Hatcher. During this time, Brodie fantasized about the life and questions of science—wondering how ideas were generated for such experiments, frequently envisioning himself as a scientist. Brodie loved this work and was, at last, motivated to work, learn, and understand. His grades soon improved from Cs to As. With help from his chemistry professor, his first paper was published in 1931 when Brodie was only 23 years old (1).

After graduating from McGill in 1931, Brodie enrolled in an organic chemistry program at New York University under R. R. Renshaw. He earned his Ph.D. in organic chemistry in 1935 from NYU, and took a position in the Pharmacology Department with George B. Wallace for $800 per year. In 1937, Brodie published three papers that foretold his future direction: two were published with Wallace in the *Journal of Pharmacology and Experimental Therapeutics* on the distribution of administered iodide and thiocyanate in relation to chloride under both normal and pathological conditions (2, 3); and the third with M. M. Friedman in the *Journal of Biological Chemistry* on a new method to measure thiocyanate in tissues (4). Two years later, Brodie published another paper in the *Journal of Biological Chemistry* on a new method to measure bromide in tissues and biological fluids (5). He also developed assays for calcium and magnesium to determine the effects of these cations on the sleep cycle of dogs.

These papers symbolized the intellectual and experimental pattern that recurred throughout Brodie's scientific career. First, he developed a methodology to understand drug metabolism, disposition, and response. Second, the methodology served as a precursor to the generalization of fundamental principles and underlying concepts.

The ambience in the NYU Pharmacology Department was irresistible, leading Brodie to commit himself totally to a career in pharmacological research. There were two influential persons at NYU who contributed to his commitment: Wallace, a distinguished pharmacologist who had trained in Europe, and Otto Loewi, the Nobel Laureate. Coincidentally, Loewi escaped from Nazi Germany with the help of Brodie's uncle. Brodie admired Wallace and learned from him the importance of formulating a creative and testable hypothesis in well designed experiments. Brodie also learned about the critical role of reliable and accurate methods of measurement, crediting Wallace with transforming him from the organic chemist he trained to be into a pharmacologist and biologist. At that time, pharmacology was largely a descriptive science, but Brodie was singularly responsible for combining the two disciplines by using his knowledge of organic chemistry to create novel methods of drug measurement.

In 1941, the Goldwater Research Service of NYU was directed by James A. Shannon, a fine, young renal physiologist trained by Homer Smith.

Shannon had an unsurpassed gift for scientific leadership, and exhibited an uncanny ability for selecting outstanding young scientists, organizing them into an effective and productive team, and guiding them along the arduous path of difficult problem-solving. He inspired these young men and gave them freedom to express their creativity.

In 1941, Shannon's mission at Goldwater was specific: to develop more effective and less toxic antimalarial therapy. Allied troops were at great risk because the Japanese invasion had cut the world's supply of quinine. He assembled an outstanding group of young, enthusiastic investiators, including Brodie, John Burns, Julius Axelrod, Sidney Udenfriend, Robert Berliner, Gordon Zubrod, J. Tagert, D. Earle, and Murray Steele, whose job was to coordinate clinical and laboratory studies to improve the therapy of malaria. Clinicians and basic scientists were brought together for the first time to solve a fundamental therapeutic problem. This synergistic approach was novel for the academic world, and resembled the subsequent collaboration of scientists in Los Alamos who were brought together to design the atom bomb.

Atabrine had been synthesized and used as an antimalarial, but its toxicity was considerable and its effectiveness against the disease varied. A new therapy was needed, different from that which adjusted the dosage of atabrine according to clinical response. Shannon knew the distinguished pharmacologist Kenneth Marshall from Johns Hopkins University, and was familiar with his brilliant demonstration that dosages of sulfonamides could be adjusted more effectively through blood sulfonamide concentrations. This work implied that a specific narrow range of plasma sulfonamide concentration was associated with a therapeutic effect, whereas lower blood levels were clinically ineffective and more toxic. Shannon suspected that the same principle might apply to atabrine. Until that time, however, atabrine concentrations in plasma and tissues could not be accurately measured with existing methods because of interference from its metabolites.

By extraction in solvents of different polarity, Brodie and Udenfriend separated atabrine from its main metabolites (6). The assay revealed that the drug localized predominantly in liver and skeletal muscle where it was present in several hundred-fold excess over plasma. However, atabrine had to be present in the plasma to attack the blood-borne malarial parasite. The knowledge of atabrine's peculiar distribution in the body provided the foundation for devising more satisfactory dosing regimens. A very high loading dose of atabrine successfully provided effective antimalarial plasma atabrine concentrations. This loading dose was followed by smaller doses to sustain plasma atabrine concentrations at the necessary and effective antimalarial level. A plan for maximizing the benefits of atabrine had been developed by the spring of 1943, less than two years from the start of the effort, largely due to Brodie's assay (7).

Brodie recalls the Goldwater period not only as one of the most exciting in his life, but also as the real beginning of his scientific career. The personal and professional precepts developed through the Goldwater experience were successfully employed throughout different contexts at LCP: the heady exhilaration and excitement of group efforts urgently committed to solve difficult scientific problems; the need to develop new, sensitive, and reliable methods, such as those that successfully separated parent drugs from their principal metabolites; the identification of drug distribution patterns in body tissues and fluids in order to develop rational dosage regimens; and, perhaps the most important, how to foster productive collaborations among basic scientists.

In 1946, Shannon left Goldwater to become director of the Squibb Institute for Medical Research. Brodie remained at Goldwater and, in January, 1947, published a series of six papers describing the separation of drugs from their principle metabolites. These papers emerged from the atabrine work but transcended it by creating broad, general principles that showed how to accurately measure drugs and their metabolites, uncontaminated by one another (8–13).

The generalizability of these methods allowed their application to the separation of other basic drugs from their metabolites. Brodie himself used these methods to describe the metabolism and disposition of drugs common at that time: acids, as well as bases, including acetanilide, antipyrine, aminopyrine, phenacetin, atropine, phenylbutazone, procaine, procainamide, dicumarol, thiopental, and salicylic acid.

In 1949, Shannon was named scientific director of the National Heart Institute and persuaded many of his Goldwater associates, including Brodie, to join him in Bethesda, Maryland. At that time, NIH had little prestige. Many academic mentors of the young Goldwater scientists counseled against following Shannon to NIH. Those who ignored this advice were Brodie, Udenfriend, Axelrod, Berliner, and Zubrod, among others. With imaginations ignited by the Goldwater experience, they determined to succeed. This determination surpassed all dreams. During the twenty year span between 1950–1970, Brodie's LCP was extraordinarily productive; it opened exciting new vistas in remarkably diverse fields, created better methodology, and developed novel scientific paradigms. James A. Shannon, the man largely responsible for creating NIH as it exists today, stated (14):

"Starting his career with a Ph.D. in organic chemistry from New York University in the thirties, Dr. Brodie moved into biology through quantitative studies on the physiological handling of halogens. He expanded his views on the interaction of chemicals with complex biological systems incidental to his studies, which provided a quantitative base for the rational development of antimalarials during World War II. He has moved through a series of subspecialties in pharmacology and related physiology in a highly productive fashion since that time. These included the fields of analgesics, antiarthritic and antiarrhythmic agents, and the exploration of the control of CNS function, to mention a few. His activities have resulted in a profound increase in our understanding of drug action, but more importantly, have modified our views on what could or indeed should be done to clarify the complexities of a broad field of concern common to all of us. But then Dr. Brodie's contributions to medicine, and its scientific base, is reflected less in his direct contributions than in the careers of the many scientists who have been affected by his work."

Science in the LCP

This brief bibliography inadequately describes the dynamic nature of Brodie's insatiable quest to discover fundamental biological truths and mechanisms. This was his driving passion, and the commitment and inspiration he instilled in those who worked with him is what distinguished the LCP from other laboratories. This quest often led to endless sequences of experiments lasting days and weeks, punctuated by long discussions and debates on the meaning and inferences of experimental results. If the experimental conclusions supported the obvious, Brodie insisted on another perspective to aim at the unknown and the novel. He penetrated beneath the superficial and the obvious to the profound, conceptually innovative, and biologically meaningful. Thus, the laboratory traveled with him in an exciting journey beyond the existing limits of knowledge. Future experiments would generate something entirely new, something previously unsuspected. The stimulating challenge of discovery made the researchers in the laboratory more than willing to endure the many associated difficulties and discomforts.

The characteristic of relentless inquiry explains why, for Brodie, there were no simple, straightforward results. Experimental data possessed wider significance, if only one would take the time and trouble to search for the more complete meaning. Results of experiments gave hints and clues of the overall, larger picture: the important underlying biological principle. Brodie's particular gift was to detect and extract information from isolated results, which often concealed the overarching concept and the unifying idea among the data. In other words, his course was never deterred by petty detail, and he maintained that it would be unfortunate and unwise to let a good idea expire on the basis of a poor experiment or incomplete "facts".

Detail annoyed him, as, for example, undue concern about the statistical significance of results. If a scientist had to rely heavily on statistics to discover a basic, organizing pattern, he felt, something was amiss. An overreliance on statistics might encourage the vision of a new biological principle that was, in reality, nonexistent. One did not need statistics when a bona fide phenomenon emerged.

Brodie's approach to the scientific literature was similar: if it was necessary to consult the literature to determine whether an idea was new and worth pursuing, it probably was not. Any new and worthy experiment would, by definition, not have been conceived and executed previously. Therefore, lengthy literature searches were unnecessary.

While such attitudes may surprise contemporary scientists, they were justified in Brodie's era. He possessed a remarkable, if not unique, intuition and sensitivity for uncovering broad biological principles and for sensing what was biologically significant and what was not. Thus, his contributions were original, innovative, and represented major departures from contemporary dogma. His mind sought connections between diverse facts and multiple experiments, always careful to avoid superfluous detail and unimaginative, limited conclusions. Brodie continually challenged himself to synthesize disparate phenomena for the purposes of conceptual unification, in much the same way as Einstein attempted to synthesize diverse phenomena in physics for the purpose of a general field theory.

To do this, Brodie continuously asked questions, often in rapid fire succession, but always orderly and logical, progressing from the seemingly elementary to the increasingly complex. Questions flowed with animation and vivacity as though they were part of his inner nature and needed immediate release. This process was, of course, the Socratic method of discussion—a series of progressive questions, each designed to test and advance beyond the preceding answers, ideas, and insights to approach the truth.

Brodie loved, and wisely availed himself of, the flourishing intellectual environment at NIH which abounded with experts. In many discussions with the NIH authorities his continuous line of debate and questions, seemingly uninspired at times, led ultimately to new insights and the emergence of experimentally testable hypotheses.

Brodie's urge to pursue new knowledge often occurred in the early morning hours when he would invite colleagues to his house to participate in long Socratic discussions. If an erroneous hypothesis occasionally led them astray, it would self-correct when an experiment revealed the error of a particular idea and a new and better idea would replace it. As unquenchable as was Brodie's appetite for asking questions, his appetite for suggesting, developing, and performing experiments was even keener. Brodie's experimental designs were generally direct, straightforward, logical, and easily reproducible. Excuses for not performing experiments were unacceptable: in his view they were always possible. He was fond of saying that experiments would never be performed if scientists allowed themselves to be talked out of doing them.

It is indeed remarkable how many brilliant, new, unifying concepts Brodie was responsible for discovering in his career. These will be discussed in later sections, but in conclusion, a frequently quoted and misunderstood remark of Brodie's, "Let's take a flier on it", needs to be explained. This remark has been attributed to Brodie's penchant for gambling and following far-out scientific hunches. His close friends named him after the legendary bartender and gambler, Steve Brodie, who, on a bet, jumped off the Brooklyn Bridge and won by surviving. An amusing anecdote to Bernard Brodie's scientific style perhaps, but it incorrectly conveys the role of chance in his scientific career, a false impression propagated in Robert Kanigel's book *Apprentice to Genius* (15). This book, generally accurate and stimulating, deserves much credit for its insights and for being the first to describe in detail Brodie, LCP, and its rich environment. Kanigel wrote that Brodie's favorite challenge— "Let's take a flier on it"—embodied a style of science characterized by a breathtaking freedom to be wrong. On the contrary, when Brodie suggested an experiment, he was convinced that the effort would not be wasted, but rather would advance in logical, rational, and necessary steps on a path ultimately leading to truth. The intention of a scientific gamble, or flier, was to stimulate, even entice, his colleagues into viewing things from his perspective in order to perform often arduous and tedious work. Brodie's remarkable ability to perceive the big picture and his intuition for the discovery of novel scientific principles made most of his experiments a success. Because an astonishing number of his concepts received experimental confirmation and opened new areas in pharmacology, Brodie, who published approximately 400 scientific papers, is considered by many to be the father of modern biochemical pharmacology and neurochemical pharmacology.

Contributions of LCP to Drug Metabolism and Disposition

The contributions made by Brodie and LCP to drug metabolism and disposition were numerous and fundamental. Although he was the director of the LCP and indisputably the intellectually dominant figure, he acknowledged the critical role played by his close associates. The LCP team was large, its composition continually changing and constantly enriched by new members. Indeed, there were literally hundreds of scientists trained there who left for subsequent scientific positions in other institutions armed with the concepts developed at LCP. They also took with them the inspiration, excitement, and voracious appetite for novel experiments that characterized the spirit of the LCP. They too thirsted for a fresh vision of nature. Many became directors of laboratories and institutes throughout the world and are today considered international leaders in pharmacological research and in the neurosciences: a measure of their success.

We had intended to include in this reminiscence a list of such scientists, but it was too long; it kept growing as we tapped more sources and questioned more colleagues. Such a list, too, would not include those who did not serve an apprenticeship at LCP but were nonetheless profoundly influenced and even transformed by the ideas and principles generated there. A count of guest scientists from foreign countries who trained at LCP from 1950–1970 yielded 79 names from 29 different countries. Thirteen others received their doctoral degrees in pharmacology from George Washington University for research performed in LCP under Brodie's direction. The scientists there were from different disciplines and countries and provided a rich intellectual environment; without this, the enormous achievements and discoveries that emanated from LCP would have been impossible.

One notable feature of the research performed at LCP in the traditional area of drug metabolism and disposition was the extraordinary array of topics, problems, drugs, and approaches; that is, the staggering diversification of themes. The basic and applied implications and significance of these themes characterized the wide significance of LCP research. Brodie's fundamental contributions to analytical methodology were based on principles of differential extraction of drugs and their main metabolites by solvents of different polarity. These papers had such a profound influence on pharmacology and medicine that it is difficult to overestimate their importance. Present day assays of drugs and their metabolites depend on these principles; new drugs cannot be approved without first obtaining information on metabolism and disposition.

Other applications of these assays include (*a*) studies on drug interactions; (*b*) pharmacogenetics (individual differences in drug elimination due to genetic factors); (*c*) environmental and developmental factors that influence rates of drug clearance; (*d*) diagnosis and management of poisoned and overdosed patients; (*e*) assessment of patient compliance with medication regimens; and (*f*) selection of appropriate routes of drug administration, dose, and dosage intervals. This list exemplifies the principle that the more fundamental and basic a discovery is, such as the idea that drug and metabolite concentrations can be quantified through the use of solvents of different polarity, the broader the application.

Another fundamental discovery made in LCP was that of the mixed function oxidases or drug-metabolizing enzyme system. This system, also designated the cytochrome P-450-mediated monooxygenases, is located in the smooth endoplasmic reticulum in liver cells and other tissues. It requires molecular oxygen as well as NADPH and was first described in 1953 (in abstract form) by LCP's Bert La Du, who showed how it was responsible for the hepatic demethylation of aminopyrine (16). It was probed and described in several additional papers from LCP (17–23).

In 1953, Axelrod was also working in LCP on the same problem and described (in abstract form) the actions of this hepatic enzyme system on amphetamine and ephedrine (24). The significance of the enzyme system has only been fully appreciated in the last few decades with the recognition of its central role in the detoxification, and occasional activation, of many drugs and foreign organic chemicals. Brodie recognized the biological significance of this discovery when it was made in his laboratory in 1953. It was customary for him to have several different individuals or groups in LCP work on the same problem, if he considered it particularly important. He recalls that his first intimation of the universal aspects of this enzyme system came from the capacity of SKF-525A to inhibit the metabolism of markedly different substrates. The system's broad substrate specificity was not lost on Brodie, and he suspected that a fundamental principle would underlie a system functioning to detoxify such a wide variety of foreign chemicals.

He postulated a grand evolutionary scheme: aquatic creatures did not need such a system for detoxifying foreign chemicals because they could readily eliminate lipid soluble foreign compounds through their gills. But as vertebrates evolved to land-living forms, they required enzymes that catalyzed disposal of lipid soluble alkaloids present in their foods; otherwise such compounds would remain in fat depots indefinitely. Drug-metabolizing enzymes allowed plant alkaloids and other lipid-soluble environmental chemicals to be converted to more polar, and hence, more excretable metabolites. Thus, their activity increased progressively as life evolved from the creatures confined to the sea to amphibians, reptiles, birds, and mammals (25).

His evolutionary hypotheses inspired studies on species, strain, and sex differences in rats, as well as pathways of drug metabolism. Brodie shared this interest with his longtime friend, R. T. Williams, from St. Mary's Hospital, London. Perhaps the best known expression of Brodie's commitment to investigating species differences is the paper by Quinn, Axelrod, & Brodie (26).

A corollary of the research addressing the evolution of drug-metabolizing systems in vertebrates was the investigation of the activity of drug-metabolizing enzymes in the neonatal period. Brodie's ontogenetic studies revealed that the drug-metabolizing enzymes in newborn mammals are greatly reduced or even absent. These observations had profound implications and explained the long-recognized increased sensitivity of newborns to many drugs and other environmental chemicals (27). It also reaffirmed the fundamental biological principle that ontogeny recapitulates phylogeny.

The identification of large differences among normal human subjects in their rates of drug elimination was influenced by the studies on species and ontogenetic variations in drug metabolism, the development of methodology to measure drugs and their metabolites, and the discovery of the drug-metabolizing enzymes. Although individuals often differed by as much as 10-to 20-fold in the rates of drug elimination, repeated measurements on the same individual were highly reproducible. Stimulated by Brodie's probing questions, twin and family studies were conducted using a wide variety of drugs metabolized by this enzyme system. These studies revealed genetic control of large interindividual differences. In identical twins, metabolic variations in drug elimination vanished, but in fraternal twins differences were largely preserved. As genetic variation disappeared, so too did pharmacokinetic differences. Yet other studies in LCP revealed that environmental factors could play important roles in affecting rates of drug elimination. One of these was exposure to numerous inducing agents, such as phenobarbital. Response to inducing agents was not uniform; rather, it was variable and influenced by genetic factors (28). Clearly, genes and environment dynamically interact to cause interindividual, as well as intraindividual, differences among patients in their response to drugs.

Many studies emanating from LCP using Brodie's methods for measuring drugs and their metabolites involved drug distribution. Fundamental principles about the influence of pH on the passage of drugs across lipid membranes emerged from these studies. Acidic drugs under acidic conditions were shown to be lipid soluble and, hence, crossed membranes most rapidly. Conversely, basic drugs rapidly transversed lipid membranes under alkaline conditions. These fundamental principles had many practical applications in medicine as well as in the pharmaceutical industry in designing products for maximal absorption.

Two of Brodie's major contributions can be cited as examples (29, 30). The first correlated the distributional properties of thiopental with its clinical characteristics. Its rapid onset of action followed from its lipid solubility, derived from its being largely nonionic at plasma pH, and its very high partition coefficient. These factors accounted for its rapid uptake by the brain. Its short duration of action as an anesthetic, lasting approximately five minutes, initially appeared attributable to rapid hepatic metabolism. However, after a series of questions, the redistribution from brain to other tissues as plasma thiopental levels decreased was shown as a critical event. From plasma, thiopental localized in adipose tissue.

The second example concerns the role of plasma protein binding of a drug in determining its clinical properties. It is also illustrated by thiopental: its rapid uptake by the brain occurs in part because it is negligibly bound to plasma proteins. Since this is a negative instance of an important principle, we will illustrate the principle with a positive example: Brodie and his associates showed that the long duration of action of phenylbutazone and dicumarol arises from their extensive and tight binding to plasma protein.

Another fundamental principle emerging from LCP concerns the potential pharmacological interest and clinical utility of drug metabolites in their own right. Biotransformation of phenylbutazone to the pharmacologically active oxyphenylbutazone (Tandearil[R]) and to another metabolite that ultimately yielded sulfinpyrazone (Anturan[R]) exemplified this concept (31).

Space limitations prevent a description of all the major contributions by

Brodie and the LCP to the field of drug metabolism. The final example is chronologically the last in which Brodie was the main contributor and, once again, is anecdotal. On a visit to a sheep station while vacationing in Australia, Brodie was told that CCl_4 was used to deworm sheep. Almost all the sheep tolerated CCl_4 well, but an occasional one died of hepatic necrosis. It was known that pretreatment of sheep with phenobarbital greatly increased the liver toxicity of CCl_4. The hepatic drug metabolizing system was not acting to detoxify foreign compounds, but rather to activate them to produce transient metabolites, such as epoxides. Brodie perceived that the highly reactive intermediates could initiate processes leading to tissue necrosis, mutation, carcinogenesis, and teratogenesis through covalent bonding to critical cellular macromolecules (32). Many LCP researchers participated in this landmark work that covered a broad range of toxic metabolites formed from such chemicals as carbon tetrachloride, bromobenzene, and acetaminophen. Key contributors on the research team included Jim Gillette, Jerry Mitchell, David Jollow, Watson Reid, Jack Hinson, and Bill Potter. This work exposed a general principle of broad biological significance; few examples of toxicity from environmental chemicals are excluded by it. The proof involved subtle use of chemicals that act on the drug-metabolizing enzymes, specifically pretreatment with inducers such as phenobarbital that enhance toxicity or with inhibitors such as piperonyl butoxide that reduce toxicity. Again, the practical applications of this fundamental original concept were enormous. In fact, their importance unfolds to this day.

The Serotonin (5HT) and Reserpine Connection

A series of studies conducted in the LCP on the relationship of 5HT and reserpine marked the beginning of the discipline of Neurochemical Pharmacology. These studies pioneered a novel methodological strategy using drugs as tools; they were used to expose neurochemical changes of physiological relevance and then to study the relationship between neurochemical and physiological change. This strategy, developed over three decades ago, is still used repeatedly in several laboratories throughout the world, and was instrumental in uncovering the relationship between 5HT and reserpine. This drug was observed to metabolize rapidly in the body; however, its action persisted while drug plasma and tissue levels were undetectable (33). Researchers in Brodie's laboratory noticed a structural resemblance between reserpine and 5HT and examined the effects of reserpine on the urinary content of 5-hydroxyindole acetic acid (5HIAA) a major 5HT metabolite. It was found to be markedly elevated. In 1955, many laboratories speculated about the role of 5HT in brain function; with the technology to measure it (34), researchers at LCP were able to show that the 5HT content of rat and rabbit brain would become undetectable after an intravenous injection of reserpine (35). After reserpine, brain 5HT became undetectable for longer than 24 hr; this decrease persisted long after reserpine was virtually absent in tissue extracts. Brodie and his collaborators interpreted this as evidence of a reserpine-induced syndrome related to an inhibition of 5HT storage caused by undetectable amounts of the drug bound to a specific process operative in 5HT storage. From these studies, Brodie and his colleagues suggested that reserpine inactivates a mechanism essential for 5HT storage (36).

Since reserpine elicits a long-lasting symptomatology characterized by immobility, increased muscle tone, blepharospasm, salivation, chromodacryorrhea, hypotension, and other cardiovascular changes, the time course of the symptoms evoked by reserpine was discovered to be related to the depletion of brain 5HT (37). It was suggested that the depletion of brain 5HT was central to explaining the syndrome evoked by reserpine. Later, other laboratories reported that brain norepinephrine and dopamine contents were also depleted by the reserpine action on the monoamine storage process suggested by Brodie (36). This catecholamine depletion helped to explain some of the symptoms of the reserpine syndrome.

Drugs, Neurotransmitters and Function

Several different paths of research were opened by the initially exciting findings on the mechanism of action of reserpine. Following the demonstration in LCP that the long-lasting effects of reserpine were associated with the depletion of 5HT, and despite the evidence that brain catecholamines were depleted, a series of studies led Brodie to establish that the impairment of brain 5HT storage was operative in the sedative action of reserpine (38–41).

The heuristic and novel idea of explaining drug actions via their effects on transmitter function was exploited in the LCP with such drugs as monoamine oxidase (MAO) inhibitors, antidepressants, and hypotensives. As with reserpine, it was necessary to determine whether the central actions of MAO inhibitors were related to the effects on brain 5HT or norepinephrine. The pertinent studies showed a relationship between the characteristically delayed action of some MAO inhibitors to the gradual increase in norepinephrine levels (42). The resulting excitatory action of MAO inhibitors, such as iproniazid, was compared in Brodie's laboratory with the association of reserpine and a new category of drugs, tricyclic antidepressants, which included imipramine (43). Brodie recommended that 5HT rather than norepinephrine may be involved in the antidepressant action of imipramine (44, 45). These studies, combined with the classical drug metabolism approach of LCP, led to the use of desmethylimipramine in depression (an imipramine metabolite, particularly abundant in rats receiving imipramine). The early concept of the anti-5HT mechanism of the action of LSD was consistent with Brodie's demonstration of the importance of 5HT in brain regulatory mechanisms (46).

It is worth remembering that researchers at LCP initiated biochemical studies of the 'adrenergic neuron blockers'—bretylium and guanethidine (47). The demonstrated action of these drugs on the sympathetic nerve terminals was defined at LCP on the basis of their selective depletion of heart norepinephrine stores; bretylium did not change heart NE content, which is descreased by guanethidine (47, 48). Furthermore, Brodie stressed the importance of the physicochemical characteristics of these drugs acting solely at the periphery (48, 49). One important implication of these studies is their clinical application in the treatment of hypertension.

Altogether, these studies led to the development of important concepts of neurochemical pharmacology that included brain storage, uptake, and release of monoamines, and the role of MAO as a regulatory process of neuronal monoamine content. It should also be recognized that this work preceded the histochemical documentation using light and electron microcopy that showed that neuromodulatory monoamines are stored in the synaptic vesicles of specialized neuronal systems and that MAO is located in mitochondria. There are many representative publications illustrating the importance of this work (50–53).

An important aspect of these studies was their emphasis on the relationship between the pharmacodynamic and behavioral effects of drugs acting on the central and peripheral nervous system, particularly as it relates to their actions on monoamine stores. Besides the tranquilizing and sedative action, effects ranging from anticonvulsant to antihypertensive actions were described in detail to depend on their action on the dynamics of neuronal monoamine stores (54, 55). There was, of course, much interest in these studies as they emphasized the functional significance of the changes in turnover rate of peripheral and central monoamine transmitters. There are several illustrative examples of this approach (56, 57).

The interest in the functional aspects of drugs affecting brain monoamine synthesis and storage led Brodie to creatively suggest that 5HT and norepinephrine are regulatory agents of the central sympathetic and parasympathetic systems (58). An even broader view of the function of these amines could be developed along the lines of the physiological concepts earlier proposed by Hess (34). As formulated by Brodie, autonomic, extrapyramidal, motor, and psychic functions are coordinated by two opposing subcortical divisions: ergotropic and trophotropic (59). They are represented particularly in the posterior and anterior hypothalamus, respectively, and in additional diencephalic areas. In fact, the LCP accumulated behavioral, pharmacodynamic, neurochemical, and other evidence to show that free brain 5HT produces actions opposite to those of free norepinephrine, while LSD mimics the effects of the latter and antagonizes the effects of 5HT and its precursor, tryptophan. In other words, free brain 5HT is trophotropic and sedative, while free brain norepinephrine is ergotropic and excitatory. There is a certain ambiguity in this reasoning, however, because the tranquilizing and sedative drug, reserpine, is a 5HT depleter; but 5HT is a trophotropic agonist. However, Brodie distinguished between depletion on the one hand and the free 5HT, on the other, that resulted from normal or accelerated synthesis of 5HT associated with the inhibition (by reserpine) of the process that stores and protects 5HT from uncontrolled release. As a result, reserpine could cause the unregulated availability of the neuromodulator to its specific receptors. Some studies partially represent these ideas (45, 50, 59). Brodie's demonstration of the functional and behavioral role of central monoamines supported their role as central neuromodulators or neurotransmitters.

Transduction, Modulation, Turnover and Second Messengers in Transmitter Function

In the late 1950s and early 1960s Brodie was somewhat unhappy with the concept of the monotypic chemical signal in synaptic transmission, on account, perhaps, of his awareness of the biochemical heterogeneity of nervous tissue, or of the interplay between 5HT, catecholamines, and acetylcholine, the parasympathetic neurotransmitter, which had been demonstrated at the LCP. This thinking influenced his concept of the synapse (particularly those activated by the sympathetic nerve endings) as a "highly organized molecular unit . . . the neurochemical transducer . . . mediating . . . the change of one kind of energy into another" (54). In this complex functional structure of the 'neurochemical transducer', a number of processes were controlled in the presynaptic terminals, including the kinetics of the neurotransmitter turnover that kept enough transmitter in storage, even in the face of wide changes in the release rates. He also was aware of the crucial role played by the postsynaptic transduction of the synaptic signal into a specific message for the postsynaptic neurons via enzymatic processes that were subsequently indicated to produce specific second messengers activating important metabolic processes (phosphorylation) in the postsynaptic cells. This was the beginning of our current understanding of polytypic- or polytransmitter-synaptic signaling. This process, operative in the regulation of synaptic strength, is changed by the quality of the interaction of synaptic signals and can become somewhat independent from the amount of the primary transmitter that is released. This was a more innovative and dramatic concept than the classical conception of monotypic synaptic signaling of the late 1950s and early 1960s.

The kinetic studies that maintain transmitter steady state in the monoamine stores located in presynaptic endings soon led to the concept of the dynamic state of transmitter storage regulation. By measuring the oscillation of this dynamic state associated with functional changes, drug action in a given neuronal system could be assessed. This measurement, known as the evaluation of transmitter turnover, became an important procedure in the detection of indirect drug action on a given neuronal system (60). In establishing this concept at the LCP, the necessary methodological and mathematical analyses were formulated for the measurement of transmitter turnover (61, 62). These measurements included the pioneer use of radiolabeled transmitters and their precursors to assess monoamine turnover rates. At this time, it was reasoned that a measurement of transmitter turnover would allow the assessment of the degree of functional participation in a specific brain structure of a given neuronal pathway. Today, the profiles of drug action on the CNS are routinely studied by measuring the turnover of various transmitters in different brain structures, rather than measuring changes in transmitter levels. Furthermore, the methodology originally established for the measurement of the turnover of norepinephrine was adapted to the measurement of the turnover of other transmitters (e.g. 5HT, acetylcholine, and the amino acid GABA).

No less important were the advances made in the area of synaptic modulation by receptor–receptor interaction. The LCP studies that reported how specific ganglionic catecholamine stores regulated cholinergic ganglionic transmission constituted the first demonstrations of both the phenomenon of modulation and the participation of this phenomenon in the regulatory mechanism contributing to the individual diversity of brain function. As stated at the time, the results obtained with reserpine and monamine oxidase inhibitors "support the notion that norepinephrine in the ganglia modulates the action of acetylcholine" (63). It is now known that both pre- and post-synaptic modulation of transmitter receptors participate in the regulation of synaptic strength.

The unique discovery of the changes of synaptic strength through receptor–receptor interaction remained dormant for over 15 years; when revived, however, it guided the study of the mode of action of benzodiazepines. Brodie participated in these studies after he retired from NIH. Indeed, as it turned out, these studies led him to formulate the concept that a homeotypic receptor modulation was involved in explaining the role of GABA receptor modulation to understand the molecular mechanisms whereby benzodiazepines exert their anxiolytic action (64).

Other advances were made at the LCP on the concept of signal transduction at synapses and its involvement in the generation of a second messenger. Important initial studies were carried out on the second messenger role of cyclic nucleotides in the FFA mobilization by sympathetic neurons. If time had permitted, Brodie would undoubtedly have demonstrated the second messenger role of the cyclic nucleotide system in the pharmacodynamic response of 5HT and catecholamines. Many representative publications exemplify this work (65, 66).

The principles discovered at LCP are still extensively applied in many laboratories as tools to study the mechanism of drug actions on the CNS, and have become standard procedures for determining the neurochemical profiles of drug actions in the central nervous system. In homeotypic receptor regulation the process of allosteric modulation of the primary transmitter recognition site changes the probability of the primary transmitter action. In summary, under Brodie's leadership (63) the LCP has provided the seminal observation that synaptic strength can be modulated by the quality of multiple chemical signals released from nerve terminals and can thereby become relatively independent of the changes in the amount of primary transmitter released.

Two major findings at LCP have directed brain researchers to discover heterotypic and homeotypic regulation of transmitter receptor excitability. In 1961, while studying the transmission of nerve impulses in cervical sympathetic ganglia of cats and rabbits receiving reserpine, it was found that one or two hours after reserpine injection the transsynaptic excitation of nicotinic receptors were facilitated (63). Such a facilitation was possibly mediated by the reserpine-induced depletion of specific stores of ganglionic biogenic amines operative in synaptic strength modulation. Extending this mechanism to the brain, we now recognize many examples where monoamines are not primary transmitters, but exert a modulatory role on the excitability of other transmitter receptors.

Brodie's second major contribution can be seen in the work on GABA receptor homeotypic modulation (64). Through this work, it is now widely accepted that a heterotypic receptor interaction and a homeotypic receptor modulation participate in synaptic transmission variability, which probably is a component for individual brain function variability.

Hypothalamico-Pituitary-Adrenal Axis, Psychotropic Drugs and Homeostasis

Brodie suggested that reserpine stimulates the pituitary-adrenocortical system and causes a hypersecretion of ACTH (37). Ultimately, ACTH depletion ensues. Additional work established the relationship between FFA mobilization, glycolysis, thermocontrol, and catecholamines; in addition, the parasympathetic nervous system was also shown to be a necessary part of the physiological mechanism (65) involved in energy and temperature regulation. Thus, cholinergic agonists such as tremorine or physostigmine evoked the loss of body heat in animals exposed to low environmental temperature and decreased the rate of utilization of energy substrates, while catecholamines exerted the opposite effect. The effects of catecholamines, as well as of cold, on the mobilization of FFAs and glucose depended on the integrity of the hypothalamico-pituitary-adrenal axis.

Peripheral way-stations were involved in these effects as well, and some of these phenomena depended on the intactness of the sympathetic ganglia and on the activation of the adipose tissue lipase by the sympathetic nervous system (67). Altogether, these studies demonstrated that the autonomic, central, and neurohumoral parameters underlie Claude Bernard's principle of homeostasis (37, 68, 69).

CONCLUSIONS

Brodie established chemical pharmacology and developed many research trends in biochemical and neurochemical pharmacology that are still contemporary points of departure for research on drug action. He was well-loved and respected by his many colleagues, and to this day his work instills enthusiasm, dedication, and intellectual excitement in those who were fortunate enough to work with him. He created many important and distinctive fields of inquiry in our discipline that are functionally alive; his work guides us in our quest to develop new and more effective drugs for the treatment of various pathological states.

Literature Cited

1. Hatcher, W. H., Brodie, B. B. 1931. Polymerization of acetaldehyde. *Canad. J. Res.* 4:574–81
2. Wallace, G. B., Brodie, B. B. 1937. The distribution of administered iodide and thiocyanate in comparison with chloride and their relation to body fluids. *J. Pharmacol. Exp. Therap.* 61:397–421
3. Wallace, G. B., Brodie, B. B. 1937. The distribution of administered iodide and thiocyanate in comparison with chloride in pathological tissues, and their relation to body fluids. *J. Pharmacol. Exp. Therap.* 61:412–21
4. Brodie, B. B., Friedman, M. M. 1937. The determination of thiocyanate in tissues. *J. Biol. Chem.* 120:511–16
5. Brodie, B. B., Brand, E., Leshin, S. 1939. The use of bromide as a measure of extracellular fluid. *J. Biol. Chem.* 130:555–63
6. Brodie, B. B., Udenfriend, S. 1943. The estimation of atabrine in biological fluids and tissues. *J. Biol. Chem.* 151:299–317
7. Shannon, J. A., Earle, D. P., Brodie, B. B., Taggart, J. V., Berliner, R. W. 1944. The pharmacologial basis for the rational use of atabrine in the treatment of malaria. *J. Pharmacol. Exp. Therap.* 81:307–30
8. Brodie, B. B., Udenfriend, S., Baer, J. E. 1947. The estimation of basic organic compounds in biological material. I. General principles. *J. Biol. Chem.* 168:299–309
9. Brodie, B. B., Udenfriend,, S., Dill, W., Downing, G. 1947. The estimation of basic organic compounds in biological material. II. Estimation of fluorescent compounds. *J. Biol. Chem.* 168:311–18
10. Brodie, B. B., Udenfriend, S., Dill, W., Chenkin, T. 1947. The estimation of basic organic compounds in biological material. III. Estimation by conversion to fluorescent compounds. *J. Biol. Chem.* 168:319–25
11. Brodie, B. B., Udenfriend, S., Taggart, J. V. 1947. The estimation of basic organic compounds in biological material. IV. Estimation by coupling with diazonium salts. *J. Biol. Chem.* 168:327–34
12. Brodie, B. B., Udenfriend, S., Dill, W. 1947. The estimation of basic organic compounds in biological material. V. Estimation by salt formation with methyl orange. *J. Biol. Chem.* 168:335–39
13. Josephson, E. S., Udenfriend, S., Brodie, B. B. 1947. The estimation of basic organic compounds in biological material. VI. Estimation by ultraviolet spectrophotometry. *J. Biol. Chem.* 168:341–44
14. Shannon, J. A. 1971. Introductory Remarks *Ann. N.Y. Acad. Sci.* 179:910

15. Kanigel, R. 1986. *Apprentice to Genius. The Making of a Scientific Dynasty.* New York: Macmillan

16. La Du, B. N., Trousof, N., Brodie, B. B. 1953. Enzymatic dealkylation of aminopyrine and other alkylamines in vitro. *Fed. Proc.* 12:399

17. Brodie, B. B. 1956. Pathways of drug metabolism. *J. Pharm. Pharmacol.* 8:1–17

18. Brodie, B. B., Gillette, J. R., La Du, B. N. 1958. Enzymatic metabolism of drugs and other foreign compounds. *Ann. Rev. Biochem.* 27:427–54

19. Brodie, B. B., Axelrod, J., Cooper, J. R., Gaudette, L., La Du, B. N., Mitoma, C., Udenfriend, S. 1955. Detoxification of drugs and other foreign compounds by liver microsomes. *Science* 121:603–04

20. Cooper, J. R., Brodie, B. B. 1955. The enzymatic metabolism of hexobarbital (evipal). *J. Pharmacol. Exp. Therap.* 114:409–17

21. Cooper, J. R., Brodie, B. B. 1957. Enzymatic oxidation of pentobarbital and thiopental. *J. Pharmacol. Exp. Therap.* 120:74–83

22. Gillette, J. R., Brodie, B. B., La Du, B. N. 1957. The oxidation of drugs by liver microsomes: On the role of TPNH and oxygen. *J. Pharmacol. Exp. Therap.* 119:532–40

23. La Du, B. N., Gaudette, L., Trousof, N., Brodie, B. B. 1955. Enzymatic dealkylation of aminopyrine (pyramidon) and other alkylamines. *J. Biol. Chem.* 214:741–52

24. Axelrod, J., Reichenthal, J., Quinn, G. P., Brodie, B. B. 1953. Mechanism of the potentiating of beta-biethylaminoethyl dithenylpropylacepate HCI. *Fed. Proc.* 12:299–300

25. Brodie, B. B., Maickel, R. P. 1962. Comparative biochemistry of drug metabolism. *Proc. Intl. Pharmaco. Meeting. 1st. Stockholm, 1961,* 6:299–324. Oxford: Pergamon

26. Quinn, G. P., Axelrod, J., Brodie, B. B. 1958. Species strain and sex differences in metabolism of hexobarbitone, amidopyrine, antipyrine and aniline. *Biochem. Pharmacol.* 1:152–59

27. Jondorf, W. R., Maickel, R. P., Brodie, B. B. 1958. Inability of newborn mice and guinea pigs to metabolize drugs. *Biochem. Pharmacol.* 1:352–54

28. Vessel, E. S., Page, J. G. 1969. Genetic control of the phenobarbital-induced shortening of plasma antipyrine half-lives in man. *J. Clin. Invest.* 48:2202–09

29. Brodie, B. B., Mark, L. C., Papper, E. M., Lief, P. A., Bernstein, E., Rovenstine, E. A. 1950. The fate of thiopental in man and a method for its estimation in biological material. *J. Pharmacol. Exp. Therap.* 98:85–96

30. Brodie, B. B., Mark, L. C., Lief, P. A., Bernstein, E., Papper, E. M. 1951. Acute tolerance to thiopental. *J. Pharmacol. Exp. Therap.* 102:215–18

31. Burns, J. J., Rose, R. K., Goodwin, S., Reichenthal, J., Horning, E. C., Brodie, B. B. 1955. The metabolic fate of phenylbutazone (butazolidine) in man. *J. Pharmacol. Exp. Therap.* 113:481–89

32. Brodie, B. B., Reid, W. D., Cho, A. K., Sipes, G., Krishna, G., Gillette, J. R. 1971. Possible mechanism of liver necrosis caused by aromatic organic compounds. *Proc. Natl. Acad. Sci. USA* 68:160–64

33. Hess, S. M., Shore, P. A., Brodie, B. B. 1956. Persistence of reserpine action after the disappearance of drug from brain. Effects on serotonin. *J. Pharmacol. Exp. Ther.* 118:84–89

34. Bogdanski, D. F., Pletscher, A., Brodie, B. B., Udenfriend, S. 1956. Identification and assay of serotonin in brain. *J. Pharmacol. Exper. Ther.* 117:82–88

35. Brodie, B. B., Pletscher, A., Shore, P. A. 1955. Evidence that serotonin has a role in brain function. *Science* 122:968

36. Pletscher, A., Shore, P. A., Brodie, B. B. 1955. Serotonin release as a possible mechanism of reserpine action. *Science* 122:374–75

37. Brodie, B. B., Maikel, R. P., Westerman, E. O. 1961. Action of reserpine on pituitary-adrenocortical system through possible action on hypothalamus. In *Regional Neurochemistry,* ed. S. S. Kety, J. Elkes, pp. 351–61, Oxford: Pergamon

38. Brodie, B. B., Finger, K. F., Orlans, F. B., Quinn, G. P., Sulser, F. 1960. Evidence that the tranquilizing action of reserpine is associated with change in brain serotonin and not in brain norepinephrine. *J. Pharmacol. Exp. Therap.* 129:250–56

39. Brodie, B. B., Kuntzman, R. 1960. Pharmacological consequences of the selective depletion of catecholamines by antihypertensive agents. *Ann. N.Y. Acad. Sci.* 88:939–43

40. Costa, E., Gessa, G. L., Kuntzman, R., Brodie, B. B. 1962. The effect of drugs on storage and release of 5HT catecholamine in brain. Pharmacological Analysis of Cerebral Nervous Action. *Proc. Intl. Pharmaco. Meet.,* 8:43–71, Oxford: Pergamon

41. Brodie, B. B., Costa, E. 1962. Some current views on brain monoamines. In *Monamines et Système Nerveux Central,* ed. H. de Ajuriaguerra, pp. 13–49, Geneva: George & Co.

42. Spector, S., Prockop, D., Shore, P. A., Brodie, B. B. 1958. Effect of iproniazid on brain levels of norepinephrine and serotonin. *Science* 127:704

43. Shore, P. A., Brodie, B. B. 1957. LSD-like effects elicited by reserpine in rabbits pretreated with iproniazid. *Proc. Soc. Exp. Biol. Med.* 94:433–35

44. Sulser, F., Watts, J., Brodie, B. B. 1962. On the mechanism of antidepressant action of imipramine-like drugs. *Ann. N.Y. Acad. Sci.* 96:279–86

45. Brodie, B. B., Sulser, F., Costa, E. 1961. Psychotherapeutic Drugs. *Ann. Rev. Med.* 12:349–68

46. Shore, P. A., Silver, S. L., Brodie, B. B. 1955. Interaction of reserpine, serotonin and lysergic acid diethylamide in brain. *Science* 122:284–85

47. Cass, R., Kuntzman, R., Brodie, B. B. 1960. Norepinephrine depletion as a possible mechanism of action of guanethidine (SU 5864), a new hypotensive agent. *Proc. Soc. Exp. Biol. Med.* 103:871–72

48. Brodie, B. B., Chang, C. C., Costa E. 1965. On the mechanism of action of guanethidine and bretylium. *Brit. J. Pharmacol.* 25:171–78

49. Costa, E., Kuntzman, R., Gessa, G. L., Brodie, B. B. 1962. Structural requirements of bretylium and guanethidine-like activity in a series of guanidine derivatives. *Life Sci.* 1:75–80

50. Brodie, B. B. 1959. Comments on a symposium entitled A pharmacological approach to the study of the mind. In *A Pharmacologic Approach to the Study of the Mind,* ed. R. M. Featherstone, A. Simon, pp. 52–63, Springfield, Ill.: Charles C. Thomas

51. Brodie, B. B. 1959. Effects of chlorpromazine, reserpine and monoamine oxidase inhibitors on the cardiovascular system by interaction with central and peripheral neurohumoral agents. *Proc. Counc. High Blood Pressure Amer. Heart Assoc.* 82–89

52. Brodie, B. B., Spector, S., Shore, P. A. 1959. Interaction of drugs with norepinephrine in the brain. *Pharmacol. Rev.* Part II 11:548–64

53. Shore, P. A., Mead, J. A. R., Kuntzman, R., Spector, S., Brodie, B. B. 1957. On the physiological significance of monoamine oxidase in brain. *Science* 126:1063–64

54. Costa, E., Brodie, B. B. 1964. Concept of the neurochemical transducer as an organized molecular unit at synaptic nerve endings. *Prog. Brain. Res.* 8:168–85

55. Costa, E., Boullin, D. J., Hammar, W., Vogel, W., Brodie, B. B. 1966. Interactions of drugs with adrenergic neurons. *Pharmacol. Rev.* Part 1, 18:577–97

56. Prockop, D. J., Shore, P. A., Brodie, B. B. 1959. Anticonvulsant properties of monoamine oxidase inhibitors. *Ann. N.Y. Acad. Sci.* 80:643–50

57. Quinn, G. P., Shore, P. A., Brodie, B. B. 1959. Biochemical and pharmacological studies of RO 1-9569 (tetrabenazine), a non-indole tranquilizing agent with reserpine-like effects. *J. Pharmacol. Exp. Therap.* 127:103–09

58. Brodie, B. B., Shore, P. A. 1957. On a role for serotonin and norepinephrine as chemical mediators in the central autonomic nervous system. In *Hormones, Brain Function, and Behavior,* ed. H. Hoagland, pp. 161–80, New York: Academic

59. Brodie, B. B. 1957. Serotonin and norepinephrine as antagonistic chemical mediators regulating the central autonomic nervous system. In *Neuropharmacology,* ed. H. A. Abramson, pp. 323–41, Madison, N.J.: Madison

60. Montanari, R., Costa, E., Beaven, M. M., Brodie, B. B. 1963. Turnover rates of norepinephrine in hearts of intact mice, rats and guinea pigs using tritiated norepinephrine. *Life Sci.* 2:232–40

61. Beaven, M. A., Costa, E., Brodie, B. B. 1963. The turnover of norepinephrine in thyrotoxic and nonthyrotoxic mice. *Life Sci.* 2:241–46

62. Brodie, B. B., Costa, E., Dlabac, A., Neff, N. H., Smookler, H. H. 1966. Application of steady state kinetics to the estimation of synthesis rate and turnover time of tissue catecholamines. *J. Pharmacol. Exp. Therap.* 154:493–98

63. Costa, E., Kuntzman, R., Revzin, A. M., Spector, S., Brodie, B. B. 1961. Role for ganglionic norepinephrine in sympathetic transmission. *Science* 133:1822–23

64. Biggio, G., Brodie, B. B., Costa, E., Guidotti, A. 1977. Mechanisms by which diazepam, muscimol, and other drugs change the content of cGMP in cerebellar cortex. *Proc. Natl. Acad. Sci. USA* 74:3592–96

65. Maickel, R. P., Stern, D. N., Brodie, B. B. 1964. The role of autonomic nervous function in mammalian thermoregulation. *Proc. Intl. Pharmaco. Meet. 2nd.* 2:225–38, Oxford: Pergamon

66. Byus, C. V., Costa, E., Sipes, I. G., Brodie, B. B., Russell, D. H. 1976. Activation of 3',5'-cyclic AMP-dependent protein kinase and induction of ornithine decarboxylase as early events in induction of mixed function oxygenases. *Proc. Natl. Acad. Sci. USA* 73:1241–45

67. Maickel, R. P., Matussek, N., Stern, D N., Brodie, B. B. 1967. The sympathetic nervous system as a homeostatic mechanism. I. Absolute need for sympathetic nervous function in body temperature maintenance of cold-exposed adrenalectomized rats. *J. Pharmacol. Exp. Therap.* 157:111–16

68. Paoletti, R., Maickel, R. P., Smith, R. L. Brodie, B. B. 1963. Drugs as tools in studies of nervous system regulation of release of free fatty acids from adipose tissue. *Proc. First Int. Pharmacol. Meet.* 2:29–41, Oxford: Pergamon

69. Westerman, E. O., Maickel, R. P., Brodie, B. B. 1962. On the mechanism of pituitary-adrenal stimulation by reserpine. *J. Pharmacol. Exp. Therap.* 138:208–17

Julius Axelrod

Ann. Rev. Pharmacol. Toxicol. 1988. 28:1–23

AN UNEXPECTED LIFE IN RESEARCH

Julius Axelrod

Laboratory of Cell Biology, National Institute of Mental Health, Bethesda, Maryland 20892

BEGINNINGS

Successful scientists are generally recognized at a young age. They go to the best schools on scholarships, receive their postdoctoral training fellowships at prestigious laboratories, and publish early. None of this happened to me.

My parents emigrated at the beginning of this century from Polish Galicia. They met and married in America, where they settled in the Lower East Side of New York, then a Jewish ghetto. My father, Isadore, was a basketmaker who sold flower baskets to merchants and grocers. I was born in 1912 in a tenement on East Houston Street in Manhattan.

I attended PS22, a school built before the Civil War. Another student at that school before my time was I. I. Rabi, who later became a world-renowned physicist. After PS22 I attended Seward Park High School. I really wanted to go to Stuyvesant, a high school for bright students, but my grades were not good enough. Seward Park High School had many famous graduates, mostly entertainers: Zero Mostel, Walter Mathau, and Tony Curtis. My real education was obtained at the Hamilton Fish Park Library, a block from my home. I was a voracious reader and read through several books a week—from Upton Sinclair, H. L. Menken, and Tolstoy to pulp novels such as the Frank Merriwell and Nick Carter series.

After graduating from Seward Park High School, I attended New York University in the hope that it would give me a better chance to get into medical school. After a year my money ran out, and I transferred to the tuition-free City College of New York in 1930. City College was a proletarian Harvard, which subsequently graduated seven Nobel Laureates. I majored in biology and chemistry, but my best grades were in history, philosophy, and literature. Because I had to work after school, I did most of my studying during the subway trip to and from uptown City College. Studying in a crowded, noisy New York subway gave me considerable powers of concentration. When I graduated from City College, I applied to several medical schools but was not accepted by any.

In 1933, the year I graduated from college, the country was in the depths of a depression. More than 20% of the working population was unemployed, and there were few jobs available for City College graduates. I had heard about a laboratory position that was available at the Harriman Research Laboratory at New York University, and although the position paid $25 a month, I was happy to work in a laboratory. I assisted Dr. K. G. Falk, a biochemist, in his research on enzymes in malignant tumors. I also purified salts for the preparation of buffer solutions and determined their pH. The instrument used to measure pH at that time was a complex apparatus; the glass electrode occupied almost half a room. In 1935 the laboratory ran out of funds, and I was fortunate to get a position as a chemist in the Laboratory of Industrial Hygiene. This laboratory was a nonprofit organization and was set up by New York City's Department of Health to test vitamin supplements added to foods. I worked in the Laboratory of Industrial Hygiene from 1935 to 1946.

My duties there were to modify published methods for measuring vitamins A, B, B_2, C, and D so that they could be assayed in various food products that city inspectors randomly collected. Vitamins had just been introduced at that time, and the New York City Department of Health wanted to establish that accurate amounts of vitamins were added to milk and other food products. The methods used for measuring vitamins then were chemical, biological, and microbiological. It required some ingenuity to modify the methods described in the literature to assays of food products. This experience in modifying methods was slightly more than routine, but it proved to be useful in my later research. The laboratory subscribed to the *Journal of Biological Chemistry*, which I read with great interest. Reading this journal made it possible to keep up with advances in the enzymology, nutrition, and methodology. During the time I was in the Laboratory of Industrial Hygiene, I received a MS degree in chemistry at New York University in 1942 by taking courses at night. My thesis was on the ester-hydrolyzing enzymes in tumor tissues. Because of the loss of one eye in a laboratory accident, I was deferred from the draft during World War II. In 1938, I married Sally Taub, a graduate of Hunter College who later became an elementary school teacher. We had two sons, Paul and Alfred, born in 1946 and 1949.

FIRST EXPERIENCE IN RESEARCH: GOLDWATER MEMORIAL HOSPITAL

I expected that I would remain in the Laboratory of Industrial Hygiene for the rest of my working life. It was not a bad job, the work was moderately interesting, and the salary was adequate. One day early in 1946 the Institute for the Study of Analgesic and Sedative Drugs approached the president of the Laboratory of Industrial Hygiene with a problem. The president of the Laboratory at that time was George B. Wallace, a distinguished pharmacologist who had just retired as Chairman of the Department of Pharmacology at New York University. Many analgesic preparations contained nonaspirin analgesics, such as acetanilide or phenacetin. Some people who became habituated to these preparations developed methemoglobinemia. The Institute for the Study of Analgesic and Sedative Drugs offered a small grant to the Laboratory of Industrial Hygiene to find out why acetanilide and phenacetin taken in large amounts produced methomoglobinemia. Dr. Wallace asked me if I would like to work on this problem. I had little experience in this kind of research, and he suggested that I consult Dr. Bernard "Steve" Brodie. Dr. Brodie was a former member of the Department of Pharmacology at New York University and was doing research at Goldwater Memorial Hospital, a New York University Division.

I met with Brodie in February 1946 to discuss the problem of analgesics. It was a fateful meeting for me. Brodie and I talked for several hours about what kind of experiments could be done to find out how acetanilide might produce methemoglobinemia. Talking to Brodie about research was one of my most stimulating experiences. He invited me to spend some time in his laboratory to work on this problem. One of a number of possible products of acetanilide that would cause the toxic effects was aniline. It had previously been shown that aniline could produce methemoglobinemia. Thus, one approach was to find out whether acetanilide could be deacetylated to form aniline in the body. With the help and guidance of Steve Brodie, I developed a method for measuring aniline in nanogram amounts in urine and plasma. After the administration of acetanilide to human subjects, aniline was found to be present in urine and plasma. A direct relationship between the level of aniline in blood and the amount of methemoglobin present was soon observed (1). This was my first taste of real research, and I loved it.

Very little acetanilide was found in the urine, suggesting extensive metabolism in the body. Since acetanilide was almost completely transformed in the body, we looked for other metabolic products. Methods to detect possible metabolites, *p*-aminophenol and N-acetyl-*p*-aminophenol, were developed that were specific and sensitive enough to be used in the plasma and urine. Within a few weeks, we identified the major metabolite as hydroxylated acetanilide N-acetyl-*p*-aminophenol and its conjugates. This metabolite was also found to be as potent as acetanilide in analgesic activity. By taking serial plasma samples, acetanilide was shown to be rapidly transformed to N-acetyl-*p*-aminophenol (1). After the administration of N-acetyl-*p*-aminophenol, neglible amounts of methemoglobin were produced. As a result of these studies, Brodie and I stated in our paper (1), "the results are compatible with the assumption that acetanilide exerts its action mainly through N-acetyl-*p*-aminophenol [now known as acetaminophen]. The latter compound administered orally was not attended by the formation of methemoglobin. It is possible therefore, that it might have distinct advantages over acetanilide as an analgesic." This was my first paper, and I was determined to continue doing research.

Soon after Brodie and I examined the physiological disposition and metabolism of acetanilide, we turned our attention to a related analgesic drug, phenacetin (acetophenetidin). I spent some time developing sensitive and specific methods for the identification of phenacetin and its possible metabolite, *p*-phenetidine. Brodie and I soon found that in humans, the major metabolic product was also N-acetyl-*p*-aminophenol arising from the deethylation of the parent compound (2). A minor metabolite was *p*-phenetidine, which we found was responsible for the methemoglobinemia formed after the administration of large amounts of phenacetin to dogs. After the administration of phenacetin to human subjects, N-acetyl-*p*-aminophenol was rapidly formed. The speed and the amount with which N-acetyl-*p*-aminophenol was formed in the body suggested that the analgesic activity resided in its deethylated metabolite.

The laboratories at Goldwater Memorial Hospital where I began my research career were set up during World War II to test newly synthesized antimalarial drugs for their clinical effectiveness. Early in the war, the Japanese had cut off most of the world's supply of the antimalarial quinine. James Shannon, then a renal physiologist at New York University, was put in charge of this program. Shannon had the remarkable capacity to pick the right young people to carry out research in the antimalarial project. Members of the team that worked at Goldwater in addition to Steve Brodie were Sid Udenfriend, Robert Berliner, Bob Bowman, Tom Kennedy, and Gordon Zubrod. The atmosphere at Goldwater was highly stimulating, and an outpouring of important new findings resulted. It was in this atmosphere that, in a period of a few years, I became a researcher.

After completion of the studies on acetanilide and phenacetin, Brodie invited me to stay on at Goldwater to study the fate of other analgesic drugs. We received a small grant from the Institute for the Study of Analgesic and Sedative Drugs, and the Laboratory of Industrial Hygiene paid my salary. Another drug we investigated was the analgesic antipyrine. A sensitive method for the detection of this drug was developed, which has since been used by other investigators as a marker to determine the activity of drug-metabolizing enzymes in vivo. We identified 4-hydroxyantipyrine and its sulfate conjugate as metabolites of antipyrine. We also observed that antipyrine distributed in the same manner as body water. Because of this property, antipyrine has been used for the measurement of body water. Another analgesic we studied was aminopyrine. We found that this drug was de-methylated to aminopyrine and N-acetylated to N-acetylaminopyrine. Many of the drugs whose fate Brodie and I studied were later used by many investigators as substrates for the microsomal drug-metabolizing enzymes: aminopyrine for N-demethylation, phenacetin for O-dealkylation, and aniline for hydroxylation. Together with Jack Cooper, we developed a method for measuring the anticoagulent dicoumerol in plasma. In a study on the disposition of dicoumerol in humans, an exceedingly wide difference in the plasma levels of this drug was found, suggesting genetic differences in drug metabolism.

MOVE TO THE NATIONAL HEART INSTITUTE

Because I did not have a doctoral degree, I realized that I would have little chance for advancement in any hospital attached to an academic institution. I had neither the inclination nor the money to spend several years getting a PhD, so I decided to join the National Heart Institute as a research chemist. In 1949, Shannon was chosen as the director of the newly organized National Heart Institute in Bethesda, and he offered me a position. Also coming to the National Institutes of Health (NIH) at that time were many members of the Goldwater staff—Brodie, Udenfriend, Berliner, Kennedy, and Bowman.

At the National Heart Institute from 1950 to 1952, I collaborated with Brodie and his staff on the metabolism of analgesics and adrenergic blocking agents and the actions of ascorbic acid on drug metabolism. After a while, I became dissatisfied with working with a large team and was allowed to work independently. The first problem I chose was an examination of the physiological disposition of caffeine in man. Very little was known about the physiological disposition and metabolism of this widely used compound. A method for measuring caffeine in biological material was developed, and the plasma half-life and distribution were determined (3). Because of my work on analgesics and caffeine, I was delighted to be elected without a doctorate as a member of the American Society of Pharmacology and Experimental Therapeutics in 1953. K. K. Chen and Steve Brodie were my sponsors.

At that time, I became intrigued with the sympathomimetic amines. In 1910, Barger and Dale reported that numerous β-phenylethanolamine derivatives simulated the effects of sympathetic nerve stimulation with varying degrees of intensity and precision, and they coined the term *sympathomimetic amines*. Sympathomimetic amines such as amphetamine, mescaline, and ephedrine also produced unusual behavioral effects. In 1952, very little information concerning the metabolism and physiological disposition of these amines was known. Because of my experience in drug metabolism, I decided to undertake a study on the fate of ephedrine and amphetamine. In retrospect, this was an important decision.

The first amine that I studied was ephedrine. Ephredrine, the active principle of *Ma Huang,* an herb used by ancient Chinese physicians, was introduced to modern medicine by Chen and Schmidt in 1930. I soon found that ephedrine was transformed in animals by two pathways (demethylation and hydroxylation) to yield metabolic products that had pressor activity. Various animal species showed considerable differences in the relative importance of these two metabolic routes. The next sympathomimetic amines I examined were amphetamine and methylamphetamine. These compounds were shown to be metabolized by a variety of metabolic pathways including hydroxylation, demethylation, deamination, and conjugation. Marked species variations in the transformation of these drugs were also observed.

THE DISCOVERY OF THE MICROSOMAL DRUG METABOLIZING ENZYMES

When amphetamine was given to rabbits, it disappeared without a trace. This puzzled me, so I decided to look for enzymes that metabolized this drug. I had no experience in enzymology, but there were many outstanding enzymologists in Building 3 on the NIH campus where my laboratory was located. Gordon Tomkins, who occupied the lab bench next to mine, offered me good advice. Gordon had the capacity for demystifying enzymology and told me that all I needed to start in vitro experiments was a method for measuring amphetamine, an animal liver, and a razor blade. I did my first in vitro experiment with rabbit liver in January 1953. When rabbit liver slices were incubated in Krebs Ringer-buffer solutions with amphetamine, the drug was almost completely metabolized. Upon homogenization of the rabbit liver, amphetamine was not metabolized unless cofactors such as DPN (NAD), TPN (NADP), and ATP were added. I then decided to examine which subcellular fraction was responsible for transforming amphetamine. Hogeboon and Schneider had just described a reproducible method for separating the various subcellular fractions by homogenizing tissue in isotonic sucrose and subjecting the homogenate to differential centrifugation. After separation of nuclei, mitochondria, microsomes, and the cytosol, none of these fractions were able to metabolize amphetamine, even in the presence of added cofactors. However, when the microsomes and cytosol were combined, amphetamine rapidly disappeared upon the addition of DPN, TPN, and ATP. At that time Bert La Du, a colleague at the NIH, observed that the demethylation of aminopyrine in a dialyzed rat liver whole homogenate required TPN. In a subsequent experiment I found that amphetamines were metabolized in a dialyzed preparation of microsomes and cytosol in the presence of TPN, but not DPN or ATP. However, when the microsomes and cytosol were separately incubated, little or no drug was metabolized, despite the addition of TPN. I realized then that I was dealing with a unique enzymatic reaction.

Before I went further, I decided to identify the metabolic products of amphetamine produced when the combined microsomes and cytosolic fraction were incubated with TPN. One of the possible metabolic pathways might be deamination, leading to the formation of phenylacetone. After incubation of amphetamine with the above preparations, phenylacetone and ammonia were identified. These results indicated that amphetamine was deaminated by an oxidative enzyme requiring TPN either in the microsomes or cytosol to form phenylacetone and ammonia. Because of its properties and the structure of the substrate, it was apparent that this enzyme differed from another deaminating enzyme, monoamine oxidase.

Where was the enzyme located, in the microsomes or the soluble supernatant fraction? An approach that I used to locate the enzyme was to heat each fraction for a few minutes at 55°C, a temperature that would destroy heat-sensitive enzymes. When the cytosol was heated to 55°C and then added to unheated microsomes and TPN, amphetamine was deaminated. When the microsomes were heated and added to the cytosol fraction together with TPN, amphetamine was not metabolized. This was a crucial experiment, which demonstrated that a heat-labile enzyme that deaminated amphetamines was localized in the microsomes and that the cytosol provided factors involving TPN necessary for this reaction.

Bernard Horecker, then working in Building 3, prepared several substrates for the TPN-requiring dehydrogenase for his classic work on the pentose phosphate pathway. He generously supplied me with these substrates, which I could test on my preparations. I found that the addition of glucose-6-phosphate, isocitric acid, or phosphogluconate acid, together with TPN, to unwashed microsomes transformed amphetamines. A reaction common to these substrates is the generation of TPNH, suggesting that the enzymes in the cytosol fraction were reducing TPN. Incubating microsomes with a TPNH-generating system using glucose-6-phosphate and glucose-6-phosphate dehydrogenase resulted in the deamination of amphetamines. Upon incubation of chemically synthesized TPNH, microsomes, and oxygen, amphetamine was deaminated. At about the same time, I also found that ephedrine was demethylated to norephedrine and formaldehyde by enzymes present in rabbit microsomes that required TPNH and oxygen. By the end of June 1953, I felt confident that I had described a new enzyme that was localized in the microsomes, required TPNH and oxygen, and could deaminate and demethylate drugs. I reported these findings at the 1953 fall meeting of the American Society of Pharmacology and Experimental Therapeutics (4, 5).

After the description of the TPNH-requiring microsomal enzymes that deaminated amphetamine and demethylated ephedrine, several members of the Laboratory of Chemical Pharmacology at the NIH described similar enzyme systems that could metabolize other drugs by a variety of pathways, N-demethylation of aminopyrine (La Du, Gaudette, Trousof, and Brodie), oxidation of barbiturates (Cooper and Brodie), and the hydroxylation of aniline (Mitoma and Udenfriend) (6). In a study of the N-demethylation of narcotic drugs that I made soon after, it became apparent that there were multiple microsomal enzymes that required TPNH and O_2 (7). Research on the microsomal enzymes (now called cytochrome-P450 monooxygenases) has expanded enormously and has had a profound influence in biomedical sciences, ranging from studies of metabolism of normally occurring compounds to carcinogenesis. In retrospect, the discovery of the microsomal enzymes is among the best work I did.

Brodie and I were struck by the findings of investigators at Smith Kline & French that SKF525A, a compound with little pharmacological action of its own, prolonged the duration of action of a wide variety of drugs. We conjectured that the compound might exert its effects by inhibiting the metabolism of drugs. The effects of SKF525A on the metabolism of ephedrine in dogs and on the metabolism and duration of action of hexabarbital were examined. We found that SKF252A slowed the demethylation of ephedrine in the intact dog. It also prolonged the presence of hexabarbital in the plasma and the sleeping time in rats and dogs. Thus, the ability of SKF525A to prolong the action of drugs could be explained by its ability to slow their metabolism. As soon as the microsomal enzymes were described, it was observed that SKF525A inhibited this class of enzymes. Subsequently, SKF525A was widely used as an inhibitor of the microsomal enzymes.

The effect of the microsomal enzymes on the duration of drug actions was examined with the collaboration of Gertrude Quinn, a graduate student at George Washington University, and Steve Brodie. Since sleeping time of hexabarbital was easy to measure, we chose that drug to make this study. Cooper and Brodie had found that hexabarbital was metabolized by microsomal enzymes in the liver (6). The sleeping time of a given dose of hexabarbital was compared with its plasma half-life and with the activity of a liver enzyme preparation using the barbiturate as a substrate in a number of mammalian species. There were considerable differences in the plasma half-life, sleeping time, and enzyme activity among the various species (8). A high correlation was observed between the plasma half-life and sleeping time of the barbiturate. There was also an inverse relationship between the duration of action of hexabarbital and its ability to be metabolized by the microsomal enzymes.

In 1956, I reported that narcotic drugs such as morphine, meperidine, and methadone were N-demethylated by the liver microsomes requiring TPNH and O_2 (7). Differences in the rate of N-demethylation of various narcotic drugs in several species made it apparent more than one enzyme was involved in their N-demethylation. There was also a marked sex difference in the N-demethylation of narcotic drugs by rat liver microsome enzymes. Microsomes obtained from male rats were found to N-demethylate narcotic drugs much faster than those from female rats. When testosterone was administered to oophorectomized female rats, the activity of the demethylating enzyme was markedly increased. Estradiol given to male rats decreased the enzyme activity. Subsequent work by many investigators found similar sex differences in microsomal enzyme activity for many metabolic pathways.

While working on the metabolism of narcotic drugs, I observed that the repeated administration of narcotic drugs not only produced tolerance to these drugs, but also markedly reduced the ability to N-demethylate them enzymatically (9). There was also a correlation between the rate of demethylation of opiate substrates and their cross-tolerance to morphine. Opiate antagonists not only blocked the development of tolerance, but also prevented the reduction of enzyme activity. On the basis of these observations, a mechanism for tolerance to narcotic drugs was proposed. In a paper reporting these experiments, the following statement was made: "The changes in enzyme activity in morphine-treated rats suggests a mechanism for the development of tolerance to narcotic drugs, if one assumes that enzymes which N-demethylate narcotic drugs and the receptors for these drugs are probably closely related. The continuous interaction of narcotic drugs with the demethlyating enzymes inactivates the enzymes. Likewise the continuous interaction of narcotic drugs with their receptors may inactivate the receptors. Thus, a decreased response to narcotic drugs may develop as a result of unavailability of receptor sites." This hypothesis stimulated considerable critical reaction, mostly negative.

Although I had just described the physiological disposition of caffeine, demonstrated the variety of metabolic pathways of amphetamine and ephedrine, and independently described the microsomal enzymes and their role in drug metabolism, it was difficult for me to obtain a promotion to a higher rank at the National Heart Institute because I had no doctorate. I decided to get a PhD degree at George Washington University, since few courses were required if a candidate already had an MS degree. However, it would be necessary to take demanding comprehensive examinations in several subjects. Paul K. Smith, then Chairman of Pharmacology, accepted me as a graduate student in his department. He allowed me to submit my work on the metabolism of sympathomimetic amines and the microsomal enzyme for my dissertation. I took a year off to attend courses at George Washington University, and I found going back to school pleasant and challenging. A few of the medical students did better than I did in the pharmacology examinations. On one occasion a multiple-choice question on antipyrine, a compound on which I published several papers, was asked, and I gave the wrong answer. After a year's study, I passed a tough comprehensive examination, and my thesis *The fate of phenylisopropylamines* was accepted. In 1955, at the age of 42 years, I received my PhD.

SETTING UP A LABORATORY AT THE NATIONAL INSTITUTE OF MENTAL HEALTH

While studying for my PhD, I was invited by Edward Evarts to set up a Section of Pharmacology in his Laboratory of Clinical Sciences at the National Institute of Mental Health (NIMH). To get started on my new position at the NIMH I took a few afternoons off my classes at George Washington University to do laboratory work. I thought that a study of the metabolism and distribution of LSD would be an appropriate problem for my new laboratory in the NIMH. LSD was then used as an experimental drug by psychiatrists to study abnormal behavior. Bob Bowman at the NIH was in the process of building a spectrofluorometer. He was kind enough to let me use his experimental model, which allowed me to develop a very sensitive fluorometric assay for LSD. This made it possible to measure the nanogram amounts found in brain and other tissues. This instrument later became the well-known Aminco Bowman spectrofluorometer. The availability of this instrument made it possible for many laboratories to devise sensitive methods for the measurement of endogenous epinephrine, norepinephrine, dopamine, and serotonin in brain and other tissues. These newly developed methods for biogenic amines were crucial in the subsequent rapid expansion in neurotransmitter research.

Just before I left the Heart Institute, I read a report in the literature that uridine diphosphate glucuronic acid (UDPGA) was a necessary cofactor for the formation of phenolic glucuronides in a cell-free preparation of livers. Jack Strommiger, a biochemist then at the NIH, and I discussed the possible mechanism for the enzymatic synthesis of UPDGA. We suspected that it would arise from the oxidation of uridine diphosphate glucose (UDPG) by either TPN or DPN. We obtained a sample of UDPG from Herman Kalckar and did a preliminary experiment in which I measured the disappearance of morphine in guinea pig liver. When morphine was incubated with guinea pig liver microsomes and soluble fraction with DPN and UDPG, morphine was metabolized; TPN had no effect. When either DPN, UDPG, soluble fraction, or liver was omitted, the disappearance of morphine was negligible. After a period of incubation during which the mixture was heated in 1N HCl, the morphine that disappeared was recovered. These experiments suggested that morphine was enzymatically conjugated in the presence of UDPG and DPN, presumably by the formation of UDPGA followed by morphine glucuronide. I had little time to continue this problem because I was in the process of getting my PhD. Strominger and coworkers then went on to purify an enzyme UDPG dehydrogenase that formed UPDGA from UDPG and DPN.

After completion of my PhD, I returned to the glucuronide problem in my new laboratory at the NIMH. As expected from my preliminary experiment with morphine, I found that morphine and other narcotic drugs formed glucuronide conjugates by an enzyme present in liver microsomes that required UDPGA. Working together, Joe Inscoe, a graduate student at George Washington University, and I showed that glucuronide formation could be induced by benzpyrene and 3-methylcholanthrene.

The work on glucuronide conjugation led to a study on the role of glucuronic acid conjugation on bilirubin metabolism. Rudi Schmid, then at the NIH, made the interesting observation that bilirubin was transformed to a glucuronide. Schmid and I then went on to describe the enzymatic formation of bilirubin glucuronide by enzymes in the liver requiring UDPGA. This conjugating enzyme served as a mechanism for inactivating bilirubin. This led to an interesting clinical observation concerning a defect in glucuronide formation. In congenital jaundice there is a marked elevation of free bilirubin in the blood. This suggested to us that something might be wrong with glucuronide formation in this disease. The availability of a mutant strain of rats (Gunn rats) that exhibited congenital jaundice made it possible to examine whether the glucuronide-forming enzyme was defective. We then went on to demonstrate that these rats showed a marked defect in the ability to synthesize glucuronides from UDPGA (10). Glucuronide formation was also examined in humans with congenital jaundice by measuring the rate and magnitude of plasma acetominophen glucuronide after the administration of the acetominophen. A defect in glucuronide formation in this disease was demonstrated.

CATECHOLAMINE RESEARCH

When I joined the NIMH, I knew very little about neuroscience. My impression of neuroscience then was that it was mainly concerned with electrophysiology, brain anatomy, and behavior. These subjects were to me somewhat strange and esoteric and concerned with complicated electronic equipment. I believed that an investigator had to be a gifted experimentalist and theorist to do research in the neurosciences. Ed Evarts, my lab chief, assured me that I could work on whatever problem I thought would be likely to yield new information. The philosophy of Seymour Kety, then head of the Intramural Programs of the NIMH, was to allow investigators working in the laboratories of the NIMH to do their research on whatever was potentially productive and important. Kety believed that without sufficient basic knowledge about the life processes, doing targeted research on mental illness would be a waste of time and money.

Instead of working on a neurobiological problem, I thought it would be best to work on one that I knew something about, and that might be appropriate to the mission of the NIMH. I began to experiment on the metabolism and physiological disposition of LSD and the enzymes involved in the metabolism of narcotic drugs. I also worked on the enzymatic synthesis of glucuronides described above.

Although the NIMH administrators were supportive of the type of research I was doing, I still felt guilty that I was not working on some aspect of the nervous system or mental illness. Dr. Kety, in a seminar to our laboratory, gave a fascinating account of the findings of two Canadian psychiatrists. They reported that adrenochrome produced schizophreniclike hallucinations when it was ingested. Because of these behavioral effects, they proposed that schizophrenia could be caused by an abnormal metabolism of epinephrine to adrenochrome. I was intrigued by this proposal. In searching the literature, I was surprised to find that little was known about the metabolism of epinephrine at that time, in 1957. In view of the provocative hypothesis about the abnormal metabolism of epinephrine in schizophrenia, I decided to work on the metabolism of epinephrine. Epinephrine was then believed to be metabolized and inactivated by deamination by monoamine oxidase. However, with the introduction of monoamine oxidase inhibitors by Albert Zeller and coworkers, it was observed that, after the inhibition of monoamine oxidase in vivo, the physiological actions of administered epinephrine were still rapidly ended. This indicated that enzymes other than monoamine oxidase metabolized epinephrine. A possible route of metabolism of epinephrine might be via oxidation. I spent several months looking for oxidative enzymes for epinephrine without any success.

An abstract in the March 1957 *Federation Proceedings* gave me an important clue regarding a possible pathway for the metabolism of epinephrine. In this abstract, Armstrong and coworkers reported that patients with norepinephrine-forming tumors (pheochromocytomas) excreted large amounts of an O-methylated product, 3-methoxy-4-hydroxymandelic acid (VMA) (11). This suggested that this metabolite could be formed by the O-methylation and deamination of epinephrine or norepinephrine. The O-methylation of catecholamines was an intriguing possibility that could be experimentally tested. A potential methyl donor could be *S*-adenosylmethionine. That afternoon I incubated epinephrine with a homogenate of rat liver, ATP, and methionine. I did not have *S*-adenosylmethionine available, but Cantoni had shown that an enzyme in the liver could convert ATP and methionine to S-adenosylmethionine (12). I found that epinephrine was rapidly metabolized in the presence of ATP, methionine, and liver homogenate.

When either ATP or methionine was omitted or the homogenate was heated, there was a negligible disappearance of epinephrine. This experiment suggested that epinephrine was O-methylated in the presence of a methyl donor, presumably *S*-adenosylmethionine. In a following experiment, I obtained *S*-adenosylmethionine and observed that incubating liver homogenate with the methyl donor resulted in the metabolism of epinephrine. The most likely site of methylation would be on the *meta* hydroxyl group of epinephrine to form 3-O-methylepinephrine. I prevailed on my colleague Bernhard Witkop, a bioorganic chemist, to synthesize the O-methyl metabolite of epinephrine. A few days later Sero Senoh, a visiting scientist in Witkop's laboratory, synthesized meta-O-methylepinephrine, which we named metanephrine. After incubating liver and S-adenosylmethionine, the metabolite formed from epinephrine was identified as metanephrine, indicating the existence of an O-methylating enzyme. The O-methylating enzyme was purified and found to O-methylate catechols, including norepinephrine, dopamine, L-DOPA, and synthetic catechols, but not monophenols (13). In view of the substrate specificity, the enzyme was named catechol-O-methyltransferase (COMT). The enzyme was found to be widely distributed in tissues, including the brain.

Injecting catecholamines into animals resulted in the excretion of the respective O-methylated metabolites. We soon identified normally occurring O-methylmetabolites such as normetanephrine, metanephrine, 3-methoxy tyramine, and 3-methoxy-4-hydroxyphenylglycol (MHPG) in liver and brain. As a result of the discovery of the O-methylation metabolites, the pathways of catecholamine metabolism were clarified (13). Catecholamines were metabolized by O-methylation, deamination, glycol formation, oxidation, and conjugation. As as result of these findings, I then considered myself a neurochemist. This work also gave me a long-lasting interest in methylation reactions that I describe later. The metabolites of catecholamines, particularly MHPG, have been used as a marker in many studies in biological psychiatry.

A major problem in neurobiology research is the mechanism by which neurotransmitters are inactivated. At the time I described the metabolic pathway for catecholamines in 1957, it was believed that the actions of neurotransmitters were terminated by enzymatic transformation. Acetylcholine was already known to be rapidly inactivated by acetylcholinesterase. However, when the principal enzymes for the metabolism of catecholamines, catechol-O-methyltransferase and monoamine oxidase, were almost completely inhibited in vivo, the physiological actions of injected epinephrine were rapidly ended. These experiments indicated that there were other mechanisms for the rapid inactivation of catecholamines.

The answer to the question of the inactivation of catacholamines came in an unexpected way. When the metabolism of catecholamines was described, Seymour Kety and coworkers set out to examine whether or not there was an abnormal metabolism of epinephrine in schizophrenic patients. To carry out this study, Kety asked the New England Nuclear Corporation to prepare tritium-labeled epinephrine and norepinephrine of high specific activity. The first batch of ^{3}H-epinephrine that arrived in late 1957 was labeled on the 7 position, which we found to be stable. Kety was kind enough to give me some of the ^{3}H-epinephrine for my studies. I thought it would be a good idea to examine the tissue distribution and half-life of ^{3}H epinephrine in animals.

At about that time, Hans Weil-Malherbe spent three months in my laboratory as a visiting scientist, and together we developed methods for measuring ^{3}H-epinephrine and its metabolites in tissues and plasma. To our surprise, when ^{3}H-epinephrine was injected into cats, it persisted unchanged in the heart, spleen, and the salivary and adrenal glands long after its physiological effects were ended. This phenomenon puzzled us. We also found that ^{3}H-epinephrine did not cross the blood-brain barrier. Just about this time Gordon Whitby, a graduate student from Cambridge University, came to our laboratory to do his PhD thesis. I suggested that he use methods for assaying ^{3}H-norepinephrine similar to those we used for ^{3}H-epinephrine to study its tissue distribution. As in the case of ^{3}H-epinephrine, ^{3}H-norepinephrine remained in organs rich in sympathetic nerves (heart, spleen, salivary gland). These studies gave us a clue regarding the inactivation of catecholamine neurotransmitters: uptake and retention in sympathetic nerves.

The crucial experiment that established that catecholamines were selectively taken up in sympathetic neurons was suggested by Georg Hertting from the University of Vienna, who joined my laboratory as a visiting scientist. In the next experiment, the superior cervical ganglia of cats were taken out on one side, resulting in a unilateral degeneration of the sympathetic nerves in the salivary gland and eye muscles. Upon the injection of ^{3}H-norepinephrine, radioactive catecholamine accumulated on the innervated side, but very little appeared on the denervated side (13, 14). This simple experiment clearly showed that sympathetic nerves take up and store norepinephrine. In another series of experiments, Hertting and I found that injected ^{3}H-norepinephrine taken up by sympathetic nerves was released when these nerves were stimulated (15). As a result of these experiments, we proposed that norepinephrine is rapidly inactivated by reuptake into sympathetic nerves. Other slower mechanisms for the inactivation of catecholamines proposed were removal by the blood stream, metabolism by O-methylation, and/or deamination at effector tissue or by liver and kidney.

In 1961, the first postdoctoral fellow, Lincoln Potter, joined my laboratory via the NIH Research Associates Program. The NIH Research Associate Program and the Pharmacology Research Associate Program provided an opportunity for recent PhD or MD graduates to spend two or three years in Bethesda doing full-time research. Because of the number of applicants for this program, the investigators in the Intramural Program at the NIH would get the best and brightest postdoctoral fellows. During the past 25 years more than 60 postdoctoral fellows joined my laboratory to do full-time research. With one or two exceptions, most of the postdocs who worked in my laboratory went on to productive careers in research.

When a postdoc joins my laboratory I try to start him on a problem that has a good chance of success but is not trivial or pedestrian. There is an open and free exchange of ideas between my postdocs and myself, which makes it possible to try novel approaches to problems. By the time postdocs are ready to leave the laboratory, they are independent investigators. I found the interactions with bright and motivated young people stimulating and highly conducive to productive and original research.

When Linc Potter joined my laboratory, we directed our attention to the sites of the intraneural storage of norepinephrine. We suspected that ^{3}H-norepinephrine, already shown to be taken up by sympathetic neurons, would label intracellular storage sites. ^{3}H-norepinephrines was injected into rats, and their hearts were homogenized in isotonic sucrose; then the various subcellular fractions were separated in a continuous sucrose gradient. There was a sharp peak of radioactive norepinephrine in a fraction that coincided with endogenous catecholamines and dopamine β-hydroxylase, the enzyme that converts dopamine to norepinephrine. The norepinephrine-containing particles exerted a pressor response only when they were lysed. In another experiment, ^{3}H-norepinephrine was injected, and the pineal gland, an organ rich in sympathetic nerve terminals, was subjected to radioautography and electron microscopy (16). Photographic grains of ^{3}H-norepinephrine were highly localized over dense core-granulated vesicles of about 500 angstroms. All these experiments indicated that norepinephrine in sympathetic nerves was stored in small, dense core vesicles.

Subsequent studies with another postdoc, Dick Weinshilboum, showed that upon stimulation of the hypogastric nerve of the vas deferens, both norepinephrine and dopamine-β-hydroxylase were disscharged from the nerve terminals. This suggested that norepinephrine and dopamine-β-hydroxylase were colocalized in the catecholamine storage vesicles of sympathetic nerves and were then discharged together by exocytosis (17). These findings led us to the postulation that the released dopamine-β-hydroxylase would appear in the blood, which was soon confirmed. Later, our laboratory and others found abnormally low levels of plasma dopamine-β-hydroxylase in familial dysautonomia and Down's syndrome, and high levels in patients with torsion dystonia, neuroblastoma, and certain forms of hypertension.

As soon as it was found that catecholamines could be taken up and inactivated by reuptake into sympathetic nerve terminals, I and my coworkers turned our attention to the effect of adrenergic drugs on this process. We designed relatively simple experiments for this study, injecting the drug into rats and then measuring the uptake of injected ^{3}H-norepinephrine in tissues. Cocaine was the first drug we examined. It had been postulated that cocaine causes supersensitivity to norepinephrine by interfering with its inactivation. After pretreatment of cats with cocaine, there was a marked reduction of ^{3}H-norepinephrine in tissues that were innervated by sympathetic nerves after the injection of the radioactive catecholamine (18). This experiment indicated that cocaine blocked the reuptake of norepinephrine in nerves and thus allowed large amounts of the catecholamine to remain in the synaptic cleft and act on the postsynaptic receptors for longer periods of time. Using a similar approach, we observed that antidepressant drugs and amphetamine and other sympathomimetic amines also blocked the uptake of norepinephrine. In another type of experiment, using an isolated, perfused beating rat heart whose nerves had previously been labeled with ^{3}H-norepinephrine, we found that the physiological action of sympathomimetic amines, such as tyramine, was mediated by releasing the norepinephrine from sympathetic nerves (19). After repeated treatment of the isolated heart with tyramine, the heart rate and amplitude of contraction were gradually reduced, presumably by the depletion of the releasable stores of the neurotransmitters. After replenishing the isolated heart with exogenous norepinephrine, the heart rate and amplitude of contraction of the isolated heart were restored. Amphetamine also released norepinephrine, and it was later shown by others that the physiological effects of the amine were due to the release of dopamine.

Most of my early work in catecholamines was done in the peripheral sympathetic nervous system. Hans Weil-Malherbe and I had found that catecholamines did not cross the blood-brain barrier. This made it impossible to study the metabolism, storage, and release of norepinephrine in the brain by peripheral administration of ^{3}H-norepinephrine. It was Jacques Glowinski, a visiting scientist from France, who circumvented this problem. He devised a technique to introduce ^{3}H-norepinephrine directly into the brain by injection into the lateral ventrical. Subsequent experiments showed that ^{3}H-norepinephrine was mixed with the endogenous catecholamines in the brain. As in the peripheral nervous system, the ^{3}H-norepineprine was found to be metabolized by O-methylation and deamination. In a series of experiments we established that ^{3}H-norepinephrine could serve as a useful tool in studying the activity of brain adrenergic nerves (13).

After labeling the brain adrenergic neurons, Glowinski and I examined the effect of psychoactive drugs on brain biogenic amines. We found that only the clinically effective antidepressant drugs block the reuptake of ^{3}H-norepinephrine in adrenergic nerve terminals (20). This, together with the observation that monoamine oxidase inhibitors have antidepressant actions and that reserpine, a depleter of biogenic amines, sometimes causes depression, led to the formulation of the catecholamine hypothesis of depression

(21). We also found that amphetamines block the reuptake as well as the release of ^{3}H-norepinephrine in the brain. Other investigators later showed that paranoid psychosis caused by excessive ingestion of amphetamines is due to the release of the catecholamine dopamine. One of the reasons that Les Iversen came to my lab as a postdoctoral fellow was to learn about the brain and its chemistry. Iversen and Glowinski worked extensively together in my laboratory on the effects of drugs on the adrenergic system in different areas of the brain. To conduct this study they devised a method of disection of various parts of the brain that has become a classic procedure.

For several years our laboratory was concerned with adaptive mechanisms of the sympathoadrenal axis. One such mechanism, the induction of the catecholamine's biosynthetic enzyme, tyrosine hydroxylase, was observed in an unexpected manner, as often happens in research. Hans Thoenen, then working in Basel, asked to spend a sabbatical year in my laboratory. He and Tranzer had observed that injected 6-hydroxydopamine selectively destroys catecholamine-containing nerve terminals (22). I invited Thoenen to join my laboratory and bring 6-hydroxydopamine. The first experiment that Thoenen tried was to examine the effects of the destruction of peripheral sympathetic nerves on tyrosine hydroxylase. After the injection of 6-hydroxydopamine, as expected, tyrosine hydroxylase almost completely disappeared from sympathetically innervated nerves. A surprising observation was a marked elevation of tyrosine hydroxylase in the adrenal medulla. 6-Hydroxydopamine was known to cause persistent firing of nerves. We suspected that tyrosine hydroxylase was elevated in the adrenal medulla by continuous firing of the splanchic nerve innervating the adrenals. This supposition was confirmed when other drugs that caused prolonged nerve firing, such as reserpine and α-adrenergic blocking agents, also increased tyrosine hydroxylase (23). Subsequent experiments showed that increased nerve firing induced the synthesis of new tyrosine hydroxylase molecules in nerve cell bodies and the adrenal medulla in a transsynaptic manner. Similar results were obtained with another catecholamine biosynthetic enzyme, dopamine-β-hydroxylase (13).

Another regulatory mechanism for catecholamine synthesis was found by asking the right questions rather than by serendipity. The ratio of epinephrine to norepinephrine in the adrenal medulla was known to be dependent on how much of the medulla was enveloped by the adrenal cortex. In species in which the cortex is separated from the medulla, norepinephrine is the predominant catecholamine, while in species in which the medulla is surrounded by the adrenal cortex, the methylated catecholamine, epinephrine, is by far the major amine. Dick Wurtman, a research associated in my laboratory, suggested an elegant experiment to determine the role of the adrenal cortex in regulating the synthesis of epinephrine. He removed the rat pituitary, a procedure that depleted glucocorticoid in the adrenal cortex, and then measured the effect on the levels of the epinephrine-forming enzyme, phenylethanolamine-N methyltransferase (PNMT), in the medulla. I had just characterized PNMT and found that it was highly localized in the adrenal medulla. The ablation of the pituitary caused a profound decrease in PNMT in the medulla after several days (24). The administration of ACTH, a peptide that increases the formation of glucocorticoids in the adrenal cortex, or the injection of the synthetic glucocorticoid, dexamethasone, increased PNMT in hypophysectomized rats almost to normal values.

METHYLTRANSFERASE RESEARCH

After the description of catechol-O-methyltransferase, I became very much involved with methyltransferase enzymes (25). I spent most of my time at the lab bench working on methylating enzymes for many years. Soon after describing COMT, I turned my attention to the enzymatic N-methylation of histamine. A major pathway for histamine metabolism occurs via N-methylation. This prompted a search for a potential histamine-methylating enzyme. As in the case of other methyltransferases, I suspected that the most likely methyl donor would be S-adenosylmethionine. To make the identity of the histamine-methylating enzyme possible, Donald Brown, a postdoc in the lab of a colleague, and I synthesized [^{14}C-methyl]-S-adenosylmethionine enzymatically from rabbit liver with ^{14}C-methylmethionine and ATP. Because of its ability to label the O or N groups of potential substrates by the transfer of ^{3}H-methylmethionine, the availability of ^{14}C-S-adenosylmethionine led to the discovery of a number of methyltransferase enzymes. Histamine N-methyltransferase was soon found and purified and its properties described. The enzyme is highly localized in the brain, and it also has an absolute specificity for histamine. Other methyltransferases soon discovered using [^{14}C-methyl]-S-adenosylmethionine were PNMT, hydroxyindole O-methyltransferase, the melatonin-forming enzyme, a protein carboxymethyltransferase, and a nonspecific N-methyltransferase. This latter enzyme was found to convert tryptamine, a compound normally present in the brain, to N-N-dimethyltryptamine, a psychotomimetic agent.

These methyltransferase enzymes, together with [^{3}H-methyl]-S-adenosylmethionine of high specific activity were used in developing very sensitive methods for the measurement of trace biogenic amines. We were able to detect, localize, and measure octopamine, tryptamine, phenylethylamine, phenylethanolamine, and tyramine in the brain and other tissues. The methyltransferases and [^{3}H-methyl]-S-adenosylmethionine also made it possible to measure norepinephrine, dopamine, histamine, and serotonin in 130 separate brain nuclei. Because of the sensitivity of the enzymatic micromethods, my colleagues and I were able to show the coexistence of several neurotransmitters in single identified neurones of Aplysia (26). Later, Thomas Hokfelt, using immunohistofluorescent techniques, demonstrated the coexistence of neurotransmitters in many nerve tracts (27).

THE PINEAL GLAND

I was struck by an article from Aaron Lerner's laboratory, published in 1958, that described the isolation of 5-methoxy-N-acetyltryptamine (melatonin) from the bovine pineal gland, a compound that had powerful actions in blanching the skin of tadpoles (28). This compound attracted my attention for two reasons: it had a methoxy group and a serotonin nucleus. The methoxy group of melatonin had a special attraction for me. Also, at that time, serotonin was believed to be involved in psychoses because of its structural resemblance to LSD. I thought it would be fun to spend some time working on the pineal gland, an organ that was a mystery to me. The best way to start was to concentrate my efforts on aspects of the problem that I was familiar with, such as O-methylation.

Herbert Weissbach expressed an interest in collaborating with me in working out the biosynthetic pathway for melatonin. Weissbach had already made important contributions on the metabolism of serotonin. The availability of S-adenosyl-L-methionine with a radioactive methyl group provided an opportunity to examine whether the pineal gland could form labeled melatonin from potential precursor compounds. When we incubated bovine pineal extracts with N-acetylserotonin and [^{14}C-methyl]-S-adenosyl-L-methionine, a radioactive product that we soon identified as melatonin was found (29). Weissbach and I then purified the melatonin-forming enzyme, which we named hydroxyindole-O-methyltransferase (HIOMT), from the bovine pineal gland. We also found another enzyme that converted serotonin to N-acetylserotonin in the rat pineal. From these observations, we proposed that the synthesis of melatonin in the pineal proceeds as follows: tryptophan $\rightarrow$ 5-hydroxytrypotophan $\rightarrow$ serotonin $\rightarrow$ N-acetylserotonin $\rightarrow$ melatonin (30). Irwin Kopin, Weissbach, and I also found that melatonin was mainly metabolized by a microsomal enzyme via 6-hydroxylation. In a study of the tissue distribution of HIOMT we observed that the enzyme was highly localized in the pineal. This convinced me that the pineal was a biochemically active organ containing an unusal enzyme and product and was worth further study.

During 1960–1962 I spent little time doing pineal research. Most of my efforts were directed towards the biochemistry of catacholamines and the effect of psychoactive drugs. In 1962, when Wurtman joined my laboratory, I thought that he should devote most of his time to catecholamine research. As a medical student Wurtman had already made an important finding that bovine pineal extracts blocked gonadal growth in rats induced by light. Although pineal research was not a fashionable subject for research then, Wurtman and I were caught up by the romance of this organ, so we decided to spend our spare time working on the pineal. We thought that a good place to start was the isolation of the gonad-inhibitory factor of the pineal. Neither of us wanted to go through a tiresome isolation and bioassay procedure, and we decided to take a chance and examine the effects of melatonin. We soon found that melatonin reduced ovarian weight and decreased the incidence of estrus in the rat (30).

Wurtman and I turned our attention to the effects of light on the biochemistry of the pineal. We found that keeping rats in the dark for a period of time increased HIOMT activity, compared to those kept in continuous light. This experiment gave Wurtman and me a biochemical marker to study how light transmits its message to an internal organ. Ariens Kappers had found that the pineal is innervated by sympathetic nerves arising from the superior cervical ganglia. This finding suggested an experiment to determine the effects of light on the pineal by removing the superior cervical ganglia and examining the effects of light and dark on the HIOMT. When the superior cervical ganglia were removed, the effects of light on HIOMT were abolished. This experiment told us that the effects of light on melatonin synthesis were mediated via sympathetic nerves arising from the superior cervical ganglia.

In 1964, Sol Snyder joined my laboratory as a postdoc, and he too was fascinated by pineal research. Quay had just made an important observation that the levels of serotonin, a precursor of melatonin in the pineal, are high during the day and low at night. Snyder and I developed a very sensitive assay for measuring serotonin in a single pineal. This gave us the opportunity to study how the serotonin rhythm, which can serve as a marker for the melatonin rhythm, is regulated by light in a tiny organ such as the pineal. We found that in normal rats in continuous darkness, or in blinded rats, the daily serotonin rhythm in the pineal persisted (31). This indicated that the indolemine rhythms in the pineal were controlled by an internal clock. Keeping rats in constant light abolished the circadian serotonin rhythm, showing that light somehow stopped the biological clock. These experiments were the first demonstration that the rhythms of indoleamines in the pineal were endogenous and that they were synchronized by environmental lighting. We found that the circadian serotonin rhythm was abolished after ganglionectomy and also after decentralization of the superior cervical ganglion, indicating that the circadian clock for the serotonin and presumably the melatonin rhythm resided somewhere in the brain. Wurtman and I published an article in *Scientific American* in which we suggested that the pineal serves as a neuroendocrine transducer, converting light signals to hormone synthesis via the brain and noradrenergic nerves (32).

Shein, a psychiatrist at McLean Hospital, Wurtman, who was then at MIT, and I decided to see whether the rat pineal in organ culture metabolized tryptophan to melatonin, and it did. This finding provided an opportunity to examine whether the neurotransmitter of the sympathetic nerve, norepinephrine, could affect the synthesis of melatonin in pineal organ culture. The addition of norepinephrine to rat pineals in organ culture increased the synthesis of melatonin from tryptophan. Shein and Wurtman then showed that noradrenaline specifically stimulated the β-adrenergic receptor.

For two years after 1970 I did little work on the pineal until Takeo Deguchi, a biochemist from Kyoto, joined my laboratory. Because interest in receptors was beginning to grow at that time, we decided that the pineal gland would be a good model to study the regulation of the β-adrenergic receptor. The activity of the β-receptor could be determined by measuring changes in serotonin N-acetyltransferase (NAT). David Klein previously showed that pineal serotonin N-acetyltransferase had a marked circadian rhythm that was controlled by a β-adrenergic receptor (33). Deguchi and I devised a rapid assay for N-acetyltransferase and soon confirmed Klein's findings. We then found that the nighttime rise in NAT was abolished by β-adrenergic blocking agents, reserpine, decentralization, ganglionectomy, and agents that inhibit protein synthesis (30). This told us that noradrenaline released from sympathetic nerves innervating the pineal gland stimulated the β-adrenergic receptor, which then activated the cellular machinery for the synthesis of NAT. Blocking the β-adrenergic receptor with propranolol at night or exposing rats to light also caused a rapid fall of NAT. These results indicated that unless the β-adrenergic receptor is stimulated by norepinephrine at a relatively high frequency, NAT rapidly decays. We thought that the rapid synthesis and fall of NAT would provide a useful model to study the molecular events in receptor linked synthesis of a specific protein (NAT) leading to the formation of a hormone (melatonin).

The regulation of supersensitivity and subsensitivity of receptors is an important biological problem. The rapidly changing pineal NAT provided a productive approach to study the mechanism of super- and subsensitivity of the β-adrenergic receptor (30). Procedures that depleted the neuronal input of noradrenaline in the rat pineal (denervation, constant light, or reserpine) caused a superinduction of NAT when rat pineals were cultured and treated with the β-adrenergic agonist, 1-isoproterenol. When pineal β-adrenergic receptors were repeatedly stimulated by injections of 1-isoproterenol into rats, the cultured pineals became almost unresponsive to the β-adrenergic agonist. In collaboration, my postdoctoral fellows Jorge Romero and Martin Zatz and I showed that the regulation of NAT and subsequent melatonin synthesis consists of a complex series of steps involving: β-adrenergic receptor, cyclic AMP, cyclic GMP, protein kinase, specific activation of mRNA for NAT, and synthesis of NAT (30). Decreased nerve activity induced by light caused an increase in receptor number and adenylate cyclase and kinase activity. This cascade of events then explained why a small change in release of noradrenaline from nerves causes a large change in pineal NAT. With the onset of darkness, there is an increase in sympathetic nerve activity that acts on the supersensitive receptor, cyclase, kinase, etc. This, we believe, considerably amplifies the signal (norepinephrine) to cause the large nighttime rise in NAT formation. Klein later showed that norepinephrine acting on an α_1-adrenergic receptor further amplified the NAT levels.

THE LAST TEN YEARS

Because of space limitations I can only give a brief description of my research during the past ten years. Most of the research in my laboratory was concerned with how neurotransmitters transmit their specific messages. About ten years ago Fusao Hirata, a visiting scientist in my laboratory, and I observed that the occupation of certain receptors stimulated the methylation of phospholipids. On the basis of these findings we proposed a mechanism for the transduction of biological signals (34). This proposal generated considerable controversy, and the role of phospholipid methylation in signal transduction still remains to be resolved. Later, with the collaboration of several postdoctoral fellows, we reported on the interaction of stress hormones (catecholamines, ACTH, and glucocorticoids) and the multireceptor release of ACTH (35).

In 1984, I officially retired from government service at the age of 72. The NIMH allows me to keep my small laboratory and generously supports my research. With the help of postdoctoral fellows, I continue to be actively engaged in studying transduction mechanisms of neurotransmitters and hormones. We have evidence for a receptor-mediated release of arachidonic acid and its many metabolites via the activation of a GTP-binding protein linked to phospholipase A_2 (36). This pathway promises to be an active area of future research.

F. Scott Fitzgerald once stated that there are no second acts in American lives. After a mediocre first act, my second act was a smash. So far the third act has not been so bad.

Literature Cited

1. Brodie, B. B., Axelrod, J. 1948. The fate of acetanilide in man. *J. Pharmacol. Exp. Ther.* 94:29–38
2. Brodie, B. B., Axelrod, J. 1949. The fate of acetophenetidin (phenacetin) in man and methods for the estimation of acetophenetidin and its metabolites in biological materials. *J. Pharmacol. Exp. Ther.* 97:58–67
3. Axelrod, J., Reichenthal, J. 1953. The fate of caffeine in man and a method for its estimation of biological material. *J. Pharmacol. Exp. Ther.* 107:519–23
4. Axelrod, J. 1954. An enzyme for the deamination of sympathomimeticamines. *J. Pharmacol. Exp. Ther.* 110:2
5. Axelrod, J. 1982. The discovery of the microsomal drug-metabolizing enzymes. *Trends Pharmacol. Sci.* 3:383–86
6. Brodie, B. B., Gillette, J. R., LaDu, B. 1958. Enzymatic metabolism of drugs and other foreign compounds. *Ann. Rev. Biochem.* 27:427–84
7. Axelrod, J. 1956. The enzymatic N-demethylation of narcotic drugs. *J. Pharmacol. Exp. Ther.* 117:322–30
8. Quinn, G. P., Axelrod, J., Brodie, B. B. 1958. Species, strain and sex differences in metabolism of hexobarbiton, amidopyrine, antipyrine and aniline. *Biochem. Pharmacol.* 1:152–59
9. Axelrod, J. 1956. Possible mechanism of tolerance to narcotic drugs. *Science* 124:263–64
10. Axelrod, J., Schmid, R., Hammaker, L.

1957. A biochemical lesion in congenital, non-obstructive, non-hemolytic jaundice. *Nature* 180:1426–27

11. Armstrong, M. D., McMillan, A. 1957. Identification of a major urinary metabolite of norepinephrine. *Fed. Proc.* 16:146

12. Cantoni, G. L. 1953. Adenosyl methionine: A new intermediate formed enzymatically from l-methionine and adenosine triphosphate. *J. Biol. Chem.* 187:439–52

13. Axelrod, J.: Noradrenaline: Fate and control of its biosynthesis. In: *Les Prix Nobel.* Imprimerieal Royal P. A. Norstedt and Soner, Stockholm, 1971, pp. 189–208; *Science* 173:598–606, 1971

14. Hertting, G., Axelrod, J., Kopin, I. J., Whitby, L. G. 1967. Lack of uptake of catecholamines after chronic denervation of sympathetic nerves. *Nature* 189:66

15. Hertting, G., Axelrod, J. 1961. The fate of tritiated noradrenaline at the sympathetic nerve-endings. *Nature* 192:172–173

16. Wolfe, D. E., Potter, L. T., Richardson, K. C., Axelrod, J. 1962. Localizing tritiated norepinephrine in sympathetic axons by electron microscopic autoradiography. *Science* 138:440–42

17. Weinshilboum, R., Thoa, N. B., Johnson, D. G., Kopin, I. J., Axelrod, J. 1971. Proportional release of norepinephrine and dopamine-β-hydroxylase from sympathetic nerves. *Science* 174:1349–51

18. Whitby, L. G., Hertting, G., Axelrod, J. 1960. Effect of cocaine on the disposition of noradrenaline labelled with tritium. *Nature* 187:604–5

19. Axelrod, J., Gordon, E., Hertting, G., Kopin, I. J., Potter, L. T. 1962. On the mechanism of tachyphylaxis to tyramine in the isolated rat heart. *Br. J. Pharmacol.* 19:56–63

20. Glowinski, J., Axelrod, J. 1964. Inhibition of uptake of tritiated-noradrenaline in the intact rat brain by imipramine and structurally related compounds. *Nature* 204:1318–19

21. Schildkraut, J. J. 1965. The catecholamine hypothesis of affective disorders: A review of the supportive evidence. *Am. J. Psychiatry* 122:509–22

22. Thoenen, H., Tranzer, J. P. 1968. Chemical sympathectomy by selective destruction of adrenergic nerve endings with 6-hydroxydopamine *Naunyn-Schmiedebergs Arch. Pharmacol.* 261:271–88

23. Thoenen, H., Mueller, R. A., Axelrod, J. 1969. Increased tyrosine hydroxylase activity after drug induced alteration of sympathetic transmission. *Nature* 221:1264

24. Wurtman, R. J., Axelrod, J. 1966. Control of enzymatic synthesis of adrenaline in the adrenal medulla by adrenal cortical steroids. *J. Biol. Chem.* 241:2301–5

25. Axelrod, J. 1981. Following the methyl group. In *Psychiatry and the Biology of the Human Brain, A symposium dedicated to S. S. Kety,* ed. S. Matthysee, pp. 5–14. New York: Elsevier/North-Holland

26. Brownstein, M. J., Saavedra, J. M., Axelrod, J., Zeman, G. H., Carpenter, D. O. 1974. Coexistence of several putative neurotransmitters in single identified neurons of *Aplysia*. *Proc. Natl. Acad. Sci. USA* 71:4662–65

27. Hokfelt, T., Johansson, A., Ljungdahl, A., Lundberg, H. M., Schultzberg, M. 1980. Peptidergic neurons. *Nature* 284:515–21

28. Lerner, A. B., Case, J.D., Takahashi, Y., Lee, T. H., Mori. W. 1958. Isolation of melatonin, the pineal gland factor that lightens melanocytes. *J. Am. Chem. Soc.* 80:2587

29. Axelrod, J., Weissbach, H. 1961. Purification and properties of hydroxyindole-O-methyl transferase. *J. Biol. Chem.* 236:211–13

30. Axelrod, J. 1974. The pineal gland: A neurochemical transducer. *Science* 184:1341–48

31. Snyder, S. H., Zweig, M., Axelrod, J., Fischer, J. E. 1965. Control of the circadian rhythm in serotonin content of the rat pineal gland. *Proc. Natl. Acad. Sci. USA* 53:301–5

32. Wurtman, R. J., Axelrod, J. 1965. The pineal gland. *Sci. Am.* 213:50–60

33. Klein, D. C., Weller, J. L. 1970. Indole metabolism in the pineal gland: A circadian rhythm in N-acetyltransferase. *Science* 169:348–53

34. Hirata, F., Axelrod, J. 1980. Phospholipid methylation and biological signal transmission. *Science* 209:1082–90

35. Axelrod, J., Reisine, T. 1984. Stress hormones: Their interaction and regulation. *Science* 224:452–59

36. Burch, R. M., Luini, A., Axelrod, J. 1986. Phospholipase A_2 and phospholipase C are activated by distinct GTP-binding proteins in response to α_1-adrenergic stimulation in FRTL5 thyroid cells. *Proc. Natl, Acad. Sci. USA* 83:7201–5

Annu. Rev. Phys. Chem. 1991. 42: 1–22

FROM HIGH-RESOLUTION SPECTROSCOPY TO CHEMICAL REACTIONS

Eizi Hirota

The Graduate University for Advanced Studies, Yokohama 227;
The Institute for Molecular Science, Okazaki 444, Japan

KEY WORDS: large-amplitude motion, transient molecules, kinetic spectroscopy, photochemical reaction, discharge plasma, microwave spectroscopy, infrared spectroscopy

INTRODUCTION

One can learn a lot from history, and from the history of science. However, history itself is not science. In other words, we cannot carry out an experiment to prove a statement made about history; history depends on who has described the past. Many people have probably learned that history taught at school differs depending on the teacher; some teachers are very interesting, but others are boring.

The Prefatory Chapters of the *Annual Review of Physical Chemistry* have been prepared by distinguished scientists. Many of them summarize what they contributed to science; their articles are a history of science based upon their personal experience. This reflection made me hesitant to prepare this Prefatory Chapter; I am afraid that my personal experience may not invoke any interest of the readers. However, I have convinced myself that a description of the achievements of my research coworkers must be of some interest and some value at least to a few people, and finally I accepted the invitation of the Editor.

HISTORICAL BACKGROUND

In 1952, I joined Professor Morino's group as an undergraduate student of the University of Tokyo. One day, I found a short article in a Japanese journal. I forget the details, but the author was probably Masataka Mizushima. He reported structural studies of simple molecules by using a spectroscopic method. I was very surprised by the enormous number of digits he quoted for the structural parameters. At that time, we had to be content with three digits, e.g. 1.76 ± 0.02 Å for the C–Cl bond length. So I brought this journal to the laboratory and naively asked Professor Morino how they could achieve such extraordinarily high accuracy. It was already 1952, but I had only very limited information on the new development of science. Professor Morino replied that it must be done with microwave spectroscopy. At that time, Townes & Dailey (1–3) published their famous method of interpreting the nuclear quadrupole coupling data; many chemists were already engaged with this new method of microwave spectroscopy, but I did not know anything about it. The next year, I started my graduate study by using electron diffraction; rotational isomerism was the central theme of my graduate work. However, I kept the memory that microwave spectroscopy was a fascinating technique, and I wished I could use it.

In the fifth, and last, year of my graduate course, Professor Morino allowed me to change my method from electron diffraction to microwave. Fortunately, in the Physics Department, which was and still is next to the Chemistry Department, Professor Shimoda had already started microwave spectroscopy, and I could learn the necessary techniques from his collaborators. Takeshi Oka was a chemistry student, but he spent most of his graduate course in the Physics Department under the guidance of Professor Shimoda. Thus he was already an expert in microwave when I started construction of a microwave spectrometer in the Chemistry Department. He was of great help when I started research in microwave spectroscopy.

I enjoyed studying a few molecules with my newly constructed spectrometer (4, 5), but soon I began to seek directions of research that other microwave researchers had not been taking. One problem we young people in Professor Morino's group thought of was vibration-rotation interactions. This problem had mainly been attacked by vibrational spectroscopists, and microwave researchers had no intention of applying their method to this problem. One of our results on this project was obtained with sulfur dioxide (6). I planned to detect rotational spectra in the excited states of all three normal modes and hoped to determine the equilibrium structure, as well as all six third-order anharmonic potential constants. It took us a long time to observe all the vibrational satellites we needed, but finally Kikuchi, Saito, and I arrived at the goal. In this connection, I am pleased to add that Professor Morino has recently extended this study and has concluded that our earlier results are quite satisfactory.

The second direction I thought of was large-amplitude motions. In the

Chemistry Department, the late Professor San-ichiro Mizushima, Professor Morino, and their coworkers had carried out systematic studies on rotational isomerism (7). As is well known, the term "gauche" was coined by Professor Mizushima. The main theme of my graduate study constituted a small part of this project. I thought, however, that rotational isomers meant nothing but the presence of minima in the internal-rotation potential function. In other words, they merely represented some static concepts, and we should try to get a more global view of internal rotation. Specifically, we should determine the potential function itself experimentally. For this purpose, microwave spectroscopy should be extremely useful; I suspected it would provide us with various types of detailed data. To me, dynamic phenomena were and are much more interesting than static ones.

Fortunately I could start the study at Harvard University in Professor E. Bright Wilson's laboratory; I stayed at Harvard from 1960 to 1962, and amply enjoyed the pertinent instruction of Professor Wilson. There, I started on a program to determine, or "estimate," the internal-rotation potential function by using data supplied by microwave spectroscopy. The first compound I studied was n-propyl fluoride (8). The internal-rotation potential parameters I determined are poor in precision, as reproduced in Table 1. Later, Durig et al (9) and Caminati et al (10) revised my result. Although I could study only a few molecules in this way (11–13), other people, in particular Professor Durig and coworkers, have been supplying

Table 1 Internal rotation potential function of n-propyl fluoride[a]

Fourier coefficient (kcal/mol)	Hirota (8)	Durig et al (9)	Caminati et al (10)[b]
V_1	3.22(202)	0.22(10)	—
V_2	−3.05(172)	−0.683(31)	—
V_3	6.48(215)	3.660(11)	2.776(43)
V_4	−1.25(117)	—	—
V_5	0.37(34)	—	—
V_6	−0.76(54)	0.034(6)	−0.079(20)
V_7	−0.002(290)	—	—
V_{max1}[c]	10.4(44)	4.246	4.186
V_{max2}[d]	4.7(15)	3.560	3.465

[a] Values in parentheses represent estimated uncertainties and apply to the last digits of the constants.
[b] Structural relaxation has been considered.
[c] V_{max1} denotes the barrier height between the two gauche forms, namely the *cis* barrier.
[d] V_{max2} denotes the barrier height between the gauche and *trans* forms, measured from the gauche bottom.

906

a large amount of beautiful data; the number of molecules for which the potential function was derived surely exceeds 50, and probably is now approaching 100.

I was and am very interested in not only internal rotation and rotational isomerism, but also in large-amplitude motions in general. One recent example is $NaBH_4$, which Yoshiyuki Kawashima and I have been studying (Y. Kawashima et al 1990, submitted for publication). The molecule shows a very interesting spectrum, which we think is at least partly due to the internal motion of the BH_4 group.

FREE RADICAL STUDIES AT THE UNIVERSITY OF TOKYO AND KYUSHU UNIVERSITY

Chemical reactions are certainly more dynamical topics and more central to chemistry than large-amplitude motions. During my two-year stay in the US, I began to think of ways to find an access to this important problem and what I could contribute with my chosen tool, microwave spectroscopy. Finally, I arrived at a conclusion: We should try to detect reactive intermediate species by microwave spectroscopy.

I admit that we were quite influenced by Herzberg and coworkers at the National Research Council (NRC) Canada. They had already accumulated a large amount of extensive, important, and beautiful data on such transient molecules by using a combination of flash photolysis and high-resolution electronic spectroscopy in the ultraviolet to near infrared (IR) regions (14). I thought that if we could apply microwave spectroscopy to this problem, we would get more detailed information on transient molecules than we would by using electronic spectroscopy. Up to 1960, only two transient molecules were detected by microwave spectroscopy: OH by Townes and coworkers (15) and CS by Mockler & Bird (16). In the US and Canada, I visited three microwave laboratories, which had been involved in similar projects: those of Rollie Myers at Berkeley, David Lide at the National Bureau of Standards Washington, and C. C. Costain at NRC Canada. But, by the time of my visit, they had apparently given up; they told me that it was too difficult to detect transient molecules by microwave spectroscopy.

In spite of this pessimistic situation, I returned to Tokyo in 1962 hoping that we would be able to detect transient species by microwave spectroscopy and open a new field. It was fortunate for us that Takayoshi Amano joined our group when I returned; he was interested in some dynamical problems, rather than in static structure determinations. Thus, we started free radical studies by microwave spectroscopy. At first, we did not know anything about free radicals; we did not know how to generate them, we did not know how reactive they were, etc. So, we spent a few years without any positive results. However, from 1967 to 1975, we made slow but steady steps and detected SO (17), ClO (18, 19), NS (20), BrO (21), NCO (22, 23), and SF (24). I certainly admit that our progress was very slow, but I must also say that the detection of free radicals was extremely difficult at that time. We needed some breakthrough in spectroscopic techniques.

FREE RADICAL STUDIES AT THE INSTITUTE FOR MOLECULAR SCIENCE, AND SPECTROSCOPY OF THE METHYL AND SILYL RADICALS

In 1975, I was invited to join a new research institute, the Institute for Molecular Science in Okazaki. At a research institute, we could concentrate on difficult problems. As a supervisor of students, however, I would not give these problems to students, who could stay with us for only a limited period of time. When I accepted the invitation of Professor Akamatu, the first Director-General of the Institute, I immediately chose the spectroscopic study of free radicals as the main theme of my group. First, I selected good people, otherwise we would lose the game. Fortunately, I succeeded in getting able collaborators, including Shuji Saito, Chikashi Yamada, Yasuki Endo, and Kentarou Kawaguchi. We also introduced IR laser spectroscopy in addition to microwave spectroscopy as our main tools. We have successfully detected and studied about 100 transient species (25). I concentrate here on two examples, the methyl and silyl radicals.

The Methyl Radical

Herzberg (26) was the first to determine the structure of this radical in the gaseous phase experimentally. He photographed ultraviolet transitions to Rydberg states and he could resolve rotational structures for a CD_3 band at 2140 Å and bands of CH_3 and CD_3 at 1500 Å. As shown in Figure 1 (27), Herzberg took the missing R(0) line in the CD_3 $\tilde{B}-\tilde{X}$ band as an evidence that the methyl radical was planar or nearly planar in the ground electronic state. The absence of any strong hot bands led Herzberg to conclude that any potential hump should be lower than 200 cm^{-1}. Numerous other studies were also carried out on this important free radical, including the fine work by Milligan and Jacox (28, 29), but information obtained on the structure was more or less indirect. I should also mention a pioneer work of Pimentel and coworkers (30), who used a rapid scan IR spectrometer. This work and Herzberg's results indicated that the v_2 band, i.e. the out-of-plane bending mode, will appear around 600 cm^{-1}.

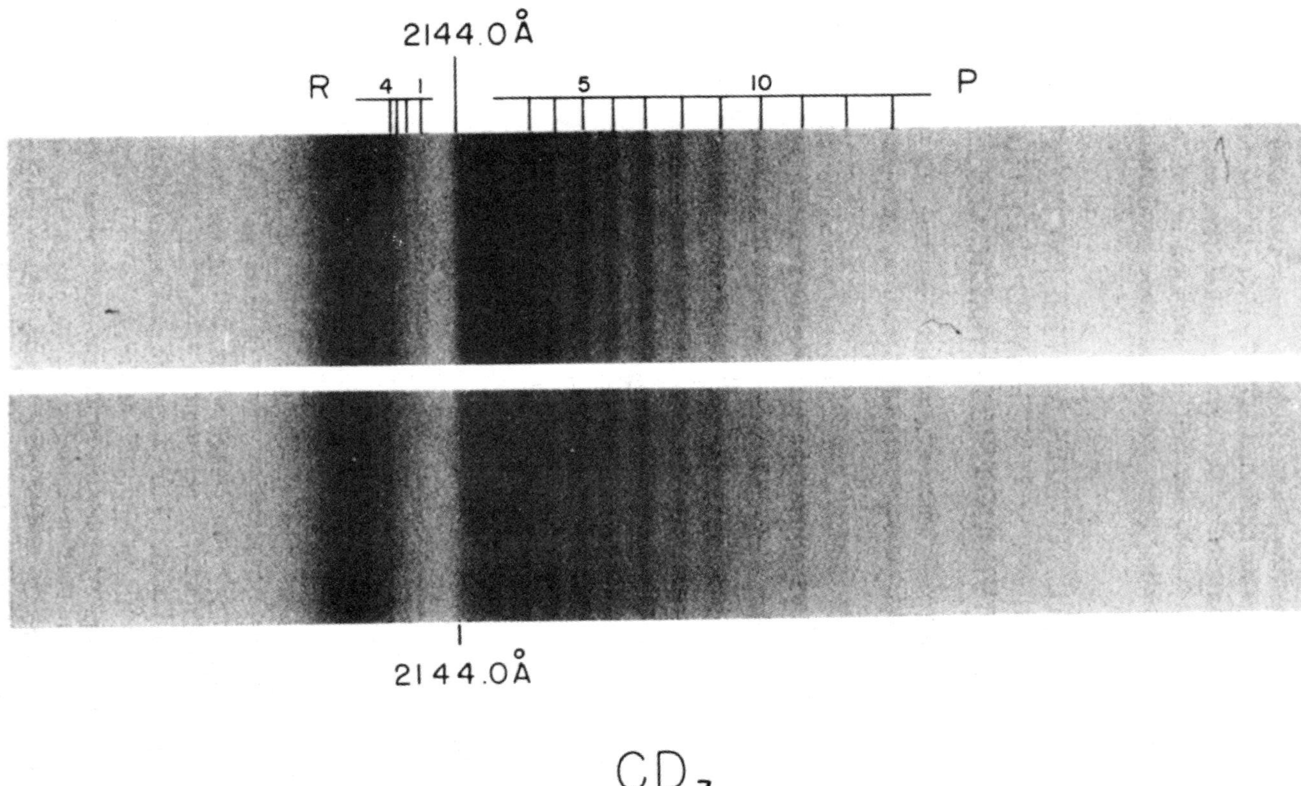

Figure 1 Spectrogram showing the parallel band of CD_3 at 2144 Å, taken by G. Herzberg (supplied by the courtesy of Dr. G. Herzberg).

Because of the fundamental importance of this radical, I decided to apply our IR diode laser spectrometer to it. Fortunately, Hiizu Iwamura, who is now in the Department of Chemistry of the University of Tokyo, was Professor at the Institute when I planned to investigate methyl. He suggested that we use di-*tert*-butyl peroxide as a precursor. Our problem was the diode laser, as was often the case; at that time, i.e. 1980, the American Government prohibited the export of IR diodes around 16 μm, because the IR spectrum of UF_6 in that region was classified. Fortunately, Fujitsu, a Japanese electric company, supplied us with several diodes of good quality, and thus Yamada and I (31) could observe the $CH_3 v_2$ band. As shown in Figure 2, the spectrum was very strong, and it was easy to identify CH_3 lines even on a scope; CH_3 lines appeared as broad lines because of large Doppler widths. We made assignments for $v_2 = 1$–0, 2–1, and 3–2 and derived the potential function, as shown in Figure 3.

In 1953, Walsh (32) published a series of papers, in which he predicted the geometry of simple molecules based on a simple molecular-orbital consideration. However, the methyl radical was a case for which his theory could not make any definite conclusion as to its planarity. As Figure 3 shows, there is no potential hump at the center, although this function has an unusually large quartic term. We suspected that this large quartic term was caused by a vibronic interaction with an excited electronic state of a_2'' symmetry. When I spoke on this result at Texas Tech University, G. Wilse Robinson questioned our conclusion. He said that he could not believe the planarity of the molecule until we got data on CD_3 consistent with our data on CH_3. Sears and coworkers (33, 33a) subsequently observed the v_2 band of CD_3 in support of our interpretation. So, I believe that now Robinson is also satisfied. We have extended the observation and have assigned the $v_2 = 4$–3 band. As I discuss below, Hermann & Leone (34) observed emission spectra of higher hot bands of CH_3, which, although of much lower resolution, are also consistent with our results.

We then observed the v_3 band, the degenerate C–H stretching band in the 3-μm region, by using a difference-frequency laser system, in collaboration with Amano, Bernath, and coworkers (35) at NRC. Because the methyl radical is planar and of D_{3h} symmetry, the v_1 band is IR inactive. Engel and coworkers (36, 36a) thus employed coherent anti-Stokes Raman spectroscopy to observe this totally symmetric band. Although their data are of lower resolution than the IR data, they could determine the vibration-rotation interaction constants. I started an analysis of the anharmonic potential function of methyl in 1979, well before our observation of the v_2 band. After we successfully detected this band, I slightly adjusted the potential constants (37). Figure 4 compares the α constants thus calculated with the observed values. The agreement is good not only for CH_3, but

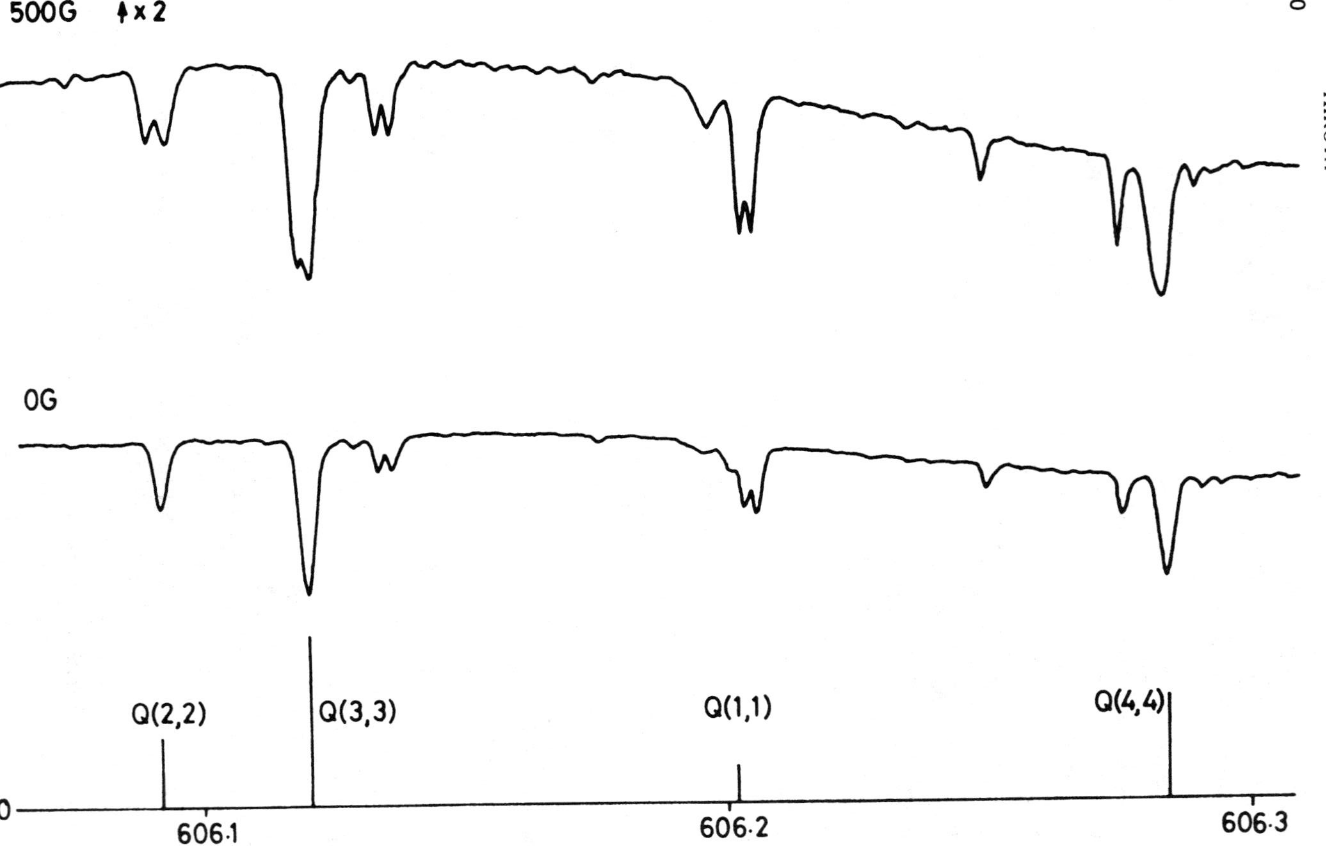

Figure 2 $^{q}Q_{K}(N)$ lines with $N = K = 1$–4 of the $CH_3\nu_2$ fundamental band. The lower trace represents zero-field spectra, whereas the upper trace is obtained with a magnetic field of 500 G (reproduced with permission from Ref. 31).

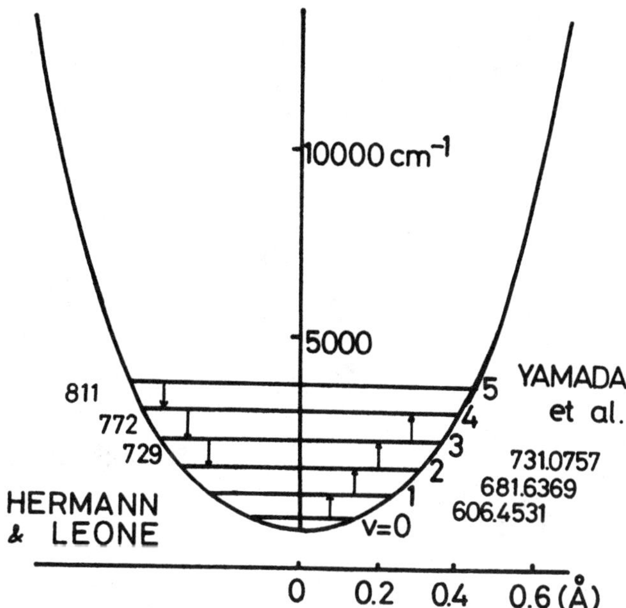

Figure 3   Potential function for the $CH_3 \nu_2$ out-of-plane bending mode.

also for CD_3. There is one normal mode left unobserved, namely the ν_4 band. We have some data on this band, and I hope to analyze this band in the near future.

The methyl radical is of fundamental importance in many fields, which also holds for the analysis of vibration-rotation interactions. We can give explicit expressions for many of the vibration-rotation interaction constants. Recently, a beautiful paper was published by Oka's group (38); they have been applying their results mainly to CH_3^+. This ion is a more ideal D_{3h} molecule; its ν_2 potential function is closer to a harmonic potential than that of CH_3. Interestingly, there is a simple rule for vibration-rotation interaction constants (E. Hirota 1988, unpublished): When n normal modes belong to one symmetry species, all vibration-rotation parameters, like the expansion coefficients of the moment of inertia in terms of normal coordinates, are fixed by $n(n-1)/2$ pieces of information in the harmonic potential function together with the atomic masses and the geometry of the molecule. The methyl radical has two one-dimensional symmetry species, a_1' and a_2'', and one two-dimensional symmetry species, e'. Therefore, only one piece of information may be derived from vibration-rotation constants of this radical, or in other words, all the vibration-rotation

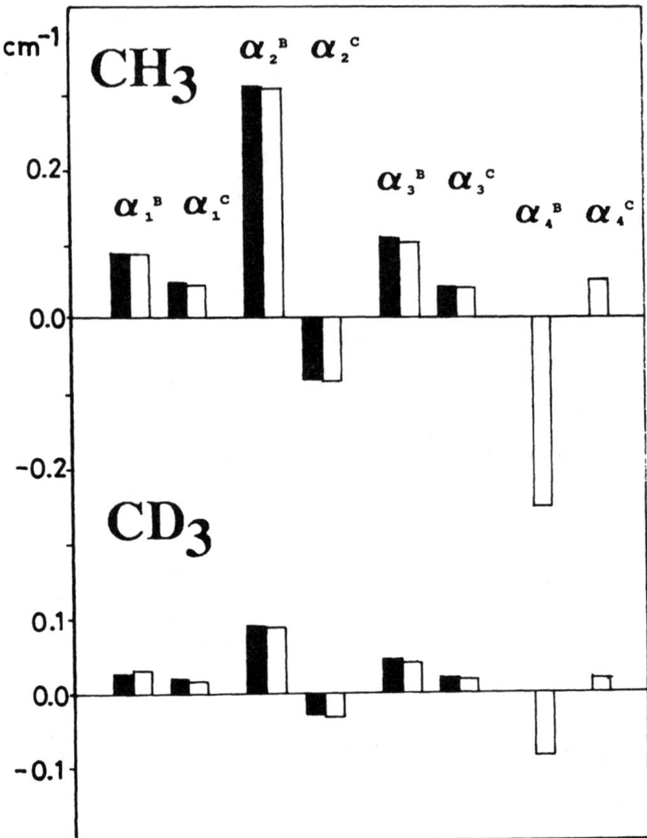

Figure 4 Vibration-rotation interaction constants of the CH_3 and CD_3 radicals. Black and white poles represent the observed and calculated values, respectively. The third-order anharmonic potential constants employed are $F_{rrr} = -33 \pm 6$ md/$Å^2$, $F_{xxx}/\sqrt{6} = -0.57 \pm 0.20$ mdÅ, and $F_{144} = -0.4$ md (1 and 4 stand for the totally symmetric and the degenerate bending internal coordinates, respectively).

interaction constants of this molecule are determined by one parameter related to the harmonic potential force field. Oka chose the first-order Coriolis coupling constant and expressed the inertial defects in terms of this parameter.

I have left one important constant unanalyzed, i.e. the ℓ-type doubling constant q, which has been determined for the v_3 band of both CH_3 and CD_3 (39). This constant should supply us with information on the anharmonic force field; it yields a linear relation between k_{333} and k_{334}

and thus should be useful in refining my potential function. The force field I employed to predict the α constants gives a value for q_3 an order of magnitude larger than the observed value. But, the two k constants are of opposite sign and the ℓ-type doubling constant corresponds to a difference between two large terms. The ℓ-type doubling constant in the $v_4 = 1$ state yields a similar relation between k_{344} and k_{444}.

The Silyl Radical

It is a natural extension of our research to go down the periodic table. This tendency surely explains why we studied the silyl radical. Yamada and I obtained such nice results on the methyl radical that we thought we could enjoy spectroscopy of silyl, as well. I also suspected that the silyl radical would afford us more than that. Although my knowledge of the fabrication of electronic devices was very limited, I thought this species might play an important role in many processes of practical importance. So we decided to try to detect this radical by IR diode laser spectroscopy.

Before we started, we knew that electron spin resonance (ESR) studies (40, 41) had already established the nonplanarity of the molecule; they estimated the HSiH angle to be about $112.7°$. Molecular orbital calculations also predicted the molecule to be nonplanar (42–44), which contrasts sharply with CH_3. Why are the two molecules so different? When I presented our IR spectroscopic results on SiH_3 at the 1986 Austin Symposium, someone asked me this question, but I could not answer. I thought it was too involved to give a simple answer, and I still think so. But, at the symposium a kind gentleman did reply to the question. He said that the energy separation between s and p orbitals is larger in Si than in C, thus sp^2 or sp^3 hybridization would become less important in SiH_3 than in CH_3, making the bonding orbitals of Si closer to a pure p orbital than those of C. This answer is probably right, and will be understandable to those who like this sort of explanation.

Yamada and I (45) started searching for the spectrum of the v_2 band of SiH_3, because the corresponding band of CH_3 was so strong. However, we found it quite difficult to observe the IR spectrum of SiH_3; the molecule is too light for IR diode laser spectroscopy and the absorption was not as strong as the v_2 band of CH_3. Furthermore, ab initio calculations predicted vibrational frequencies that were too high; we often lost our way. After a long search, we finally got the spectrum and its assignment.

We found the v_2 band to be split into two components, which were probably caused by inversion, as in the case of ammonia. The splitting of about 6.9 cm^{-1} was thus ascribed mainly to the splitting in the upper state $v_2 = 1$. Although our data were limited, we could estimate the barrier height to the inversion to be about 1900 cm^{-1}, and the splitting in the ground state to be about 0.1 cm^{-1}. The band origin we determined is 727.9438 and 721.0486 cm^{-1} for $v_2 = 1^- \leftarrow 0^+$ and $1^+ \leftarrow 0^-$, respectively, which are 50–60 cm^{-1} lower than ab initio estimates. We also obtained approximate values for the Si–H length and the H–Si–H angle to be 1.456–1.468 Å and 108.5–110.5°, respectively.

DEVELOPMENT OF HIGH-RESOLUTION KINETIC SPECTROSCOPY AND ITS APPLICATIONS TO CHEMICAL REACTIONS

As we accumulated the spectroscopic data on transient molecules, I started to dream of using these data in related fields. For example, I wished we could provide more dynamical information on chemical reactions by using our own spectroscopic data and our own high-resolution spectroscopic methods. I thought we should maintain the high resolution of our methods, otherwise we would lose our identity.

Figure 5 illustrates what I dreamed of: I hoped to record two-dimensional spectra, one axis being the ordinary frequency or wavelength scale at high resolution and the other representing the time axis. The uncertainty principle of Heisenberg prohibited us from getting very detailed data beyond a certain limit in this two-dimensional area, but our method is still far from this limit. For example, a frequency resolution of 1 MHz can be achieved only when the time resolution is worse than 0.17 μs. Our signal detection circuitry had a time response of about 0.3 μs. Thus, it was still quite far from the Fourier limit imposed by the uncertainty principle, in

Time-resolved high-resolution spectrum

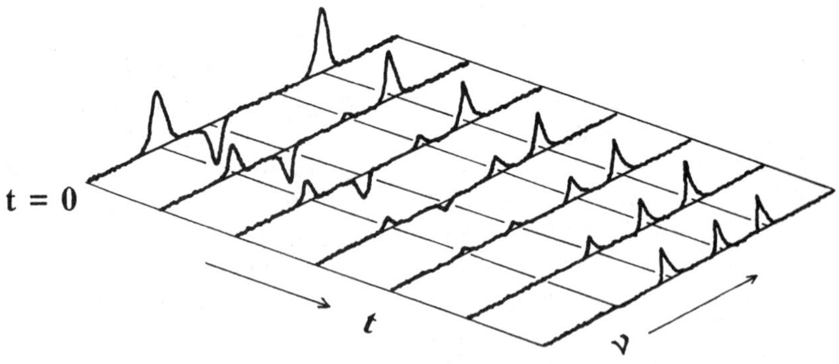

Figure 5   Observation of time-resolved high-resolution spectra.

particular in the case of IR spectroscopy where the line width is normally determined by Doppler broadening (several tens of MHz being a typical linewidth).

It was again extremely fortunate for us that we got the right people to start this new project. In 1981 or 1982, I received a letter from Jim Butler of the Naval Research Laboratory (NRL). He asked to stay in my laboratory for a year or so, as he planned to learn the technique of IR diode laser spectroscopy from my group. He could get support from the NRL. Shortly after he joined my group, we started to examine the possibility of making our IR spectrometer time-resolved to observe time-dependent phenomena in real time. Two able people, Hideto Kanamori and Tatsuya Minowa, soon joined us; they started setting up our kinetic spectroscopy system. They introduced a transient digitizer in the signal detection and recording circuitry and controlled it by a personal computer (46, 47). We thought that the photochemical decompositions of molecules would be the simplest systems to study first with our new method and we could induce these processes by irradiating appropriate precursors with an excimer laser. The output of an excimer laser would be large enough to produce transient molecules at concentrations sufficient to detect their vibration-rotation spectra with our IR spectrometer (48).

Before we started the new project, I was worried about electrical noise that the excimer laser might cause. Our high-resolution spectrometers are delicate, home-made machines; if they are placed in a noisy environment, we would be unable to record the faint signals of photofragments generated at very small concentrations. Fortunately, I did not need to worry about noise; the excimer laser we got was much quieter than I had expected, and our spectrometer worked well.

Attempts to detect photolysis-produced transient molecules by using high-resolution spectroscopy are by no means new. Pickett & Boyd (49), for example, employed microwave spectroscopy to observe the rotational spectra of HCO generated by the photolysis of acetaldehyde. But in those early days, the light sources for photolysis were weak, and thus there were few reported examples of success. The low intensity of the light source yielded transient species at too low concentration to be detected by high-resolution spectroscopy, like microwave spectroscopy. Advances in laser technology have made it routine to use high-power excimer lasers as photolysis sources. An excimer laser, for example, emits a light pulse of 100 mJ with a few ns duration at 10 Hz repetition rate. This light pulse contains about 10^{17} photons, which is surely an enormous number. Even if the quantum yield is as small as 10^{-4}, we still get 10^{13} photofragments, a number large enough to observe their vibration-rotation spectral lines with our IR diode laser spectrometer. We need 10^8 molecules to record their IR spectra in the most favorable cases, and we should be able to detect 10^{13} molecules without too much difficulty in most cases.

Endo, who was in my group, transferred our experience with time-resolved IR spectroscopy to microwave spectroscopy (50). It was fortunate for us that, when Endo started this modification, Soji Tsuchiya of the University of Tokyo joined the Institute for Molecular Science as an adjunct professor. He had been very active in the field of molecular dynamics and chemical reaction, and he proposed that we study a reaction system that was very appropriate for my group: the reaction of ethylene with oxygen atom (51, 52). I will not go into the details of this study, but our results have changed the view widely accepted in the field of chemical reactions. Namely, the $O + C_2H_4$ reaction proceeds mainly in two channels: vinoxy (better named formyl methyl) + H [A] and $CH_3 + HCO$ [B]. People had believed that channel [B] was opened only under high pressure conditions, i.e. [B] was essentially a collision-induced process, and only [A] occurred at low pressures. We demonstrated clearly that this conclusion was erroneous, as the branching ratio of [A] and [B] was almost independent of the sample pressure.

THE PHOTOLYSIS OF METHYL IODIDE AT 248 nm

We have applied our time-resolved spectroscopic system to a few reactions. As an example, I describe in some detail the photolysis of CH_3I at 248 nm. This reaction is one of the most important photochemical reactions and thus has been studied many times with time-of-flight (TOF) mass spectrometry (53–55) and IR emission spectroscopy (56). Although there is a potential crossing, most of these studies have agreed that the process leads to a methyl radical with an iodine atom mainly in the excited fine structure level $^2P_{1/2}$, referred to as the I* channel; its fraction is about 70–80%. In addition, because methyl iodide is excited by the 248 nm excimer laser light to a Franck-Condon state, where the CH_3 group keeps a nearly tetrahedral configuration, the photofragment CH_3 is highly excited in the v_2 mode. However, no one had ever resolved the spectra, either TOF or others, into components that correspond to the vibrational states of the methyl radical. Most people have simply assumed that CH_3 would be distributed over the v_2 mode in a way they anticipated. It is really strange that, once someone reported such a distribution of "reasonable shape," subsequent studies followed similar steps and found nearly identical distributions. No one really questioned how such a vibrational distribution of CH_3 radical could have been derived from unresolved data.

When we completed our time-resolved spectroscopic system, we chose the CH_3I photolysis to test our method (57), because we have extensive data on the v_2 mode (31). Because an inverted distribution was reported in the literature, we expected to observe induced emission for the v_2 band, as in our study on SO generated by the 193-nm photolysis of SO_2 (48). To our surprise, we did not observe any emission; all the vibration-rotation lines of this band appeared as absorptions. It was still an early stage of our study, so we were not entirely confident of our technique; some members of my group seriously criticized our results and proposed an unusual explanation. I could not accept this explanation and did not see any firm ground for their criticism. However, we postponed the publication of our data for a long time; I was afraid there might be something wrong in our method that we could not detect. In the meantime, Steve Leone stayed with us for a few weeks and joined Kanamori to repeat the observation (58). I hoped that he would also agree with our findings.

Our results gradually prevailed, and people started to question the previous result, an inverted vibrational population of the methyl radical. Among others, Sears and coworkers (59) followed us by studying the fully deuterated species CD_3. They also found that the v_2 vibrational population was not inverted, as I expected. The original TOF experiment also seemed to have been revised (Y. T. Lee 1989, private communication), although the details have not been published; the results now support our conclusion.

About two years ago, Toshinori Suzuki joined me. He accepted my proposal that we repeat the observation of methyl radicals produced by the 248-nm photolysis of methyl iodide. He carried out the experiment very carefully (T. Suzuki et al 1990, submitted for publication). One of his accomplishments was to explain satisfactorily the spike that appeared in an early stage of the time profile of the signal. As shown in Figure 6, all four bands show a very similar spike, which indicates that all methyl radicals produced by the photolysis behave more or less in the same way.

The nice thing about our IR experiment is that, because we are observing a particular vibration-rotation transition and we know in which electronic states our photofragments are prepared, we can calculate the translational velocity of the fragments. The velocity of CH_3, thus calculated, is very large, something like 4100–4400 ms^{-1}, and it takes only a few μs for methyl radicals to cross the IR beam. Therefore, the spike corresponds to the absorption signal of initially prepared CH_3, which escapes from the IR beam within a few μs. We examined the lineshape within the time interval of the spike and confirmed that the spectral width, namely the Doppler width did not change at all. This means that no vibration-rotation relaxation took place within this time interval, and our observation is independent of such relaxation. This relaxation has been referred to in the

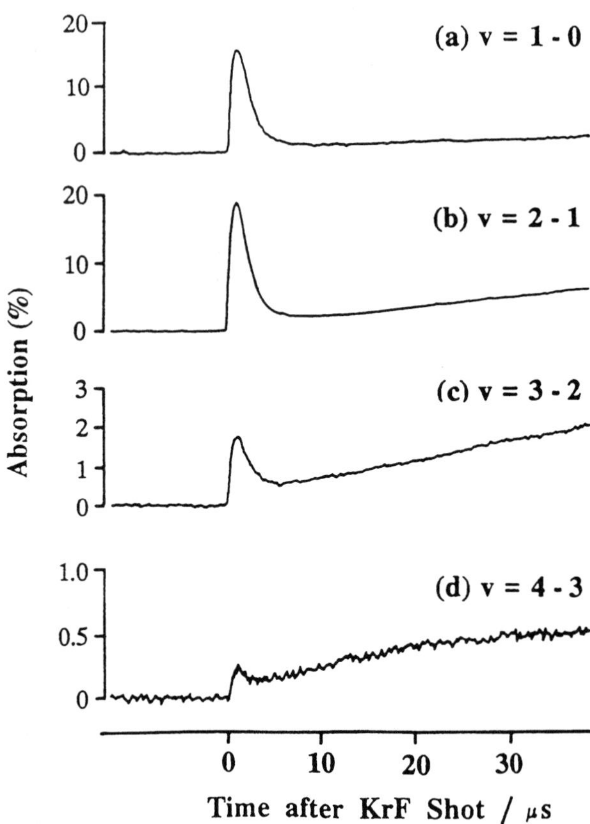

Figure 6   Time profiles of the $^qQ_3(3)$ lines of the $v_2 = 1$–0 to 4–3 bands of CH_3 prepared by the 248-nm photolysis of CH_3I.

criticism of our data, but Suzuki's analysis has clearly shown that this criticism does not hold for our results.

The vibrational distribution of the methyl radical Suzuki obtained is compared with those of previous studies in Figure 7. As I discussed earlier, the discrepancy is striking, but it is not unexpected; previous studies did not obtain any data that directly gave the vibrational distribution. Also, Sears et al (59) did not observe the spike for CD_3; I do not know why. The velocity of CD_3 is surely lower, but not much. If the sample pressure is too high, the spike may be smeared out, and then we need to worry about the effect of relaxation.

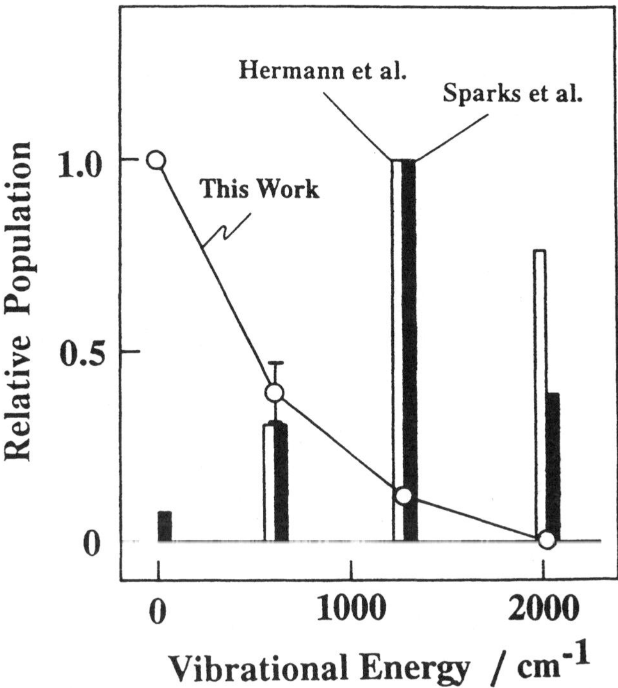

Figure 7 Vibrational distribution of CH_3 generated by the 248-nm photolysis of CH_3I. Our result designated by open circles is compared with those of Refs. 53 and 56.

DIAGNOSIS OF THE SILANE DISCHARGE PLASMA WITH IR DIODE LASER SPECTROSCOPY

There must be numerous applications of high-resolution spectroscopy of transient molecules. Development of the time-resolved observation technique has certainly widened the potential of high-resolution spectroscopy. A few years ago I was invited to speak at a symposium on plasma processing. The audience was mainly associated with the development of electronic devices. Toshio Goto, of the Department of Electronics at Nagoya University, also attended this symposium. He was interested in applying our IR diode laser spectroscopy to diagnose the silane discharge plasma employed to fabricate amorphous silicon. We thus started a collaborative study. From my group, Yamada, who observed and analyzed the v_2 band of SiH_3 (45), took the initiative, in collaboration with Goto's graduate students Naoshi Itabashi, Nobuki Nishiwaki, and others, in observing the

IR spectra of the three silicon hydrides—SiH, SiH_2, and SiH_3—that were expected to be important intermediates in the silane plasma.

There was controversy as to the role of SiH_2 and SiH_3 in the plasma. Some groups thought SiH_2 to be the most important intermediate; others believed SiH_3 to play a crucial role in the process. But no one was able to monitor these species in the discharge media, and thus no clear-cut conclusion had been obtained.

I searched the literature to find data on the IR spectra of these three species. We knew that our result on the v_2 band was only the IR data on the SiH_3 radical. To our surprise, no IR spectrum had been observed for the SiH_2 molecule in the gaseous phase, although its electronic spectrum in the visible region had been well known and the observation had already been carried out extensively. So, we decided to observe the IR spectrum of SiH_2 by using our IR diode laser spectrometer. We found the photolysis of phenylsilane at 193 nm to be a most convenient way of generating SiH_2. First, we investigated the v_2 band, because this band is isolated from the rest and thus free of any perturbations. We found that the spectrum was weak, and it took us some time to complete the assignment (60). Although we got an assignment for the v_2 band in this way, the band was a bit too weak to carry out a quantitative analysis by using this spectrum; we could obtain some qualitative information from this band.

We thus extended the measurements to the 5-μm region, where we expected to observe three bands, v_1, v_3, and $2v_2$; these bands were supposedly several times stronger than the v_2 band. Unfortunately, they interacted with each other: v_1 and $2v_2$ coupled with each other by Fermi interaction through an anharmonic potential term k_{122}; v_1 and $2v_2$ both interacted with v_3 through Coriolis terms. We have already observed more than 100 lines in the 1900 cm^{-1} region and it is my job to make assignments for these lines. So far, however, I have been unable to get a solution because of many commitments. When I succeed, we will provide more detailed data on the concentration of SiH_2 in the discharge plasma. For the third molecule, SiH, extensive observations had already been carried out in the IR region, so we had no difficulties in monitoring this species in the plasma.

We switched the discharge, either direct current or radiofrequency, on and off and observed the change in the IR absorption. We employed the software (46, 47) that we developed for time-resolved spectroscopy combined with the photolysis. We readily found that the lifetimes of SiH and SiH_2 were an order of magnitude shorter than that of SiH_3 (61, 62). This result may be ascribed to the fact that both SiH and SiH_2 react with the precursor SiH_4 very rapidly, whereas SiH_3 is eliminated mainly by a bimolecular recombination reaction. We followed the profile of the SiH_3

signal in some detail, and found that the decay curve could be reproduced only when we considered the contributions of both the bimolecular recombination and the diffision. The former was proportional to the square of the SiH_3 concentration, whereas the latter was linearly dependent on the concentration; thus, the two contributions could be separated (63). Because of the difference in lifetime, the SiH_3 radical was an order of magnitude or so more abundant than SiH and SiH_2.

One serious drawback of the IR diagnosis is that, because it is based upon weak absorptions of molecules, we need a long path length to increase the signal intensity. This means that we can get enough sensitivity only at the expense of spatial resolution. However, amorphous silicon is ordinarily produced in a cell with parallel plate electrodes (circular plates) for discharge. For this type of cell, we may determine the concentrations of fragment molecules as functions of the distance from the surface of the base electrode on which amorphous silicon grows. This observation, combined with the diffusion coefficient we determined, led us to conclude that the main process of producing amorphous silicon took place through the SiH_3 radical (64).

Our conclusion may be of some significance for fabrication of amorphous silicon in factories. Only a trial-and-error approach has been employed, but, if we can diagnose the molecular processes taking place in the plasma in real time, it would become much easier to find better conditions. When such conditions are established, the spectra of fragments, such as SiH_3, may be employed to control the process.

I have learned a lot from this collaborative work. At one time, I thought that the study of my group could be done only within the framework of so-called fundamental science, and that our results were not expected to be of practical significance. However, any important research in the fundamental sciences will strongly influence many fields, including practical applications. The relationship between fundamental science and practical applications has been discussed many times in many cases. I do not deny that there is a strong correlation between them, and it would certainly be profitable for both sides to appreciate this relationship. However, this fact does not imply that all fundamental science must have some practical application. Its value must be judged by how much it affects other intellectual activities in both fundamental and applied areas.

FURTHER REMARKS AND THE FUTURE

We have shown that high-resolution spectroscopy, like infrared and microwave, can be of much more potential use in wide research areas by introducing a function of time-resolved observation in it. We have applied it to a few chemical reactions, and in almost all cases we found something fundamentally new. In a few cases, we completely revised the opinion hitherto widely accepted.

I might add that the photolysis method we developed is also of some use to generate new transient species that are difficult to obtain by earlier methods like electric discharge. One good example is the vinyl radical, which plays an important role in some reactions of organic molecules. Kanamori et al (65) successfully produced this molecule by the photolysis of vinyl halides. He also tried to obtain this transient molecule by discharge, but failed; the molecule seems to be too fragile to generate by such severe methods as discharge.

Recently, resonance-enhanced multiphoton ionization spectroscopy (REMPI) has been advanced as a powerful means of studying transient molecules. In fact, it has been applied to the two free radicals, CH_3 (66–71) and SiH_3 (72). The sensitivity of this method seems to be extremely high, because it counts ions generated by ionization; the spectra reported show exceptionally good signal-to-noise ratio. This method will supply us with much information on transient molecules and the chemical reactions that they are involved in, information not obtainable with methods hitherto employed. However, I am skeptical about its applications to quantitative analyses. It will be difficult to extract information on the concentration of transient molecules; generally, more than one process is induced by several laser photons of different colors. Furthermore, it is almost impossible to apply this technique to discharge plasma; there already exist an enormous number of charged particles in the plasma, and REMPI will add too small a number of ions to record.

Astronomy offers another important application of high-resolution spectroscopy. The interplay between astronomy and molecular spectroscopy may be traced back even to the last century. A new trend arose in the 1960s. Thanks to developments in computer and other technology, it became possible to observe microwave spectra emitted from interstellar molecules by using a radio telescope. Up to the present, about 80 molecules have been identified, and about half of them are transient molecules in terrestrial conditions. Here, again, we have an example of nice interplay between astronomical and laboratory observations.

Up to this point I have been concerned with the past, i.e. my personal history, which is symbolized by the phrase from high-resolution spectroscopy to chemical reactions. This project will continue; although I do not know who will continue it, I am certain the activity in this area will even expand in the future. It is very tempting to think of future developments, and I take here the privilege of stating what I plan to do in the future, if I am allowed to continue research.

High-resolution spectroscopy of transient molecules and ions is a fascinating subject, and, although it is not easy to carry out, it is still very rewarding. We can easily mention candidate molecules which we may detect, and determine their structures by using high-resolution spectra. For example, let us consider hydrocarbon radicals that contain three or four carbon atoms, C_3H_n and C_4H_m, where $n = 1$–7 and $m = 1$–9. It is striking to see that we know only a few of them. The situation is even premature for ions. It would not be absurd to think that some of these "unknown" molecules have properties that we have not even thought of. We must develop new high-resolution spectroscopic techniques to expand our scope.

ACKNOWLEDGMENTS

I thank Jon T. Hougen for a critical reading of the present manuscript. My coworkers and I are very indebted to him for clarifying various aspects of our research.

Literature Cited

1. Townes, C. H., Dailey, B. P. 1949. *J. Chem. Phys.* 17: 782–96
2. Townes, C. H., Dailey, B. P. 1952. *J. Chem. Phys.* 20: 35–40
3. Dailey, B. P., Townes, C. H. 1955. *J. Chem. Phys.* 23: 118–23
4. Hirota, E., Oka, T., Morino, Y. 1958. *J. Chem. Phys.* 29: 444–45
5. Hirota, E., Morino, Y. 1960. *Bull. Chem. Soc. Jpn.* 33: 158–62
6. Morino, Y., Kikuchi, Y., Saito, S., Hirota, E. 1964. *J. Mol. Spectrosc.* 13: 95–118
7. Mizushima, S. 1954. *Structure of Molecules and Internal Rotation.* New York: Academic. 254 pp.
8. Hirota, E. 1962. *J. Chem. Phys.* 37: 283–91
9. Durig, J. R., Godbey, S. E., Sullivan, J. F. 1984. *J. Chem. Phys.* 80: 5983–93
10. Caminati, W., Fantoni, A. C., Manescalchi, F., Scappini, F. 1988. *Mol. Phys.* 64: 1089–1103
11. Hirota, E. 1965. *J. Chem. Phys.* 42: 2071–89
12. Hirota, E. 1968. *J. Mol. Spectrosc.* 26: 335–50
13. Meakin, P., Harris, D. O., Hirota, E. 1969. *J. Chem. Phys.* 51: 3775–88
14. Herzberg, G. 1966. *Molecular Spectra and Molecular Structure III. Electronic Spectra and Electronic Structure of Polyatomic Molecules.* Princeton, NJ: Van Nostrand. 745 pp.
15. Dousmanis, G. C., Sanders, T. M. Jr., Townes, C. H. 1955. *Phys. Rev.* 100: 1735–54
16. Mockler, R. C., Bird, G. R. 1955. *Phys. Rev.* 98: 1837–39
17. Amano, T., Hirota, E., Morino, Y. 1967. *J. Phys. Soc. Jpn.* 22: 399–412; 1968. *J. Phys. Soc. Jpn.* 23: 300
18. Amano, T., Hirota, E., Morino, Y. 1968. *J. Mol. Spectrosc.* 27: 257–65
19. Amano, T., Saito, S., Hirota, E., Morino, Y., Johnson, D. R., Powell, F. X. 1969. *J. Mol. Spectrosc.* 30: 275–89
20. Amano, T., Saito, S., Hirota, E., Morino, Y. 1969. *J. Mol. Spectrosc.* 32: 97–107
21. Amano, T., Yoshinaga, A., Hirota, E. 1972. *J. Mol. Spectrosc.* 44: 594–98
22. Saito, S., Amano, T. 1970. *J. Mol. Spectrosc.* 34: 383–89
23. Amano, T., Hirota, E. 1972. *J. Chem. Phys.* 57: 5608–10
24. Amano, T., Hirota, E. 1973. *J. Mol. Spectrosc.* 45: 417–19
25. Hirota, E. 1989. *Int. Rev. Phys. Chem.* 8: 171–205; 1985. *High-Resolution Spectroscopy of Transient Molecules.* Berlin: Springer. 233 pp.
26. Herzberg, G. 1961. *Proc. R. Soc. London Ser. A* 262: 291–317
27. Herzberg, G. 1966. See Ref. 14, p. 227
28. Milligan, D. E., Jacox, M. E. 1967. *J. Chem. Phys.* 47: 5146–56

29. Jacox, M. E. 1977. *J. Mol. Spectrosc.* 66: 272–87

30. Tan, L. Y., Winer, A. M., Pimentel, G. C. 1972. *J. Chem. Phys.* 57: 4028–37

31. Yamada, C., Hirota, E., Kawaguchi, K. 1981. *J. Chem. Phys.* 75: 5256–64

32. Walsh, A. D. 1953. *J. Chem. Soc.* 2260–2331

33. Frye, J. M., Sears, T. J., Leitner, D. 1988. *J. Chem. Phys.* 88: 5300–6

33a. Sears, T. J., Frye, J. M., Spirko, V., Kraemer, W. P. 1989. *J. Chem. Phys.* 90: 2125–33

34. Hermann, H. W., Leone, S. R. 1982. *J. Chem. Phys.* 76: 4759–65

35. Amano, T., Bernath, P. F., Yamada, C., Endo, Y., Hirota, E. 1982. *J. Chem. Phys.* 77: 5284–87

36. Holt, P. L., McCurdy, K. E., Weisman, R. B., Adams, J. S., Engel, P. S. 1984. *J. Chem. Phys.* 81: 3349–51 (CH_3)

36a. Miller, J. T., Burton, K. A., Weisman, R. B., Wu, W.-X., Engel, P. S. 1989. *Chem. Phys. Lett.* 158: 179–83 (CD_3)

37. Hirota, E., Yamada, C. 1982. *J. Mol. Spectrosc.* 96: 175–82

38. Jagod, M.-F., Oka, T. 1990. *J. Mol. Spectrosc.* 139: 313–27

39. Fawzy, W. M., Sears, T. J., Davies, P. B. 1990. *J. Chem. Phys.* 92: 7021–26

40. Morehouse, R. L., Christiansen, J. J., Gordy, W. 1966. *J. Chem. Phys.* 45: 1751–58

41. Jackel, G. S., Gordy, W. 1968. *Phys. Rev.* 176: 443–52

42. Bunker, P. R., Olbrich, G. 1984. *Chem. Phys. Lett.* 109: 41–44

43. Ellinger, Y., Pauzat, F., Barone, V., Douady, J., Subra, R. 1980. *J. Chem. Phys.* 72: 6390–97

44. Marynick, D. S. 1981. *J. Chem. Phys.* 74: 5186–89

45. Yamada, C., Hirota, E. 1986. *Phys. Rev. Lett.* 56: 923–25

46. Kanamori, H., Butler, J. E., Minowa, T., Kawaguchi, K., Yamada, C., Hirota, E. 1985. *Laser Spectroscopy VII*, pp. 115–17. Berlin: Springer. 419 pp.

47. Kanamori, H., Butler, J. E., Kawaguchi, K., Yamada, C., Hirota, E. 1985. *J. Mol. Spectrosc.* 113: 262–68

48. Kanamori, H., Butler, J. E., Kawaguchi, K., Yamada, C., Hirota, E. 1985. *J. Chem. Phys.* 83: 611–15

49. Pickett, H. M., Boyd, T. L. 1978. *Chem. Phys. Lett.* 58: 446–49

50. Endo, Y., Kanamori, H., Hirota, E. 1987. *Laser Chem.* 7: 61–77

51. Endo, Y., Tsuchiya, S., Yamada, C., Hirota, E., Koda, S. 1986. *J. Chem. Phys.* 85: 4446–52

52. Koda, S., Endo, Y., Hirota, E., Tsuchiya, S. 1987. *J. Phys. Chem.* 91: 5840–42

53. Sparks, R. K., Shobatake, K., Carlson, L. R., Lee, Y. T. 1981. *J. Chem. Phys.* 75: 3838–46

54. Van Veen, G. N. A., Baller, T., De Vries, A. E., Van Veen, N. J. A. 1984. *Chem. Phys.* 87: 405–17

55. Barry, M. D., Gorry, P. A. 1984. *Mol. Phys.* 52: 461–73

56. Hermann, H. W., Leone, S. R. 1982. *J. Chem. Phys.* 76: 4766–74

57. Butler, J. E., Kawaguchi, K., Kanamori, H., Yamada, C., Hirota, E. 1983. *IMS Ann. Rev.* II-B-4, Inst. Mol. Sci., Okazaki, Japan

58. Kanamori, H., Leone, S. R., Hirota, E. 1988. *IMS Ann. Rev.* II-A-9, Inst. Mol. Sci., Okazaki, Japan

59. Hall, G. E., Sears, T. J., Frye, J. M. 1989. *J. Chem. Phys.* 90: 6234–42

60. Yamada, C., Kanamori, H., Hirota, E., Nishiwaki, N., Itabashi, N., et al. 1989. *J. Chem. Phys.* 91: 4582–86

61. Itabashi, N., Kato, K., Nishiwaki, N., Goto, T., Yamada, C., Hirota, E. 1988. *Jpn. J. Appl. Phys.* 27: L1565–67

62. Itabashi, N., Nishiwaki, N., Magane, M., Goto, T., Matsuda, A., et al. 1990. *Jpn. J. Appl. Phys.* 29: 585–90

63. Itabashi, N., Kato, K., Nishiwaki, N., Goto, T., Yamada, C., Hirota, E. 1989. *Jpn. J. Appl. Phys.* 28: L325–28

64. Itabashi, N., Nishiwaki, N., Magane, M., Naito, S., Goto, T., et al. 1990. *Jpn J. Appl. Phys.* 29: L505–7

65. Kanamori, H., Endo, Y., Hirota, E. 1990. *J. Chem. Phys.* 92: 197–205

66. Loo, R. O., Hall, G. E., Haerri, H.-P., Houston, P. L. 1988. *J. Phys. Chem.* 92: 5–8

67. Loo, R. O., Haerri, H.-P., Hall, G. E., Houston, P. L. 1989. *J. Chem. Phys.* 90: 4222–36

68. Black, J. F., Powis, I. 1988. *Chem. Phys.* 125: 375–88

69. Black, J. F., Powis, I. 1988. *J. Chem. Phys.* 89: 3986–92

70. Black, J. F., Powis, I. 1988. *Laser Chem.* 9: 339–58

71. Powis, I., Black, J. F. 1989. *J. Phys. Chem.* 93: 2461–70

72. Johnson, R. D. III, Tsai, B. P., Hudgens, J. W. 1989. *J. Chem. Phys.* 91: 3340–59

Sam Weissman

Annu. Rev. Phys. Chem. 1990. 41: 1–13

THE WAY IT WAS

Samuel Weissman

Department of Chemistry, Washington University, St. Louis,
Missouri 63130-4899

My account of the Way it Was begins with the years 1933–1941, skips the
war years 1941–1945, and ends in the mid 1950s.

The first volume of the *Journal of Chemical Physics* appeared in 1933,
the year of my beginning graduate work in physical chemistry at the
University of Chicago. Many of the great contributors to physical chemistry—Langmuir, Debye, Lewis, Cross, Pauling, Libby, Urey, Kistiakowsky, Van Vleck, Wilson, Eyring, Mayer, Kirkwood, Hildebrand, James,
and Coolidge—published in that volume. But in 1933, I had heard in my
undergraduate work only of Lewis, Langmuir, and Debye.

The department, under the chairmanship of the organic chemist Julius
Stieglitz, retained vestiges of Teutonic organization. The senior professor
in each field had as his assistant a younger person with a nominal faculty
appointment who helped run his research program and pretty much took
orders from his boss. The senior physical chemist was W. D. Harkins;
with his assistant David Gans he maintained active research programs in
nuclear chemistry and surface chemistry, with occasional excursions into
other fields, including Raman spectroscopy and synthesis of organic compounds in electrical discharges. The physical chemists not tied to Harkins
were T. F. Young, T. R. Hogness, and Simon Freed, all graduates of the
chemistry department at Berkeley. John Kirkwood joined the department,
but departed after only one year.

In his research, Young concentrated on precision measurements of
the thermodynamic properties of electrolyte solutions. I recall only one
departure from his beloved heats of dilution and freezing point
depressions—an imaginative attempt to measure the photoelectric properties of solutions of metals in liquid ammonia. Hogness was a pioneer in
chemical applications of mass spectrometry. Several of his graduate students worked in that field, others in spectroscopy and a few other associated fields. He later turned to enzymology. Freed, fresh out of Berkeley

and work with G. N. Lewis, was the youngest and, in my view, the most imaginative member of the department. He worked on magnetism, spectroscopy of rare earth ions, and some other projects not readily categorized. His work in magnetism was concentrated on measurements of the concentration and temperature dependence of the magnetic susceptibilities of solutions of metals in liquid ammonia. He thought of the solutions as electron gases to which the recent developed theories of electrons in metals could be applied. Another conception that he hoped would be accessible to measurement was Landau's that, owing to a diamagnetic correction, the effective magnetic moment of electrons in a dilute electron gas is two thirds of a Bohr magneton.

Harry Thode, Richard Metcalf, and Nathan Sugarman did the metal ammonia work. For their measurements of the susceptibilities by the Guoy method, they had to contend with unstable solutions, an inadequate magnet, a succession of fragile and frequently broken home-built Dewar flasks necked down to fit into the narrow gap of the magnet, and a temperamental microbalance borrowed from the Universal Oil Company. Constant demands by the Universal Oil people for return of the balance were not helpful. Nevertheless, they succeeded in getting reliable measurements over a considerable range of concentrations. Freed was able to account for the data by treating the system as a gas of fermions, but his interpretation seemed not to gain wide acceptance after Pauling showed that the data could be accounted for by an equilibrium between paramagnetic monomers and diamagnetic dimers. Pauling's language was probably more appealing to chemists than Freed's, but I suspect that the two are equivalent.

I recall one magnetic measurement of Freed's that might now be considered childishly naive: the magnetic susceptibility of ethylene. (Mulliken, a member of the department of physics at the university, helped instigate the experiment.) Might not ethylene, since it is isolectronic with dioxygen, be paramagnetic? Everyone now knows the answer.

At the time that I was hunting around for a project to work on, Freed had become interested (fascinated is probably the better word) in the high rotational speeds that could be reached with gas-driven, gas-supported rotors. He thought that new phenomena were waiting to be uncovered at the high centrifugal fields, greater than 10^6 g, that had become accessible.

I decided to take a flyer with Freed in the top-spinning business. (Harkins averred that Freed was doing secret work with spinning tops.) Probably the times had something to do with my decision. The world was in turmoil—depression in the United States, Hitler in power in Germany, the future looking bleak—so what the hell, take a chance on a long shot that probably would not pay off but might be fun for a while.

As to the results of my experiments with the spinning tops, there is little to say. I did learn how to work a lathe and a milling machine. (I built the damned things myself.) I looked for three different effects: spectroscopic shifts or splittings from a source in a high centrifugal field, rotation of the plane of polarization of linear polarized light propagating along the axis of a rotating cylinder, and development of an electrostatic potential difference between center and periphery of a rotating metallic disk ("centrifuging electrons"). The attempt to centrifuge electrons has since revealed to me the conceptual errors of which I was guilty. I calculated the free energy difference between electrons in an electron gas at the center and periphery of the disk and assumed that the corresponding electromotive force would appear as an electrostatic potential difference. As Bridgman showed in his book, *Thermodynamics of Electrical Phenomena in Metals*, it is only under special conditions (not fulfilled in my experiment) that the two are equal. The measured differences were much larger than I had calculated, I couldn't explain them, and gave up. The experiment on spectral shifts gave null results, and the one on rotation of the plane of polarization I could not do.

The department mercifully gave me a degree for a good try. I am totally amnesic about my PhD oral examination. Maybe there wasn't one.

[Two of the experiments have since been carried out successfully. R. V. Jones showed that the plane of polarization of linearly polarized light propagating along the axis of a transparent rotating cylinder is in fact turned, and Jesse Beams carried out the centrifuging of electrons. His experiment was instigated by Fairbanks' attempt to find out whether the gravitational interaction between matter and antimatter is repulsive or attractive. I don't have space to go into the connection.]

Both in physical chemistry and physics, much of the graduate students' time was spent in building apparatuses. But we did find some time for thinking. Freed, recognizing the importance of symmetry and its mathematical realization in group theory, organized a tutorial in the subject. We (Freed, Norman Davidson, George Boyd, Martin Kamen, Eugene Rosenbaum, Ruth Comroe, I, and a few others) met in the evenings to teach ourselves group theory. We studied Speiser's book on finite groups and Wigner's newly published *Gruppentheorie und Ihre Anwendung auf die Quanten Mechanik der Atomspektren*. It was great fun—all those new ideas. We finally were able to work through Bethe's 1929 paper on the splitting of levels in fields of various symmetries. It was all wondrously new to us then. Now it's old, taught, after a fashion, in undergraduate courses in inorganic chemistry.

The university was caught up in the political, philosophical, and scientific ferment of the times. Stalinists, Trotskyites, Norman Thomas socialists,

New Dealers, far right Republicans were having at each other. Professor Henry Gordon Gale, Dean of the division of physical sciences, described by the *Chicago Tribune* as a "soldier scientist," endorsed Alf Landon for the presidency in 1936. At the physics colloquium on the day after Gale's endorsement, almost all the graduate students and some of the faculty turned up wearing large Roosevelt buttons.

The Aristotelians, led by Mortimer Adler, were deriding modern science (not only had medicine made no progress since Galen, it had retrogressed owing to its ignoring his teachings); the scientists, led by Anton J. Carlson, responded vigorously and not too politely.

In the physical sciences, a group of operationalists, converted by Bridgman's book, *The Logic of Modern Physics*, had become more Catholic than the pope. One of them, a graduate student in physics, confronted one of his fellow students with the question, "What are you doing?" The answer: "I'm measuring the cross section of beryllium for alphas in the alpha-n reaction." Operationalist: "Don't be childish. What scale readings are you making and what correlations are you finding between them?" At their next encounter, before the operationalist could repeat his question he was told, "I'm sawing this fucking piece of brass."

But some aspects of the time were more grim than polemics over the philosophy of science. The senior machinist in the chemistry department was a Nazi. Daily at noon he read aloud to his young assistant a selection from *Mein Kampf*. I had frequent dealings with him. We got along well when the issue was how to machine the flutings in a turbine, not so well when the issue was fascism. I finally goaded him into a violent outburst. "It used to be that every Jew and Socialist could shit out his guts on Germany, but no more."

Finding jobs, particularly for Jews, was not easy. Recruiters from the chemical companies visited the department regularly, even during the depression years. One blue eyed blond student reported that his interview with the duPont recruiter went well until he revealed that he was Jewish. The interviewer slammed down the papers that he had been filling out and exclaimed, "Damn it, you can't tell a Jew by looking at him anymore."

James Franck had joined the department in 1938. A naturally exuberant man, he was constantly concerned and oppressed by the Nazi atrocities in his native Germany. On the morning after Kristallnacht, he came to the laboratory distraught and ashen faced, muttered a few words in German and then left. Franck, a decorated veteran of the First World War, had no illusions about the Nazis.

Not all the responses to Nazism and war were totally grim. Roosevelt's attempts to get the country out of the depression included provision of assistants to university laboratories. Freed was allotted Ben J. Ben (he said that wasn't his real name; his real name couldn't be pronounced by Americans). His task, he said, was to "streamline" himself for participation in the war that the United States would soon be waging against the Nazis. He managed to distill a few solvents and wash a few dishes. Perhaps he should have been assigned to the State Department. He explained, following the Nazi-Soviet pact, that Hitler would say to Stalin, "Joe, we got pact, no? How's about you give me Ukraine (pronounced ookrayeen)?" Stalin gonna say "No, then will be invasion of Russia."

I hung on at the University of Chicago until the spring of 1941. I worked for a few months as a technician with Franck's group on a photosynthesis, and for the rest of the time with Freed on spectroscopy of rare earth ions, particularly Europium. During my stint with Franck's group, I built a photometer designed to record "rapid" (about 0.1 second) changes in fluorescence intensity from photosynthesizing leaves (spinach was the good stuff) when the composition of the atmosphere in which they were bathed was changed. The photometer included a vacuum tube amplifier. Franck was pleased with its performance but confessed, "These vacuum tubes I do not trust. Electrometers I trust."

Of the many colloquia that I attended at the University of Chicago only a few remain in my memory. (Perhaps I should use more modern language and say that I can access only a small portion of what may possibly be stored in my ROM.) I recall Pauling's talk on the entropy of ice not only for its scientific virtuosity but for his light-hearted twitting of the humorless Professor Harkins. "All of this hearkens back to—hearkens, harkins, anybody every heard that word around here?"

Another memorable colloquium was by Phipps, a report of the experiments that he had done during a sabbatical stay in Otto Stern's laboratory. He isolated one spin-selected beam from a Stern-Gerlach sorter and ran it into a second sorter whose magnetic field was rotated relative to the first. Instead of getting the expected two beams with relative intensities $\cos^2 (\theta/2) \sin^2 (\theta/2)$ where θ is the angle between the fields in the two sorters, he got only one beam. Analysis of the experiment revealed that the beam velocities and field gradients were such that the spins in the selected beam followed the ambient field adiabatically. The notion of adiabaticity remained murky to me until many years later when I studied Abragam's book on magnetic resonance.

In the physics department there was great, but short-lived, excitement after the colloquium in which Robert Shankland, one of A. H. Compton's students, presented the results of his experiments designed to find out whether the electron and scattered photon in Compton scattering appeared simultaneously. He set up two coincidence counters, one to detect electrons, the other photons, and recorded the dependence of coincidence rate on relative orientation of the two counters. He found no more coincidences at the Compton angles than at the other angles and concluded that the scattered electron and photon did not appear simultaneously. Carl Eckart announced to his class in quantum mechanics that everything he had been teaching was wrong. Repetition of the experiment in other laboratories and by Shankland ultimately revealed that the scattered particles did appear simultaneously. A subtle artifact had crept into Shankland's experiment. Quantum mechanics was saved.

Finally, Carl Anderson's report on his discovery of the positron remains vivid. His data were unambiguous. There was one question over naming the new particle, and some highbrow types suggested Oreston after Electra's brother.

In 1941 I moved on to Berkeley to work as a National Research Council fellow under G. N. Lewis. I reported to Miss Kittridge, boss of the department office (and alleged by some to be boss of the department), and was directed to Lewis' laboratory. There I found him shining a flashlight on a sample cooled by liquid air in an unsilvered Dewar. He tucked the Dewar under his sport jacket and observed the afterglow. "I would say the half life is about one second, wouldn't you? And the color I maintain is yellowish green despite my associate Dr. Lipkin's stubborn insistence that it's greenish yellow." Thus my introduction to the phosphorescent state, later identified by Lewis as the triplet state.

At Berkeley, I felt for the first time that here I was where the physical chemistry that counted was being done. Lewis, Hildebrand, Latimer, Rollefson, Eastman, Olson, Bray, Gibson, Giauque, Branch, and the bright young men Pitzer, Seaborg, Kennedy, Calvin, Rubin were doing it. There was a sense of excitement and adventure that I had not experienced at Chicago. There was a most able group of graduate students, among them Gwinn, Connick, Duffield, Wilmarth, Campbell, Wahl, Brewer, Gofman, Hill, Bigeleisen. David Lipkin was Lewis' research assistant. We have been colleagues and friends ever since.

Everyone, it seemed to me, was doing important work and doing it well. Gwinn and Pitzer were completing calculations on the contributions of internal molecular rotation on thermodynamic properties. (Questions about the barrier to rotation in ethane had not yet been resolved.) Campbell, working with Hildebrand, was looking at the structure of liquid mercury by x-ray diffraction. Terrel Hill, then nominally an organic chemist, worked with Branch on relations between color and structure of organic compounds. Art Wahl, in a feat comparable to Madame Curie's extraction of radium from pitchblende, got at nuclear and chemical properties of plutonium from the few micrograms of it that he fished out of hundreds of pounds of irradiated uranyl nitrate.

I was given laboratory space in Gilman Hall in a room that had a sunlit balcony. The extraordinary resources of the laboratory (by Chicago standards) soon became apparent. I wanted to work on a phenomenon that I had stumbled across just before leaving Chicago. To photograph the absorption spectrum of an organic chelate of europium I mounted the sample close to the slit of a spectrograph and focussed the filament of a tungsten lamp on it. The light entering the spectrograph showed the expected europium absorption bands, but on the intense red background from the unfiltered light of the lamp there appeared the brighter fluorescence lines of the europium ion. The blue light of the lamp was absorbed by the ligand and the excitation transferred with high efficiency to the europium ion. To study the excitation spectrum I wanted a continuous light source, a monochromator, and a photometer. No problem—here they are. What kind of light source? Here's a heliostat. I set up the experiment on the sunlit balcony with the light produced by hot fusion 93 million miles distant.

The europium that I was using had been given to Freed by Herbert McCoy, who had devised a scheme for its separation from the other rare earths. It was rare stuff.

McCoy had served on the University of Chicago faculty from 1901 to 1911. He made fundamental observations in radioactivity and is credited with having been the first to recognize the existence of isotopes. In 1919 he became vice president of the Lindsay Light and Chemical Company. The company extracted products that it could sell from monazite, a mineral containing thorium and rare earths. The thorium went into thorium dioxide, an ingredient of Welbach mantles, the radiating element in gas lights. So McCoy had a supply of raw materials from which he extracted rare earths. He retired to Pasadena, set up a laboratory complete with laboratory bench and grand piano. He played Mozart and produced pure europium. He exploited stability of the $(4f)^7$ configuration that permitted reduction of $Eu^{3+}(4f)^6$ to $Eu^{2+}(4f)^7$, in which oxidation state it behaves like an alkaline earth and is easily separated from the other tripositive rare earth ions.

Freed had given me about 150 milligrams that I used over and over again as I prepared various chelates. McCoy was a good friend of Lewis' and turned up to visit him at a time when all my europium was in a few crystals of some chelate or other that transferred the excitation with high efficiency. In daylight the crystals glowed bright red. Lewis brought McCoy to the laboratory where I showed him the glowing crystals. "Could you use some more europium?" he asked. Before I could answer he went on, "Got a weighing bottle?" I produced one and he took from his briefcase a large bottle of pure europous carbonate—the world supply of pure europium. He dumped about five grams into the weighing bottle, spilled a few hundred milligrams, blew them away before I could mop them up, and said "There you are." As I began an elaborate speech to thank him, Lewis returned and said, "Herbert, would you like a drink?" They disappeared into Lewis' office.

During the two years 1941–1943 that I spent in Berkeley I was privileged to witness Lewis at work. He was constantly probing in areas where he felt that new concepts were needed and was impatient with those who worked where he felt that the concepts were well established. When one of his colleagues came into the laboratory to describe yet another elaboration of an old method for measuring activity coefficients, Lewis listened with obvious annoyance and then said as the colleague was barely out of earshot, "Well, you can't stop a man from doing that sort of thing, but you certainly shouldn't pay him for it."

During my stay Lewis had as coworkers first David Lipkin, then Jacob Bigeleisen. Much of their work dealt with spectroscopy and photochemistry in rigid media. ("A rigid medium is just like a vacuum," Lewis once said, looking around for someone to have an argument with.) Lewis carried out many of the experiments himself. He seemed to enjoy working at the vacuum line. Once something went wrong, and he filled the line with an opalescent smoke that he couldn't get rid of. I diagnosed his problem. "Professor, you've filled the line with impalpable motes." He thumped me on the chest and said "Young man, did you ever palp a mote?"

In the summer of 1941, Lipkin and I, responding to the ever more ominous developments in the European war and to the growing preoccupation at Berkeley with the possibility of making nuclear explosives, tried to separate the uranium isotopes photochemically. Our attempt was provoked by a paper on the effect of nuclear spin on radiative rates in forbidden transitions. The atomic transition $^3P_0 \leftrightarrow {}^1S_0$ is forbidden as a one-photon process for all multipole orders. But the $^3P_0$ state of the odd isotopes of mercury does radiate to $^1S_0$ with a lifetime of about 0.1 second owing to coupling to the nuclear spin angular momentum. We played the long shot that the long-lived (about 1 millisecond) excited state of uranyl ion might be shorter lived in $^{235}UO_2^{2+}$ than in $^{238}UO_2^{2+}$. We did not do the obvious first experiment, a direct comparison of the lifetimes (much more easily done now than then), but plunged right into a photochemical experiment—competition between photochemical reduction and fluorescence. Lipkin devised a mixture of phosphoric and hypophosphorous acids that met our requirements, but we got no isotope separation. We discovered by a tracer experiment that out chemical separation following photochemistry would have rescrambled the isotopes. We don't know to this day whether we had any isotopic enrichment in the photochemical process.

At first we had no formal connection with any organized project—we just prepared our samples and gave them to someone in the Radiation Laboratory for isotopic analysis. Later Lipkin resigned his assistantship and I my fellowship to become full time employees of the Radiation Laboratory. We worked on the chemistry of source materials for E. O. Lawrence's electromagnetic separations.

While working on isotope separation we did manage to moonlight an experiment on the phosphorescent state. There had been speculation that owing to some unspecified cause, long-lived phosphorescence proceeded not by an electric dipole mechanism but by a higher order multipole. We determined the mechanism by observation of wide angle interference of the beta phosphorescence of fluorescein—one of Lewis' favorite phosphorescences. The experiment was finished in one week; the answer, pure electric dipole. As we were setting up the experiment I managed to get beautifully bright fringes from a fluorescent source. I tried to get Lewis to look at them. At first he refused: "Every time someone asks me to look at something through an eyepiece, I don't see a damned thing." "So try already, professor." Finally, he tucked his cigar behind him, looked through the eyepiece and exclaimed, "By god, they're there."

Following the work of the war years at Berkeley and Los Alamos, Joe Kennedy, the newly appointed chairman of the chemistry department at Washington University, brought with him five of his colleagues, Helmholz, Lipkin, Potratz, Wahl, and me. The appointments were made without benefit of advertisements in the *Journal of Higher Education* and approval by an Affirmative Action Committee.

Resuming academic work was relatively easy. We no longer had the unlimited support that we had experienced in the Manhattan Project, but we had the encouragement of the university (but not much money) and a fine group of graduate students, most of whom had worked at Los Alamos. Among the undergraduates were many veterans who had come to the university under the G.I. Bill, some older than we, and almost all with the irreverence that their years in the service had produced. I almost lost a class when I first wrote on the blackboard, $G = H - TS$. TS (tough shit in army language) did it. Thereafter, it was $G = H - ST$. It's a good thing that S and T commute.

An experience of Lipkin's in our first year at Washington University was an augury of things to come. Los Alamos volunteered financial support of his work. The university demanded substantial overhead. Overhead, what's that? It was a new concept. The university got its overhead but not as much as it had asked for.

My first project at Washington University was a study of phosphorescence. By 1943 Lewis had become convinced that the "phosphorescent state" was a triplet state. With Calvin he then demonstrated its paramagnetism by a difficult, but straightforward, observation of its magnetic susceptibility. Phil Yuster, my first graduate student, and I demonstrated perturbation of phosphorescence by paramagnetic ions. As we were getting started on that work, Bloch and his coworkers at Stanford and Purcell at Harvard with his coworkers announced their observations of nuclear magnetic resonance in condensed phases. (Purcell emphasizes that NMR had been seen years earlier by Rabi in molecular beams.) Zavoisky's discovery of electron paramagnetic resonance (EPR), published in 1945 in the Soviet Union, became known in the West at about the same time. With little understanding of how magnetic resonance works, I thought that EPR would be the way to look at the triplet state. The importance of NMR for chemistry was appreciated almost immediately after its discovery by a physicist, Philip Morrison. Lipkin asked him whether he wasn't excited over this new way for measuring nuclear magnetic moments. Morrison allowed that yes, that was pretty nice, but the real use of it would be in chemistry.

I tried to arrange a summer visit to the newly established Brookhaven Laboratory for the summer of 1948. Bill Cohen had an EPR spectrometer; I would go there and look at the EPR of triplet states. It was five years before I could go to Brookhaven—a dismal but not uncommon business about clearance—and ten years before Clyde Hutchison did see the EPR of a triplet state.

As I tried to learn about magnetic resonance I realized the inadequacy of the concepts that I had used in my earlier work in optical spectroscopy. I listened to a lecture by Bloch in 1947. What's this crossed coil business? You irradiate with one set of coils and pick up a signal with the other. Is it like resonant fluorescence? What's all this about coherence? Spins can point up or down but not sideways as he alleges. And in Purcell's early papers, this business about motional narrowing comes up. Fast molecular reorientations, the faster the better, produce narrowing of spectral features and improved resolution. That's contrary to all my previous experience in optical spectroscopy, where one always tried to restrict molecular motions to improve resolution. Great new stuff to try to understand.

My entry into magnetic resonance was made easy by the arrival of George Pake at Washington University in the fall of 1948. His doctoral work, with Purcell on the NMR of a two-proton system, water, in calcium sulfate dihydrate (the "Pake Doublet"), had in it most of the ideas that were later involved in dealing with the EPR of electronic triplet states.

The vigorous research program in magnetic resonance that George Pake initiated at Washington University has grown over the years. Norberg, Conradi, and Fedders in physics, Ackerman, Schaefer, and Lin in chemistry are currently active in the field. Among the early advances were Lowe, Norberg, and Kessemeyer's exploitation of magic angle spinning, and Lowe and Norberg's fourier transform spectroscopy. Lowe and Norberg recorded oscillatory free induction decays in the fluorine resonance of calcium fluoride. They realized that they were seeing the fourier transform of the cw spectrum. Norberg engaged an undergraduate to do the fourier transformations on a mechanical computer. I was reminded of Michelson's "harmonic analyzer" that he had used for converting his optical interferograms to frequency domain spectra.

Not long after Pake's arrival at Washington University, Professor Immanuel Estermann, then on loan to the Office of Naval Research (ONR), turned up at the University with the proposal that ONR support us in work on EPR. Perhaps a sensitive magnetometer for detection of submarines could be developed. ONR would support a few laboratories in the work. No reports or proposals—we would get together every few months and tell each other what we had been doing. We'd never had it so good.

So Pake, Jonathan Townsend, and I began work on EPR. My first observations of an EPR signal was done on a 14 MHz cw instrument designed and constructed by Nick Shuster, one of Pake's students. I prepared a polycrystalline sample of an organic free radical (tris p-nitrophenylmethyl) from starting material that Dave Lipkin had synthesized. We got a huge signal, only a few tenths of a gauss broad, at an applied field of 5.0 gauss. (I had underestimated the expected signal strength by a mere factor of 5×10^5.) Much of the early EPR work on solid organic free radicals dealt with line breadths and g values of the pure solids. Townes had discovered that solid diphenylpicrylhydrazyl (DPPH), despite the expected dipolar breadth of several hundred gauss, had a line breadth of only one or two gauss. Van Vleck explained the phenomenon: The spin-spin exchange interaction conserves the second moment but increases the fourth moment of the spectral distribution, thus a narrow central component and broad weak wings. Studies of the solids would perhaps reveal something about the exchange interaction. One skeptic said to me "So what are you guys going to find out? Some g-values and some line breadths. Who cares?"

But we persisted. I am not sure whether it was the easy money from ONR or burning interest in exchange interactions that kept us at it. Jonathan Townsend constructed our first X-band spectrometer from a discarded mass spectrometer magnet, war surplus wave guides, and klystrons and home built electronics. (Surplus 2K25 klystrons cost less than one dollar. There were bins of them in war surplus stores.)

I tried to make a paramagnetic ionic solid in which both anions and cations are paramagnetic. The anion was to be Fremy's ion nitrosyl-disulfonate, the cation Wurster's blue. Mixing approximately stoichiometric amounts of the two produced a solid. In my haste to look at the solid, I failed to remove the liquid that stuck to it and got a spectrum of three lines, centered close to $g = 2$ with separation 13.0 gauss between lines. We soon found that the crystals were diamagnetic and the three-lined spectrum came from the liquid that was sticking to them. It was our first hyperfine splitting in dilute liquid solution and came from the paramagnetic anion $(SO_3)_2NO^{2-}$. We put solids and triplet states aside and concentrated on our three-lined bonanza. (Jack Townsend called it our meal ticket.) At the next meeting of Estermann's group, Clyde Hutchison and I almost simultaneously informed each other "Hey, we've seen nitrogen hyperfine splitting in solutions." Clyde had looked at dilute solution of DPPH and had seen the splitting by two ^{14}N nuclei.

The happy arrangement with ONR that Estermann had managed was too good to last. The Way it Was evolved into the Way it Is: reports, proposals, grantsmanship, etc. We continued work with Fremy's salt. We found that in liquid solution the spectrum was accurately accounted for by a scalar interaction $A\mathbf{I} \cdot \mathbf{S}$. We measured the spectra over a range of fields from zero to several hundred gauss and found that everything fit the Breit-Rabi formula. It was astonishing, to me at least, that the same formula, properly scaled, that accounted for the magnetic field dependence of the spectra of deuterium atoms worked for an 81-electron polyatomic system tumbling about in a dense liquid, and that the only part of the hyperfine interaction that survives the buffeting in a liquid is the Fermi contact term. (Not quite true; the rest of the hyperfine interaction that is allegedly "averaged out" still affects relaxation processes. The averaging at zero field demonstrates the role of non-adiabaticity. Thus spins, owing to their gyroscopic properties, don't follow the molecular axes as the molecule tumbles.)

With improvements in sensitivity and resolution came the identification of many new radical species and the uncovering of detailed features of structure and dynamics. Our earliest observation of the EPR of tri-phenylmethyl in liquid solution revealed only a single line about 15 gauss broad. Even with the poor resolution then available to us, we could still see the 26 gauss ^{13}C hyperfine splitting in a sample enriched with ^{13}C in the methyl posiion. Howard Jarrett at duPont soon looked at tri-phenylmethyl in a good spectrometer that he had constructed; he got 196 lines—the complete proton hyperfine splitting was resolved. A rich lode of free radicals turned up following Dave Lipkin's suggestion that the adducts of alkali metals to aromatic hydrocarbons discovered by Scott, Walker, and Hansley in 1936 were anion-free radicals. He was right and we, finally having equipment capable of resolving proton hyperfine splittings, had a new batch of free radicals to play with.

The origin of the Fermi contact hyperfine splittings by protons in planar aromatic radical ions (cations were soon included) was a matter of some interest for a while. The simple orbital picture of "the" odd electron occupying a pure pi orbital with a node at the equilibrium position of the protons is clearly inadequate. McConnell's theory, later elaborated by others, accounts for the phenomena. To McConnell's annoyance, I pointed out that the theory had really been done in the 1930s by Fermi in his explanation of the contact interaction in a p state of thallium. The single configuration picture, with individual orbitals classified as pi or sigma, is simply inadequate. For odd alternate radicals such as triphenylmethyl, the question, "where is the odd electron?" clearly became inapplicable. Spin is distributed all over the molecule with alternating sign at adjacent carbon atoms. No single simple molecular orbital configuration is adequate; in the valence bond description, negative spin density turns up in the cross term of the hyperfine operator between pairs of valence bond structures.

In measuring rates of chemical processes and tracing out details of mechanisms via analysis of magnetic resonance spectra I found one of the most exciting features of the field. The earliest experiments had been done by Gutowsky & Saika and by Piette & Anderson. How do you get a rate from a continuous wave spectrum of a system at equilibrium? Does the observation disturb the equilibrium? Although the questions might now seem ridiculously naive, I worried about them nevertheless. To the question concerning disturbance of the equilibrium by the measurement, the answer is that it does. The measurement process induces coherences in the system; stochastic chemical events tend to destroy the coherences, and it's the destruction of the coherences that turn up in the spectra.

It's all pretty well illustrated by an encounter I had with the late V. V. Voevodskii, a Russian pioneer in chemical applications of EPR. He measured rates by preparing systems far off equilibrium and determining the time evolution of the composition of the mixture by EPR. He visited our laboratory during one of the periods of détente. After dinner, his Marxist zeal stimulated by capitalist Jack Daniels, he undertook to give me a lesson in dialectical materialism: "Colleague Weissman, I must tell you, in your country are the experiments static, in ours dynamic." My response: "It's OK, doc, we just take the Fourier transform."

Robert H. Cole

Annu. Rev. Phys. Chem. 1989. 40: 1–28

DIELECTRICS IN PHYSICAL CHEMISTRY

Robert H. Cole

Department of Chemistry, Brown University, Providence,
Rhode Island 02912

INTRODUCTION

When I was invited to contribute this prefatory chapter, I felt both honored
and concerned about my qualifications, the latter when I remembered
some questionable credentials and a piece of advice from P. Debye.

The credentials in question or lacking go back to the 1930s. My undergraduate training included only a single course in chemistry. This was not
entirely my fault: I had thought of taking physical chemistry at Oberlin,
if only for cultural reasons, but had neither time nor inclination to meet
the inviolable prerequisite of qualitative analysis while concentrating in
physics and mathematics. I managed only a little better, and not for the
record, when I was a graduate student in physics at Harvard, as my only
instruction in chemistry then was auditing G. B. Kistiakowsky's course in
thermodynamics.

As accounts that follow suggest, my developing research interests and
activities increasingly exposed me to physical chemistry and physical
chemists. So I was not entirely taken aback when one of them I had come
to know well in World War II years, Paul Cross, was able with support
from Donald Hornig and J. S. (Spike) Coles to offer me a position in the
chemistry department at Brown. Not long after accepting and beginning
to learn more by teaching and doing, C. A. Kraus told me I should join
the American Chemical Society. When I innocently asked why, I was
merely given a form to fill out and expose my pitiful lack of qualifications.
The application was summarily rejected. When I reported this to the King,
as Kraus was described by his students, he called Alden Emery, then long-
time ACS executive secretary, and told him to admit me forthwith, which
he did.

0066–426X/89/1101–0001$02.00

The directions of research that I hoped would give me further respectability were then as now largely in the field of dielectric behavior, perhaps more generally and better called "electrophysical chemistry." This is not common usage, but there is a distinguished precedent: the entry for Lars Onsager in *American Men and Women of Science* (12th edition, 1972) listed a primary interest as "electrophysics."

I hope the second description is apt enough to make these recollections and thoughts of sufficiently general interest, but I have occasional qualms when I remember some fatherly advice from Debye in 1963. I had helped to organize an ACS symposium in his honor and had contributed a somewhat tedious talk on electric dipole interactions to the occasion, but at a party afterward he took me aside and suggested that I think about something more worthwhile, like NMR. This shook me a bit, and I continued to remember it whenever a current line of interest loses some of its initial zest, but something new and exciting has always come along to which I could relate without major disruption. May this account and future developments justify my belief that there is now as ever much of interest and much to be learned.

THE INFERNAL FIELD PROBLEM

The traditional problem in dielectrics has long been to account for experimentally observed relations between the macroscopic polarization or electric moment density P and macroscopic electric field E. A stumbling block of equally long standing has been the problem of how adequately to describe the underlying fields at microscopic and molecular levels in tractable ways. As a substitute for many body problems with long-range orientation-dependent forces, local or mean field approximations derived from macroscopic quantities have of necessity been introduced, debated, and often found wanting, to the point that they have been called "infernal fields" (by a research student at Oxford in 1962).

A mainstay for many years and still useful in its proper place is the Lorentz local field E_{LOR}. In 1909, Lorentz (1) considered isotropically and linearly polarizable atoms on a cubic lattice or in an isotropic continuum, and showed that for a spherical sample in a field E_0 external to it, the sum at any one of the induced dipole fields of all the others vanishes and the field at any site is $E_{LOR} = E_0 = E + (4\pi/3)P$, where E is the macroscopic average in the sample and the relation $P = (\varepsilon - 1)E/4\pi$ if the problem is linear gives $E_{LOR} = (\varepsilon + 2/3)E$ where E is the dielectric constant (or relative permittivity). This led to the appearance of the ratio $(\varepsilon - 1)/(\varepsilon + 2)$ in many subsequent developments to relate polarization and field to charge displacements at the molecular level, with disastrous results for polar molecules in liquids and solids.

Debye's classic work (2, 3) in relating dielectric constant to polarizability α and permanent dipole moment μ of polar molecules opened up the whole field of dielectric studies for chemical purposes. His formula in simplest form is

$$(\varepsilon-1)/(\varepsilon+2) = (4\pi/3)\,(N/V)\,[\alpha+\mu^2/(3kT)] \qquad\qquad 1.$$

in which N/V is the number density of molecules and the Lorentz local field, or Clausius-Mossotti function, has been incorporated. The disaster for polar molecules just mentioned, which Van Vleck (4) called the $4\pi/3$ catastrophe, is evident from this formula predicting an infinite value of ε for the right-hand side equal to unity. This is easily realized for only moderate values of μ and T, in complete contradiction to experiment as shown schematically in Figure 1, where $\varepsilon-1$ is plotted against $(N/V)\,(\mu^2/3kT)$ for a variety of polar liquids at temperatures in the normal liquid range. The figure also shows a common pattern of behavior, with the only major deviations being for hydrogen bonding, or otherwise associated, liquids.

Much work since the 1920s has been devoted to modifying or embellishing the basic Lorentz field picture, with little or no success for even moderately polar media, and a variety of empirical functions have been proposed to describe the pattern of behavior suggested by Figure 1. One of the latter, proposed by Jefferies Wyman in 1936 (5), was apparently an incentive for the breakthrough in Onsager's famous 1936 paper (6), which was soon to be followed by Kirkwood's equally famous 1939 paper (7).

The gist of Onsager's contribution was in the observation that the electric field at a polar molecule was in part a reaction field induced by its own dipole moment and as such could have no role in reorienting it, as tacitly supposed in using the Lorentz field. The result of evaluating both this field and the field in a cavity volume occupied by the molecule for an electrostatic continuum model was Onsager's equation. The prediction of the equation for an average value of the contribution from polarizability is shown as the *solid curve* in Figure 1. Thus the $4\pi/3$ catastrophe from the bootstrap effect in the Clausius-Mossotti function was averted and a remarkably good account given of average behavior in the absence of specific short-range orientation effects not accounted for by the simple model. The Kirkwood treatment based on separation of specific short-range effects and correlations from long-range dipolar (Coulomb) forces, with only the latter treated macroscopically, was to appear shortly. With this, a basis became available for characterization and modeling of hydrogen bonding or other associative effects, molecular shapes, and the like to

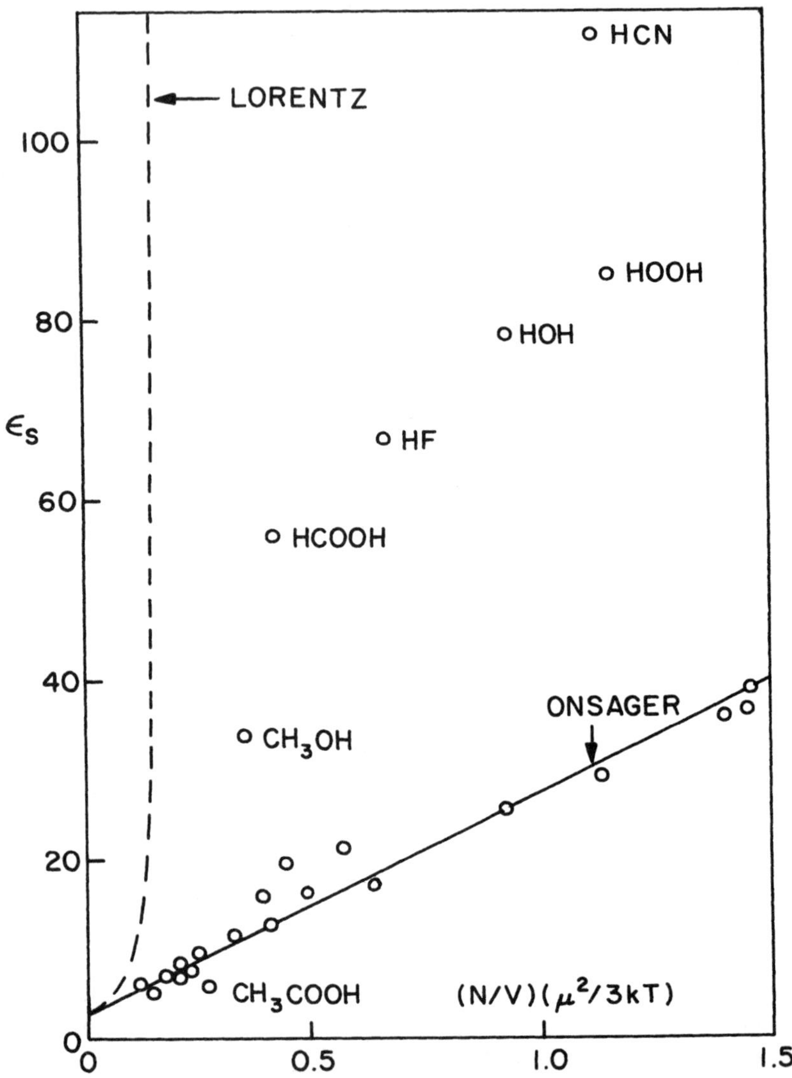

Figure 1 Permittivities of polar liquids at or near 25°C plotted as a function of relative dipole energy $(N/V)(\mu^2/3kT)$. *Points* are experimental, *curves* are from Lorentz and Onsager expressions for nominal permittivity $\varepsilon_\infty = 2.25$ of induced polarization.

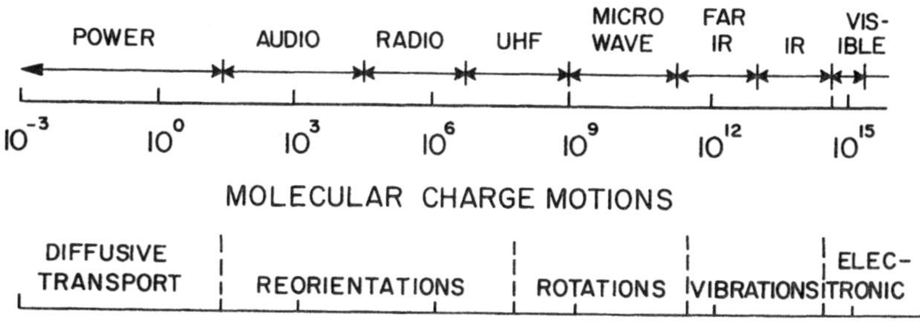

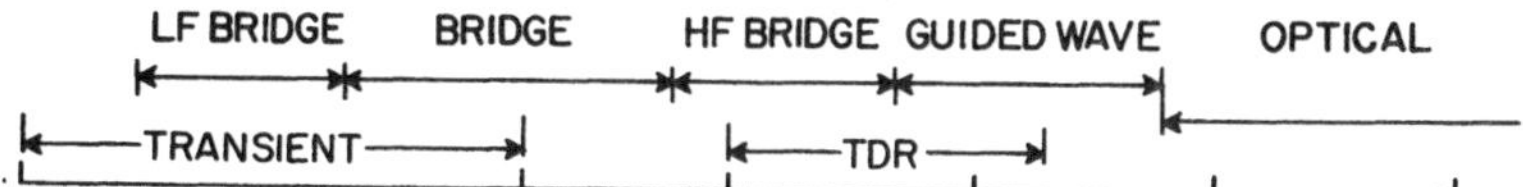

Figure 2 The dielectric regions of the electromagnetic spectrum on a logarithmic frequency scale, with representative motions of charges and experimental methods indicated.

account for gross differences and lesser deviations from the Onsager result in Figure 1. These predictions and previsions were yet to be tested, questioned, and refined, but new approaches to old problems were at hand.

These were exciting times for me. I had served a rewarding apprenticeship for several summers with my older brother, Kenneth Cole, and learned about electrical measurements and the state of physics in understanding impedance behavior of biological systems. The dielectric aspects particularly intrigued me as I worked with my brother to survey experimental evidence and theories of relaxation. I was starting graduate research with J. H. Van Vleck at Harvard as my mentor when Onsager's paper generated considerable excitement because of its obvious relevance to problems of dipole coupling and magnetic susceptibilities. It also changed my thesis problem from magnetic susceptibilities of single crystals of Tutton and (I had hoped) rare earth salts to the very different one of measuring dielectric constants of liquids and interpreting both these static results and relaxation data in the light of the new theoretical advances.

A particularly memorable occasion was a New York Academy of Sciences conference in April of 1939 (8), which featured papers by Kirkwood (9) and Van Vleck (4) among others. Van Vleck's paper summarized the state of local field treatments and discussed results of his calculations of dipole-dipole interactions on rigid lattices. The latter had considerable bearing on several theoretical and experimental problems of interest, discussed more below. Kirkwood's contribution was an outline of his statistical mechanical treatment and a first crude model for dipole correlations and the Kirkwood g-factor of liquid water that led to a static dielectric constant of 67 rather than the experimental 78 at 25°C. This was a major achievement at the time, and an incentive both for devising models of hydrogen bonding in other liquids, such as aliphatic alcohols, HCN, and several acids, and for better calculations to come as more realistic theories of liquids were developed and computer simulations became feasible. An interesting exchange in discussion, not recorded in the published account of the meeting, was between Kirkwood, insisting on spherical samples, and Onsager, equally determined to discuss a sample of arbitrary shape provided only that it was covered with tin foil. Although it was not clear to me at the time, this may have been the first mention of the Onsager "tin foil theorem," never published by him but derived much later by Felderhof (10).

All these developments gave me much to think about and also were the beginning of my conversion from a physicist to a physical chemist, at least in name and as much by accident as by design. As World War II approached in 1941, Kirkwood was commissioned by E. Bright Wilson to recruit someone who might know enough electronics to help develop instrumentation and methods for study of shock waves and other underwater explosion phenomena. Apparently on the strength of having talked with me about mutual interests and having seen me face to face with an oscillograph when visiting Harvard, Jack was responsible for shifting my activities to the National Defence Research Council (NDRC) project on explosives under G. B. Kistiakowsky as part of the war effort. It was not until 1947 that I got back to academic research, thanks to Paul Cross.

The more recent developments from theory of liquids and of dipolar crystals would require at least a chapter to do them justice, a task that I attempted in part not long ago (11). Here it is worth mentioning results of some relevant calculations that seem not to be widely known.

The first are from papers of M. Lax and co-workers (12, 13), in which an analytic theory of coupling of dipoles on a rigid lattice in an applied electric field is developed. The partition function is expressed in terms of dipole wave sums and can be evaluated to give the polarization and permittivity if the spherical model approximation is made of replacing the N constraints $\mu_i \cdot \mu_i = \mu^2$ of orientations of N dipoles μ_i by the single constraint $\Sigma \mu_i \cdot \mu_i = N\mu^2$, a procedure that has worked reasonably well for other models of cooperative interactions.

The results are of interest in several respects. First, when the discrete lattice sums for rigid dipoles are replaced by continuum integrations in the long wavelength limit, Onsager's equation is obtained from the seemingly unreasonable choice of lower boundary radius a to satisfy $a^3 = 3V/4\pi N$, i.e. the cavity volumes fill the sample volume. For face and body centered cubic lattices, however, somewhat larger permittivities are predicted (corresponding to a Kirkwood g-factor greater than one) with a phase transition predicted for a temperature $T_c \simeq 0.1(4\pi N/3V)\mu^2$, i.e. a factor 10 smaller than for the Lorentz field. A further feature is that induced moments can be included if they are represented by harmonic oscillators. The principal effect is to multiply the rigid dipole moments μ by a factor $(n^2+2)/3$, where n is the refractive index of induced polarization. A similar conclusion seems to be indicated by Wertheim's treatments for liquids using graph theory methods (14), so it may be a useful result for correcting other calculations with rigid dipoles for comparison with liquids of real polarizable dipoles. These and other features all seem to deserve further study (despite Debye's advice to me mentioned in the introduction).

Computer simulations are proving to be valuable in assessing the magnitudes and range of specific orientational correlations of simple molecules as they affect dielectric properties and serve to test the Kirkwood formulation. A stumbling block discussed in several places (14, 15) has been proper use of cyclic boundary conditions when long-range dipole interactions are involved. The calculations by Claude Brot and co-workers (16) are illuminating because they avoid the problem by using a spherical potential well or "box" to confine the simulation particles and by evaluating the dipole correlations for both an inner Kirkwood sphere of specific interactions and the entire volume.

For interactions of point dipoles and parameters corresponding roughly to CH_3F at 206 K, the results indicate that Kirkwood's formulation is approximately correct for inner sphere diameters as small as 3σ, where σ is the Lennard-Jones potential parameter used for the calculations. At the same time, however, the calculated Kirkwood g-factors are of order 2.5 whereas the experimental value is nearly unity; but this discrepancy is greatly reduced by using either off-center dipoles or quadrupole moments, indicating the importance of using more realistic descriptions of molecular charge distributions than merely point dipoles.

Obviously many points deserve further study before equilibrium properties of dipolar materials are understood really well. At the same time, the progress to date is encouraging, and, in any case, even quite simple applications of the various formulations often give semiquantitative results of respectable accuracy.

Finally in this section, it would not do to omit any mention of the very fundamental formal treatment pioneered and being implemented by Robert Fulton. In a sense, much of the controversy about local fields over the years has been confused by failures to realize that the electric field

948   E of the macroscopic Maxwell equations is, like the polarization P, a macroscopic average quantity. Fulton's approach (17, 18) is to recognize this at the outset and to obtain expressions for both in terms of quite arbitrary charge and current density distributions as external sources. In this way, dielectric response functions are expressed in terms of functional derivatives with respect to source fields and macroscopic E without introducing molecular cavities and the like. The general results have been shown to reduce to the Kirkwood-Fröhlich equation and Onsager-like static and relaxation functions by appropriate specialization, as well as providing a basis for evaluation of nonlinear effects. These developments have received less attention than it seems to me they deserve, no doubt in part because of formal quantum electrodynamic methods formidable to the uninitiated. The extent to which these and other approaches in the same spirit can be put together with adequate molecular dynamics to give further results remains to be seen, but the signs are encouraging.

THE RANGE AND CAPABILITIES OF DIELECTRIC SPECTROSCOPY

The term *dielectric spectroscopy* has a variety of connotations depending on one's interests and point of view, as suggested by such other terms as *impedance* or *immitance spectroscopy*. Here we take it to mean the study of relaxation processes, primarily in condensed phases, of interest to physical chemists. Defining a boundary or transition in the electromagnetic spectrum between such spectroscopy, using electronic methods primarily, and resonance spectroscopy by optical methods is arbitrary but is reasonably taken to be the region between 100 and 1000 GHz.

The distinction between the two is fairly clear-cut in terms of experimental methods: Microwave styles of measurement are currently limited to wavelengths greater than about 1 mm (frequency < 300 GHz), and this is the same as the 10 cm^{-1} lower limit commonly quoted for far IR spectrometers. The change in character from IR resonance lines or bands to regions of relaxation dispersion and absorption, with line widths of a decade or more in frequency, is less well defined either conceptually or experimentally, as there is a comparative dearth of measurements in the difficult transition region.

As suggested by the schematic representation in Figure 2, the dielectric region of the spectrum extends over some 14 decades in frequency, with only a little further space on a logarithmic scale needed to accommodate the whole of resonance spectroscopy. (This obviously distorted viewpoint can of course be reversed by using a linear frequency scale to reduce the whole dielectric region to little more than a dumping ground for products

of resonance excitations.) Some of the changes in character of molecular   949
charge motions responding in different regions are suggested in Figure 2,
together with an equally sketchy indication of kinds of methods available
for their study.

This available range is not as enormous as it might seem to be without
considering the natural line widths and shapes for relaxation processes
and the changes of their characteristic relaxation true scales with temperature. The prototype of relaxation functions is of course a simple
exponential approach of polarization to a new equilibrium state with time,
$\exp(-t/\tau)$, after an instantaneous change in applied field. This is shown
in Figure 3, together with the corresponding steady state a.c. response to
applied sinusoidal field of frequency $f = \omega/2\pi$, as represented by the
real (dispersive) ε' and imaginary (absorptive) ε'' parts of the complex
permittivity $\varepsilon^* = \varepsilon' - i\varepsilon''$. This function by Laplace transformation of
$\exp(-t/\tau)$ is

$$\varepsilon^* = \varepsilon_\infty + \frac{(\varepsilon_s - \varepsilon_\infty)}{1 + i\omega\tau} \qquad\qquad 2.$$

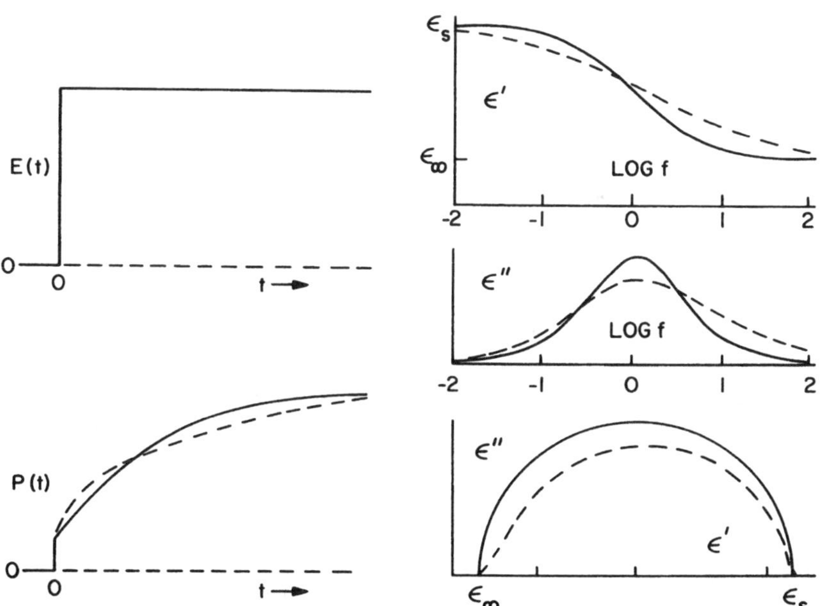

Figure 3   *Left side*: Response of polarization $P(t)$ to an applied step field $E(t)$. *Right side*:
Dispersion ε' and loss ε'' as a function of log frequency and in the complex plane. *Solid curves*
are for exponential (Debye) relaxation, *dashed curves* are deviations found experimentally.

where ε_∞ is the somewhat ambiguous high frequency limit of the relaxation and ε_s the low-frequency limit.

It is to be noted that the dispersion and absorption curves require at least a decade in frequency, and preferably two or more, to be reasonably well defined. Coupled with the facts that one is usually dealing with thermally activated processes with an Arrhenius or stronger temperature dependence of relaxation time scale and that observed relaxation processes are commonly more spread out in frequency or time, as indicated by the dashed curves in Figure 3, the need for broadband coverage is only too apparent.

A good deal of my research effort over the years has of necessity been devoted to developing or taking advantage of newer and better methods to meet the need. By all odds the easiest part of the frequency range is from ca. 20 Hz to 200 kHz or more, particularly by use of transformer bridges. The first of four such, built in 1948, got me into dielectrics again after World War II, and the last in 1963 is still serving faithfully and well, even though it lacks such newer amenities as automatic balancing and programmed frequency sweeping at a series of temperatures to let the equipment do all the work by itself overnight.

With such equipment, one could readily do quite definitive studies of relaxation in polar liquids of large molecules or at low temperatures, molecular crystals, and amorphous solids, as well as very precise measurements of static permitivities of gases. At the same time, however, the studies usually underscored the need to extend the range of measurement to both lower and higher frequencies; for example, on the one hand to follow further the approach of liquids to glassy behavior and on the other to study properly the dielectric aspects of dynamics of small molecules and local motions in larger ones at ordinary temperatures. The former has been taken nearly to the limits of patience or thermal noise by ultralow frequency steady state (19) and transient response methods (20, 21).

For a long time, dielectric measurements at megahertz and gigahertz frequencies were not to be undertaken lightly, but in the past 25 years or so measurements of modest accuracy at a few frequencies with any one instrument have been replaced at an increasing rate to ever higher frequencies by broad band techniques, made available largely by response to the needs of communication and information technology.

Impedance and network analyzers with a gamut of programming and data processing capabilities are becoming commonplace to frequencies upwards of 10 GHz, as are time-domain picosecond response methods using fast pulse generators and sampling oscilloscope detection. My own efforts to approach this range for looking at simpler dynamics of reasonably small molecules began as modest forays that resulted in some progress and more frustration in trying to extract molecular response functions from the obscurity of transcendental functions relating them to observables with equipment based on 50 ohms as the point of reference. This all changed for the better in 1970 when I heard Fellner-Feldegg talk about time domain reflectometry (TDR), made possible by then new developments of fast pulse and pulse sampling techniques. His somewhat deceptively simple methods (22) of getting dielectric information from a few picoseconds to nanoseconds or more all at once helped set me off on several new courses (23), about which more follows.

In the last year or two at the time of this writing, a number of new instruments have been announced or rumored that promise to open up the high-frequency range considerably. These include network analyzers to 40 GHz, time domain pulse generators and sampling systems to as high as 100 GHz, and time domain opto-electronic equipment to 350 GHz. I lack the temerity at this stage to consider very seriously the considerable problems in using these and other developments in prospect for dielectric measurements, but there are good reasons to think that such efforts will be rewarded by important results. A noteworthy example has already been provided by the work of Alain Gerschel and his Polish colleagues (24) for such liquids as halogenated methanes and chlorobenzene in the range 50–600 GHz from millimeter wave interferometer measurements that join up with far IR data above 450 GHz. These show deviations from extrapolated lower frequency patterns of behavior that are far from understood, but with various aspects attributed to correlated rotations and librations (25), ordering by short-range dipolar interactions, and dipolarons or other collective modes.

More such experiments are needed with better sources and detectors than were available for the original work. Although it is unlikely that any single approach alone will clarify the underlying dynamics, I can also believe that dielectric evidence will prove to be an important part of what is needed.

DIELECTRIC VIRIAL COEFFICIENTS OF GASES

As part of this somewhat random walk through past and present, a digression here on equilibrium dielectric properties of less than ideal gases is in keeping, as it shows how a new direction of research can develop from a combination of circumstances rather than by design and with some unexpected results.

I had begun to wonder early on about the adequacy of point dipoles, permanent or induced, for describing real molecules and their dielectric interaction effects in local field problems, Kirkwood correlation g-factors, and the relation of dipole coupling to phase transitions and relaxation in orientationally disordered molecular crystals, notably HCl, HBr, and HI. These last intrigued me because it was hard to think of any real molecules that more closely resembled the unreal spherical objects with point dipoles assumed in theory and because available evidence indicated that the crystal structures were all simple face-centered. A literature search yielded only wildly discordant information about equilibrium, let alone relaxation, dielectric behavior, and some calculations by Krieger & James (26) to suggest that interactions of molecular quadrupole moments could help to explain the multiplicity of solid phases. This was enough to start a series of investigations of the solid forms, discussed more in the next section.

My interest in gases was aroused when papers by Zwanzig (27) and Buckingham & Pople (28) appeared that pointed out collision-induced effects in gases from the field of one molecular quadrupole that induced dipole moments in neighbors during collisions. Their predictions were impossible to test and use for my purposes, however, because magnitudes of molecular quadrupole moments were virtually unknown, and the few available measurements of gas imperfections, as expressed by dielectric second virial coefficients of pair interactions, were highly uncertain about the magnitude and even the sign, let alone temperature dependence.

By this time, my research students and I had developed transformer bridge measurements to a high degree of precision for other purposes, so with realization that much more could be done it seemed entirely feasible to make the necessary electrical measurements of dielectric constants with a precision of the order of parts per million. This proved to be possible without undue difficulty, the more demanding problems proving to be in design and use of sufficiently stable capacitance cells and in sufficiently accurate determinations of gas densities.

The density problem comes about because theory leads to a virial expansion of the Clausius-Mossotti function in powers of number density N/V.

$$\frac{\varepsilon-1}{\varepsilon_2} \cdot \frac{V}{N} = A_\varepsilon + B_\varepsilon \frac{N}{V} + C_\varepsilon \left(\frac{N}{V}\right)^2 + \cdots \qquad 3.$$

where A_ε contains the single molecule contributions from polarizability and permanent dipole moments as in Eq. 1 and the second virial coefficient B_ε, the effects of pair interactions. Determination of densities from measured pressures was undesirable, because the conversion of pressure to density leads to a modified virial expansion with B_ε replaced by $B_\varepsilon - A_\varepsilon B_p$ where B_p is the second pressure virial coefficient for expansion of pressure P in powers of density. The correction term $A_\varepsilon B_p$ is typically 10–20 times larger than B_ε, and literature data for B_p of gases of interest often lacked the necessary accuracy. While recoiling from the prospect of major efforts in PVT measurements, I remembered the Burnett expansion method of generating a series of decreasing densities in constant ratio that could be determined by a modest number of measurements at low pressures. The basic scheme (29) was later elaborated into a sequence of expansions from one cell to two others (30), which provided valuable checks on consistency and clues to sources of error, and Roland Orcutt among others and I were finally in business.

Some early results were unexpected. Although the principal interest at the time was in quadrupolar molecules, we decided to measure some atomic gases, primarily as a test of the apparatus, as we supposed that the expected small positive B_ε values could be calculated with sufficient accuracy from Kirkwood's early theory (31) of induced dipole fluctuation effects. The experimental positive values of B_ε for argon were somewhat less than calculated, while more difficult measurements of the much smaller effect in helium (for which ε deviates from unity by 60 ppm at one atmosphere) gave negative values of B_ε. This caused us considerable concern, as it suggested the presence of systematic errors in the measurements. We could find no likely reason, and happily I heard a report of a similar anomaly in pressure-induced spectral line shifts in helium. With these reassurances, we reported these and other results (29). The effect for helium presented an interesting challenge to theorists, as one is dealing with the calculation of the polarizability of a pair of helium atoms as a function of their separation. As it has finally turned out, even such a seemingly simple problem requires very accurate wave functions, and a variety of calculations were made, not without controversy about their methods, before the matter was settled.

Meanwhile, my own work with a series of students went on to study the effects for molecules of increasing complexity. We took CO_2 as a prime example of a quadrupolar molecule, and from measurements by Tapan Bose to 150 atm of both the pure gas and mixtures with argon (32) in conjunction with theory developed by Buckingham & Pople (28) we were able to derive a rather large quadrupole moment $Q = 4.3 \times 10^{-26}$ esu cm^2. Some earlier estimates had given considerably larger values which we found to be the result of neglecting quadrupole-quadrupole interactions in the pure gas, as verified by the experiments in the mixtures with argon from which the interaction of $CO_2 - Ar$ pairs lacking the $Q - Q$ interaction could be extracted. The results were gratifying, as they demonstrated the importance of both quadrupole moment fields inducing dipoles in neighbors and their interactions with each other at fairly short range. A further satisfaction was the agreement of our value with ones from other effects, notably the one from the elegant but difficult Buckingham-Disch experiment (33) in which the anisotropy of orientations of CO_2 molecules by quadrupole interaction with a strong external field gradient was measured by the small but cumulative induced birefringence for an optical path length of ca. 1 meter.

After these and other studies of quadrupolar and octapolar molecules, our attention turned to simple polar molecules for which quadrupolar effects could also be expected, albeit added to larger effects of dipolar interactions, and such molecules as CF_3H, CHF_3, $CClF_3$, $CClH_3$, and $CClF_2H$ were chosen.

The values of B_ε obtained by Sutter (34) varied widely in magnitude and in sign as a combined result of effects of dipole and quadrupole interactions together with shape-dependent short-range repulsive forces, but obtaining consistent fits without including quadrupole effects proved impossible. Uncertainties as to best values of other parameters made derived values of quadrupole moments very approximate, but the magnitudes were in reasonable agreement with other estimates. The principal value of the work was thus to establish both the importance of quadrupole effects and the high sensitivity of dielectric effects to orientational forces, with pressure virial coefficients much less sensitive, and the temperature dependence of viscosity scarcely sensitive at all, as previously pointed out by Spurling & Mason (35).

More selective evidence as to the magnitude of quadrupole-induced electric moments during collisions is provided by the absorption spectra, in which one expects rotational lines with selection rules $\Delta J = 0, \pm 2$ rather than $\Delta J = \pm 1$ for dipole transitions. Ordinarily these are in the far infrared and too broadened because of short durations of collisions to be readily resolved, but HCl and HBr are exceptions: Because of their small moments of inertia, the quadrupole lines are widely spaced and in the near infrared. From this, Shmuel Weiss and I (36) were able to find and identify the predicted lines quite precisely and from estimated intensities obtain a quadrupole moment $Q = 5.5 \times 10^{-26}$ esu cm^2 for HBr. This was convincing evidence for the reality of quadrupole effects of sufficient magnitude to play a significant role in the structure of orientational phases and phase transitions in hydrogen halides, thus helping to resolve questions about interpretations of dielectric effects in these solid phases that had led to the studies of dielectric virial coefficients of gases in the first place.

RELAXATION OF DIPOLAR LIQUIDS AND CRYSTALS

When several students from the World War II generation and I started studies of dielectric relaxation in 1947, the initial instrumentation for audio and radio frequencies together with questions raised by studies of earlier literature suggested looking at solid phases with orientational freedom of small dipolar molecules. Two prime examples were hydrogen bromide, which we hoped was representative of simple molecules in well-defined structures, and ice, important for understanding "water substance" in any of its forms. In both cases, the experimental data from previous work were meager, with no agreements between results of different authors. We were soon to learn the hard way about the difficulties of measurements on any solids, and especially ones of high permittivity at low temperature.

The mundane problems of preparing and maintaining samples free of cracks and voids and the lack of adhesion of electrodes appeared in extreme form for HBr, with Norman Brown the chief sufferer. Solid samples grown from the liquid in an elegant coaxial cylinder cell invariably developed cracks and voids in cooling, with catastrophic results as peak permittivities near the order-disorder transition at 89 K were approached, but reproducible values were finally obtained by resorting to applying compression to parallel plate electrodes after a sample initially free of any visible defects was frozen in by judicious temperature cycling. The result was a spectacular and reproducible peak in static permittivity at $\varepsilon_s \cong 200$ (rather than 10–35 from earlier work), followed by not one but two distinct relaxations at lower temperatures of the partially ordered crystal (37). The slower and better-defined one was a beautiful depressed circular arc in the complex plane plots of $\varepsilon^* = \varepsilon' - i\varepsilon''$ expressed as $\varepsilon^* - \varepsilon_\infty = (\varepsilon_s - \varepsilon_\infty)[1 + (i\omega\tau)^{1-\alpha}]^{-1}$, so here there was a seemingly clear-cut example of the behavior my brother and I had uncovered years before (44). At the time, we did not realize that the low-temperature phase was ferroelectric rather than antiferroelectric as I supposed, with the slow relaxation due to the presence of domains. This was discovered only some 20 years later when Japanese workers reported hysteresis loops (38), which Stan Cichanowski and I confirmed and studied further (39). Other students and I in the meantime had been able to define the dielectric properties of hydrogen halides much more extensively (40, 41). The results sufficed to map some common patterns of behavior but equally brought out marked differences among HCl, HBr, and HI; these differences both dashed my hopes for any simple understanding and emphasized the need for much more refined treatments of pair and many body interactions in these supposedly simple systems.

By way of contrast, the early dielectric studies of ice that I and Bob Auty (42) performed gave very simple, clean results once the problems of void formation, electrode contacts, and electrode polarization had been diagnosed and dealt with reasonably well. The relaxation of the polycrystalline samples was accurately fitted by the simple Debye Eq. 2, with a relaxation time of 40 μs at 0°C increasing to 2 ms at -90°C and an activation energy at 13 kcal/mole. Numerous studies since have been extended to single crystals and the various polymorphic forms, but it has been gratifying to see how well the original values have been confirmed (the more so after Lars Onsager in a lecture on Bjerrum fault sites as the agents for dipole reorientations referred to me as an experimentalist whose results could be trusted).

Below $-40°C$, the relaxation in ice was too slow for measurement by available bridge methods, so with Auty and Donald Davidson, I set out to develop transient methods (43) for this and other relaxations at longer times. I wanted particularly to look for good examples of the $t^{-\alpha}(\alpha < 1)$ time dependence so often found in the literature and deviations from it at long times (44), and as a test of the new instrumentation picked glycerol for measurements by both bridge and transient measurements over a range of temperatures.

Glycerol was chosen partly because analysis of earlier measurements had seemed to show that the relaxation was described by the circular arc (Cole-Cole) relaxation function (regardless of whether glycerol could be considered representative of anything but itself), but more from the convenience that it was as slow as molasses in January and could be supercooled indefinitely with ease. This was the genesis of the skewed arc (Cole-Davidson, C-D) relaxation function (45).

When Don Davidson showed me his first results with a skewed rather than the expected symmetric circular arc in the complex plane representation, I asked him to try again with more attention to avoiding possible sources of error. After he forced me to take him seriously, I remembered the impedance characteristic of an infinite leaky cable (with capacitances charged through resistors) described by a relaxation function of the form $(1 + i\omega\tau)^{-1/2}$. Changing the exponent from $-\frac{1}{2}$ to a variable parameter β with $0 < \beta < 1$ gave the skewed arc function $\varepsilon^* - \varepsilon_\infty = (\varepsilon_s - \varepsilon_\infty)[1 + i\omega\tau]^{-\beta}$. For β of order 0.6–0.8 this gave beautiful fits except at the very highest frequencies with $\omega\tau \gg 1$ and set me thinking of such things as coupled diffusive mechanisms and diversity of hydrogen bonding. Going on to propylene glycol with two OH groups gave similar results except for β values nearer to value unity, whereas 1-propanol with a single OH gave the $\beta = 1$ characteristic of Debye relaxation except for evidence of small but distinct secondary relaxations at much higher frequencies. All three liquids could be supercooled quite readily to very low temperatures and long times, and all showed characteristic slowing down of relaxation time on approaching the glass transition. This could be described by the familiar non-Arrhenius law $\ln \tau = A \exp B/(T - T_0)$, with T_0 some 20–30 degrees below the glass transition temperature T_g, which I attributed to Tammann and is now commonly known as the Vogel-Tammann-Fulcher (VTF) equation.

To test whether such striking results were to be found in other than associated liquids, I naturally looked first in C. P. Smyth's book (46) with its summaries of his many pioneering exploratory studies. This soon suggested alkyl halides for further investigation, as they were normal polar liquids by usual criteria, including that of reasonable conformity with Onsager's equation for their static permittivities, and several likely candidates could be supercooled quite readily. Donald Denney had recently finished a PhD thesis on alcohols and agreed to return for a year for work on such problems and to help other students while I was on sabbatical in 1955–1956.

Any anxiety about progress in the lab during my absence was quickly dispelled by reports of very similar behavior in i-butyl chloride, i-butyl, and i-amyl bromide (47). The last was studied even more extensively in the next few years, thanks to the development of a microwave bridge method by Sivert Glarum (48) and other somewhat crude but usable methods to cover the range from 1 to 900 MHz, with the result that the relaxation spectrum of i-amyl bromide was reasonably well defined over 13 decades of frequency or time. At the same time that we were engaged in these and related studies, there were increasingly frequent reports of other examples of similar relaxation line shapes and non-Arrhenius temperature dependence in liquids and amorphous solids, including dielectric results for a variety of polymers and other systems, viscosity, and viscoelastic effects. Much of the dielectric work has been ably reviewed by Graham Williams (49), including the extensive work of his own research group, and a notable contribution of his was the finding of an empirical relaxation function variously known as the Williams-Watts, Kohlrausch-Williams-Watts (KWW), or "stretched exponential." (The addition of Kohlrausch is to recognize that he had used the same function many years before to describe time dependence of viscoelastic behavior.)

The term "stretched exponential" refers to the (macroscopic) relaxation function of the form $\gamma(t) = A \exp\left[-t/\bar{\tau}\right]^{\bar{\beta}}$ with $0 < \bar{\beta} < 1$, from which the complex permittivity ε^* can be obtained by the relation $\varepsilon^* - \varepsilon_\alpha = (\varepsilon_s - \varepsilon_\infty)$ $[1 - i\omega L\gamma(t)]$, where L denotes the Laplace transform. Although no solution in terms of known functions exists, extensive numerical evaluations have been made and used to show that the calculated frequency dependence, although very similar to that of the skewed arc (Cole-Davidson) function above, is less asymmetric in the ε'' absorption line shape for values of β and $\bar{\beta}$ required to produce good average agreement. Numerous examples have been, or could be, cited for preferring one, the other, or neither, but while mentioning a few of these (50) I expressed my opinion that the similarities are more important than the differences. In any event, the KWW function has gained widespread acceptance and use, to the extent that it is often referred to as a universal function for relaxation processes, but so have others.

Whether or to what extent any function can be called universal is at least partly a matter of semantics. Dictionaries provide a multitude of definitions, ranging from the uncompromising "applicable to all cases" (Concise Oxford) to "any general or widely held principle, concept, or notion" (American Heritage). By the first definition, I know of no single relaxation function that is universal for liquids, let alone solids as well, whereas the second is more accommodating.

Theoretical interpretation of the widespread occurrence of characteristic patterns of behavior has until quite recently lagged far behind the accumulation of experimental evidence and empirical descriptions. A common practice has long been to beg the real questions by invoking distributions of relaxation times in a sum of individually exponential processes as an explanation. Ever since I first looked at the necessary form of such distributions to fit the data, I have been arguing that such a description may be both unnecessary and inappropriate, and I had accumulated a few examples to support the thesis drawn from electrolyte diffusion and extended electrical network theory. A more satisfying example became available when Sivert Glarum as part of his PhD thesis produced entirely on his own his "defect diffusion" model (51) of a linear chain of molecules relaxing exponentially unless and until the arrival of a defect, represented by random walk between sites, resulted in complete loss of orientational correlation. The resulting relaxation function $\gamma(t)$ is a sum of the two effects of an exponential and a cooperative diffusive process characterized by a complementary error function of time. Different ratios of the two characteristic times gave Debye relaxation in one limit, with the circular arc with $\alpha = 0.5$ in the other, and the skewed arc with $\beta = 0.5$ for equal times, with a continuous variation of breadth and asymmetry between the extremes.

Glarum's model was admittedly a highly simplified approach, and difficulties were encountered in attempting to elaborate on or modify his assumptions, but it also was instructive in showing how characteristic observed behavior could emerge from treatment of relaxation as a cooperative process. It has also evidently been seminal for recent developments: by Shlesinger & Montroll (52a,b) on non-Markovian distribution of pausing times between sequential reorientational steps, and by Anderson, Palmer, et al (53) on hierarchically constrained dynamics. Both lead for some choices of the distribution or constraints to relaxation functions essentially of the Williams-Watts form and also to predictions or inferences about their temperature dependence. Space here permits only two general comments about these very interesting developments.

The first comment is that although avowedly molecular in concept, there is no evident place in these theories for individuality of molecules, and the treatments are stochastic in nature. This comment is not meant to be unduly critical: treatment of many body dynamics at the molecular level remains formidable and, as Montroll once remarked to me, one adopts stochastic methods when the molecular going gets too rough. The second comment is that although these and other theories make connections between the onset of broadened relaxation line shapes and non-Arrhenius temperature dependence of characteristic times, there are several examples from experiment that one can perfectly well have one without the other. One already mentioned is of alcohols, which have both a principal relaxation of Debye form and relaxation time described by the VTF equation over considerable ranges of temperature. A second is of so-called conductivity relaxation in ionic glasses, where the converse is true.

A number of more specifically molecular models have also appeared, of which two for polymer chain dynamics are of interest in the present context because of their formulation and predictions. The first is the Shore & Zwanzig calculation (54) for transverse dipoles attached to polymer chain segments that are both coupled by torsional forces between segments and subject to rotational diffusion forces of the surroundings. Depending on the strength of the coupling forces relative to thermal energy kT, number of chain segments N, and number of these carrying interacting dipoles, the resulting relaxation spectra show deviations from simple Debye relaxation that are variously like the C-D, KWW, or Havriliak-Negami (55) function. (The last, which is an empirical combination of skewed and symmetrical circular arc functions, often fits polymer relaxation data better than the other functions with one less adjustable parameter.) The original paper should be consulted for further details; the above is no more than a sketch of a previously attempted summary (11) of the variety of conclusions that can be drawn about the relative importance of the assumed forces in chain dynamics.

A second kind of development by Skinner (56) invokes solitons as descriptions of localized conformational misalignments, or spins. These can perhaps be regarded as an example of the "defects" assumed by Glarum without discussion of such possible origins as localized fluctuations generated by the heat bath. Results for different constraints on the soliton dynamics also gave in this model relaxation functions of both Williams-Watts and Cole-Davidson form.

All these developments seem to show real promise for better understanding, but there are some awkward points from earlier work that have been brought back to mind by some recent preliminary results. Glarum

(51) found, for example, that experimental data for i-amyl bromide at room temperature and a limited range of microwave frequencies sufficed to show that the skewed arc representation of data at lower temperatures and frequencies was not a good fit. Rather, the absorption curve suggested the presence of two reasonably distinct peaks. These could plausibly be attributed to intramolecular conformational effects brought out by increased thermal energy, and some unpublished exploratory measurements by Elpidio Tombari using time domain (TDR) methods with high resolution to 10 GHz show what may be similar effects. These are shown in Figure 4 as complex plane loci for 1-bromopentane at $-96°C$ and 1-iodoctane at $-40°C$, together with data for 3-bromopentane at $-96°C$, known from previous TDR measurements (50) to be accurately fitted by the skewed arc function, with $\beta = 0.88$ giving the *solid curve* in the figure. The deviations from similar behavior for the other two substituted alkanes are evident, but no attempt has yet been made to establish better representations.

A further interesting finding is that when the neat liquids are diluted with hexane the relaxation spectra shift to considerably higher frequencies but without any significant changes in shape. The choice of hexane as a nonpolar solvent limits the temperature range of dilution studies, but these limits are being explored by John Berberian using 3-methylpentane, known from his work to be a readily supercooled solvent. Just to tantalize me and any interested reader, he has told me of rather startling preliminary results at low temperatures that are as yet too fragmentary to be presented, much less rationalized, at this stage. Only one conclusion is clear: The old truism continues to hold that further investigations can, and often do, raise as many questions as they seek to answer and call for further work.

ELECTROLYTES AND CONDUCTING GLASSES

Although the Kraus tradition was strong in chemistry when I arrived at Brown in 1947, and I helped out a little with research in progress then, my first serious involvement with electrolytes was as usual in an unexpected and unplanned way. Ronald Gillespie wrote in 1953 to ask whether he could work in the laboratory to determine dielectric properties of neat sulfuric acid, which he needed to know better for his classic studies of ionic equilibria in this solvent. The extraordinarily high specific conductance, about that of 0.1 m KCl in water, gave us great difficulties with electrode polarization effects even at the highest frequency (< 1 MHz) then available, but we did manage to obtain static permittivities of order 100, together with slight indications of dispersion at higher frequencies (57). When Stuart Lovell and I had developed methods for measurements to 250 MHz soon

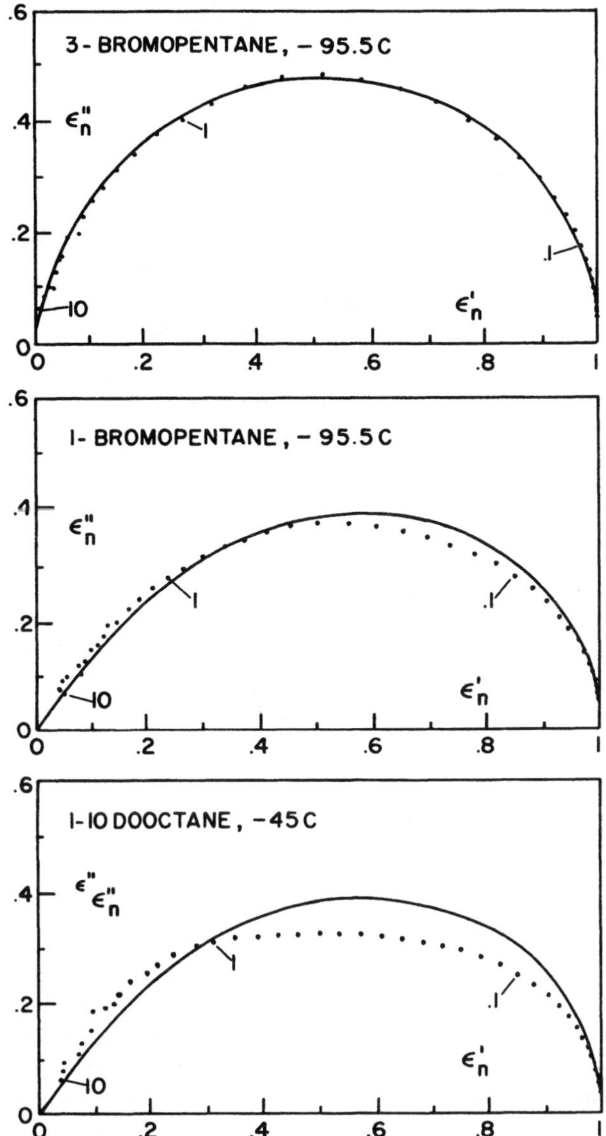

Figure 4 Complex plane plots of permittivity for three substituted alkanes. Indicated frequencies of experimental points are in GHz, *solid curves* are skewed-arc (CD) functions with $\beta = 0.88$ (3-bromopentane) and 0.60 (1-bromopentane, 1-iodo octane). The numbers along the curves are values of the frequency in GHz.

thereafter (58), sulfuric acid revisited gave some very strange results: Although the static permittivity was indeed very large as previously found, dielectric relaxation of Debye form as best we could tell was centered extraordinarily low frequency, and even slight deviations from the stoichiometric $H_2O \cdot SO_3$ composition produced large decreases of permittivity. I suspected instrumental errors, possibly as a result of higher conductivities from shifts of ionic equilibria, and refrained from publishing the results pending some grounds for more confidence in them. This finally came in 1974 on hearing a talk by Lars Onsager at Yale when Raymond Fuoss retired. My ears perked up when sulfuric acid was mentioned in the context of rate-limiting steps in proton transfer and kinetic depolarization by delayed dielectric response to changing ionic fields. I realized that the latter could explain our findings in terms of the very long solvent relaxation time, ca. 400 ps as compared to 8 ps of water and 50 ps of methanol, for example, and our results agreed with the Hubbard-Onsager prediction to the accuracy of the rather suspect measurements.

Soon thereafter, I was able to catalyze an interaction of Onsager and Hubbard with W. M. Van Beek and Michel Mandel at Leiden (59), as I had heard there about the latter's puzzlement of finding decreases of permittivity of salts in H_2O and MeOH that correlated better with conductance regardless of salt type than with theories of ion solvation. At the time, I had been developing time domain (TDR) methods for higher frequency dielectric measurements, and with David Hall returned to sulfuric acid as a prime example of dominant kinetic effects, because of the long relaxation time. The results (60), by a new method in an early stage, were gratifying as they confirmed and extended the previous results and also gave near quantitative agreement with the theory if "stick" boundary conditions for ions and solvent molecules were used in the hydrodynamic continuum model.

A series of studies of Paul Winsor of ions in other solvents (61) showed that for more representative electrolytes both the kinetic (depolarization) and static (solvation) effects had to be considered in comparison with theories of the two kinds of effects on dielectric polarization, and also on ion mobilities in conduction (62). A sticking point in comparing these and other experiments with theory is uncertainty about the roles of various complicating factors. Experimentally, for example, the classic Debye-Falkenhagen effect of ion atmosphere relaxation has yet to be observed directly and unambiguously as far as I know, and the problem of properly defining a "static" dielectric constant in a conducting medium is not trivial. Some of the theoretical difficulties have been stated succinctly by Hubbard, Colonomos & Wolynes in their paper on molecular theory of ion dynamics (63): "All equilibrium or static effects have been treated independently one from another; a complete theory should include all these effects in a self consistent form." I would also add *a fortiori*—and so should a complete dynamical theory. Ramifications as they appeared in 1980 have been ably reviewed by Wolynes (64).

Problems with the interplay of ionic charge transport and displacements are also evident in conducting glasses and the various interpretations of dielectric and conductive aspects of their behavior that have been proposed. These are suggested schematically in Figure 5 by plots of measured conductance and permittivity as a function of frequency. At low frequencies, the apparent conductance typically has a plateau taken to be a d.c. or steady state value, but with a decline at still lower frequencies accompanied by enormous increases in permittivity. These are ordinarily attributed to space charge effects at electrodes or other boundary layers rather than to intrinsic properties of the solid, but unambiguous separation of the two is even more difficult for solids than for conducting liquids.

My own attitude toward the low frequency problem over the years had been to avoid it as much as possible, but there was no escaping it entirely

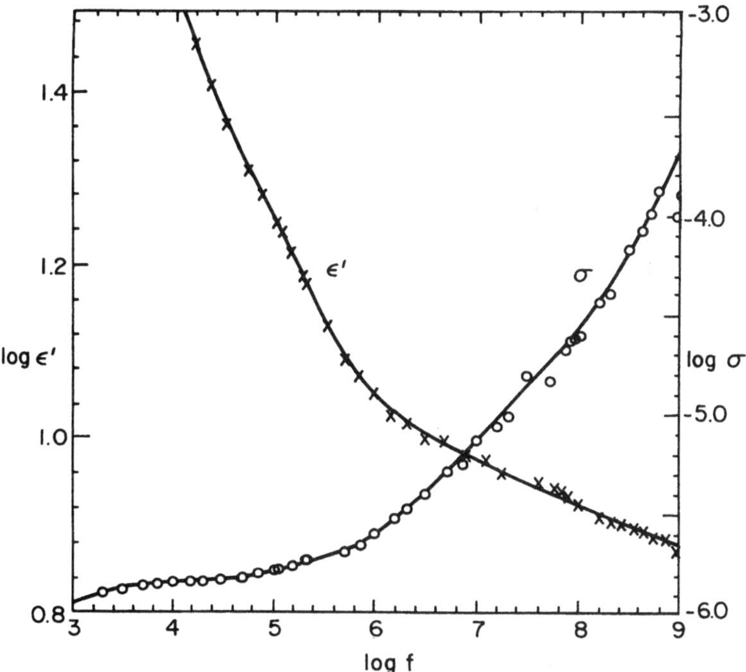

Figure 5 Logarithmic plots of permittivity ε' and conductivity σ' (mho/cm) for sodium trisilicate glass of 168°C.

when Bill Risen asked me and I agreed to look into what might be learned about fast ionic conducting glasses over the dielectric spectrum, and in particular at higher frequencies in the gap between bridge measurements below 1 MHz and far IR absorption well above 10 GHz. That there are thermally activated relaxation processes in addition to simple ion mobilities and glass network charge displacements complicated by space charge effects below 1 MHz was clear enough from the considerable and growing literature of low frequency results as typified by the plots in Figure 5. The region above 1 MHz was of particular interest to us because there seemed to be little or no indication of either conductance or permittivity approaching high frequency plateau values expected from applying either strong or weak liquid electrolyte theory to the solid state problem.

In progress to date on the problem, it has been possible by using high frequency bridge and TDR methods to show that for sodium trisilicate and lithium fluoroborate glasses conductance continues to rise and permittivity to decrease roughly as fractional powers of frequency to at least 3 GHz, and that the former may well join smoothly and continuously with the low frequency wing of far IR absorption, as fearlessly interpolated through four decades of frequency by Wong & Angell (65) in their log-log plots of absorption over 13 decades in magnitude and 12 decades in frequency. The excess of absorption seems, as Moynihan has remarked (66), "to be endemic to the solid state." Another feature that may also be common is the virtual independence of overall shape on temperature over a considerable range, despite shifts over several decades of the characteristic time scale.

All this and other evidence raises more questions than it provides answers. Are the patterns of behavior in these and other systems to be understood in terms of distinct low and high frequency processes or are they aspects of a continuum of behavior? Are they best interpreted as conductivity or polarization relaxations or (again) as aspects of a mingling of the two?

NONLINEAR DYNAMICS—SECOND AND HIGHER ORDERS

The appearance of Kubo-Green linear response theory and correlation function methods in the 1950s was another landmark for me, as it put into better perspective both the time evolution of dynamical processes generally and the relation of dielectric observables to other mechanical and thermal transport processes in particular. After Sivert Glarum introduced me to the subject as part of his thesis, I was able to use the new formalism to derive a number of earlier results obtained in other ways and to add a few new ones (67, 68). As a result, I was receptive to a proposal by Graham Williams that the possibility be explored of applying response theory and correlation function methods to analysis of the dynamics of Kerr effect birefringence more generally than for the rotational diffusion model of the classic work by Benoit (69, 70).

In Kerr effect experiments, anisotropically polarizable molecules are partially oriented by an external bias field E_b. The degree of orientation is detected by the induced birefringence of initially linearly polarized light passing through the sample at right angles to E_b. The dynamic effect as a measure of molecular response most simply consists in observing the changes in birefringence as a function of time after applying or removing a step-like pulse with sufficiently short transition time. For nonpolar molecules, only the first-order response proportional to E_b^2 is needed, and for axially symmetric molecules the first-order response is a measure of the correlation function for $P_2 (\cos \theta)$, where $\theta = \theta(t)$ is the angle between the axis and field directions and P_2 is the second Legendre polynomial. For molecules with permanent dipole moments, the analysis becomes more interesting, as dipole orientation energy proportional to E_b and $P_1 (\cos \theta)$ is also involved. This involvement changes the magnitude without affecting the P_2 character of response after E_b is removed, but it complicates the problem for response after E_b is applied because the second-order effect of dipole torques ($\sim E_b$) is required to obtain the overall response of order E_b^2.

There were intriguing differences between observed field-on and field-off responses when compared to each other and to P_1 responses from conventional dielectric measurements. Graham Williams had made me aware of these in his beautiful work on large molecules and supercooled liquids, which he has ably reviewed elsewhere (71), and I also knew of the extensive studies of biopolymers, notably by O'Konski and co-workers (72). I had supposed at the outset that a simple extension of response theory perturbation expansion in powers of E_b to second order in E_b^2 would readily yield simple expressions in terms of P_1 and P_2 correlations to account for the observations and thus increase their usefulness. The problem proved to be nontrivial, however, and it was only after considerable floundering and frustration that I was able to disentangle the nested convolutions of self and joint correlations in the field-on response to achieve the hoped for results (73).

If nothing else, this modest excursion into nonlinear dynamics gave me simple examples of several truisms. Two are that probes of two different aspects can be much more revealing than either one alone, and that higher order effects are much more than a simple extension of first order ones. A simple demonstration is that the relation of P_1 and P_2 responses is sensitive to the type of orientational dynamics, as the time constant τ_1 of P_1 response is three times τ_2 of P_2 response for small-step diffusional motion, whereas the two are the same for strong collisional loss of correlations at the other extreme, and that effects of both are observable by comparison of Kerr responses for field-on and off.

The construction of theories of nonlinear dielectric effects for molecules of arbitrary symmetry, with joint correlations, and to higher orders seems likely to encounter considerable difficulties, but at least two recent developments may help in overcoming them. The first is the approach of Fulton mentioned in the first section, the second are formal solutions, by Morita (74) and Morita & Watanabe (75), of the time evolution to arbitrary order (and increasing complexity) in powers of field strength, together with generalization of Kramers-Kronig and other relations among transient and steady state solutions.

Experimentally, new capabilities from laser techniques and high power fast electronic devices should make it possible to study the dynamics of smaller molecules and of larger ones to much higher frequencies.

CONCLUSION

The writing of any report or personal review such as this one invariably recalls unresolved questions put aside for one reason or another, brings some recognition of new ones in the light of recent developments, and adds to the store of things to think about. A "short list" of some of these, more or less in order of appearance as this is written, includes the following:

1. Small molecule dynamics and high frequency processes.
2. The roles of dipolar and shorter range forces in relaxation.
3. Unified theories of polarization and conduction.
4. Interfacial Polarization; Debye-Falkenhagen effect.
5. Nonlinear dynamics.

I am sanguine enough to believe that these and other dielectric problems, as aspects of much larger problems, will be as challenging and rewarding in the future as in the past. As Wolynes put it (64): "How a charge moves from place to place is a question that is likely to remain relevant for some time to come."

ACKNOWLEDGMENTS

I owe thanks to many friends and colleagues for help over the years. Of them, I should mention those who have been especially helpful as I was thinking about and writing this chapter: John Berberian, Satoru Mashimo,

Bo Gestblom, Giorgio Chryssikos, Elpidio Tombari, and Andrew Burns. Support from the National Science Foundation through Grant CHE-8518365 is gratefully acknowledged.

Literature Cited

1. Lorentz, H. A. 1914. *Vortrage über die Kinetischen Theorie der Materie und der Eleckrizitat*, pp. 167–92. Leipzig: Teubner
2. Debye, P. 1912. *Physik Z.* 13: 97–100
3. Debye, P. 1929. *Polar Molecules.* New York: Chemical Catalog. 172 pp.
4. Van Vleck, J. H. 1940. *Ann. NY Acad. Sci.* 40: 293–313
5. Wyman, J. 1936. *J. Am. Chem. Soc.* 58: 1482–85
6. Onsager, L. 1936. *J. Am. Chem. Soc.* 58: 1486–93
7. Kirkwood, J. G. 1939. *J. Chem. Phys.* 7: 911–19
8. Conference papers. 1940. *Ann. NY Acad. Sci.* 40: 289–482
9. Kirkwood, J. G. 1940. See Ref. 8, pp. 315–20
10. Felderhof, B. U. 1979. *Physica A* 95: 572–80
11. Cole, R. H. 1984. In *Molecular Liquids, Dynamics and Interactions*, ed. A. J. Barnes, W. J. Orville-Thomas, J. Yarwood, pp. 59–110. Dordrecht: Reidel. 590 pp.
12. Lax, M. 1952. *J. Chem. Phys.* 20: 1351–59
13. Toupin, R. A., Lax, M. 1957. *J. Chem. Phys.* 27: 458–64
14. Wertheim, M. S. 1979. *Annu. Rev. Phys. Chem.* 30: 471–501
15. Alder, B. J., Pollock, E. L. 1981. *Annu. Rev. Phys. Chem.* 32: 311–30
16. Brot, C., Bossis, G. 1980. *Mol. Phys.* 40: 1053–72
17. Fulton, R. L. 1975. *Mol. Phys.* 29: 405–13
18. Fulton, R. L. 1983. *J. Chem. Phys.* 78: 6856–84
19. Berberian, J. G., Cole, R. H. 1969. *Rev. Sci. Instrum.* 40: 811–17
20. Davidson, D. W., Auty, R. P., Cole, R. H. 1951. *Rev. Sci. Instrum.* 22: 678–82
21. Mopsik, F. I. 1984. *Rev. Sci. Instrum.* 54: 79–87
22. Fellner-Feldegg, H. 1972. *J. Phys. Chem.* 76: 2116–22
23. Cole, R. H., Berberian, J. G., Mashimo, S., Chryssikos, G. D., Tombari, E., Burns, A. 1989. *J. Appl. Phys.* July
24. Gerschel, A., Grochulski, T., Kisiel, Z., Pszczolkowski, L., Leibler, K. 1985. *Mol. Phys.* 54: 97–117
25. Gerschel, A. 1984. See Ref. 11, pp. 163–99
26. Krieger, T. J., James, H. M. 1954. *J. Chem. Phys.* 22: 796–814
27. Zwanzig, R. W. 1956. *J. Chem. Phys.* 25: 211–16
28. Buckingham, A. D., Pople, J. A. 1955. *Trans. Faraday Soc.* 51: 1029–35; 1957. *J. Chem. Phys.* 27: 820–21
29. Johnston, D. R., Oudemanns, G. J., Cole, R. H. 1960. *J. Chem. Phys.* 33: 1310–17
30. Orcutt, R. H., Cole, R. H. 1967. *J. Chem. Phys.* 46: 697–702
31. Kirkwood, J. G. 1936. *J. Chem. Phys.* 4: 592–601
32. Bose, T. K., Cole, R. H. 1970. *J. Chem. Phys.* 52: 140–47
33. Buckingham, A. D., Disch, R. L. 1963. *Proc. R. Soc. London Ser. A* 273: 275–89
34. Sutter, H., Cole, R. H. 1970. *J. Chem. Phys.* 52: 132–39
35. Spurling, T. H., Mason, E. A. 1967. *J. Chem. Phys.* 46: 322–26
36. Weiss, S., Cole, R. H. 1967. *J. Chem. Phys.* 46: 644–49
37. Brown, N. L., Cole, R. H. 1953. *J. Chem. Phys.* 21: 1920–26
38. Hoshino, S., Shimaoka, K., Nimura, N. 1967. *Phys. Rev. Lett.* 19: 1286–88
39. Cichanowski, S. W., Cole, R. H. 1973. *J. Chem. Phys.* 59: 2420–26
40. Cole, R. H., Havriliak, S. Jr. 1957. *Discuss. Faraday Soc.* 23: 31–38
41. Groenewegen, P. P. M., Cole, R. H. 1967. *J. Chem. Phys.* 46: 1069–74
42. Auty, R. P., Cole, R. H. 1952. *J. Chem. Phys.* 20: 1309–14
43. Davidson, D. W., Auty, R. P., Cole, R. H. 1951. *Rev. Sci. Instrum.* 22: 678–82
44. Cole, K. S., Cole, R. H. 1941. *J. Chem. Phys.* 9: 341–51; 1942. *J. Chem. Phys.* 10: 98–105
45. Davidson, D. W., Cole, R. H. 1951. *J. Chem. Phys.* 19: 1484–90
46. Smyth, C. P. 1955. *Dielectric Behavior and Structure*, pp. 99–131. New York: McGraw-Hill. 441 pp.
47. Denney, D. J. 1957. *J. Chem. Phys.* 27: 259–64
48. Glarum, S. H. 1958. *Rev. Sci. Instrum.* 29: 1016–19

49. Williams, G. 1984. See Ref. 11, pp. 239–74
50. Berberian, J. G., Cole, R. H. 1986. *J. Chem. Phys.* 84: 6921–27
51. Glarum, S. H. 1960. *J. Chem. Phys.* 33: 639–43
52a. Shlesinger, M. F., Montroll, E. W. 1984. *Proc. Natl. Acad. Sci. USA* 81: 1280–83
52b. Bendler, J. T., Shlesinger, M. F. 1987. *J. Mol. Liq.* 36: 37–46
53. Palmer, R. G., Stein, D. L., Abrahams, E., Anderson, P. W. 1984. *Phys. Rev. Lett.* 53: 958–65
54. Shore, J. E., Zwanzig, R. W. 1975. *J. Chem. Phys.* 63: 5445–58
55. Havriliak, S. Jr., Negami, S. 1966. *J. Polym. Sci. Polym. Symp.* C14: 97–117
56. Skinner, J. L. 1983. *J. Chem. Phys.* 79: 1955–64
57. Gillespie, R. J., Cole, R. H. 1956. *Trans. Faraday Soc.* 52: 1325–31
58. Lovell, S. E., Cole, R. H. 1959. *Rev. Sci. Instrum.* 30: 361–62
59. Hubbard, J. B., Onsager, L., Van Beek, W. M., Mandel, M. 1977. *Proc. Natl. Acad. Sci. USA* 17: 401–4
60. Hall, D. G., Cole, R. H. 1981. *J. Chem. Phys.* 85: 1065–69
61. Winsor, P. IV, Cole, R. H. 1982. *J. Phys. Chem.* 86: 2486–94
62. Evans, D. F., Tominaga, T., Hubbard, J. B., Wolynes, P. G. 1979. *J. Phys. Chem.* 83: 2669–77
63. Hubbard, J. B., Colonomos, P., Wolynes, P. G. 1979. *J. Chem. Phys.* 71: 2652–61
64. Wolynes, P. G. 1980. *Annu. Rev. Phys. Chem.* 31: 345–76
65. Wong, J., Angell, C. A. 1976. *Glass, Structure by Spectroscopy*, Chapter 11. New York: Dekker. 864 pp.
66. Moynihan, C. T., Boesch, L. P., Laberge, N. L. 1973. *Phys. Chem. Glasses* 14: 122–25
67. Cole, R. H. 1965. *J. Chem. Phys.* 42: 637–43
68. Cole, R. H. 1981. In *Physics of Dielectric Solids*, ed. C. H. L. Goodman, Conf. Ser. No. 58, pp. 1–21. Bristol/London: Inst. Physics. 151 pp.
69. Benoit, H. 1951. *Ann. Phys. Paris* 6: 561–609
70. Benoit, H. 1952. *J. Chim. Phys.* 49: 517–21
71. Williams, G. 1984. See Ref. 11, pp. 237–74
72. O'Konski, C. T., ed. 1976. *Molecular Electro-Optics*, Vol. 1. New York: Dekker
73. Cole, R. H. 1982. *J. Phys. Chem.* 86: 4700–3
74. Morita, A. 1986. *Phys. Rev. A* 34: 1499–1504
75. Morita, A., Watanabe, H. 1987. *Phys. Rev. A* 35: 2690–96

Ann. Rev. Phys. Chem. 1988. 39: 1–37
Copyright © 1988 by Annual Reviews Inc. All rights reserved

50 YEARS OF PHYSICAL CHEMISTRY, A PERSONAL ACCOUNT

Sidney W. Benson

Donald P. and Katherine B. Loker Hydrocarbon Research Institute, Department of Chemistry, University of Southern California, University Park, Los Angeles, California 90089–1661

I have always been interested in history, ancient history, modern history, chemical history, and even just plain gossip. However, I have never had either the time or motivation to learn any of these very thoroughly. What knowledge I may have in these areas is a somewhat random collection of memories, not necessarily accurate, integrated over a lifetime of episodes. Thus it was with some pleasure and numerous misgivings that I considered the invitation from the Editors to write this chapter. I have been actively involved with chemistry for over half a century, and my activities have always seemed to me very personal and possibly idiosyncratic. It was not very clear to me that I had any profound messages for my colleagues or the current generation of chemists. So I shall not attempt to invent any here. What may be of interest to the reader are my personal recollections of what has turned out in retrospect to have been a very exciting period both in human history and in chemistry. They are intimately related.

There are five billion of us on this planet, of whom about 250 million are located in the United States. Of that latter number, about 200,000 are currently involved in the diverse activity we call chemistry. Taken together with their families, chemists today represent a no-longer trivial fraction of the domestic American scene. Since 1900, their numbers have increased about 20-fold. These numbers are testimony to the profound and rapid changes that have occurred in the western world during the twentieth century and that will probably point the direction for what we call today the Third World.

George B. Kistiakowsky was fond of saying to his graduate students,

"The days of the gentleman chemist are over." As a 20-year-old member of that group, reared on the streets of New York City, much of the meaning of that remark escaped me. Few, if any, of the motley crew of graduate students in chemistry entering Harvard in the Fall of 1938 looked much like the Hollywood version of a gentleman, and few of us seemed to cherish that image as an immediate or distant goal. All of us had been reared in the background recollection of the bloodiest period in human history, World War I, and we were keenly aware, struggling as the United States still was with the effects of the Great Depression, that we were on the threshold of still another, possibly bloodier period—World War II.

Curiously enough, this social setting did little to change the interest or intensity with which we entered into graduate classes and research activities. There is nothing more immediate than the present, and the world of science and chemistry was witnessing its own rapid developments. Quantum mechanics was a relatively new discovery just beginning to make impacts on chemistry. Special and General Relativity were still exciting philosophers and physicists. Having been myself reared in the classical, rational world of Newtonian mechanics and thermodynamics, these were not particularly welcome changes. The Uncertainty Principle was not of great comfort to a student raised in a causal universe.

Many of my classmates were looking forward to careers in chemical industry. My own horizon was somewhat cloudy but, even more than I was then aware, I had already been infected by the bug of research. It was to be a chronic disease from which I have still not recovered.

At 20, I believed quite fervently that life was goal-oriented. One did research to learn something. The research was only a means. Was there life after finishing a PhD degree? Well it was clearly training for some more ambitious goal. But what? The future was, indeed, cloudy. If I have learned anything since then it is that while the ends may justify some means, life itself consists mainly of means, the things one does from day-to-day. If these are not on the whole satisfying, the end is not terribly important. For those who like it, research can be a very satisfying way.

My earliest interest in chemistry dates back to when I was 11 years old and had acquired my first chemistry set. I was fascinated by turning red water into colorless water and then back-again with a "magic wand" that had a capsule of sodium carbonate embedded in one end and crystals of tartaric acid in the other. Even more fascinating were the recipes for making gunpowder and colored gunpowders. Carbon, sulfur and potassium nitrate were the basic ingredients. The arcane world of strontium, calcium, barium, and sodium salts to achieve a color spectrum was able to provide fascinating hours of uninterrupted experiments.

However, one particular set one summer evening in the Bronx was interrupted by my frantic mother together with some neighbors, who burst into the kitchen, moving through a fog of fumes to find me, quite oblivious, concocting another potion to ignite. I had earlier noticed a very strong smell of burning sulfur and nitric oxides, and I had opened the doors and windows to create some ventilation, but neither the gases nor the smoke seemed to disturb me. I was quite surprised by their agitation. But that was the end of fireworks in the kitchen.

In the Fall of 1938, the world of chemistry was quite small. In fact the entire world of science was quite small. A few months after I had entered Columbia College as a freshman in the Fall of 1934, Harold Urey gave a lecture to which interested chemistry students were invited. He had just won the Nobel Prize for his separation of heavy water, and he demonstrated the properties of this exotic substance to the audience of about 250 students and faculty who had assembled in the then Victorian Havemeyer Hall. Most of the chemistry faculty of about eight were there, as were such notables as Professors G. Pegram and I. I. Rabi from the physics department. This distinguished event was still being eclipsed in the local press by the victory of the Columbia College football team over Stanford University in the Rose Bowl earlier that year. An ardent sports fan, I equally enjoyed the excitement of both occasions. Looking back at myself now, I am amused by the blasé casualness with which I found myself participating in these events. Part of it certainly must be attributed to the usual arrogance of being a native New Yorker, which was clearly identified with being in the center of the known universe. Provincialism may exist in the provinces, but I suspect it was exported from the big city centers.

My first serious interest in chemistry started with High School. I had always been interested in doing things with my hands and learning how things worked. Robinson Crusoe was one of my childhood heroes. When I finished public school, I knew that I wanted to do something in science or engineering. Through an older friend who had already graduated, I was encouraged to apply to the New York City Science High School of the time, Stuyvesant. My parents took a very dim view of it. We lived 20 miles away in the then, still rural West Bronx. Stuyvesant was located in a somewhat decrepit brick building in a slum area of the lower East side of New York, sandwiched in between an emergency hospital which forever reeked of ether and some equally foreboding dirty brown-stone, cold water, tenement buildings. It took a 30–40 minute ride in the subway followed by about a three quarter mile walk to reach it.

My three and a half years in this most unengaging environment were to turn out to be among the most exciting, exhilarating, and memorable of my early life. This unprepossessing exterior contained a group of the most dedicated teachers I have met anywhere, together with the most extraordinary facilities I could imagine at this time. Following a fairly uneventful first year in what was termed "the afternoon session" (due to a lack of space), I was promoted in my second year to the "morning session." We started classes at 8:00 A.M. and finished by 12:30 P.M.,with only a six minute break between five solid classes. A quick lunch followed in the school auditorium, which doubled as a makeshift, fast food lunchroom. But then, for this privileged "morning session," the school resources were made available. Shops and facilities were open to interested students. There were clubs for every taste from Latin to math and chemistry, with enthusiastic faculty in charge. For me it was like entering an enchanted world. I seldom arrived home before 6:30 P.M.!

For the first time I met many boys (no girl students then) my age with similar interests in science. Some had labs at home—many very sophisticated, with chemicals I had never heard of and some with then newly discovered "electronic" gear. There was a wonderful young math teacher, Hyman Marcus, who first got me interested in mathematics and then later in fencing. He had just graduated from Columbia and, although I didn't realize it at the time, that pointed me in my future direction. Another wonderful chemistry teacher, Abraham Kerner, with whom I maintained contact until his death some ten years ago, opened up the chemistry labs to me. Perhaps most important was a dimunitive English teacher, Hyman Mostow, seemingly possessed of infinite sources of energy. He assumed planning my career as well as introducing me in his extraordinary course in English to Shakespere and English poetry. He had about eight other students in his menage who he also guided with a gentle but iron hand. In his spare time he coached the somewhat nebulous basketball team.

Leading this most unexpected coterie of teachers and mentors was the even more unlikely Principal of the High School, Dr. Earnest R. von Nardroff. Each semester he filled the 2000 seat auditorium for a special lecture, marvellously demonstrated, on some topic in science. They were extremely well-done, effective and for me a constant example of how science should be presented. I can still remember most of them, so vivid were they! Most dramatic was a lecture titled "The Spheroidal State." The large lecture stage was lined with a sand trough and half-way through the discourse on the cooling effects of evaporation and the pressure produced by gases, several students in asbestos aprons and boots, wearing asbestos gloves appeared, carrying a large, white hot crucible containing molten iron. They proceeded to the podium where they tilted the crucible into the trough, producing a small stream of white hot molten iron. Dr. von Nardoff, having earlier explained his protocol, carefully washed his hands and then passed his open palm through the stream, scattering a fountain of incandescent, sputtering globules all over the stage. The audience gasped the loudest gasp of astonishment I've ever heard, while three assistants with pails of sand extinguished flames from potential fires started where globules had penetrated the sand to the wooden floor below. Thunderous applause then elicited a repetition of the demonstration!

Stuyvesant provided more than a fertile soil for future scientists. It opened a view onto the intellectual world of ideas of which I had never been aware. My classmates came from many diverse social, ethnic, and political backgrounds, and passionate discussions could be heard and attended on politics, economics, sex, as well as on science. The once large family and neighborhood world of the Bronx and Brooklyn in which I had been raised steadily diminished in size and importance during those exciting years. I also discovered the incredible used book stores of lower Manhattan where one could browse for hours through books on subjects I had never heard of. Equally exotic was the chemical supply house of Eimer and Amend on Third Avenue, where a trip through their apparatus stockroom could intoxicate the imagination for weeks trying to guess the purpose of the various metal and glass constructions. Their chemical catalog was no less evocative. Only my mother expressed some dismay at the increasing arsenal of chemicals residing in her once splendid, now slowly invaded glass china cabinet. Weekend experiments had become less visible, more sophisticated, but no less odorous. When I began to explore qualitative analysis, the schemes for separation of metals based on solubilities became an adventure in devising new routes and new combinations of metallic groups. I found hydrogen sulfide a socially deplorable reagent and hunted for more acceptable means of separation, some of which would turn out later in college to save me many hours in the laboratory.

No experiences at Stuyvesant, however, were to prove as tense, anxious, as well as exciting as those with the Math Team. While I might have preferred a role in the somewhat pathetic but nevertheless, to me, glamorous football team, nature had not designed me in size, mass, or speed for a role in American football. The Math Team turned out to be a more than adequate substitute. Invited at first by classmates, and having to overcome an initial unfamiliarity and just plain ignorance, I discovered after a few months an unexpected facility in problem solving. Then encouraged, despite numerous failures, by an enthusiastic Mr. Marcus, I quickly began to develop a real interest in algebra, geometry, and mathematical puzzles. Most remarkable, I discovered books in the school library on mathematics that I could read and, to my amazement, from which I could learn. It was a heady experience. By my fourth semester, I was a regular on the school math team, and in my last year I was elected captain.

The New York City Official Math Meets were something else. I don't remember ever being able to sleep the night before a Meet. And during the actual 40 minutes of the two- or three-school competition my teeth clattered so hard I had to put a pencil between them to keep from attracting attention. Problems were written on a blackboard. We had one minute to read them and then, depending on their complexity, from one to a maximum of ten minutes to solve them. Only the answer counted. The team would spend hours practicing on all the available examples from texts, puzzle books, and examples from previous meets. It was an altogether nerve-wrecking event softened only by the more relaxed post-mortems that followed.

If Stuyvesant was an introduction to the arena of science and ideas, Columbia was the big time. Unlike Stuyvesant it did not emphasize science. Rather its main thrust was the humanities. However, it did have very active and respectable chemistry, physics, and mathematics departments. In 1934, I would guess that perhaps 20 American universities carried on serious research efforts in the natural sciences. Perhaps another 20 supported token research efforts. There were then only three American journals of importance in physical chemistry, one of which, *The Journal of Chemical Physics*, had just begun publication in 1933, while another, *The Journal of Physical and Colloid Chemistry*, was considered an eccentric enterprise started by Professor Bancroft of Cornell. If we add to these the three or four major foreign journals, one could easily read all new significant articles in any given area of physical chemistry and still have time for the Sunday comics. Deciding to become a research chemist was very much like joining a rather small, select club.

The once miniature American chemical industry was just beginning to grow. With a huge domestic market and a very rich material base in petroleum, gas, coal, and iron, the stage had been set for a huge expansion. DuPont advertised "Better Living Through Chemistry." Vitamins had been recently discovered and plastics such as nylon were about to come into the market place. Despite the use of poison gas in World War I, chemicals did not have the ominous significance they were later to gain starting about 1950. For the neophyte chemist, chemistry was clearly a potentially hazardous profession, but that, if anything, gave it a romantic aura. The reality of chemistry classes, however, was something different.

After a somewhat neutral and repetitious first year my second year nearly ended my interest in chemistry. I have never found anything more excruciatingly dull than quantitative analysis as it was then practiced. A full two semesters were required. The intellectual content was provided in about half a semester. The remainder was a grinding drill, which even the instructor found boring.

Quantitative analysis is the bread and butter of chemistry, the necessary foundation for about any chemical procedure, but it is not the most stimulating way to spend a year. Only chemists who have experienced the tedium of counting swings on a beam balance and have spent two weeks calibrating weights can truly savor the sense of delivery brought about by automation and instrumental methods of analysis. It took great self-discipline to survive two semesters of "wet" analysis, and it was to be several years before I again had any pleasure in walking into a chemistry laboratory.

The interesting subjects during this period were in physics and mathematics. Newtonian mechanics, electricity, magnetism, probability theory, and theory of functions were challenging and exciting.

My introduction to research came unexpectedly enough in the second semester of quantitative analysis. Our laboratory assistant, James "Spike" Coles, suggested doing an original research problem, the determination of the bicarbonate content of sodium carbonate using a titration technique. It was easy enough to calculate that titration with a strong acid would produce a weak end-point at the pK of sodium bicarbonate, about 8.4, corresponding to the carbonate content, while further titration would yield a sharp end-point at pH 6, corresponding to the sum of both. After much experimenting with pH-sensitive dyes, I finally found one that gave reasonably sharp color changes at both pH's. Then began my first contact with the real world with the discovery that so-called solid sodium carbonate would if exposed to the laboratory atmosphere pick up CO_2 and moisture and perversely change its composition from day to day! A fairly elementary finding—but it took about six weeks of very frustrating efforts.

Louis P. Hammett was my advisor. I was but dimly aware of his growing reputation. He was an extremely shy person who was at first meeting rather taciturn. His colleagues many years later told me that it took two martinis to really make him relax. I found him kind and helpful if not loquacious. Columbia had a wonderful policy of waiving course requirements in any area upon consent of the faculty and the passing of a written examination in the course. He allowed me to take a special exam in physical chemistry, for which I studied in the Summer and passed in the Fall. That opened to me a huge menu of electives in my junior year which renewed my waning interest in chemistry. Very memorable was an encyclopedic course in Chemical Kinetics given by Victor K. La Mer. The idea of connecting time and chemistry in a quantitative form was new and stimulating and this course evoked an interest which has never ceased. Another course by Ray Crist on the new field of photochemistry gave further fuel to this interest in kinetics. It also for the first time made me aware of bond strengths, and the significance of energy in elementary reactions.

By my final year, I was entirely relaxed and enjoying the "intellectual" life. A $400 math prize in my junior year had made it possible for me to purchase my first automobile, an incredible luxury. I had finished all the required courses for my degree and could take any elective courses I wished. I took a few, notably Professor Harold Urey's courses in Quantum Chemistry and Thermodynamics and Professor Biot's course in Vector and Tensor Analysis. These and a very demanding Physical Chemistry laboratory were not enough to keep me from weekend and holiday drives outside of New York with my friends. I also had a job with the newly founded "Young Students Administration," organized by then-President Roosevelt to assist needy college students. It paid the magnificent wage of 50 cents an hour for 30 hours a month. That was enough to pay for most of my meals in the Columbia dormitory cafeteria. Originally assigned to dusting animal and human skeletons in the Schermerhorn Museum of Anthropology, I succeeded in getting Professor Urey to request my transfer to his laboratory, where I assisted in maintaining the one and only mass spectrometer.

My major job was to maintain the distilled water level in some 200 storage batteries used as a constant voltage source for his home-made magnetic sector instrument. Fortunately, the current drain was so small that the water level needed only occasional attention. When Urey found out I had taken courses in mechanics and electricity, he asked me to investigate the focusing quality of the magnetic field for different speed ions coming in at a variety of angles. I worked on this for awhile but soon decided it was not soluble in closed form. I didn't have then the background in the real essence of mathematical physics, namely, the fine art of approximation.

During most of my four years at Columbia, I had only a very vague idea of what I would do upon graduation. Teaching seemed the most attractive, but it clearly required a PhD at the University level. Jobs in the sciences were very few and, in the absence of any concrete ideas of how professional chemists spent their time, not very real. Reality started to intrude one Fall day in my senior year when Professor Hammett suggested that if I was interested in a PhD I should start writing to graduate schools. I asked about continuing at Columbia and he replied that it was best for me to go elsewhere; also, it was also against their policy to retain undergraduates. This produced a real shock! It had never seriously occurred to me that I would live anywhere else but New York City and the thought of a three or four year period away seemed like permanent exile. A follow-up audience with Professor Urey elicited the same advice with the additional suggestions that I write to Cal Tech, Princeton, and Harvard. I was fortunate enough to receive offers at each of them. Much soul searching and conversations with anyone who would listen finally led to my choosing Harvard, based on their fellowship offer that would give me the freedom to use my own time as I saw fit. Later that Spring, when I celebrated my graduation by driving with my mother, sister, and brother for a vacation and visit to some friends in California, I visited Cal Tech and got to know California. It was love at first sight. Although committed to Harvard, I made the decision there that when I finished my studies, I would try to come back to California. About six years later, in the most unexpected circumstances, I did.

Graduate school is a psychological quantum step removed from college. In college, I was busy learning, living, and observing. In graduate school, I was suddenly aware of doing things—research, and also preparing for a professional career of some kind. Columbia represented an existence, exciting and stimulating, since at that age so many experiences are new. It carried with it the feeling that it would go on forever. Graduate school and Cambridge, while no less interesting, were clearly transient states. Life had suddenly become more serious.

My introduction to Harvard started dramatically. My cousin, who was entering law school, and I set out on a sunny day for the five hour drive. It proved to take two days. The day before we started New England had its first hurricane of the century and we spent a night stranded at the Connecticut-Massachusetts border.

Although he doesn't truly realize it at the time, the most significant decision a graduate student will make is to choose his research director. In my own case it was very simple. None of the research being done by the physical chemists on the faculty then seemed terribly interesting to me. George B. Kistiakowsky had the practice of giving parties about once a month during the year for his graduate students and post-doctorates. They were interesting parties with good food to eat, a variety of interesting guests, and real liquor. My choice was made very easy. Choosing a research problem was much more difficult.

Kisty, as we called him, was a great lecturer, a very easygoing and gregarious person and an extraordinarily gifted experimentalist. He was a skilled machinist and a talented glass blower able to work with quartz, pyrex, soft glass or any grade in between. He had learned from Max Bodenstein, in Berlin, all the techniques for doing gas phase kinetics. I learned to construct the famous greaseless quartz helical manometers and the glass bellows–activated, greaseless, Bodenstein glass stopcocks. Kisty required his students to construct all their own equipment, except for these last two. After I completed my thesis, my last act in the laboratory was to destroy my vacuum system, first salvaging the mercury diffusion pump, the ubiquitous McCloud Gauge, and the several stopcocks. The next student came to a bare wooden frame!

Although I had a general interest in kinetics from my experiences at Columbia, I had little knowledge of the important or interesting problems. During the first semester, a course by Paul Bartlett, then a young Assistant Professor in physical organic chemistry, inspired my interest in mechanism with examples of how one could draw conclusions about mechanism from kinetic studies. Free radicals were coming into prominence at about that time, and I was intrigued by the Paneth experiments on pyrolysis using metallic mirrors as radical scavengers. F. O. Rice and K. Herzfeld had just published their classic paper on chain pyrolysis of organic molecules, and F. O. Rice and K. K. Rice had just written a very readable monograph, *The Aliphatic Free Radicals*, which summarized the known data. Although many chemists questioned the existence of free radicals in any given pyrolysis or photochemical study, the evidence to me seemed overwhelming for their existence. By this stage, I had become an avid reader, both of monographs and research articles. An article by Pearson and Purcell on biradicals had a special appeal. How to prepare these exotic species, then demonstrate their existence, and finally measure some of their properties seemed a real challenge.

The photochemical decomposition of acetone and higher ketones had already been studied. This was described in great detail by another just-published classic book by W. A. Noyes, Jr. and P. Leighton on *Photochemistry*. Absorption of a photon more energetic than 90 kcal appeared sufficient to cleave the carbon–carbon bond in acetone or ketones generally and produced two free radicals. What if the ketone were a cyclic molecule like cyclopentanone? Would photochemical excitation at 2900 Å produce a biradical with free valences at the two ends of the chain? W. G. R. Norrish and a student had performed such studies in 1938 and found good evidence for CO production from cyclopentanone, cyclohexanone, cycloheptanone, and cyclobutanone. This seemed to me unequivocal evidence for ring cleavage and biradical production. Thus, in the case of cyclopentanone a possible reaction sequence might be:

$$\begin{array}{l}
\underset{\displaystyle CH_2-CH_2}{\overset{\displaystyle CH_2-CH_2}{\Big\rangle}} C{=}O \;\xrightarrow{\;h\nu\;}\; \cdot\overset{\displaystyle O}{\overset{\|}{C}}-CH_2-CH_2-CH_2\dot{C}H_2 \\[2mm]
\hspace{5cm}\downarrow \\
\hspace{3cm} CO + \dot{C}H_2-CH_2-CH_2-\dot{C}H_2 \\
\hspace{5cm}\downarrow \\
\hspace{3cm} CH_2{=}CH-CH_2-CH_3.
\end{array}$$

It would appear as though the products, butene-1 and CO, must arise from a biradical. Similar products were found with all cyclic ketones. The sophisticated reader needs to understand that many chemists at the time believed that the many steps outlined above could all take place in one concerted act.

Not wanting to repeat someone's experiment, I spent a semester trying to synthesize 1,2-diazacyclopentane, better known as pyrazoline. The ring cleavage should produce N_2 and a trimethylene bi-radical.

Extrapolating from Norrish's results, one might suppose that the biradical might end up as propylene or dimerize to cyclohexane. In any case, the reaction should be very clean and free of side products, one hopes.

It proved to be a very frustrating experience, smelly and hazardous as well. Acrolein (CH_2=CHCHO), an extremely potent lachrymator, and hydrazine (N_2H_4), a very corrosive and metastable substance, were the starting reagents. There was no trick to following the published recipe, but several months' work went into futile purifications before I realized that pyrazoline was in tuatomeric equilibrium with its ene-imine:

The photochemistry of the tautomer was totally unrelated to that of pyrazoline. So back to the cyclic ketones! Kisty was kind enough to hire an organic student during the Summer to prepare cyclobutanone. It was a nearly disastrous experience. The starting materials were the very toxic and explosively unstable diazomethane (CH_2N_2) and the very labile ketene. One major explosion nearly lost us our organic student. He fortunately suffered only superficial injuries and was finally able to provide the magic ketone.

By the end of my first year, I was beginning to enjoy graduate life. I finished all my formal course work, was well begun in research, and had become a reasonably skillful glassblower, mostly by virtue of having broken several pieces of equipment in the lab and then having had to repair them. It had been necessary for me to do tutoring to supplement my fellowship, and Kisty generously offered to hire me to do some research for him over the Summer while he made a trip to Europe. It was my first real job and paid the unbelievable salary of $40 per week. That came to about 70 cents an hour, but it added up to nearly $180 a month (no withholding!) More importantly, three months would guarantee me full-time eating the following year.

If the New England hurricane of 1938 had been dramatic, it was minor compared to what was happening in Europe. The Chamberlain–Hitler pact in 1938 and the subsequent invasion of Czechoslovakia was a grim prelude to World War II. We had already witnessed the Italian invasion of Ethiopia in 1935, and the prolonged and bloody civil war in Spain, which only ended in the Spring of 1939. Many of my friends had been active in the loyalist cause and two had joined the famous Lincoln Brigade. Kisty, when he left, had felt that this might be the last chance to visit Europe. He proved to be correct. He managed to return on the last boat to sail from England just after the war was declared in August 1939. To those of us who, like myself, had been sympathetic to the communists and the Soviet Union, the Stalin–Hitler Pact was both a political bombshell and a profound disillusionment. The rapid dismemberment of Poland, followed only months later by the invasion of Finland by the Soviets, was a last cynical disenchantment.

My generation had been brought up on the profound disillusionment that followed World War I. The First World War had appeared a totally pointless war, and I had never understood, until quite recently from reading Barbara Tuchman's *The Zimmerman Telegram*, just how the United States had become involved.

The aftershock from the Stalin–Hitler pact was not long in coming. The invasion by Germany of Holland and Belgium and then, in May 1940, the Fall of France provided a watershed in American thinking. Suddenly, Hitler and his prophesised 1000-year Reich started to look very real and the European scene very grim. I had been very anti-war and active in the large anti-war student group on campus during 1939–1940. The events in Europe began to look like an inescapable tragedy approaching.

Roosevelt had immediately cast his lot with England and the now fallen France and many Harvard faculty had already responded to the Washington initiative to become active in what might follow. From the day after his return until the Fall of France in 1940, Kisty had begun to take part in some government-sponsored research projects that were just being organized. He spent about half his time in Washington during that year and much more after Pearl Harbor. He was kind and thoughtful enough, even though we did not quite agree politically, to arrange for me to be hired for the Summer of 1940 at the famous General Electric labs at Schenectady. As he put it, "They owe me one!" It was to be a marvellous experience.

The General Electric labs were predominantly a "physicist's" laboratory, presided over scientifically by Irving Langmuir, who was perhaps America's first chemical physicist. He was then still pursuing with Katherine Blodgett an active program on surface films, for which in 1932 he had been awarded the Nobel Prize in chemistry. He had earlier in his career endeared himself to General Electric by his work on tungsten filaments, atomic hydrogen (and welding), and vacuum tube getters. Chemistry was distinctly a minor actor in the General Electric script, but it had a very prestigious staff. Abe (Lincoln) Marshall headed the chemistry lab. Ray Fuoss, Lewis Tonks, and Saul Dushman were senior staff members along with Gene Liebhafsky. Eugene Rochow was just at the beginning of his famous researches on silicon chemistry, and the silicone polymers were then in the experimental stages. Marshall assigned me to a laboratory and a project related to work on alkyl polymer resins. A few weeks after I started, a summer visitor, John Kirkwood, agreed with some misgivings to share the office space located in my small laboratory. He never quite appreciated the noise of the vacuum pumps I used or the way in which my equipment progressively cluttered the available space.

Weekly seminars in any area of interest to the laboratory were presided over by Langmuir, who seemed to have an uncanny insight into whatever topic was being presented. My own seminar on photochemistry was explored by him in enervating detail. Although I succeeded in answering most of his questions, I breathed a sigh of relief when he had finished, and I had no difficulty falling asleep early that night, despite my usual custom of reading or writing until about 1:00 A.M.

The experience at General Electric was an important and interesting one, giving me my first insight into how industrial laboratories operated and how they selected their problems. Coolidge (he of the x-ray tube) was nominal head of the lab and an extremely gracious man. Although in principle all work was project-related, Fuoss was working on electrolyte theory, Tonks was exploring the statistical mechanics of two-dimensional gases, and Dushman was experimenting with the semiconductor properties of lead sulfide crystals. What I found fascinating in talking to them was their ready ability to relate what seemed like abstract research to the practical problems of the company. My experience at General Electric was also to be a practical introduction to big business and bureaucracy.

Once my problem had been selected and mutually agreed upon, I outlined my approach and the equipment to be used. I set about building the equipment myself. It took me about three weeks to finish it, whereupon a visit from the head electrician produced the incredible (to me) instruction that it didn't meet code specifications and would have to be redone by their department. No pleading or cajoling could alter that decision, and for the next two weeks I watched a perfectly functioning apparatus disassembled and reconstructed in ponderous B-X cable and conduit. It took two weeks of precious time because the labs opened at 8:00 A.M. and closed at 4:00 P.M.! Executive orders were required to stay late and weekend work was unheard of. It took me some time to reconcile myself to this totally absurd schedule. Research is not something like bread that is cut in so many slices and then consumed. Few things are more frustrating than to begin an experiment you know will take at least four hours only to realize that at 2:00 P.M. it is two hours too late. How the famous laboratories of Edison and Steinmetz reached this state of organization was never made clear to me.

During my last month, a photochemical project, the first in the laboratory, was begun. It was a project on the photochemistry of uranium salts. I didn't know it, but the nuclear age had begun.

Both chemistry and physics were usually represented in my high school and college courses in a very abstract and formal manner. Industrial processes were dragged in as an example of the application of chemistry or physics to some practical goals, but the real motivations were never explored. Outside of the alchemists' dreams of changing lead into gold—which always seemed to me a total waste of time—there appeared very little connection between science and everyday life. Scientists appeared to be a brand of medieval monks exploring some self-appointed tasks in natural phenomena. The lively give and take between hypothesis and laboratory, the scaffolding of science, was hygienically removed, leaving what seemed like a most improbable structure—the final result. The infrastructure of scientific research, how research was paid for or motivated, was also totally missing. At Columbia and Harvard one was aware of the Rockefeller Foundation and a few benevolent companies or individuals who made gifts. At Harvard, the unexplained benevolence of the Mallenkrodt Chemical Company made available a fund from which every graduate student was allowed $50 a semester for chemicals or equipment— a handsome sum actually, which was seldom exceeded by physical chemists. The university also made a research allowance (about $2000 a year) to each professor doing research—a very different system from our current federally supported research.

Years later when I started on the faculty of the University of Southern California, it was understood that the school would make funds available to me for my research, at a modest level of course. Grants didn't exist then, but I was asked to write a short summary of the work I wanted to do and include some statement as to why it was of importance. I never learned who read these summaries or what response they triggered.

When I returned to Harvard in September for my third year I had decided to devote myself to making it my final year. General Electric glass and machine shops had been very helpful. In exchange for teaching their glassblowers how to make the Bodenstein helical quartz manometers they built me a beautiful Blacet–Heldman apparatus for doing micro-analysis of gases. It was a technique of the time for quantitatively analyzing as little as 0.1 cc STP of a gas (about 4 micromole) and was to speed my research enormously. The military draft approved in the Summer of 1940 began in October. My best friend at the time, Milton Soffer, an organic chemist, had the first number picked by lot. Ironically he had been President of the very large Harvard Anti-War Society. He became very upset at his "luck" in the drawing; his state of mind was not helped by the group of us listening to the radio, laughing uproariously at his being the first chosen. An hour later the second number was announced. It was mine! I had been secretary of the same society. I have always had a somewhat fatalistic approach to such events. In any case nothing imminent was about to occur in October 1940. We were not at war, and students were given automatic deferments. However, the sudden turn of events reinforced my decision to graduate quickly, inaugurating a prolonged period of what was to be the hardest, most sustained working period in which I have ever been involved.

For the next seven months, I worked in the laboratory six days a week, usually for ten to 12 hours. Each days' work was outlined the night before and the end of the day was determined by the experiment at hand. By this time, the work had become routine, punctuated only by accidents and breakage of equipment. Accidents, I quickly observed, came at night generally, when I was tired and trying to hurry. It took me a long time to realize that at some point in a working day the probability of a destructive event began to exceed the probability of doing something useful. Fortunately, I survived the learning, but there were more than a few scares along the way.

I used large quantities of liquid refrigerants in my gas handling and analyses, particularly dry ice/acetone slurries, which I kept in a large Dewar container. A similar Dewar was used for shaved ice. One day I reached into what I thought was the ice Dewar only to find I had immersed my hand in dry ice/acetone at $-80°C$. I withdrew my numbed hand from the Dewar, staring at it with horror. I had no sensation in my fingers, which were still curled about some dry ice. The skin had turned yellow, and I was sure I had lost all my fingers. Instinctively I plunged my hand into my mouth to warm it and then as sensation slowly returned, put it under the faucet. Within about ten minutes everything was back to normal but my fright lasted for some time. Thereafter, I was very cautious about where I put my hands.

On another occasion, in the late evening at about 1:00 A.M. after an unusually long day, I managed to break part of my vacuum line. I stopped the experiment and began the laborious reconstruction using the lab oxygen torch for the glassblowing. In the middle of the repair I hung it on the vacuum rack, the flame still going strong, not realizing that it was pointing at my lab notebook on the other side of the metal frame. Suddenly, I smelled smoke and glanced up to observe, in awed disbelief, large flames coming from my lab notebook, containing all my results!—six months of arduously won information. I didn't think. I just picked it up and smothered the flames with my arms and lab coat. Again, I was lucky. No serious damage was done to my precious data. The book cover and edges were badly charred but the contents had been spared. I only had minor burns on my fingers, but I can still recall the anxious moment after as I scanned through the precious pages.

I succeeded in completing my studies on cyclohexanone and cyclopentanone photolysis with very gratifying results. Contrary to the earlier published reports, the major products were cyclopentane + CO from the former and cyclobutane, ethylene + CO from the latter. The earlier workers had misidentified the hydrocarbons. It looked like solid evidence for biradicals with a sequence:

In the case of cyclopentanone, the tetramethylene biradical had two competing paths that appeared to be temperature independent and hence with little or no activation energy:

Kisty induced the Cabot Corporation to patent the method as a cheap praparation of the very rare cyclobutane. He thought it might be a useful anesthetic. It did become a standard research synthesis but found no greater use.

While doing cyclobutanone I seemed to have found a similar result with a gas product whose molecular weight was 42, as expected for cyclopropane or propylene, and whose boiling point was also about right, $-48°C$. However, attempts at analysis were totally disconcerting. Gas volumes were measured by isolating a sample in a 2 mm capillary between mercury threads. This gas burette had worked well with everything else I had examined, but with the presumed cyclopropane/propylene the gas volume kept shrinking in the capillary. After three days I gave up and went to the movies, convinced I was going to spend another year at Harvard. I didn't come back to the laboratory over the weekend. Monday I returned to consider my plans.

It was my custom to save all condensable gas samples by freezing the samples in small capillaries and sealing them off under vacuum. This morning, I looked at my collected samples, gathered in a large beaker, and saw something totally unexpected; the usual colorless hydrocarbons in three of them were a brilliant, burgundy red. Eureka! What I thought was cyclopropane was something else, ketene as it turned out. Ketene is notoriously reactive, and it didn't take long in the library to confirm a red dimer. Cyclobutanone, like cyclopentanone under photolysis, gave two sets of products:

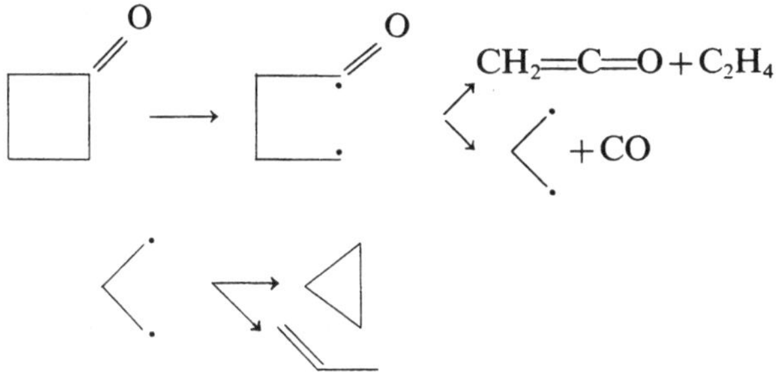

$$CH_2{=}C{=}O + C_2H_4$$
$$+ CO$$

I completed my German language exam a week before the December deadline required for June graduation. My thesis was accepted three days before its April deadline and my final oral exam two days before the May deadline. I didn't wait for June. In May 1941, I left Harvard, I thought, forever. Four months later I returned.

During my race to complete the PhD requirements I gave little thought to a job. In April 1941, I had become qualified, as a result of my employment at General Electric the summer before, for unemployment insurance, one of the reforms introduced under Roosevelt's New Deal. The \$15.00 per week I was able to receive made the last month at Harvard and the subsequent transition to civil life much easier once I returned to my beloved New York City.

I set about in earnest to write to various schools, to consult the want ads, and to visit the few employment agencies in downtown Manhattan that specialized in scientists. Jobs were still very scarce. The "Great Depression" had not yet abated and a PhD in chemistry was of questionable significance in the New York job market. Finally, just as my last thirteenth weekly unemployment check was spent I received a good offer doing classified, war-related work at the Rockefeller Institute in midtown Manhattan under the direction of a very distinguished physical chemist, Duncan MacInnes. My acceptance was made with very mixed feelings. Although I was not a pacifist, the words "War" and "War Work" evoked very negative reactions in me. The idea of spending my efforts in creating or improving destructive instruments was not part of my self-image. Even less appealing, our assignment was to improve the technique of dispersion of poison gases, nothing to proudly discuss with friends or family.

Rockefeller Institute was already a famous place. Wendell Stanley had done his classical work on viruses there, and the scientific staff were very interesting. After about six more weeks of a relatively pleasant readjustment to New York life my equilibrium was suddenly shattered by an early morning telegram from MacInnes saying that my services were no longer required. A special delivery letter followed, containing my paycheck and unexpectedly an equal amount from MacInnes. No reasons for the sudden decision were given, and MacInnes refused to talk to me on the telephone. Reflecting as calmly as I could on the circumstances, I concluded that my "political" activities were responsible. All war work required security clearance and evidently I had not been cleared.

Let me explain. After the Stalin–Hitler pact I had become disillusioned, critical of my communist-leaning friends, but also more interested in politics. I met and became friendly with some of the Socialist and Trotskyite students on campus. My anti-war activities brought me in even closer contact with them, and an off-campus incident in the Spring of 1940 in which I had been active in organizing a Cambridge protest against a particularly corrupt, local politician with the picturesque name of "Ratty" Hamilton brought me to the attention of the local police. Boston politics in 1940 were only a bit more covert than that of the notorious "Boss Tweed" of old Manhattan. Once returned to New York, I expanded my circle of Trotskyite friends. All of this I decided must not have passed FBI muster. Swallowing much of my by now battered pride I wrote to Kistiakowsky, asking if I could do some post-doctoral work for a while. He was extremely kind. He evidently had learned of my situation and offered a job at Harvard on the condition that I not engage in any political activity. He was by now spending about two thirds of his time in Washington DC on war work and he felt somewhat ambiguous about my political sentiments.

It was a moment of great soul-searching from which emerged the realization that my scientific training and underriding skepticism were more important to me than transient commitments to political opinions. I accepted his offer. I have never regretted the choice.

The post-doctoral year was a tranquil contrast to the preceding graduate student existence. I enjoyed the great luxury of a regular salary, time to read and attend seminars, and good laboratory facilities. Although Kisty had made me the offer, Professor George Shannon Forbes was my sponsor. Forbes was the very epitome of the stereotypical, Victorian, university Professor. From his, even by 1940 standards, somewhat archaic, celluloid collar and bow tie to his trimmed moustache, polished shoes, and very formal lecture style, he was the very soul of decorum. He had been one of the American pioneers in physical photochemistry and an extremely careful and meticulous experimenter. He was exceptionally shy in conversation, but one of the kindest men I have ever met. His hobby was photography, and he would spend each summer out west in either Colorado or California taking black and white pictures of the glorious views, mountains, and forests. One of these would be carefully selected, enlarged, mounted on a white mat, a small calendar attached and a copy sent to each of his students for the New Year. I received one faithfully every year thereafter until a year before his death at the age of 93. The last few years, he accompanied the calendars with apologies for the fact that the pictures were old since his health prevented him from taking hikes.

I succeeded in completing what I considered to be a thorough study of the photolysis of acetone at 2560 Å and proved that every photon absorbed produced CO + two methyl radicals. However, the effort invested in these modest conclusions was so great that I decided then and there that if I pursued research it would not be in photochemistry.

After the attack on Pearl Harbor in December 1941, everything suddenly took on a different significance. I received a letter from one of the schools I had written to the year earlier. New York City College offered me a job as instructor in chemistry at the magnificent salary of $2300 for nine months. It was a dream come true!

The reality was to be less than the dream. CCNY, as it was known, had the pick of the brightest students in the New York area. I enjoyed teaching very much. However, the students proved much less sanguine. Despite their intelligence and grades they were not as eager to learn all the chemistry I wanted to teach them. I suffered the usual disillusion of the beginning teacher, but it was quickly absorbed. Less attractive were the very poorly equipped and crowded laboratories and the one half desk that I shared in shifts with another instructor. Most disillusioning was the rapid realization that I could not live in New York on my salary. During Christmas and Easter I was reduced to taking a job in a store selling shoes. So much for academic "free" time. Before the second semester had begun, I was back interviewing for jobs. I was offered one by the Kellex Corporation of New Jersey. It was a subsidiary of the H. W. Kellog Company, a well known chemical engineering firm. They had been newly formed to perform special war work for the government. The salary was almost triple what I was making, and the people I spoke to, Manson Benedict and Charles Johnson, were scientifically impressive. I explained my history with war work. They felt it wouldn't matter and a month later when I next heard from them I was told that I had passed security clearance. I accepted the job with great enthusiasm, and as soon as my last spring lecture was over in early May I began my new work on what was to become better known as the Manhattan Project.

My announcement to my colleagues at CCNY that I would leave at the end of the semester was greeted first with stark disbelief and then with incredulity at my naive behavior. "No one had ever resigned from the department. Some people had waited 14 years at the Assistant Professor rank in just the hope of gaining tenure! Why would I throw away so promising a career?" All of this helpful advice gave further support to the wisdom of leaving.

The discovery of atomic fission had appeared in the scientific literature without inspiring much publicity, and I was but dimly aware of it. When I learned what we were engaged in and the magnitude of the energy available in an atomic bomb I was stunned. It was a difficult time. It took several days to absorb the facts and the potential implications. For many months I softened the impact by trying to see how it might work and decided that it was an impossible effort. I had no knowledge of neutron cross-sections and no idea of how, eventually, implosions would be used to achieve critical mass, so my skepticism seemed well founded. It was to be reinforced in the months ahead as our research proceeded.

The Kellex Corporation was responsible for the construction of the Gas Diffusion Plant to be built at Oak Ridge. U^{235} was to be separated from U^{238} by diffusion of UF_6 (a gas at room temperature) through a thin, porous nickel membrane. The lighter molecules of $U^{235}F_6$ had a higher average velocity by about 0.5% than $U^{238}F_6$, hence at each stage its concentration would be enriched by this amount. To go from a natural abundance of U^{235} of about 0.7% in uranium to, let us say, 25% would require about 7000 stages of diffusion if each one worked perfectly. Even small leaks in the barriers in the early stages would lower the efficiency to the point where the system would not operate. Of great concern was the fact that UF_6 was potentially reactive with nickel. Such a reaction would plug the pores with non-volatile UF_4 and NiF_2 and also close the system down. Since the system acted in a cascade fashion, no stage could begin until the earlier stages had begun to operate and had reached steady state. A failure in a few of the many units could shut the entire system down.

My research problem was to study the stability of the barriers to corrosion and plugging. At the time I began work, in early 1943, the barriers had not been developed, so only experimental samples were available to us. When Kellex had begun work on the plant design only 18 months earlier, Benedict and Johnson had been assigned to do whatever research might be needed! Within six months they concluded that a large effort was needed, and within the incredibly short time of six months, a swamp in New Jersey had been drained and a 20,000 square foot laboratory was built. This was at the height of the war effort when materials were tightly rationed and trained manpower had become scarce. Within a month after I was hired, I was promoted to group leader and our group was expanded from six to 24 people working three shifts around the clock.

For three months my two other group leaders and I took turns on the different shifts. I found the graveyard shift, 12:00 midnight to 8:00 A.M., most difficult to adjust to but it was an educational experience in a different style of living. It was also a period of learning for me—learning an engineering approach to scientific problems, learning to work with metal apparatus instead of my accustomed glass, and finally learning to direct research.

Because of the urgency of the project we had the highest AAA-1 priority in buying scarce materials, and every effort was made to facilitate our work. When technicians were unavailable, drafted personnel from the Army with scientific training were reassigned to us. However, the Army bureaucracy still exercised visible control. All capital equipment in excess of $100 had to be purchased by competitive bidding. Our solution to this was to relabel all items "supplies" and purchase them with petty cash by distributing the cost among as many vouchers as were needed to keep the total under $100 per item. When our petty cash account went suddenly from about $200 a month to about $10,000 we had a visit from the Army auditors and our purchasing scheme was terminated. We replaced it by one in which we purchased capital items in pieces, each a part of the whole, for less than $100.

At one point General Groves and the scientific directors, including Professor Urey, were to visit the labs. Incredibly, all work was brought to a halt while we spent two days cleaning up the lab for the visit. After a half-day show our visitors left and we plunged back into our usual disorderly routine.

It was an exciting period because of the urgency of our work and the understanding that we were also possibly in a race with the Nazis. It also carried with it many of the frustrations of mission-oriented research. A number of very intriguing observations we made seemed to cry out scientifically for further research and understanding. Because they were peripheral to our goals they had to be ignored.

There were the usual mishaps, accidents, and one very dramatic fluorine fire involving a large tank containing F_2 gas at 600 pounds pressure. Within seconds the lab was crossed by an intense 20-foot blue flame issuing from the tank nozzle. It was hard to understand how such a flame of F_2 in air could be maintained with no evident fuel but the moisture in the air. The entire lab was evacuated in less than a minute until we realized that the ventilation system was making the air outside heavy with HF and much more dangerous than the labs and offices. We quickly returned despite warnings from the firemen.

We made one major contribution, as usual in my experience, quite by accident. One day, opening up one of the copper lines used in our experiments for handling UF_6 and F_2 at temperatures up to 500°C, we discovered it to be bright and uncorroded by the usual green CuF_2. Following up on its history led to a method for passivating metals like Ni and Cu against corrosion by F_2 and UF_6. It is one of the few findings made on the Manhattan Project that has still, inexplicably, not been declassified. The discovery made possible a huge cost savings in the diffusion plant.

Near the end of the year, we had successfully completed our assigned work, and it was announced that the labs were to be moved to Oak Ridge. A few months earlier, in a response to an advertisement in *Chemical and Engineering News*, I had received a letter from Anton Burg at the University of Southern California (USC) in Los Angeles. He offered me a job as Assistant Professor at the salary of $2300 per year. I turned it down quickly, explaining my difficulties with this princely sum in New York. Three months later a return letter brought an offer of $2600, "In Los Angeles," I was assured, "an excellent salary." Discussions with Kellex elicited the counter-proposal to go either to Oak Ridge—I couldn't find it on the map and would go only if carried—or to Site X in New Mexico. Site X was somewhere in the mountains near Albuquerque. It sounded even less appealing. But I didn't appear to have much choice. Fate intervened most unexpectedly. I have never done well in New York winters and the Winter of 1943 had been most severe. I had come down with severe sinus infections and a chronic cold. Visits to the doctor brought the unhelpful suggestion that I move to a warmer, drier climate. Now armed with a letter from the doctor, I was able to obtain a release from the Manhattan Project. So near the end of 1943, I piled all my clothing, books, and furniture in the back of my once new 1939 Plymouth and set out again to California.

Los Angeles was beautiful, gleaming white in the brilliant Fall sun, composed of mainly single-story homes and low office buildings, small in population and vast in area. In strong contrast to the East and Midwest, it was spanking clean and surrounded by high mountains just beginning to be covered with snow. Smog hadn't yet made its appearance, and from the top of any building it seemed you could see forever.

USC was a private and almost rural University then, small in size and known mainly for its football team and professional schools. A former classmate from Harvard, Ron Brown, had just started as assistant professor in organic chemistry. UCLA, its neighbor, was the recently formed southern branch of the state university and was still struggling to establish itself. Two UCLA faculty members I had known while at Harvard, Tom Jacobs and Saul Winstein, were just starting their research programs. Best known to me by reputation was Francis Blacet, well recognized for his researches on photochemistry. Cal Tech was much better known. Linus Pauling had already become one of the dominant figures in American physical chemistry, and Don Yost and R. Badger were figures on the national chemical scene.

Although I had been hired (at the prompting of four younger, ambitious members of the six-man chemistry department) for my research promise, a 12-hour assignment in teaching relegated that to a hope more than a reality. In the winter semester, still inspired by experiences on the Manhattan Project, I organized for the evening program what I believed may have been the first course in Instrumental Analysis. In attendance were scientists from the local industry ranging in age from 40 to 78. They were a great class, motivated, knowledgeable, and attentive. My college program was largely freshman chemistry, two sessions each of about 200–250 students. They also were well-motivated and attentive, but not nearly so knowledgeable. Within six months, I was moved into war work once again. A government program was directed by Anton Burg, chairman of the chemistry department and on his way to becoming world famous for his research on boron chemistry. It was supported by Division 10 of the National Defense Research Council and was engaged in an intense effort to find a chemical stabilizer for cyanogen chloride (CNCl), a newly explored poison gas, discovered by the Germans, but not known to the Japanese. It had "great" lethal properties and could pass through the Japanese poison gas cannisters. Enormous quantities of 500 and 1000 pound bombs of the liquid were stockpiled in the Pacific Islands and in the West in preparation for the invasion of Japan to begin in 1945. It had the unfortunate property of polymerizing very exothermically and explosively to a dense plastic (melamine) or to an aromatic ring trimer.

This was for me a somewhat unwelcome digression and not exactly in my strongest areas of research. But it was again a useful learning period. I picked up some very ingenious techniques from Burg on the handling of sensitive, volatile materials in greaseless systems. One of them, using liquid propane as a refrigerant in gas separations, was nearly disastrous. Liquid propane was obtained from a large tank by allowing the gas to bubble into a Dewar of dry ice. As the gas liquified the dry ice evaporated. In about 15 minutes one could obtain 100 cc of liquid propane. While waiting for this to happen one day I stopped to make some calculations on a lab table in the middle of the room. Suddenly I became aware of a hissing sound and looking up saw a sheet of thin yellow flame approaching me. It stretched from the floor to the ceiling and came from near the lab door. I had just time to see it approach. It bifurcated in front of me, one sheet crossing the sheet of yellow paper on which I had been writing and burning a neat line across it. The other sheet went around my back. they joined on the other side and continued toward the casement window where the tank of propane was standing. Finally there was a "VROOM" and a roar. The door was blown shut, the window was blown open and several large bushes outside were blown absolutely flat on the ground. The dry ice Dewar was converted to tiny fragments, but strangely enough no other damage occurred. Some of my colleagues and students came in to see what had happened but I, standing stunned but unharmed, was relatively non-committal. A lady passing by in the street outside complained to the university authorities and Anton Burg was questioned about the damaged foliage. I never told him of the explosion. I was lucky and relieved to come out in one piece.

Shortly after April 1945, and the end of the War in Europe, work on our CNCl project was intensified. Francis Blacet at UCLA was engaged in a parallel program. Two students in Army uniform were helping him at the time, Jim Pitts and Jack Calvert. After the war they completed their PhDs with Blacet. We had pleasant but intermittent contact with each other. With the shut-down of the European theatre of operations, Professor Morris Kharasch of the University of Chicago was assigned to our program, and within three months the problem was dramatically solved.

The Army had imposed some interesting boundary solutions on our problem. Because of the scale of the problem they decided that the CNCl bombs had to be stabilized in situ by means of some additive. Because an Army sergeant and privates would effect the remedy no solid additives could be used. The reason was that in opening the containers, a large screw closure had to be removed and then tightly resealed. It was unlikely, the Army decided, that available personnel could be relied upon not to get solid granules or powder into the exposed threads, thus jeopardizing the tightness of the seal when the screw was replaced. A loose seal was as hazardous as the potential explosions, ergo no solid additives.

Kharash in typical fashion did some quick and dirty experiments, sufficient to convince him he was dealing with a general acid, metal ion–catalyzed reaction. Chelating agents such as phosphates were shown to suppress this, so a good dose of solid sodium metaphosphate was the answer. There was no scientific problem remaining—only a personnel problem. The case was summarily closed and the ensuing discussion with Army brass led to a change in guidelines! It was frustrating to Burg, the thorough scientist, but illuminating to me.

World War II had catalysed transformations in American life that were little appreciated at the time. The sudden release to civilian status of millions of enlisted men, together with the GI Bill giving them educational benefits, sent a crashing wave of mature, motivated, hard-working students into the universities. It was a golden age in higher education. The GIs not only influenced the perceptions of the faculty as to how well students might perform, they also set a standard for their younger peers. This was to continue with the Korean War in 1949–1952, until almost 1960. For research scientists in physics and chemistry it was a golden age in a more material sense. For the first time, under the stimulus of the Office of Naval Research, government funds were made available to university professors. Summer employment could be paid and stipends made available for graduate students and post-doctoral research helpers. A new world had started! Within 20 years the US had changed from a minor to the major center of scientific research on the world scene. For me, 1945 started a process of examining for the first time what I wanted to do and how I might do it.

An assistant professor starting at a research university has five years in which to establish a research program and publish significantly. There is a very strong compulsion to continue work along the lines of a just completed post-doctorate research or the earlier PhD dissertation. The cost-benefit analysis of continuing research on photochemistry in 1946 was for me very negative. Too much input, very little output. Theoretical work, particularly in statistical mechanics, had a strong appeal for me. On the other hand, if I expected to work with students I needed an experimental program. Biochemistry seemed a relatively new field where physical chemistry might provide useful insights. A number of students expressed an interest in understanding catalysis and surface phenomena. Thus, I began a period of reading, learning, and reorientation. Kinetics was abandoned and I began some work on the theory of liquids and surfaces.

This kind of change in interest was not uncommon at the time. I had a number of current examples to inspire me. Linus Pauling at neighboring Cal Tech, after establishing a brilliant reputation in quantum chemistry and in x-ray and electron diffraction studies of solids, was now wholly involved in studying protein structure. John Kirkwood, just newly at Cal Tech, who had done such elegant work in statistical mechanics, had shifted to studying the theory of electrophoresis of protein molecules.

This was the prelude to what was to turn out to be for me a ten-year program on the structure of proteins explored by examination of their surface properties and their adsorption of polar and nonpolar gases. It was to be a very interesting program for me. In a few years, I was able to attract generous support from the National Institute of Health. Ironically, I was eventually forced to abandon it when it proved increasingly difficult for me to interest physical chemistry students in biochemical problems.

Another new research program began quite accidentally. During one of my lectures in freshman chemistry I had explained to the class how small polar water molecules could interact strongly enough with the ions in solid ionic salts to overcome the strong lattice forces completely and dissolve them. After the lecture one of the students approached me with the very innocent question, "Why doesn't steam dissolve salts?" I wrestled with that for some time after he had left, apparently satisfied with my facile answer, "There aren't enough H_2O molecules in steam." Finally, I decided that at high enough densities steam must be capable of dissolving salts. Some work in the library showed that no real information was available. A few engineering studies on steam turbines showed that, indeed, salt fouling of turbine blades could be a significant problem. In discussing the question further with a colleague, Charles Copeland, we conceived a simple experiment to resolve the question. We built a heavy-wall, stainless cell fitted with an auto spark plug. We filled it about one third full of a 0.1 Molar NaCl solution, tightened the spark plug, and breathlessly watched for a current flow as we heated it in an oven to the critical temperature of water, $374°C$. At about $340°C$ we saw a small current and by $374°C$ it was much larger! A grant from the Research Corporation on the basis of these observations launched a more controlled study of supercritical salt solutions. Subsequent support by the Office of Naval Research made possible a fascinating study of supercritical solutions, which we extended to alcohols, SO_2 and NH_3. It was a nice field to work in, since we were pioneers and for many years the sole workers.

What was to be my last excursion into politics, in this case, very local, school politics, came just after the war ended in 1945. USC was then a very small school with about 6000 students and a faculty of about 350. Within a few months after arriving I had met most of my fellow faculty. The faculty lunch room had a very large round table where up to 16 at a time could be served. It was a real public forum. Very quickly I became aware that my salary was one of the highest in the school. Salaries hadn't changed since 1939 although the cost of living had slowly climbed during the war and then very rapidly after. Dissatisfaction with salaries was widespread and vocal. A Fall meeting called by President Rufus von Kleinschmidt to greet incoming faculty and boost morale turned into a fiasco when one of the older professors, Fred Weirsing, disappointed by what had turned into an empty pep talk, stood up and asked why, in the face of mounting income from all the new students, no changes had been made in salaries. Spontaneous applause from all sides suddenly made the mood very tense. V.K., as he was popularly known, a tall, stately, fatherly figure with a flowing white mane of hair, tried first to dissipate the mounting feeling with some cursory statements and then finally adjourned the meeting.

As a relative newcomer to the faculty I had a somewhat dispassionate reaction to the proceedings, but months of lunchtime discussions had made me keenly aware of widespread injustices and a very high-handed and dictatorial management. I decided that the moment was ripe for stronger action. I spent the next few hours in conversations with some friends who I believed had the support and the credibility of their colleagues and agreed to convene a rump meeting of the faculty. The response was overwhelming. Two days later about 100 faculty met to elect an executive committee to draw up a program for action. My experiences at Harvard stood me in good stead, and my facility at mental arithmetic established me as one of the active members. Not much in the way of arithmetic was required to calculate from the number of new students and the tuition the additional revenue and, even further, to estimate the salary budget. There could be no financial reason for not adjusting salaries! In a short time what became the largest single branch of the Association of University Professors had organized and become a virtual trade union. It was a real faculty revolution. Within six months over 95% of the faculty had deposited signed pledges saying that they would not sign their next year's contracts before holding a joint meeting in the Fall. I was given charge of these pledges, which also unprecedentedly contained the first revelation of the faculty salaries. Now I was able to do a very precise estimate of the faculty budget. Salaries turned out to be much lower than anyone had imagined!

Some very stormy meetings of our executive committee with the President and finally with the Board of Trustees led to meeting all of our demands. The revolution had succeeded! It was an exhilarating experience, but a few disagreeable events along the way convinced me that my future efforts would be better spent doing science. In 1949, I was elected Vice-President of the newly organized Faculty Senate, and six months later I resigned.

Before the war a serious American scientist could expect to spend one to two years as a post-doctoral in England, Germany, or France. Europe was clearly the world center for all major sciences. Although the war had completely altered that situation, my psychological outlook had not changed. A sabbatical year in Europe was still a fond goal. I had studied French and German in high school and had continued French in college. France had become a romantic dream. Guggenheim and Fulbright Fellowships in 1950 turned the dream into reality. In September, I arrived in Paris, where I had been offered space at the Laboratoire de Chimie Physique. Professor E. Bauer, a distinguished scientist of the old school, presided over physical chemistry, and Professor Michel Magat was actively in charge of the programs, principally in kinetics and polymer chemistry. European science was a relatively compact, tight community. Everyone knew everyone else, not only in their own fields, but in neighboring fields. My visit was propitious. Victor Weiskopf, a physicist from MIT, was doing a year's sabbatical in the next building, and Sam Eilenberg from Columbia was a guest of the local mathematicians. Paris was a central point in Europe, and it was a rare week that didn't bring from three to five distinguished visitors to the laboratory.

It was a stimulating year. For the first time in many years I had no fixed schedule to meet and no obligations. I could read, discuss and think. My project was to write a book on chemical kinetics. I had taught it for five years at USC at the graduate level, my notes were fairly extensive, and I thought I could finish it in one year. As it turned out I wrote about half in France and then spent the next eight years, mainly summers, completing it. Had I known how long it would take I would probably never have started. My decision to write the book stemmed from some recent advances in kinetics that seemed to have profound implications.

In 1950 Noel Slater, a mathematician at Manchester, used a theorem discovered by Mark Kac on the roots of trigometric equations to solve a very important problem in statistical mechanics. Treating a molecule as made up of coupled harmonic oscillators, he had derived the result of unimolecular rate theory on the frequency of passage through a given energy distribution. It yielded an A-factor in the range 10^{13}–10^{14} sec^{-1} for all unimolecular reactions and related it to the actual molcular frequencies. This had rekindled my interest in kinetics. In addition, Rudy Marcus, a post-doctorate student working with Oscar Rice, had been able to unify Eyring's Transition State Theory with the Rice–Ramsperger–Kassel theory of unimolecular processes to provide a quantitative model for all such reactions. I felt the time was ripe to review the relatively small amount of experimental data and reestablish chemical kinetics on a firmer base. The historical background needs some elaboration.

Chemical kinetics was a relatively recent science. Gas phase kinetics had started with Bodenstein in 1890 and the theoretical basis for it rested on the Arrhenius formulation from his famous 1888 paper plus the prior background in statistical mechanics. Chain reactions were discovered in 1916 by Nernst, and branching chains were discovered by Semenov in 1923. In 1923 Lindemann proposed the collisional mechanism for energy transfer, and within three years Rice and Ramsperger and then Kassel showed how this could account for the pressure dependence of the apparent rate constant for unimolecular reactions at low pressures. This was followed by a deluge of activity in the 1920s and 1930s led by Hinshelwood in England and Steacie in Canada to characterize all pyrolysis reactions as having rates determined by unimolecular fission processes. The development in 1934 of "Absolute Rate Theory" by Eyring and his colleagues seemed to complete the molecular framework of chemical kinetics.

The picture slowly unravelled after the war. By 1948 every reaction considered unimolecular before had been shown to be a complex chain process except for two, the isomerization of cyclopropane to propylene and the decomposition of nitrogen pentoxide:

$$N_2O_5 \rightarrow N_2O_4 + \tfrac{1}{2}O_2.$$

However, the latter fit none of the criteria of unimolecular rate theory (RRK) whereas the former had an abnormally large Arrhenius A-factor. Richard A. Ogg was able to resolve the N_2O_5 dilemma by showing that it must be a complex radical process. Thus by 1950 kineticists had an elegant theory and no reliable data. The stage seemed ready for a second effort to do some selected experiments in gas kinetics.

The year in Europe was an unparalleled opportunity for me to visit all the active laboratories. I went to London, Manchester, Birmingham, Oxford, and Cambridge. M. C. Evans and M. Polanyi were the leading figures in British kinetics at this time. Hinshelwood had just turned to biochemical kinetics of cell growth and, with the recent third edition of his famous text, seemed to have lost all interest in gas phase reactions. He was still directing a few graduate students when I visited him but not with much real interest. Evans was at the peak of his career at Manchester, while Polanyi had just resigned from chemistry to go into economics. Evans's premature death of cancer a few years later and Polanyi's retirement were an enormous loss to physical chemistry. However, Porter and Norrish were just completing their classical studies on flash photolysis at Cambridge, while a young Fred Dainton was establishing a reputation at Oxford. Christopher Ingold and colleagues at University College, continuing their elegant work on solution kinetics, were at their peak. H.

Melville was completing a dazzling career in kinetics and polymer chemistry at Birmingham and would soon become an elder statesman.

Eric Rideal, already elder statesman, was at the Royal Institution but still active in studying catalysis and surfaces, while Dennis Riley was doing his so-important x-ray studies of DNA. The Faraday Society Symposia that I was able to attend were a thrilling introduction to how scientific meetings could be run. Authors' manuscripts had been pre-distributed, they were given five minutes to summarize their findings, and then the fun began. The Discussions were just that—real discussions! Guggenheim, then at Reading, was his famous caustic self and woe to the unwary scientist who might draw his scathing attention!

In France the leading kinetics group was that of Letort at Nancy. Sadron was just starting polymers at Strasbourg. Lefort at Lyons had an active program in catalysis. Best known on the European mainland was the Brussels' group, with Prigogine in theoretical work and Goldfinger in kinetics. In Germany, Jost and Bonhoeffer were just beginning to reconstruct their laboratories and Eigen was becoming known for his work on fast reactions. I was not prepared for the warmth with which I was received on my visits. Even more surprising was the respect and even deference shown. European science was organized on very authoritarian lines. The professor of a department usually had lifetime tenure and total command of a state-provided budget. He could literally hire and fire at will. I was a recently promoted associate professor at USC. The title didn't exist in Europe. My hosts everywhere treated me like a European professor, even those who knew the plethora of assistant associate and full professor titles in the States. They couldn't divorce the title Professor from the European image.

We had quite a number of American visitors to Paris. These visits gave me an opportunity to get better acquainted with some of my compatriots. These included Jack Aston of Pennsylvania State University, then on sabbatical in Leyden, Ken Pitzer of Berkeley, then Commissioner of the Atomic Energy Commission, and Gerald Oster, a post-doctoral at the lab. Stan Ulam, a former colleague in mathematics briefly at USC and a once-active participant in the Manhattan Project, most recently from Los Alamos, made a surprise visit in the Spring of 1951. He cheerfully informed me at lunch one afternoon that the US had detonated an H-bomb and made a Pacific Island vanish. It was still secret and he didn't tell me about the role he had played in designing it. This was in the heyday of the McCarthy Committee and during the Korean War, a time punctuated by scandalous accounts on all sides of American and British spies serving the Soviet interests. I asked how he managed to come to Europe with his information. He directed my attention to the next table (at the Cafe de la

Paix where we were lunching). "They go everywhere with me, those two men. I'm sure if I talked to the wrong people I might have an accident." It didn't seem to perturb his usual good cheer.

I returned from Europe with plans for beginning a new program in kinetics. It had become clear to me during my sabbatical that one of the basic problems of kinetics was mechanism. It was pointless to speculate on the transition state of a chemical reaction until one was very certain as to what reaction one was observing. I searched long and hard for a simple system as free from chemical complexity as possible. The decomposition of N_2O to give $N_2 + O_2$, which was well studied, seemed too complex. Small amounts of NO were produced and it was not clear what role they might play, since NO was a known catalyst for the decomposition. Among the simply systems only ozone, O_3, fit the requirements. Only three species existed in the system, O, O_2, and O_3. Only one of these was reactive, O atoms. Arthur Axworthy, a new student, was interested and we began what became my most hazardous research. Our first experiment with pure O_3 at 100°C exploded, destroying our beautifully constructed equipment and leaving only fine particles of pyrex. Axworthy, miraculously, came out of it with only minor cuts. Over the next three years about one out of three experiments exploded. But we were well prepared with explosion barriers, so only a day was required to rebuild. Axworthy became a superb glassblower.

I describe the ozone work because it provides an excellent example of the connectedness of science and technology. We began it with no thought of any practical applications or idea that ozone might be of more general interest. I had no idea of the stratospheric ozone layer in 1952, let alone its significance. By 1957, when we published our findings, we had become "world experts" on a material that had become important enough to have an International Symposium devoted to it. We were able to verify the original two-step Chapman mechanism (1906) for its decomposition, prove that the 1934 16-step mechanism was incorrect, and provide the first detailed rate constants for the two steps. We also showed the significance of excited $O(^1D)$ in the photolysis and explained the surprising catalytic effect of H_2O on the photolysis.

This experience was typical of the connections that have occurred over and over in my own work and in the work of many of my colleagues. When photochemical smog was discovered in Los Angeles in 1946 and later ozone was shown to be one of its active agents, I was pulled into work on air pollution that has continued over 30 years. This work on air pollution excited an interest in oxidation and peroxides and this in turn has involved me in combustion research since 1964.

The decision to go back into chemical kinetics was a fortunate one. It was a propitious moment in the development of the field, and some of my most important discoveries were to be in kinetics and its closely coupled field, thermochemistry. Perhaps one of the best known, Group Additivity, began very innocently in 1957 with an analysis of elementary radical-molecule reactions. One of my very dedicated graduate students at this time was Jerry Buss. He had the most important gift of learning—thoroughness. We had many protracted discussions during the work on his degree. I never felt certain about an idea until I had succeeded in convincing Jerry of its validity. When he came up with a new idea it was almost certain to be important and insightful. His self-critical ability was unusual.

One bright winter day, just months before his thesis was completed, Jerry approached me with some intriguing observations. We were interested in the C–H and C–C bond strengths in some alkanes, ketones, and aldehydes. In almost every case the available data were insufficient for a comparison. However, closely related thermochemical data on heats of formation ($\Delta_f H°$) were available, and Jerry had noticed that $\Delta_f H°$ for CH_3CHO was equal to the average of CH_3COCH_3 and CH_2O. Then followed an exciting two weeks of scouring the thermochemical literature and finding similar results for $ClCH_2CH_3$ and the related symmetrical pair $ClCH_2CH_2Cl$ and CH_3CH_3. The obvious generalization was that in disproportionation reactions of the type

$$RAR + SAS \rightleftharpoons 2RAS$$

$\Delta H \simeq 0$.

We quickly discovered that it worked for molar heat capacities, for intrinsic entropies, and for virtually all molecular properties. Two further months of euphoric discoveries were finally distilled into a series of five short notes to the Editor of the *Journal of the American Chemical Society* (*JACS*). Jerry graduated with honors and went on to his first well-paid job, with A. D. Little in Boston. I started a senior post-doctorate year at Cal Tech, ostensibly to complete my exhausting *Foundations of Chemical Kinetics*. I've never had a busier year. My son Nick was born, my wife started law school, and I commuted to Cal Tech.

In August a deflating letter from the Editor informed me that the notes could not be published. They were not that new, they should be consolidated, and they should be related to earlier reported and similar observations. Thinking wicked thoughts about the long-since departed but sorely missed Jerry, I put aside my writing and spent a month in reflection and rewriting. The result was the discovery of what I like to think of as Raoult's law for molecules, or a generalization of all laws of additivity of molecular properties. Group Additivity was the second order level and the most accurate. It has since received widespread acceptance in physical chemistry and engineering.

Jerry Buss returned to Los Angeles, made an outstanding record in Aerospace Research, first with Douglas Aircraft and later with STL (now Aerospace). He soon decided that the challenges in aerospace were neither sufficiently interesting nor rewarding. He went instead into real estate and within ten years became an American success story. Today, he is a popularly recognized sports figure, and current owner of the world-famous Los Angeles Lakers basketball team. Most sports audiences don't realize that Dr. Buss, as he is popularly referred to, has his degree in physical chemistry.

In 1964, just a year after I had begun research in the newly installed department of Chemical Kinetics and Thermochemistry at the Stanford Research Institute (SRI), I had a similar experience. I had received a short paper to review from *JACS* on the B–B bond strength in diborane (B_2H_6). Since I had just published a paper on the subject that I felt fixed the bond at 35 ± 1.5 kcal/mole, I was somewhat outraged to read an experiment that raised it to over 54 kcal. My first impulse at such moments was simply to stop reading—the result was certainly in error! However, I had a very persistent, bright young post-doctorate working for me, one David Golden. David was to be one of my closest friends and colleagues during my 13-year stay at SRI. He proved a wonderful counterpoise for my impatience (impetuosity?). He was and is a meticulous and careful research worker, bright, energetic and indefatigable. When David felt some piece of research was done, it was really done; sometimes, in my opinion, overly done. I never felt comfortable with a result until it had Dave's blessing.

In this instance of diborane he kept nagging me. Why was the experiment wrong? Finally convinced that it deserved some thought, I pondered over it, eventually realizing that the experiment had been done at such low pressures (~ 1 mtorr) that the unimolecular rate had fallen off many powers of ten from its high pressure limit. The very low rate constant could only be rationalized by an apparently high activation energy, hence bond strength. This reflection was to be very rewarding. No one who has worked in close proximity to Anton Burg is unaware of the fact that boron hydrides are extremely sensitive to surfaces and impurities. In this case, I was astonished at the apparent stability of B_2H_6 (up to 700°C) in the presence of a very catalytic stainless steel reactor. Under normal conditions (760 torr) it decomposes rapidly in glass at 100°C. Surface-catalyzed reactions usually slow down at high temperatures and low pressures, since adsorption equilibrium favors the gas phase under these conditions. This suggested an experiment, and the technique of Very Low Pressure Pyrolysis (VLPP) was born. Another colleague, Neil Spokes, and I designed it.

In VLPP a reactant was allowed to flow into a Knudsen Cell (300–

1100°C) at pressures less than 1 mtorr. Under these conditions all uni-molecular reactions tend to be at their low pressure limits, and all collisions are with the walls. A molecular beam exiting the reactor orifice (1–5 mm) is analyzed for reactants and products by a mass spectrometer. High pressure limiting rate constants can be estimated using RRKM Theory with the wall as a strong collider. The technique has given valuable insight into unusual fission reactions and has been adapted to studying both bimolecular and surface reactions. A variant technique developed in the past ten years at USC, the Very Low Pressure Reactor, has been used to get very accurate rate and equilibrium constants in systems such as:

$$RH + X \rightleftharpoons R + HX.$$

Under VLPR conditions, concentrations of radicals and atoms can be so low that secondary rections do not occur.

I was very grateful to Dave for his nagging. An important incentive for his persistence was the fact that Fred Stafford, the author of the B_2H_6 note, was an old friend. It later turned out that Fred was an alumnus of Camp Rising Sun, a small, unusual, summer scholarship camp in New York. A generation earlier, in 1934, I had attended the camp. It had had a strong influence on my life and attitudes, which I still recollect with fondness. Science is a small world!

World War II sponsored a vast mobilization of scientific and technical manpower, mainly on the home front. The Cold War that followed led to a revolution in scientific research both in its conduct and its funding. Starting in the physical sciences with the Office of Naval Research, then Air Force, the Army, and the Atomic Energy Commission (AEC), a parallel explosion in the life sciences was supported by the Public Health Services (PHS) (now the National Institute of Health). This culminated in 1952 with the installation of the National Science Foundation (NSF). By 1957 the scientific population had more than quadrupled. With Sputnik in 1957 and the development of the National Aeronautical and Space Administration (NASA) from the early NAACA, scientific funding in American universities had become predominantly federal.

In this same period, the Chemistry Department at USC grew from a 7 to a 12 to a 17-member department. Its members were young, almost uniform in age distribution, enthusiastic, and devoted to research. Although living in the shadow of Cal Tech in Pasadena and the more distant shadow of UC Berkeley, physical chemistry was surprisingly strong. A strong surface science group was possibly unique in the US, consisting of Robert and Marjorie Vold, Todd Dosher, Karol Mysels, and Arthur Adamson. Physical chemistry had, in addition to Charles Copeland (ionic solutions) and myself, Harold Friedman, Harry Frisch, Jerry

Donohue (X-rays), and Robert Simha. These together with Anton Burg, Wayne Wilmarth, and Jim Warf in inorganic chemistry and Jerry Berson and Norman Kharasch in organic chemistry, managed to attract enough research support to be counted among the top ten departments in research productivity (measured in papers/faculty). We were also fortunate in attracting a good, talented, hard-working graduate student enrollment. In retrospect I would be inclined to say anomalously good.

For me, the postwar period was a very happy and productive one. I could indulge my scientific tastes, collaborate with many of my colleagues, and enter the world of consulting. With support from Goodyear, we did work on polymers. NSF and AEC supported free radical studies. Funding from ARO allowed us to do ozone and chlorine kinetics. PHS continued supporting our protein work while the Office of Naval Research (ONR) supported our work on supercritical salt solutions. Consulting work at Douglas Aircraft allowed us to initiate some of the pioneering studies of collisional energy transfer by computer simulation.

When in 1963 my wife's illness forced us to move from Los Angeles to the Stanford Research Institute (now SRI International), very little of this changed, including the consulting, which extended by then to the Jet Propulsion Lab (JPL) and the Aerospace Corporation. This is probably a result of the federal nature of research funding.

SRI was an interesting introduction into the world of sponsored research, both federal and industrial, and provided a wonderful opportunity for me to see how the chemical industry operated. Most importantly, it gave me immediate and broad contact with a large and extraordinarily talented group of scientific and engineering professionals. When it finally severed its close ties to Stanford after 1970, it became more income-driven, and I decided after a few years with many mixed feelings to return to academia. In 1976, I returned to USC.

Fifty years have now passed since I started my first graduate studies at Harvard—a period of extraordinary development in physical chemistry. The explosive growth in electronics, optical methods of analysis, gas chromatography, mass spectrometry, and most recently computers has given us the facility to ask questions of nature we would never have dreamed of asking even 20 years ago. As a veteran of this period, I am not sure that I want even to know the answer to some of the questions my junior colleagues are now asking, but I cannot help but recognize, admire, and salute their expertise and enthusiasm. In the 60 years since Dirac added the last piece to the conceptual background of atomic science, chemistry has exploded. This explosion might be described as being application-driven. In this sense it is amusing to me to hear the endless discussions that go on among scientists concerning the various virtues of applied or basic research. No research is funded today without the author describing succintly the possible results of that research and their potential importance. As beauty is in the eye of the beholder so does importance lie in the ear of the importuned.

More clearly than ever before scientific research is a societal endeavour. It draws its support and its ambiance from society. If it is not responsive to the needs of that society it will not long endure. This is not to say that all research must be practical. Society has an endless need to be amused or entertained, to indulge in fantasy, and this is only slightly less to be addressed than more pressing problems such as hunger, disease, and war. A large amount of research in astrophysics and particle physics I would place in the entertainment category, but that should not diminish the consideration it is given.

A different problem affects research today, namely, the resources devoted to its performance. Big and small science are no longer negligible items in the national (or international) budgets. Science has been so successful that it has grown huge over the last 50 years. How much support can it expect? I do not know of any simple answers to that question. How many symphony orchestras should there be? How many hospitals?

As I approach my fiftieth year of research and teaching, I rediscover a typical phenomenon that touches teachers. Everyone around me seems to be the same age as when I started teaching but my reflection in the morning mirror has unaccountably aged. My feelings of excitement about science and discovery are just as strong as they always were. Periodically, just when I feel that no new findings can be as important as the ones already made, I find myself looking at phenomena from a totally new perspective and getting as agitated as I did at 20.

Things are easier for me now. I have worked in almost every area of physical chemistry so that most problems seem familiar. Pondering the more difficult, still unsolved problems is fun. Nonbonded interactions in large molecules, the structure of H-bonded strongly polar liquids, intrinsic activation energies, metallic bonding, the energetics of bonds on catalyst surfaces, correlation of electron motions in atoms and molecules, how big is a photon—these are some of the more interesting of the unsolved problems. Every now and then we stumble on some totally unexpected discovery. Within the last three years, one of my more enthusiastic and hard-working post-doctorates, Mohammed Hisham, succeeded in finding a set of simple empirical relations that permit us to estimate the heats of formation of any salt to within about 2 kcal/mole. We had started by trying to understand the thermochemistry and kinetics of some common catalytic process and unexpectedly ended predicting the thermochemistry of solids. We also feel we have discovered some new insights in catalysis and metal bonding. Also unexpectedly, out of this has come a new and economical process for the chemical conversion of HCl into Cl_2.

In the last year, I have been amusing myself by going back to quantum mechanics and reexamining the way the wave equation is solved. My intuitive feeling is that electrons are localized relative to each other in both atoms and molecules—not only in radial shells but angularly. I do not believe, for example, that the two electrons in the ground state He atom ever find themselves on the same side of the nucleus. Our best wave functions say they do. I think they are wrong. They have too much electron-electron repulsion and compensate by decreasing the electron-nucleus separation to give a good energy—but possibly a bad wave function. Can I do better? Should I even try? I do not know, but it is fun to think about.

Robert L. Post

Annu. Rev. Physiol. 1989. 51:1–15
Copyright © 1989 by Annual Reviews Inc. All rights reserved

SEEDS OF SODIUM, POTASSIUM ATPase

Robert L. Post

Department of Molecular Physiology and Biophysics, Vanderbilt University Medical School, Nashville, Tennessee 37232

> For who "can look into the seeds of time,
>
> And say which grain will grow and which will not"?
>
> Adapted from Shakespeare: *Macbeth* I,iii
>
> "There are more things in heaven and earth, Horatio,
>
> Than are dreamt of in your philosophy."
>
> Shakespeare: *Hamlet* I,v

HISTORY

Comroe (7) examined the history of medical discovery to find out how it is done and how to do it better. I describe part of the natural history of one discovery as a correlation and adjustment of interlocking ideas until new priorities and more effective combinations are found.

My Background

My father was Professor of Greek at a small Quaker college in Haverford outside Philadelphia. He believed that Western Civilization began with the ancient Greeks. The fundamental problems and solutions started there; a perspective of more than 2,000 years was his frame of reference for current events.

0066-4278/89/0315-0001$02.00

He sent me to the best schools he could afford. At Harvard College, even though I had studied Latin for six years and Greek for three years, I was more attracted to science. Isaac Newton was my boyhood hero; Newton's Laws of Motion expressed more understanding in fewer words than anything I had encountered. I did an experimental thesis in Biochemical Sciences under J. T. Edsall and Jeffries Wyman. After college I wanted contact with "real" life and chose Harvard Medical School. In my second year, George Acheson (later Chairman of Pharmacology at University of Cincinnati) asked if I would like to investigate the mechanism of digitalis: I could isolate an endogenous factor from the plasma that would mimic its effects. Fortunately I declined; that problem is still unsolved! After medical school, a rotating internship nearly finished me. The necessity to stay awake required constant attention; we had only every other night off duty. I knew I did not understand what I was doing, so I devised a shorthand and kept a record of every patient I saw. If I had enough of these and organized them, I thought I could make sense out of the system. At its lowest level, the practice of medicine requires only that one do what any other physician would do next.

> Life is short
>
> And the art long
>
> The occasion instant
>
> Experiment perilous
>
> Decision difficult.
>
> An inscription at Harvard Medical School

The problems come to you, regardless of your understanding. A major advantage of a career in research is that one can choose one's problem.

After my internship, an opening in the Department of Physiology at the University of Pennsylvania under H. C. Bazett led me to research on the circulation of the blood. A later opening at Vanderbilt under H. J. Curtis introduced me to electrophysiology. When Howard Curtis left, my new chairman, Rollo Park introduced me to membrane transport.

Rationalization of Concentration Differences Between Cytoplasm and Extracellular Fluid in the 1950s

MEMBRANE TRANSPORT   At this time there was a well-developed understanding of the plasma membrane as a lipid barrier through which specialized transport, including active transport, could take place. Active transport was considered net movement away from thermodynamic equilibrium. It was also defined as dependent on metabolism. The difference between the two definitions was often neglected. Kepner has collected key papers on these ideas (24).

SORPTION   However, some investigators thought that the plasma membrane did not do anything, even if it did exist. (The electron microscope had only recently been invented.) Troshin took coacervates of gelatin and gum arabic, for instance, as a model of protoplasm; both systems took up solutes from their surrounding medium in a nonlinear fashion. He made a scholarly and complete survey of the literature arguing for "sorption" (49). Ling, who invented the intracellular microelectrode, found that the membrane potential of muscle cells persists in the presence of metabolic poisons and developed an elaborate a priori physicochemical model, his "association-induction hypothesis" (27), to explain an assumed selective binding of potassium to cytoplasm and the basis of action potentials.

OUTCOME   The membrane and sorption models are not contradictory except in the minds of their advocates. Both models postulate binding sites that are variably selective for sodium or potassium. The sorption model envisions a great many of them attached to macromolecules of the cytoplasm; membrane transport envisions a few embedded in the membrane and varying their selectivity as they move sodium or potassium from one side of the membrane to the other. Sorption keeps cell components from separating by forces between the components; membrane transport confines the components by means of a double Donnan equilibrium across the membrane. Robinson compared this controversy to that over the nature of combustion in the eighteenth century (40). He likened the membrane theory to oxygen and the association-induction theory to phlogiston. But proteins do bind water and can bind some ions (8), but not to the extent imagined in the sorption theory.

Counter-Transport of Glucose

In those days, "active transport" usually meant, as now, transport against an electrochemical gradient, away from equilibrium, "uphill" (8a) although some writers did not distinguish this from carrier-mediated transport. Carrier-mediated transport showed saturation kinetics, chemical specificity for the transported substrate, and competitive inhibition by structurally related molecules. The imagined carrier was a small lipid-soluble molecule in the membrane that formed a molecular complex with the transported substrate and allowed it to move through the hydrophobic barrier of the membrane. Today an ionophore, valinomycin, embodies this model for transport of potassium ion.

Rollo Park aimed to show that insulin regulated a glucose carrier in the plasma membrane of rat diaphragm. First he studied rabbit erythrocytes, and I was his consultant. After a while we realized that we could distinguish a mobile carrier from a selective channel if we could drive glucose uphill by transporting a competitive substrate downhill in the opposite direction, as predicted by Widdas (24). The experiments worked (34). Rosenberg & Wilbrandt (41) showed the same phenomenon in the more difficult human erythrocyte system and named it counterflow. They also reasoned that active transport implied a "mobile carrier" model rather than a fixed binding site in the membrane.

Active Transport of Sodium and Potassium

In 1954 active transport of sodium, with or without potassium, had been demonstrated in erythrocytes, skeletal muscle, giant axons, and across frog skin. Sodium transport across frog skin occurred between aqueous solutions at equilibrium and carried an electric current (51). This active transport was proportional to an increment in oxygen consumption.

CARDIOACTIVE STEROIDS   In 1953 Schatzmann published a most significant paper (42). He showed that the active transport of sodium and potassium in erythrocytes was inhibited by cardioactive steroids. This project was based on work done elsewhere in the 1930s showing that digitalis provoked a loss of potassium from muscle. Schatzmann's mentor, Wilbrandt, thought that adrenal mineralocorticoids might be the putative lipid-soluble sodium carrier and that the lactone ring of the cardioactive steroids might displace the carrier.

DEPENDENCE OF ACTIVE TRANSPORT SPECIFICALLY ON ATP   In 1954 in Hungary, Gárdos, using F. B. Straub's new technique of reversible hemolysis of erythrocytes, incorporated extracellular ATP into these cells and showed that this ATP supported potassium uptake (13). In 1960 in Cambridge, England, Caldwell et al (5) stimulated radioactive sodium efflux and potassium influx in the squid axon by injecting arginine phosphate into the cytoplasm. Injection of ATP stimulated only sodium efflux.

LINKAGE OF ACTIVE SODIUM AND POTASSIUM TRANSPORT   In 1956 Glynn (13a) gave an elegant and scholarly analysis of kinetics of radioactive tracer fluxes in erythrocytes. He related his findings to a cyclical carrier hypothesis for active transport as follows.

> It is supposed that K^+ and Na^+ can only cross the membrane in combination with the carriers X and Y, X being K^+ specific and Y being Na^+ specific. . . . X and Y are, by hypothesis, interconvertible and are in equilibrium at the outer surface of the membrane; at the inside surface, X is converted into Y with the expenditure of metabolic energy.

STOICHIOMETRY OF ACTIVE SODIUM AND POTASSIUM ION TRANSPORT IN ERYTHROCYTES   In 1954 I chose to work with human erythrocytes because they are simple, structurally and metabolically, and easy to get (from myself, at times!), and they permit aliquots. A new instrument, the flame photometer, made the analyses convenient. Incubation for a week or two in the cold room moved ion concentrations toward equilibrium, increasing cell sodium and reducing cell potassium. Subsequent warming with added metabolic substrates restored the normal composition in a few hours, moving both ions away from equilibrium and demonstrating active transport. In order to estimate cell content rather than concentration, I measured the ratio of the cations to hemoglobin. I avoided radioactive tracers because I thought that net transport was more fundamental. In 1957 Post & Jolly reported a stoichiometry of 3 Na^+ outward per 2 K^+ inward for the component of transport inhibited by a cardioactive steroid; "The discrepancy in the pumping rates for sodium and potassium would have a considerable effect on the membrane potential due to the net movement of positive charge if it were not for the fact that the erythrocyte membrane is freely permeable to small anions . . ." (36).

AN ELECTROGENIC SODIUM PUMP   In 1957 Ritchie & Straub carefully characterized posttetanic hyperpolarization in small nerve fibers, showing that it was produced by the sodium pump. They hesitated to conclude that the pump itself was electrogenic (39). Unfortunately, I was not aware of this work at that time.

Na,K-ATPase

In 1954 Na,K-ATPase, was unknown. In 1945 Racker & Krimsky (38) reported that sodium ions inhibited glycolysis in a mouse brain homogenate. In further work on inhibition by sodium, Utter (52) wrote in 1950, " . . . the dephosphorylation of ATP and ADP was stimulated. The stimulation of Apyrase was shown . . . at low levels of Na^+ and to be dependent upon . . . Mg^{++}." " . . . one or more of the enzymes affected by Na^+ is removed by centrifugation."

In 1957 Skou (46) noted in the introduction to his landmark paper that metabolic poisons inhibited active transport of sodium ions out of giant axons (21) and reduced their content of energy-rich phosphate esters (4). "A further study on the ATPase in nerves and its possible role in the active outward transport of sodium ions seems warranted" (46). He based his experiments on those of Abood & Gerard (1), who found ATPase activity in the light microsomal fraction of a homogenate of rat peripheral nerve. From this fraction of crab nerve, Skou obtained and characterized an ATPase activity highly specific for sodium ions. He summarized,

> Leg nerves from the shore crab *(Carcinus maenas)* contain an adenosine triphosphatase which is located in the submicroscopic particles. . . . Sodium ions increase the activity when magnesium ions are present. Potassium ions increase the activity when the system contains both magnesium and sodium ions. . . . This observation, as well as some other characteristics of the system, suggest that the adenosine triphosphatase studied here may be involved in the active transport of sodium from the nerve fibre (46).

In 1957 it was hardly credible that active sodium transport could be generated by anything so simple as a specific sodium and potassium ion-transport ATPase embedded in the plasma membrane of almost every animal cell. When Skou's paper appeared, it aroused little immediate attention. Nevertheless, before his paper the rain of evidence trickled in diverse streams or sat in puddles. After his paper, it flowed in a progressively larger and broader river. Input came from both membrane transport and enzymology. My colleagues and I were fortunate to contribute seven drops of convergent evidence that the Na^+-K^+ pump and the Na,K-ATPase were a single entity (37).

Calcium Ion-Transport ATPase from Sarcoplasmic Reticulum

Na,K-ATPase was soon joined by the calcium ion-transport ATPase of sarcoplasmic reticulum of skeletal muscle. A particulate ATPase of W. W. Keilley and O. Meyerhof in 1948–50 and B. B. Marsh's "relaxing factor" of muscular contraction in 1952 converged in the work of S. Ebashi in 1958 and became a calcium pump in the work of W. Hasselbach and M. Makinose in 1960 and of S. Ebashi & F. Lipmann in 1962 (18). It was a complicated system. Much effort was expended before it was realized that free calcium ion in the cytoplasm catalyzed contraction and that resealed vesicles of sarcoplasmic reticulum pumped calcium into their interior to produce relaxation.

THE CURRENT SCENE

I apologize to all those authors who might reasonably expect to find their work cited here and who do not find it, especially those who sent me reprints. I apologize to readers who might reasonably expect to find a better coverage. By the time I warmed up to my task, there was no more time.

Reviews

Glynn has written a comprehensive review (14). Nørby contributed further informaton and perspective (32). Jørgensen reviewed the function and regulation of the enzyme in the kidney (22). Fambrough et al reviewed regulation of the pump from the perspective of a molecular biologist (11). Skou has written a very nice overview (47).

Subunit Interactions

The formation of two-dimensional crystals in the lipid bilayer shows that subunit interactions by lateral association of $\alpha\beta$-protomers can occur (30). Retention of many functions by solubilized protomers shows that protomer interactions are not necessary (53). Although many partial answers have been offered, the question whether protomer interactions modify function in the native membrane-bound enzyme remains open. There is at least one report of retention of some function by an isolated α-subunit (29). Askari has written a mini-review (2). See also a contribution from Schoner's laboratory (43).

Endodigin

Endodigin is Arnold Schwartz's name for endogenous digoxin, a putative regulatory substance of the pump in animals, reflected in the high specificity of cardioactive steroids for inhibition of this enzyme. It may be involved in hypertension.

A major stumbling block to finding this substance is the necessity of a major commitment to a tissue source, a means of extraction, and an assay system before knowing if the right ones were chosen.

If endodigin is a regulatory substance in multicellular organisms, then Na,K-ATPase in unicellular organisms such as protozoa should not be sensitive to cardioactive steroids.

For recent articles, see the book edited by Mulrow & Schreier (31) and two reviews (15, 17). Inagami's laboratory reported a substance with properties very close to those of a cardioactive steroid (48).

Na,K-ATPase AS AN OSCILLATOR

Qualifications

For an enzyme to qualify as an oscillator, its reaction sequence should consist in a single cycle that includes at least two reactive states. Two of these reactive states should exhibit maximum and minimum values of some parameter, such as the volume, the dipole moment, or the fluorescence intensity of a probe attached to or part of the enzyme. I refer to these reactive states as principal reactive states. The turnover number of the enzyme, thus, corresponds to the frequency of the oscillator. The frequency depends not only on the enzyme but also on the reaction mixture in which it functions, particularly on the concentrations of substrates and products.

Experimental Limitations

A major limitation to detecting a signal from such an enzyme is the necessity to synchronize the molecules and to keep them synchronized. Even if the enzyme molecules all oscillate at the same frequency, no signal can be detected if the molecules have a random phase relationship to each other. In principle, they can be synchronized by varying a force, such as pressure or an electric field, if the principal conformations of the enzyme differ in a corresponding displacement such as volume or dipole moment respectively. For instance, if the volume of the enzyme is different in the principal conformations, as each molecule turns over, it will emit a tiny sound wave at a frequency corresponding to its turnover number. Now, if a sound wave is applied to the reaction mixture from an external source at the frequency of the turnover number, those enzyme molecules that are out of phase with the applied wave will be brought into phase with it. The ones that are ahead of the applied wave, leading the phase, will be slowed slightly and the ones that are behind, lagging the phase, will be speeded up slightly until all the molecules are in phase with the applied sound wave. Most of the energy is supplied by the splitting of the substrate of the enzyme; the sound energy is needed only for synchronization. [Emission of sound produced by flashes of light was used to estimate enthalpy changes in the reaction sequence of bacteriorhodopsin (26a)].

An Acoustic Oscillator

EXPERIMENTS John Wikswo and Pablo Vasallo helped me with this project. We used Na,K-ATPase from the outer medulla of dog kidney prepared by Jørgensen's zonal method (23). For the output signal we chose the fluorescence intensity of a fluorescein isothiocyanate molecule attached to lysine residue number 501 of the α-subunit (sheep kidney sequence). Its intensity is 20% greater in the presence of Na^+ than in the presence of K^+. An SLM-Aminco DW-2c spectrophotometer was available with a fluorescence attachment. This instrument can respond to frequencies as high as 250 cycles per second. For the input parameter we chose pressure since it was simple to apply sound vibrations to the enzyme by coupling a loudspeaker driver to a cuvette in the spectrofluorometer. For a substrate we chose acetylphosphate since the enzyme derivatized in this way shows no response to ATP. In using acetylphosphate we had to separate the Na^+,K^+-dependent activity with this substrate from the K^+-dependent activity. We did this by choosing Na^+ and K^+ concentrations that permitted the Na^+,K^+-dependent acetylphosphatase activity and prevented K^+-dependent p-nitrophenylphosphatase activity. We obtained a turnover number of 17 per second for the acetylphosphatase activity at 20°C. [This was an upper limit since the Na^+,K^+-activity can stimulate a little K^+-dependent activity; when K^+ gets on the enzyme, it tends to come off slowly (35).]

In order to improve the ratio of signal to noise we used a lock-in amplifier. This device filters a noisy input signal. It also accepts a signal from a reference oscillator and reports the amplitude and phase of the component of the noisy signal that is at the frequency of the reference signal. We fed it the output from the spectrophotometer and a reference signal from the oscillator that was driving the loudspeaker. If the enzyme was synchronized to the sound waves, we thought we would detect a modulation of the fluorescent intensity at the same frequency and at a constant phase relationship to the driving signal.

When we got the derivatized enzyme, the reaction mixture, and the apparatus all together, we varied the frequency of the applied sound wave from 5 to 100 cycles per second and looked for a response from the lock-in amplifier. There were several sharply tuned responses. But the inhibitor ouabain, which stops the acetylphosphatase activity, had no effect. All the responses were artifacts produced by vibrations of the mirrors of the spectrophotometer caused by the vibrations of the loudspeaker driver.

SIGNIFICANCE   If we had obtained a usable signal, we could have determined whether the sodium form, E_1 conformation, or the potassium form, E_2 conformation, of the enzyme had the smaller volume. For instance, if the potassium form had the smaller volume, then the minimum of the fluorescence intensity should have coincided with the compression of the sound wave and not with the rarefaction. In the calcium ATPase of sarcoplasmic reticulum, Blasie et al found that the protein of that enzyme moved from a position outside of the membrane slightly into the membrane when that enzyme went from an E_1 to an E_2 conformation (3). This motion does not necessarily imply a change in volume, of course.

EPILOGUE   We learned later by a personal communication from George Fortes that a pressure of about 1,000 atm is needed to inhibit Na,K-ATPase. Since out apparatus produced a peak-to-peak change of pressure of a little more than one atm, we were applying too little pressure to have an effect.

An Electric Oscillator

This was not the end of our experiments, however. The enzyme might be an electric oscillator, if not an acoustic oscillator. Na,K-ATPase is an electrogenic pump, characteristically transporting 3 Na^+ outward per 2 K^+ inward per cycle and thus generating a net movement of one positive charge per cycle, a small electric current. Furthermore, it is the outward sodium stroke of the cycle that moves the charge (9).

Even apart from sodium transport, the enzyme may respond to and therefore generate oscillating electrical fields. There are occluded conformations for both the sodium phosphoenzyme and the potassium dephosphoenzyme. In an occluded conformation, the enzyme holds a transported substrate (ion) without allowing it access to either the solution of origin or the solution of destination. Such an enzyme must have two gates, one on either side of the active center or cavity of occlusion. In the occluded state both gates are shut. Furthermore, in the course of the transport step only one gate may be open at one time. If both gates open simultaneously, the pump becomes a channel and the transported ion rushes across the membrane toward equilibrium, impairing the function of the pump.

In the voltage-gated sodium channels of electrically excitable membranes, the gates flicker transiently open and shut. The fraction of time in which they are open is regulated by the electrical potential across the membrane. This regulation implies a "sensor," a dipole in the protein that moves in response to the electric field across the membrane. The motion of the sensor generates in turn a transient electric current, a gating current, that has been demonstrated experimentally (20). Such flickering gates have been reported in preparations of Na,K-ATPase incorporated into artificial lipid bilayers (25, 26, 29). Forbush has proposed a flickering gate model to explain transient kinetics of rubidium release from the enzyme (12). In the secondary structure of proteins there are alpha helices. An alpha helix of more than two turns has a dipole moment equivalent to one-half charge at either end, even in the absence of charged residues. The amino-terminal end is positive (28). Thus, even the secondary structure of a protein can have a dipole moment. Gresalfi & Wallace reported that the secondary structure of Na,K-ATPase differs greatly between the sodium and the potassium forms (16) although this conclusion has been denied elsewhere (6, 19).

Finally, consider the experiments of Dux & Martonosi (10) on the related calcium ATPase of sarcoplasmic reticulum. They crystallized this enzyme in its native vesicles by adding vanadate or decavanadate and observed the rate of crystallization. The enzyme crystallizes preferentially in the presence of E_2 ligands, namely inorganic phosphate or vanadate, magnesium ion, and ethyleneglycol-bis-(β-aminoethyl ether)N,N,N',N'-tetraacetic acid (i.e. in the absence of calcium ion). Dux & Martonosi studied the effect of membrane potential on the rate of crystallization. They varied the membrane potential by applying suitable salt concentration gradients across the membrane. The rate of crystallization was faster when the cytoplasmic side of the membrane was negative. When it was positive, crystals that had already formed broke up. Thus the E_2 conformation was stabilized by a negative membrane potential, and the E_1 conformation was stabilized by a positive membrane potential. This stabilization implies a difference in the orientation of a dipole in the enzyme.

All these indirect considerations suggested that it might be possible to make the enzyme vibrate between sodium (E_1) and potassium (E_2) conformations by applying an oscillating electric field across the membrane. The alternating electrical field would move a dipole, which would alter the conformation of the enzyme. The change in conformation would change the intensity of fluorescence of the probe.

The membranes containing Na,K-ATPase in the Jørgensen preparation are discs about 0.2–0.4 μm in diameter. To apply an electric field uniformly, all the discs must first be oriented in the same direction at right angles to the electric field. Fortunately, Ormos et al oriented purple membranes of bacteriorhodopsin by applying a steady electrical potential (33). We applied 30 V along 3 cm of a cuvette. To this potential we added an alternating sine wave potential of 50 V peak-to-peak and varied its frequency. The derivatized enzyme was suspended in 4 mM imidazole/3-(N-morpholino) propanesulfonic acid buffer at pH 7.1 at 23°C. At 5 cycles per second, the lowest frequency available, we obtained a signal about 0.5% as large as the maximum change in fluorescence intensity produced by changing sodium and potassium ion concentrations. This signal was three to five fold larger than the noise level. As we increased the frequency, the phase of the response lagged and the amplitude decreased, disappearing at about 10 cycles per second. The response required about 10 seconds to reach its maximal level. Both the steady and the sine wave components of the stimulus were required. Addition of the inhibitor, ouabain, did not affect the response. Ouabain locks the enzyme in an inhibited conformation. Thus the response was probably not due to a conformational change. The membrane fragments containing the enzyme visibly migrated toward the anode in this system. We speculated that they were tumbling at the frequency of the alternating potential. As they tumbled, they could variably shadow the fluorescent probe.

In a different reaction system, Serpersu & Tsong stimulated the pump electrically. They stimulated ouabain-inhibited net rubidium uptake into human red blood cells with an alternating electric field (44, 45). Their optimum stimulation was a sinusoidal potential of 12 mV across the membrane at 1,000 cycles per second. The temperature coefficient of the stimulated uptake was less than that of the unstimulated, metabolically driven uptake so that the stimulated uptake was about fourfold greater than the unstimulated uptake at 6°C and about one half as great at 26°C. The stimulated uptake was an increment above the unstimulated uptake. Efflux of rubidium or potassium and influx or efflux of sodium were not stimulated correspondingly. The kinetics with respect to extracellular rubidium concentration were similar for the unstimulated and the stimulated uptakes at 26°C. Intracellular sodium was required with about the same affinity for both uptakes ($K_m \simeq 8$ mM). The stimulated uptake was less dependent on the ATP concentration and was less inhibited by vanadate than was the unstimulated uptake. From a mathematical model they concluded that a net or cyclic charge translocation across the membrane and an asymmetric stability of the enzyme states could account for the results (50). Since the unstimulated potassium transport step is electrically neutral (9), an outward movement of an empty carrier, presumably carrying two negative charges, is required for the stimulated uptake. Since there was no stimulation of Na^+ efflux, their stimulation must have displaced the cytoplasmic Na^+, which was required for the Rb^+ influx, to permit the empty charged transport site to gain access to the extracellular face of the membrane. They estimated a turnover number of their enzyme at 10 cycles per second at a stoichiometry of 2 Rb^+ per cycle (45). Thus, their stimulation frequency of 1,000 cycles per second was far above the turnover number. They were probably not vibrating the whole enzyme but only a part of it, possibly the "lobe" that extends from the massive "body" of the enzyme in the parts farthest from the membrane (30). Presumably, they were not only driving cytoplasmic Na^+ out of the transport site but were also making one or both of the gates of the empty occluded transport site flicker.

ACKNOWLEDGMENT

This work was supported by a grant, R01 HL01974, from the National Heart, Lung, and Blood Institute of the National Institutes of Health.

MAXIMS FOR A YOUNG INVESTIGATOR

If we could first know where we are and whither we are tending, then we could better judge what to do and how to do it.

Abraham Lincoln

Whenever a new discovery is reported to the scientific world, they say first, "It is probably not true." Thereafter, when the truth of the new proposition has been demonstrated beyond question, they say, "Yes, it may be true, but it is not important." Finally, when sufficient time has elapsed to fully evidence its importance, they say, "Yes, surely it is important, but it is no longer new."

Montaigne

One of the principal objects of theoretical research in any department of knowledge is to find the point of view from which the subject appears in its greatest simplicity.

J. Willard Gibbs

THINK BIG!

Sidney Colowick

Do it right the first time.

C. R. Park

Most things don't work. Confirm anything that does.

Murphy

Put yourself in a position to be lucky. Be lucky. Know that you have been lucky. Persuade others that you were lucky.

R. L. Post

Literature Cited

1. Abood, L. G., Gerard, R. W. 1954. Enzyme distribution in isolated particulates of rat peripheral nerve. *J. Cell. Comp. Physiol.* 43:379–92
2. Askari, A. 1987. (Na$^+$ + K$^+$)-ATPase: on the number of the ATP sites of the functional unit. *J. Bioenerg. Biomembr.* 19:359–74
3. Blasie, J. K., Herbette, L. G., Pascolini, D., Skita, V., Pierce, D. H., Scarpa, A. 1985. Time-resolved x-ray diffraction studies of the sarcoplasmic reticulum membrane during active transport. *Biophys. J.* 48:9–18
4. Caldwell, P. C. 1956. The effects of certain metabolic inhibitors on the phosphate esters of the squid giant axon. *J. Physiol.* 132:35P
5. Caldwell, P. C., Hodgkin, A. L., Keynes, R. D., Shaw, T. I. 1960. The effects of injecting 'energy-rich' phosphate compounds on the active transport of ions in the giant axons of *Loligo*. *J. Physiol.* 152:561–90
6. Chetverin, A. B., Brazhnikov, E. V. 1985. Do sodium and potassium forms of Na,K-ATPase differ in their secondary structure? *J. Biol. Chem.* 260:7817–19
7. Comroe, J. H. 1977. *Retrospectroscope. Insights into Medical Discovery*, p. 186. Menlo Park, Calif.: Von Gehr
8. Creighton, T. E. 1984. *Proteins. Structures and Molecular Principles*, pp. 153, 243, 251–52, 377, 387. New York: Freeman
8a. Davson, H. 1951. *A Textbook of General Physiology*, pp. 274–77. Phila. Blakiston
9. De Weer, P., Gadsby, D. C., Rakowski, R. F. 1988. Voltage dependence of the Na-K pump. *Ann. Rev. Physiol.* 50:225–41
10. Dux, L., Martonosi, A. 1983. The regulation of ATPase-ATPase interactions in sarcoplasmic reticulum membrane. II. The influence of membrane potential. *J. Biol. Chem.* 258:11903–7
11. Fambrough, D. M., Wolitzky, B. A., Tamkun, M. M., Takeyasu, K. 1987. Regulation of the sodium pump in excitable cells. *Kidney Int.* 32(Suppl. 23):S97–S112
12. Forbush, B. 1987. Rapid release of ^{42}K or ^{86}Rb from two distinct sites on the Na,K-pump in the presence of P$_i$ or vanadate. *J. Biol. Chem.* 262:11116–27
13. Gárdos, G. 1954. Akkumulation der Kaliumionen durch menschliche Blutkorperchen. *Acta Physiol. Hung.* 6:191–99
13a. Glynn, I. M. 1956. Sodium and potassium movements in human red cells. *J. Physiol.* 134:278–310
14. Glynn, I. M. 1985. The Na$^+$,K$^+$-Transporting Adenosine Triphosphatase. In *The Enzymes of Biological Membranes*, ed. A. N. Martonosi, pp. 35–114. New York: Plenum. 2nd ed.
15. Graves, S. W., Williams, G. H. 1987. Endogenous digitalis-like natriuretic factors. *Ann. Rev. Med.* 38:433–44
16. Gresalfi, T. J., Wallace, B. A. 1984. Secondary structural composition of the Na/K-ATPase E1 and E2 conformers. *J. Biol. Chem.* 259:2622–28
17. Haber, E., Haupert, G. T. 1987. The search for a hypothalamic Na$^+$,K$^+$-ATPase inihbitor. *Hypertension* 9:315–24
18. Hasselbach, W. 1964. Relaxing factor and the relaxation of muscle. *Prog. Biophys.* 14:167–222
19. Hastings, D. F., Reynolds, J. A., Tanford, C. 1986. Circular dichroism of the two major conformational states of mammalian (Na$^+$ + K$^+$)-ATPase. *Biochim. Biophys. Acta* 860:566–69
20. Hille, B. 1984. *Ionic Channels of Excitable Membranes*, pp. 345–47. Sunderland, Mass.: Sinauer
21. Hodgkin, A. L., Keynes, R. D. 1955. Active transport of cations in giant axons from *Sepia* and *Loligo*. *J. Physiol.* 128:28–60
22. Jørgensen, P. L. 1986. Structure, func-

tion and regulation of Na,K-ATPase in the kidney. *Kidney Int.* 29:10–20

23. Jørgensen, P. L. 1988. Purification of Na⁺,K⁺-ATPase: enzyme sources, preparative problems, and preparation from mammalian kidney. *Methods Enzymol.* 156:29–43

24. Kepner, G. R. 1979. *Cell Membrane Permeability and Transport. Benchmark Papers in Human Physiology.* Stroudsburg, Penn.: Dowden, Hutchinson Ross

25. Kumazawa, N., Tsujimoto, T., Fukushima, Y. 1986. Influence of voltage and ATP on ion-channel of (Na,K)ATPase incorporated into solvent-free phospholipid planar bilayers. *Biochem. Biophys. Res. Commun.* 136:767–72

26. Last, T. A., Gantzer, M. L., Tyler, C. D. 1983. Ion-gated channel induced in planar bilayers by incorporation of (NA⁺,K⁺)-ATPase. *J. Biol. Chem.* 258:2399–2404

26a. LeGrange, J., Cahen, D., Caplan, S. R. 1982. Photoacoustic calorimetry of purple membranes. *Biophys. J.* 37:4–6

27. Ling, G. N. 1962. *A Physical Theory of the Living State: The Association-Induction Hypothesis; With Considerations of the Mechanisms Involved in Ion Specificity.* New York: Blaisdell

28. Matthew, J. B. 1985. Electrostatic effects in proteins. *Ann. Rev. Biophys. Biophys. Chem.* 14:387–417

29. Mironova, G. D., Bocharnikova, N. I., Mirsalikova, N. M., Mironov, G. P. 1986. Ion-transporting properties and ATPase activity of (Na⁺ + K⁺)-ATPase large subunit incorporated into bilayer lipid membranes. *Biochim. Biophys. Acta* 861:224–36

30. Mohraz, M., Simpson, M. V., Smith, P. R. 1987. The three-dimensional structure of the Na,K-ATPase from electron microscopy. *J. Cell Biol.* 105:1–8

31. Mulrow, P. J., Schreier, R. 1987. *Atrial Hormones and Other Natriuretic Factors,* pp. 127–71. Bethesda, Md.: Am. Physiol. Soc.

32. Nørby, J. G. 1987. Na,K-ATPase: Structure and kinetics. Comparison with other ion transport systems. Nobel Symposium 66 Membrane Proteins: Structure, Function, Assembly. *Chem. Scr.* 27B:119–30

33. Ormos, P., Dancsházy, Z. S., Kesthelyi, L. 1980. Electric response of a back photoreaction in the bacteriorhodopsin photocycle. *Biophys. J.* 31:207–13

34. Park, C. R., Post, R. L., Kalman, C. F., Wright, J. H., Johnson, L. H., Morgan, H. E. 1956. The transport of glucose and other sugars across cell membranes and the effect of insulin. *Ciba Found. Collog. Endocrinol.* 9:240–60

35. Post, R. L., Hegyvary, C., Kume, S. 1972. Activation by adenosine triphosphate in the phosphorylation kinetics of sodium and potassium ion transport adenosine triphosphatase. *J. Biol. Chem.* 247:6530–40

36. Post, R. L., Jolly, P. C. 1957. The linkage of sodium, potassium, and ammonium active transport across the human erythrocyte membrane. *Biochim. Biophys. Acta* 25:118–28

37. Post, R. L., Merritt, C. R., Kinsolving, C. R., Albright, C. D. 1960. Membrane adenosine triphosphatase as a participant in the active transport of sodium and potassium in the human erythrocyte. *J. Biol. Chem.* 235:1796–1802

38. Racker, E., Krimsky, I. 1945. Effect of nicotinic acid amide and sodium on glycolysis and oxygen uptake in brain homogenates. *J. Biol. Chem.* 161:453–61

39. Ritchie, J. M., Straub, R. W. 1957. The hyperpolarization which follows activity in mammalian non-medullated fibres. *J. Physiol.* 136:80–97

40. Robinson, J. D. 1982. The sodium pump and its rivals: an example of conflict resolution in science. *Perspect. Biol. Med.* 25:486–95

41. Rosenberg, T., Wilbrandt, W. 1957. Uphill transport induced by counterflow. *J. Gen. Physiol.* 41:289–96

42. Schatzmann, H.-J. 1953. Herzglycoside als Hemmstoffe für den aktiven Kalium- und Natriumtransport durch die Erythrocytenmembran. *Helv. Physiol. Acta* 11:346–54

43. Scheiner-Bobis, G., Fahlbusch, K., Schoner, W. 1987. Demonstration of cooperating α subunits in working (Na⁺ + K⁺)-ATPase by the use of the MgATP complex analog cobalt tetrammine ATP. *Eur. J. Biochem.* 168:123–31

44. Serpersu, E. H., Tsong, T. Y. 1983. Stimulation of ouabain-sensitive Rb⁺ uptake in human erythrocytes with an external electric field. *J. Membr. Biol.* 74:191–201

45. Serpersu, E. H., Tsong, T. Y. 1984. Activation of electrogenic Rb⁺ transport of (Na,K)-ATPase by an electric field. *J. Biol. Chem.* 259:7155–62

46. Skou, J. C. 1957. The influence of some cations on an adenosine triphosphatase from peripheral nerves. *Biochim. Biophys. Acta* 23:394–401

47. Skou, J. C. 1988. Overview: the Na,K-pump. *Methods Enzymol.* 156:1–25

48. Tamura, M., Lam, T.-T., Inagami, T. 1988. Isolation and characterization of a specific endogenous Na$^+$,K$^+$-ATPase inhibitor from bovine adrenal. *Biochemistry* 27:4244–53
49. Troshin, A. S. 1966. *Problems of Cell Permeability*. London: Pergamon
50. Tsong, T. Y., Astumian, R. D. 1988. Electroconformational coupling: how membrane-bound ATPase transduces energy from dynamic electrical fields. *Ann. Rev. Physiol.* 50:273–90
51. Ussing, H. H., Zerahn, K. 1951. Active transport of sodium as the source of electric current in the short-circuited isolated frog skin. *Acta Physiol. Scand.* 23:110–27
52. Utter, M. F. 1950. Mechanism of inhibition of anaerobic glycolysis of brain by sodium ions. *J. Biol. Chem.* 185:499–517
53. Vilsen, B., Andersen, J. P., Petersen, J., Jørgensen, P. L. 1987. Occlusion of ^{22}Na$^+$ and ^{86}Rb$^+$ in membrane-bound and soluble protomeric $\alpha\beta$-subunits of Na,K-ATPase. *J. Biol. Chem.* 262:10511–17

Andrew Huxley

Ann. Rev. Physiol. 1988. 50:1–16

PREFATORY CHAPTER: MUSCULAR CONTRACTION

Sir Andrew Huxley

The Master's Lodge, Trinity College, Cambridge CB2 1TQ, England

MUSCULAR CONTRACTION

Since 1950 each issue of the *Annual Review of Physiology* has opened with an article by a well-known physiologist reviewing aspects of the history of the subject or of his own life and work. The Editor invited me to write this year's prefatory chapter but proposed that this, the fiftieth issue, should be the first in which there should be a different emphasis. He proposed that the new format "would present perspectives on topics by experts in the field . . . We would hope that the author would summarize the key background material, provide a state-of-the-art view of the subject, pose the important unanswered questions (as well as the approaches to possible solutions), and speculate a bit about future developments." He asked me to write on skeletal muscle, adding in a parenthesis "you could include cardiac and smooth muscle too, if you so desire." I do not so desire, for two good reasons: First, I am not familiar enough with either of those rapidly expanding fields, and second, to do so would expand the topic far beyond what could be dealt with in the space I was offered. For the same two reasons, I shall not aim to cover the whole field of "skeletal muscle" but shall restrict myself to an aspect on which I have myself been engaged, namely, the actual mechanism by which force and shortening are brought about. I shall not touch on the metabolism, excitation, excitation-contraction coupling, structure, development, or disease of muscle. Again, for the same two reasons I shall not aim to be comprehensive but will take this as an opportunity to present a personal viewpoint, concentrating on experimental work on the intact contractile system rather than biochemical or structural studies or in vitro experiments on separated filaments or other reconstituted systems. There will inevitably be huge omissions (not least because posts I have held in the last few years have prevented me from keeping up to date with the literature in the way that I could have wished), and no one must be offended because his or her work has chanced to be among these omissions.

Revolutions in Muscle Theory: ATP, Actin, and Myosin

During the last sixty years, theories of muscle contraction have undergone a series of revolutions. It was A. V. Hill (29) who first used the word "revolution" in this context, in his article of 1932 in *Physiological Reviews*, describing the overthrow of the lactic acid theory by Lundsgaard's (52) experiment of 1930 in which a muscle poisoned with iodoacetic acid was found to be capable of performing many contractions without producing any lactic acid. The breakdown of "phosphagen" (phosphoryl creatine) had already been demonstrated in 1927 both by the Eggletons (12) and Fiske & Subbarow (16) and was therefore ready to replace lactic acid formation as the "primary energy-producing reaction." A second revolution which replaced phosphoryl creatine by adenosine triphosphate (ATP) took place a few years later, though it was not finally clinched until 1962 when Cain et al (8) found a poison (fluorodinitrobenzene) that relegated the splitting of phosphoryl creatine (PC) to the recovery processes, just as iodoacetate had relegated lactic acid formation. The key role of ATP is now universally accepted. The next two breakthroughs do not in my view rank as "revolutions" since they did not overthrow views previously accepted, but they were of major importance: the discovery by Engelhardt & Ljubimowa (13) (1939) in Moscow that myosin is itself an ATPase, and the discovery by Straub (60) in Hungary that "myosin" was a complex of two proteins, actin and what is now known as myosin, together with the recognition by Albert Szent-Györgyi (61) that the dissociation of these two proteins was brought about by ATP.

Sliding Filaments

The next real revolution—and the last up to the present—was the discovery that shortening of a muscle fiber takes place by relative sliding motion of two sets of filaments in each half-sarcomere rather than by shortening of continuous protein filaments as had previously been universally supposed. The key pieces of evidence were, first, that the myosin and actin were differently localized in relation to the striation pattern with myosin in the A band (26, 28), second that transverse sections under the electron microscope showed a double array of overlapping filaments (42), and third, that the width of the A band did not change during shortening [interference microscopy of intact living muscle fibers, A. F. Huxley & Niedergerke (38, 39); phase microscopy of separated myofibrils, H. E. Huxley & Hanson (44), Harman (27)]. The original observations in each case were made in 1953 and were made entirely independently by the different groups; the first published suggestion of shortening by sliding was made by H. E. Huxley (42).

For the purposes of this article, I shall assume that changes of length of striated muscle take place entirely by sliding, without any shortening or lengthening of the individual filaments. There may of course be yet another revolution in store for us, in which perhaps it might be shown that small cyclical length changes in one of the filaments are converted into continuous motion by some ratchet mechanism. Such a process is indeed to be found among the many proposals for specific mechanisms that have been made during the last thirty years [see my review of 1974 (36)], but I do not know of any evidence for it. There are strong hints that in striated muscle of the horseshoe crab (*Limulus,* an arachnid) the thick filaments shorten when a whole fiber shortens. This finding needs to be pursued, but there is at present no comparable evidence for vertebrate or other striated muscles. Some kind of shortening—perhaps crumpling of the ends—must take place in thick filaments during extreme shortening when contraction bands are formed at the level of the Z line. This again calls for further investigation but is outside my scope in this article.

Independent Force Generators

As regards the mechanism that generates the sliding motion, a rather general proposition that is widely, though not universally, accepted is that a relative force between adjacent filaments of the two types is generated more or less independently at each of a number of sites uniformly distributed along that part of their lengths where they overlap each other. This was proposed in our 1954 article (38) on sliding filaments on the basis of evidence that was already more than a decade old, namely the observation by Ramsey & Street (58) that active tension declines linearly as an isolated muscle fiber is stretched (prior to stimulation) beyond an optimum near its slack length and reaches zero at about double that optimum length. According to the independent-force-generator version of the sliding-filament theory, this decline in force would be a direct consequence of the reduced amount of overlap of thick and thin filaments and hence the smaller number of sites within each overlap zone contributing to the total force.

However, those measurements of force at different muscle lengths were complicated by the variation of sarcomere length (and hence amount of overlap) along the fiber found by A. F. Huxley & Peachey (40). Most importantly, in a stretched fiber the sarcomere length was often substantially less in the few tenths of a millimeter at each end of the fiber than in the middle, and when the fiber was stimulated these regions would undergo extreme shortening with consequent stretching of the middle parts and an upward "creep" of tension. To avoid this complication, my colleagues and I devised a servo control system which, during a contraction, holds constant the length of a selected segment of the fiber. With this equipment we found (25) that the developed force was closely proportional to filament overlap, estimated from sarcomere length within the segment and electron microscopic measurements of filament length. More particularly, force was proportional to the amount of overlap of each thin filament with that part of the thick filament that carried the projections known as cross-bridges, demonstrated in 1957 in the elegant electron micrographs of longitudinal sections by H. E. Huxley (43), greatly strengthening his suggestion that these projections were themselves the force generators.

This interpretation of the length-tension relation was strongly challenged about ten years ago by Pollack and his collaborators (56, 63, 64), who found much less decline of force with increasing length than in our results. However, they always recorded tension in fibers held by stationary hooks in the tendons at the two ends and never used servo control of segment length, a precaution that I regard as essential in experiments of this kind. Indeed, we did not attempt such experiments until we had developed our "spot follower" and servo system. In tetanic contractions at a length above the optimum, with the fiber ends held stationary, the rise of tension during the creep is often very substantial, and it is this raised value that was taken by Pollack. The creep was much reduced in our experiments using servo control of segment length, and since we attributed the residual creep to progressive increase of nonuniformity of striation spacing, we took the tension value that was reached during the initial quick rise (25). This explanation of the creep has since been confirmed by several groups who have used servo control of length in shorter and more uniform segments and obtained flat-topped tetani, with no detectable creep (4, 11, 47).

Pollack claimed that his measurements (by light diffraction) of sarcomere length along his fibers did not show a large enough increase of variability during a contraction to account for the creep. However, this evidence was not satisfactory because the ends of frog fibers, where sarcomeres are shortest and great shortening occurs during a tetanus, are commonly obscured by overlying tendon, which would defeat sarcomere length measurement by light diffraction; I have never seen a fiber free at *both* ends from such obscuration. In a more recent paper (2), however, Altringham & Pollack used straightforward photomicrography to record the local shortening of the ends of fibers that Peachey and I had reported twenty years before (40). They concluded that the creep is indeed explained by this local shortening, and they withdrew the objection that Pollack had previously raised against our interpretation (25) of the length-tension relation. More detail is presented in the paper of Altringham & Bottinelli (1), who also introduced a new and ingenious way of avoiding the complications due to local length changes. They recorded continuously by light diffraction the striation spacing at a position within the central region of a fiber. At the onset of stimulation the spacing first falls (the fiber shortens as it stretches the tendon attachments) and later begins to rise as the fiber ends, which have shorter striations, generate more force and shorten and stretch the middle of the fiber. Tension was recorded at the time when striation spacing was stationary at its minimum and was plotted against this minimum value. The resulting length-tension relation agreed closely with that found by Gordon et al (25). Good agreement was also found in measurements on skinned fibers when force was plotted against the shortest, rather than the average, sarcomere length within the segment (48).

The following points emerge from more recent determinations of length-tension curves. Akster et al (68) compared the curves from two muscles of a fish and found a shift agreeing with a difference in thin-filament length determined by electron microscopy. Bagni et al (4) found a similar but smaller shift between fibers from tibialis anterior and semitendinosus of the frog. Edman & Reggiani (11) found a more rounded relationship than other observers, suggesting greater variability of overlap. The latter two points need to be explored with the electron microscope.

If all cross-bridges are identical and act independently, the speed of shortening under zero load ought not to be affected by the amount of overlap, since even a very small number of cross-bridges would be able to cause sliding motion at a speed limited by their own intrinsic properties, and having a larger number of cross-bridges working in parallel would not increase this speed. This result was found by us experimentally (25) in isotonic contractions without after-load. A more thorough investigation of speed of shortening of completely unloaded fibers was made by Edman (10), who found a constant speed of shortening at lengths from substantially below that at which overlap is just complete up to that at which appreciable resting tension appears. At still greater lengths the speed was actually increased (up to well over double that at full overlap), and he attributed this to shortening being aided by the passive tension. It is not known whether in this situation the myofibrils remain straight or whether they are thrown into waves by the longitudinal compressive force, as occurs in fibers passively shortened below their slack length (7, 24).

For these reasons I believe—and I shall assume throughout this article—that the force generators are the cross-bridges between thick and thin filaments seen with the electron microscope (43) and that they act more or less independently of each other to produce more or less equal contributions to the total force.

Energy Utilization and Cyclic Action of Cross-Bridges

An essential piece of background to all thinking about the mechanism of contraction is the 1938 paper of A. V. Hill (30), which established relationships between load, speed of shortening, and rate of energy liberation (heat plus work) during contraction. These relationships, expressed by simple equations, made quantitative the discovery by Fenn (14, 15) [made in Hill's laboratory soon after World War I and foreshadowed by experiments of Heidenhain and Fick half a century earlier (cited in 31)] that the total energy liberated is greater if a stimulated muscle is allowed to shorten, thus doing mechanical work, than if it is held at constant length. This clearly implies that the rates of the important chemical reactions are governed by the mechanical conditions, a fact not sufficiently recognized by many biochemists even now.

A very perceptive point made by Dorothy Needham (54) in 1950 (well before any idea of sliding filaments) was that A. V. Hill's relation between rate of energy liberation and speed of shortening had analogies with the Michaelis-Menten relation between rate of reaction and substrate concentration in enzyme-catalyzed reactions. This led her to propose that the contractile elements act cyclically, as enzymes do. This idea did not catch on at the time because it is difficult to imagine *cyclic* action of elements in a continuous contractile filament: Creation of new folds would have a *cumulative* effect in reducing the length of the filament. This difficulty disappears if one thinks in terms of sliding filaments. Indeed, if one accepts the idea of myosin molecules acting as independent force generators, it is necessary to suppose that they act cyclically since the total length of each molecule is only a fraction of the range over which sliding of adjacent filaments can occur in a single contraction. Several noncyclic mechanisms for generation of the sliding movement have been proposed (see 36), but they have not been satisfactorily reconciled with the length-tension relation or with the relation between shortening speed and energy liberation, so I shall not consider them further here.

The almost universally accepted idea at present, which I shall assume to be true for the rest of this article, is that during shortening each cross-bridge, consisting of the head(s) of a myosin molecule, attaches to actin molecule(s) of the adjacent thin filament, exerts force until it detaches after a certain amount of sliding has occurred, and begins the cycle afresh by reattaching further along the thin filament. The proposition that actual attachment occurs is not universally accepted, but I regard it as being sufficiently strongly supported by (*a*) the great increase in stiffness that occurs when a muscle is activated and (*b*) the observation of Sheetz & Spudich (59) that a myosin-coated polystyrene bead ceases to undergo Brownian movement when it comes in contact with, and begins to move along, an actin bundle.

The Cross-Bridge Cycle

The first mathematical theory of contraction based on sliding filaments with cyclic action of cross-bridges was worked out by me in the couple of years after the discovery of sliding filaments and published in 1957 (34). It made admittedly primitive assumptions about the rate constants for attachment and detachment (by utilization of a "high-energy phosphate" molecule) of cross-bridges as functions of the relative positions of the thick and thin filament sites, which necessarily move relative to one another as the filaments slide. I was able to get an adequate fit to A. V. Hill's 1938 relations between load, shortening speed, and rate of energy liberation during shortening. As regards lengthening, the discontinuity in the force-velocity curve between shortening and lengthening (50) was immediately explained, and the very low rate of energy liberation (33, 46) was explained by the additional postulate that overstretched cross-bridges can break without utilization of energy from chemical reactions. Fresh experiments by A. V. Hill (32) a few years later showed that the rate of energy liberation did not (as in his 1938 equations) increase indefinitely with shortening speed but passed through a maximum. I was able to accommodate this new evidence (35) by postulating that attachment takes place in two stages and that the first, but not the second, is easily reversible without utilization of chemical energy.

Even with the last-mentioned addition, it is now clear that the theory is incomplete. It cannot explain the length transients seen in response to a sudden change of load, reported in 1960 by Podolsky (55), or the tension transients following a sudden change of length, which my colleagues and I have studied since the late 1960s (17–20), without postulation of additional steps that occur on the millisecond time scale while the cross-bridge is attached. Also, the functions chosen to represent the rate constants for attachment and detachment were only first guesses and cannot be expected to be better than rough approximations to the truth even if, as I believe, the general outline is correct, i.e. that the cross-bridge is elastic, that its rate constant for attachment is moderate if the relative positions of the thick and thin sites are such that positive force would be exerted if attachment occurrs, but that its rate constant for detachment is low until shortening by sliding brings the cross-bridge to the position where it exerts zero force and becomes high at positions where the force exerted is negative. I will assume this much for the purposes of this article. It is a skeleton theory: Plenty of questions remain, not least because the theory contains no specific statements about the nature of the actin-myosin bond, the nature or location of the elastic element, or the mechanism for breaking the bond at the right time.

In my 1957 article I denoted the "high-energy phosphate" molecule by XP and avoided committing myself to the assumption that it actually was ATP. Also, I assumed that it caused dissociation by becoming bound to actin, not to myosin. What was in my mind was that myosin was the enzyme and that XP bound to actin was the substrate. A related possibility that I was thinking of, but did not incorporate in the theory, was that adenosine might be permanently bound to actin as a prosthetic group and be rephosphorylated from the diphosphate to the triphosphate by reaction with PC in solution in the sarcoplasm; in that case, XP would have been PC, not ATP. These possibilities evaporated when, five years later, Cain et al (8) showed that muscle (appropriately poisoned) can contract more or less normally with utilization of ATP but not of PC, and many in vitro studies showed that ATP could both bind to, and be split by, myosin without any involvement of actin. My theory therefore needed to be modified by (a) specifying that XP was in fact ATP and (b) that it caused dissociation by binding to the myosin side of the cross-bridge, not the actin side. The kinetics, which are the interesting part of the theory, are not altered by this change in postulates.

The general outline of the 1957 theory has recently been supported, as regards overall events in steady shortening, by the experiments of Brenner & Eisenberg on skinned muscle fibers contracting under conditions of altered ATP concentration (6).

Transient Responses

I have already mentioned that the initial changes of length when load on stimulated muscle is changed (isotonic transient) and the initial changes of force when length is suddenly changed (isometric transient) require additions to my 1957 theory. The two types of transient are clearly alternative expressions of the same underlying properties of the muscle. If one assumes linearity for small changes of load or length, it is easy to calculate the expected time course of the isotonic transient given the time course of the isometric transient. This was done by C. M. Armstrong, F. J. Julian and myself in 1966; we obtained agreement as good as could be expected in view of the nonlinearities of the phenomena, but in our published note (3) we only stated the conclusion and did not present the results. I shall assume that the underlying phenomena are the same in both cases. Since the isotonic transient is a damped oscillation while the isometric transient has only a finite number of phases (three components decaying approximately exponentially after the initial tension change), it is probable that the directly relevant independent variable is length change rather than force. For this reason we have concentrated on the isometric transient (force response to step change of length) both in our experiments and in our attempts to explain the results.

Measurements of the isometric transients at different muscle lengths (18) led to the conclusion that the phenomenon is a property of the cross-bridges alone; the filaments are so stiff that their compliance contributed only a small fraction to that which was measured. It has to be admitted, however, that it was assumed tacitly that cross-bridge stiffness was not affected by change of lateral spacing between filaments; if this turns out to be an important factor (23), the conclusion about filament compliance may have to be reconsidered. It is most unlikely that the conclusion that each cross-bridge contains an elastic element almost free from damping and obeying Hooke's law closely could be upset. According to our present estimates, the compliance of this elastic element is such that force is reduced to zero from the isometric value by ~4 nm of relative sliding movement of the filaments.

In 1971 R. M. Simmons and I (41) gave a tentative explanation of the isometric transient. We postulated a sequence of attached states with progressively increasing affinity and with progressively increasing stretch of the elastic component of the cross-bridge. This provided a semiquantitative explanation of two striking nonlinearities in the phenomenon. My present view of that theory is that it is still at least a valuable working hypothesis with a substantial chance of turning out to be correct in essentials, but I do not claim that it would be justifiable at present to place much reliance on its correctness. A testable prediction from the theory is that even during isometric contraction each cross-bridge should be in an equilibrium, switching between adjacent attached states on a time scale related to the time constant of the early tension recovery phase of the isometric transient (a few tenths of a millisecond in frog muscle at 0°C). This time scale is too long for investigation with most of the probes of cross-bridge orientation that have so far been extensively used [fluorescence and electron paramagnetic resonance (epr)], but there is hope for solution of the problem with probes using phosphorescence or saturation transfer epr.

Stepwise Shortening

Reports that length changes in muscle often take place in a stepwise or oscillatory manner have given rise to a good deal of controversy in the last few years. The situation was presented recently in a pair of articles (37, 57). My own view, presented in one of those articles (37), is that some of the phenomena are instrumental artifacts and that those which are genuine are expressions of properties that are more effectively studied as "isotonic transients" (55) or better still as "isometric transients" (17–20).

UNANSWERED QUESTIONS

In my 1974 review (36) I finished by listing seven unknowns, as follows:

1. which structure is the elastic element in the cross-bridge;
2. what is the nature of this elasticity;
3. what structure undergoes the stepwise change;
4. whether the attachment of the myosin head is to a single actin monomer or to two (or more) monomers within the thin filament;
5. what kind of bonds hold the myosin head to the thin filament;
6. how binding of ATP causes myosin to dissociate from actin; and
7. what is the significance of the fact that each myosin molecule has two heads.

None of these questions has yet been answered, and I still regard them as among the most important. Question 3 was formulated on the provisional assumption that the essentials of our 1971 theory (41) are correct, i.e. that each attached cross-bridge has more than one possible state and that it is in fairly rapid equilibrium between two or more of these states.

Since these questions are not yet answered, it might be supposed that little progress has been made in recent years. This is not the case because several fresh approaches have been made to the problem in the last few years and have begun to yield a new crop of information bearing not only on the questions I have listed but on other aspects of the problem. It is still too early to expect decisive answers to have emerged.

NEW APPROACHES

Up-to-date reviews of several aspects of the molecular mechanism of muscle contraction were published in the 1987 issue of *Annual Review of Physiology* (Volume 49) as a Special Topic section. Those articles cover most of what I might have said in this section in much greater detail than would be possible in the space allotted to me. The following paragraphs therefore concentrate for the most part on a few aspects not covered in that group of reviews.

Molecular Genetics and Myosin Structure

Watson & Crick proposed the double helix structure of DNA in 1953 and it was soon a commonplace to say that it was revolutionizing every branch of biology. However, it was some 30 years before it even began to make an impact on our understanding of muscle. In 1983 Karn et al (49) sequenced a myosin gene from the nematode *Caenorhabditis elegans,* and since then many sequences, partial or complete, either from DNA or from the protein itself, have been obtained for myosins from other organisms, including *Drosophila.* Several highly conserved regions have been recognized, but their respective functions have not yet been identified. High-quality crystals of the head (S-1 fraction) of myosin were obtained a few years ago (66), and great progress is to be expected when the tertiary structure has been worked out from these crystals by x-ray diffraction and has been correlated with the sequence. It is to be hoped that this will lead to identification of the site(s) for attachment to actin and to elucidation of the allosteric process by which binding of ATP to myosin reduces its affinity for actin.

Time-Resolved X-Ray Diffraction

Low-angle x-ray diffraction has been the principal method for obtaining ultrastructural information from living muscle, but until recently the exposures required were so long that variations with time could not be studied except by stroboscopic methods on the oscillatory activity of asynchronous insect flight muscles. The advent of the synchrotron as an x-ray source of much higher intensity than that of normal x-ray tubes, together with digital recording to allow averaging of results from several contractions, has made a time resolution of the order of 1 ms possible; several groups are exploiting this technique.

The most interesting results refer to two situations, the onset of contraction (51) and the isometric transient (response to small stepwise shortening) (45). As regards the former, changes in the relative intensities of the equatorial reflections indicate a shift of material from the surface of the thick filaments towards the thin filaments. It is debated whether this directly indicates actual attachment of cross-bridges or merely outward movement of cross-bridges to permit subsequent attachment. Since the change is substantially earlier than the rise of tension, the former interpretation if correct is usually taken to imply that there is a lag (several milliseconds in frog muscle at low temperature) within the cross-bridge cycle between attachment and contribution to force. An alternative explanation could be that during the rise of activation there is a phase when cross-bridges attach but do not exert force, or exert less force than during full activation.

The other very striking result I wish to draw attention to is the large drop in intensity of the meridional reflection at 14.3 nm observed by H. E. Huxley and his collaborators when the muscle is suddenly shortened by say 10 nm per half sarcomere (45). This change appears to lag behind the length change itself by about 0.5 ms (frog muscle at 5°C) and is therefore likely to be related to Phase 2 of the tension transient (early tension recovery) rather than Phase 1 (tension drop simultaneous with the shortening). It may seen ungrateful to respond to such a technical triumph by asking for more, but I would dearly like to see this experiment repeated with a quicker step and with finer time resolution in the recording, to see whether the time course of the drop of intensity is similar to that of the early tension recovery in the tension transient. Several explanations for the drop of intensity are possible. H. E. Huxley et al (45) suggest either unequal longitudinal displacements of the ends of myosin heads closest to the thick filament shaft or an increased obliquity of the myosin heads, which might be nearly perpendicular to the fiber axis during isometric contraction. Another possibility is that only one of the two heads of each myosin molecule is attached and generating force during isometric contraction and that the other head is able to attach rapidly after the shortening step at a position roughly midway between the positions of the already-attached bridges. This mechanism would require a much higher rate constant for attachment of the second head than for a head whose twin was unattached. This supposition is not unreasonable since the attached first head might hold the second head in a favorable position. Another difficulty for this idea is that no increase in stiffness accompanies the early rise of tension; this could be explained if the two heads share a single elastic element (e.g. in S-2).

This method clearly has great promise for the future. It would be exciting to see the results of applying it to muscle of teleost fishes, in which Luther & Squire (53) have shown that all the thick filaments in a sarcomere are similarly oriented (in contrast to the situation in other vertebrates), so that the diffraction pattern is much more informative.

"Caged" ATP and Other Substances

The phrase "caged X" refers to a substance that is decomposed by a flash of light into X plus another (preferably inert) substance; the original caged X does not share in the characteristic actions of X. Thus a skinned muscle fiber loaded with caged ATP will undergo a sudden increase in the ATP concentration when exposed to an appropriate flash. The most thorough development of this technique in its application to muscle has been made by Goldman and colleagues (21, 22). From rigor (zero ATP) in the absence of calcium, the liberated ATP is expected to dissociate the cross-bridges and produce a relaxed condition. This was indeed found to happen, but only after a short-lived rise of tension attributable to active contraction, i.e. the system was turned on by the presence of rigor bridges (5) despite the absence of calcium. A refinement of this experiment involved comparison of the tension time courses observed with different amounts of initial tension in the rigor fiber. This made it possible to separate the component due to preexisting cross-bridges from that due to cross-bridges formed after the release of ATP.

This method is also being used with caged calcium to produce sudden activation of a skinned fiber. This approach promises to greatly increase the value of the skinned fiber preparation since the chief drawback of the latter has lain in the slowness of activation by inward diffusion of added calcium, which causes nonuniform contraction and consequent disorganization of the fiber structure.

Birefringence

Muscle has been studied with polarized light for a century and a half [e.g. in 1858 von Brücke (65) deduced correctly that the birefringence was due to rodlets that were not individually stretched when a muscle fiber as a whole was stretched], and records of the fall of birefringence on stimulation were made in several laboratories before World War II. Nevertheless this technique was little used except in the elegant study by Eberstein & Rosenfalck (9) on living isolated fibers, until the mid-1970s. Two investigations (62, 67) then showed that the strength of birefringence is reduced in rigor as well as during contraction and that the amount of the reduction is proportional to filament overlap. These results imply that the change occurs only in the overlap zone.

The birefringence technique, like x-ray diffraction, has the advantage of providing ultrastructural information on living tissue, at rest or in contraction. It is less specific than x-ray diffraction but is capable of much better time resolution. The major changes observed are presumably due chiefly to changes in the orientation of parts of the myosin molecules, though some of the small changes soon after stimulation are probably due to membrane events and are thus outside the scope of this article. This technique deserves to be more widely used. Observations by Dr. Malcolm Irving on changes in birefringence in response to quick changes in length are due to be published soon.

CONCLUSION

In this article I have restricted myself almost entirely to aspects of the mechanism of force development and shortening that can be elucidated by experiments on intact, working muscle. Recent experiments of this and other kinds have tended to confirm the version of the sliding-filament theory that came to be widely accepted within a dozen years after the first proposal of sliding filaments in 1953: Force is generated by cross-bridges that consist of the heads of myosin molecules projecting from each thick filament and that act cyclically and more or less independently of one another. They attach to an adjacent thin filament, generate force, and detach as a consequence of binding of ATP. The rate constants for attachment and detachment are functions of the displacement of the myosin site relative to the actin site that occurs as the filaments slide past each other. It is this last feature that makes experiments on the intact contractile structure a continuing necessity: In vitro it is impossible to apply a load or to control the sliding motion, even if the contractile proteins are in their filamentary form. Although experiments on intact muscle can give no more than hints about the intimate details of the contraction process, understanding at this deeper level is advancing fast not only through in vitro biochemical investigations and structural studies but especially through techniques that are partly physiological and partly biochemical in nature. These include methods that depend on removing the surface membrane of muscle fibers so as either to change the concentrations of key solutes or to attach probes to the working protein molecules and include methods that follow movements of separated thick or thin filaments sliding over stationary actin or myosin, respectively. Molecular genetics too has begun to make an impact. It is nowadays necessary for those interested in muscle contraction, whether their own work is physiological, biochemical, structural, or genetical, to try to keep abreast with the advances made by the other disciplines: New observations can be interpreted only within the complex framework provided by the many different approaches now being used.

Literature Cited

1. Altringham, J. D., Bottinelli, R. 1985. The descending limb of the sarcomere length-force relation in single muscle fibres of the frog. *J. Muscle Res. Cell Motil.* 6:585–600

2. Altringham, J. D., Pollack, G. H. 1984. Sarcomere length changes in single frog muscle fibres during tetani at long sarcomere lengths. In *Contractile Mechanisms in Muscle*, ed. G. H. Pollack, H. Sugi, pp. 473–93. New York/London: Plenum

3. Armstrong, C. M., Huxley, A. F., Julian, F. J. 1966. Oscillatory responses in frog skeletal muscle fibres. *J. Physiol. London* 186:26–7P

4. Bagni, M. A., Cecchi, G., Colomo, F., Tesi, C. 1986. The sarcomere length-tension relation in short length-clamped segments of frog single muscle fibres. *J. Physiol. London* 377:91P

5. Bremel, R. D., Weber, A. 1972. Cooperation within actin filament in vertebrate skeletal muscle. *Nature New Biol.* 238:97–101

6. Brenner, B., Eisenberg, E. 1986. Rate of force generation in muscle: Correlation with actomyosin ATPase activity in solution. *Proc. Natl. Acad. Sci. USA* 83:3542–46

7. Brown, L. M., González-Serratos, H., Huxley, A. F. 1984. Structural studies of the waves in striated muscle fibres shortened passively below their slack length. *J. Muscle Res. Cell Motil.* 5:273–92

8. Cain, D. F., Infante, A. A., Davies, R. E. 1962. Chemistry of muscle contraction. Adenosine triphosphate and phosphorylcreatine as energy supplies for single contractions of working muscle. *Nature* 196:214–17

9. Eberstein, A., Rosenfalck, A. 1963. Birefringence of isolated muscle fibres in twitch and tetanus. *Acta Physiol. Scand.* 57:144–66

10. Edman, K. A. P. 1979. The velocity of unloaded shortening and its relation to sarcomere length and isometric force in vertebrate muscle fibres. *J. Physiol. London* 291:143–59

11. Edman, K. A. P., Reggiani, C. 1987. The sarcomere length-tension relation determined in short segments of intact muscle fibres of the frog. *J. Physiol. London* 385:709–32

12. Eggleton, P., Eggleton, G. P. 1927. The inorganic phosphate and a labile form of organic phosphate in the gastrocnemius of the frog. *Biochem. J.* 21:190–95

13. Engelhardt, W. A., Ljubimowa, M. N. 1939. Myosine and adenosine-triphosphatase. *Nature* 144:668–9

14. Fenn, W. O. 1923. A quantitative comparison between the energy liberated and the work performed by the isolated sartorius muscle of the frog. *J. Physiol. London* 58:175–203

15. Fenn, W. O. 1924. The relation between the work performed and the energy liberated in muscular contraction. *J. Physiol. London* 58:373–95

16. Fiske, C. H., Subbarow, Y. 1927. The nature of the "inorganic phosphate" in voluntary muscle. *Science* 65:401–3

17. Ford, L. E., Huxley, A. F., Simmons, R. M. 1977. Tension responses to sudden length change in stimulated frog muscle fibres near slack length. *J. Physiol. London* 269:441–515

18. Ford, L. E., Huxley, A. F., Simmons, R. M. 1981. The relation between stiffness and filament overlap in stimulated frog muscle fibres. *J. Physiol. London* 311:219–49

19. Ford, L. E., Huxley, A. F., Simmons, R. M. 1985. Tension transients during steady shortening of frog muscle fibres. *J. Physiol. London* 361:131–50

20. Ford, L. E., Huxley, A. F., Simmons, R. M. 1986. Tension transients during the rise of tetanic tension in frog muscle fibres. *J. Physiol. London* 372:595–609

21. Goldman, Y. E., Hibberd, M. G., Trentham, D. R. 1984. Relaxation of rabbit psoas muscle fibres from rigor by photochemical generation of adenosine-5'-triphosphate. *J. Physiol. London* 354:577–604

22. Goldman, Y. E., Hibberd, M. G., Trentham, D. R. 1984. Initiation of active contraction by photogeneration of adenosine-5'-triphosphate in rabbit psoas muscle fibres. *J. Physiol. London* 354:605–24

23. Goldman, Y. E., Simmons, R. M. 1986. The stiffness of frog skinned muscle fibres at altered lateral filament spacing. *J. Physiol. London* 378:175–94

24. González-Serratos, H. 1971. Inward spread of activation in vertebrate muscle fibres. *J. Physiol. London* 212:777–99

25. Gordon, A. M., Huxley, A. F., Julian, F. J. 1966. The variation in isometric tension with sarcomere length in vertebrate muscle fibres. *J. Physiol. London* 184:170–92

26. Hanson, J., Huxley, H. E. 1953. Structural basis of the cross-striations in muscle. *Nature* 172:530–32

27. Harman, J. W. 1954. Contractions of skeletal muscle myofibrils by phase microscopy (motion picture). *Fed. Proc.* 13:430

28. Hasselbach, W. 1953. Elektronenmikroskopische Untersuchungen an Muskelfibrillen bei totaler und partieller Extraktion des L-Myosins. *Z. Naturforsch.* 8b:449–54

29. Hill, A. V. 1932. The revolution in muscle physiology. *Physiol. Rev.* 12:56–67

30. Hill, A. V. 1938. The heat of shortening and the dynamic constants of muscle. *Proc. R. Soc. London Ser. B* 126:136–95

31. Hill, A. V. 1959. The heat production of muscle and nerve, 1848–1914. *Ann. Rev. Physiol.* 21:1–18

32. Hill, A. V. 1964. The effect of load on the heat of shortening of muscle. *Proc. R. Soc. London Ser. B* 159:297–318

33. Hill, A. V., Howarth, J. V. 1959. The reversal of chemical reactions in contracting muscle during an applied stretch. *Proc. R. Soc. London Ser. B* 151:169–93

34. Huxley, A. F. 1957. Muscle structure and theories of contraction. *Prog. Biophys. Biophys. Chem.* 7:255–318

35. Huxley, A. F. 1973. A note suggesting that the cross-bridge attachment during muscle contraction may take place in two stages. *Proc. R. Soc. London Ser. B* 183:83–86

36. Huxley, A. F. 1974. Muscular contraction. Review lecture given at the meeting of the Physiological Society at Leeds University on 14–15 December, 1973. *J. Physiol. London* 243:1–43

37. Huxley, A. F. 1986. Comments on "Quantal mechanisms in cardiac contraction." *Circ. Res.* 59:9–14

38. Huxley, A. F., Niedergerke, R. 1954. Interference microscopy of living muscle fibres. *Nature* 173:971–73

39. Huxley, A. F., Niedergerke, R. 1958. Measurement of the striations of isolated muscle fibres with the interference microscope. *J. Physiol. London* 144:403–25

40. Huxley, A. F., Peachey, L. D. 1961. The maximum length for contraction in vertebrate striated muscle. *J. Physiol. London* 156:150–65

41. Huxley, A. F., Simmons, R. M. 1971. Proposed mechanism of force generation in striated muscle. *Nature* 233:533–38

42. Huxley, H. E. 1953. Electron microscope studies of the organisation of the filaments in striated muscle. *Biochim. Biophys. Acta* 12:387–94

43. Huxley, H. E. 1957. The double array of filaments in cross-striated muscle. *J. Biophys. Biochem. Cytol.* 3:631–48

44. Huxley, H. E., Hanson, J. 1954. Changes in the cross-striations of muscle during contraction and stretch and their structural interpretation. *Nature* 173:973–76

45. Huxley, H. E., Simmons, R. M., Faruqi, A. R., Kress, M., Bordas, J., Koch, M. H. J. 1983. Changes in the x-ray reflections from contracting muscle during rapid mechanical transients and their structural implications. *J. Mol. Biol.* 169:469–506

46. Infante, A. A., Klaupiks, D., Davies, R. E. 1964. Adenosine triphosphate: Changes in muscles doing negative work. *Science* 144:1577–78

47. Julian, F. J., Moss, R. L. 1980. Sarcomere length-tension relations of frog skinned muscle fibres at lengths above the optimum. *J. Physiol. London* 304:529–39

48. Julian, F. J., Sollins, M. R., Moss, R. L. 1978. Sarcomere length nonuniformity in relation to tetanic responses of stretched skeletal muscle fibres. *Proc. R. Soc. London Ser. B* 200:109–16

49. Karn, J., Brenner, S., Barnett, L. 1983. Protein structural domains in the *Caenorhabditis elegans unc-54* myosin heavy chain gene are not separated by introns. *Proc. Natl. Acad. Sci. USA* 80:4253–57

50. Katz, B. 1939. The relation between force and speed in muscular contraction. *J. Physiol. London* 96:45–64

51. Kress, M., Huxley, H. E., Faruqi, A. R., Hendrix, J. 1986. Structural changes during activation of frog muscle studied by time-resolved x-ray diffraction. *J. Mol. Biol.* 188:325–42

52. Lundsgaard, E. 1930. Untersuchungen über Muskelkontraktionen ohne Milchsäurebildung. *Biochem. Z.* 217:162–77

53. Luther, P. K., Squire, J. M. 1980. Three-dimensional structure of the vertebrate muscle A-band. II. The myosin filament superlattice. *J. Mol. Biol.* 141:409–39

54. Needham, D. M. 1950. Myosin and adenosinetriphosphate in relation to muscle contraction. *Biochim. Biophys. Acta* 4:42–49

55. Podolsky, R. J. 1960. Kinetics of muscular contraction: The approach to the steady state. *Nature* 188:666–68

56. Pollack, G. H. 1983. The cross-bridge theory. *Physiol. Rev.* 63:1049–1113

57. Pollack, G. H. 1986. Quantal mechanisms in cardiac contraction. *Circ. Res.* 59:1–8

58. Ramsey, R. W., Street, S. F. 1940. The

isometric length-tension diagram of isolated skeletal muscle fibers of the frog. *J. Cell. Comp. Physiol.* 15:11–34

59. Sheetz, M. P., Spudich, J. A. 1983. Movement of myosin-coated fluorescent beads on actin cables in vitro. *Nature* 303:31–35

60. Straub, F. B. 1943. Actin. *Stud. Inst. Med. Chem. Univ. Szeged* (1942) 2:3–15

61. Szent-Györgyi, A. 1942. Discussion. *Stud. Inst. Med. Chem. Univ. Szeged* (1941–42) 1:67–71

62. Taylor, D. L. 1976. Quantitative studies on the polarization optical properties of striated muscle. 1. Birefringence changes of rabbit psoas muscle in the transition from rigor to relaxed state. *J. Cell Biol.* 68:497–511

63. ter Keurs, H. E. D. J., Iwazumi, T.. Pollack, G. H. 1978. The sarcomere length-tension relation in skeletal muscle. *J. Gen. Physiol.* 72:565–92

64. ter Keurs, H. E. D. J., Iwazumi, T., Pollack, G. H. 1979. The length-tension relation in skeletal muscle: Revisited. In *Cross-Bridge Mechanism in Muscle Contraction,* ed. H. Sugi, G. H. Pollack, pp. 277–95. Tokyo: Univ. Tokyo Press

65. von Brücke, E. 1858. Untersuchungen über den Bau der Muskelfasern mit Hülfe des polarisirten Lichtes. *Denkschr. Akad. Wiss. Wien Math.-Naturwiss. Kl.* 15:69–84

66. Winkelmann, D. A., Mekeel, H., Rayment, I. 1985. Packing analysis of crystalline myosin subfragment-1: Implications for the size and shape of the myosin head. *J. Mol. Biol.* 181:487–501

67. Yanagida, T. 1976. Birefringence of glycerinated crab muscle fiber under various conditions. *Biochim. Biophys. Acta* 420:225–35

ADDED IN PROOF:

68. Akster, H. A., Granzier, H. L. M., ter Keurs, H. E. D. J. 1984. Force - sarcomere length relations vary with thin filament length in muscle fibres of the perch (Perca fluviatilis L.). *J. Physiol. London* 353:61P

PLANT PHYSIOLOGY AND PLANT MOLECULAR BIOLOGY

Annu. Rev. Plant Physiol. Plant Mol. Biol. 1991. 42:1–20

SHORT STORY OF A PLANT PHYSIOLOGIST AND VARIATIONS ON THE THEME

Erasmo Marrè

Department of Biology, University of Milan, Italy

KEY WORDS: plant physiology, biology, vision of the world

CONTENTS

This is the use of memory: for liberation

T. S. Eliot, Little Gidding

The invitation to write this prefatory chapter is a great honor. However, what to write? Not much in my experience seems worth the telling.

While pondering my task, I happened to read the verse of T. S. Eliot cited above, which suggested an approach. I thus write this chapter with the aim of obtaining, from a recollection and a reconsideration of my life, a better understanding of my actions and emotions, of their interrelationship, and of my interactions with others. I reflect, in other words, on how I have played my part in this world. If Eliot is right, these recollections of my history, seen as a whole and as objectively as possible, can lead to a liberation from futilities and self-attachment, and thus to a deeper understanding of the importance of events and persons in shaping my life.

Here I treat also of my activities and ideas in Plant Physiology. I feel no conflict between this "official" reportage and the more personal approach, for biology in general, and plant physiology in particular, have played a large part in my life and in my mental development.

BIOGRAPHICAL

Early Years

When very young—say between 5 and 15—I lived with few external restrictions in a small villa close to Genoa and the sea, a villa surrounded by gardens and orchards. Here I developed the wild side of my nature. Through fishing, through capturing and trying to tame small animals the seed of my interest in the functioning (the physiology!) of living beings began to germinate. Those who are strongly opposed to hunting forget that perhaps its deepest root is a kind of love: a longing to take into one's hands (to possess) the object of love. This distorted form of love can be a primitive stage of a less egoistic, more enlightened feeling for the beauty of nature and for our deep involvement in it. Fishing and hunting imply knowing the names and habits of fishes, birds, and squirrels; constructing traps stimulates technical ingenuity; taming the captured animals involves principles of both physiology and ethology. I gave up hunting at about the age of 20, when I realized that killing or caging did not give me what I sought—namely, an intimate contact with wild creatures.

I had access to a large and varied collection of books. My schoolwork took up only a small part of my time, and I dedicated much of the remainder to reading the classics: Shakespeare and Chesterton, de Kruif and Pascal, Fabre and Melville. This omnivorous reading increased in my high school years, when my family moved to Genoa. I became seriously interested both in literature and in the natural sciences. Thus I discovered the great theories and controversies related to biological evolution; I became familiar with the works of Lamarck, Darwin, Haeckel, and Huxley, and with the opposition of the Catholic Church. Seeking a bridge between biology, the human sciences, and religion, I studied Spencer, Bergson, and Theilard de Chardin.

In 1938 I entered the university as a student of medicine and immediately started laboratory work in anatomy and histology. Physiology became part of my studies almost immediately, inasmuch as my anatomy professor, Ganfini, was interested in the relationship between body weight and the cytological features of the hypophysis. This first contact with the problem of biological regulation informed my entire scientific life.

Having realized that I was interested in basic rather than in applied biology, and in dynamic rather than in descriptive biology, I switched from Medicine to Natural Sciences. Greatly influential in this decision was Ettore Remotti, professor of Comparative Anatomy and Physiology. Remotti was a man of modern scientific views who moved with equal ease and acuity in the fields of morphology and physiology. Remotti introduced me to the wonderful world of embryology, and my first published paper concerned the development of the embryo of a small viviparous fish and (of interest in the light of my later work) the mechanisms of nutrient exchange between mother and embryo.

The intensity of my dedication to laboratory work in those years left little time to other occupations. I was still reading very widely: philosophy (Spinoza, Kant) and novels (I discovered Poe, Dostoevski, Gogol). My social life was limited to contact with students attending the biological laboratories and to the membership of a Catholic association of students. In the latter organization I met two most enlightened priests, E. Guano and F. Costa, who had a profound influence upon me, helping to develop my feeling that a man's highest aim should be to understand human life in its wholeness. For some time I seriously considered entering a monastery. At about the same time, however, I was greatly attracted to a girl I met at the university. During vacation time, we mountaineered in the Alps and explored the Ligurian caves. In 1945 she agreed to become my wife; we still share the good and the bad, but that is another story.

The War

Three different tendencies (a passion for science, a fascination with philosophy, and love for a girl) coexisted in me in a largely unconscious but deep conflict, when the events of the war abruptly spurred me into action. The war until then had little affected my life; my feelings were definitely against nazism and fascism, but I had seen few opportunities to translate these feelings into action. However, July of 1943 saw the fall of Mussolini, and in October Italy surrendered to the Allies. The Germans invaded northern Italy, and the official government in southern Italy (occupied by the Allied armies) soon found itself in open conflict with its former ally. In northern Italy the fascists established a republican government, largely controlled by the Germans.

As a medical student I had been exempted from military service. However, in October 1943 I was informed that my call-up was imminent, and so, with three friends, I decided to leave Genoa for the south to join the Italian troops being organized to fight on the side of the Allies. After an adventurous trip through central Italy we crossed the line close to Termoli and, to cut a long story short, a month later we joined a group of some 30 people organized in Naples by the Italian National Liberation Committee in close connection with the American Vth Army. The purpose of this group was the dispatch to northern Italy of small teams of four men and a radio in order to establish contact and collaborate with the partisans operating against the fascists and the Germans.

After three months' training, including instruction in sabotage techniques and a parachutist course at Brindisi, I was parachuted into a partisan-occupied valley in the Alps, and moved to the mountains north of Genoa, in the area occupied by the partisans of the VI zone. This partisan group was politically controlled by the communists, but most of the military commanders were independent of political parties and were merely collaborating with the Communists in the fight against fascism. After some months of loyal and active collaboration with the partisans of the VI zone (organizing an information service for our base in South Italy, receiving air drops of weapons, sabotaging electrical lines, railways, etc), I was led by circumstances to assume a different role. A partisan brigade (the Arzani) had been dispersed by the Germans and its commander captured in December 1944. I was asked to take command of this brigade and to reorganize it. I spent the last four months of the war doing this, I think satisfactorily.

I dwell on this part of my story because I am convinced that these 18 months of "atypical military service" formed a crucial part of my education. Even if all my forces were concentrated on fighting fascism, my past experience conditioned my actions and decisions, and much of what I learned through action and experience was indelibly to influence my subsequent development. Thus, my earlier hunting, fishing, and mountain climbing made it easier than it would otherwise have been for me to move in the wilderness of the Apennine mountains; my training in experimental science helped me to see military problems in terms of final goals (survival, victory), to consider the means available before making plans, and, above all, to be aware of the difference between abstract plans and the concrete possibility of putting them into practice, when most of the parameters were unpredictable or uncertain. On the other hand, the human context in which I was acting was absolutely new to me. I had suddenly found myself part of an extremely mixed population in which communist idealists with years of imprisonment behind them, antifascist officers from the disbanded Italian army, liberal-minded intellectuals like myself (few indeed), socialist workers, and individualistic peasants moved to defend their rights all struggled to find a ground for unitary action under the pressure of circumstances. Understanding others and making myself understood in sometimes very difficult discussions, participating in actions, and confronting the emotions of such various people was an exceptionally important experience; it confirmed or corrected the vague concepts I had derived from books. As a result I formed a number of convictions that I still consider valid. First, although adherence to dogmatic interpretations of life may affect a person's objectivity and rationality, no one should be judged on the basis of ideology alone. Second, events (political and historical processes) are determined by internal, indefinite forces, arising from the depths of individual consciences, rather than by the conscious will and rationality of those holding official power (see Tolstoi's *War and Peace*). The behavior of each individual in a given moment is conditioned by impulses springing from the subconscious levels of personality, as shaped by genetics, environment, and previous choices. These impulses can be analyzed and modified at the conscious level where information of all kinds (i.e. culture in the widest sense) are integrated and utilized for practical action. Thus the development of rationality and knowledge serves as a filter for impulses, and both an artistic and a scientific education play a determinant role in this development. In this perspective, the contribution of the study of experimental sciences to the formation of personality can be ordered according to a hierarchy of values, rising from those studied in physics to those investigated in ethology and psychology, where a connection is established with subjects treated in the human sciences and arts.

I think I was born with a propensity to influence the events around me. This tendency developed further when I found myself the head of some 200 persons, with the power of decision in all matters concerning their military activity. Authority is responsibility first toward the objective justifying it and second toward the others working to accomplish this objective. What I rapidly realized was that authority is only legitimated—and accepted—when it is based on example and on the capacity to understand the needs of those who are subjected to it, to make good use of the abilities of others, and to win esteem rather than affection. Such qualities, developed while with the brigade, were to be useful long after the war, when I became the leader of a research group.

The war parenthesis ended in April 1945 with the surrender of the German army in Italy. My experience with political parties in the previous months caused me to decline invitations to enter politics. Instead I concentrated on my two main aims at that moment: to marry the girl I loved, and to find an acceptable job that would enable me to maintain a family. Of the few appointments offered to me, I chose an assistant professorship in Botany at the University of Genoa, which corresponded to my interests. At the beginning of 1946 I found myself married and fully involved in plant biology.

On the Way to Plant Physiology

At the end of the war the general situation of the biological sciences in Italy was far from satisfactory, for ancient as well as recent reasons. In spite of some great names (Malpighi, Redi, Spallanzani, Golgi—isolated peaks), the Italian cultural tradition after 1600 had favored speculative philosophy over experimental biology. The experimental investigation of the mechanisms of life was discouraged, the theory of evolution being considered an attack on religious dogmas. Thus most biological schools and studies were concerned more with description than with the elucidation of mechanism. This unfavorable situation had been worsened by fascism, with its idealistic roots and autarchic (cultural as well as economic) policy. As a consequence of these historical, cultural, and political conditions, the funds for experimental research (books, instruments, jobs) were scarce.

In 1945 the situation of experimental botany was particularly bad. But while most Italian botanists were then involved in descriptive work, in either taxonomy or morphology, some groups were developing more modern and dynamic approaches. In Florence, Chiarugi and collaborators, among whom was Francesco D'Amato, were carrying out good work in caryology, embryology, and cytogenetics, and some of their younger collaborators were initiating a fruitful genetic approach to plant biology. In Pavia, Ciferri was taking the first steps from plant pathology towards physiology, sending one of his best assistants, Felice Bertossi, abroad (to Gautheret's laboratory) to learn methods of tissue culture. Finally, and most important for my future work, in Padua a highly enlightened and dedicated scientist, Giuseppe Gola, was establishing the basis of a really modern Italian plant physiology, with regard both to problems and to methods. (As pointed out by Robert Hill, Gola was the first to recognize the role of a cytochrome in plant oxidation-reductions— in 1915.) The torch of plant physiology lit by Gola passed into the hands of some of his collaborators, in particular Sergio Tonzig, who in 1940 had moved to Milan, where he had started to set up a modern laboratory of plant physiology. The laboratory of Botany in Genoa, where I had found a job in 1946, was then ruled by Giuseppina Zanoni, an intelligent woman sincerely dedicated to biology. At that time, as the new full professor in Botany (having succeeded a strict taxonomist), she was just starting to organize a modern plant biology laboratory, building it up literally from its very foundations (the old building had been almost completely destroyed during the war).

For me this situation had advantages as well as disadvantages. Laboratory equipment and facilities were nonexistent, contacts with other botanists were scarce, and there was no established scientific tradition; but these disadvantages enabled me to start my scientific activity with a minimum of conditioning and a maximum of freedom in choosing a line of work.

Choice of Regulation as the Central Problem

A lucky coincidence of my interests and those of Professor Zanoni facilitated the choice of the field in which I could work. From the beginning of my studies I had been strongly attracted by the problem of hormonal correlations.

I was interested in the mechanisms that integrate the various life processes and allow each organism to behave as a unity, which I regarded as the most amazing and splendid mystery of life. Thus work on certain modest aspects of the problem of unity in a plant absorbed me.

My interest in hormonal correlations fitted in well with the views of Zanoni, who had for some years been conducting original research on the possible role of the sexual organs of plants in the control of differentiation, a role such organs were known to play in animals. While collaborating in her research on the specific actions of anthers and pistils, I concluded that these organs markedly influenced the growth of other plant parts only very early in development and that the removal not only of young anthers or ovaries but even of the growing apexes of shoots resulted in the disappearance of starch from the neighboring territories, an effect prevented by substituting auxin for these growth centers. Thus auxin—and possibly other plant hormones—could influence metabolism. This observation introduced me abruptly into the magic world of biochemistry, seen as a tool with which to investigate the mechanism of hormonal correlations. My new biochemical education was strongly stimulated by the sympathy and advice of Arturo Bonsignore, professor at Genoa and head of one of the best Italian schools of biochemistry, a physiologically minded biochemist of great imagination and technical skill. With his help, a lot of enthusiasm, and little knowledge, I set out to analyze my plant tissues for enzymes, substrates, and respiration. I like to remember, among my more instructive experiences in that period, the successful efforts I made to prepare the phosphate esters required to measure phosphorylase and other enzymes of carbohydrate metabolism.

This almost solitary fight against technical difficulties, this effort to enter a completely new world of information and methods, profoundly shaped my further development. Moreover, entering the world of biochemistry, then almost unknown to Italian botany, led to a series of friendly contacts with many other young Italian biochemists, among whom were Carlo Ricci and Sandro Pontremoli, two brilliant collaborators of Bonsignore's. Here I should also mention Enzo Boeri, whose talk in Siena on the interaction between large and small molecules opened my eyes to the control of protein activities. In the same years, circumstances led me to study the regulation of respiratory metabolism. As I was beginning to appreciate the importance of parameters such as temperature, pressure, and agitation with regard to quantitative O_2 and CO_2 measurements, a good friend of Zanoni's, Professor Vernoni, told her that I would benefit from a familiarity with what was already known! Thus, after some hurried reading of papers by Thunberg, Warburg, Krebs, and other luminaries of the field, I spent a week learning of the marvelous discoveries one can make by the correct use of a respirometer. My mentor in this field was Massimiliano Aloisi, now a leading figure in Italian pathology, to whom I feel bound by a warm and enduring gratitude.

The USA Experience

As I was still engaged in refining my biochemical and physiological tools with a view to attacking the problem of regulation, my research—until then almost autodidactic—was abruptly shifted to the open air and the international field of plant biology. In response to my request, Professor Zanoni arranged for me to study with Andrew Murneek, a distinguished and physiologically minded professor of Horticulture at Columbia, Missouri. Having obtained a Fulbright fellowship, I left for the United States in September 1951. On the way to Columbia, I stopped for a few days in Boston, where, in spite of my almost incomprehensible English, I had long talks with Kenneth Thimann—one of the most important events in my scientific life. These few hours revealed to me the extent of the difference between the Italian and the international scientific environment. In Thimann I found the closest approximation to my ideal of the scientist: a range of biological interests extending far beyond the boundaries of plant life; a capacity to apply the modern developments in biochemistry and biophysics to physiological problems; a sincere and intense interest in the development of science all over the world; the capacity to establish a warm, friendly relationship with younger people; a wide humanistic culture; and a faith in the value of scientific research accompanied by an awareness of the intrinsic limitations of this type of knowledge. I was prepared for this aspect of Thimann's personality by the quotation at the beginning of his book on plant hormones ("Peas, beans, oats and barley grow / But can you, or I, or anybody know / why peas, bean, oats and barley grow?"). Besides discussing and encouraging my plans and ideas with regard to regulation, Thimann gave me a precious overall picture of the enormous effort of renewal going on in plant biology, and of the impact on it of general biochemistry: People such as Kalckar, Lipmann, Krebs, Lynen, Ochoa, Kornberg, and many others were rapidly developing a far-reaching synthesis of the mechanistic analysis of processes and their physiological interpretation. As a consequence, plant physiology was then moving from the identification of forms and functions to the attempt to recognize and reconstruct the deterministic cause-effect chains of events mediating the life processes, the relationships between stimuli and responses, and (my chosen problem) the integration of the various processes in the unit of the organism.

Andrew Murneek, my Fulbright mentor, was a kind and intelligent man, interested in hormonal correlations in plants and, in particular, in the impact of hormones produced by the young flowers and fruits on photosynthesis, translocation, and productivity. It was easy to work out together a plan for my work at Columbia. Moving from Murneek's experience of auxin-induced parthenocarpy and my findings concerning the effects of growth centers on carbohydrate metabolism, I was able to confirm the hypothesis of an effect of growth centers on the activation of sugar metabolism and reserve accumulation in the neighboring tissues. In this, and on more general methodological issues, I was greatly helped by Nobel Prize winner Carl Cori, then teaching at Washington University, St. Louis, who kindly gave me some of his time, the opportunity to discuss my results, and some very precious advice. I also still remember with pleasure the long hours spent with biochemistry books, in the hard but exciting and rewarding effort to bridge the gap between my background and that of my young colleagues in the lab. With two of my fellow investigators I was to establish a lasting friendship and collaboration: Fred Teubner (who died prematurely some years later) and Bob Goodman, now a distinguished professor of Plant Pathology at the University of Missouri.

I was deeply immersed in my laboratory work and reading at Columbia when quite unexpectedly something happened that was to be of the utmost importance for my future. One of the leading Italian botanists, the already mentioned Sergio Tonzig, was then organizing his new laboratory in Milan. He had accepted the invitation of the Rockefeller Foundation to spend three months in the United States, to visit a series of modern botany laboratories, among which was Murneek's. I had never met Tonzig, and thought of him as one of the figures then ruling the universities in Italy by means of much politics and little science. But how wrong I was! The man immediately fascinated me with his subtle sense of humor and his passion for discussing the problems of Italian scientific research in general and the development of plant physiology in particular. I was also impressed by his willingness to discuss any problem, including those relating to his or my experimental work, on a plane of equality, despite the great difference in our age, experience, and academic status.

Tonzig amicably reproached me for the way I was spending my time in the United States, "stuck in a hole like a mouse who has found some cheese," instead of seeing that splendid country and meeting people who could give me new ideas. He persuaded me to accompany him on his visit to some plant physiology laboratories in California. The two weeks we spent together moving slowly west on Greyhound buses were very pleasant; short visits to famous beauty spots alternated with instructive visits to famous laboratories and discussions with some of the most eminent scientists in our field—from James Bonner to Went and Haagen Smit at Pasadena, Van Niel and Arnon at San Francisco, and many others. Tonzig's conversation touched on almost every subject: the *Divina Commedia* (Tonzig knew many passages by heart), the philosophy of science (plant sciences in particular), how to organize a modern plant physiology laboratory in Italy, and the importance of particular research areas.

Back at Columbia, much richer in information and ideas, I spent the time that remained completing my work on the influence of fertilization and of auxin-induced parthenocarpy on the levels of carbohydrate and phosphate esters and on the enzymes of the pentose phosphate pathway. Before returning to Italy, I spent some weeks in Thimann's laboratory at Harvard University. I was able to establish a precious and lasting contact with him, and friendly relations both with Carl Price, then beginning his attempt to clarify the mechanism of the primary action of auxin (a still unsolved problem), and with David Hackett, a promising plant biochemist who died tragically a few years later.

Contacts between European and American scientists are fundamentally important, not only because they advance experimental science but also because they help to develop the awareness that researchers are linked by common aspirations and interests irrespective of differences of nationality or age. The great hospitality offered by the United States to so many young people then fighting to find their way in an economically exhausted but intellectually still vital Europe was the main driving force for the development of a worldwide scientific community, in which discoveries and theories are exchanged, and friendships and collaborations formed without impediment. From the point of view of human progress (or cultural evolution) the development of this cooperation was more important than even the most brilliant achievements of science and technology.

Milan: a Research Team

Soon after my return to Genoa, I found myself in conflict with Professor Zanoni owing to my firm intention to maintain contacts and collaboration with other laboratories—the one directed by Tonzig in particular. Zanoni, like many other Italian botanists, had a restrictive concept of the identity of her laboratory and the importance of hierarchy. Ultimately I accepted Tonzig's invitation to move to his laboratory in Milan, where he offered me "nothing but every possible help to do good work," a promise he fulfilled far beyond any reasonable expectation. In a short time I had a reasonably well-equipped laboratory and, even more important, some young collaborators—the beginning of a small, fully independent research group specifically dedicated to plant physiology. Among the members of this group who made important contributions, were Giorgio Forti, Oreste Arrigoni, and, some time later, Renato Bianchetti. Our progress in the ensuing years was the fruit of a real and spontaneous unity of scientific purposes, our complementary tendencies and personalities, and our capacity to combine a lively and free discussion of the objectives to be pursued with unity of effort in carrying out the laboratory work. Thus Arrigoni (now Professor of Botany at Bari) contributed with his bright, sometimes poetic, imagination and enthusiasm, Forti with his outstanding capacity to sense the importance of new developments in the scientific world, and Bianchetti with his talent for lucid criticism and clear understanding of the key aspects of a problem. In the period from 1952 to 1959 the general philosophy of our research was substantially the one I had developed previously, namely to approach the problem of the regulation and integration of processes by seeking knowledge about the particular processes present in the organism, such as water relations, biosynthetic and energetic metabolism, transport of solutes, and so on.

The birth and growth of this group would not have been possible without the generous assistance of Tonzig, who showed a constant interest in our work, used his great authority in the Italian scientific world to obtain recognition for it, contributed to it with intelligent criticism and advice, and helped us to establish contacts with other laboratories, without ever restricting our freedom. As a result of Tonzig's efforts, in 1956 the Italian National Council established in our laboratory the first Center of Research in Plant Physiology. The institution of this Center (directed at first by Tonzig, then by me, and now by Giorgio Forti) was of decisive importance. It provided a stable budget, increased the number of research positions, and ensured a substantial scientific continuity in the group's work between 1953 and the present day.

Among our contributions of some importance in those years one can mention research on the role of ascorbic acid, on the extramitochondrial respiration pathways, on the metabolic changes associated with the transition from rest to growth, on the mechanism of thermoresistance, on the biochemistry of seed germination, and finally on photosynthesis.

Our work on ascorbic acid and on nonmitochondrial respiration sprang from the confluence between (a) Tonzig's observations of the effects of ascorbic acid on growth by elongation, plasma viscosity, and resistance to plasmolysis and (b) my hypothesis, developed while in the United States, that glucose-6-phosphate oxidative metabolism might be involved in auxin action and the regulation of growth. We started working on soluble dehydrogenases and oxidases involved in the transfer of electrons from glucose-6-phosphate to oxygen through pyridine coenzymes, glutathione, ascorbate, and polyphenols, as well as on the effect of auxin on the oxidation-reduction state of these factors. We also obtained evidence that auxin, as well as conditions breaking dormancy in resting tuber slices, reduced the activity of the pentose phosphate pathway and shifted the NADP, glutathione, and ascorbate systems toward reduction. Although far from definitive, these results pointed to an important role of this pathway in the regulation of cell activity. Work to substantiate this notion (mainly centered on the importance of thioredoxin) is still going on (or planned).

Our research in this region of metabolism immediately and quite naturally opened the way to the study of other aspects of respiration, in particular the possible involvement of the state of energization of the adenylate system in the action of hormones on growth and respiration. These studies, although they did not show such involvement definitively, were of great importance in increasing our direct experience with regard to respiratory and energetic metabolism and preparing the group for a new and exciting adventure, this time in the field of photosynthesis.

In 1952 Daniel Arnon had shown me how to prepare clean chloroplasts. Interested as we were in NADP function and curious about how this cofactor was manipulated in photosynthesis, we decided to look into this related problem. Working first with Servettaz, then with Forti, we were among the first to show that NADPH was indeed a link in the energy-yielding photosynthetic cycle. We obtained indications that a diaphorase and a cytochrome of the c type (cyt f) played roles in the cycle. (The nature of these roles was later independently worked out much more extensively by Giorgio Forti after his long and fruitful visit to André Jagendorf's laboratory.)

The Academy of Cats and the Birth of the Italian Society of Plant Physiology

In the same period, a considerable part of my time was devoted to social-political activity concerning the development of plant biology in Italian universities. The situation of young researchers in most of our botany laboratories (with very few exceptions, such as mine in Tonzig's laboratory) was then far from satisfactory. Old, politically powerful professors ruled the field in an autocratic way, severe and nondemocratic rules of hierarchy called for blind obedience on the part of assistants, and distances between the various academic ranks marked. Independent work was discouraged; most laboratories functioned as closed systems, and contacts among young researchers were discouraged or even prohibited. This situation seemed to me hardly tolerable. In particular, the hostility toward direct contact between young scientists belonging to different laboratories (regardless of the relations of their superiors) appeared to me an unbearable offense to dignity, and also a great impediment to the development of plant research in Italy. To these feelings and conclusions I was moved both by my experience during the partisan war and by what I had seen in the more liberal and cooperative scientific world of the United States. A few friends and I therefore started to establish a kind of mutual-help organization for young botanists, with the enthusiastic cooperation of several colleagues at various Italian laboratories. Notable among these were Felice Bertossi, an excellent plant physiologist at Pavia, Tullio Dolcher at Padova (who was to become my closest friend), Aldo Merola, a very good and complete botanist (ranging from taxonomy to ecology to physiology) at Naples, and Mario Orsenigo, a good plant pathologist at Padova; also included were Forti, Laudi, Arrigoni, Honsell, Pignatti, Meletti, Sarfatti, and Orio Ciferri. With these and some other younger researchers we gave birth to the "Academy of Cats" (cat signifying independence) with elaborate rules, the first of which was that no full professor could be admitted. This group operated efficiently (with periodic, semi-secret meetings) for some years, after which it was absorbed—with most of its original spirit—by the Italian Society of Plant Physiology (formed in 1961). Nearly all the members of the Academy of Cats are now full professors of either Plant Physiology or Botany and today play an important role in the development of plant sciences in Italy.

In 1959 Tonzig obtained for me a full professorship in Plant Physiology at Milan. This did not much alter the previous situation as far as relationships with Tonzig and his Institute of Botany were concerned: The old institute immediately offered hospitality and research facilities of all kinds to the new. My promotion, however, coincided with important changes in the organization of my research team. Most of my former collaborators had by this time reached a level of full independence in research. Some, including Arrigoni, Laudi, and Servettaz, now occupied stable positions in Botany with good prospects of advancement in their academic careers, and my closest collaborator, Giorgio Forti, then at André Jagendorf's laboratory at the Johns Hopkins University in the United States, had already started an independent line of work in photosynthesis. Only Renato Bianchetti (the youngest of my former associates) was still very efficiently cooperating in my research on the regulation of metabolism. I was therefore pleased to be allocated some research positions, which were soon occupied by three persons who were to be invaluable in carrying on my old line of work, modified according to the new circumstances and the advances of plant research in the world. These were Lilia Alberghina, Piera Lado, and Sergio Cocucci. I wish here to express my gratitude for their enthusiasm, spirit of collaboration, and dedication in the years they spent with me before moving on to independent positions, responsibilities, and areas of research.

The organization of the team coincided with certain changes in our research orientations. I had been asked by the *Annual Review of Plant Physiology* to write an article on "Phosphorylation in higher plants" and had been invited to give a lecture on respiratory metabolism in plants at the Botany Congress in Montreal. To prepare these papers I had to work hard to bridge the gaps in my biochemistry. Taking advantage of a grant from the Rockefeller Foundation, I spent three months visiting American laboratories, especially the laboratory of Harry Beevers, whose excellent work was then raising plant biochemistry to a level comparable to that of animal and microbial biochemistry. Harry was then engaged in the utilization of radioactive ^{14}C to determine the relative importance and physiological significance of the different metabolic pathways. It is difficult for me to find words to thank Harry for all the things I learned in those few weeks: from general ideas to specific techniques, from intellectual modesty to rigor in the interpretation of results.

Meanwhile the institution of full professorships in Plant Physiology in Italy had started a kind of autocatalytic process in this rather neglected area. The second chair in Plant Physiology had been given to Felice Bertossi, an intimate friend whom I have already mentioned as one of the founding members of the Academy of Cats. We, with most of the other members of that Academy, decided that the time had come to organize an Italian Society of Plant Physiology. This new, very democratic, and very active association was officially formed in 1961 and immediately began to promote contacts, collaborations, scientific meetings and various cultural activities. Particularly consequential was the initiation of three courses for researchers in Plant Physiology. These courses, organized by the Milan group with the active cooperation of Bertossi and of an excellent and open-minded geneticist, Giorgio Morpurgo, aimed to make our plant physiologists aware of the enormous advances that had been made in the fields of biochemistry, physiology, genetics, embryology, molecular biology, general physiology, chemistry, and physics. These courses, lasting two weeks and attended by some 40 young researchers, benefited from the contribution (as teachers and debaters) of most of the best scientists then available in Italy: from Falaschi to Tocchini Valentini for molecular biology; Monroy and Giudice for embryology; Scoffone for organic chemistry; Magni, Morpurgo, and Calef for genetics; Scarano and Guerritore for biochemistry; Ceccarelli for physics, and many others. Characteristic of these courses was the warm atmosphere of friendship created by the continuous informal contacts among relatively young people moved by a common purpose. Their influence on the development in Italy of a group of open-minded, modern plant physiologists was considerable.

The research activity of my new group in those years was initially concentrated on the relationships between growth stimulation and respiratory metabolism, with particular reference to the pentose phosphate pathway and to the NADP reduction state. By means of metabolite determinations and the ^{14}C techniques learned in Harry Beevers's laboratory, we were able to demonstrate that the auxin-induced stimulation of growth was associated with a marked decrease of activity of the pentose phosphate pathway and an increase of the levels of glucose-6-phosphate and reduced NADP, which tallied with our previous observations of the increase of the reduction state of glutathione and ascorbate in auxin-treated tissues.

From Protein Synthesis to Transport

This research was in full course when our attention was at least partially diverted to a new aspect of growth regulation, namely nucleic acid and protein synthesis. The need to include such an approach in our investigation of the problem of regulation was clearly suggested by certain dramatic successes in molecular biology—in particular, by the demonstration of specific mechanisms of regulation at the transcriptional level. Our entry into this field was efficiently catalyzed by the friendly, generous advice and collaboration of

Alberto Monroy, then carrying out excellent work on the changes in RNA and protein metabolism associated with sea urchin egg fertilization and embryonal development. Using Monroy's experimental approach as a guide, we started investigating the somewhat analogous transitions of macromolecular patterns in the three phases of seed development (namely growth and maturation, fall into dormancy and dehydration, and germination). Most of the forces of our laboratory came to be engaged full time in this new field. Among the main results obtained were the demonstration of the inducibility of isocitrate lyase and galactose kinase, of the reversible inactivation and reactivation of soluble and mitochondrial enzymes during the seed cycle, and, even more interesting, of the fundamental role of water availability (turgor?) in regulating protein synthesis.

We were slowly proceeding with these problems when, once again, circumstances (the most important, as usual, being contacts with clever people working in neighboring areas) modifed the main course of our research by introducing the physiology of transport as an important parameter of the general problem of regulation. Prompting us to start along this new course was the discovery of the fungal toxin fusicoccin and of its peculiar physiological action by two good friends: Antonio Graniti, who isolated the toxin and described its capacity to induce stomata opening and wilting, and Alessandro Ballio, who had worked out the structure of fusicoccin and had started investigating its mechanism of action. Collaborating with Ballio, we soon found that this toxin had a growth-promoting activity similar to that of auxin and could be used as a tool to elucidate the action of auxin. The excellent work of Cleland, Rayle, and Evans in the United States, and of Hager, Lüttge, and Haschke in Germany, had suggested that the effect of auxin on cell enlargement might depend on the stimulation of acid secretion. Using fusicoccin, we were able to demonstrate that growth stimulation was indeed correlated with acid secretion in the case of fusicoccin as well as in that of auxin. The investigation of the mechanism of acid secretion led us into the area of ion transport. In collaboration with Arnaldo Ferroni, a distinguished electrophysiologist at our University, we soon found that the fusicoccin- (and auxin-) stimulated acid secretion was associated with a hyperpolarization of transmembrane potential and with an increase of K^+ uptake, which suggested the operation of a proton pump. Because such work was completely outside our previous experience, our progress depended, once again, on the generous help and advice of several outstanding scientists then engaged in the biophysical and physiological aspects of ion transport in plants. Since 1968, this group of researchers had organized periodical workshops on transport. Characterized by a rare spirit of collaboration, this community of scientists played an extremely important role in the development of modern plant physiology and in the integration of physiology with biophysics, biochemistry, and molecular biology. I am still grateful to Noe Higinbotham for encouraging me to join this group. Thereafter, our work in plant transport proceeded in close contact with many "transportists," among whom I remember Jack Dainty, Clifford Slayman, Tom Hodges, Michael Pitman, Widmar Tanner, Ulrich Lüttge, John Raven, and Anton Novacky with particular gratitude.

Though we were now concerned with the mechanism of proton transport, our main problem remained that of the regulation of transport and the physiological consequences of changes in activities of transport processes. Here again precious help came from two distinguished scientists: Rainer Hertel of Freiburg, with whom we were able to give a first demonstration of a fusicoccin receptor in the plasma membrane, and Jean Guern, whose competence in the use of ^{31}P-NMR spectroscopy was of great value in establishing that fusicoccin-induced stimulation of the proton pump was indeed associated with an increase of cytosolic pH.

An interesting result of my effort to combine hormonology, biochemistry, and biophysics in the study of the integration of processes in plants was a workshop on "Hormonal regulation of transport" organized by our group at Pallanza in 1976, which was attended by many of the most representative scientists in these fields. The workshop was important in developing awareness of the complementarity of three fields: the physiology of transport, biochemistry, and hormonal regulation. It also gave rise to some unforeseen collaborations between people involved in quite different fields, such as between the electrophysiologist Noe Higinbotham and the hormonologist Bob Cleland.

The attention of our group since 1970 has been centered on proton transport; we have studied its regulation in intact cells and in membrane vesicles, its effects on intracellular pH, its consequences on metabolism, and its relationships with other proton transport systems such as the plasmalemma redox system; such work continues in the present. Some widening of our perspectives and programs has been prompted recently by advances in plant genetics and molecular biology as well as by the increasing tendency of Italian and international organizations to favor applied over basic research. Our recent programs thus tend towards the use and investigation of mutants in transport processes, with the twofold aim of obtaining further information about the mechanisms and regulation of transport, on the one hand, and of investigating the relationship between transport and productivity-limiting conditions (such as water, salt, and temperature stresses), on the other.

CONCLUSIONS

The recollection of one's past life tempts one to draw conclusions. Allow me here to reflect on certain of my firm beliefs concerning research, plant physiology, and the practice of experimental biology in relation to a general vision of the world.

Motivations for Research

At the root of my choice of research and teaching as a profession I find two distinct motivations: a thirst for self-fulfilment and a desire to enjoy the order and beauty of reality, and to share this enjoyment with others. While the impulse towards self-fulfilment has a well-defined physiological meaning (it helps human beings to develop by accepting challenges—e.g. climbing a mountain, solving a problem, overcoming laziness and discouragement) it also entails competitive confrontation with other members of the human community. On the other hand, the desire to enjoy the world and to communicate this pleasure gives rise to some of the highest manifestations of human nature, such as solidarity and friendship. The positive role of the spirit of competition is emphasized and encouraged by the official organization of science much more than the spirit of collaboration, possibly because of the political-social demand for practical results. I am deeply convinced, however, that the feeling of being associated in a common cause—which favors collaboration—is a much more important condition for scientific progress. Competition leads to the exploitation of our colleagues, considered merely as sources of information and ideas, whereas collaboration—when understood in the full meaning of the word—leads to solidarity and potentially to friendship, and is moved by the desire to help others and to share with others anything good achieved in our (or their) work. My conviction that friendly collaboration is far more favorable to scientific progress than competition is born of my experience. My scientific career was decisively influenced by collaborative contacts with other scientists, notably Thimann, Tonzig, Boeri, Monroy, Beevers, Higinbotham, Hertel, Guern, Tanner, Dainty, and many others. I am convinced that the positive contribution to science arising from their unselfish dedication in helping colleagues and students was even greater than that of their outstanding scientific discoveries.

The Political Trend towards Application and the Necessity of Integration of Plant Sciences

As in any other scientific area, the objective of research in plant physiology is twofold: knowledge for its own sake, and application to practical needs. Obviously enough, these two aspects are strictly interlinked: Good basic research produces tools for application, and application pressure stimulates research on basically central topics or problems.

Recently in Italy, and more or less all over the world, governments, under the pressure of expanding economic and social needs, have started to favor applied over basic research. Many of us feel that this political trend has something intrinsically wrong, demagogic, and illogical about it. Knowledge must precede application, and the progress of knowledge has its rules, among which intellectual freedom is the most important.

Some aspects of the trend toward greater support of applied science might be usefully exploited to increase awareness of the need to integrate the different scientific areas. The application of biological science—in our case of plant science—is a highly synthetic activity. It implicates a capacity to predict the behavior of an organism in given environments—behavior that results from the integration of all its activities. Knowledge of the mechanisms and knowledge of the integration of these processes are thus the two main bottlenecks that limit application. Our efforts to remove the "integration" bottleneck helps to identify the missing links in our knowledge about each mechanism, just as any increase in mechanistic knowledge contributes to an understanding of the integration.

From this point of view, the negative consequences of modern pressure in favor of application and technology are much greater in plant research than in animal and microbiological biology, where basic research is so advanced that the interactions among basic research areas and their impact on application have become obvious. In contrast, basic knowledge concerning plants is much poorer, and the need for closer interactions among the different areas (e.g. genetics, molecular biology, biophysics, systematics, plant physiology in its different branches, etc) has only recently become clear. An effort of interdisciplinary integration is urgently needed to enable us not only to understand better the phenomenon of life in its higher expressions, but also to establish a closer, more rational relationship between basic research and application.

Plant Physiology, Biology, and the Unity of Life

It seems there are two different ways to interpret one's role (significance) in life. One way is to accept—as a working hypothesis—that in the framework of a general design of the universe (difficult to rationalize and open only to intuitive understanding) each individual has a definite role, that individual roles aggregate to serve a primary purpose, and thus that all activities are organized to accomplish that purpose (just as the organization of physiological processes makes possible the survival and the functioning of an organism). Alternatively, one can see life as an incomprehensible puzzle, forget about the problem of unity, and allow oneself to behave according to one's impulses. Very near the end of a rather complex life, and having personally chosen the first alternative, I now wonder how much unity I have achieved, and how important it was to have dedicated so large a part of my time and thought to plant physiology.

Research has heavily influenced both my way of thinking and my vision of the world. Physiology is the study of the intimate nature and the dynamism of the most complex system existing in the universe, from its most microscopic to its most macroscopic manifestations. It is a study carried out by a well-defined method, the experimental method, which has been developed over hundreds of thousands of years, in a process that started with the appearance of *Homo sapiens,* gained rationality from the work of such men as Galileo, Newton, and Claude Bernard, and is still being refined today. Even with the limitations implicit in the circumstantial and hypothetical nature of scientific knowledge, the correct application of the experimental method remains necessary for the description and interpretation of the phenomena occurring in the world external to our consciousness, including the biological aspects of human history and behavior. A long-lasting dedication to the use of this method, as required for good work in physiology, tends to refine one's capacity to analyze, critically evaluate, and interpret any type of information received from the environment (including ideological, political, and historical information).

How has the study of biology influenced my vision of the world, and my attempt to organize it? My conviction that it has is difficult to explain, but I can identify some of its underlying aspects. The observation and analysis of life confronts us with the highest manifestation of complexity and order, thus nourishing our innate hunger for both truth and beauty (as C. R. Stocking so nicely put it in his prefatory chapter to Volume 35 of this *Annual Review*). This nourishment establishes an intimate and strong interaction between biology, as an experimental science, and all other forms of knowledge, including philosophical and artistic knowledge, which combine to form a single whole in the individual consciousness. In this regard, certain properties of physiology are of particular interest. The aim of physiology is to describe and understand the nature and integration of life processes. Even if we exclude a naive finalism from our effort to interpret reality, we are constantly led to recognize the fact that almost all the components of the biological machinery we investigate play an essential role in enabling it to achieve a final unitary result: the existence of life. As a consequence of this observation, the discovery of new processes is constantly followed by an effort to understand their physiological relevance. A similar situation is found when we move from organismic physiology to the analysis of the biological system as a whole, considering evolution as a process to be studied by the same method as we use for organisms, or parts of them. Can we identify a general design, and possibly a unitary aim, in the appearance of life and in its evolution to extreme manifestations of coordinated complexity? The recent attempts to reproduce in nonbiological systems the synthesis of biological energy-rich compounds and macromolecules, such as ATP, nucleic acids, and proteins, indicate that both physicochemical and historical restrictions may have played a decisive role in the origin of life, just as in the previous evolution of the universe. If this is the case, then the appearance of life is an improbable event privileged by the intrinsic nature of the matter-energy context. One might easily speculate that similar (although still unexplored) restrictions have channeled biological evolution in its progress towards coordinated complexity and consciousness. To accept this view implies the interpretation of evolution as the development of a design. It also implies that every element in the evolving system—myself included—has a definite role to play. Our responsibility is to accept our roles (insofar as we can understand them) in a way that conforms to the design.

I am well aware that a vision of the world based on the recognition of both a general design and a personal role implies a choice belonging to the domain of faith rather than to that of science. However, I am glad to avail myself of the present opportunity to state that in making this choice I see a confluence of my scientific experience with what knowledge I have been able to obtain from other sources. To paraphrase a famous sentence of Kant's, I like to say that two things in the world fill me with unlimited wonder: the evolution of the universe outside me, and the moral conscience within me. As a physiologist, I accept the challenge to seek a unitary interpretation of life.

Jack Dainty

Annu. Rev. Plant Physiol. Plant Mol. Biol. 1990. 41:1–20

PREFATORY CHAPTER

Jack Dainty

Department of Botany, University of Toronto, Toronto, Ontario, Canada M5S 3B2

This is the story of how I became and what I have done as a plant biophysicist, and what I consider to be the function and importance, if any, of such people in plant physiology.

I was born and grew up in a depressed area of South Yorkshire, England. We lived on the edge of D. H. Lawrence country, surrounded by miserable, poverty-stricken, coal-mining towns, near the steel-making city of Sheffield. The waters and soil were heavily polluted, and the unpleasant smell of industrial gaseous waste hung in the air. It was not then, nor is it today, a place anybody would choose to live. The pollution has greatly decreased because many of the coal mines and all the steel works have been shut down, with resulting heavy unemployment.

The small mining towns were practically contiguous and equally depressing. "Deprived" is the word that best describes them, a characterization I first became aware of only after I had gone away to university. Nearly everybody was poor; many—like us—were very poor. The "aristocracy" of the towns consisted of the small shopkeepers and school teachers. The vast majority of us voted Labor (socialist)—indeed the area produced the largest Labor majorities in the country. I grew up, and have remained, a socialist.

My father having died young (when I was three), my brother, my mother and I lived with my grandparents in poverty. We often received "welfare"— an even more distressing experience then than now. Inevitably, the atmosphere of home was devoid of culture, literature, and learning. Nobody in the immediate family had been to school beyond the age of 14, and my many cousins, uncles, and aunts had the same background. By some quirk, however-er, I was precocious, learning to read by the time I was three. I thus early discovered the world of books—still the world in which I like to live.

At that time (the 1920s and 1930s), a child such as I entered the educational system through the so-called elementary school; at the age of 10 or 11 one took an examination for entry to the high school—in my town of Mexborough it was called the secondary school. Fewer than one child in ten passed this scholarship examination; those who did not pass were condemned to stay in the elementary school until age 14. I passed, but to keep going in the secondary school I needed the charity of free school meals, clothes (hated) given by the local Rotary Clubs, and the encouragement of my mother.

At the end of the first four years, during which we studied a broad range of subjects—French, Latin, English literature, mathematics, physics (no chemistry or biology), geography, history, art, music, workshop—there was a national examination. If one was reasonably successful in this, one could pass on to a specialized program for another two years. This was a crucial point in the educational system; it involved a narrow choice of specialized subjects: not merely whether to do arts (i.e. humanities) or science but what kind of arts or sciences. I chose science because I believed I would rank high up in the class, and I chose to do mathematics, physics, and chemistry rather than botany, zoology, and chemistry (the only two options offered by the school) both because I thought I was quite good at mathematics and because boys did mathematics while girls did Biology! Such are the determining factors in a situation such as mine was, with good advice unavailable.

I was not uninterested in biology. In fact, from an early age I had been a dedicated naturalist, spending much of my time going for long, lonely walks in the neighboring countryside (which despite the pollution was not unattractive) and on the moors and hills of Derbyshire and Yorkshire. I read deeply in the English naturalist literature: Belt, Wallace, Bates, Waterton, Gilbert White, Darwin, and others. It did not occur to me to translate this great interest into the formal study of biology, and there was nobody to suggest such a strategy. However, my much later move from physics to biology was based on this early interest in natural history.

Thus I was set, at the age of 15, on the mathematics-physics-chemistry track in the early specialization characteristic of English higher education. After two years of such specialization there was another national examination, and in this I did well in all three subjects. Throughout my education, in fact, I was an accomplished examinee. (This trait was of course useful, but it is not, I believe, closely related to ability in scientific research, which needs different qualities.) After this success it was suggested to me that I might consider the great heights (for Mexborough) of the University of Cambridge and that I should stay at the school for another year in order to take the scholarship examination for university entry. I spent this final year doing mathematics only and was successful in getting a scholarship.

I went to Cambridge to study mathematics with more than adequate financial support from scholarships. It seemed—it *was*—an extraordinary place. The faculty were outstanding. Most of the students were male, the vast majority from well-to-do families. They had gone to private schools where the standards of education had often been excellent. Both their families and these prior schools had influence in the Cambridge colleges, and there were long traditions behind the entry of these students into the University. Many seemed arrogant and rather stupid; some were brilliant; almost none would fail: the system did not allow it. The percentage of students who had arrived there by my route was small. I felt thoroughly out of place, with my Yorkshire accent and lack of social know-how—a feeling that certainly did nothing to weaken my socialist sympathies.

I settled down, made one or two friends, and had some success as a football (soccer) player; indeed, as I had also done at the secondary school, I spent rather a lot of time playing football. I was "reading" mathematics, which meant doing only mathematics. At the secondary school in the backwoods of South Yorkshire, I had thought myself rather good at mathematics, but I discovered in Cambridge that many were much better than I. Both because I did not like this and because I felt that mathematics only was too narrow, I switched to physics. By this time the Second World War was starting, but I was allowed, indeed instructed by the military authorities, to finish my physics undergraduate degree at Cambridge. By the end of this course and the final examinations it was June 1940 and Hitler's armies were overrunning France, the Netherlands, Luxembourg, Belgium, Denmark, and Norway. Almost all the rest of the graduating physics class had been recruited to work on radar and other urgent applications of technology to war. The recruiters must have thought little of my technical competence for I was not asked to take part in this important work. However, after a short time I was asked to return to Cambridge to join one of the two small teams (the other being at Liverpool University) being then set up to work on nuclear fission—i.e. to carry out studies leading to the eventual construction of an atomic bomb.

We were a mixed lot in the Cambridge team: two French physicists who had escaped from France with a large stock of heavy water, a Swiss who was working in Cambridge, two or three German and Austrian refugees, and a handful of Britishers who had not been swept up by the radar net. We soon realized that the making of an atomic bomb was more a technological and industrial problem than an intellectual one—that it involved either the separation of large amounts of ^{235}U or the construction of large reactors for the production of fissionable plutonium. Under wartime conditions, only the United States could set aside sufficient industrial power to make such a bomb, but we carried on with our fundamental studies of nuclear fission in Cambridge. The team was greatly reduced in number when some were sent to help the Americans in Los Alamos and others to Montreal to join with Canadians in initiating the Canadian Atomic Energy effort, the latter being solely concerned with peaceful applications of the release of nuclear energy.

I remained with the Cambridge group, indeed became head of the cyclotron team, and spent the war carrying out rather academic nuclear research on fission. Because of the shortage of faculty, I also did a great deal of teaching during those years. I gave an advanced course of lectures on atomic and nuclear physics, instructed in the practical classes, and did a lot of individual tutorial teaching, according to the Cambridge system. This research and particularly this teaching strengthened my understanding of physics, which was of great importance later when I switched to biological problems. I stayed in Cambridge until the fall of 1946. I retained my interest in natural history, chiefly birds at this time, and improved my understanding of some aspects of biology, chiefly that of evolution, by study of the works of Darwin, Huxley, Mayr, Dobzhansky, Haldane, and others. I also read the little book *What is Life* by Schrödinger. My love of literature developed greatly during these Cambridge wartime years, a love that has strengthened as the years pass.

In 1946 I made a half-hearted attempt to move from nuclear physics to biology by joining the Division of Biology and Medicine of the Canadian Atomic Energy Laboratories at Chalk River, Ontario; but the problem I worked on was really pure nuclear physics. I set out to develop a method of measuring the dose (energy per gram) delivered to biological tissue by the fast neutrons in, for instance, a nuclear reactor where the radiation would be a mixture of fast neutrons and gamma rays. The solution, differential ionization chambers, one with walls of polystyrene [$(CH)_n$, approximately] filled with acetylene and the other with walls of polyethylene filled with ethylene, was relatively simple, but I spent about two years on it. Again I carried on with my natural history (largely birds) and my reading of the classics of evolution. Deep River, the newly built town on the banks of the Ottawa River, was then a stimulating place for an amateur naturalist.

In early 1949 I returned to the United Kingdom to join the faculty of the Department of Natural Philosophy (physics) at the University of Edinburgh. I had finished, or so I thought, with any dabbling in biological problems, and was going back to "proper" (nuclear) physics. And so I did for two or three years. I lectured on relativity, quantum theory, and the theory of errors; taught in practical classes; and built up a laboratory around a small accelerator. I supervised graduate students in nuclear physics.

Around 1952 the department asked me if I would take on the onerous task of teaching elementary physics to a joint class of 300 medical, dental, and veterinary students. (In the United Kingdom such students are recruited directly from high school. In Scotland at that time all had to spend their first year on physics, chemistry, and biology). Initially I refused. In order to handle such a body of students, who refused to take physics seriously no matter how relevant one made it to physiology and medicine, one needed the qualities of an army sergeant rather than those of a university lecturer. But the head of the department, the nuclear physicist Norman Feather, knew of my interest in biology and got the university to promise that I could develop a department of biophysics if I would undertake this teaching task. I finally agreed, with considerable misgivings arising not only from the nature of the teaching task but also because I was comfortable in (nuclear) physics. I knew the literature, many of the techniques, and many of the leading people involved. After all, I had effectively been a nuclear physicist for 12 years or so.

Thus I moved into biology almost by accident. I'd had a long interest in certain aspects of the field, but not in those where I felt I could use my knowledge of physics. I had been interested not in how living organisms work but in their diversity and evolution. Therefore I had first to find out what I could do. Many people in Edinburgh were prepared to help. Since I was teaching medical students, some clinicians wanted my help, as a kind of high-grade technician, with the more physical aspects of their problems. C. H. Waddington, the geneticist and embryologist, tried to interest me in the physical problems of morphogenesis, and I carried out some experiments on early *Xenopus* embryos. The zoologists also had me doing experiments on heat production as a function of the cell cycle in developing sea urchin eggs. But the work that really started me moving in the direction I was ultimately to take was a collaborative study in 1953, with K. Krnjevic of the physiology department, of the exchange of sodium across the membranes of nerve bundles of the cat. It was this that led me to the work of Hodgkin, Huxley, Katz, Cole and others on membrane transport, particularly that on the squid axon. I did not exactly enjoy working with the nerves of cats, but membrane transport seemed to me a physiological process that one had to think of in terms of physics, at least in its overall aspects. In choosing a membrane system to investigate, I certainly did not want to compete with Hodgkin et al. It soon struck me that almost nobody was looking at membrane transport in plant cells from a biophysical point of view. And plants were not cats!

I was fortunate that at about this time a very good physics student, Enid MacRobbie, wanted to do biophysical research with me. The early work on plant cells was a good collaboration between the two of us. Our ideas were guided—perhaps too much, as will become clear later—by the work of animal physiologists on the squid axon. We had no botanical contacts, and we decided that it would be best to work with a marine alga—i.e. with cells immersed in a well-defined medium of fairly high concentration: sea water. Mostly in the 1930s and predating the squid axon work, excellent work on algal cell membranes had been done by Osterhout and Blinks on *Valonia*, *Halicystis*, and *Nitella*. It seemed to us that this work had been forgotten by plant physiologists.

We did not at first take up *Nitella* and the Characeae in general because they are mostly freshwater species and we had this idea that we needed a concentrated bathing medium, sea water. *Valonia* and *Halicystis,* being tropical, were hardly likely to occur on the icy coasts of Scotland, so we chose to work with a red alga, *Rhodymenia palmata,* and to a lesser extent with a green alga, *Ulva lactuca.* In these early experiments we measured concentrations of Na, K, and Cl in the cells and, using radioactive isotopes, the fluxes out of and into the algal discs we were using. What we did not realize (and did not even look at sections to find out—such a poor biologist was I) was that we were dealing with complex tissues and a very inhomogeneous cell population. We could not easily express our fluxes on a meaningful moles per square meter (square centimeter in those days) per second basis and could not move forward in any biophysical sense. Our minds were thus focused on finding a homogeneous cell population or, better, a large single cell of simple geometry. And since we had no *Valonia* in Scotland, we looked at the Characeae. We were still obsessed by the notion of a concentrated, well-defined external medium for our cell. We came across some work being carried out in Finland by Collander & Wartiovaara on *Nitellopsis,* an ecorticate Characean species living in the Baltic at a salt concentration of about 25 mM. This was pretty dilute sea water, but it was well defined and concentrated enough for our purposes. *Nitellopsis obtusa* therefore became our first real "guinea pig," and ion transport in this organism was the object of Enid MacRobbie's PhD thesis, the first one produced at the Biophysics Department at the University of Edinburgh.

The *Nitellopsis* studies were basically kinetic, aimed at measuring the influx and efflux of Na^+, K^+, and Cl^- across the plasmalemma and tonoplast. We sought to interpret the measurements in the light of the concentrations of these ions in the protoplasm and vacuole, taking into account some early measurements of the electric potential between vacuole and external solution by our colleagues E. J. Williams and R. Johnston. Although hindsight reveals that the compartmental analysis of the efflux curves (which seemed to split nicely into two exponentials and a fast diffusion curve) was somewhat flawed, the general conclusion was valid: There had to exist a sodium extrusion pump at the plasmalemma and an inwardly directed chloride pump, which we located at the tonoplast. (We missed the point that there must also be one at the plasmalemma.) We also deduced that most of the potential difference between vacuole and tonoplast was likely to be across the plasmalemma. This agreed with earlier measurements by Alan Walker in Australia of a low, positive potential across the tonoplast and a high, negative one across the plasmalemma. We located the fast-diffusion so-called free-space fraction exclusively in the cell wall and thus came down against the then prevalent idea, put forward by Briggs, that the protoplasm was part of the apparent free space of the cell. This latter idea then, I think, began to die. We did not consider the hydrogen ion concentrations and any possible proton pump; this came a decade or more later.

While we were developing our biophysical approaches to plant membrane transport during the 1950s in Edinburgh, somewhat similar ideas were being followed in Australia, particularly by Alex Hope. Alex had been a member of a remarkable group in the Physics Department of the University of Tasmania at Hobart. This group was the brainchild of Lester MacAulay, the head of the department, who had an interest in how plants function and encouraged some graduate students to share it. Bruce Scott, Alex Hope, Alan Walker, Geoff Findlay, and Ian Newman are all from this school and have helped to make plant biophysical physiology eminent in Australia. In 1958 Alex was in the C.S.I.R.O. Plant Physiology Unit at the University of Sydney, which was headed by the Australian plant physiologist R. N. (Bob) Robertson, who himself had a major interest in membrane transport but was more biochemically oriented than Alex and me. Robertson was the ideal person to head this unit and to encourage a biophysical approach.

I did the obvious thing and applied for four or five months of research leave in 1958 to work with Alex in Sydney. This turned out to be an extremely fruitful collaboration. We came up with what I believe to be one or two important concepts from the experiments we carried out on the water permeability of *Chara australis* (now *corallina*) and on the ion exchange properties of the *Chara* cell wall.

The water permeability of a cell can be measured, indeed specified, in two ways: The net flow of water across a membrane separating two solutions of different osmotic pressure gives, for a semipermeable membrane, the hydraulic conductance, L_p; the exchange flux of labeled water under conditions of equilibrium gives the so-called diffusional permeability coefficient, P_d. If these two measures are expressed in the same units and compared, then if there is no equality, or specifically if $L_p > \bar{V}_w P_d/RT$, where $\bar{V}_w$ is the partial molar volume of water, this indicates that water crosses the membrane via aqueous pores; the relative magnitudes give some idea of the diameter of these pores. We were inspired to make water permeability measurements in these two ways on *Chara australis* by similar work on animal cells, in particular that of A. K. Solomon and his associates on red blood cells. Measuring L_p was no problem using the method of transcellular osmosis invented by Kamiya & Tazawa a year or two previously. Measuring P_d likewise appeared to be straightforward, using heavy water as the tracer and following the exchange of D_2O for H_2O by a simple weighing technique. However, when we repeated the P_d measurements on a cylinder of agar of exactly the same dimensions, the exchange kinetics were precisely the same; the agar cylinder had of course no membrane(s), and thus we could only conclude that the water exchange was rate controled by diffusion in the unstirred layers of water external (and to some extent internal) to the cell membrane! We thus rather dramatically "discovered" the importance of the unstirred layer in membrane transport, although of course it had long been well known to chemical engineers.

We were able to point out that P_d is always underestimated to a greater degree than L_p, because of the unstirred-layer effect, and thus to throw much doubt on some of the estimates of membrane pore size made by animal physiologists. We also used the unstirred layer to "explain" the difference between the L_p values when water is flowing into a cell and those when it is flowing out. It was impossible to correct for the unstirred layer in the P_d experiment with *Chara australis* because the effect was so dominant. No estimate of pore size could be made. In fact only one good experiment has been carried out on L_p and P_d for a plant cell: In my laboratory some years later John Gutknecht found in *Valonia* that after correcting P_d for the unstirred layer effect, $L_p = P_d \bar{V}_w/RT$. The osmotic and diffusional permeabilities when expressed in the same units were the same. Thus there are *no* aqueous pores in the plasma membrane of *Valonia;* water molecules must go through the membrane one by one.

I consider the little experiment on *Chara* one of the most important with which I have been involved, for it brought to the notice of physiologists in general, even if it took a long time to be recognized, the potentially great importance of the unstirred layer in transport processes.

We met the unstirred layer again in our studies in Sydney on ion exchange with the plant cell wall. I think these studies were initiated to resolve definitely the question of whether any part of the apparent free space (AFS) was in the protoplasm or whether it was all in the cell wall. The experiments certainly resolved this question: The AFS is all in the wall. In addition it was shown that the exchanges of Na^+ in the wall for Na^+ in the bathing solution and of Ca^{2+} in the wall for Ca^{2+} in the bathing solution were rate controled by diffusion in the unstirred layers of solution adjacent to the wall. Indeed, calculations of the thickness of the unstirred layer from the two separate experiments (Na^+ exchange and Ca^{2+} exchange) gave similar values of about 100 μm, a reasonable value under the conditions of stirring and cell wall size in the experiments. Naturally we were delighted with this reinforcement of the importance of the unstirred layer in a quite different example.

This cell wall work was also important in several other respects. For instance, we showed that the negative fixed charge of about 1.3 meq/g dry weight probably arose from the uronic acid groups on the pectins with a pK of about 2.2. Mobile anions such as Cl^- or I^- were not excluded from the wall water as much as expected by such a high fixed-charge density, if the latter were uniform. Thus a reasonable simple-minded view would be that this heterogeneous wall could be divided into a water free space (WFS), not under the influence of the negative fixed charges, and a Donnan free space (DFS) in which the fixed charges were uniformly distributed and the Donnan theory applied. We made this view more plausible (to us) by imagining the fixed charges giving rise to electrical double layers in the wall water. Such a picture "explained" rather naturally the idea of WFS and DFS, although it has been recently superceded by the rather *more* natural picture of linear chains of charges.

This experience of working in Australia with Alex Hope was of the greatest benefit to me, not only owing to the ideas we had and the experiments we did. It was the first time I had worked in a biological milieu, and I believe I derived much from my close contacts with Bob Robertson, Joe Wiskich, and others who were more biochemically inclined than I and were trained as biologists, not physicists.

Back in Edinburgh later in 1958 I started something that turned out to be useful. We developed a series of graduate courses in biophysics designed to attract physics graduates. We hoped to orient such students as well as possible toward biology without preventing them from thinking as physicists. For a year's set of courses plus a small three- or four-month research project they were awarded a Diploma in Biophysics. We had five or six students each year, and the best of them stayed on to do a PhD. I think the program was rather successful, attracting such students as Peter Anderson, Roger Spanswick, Regis Kelly, Julian Collins, Randall House, Peter Kohn, David Aikman, and John Sinclair. One part of the course not so well appreciated by the students was a trip to the top of a Scottish mountain, Ben Lawers, to see the alpine flora there.

Early in the 1960s I was asked to write a review on "Ion transport and electrical potentials in plant cells" for the *Annual Review of Plant Physiology*. For me this was an important means of introducing to plant physiologists a biophysical approach to plant membrane transport, and it became a piece of pedagogy rather than strictly a review. Rereading it now I am struck by the excessive influence of the work of such animal physiologists as Hodgkin and Katz; indeed, I think I subconsciously looked upon single cells of the Characeae as green squid axons. Thus although I discussed the possible electrogenicity of ion pumps, I preferred to consider them neutral, as they at that time appeared to be in animal cells, despite evidence from Noe Higinbotham's group of the electrogenic nature of some ion pump(s) in higher plant cells. And again, although I discussed in some detail the great discrepancy between the resistance of the *Chara* or *Nitella* plasmalemma as calculated from the ion fluxes of Na^+, K^+ and Cl^- and as measured directly, I tried to interpret this in terms of some kind of ion interaction. It did not occur to me at that time to consider the hydrogen ion, presumably because the animal physiologists had not thought about its possible importance in membrane transport. Nevertheless I think this "review" did play some part in familiarizing plant physiologists with a more biophysical understanding of transport.

Immediately after this paper on ion transport, I wrote another "review" article on "Water relations of plant cells" for the first volume of *Advances in Botanical Research* in 1963. This was perhaps an even more didactic piece of work than the ion transport "review" and was, rather deliberately, quite heavily theoretical. At the outset I strongly supported the suggestions made a year or two previously by Taylor & Slatyer that the ancient concepts of "suction pressure" (European) or "diffusion pressure deficit" (American) should be replaced by a more thermodynamic approach to the driving forces of plant water relations; these old concepts should be replaced by that of water potential, which is a rough expression for the free energy per unit volume of water in a system. I regret now that I did not go further than Taylor & Slatyer (who, I believe, wished to keep some continuity with the past) and recommend that we use not water potential but the chemical potential of water—i.e. the free energy per *mole* of water, in water relations considerations.

I used this article also to introduce to plant physiologists the theory of irreversible thermodynamics as it applies in water transport, and in particular to derive the two equations for water flow and for solute flow in a real, nonsemipermeable membrane. These equations bring out the importance of the three basic parameters of water transport—the hydraulic conductance, L_p; the solute permeability, ω; and the reflection coefficient, σ. The latter parameter, the reflection coefficient, is the "new" solute-water interaction coefficient brought out by the irreversible thermodynamic formulation. I discussed it, following some work Benz Ginzburg and I had done together, in terms of a frictional model of membrane transport in which a specific frictional solute-water interaction is invoked. Perhaps this is no longer important. Certainly little is heard now of another application of irreversible thermodynamic theory I discussed: electro-osmosis, where the solute (ion)-water interaction is partly electrical. My interest in this, at that time, arose from the visit to Edinburgh of David Fensom and also from the suggestion by Bennet-Clark and others that active transport of water could take place through some kind of electro-osmotic "pump." I think we were able to show by both experiment and theory that electro-osmosis was unlikely to play any great physiological role.

I took advantage of this water-relations "review" to discuss two of my hobbyhorses: the mechanism of osmosis and the importance of unstirred layers.

I have already discussed my realization with Alex Hope in Australia that membrane transport processes could be seriously affected by diffusion in the unstirred layers of solution adjacent to the membrane. I took the opportunity in this article to set out the theory of unstirred layers and their effects on measurements of solute and water permeability, of reflection coefficients, and of their unequal effects on measurement of L_p depending on whether the water was entering or leaving a cell. (In the ion transport review I had also discussed unstirred-layer effects on membrane potentials, particularly in nonequilibrium situations.)

The mechanism of osmosis has long been an interest of mine, stimulated from time to time by discussions with and articles by Alex Mauro, Peter Ray, Patrick Meares, Alan Walker, Jack Ferrier, and others. From a thermodynamic point of view the mechanism presents no problem, as Alex Mauro has demonstrated. In the pores of a porous but semipermeable membrane separating two solutions at the same hydrostatic pressure but of different concentration there must be a hydrostatic pressure gradient that is the actual driving force on the water within the membrane. The hydrostatic pressure difference within the pores must be equal to RT multiplied by the concentration difference in the solutions. This can be shown from the necessity that the chemical potential of the water, which depends on both pressure and solute concentration, must be continuous (i.e. exhibit no sharp jumps) from one solution to the other. A molecular picture of osmosis, however, is more difficult to envisage and has proved more contentious. In the water relations article I put forward an "explanation," developed from an idea of Peter Ray, based on the unbalanced molecular jumping of water across the mouth of the pore, leading to a sharp drop in hydrostatic pressure across this mouth. (I return to the mechanism of osmosis below.)

We had academic visitors to the small Biophysics Department in Edinburgh. I have already mentioned David Fensom, who made us think about electro-osmosis and helped us demonstrate that it seems unlikely to be of major physiological importance. (We missed an important unstirred-layer effect associated with electro-osmosis, later demonstrated by Peter Barry and Alex Hope: The inequality between the transport numbers of ions in the membrane and in the bathing solutions leads to the creation of a local osmotic pressure difference, and hence osmotic flows, across the membrane.) Other important visitors were Karel Janáček from Czechoslovakia, Jan Stolarek from Poland, Bud Etherton, who brought with him insights obtained in the laboratory of Noe Higinbotham at Washington State University in Pullman, and Benz Ginzburg from Jerusalem.

Through Benz Ginzburg I came to know rather well Aharon Katchalsky, whose life was cut short by terrorists in the airport at Tel Aviv. It was Aharon, of course, together with his colleague Ora Kedem, who developed the basic membrane transport equations from Stavermann's formulation of the theory of irreversible thermodynamics. Benz and I embarked on some experiments and theorizing on transport across the membranes of single cells of *Chara australis* and *Nitella translucens,* inspired by the approach of Kedem & Katchalsky. Our exercise in basic theory was to calculate the membrane transport parameters L_p and σ for a lipid/pore model of a cell membrane in terms of frictional interactions between water and solute in an aqueous pore, between water and pore wall, between solute and pore wall, between water and the lipid part of the membrane, and between solute and the lipid part of the membrane. We were successful in obtaining suitable equations for L_p and σ, compatible with other equations derived by Katchalsky & Durbin, and we were able to make use of them in our experimental work. This experimental work in part consisted of showing that the hydraulic conductance, L_p, and the permeability of the plasmalemma of *Nitella* to urea both decreased markedly as the ambient osmotic pressure, due to sucrose, increased. We suggested that a kind of dehydration of the membrane was responsible for this permeability decrease. The other experiments were concerned with the somewhat difficult task of measuring permeabilities and reflection coefficients of *Nitella translucens* for rapidly penetrating solutes such as methanol, ethanol, and isopropanol. In these cases the unstirred layer was of great significance and we had to handle a situation in which the concentration profile of the solutes in the unstirred layer varied with time. It was an interesting exercise; we were also able to consider the reflection coefficients in the light of our frictional model equation for σ.

Benz Ginzburg left Edinburgh in 1962, I think, to spend some time in Thimann's laboratory at Harvard. Ron Poole came to Edinburgh from Harvard, but unfortunately we did not overlap much because in 1963 I was on my way to taking up a position at the new University of East Anglia, Norwich.

Before leaving the topic of Edinburgh, so important to me, there are one or two more things to say. Peter Mitchell was at Edinburgh, in the Zoology Department, during the time I was there. He was then developing his ideas on chemiosmosis and spent much time discussing various aspects of them with me. He would come to me to try out the physical and physicochemical aspects of his hypothesis that a proton gradient across the inner mitochondrial membrane drives a vectorial chemical reaction and results in the synthesis of ATP. I really missed a great chance here of realizing very early the significance of these momentous developments; then I might have enlarged my research to become more biochemical. But I was still too much an isolated physicist, working with physicists, and far too ignorant of biochemistry, for which I had no feeling. Thus I did not sufficiently realize the great importance of what Peter Mitchell was doing!

An event of considerable importance took place in the summer of 1960: the first (biophysical) membrane transport meeting held in Prague. There were about 50 invited delegates, all except me working on animal or microbial systems. I found it a great stimulus to meet the great names in the field. Peter Mitchell there gave the first (I think) exposition of his ideas. It was a notable event in the history of membrane biophysics. For me it underlined the influence of the animal biologists—not necessarily too good a thing.

The University of East Anglia (UEA) was a new University. It began operation in 1963, in wooden huts. It was to be a strong center for biology, an emphasis to be stressed by having a School of Biological Sciences—a name then new, I believe, at least in the United Kingdom. Thomas Bennet-Clark, a distinguished British plant physiologist, was made first head (called "Dean") of the School and given the task of building it up. There were to be four "chairs"—botany, zoology, biochemistry and biophysics—and four corresponding sections of the school. It was to be a loose arrangement with room for section-spanning disciplines such as genetics and developmental biology. B-C, as we called Bennet-Clark, asked me if I would take the chair of biophysics. Despite the fact that I had been able to build up a nice small department in Edinburgh, I accepted the chance to start again at UEA. I was excited by the concept of the School, by the promise that biology would be the strong science, and particularly by the thought that I was going to be part of a biology environment, involved in the teaching of undergraduate biologists. Another attraction was the pleasant city of Norwich and the surrounding county of Norfolk—particularly beloved by naturalists, among whom I still counted myself.

I took with me to UEA some of the people, including graduate students, from Edinburgh and rapidly built up a small group that included Alan Walker for four years or so before he decided to go back to Sydney. We had excellent financial support from what is now the Science and Engineering Research Council and from the Nuffield Foundation. We taught physics and mathematics to biology undergraduates through biology (i.e. using biological examples) and offered a third year (final year) option in biophysics. At the graduate level we carried on with the Edinburgh tradition and gave a series of courses in biophysics. These plus a small research project led to the degree of MSc in Biophysics.

The years at UEA in Norwich were good ones. There were excellent MSc and PhD students such as Julian Collins, John Sinclair, David Aikman, George Duncan, Mel Tyree, Ed Tarr, and Jim Barber. We had a small research group on root physiology led by Randall House and Peter Anderson. And we had many academic visitors and postdoctoral fellows—e.g. Alex Hope, Geoff and Nele Findlay, John Gutknecht, David Clarkson, John Cram, Jack Hanson, Bob Lannoye, and Tom Crossett.

This period 1963–1969 still saw relatively little research—at least from a biophysical point of view—on plant cell membranes. Perhaps the dominant transport idea at that time among plant physiologists was that of Emanuel Epstein—i.e. that of comparing the kinetics of uptake of a solute (by a plant root), via a carrier, with enzyme kinetics. This was a concept that we biophysicists—wrongly—did not take seriously, for it seemed to ignore any possible passive movement of solutes, the effects of membrane potential and cell inhomogeneities. However, we should have taken this idea—as Clifford Slayman, Dale Sanders, Peter Hansen, and others later did—and elaborated it into a proper kinetic model of the carrier-mediated transport of a solute, for I am now convinced that such a model will greatly enhance our understanding of this aspect of membrane transport. This was a typical error of judgment on my part; I was still too much under the influence of the animal physiologists, who were working with systems in which passive ion transport was more important than it is in plant systems.

The kinds of studies we made in Norwich involved in general this biophysical approach. Julian Collins tackled the problem of the ion-exchange capacity of plant cell walls using *Sphagnum* as material. John Sinclair studied the ionic relations of a moss, *Hookeria lucens*, chosen for its large cells and relatively large leaves, one cell thick. David Aikman studied the ionic relations of *Valonia ventricosa*, and Jim Barber those of *Chlorella*. (Jim Barber, now of photosynthesis fame, was the first to measure the membrane potential of *Chlorella* using a microelectrode.) Alan Walker carried on his electrophysiological studies of the Characeae. Some nice experiments on water and ion uptake by maize roots were carried out by Randall House and Peter Anderson. Our postdoctoral visitor, John Gutknecht, did the important experiment, to which I referred earlier, of measuring the two water permeabilities, L_p and P_d, of *Valonia;* after correcting for unstirred layers he showed that the two permeabilities were equivalent and thus that there was no evidence that water-filled pores in the membrane(s) of *Valonia* were an important route for the movement of water across the membrane(s). Another visitor, Bob Lannoye, together with Ed Tarr did some good work on the resting and active electrophysiological properties of *Chara corallina*.

I remember with considerable chagrin that at this time ($\sim$1964) Alan Walker and I tried to see if we could pick up any net movement of protons in or out of a *Chara* cell under conditions that I can no longer remember. The reason for such an experiment was our constant worry about the great discrepancy between the electrical conductance of the plasmalemma of *Chara* as calculated from the ion fluxes (of Na^+, K^+ and Cl^-) and as directly measured. The latter was 10–15 times that of the former! Was there a missing moving ion? If so, it could only be H^+. But for reasons we cannot now understand we could detect no change in the pH of the medium, and decided that H^+ did not play any significant role in the ionic relations of *Chara*. Thus the idea of the H^+ extrusion pump, developed from the work of Kitasato in 1968, and the electrogenic nature of some pump, was missed by us.

Towards the end of the 1960s I was persuaded to move from the University of East Anglia to the University of California, Los Angeles—at first to the Laboratory of Nuclear Medicine and Radiation Biology and then to the Department of Botany. As it turned out, I did not stay long at UCLA, but moved after two years to become chairman of the Department of Botany at the University of Toronto. My short time at UCLA was certainly pleasant, thanks to George Laties, Park Nobel, Jacob Biale, and others.

From the point of view of my own scientific development it was a period in which I asked myself whether I could make any contributions to plant ecology. In part this query arose because when I arrived Hal Mooney had just moved from UCLA to Stanford, leaving behind one or two graduate students and the course in plant ecology. I volunteered both to look after the graduate students and to give the course. This was truly "the blind leading the blind," but I think we all enjoyed it—even though I put a biophysical slant on everything. Since my early interest in biology had been as a naturalist, I wondered whether I could fruitfully tap that early enthusiasm and combine it with my training as a physicist and my experience as a plant cell membrane biophysicist. I tried, but it didn't work. I didn't know what to do. I couldn't think of the right questions to ask. Ecology was too complex, quite the wrong milieu for a physicist who had to work with simple things for his approach to be fruitful. (Park Nobel, on the other hand, has made important biophysical contributions to plant ecology.) In any case, after moving to Toronto in the summer of 1971, I went back to being a biophysical plant physiologist.

Being chairman of what I believe is the largest botany department in North America leaves little time for one's own research. But I was fortunate in that Mel Tyree joined the faculty in Toronto at the same time as I did. He initiated some fundamental research on the water relations of plants as investigated by the so-called pressure bomb. (This device had been reinvented by Scholander & Hammel after Dixon, who had invented it around 1920 but made it of the wrong material—glass.) Our chosen plant was hemlock, *Tsuga canadensis*. I cooperated to a limited extent in this work, which was quite extensive and exposed, or so we thought, the advantages and the limitations of the technique. At least with the material we used, the technique seemed reasonable for measuring such "static" parameters as water potential as a function of volume of water expressed from a shoot and the more "indirect" parameters such as turgor pressure, osmotic pressure, and volumes of water both in the cells and in the apoplast. But our attempts to obtain "kinetic" parameters such as hydraulic conductance, L_p, were unsatisfactory. Perhaps the best thing about this research was that it initiated Tyree's interest in the physiology, and particularly the water relations, of trees, which he still pursues with great success.

Another interest of Tyree's at this time was phloem translocation. He theorized that the flow might be driven by hydrostatic pressure gradients—i.e.

he helped develop a mathematical model of flow according to the Munch hypothesis. In this work he became associated with a physicist from Ohio State University, Jack Ferrier. Jack Ferrier joined us in Toronto, initially as a postdoctoral fellow. (For the last few years he has been with a group in the University of Toronto Dental School, where he works on the electrophysiology of cells important in bone.) During his time in the Department of Botany, Jack Ferrier made a number of important biophysical contributions. At first he worked on developing this computer model of phloem transport. Later he became interested in plant water relations and we tried to develop a so-called external force method for measuring hydraulic conductivities and elastic coefficients for higher plant cells. (The idea involved squeezing water out of cells.) The theoretical aspects of this method proved horrendous, and we eventually dropped it.

We had a momentous visitor, first as a postdoctoral fellow and then as research associate, in Bill Lucas, who was a very good experimenter indeed. He worked at this time on the spatial distribution of the apparent H^+ pumps, OH^- carriers, and CO_2 or HCO_3^- uptake regions of the *Chara corallina* plasmalemma. Bill Lucas and Jack Ferrier (and I to some extent) became associated through this work, Jack Ferrier in particular developing realistic models for the flow of ions externally between the acid and alkaline bands.

Claudine Morvan came to us from Rouen on a year's postdoctoral fellowship. She had worked in Rouen on the plant cell wall, but in Toronto, working with Jack Ferrier and Bill Lucas, she helped us initiate studies on electrical noise, both voltage and current fluctuations, in the plasma membrane of *Chara corallina*. At the time we were not able to say exactly what was the origin of the noise: within (passive) channels, the fluctuations in the numbers of open channels or active transport fluctuations? This study of electrical noise in *Chara* was later taken up by a graduate student, Stephen Ross, who managed to convince himself that the low-frequency noise was indeed attributable to active transport, probably the H^+ pump. By injecting white noise current and measuring the consequent voltage noise Stephen Ross also measured the impedance of the *Chara* plasma membrane, obtaining the apparently strange result that at very low frequencies (<0.5 Hz) the membrane capacitance becomes large and negative. We (Stephen Ross, Jack Ferrier, and I) recognized that this phenomenon is related to the diffusion-driven movement of protons in the unstirred layer. This is perhaps an amusing rather than an important effect, but it illustrates the necessity of a sound physical understanding in some aspects of membrane electrophysiology.

Jack Ferrier has recently had a very nice idea about the mechanism of osmosis and we have worked on it a little together. If a porous semipermeable membrane separates, say, pure water from a not-too-concentrated solution, then only occasionally, and for a brief time, would there be a solute molecule opposite the pore mouth on the solution side. Only at such an instant would the water in such a pore "realize" it was facing a solution—and not pure water; and only at such instants would there be a force, between solute molecule and water molecule, pulling water through the pore from the pure-water side. (The solute molecule could not move: It would be stopped by the membrane.) The theory, made quantitative by means of some simple physical ideas, *works*—i.e. gives the right answer that the equivalent pressure driving the water through the membrane is RT multiplied by the solute concentration. Even though the concept is still in rough form, I feel we have an adequate theory of osmotic flow at last. Note that although a pressure difference of $RT\Delta C$ will produce exactly the same amount of flow as a concentration difference of ΔC, the pressure difference would give a smooth flow while the flow produced by concentration difference would comprise spurts—i.e. would be noisy. It seems unlikely that these different flow types could be demonstrated experimentally. Such is a "hobbyhorse" of mine—of no real importance, but intellectually challenging and satisfying.

During the past 7 or 8 years I have returned to another old interest of mine: the ion exchange properties of the plant cell wall. Marie Kleinová, who worked with me for two years, started the work on the isolated cell walls of *Chara* and of some higher plants. She also carried out studies on the water sorption properties of such walls, an important aspect of their physical chemistry. The ion exchange work was taken over by my graduate student, Conrad Richter, who used cell wall material from the moss *Sphagnum*. He was able to interpret the ion exchange properties in terms of the Manning model of linear systems of fixed negative charges comprising the pectin fraction of the wall.

But now the plant biophysics group in Toronto, as such, is practically finished. Mel Tyree left 3 or 4 years ago. Jack Ferrier, although still in Toronto, devotes his time to his bone cell studies. I have retired, to play only a minor role in the research of the department. However plant membrane studies are more vigorous than ever, although now permeated with a more biochemical approach, led by my successor, Eduardo Blumwald. It is appropriate that there should now be an integrated approach—biophysical, biochemical, cell physiological, molecular biological—to plant membrane transport; and this is what is happening in Toronto.

Of course, my activities in Toronto were not confined to encouraging biophysical membrane research. Teaching is the other pillar of a university, and I wanted to inculcate in at least a few students some feeling for the biophysical approach to biology—to plant physiology in particular. After all, living organisms obey the laws of physics (and of chemistry, which is also physics).

As an administrator I did not repeat the system I had developed in Edinburgh and Norwich of offering a MSc in Biophysics that involved a lot of course work plus a little research. Instead I attempted to inject my ideas, my approach, into the undergraduate teaching program. For example, I offered a course to first-year biologists on the "Design of Organisms" in which we discussed questions of size and shape—how these were determined by the simple physics of diffusion, circulation of fluids, and/or mechanical constraints. The textbook for this course was *Gulliver's Travels!* After their shocked discovery that the course treated not only physics but also English Literature, most students quite enjoyed it; but whether it had any subsequent influence on their biology, I very much doubt. I also handled the more biophysical aspects of ionic and water relations in the basic plant physiology course, taken in the second year in Toronto. Students clearly found this difficult, despite the fact that they had all had physics, chemistry, and mathematics during their first year. They seem to keep their individual courses in little boxes which, once closed upon completion of the term are not opened and looked into ever again. The North American System involving frequent assessment of progress in individual courses, examinations confined to the content of each course in isolation, and a transcript of courses taken with grades given may encourage this tendency to keep one's knowledge in separate compartments. The other course I gave, for advanced undergraduate and graduate students, was entitled "Cellular Transport Processes"—really a course on the biophysics of membrane transport in plant cells. I shared with Mel Tyree advanced and graduate courses on "Physical Environment of Plants" and "Plant Biophysics." The former treated the interactions among heat, light, humidity, etc, and the plant; the latter contained more membrane biophysics, some soil physics, and a lot on translocation.

Did these courses have much impact? I confess that I feel they had little. In some ways most biology students have preselected themselves as feeling hostile to mathematics and the physical sciences; they can almost be described as refugees from the physical sciences. Therefore the impact of one or two biophysical courses on their general outlook is negligible. All their other courses tend to be given by people who share their distaste for, and incompetence in, physics and physical chemistry. This is a somewhat depressing conclusion from so many years of attempting to explain, to put over, a physical viewpoint that I fervently believe is essential. My greatest success in teaching, and perhaps the only thing that has justified my academic career, was in bringing people from a background in physics or physical chemistry into biology through the Diploma in Biophysics at Edinburgh and the MSc in Biophysics at East Anglia. Does this mean that the best hope for plant biophysics, if it is deemed of some importance, is to attract students already trained—or perhaps training—in physics and/or physical chemistry? Maybe this will prove to be the only way; perhaps the demand for plant biophysicists is not great enough to encourage what is really a ridiculous idea—revolutionizing the training of biologists.

My years in Canada have coincided with a great change in the study of plant membrane transport. From being a field in which only a handful of people worked, particulary from a biophysical point of view, it has developed into a sophisticated and rapidly changing subject. No longer is it a poor relation of the study of animal and microbial cell membrane transport. One important coordinating feature of this development has been the series of International Workshops on Plant Membrane Transport we have been holding every three years or so for the past 20 years. These started in 1968 at Schloss Reinhardsbrunn in East Germany but perhaps really got going at the Workshop organized by Peter Anderson in Liverpool in 1971. The third workshop was held in Jülich, West Germany, in 1974, organized by Ernst Steudle and Ulrich Zimmermann, and the fourth in 1976, in Rouen and Paris, the organizer being Michel Thellier. The fifth took place in Toronto in 1979; then came Prague in 1983, Sydney in 1986, and this year, 1989, Venice. These workshops have been attended by essentially all workers in the field. The proceedings of each of them have been published and given a clear record of the rapid increase in our understanding of transport processes across plant cell membranes. The International Workshops have acted both as a periodic focus of interest and as an important stimulus to work in this field. I think it fair to say that from the outset the guiding approach was biophysical.

In addition to these Workshops I have enjoyed being an Associate Editor of the journal *Plant Physiology* for two periods of two or three years, handling the more biophysical papers—especially but not exclusively those on membrane transport. As has been said many times the ASPP and the journal owe a tremendous debt to Martin Gibbs, who has now been Chief Editor for more than 25 years. When Marty asked me if I would succeed Noe Higinbotham as an Associate Editor, I didn't hesitate to accept the important job. I thus succeeded the person who did more to move forward the field of plant membrane transport, particularly in North America, than anyone else I can think of. I found being an editor of scientific papers a difficult but rewarding task. Accepting a paper rarely involved a straightforward decision, for the assessments of referees often disagreed. Sometimes a paper had to be declined, never a pleasant decision to take, but I never had an angry "customer"—at least authors never vented any anger at me. I look upon this associate editorship as one of the most useful things I have done, always made pleasant by the cordial relations with Marty Gibbs and his secretary.

Now 70, retired, and able to spend more time reading and studying my beloved literature (not scientific), how can I summarize what I think I have done for plant physiology? Whatever I have done, however it may be judged, I see it purely as an accident or series of accidents, as should be clear from what I have so far written: the accidents of my year of birth, my study of physics, my change from physics to biology (because the University of Edinburgh wanted me to teach physics to medical students!); and then the taking up of the study of membrane transport in plants for reasons that certainly did not include a burning desire to do so. At that time the field was nearly empty. Whatever one did—particularly from a biophysical point of view—was new and exciting. One could not help but make some progress, succeed in a kind of way. I myself have made only one or two small contributions, nothing that justifies my writing about them as I have done here. My main contribution has been, as I see it, the introduction to the field of plant biophysics of a number of outstanding people, among them Enid MacRobbie, Roger Spanswick, and Jim Barber. They and their students have made *major* contributions. Somehow or other the biophysical approach to plant physiology has become respectable, and for this I have shared with others some responsibility.

Annu. Rev. Plant Physiol. Plant Mol. Biol. 1989. 40:1–18

MY EARLY CAREER AND THE INVOLVEMENT OF WORLD WAR II

Noburô Kamiya

National Institute for Basic Biology, Okazaki, 444 Japan

CONTENTS

It is an honor to be asked to contribute a prefatory article to the *Annual Review of Plant Physiology and Plant Molecular Biology,* and I accepted the invitation without deeply considering what I should write. Since, however, I underwent unforgetable experiences in my youth—especially before and during World War II—I allow myself to record here my early personal history, remembering especially my teachers and others to whom I owe much.

CHILDHOOD

Born on July 23, 1913 as the third son of Ichiro Kamiya in the uptown district of Tokyo, I spent a calm childhood in a peaceful family. The event most shocking to me during my childhood was the great earthquake that hit the

1040-2519/89/0601-0001$02.00

Tokyo and Yokohama area on September 1, 1923. At the onset of the sudden quake around noon of that hot day, I jumped out into the garden, but owing to the violent horizontal and vertical vibrations of the ground I could hardly walk. Fortunately our single-story wooden house did not collapse, but our fire-proof two-story storehouse with its thick mortar wall was severely damaged. A few hours after the earthquake, gigantic cloud columns rose up in the southeastern sky: Fires that lasted until the next day reduced practically all of downtown Tokyo to ashes.

After a few days, accompanied by my father I took a look at the fire-ravaged areas. I could hardly bear to look at the disaster's consequences. Thousands of charred bodies were piled up along the roadsides; drowned bodies were scattered in moats and riverbeds from which victims, surrounded by the fire and smoke, had been unable to escape. To a boy of 10 years old this sight was ghastly. I understood that if our house had been in the downtown distict, I might have encountered the same fate. Rome was not built in a day, but Tokyo was destroyed overnight by a natural calamity. This miserable experience must have changed my view of life. It brought home to me firmly what a blessing it is simply to be alive and how wonderful is a peaceful existence, a keen awareness that has stayed with me ever since.

This terrible blow aside, I spent a happy boyhood playing with brothers and friends. I was absorbed in making various scientific toys and models such as small electric motors and steam engines, devoting less effort to the homework assigned by my schoolteachers. Some boys were enthusiastic about collecting insects or plants, but I was not. Probably I was not a good naturalist. As a high school junior and senior I thought there was no choice for me but to become an engineer. Thus I took a preparatory curriculum in which less weight was laid on biological sciences and in which English was the foreign language emphasized.

BOTANY AS A MAJOR

Being intrigued by the mystery of life, however, I had a latent interest in biology. What dissuaded me from majoring in basic biology was the fact that it was difficult to get a proper job in the university or elsewhere that permitted research in pure biology. The jobs available to biology graduates were mostly posts as high school teachers, whose main duty was the education of junior or senior high school pupils. The graduates from the engineering school had no problem finding good jobs and salaries.

After repeated consideration, however, I concluded that one should select one's profession not according to the convenience and profit but in response to an awareness of one's calling. Getting a good job is of secondary importance. My teacher of botany, the late Dr. Yoshitaka Imai, an internationally known geneticist at what is now Tokyo Metropolitan University, encouraged me in my resolution. To study cell biology or general physiology, however, we had to select either plant science or zoological science—there were at that time no departments of what we now call cell biology, biophysics, molecular biology, and so on. Without hesitation I selected plant sciences, because I did not like to kill animals, not even small insects.

Before World War II, the systems of education in the United States and Japan were somewhat different. In Japan, we had adopted the so-called 6-5-3-3 system [6 years of primary school, 5 years of middle school (junior high, or lower-secondary school), 3 years of high school (senior high, or upper-secondary school), and 3 years of undergraduate study in a university]. I was allowed to enter the Department of Botany, Faculty of Science, University of Tokyo in 1933. Although I found myself in a highly traditional environment, I gleaned enjoyment from each lecture. Plant taxonomy, morphology, genetics, physiology, ecology, microbiology, biochemistry and their class works—everything was fresh to me.

The lectures, however, were disconnected from each other. Morphology, for instance, was pure morphology, including fossil plants; it did not take into consideration the physiological function or meaning of plant structures. I, on the other hand, was keener on interdisciplinary fields, especially the functional aspects of the plant cell. But cytology was a part of morphology, and the class work of cytology was concerned mostly with the morphology of a variety of plant cells. To study chromosomes, whether in connection with taxonomy or genetics, we had to fix the cell with an appropriate fixative and stain with a dye. Few plant physiologists in Japan at this time were interested in either the physical chemistry of the living cell or its dynamic activities. Soon, before the third year of the undergraduate course, I asked my teacher of cytogenetics, Dr. Yoshito Sinotô, to let me work on living cells for a Bachelor's thesis. Dr. Sinotô was broad-minded enough to accept me in his laboratory. Although he was not himself a specialist in this field and could not guide my research, he encouraged me to open up this potentially important field in Japan through my own efforts.

As a happy new member of Dr. Sinotô's laboratory, I immediately faced a serious problem. Most of the important papers in my field of interest were written in German. I was only poorly educated in German, because German was not then as important to prospective engineers as it was to medical and biology students, for whom it seemed almost a prerequisite. I made up my mind to learn the German language all over again. I went to a German-training school and took private lessons from a German woman while I was a university student. In a year or two, I made fairly good progress. I could read German technical papers, and I wrote my Bachelor's thesis in (not very expressive) German. The thesis was on the effect of direct current on mitosis in the stamen hair of *Tradescantia* (1). It reported that the mitotic apparatus moved as a unit mass toward the anode, reversibly in the presence of a comparatively strong but short-lasting electric current. The cell was not killed by this treatment. I did not regard this phenomenon as electrophoresis of the chromosomes having a negative charge but believed it was caused by a temporary hydration and dehydration of the cytoplasm brought about by the ionic imbalance on the anodal and cathodal sides of the cell. I believe this interpretation is still thought to be correct.

In 1937, the year I entered the Graduate School of the University of Tokyo, the International Congress of Education was held on its campus. At that time, not only English (as in recent years) but also German and French were the official languages of the Congress. The office of the Congress therefore recruited translators of German and French from the University of Tokyo to help participants in the Congress from non-English-speaking countries. Although my German was still not fluent enough, I applied for the job and was accepted as one of the German translators. The mission of the translators was not to translate lectures presented at the Congress, a task too difficult for a student from a different discipline, but to accompany distinguished guests from abroad. I was assigned to guide a world-famous German philosopher, Eduard Spranger, who stayed in Japan as a guest professor for some time before and after the Congress. Regardless of his high scholarship and academic reputation, he was always courteous and friendly to me. When the Congress was over, he gave me a beautiful German photo album with a dedication in token of his gratitude. I think his influence helped me to be selected as a Germany-Japan exchange student the next year, providing an opportunity to study in Germany what it was then difficult to study in Japan.

VISITING ERNST KÜSTER AT THE UNIVERSITY OF GIESSEN

My destination in Germany was the Botanical Institute of the University of Giessen, presided over by Professor Ernst Küster, then the world authority on plant cell biology and plant cell pathology. Giessen was a pretty, quiet, garden city in western Germany with a population of only 35,000. It was situated on the east side of the Lahn river, a branch of the Rhein.

I left Japan for Germany in the early part of April, 1938, on board a Japanese freighter. Cherry blossoms were then at their best in Japan. When I reached Giessen two months later, in early June, this cosy city was basking in the fresh verdure of linden trees. A dignified, scholarly looking old gentleman greeted me warmly. Küster was then 63 years old and I was 24. He might well have been curious about me, because there were no Oriental people in Giessen in those days. I was the first (and last) Japanese student ever to work in his laboratory. On meeting me for the first time, he told me he would like to make my stay in his laboratory as fruitful and effective as possible. The professor soon guided me to a chair before his grand desk, just as he would have done had he been giving an oral examination to one of his own, German students. He needed to determine how far my studies had progressed in Japan. I think I answered the first questions pretty well, because they concerned osmotic phenomena in plant cells; but I was shaky on the second group of questions. He asked me if I knew the names of several algae and fungi and their characteristic features. This was the weakest point in my botanical knowledge. What I had learned in systematic botany at the University of Tokyo primarily concerned plants higher than ferns. I knew popular class-work materials such as *Spirogyra, Nitella, Phycomyces,* etc, but otherwise my answers were very unsatisfactory. Professor Küster, however, did not dwell on my ignorance in the matter, but said "Here we are doing experimental work with special attention to lowly plant cells and protophytes. Are you not interested in research along this line?"

I was delighted to hear that I could study there, from that time on, the subjects about which I had little knowledge but that interested me keenly. My life in Giessen thus began. Looking back upon my student days, when I was not yet an independent worker, I think in Giessen I spent the most enriching time in training myself. My stay there was to be cut short at one year and three months by the outbreak of the second European War, a subject to which I shall return below.

DAYS SPENT IN GIESSEN

In the summer semester when I arrived in Giessen, Küster's lecture on General Botany was to be heard every day. I decided to attend this class not only to revise my knowledge in General Botany but also to train my ear for German. This lecture was intended primarily for premedical students, and its level was about that of the famous Strasburger's *Allgemeine Botanik*. Küster spoke without a manuscript (a tiny memo if anything at all), yet the content was condensed and perfect. He said everything necessary and important but nothing unnecessary or trivial for the students, demonstrating simple experiments and showing many specimens arranged on the large lecture table. The lecture was so perfect that one thought the stenographed transcript would make an excellent textbook.

One thing, however, disturbed me. For some months after I had reached Giessen, I stayed in a student dormitory, where the dining hall opened at 7:30 a.m. Küster's lecture on General Botany started punctually as early as 7:00 a.m. every morning and ended at 8:00. The lecture was followed immediately by research work, usually until 12:30. Therefore, I had no chance to take breakfast and had to work the whole morning on an empty stomach! I gradually became accustomed to this habit, and I rather enjoyed every minute of my laboratory work.

It was not Küster's way to use sophisticated new measuring apparatuses or to try to develop a new gadget for a specific purpose. Instead he observed the variety of living plant cells in nature, mostly those of lower plants, and induced abnormal cell structure or cell behavior experimentally. During my sojourn in Giessen, I wrote five papers, of which four were on protophytes [*Melosira* (a diatom)(2), *Euglena* (4), *Spirogyra* (5), and *Lyngbya* (a blue-green alga)(6)] and only one was on onion scale epidermis cells (3). I followed no predetermined plan of study and therefore sought no particular connection among my various observations. My main purpose was to familiarize myself thoroughly with as wide a variety of lower-plant materials as possible. We did not use complicated methods; as a means of experimental study we frequently applied plasmolysis, vital staining, centrifugation, etc.

There were 10 or so research people working in the Botanical Institute, including three postdoctoral assistants, several doctoral candidates, and several guest research workers from other universities. Just before I arrived at Giessen, Walter Michel (19) had succeeded in bringing homospecific as well as heterospecific plant protoplasts into fusion. At that time there were no suitable enzymes strong enough to dissolve the cell wall quickly, and we hence had to resort to plasmolysis followed by the opening of cells in order to obtain the intact plant protoplasts. Though the efficiency was low, it was nevertheless possible to get naked plant protoplasts by this means. After repeated trials and errors, Michel finally fused two protoplasts, a memorable achievement in experimental plant cell research. Who could then have expected that protoplast fusion would become such a popular and important method in plant biotechnology?

It was Küster's daily habit to visit each research worker every other hour, asking "Now, do you have anything new?" Having to listen to such a "greeting" several times a day was certainly too much for us. But if someone was able to exhibit noteworthy phenomena under the microscope, Küster became excited, saying "Sehr interessant! Sehr interessant! Kolossal!" Occasionally Küster would unceremoniously displace a young researcher at his bench and proceed to describe what he observed, while the host of the preparation, sitting to one side, wrote down quickly what Küster dictated. The transcript would be a part of the paper to be published later. It was a good idea to write such descriptions on the spot, especially for phenomena that occurred only rarely. Our memories may so soon fade.

While Küster paid close attention to the research progress of each of his students, he himself wrote more than ten original papers each year. He was the author not only of *Die Pflanzenzelle* (1st ed. 1935; 2nd ed. 1951; 3rd ed. 1956, after his death), his magnum opus, but also of many important monographs. He also acted as a chief-editor of *Zeitschrift Für Wissenschaftliche Mikroskopie und für mikroskopische Technik*. I wondered how so much could be achieved by only one man.

It was not long after I had arrived at his laboratory that I saw a typist entering his big office and taking a seat in front of a large table set at the center of the room. Then the door was closed. Before long, I heard Küster's footsteps as he walked slowly around the central table; he then began dictating to the typist, and there followed the sounds of continual typewriting. Though of course we were not allowed to enter the room then, I learned from one of the professor's assistants that Küster used nothing but a small memo when he dictated. Simply through concentration, he found the proper layout, context, and sentences for a scientific paper while walking at a slow pace around the table. After a few hours, bunches of typewritten sheets, say 30 or so, were finished. Küster took a look at them and made corrections, insertions, or deletions with a pen; but the manuscript was not retyped; it was sent out to a publisher in a day. It must have required very special talent to write a scientific paper in such a way, even in one's mother language.

These recollections of my Giessener days are already half a century old. One of my German colleagues noted that Küster was among the last university professors of an old German type.

Another memorable experience occurred one day when we took dinner together with a few colleagues in a Giessen restaurant. At a table next to ours sat Robert Feulgen, who had established the Feulgen method of staining DNA. Since I thought Feulgen a great figure in the history of cytochemistry, it delighted and surprised me to see him near at hand, taking dinner with a glass of beer just as any other man might.

Things have been much changed since World War II. Nowadays many scientists whose mother tongues are not English (including German scientists) write their papers in English, the most popular and generally accepted language internationally. Therefore, learning English and having to write papers in English are inevitable demands upon internationally minded research workers. This is a big handicap for Japanese scientists. Not a few of them believe that they are the people poorest at learning Western languages, probably because of the basic difference in construction between their mother language and the Western ones. As a matter of fact, it is not rare for us, and I don't hesitate to include myself, to spend more time and energy in preparing the manuscript in English than in performing experiments. Preparing manuscripts of scientific papers by dictation, as Küster did, is difficult enough in one's mother language, let alone in a foreign one. Such expertise in English is at present a dream for most Japanese scientists.

HIDDENSEE

A small marine biological laboratory stood on a tiny island called Hiddensee, just off Rugen Island in the Baltic Sea. The island belongs now to East Germany. It was a custom for Küster and his students, about 10 altogether, to spend three summer weeks at Hiddensee. I had the good fortune to spend two summers there, in 1938 and 1939. It was an enjoyable time for all of us, combining research and recreation. We collected various algae there, finding interesting materials for observation and experiments. If our experiments appeared promising or needed longer to finish up, we took the material back to Giessen. In 1938 we were invited by Professor Boysen-Jensen, a great figure in the field of plant growth physiology, to visit his laboratory in Copenhagen on the way back to Giessen. This was a pleasant and instructive experience for me.

The next summer, 1939, Hiddensee Island was as quiet and peaceful as it had been the year before, but European international relationships were growing more and more strained. On August 25, close to noon, as I was working at the microscope as usual, I got an unexpected long-distance telephone call from Berlin. A Japanese friend of mine who knew my where-abouts transmitted urgent advice from the Japanese Embassy. I was to prepare immediately to depart Germany. I was to board a Japanese boat, the *M. S. Yasukunimaru,* that happened to be in Hamburg and was scheduled to leave the next evening at 8:00 p.m.

I immediately informed Professor Küster of this urgent message from the Japanese Embassy. He was astonished to hear it and told me earnestly, "I feel very sorry to have you leave us. If the reason for leaving Germany is simply a financial problem, don't be bothered with it. As long as I am alive, I think I can help you financially, even if war should break out. When you go to Berlin and meet the officials of the Embassy, tell this to them so that you will be allowed to stay on in Giessen. Giessen is a comparatively safe place if the war begins. If you are allowed to stay in Germany, you need not come back here again, as we shall not stay here long anyway. We will see you in Giessen again. We will take your material (the blue-green alga *Lyngbya*) with us to Giessen. . . ."

Mrs. Küster gave me as a souvenir a photo album of Hiddensee together with a sheet of paper on which she wrote a beautiful poem by the German poet Isolde Kurz entitled "The sun going down in Hiddensee." Underneath this poem she wrote "To the memory of pleasant hours on Hiddensee and Giessen absorbed in the wonder of protists," signing it "Frau Dr. Gertrud Küster." Mrs. Küster and a colleague made me sandwiches, each wrapped separately, and added fruits and chocolate decorated with a branchlet of the beautiful wild flowers of Hiddensee. This would be my snack on the voyage from Hiddensee to Stralsund and in the train from Stralsund to Berlin. The time hurried past. I had to catch a boat bound for Stralsund at 4:30 p.m. The summer sun was still high in the sky. The sight of Professor and Mrs. Küster seeing me off on the pier of Kloster harbor on Hiddensee still lives in my mind. Two of my colleagues accompanied me to the next harbor, Vitte, on the same island. I arrived in Berlin late that evening.

Next morning I met in Berlin several Japanese people and an official of the D.A.A.D. (German Academic Exchange Agency). Although we had received not an evacuation order but only advice, all the Japanese I met in Berlin strongly opposed my going to stay in Giessen. They urged me to go immediately on board *M. S. Yasukunimaru,* a refugee boat. This boat would sail first for Bergen, Norway, where she would stay for some time to see how the international situation developed; if war did not break out, she would return to Hamburg.

It was difficult to anticipate the fate of a foreign student or the possibility of his performing research work in wartime, even if he was not in an enemy country. I decided finally to go to Hamburg, in spite of Küster's kind advice, to board the refugee boat with faint expectation of the ship's returning to Hamburg. If I was to catch the boat, I could not go back to Giessen to gather my belongings from the laboratory and the boarding house. What I had with me was only a small suitcase taken from Giessen to Hiddensee, containing mainly the clothing necessary for a three weeks' stay.

With about 180 refugees aboard, the boat left on time at 8:00 p.m. on the moonlit night of the 26th of August, 1939. Hearing Japanese spoken all around me, I experienced an illusion of being suddenly back in Japan. Most of the refugees were businessmen and their families. Among the few scientists were two physicists, Drs. H. Yukawa and S. Tomonaga. Their names were not so well known then, but they both won the Nobel Prize later. Since Tomonaga was also one of the exchange students of the D.A.A.D., he and I became good roommates in the same second-class cabin.

Two days after embarkation, the refugee boat arrived at Bergen and anchored there for one week, watching for changes in the international situation. While we were in Bergen, Germany invaded Poland and World War II broke out. This deprived me of my last hope of going back to Giessen. I wrote a letter to Küster thanking him for his willingness to have me continue my research and for all he had done for my sake. It was announced that the boat would leave Bergen at noon on the 4th of September, 1939, for Yokohama by way of New York, the Panama Canal, and San Francisco.

In this situation, it occurred to me that it might be possible to continue my work in the United States, since I earnestly hoped to study further before returning home. I knew the name of William Seifriz at the University of Pennsylvania, as I had been a devoted reader of his book *Protoplasm* (1936) when I was in Japan. But I was of course not prepared for landing in the New World. I had no proper visa, no letter of introduction, no acquaintances in the United States. I could read and write English to some extent, but I had had no practice in speaking it. What I had was barely enough money to transit the American continent by train.

When the refugee ship was still in Bergen, I got a "transit" visa, good only for one month. With it I was able to land in the United States. In Bergen I bought a book on English conversation. It was written in Norwegian, but I could pick up English expressions in it and try to memorize them during the voyage to New York. Such hasty efforts did not help much toward speaking English, however. Before the boat reached New York, I wrote a long letter in German to Professor William Seifriz explaining the situation and asking about the possibility of working in his laboratory for a time. Upon my arrival in New York I posted the letter, appending my curriculum vitae and a list of my publications.

VISITING WILLIAM SEIFRIZ AT THE UNIVERSITY OF PENNSYLVANIA AT THE OUTBREAK OF THE EUROPEAN WAR

Taking a far northern route in the Atlantic Ocean, *M. S. Yasukunimaru* reached New York ten days after its departure from Bergen. I think it was I alone who got off the ship at New York.

The first thing I did in New York was visit the Japan Institute. Utterly independent of politics and economy, it had been founded for the purpose of introducing Japanese culture to Americans. Since I had no letter of introduction, I introduced myself to a young official there. Soon another official joined us, and eventually I was guided to the office of the director of the Institute, Tamon Mayeda. (Mayeda was to become the first Minister of Education after World War II.) He must have wondered at the queer young Japanese refugee who had appeared on his Institute's doorstep. I told him my whole story and of my desire to continue to study with Professor Seifriz at the University of Pennsylvania. Mayeda kindly tried to contact Professor Seifriz by phone: "There is a young Japanese scientist now in New York evacuated from Germany on account of the European war. He earnestly hopes to see you and, if possible, to work with you for some time. Would you be good enough to meet him?" Thus Tamon Mayeda made me an appointment with Seifriz on the 18th morning of September, 1939.

It was a clear autumn day with deep blue skies, but my innermost feeling was a complex mixture of aspiration and anxiety. If Seifriz did not speak German, what could I do? Still, I had no choice but to visit him. The letter I had written from the ship must already have been in his hands. Having bought a city map of Philadelphia at the terminal station, I walked to the University campus. I asked the way several times and finally got to the laboratory of Seifriz. It was a rather old fashioned, detached house called Botany Annex.

To my great surprise, the person I met at the entrance of the laboratory was a young Japanese woman, a graduate student of Seifriz. She said "Are you Mr. Kamiya? I am so glad to see you. Dr. Seifriz is waiting for you. He loves to speak in German!" It was a further surprise that she already knew my name.

Guided by her, I stepped upstairs. When I came to the entrance of Seifriz's office, the door of which was open, he walked toward me and shook hands, saying in German that he was very pleased to have me there. He had read my letter already. The young lady, whose name was Masa Uraguchi, seemed to have been waiting downstairs for my coming. It was indeed very fortunate for me that Seifriz spoke German as fluently as his native language. Although he had been born in Washington D.C., near the Capitol, his parents were German and he had studied colloid chemistry for some time with Freundlich in Berlin.

At that time Seifriz was interested in the slime mold *Physarum polycephalum*. It was an excellent experimental material for the study of protoplasm, especially regarding the rhythmicity of its protoplasmic streaming. He was studying the pulsation of the slime mold by taking time-lapse cinefilm. My small paper on the rhythm of euglenoid movement (4) was a hastily written report of a simple observation done in Giessen, but Seifriz seemed to have paid some attention to it, probably because it dealt with a biological rhythm. Since we shared a close interest in this topic, our conversation was lively. A room in front of Seifriz's office, just then vacant, was allotted to me for my laboratory.

It was unusual for Professor Seifriz to have lunch off-campus, but on this day he took Masa Uraguchi and me in his car to a fancy restaurant somewhere near the university. It was funny that we had no common language among us. Seifriz and Masa spoke in English, Seifriz and I spoke in German, and Masa and I spoke in Japanese. But we shared a delightful mood.

On the same day, a modest room was found for me to stay in near the campus. Thus it became possible for me to settle down in Philadelphia at least for a short period. I returned to New York to fetch my suitcase and to report the matter to Tamon Mayeda.

DAYS SPENT AT SEIFRIZ'S LABORATORY

The extension of my transit visa was kindly taken care of by the Chairman of the Department, Professor J. R. Schramm. Although the problem of monetary support was still pending, it had thus become possible for me to stay in Philadelphia and work on *Physarum* with Seifriz as far as my finances would permit. There in Seifriz's laboratory I saw for the first time the beautiful culture of the plasmodium of *Physarum polycephalum*. An important merit of this material for experimental work was that it could be easily cultivated in the laboratory simply by feeding it with oats. Unlike Küster, Seifriz did not suggest what theme I should study or what material I should use. There were a variety of interesting experimental materials in the well-tended greenhouse, but I was attracted to this golden yellow slime mold.

Seifriz's lecture on "The Physics and Chemistry of Protoplasm" began on the 1st of October 1939. However, this was the first rainy day after my arrival in Philadelphia. I had neither umbrella nor raincoat, and I had to run to the laboratory in the rain. Looking at my wet clothes, Seifriz gave me his own English raincoat, saying that he had no need of it because he commuted in his own car.

In his lecture Seifriz gave not merely dry textbook facts, but also the background of the subject and lines of thought by which the discoveries had come about. As was not the case at Küster's lecture, students here had ample opportunity to ask questions or discuss whatever subjects lay within the scope of the lecture. The atmosphere of the class was quite different from that in Giessen. As lecturers the only point Seifriz and Küster had in common was that neither needed elaborate notes.

Seifriz, who was also known as a plant geographer, gave another lecture on "Plants and Climate," with many beautiful slides he had taken during botanical trips all over the world, from the tropical to the arctic zones. These slides, shown more than 50 years ago, were large glass monochrome plates.

Seifriz's main objective in the laboratory was to gain insight into the submicroscopic structure of the protoplasm. Because electron microscopes were not yet available, he used physical properties as a means of looking into the structure of protoplasm. To investigate its physical properties, he liked to resort to the delicate technique of microdissection, which had hitherto not been in general use. In the early days of Seifriz's scientific career, when the now popular apparatus known as the micromanipulator was not available, he used the Barber pipette holder, following Kite & Chambers for microdissection. He tried to estimate such physical properties as the viscosity and elasticity of protoplasm with a glass needle.

Seifriz emphasized that protoplasm is non-Newtonian and generally elastic and insisted that these physical properties are explained by the existence of long-chain molecules attached to one another in one way or another. At a time when the idea of the structure of protoplasm was a confused mixture of fact and theory, Seifriz took the lead in the modern view. The image he had 50 years ago matches well our contemporary knowledge about the cytoskeleton. As to the unique scholarship and personality of Seifriz, I have already written a memorial article elsewhere (9).

For some time after I came to Seifriz's laboratory, I continued to observe the behavior of the slime mold under various experimental conditions, especially its response to mechanical stress. A slight, locally applied mechanical pressure modified the speed and direction of flow of the endoplasm sensitively, and yet produced no injury to the plasmodium. This behavior suggested that the normal flow of the endoplasm, too, was caused by the internal local difference in pressure. Curiosity about such behavior of endoplasmic flow led me to perform further experiments in which air pressure was used instead of mechanical pressure. The double-chamber method for measuring the difference in local internal pressure (i.e. the motive force responsible for protoplasmic streaming) was thus developed (7, 8). The greatest difficulty in preparing the double-chamber lay in separating the two protoplasmic blobs connected by a single vein into two airtight compartments. This was because the vein in which endoplasm flows back and forth was so soft and delicate that it was readily collapsed by mechanical disturbances. The problem was solved using semi-fluid, lukewarm agar.

It took about three months after I first saw the *Physarum* plasmodium to establish this simple technique—until the end of 1939. When we plotted the motive force responsible for streaming against time using this method, we got undulating curves with various wave patterns. It was found, by means of waveform analysis, that all the curves thus obtained could be reconstructed neatly with only two or three sine waves having different periods and amplitudes. Seifriz was very pleased with this result, because it measured the motive force responsible for protoplasmic streaming for the first time and showed that the visible pulsation was composed of more than a single rhythm.

Although I was blessed with good luck in my work, the problem of a source of living expenses had still not been solved. Not only Professor Seifriz but also the staff of the Department were good enough to seek funds for me. However, public opinion in the United States was then set severely against Japan. To the United States at that time Japan was already a so to speak quasi-enemy country. There were not a few foreign students and post-docs, but I met no Japanese students supported by American fellowships. Masa Uraguchi's was an exceptional case: She had come to the United States on a Friends (Quaker) fellowship.

One day in April, Seifriz asked me to accompany him to the dinner of Mrs. Curtin Winsor. I did not know it then, but later I learned that she was a daughter of the President of DuPont and had once been the wife of a son of President Roosevelt. Unaware in what connection Seifriz knew her, I supposed she was his former student. At any rate, Seifriz took me, along with a cine projector and a few reels of cinefilm, to Winsor's home in the Main Line, the most prestigious residential area west of Philadelphia. Mrs. Winsor seemed to be still young, probably in her 30's, a refined and gracious lady. Seifriz introduced us. After dinner was over, Seifriz persuaded her of the importance of basic science and showed the film of the slime mold, which he had taken himself. Seifriz asked her if it would be possible for her to help with my living expenses for one year for the sake of progress in basic science. Mrs. Winsor gave her consent without hesitation. Thus thanks to Seifriz's enthusiasm and Mrs. Winsor's philanthropic support, my financial problems were solved for a year to come. As we rode back toward my boarding house in Seifriz's car, the city of Philadelphia shrouded in spring rain appeared to me a dreamland.

Owing to the efforts of Professor J. R. Schramm, the chairman of the

Department of Botany, I obtained my formal student visa. The problems that had been pending since I landed in New York were thus all solved through the goodwill and efforts of the people around me.

During the Christmas season of the same year, 1940, the annual meeting of the AAAS was held in Philadelphia. As part of the program of the American Society of Plant Physiologists, Seifriz organized a symposium on "The Structure of Protoplasm." Invited to participate, I presented the data so far obtained on *Physarum*. This was my first lecture in English to an audience of about 300. Seifriz chaired the symposium with great charm and humor. The record of the event was published as a book (20) in the early part of 1942—that is, soon after the outbreak of the Pacific War.

MRS. WILLIAM H. COLLINS

Shortly before Miss Masa Uraguchi left for Japan (having finished her thesis), a pious elderly Quaker lady, Mrs. William H. Collins, invited Masa, her close friend Miss Miyeko Mayeda (the first daughter of Tamon Mayeda), and me to a dinner at her home on the campus of Haverford College not far from Philadelphia. Soon after that, Mrs. Collins asked me whether I might be interested in staying in her home, where everything would be taken care of by herself and her able American-Indian butler. My only duty would be to take breakfast and dinner with Mrs. Collins and to accompany her to the Friends Meeting House every Sunday morning. She was the only daughter of a world-renowned 19th-century paleontologist, Edward D. Cope, and the wife of Dr. W. H. Collins, Professor of Astronomy at Haverford College, who had died a couple of years before. She was a graceful Japanophilic lady more than 70 years old and without children.

I gladly accepted her kind invitation. There was a cosy bedroom and a bathroom for my own use. Further, her late husband's study was allotted to me for my work. Delicious meals were served by her butler, an excellent cook. She treated me as if I were her son-in-law. I not only paid no board, but she gave me a coat to wear at home, ties, underwear, etc.—even a golden wristwatch when I somehow lost mine. Occasionally I told her that I felt sorry to be given so much by her while I had nothing to give in return. Then her answer was always "Oh no, Noburô! You give me friendship!" The only inconvenience to me was that I had to return to the house by 6:00 p.m. to take dinner with her, when it had been my habit to work late in the laboratory. At any rate, the blessed encounter with Mrs. Collins changed my life-style.

OUTBREAK OF THE PACIFIC WAR

The international relationship between the United States and Japan worsened day by day, and many people in the States had a foreboding of war with Japan. At the very beginning of December, 1941, the Japanese Government sent a refugee boat, *M. S. Tatsutamaru*, to repatriate Japanese citizens in the

United States. I prepared to go home on board that boat and packed personal articles so that I would be ready to leave Haverford for San Francisco by train on the 8th of December. But alas, when everything was ready for departure and I waited, sad at the thought of bidding Mrs. Collins farewell soon, the radio announced in the early afternoon of December 7th, 1941, the shocking news of a sudden attack on Pearl Harbor by the Japanese Navy. My plans for going back to Japan had all been in vain. From that moment I became an enemy alien. Mrs. Collins was of course much shocked by this drastic news, but her attitude toward me was as good as before; and seemed even gentler, reading my mind.

After that I was visited often by FBI and Naval officers. Once I was asked to appear in person at the Attorney General's Office in downtown Philadelphia. Mrs. Collins, in spite of her age, accompanied me, testifying that I was not a dangerous person at all. Japanese in important positions were sent to the concentration camp on Ellis Island in New York Harbor. As a student I was left more or less free, except that I was not allowed to go farther than 3 miles from my residence without permission. I remained at Mrs. Collins's home under her protection. Since I could not go to the University of Pennsylvania, Seifriz made it a rule to visit me twice a week, regularly.

Surprisingly enough, I got a message a few months after the outbreak of the Pacific War from the Chairman of the Department of Biology of Haverford College, Professor H. K. Henry, granting me the privilege of using a laboratory and library there. I was touched by this generosity of the Quaker College in that most serious international situation. As the building of the Biology Department was a 10 minutes' walk from the Collins home on the same beautiful campus, it was very convenient for me. I gratefully accepted his offer.

Although there was nothing in the laboratory but a large table, a stool, and a lot of bottled snakes preserved in formalin, it was at least possible to cultivate the slime mold there with oats in soup dishes and wet sheets of paper towel. I inverted other soup dishes on the top of the culture to cover it. Having carefully observed the slime mold culture in the soup dishes with the naked eye, I noticed a tendency of the slime mold strands to twist themselves when they came into contact with the water that half-filled the dish.

I then asked Seifriz to bring several small glass cylinders with bottoms and several glass plates of about 2 in^2 to cover them. Pouring a small amount of water in the bottom of the cylinder to make the air inside moist, I tried to hang a 3–5-cm-long strand (vein) segment cut out from the stock culture at the center of this tube from the underside of the flat glass cover. A delicate inverted T-form glass needle was attached to the lower free end of the strand to see how it rotated. If we placed a handmade 360° protractor underneath the bottle and observed the inverted T-form glass needle from the top, we could follow, approximately, even with the naked eye, how the strand segment twisted itself with time. In the first 20–30 min, torsion of the strand was modest and irregular, but it gradually tended to twist regularly and alternately in clockwise and counterclockwise directions in a period of about 2 min, often with a swing angle of more than 180°. But curiously enough, this torsional movement was asymmetrical. It twisted in one direction more each time than in the other. Statistically, I encountered more often the case in which counterclockwise rotation was dominant. At any rate, by repeating the asymmetric movement, the strand twisted itself substantially in one direction many times, practically endlessly. This simple observation made with this simple setup was repeated after the war was over; but its implication in relation to the cytoskeleton and the mechanism of synchronization of local irregular rhythms into one regular, larger rhythm remain interesting problems today (14).

BACK TO JAPAN DURING THE WAR

While I was spending quiet days at Haverford under the protection of Mrs. Collins, sometime in May of 1942, I suddenly got a telegram from the Department of State in Washington, D.C. notifying me of the possibility of repatriation by means of exchange boats. The arrangement had been made by the Red Cross in Switzerland to exchange American civilians in Japan and Japanese civilians in the United States. The telegram said that if I wished to go back to Japan, I should come to the Hotel Pennsylvania in New York by the 10th of June, 1942. The volume of allowable luggage was less than 30 cubic feet. The telegram requested me to answer as soon as possible whether or not I wished to go back. Of course I wanted to take advantage of this opportunity to return home. Since the volume of my luggage was far less than 30 cubic feet, and no restrictions on the contents of the luggage were noted, I went to the Hotel Pennsylvania on the designated date, taking all my belongings. On parting from Mrs. Collins, I expressed to her my hearty gratitude for all she had done for my sake during the previous two years.

At the Hotel Pennsylvania in New York I found many Japanese of various professions. We were all leniently confined on specified floors of the hotel for about one week, and the meals and services of the hotel were good enough for enemy aliens. During the wait all our luggage was inspected severely. When my turn came, not only all the experimental data, but even newspaper sheets used for wrapping were confiscated. When the inspector found several sheets of yellow sclerotia of *Physarum polycephalum* spread on dried sheets of paper towel and filter paper, he asked me with perplexed eyes what they were. In spite of my desperate efforts to explain that it was a dried specimen of an absolutely harmless organism called slime mold, he confiscated all of them mercilessly, saying nothing. In other words, we were allowed to carry only our clothes. For an experimental scientist, loss of his experimental data is grievous. All my efforts of the past few years had thus come to nothing. I was stunned by this result; I felt heartbroken.

We left New York on board a Swedish-American Liner, *S. S. Gripsholm,* on the 18th of June, 1942, if my memory is correct. The then ambassabor, Normura, extraordinary plenipotentiary ambassador Kurusu, and the director of the Japan Institute, Tamon Mayeda, whom I have already mentioned, were all on board. Since the Panama Canal was closed, *S. S. Gripsholm* took a southern route and touched in at Rio de Janeiro for a few days, probably for refueling and reprovisioning. We were not allowed to disembark but we enjoyed the beautiful landscape surrounding the harbor. Then the boat left for Lourenço Marques on the east coast of Africa, rounding the Cape of Good Hope.

The day after we arrived at Lourenço Marques, a Japanese boat, *M. S. Asamamaru,* with large, white crosses on her sides, reached the same port. She carried American citizens from Japan. During our week's stay there, passengers and luggage were exchanged. We had plenty of time to take walks in the city and the suburbs. As a minor botanist, I observed with special curiosity the flora, which was quite different from that of the northern hemisphere. At last all of us were on board the *M. S. Asamamaru* and we left for Japan across the Indian Ocean. The boat touched in at Singapore for a while and reached Yokohama on the 20th of August, 1942. The long voyage from New York to Yokohama had taken more than two months.

I was not drafted until the end of the war. In April, 1943, I was nominated as lecturer at the Botany Department of the University of Tokyo and taught plant cell physiology to the students until 1949. During the war, Tokyo was severely attacked from the air. My house was reduced to ashes, but fortunately the buildings of the University of Tokyo remained almost intact. After the war ended in August of 1945, I tried to restart my experiments on *Physarum*. Although the laboratories were not in good order right after the war, it was not impossible to do some experiments, because my work on the slime mold did not require much money. I constructed myself a queer gadget that enabled us to record the motive force of protoplasmic streaming semiautomatically.

I also started to work on protoplasmic streaming in giant algal cells, especially in the intermodal cell of *Nitella*. Such work was feasible in a devastated, poorly equipped laboratory after the war. Protoplasmic streaming was an intriguing phenomenon, the mechanism of which was hardly known at that time. Whether it was shuttle streaming in *Physarum* or rotational streaming in *Nitella*, protoplasmic streaming was thought to visualize the mystery of cellular activities. As a matter of fact, neither the nature of the force driving the protoplasm nor the site where it was produced in the cell was known.

In 1949, a new Department of Biology was established in Osaka University and the post of professor in charge of cell physiology was offered to me. The next year, in 1950, I was invited again to the University of Pennsylvania to work with Seifriz for 9 months. I continued then the work on torsion of the protoplasmic strand started at Haverford College during wartime but interrupted by my repatriation (17). The subsequent progess in the field of cytoplasmic movements, mostly in the slime mold and giant algal cells, was reviewed by the author in the *Annual Review of Plant Physiology* (11, 13) and elsewhere (10, 12, 14–16).

Looking back on youth, especially the several years I spent in Germany and the United States before and during the Second World War, I would like to believe that an invisible thread of destiny enabled me to meet Professors Küster and Seifriz, Mrs. Collins, and the many benefactors to whom I owed so much. Who could have expected that Dr. Tamon Mayeda, whom I had met in New York for the first time without any letter of introduction, would become my father-in-law after the end of the war. Both Küster and Seifriz had outstanding (and quite different) personalities. Both were most precious teachers in my scientific career, and I can hardly estimate their influence upon my subsequent life.

I would like to express my cordial thanks to Mrs. Judith Fujimoto for her kind help and advice in overcoming the linguistic difficulties I had to cope with here.

Literature Cited

1. Kamiya, N. 1937. Untersuchungen über die Wirkung des electrischen Stromes auf lebende Zellen. I. Das Verhalten der mitotischen Figur unter der Wirkung des Gleichstromes. *Cytologia* (Fujii-Jubiläumsband):1036–42

2. Kamiya, N. 1938. Über Doppelschalen bei *Melosira*. *Arch. Protistenkunde* 91: 324–42

3. Kamiya, N. 1939. Zytomorphologische Plasmolysestudien an *Allium*-Epidermen. *Protoplasma* 32:373–96

4. Kamiya, N. 1939. Die Rhythmik des metabolischen Formwechsels der Euglenen. *Ber. Dtsch. Bot. Ges.* 57:231–40

5. Kamiya, N. 1939. Beiträge zur Pathologie der Zellteilung und der Querwandbildung. *Protoplasma* 33:427–39

6. Kamiya, N. 1940. Parasiten in Oscillatoriaceen. Beiträge zur Pathologie der Cyanophyceenzelle). *Arch. Protistenkunde* 94:201–11

7. Kamiya, N. 1940. The control of protoplasmic streaming. *Science* 92:462–63

8. Kamiya, N. 1942. Physical aspects of protoplasmic streaming. See Ref. 20, pp. 199–244

9. Kamiya, N. 1956. In memoriam William Seifriz. *Protoplasma* 45:513–24

10. Kamiya, N. 1959. Protoplasmic streaming. *Protoplasmatologia*, 8, 3 a. Vienna: Springer. 199 pp.

11. Kamiya, N. 1960. Physics and chemistry of protoplasmic streaming. *Annu. Rev. Plant Physiol.* 11:323–41

12. Kamiya, N. 1962. Protoplasmic streaming. *Handb. Pflanzenphysiol.* 17(2): 979-1035

13. Kamiya, N. 1981. Physical and chemical basis of cytoplasmic streaming. *Annu. Rev. Plant Physiol.* 32:205–36

14. Kamiya, N. 1985. Some motility characteristics of living cytoplasm: from observations on *Physarum*. In *Cell Motility: Mechanism and Regulation*, ed. H. Ishikawa, S. Hatano, H. Sato, pp. 577–85. Tokyo: Univ. Tokyo Press, 629 pp.

15. Kamiya, N. 1986. Cytoplasmic streaming in giant algal cells: a historical survey of experimental approaches. *Bot. Mag. Tokyo* 99:441–67

16. Kamiya, N., Allen, R. D., Yoshimoto, Y. 1988. Dynamic organization of *Physarum* plasmodium. *Cell Mot. Cytoskel.* 10:107–16

17. Kamiya, N., Seifriz, W. 1954. Torsion in a protoplasmic thread. *Exp. Cell Res.* 6:1–16

18. Kamiya, N., Yoshimoto, Y., Matsumura, F. 1982. Contraction-relaxation cycle of *Physarum* cytoplasm: concomitant changes in intraplasmodial ATP and Ca^{++} concentrations. *Cold Spring Harbor Symp. Quant. Biol.* 46:77–84

19. Michel, W. 1937. Über die experimentelle Fusion pflanzlicher Protoplasten. *Arch. Exp. Zellforsch.* 20:230–52

20. Seifriz, W., ed. 1942. *The Structure of Protoplasm.* Monogr. Am. Soc. Plant Physiol. Ames, Iowa: Iowa State Coll. Press. 283 pp.

Ralph O. Erickson

Ann. Rev. Plant Physiol. Plant Mol. Biol. 1988. 39:1–22

GROWTH AND DEVELOPMENT OF A BOTANIST

Ralph O. Erickson

Department of Biology, University of Pennsylvania, Philadelphia, Pennsylvania
19104

FOREWORD

I suppose that my earliest interest in plants stems from my childhood in
northern Minnesota and Michigan. Winters there are severe but the long
summer days among trees and lakes are delightful. Our family was close to
nature in many ways. We often spent days picking wild blueberries and
raspberries which my mother canned for desserts through the year. Pin
cherries and wild strawberries made excellent jelly. We used our pocket
knives to make whistles of poplar twigs, and selected symmetrical maple
crotches for sling shots. Later, when my commitment to botany was firm, I
was drawn to studies of plant growth and development from a variety of
educational and research experiences, but with the strong influence of Edgar
Anderson and David R. Goddard. I have had little conventional training in
plant physiology, and it may be interesting to try to trace the development of
my teaching and research interests from such diverse fields as plant taxonomy, plant anatomy, cytology and cytogenetics, evolutionary theory and
ecology, and an interest in statistics and numerical analysis. I can probably be
accused of dilettantism.

Discussing this theme in a personal vein, with a bit of apprehension, has
resulted in a sort of selective autobiography. One author (24) has written that
"autobiography is a most peculiar genre or form....[It] presuppose[s] a particular kind of arrogance, a conviction that one's life is in some serious way
exemplary...." However, I may hope that frequent use of the first person
pronoun will not be taken as conceit, but as the candor that I intend.

ROOTS

My grandparents were emigrants from Sweden in the 1870s, a decade or so after the Sioux massacre of settlers in the Minnesota River valley in 1862. My mother's parents were from Småland and Västergötland and homesteaded near Bernadotte in southern Minnesota, where they raised ten children. My father's parents came from the Åland Islands and settled near St. Hilaire in northern Minnesota, also as homesteaders. My father, Charles, the second of four children, left the farm at age 19 or 20, against his father's wishes, to attend high school, went on to Gustavus Adolphus College, and graduated in 1913. My mother, Stella Sjostrom, and he were married in 1913, and after two years in Duluth, Minnesota, moved to Rock Island, Illinois, where my father attended Augustana Seminary, graduating in 1918 with a B. D. degree. He was then a Lutheran pastor in Clearbrook, in northern Minnesota. I was born in Duluth, 27 October 1914, the first of six children. When I was five, my mother suffered a severe "nervous breakdown," and the family, then of four children, were dispersed. I was sent to my grandparents Sjostrom, who had retired to St. Peter, Minnesota, and there I attended kindergarten. I have virtually no memory of these years, but I am told that they were unhappy.

My father left the church in Clearbrook for nearby Leonard, a town of 75 people about 15 miles from the Red Lake Indian Reservation (Chippewa). He cleared land beside a small lake and single-handedly built a four-room house for the family, which is still lived in. He was pastor of the church in Leonard, and principal of the three-room school, where I advanced through the first six grades in four years. In sixth grade I heard the seventh and eighth grade recitations, since they were in the same room. At the end of the year I was given, and passed, the State Board examinations for graduation from eighth grade and was then entitled to enter high school when I was not quite eleven.

SI QUAERIS PENINSULAM AMOENAM...

My father was called to a church in Iron River in the upper peninsula of Michigan, which in 1925 was a fairly prosperous iron mining town. The school authorities were dubious about my starting high school, and instead I entered eighth grade. Some of my classmates were the children of the immigrant Italian, Polish, and Finnish miners. It was a rough town. I suppose that my school experience was unusual, since I was two or three years younger than my classmates, and in one's early teens this is important. In high school it meant that I could not join in sports such as football, basketball, or hockey, nor could I join a gang. Also the puritanical character of midwestern Lutheranism at that time meant that, as preacher's kids, we were not allowed to do such "sinful" things as see movies or go to dances.

After classes, piano practice, band rehearsal, and delivering newspapers, I had time to read, and I read widely. In addition to pulps, Jules Verne, and H. G. Wells, I tackled the Bible, and such books as Dostoyevski's *The Brothers Karamazov*. My father's library included, in addition to homiletics and concordances, "Dr. Eliot's Five-foot Shelf of Books" (*The Harvard Classics*), and I believe that I read the entire collection, with or without comprehension. I recall reading Darwin's *Origin of Species* and *Voyage of the Beagle*. I also spent much time with the *Encyclopedia Americana*. At least I learned some words. On entering college, I scored above the median for college graduates on a standardized vocabulary test.

During the summers, we spent much of our time at nearby Fortune Lake where my father organized and built a summer Bible Camp. Having grown up on a farm, he knew carpentry. He also had a woodworking shop at home, which I could sometimes use under strict supervision. At the camp he, my brother and I, and occasionally others cleared brush, built roads and paths, erected buildings, built tables and benches, even rowboats, painted, and installed electric wiring—whatever was required. After a time, I realized that I could cut rafters or hang a door as accurately as a professional carpenter who worked at the camp for a while. We were not paid and so could spend a part of our time at tennis, swimming, diving, and boating. We longed for a canoe. My fond memories of Fortune Lake include recollections of loons, ducks, herons, woodchucks, and deer. I did not care to fish, but heard stories about the bass, trout, and pickerel. Fortune Lake was one of a chain of five or six, and occasionally I would row the whole length to check the beaver lodges or perhaps look for lady's slippers. I remember some irritation that I could not easily learn the trees and other plants. Names people used were contradictory, and I suppose I did not realize then that I needed a manual.

My father also taught us photography, since he had earned a part of his college expenses by photographing barns, livestock, and families and selling the picture postcards he made to the farmers.

GUSTAVUS ADOLPHUS COLLEGE

When the stock market crashed in 1929, the mines closed almost immediately. With widespread unemployment, the church was unable to pay my father's full salary and, when I graduated from high school in 1930, college seemed out of the question. During the next year I had a part-time job and took a course or two at the high school. The following year, my father announced that I could go to Gustavus Adolphus College, St. Peter, Minnesota. As I recall, tuition was about $75 per semester and I lived with relatives for two years, then in a dormitory as a proctor. I worked for meal tickets as an attendant in the library (more opportunity to read), and as a reader for a blind classmate.

At college I was still two years younger than my classmates and felt exceedingly shy. I was regularly permitted to take one or two courses beyond the required four per term. This entailed some scattering of effort, but I graduated magna cum laude. My major was biology, but I also took most of the math courses offered and not quite enough chemistry for a second major. In addition to zoology, comparative anatomy, human physiology, etc, there was one botany course, taught by a zoologist. One of the assignments was to turn in dried specimens of 10 plants. I did many more. My roommate for two years was a born naturalist, who kept plants and tropical fish in our room. The two of us spent many extra hours in the biology lab and in the field, collected material for use in the biology course, frog eggs in season, and many other things. He became a high school biology teacher in St. Peter and we continued our joint biology ventures for some time.

Music was important at Gustavus; the a capella choir was nearly as important as the football team. I took music seriously. I had voice and piano lessons one year (even played a recital), played clarinet in the band and bassoon in the orchestra, and sang in the choir. I also wanted some music theory. In high school the music teacher, remarkably, gave a course in harmony to a few of us, and in college I was the only person who wanted the course in advanced harmony and counterpoint. One term Dr. Nelson agreed to meet with me once a week at the piano, to play and correct exercises I had written. This was one of my most demanding and satisfying courses. Shortly before the spring choir tour my senior year, Dr. Nelson was ill for about three weeks, and it fell to me to direct the choir in rehearsals and the first concert, which included the Bach motet, *Singet dem Herrn*. This was nearly disastrous, since I felt I had to devote my full time to studying the music, instead of attending classes. In later years my performing skills have atrophied but music continues to be an important part of our family life.

At graduation from college, only two of my classmates had job prospects. I was qualified for certification as a high school teacher, but there were no opportunities. I spent the summer at Fortune Lake wondering what I might do. Late in August a letter arrived from the president of Gustavus offering me the position as assistant (really instructor) in biology, which had just become vacant. The salary was very low even by 1935 standards, but I immediately hitchhiked to St. Peter. The biology faculty consisted of Dr. J. A. Elson and me. I spent four years teaching there, in sole charge of the elementary biology labs and the botany course, 24 contact hours per week. In my fourth year I introduced a course in genetics, using *Drosophila* and segregating ears of corn in the lab.

Funds for lab material were limited; for instance, gophers caught in nearby fields were dissected in the zoology course, instead of specimens bought from a supply house. The microscope slide collection was inadequate, but the department had a microtome and with an improvised paraffin oven I prepared slides of stem, root, and leaf sections, shoot apices, slides for animal histology, even parasitic flat worms. I converted an ice-box into an incubator, prepared whole mounts of blastoderms and introduced lab exercises on chick embryology. I also made many 2 × 2 inch lantern slides. At times student volunteers helped with this work. I had read Cooper's article on embryo sac development in lilies (7) so I bought some Easter lily plants with flower buds of various ages at a local greenhouse and prepared sections of anthers and ovaries. Since I was into mass production, I could select choice slides for myself of the crucial stages of pollen and embryo sac development, which have continued to be useful in teaching for many years, and I recently learned that some of my slides are still in use at Gustavus.

Something more should be said about Gustavus. The announced mission of the college was, and is, training for Christian leadership. A daily chapel service was compulsory, there were evening prayer meetings, etc. In my third year of teaching, as I recall, a decision was made to ordain the Gustavus professors into the Lutheran ministry. I took little part in the religious life of the college because my interests were elsewhere. The faculty at Gustavus were dedicated teachers, but it occurred to me later that I knew of none who were engaged in scientific research or any other scholarly work. I graduated with only a slight understanding of academe in the wider sense.

SUMMER SCHOOL

After my first year of teaching, I attended a summer session at the Douglas Lake Biological Station of the University of Michigan, taking systematic botany and plant anatomy. The former course was devoted to the local flora and consisted of all-day field trips in which we filled our vasculums, then sat down in some pleasant place to key out our specimens with Gray's *Manual*. It pretty well satisfied my desire to be able to identify plants. C. D. LaRue's plant anatomy was less cut and dried. LaRue was a challenging and entertaining lecturer and he taught us the paraffin technique. I recall some dissatisfaction with the static descriptions in Eames & McDaniels, our text. When I asked how fast the onion root grew and how rapidly cells in the meristem divided, I found no answers. To say that I could not imagine how the root tip could grow so as always to look the same in sections, is perhaps invoking too much hindsight. It was interesting that George Avery shared LaRue's laboratory that summer, working up his sections of *Aesculus* shoots to explore the possible role of auxin in the initiation of cambial activity in the spring (6). This was my first view of research in progress.

The following two summers I had courses at the Lake Itasca Forestry and Biological Station and at the Minneapolis campus of the University of Minnesota. Among them was a field course on the ecology of Itasca Park, a course in genetics, an elementary plant physiology and a seminar course concerned with the structure of chlorophyll and the physiology of photosynthesis, for which I was not prepared. The structure of grana was not then known. This is the extent of my formal training in ecology, genetics, and plant physiology.

SHAW'S GARDEN

Summer experiences and extracurricular fooling around in the laboratory at Gustavus reinforced my desire for graduate study in biology. I had made unsuccessful inquiries about the possibility of doing full-time graduate work in the botany department at Minnesota, and during my fourth year of teaching, I resolved to make a more serious effort to get into graduate school, realizing by that time that my chances of being accepted were slim. My academic record at Gustavus was good, but I was sure that my grades and my recommendations would be discounted. With the advice of people in botany at Minnesota I applied to 12 schools, mainly Ivy League and State Universities, at which work in plant cytology was going on. I got rejections, or no word, from all but one.

Edgar Anderson wrote that he was impressed with my application, that there were no opportunities for support at the time, but that he would "by hook or by crook" see that I could come to Washington University. It is still a mystery to me what merit he could see in my application. Shortly before the start of classes in September 1939, I was awarded a University Fellowship. I hitchhiked to St. Louis, found a boarding house near the Missouri Botanical Garden, and became a graduate student in the Henry Shaw School of Botany. When I had paid tuition, room, and board, I had ten dollars per month. I took Jesse Greenman's course on the flowering plants, which was quite another thing than a local flora. Greenman had studied at Berlin with Adolph Engler and his course was a *grosses Praktikum*, intended to acquaint us with virtually all the plant families, through lectures, and dissection and drawing of boiled-up specimens of dried flowers, fruits, etc., filched from herbarium sheets. I enjoyed it and still value it greatly. During my three years at the Garden, I made a point of walking through the conservatories at least once a week. At the main campus of the University, a course in physical chemistry fulfilled my old intention to major in chemistry, and I had a cytology course, a seminar course in animal embryology, and others.

I chose to do a taxonomic problem for a master's degree, which was narrowed down to a revision of the *Viorna* section of *Clematis*, under Greenman's direction. In retrospect, I can perhaps see Edgar Anderson's hand in this choice. (By a curious coincidence, I was given a cordial welcome to the Academy of Natural Sciences in Philadelphia and the botany department of the University of Pennsylvania in 1942, when I made a bus trip to Eastern herbaria to study *Clematis* specimens.) My thesis was published (9) as my thickest paper to date, and it earned me several pages of testy criticism from M. L. Fernald in *Rhodora*. However, I am cited in Fernald's *Manual* as author of one variety of *Clematis*, so I can claim to have had a bit of taxonomic competence.

The most valuable part of my experience in St. Louis was association with Edgar Anderson. It was strenuous. From the day I arrived he subjected me to a continual barrage of discussion, wisecracks, pithy anecdotes about other biologists, genetic questions intended to stump me, a continuing contest of wits. I often went with him on field trips, and *Faltboot* trips on Ozark rivers; taught his wife, Dorothy, and him to play recorders; accepted many invitations to the barn at the Gray Summit Arboretum, which he had converted to a summer place; and listened to a certain amount of advice on how I should conduct my life. I also learned a great deal about species of *Iris, Tradescantia, Acer*, and other genera: their geographical distribution, morphological variation, cytology and, introgressive hybridization. Anderson paid me to help with his study of F2 segregation in a species cross between *Nicotiana alata* and *N. langsdorfii*, photographing and measuring flowers and leaves, and making preliminary extractions of tissue for auxin analysis in F. W. Went's laboratory. Anderson had published (2) on the hindrance to recombination imposed by linkage, and in these studies he wished to document a further hindrance apparently imposed by developmental constraints. I do not believe that this work was published. Anderson was at that time beginning his survey of the indigenous varieties of maize. One summer he paid me to plant, hoe, and self-pollinate a number of strains of maize from the Hopi and other southwest Indians, from Mexico and Guatemala. The latter grew to about 20 feet and pollinating them required a ladder when they tasseled in September, contrasting with an 18-inch Hopi strain. I learned a great deal about the diversity of *Zea*.

Anderson gave a great deal of thought to methods of analyzing and representing variation in natural populations. Although he was far from naive in statistics and mathematics, he preferred graphical methods, such as pictorialized scatter diagrams (3), rather than formal statistical analyses of his data. I was impressed with the outcome of his association with R. A. Fisher, whose discriminant function (23) was worked out using Anderson's data on three species of *Iris* as an example, and has become a useful technique in multivariate analysis. One might argue that Anderson's ideographs (1, Plate 23) visually demonstrate the relationships among the three species as well as Fisher's Figure 1 does. I was also taken with his graphs of internode length vs number (4) and made similar plots of growing *Clematis* vines. I joked that he had plotted the first derivative of the plants.

My doctoral problem grew out of these discussions, or perhaps it was tactfully assigned to me. The idea of making a thorough field study of the glade leather leaf (*C. fremontii* var. *riehlii*) was roughly formulated in the spring of my first year. I was able to buy a used Model A Ford and a sleeping bag that summer (from hoeing corn), and spent a large part of my time on the Ozark glades (dolomitic barrens with an attractive endemic flora), during all seasons for the next two years, and a lesser part for another two years. At Anderson's urging I took a microscope to the glades and made squash preparations of anthers to look for irregularities in meiosis, which might indicate introgressive hybridization with another *Clematis*. I found none and went on to a study of the ecology, reproduction, and natural variation of the population. This constituted a major in botany and a minor in trespassing. I had found that walking up to the door of a farmhouse to ask the farmer's wife for permission to look at a glade tended to frighten her. A farmer once found some plants with bags over them on his glade and called the State Police, thinking that someone was growing hashish The Police brought one of the bagged plants to the Gray Summit Arboretum and I was called in for an explanation. The specimen, which I had bagged to find out if it would self-pollinate, was then annotated and deposited in the herbarium at Shaw's Garden.

I wrote up my work on the glades as a dissertation and received my Ph.D. degree in 1944. This was well enough regarded to be reprinted (10). In later years, I have had intentions of continuing field work. I made a few collections of two species of *Uvularia* in western New York state, with the idea of studying their relationship, but did nothing with them. Later there were several trips with Robert B. Platt to shale barrens of Virginia and West Virginia, where I learned to know the *Clematis* species, closely related to *C. fremontii*, which are restricted to the barrens. The *Clematis* leaves that I collected have served as samples for the analysis of variance by many biometry classes, but nothing else has come of these efforts. On several occasions I have taken a break from other things and driven to Missouri to revisit the glade *Clematis*, often with one or two students.

WESTERN CARTRIDGE COMPANY

My going to St. Louis in September 1939 nearly coincided with the German invasion of Poland. A year later the Selective Service Act was passed. I was classified 1-A, appealed, and was granted deferment as a student. In the spring of 1942 student deferments were abolished, and through Anderson's acquaintance with the research director at a defense plant, who was an amateur botanist with a master's degree in botany, I was offered a job as a chemical microscopist. Western Cartridge Co., East Alton, Illinois (later a part of Olin Industries) manufactured small arms ammunition, including smokeless powder. The lab to which I was assigned was mainly concerned with problems of variability of the powder charge in cartridges, and with a polarizing microscope and the guidance of Chamot's & Mason's *Chemical Microscopy*, I was able to learn something about the composition of powder grains (nitrocellulose, nitroglycerin and a plasticizer) from thin sections. I boned up on the processing of wood pulp and on the chemistry of cellulose, nitrocellulose, and polymers generally, all with a crowd of chemical engineers. It seemed to a friend and me that some of the variability of the ballistic tests might arise in the blending of various batches of powder, and we made a statistical test. We had a barrel of marked powder poured through the blending tower with many unmarked barrels, then counted marked grains in samples of the output. Our statistics showed that the blending was very poor, but so far as I know nothing was done about it. After a year I was put in charge of a laboratory to study dry cells (flashlight batteries). So now the topic was electrochemistry, specifically of the Leclanché cell. With a punch press and other equipment, we did not succeed in a year's time in making experimental cells that equalled production cells in their service life. At least I learned some more chemistry, a bit of chemical engineering, and broadened my understanding of microscopy.

ROCHESTER

A position at the University of Rochester became available in the spring of 1944, and Anderson suggested my name to David R. Goddard. He planned to be in Terre Haute, Indiana, on a consulting job and asked if I could meet him there. I played hooky from my job, had the first of countless exhilarating discussions with Goddard and in effect was promised the job then. I now needed approval from the War Manpower Board to change jobs, but that turned out to be a breeze, since I was to be an instructor, teaching in the Navy V-12 program at Rochester. My superiors at Western Cartridge offered me a handsome raise and painted a rosy picture of my future there. When I pointed out that I expected a much lower salary at Rochester, that discussion ended, as did my career in industry.

The biology group at Rochester was largely assembled by Benjamin Willier several years before I came. It was a congenial and exciting group. Curt Stern was a great geneticist who had the collaboration of Ernst Caspari, Warren Spencer, and others on classified work for the Manhattan Project. There were many luncheon discussions of genetics and many other topics, as well as a journal club and "Festschrift." Sherman Bishop, A. W. Küchler, and I organized a memorable seminar on biogeography. I am indebted to Donald R.

Charles for patiently guiding me through an analysis of covariance and the solution of a discriminant function, as well as introducing me to *The Calculus of Observations* (39). I sat in on Dave Goddard's course in plant physiology, and for the first time heard critical lectures on thermodynamics and metabolic cellular physiology. He and I had many free-wheeling discussions of science. I particularly remember that we both felt that the nucleic acids richly deserved study, a few years before Watson & Crick.

I had told Goddard that I wanted to do research in experimental cytology, having only a vague idea of what that meant. Having been fascinated by Darlington's (8) speculations about the evolution of sexuality, and the contrast between mitosis and meiosis, I thought that it would be interesting to try to do something that might illuminate the difference between the latter two processes. This narrowed down to a plan to study the respiration of microsporocytes, microspores, and pollen. Thinking that it would be good to select a plant that had large anthers and was easy to manage, I made a little survey, and concluded that it would be hard to beat the Easter lily, *Lilium longiflorum*. I ordered some bulbs to set out in the greenhouse on a staggered schedule, and Goddard taught me the ins and outs of using the Fenn microrespirometer. This resulted in papers on the respiration of anthers (11), on growth of the flower bud and its parts using log bud length as a developmental index (12), and later, on nucleic acids in the anthers (33). Lilies have now been used by other workers, notably H. Stern (37), in a number of important studies of biosynthetic aspects of microsporogenesis, and Moens (32) for a study of the synaptinemal complex in meiosis.

My wife, Elinor Borgstedt, and I were married after my first year in Rochester. We have two daughters and two granddaughters. Elinor had been a secretary to the research director at Western Cartridge for a year, and was a music student at the University of Illinois at Urbana. She transferred to the Eastman School of Music at Rochester and earned her B. Mus. degree there. She has had a rewarding career as an organist and choir director, and now devotes her efforts to the piano. When long-playing records were announced in 1947, we bought six from the very first list, assembled an amplifier kit from war surplus parts, connected a turntable and speaker, and were both overjoyed with the music. We have been audio fans ever since.

PENN

During my second year at Rochester, Goddard accepted a professorship at the University of Pennsylvania and was succeeded at Rochester by F. C. Steward. With two colleagues at Penn, Goddard obtained research grants from the National Cancer Institute and the American Cancer Society for studies of cell division in plants. After three years at Rochester, I had been promoted to an assistant professorship, but as it turned out I did not serve in that capacity, since Goddard offered me a position as a research associate in his program. After some soul searching and negotiation, we moved to Philadelphia. Maurice Ogur, a biochemist with particular interests in nucleotides and nucleic acids, joined the group as a research associate, as well as three excellent technicians, Kathie Sax, Gloria Rosen, and Connie Holden.

Goddard and I had reasoned that root meristems, as well as anthers, were favorable for studying cell division, and we began experiments with the primary roots of corn seedlings. They were grown in the presence of certain alkaloids, which were candidates for cancer therapeutic agents. The control and poisoned roots were fixed, sectioned, stained, and examined for mitotic abnormalities. After many weeks of study of the slides, I was frustrated at trying to imagine what had happened, for instance, to nuclei that had surely been in metaphase and after treatment looked something like interphase nuclei. I proposed that we abandon this traditional approach and try first to learn something about how roots and their cells grow. I knew in some detail about the work which Richard H. Goodwin, my predecessor at Rochester, and William Stepka had done in describing the growth pattern of *Phleum roots* (25). (In a footnote they acknowledge the assistance of Don Charles.) Their microscopic method of studying the minute grass roots was not feasible, since we had chosen to work with the much larger roots of *Zea* in anticipation of getting biochemical and metabolic data. At a meeting of our group, I suggested the kinds of data we should try to get and what I had in mind for a growth analysis. I put together a special camera rig to automatically record the displacement of marks placed on the roots, we worked out methods of counting cells, etc. I also supposed that I could handle the math involved in coordinating and interpreting the data. This is the rashest statement I have ever made. It took about 18 months of study to arrive at the analysis presented in the first papers (18, 21, 22), and it is apparent from recent publications by several authors that much remained to be done. In addition to the root work, we made the study of nucleic acids in *Lilium* anthers referred to above (33). The collaboration with Ogur was a great education in biochemical principles and methods of analysis.

After two years Ogur left Penn and a year later I was appointed associate professor. I had participated with John Preer in a biometry course (mostly statistics) for two years, but I now had additional teaching and some administrative duties. Goddard had great talents as an administrator as well as a teacher and researcher, and it was by his efforts that the departments of botany, microbiology, and zoology were merged in 1954 into a greatly strengthened division of biology. In 1961 he became provost of the university and served Penn eminently throughout the turbulent 1960s. Unfortunately, however, the analytical (chemical) work on roots was published only in summary form (18). The work that had been done on respiratory metabolism of root segments would have yielded estimates of the energetic requirement for growth, but this could not be analyzed or interpreted without Goddard's participation.

With the assistance of Roman Maksymowych, the root studies continued at a reduced scale. We undertook to explore the effects of inhibitors of root growth and cell division, based on the growth analysis that had been worked out on a descriptive basis. However it seemed too laborious to work out elemental growth rates, so we photographed the growing roots with an automatic camera, and on the basis of hourly readings of total root length and measurements of mature cell lengths in sections of roots fixed at the end of each run, were able to calculate average rates of elongation and of cell production. A variety of substances were tried and three distinctive patterns of inhibition were found. Metabolic inhibitors such as cyanide and azide inhibit elongation in a manner suggestive of enzyme inhibition, with no effect on the rate of cell production. Substituted nitrogen bases are potent inhibitors of cell division with no effect on elongation for many hours. Several alkaloids depress both processes. Unfortunately, these results have not been adequately analyzed nor published, since there are certain points that I have not fully understood until recently. Another study was based on data on growth and cell division in *Phleum* roots, which Goodwin kindly provided. Reanalysis of the data for cell division rates showed that all the cells in the apical part of the meristem divide, whereas in the basal part, the proportion of cells that divide falls progressively to zero (13).

The studies of lily anthers and of the corn root were motivated by the idea of doing "experimental cytology." In both cases, however, my interest shifted from the cells per se, to questions of how the organs, the flower and anther, or the root, grow. I was impressed in both cases by the great regularity and coordination of cellular processes into a predictable morphogenesis. If there is a question of whether growth should be modeled as a stochastic process or a deterministic one, I would certainly argue for the latter. While there seemed to be sufficient opportunities to devote a career of research to either anthers or roots, I began to wonder whether the same regularity would be found in other developing systems, such as shoot apical meristems, and set out to obtain growth data.

Zygmunt Hejnowicz joined my lab in 1963–64 and collaborated in root studies. He worked out a method of recording growth using fluorescent marks illuminated with near-UV light, with the idea of applying it to a study of gravitropic curvature of roots, and made a study of the inhibition of root growth by auxin (27). I owe a great deal to him for many animated discussions of growth problems, particularly their mathematical and physical aspects, and we have had the pleasure of visiting him in Poland.

Hejnowicz suggested that we use celery, *Apium graveolens*, for studies of the shoot apex, since it has a large and relatively flat apex. After dissecting out a few young leaves from an otherwise intact potted plant it was possible to focus an Ultropak objective on the apex, and with an automatic camera to obtain photographs with cellular detail. After a great deal of effort we gave this up because of inadequacies in the Ultropak image, based as it is on shifting light reflections from the cell surfaces. We had also attempted to photograph shoot apices of *Xanthium*, chosen because we could assure vegetative growth by keeping these short-day plants on a non-inductive light schedule. This photography was similarly unsuccessful and I began to think of the possibility of an indirect approach, like using log bud length as an index of the development of lily flower buds. Recalling the plots of internode lengths of growing *Clematis* plants which I had made long before, I began daily measurements of internodes of *Xanthium* plants, and saw only that they were quite variable in length and apparently erratic in their growth. There had been some discussion of leaf growth with graduate students, and this led to daily measurements of leaf length in *Xanthium*. One day during a discussion with Mike Michelini of a semilog plot of this data, I found myself writing the formula for the plastochron index on the blackboard, as if by an inspiration (20). The plastochron index has now been used by many authors, including ourselves, in a variety of ways.

Richards & Kavanagh (34) published their analysis of Avery's (5) data on the growth of a tobacco leaf, marked with a grid of points, in 1943. As it happened, I read their paper when that issue of *The American Naturalist* appeared in the current literature box at Western Cartridge. I did not understand it fully then but was convinced that it was important, since it dealt with differentials of spatial dimensions as well as time. Later at Penn, I felt that their analysis could be repeated more satisfactorily with *Xanthium* leaves, and eventually I was able to complete it (14). This two-dimensional analysis of elemental growth rates was then applied to younger *Xanthium* leaves, to a fern prothallus (Mae Chen's Master's thesis), and to the thallus of *Marchantia*. The intention also was to use this analysis for data on shoot apical meristems, but as I indicated, this did not work out. This kind of analysis has scarcely been followed up by other workers but it has been important in our thoughts about the nature of plant growth, and growth analysis, including root growth analysis.

Of the many courses I have taught at Penn, the one I most enjoyed was developmental plant morphology, given occasionally to a class of graduate and undergraduate students, sometimes with the collaboration of a colleague or a visiting botanist. It was a mix of talks by members of the class and myself, with small projects in the lab. The first class, in fall 1952, was a remarkable group, some of whom are now professional botanists. Paul B.

Green took this course as an undergraduate and, in the next term, he enrolled for an independent study course, made a study of the growth of the *Nitella* internode cell, and published it. He went on to graduate school and, a few years later, returned to Penn as a member of the biology department. During his years at Penn, Green was a stimulating colleague. While our formal collaboration (in print) has been minimal, we shared discussions continuously. I trust that I was helpful to Green in some technical matters and his viewpoint has certainly had, and continues to have, a great influence on my thinking.

Wendy K. Silk came to Penn as a graduate student in 1969, with a degree in biomathematics, and for personal reasons stayed only a year. When she had completed her graduate work at the University of California, Berkeley, she came to my lab as a postdoctoral fellow. She had made a compartmental analysis of the uptake and release of gibberellic acid by excised hypocotyl segments of lettuce, *Lactuca*, and wished to analyze growth of the segments in detail. This analysis did not work out well and we began to talk about the curvature of the hypocotyl hook. She set up a time-lapse camera to photograph intact lettuce seedlings and made a thorough analysis of the kinematics of hook maintenance in the growing hypocotyl, which required rather deep study of continuum mechanics. I then suggested that there should be a general article on the kinematics of plant growth (36). In my view, this work has provided a sort of capstone to the growth studies, answering questions that I had only dimly perceived. It has been followed by a number of theoretical papers on plant growth by other authors.

PHYLLOTAXIS AND OTHER THINGS

A topic that I found confusing at first was phyllotaxis, as discussed by taxonomists, morphologists and plant anatomists, and at one point I decided to look into the copious classical literature. I felt that it must somehow be important to understanding plant morphogenesis, implying as it does a very close regulation of the process of leaf initiation at the shoot apex. Aristid Lindenmayer was then a member of the Penn botany department, and in many luncheon discussions, he was very helpful in my early puzzlement about phyllotaxis. (It may be that these discussions played some part in Lindenmayer's later formulation of the cellular automata known as L-systems.) I read Church's 1904 monograph, failing to understand his emphasis on orthogonal parastichies as implying some mysterious physical analogy. I also studied F. J. Richards's papers and quite a few others. Van Iterson's thesis of 1907 was far more difficult since it is in German and bristles with equations, but it impressed me as a much more comprehensive work than Church's. After some time, I came to feel that the supposed conflict between Church's and van Iterson's models was minor, and could be resolved by rather simple notational changes in the equations used. When I felt that I understood this, a Dutch friend suggested that I write to van Iterson. He responded, generously sending me a copy of his monograph, which I have had bound, and treasure. I have recently completed a review of phyllotaxis (17).

This concern with phyllotaxis has had some unexpected consequences. Maksymowych at Villanova University had described striking changes in the morphology and the growth pattern of vegetatively grown *Xanthium* shoots, as the result of a single application of gibberellic acid. Among other things he made transverse sections of the shoot apex and young leaves of control and treated plants. When I saw them, I exclaimed that the treatment had altered the phyllotaxis. I proposed that we make a careful analysis of the arrangement of leaf primordia at the apex in these plants. We found that the normal pattern had been changed to a stable higher-order pattern (31) and speculated that the effect had some similarity to changes that occur on photoperiodic floral induction. Roger Meicenheimer (19) then induced *Xanthium* plants to flower and found that indeed the shoot apex underwent an identical but transient change in its phyllotaxis. The gibberellic acid effect is one of the few instances of an experimental modification of leaf arrangement in plants.

A second development is at the molecular level. Arthur Veen, a student with Lindenmayer at Utrecht University, had written a computer program to simulate growth of a shoot with the initiation of leaves in phyllotactic patterns (38). Their model was developed on a cylindrical surface, and I was sufficiently interested to write a preliminary program to carry out the simulation in a plane. Veen came to my lab to work with me on it. At that time, Lewis Tilney had developed an elegant technique of high-resolution electron microscopy of negatively stained microtubules. Lewis Routledge, working with Bernard Gerber, was using the technique for studies of bacterial flagella. When Tilney and Routledge showed me their pictures and asked how to analyze the obviously helical arrangement of subunits, my immediate impression was that they resembled certain of van Iterson's models. I suggested measuring distances between the units and certain angles, and constructing cylindrical models. After further discussion, I proposed to do the analysis if they would provide micrographs, references, etc. Veen and I were then deeply involved with computer modeling of phyllotaxis, so that the mathematical work and computations went quickly. In little more than a month the manuscript on tubular packing of spheres was completed (15). I had not been aware of the closely related work on cylindrical crystals by William F. Harris until our manuscript was completed, but I then sent him a copy. This led to voluminous correspondence, a visit by him to my laboratory, a visit to his at the University of the Witwatersrand, and to a far more rigorous analysis of tubular packings, from a crystallographic point of view (26).

The *Science* article on tubular packings (15) attracted the attention of others than biologists. At Penn I was one of the founders of a discussion group of people with varied interests in "form." For about five years this "Form Forum" met monthly, with wine and cheese, for talks and discussions of a remarkably wide range of topics, usually with demonstrations of paintings, sculptures, architectural renderings, tilings, polyhedra, electron micrographs, computer simulations, music, and poetry. I have also participated in two mathematical conferences on polyhedra. It is my hope that this broad approach to the geometry of form may be valuable in biological problems, such as the analysis of cellular patterns.

CALIFORNIA

I have taken three sabbatical leaves from Penn, and in each case have chosen to go to a California institution. In 1954–1955, with the award of a Guggenheim Fellowship, I worked at Frits W. Went's phytotron at the California Institute of Technology where I was incarcerated daily with Lloyd T. Evans, Harry R. Highkin, William S. Hillman, Margaretta G. Mes, Paul E. Pilet, Roy Sachs, and Went, when he was not travelling. There was much discussion of plant physiology, with a slant toward problems of floral induction. My idea was to grow *Xanthium* and perhaps other plants in a range of environmental conditions, using the plastochron index to assess temperature and light effects on the development pattern of the plants. I shared Anton Lang's dissatisfaction with the Cal Tech phytotron (30). My complaint was that growing conditions were not under control! Because carts were moved twice-daily to meet the schedules all of us had requested for our plants, there was no way, short of being an outright stinker, of knowing on a given day whether one's plants would be next to a flat of oat seedlings, or under the shade of a coffee bush. It seemed to me that the temperature coefficients for leaf growth and for the rate of leaf initiation (inverse of the plastochron) were nearly three, implying that the ratio of relative elongation rate to initiation rate (that is, the relative plastochron rate) was nearly constant over a broad range of temperatures. I took this to be an evidence of temperature regulation of morphogenesis, as to leaf initiation and growth, such that plants grown at different temperatures look much the same. Horie et al (29) have since described similar findings with cucumber plants.

But I was unable to get respectable data to support these ideas, and did not publish them. Looking around for other things to do, I set up lights and a 16-mm movie camera, which I had brought from Philadelphia, in a machinery room, which it turned out was air-conditioned. Time-lapse equipment was not then easily available or affordable, but I had brought home-made timers and mechanisms to operate cameras. There was an excellent machinist at the phytotron who, understandably, did not welcome others to use his shop. But when I had satisfied him that I was not likely to abuse his machines and tools, he gave me nearly free rein, excellent instruction in shop practices, and good advice about getting my gadgets to work. Several striking scenes of *Xanthium* growing from seed to plastochron 15 or so, in continuous light or with a non-inductive dark period, resulted, and some footage on *Coleus*, before the year was out. I also rigged up an automated 35-mm camera, and made the photographs of *Xanthium* leaves, which were analyzed much later (14). Despite my dissatisfaction at the time, it was a profitable year.

Our second California trip was to La Jolla in 1966–1967, where Herbert Stern had welcomed me to his lab at the new campus of the University of California, San Diego. This might have been an opportunity to learn modern biochemical techniques at the bench. However, Stern and I got to talking about some published work on the inhibition of cell division in roots of the broad bean, *Vicia faba*, by a thymine analog, and I decided to resume inhibition experiments with *Zea* roots. I had again brought camera equipment with me, and a student who wanted to learn the paraffin technique assisted. We could find no mitotic figures in sections of roots grown in the presence of purine and pyrimidine analogs, although the overall rate of elongation was normal for as long as 18 hours, during which time the meristem appeared to be "used up." To the question of where these inhibitors were incorporated, Yasuo Hotta suggested the simple expedient of using radioactively labeled inhibitors, putting root segments into the cocktail of scintillation vials and counting them. To our slight surprise, the label appeared not only in the former meristematic region but also in cells quite some distance behind it; but time ran out before this result was reconciled with the growth data. I also set up a time-lapse movie camera to photograph growing thalli of *Marchantia* that a postdoctoral fellow provided, and later analyzed them for the pattern of growth in area. I was enamored of the CDC 6600 computer at UCSD, far more satisfactory than the IBM machine at Penn, and spent a part of my time at the computer center working on problems such as the numerical solution of differential equations.

At Stanford University, in the fall of 1978, we had the pleasure of renewing our long acquaintance with Paul and Margaret Green. Following some discussions with Paul, I began calculations of the effect of growth deformation on the multi-net pattern of cellulose microfibrils in cell walls. This resulted quickly in a geometrical and statistical model of changes in the microfibrillar pattern, which agrees satisfactorily with experimental data on the walls of growing *Nitella* internode cells (16).

In January, we moved to the University of California, Davis, where I had a visiting professorship for the spring term. I collaborated with Wendy Silk in a graduate seminar on plant growth analysis and gave a few other lectures. We set up an automatic camera to record the growth of marked *Avena* coleoptiles, with the intention of following changes in phototropic curvature (curvature in the mathematical sense.) These preliminary experiments were not entirely satisfactory, but we did get some promising photographs.

One can conclude from these experiences that experiments do not always work out in a new laboratory with a time limitation. By and large, however, the scholarly leave is a valuable institution. The new viewpoints, acquaintances, and intangible benefits that result are ample justification for the inconvenience of moving and the frustrations one encounters. I wish I had taken leaves more often.

COMPUTERS AND CALCULATORS

The history of the computer has been written more than once (28, 35) but there may be some interest in my personal experiences as a user. The situation before the computer revolution is well stated in the preface of Whittaker & Robinson (39), written in 1924 "Each student should have a copy of Barlow's tables of squares, etc...a stock of computing paper (i.e., paper ruled into squares...), and...computing forms for...Fourier analysis... With this modest apparatus nearly all the computations hereafter described may be performed, although time and labour may often be saved by the use of multiplying and adding machines when these are available." As a boy I was impressed by the facility at mental arithmetic of one of my uncles, a bookkeeper, but I was all for machines. In high school I bought a cheap slide rule, possibly the only one in the school, and later I had a simple adding machine with dials to be turned with a stylus, which I could laboriously multiply with. I wore them both out. At Washington University and later, I was usually able to find a mechanical desk calculator of some sort, but when I came to Penn in 1947 there was no calculator in the botany department, only an ancient Monroe in the zoology office. I immediately ordered a Marchant, and soon additional machines were obtained for the biometry course and research. When I could afford it, I bought a Curta hand-held calculator to use at home and was delighted with the watch-like precision of its construction. In the summer of 1951, when the analysis of our root growth data needed to be done, a few of us moved to the Morris Arboretum with Marchant calculators and, on a pleasant terrace, punched the machines every day for about three weeks.

In 1972 Hewlett-Packard announced their first pocket scientific calculator, the HP-35, and I of course bought it for myself and the lab. As improved models appeared, I acquired and used several, including the current HP-28C, which has memory exceeding that of the IBM mainframe computer at which I learned FORTRAN.

By a curious coincidence, at the first scientific meeting I attended, the AAAS Christmas meeting at Richmond in 1938, I saw an exhibit of computing equipment from the Moore School of Electrical Engineering at Penn. It undoubtedly had to do with Weygandt's differential analyzer, an analog computer (35). So I was not as surprised as I might otherwise have been when I learned of the first digital computer, the ENIAC built at the Moore School. Penn had a computer lab when I came, which at one time housed a Univac, and later other machines. Occasionally I visited this lab and found that the staff were friendly enough, but the computers were not. One needed to know a great deal about the machines to use them. In January 1964, Penn acquired an IBM 7040, which was one of the first computers with an operating system designed to accommodate ordinary users. A full-scale computer center was quickly organized and I immediately took the FORTRAN course. At the end of the course each student was to write a program and run it. The instructor suggested a payroll problem, but I wrote my own program to calculate Fibonacci numbers and plastochron ratios. It ran on the first submission. I then proceeded to program the computation of elemental rates of growth in area of the *Xanthium* leaf, and completed the analysis for the 1964 Edinburgh Congress (14). I have now had experience with three or four mainframe computers, as well as two desktop computers.

When microcomputers appeared on the market we were in Silicon Valley, and I spent some time learning about the first Apple and other micros, skeptical at first about their usefulness for serious work. I bought an AIM 65 and set it up in our bedroom in Davis. Soon I had wired up a small speaker and written a program to play tunes through one of the output ports: *Papa Haydn's Dead and Gone*, Brahms's *Lullaby*,…. The AIM was perhaps the least expensive, and one of the most educational of the early machines, with provisions for expansion of the hardware and the monitor program. I brought it back to Penn and used it to good effect for about five years. The machine at which I am writing this text is a far more competent personal computer, an MTU-130, which in its turn is about five years old….

LOOKING FORWARD

The years I have written about have of course seen a great revolution in biology, with the development of molecular genetics and many other advances. Reflecting on the state of what might be called the biometry of growth, say in 1936, one realizes that this received very scanty treatment in plant physiology textbooks, and none at all in plant anatomy. One learned of the "grand period of growth," of auxanometers, and of Julius Sachs's root-marking experiments of the 1860s. Plant anatomists wrote of gliding growth, and used the word plastochron in a purely descriptive sense. In the 1870s Kreusler et al in Germany had studied the growth of *Zea* plants, but in the English-speaking world serious consideration of plant growth seems to have begun with defining of the efficiency index, or relative growth rate, by Blackman and by Briggs, Kidd & West in the early 1920s. Robertson's ideas about a master reaction in control of growth were widely quoted, and Huxley's allometric coefficient was fashionable. Biologists had a definite prejudice against mathematics and statistics and for many years I felt that my work was mostly quietly ignored. At the first presentation of our root work at a meeting, an older botanist took me aside and offered me the fatherly advice to soft-pedal the math.

In the intervening years the analysis of the growth of whole plants and plant organs has advanced considerably, particularly in connection with agricultural research. Statistics such as the absolute and relative growth rates, unit leaf rate, and leaf area ratio in the analysis of growth of individual plants, and related quantities for the growth of crops, appear to be firmly established, with general agreement about how they are to be estimated. Methods for fitting of growth curves, particularly the F. J. Richards function, have been highly developed. There is much activity in mathematical modeling of the growth of plants and crops.

The concept of elemental growth rates was introduced by Richards & Kavanagh in 1943 (34) and much of what I have discussed above, such as root growth analysis and analysis of growth of the *Xanthium* leaf, is related to this idea. This has led to the consideration of the kinematics of plant growth (36) in the context of continuum mechanics. The importance of distinguishing between material and spatial specifications has become clear. Kinematics deals only with motion without consideration of mass and force, but with kinematics as a basis we can look forward to the development of the dynamics of plant growth. Since a large part of plant physiology has to do with growing tissues, it will be important to deal effectively with the kinematics of growth, in order to make valid estimates of biosynthetic rates, for instance. Studies of the role of water and solute transport in tissue growth will have to take account of the kinematics of the growing tissue, as will considerations of the energetics of tissue growth. It is also true that morphogenesis, the development of the form of plant organs from meristematic tissue, is to a large extent a matter of tissue deformation, and it will be necessary to consider the forces that give rise to kinematic displacements and their origin. Work in these directions is under way, and we can look forward to further advances in the empirical analysis of plant growth processes, and in theoretical treatments of plant growth.

Literature Cited

1. Anderson, E. 1936. The species problem in *Iris. Ann. Mo. Bot. Gard.* 23:457–509
2. Anderson, E. 1939. The hindrance to gene recombination imposed by linkage: an estimate of its total magnitude. *Am. Nat.* 73:185–88
3. Anderson, E. 1967. *Plants, Man and Life*. Berkeley: Univ. Calif. Press
4. Anderson, E., Schregardus, D. 1944. A method for recording and analyzing variations of internode pattern. *Ann. Mo. Bot. Gard.* 31:241–47
5. Avery, G. S. Jr. 1933. Structure and development of the tobacco leaf. *Am. J. Bot.* 20:565–92
6. Avery, G. S. Jr., Burkholder, P. R., Creighton, H. B. 1937. Production and distribution of growth hormone in shoots of *Aesculus* and *Malus* and its probable role in stimulating cambial activity. *Am. J. Bot.* 24:51–58
7. Cooper, D. C. 1935. Macrosporogenesis and development of the embryo sac of *Lilium henryi. Bot. Gaz.* 97:346–55
8. Darlington, C. D. 1937. *Recent Advances in Cytology*. Philadelphia: Blakiston. 2nd ed.
9. Erickson, R. O. 1943. Taxonomy of *Clematis* section *Viorna. Ann. Mo. Bot. Gard.* 30:1–62
10. Erickson, R. O. 1945. The *Clematis fremontii* var. *riehlii* population in the Ozarks. *Ann. Mo. Bot. Gard.* 32:413–60. Reprinted in Ornduff, R., ed. 1967. *Papers on Plant Systematics*. Boston: Little, Brown. Pp. 319–66
11. Erickson, R. O. 1947. Respiration of developing anthers. *Nature* 159:275
12. Erickson, R. O. 1948. Cytological and growth correlations in the flower bud and anther of *Lilium longiflorum. Am. J. Bot.* 35:729–39
13. Erickson, R. O. 1961. Probability of division of cells in the epidermis of the *Phleum* root. *Am. J. Bot.* 48:268–74
14. Erickson, R. O. 1966. Relative elemental rates and anisotropy of growth in area: a computer programme. *J. Exp. Bot.* 17:390–403
15. Erickson, R. O. 1973. Tubular packing of spheres in biological fine structure. *Science* 181:705–16
16. Erickson, R. O. 1980. Microfibrillar structure of growing plant cell walls. *Lect. Notes Biomath.* 33:192–212
17. Erickson, R. O. 1983. The geometry of phyllotaxis. In *The Growth and Functioning of Leaves*, ed. J. E. Dale, F. L. Milthorpe, 3:53–88. Cambridge: Cambridge Univ. Press
18. Erickson, R. O., Goddard, D. R. 1951. An analysis of root growth in cellular and biochemical terms. *Growth, Symp.* 10:89–116
19. Erickson, R. O., Meicenheimer, R. D. 1977. Photoperiod induced change in phyllotaxis in *Xanthium. Am. J. Bot.* 64:981–88
20. Erickson, R. O., Michelini, F. J. 1957. The plastochron index. *Am. J. Bot.* 44:297–305; 47:350–51
21. Erickson, R. O., Sax, K. B. 1956. Elemental growth rate of the primary root of *Zea mays. Proc. Am. Philos. Soc.* 100:487–98
22. Erickson, R. O., Sax, K. B. 1956. Rates of cell division and cell elongation in the growth of the primary root of *Zea mays. Proc. Am. Philos. Soc.* 100:499–514
23. Fisher, R. A. 1936. The use of multiple measurements in taxonomic problems. *Ann. Eugen. London* 7:179–88
24. Gilman, R. 1987. Noted with pleasure. *N. Y. Times Book Rev.* 1 Feb.
25. Goodwin, R. H., Stepka, W. 1945. Growth and differentiation in the root tip of *Phleum pratense. Am. J. Bot.* 32:36–46
26. Harris, W. F., Erickson, R. O. 1980. Tubular arrays of spheres: geometry, continuous and discontinuous contraction, and the role of moving dislocations. *J. Theor. Biol.* 83:215–46
27. Hejnowicz, Z., Erickson, R. O. 1968. Growth inhibition and recovery in roots following temporary treatment with auxin. *Physiol. Plant.* 21:302–13
28. Hodges, A. 1983. *Alan Turing: the Enigma*. New York: Simon & Schuster
29. Horie,T., de Wit, C. T., Goudrian, J., Bensink, J. 1979. A formal template for the development of cucumber in its vegetative stage. *Proc. Kon. Nederl. Akad. Weten.* 82:443–79
30. Lang, A. 1980. Some recollections and reflections. *Ann. Rev. Plant Physiol.* 31:1–28
31. Maksymowych, R., Erickson, R. O. 1977. Phyllotactic change induced by gibberellic acid in *Xanthium* shoot apices. *Am. J. Bot.* 64:33–44
32. Moens, P. B. 1968. The structure and function of the synaptinemal complex in *Lilium longiflorum* sporocytes. *Chromosoma* 23:418–51
33. Ogur, M., Erickson, R. O., Rosen, G. U., Sax, K. B., Holden, C. 1951.

Nucleic acids in relation to cell division in *Lilium longiflorum*. *Exp. Cell Res.* 11:73–89

34. Richards, O. W., Kavanagh, A. J. 1943. The analysis of relative growth gradients and changing form of growing organisms: illustrated by the tobacco leaf. *Am. Nat.* 77:385–99

35. Shurkin, J. 1985. *Engines of the Mind, a History of the Computer.* New York: Washington Square Press

36. Silk, W. K., Erickson, R. O. 1979. Kinematics of plant growth. *J. Theor. Biol.* 76:481–501; 83:701–3

37. Stern, H., Hotta, Y. 1977. Biochemistry of meiosis. *Philos. Trans. R. Soc. London Ser. B* 277:277–94

38. Veen, A. H., Lindenmayer, A. 1977. Diffusion mechanism for phyllotaxis, theoretical, physico-chemical and computer study. *Plant Physiol.* 60:127–39

39. Whittaker, E., Robinson, G. 1944. *The Calculus of Observations. An Introduction to Numerical Analysis.* New York: Dover Publications. 4th ed., reprint

SOCIOLOGY

Matilda White Riley

THE INFLUENCE OF SOCIOLOGICAL LIVES: PERSONAL REFLECTIONS

Matilda White Riley

National Institute on Aging, National Institutes of Health Building 31C, Room 5C32, Bethesda, Maryland 20892

KEY WORDS: influence, lives, age, gender, dynamic systems, sociological practice

Abstract

In contrast to the well-established influence of social change on sociological lives, the author develops a theory of the influence exerted *by* the lives and experiences of sociologists on social and intellectual structure and change, both in sociology and in society as a whole. Writing in a semi-autobiographical vein, she uses as examples of this influence fragments from her earlier writings in four areas of current sociological concern: sociological practice, gender, age, and dynamic social systems. These fragments are interwoven with anecdotal accounts of experiences from the lives of well-known sociologists and the author herself, spanning much of the twentieth century. While giving a flavor of her own contributions and also several thwarted attempts, these reflections illustrate how the degree of sociological influence depends on the mesh between the attributes of particular lives and the opportunities afforded at the time by the state of the discipline and of society. Several types of historical structures and changes are identified as either facilitating or hindering the flow of influence, including trends in sociological thought and methods of research, ideologies and values paramount in the discipline, and social and cultural changes in society as a whole.

For the future, the question raised is how, in a rapidly changing world, to identify channels for exercising influence on sociology as well as on public policy and professional practice. The concluding hope is that other sociologists may be sensitized to a self-conscious awareness of the special opportunities for influence available in the unique historical era in which their lives unfold.

INTRODUCTION

In the summer of 1949, while my husband and two children were setting up camp in Grand Teton National Park, I sat on a huge boulder writing a grant application. My proposal was to describe and show the relevance for sociology of numerous research designs developed during my years of directing research at the Market Research Company of America. Encouraged by our Chicago colleagues, Clyde Hart and Everett and Helen Hughes, the proposal was designed to influence sociological work by making available from my own life experience examples of the use of methods that were new at that time (as in cross-section surveys, intensive interviewing and observation, panel studies, probability sampling, and small group interactions).

That proposal failed. Yet analysis of the failure is instructive here for exploring the topic of this essay: *the influence exerted through the lives of sociologists on sociology and on society.* The experience illustrates some of the major obstacles that block channels of influence at particular periods of history: in this instance, the difficulties of securing funding (that application was for $5000, a modest sum even in those days); the paucity of vehicles for communicating potentially useful research approaches; the marginal level of sociological interest in methodological advances; and the consequent lag in influence as many of these approaches became the stuff of sociological research only many years later, if at all.

Now, 40 years later, this time on the shores of Maine's Casco Bay, I look back on the implications of this and other personal experiences for the concept under consideration here: the influence of sociological lives. My reminiscences are prompted by the incipient development of such a concept in the introductory "Notes" (1988a) to *Sociological Lives* (Riley 1988b). Those "Notes" examined the experiences of the eight sociologists contributing to that volume. Their essays, though only indirectly focused on the topic, nevertheless suggest numerous ways in which the influence of their lives as sociologists was felt by the discipline. From just these few life histories it becomes clear that the degree of sociological influence depends on the mesh between the attributes of particular lives and the concurrent opportunities afforded by the state of the discipline and of society. That introductory chapter concluded with these thoughts:

Collectively, the essays begin to show how sociologists living at particular moments in history have been influencing social policies, practices, and the shape of social structures; and how they have been influencing the development of sociology, its content and goals, and its intellectual and organizational arrangements. Perhaps the concept of "the influence of sociological lives," like the concept of "sociological autobiography,"[1] may enhance our understanding of the interplay between social structures and human [in this instance sociological] lives. (p. 40)

Intrigued by the potential of such a concept, I want now to pursue it further by adding to the experiences reported in *Sociological Lives* a few experiences and fragmentary writings from my own life and the lives of close associates. My emphasis throughout is not on the well-established influence of social and intellectual structures *on* sociological lives. Rather, the focus is in the other direction of the interplay: on the influence exerted *by* the lives and experiences of sociologists on social and intellectual structure and change in sociology and in society as a whole. The concern with lives in historical context is distinct from evaluation of particular ideas, exegesis of particular works, or particular continuities in social research.

This attempt to learn more from my own life about how influence operates begins with some clues from *Sociological Lives* and then reviews fragments from my earlier writings and associated experiences in four areas of current sociological concern: sociological practice, gender, age, and dynamic social systems. The essay ends with the thought that other sociologists might be sensitized by a self-conscious awareness of the special significance of the unique historical era in which their lives unfold.

AFTERTHOUGHTS ON *SOCIOLOGICAL LIVES*

What, then, is this concept of the influence of sociological lives? Quite inadvertently, the autobiographical essays in *Sociological Lives*[2] raise important questions about how structural changes evolve from the experiences of successive cohorts of individual sociologists, and most particularly from the lives and work of individuals who reach positions of leadership at particular moments in history. For example, how does it happen that the intellectual growth of William Julius Wilson is a powerfully active force in influencing both sociological thought and public policy—rather than merely a passive reflection of structural change and academic controversy? About other sociologists, whether born earlier or later than he, such general questions are

---

[1]As in the opening chapter by Robert K. Merton of the volume on *Sociological Lives*.

[2]*Sociological Lives*, to which frequent reference is made here, is the companion volume to *Social Structures and Human Lives* in the Presidential Series of the American Sociological Association (ASA).

raised as: How does sociological influence operate? What "channels of influence" open or close as social and intellectual structures change? What can block the channels, and how are barriers removed? What attributes of lives serve as sources of influence under particular historical conditions? In short, how do lives of sociologists in successive cohorts mesh with the changing channels of opportunity?

Partial answers to these questions are suggested by the essayists, who represent cohorts from four decades of this century (their birth dates range from 1909 to 1947) and who thus confronted widely differing historical structures and exigencies. Among these answers, *sociotemporal location* emerges as a central force, affecting both the lives of sociologists as sources of influence and the channels that provide access to influence. On the one hand, whether or not a sociologist's work can spark enthusiastic response depends on the historical context. On the other hand, whether or not the intellectual and social climate is open for positive response also depends on the historical context. Over the course of history, the succession of cohorts provides the linkages between the influencer and the influenced, between sociological lives and changes in the discipline and in society.

Sources of Influence

Several essays illustrate how sociologists living under particular sociotemporal conditions can exert influence: through the focus or style of their work, the nature of their experiences, or according to such attributes as gender or race. For example, Rosabeth Kanter, referring to the stormy 1960s, describes the mesh between phases of discontinuity in her unfolding individual career and the changing society; thus she points to changes in her own life encouraging her to seize timely opportunities for influence. Tad Blalock complains that, with few exceptions during the 1950s and 1960s, his mathematical modelling as a form of theorizing was entirely foreign to his sociological contemporaries; thus his work was ahead of its time, producing a lag in its impact on the discipline. By contrast, Bernice Neugarten's career was precisely "on time" for influencing studies in human development at the University of Chicago, though here again there has been a lag in widespread sociological recognition.

Other essays illustrate how influence is affected by such attributes of sociological lives as gender or race, depending upon the attitudes prevalent at the time. Thus, Alice Rossi describes how, during the 1950s, sex discrimination and antinepotism rules greatly reduced the educational and occupational openings for academic women in her cohort—forcing many to lower aspirations of ever becoming influentials, or even to withdraw from the labor force into fulltime motherhood. (Only much later, following the women's movements, did the topic of gender become central to sociological thinking, thus opening channels of influence.) The essays tell little about the undoubtedly significant implications for influence of differing statuses in the world of sociology, because all these authors are similarly active and visible in the field. Nor do the essays provide many clues to the importance of location in the age structure, since none of the authors have yet completed their life course. (Left for future analysts are questions of how age affects the potential for influence of sociologists born recently—like Bill Wilson or Theda Skocpol—or even those, like Bernice Neugarten or myself, who were born much earlier.)

Opportunities and Obstacles

While individual sociologists in successive cohorts are pursuing their careers, the historical milieu is changing, and with it the channels through which their influence may be felt. My "Notes on Influence" suggest various ways in which the opportunities and obstacles encountered at particular moments in a person's life reflect the coexisting exigencies in the discipline and in society:

> Whether or not channels of influence are accessible during the lifetimes of particular sociologists seems to depend on broad social and intellectual trends and structural changes. Changes affecting the flow of influence include those in the state of the art and the organization of research; in the effectiveness of communication; and, perhaps most important, in the goals and interests of the discipline. In sociology, some of these goals and interests seem to derive from trends in thought; others from emergence of immediate social problems, disruptions, or controversies; and still others from ideologies and values paramount at the time. Sometimes adherence to vested ideas and concerns produces outright resistance to new influences. Sometimes influences from sociology diffuse outward to other disciplines or to policies and practices. Moreover, channels blocked at one time may be opened at a later time, producing a "lag" in the cumulation of scientific work (Riley 1988-a:36).

As a starting point, then, for further conceptual exploration here, at least five types of historical structures and changes can operate either to facilitate or to hinder the flow of influence: (*a*) social and cultural changes in the society as a whole; and trends in (*b*) sociological thought, (*c*) methods of sociological research, (*d*) the organization of sociological research, and (*e*) ideologies and values related to the discipline. My personal exploration relates to the conjunction between particular efforts in my lifetime and such types of period-specific opportunities or barriers to influence at those particular times. Dates are essential for linking lives to history (especially when only a single cohort—rather than the (preferable) succession of cohorts—is available for identifying the linkages).

To broaden the base beyond my own cohort, I draw freely on the work of the essayists in *Sociological Lives*, of other sociologists whose experiences touch upon my own, and of my significant collaborators and colleagues. In particular, my own recollections cannot be separated from those of my husband, Jack (properly, John W. Riley, Jr.). Our joint lives have been intertwined for nearly 60 years of marriage and colleagueship. As he and I often say, our collaboration as sociologists has spanned a large part of the twentieth century and almost our entire life course: we first published empirically based articles on contraception; then sequentially on children, adolescents, and midlife careers; then on old age and death; and now we plan to turn back to leisure, the topic which, in our earliest and most idealistic years, we had dreamed of being one day "mature" enough to tackle!

SOME AREAS OF POTENTIAL SOCIOLOGICAL INFLUENCE

This pursuit of reminiscences and introspections that might identify and clarify factors opening or blocking channels of influence is a nostalgic affair. It touches the nerve involved in steering between self-blame and self-congratulation. It requires searching through many dusty pages of books, reprints, and unpublished manuscripts. These manuscripts span five decades of my own sociological biography.[3] They are full of surprises (how often, by forgetting, we fail to be influenced by our own earlier efforts!). Yet they provide fertile ground for asking why some seem to have exerted a degree of influence (though often marked by considerable time lag), whereas others have evoked strong resistance or—worse still—no response whatsoever.

Hoping to avoid a particularistic evaluation of certain topics as more or less "significant" than others, I have selected as case examples a few fragments and incidents relevant for each of four areas of sociological concern today: sociological practice, gender, age, and dynamic social systems.[4] Nonetheless, my biases show through. I cannot resist an overarching emphasis on the fourth area, "dynamic social systems" (variously involving "multilevel" or "macro-micro" analysis), especially as it evokes the central goal of improving the mesh between theory and research.

This dual focus (though with reference only to measurement, not to analysis) appears in our work as early as 1954, in *Sociological Studies in*

---

[3]In still an earlier decade, as a junior at Radcliffe, I coauthored a book on *Gliding and Soaring* (1931. P. & M. White. New York: Whittelsey House) which, though scarcely relevant for sociological influence, provided an invaluable introduction to the culture of preparing and publishing a book.

To shorten the "literature cited" for this essay, only immediately relevant works are cited; a full bibliography may be obtained from the author.

[4]This arrangement by substantive areas rather than by chronology reflects the conceptual, as opposed to any autobiographical, focus of this essay; thus the discussion skips back and forth over the course of my life.

Scale Analysis, by Riley, Riley & Toby. Jack introduced the first chapter of that book as follows:

> [*Fragment*] This book attempts to codify, in methodological terms, one aspect of the interplay between theory and research . . . It is the reconstructed story of how a recently contrived research method, scale analysis, was bent to the task of translating vague theory into precise variables.
>
> . . . the study of the single variable is not a simple task . . . On the one hand, many concepts involve the notion, not of an isolated act or attitude, but of *patterns of action,* of the 'uniformities in the sayings and doings of men,' to use Talcott Parsons' phrase. On the other hand, sociological concepts, insofar as they reflect the collective acts or attitudes of the members of a group, imply a *structure among the members* of the group who act in complementary roles with reference to one another. (pp. 5, 8)

Sociological Practice

In the area now called "sociological practice" (or, as Paul Lazarsfeld called it as the theme of his Presidential volume, *The Uses of Sociology,* 1967), the American Sociological Association (ASA) has just launched the new *Sociological Practice Review*—a considerable "lag" indeed from my effort in 1949 to specify for sociologists the relevant "practical" methods of market research.

EARLY EXPERIENCES These experiences of my early life were part of a persistent effort to link theory with research methods. I was firmly imbued with the sociological perspective even before entering the "practice" of market researcher. (As a wife I then simply accepted whatever occupational opportunities arose.) When I married Jack in 1931 he had just entered the newly formed Harvard Department of Sociology, and we shared the high excitement of discovering the convergence between his studies with Pitirim Sorokin on social change and its meaning, and my studies with Irving Babbitt on the universals in human thought that persist across time and place. Through Jack, I learned the basics of sociology. As the first graduate assistant in the Department, I had the privilege of working with Carl Zimmerman on an analysis of the Le Play studies of family budgets in Europe. This task entailed first translating from many languages and then analyzing the detailed case studies of households in diverse cross-cultural settings, in order to test certain Engelian "laws" (concerning the inelasticity of demand for food, in contrast to other budgetary categories). From this endeavor I derived an early comprehension of the family group as a solidary system of interdependent members.

While not in the forefront of my mind when I joined the Market Research Company of America, this exposure to sociology surely informed the concepts as well as the methods of the numerous and varied studies we designed during the late 1930s and 1940s. Often advised by such experts as Paul Lazarsfeld, W. Edwards Deming,[5] or Raymond Franzen, we gradually solved many practical problems. For example, we used door-to-door interviewers to conduct classical experiments in the household, in order to examine consumption of new products (White & White 1948). We devised national surveys of contraceptive behaviors and attitudes, leading to improved estimates of fertility and to the development of "the pill." We were the first, outside the Federal government, to design and conduct a national probability sample of the United States. We invented a gadget (called the "Chronolog") for obtaining private information from different members of a household. Along the way we were forced to learn from many false starts and mistakes. It was my enthusiasm for communicating to sociologists the practical implementations of these and other methodological tools which prompted my unsuccessful early foray into grantsmanship.

CHANNELS OF INFLUENCE   That this attempt failed in 1949 is not surprising. The incident is one example of abandoning the effort to communicate systematically what I had learned as a sociological practitioner. (Instead, I accepted the first Executive Officership of the then American Sociological Society.) How many such efforts are abandoned and hence never recorded in the literature? (I seem to remember Bob Merton once referring to his files of unfinished projects as "little dead children"—though he now says my memory does not "resonate" with his.)

Further contributing to my failure here was the general resistance at that time to "commercial" research—even though Paul Lazarsfeld, on his first visit to this country as a Rockefeller Fellow in the 1930s, found market research in some respects more advanced than academic research. Jack has never forgotten Sorokin's ire when, as a graduate student asked to design some research on the effectiveness of advertising, Jack came up with a plan drawn from market research. Far more sophisticated than the academic procedures of the day, Jack's plan was scoffed at as "tainted." Only later did the innovative experiences of market research take firm root in sociology, after they had been legitimated through the work of prestigious sociological researchers, and through such organizations as the Bureau of Applied Social Research at Columbia and the National Opinion Research Center at the University of Chicago. Even Hans Zeisel's *Say It With Figures* (first published during World War II, when Hans was my colleague in the Market Research Company of America, and now in its sixth edition, 1985) has perhaps been less influential among his fellow sociologists than among

---

[5]The mathematician who used his methods of quality control, first, to invent area sampling and, more recently, to guide Japanese manufacturers in procedures for which they hold him in the highest esteem.

researchers across the full range of scholarly, professional, and practical affairs.

These experiences also illustrate the outcome of sociological influence for the little-developed midcentury organizational arrangements for funding and conducting sociological research. They supplement the report by William Sewell in *Sociological Lives* of the long lag before his final successes in enlarging the scale, scope, and funding of sociological research and graduate training, both in his own University of Wisconsin and at the national level (see also Sewell 1989).

CONTINUING EFFORTS   Not deterred by my own initial setback, I have made continuing efforts to apply "practical" market research experiences to sociological work. Jack and I reported the market research survey of contraception and published it in an early issue of the *American Sociological Review* (and signed it as Riley & White, 1941—note the use of my "maiden name" even when collaborating with my husband). Though largely atheoretical, the findings startled the demographers who had heretofore assumed that most Catholics used only "natural" contraceptive methods (despite the name Riley, we are not Catholics). That study became the baseline for subsequent series on fertility. This and other experiences with market research based on groups (such as families or companies) led to my long struggle against studies which inappropriately treat groups or societies as mere aggregates of individuals.

Another market research example from these early writings is found in a conceptual chapter in *Sociological Studies in Scale Analysis*. Here we provided for use in sociological measurement an analogue of Durkheim's "organic solidarity" (the changes since 1954 in family buying behaviors are striking):

[*Fragment*]   Lacking any systematic theory, market research has often made use of the group, recognizing it, on a common-sense basis, as an interpersonal, frequently differentiated unit. Conscious use of the group in market research rests . . . upon the obvious fact that the market for many products is made up, not of individuals as such, but of households. In some instances, to be sure, the demand for a product may come from only one member of the family: the housewife, not the whole family, may be the sole user and purchaser of a household cleanser.

In other instances, however, the demand is a composite of the demands of the several family members, as in the case of most food products. Hence, if a new food product is to be tested for market acceptance, usually it is not enough merely to interview the housewife, ask her to taste it or try it, and to report her reactions. When, instead, samples of the product are left in the home for use under normal conditions, it is often the case that the several members have differing reactions: the housewife may like the taste but find it difficult to prepare; another may dislike the product but wish to collect the jars it comes in; another may like it but prefer a substitute which he is more in the habit of eating, and so on. The researcher recognizes it as his task to obtain the synthesis of all such reactions, the net which can be used to predict whether or not this household would purchase this new product if it were put onto the market. The new product researcher has effective operating techniques for just this purpose. After the period of trial use of the product [in which actual consumption by all family members is observed], he offers the housewife her choice of a supply of the new product or a supply of the old one. Her choice is then taken to represent the composite family demand, in the same sense that her purchase at the grocery store reflects not only her wishes but also those of the other family members taken as a whole. (Riley, Riley & Toby 1954, pp. 193–194)

In market research, as in all sociological practice, the channels of influence run in both directions: sociologists use sociological tools in nonacademic settings, just as they also, in turn, bring the results of this work back into sociology. In the 1950s, during my early days as Executive Officer of the ASA, I worked with Donald Young, then President of Russell Sage Foundation, on a series of ASA-Russell Sage Bulletins on Sociology and the Practicing Professions. In another instance, Jack Riley, shifting his career in 1960 from the university to the insurance industry (whose business was viewed as involving the lives, hopes, and problems of *people*), led the way in applying sociological principles to the world of corporate affairs. Today, prestigious sociologists teach in schools of business and other applied areas; in *Sociological Lives*, for example, Rosabeth Kanter's autobiography shows how her sociological influence is exercised through teaching tomorrow's economic and political leaders, working with corporations, and publishing widely read books. Today sociology, though by no means universally accepted, is utilized throughout the business world, just as knowledge and approaches developed in business and industry have become essential to sociology.

OTHER EXAMPLES   In another domain of sociological practice, military research, Jack's experience (but only indirectly my own) illustrates how channels of influence become open or closed with historical changes in values and ideologies. During World War II, while Samuel Stouffer and his colleagues were producing their monumental studies on *The American Soldier*, Jack's studies of French civilian attitudes toward the Allies (e.g. J. Riley 1947) not only provided essential facts to Eisenhower's military headquarters, but also fed directly into the sociological literatures on war and on mass communication. During the Korean war, he again participated in surveys that were useful not only in advising generals in the field, but also in informing scholars about Communist methods of sovietizing populations under their control; *The Reds Take a City*, by J. Riley & Schramm 1951, has been translated into some 15 languages (see also Schramm & J. Riley 1951). The Vietnam war, however, with its contributions to newly emerging values and life-styles, dampened much of the influence of military sociology, with the personal consequence that for many years neither Jack nor I have cited his war studies. With current changes in public attitudes and the recrudescence of sociological analysis of the impact of war on human lives, however, this early work takes on new significance. The recent review by Robin Williams (1989) of *The American Soldier* "several wars later," which dramatizes the impact of "shifts in attention, emphasis, and evaluation," points to renewed interest in "military sociology" after its long separation from "peace studies" (pp. 166, 170; see also Mayer 1988, and Elder & Clipp 1988).

Sociology of Gender

The trends in ideologies and values that have often operated to open or close channels of influence in sociological practice take a different form in research on gender. Here the channels are affected by massive social and cultural changes in society as a whole and also by trends in sociological thought. During the 1950s at Rutgers University, though the changes were already swirling around us, we had little interest in gender. We spent years studying adolescents, focusing on such topics as socialization, interpersonal influence, and intergenerational relationships. But the early findings were frustrating. Why? Because, as we repeatedly said to one another, "They only show that boys differ from girls. And everyone already knows that!" Again, we failed to publish—this time because we expected little sociological response.

THE ERA OF "INNOCENCE"   In the first half of this century, as the autobiographies in *Sociological Lives* show, gender was scarcely a central concern of sociology, and hence it was not a source of influence. Even Alice Rossi, talking of a time prior to her many influential writings on gender, speaks of her early "innocence" of the invidious position of women. Bernice Neugarten reports *no* instance in which her gender worked to her advantage or her disadvantage in her research career! For myself, my daughter rebukes me for passive acceptance when, in my college years, the publisher of my book on gliding and soaring changed my name to "Mat" because "no one will read a book on that subject written by a girl;" or when, in 1931, I was turned down for a teaching assistantship because, I was told, "as a woman you will not continue a career"! It was only later, as sex discrimination became a social issue, that gender became a major area of sociological inquiry. Thus, Theda Skocpol, representing the most recent cohort in *Sociological Lives*, writes of her plans to "join the many others in our discipline who are already drawing on the travails of changing gender relations to enrich the sociological imagination."

WOMEN'S CHANGING OCCUPATIONAL ROLE   Yet well before the women's movements of the 1960s, the long-term increases in women's participation in the labor force had begun subtly to redirect the course of sociological influence. In line with this change, we finally turned attention to the gender differences observed in the Rutgers research, as reported in a piece on "Woman's Changing Occupational Role: A Research Report" (Riley, Johnson & Boocock, 1963). The research objective here was to discern the factors that dispose some members of the oncoming cohort of young people to favor, and others to frown upon, the norm of women's employment. Beginning with data collected in 1961 from a sample of New Jersey high school students, the study went on to test its predictions through a special analysis of Census data for adult women in the United States as a whole. These predictions were that the economic status of the family, traditionally the major determining factor in whether or not a wife is in the labor force, was giving way to the rising power of education, and to the redefinition of women's paid work as contributing not only to the family income, but also to self-actualization and the "good life." In retrospect, the study is historically interesting for its microlevel contributions to macrolevel changes in a period of social transition, as illustrated in the following excerpt:

[*Fragment*]   The study reveals the unanticipated finding that the family's economic status (as measured by the father's occupation)—often so powerful a factor in sociological analysis—proves to bear little relationship to the career attitudes of these adolescents.

This surprising finding . . . proves, upon further investigation, to reflect two quite different tendencies in the data. In the lower status families, the mothers are more likely to work. And since acceptance is relatively high among adolescents whose mothers work, many girls from lower status families will expect to work themselves—even for economic reasons. On the other hand, the higher status families are more likely than the lower to provide college or graduate educations for their offspring. And this fits into the emerging tendency for adolescents planning advanced education to want to work for primarily non-economic reasons. Thus the "effect" of father's lower economic status appears to be mediated through the experience of having a mother who works; whereas the "effect" of father's higher economic status seems mediated through education and a good life view. High status and low status families produce similar acceptance, but by quite different paths.

This process among adolescent girls, by which career acceptance becomes relatively independent of an economic base, seems to reflect the situation [as seen in Census data] in the country as a whole. At a given point in time, two factors (among others) appear to exercise marked effects upon the labor force participation of the wife—but in opposite directions. Her participation tends to *increase* as her education rises (reflected in our survey data by the association between planned education and the adolescent girl's disposition to work). Yet, at the same time, her participation tends to *decrease* as her husband's income rises (consistent with the association between mother's working and family status in our sample). Moreover, the better educated women tend to marry the financially more successful men. Thus, at this stage of the transition, many women are under the marked crosspressures of high education, which appears to facilitate their working, and high income, which appears to inhibit it.

There are some indications that over time this conflicting influence of husband's income and wife's education may be resolved. Since more and more women are going on to higher education, one might expect the decline to occur in the importance of the economic, rather than the educational, factor. And, indeed, the rather scanty trend data available suggest that . . . the husband's income seems to be having a decreasing effect upon the labor force activity of his wife.

> What is the broader meaning of the loosened ties between the woman's occupational role and its traditional economic base? Perhaps, to be the husband-of-the-working-wife . . . may no longer constitute either a stigma or a threat. A job for the individual woman may become an extension of her traditional primary role as wife and mother within the family . . . Thus redefined, the woman's job may come to strengthen both the differentiation and the integration of the roles of husband and wife within the family (Riley, Johnson, & Boocock (1963).

Though timidly stated and now outdated, these observations foreshadowed some of the subsequent changes both in popular thought and in sociology. There were also intriguing findings on socialization for the future: Parents of our respondents, while favoring a career for their daughters, wanted their sons to marry a wife who would stay home and look after household and children! Provocative as such findings were, however, they were ahead of their time. Talcott Parsons, who often brought into our Rutgers research his eminent common sense on concrete issues, made somewhat similar remarks about the changes in women's occupation and education (Parsons & White 1961, p. 120 n.), though the remarks did little to change his image as a "traditionalist" in matters of gender role. The trends in thought had not yet caught up with social changes.

Today, sociology's major focus on gender is exemplified in many ways: e.g. the launching of a new journal on *Gender and Society,* the 1989 address by Beth Hess as President of the Eastern Sociological Society, or the ASA Presidential volume on *Gender and the Life Course* by Alice Rossi (1985). In one of its aspects, my own current work comports with this focus (e.g. as one contributor to the Rossi volume, author of several pieces on older women, and sponsor of a forthcoming volume on *Gender, Health, and Longevity*—Ory & Warner, 1990).

STUDIES OF SEXUAL ATTITUDES AND BEHAVIORS   One other current experience has disturbing implications for sociological influence. Half a century ago (as indicated above) our studies of contraception were successful in developing nonthreatening research approaches that could encourage women respondents to talk freely and in detail about their sexual behavior. Today at the National Institutes of Health, together with other government agencies and with such sociological colleagues as Wendy Baldwin, Edward Laumann, and James Coleman, we have been making elaborate preparations for a cross-sectional survey of health and sexual behavior, with special reference to prevention of AIDS. However, the topic has proven so politically divisive that Federal officials are refusing to fund "a project that would seek on a national scale to inquire in great detail about the most private aspects of people's sex lives" (*Washington Post,* 7/26/89). Even under threat of mortal epidemic, then, ideologies can be politicized to create major obstacles to the flow of sociological influence (cf. Riley, Ory & Zablotsky 1989).

Sociology of Age

If our early work on gender illustrates failures of influence because of intellectual timing and problems of communication, our work on age has also been somewhat ahead of the major trends in the society, in sociological thought, and in both the prevailing styles and the capacities of research methods. Because much of our recent writing documents the development of a sociology of age (e.g. Riley, Foner, & Waring 1988; Riley Huber, & Hess 1988; Riley 1987), the topic needs scant attention here. I shall merely indicate some of the ways in which age is now gradually attracting systematic sociological attention—both positive and negative.

EXPERIENCES OF A DEVELOPING FIELD   One might have expected a sociology of age to develop early. Aging and cohort flow are powerful and universal processes, and the roles and institutions of every society are structured to accommodate people who differ in age. Indeed, many sociologists in previous cohorts had already dealt with isolated aspects of the topic; but the work was not cumulative. What was missing was a model that could integrate these aspects and could embrace the dynamic interplay between individuals who are growing older and sociocultural structures that are changing.

In the early 1970s, we formulated one such model, referring to it as an "age stratification system" (Riley, Johnson & Foner, 1972). In terms of sociological influence, our strategy for specifying and implementing this model, as I now look back on it, should have been more immediately effective than it was, for we involved an entire working group of influential scholars, including such sociologists as Robert Merton, Talcott Parsons, John Clausen, Harriet Zuckerman, Orville Brim, Gerald Platt, and Harris Schrank.

In fact, however, during the subsequent decade, the model seemed to prove useful to only a few sociologists, while it also elicited considerable criticism and a great deal of confusion. Despite our efforts to explicate and simplify, the complexity of a dynamic multilevel system of this kind eludes easy grasp. The discipline appeared unready conceptually and methodologically to handle such complexity, or to deal with concepts like the "asynchrony" between aging processes and social change, or the "structural lag" that occurs in society because few useful or esteemed roles are available for the growing numbers of long-lived older people. Ironically, our efforts to simplify in themselves brought complaints about our parsimonious use of familiar terms and overly elliptical statements.

In the early 1970s, age had not yet become a central focus of sociological concern. It had not generated the social issues that Charles Willie (in his Epilogue to *Sociological Lives*) describes as "an abiding and stimulating force in the careers of many sociologists." Only recently have several social changes begun to unblock the channels of influence. Among these changes are: a sudden awakening to the fact that we now live in an aging society in which most people survive into old age; a new appreciation of the succession of cohorts, most particularly the "baby boom" cohort; and a recent politicized controversy over "intergenerational equity." This controversy, brought to sociological attention by Samuel Preston (1984, 1988), further documents the importance of current ideological issues in opening (and, in other instances, closing) channels for sociological influence.

AGE-BASED INEQUALITIES   Much of the confusion blocking acceptance of the age stratification model has centered on issues of inequality. The confusion arises from use here of the term "stratification," which is often mis-interpreted as class stratification. This misinterpretation misses the larger points about age, of which inequality is but one. Anne Foner, co-author of much of my work on age, has attempted to set the record straight by repeatedly demonstrating that age is itself an independent basis for inequality, distinct from social class but interrelated with it (e.g. Foner 1974, 1988).

Foner argues that the dynamics underlying *cohort succession* and *aging* produce age differences in economic status and power *within* social classes, thereby tending to undermine class solidarity. In contemporary America, regardless of class, both young and old are on the average less advantaged than are the middle-aged—in part, because the young and old belong to cohorts with very different experiences and different adaptations. Many members of the cohorts now young have suffered from prolonged periods of unemployment—and are thus likely to feel excluded and alienated from family and community. In contrast, members of cohorts now past age 65, though also disadvantaged, have lived through an era marked by improvement in material welfare—which may help explain their relatively high levels of life satisfaction and their failure to act like other disadvantaged groups.

The processes of aging also affect relationships among older, middle-aged, and younger adults within social classes. Those now of working age have supported government programs that provide pensions and other benefits for the old, expecting that they themselves in turn will receive benefits from these programs when they grow old. Such anticipations of one's own aging may help explain why prolonged or serious age-based conflicts over economic issues have not erupted in the past and do not appear likely in the future (here Foner's estimate departs from Preston's).

Social class, so long a central focus of sociology, is only one of several bases for social inequalities and cleavages (cf. Foner 1979). Moreover, as Sorokin (1968) put it, the several lines of social differentiation are continually changing, so that social class in its modern sense "existed before the eigh-teenth century only in rudimentary form, and only since then has it grown into one of the most powerful social groups in the Western world" (14:488).

As so often in the past, Sorokin's words may be prophetic: given the contemporary changes in social structures in this country, age, gender, race, and other bases of inequality may well be undermining the overarching power of social class.

AGE AND VALUES   In addition to issues of inequality, other assumptions about basic values are implicit in the sociology of age. Such assumptions, not immediately attuned to current thinking about stratification, are only occasionally brought into question by sociologists. Yet work in the sociology of age—with its encompassing concern with human lives and sociocultural structures—should ultimately recall attention to underlying values, perhaps thereby influencing sociological thought.

One set of questions concerns the value attached to the prevention of disease, a central goal underlying much work in gerontology. The "medical model," with its overemphasis on disease or physical health is currently supported by legislators, lobbyists, and vested pharmaceutical interests, and threatens to override federal funding for research based on broader models of aging (cf Estes & Binney 1989). Sociologists, many of whom tend to reject biological emphases, would undoubtedly agree on the value of maintaining health and effective functioning far into old age. But what about longevity? Do we prize long life in itself, even without those elements of "life satisfaction" that received so much attention in earlier research on aging?—longevity, even without a sense of self-esteem or personal efficacy?

Quite another question concerns the value currently placed, in modern capitalist societies, on achievement—on work, material gain, success (a question recently explored in depth by Amitai Etzioni in *The Moral Dimension,* 1988). Yet changes in achievement values may be needed if large numbers of older individuals increasingly undertake new assignments—paid or voluntary—at lower than their former levels of pay or prestige. Is it likely that new societal values may be forged by current cohorts of older people, as they spend many years in retirement or, alternatively, undertake new careers with small financial reward?

Even though most might agree in principle that good health, combined with adequate income, ought to be an essential floor for the good life of older people, what account should be taken (reverting to W. I. Thomas) of recognition, esteem, the chance for new adventure, love? Perhaps these questions are moot, but is it not possible that a focus on age could redirect sociological inquiry along such lines?

Dynamic Social Systems

Studies of age, because they involve dynamic social systems, also raise the pertinent conceptual and methodological issues running throughout my sociological life. I was strongly influenced during the 1950s by continuing communication with Talcott Parsons, Freed Bales, Edward Shils, and many of their collaborators and students, whose notion of groups as systems composed of interdependent parts began to pervade my own sociological efforts. A "social system (or multi-level) approach" became the hallmark of much of my work, particularly in efforts to link theory and research method. The wide applicability of this approach is documented by its use as the framework for analyzing the broad range of studies examined two-and-a-half decades ago in *Sociological Research* (Riley et al 1963). That Joan Huber chose the "macro-micro" theme for the 1989 meetings of the ASA speaks to the revitalized concern with aspects of a social system approach.

Three examples of this pervasive approach illustrate how the flow of influence can be blocked by countervailing trends in thought and in methods of research.

USE OF THE APPROACH IN RESEARCH   In our early use of scale analysis to study informal groups of adolescents (Riley, Riley & Toby 1954), empirical regularities were observed in the "collective data" which did not appear in the individual data. Buried in our discursive explorations of the meaning of such a finding is the following simplistic example:

[*Fragment*]   A relevant single case is provided by William Foote Whyte in which Long John's 'friendship with the three top men [Doc, Mike and Danny] gave him a superior standing' in the gang. In Doc's words, 'We give him so much attention that the rest of the fellows have to respect him.' [Or, as restated schematically] The respect of the 'other fellows' (B) for Long John (C) is found to depend upon the fact that the gang leaders, Doc, Mike and Danny (A), like Long John. This is analogous to our statistical data which suggest that whenever one person, B, follows C as a leader, then another person, A, must like C as a friend." (pp. 248–249)

It was in this connection that, as elsewhere in the same book, we applied Durkheim's schema of organic vs mechanical solidarity to analysis of small social systems:

[*Fragment*]   [The problem] requiring clarification deals with the distinction between a social system, on the one hand, in which the acts of individuals are interdependent, and a mere aggregate of individuals, on the other, who are not "interacting" with reference to the dimension under scrutiny. (p. 189)

After numerous empirical experiments with the data, and sophisticated mathematical support from our colleague, Richard Cohn, we formulated two key elements in a social system approach: (*a*) That the group, not the individual, should be used as the research case, but that (*b*) the contributions of the individual members within the group must be identified (pp. 229, 238). This formulation gradually became clearer in our later writings and was adapted to include analysis of process and change. These two elements proved to be central to the approach, yet extremely difficult to implement in large-scale studies where multiple levels—from individual to group to society—are all intertwined. Indeed, how can the researcher examine changes in a sample of groups, and at the same time keep track of the contributions to those changes from particular members of these groups? Although my most complete discussion of these issues is in *Sociological Research* (Riley et al 1963, Unit 12), other writings are cited here as instructive for the topic of sociological influence.

USE OF THE APPROACH IN INTERPRETATION   In 1963, in an interpretation of "Sorokin's Use of Sociological Measurement," Mary Moore and I explored how a social system approach might be used with large systems where, in contrast to small peer groups or street gangs composed of interdependent individuals, whole societies are viewed as composed of classes as the interdependent parts. I quote at length from this essay (which, incidentally, we had the privilege of discussing with Sorokin himself) because it attempts to spell out the details:

[*Fragment*]   The sociologist of today often seeks measures which will fit together two types of indicants: those which refer to collective acts or attitudes and their underlying meaning, and those which refer to the actors as parts of the group. Much present confusion over this problem has arisen because one major line of development in recent years has concentrated on measures of the *individual,* thus isolating for exclusive attention the component of actions and attitudes. And when this line of development is carried over to the group, it often leads to treatment of the *group* as an undifferentiated entity which is formally analogous to the individual. Such a "sociologistic" treatment (to borrow Sorokin's term) might classify orchestral performances, first, according to the pattern of their notes (acts) and, second, according to the proportions of the members (actors) who play strings, woodwinds, brasses, or percussion instruments—all without any fitting together of actors and their respective acts, without taking cognizance of the orchestration involved.

Yet these sociologistic measures may fail to match the concept which the sociologist has in mind. He often wishes to represent the group, not as an undifferentiated entity, but as a system of parts . . . Because of growing awareness of such problems, there is an urgent present search for *social system measures*—as distinguished in this sense from sociologistic measures. One solution which has been recently suggested lies in *identifying the actors* in respect to their contributions to the over-all pattern. Such social system measures would indicate *which* members of the orchestra contribute *which* notes to the symphonic pattern . . .

Our re-analysis of Sorokin's research suggests that some portions of it anticipate exactly this kind of solution to the problem. For example, in his appraisal of the economic well-being of countries, he starts out by rating each society as a whole according to its changing prosperity. Then he goes beyond this . . .: he also shows the prosperity ratings of each of the main *classes* within the society. As he says, "even though the country as a whole is on the upward trend, this does not mean, necessarily, that the economic situation of all its classes follows the same trend." Thus he accounts for the over-all trend in terms of the trends for the various classes (clergy, nobility, etc,). That is, in our formulation, his group measure of prosperity identifies the various parts (classes) within the group with respect to their prosperity, and follows each of the parts through the over-all process. (pp. 217–18).

RESISTANCE TO THE APPROACH   Most instructive for the topic of influence are the extraordinarily extensive and abortive discussions that ensued from what we felt was a relatively innocuous formulation of a social system approach. While sociological theorists of the day were receptive, some leading methodologists, including Samuel Stouffer and Paul Lazarsfeld simply could not accept it. To them, empirical regularities and patterns could be explained only in terms of processing through the minds of *individuals*— although, to be sure, correlated aspects of *groups* could be observed, if not explained. The interdependence between parts and whole eluded them. (There is a parallelism here to Sorokin's "sociologistic" vs "psychologistic" reductionism.) Indeed, Lazarsfeld once confessed privately that he had spent one long night attempting to derive our social system findings for *groups* through random combinations of *individual* data.

Back in 1957, the ideal opportunity arose for countering the resistance to a social system approach. Invited by Lazarsfeld to contribute a chapter to his book on panel analysis, I seized upon a problem he himself had defined, and labored over a long manuscript intended to clarify the approach by playing Hamlet to his Hamlet. His problem as reformulated, as well as my opportunity to confront it, has been described as follows:

[*Fragment*]   . . . Difficulties in relating system levels appeared in the study of friendship process (Lazarsfeld & Merton 1954) in which Lazarsfeld's scheme for panel analysis could not handle Merton's detailed account of the "patterned sequences of interactions." The Lazarsfeld scheme (his well-known 16-fold table), designed to study individuals but here transferred to friendship groups, could show only how many members liked or agreed with others, but not which members. The effect was to reify the group by obscuring any internal "division of labor" whereby individual-level changes in affect or attitude might affect the group-level formation or dissolution of friendships."[6] (Riley 1987, p. 6; see also Riley et al 1963, pp. 562, 728 ff)

That this chapter was never published (but merely delivered at the 1959 ASA meetings) tells us something about how channels of influence can be

---

[6]I remember discussing the problem at length with Bob Merton, whose response was, "I wish you luck! Paul and I communicate every day on many topics, but this is the one topic on which our communication founders." To be sure, Lazarsfeld wrote extensively about group indices, and about group properties as contextual characteristics of individuals. Without belaboring the point (fuller discussion appears in Riley et al 1963, pp. 700–739, and elsewhere), these are *partial* approximations of the system approach, as are sophisticated analyses of the 1950s and 1960s by such sociologists as James Davis, Herbert Menzel, or James Coleman.

blocked. Lazarsfeld once told me that he withheld publication of the entire book because, challenged by our approach, he was trying to subsume it within his own developing methodological framework. Our consequent failure to reach the relevant methodologists is summed up—quite appropriately in a volume in honor of Talcott Parsons—as follows:

> [*Fragment*] Despite explicit warnings over a decade ago against applying the scheme for individual panel analysis directly to social system models, experts on panel analysis today still claim—entirely disregarding the internal shifts often involved in system change—that "the determinants of organizational change could be analyzed by the same methods that are used in analyzing panels of individuals" (Barton, 1968). Once again the refinements developed for research on individuals require careful scrutiny and many modifications if their benefits are to be transmitted to sociological research on social systems. (Riley & Nelson 1971, pp. 42–43)

In sum, numerous obstacles block continuing research on dynamic social systems composed of identifiable and interdependent parts: These complex systems are difficult to comprehend and even more difficult to translate into empirical operations. In a recent review Robin Williams (1989:160), citing the work of Merton & Kitt (1950), bemoans the 40-year failure to develop their recommended indices "*both* of social structure and of the behavior of individuals situated within that structure." Furthermore, as Paul DiMaggio emphasized at the 1989 ASA meetings, "grand theories" must also incorporate the "meso-level" into their macro-micro analyses, if both social relationships and culturally institutionalized patterns are to be taken properly into account.

Mathematical difficulties have constituted a major obstacle to research on such sweeping theories, especially theories that emphasize the systemic interdependence of lower-level parts and the centrality of process and change. Much of the credit for our own—though elementary—early innovations in research design goes not to the conventional statisticians of the day, but to our mathematical colleagues, especially Richard Cohn, who worked almost daily with the Rutgers research group, and to Frederick Mosteller at Harvard, who often advised us (see, e.g., Riley et al 1954, pp. 720 ff; Cohn et al 1960). Among the few sociologists at that time to comprehend the special mathematical demands was James Coleman. At a Johns Hopkins seminar arranged by Sarane Boocock to discuss the Rutgers work, he differed from his colleagues in envisioning models beyond the conventional regression analyses! Recently, however, powerful mathematical and computer-aided advances are beginning to be applied to issues of social structure and change (as by Coleman, Charles Tilly, and Harrison White, among others), and sociologists are now far better equipped for continuing study of the complex interplay between changing societies and the successive cohorts of interacting and interrelated individuals.

PROBLEMS OF COMMUNICATION

No discussion of channels of influence can end without taking note of the sociologists' efficacy (or lack of it) in communicating their own work. As set forth in *Sociological Lives,* the life of Lewis Coser stands as one model of how sociological influence can be disseminated. Describing his "double career" over the past 35 years as both sociologist and journalist, he shows the benefits of having "cultivated a kind of double vision, a dual set of premises of pure sociological analysis and impure social and moral partisanship." While avoiding the potential role conflict between scientist and partisan, Coser has become a sophisticate in the lessons of scholarly publication. In my own experience, unlike Coser's, few of these lessons have been heeded, and an account of their neglect may serve as warnings to sociologists working today.

For one thing, the placing of scholarly articles affects the attention they attract from sociologists. Several of our writings, such as "Woman's Changing Occupational Role" described above, were not published in mainstream journals, nor included in any standard volume of abstracts. They stand in contrast, for example, to our abstracted piece in the *Public Opinion Quarterly* on "Aging and Cohort Succession: Interpretations and Misinterpretation" (Riley 1973), which has frequently been cited. Many of our other publications were prepared as contributions to edited volumes which, though they often provide useful background, tend to attract specialized or narrow audiences. Still other articles, because they appeared in interdisciplinary publications (such as handbooks of gerontology, or publications of the Institute of Medicine or the American Philosophical Society) predictably were overlooked by sociologists. Although transcending disciplinary boundaries should be a high scientific priority today, it often impedes the cumulation of a unified body of knowledge in a single discipline.

My most important contributions to sociology, I have always felt, are in the two-volume textbook on *Sociological Research* (Riley et al 1963). Though I signed it as the major author, it contains the best thinking of a host of advisors, colleagues, and students. With his characteristic acumen, Robert Merton, as editor, scrutinized and made suggestions on every page. The book was long and widely used in teaching here and abroad. Yet, as with most textbooks, its original scientific contributions are not recognized as such. Had they been separately published in refereed sociological journals, they might have been more often cited and more nearly incorporated into the developing body of sociological thought.

To be sure, the influence exerted by a textbook is invisible. Many sociologists, whose early thinking was shaped by materials read during their student days, later join the ranks of "invisible influentials." These "influentials," whether teachers or practitioners, make use of numerous and often hidden channels that proliferate throughout sociology and society, thereby imbuing students and others with sociological knowledge and perspectives.

Moreover, styles of publishing follow trends of their own that can affect sociological influence, as Edward Nelson and I learned to our dismay when we published *Sociological Observation* (1974) as part of a series sponsored by the ASA. We reprinted 28 pieces of research by leading sociologists, analyzed each piece as a case study, and wrote surrounding text to demonstrate the largely unexploited relevance of methods of observation for examining a wide range of sociological theories. The book, as a collateral text, was expected to be largely self-teaching, and several of its contributors feel it would still be useful today. Yet the entire series was quickly discontinued because the publisher found "little market at that time for books of readings." Thus another effort to strengthen theory and method was aborted.

Quite apart from influence, every author (at least since Shakespeare) hopes to leave behind a rounded body of work. Jack and I have on our agenda for retirement a volume that brings together in one place some of the activities of our intertwined sociological lives. So much we have learned from this review!

NOTES FOR THE FUTURE

In these selected reflections, I am again impressed by the power of sociotemporal location as a significant factor in sociological influence. The lives of sociologists and their consequences are linked to history in diverse ways; and the particular channels opened or closed during their careers, their successes and failures in seizing opportunities and handling obstacles, the responses of their successors—all are affected by the historical era in which their lives are embedded. As one recent example, Charles Carmic's 1989 reassessment of Parsons' *Structure of Social Action* suggests how authors in one cohort, by attacking the assumptions of scholars in earlier cohorts, often impede understanding of their own work by members of the oncoming cohorts who begin with entirely different assumptions. Camic shows how Parsons, in defending sociology against the then-established power of neoclassicism, behaviorism, and the biological sciences, has been judged by many successors as thereby *over*emphasizing the normative slant on subjective attitudes, values, and symbols.

The rudimentary clues to the operation of influence gleaned here from my own autobiography are mere afterthoughts specifying themes drawn from the eight autobiographies in *Sociological Lives*. These clues indicate how any serious study of the influence of sociological lives would be required to go far beyond such reminiscences as mine.

Among the critical lacunae here is the question of age, or location within the life course. What are the links between the ages at which particular individuals are doing their work and the historical eras experienced by their cohorts? Others have studied the influence *on* scientists of early-career experiences, or of the changes in power and esteem occasioned by midcareer advances or by long-term retirement (e.g. Zuckerman & Merton 1972; Messeri 1988). But how do such life events affect the influence exerted *by* sociologists? To answer such questions would require intensive scrutiny of many diverse sociological lives drawn from many different cohorts.

Another obvious omission here concerns the meaning of influence, and its possible measures for use in research. A large body of sociological work is available as background here. This work ranges from Sorokin's use of citations to measure influence; to the early studies of influentials by Paul Lazarsfeld, Robert Merton, and Elihu Katz, among others; to more recent developments in network analysis by such scholars as Edward Laumann, James Coleman, and Ronald Burt. Long ago Jack and I described in detail how the flow of communications involves mutual interchanges between the communicator and the recipient of the message, each influenced by a surrounding network of primary groups, and both encompassed within the same wider society and the same secular trends (J. & M. Riley 1959). Within so complex a system, new and special approaches are needed for tracing the flow of sociological influence.

For the immediate future, however, the question is how, in a rapidly changing world, to identify the appropriate channels of opportunity for exercising sociological influence—both on the development of the discipline and on issues of public policy and professional practice. Perhaps the slight answers suggested in these personal reflections may serve as useful guides.

ACKNOWLEDGMENTS

I am indebted to the many colleagues and friends who took part in these recollected experiences, though none of them is responsible for the reconstructions, which are entirely my own. Helpful comments and suggestions were provided on early and late drafts by my life-long colleague and husband, John W. Riley, Jr.; my life-long editor and friend, Robert K. Merton; my long-time associates, Anne Foner, Beth Hess, Joan Waring, and Dale Dannefer; and my new associate, Katrina Johnson. I also thank Robin Williams, who made thoughtful proposals for reworking my 1988 "Notes on the Influence of Sociological Lives;" Richard Scott, who performed yeoman service as editor of the Annual Review of Sociology; Bernard Barber, W. Edwards Deming, Bernice Neugarten, David Riesman, Charles Tilly, and several other thoughtful readers.

Literature Cited

Barton, A. H. 1968. Organizations: methods of research. In *International Encyclopedia of the Social Sciences*, ed. D. L. Sills, vol. 11:341. New York: Macmillan

Carmic, C. 1989. *Structure* after 50 years: the anatomy of a charter. *Am. J. Sociol.* 95:38–107

Cohn, R., Mosteller, F., Pratt, J. W., Tatsuoka, M. 1960. Maximizing the probability that adjacent order statistics of samples from several populations form overlapping intervals. *Ann. Math. Statistics* 31:1095–1104

DiMaggio, P. 1989. *The micro-macro dilemma in organizational research: implications for role-system theory.* Presented at Meet. Am. Sociol. Assn., 84th, San Francisco

Elder, G. H. Jr., Clipp, E. C. 1988. War experiences and social ties: influences across 40 years in men's lives. In *Social Structures and Human Lives*, ed. M. W. Riley, B. J. Huber, B. B. Hess. Newbury Park: Sage

Estes, C. L., Binney, E. A. 1989. The biomedicalization of aging: dangers and dilemmas. *The Gerontologist.* 29:587–96

Foner, A. 1974. Age stratification and age conflict in political life. *Am. Sociol. Rev.* 39:187–96

Foner, A. 1979. Ascribed and achieved bases of stratification. *Annu. Rev. Sociol.* 5:219–42

Foner, A. 1988. Age inequalities: are they epiphenomena of the class system? In *Social Structures and Human Lives*, ed. M. W. Riley, B. J. Huber, B. B. Hess. Newbury Park: Sage

Hess, B. B. 1989. *Beyond dichotomy: making distinctions and recognizing differences.* Pres. Mtgs. Eastern Sociol. Soc. Baltimore, Maryland

Huber, B. J. 1988. Social structures and human lives: variations on a theme. In *Social Structures & Human Lives*, ed. M. W. Riley, B. J. Huber, B. B. Hess. Newbury Park: Sage

Lazarsfeld, P. F., Sewell, W. H., Wilensky, H. L. eds. 1967. *The Uses of Sociology.* New York: Basic

Lazarsfeld, P. F., Merton, R. K. 1954. Friendship as social process: a substantive and methodological analysis. In *Freedom and Control in Modern Societies*, ed. M. Berger, T. Abel, C. H. Page, pp. 21–54. New York: Van Nostrand

Mayer, K. U. 1988. German survivors of World War II: the impact on the life course of the collective experiences of birth cohorts. In *Social Structures and Human Lives*, ed. M. W. Riley, B. J. Huber, B. B. Hess. Newbury Park: Sage

Merton, R. K. 1988. Some thoughts on the concept of sociological autobiography. In *Sociological Lives*, ed. M. W. Riley. Newbury Park: Sage

Merton, R. K., Kitt, A. S. 1950. Contributions to the theory of reference group behavior. In *Continuities in Social Research: Studies in the Scope and Method of The American Soldier*, ed. R. K. Merton, P. F. Lazarsfeld. Glencoe, Ill: Free Press

Messeri, P. 1988. Age, theory choice, and the complexity of social structure. In *Social Structures and Human Lives*, ed. M. W. Riley, B. J. Huber, B. B. Hess. Newbury Park: Sage

Ory, M. G., Warner, H. R., eds. 1990. *Gender, Health, and Longevity.* New York: Springer. In press

Parsons, T., White, W. 1961. The link between character and society. In *Culture and Social Character: the Work of David Riesman Reviewed*, ed. S. M. Lipset, L. Lowenthal. Glencoe Ill: Free

Preston, S. H. 1984. Children and the elderly: divergent paths for America's dependents. *Demography* 21:435–57

Preston, S. P. 1988. Age-structural influences on public transfers to dependents. In *Social Structures and Human Lives*, ed. M. W. Riley, B. J. Huber, B. B. Hess. Newbury Park: Sage

Riley, J. W. Jr. 1947. Opinion research in liberated Normandy. *Am. Sociol. Rev.* 12:698–703

Riley, J. W. Jr., Riley, M. W. 1959. Mass communication and the social system. In *Sociology Today: Problems and Prospects*, ed. R. K. Merton, L. Broom, L. S. Cottrell, Jr. New York: Basic

Riley, J. W. Jr., Riley, M. W. 1940. The use of various methods of contraception. *Am. Sociol. Rev.* 5:890–903

Riley, J. W. Jr., Schramm, W. 1951. *The Reds Take a City.* New Brunswick: Rutgers Univ. Press

Riley, M. W. 1973. Aging and cohort succession: interpretations and misinterpretations. *Public Opinion Q.* 37:35–49

Riley, M. W. 1987. On the significance of age in sociology. *Am. Sociol. Rev.* 52:1–14

Riley, M. W. 1988a. Notes on the influence of sociological lives. In *Sociological Lives*, ed. M. W. Riley. Newbury Park: Sage

Riley, M. W. ed. 1988b. *Sociological Lives.* Newbury Park: Sage

Riley, M. W., Cohn, R., Toby, J., Riley, J. W. Jr. 1954. Interpersonal orientations in small groups: a consideration of the questionnaire approach. *Am. Sociol. Rev.* 19: 715–24

Riley, M. W., Foner, A., Waring, J. 1988. Sociology of age. In *Handbook of Sociology,* ed. N. Smelser. New York: Sage

Riley, M. W., Huber, B. J., Hess, B. B., eds. 1988. *Social Structures and Human Lives.* Newbury Park: Sage

Riley, M. W., Johnson, M., Boocock, S. S. 1963. Women's changing occupational role. *Am. Behav. Scientist* 6:33–7

Riley, M. W., Johnson, M., Foner, A. 1972. *A Sociology of Age Stratification.* New York: Russell Sage

Riley, M. W., Moore, M. E. 1963. Sorokin's use of sociological measurement. In *Pitirim A. Sorokin in Review,* ed. P. J. Allen. Durham: Duke Univ. Press

Riley, M. W., Nelson, E. E. 1971. Research on stability and change in social systems. In *Stability and Social Change: A Volume in Honor of Talcott Parsons,* ed. B. Barber, A. Inkeles, pp. 407–49. Boston: Little, Brown

Riley, M. W., Nelson, E. E. 1974. *Sociological Observation: A Strategy for New Social Knowledge.* New York: Basic

Riley, M. W., Ory, M. G., Zablotsky, D., eds. 1989. *AIDS in an Aging Society: What We Need to Know.* New York: Springer

Riley, M. W., Riley, J. W. Jr., Cohn, R. M., Moore, M. E., Johnson, M. E., Boocock, S. S., Foner, A. 1963. *Sociological Research.* New York: Harcourt Brace & World

Riley, M. W., Riley, J. W. Jr., Toby, T. 1954. *Sociological Studies in Scale Analysis.* New Brunswick: Rutgers Univ. Press

Rossi, A. S., ed. 1985. *Gender and the Life Course.* New York: Aldine

Schramm, W., Riley, J. W. Jr. 1951. Communications in the sovietized state, as demonstrated in Korea. *Am. Sociol. Rev.* 6:757–86

Sewell, W. H. 1989. Some reflections on the golden age of interdisciplinary social psychology. *Annu. Rev. Sociol.* 15:1–16

Sorokin, P. A. 1968. Social differentiation. In *International Encyclopedia of the Social Sciences,* ed. D. L. Sills, vol. 14. New York: Macmillan

White, P., White, M. 1948. *New Product Development.* New York: Funk & Wagnalls

Williams, R. M. Jr., *The American Soldier:* an assessment, several wars later. *Public Opin. Q.* 53:155–74

Zeisel, H. 1985. *Say It With Figures.* New York: Harper & Row. 6th ed.

Zuckerman, H., Merton, R. K. 1972. Age, aging, and age structure in science. In *A Sociology of Age Stratification: Volume III. Of Aging and Society,* M. W. Riley, M. Johnson, A. Foner. New York: Russell Sage

Annu. Rev. Sociol. 1989. 15:1–16

SOME REFLECTIONS ON THE GOLDEN AGE OF INTERDISCIPLINARY SOCIAL PSYCHOLOGY

William H. Sewell

Department of Sociology, University of Wisconsin, Madison, Wisconsin 53706

Abstract

In the 25 years or so that began with World War II, there was a great wave of enthusiasm for interdisciplinary social psychology which resulted in the establishment of interdisciplinary social psychology training and research programs in some of the major universities in the United States. By the mid-1960s however, this seeming Golden Age had largely vanished. This article, by one of the participants in this movement, is devoted to an elaboration of how this Golden Age came about and the forces that led to its demise. Its origins are traced to the World War II experiences of social psychologists in interdisciplinary research on the adjustments of the American soldier under the leadership of Samuel Stoffer and with Rensis Likert on the US strategic bombing surveys in Germany and Japan. Many of the participants in this research were greatly impressed by the fruitfulness of interdisciplinary collaboration and were determined to establish interdisciplinary social psychology programs on their return to their universities. Several of these programs were very successful for a number of years, especially those at Harvard and Michigan, but failed to survive and become integrated into the institutional structure of the American university. The reasons for their failures are complex but at least four factors seem to have been important. First, the threat of these programs to the traditional departmental structure of the university—particularly in light of the relatively weak position of the social sciences in that structure. Second, the lack of adequate and appropriate funding from either university or federal sources. Third, the lack of a major breakthrough in

0360-0572/89/0815-0001$02.00

social psychological theory. Fourth, advancements in research methods did not produce greatly increased understanding of social psychological phenomena. These factors are examined and contrasted with the situation in the natural sciences, particularly with molecular biology.

Introduction

In a perceptive article, "The Three Faces of Social Psychology," James S. House (1977) pointed out that in the 25 years or so that began with World War II, there was a great wave of enthusiasm for interdisciplinary social psychology. This led to the establishment of several significant interdisciplinary social psychology training programs and research centers in some of the major universities in the United States. By the mid-1960s, however, this seeming Golden Age of interdisciplinary social psychology had largely vanished. By the mid-1970s, it had been almost completely replaced by three separate and largely isolated divisions of social psychology: psychological social psychology now focuses on individual psychological processes as related to social stimuli, emphasizing the use of laboratory experimental methods; symbolic interactionism concentrates on face-to-face social interaction processes, using participant observation and informal interviewing in natural settings; and psychological sociology focuses on the reciprocal relation between social structure and individual social psychological behavior, relying mainly on survey methods. House further asserted that this fractionation of social psychology grew out of the institutional and intellectual contexts in which social psychology originally developed, that the three factions have grown further apart over the last two decades, and that there was great need for more interaction between them, if a vital and well-rounded social psychology were to develop.

For the most part I agree with House's formulations and conclusions, although I still prefer to describe what most of us do with the traditional label, social structure and personality, rather than with his term, psychological sociology. Like Sheldon Stryker (1987), I see a somewhat less clear distinction between the present stance of symbolic interactionism and social structure and personality than House did a decade ago. This is particularly true now that many symbolic interactionists are using formal observation, sample surveys, and multivariate analysis in their research.

This brief review of House's article serves as background for my own reflections on what some call the Golden Age of interdisciplinary social psychology. I wish to elaborate on how it came about and to describe the forces that led to its demise. I agree with House that the intellectual and institutional contexts in which each faction developed probably predetermined its return to its original disciplinary moorings once the interdisciplinary arrangements faltered. I wish to reflect on the circumstances that may have contributed to the failure of these programs to become part of the institutional structure of our universities, while several post-war interdisciplinary programs in the natural sciences succeeded.

Background

I am neither a qualified historian of science nor a sociologist of knowledge, but I was one of the many actors in the movement and I participated in almost every aspect of it, its successes and its failures. Thus, I feel emboldened to share my reflections on it. Like most other participants, I had completed my graduate training in sociology before World War II, with a major interest but inadequate training in social psychology. There were few places where one could obtain much in the way of training in social psychology in the mid-1930s, and Minnesota, where I did my PhD in sociology, was not one of them.[1] By the time I was called to service in World War II as a reserve officer in the US Navy, I was already a fairly well-established sociologist. I had read widely in social psychology and had done research and teaching in the field. On entering active military service, I was assigned to the staff of the Research Division of the National Headquarters of Selective Service, where along with other social scientists I did research on civilian and military manpower.[2] During this period, through contacts with Samuel A. Stouffer and members of his staff, I became well acquainted with the interdisciplinary research program of the Information and Education Division of the War Department, most of which involved studies of the adjustment of soldiers to military life during World War II. Much of this research was published, under Stouffer's leadership, in the famous four-volume work *The American Soldier* (1949). I also became acquainted with Rensis Likert and several of his colleagues who directed the Program Surveys Division of the Department of Agriculture which had been doing social psychological studies of the civilian population for several departments of the Government during the war years.

After the surrender of Germany, Likert asked me to join a group that was

[1] At that time the leading centers for social psychology training were Chicago, Columbia, and Harvard, but even at these institutions the offerings were not extensive. At Minnesota I had reading courses with Clifford Kirkpatrick and sat in on a course in social psychology in the psychology department. This course was devoted largely to group differences in ability and attitudes and gave some attention to collective behavior. Fortunately, as an undergraduate, I had had courses in sociology and philosophy at Michigan State in which I had read much of Dewey and Cooley and some of Mead—at that time *Mind, Self and Society* (1934) had not yet been published.

[2] Others involved in this research were Kenneth McGill, Raymond V. Bowers, C. Arnold Anderson, Harold Faulk, Robert N. Ford, J. Mapheus Smith, and Louis Levine.

making preliminary plans for a study of the influence of strategic bombing on Japanese civilian morale. (Likert had directed a similar study in Germany; see US Strategic Bombing Survey, 1946). This group included some members of the team that had conducted the German survey and several of the social scientists who were to conduct the survey in Japan. We drew up preliminary plans for the Japanese survey, including a clear conceptualization of the aims of the survey, a list of the major components of morale, and a series of questions designed to elicit these.

Within days after the surrender, the interdisciplinary team of psychologists, sociologists, anthropologists, political scientists, a psychiatrist, and sampling statisticians, who were to carry out the study had assembled in Tokyo.[3] We immediately began to review the purposes and design of the survey and made many important revisions in both the conceptual guides for the study and the content of the survey instrument. We then pretested the interview schedule on Japanese civilians, using Japanese-American interviewers. These interviewers had participated in many of our meetings and were well acquainted with the purposes of the research. The survey directors and the interviewers then participated in the final revision of the interview schedule.

Meanwhile, our sampling experts had designed and drawn a probability sample of the Japanese adult civilian population consisting of approximately 3000 persons. We then took our teams of interviewers into the field and completed the interviewing in a period of three more months. In another month or so, after returning to Washington DC, we had developed a coding scheme for the interviews, coded the materials, and completed the statistical processing of the data. By then several of our number had been released from service and returned to their academic posts. Those of us who remained, with assistance from some of our departed colleagues, wrote the final report which was then sent to the printers. All of this was accomplished within less than a year after our arrival in Japan. The report was published by the Government Printing Office in 1947 (US Strategic Bombing Survey, 1947).

Throughout this endeavor I was very much impressed with the fruitfulness of interdisciplinary collaboration among bright and willing social scientists. In general the most innovative and insightful ideas were generated as a result of group discussions in which little regard was paid to the disciplinary origin of the idea. I was greatly impressed also with the ability of an interdisciplinary team to mount a study of this complexity and to move it to completion so

---

[3]The group included David Aberle, Conrad Arensberg, Jules Henry, and Fredrick Hulse (anthropologists); Donald Adams, Edgerton Ballachey, and Horace English (psychologists); Raymond Bowers, Burton Fisher, and William Sewell (sociologists); Morris Hansen and Harold Nisselson (statisticians); David Truman and Harold Nissen (political scientists), and Alexander Leighton (psychiatrist).

expeditiously.[4] My colleagues on the Bombing Survey, as well as those who had participated in other wartime interdisciplinary social psychology research projects, were equally impressed with their experiences and were determined to promote interdisciplinary training and research programs in social psychology on return to academic life.

Moreover, the private foundations, especially Ford, Rockefeller, Carnegie, and Sage, along with the Social Science Research Council, were stressing interdisciplinary social psychology. The Office of Naval Research, the National Institutes of Health, and later the National Science Foundation also were supportive of interdisciplinary research and training programs in social psychology. I served on and was chairman of several research grant and training committees during the period of expansion of interdisciplinary social psychology. Through this activity I came to know most of the leaders in this movement, and throughout the period I was involved with them in the promotion of interdisciplinary social psychology on the national level, as well as with others at the University of Wisconsin.[5]

As a result of all of this enthusiasm, activity, and support, interdisciplinary programs for graduate training were developed at Michigan, Harvard, Yale, Cornell, Berkeley, Columbia, Minnesota, Wisconsin, and other leading universities. In addition the National Opinion Research Center was moved to the University of Chicago, with a broadly expanded program under the direction of Clyde Hart; a new national research center, the Institute for Social Research, was established at the University of Michigan under the direction of Rensis Likert; and the Bureau of Applied Social Research was established at Columbia under the leadership of Paul F. Lazarsfeld. More locally oriented survey research centers were developed at Harvard, Yale, Princeton, Berkeley, UCLA, Illinois, Minnesota, and Wisconsin, to mention only the

---

[4]I wish it were possible to produce actual illustrative examples of how the work of the team was influenced by its interdisciplinary composition. I am sure that it was, but my memory of the instances is no longer reliable. I took no notes on our meetings and conferences, nor do I remember that anyone else did except Edgerton Ballachey. His notes on the Tokyo meetings, and those of David Krech on the earlier meetings in Washington, formed the basis for their syllabus, *A Case Study of a Social Survey,* (1948), but they do not discuss the disciplinary sources of ideas, the definitions of concepts, the hypotheses tested, or the analytic strategies of the study. Again, what I remember most clearly is that some of the best suggestions came from persons who were not directly identified with the discipline with which one would normally associate the idea. This is not too surprising because all of us had a strong commitment to social psychology regardless of our disciplinary identifications.

[5]Among the national leaders not otherwise mentioned in this paper were: Gardner Murphy, Richard Crutchfield, Clyde Coombs, Ernest Hilgard, Charles Osgood, Harold Kelley, Goodwin Watson, Muzafer Sherif, Urie Bronfenbrenner, M. Brewster Smith, David Riesman, Leonard Cottrell, Robin Williams, Arnold Rose, Fred Strodtbeck, Angus Campbell, Herbert Simon, Ralph Linton, Fredrick Redlich, August Hollingshead, Warren Dunham, and Robert Faris.

more prominent ones. The primary commitment of these centers was to interdisciplinary research on social psychological topics using sample survey research methods. The Armed Services established similar research centers to investigate problems related to their military mission.[6] During this period too the interdisciplinary research program in social psychology was developed in the Laboratory of Socioenvironmental Studies of the National Institute of Health under the direction of John Clausen and, later, Melvin Kohn.

There certainly was no lack of interest in the interdisciplinary graduate training programs on the part of social psychologists and graduate students. Social psychology was a challenging intellectual field and students were anxious to learn more about it, including what other disciplines than their own had to contribute to its theory and methods. Moreover, there was a backlog of mature graduate students whose education had been interrupted by military service and who could qualify for financial support for their graduate training under the GI Bill. Still others could be supported by training grants from the National Institute of Mental Health, National Institute of General Medicine, the National Science Foundation, and other agencies interested in increasing the supply of persons trained in social psychology. Thus, it seemed that all conditions were right for sustained growth of a new interdisciplinary field of social psychology.

For a decade or so great progress was made, particularly in the interdisciplinary training program at Michigan, under the leadership of Theodore Newcomb. The faculty included Angus Campbell, Dorwin Cartright, J. R. P. French, William Gamson, Daniel Katz, Robert Kahn, Herbert Kellman, Helen Peak, Albert Reiss, Guy E. Swanson, and Howard Schuman. At Harvard, the new Department of Social Relations was headed by Talcott Parsons, with a faculty that included Gordon Allport, R. Freed Bales, George Homans, Alex Inkeles, Clyde and Florence Kluckhohn, Gardner Lindzey, Frederick Mosteller, Richard Solomon, and Samuel Stouffer.[7] The growth was less spectacular in other universities but was by no means insignificant. With such a good start why did the interdisciplinary programs in social psychology all but vanish by the late 1960s, without ever becoming established in the institutional structure of American universities?

---

[6]For information on the work of these and related research centers see Bowers (1967).

[7]To my knowledge only the Harvard program was set up as a separate department. The usual pattern was the one followed at Michigan where the social psychology program resulted from the joint sponsorship at the departments of sociology and psychology; it never had the status of a separate department. I suspect the separate department at Harvard resulted not only from Talcott Parson's interest in social relations but also from his desire to escape Pitirim Sorokin's administrative control.

Why Did Interdisciplinary Programs In Social Psychology Fail?

The reasons for the failure of interdisciplinary programs in social psychology are complex and not entirely apparent. I believe one reason is the traditional institutional structure of American universities and the place of the social sciences in that structure. Another factor closely related to this is the system for funding science that has developed and become institutionalized in the United States and the unfavorable position of the social sciences in this system.[8] Other reasons may be found within social psychology itself, particularly in the condition of social psychological theory and methods.

The Threat to the Departmental Structure

I turn first to the traditional institutional structure of the American university and the relatively weak position of the social sciences there. The physical and biological sciences, in both their pure and applied branches, are in a position superior to the social sciences and humanities in most of our universities. This is true with respect to the funds allocated to research, to buildings, to equipment, and to salaries, but particularly to new ventures such as research centers and training programs. Thus, the existing social science departments have to defend their turf in the face of new interdisciplinary programs that might threaten their claim on the universities resources. This is much less true in the natural sciences where more ample funds are available from both local and national sources. Thus, social science departments tend to be much less supportive of interdisciplinary programs, unless additional funds for them can be brought in from the outside. This is particularly so when the program is likely to require faculty, scholarships, space, equipment, and operating funds, always in short supply, that may draw faculty and students away from the parent departments. To be sure several universities were willing to give limited support to interdisciplinary training programs in social psychology, but for the most part the faculty were part-time in the program and were budgeted to their original departments. Funds for subsidizing graduate students, faculty research, secretarial and clerical staff were expected to come from outside grants. Federal funds to meet these costs, although available, were usually inadequate and were granted for a relatively short period—usually three to five years—with no assurance they would be renewed. Given these conditions it is not surprising that the parent departments found interdisciplinary social psychology programs threatening.

Wisconsin provides a classic example of the point I have been making. Early on, I obtained funds from the Social Science Research Council to set up a faculty seminar made up of psychologists, anthropologists, and sociologists to draw up plans for a graduate interdisciplinary training program in social

[8]For further discussion of this topic see Sewell (1988).

psychology. We met for several months and developed a program consisting mainly of courses in social psychology already being taught in the departments, plus two new seminars: one on current social psychological theory and the other on current research methods in social psychology. The dean of our college supported the plan, subject to the approval of the departments involved, with the understanding that the departments would provide the faculty from their current budgets. It was further assumed that the group would seek outside funds for the subsidy of graduate students and for other requisite costs. When the plan was presented to the departments, neither sociology nor psychology would approve of our request for a joint major in social psychology. The best either department would settle for was an interdepartmental minor, with psychology requiring that students who were not psychology majors take the psychology proseminar, and sociology stipulating similar requirements for psychology students. The Graduate School approved of these arrangements, but the graduate students did not find these requirements attractive and within a few years the program ceased to exist. This did not mean that thereafter no interdisciplinary training in social psychology could be found at Wisconsin; the sociological social psychologists encouraged their students to take courses from psychological social psychologists, and vice versa, but nothing like a true interdisciplinary program emerged. I do not claim that the Wisconsin experience was typical of other programs, but I do know that several others suffered from less than adequate support by their departments and deans. In some institutions interdisciplinary programs prospered only so long as their enthusiastic and powerful founders participated in the program and strongly supported it in their departments. When they were replaced, their successors tended to lack the enthusiasm and often the power and organizational skills of the founders. Conflicts arose; the departments and college administrations withdrew their support, and the programs were soon abandoned.

Lack of Adequate Funding

Another factor in the decline of interdisciplinary social psychology programs was that they never received adequate funding from federal sources. This may seem paradoxical because it was during the period of their ascendency that funding for social science research and training became institutionalized as a part of the program for the support of science in the United States, particularly in the National Institutes of Health and later in the National Science Foundation.[9] Social psychology was especially favored in the research grant and training programs of the National Institute of Mental Health. However,

---

[9]For a brief discussion of the early development and importance of these sources of support, see Sewell (1988).

the funds going to the support of social sciences in these agencies never were more than 10% of their research budgets, and social psychology got far less than other branches of psychology and less than some other subfields of sociology. During this period funds became available for research and training in medical sociology, social problems, urban problems, juvenile delinquency, substance abuse, and aging. All of these involved social psychological research but still were competitive with programs for interdisciplinary social psychology research and training.

In any event, the major source of funds available for social psychology was NIMH. These funds had to be justified on the basis of mental health relevance, were modest in amount, limited in duration, provided for only a small number of research assistants or trainees, did not provide facilities, and usually supported only a limited portion of the salary of the principal investigator or director of the program. The consequence was that these programs were inadequately supported by either the universities or the federal agencies. This is in rather sharp contrast to the ever-increasing funds available to interdisciplinary research and training programs in the natural sciences during this period.

Unfortunately, the national survey research centers at Michigan and Chicago were somewhat underused as a source of interdisciplinary research training in social psychology. These not-for-profit organizations were only loosely connected with the universities, had their own staff, received limited financial assistance from the universities, and had to raise their own funds by doing contract work for government and private business. I do not mean to imply that they provided no support for social psychology programs. A limited number of their members participated in training programs, and both organizations provided part-time employment for a number of graduate students. They also made their research facilities available to faculty members who wished to subcontract with them for data gathering, data processing, and related services. Generally this was possible only when faculty members had outside grants for these purposes. Much the same situation held at other university sponsored survey research centers with the exception of Columbia's Bureau of Applied Social Research, where graduate students received coordinated training in Lazarsfeld's research methods seminar along with first-hand experience on the Bureau's ongoing social research projects.

Modest Advances in Theory

In speculating about the fate of interdisciplinary social psychology, I must point out that there have been no powerful theoretical breakthroughs that might have served as a stimulus to exciting new theoretical developments or new research areas during this period, or for that matter since. Advances in social psychological theory did occur, but they were modest. Although some codification took place, nothing approaching a unified body of social psychological theory emerged. Rather, there were improvements in somewhat iso-lated bodies of special social psychological theories, such as role theory, field theory, attitude theory, socialization theory, theory of interpersonal relations, communication theory, theory of collective behavior, and theory of small group processes. Perhaps the most exciting advances were made in our knowledge of small group processes, with fruitful work being done on group structure, cohesiveness, communication flow, leadership, productivity, de-viance, and the construction of social reality. Unfortunately most of the bodies of special theories mentioned above (to paraphrase Robert K. Merton; 1949:85–87) consisted of general orientations toward problems and types of variables to be taken into account, rather than verifiable statements of rela-tionships between sets of specified variables. There was little or no consolida-tion of these special theories into a general conceptual scheme for social psychology. The fate of social psychology in this regard was no different from that of the social sciences generally. In fact it could be argued that none of the social sciences made spectacular progress in developing general theory during these years.

What did take place was a great burst of research activity on a large number of social psychological topics, often with a view toward shedding light on problems of social behavior, rather than toward theory construction or testing. This is clearly reflected in the articles selected for the influential book, *Readings in Social Psychology,* edited by Theodore Newcomb and Eugene Hartley in 1947 and revised in 1952 and 1957. This book was sponsored by the Society for the Psychological Study of Social Issues. Its editors did not attempt to provide an overall framework for social psychology; instead they stressed three requirements for research in social psychology: It must adhere to rigorous canons of scientific procedure; it must draw hypotheses from the relevant psychological and social sciences; and it must bring these hypotheses to bear on systematic research on problems of human importance (Likert 1947 V). This final requirement characterized much of the research done during this period. It must be remembered that a great volume of research was done, enough to fill professional journals in the field and, for that matter, hundreds of pages in the more general journals of the various social sciences. Most of the pages in the five volumes of *The Handbook of Social Psychology* (1969), edited by Gardner Lindzey and Elliot Aronson, are devoted to summarizing this research record. But only a small fraction of this research resulted from interdisciplinary efforts.

Unfortunately, little of this outpouring of research resulted in powerful ideas that could stimulate further development of theory or research in social psychology. Rather, with few exceptions, the explanatory power of the theories and models in social psychology remained quite modest—often providing only small, though statistically significant, results. This is not the stuff that makes for a stimulating new interdisciplinary field; and it certainly does not command the long-term, high level of financial support that interdisciplinary programs need to be successful.

Advances in Research Methods

During this period of great research activity much effort was devoted to the improvement of research methods, particularly in sampling, interviewing, questionnaire construction, index and scale development, observational techniques, and statistical methods for the analysis of survey data. Because of the rapid adoption and use of sample survey methods during and after the war, the government and the public were raising questions about the adequacy and dependability of existing methods of sampling, interviewing, and data analysis. Consequently, the National Research Council and the Social Science Research Council jointly sponsored a committee, under the chairmanship of Samuel A. Stouffer, to investigate these questions. The committee in turn commissioned a study on interviewing under the direction of Herbert H. Hyman and one on sampling under the direction of Fredrick F. Stephan and Phillip McCarthy. The results were Hyman's *Interviewing in Social Research* (1954) and Stephan and McCarthy's *Sampling Opinions* (1958). Hyman's book not only brought together what was then known about interviewing, it also reported on a series of careful experimental and observational studies of sources of error in interviews, and their control. This book had a great influence on the work of survey agencies and on the teaching of survey research methods. Hyman also produced another influential book, *Survey Design and Analysis* (1955), that presented a series of detailed case studies of problems encountered in social research. This book grew out of Paul F. Lazarsfeld's well-known project at Columbia, designed to produce materials suitable for advanced training in social research. An equally important book from the Columbia project was Paul F. Lazarsfeld and Morris Rosenberg, *The Language of Social Research* (1955), which emphasized the use of partialling to control for the influence of intervening variables in studying causal relationships, and of contextual analysis to separate individual and group effects (see also Kendall & Lazarsfeld 1950 and Lazarsfeld & Menzel 1961).

The Stephan & McCarthy book (1958) describes the relationship between sampling and other components of survey design, the problems raised when methods do not conform to the underlying mathematical theory, and finally, the problems encountered in actually designing a sample survey and putting it into operation. This book was by no means a primer; it was quite influential in survey research operations and was widely used in survey research methods courses. Other important books on sampling during the period were William Edwards Deming, *Some Theory of Sampling,* (1950) and Morris H. Hansen, William Hurwitz, and William Madow, *Sample Survey Methods and Theory,* (1958).

Mention should also be made of the book by Marie Jahoda, Morton

Deutsch, and Stewart Cook, *Research Methods in Social Relations* (1951), which covered research design, observational techniques, survey methods, content analysis, measurement, and data analysis. This book, sponsored by the Society for the Psychological Issues, was widely used in introductory courses in research methods in social psychology. A more advanced text covering much the same subject matter was *Research Methods in Social Sciences* (1953), edited by two prominent social psychologists, Leon Festinger and Daniel Katz. A book used by sociologists teaching research methods in social psychology was *Sociological Studies in Scale Analysis* (1954) by Riley et al. Another important book is R. Freed Bales's *Interaction Process Analysis* (1950) which provided sociological social psychologists with a system that enabled them to observe and rate the behavior of members of small groups. This system was widely adopted by younger sociologists and produced a generation of social psychologists who continue to work on important problems of individual and group behavior.

Considerable progress was also made in the measurement of social psychological variables during this period. Of course, even before the period began, L. L. Thurstone (1928), Rensis Likert (1932), and others had developed useful techniques for scaling attitudes, opinions, and similar social psychological constructs. But early in this period (during World War II) Louis Guttman (1944, 1950) developed Scalegram Analysis for determining rank order. This technique came to be known as Guttman Scaling. Scalegram Analysis, which is easy to accomplish and produces readily understandable results, was widely adopted. It doubtless was a great stimulus to research on attitudes and to studies of attitude change, which was a popular topic during this time of great concern with intergroup relations. Guttman Scaling soon replaced the earlier techniques for scale and test construction and was used to measure a wide range of social science variables. Mention must also be made of Paul F. Lazarsfeld's development of Latent Structure Method (1950) by which the manifest relations between any two items in a questionnaire can be accounted for by a simple set of latent classes, and only by this set. This was an important contribution to scaling theory although it did not gain widespread use by social psychologists.[10]

---

[10]During the later part of this period social psychologists began to develop scales and indexes by the use of factor analytic methods. Computer programs were developed to factor analyze and assign factor weights to a large number of items with great speed; this made scaling a quick and cheap process. Scales were produced to measure almost any social psychological variable that anyone could wish for. Unfortunately, according to Otis Dudley Duncan (1984a: 119–155) who provides a number of examples, most of these scales do not meet even minimum measurement requirements and hence they produce misleading results. The matter is made still worse when the numbers produced by these scales are subjected to complex statistical analysis. Duncan concludes that until we are ready to do the hard thinking and careful analysis that the methods developed by Georg Rasch (1968; Perline, et al 1979; Duncan 1984b) require, there is little hope for adequate measurement scales in social psychology or other social sciences for that matter.

Probably the greatest area of advance in this period was in the use of computer technology and methods. We started the period using counting sorters for most of our research. I remember the hours that my wife and I spent in 1938 feeding IBM cards into a counting sorter to get the numbers I used in the analysis of the 123 items from which I selected the 36 most diagnostic ones that finally comprised the farm family socioeconomic status scale (Sewell 1940). By the early 1950s, however, we had computers that, although crude by present day standards, had sufficient speed and storage capacity to enable us to do some quite complicated multivariate statistical analysis. For example, my colleagues and I were able to do a factor analysis of a set of 38 child-training practices to test hypotheses concerning the psychoanalytic claim that the mother's child-training practices reflect her unconscious acceptance or rejection of her child (Sewell et al 1955). Incidentally, the results failed to confirm the hypothesis. The computer proved to be very useful to social psychologists in multivariate cross-tabular analysis based on large samples, such as those my colleagues and I used to parse out the influence of social background variables on educational and occupational aspirations (Sewell et al 1957). Improvements in computer technology also made it possible for social psychologists to begin large-scale longitudinal and panel surveys. One was the Wisconsin study of social and psychological factors in the educational and occupational aspirations and attainments of over 10,000 students who graduated from high school in 1957. (See Sewell & Hauser 1975 for a summary of the early work on this project). Finally, by the time this period came to a close many new computer programs enabled social psychologists to use quite advanced mathematical statistical models in their research. Of course, social psychologists were not the major contributors to these mathematical and statistical techniques or to computer technology but, like other scholars, they were quick to adopt such techniques once they became available.

In closing I must say that although there were important improvements in the research methods used during this period, their main effect was to increase the reliability of our observations rather than to extend our powers of observation. New computers and computer programs did help us to sort out some of the complexities of social psychological behavior that would been almost impossible with earlier techniques. However, none of this was sufficient to bring about major theoretical breakthroughs to fuel great advances in social psychology.

Summary and Conclusions

Unfortunately, the rather modest developments that took place in social psychological theory and methods during its Golden Age were not sufficient to serve as the basis for a new interdisciplinary field. This was particularly true because of social psychology's weak position in the university structure and the inadequate funding available from university and federal sources. Contrast this with the success of the interdisciplinary programs in the natural sciences—particularly with those in molecular biology, where tremendous theoretical breakthroughs, stemming from the work of Watson, Crick, and Wilkins on the structure of DNA, provided the stimulus for a whole new approach to biological studies.[11] This, plus the perfection of powerful new instruments for observation and measurement (such as the election microscope and several complex devices and techniques for studying large and small molecules) spawned complex research problems that could only be solved by bringing together the skills and knowledge of physicists, chemists, geneticists, bacteriologists, zoologists, and botanists. Usually it was the younger scientists in these fields who were willing to engage in this joint effort and to learn the new techniques necessary for success in solving new problems. In the early years of the new programs, most of the scientists involved maintained their departmental connections but did their research in molecular biology teams. The level of cooperation of the parent departments with the interdisciplinary programs was not uniformly high in these years, but there was no great departmental resistance because adequate funding was available to permit other scholars in the departments to continue their established research programs. At the same time there was plenty of new money for the support of molecular biology. In fact NIH and NSF were so anxious to promote interdisciplinary programs in molecular biology that they were willing to provide funds for new buildings, laboratories, and equipment as well as salary support for faculty members and ample stipends for pre- and postdoctoral trainees. Over the years support has continued at high levels for both training and research in molecular biology. In several instances molecular biology has been granted full departmental status and in all instances has had the power to set its own graduate requirements and to grant its own PhD degrees.

One cannot help but wonder what would have happened if generous support from the universities and the federal funding agencies had been available to

[11]My comments on molecular biology are based on an interview with Robert M. Bock, professor of biochemistry and molecular biology and dean of the graduate school, University of Wisconsin-Madison. The Wisconsin program in molecular biology (now called Cellular and Molecular Biology) began in 1952 with five professors from biochemistry, genetics, and physics, with support from the Graduate School, NIH, and NSF. Funds for a new ten-story building with completely equipped laboratories and offices were provided from federal sources. The program now includes 82 professors from 17 departments. All of its 100 or more graduate students are guaranteed three years support from University fellowships, or NIH traineeships or Research Assistantships. Teaching costs are now borne by the University. Research support comes mainly from federal sources. The program has never sought departmental status and maintains great flexibility in its research and teaching functions.

interdisciplinary programs in social psychology. Probably, no great new theoretical breakthroughs would have occurred or powerful new instruments or techniques of research would have been developed. The nature of social psychological phenomena makes such developments very difficult. But adequate and appropriate funding would have increased the probability of important developments in these areas and would most certainly have made greater progress in the improvement of social psychological theory and methods. I am confident that we would have been turning out more good research and more well-trained social psychologists had sociologists been collaborating with psychologists and other social scientists in interdisciplinary social psychology research and training programs. We might have been making much greater contributions to the understanding of some of the common social problems of our time.[12] This in turn probably would have led to greater support for research and training in social psychology.

ACKNOWLEDGMENTS

Remarks on the occasion of the presentation of the Cooley-Mead Award, Section on Social Psychology, American Sociological Association, August 24, 1988. The writer wishes to acknowledge the helpful comments of Archie O. Haller, Robert M. Hauser, David R. Heise, Robert Kahn, H. Andrew Michener, W. Richard Scott, and William H. Sewell, Jr.

Literature Cited

Bales, R. F. 1950. *Interaction Process Analysis: A Method for the Study of Small Groups*. Reading, Mass: Addision-Wesley

Bowers, R. V. 1967. The military establishment, In *The Uses of Sociology*, ed. P. F. Lazarsfeld, W. H. Sewell, H. L. Wilensky, pp. 234–74. New York: Basic Books

Deming, W. E. 1950. *Some Theory of Sampling*. New York: Wiley

Duncan, O. D. 1984a. *Notes on Social Measurement: Historical and Critical*. New York: Basic Books

Duncan, O. D. 1984b. Rasch measurement in survey research: Further examples and discussion. In *Surveying Subjective Phenomena*, ed. C. F. Turner, E. Martin, New York: Russell Sage Found.

Festinger, L., Katz, D., eds. 1953. *Research Methods in Social Science*. New York: Dryden

Guttman, L. 1944. A Basis for Scaling Qualitative Data. *Am. Sociol. Rev.* 9:139–50

Guttman, L. 1950. The basis for scalegram analysis. In *Measurement and Prediction: Studies in Social Psychology in World War II*, S. A. Stouffer, et al, 4:51–65. Princeton, NJ: Princeton Univ. Press

Hansen, M. H., Hurwitz, W. N., Madow, W. G. 1953. *Sample Survey Methods and Theory*. New York: Wiley

House, J. S. 1977. The three faces of social psychology. *Sociometry* 40:161–77

Hyman, H. H. 1954. *Interviewing in Social Research*. Chicago: Univ. Chicago Press

Hyman, H. H. 1955. *Survey Design and Analysis Principles, Cases & Procedures*. New York: Free Press

Jahoda, M., Deutsch, M., Cook, S. W. 1981. *Research Methods in Social Relations, Parts 1 and 2*. New York: Dryden

---

[12]For a helpful discussion of some of the important social problems that might be better studied by interdisciplinary effort, see House 1977:172–74

Kendall, P. L., Lazarsfeld, P. F. 1950. Problems of survey analysis. In *Studies in Scope and Methods of the American Soldier.* ed. R. K. Merton, P. F. Lazarsfeld, pp. 133–96. Glencoe, Ill: Free Press

Krech, D., Ballachey, E. 1948. *Study of a Social Survey: U.S. Strategic Bombing Survey (Japan).* Silabus Series T. G. Berkeley: Univ. Calif. Press

Lazarsfeld, P. F. 1950. The logical and mathematical foundations of latent structure analysis. In *Measurement and Prediction: Studies in Social Psychology in World War II,* ed. S. A. Stouffer et al, 4:362–412. Princeton: Princeton Univ. Press

Lazarsfeld, P. F., H. Menzel 1964. On the relation between individual and collective properties." In *Complex Organizations: A Sociological Reader,* ed. A. Etzioni. pp. 426–35. New York: Holt, Rinehart & Winston

Lazarsfeld, P. F., Rosenberg, M. eds. 1955. *The Language of Social Research.* Glencoe, Ill: Free Press

Likert, R. 1947. Foreword v. In *Readings in Social Psychology,* ed. T. M. Newcomb and E. Hartley. New York: Holt

Likert, R. 1932. A technique for the measurement of attitudes. *Arch. Psychol.* No. 140

Lindzey, G., Aronson, E. eds. 1968–1969. *The Handbook of Social Psychology.* 5 volumes. Reading, Mass: Addison-Wesley. 2nd ed.

Merton, R. K. 1949. *Social Theory and Social Structure: Toward the Codification of Theory and Research.* Glencoe, Ill: Free Press

Morris, C. W. ed. 1934. *Mind, Self and Society: From the Standpoint of a Social Behaviorist.* Chicago: Univ. Chicago Press

Newcomb, T. M., Hartley, E. L. eds. 1947. *Readings in Social Psychology.* New York: Holt

Perline, R., Wright, B. D., Weiner, H. 1979. The Rasch model as additive conjoint measurement. *Appl. Psychol. Measur.* 4: 1–7

Rasch, G. 1968. An individualistic approach to item analysis. In *Readings in Mathematical Social Science,* ed. P. F. Lazarsfeld, N. W. Henry, pp. 89–108. Cambridge, Mass: MIT Press

Riley, M. W., Riley, J. W. Jr., Toby, J. eds. 1954. *Sociological Studies in Scale Analysis.* New Brunswick, NJ: Rutgers Univ. Press

Sewell, W. H. 1940. *The Construction and Standardization of a Scale for the Measurement of the Socio-Economic Status of Oklahoma Farm Families.* Stillwater, Ok: Agric. Exp. Stat.

Sewell, W. H. 1988. *The changing institutional structure of sociology and my career.* In *Sociological Lives: Social Change and the Life Course,* ed. Matilda W. Riley, 2:119–43. Newbury Park, Calif: Sage Publ.

Sewell, W. H., Haller, A. O., Straus, M. A. 1957. Social status and educational and occupational aspirations. *Am. Sociol. Rev.* 22:67–75

Sewell, W. H., Hauser, R. M. 1975. *Education Occupation and Earnings: Achievement in the Early Career.* New York: Academic Press

Sewell, W. H., Mussen, P. H., Harris, C. W. 1955. Relationships among child training practices, *Am. Sociol. Rev.* 20:137–48

Stephan, F. F., McCarthy, P. J. 1958. *Sampling Opinions: An Analysis of Survey Procedure.* New York: Wiley

Stouffer, S. A. et al. 1949. *The American Soldier: Studies in Social Psychology in World War II* 4 vols. Princeton, NJ: Princeton Univ. Press

Stryker, S. 1987. The vitalization of symbolic interactionism. *Soc. Psychol. Quarterly* 50:83–94

Thurstone, L. L. 1928. Attitudes can be measured. *Am. J. Sociol.,* 33:529–54

US Strategic Bombing Survey. 1946. The Effects of Strategic Bombing on German Morale. Washington, DC: USGPO

US Strategic Bombing Survey. 1947. The Effects of Strategic Bombing on Japanese Morale. Washington, DC: USGPO

David Riesman

Ann. Rev. Sociol. 1988. 14:1–24

ON DISCOVERING AND TEACHING SOCIOLOGY: A MEMOIR

David Riesman

Department of Sociology, William James Hall, Harvard University, Cambridge, MA 02138

Abstract

The focus of the article is on the author's experience as learner and teacher. The task of preparing law students for their professional life is compared to that of teaching undergraduates who, in the uniquely strenuous College of the University of Chicago, are interested in books and ideas and are not immediately job-driven. Whereas in the law school setting the author, like other law professors, worked alone, in the College he worked with a staff of colleagues who shared in the choice of readings and in lectures to the entire student body—altogether, in the author's view, a more demanding task for the teacher. Graduate education in sociology is again different: the student is an apprentice professional, but there is no clear road toward professional practice; the faculty member as mentor on the dissertation helps pilot the apprentice through the cross-currents of conflict in the field and in the particular department.

Teaching undergraduates in Harvard College, the author reports, involved tasks different from those faced in the College of the University of Chicago. Rather than working, as at Chicago, with a staff of presumptive equals, at Harvard the author recruited advanced graduate students and junior (and on occasion, senior) faculty whom he treated as colleagues, but whom Harvard undergraduates tended to deprecate as mere "section men," i.e. teaching assistants. The author details efforts to overcome what was 30 years ago the fabled yet quite real pose of "Harvard indifference"—a lack of engagement with subject matter and with noncelebrity faculty. The Harvard colleague group worked to get students to do small pieces of fieldwork, not merely as exercises, but out of interest in a topic and an idea. When with the rise of a

0360-0572/88/0815-0001$02.00

radical student-faculty "movement" many undergraduates became less "cool," the problems of teaching subtly changed. Always, the author notes, teaching is contextual and often needs to work against the prevailing tide.

ON DISCOVERING AND TEACHING SOCIOLOGY: A MEMOIR

I propose in this essay to say a bit about the lucky accidents that brought me to Chicago to begin my second career as a Visiting Assistant Professor of Social Science in the College (I did not become a member of the graduate Department of Sociology until eight years later). Like a perpetual graduate student, I have been learning sociology as I have been teaching it.

Beginnings

To someone looking at my career from the outside, it would seem quite appropriate that I should become an academic man. My father, David Riesman, M.D., was an eminent professor of clinical medicine and later of the history of medicine at the University of Pennsylvania, a man of insatiable and admired erudition, who joined my mother in learning Italian late in life so as to be able to read Dante in the original. Known here and abroad as a diagnostician, he wrote constantly: articles on clinical observations, historical essays (e.g. Irish Clinicians of the Eighteenth Century), and late in his life, at a time when I could be of some help to him, *Medicine in Modern Society* (1939). In 1908, at age 42, he married Eleanor Fleisher, like himself of German Jewish nonreligious background, of a family established in Philadelphia for three generations. My mother, an elegant, introverted bluestocking, had led her class at Bryn Mawr College and was encouraged toward an academic career by the offer of the European Fellowship. But she declined the Fellowship, inhibited less by her family's fear that an academic career would lead to spinsterhood than by her own self-deprecation; for her, intellectuality without creativity—defined as high art—was sterile. As against my father's gregariously sanguine temper, his liberal acceptance of things and people as they were, my mother was a skeptic to the point of cynicism, an early modernist scornful of mercantile industrious and professional America.[1]

In Harvard College, I accepted my mother's definition of me as also uncreative, and so I went out not for the literary magazine but for the *Harvard Crimson*. I made the University my beat. I roamed the Business, Medical, and Law Schools for news, and devoted a full page to Henry A. Murray's then

[1] I have sketched my mother's outlook for a *Daedalus* conference on the woman in America (Riesman 1964). To shorten the "literature cited" coda to this essay, I shall include only work of my own whose bearing on the text is not obvious, or which is not found in the bibliography in Gans, Glazer, Gusfield, and Jencks (1979).

almost unknown psychological clinic. With a *Crimson* colleague I discovered President A. Lawrence Lowell's House Plan and kept the secret until land assembly was complete; we then published an entire *Crimson* issue devoted to the Plan, of which I was among my friends the sole supporter. (I benefited personally from the House Plan, spending my senior year in newly built Dunster House.)

One of the courses I most enjoyed was taught by Charles Kingsley Webster, a one-term visitor from Aberystwyth, on British diplomatic history in the years immediately after 1815. Webster had devoured and made sense of immense archival materials. I realized I was much too slow a reader seriously to consider such a pursuit. What I particularly appreciated about Webster's course was the sense that for this period, with its bearing on the origins of the Monroe Doctrine, the excavations seemed virtually complete, the intricacies illuminated. Few students discovered Webster, but some three hundred registered for the famous course in which Irving Babbitt denounced Romanticism; he did not ruffle or excite them. He intrigued me, and I wrote for him a lengthy, overly ambitious essay on the educational theories of Goethe and Rousseau.

I did not discover sociology, but I did learn that Pitirim A. Sorokin was a spirited fellow; I invited him to talk to Dunster House undergraduates in my senior year (1930–1931). People who knew only the later, pacifist Sorokin may be astonished to learn that he told us, "Before I teach my son to think straight, I teach him to shoot straight!" Carl J. Friedrich was a Tutor in Government in Dunster House, and my friendship with him changed my life. Friedrich, born in 1900, had received his doctorate under Alfred Weber in Heidelberg. I owe my introduction to the social sciences to Friedrich. But I owe much more than that. Friedrich, who later wrote, among many other books, *The Age of the Baroque,* could talk on equal terms with my parents concerning Renaissance painting or almost any other topic of high clulture. Unlike my friends, who envied me for having a mother with whom one could discuss Eliot, Faulkner, or Proust, Friedrich was not awed by my parents, though he respected them. More important, he did not share my mother's judgment of me, but regarded me as a person of potential originality.

I entered Harvard Law School aimlessly, wanting to stay on in Cambridge and to maintain my association with Friedrich. By my choice of roommates, I took care to minimize what many of us who went from the amiable amateurism of the College to the fierce competitiveness of the Law School experienced as culture shock. Having met Alexander Meiklejohn's son Donald, who was coming to Harvard as a graduate student in philosophy, I arranged to room with him. Another roommate was James Henry Rowe, Jr., a charming, cavalierly indolent literary man from Butte. Rowe went on to become Oliver Wendell Holmes's last law clerk, an early energizer of the New Deal, and one of Franklin D. Roosevelt's "anonymous assistants." We lived at the Brattle Inn, along with several other lively would-be philosophers. Elton Mayo also lived in the Brattle Inn, where he and I, when we were having dinner together, were occasionally joined by Lawrence J. Henderson. I was fascinated by Mayo's work in the Fatigue Laboratory of the Harvard Business School, and it was through him that I learned about Lloyd Warner's research, then under way in what became the Yankee City series. Mayo's combination of research strategies and wide social concerns was as remote from law school as I could have wished. Moreover, I felt that there was something mysterious about him (Trahair 1985, Smith 1987, Homans 1984).

Virtually all casebooks at that time were prepared by the professors themselves and were unannotated selections of appellate opinions from whose *obiter dicta* we were supposed to extract the holding—a figure-and-ground exercise which many of us quickly learned. I enjoyed the cases for their details, though these were of course filtered through what the judges regarded as important. When legal rules led to what were clearly untoward outcomes, the legal system was upheld by most of my professors and eventually by fellow students on the ground that allowing "each side" its day in court and its opportunities for argument would in the main lead to the discovery of the relevant facts, with the debate among the opposing counsel eventuating in their correct interpretation. I found this debaters' outlook inadequate. But I was even less attracted to iconoclastic Thomas Reed Powell, who taught Constitutional law in what was already known as the Yale realist style: a naive cynicism that saw judges manipulating precedents to arrive at decisions satisfying to their egos, their interests, or their whims—for Powell and his followers, it was enough to debunk the law.

Even for my slow pace of reading, the case assignments were not heavy; the eager law students reread and annotated them and joined small groups to dissect them. I had resolved that in my second Law School year I would try to work harder, along this fashion, deserting the comfortable Brattle Inn and seeing more of my fellow law students. Through Phillips Brooks House, Harvard College's social service center, I had taken a job as an unpaid volunteer laborer for the summer with a Grenfell Mission station in Labrador. There I was astonished to receive a cable telling me that I must return at once to Cambridge to take up my duties on the Law Review.

I enjoyed those duties. The Law Review was a diurnal affair. Few of the editors worried about their Law School classes. I remember editing an article by the man I most admired in the law, Columbia's Karl Llewellyn. I took part, with fellow editors, in an assessment of the constitutionality of the early New Deal legislation, my assignment being Section 7a of the Wagner Act, whose vague language was seized upon by John L. Lewis and his cadres of organizers to encourage workers to join unions.

I had taken no social science courses as an undergraduate; the closest I came was Arthur Schlesinger, Sr.'s course in American history. Law school provided scintillas of economics from an elective third-year course in antitrust laws and another elective on corporate finance—this latter, taught by Ralph Baker, who had been at the Harvard Business School, devoted an entire term to the examination of a single public utility rate-making case and delighted me by the meticulousness of his concentration.

I was chosen by Felix Frankfurter to become Brandeis's law clerk, a privilege which carried with it the opportunity of a postgraduate fellowship year at the Law School. I spent this time primarily under Friedrich's mentorship, helping with the preparation of his book, *Constitutional Government and Politics* (1937). Friedrich looked at America with an enthusiastic newcomer's eye. The political philosopher in him introduced me to some classics of political thought, and in time to some of his refugee friends, of whom one of the most exotic and stimulating was Eugen Rosenstock-Huessy. During the summers, the practical Vermont farmer he sought to become (with amateur help from me) led us to discover the Agricultural Extension Services and the county agent system as an illustration of American federalism. Friedrich and I also became interested in George Gallup's recently begun public opinion polling.

After my year's clerkship with Brandeis, I did not pursue invitations to a job in the New Deal, but chose instead to join a small Boston firm, where I hoped to learn something about trial practice; thus, I was able also to continue my association with Friedrich, whose seminars I attended.

Law Teaching

Before that year was up, I was asked to become professor of law at the University of Buffalo Law School. Samuel Capen, the President of the then-private University, had decided to transform what had been a practitioners' law school into a scholarly one on the Harvard model; I was one of four recent Harvard Law School graduates who formed the new faculty. I taught required first-year courses in property and in criminal law. I also offered an elective third-year seminar devoted to the ordinances of the City of Buffalo. I was unsure of my capacities as a teacher (an anxiety that has not completely left me even now), and I practiced my early lectures before my severest critic, Evelyn Thompson Riesman, herself a writer and editor.

Both the way I taught the first-year courses and my seminar reflected my acceptance of the primarily vocational interests of the students, almost invariably the first in their families to go to college and on to postbaccalaureate study. The new Dean, Frank Shea, also Harvard Law, and my Harvard-trained colleagues had high aspirations for the students, hoping to send them on to work in New Deal jobs in Washington. A few students were disoriented by the mismatch between my colleagues' high aspirations for them and the inability of any one of them to get a job in the New Deal in competition with the graduates of the major law schools. Buffalo was not yet, though it later became, a "national" law school, and when I taught property law, I included some very concrete things about landlord and tenant law in New York State—for example, whether a refrigerator went with the property when it was sold in the way that a furnace would. I responded at the margin or privately to those with wider curiosities.

Nor did I share with the students the research on which I was engaged. Some of my research did deal with the constitutional and civil liberties issues which were of central concern to my colleagues; I examined, for example, whether or not Americans had or should have a constitutional right to travel abroad and to enlist in foreign wars. One bit of research published as "Possession and the Law of Finders" dealt with the law of property. This was an empirical study which compared legal decisions about the rights of finders of lost property ("finders-keepers") with the actual practices of department stores, hotels, transportation companies, and others. Those who in the course of business acquired large amounts of lost property and held it for the owner made efforts where possible to return it, no matter what the cases held concerning rights of finders in a particular jurisdiction. At that time, studies comparing what the cases said with what actually happened in the world were uncommon.

Studies comparing cases in one country with those in another were even less common. I undertook a comparative study of libel and slander litigation. In England, Austria, Argentina, and other countries of Western European tradition, individuals on the receiving end of defamation, oral or written, frequently brought successful suits. Here seemed to be an example of American exceptionalism. The relative lack of hierarchy in the United States, the egalitarian spirit, had the consequence that everyone, public figure or otherwise (quite apart from issues raised by the First Amendment), was supposed when assaulted in speech or in the press to be able to "take it." In several essays under the heading "Democracy and Defamation: Fair Game and Fair Comment," I made cross-cultural comparisons of judicial support for or barriers against libel litigation; I also examined regional and historical patterns in the American case law. (Currently, these issues in the United States have become far more salient: we have seen the growth of a libel "malpractice" bar, and the targeting of those media regarded as liberal, as in the suits by Generals Ariel Sharon and William Westmoreland.)

The Riesmans were welcomed by the cosmopolitans in Buffalo, of whom there were relatively few. We met patrons of art, amateur chamber music players, and, to lifelong effect, Reuel Denney, Dartmouth graduate and poet, then teaching English in a technical high school. I joined the Board of the

Foreign Policy Association, lectured to local civic groups, and taught my first social science classes to two groups at the YWCA: one, "Working Girls"; the other, "Business and Professional Women," who would today be termed pink-collar or white-collar. I roamed the city with Denney, seeking to understand its large Polish and other "ethnic" working-class populations. Exploring the University, I found colleagues and friends, among them the economist Fritz Machlup and the city planner Walter Kurt Behrendt.

In 1939, at my mother's urging (she was in analysis with Karen Horney), I began an unorthodox psychoanalysis with Erich Fromm, for which I commuted from Buffalo on alternate weekends. Like Friedrich, Fromm was not awed by my parents and helped free me from their verdicts on me. When I first met Fromm, I noticed the collected works of Marx and Engels, of course in German, in his library, and remarked that he could not expect to convert me to Communism; Fromm laughed at the notion that he, an anti-Stalinist admirer of Trotsky, would try such an unprofessional thing!

Searching for a New Vocation

Harvard's Law Dean when I was a student, Roscoe Pound, had called for a sociological jurisprudence. Karl Llewellyn, publicly critical while privately friendly with Pound (Hull 1988), had come at my invitation from Columbia Law School to give an address at Harvard Law School on legal education; he was then the witty and imaginative leader of the realists. In my criminal law course at Buffalo, I used an experimental "class book," a departure from casebook traditions, put together by Jerome Michael and Herbert Wechsler, also of Columbia. Even this work, however, was not empirical enough for me; these men and others used social science research but did not engage in it directly. Although restless in this company despite its friendliness and intelligence, I did not respond to Friedrich's urging that I should take a PhD at Harvard in the Government department.

Practical considerations aside, I had begun to believe, no doubt mistakenly, that "government," Harvard's term for political science, was almost as abstract and unempirical as I found the law. Correspondingly, I turned down chances for joint appointments in law and in political science at Wisconsin, Ohio State, and Oregon. I knew I had to learn political science, or any social science, before I could teach it, and I did not think I could do that while also teaching in a law school. After three years of law teaching, I decided to take a year off to devote myself to a piece of social research. I accepted a fellowship at Columbia Law School, planning to continue my comparative study of the law of defamation.

The Columbia location would help put me in touch with people in the social sciences who, I was beginning to realize, were my truer colleagues. I had been excited by reading *Middletown* and later *Middletown in Transition* (Lynd

& Lynd 1929, 1937). The Columbia year gave me a chance to become acquainted with Robert and Helen Lynd, and at some point I discussed with Robert Lynd the prospect of working with him on another community study. I envisaged as the area of my own concern the terrain jointly occupied by sociological fieldwork and cultural anthropology. (I was already in touch with Ruth Benedict and Margaret Mead.)

Being in New York had the additional advantage of making it easier to continue my psychoanalysis. I began attending seminars for psychoanalysts in training conducted by Fromm and by Harry Stack Sullivan at the William Alanson White Institute of Psychiatry; also I audited the lectures of Ernest Schachtel at the New School, dealing inter alia with the use of projective tests in social research.

As a law student, I had been critical of the Fifth Amendment, regarding it as an overreaction to earlier British tyranny, and recognizing that other countries, at least as wedded to justice as the United States, allowed defendants to be cross-examined and denied any privilege against self-incrimination. Such a view was abhorrent to my civil libertarian colleagues and friends. I brought to the questions of civil liberties both comparative perspective and my continuing studies of public opinion. I put some of my reconsiderations into a long essay, "Civil Liberties in a Period of Transition," finished that year at Columbia; it appeared in the 1942 volume *Public Policy* edited by Friedrich and Edward Mason. The essay came to the attention of the omnivorous Edward Shils, who used it as a reading in the third-year course ("Soc 3") of the three-year undergraduate Social Science program in the College of the University of Chicago. Discovering that its author was alive, Shils persuaded the Dean of the College and the Associate Dean (who was familiar with my name, having read an article of mine in the *University of Chicago Law Review*) to bring me out for an interview with the staff of the course; this led to my appointment as Visiting Assistant Professor of the Social Sciences. My father had died in 1939, and with my mother's death in 1945, I inherited some money passed on to her by her parents. This made it easier to undertake the risks of a new career.

Starting Over

As President of the University of Chicago, Robert M. Hutchins never achieved in the undergraduate College there what he admired in the program at St. John's College, where students and tutors engaged in the Great Conversation with the texts, in the designed absence of historical context. Still, at Chicago the College was largely autonomous, with its own faculty teaching sequences in the natural sciences, biological sciences, humanities, and social sciences—resembling but not directly related to the four Graduate Divisions of the University.

Although some College faculty were in fact pursuing PhDs in the Graduate Divisions, all were full members of the four College staffs; there were no teaching assistants. The general temper of the social science staff, while rejecting what was regarded as the overspecialization of the graduate departments, was in varying degree at odds with the more insistently theoretical and philosophical orientation of the other staffs. Thus, while Soc 3 was on the whole compatible with Robert Hutchins's distaste for "mere" empiricism, it did not reject contemporary work—otherwise, something as contemporary as my essay "Civil Liberties in a Period of Transition" would not have been made a required reading.

For me, it was an extraordinarily fortunate opportunity to begin discovering sociology within the context of the other social sciences. With no time to prepare, I found myself teaching both the second term of Soc 3, begun the previous fall, and a speed-up Soc 3 for veterans which began, as I did, in January, 1946. In both courses, I stayed barely one step ahead of my students. The course, like the entire program in the College, was ambitious in its interdisciplinary reach, spanning economics (some Keynes, lectures and readings by Frank Knight) and some of the Great Books of social theory, including works of Malthus, Marx, Maine, and John Stuart Mill, Max Weber's *The Protestant Ethic*, Freud's *General Lectures on Psychoanalysis* and also *Civilization and Its Discontents*, Durkheim's *The Elementary Forms of the Religious Life*, and Thorstein Veblen's *The Theory of the Leisure Class*.

Soc 3 (and its predecessor course, Soc 2) had a pattern of two section meetings a week, and two lectures a week given in turn by each staff member to the entire class. The lectures, which the staff also attended, freed students in some measure from the idiosyncrasies of their particular section leaders, and allowed us, the staff, to educate each other in our specialties. The staff also had weekly meetings in which we argued, often vehemently, discussing plans for lectures and sharing thoughts on the prescribed texts. The sections met around seminar tables. Since the six-hour comprehensive examination at the end of the year was prepared by the College Examiner, students shopped around among section leaders not in terms of grades, but rather in terms of who was the more provocative and charismatic; this created intense and often unhappy competition among the staff. However, since students were not graded by their section leaders, and of course not on class performance, they could engage in discussion, whether timidly or assertively, free of worry about grades.

Problems in Teaching

As lecturer and section leader, I sought to enlist students' curiosity concerning society and some of its most significant interpreters. Although I found myself more engaged by Tocqueville than by Veblen, and more by Weber than by

Durkheim, I encouraged students to approach each reading at once sympathetically and critically. When it came my turn to lecture (four in a row on Freud, later revised and published [1950]), I wrote them out, concerned with density and pace, hoping to keep them moderately accessible to the less adept while holding the interest of the "whiz kids." (The temptation was always there among the staff to lecture to one another.) What I found anxiety-provoking, and still find difficult, is how to teach evocatively by discussion in seminar-size groups. At the University of Buffalo, I used the case method despite my awareness of its limitations. That procedure allows for a professorial routine of asking students at random to state a case and to define its holding, and then asking other students to criticize that synopsis.

Some of my Chicago colleagues used a comparable method they termed Socratic: they would pick a sentence out of a required text and ask a student what it meant, and then another student whether that student agreed or not. By eschewing concern with larger ideas, this formal approach could become monotonous. My own inclination was to approach each reading in terms of its particular intellectual specific gravity, questions of method that it raised, and potential contemporary significance. I hoped in discussion to relate the readings to each other and also to what had been said in earlier discussions—I kept notes of who spoke, and what was said. I learned not to be too afraid of silence.

In the vigorous intellectual milieu of the College in those first post–World War II years, the majority of students appeared to accept the authority of avant-garde ideas and of the "Great Books" without testing them against their own experience. Just as in a somewhat later era to attack Karl Marx made one out to be a despicable bourgeois, so to attack Freud among our undergraduates was to exhibit "resistance"; Freud and psychoanalysis had overwhelming authority. One way that I used to relate *Civilization and Its Discontents* to everyday life, and also to draw out some of the shyer students, was to ask those in the class who had been babysitters to pay particular attention to Freud's observation that the death instinct must be innate, because children of parents who are kindly nevertheless get enraged at their parents without reason—hostility hence must be inborn. The sanctity about the readings and especially about Freud made it difficult for an 18-year-old student to contradict Freud on the basis of merely personal, happenstance observation. I would note that just such observation was necessary to test the generalizations of Freud or those of anyone else we read in the course. Then the stories would pour out illustrating, for example, ways in which parents who appeared kindly and permissive were actually self-indulgent, not wishing to undertake the burden of disciplining their children; the parents would not declare formal rules, but would in their manner show disapproval if unspoken hopes were disobeyed. In the course of discussion there turned out to be ample reason why children of such parents might get angry with them; one did not need to posit a "death instinct" to explain children's rage.

This pedagogic device often gave the moie diffident women students confidence that in other areas they could draw upon their own experience. Once having broken the ice of speaking in a section, they found it easy to continue, particularly since I could refer in a later session to particular comments. To a certain degree, even at the University of Chicago, which had been coeducational since its founding, the classroom culture favored what might be termed "boys' games" of assertiveness and subsumesmanship.[2] I did not intend to suggest that our readings could be dismissed by students because they did not jibe with their own experiences of daily life. Rather, I intended to have students take seriously both the reading and their own experience, and to make efforts to interpret and reconcile the two.

One purpose of such teaching was to help students learn to read more carefully, and to approach heated issues with a combination of concern and detachment. Policy questions were central to Soc 3, and the focus was more on theory than on data. Perhaps for that reason and because it was the "capstone" course, Soc 3 had great prestige. I concluded that I would prefer to teach in the second-year course, Soc 2, which was also interdisciplinary but somewhat less oriented to theory and policy. With new veterans arriving and staffs expanding, I worked with Milton Singer, a philosopher turned anthropologist, to move to Soc 2, where we would have an opportunity to introduce more contemporary work. We read *Elmtown's Youth, Who Shall Be Educated?*, and other community studies, mainly of the Lloyd Warner persuasion. But I was unsuccessful in my effort to introduce "unprocessed" materials from survey research or life history studies.

Everett C. Hughes, whom I met soon after arriving in Chicago, exemplified these ramifying interests, and he became an influential mentor in the field, namely, sociology, where I had come to feel most at home (Riesman 1983). I began auditing his courses, especially one on fieldwork and participant observation, and meeting his graduate students and others who were working on race relations in industry and looking at work and occupations more generally.

Among the more empirical materials read in Social Sciences 2 was Gunnar Myrdal's *An American Dilemma*. The students in my sections found Myrdal's

---

[2]The women we recruited for our staff (sociologists Alicja Iwanska, Helen Mims, and later Sally Cassidy and Paule Verdet, anthropologist Rosalie Hankey Wax, economic historian Sylvia Thrupp) altered by their presence marginally, if at all, what is sometimes referred to as "the chilly climate for women" in discussion sections.

approach so reasonable and persuasive that they readily assumed that all white Americans regarded American race relations as a dilemma and suffered guilt for failure to live up to the ideals of equality. I would ask my students to make the experiment of thinking how a fatalistic Catholic from Spain or Italy would view American race relations. I would note that Gunnar Myrdal was almost ultra-American in his energetic rationality and belief that problems, once delineated, could be solved. I initiated such a perspective tentatively, and made clear that my purpose was not to denigrate Myrdal but to lead the students to examine the empirical questions he raised.

The collegial teaching, notably as Soc 2 developed, depended upon the partial suspension of disbelief among staff members, who had to be willing to read and then teach works in fields in which they did not feel fully at home, subject to the not-always-trusted guidance of those who were somewhat more expert. That guidance could be intimidating or garrulous, overly synthetic or overly parochial; but for me, seeking to learn my trade, the staff discussions were invaluable. The students, a large proportion of them veterans, did not come to Chicago for "college life" (Horowitz 1987), nor for the most part as rebels; many were prepared to make exemplary efforts toward integration of the offerings, while for less talented and exploratory ones the course may have been no more than the sum of its parts.

What drove the program was the contagious excitement among students and staff, the relentless efforts to understand, to come to terms with society from more than one perspective. Other virtues of the Social Science program and its willing devotees become most apparent when compared with what occurs in other institutions where students are taking a large variety of courses. At Chicago what happened in the curriculum could and did spill over into conversations among students outside the classroom. Moreover, when we took our stint as lecturers, we could take for granted what students had read and presumably learned. In contrast, at most other institutions, including, in my experience, Harvard College, one is dealing with students about whose basic knowledge one could take nothing for granted.

Graduate Teaching, Colleagueship, and Research

With the omnicompetent arrogance of some bright lawyers, Robert M. Hutchins had contempt for specialized research. Many members of the College faculty, a bit more proud than defensive, looked down on the supposedly overspecialized graduate faculty. I was at odds with this outlook of those I termed the "College Patriots," recognizing the danger, in myself and others, that if the defenders of specialties can become tyrannical or arid, the defenders of the interdisciplinary can become slovenly (cf. Trow 1976). On their side, professors in the Graduate Divisions looked upon the College faculty as grandiose and pretentious people, given to lofty overgeneralization and lacking in rigor. Furthermore, when undergraduates went on into a graduate department, they were apt to bring with them the alternately charming and tiresome hubris these intense and often highly talented young people seem to have had from birth and then enhanced as undergraduates.

As the College grip on the undergraduate curriculum became total, the antagonism of many of the faculty in the Graduate Divisions toward the College intensified. But there were others who fully supported the College and interdisciplinary work in general. Among them were Robert Redfield, who taught part-time in the College and became for me an influential colleague, and Everett Hughes, who encouraged graduate students working with him to teach in the College. Moreover, the successive Deans of the Social Science Division were supportive of the College, and they responded to Hutchins's encouragement to form new nondepartmental committees (it seemed to me for much the same reason that Franklin D. Roosevelt, when blocked by the regular Cabinet departments, created new agencies). Redfield was one of the creators of the Committee on Social Thought, which continues to attract philosophically and speculatively minded graduate students from all over the country. I had informal ties with the Committee on Planning. I was a charter member of a short-lived Committee on Communication, directed by Bernard Berelson. Most consequential for me personally was the Committee on Human Development, of which Hughes and Lloyd Warner were members, as well as Allison Davis, whose community studies in the Deep South I had admired (Davis & Dollard 1940, Davis, Gardner & Gardner 1941). The Committee on Human Development included social psychologists such as, (besides Allison Davis) Robert J. Havighurst, and Bernice Neugarten. I became a member of the committee, attracted by the colleagueship it offered.

During my first year at Chicago, I had been able to recruit Reuel Denney from Buffalo to teach in both the Humanities and the Social Science College sequences. He and I were beginning informal explorations of American popular culture. However, my teaching left me no opportunity to develop sociological research. Thus when the Yale Committee on National Policy in the summer of 1947, after my first full year of teaching at Chicago, invited me to come to Yale for two years of work on any project I chose, I was greatly tempted. The invitation came at the initiative of two members of the Committee: Harold D. Lasswell, a pioneer in integrating psychoanalysis with political science, whom I had long admired and met soon after coming to Chicago, and Eugene V. Rostow, then Dean of Yale Law School, whom I had known in my days as a law professor. I was able to work out a compromise with the College at Chicago in which I would teach the fall quarters of 1947 and 1948, helping get the course under way at the time of heaviest student load.[3]

In addition to my own stipend, I would have a small fund for research. I had been reading with admiration the work of Nathan Glazer in his "Study of Man" essays in *Commentary* and had learned from Daniel Bell, then on the Soc 2 staff, that Glazer was a graduate student in sociology at Columbia. Glazer's essays made clear that he was interested in survey research, as I had been; I was fortunate to be able to recruit him to work with me part-time at Yale. We started out in the Eastern office of NORC by looking at questionnaires dealing with political issues; we puzzled over respondents' willingness to venture opinions, with only a tiny minority responding, "Don't know." What accounted for this opinion-proneness? We were less interested in the opinions people expressed than in the evidence of an egalitarian belief that each American (whether a voter or not) was entitled to and should have an opinion on just about everything.

We then turned from reading interviews to doing them. I was familiar with the interviews Erich Fromm and Ernest Schachtel had done in the 1920s with German workers, which they interpreted projectively to indicate tendencies toward authoritarianism. We took pretty standard NORC-type questions and gave them to high school and college students. Moreover, Glazer had been taking a course with C. Wright Mills, who had offered his students the chance to examine the long interviews he had had done for what became his book *White Collar*. We sought to interpret Mills's interviews and our own "projectively," creating a portrait of the interviewee in terms of the *gestalt* of the entire protocol—an enterprise illustrated in *Faces in the Crowd*. (*Faces* also included a minuscule community study of a Vermont town for which I had recruited Martin Meyerson from Chicago's Committee on Planning and Margy Meyerson, a graduate student working with Everett Hughes.) Glazer and I thought we could roughly sort interviews with upper-middle-class young people and some older ones in terms of a tentative dichotomy of inner-directed and other-directed, with a few working-class and rural people as reminders of the residual category of tradition directed. In the academic year 1946–1947, with help from Milton Singer and Daniel Bell, I had reshaped the Soc 2 course around "culture and personality," and I brought this approach to the work that became *The Lonely Crowd*. However, my interest was in

---

[3]Chicago has operated on the quarter system, a calendrical pattern which facilitated securing unpaid leaves of absence; throughout my career, my research has primarily been done in summers and in sabbaticals and other leaves of absence. I recruited C. Wright Mills to fill in for me at Chicago during my second stint at Yale. Mills by no means granted "favored nation" status to all assigned readings: when he came to Ruth Benedict's *Patterns of Culture*, so students later reported, he threw the book on the floor and said he would be damned if he was going to spend time on a bunch of tribes!

specific shifts we thought we found in the metropolitan and cosmopolitan upper-middle class, rather than in seeking to delineate an American national character—a doubtful approach to a country as large and diverse as the United States.

When Glazer and I were reflecting on the NORC and other survey data, we found we shared similar attitudes about repeated efforts to get Americans to vote and to become more politically active. We did not believe that political engagement per se was necessarily a good thing, in the light of the opinionatedness of uninformed Americans. Part II of *The Lonely Crowd,* "Politics," sought to make a case that the apathetic should not be suddenly disturbed, and that a politics of veto groups (perhaps heirs of Madison's "factions") might be some protection in a country where liberty has always been at risk. Concurrently, Reuel Denney and I drew on some material from popular culture, then only beginning to excite sociological attention, for themes and illustrations of shifting modes of child-rearing, work, and leisure. In a more "innocent" time for society and for social science (Riesman 1987), we brought these many notions tentatively together in *The Lonely Crowd.*

Community Study in Kansas City

Back in Chicago in the fall of 1949 I still wanted the experience of taking part in a community study. After the Newburyport series, W. Lloyd Warner and associates had published three studies of social stratification, particularly in the school system of the small town of Morris, Illinois (Warner et al 1944, Hollingshead 1949, Warner & Hollingshead 1949). When members of the Committee on Human Development discussed another community study, this one focused not on youth, schools, and social class, but on aging, I suggested the possibility of tackling a larger but still nonmetropolitan city—not a small town such as Morris nor even one the size of Muncie, and also beyond the scale of James West's *Plainville USA* (1945) and Everett C. Hughes's Quebec locale for *French Canada in Transition* (1943). How big a place? I proffered as a tentative definition a community with a single influential newspaper, for which 40 or 50 leading citizens could make major decisions, legitimated by the paper's support. After brief visits, Racine and Springfield seemed unsuitable. At this point we were approached by Community Studies, Inc., of Kansas City, Missouri, which offered a base for University of Chicago researchers and a panoply of already gathered data on the city. Kansas City appeared to meet the definition and, after visits and discussion, the decision was made to launch a study there.

In 1951 a research group of four graduate students directed by Martin Loeb set up shop in the Community Studies quarters, under a directorate of Havighurst, Hughes, Warner, and me. I joined the group in Kansas City from

January through June, 1952 (when I also gave an evening course on sociology at the University of Kansas City). However, Loeb was lackadaisical and the study was dilatory in getting under way. I focused on middle- and upper-middle-class life in a city very different from Buffalo in its social, ethnic, and religious composition, though it was about the same size. The Protestant churches and sects were of particular interest to me. I contributed several essays to the ongoing work in Kansas City, but learned less than I had hoped about how to conduct a community study.

Studies of Academia

I grew up with an interest in education. My father did not bring his medical problems home with him, but he did talk about the proper education of physicians. My parents were acquainted with John Dewey, and my mother's interest in progressive education had led her to try to interest me in attending Antioch College. At Harvard College, not only did I cover the University for the *Crimson*, but I brought innovative college presidents (among them, Alexander Meiklejohn, then President of the University of Wisconsin's Experimental College) to talk to small groups of students. My particular interest in women's colleges was partly due to my mother's involvement as a graduate of Bryn Mawr ('03) and also to Evelyn Riesman's experience (Bryn Mawr '35) as well as my later introduction to Sarah Lawrence College through Helen Lynd.

Teaching at the University of Buffalo had been my first opportunity to get beyond the selective educational institutions of the eastern seaboard, and I sought in the early years at Chicago to become something more than a tourist, if less than an ethnographer, vis-à-vis colleges and universities. If the Midwest meetings were in Bloomington, I would go a day ahead to visit sociologists at Indiana University and to learn more about the place. I welcomed the chance to speak at Southern Illinois University, or wherever opportunities to visit presented themselves. Preparing to write about Thorstein Veblen and particularly about his bitter satire, *The Higher Learning in America: A Memorandum on the Conduct of Universities by Businessmen* (1918), I visited Carleton College, where he had been an undergraduate, and the University of Missouri, where he had taught after his dismissal from Stanford. Of course in the orbit around Robert Hutchins there was continuous debate about higher education, its direction and its leadership.

Everett Hughes encouraged his students to reflect on their own school and college experiences and to study teachers and schools. I had the opportunity to read the interviews and field notes of the study of the University of Kansas Medical School that became *Boys in White* (Becker et al 1961) as well as those of the Kansas undergraduate culture that led to Becker, Geer & Hughes, *Making the Grade* (1968). Regarding academic institutions as what I termed a "snake-like procession," following what they thought was the model of those of higher prestige, I wanted to understand better all the turns (and also discontinuities) in that "procession." I found the opportunity, like so much else in this account, fortuitously. Hutchins, by then President of the Fund for the Republic, had in 1954 commissioned Paul F. Lazarsfeld to conduct a survey of social scientists on the state of academic freedom in the wake of McCarthyism. Lazarsfeld had pretested an elaborate questionnaire among his fellow scholars spending a year at the Center for Advanced Study in the Behavioral Sciences at Stanford. When in the winter of 1955 the survey of 2451 social scientists went into the field, with the interviewing divided between interviewers from NORC and from Elmo Roper and Associates, there were some vehement reactions from institutions and from interviewees. At some schools, Hutchins and the Fund for the Republic were regarded with suspicion as Left Wing. In contrast, a group of Smith College faculty members sent a telegram to Hutchins objecting to the notion that a mere conventional Roper interviewer could fairly report their subtle reactions to prevailing climates of academic freedom. There was widespread doubt that any survey, and particularly this one, could accurately capture the degree of intimidation of faculty members in presumably the most vulnerable fields by the local and national Radical Right. In assigning the interviewing to two different agencies, Lazarsfeld was following a strategy earlier pioneered by Samuel Stouffer, who had used and then compared interviewers from NORC and Gallup's American Institute of Public Opinion for his surveys that became *Communism, Conformity, and Civil Liberties* (1955). Lazarsfeld's sample was stratified by type of baccalaureate-granting institution.

In response to the avalanche of attacks and criticisms, I was asked by Hutchins and Lazarsfeld if I would do a survey of the survey, in effect to assess the validity of the responses and their significance for the topics in hand. I agreed. In addition to having written on civil liberties and intellectual freedom, I had been an active member of the American Association for Public Opinion Research, which includes the leading pollers as well as academic and market researchers. Moreover, I had recruited Mark Benney, an Englishman who had worked as a survey researcher for Mark Abrams's polling organization in London, to teach in the College. We were collaborating on secondary analyses of both an NORC survey and one done by Michigan's Survey Research Center. Furthermore, I had made evident my admiration for the Bureau of Applied Social Research under Lazarsfeld's direction.

To accept the invitation to assess the survey would interrupt the work of the Center for the Study of Leisure I had created with Ford Foundation support, particularly a study of sociability which Nelson Foote and I had organized that fell wholly to me when Foote, not gaining tenure, left Chicago. However, the idea of what Everett Hughes termed "firehouse research" appealed to me. By that he meant a readiness to go down the greased pole and out into the field at a moment's notice. (The same spirit of urgency inspired Robert K. Merton's study of Kate Smith's war bond drive, brilliantly analyzed in *Mass Persuasion* [1946].) Moreover, I recognized that the survey would provide the immersion in the full range of postsecondary institutions that was impossible for a lone intermittent fieldworker.

In the spring of 1955, I moved to New York and read all the interviews and the interviewers' reports at the Bureau. With Mark Benney's help, I developed a mailed questionnaire that we sent to a sample of the respondents at 45 colleges, asking whether the interviewee believed that "a generally fair and true image of your attitudes to academic freedom was elicited by the [Lazarsfeld] questionnaire," and whether "the interviewer got down a generally fair and true record of your attitudes." In addition, I went, sometimes in company with Benney, sometimes alone, to visit 55 institutions in a geographic range from the Wisconsin state colleges at Superior and Whitewater to one of the "black land grant" institutions, North Carolina Agricultural and Technical State College at Greensboro. I interviewed about a quarter of the 212 interviewers who had worked on the survey, focusing on those who had done the survey at the colleges I visited. Being able in many cases to read the interview, interview the respondent, and interview the interviewer permitted "triangulation" of what got transmitted and what might have gotten left out or distorted. In general, my already high respect for survey interviewers' fortitude and intelligence increased. The interviewing staffs did indeed differ a good deal: NORC interviewers were somewhat better educated, more intellectual, more probing; they on the whole produced more free answers, more words altogether, but this did not mean that the summary judgments that Lazarsfeld analyzed differed depending on which staff had been assigned to which college. Sophisticated respondents, especially in the colleges of high prestige, snobbish often toward survey research and especially toward the Roper interviewers, understandably chafed at the boundaries of the questions, but more of their assessments concerning their institutions' climates got through than many had appreciated.

At the end of the summer's interviewing, I could present a preliminary report. Then with some graduate student assistants I spent several years puzzling over the materials of the survey and of my own restudy and submitted a fuller report (Lazarsfeld & Thielens 1958). This work on the survey gave me the materials and the confidence for my first book on education and convinced me that the congruence of my interests, my talents, and my limitations as a fieldworker made the study of academic institutions an appropriate agenda for me.[4]

Graduate vs Undergraduate Teaching

In 1954, with the support of Everett Hughes as Chairman, I became a member of the Department of Sociology at the University of Chicago, while retaining my affiliation with the Social Science Program of the College. Hughes himself had done graduate work at Chicago at a time when sociology and anthropology were joined (cf Helen MacGill Hughes in Merton & Riley 1980). However, it will not surprise readers to learn that the climate of the Sociology Department was not uniform. Working with Hughes or me or others who were regarded in a stereotypical way as merely qualitative, graduate students were anxious lest their theses not pass muster. With a not uncommon paranoia, they concluded that no thesis without tables in it would receive the imprimatur of the Department. I would urge them to go to the library to look at the dissertations recently done in which there were no tables. Of course I had myself no animosity against tables!

Undergraduate students are for the most part not dependent on any one instructor, for it is their overall record, their GRE, LSAT, or MCAT scores, and only to a lesser degree the verdict of their professors in their major field, that can influence their later fates. A graduate student, anxious about a thesis committee and later recommendations for jobs, depends much more heavily on mentors. At Chicago and later at Harvard, I had always to consider how a graduate student would find an "umbrella" to protect against hostile showers, and association with me might or might not help. Graduate students, like law students, are apprentice professionals, even though sociology invites many of exploratory bent. These problems relate to the dissertation, not to the graduate seminars I offered on subjects I chose—for example, one on the interview; one on mass communications taught with Elihu Katz; another on the study of leisure with Rolf Meyersohn.

Graduate seminars at Chicago (and later at Harvard on the sociology of higher education) brought me acute and engaging students and allowed me to

---

[4]My work led to my being recruited for task forces and commissions concerned with education, particularly higher education. I was a member of Nelson Rockefeller's Task Force on Education, then in 1965 and thereafter of Lyndon Johnson's Task Force on Education and of the Carnegie Commission on Higher Education directed by Clark Kerr. I used these responsibilities to contribute what I could and also to learn what I might, especially when I helped persuade the Carnegie Commission to hold its meetings all over the academic landscape of America. The Carnegie Commission brought me into close collaboration with Martin Trow, whom I had first met when he was finishing his graduate work at Columbia, and from whose work on higher education, alone and with Burton Clark (e.g., Clark & Trow 1966), I continue to profit.

draw on my contemporary research interests. Nevertheless, I gradually came to realize that I preferred to teach undergraduates, even though this meant, as in my years as a law teacher, a considerable divorce between the areas of my teaching and my research. However, there was for me a great difference between teaching law and teaching sociology: readings in law school consist primarily of cases, most of them written by quite ordinary judges, whereas in my teaching of sociology I never tired of the classics, nor of the assigned, more contemporary work, in which on rereading I would find something new.

Teaching General Education in Harvard College

When in the summer of 1954 I taught sociology at Harvard Summer School, my already considerable admiration for the constellation making up the Department of Social Relations increased, but without hesitation I turned down an inquiry as to whether I could be attracted to Harvard. Several years later, when Philip Hauser, after a bitter fight, ousted Hughes as Sociology Department Chairman, the intellectual intensity characteristic of Chicago appeared at moments to be more claustrophobically ferocious than stimulating. Accordingly, when in 1957 Dean McGeorge Bundy again approached me with the offer of a special chair in the social sciences focused primarily on the teaching of undergraduates, Harvard's more dispersed environment with its wider orbit of distractions for its faculty, in addition to one another, appeared equable as well as vigorous; I was persuaded. I would become a member of the Sociology wing of Social Relations. I liked the Department's particular interdisciplinary mix of sociology and cultural anthropology along with personality, clinical, and social psychology. The combination and its distinguished representatives attracted first-rate undergraduates and, at the graduate level, adventurous students and junior faculty who specialized in one wing of the joint Department, but were exposed in courses and in intellectual commerce to all the wings.

My appointment was intended to strengthen the General Education program whose often very large lecture courses were taught by leading faculty members assisted by "section men"; the latter met weekly with those students who chose to come to sections, and they also served as graders. Undergraduates, sometimes snobbish toward "mere" graduate students, often deprecated their fellows who took too active a part in section meetings (cf Christensen & Hansen 1987). I would be given the opportunity to recruit a group of junior colleagues and graduate students to recreate something of the collective spirit of Soc 2 and to provide another model for General Education (one had already been illustrated by Samuel Beer in his famous course, "Western Thought and Institutions"). Michael Maccoby, a graduate student in Social Relations and an assistant to Bundy, had spent a year at Chicago teaching Soc 2; in the year before I came, he was one of those I asked to help build a staff for a course that would be called Soc Sci 136, "Character and Social Structure in America." These section leaders were to become my "true colleagues"; they were primarily graduate students and young faculty from Social Relations, Government, and History. Occasionally, for the sake of colleagueship, one or several senior faculty members would also take part. The group of us (never more than a dozen, the maximum number which could be seated around the Riesmans' dining room table) met weekly to plan the course. We talked at dinner about our more-or-less successful efforts to engage undergraduates in serious discussion of the readings, and we also discussed student paper topics. After dinner, we took turns leading discussion of the coming week's readings. When advanced graduate students took part, I was careful lest the special demands of teaching in the course delayed completion of the dissertation; in general it was my experience that the colleagueship of the course helped counterbalance the frustrations many graduate students experience when they work in isolation.

One of my major aims, shared by the staff, was to get as many students as possible to do a manageable piece of fieldwork. We read William F. Whyte's *Street Corner Society* with a particular emphasis on the methodological appendix, and we brought a similar focus on method as well as content to Herbert Gans's *The Urban Villagers.* When we read the community study *Good Fortune: Second Chance Community* (Iwanska 1958) by a Polish emigrée, we invited comparison with Tocqueville as ethnographer. Robert Lane's detailed interview study of 15 men in a New Haven housing project, *Political Ideology,* helped make clear to students how much one could learn through repeated interviews concerning the explicitly Tocquevillean themes of equality, class envy, and resentment, along with the political participation or noninvolvement of the interviewees.

Much of my effort directly and, through the section leaders, indirectly went into persuading undergraduates that they might have something to contribute, something they could discover, report, and interpret in their papers. In the early years, it was difficult to persuade students that they could venture out on their own, do some interviews, or engage in participant-observation, when they were not yet trained social scientists. Even so, many who were willing to take a try at fieldwork of some sort came with the misapprehension that empirical work required a large number of survey-type interviews. We made clear that Soc Sci 136 was not an introduction in a formal way to "sociological methods," as these might be taught in a course for prospective majors. Still, some might come to appreciate the kinds of intellectual craftsmanship C. Wright Mills evokes in the Appendix to *The Sociological Imagination* (1959).

Students could draw from their earlier experience or their present milieu or their voluntary work for Phillips Brooks House, Harvard's social service organization, themes that they could explore in term papers that could be tied in to one or another of their readings. Erik Olin Wright's paper comes to mind, in which he examined the social class backgrounds of Harvard undergraduates who took, as a fifth of the students did, a leave of absence for a term or year. He had initially assumed that these would be students from more privileged backgrounds who concluded they could afford such an interruption; instead, he discovered that in that more propitious economic period of the early Sixties, nonaffluent students were quickly socialized to the prevailing norms. Many students wrote about educational settings; others explored ethnicity; still others, popular culture. One splendid paper examined political attitudes of the small businessmen of Keene, New Hampshire.

At the outset we published three volumes of student papers, by no means always finished or craftsmanlike, to encourage students to be venturesome. Every student received extensive written criticisms of the paper from his section leader and in the early years also a letter from me—one that I shared with the section leader. Later, slow reader that I am, I had to give up the pleasure and task of reading and responding to all the papers and read only a few selected ones.

In the later 1960s, with the development of student and faculty protests and the counterculture, the course found itself in an altered moral and academic climate. The papers students chose to write changed considerably. It no longer took effort to get students to go out into the field. Many preferred the field to the library. In the first years of the course, the curiosity of students was limited by inhibition and self-mistrust. But by the late 1960s, some students seemed to be inhibited by a cynical or nihilistic attitude toward society and its institutions: everything was corrupt, and since we already knew that, what was there to find out about it? Students did not invent these interpretations but were gullible to them when presented by some charismatic senior scholars and younger faculty. Extramurally, I had been involved at Harvard as earlier at Chicago in the effort to control the nuclear arms race, and in 1960 began what became a journal, *The Correspondent: Critical Dialogue and Research on Home and Foreign Affairs*, to which several of the staff contributed. We sought to divorce our political from our pedagogic tasks; staff members came, to use the outdated terms, from Left, Right, and Center. Most of us opposed the Vietnam involvement from the beginning, and most of us also opposed making the University our target. There were of course tensions within and among us, but in general we sought to overcome the closing off of inquiry by newfound radicalism as we had worked in an earlier time to overcome a more naive diffidence. These changes in the temper of students were of course not universal—one had to be careful that one did not take the vocal students as the voice of "the students." We sought to encourage a willingness to be surprised, as against a temptation (somewhat stronger at Harvard than at Chicago) to assume a blasé attitude.

As just indicated, we changed in some respects to respond counter-cyclically to changes among the most salient students. Did we change the students? We have no measures of "value added" through our several and combined efforts. We were pleased when undergraduates chose to pursue academic careers as, in the earlier years particularly, many did, but we avoided forming anything like a "school." I have found collaborative teaching, and in some measure all teaching, unfailingly stressful, and at the same time one way to fulfill my obligations while continuing the unfinished task of discovering sociology.

ACKNOWLEDGMENTS

I thank colleagues for their thoughtful reading of early and late drafts: Daniel Bell, Lewis Dexter, Herbert J. Gans, Howard Gardner, Gerald Grant, Wendy Griswold, Joseph Gusfield, George Homans, Alicja Iwanska, Steven Klineberg, Michael Maccoby, Edward McDonagh, Barbara Norfleet, Charles H. Page, Michael Schudson, W. Richard Scott, Murray Wax, Robert S. Weiss, Harvey Wheeler, Milton Yinger. I acknowledge financial support from the Exxon Education Foundation and Douglass Carmichael's grant to the Project on Technology, Work, and Character.

Literature Cited

Becker, H. S., Geer, B., Hughes, E. C. 1968. *Making the Grade: The Academic Side of College Life*. New York: Wiley

Becker, H. S., Geer, B., Hughes, E. C., Strauss, A. 1961. *Boys in White: Student Culture in Medical School*. Univ. Chicago Press

Christensen, C. R., Hansen, J. 1987. *Teaching and the Case Method: Texts, Cases, and Readings*. Boston: Harvard Bus. Sch. Press

Clark, B. R., Trow, M. 1966. The organizational context. In *College Peer Groups: Problems and Prospects for Research*, ed. T. M. Newcomb, E. K. Wilson. Chicago: Aldine

Davis, A., Dollard, J. 1940. *Children of Bondage*. Washington, DC: Am. Council Educ.

Davis, A., Gardner, B. B., Gardner, M. S. 1941. *Deep South*. Chicago: Univ. Chicago Press

Friedrich, C. J. 1941. *Constitutional Government and Democracy*. Boston: Little Brown

Gans, H. *The Levittowners*.

Gans, H. *The Urban Villagers*.

Gans, H., Glazer, N., Gusfield, J., Jencks, C. 1979. *On the Making of Americans: Essays in Honor of David Riesman*. Philadelphia: Univ. Penn. Press

Hollingshead, A. B. 1949. *Elmtown's Youth: The Impact of Social Classes on Adolescents*. New York: Wiley

Homans, G. C. 1984. *Coming to My Senses: The Autobiography of a Sociologist*. New Brunswick, NJ: Transaction

Horowitz, H. L. 1987. *Campus Life: Undergraduate Cultures from the End of the Eighteenth Century to the Present*. New York: Knopf

Hughes, E. C. 1943. *French Canada in Transition*. Chicago: Univ. Chicago Press

Hughes, H. M. 1980. Robert Ezra Park: The Philosopher-Newspaperman-Sociologist. In *Sociological Traditions from Generation to Generation*, ed. R. K. Merton, M. W. Riley, pp. 67–80. Norwood, NJ: Ablex.

Hull, N. E. H. 1988. Some realism about the Llewellyn-Pound exchange over realism, 1927–1931. *Wisc. Law Rev.* In press

Iwanska, A. 1958. *Good Fortune, Second Chance Community*. Wash. Agri. Exper. Stat., Bull. 589

Lane, R. *Political Ideology*. New York: Free Press

Lazarsfeld, P. F., Thielens, W. 1958. *The Academic Mind: Social Scientists in a Time of Crisis*. With a Field Report by David Riesman. Glencoe, Ill: Free Press

Lynd, H., Lynd, R. 1929. *Middletown*. New York: Harcourt Brace

Lynd, H., Lynd, R. 1937. *Middletown in Transition*. New York: Harcourt Brace

Merton, R. K. 1946. *Mass Persuasion*. New York: Harper

Mills, C. W. 1959. *The Sociological Imagination*. Oxford Univ. Press

Mills, C. W. 1951. *White Collar: The American Middle Classes*. New York: Oxford Univ. Press

Riesman, D. 1939. *Medicine in Modern Society*. Princeton, NJ: Princeton Univ. Press

Riesman, D. 1987. Balancing teaching and writing. Symposium in *J. Harvard-Danforth Ctr. Teach. Learn.* 2:10–16

Riesman, D. 1983. The Legacy of Everett Hughes. *Contemp. Sociol.* 12 (5)

Riesman, D. 1980. *On Higher Education*. San Francisco: Jossey-Bass

Smith, J. H. 1987. Elton Mayo and the hidden Hawthorne. *Work, Employ. Soc.* (1,1)

Stouffer, S. 1955. *Communism, Conformity, and Civil Liberties*. New York: Doubleday

Trahair, R. 1985. *The Humanist Temper: The Life and Work of Elton Mayo*. New Brunswick, NJ: Transaction Books

Trow, M. 1976. Higher education and moral development. *AAUP Bull.* 62 (1):20–27

Veblen, T. 1918. *The Higher Learning in America: A Memorandum on the Conduct of Universities by Businessmen*. New York: D. W. Huebsch; Stanford Univ. Press: Academic Reprints (1954)

Warner, W. L., Havighurst, R. J., Loeb, M. B. 1944. *Who Shall Be Educated? The Challenge of Unequal Opportunities*. New York: Harper

Warner, W. L., Hollingshead, A. B., Neugarten, B., et al. 1949. *Democracy in Jonesville: A Study in Equality and Inequality*. New York: Harper

West, J. 1945. *Plainville USA*. New York: Columbia Univ. Press

Whyte, W. F. 1943. *Street Corner Society*. Chicago: Univ. Chicago Press

APPENDIX: Historical and Prefatory Chapters Published in All *Annual Reviews* Series, 1950–1991, and in *The History of Entomology*, 1973

Adolph, E. F. 1968. Research provides self-education. *Annu. Rev. Physiol.* 30:1–14 [Reprinted in *Exc. Fas. Sci.* Vol. 2]

Ainsworth, G. C. 1969. History of plant pathology in Great Britain. *Annu. Rev. Phytopathol.* 7:13–30

Akai, S. 1974. History of plant pathology in Japan. *Annu. Rev. Phytopathol.* 12:13–26

Alexander, C. P. 1969. Baron Osten Sacken and his influence on American dipterology. *Annu. Rev. Entomol.* 14:1–18

Ambartsumian, V. A. 1980. On some trends in the development of astrophysics. *Annu. Rev. Astron. Astrophys.* 18:1–13

Anastasi, A. 1986. Evolving concepts of test validation. *Annu. Rev. Psychol.* 37:1–15

Anderson, Thomas F. 1975. Some personal memories of research. *Annu. Rev. Microbiol.* 29:1–18

Andrewes, C. H. 1978. Fifty years with viruses. *Annu. Rev. Microbiol.* 27:19–32 [Reprinted in *Exc. Fas. Sci.* Vol. 2]

Anichkov, S. V. 1975. How I became a pharmacologist. *Annu. Rev. Pharmacol. Toxicol.* 15:1–10

Arensberg, C. M. 1972. Culture as behavior: structure and emergence. *Annu. Rev. Anthropol.* 1:1–26

Askonas, B. A. 1990. From protein synthesis to antibody formation and cellular immunity: a personal view. *Annu. Rev. Immunol.* 8:1–21 [Reprinted in *Exc. Fas. Sci.* Vol. 4]

Austin, James (with Catherine Overholt) 1988. Nutrition policy: building the bridge between science and politics. *Annu. Rev. Nutr.* 8:1–20

Axelrod, J. 1988. An unexpected life in research. *Annu. Rev. Pharmacol. Toxicol.* 28:1–23 [Reprinted in *Exc. Fas. Sci.* Vol. 4]

Axelsson, J. 1971. Catecholamine functions. *Annu. Rev. Physiol.* 33:1–30

Ayala, F. J. 1976. Theodosius Dobzhansky: the man and the scientist. *Annu. Rev. Genet.* 10:1–6

Bailey, D. L. 1966. Whither pathology. *Annu. Rev. Phytopathol.* 4:1–8

Bak, T. A. 1974. The history of physical chemistry in Denmark. *Annu. Rev. Phys. Chem.* 25:1–10

Baker, K. F. 1982. Meditations on fifty years as an apolitical plant pathologist. *Annu. Rev. Phytopathol.* 20:1–25 [Reprinted in *Exc. Fas. Sci.* Vol. 3]

Baker, W. O. 1976. Role of science and engineering in human use of materials. *Annu. Rev. Mater. Sci.* 6:35–52

Baker, K. F. (with George W. Fischer) 1983. Pioneer leaders in plant pathology: F. D. Heald. *Annu. Rev. Phytopathol.* 21:13–20

Bard, P. 1973. The ontogenesis of one physiologist. *Annu. Rev. Physiol.* 35:1–16

Bardeen, J. 1980. Unity of concepts in the structure of matter. *Annu. Rev. Mater. Sci.* 10:1–18

Barker, H. A. 1978. Explorations of bacterial metabolism. *Annu. Rev. Biochem.* 47:1–33

Barker, K. R. (with E. Mae Noffsinger & G. D. Griffin) 1981. Pioneer leaders in plant pathology: Gerald Thorne. *Annu. Rev. Phytopathol.* 19:21–28

Bawden, F. C. 1970. Musings of an erstwhile plant pathologist. *Annu. Rev. Phytopathol.* 8:1–12 [Reprinted in *Exc. Fas. Sci.* Vol. 2]

Beadle, G. W. 1974. Recollections. *Annu. Rev. Biochem.* 43:1–13 [Reprinted in *Exc. Fas. Sci.* Vol. 2]

Beals, R. L. 1982. Fifty years in anthropology. *Annu. Rev. Anthropol.* 11:1–23 [Reprinted in *Exc. Fas. Sci.* Vol. 3]

Bean, W. B. 1982. Personal reflections on clinical investigations. *Annu. Rev. Nutr.* 2:1–20 [Reprinted in *Exc. Fas. Sci.* Vol. 3]

Beier, Max. 1973. The early naturalists and anatomists during the Renaissance and seventeenth century. *Hist. Entomol.*, pp. 81–94

Benacerraf, B. 1991. When all is said and done.... *Annu. Rev. Immunol.* 9:1–26 [Reprinted in *Exc. Fas. Sci.* Vol. 4]

Bennett, C. W. 1973. A consideration of some of the factors important in the growth of the science of plant pathology. *Annu. Rev. Phytopathol.* 11:1–10

Bennett, D. 1977. L. C. Dunn and his contribution to T-locus genetics. *Annu. Rev. Genet.* 11:1–12 [Reprinted in *Exc. Fas. Sci.* Vol. 3]

Benson, S. W. 1988. 50 years in physical chemistry: a personal account. *Annu. Rev. Phys. Chem.* 39:1–37 [Reprinted in *Exc. Fas. Sci.* Vol. 4]

Beritashvili (Beritoff), J. S. 1966. From the spinal coordination of movements to the psychoneural integration of behaviour. *Annu. Rev. Physiol.* 28:1–16

Bernardi, B. 1990. An anthropological odyssey. *Annu. Rev. Anthropol.* 19:1–15 [Reprinted in *Exc. Fas. Sci.* Vol. 4]

Bethe, H. A. 1988. Nuclear physics needed for the theory of supernovae. *Annu. Rev. Nucl. Part. Sci.* 38:1–28

Beyer, K. H. Jr. 1977. A career or two. *Annu. Rev. Pharmacol. Toxicol.* 17:1–10

Biale, J. B. 1978. On the interface of horticulture and plant physiology. *Annu. Rev. Plant Physiol.* 29:1–23 [Reprinted in *Exc. Fas. Sci.* Vol. 3]

Binnie, A. M. 1978. Some notes on the study of fluid mechanics in Cambridge, England. *Annu. Rev. Fluid Mech.* 10:1–10 [Reprinted in *Exc. Fas. Sci.* Vol. 3]

Birch, F. 1979. Reminiscences and digressions. *Annu. Rev. Earth Planet. Sci.* 7:1–9 [Reprinted in *Exc. Fas. Sci.* Vol. 3]

Birch, L. C., Andrewartha, H. G. 1973. The history of insect ecology. *Hist. Entomol.*, pp. 229–266

Birdsell, J. B. 1987. Some reflections on fifty years in biological anthropology. *Annu. Rev. Anthropol.* 16:1–12 [Reprinted in *Exc. Fas. Sci.* Vol. 3]

Bishop, G. H. 1965. My life among the axons. *Annu. Rev. Physiol.* 27:1–18 [Reprinted in *Exc. Fas. Sci.* Vol. 1]

Bitancourt, A. A. 1978. Phytopathology in a developing country. *Annu. Rev. Phytopathol.* 16:1–18 [Reprinted in *Exc. Fas. Sci.* Vol. 3]

Black, L. M. 1981. Recollections and reflections. *Annu. Rev. Phytopathol.* 19:1–19 [Reprinted in *Exc. Fas. Sci.* Vol. 3]

Blaschko, H. K. F. 1980. My path to pharmacology. *Annu. Rev. Pharmacol. Toxicol.* 20:1–14 [Reprinted in *Exc. Fas. Sci.* Vol. 3]

Bloch, K. 1987. Summing up. *Annu. Rev. Biochem.* 56:1–19 [Reprinted in *Exc. Fas. Sci.* Vol. 3]

Boothroyd, Carl W. 1982. Charles Chupp: extension plant pathologist. *Annu. Rev. Phytopathol.* 20:41–47

Brooks, Harvey 1978. Resources and the quality of life in 2000. *Annu. Rev. Mater. Sci.* 8:1–19

Brown, W. 1965. Toxins and cell-wall dissolving enzymes in relation to plant disease. *Annu. Rev. Phytopathol.* 3:1–18

Brown, S. W. 1973. Genetics—the long story. *Hist. Entomol.*, pp. 407–432

Bruehl, G. W. 1980. James G. Dickson: the man and his work. *Annu. Rev. Phytopathol.* 18:11–18

Bullard, E. 1975. The emergence of plate tectonics: a personal view. *Annu. Rev. Earth Planet. Sci.* 3:1–30

Bünning, E. 1977. Fifty years of research in the wake of Wilhelm Pfeffer. *Annu. Rev. Plant Physiol.* 28:1–22

Burgers, J. M. 1975. Some memories of early work in fluid mechanics at the technical university of Delft. *Annu. Rev. Fluid Mech.* 7:1–11

Burn, J. H. 1969. Essential pharmacology. *Annu. Rev. Pharmacol. Toxicol.* 9:1–20 [Reprinted in *Exc. Fas. Sci.* Vol. 2]

Burton, A. C. 1978. Variety—the spice of science as well as of life: the disadvantages of specialization. *Annu. Rev. Physiol.* 37:1–12 [Reprinted in *Exc. Fas. Sci.* Vol. 2]

Busemann, A. 1971. Compressible flow in the Thirties. *Annu. Rev. Fluid Mech.* 3:1–12

Cameron, J. W. M. 1973. Insect pathology. *Hist. Entomol.*, pp. 285–306

Campbell, C. L. 1983. Erwin Frink Smith—pioneer plant pathologist. *Annu. Rev. Phytopathol.* 21:21–27

Carlsson, A. 1987. Perspectives on the discovery of central monoaminergic neurotransmission. *Annu. Rev. Neurosci.* 10:19–40 [Reprinted in *Exc. Fas. Sci.* Vol. 3]

Chailakhyan, M. 1968. Internal factors of plant flowering. *Annu. Rev. Plant Physiol.* 19:1–36

Chandler, W. H. 1959. Plant physiology and horticulture. *Annu. Rev. Plant Physiol.* 10:1–12

Chargaff, E. 1975. A fever of reason: the early way. *Annu. Rev. Biochem.* 44:1–18

Chase, M. W. 1985. Immunology and experimental dermatology. *Annu. Rev. Immunol.* 3:1–29 [Reprinted in *Exc. Fas. Sci.* Vol. 3]

Chen, K. K. 1981. Two pharmacological traditions: notes from experience. *Annu. Rev. Pharmacol. Toxicol.* 21:1–6 [Reprinted in *Exc. Fas. Sci.* Vol. 3]

Chynoweth, A. G. 1975. Science-engineering coupling and some priorities in materials research. *Annu. Rev. Mater. Sci.* 5:27–42

Clark, G. 1979. Archaeology and human diversity. *Annu. Rev. Anthropol.* 8:1–20

Clark, W. M. 1962. Notes on a half-century of research, teaching, and administration. *Annu. Rev. Biochem.* 31:1–24 [Reprinted in *Exc. Fas. Sci.* Vol. 1]

Clarke, H. T. 1958. Impressions of an organic chemist in biochemistry. *Annu. Rev. Biochem.* 27:1–14 [Reprinted in *Exc. Fas. Sci.* Vol. 1]

Clifton, C. E. 1966. Microbiology—past, present and future. *Annu. Rev. Microbiol.* 20:1–12

Cole, K. S. 1979. Mostly membranes. *Annu. Rev. Physiol.* 41:1–24 [Reprinted in *Exc. Fas. Sci.* Vol. 3]

Cole, R. H. 1989. Dielectrics in physical chemistry. *Annu. Rev. Phys. Chem.* 40:1–28 [Reprinted in *Exc. Fas. Sci.* Vol. 4]

Colson, E. 1989. Overview. *Annu. Rev. Anthropol.* 18:1–16 [Reprinted in *Exc. Fas. Sci.* Vol. 4]

Coon, C. S. 1977. Overview. *Annu. Rev. Anthropol.* 6:1–10

Cori, Carl F. 1969. The call of science. *Annu. Rev. Biochem.* 38:1–20

Cosman, M. P. 1983. A feast for Aesculapius: historical diets for asthma and sexual pleasure. *Annu. Rev. Nutr.* 3:1–33

Costa, E., Karczmar, A. G., Vesell, E. S. 1989. Bernard B. Brodie and the rise of chemical pharmacology. *Annu. Rev. Pharmacol. Toxicol.* 29:1–29 [Reprinted in *Exc. Fas. Sci.* Vol. 4]

Cowling, T. G. 1985. Astronomer by accident. *Annu. Rev. Astron. Astrophys.* 23:1–18 [Reprinted in *Exc. Fas. Sci.* Vol. 3]

Crane, E., Townsend, G. F. 1973. History of apiculture. *Hist. Entomol.*, pp. 387–406

Crow, J. F. 1987. Population genetics history: a personal view. *Annu. Rev. Genet.* 21:1–22 [Reprinted in *Exc. Fas. Sci.* Vol. 4]

Cummins, G. B. 1978. J. C. Arthur: the man and his work. *Annu. Rev. Phytopathol.* 15:19–30

Dagley, S. 1987. Lessons from biodegradation. *Annu. Rev. Microbiol.* 41:1–23 [Reprinted in *Exc. Fas. Sci.* Vol. 3]

Dainty, J. 1990. Prefatory chapter. *Annu. Rev. Plant Physiol. Plant Mol. Biol.* 41:1–20 [Reprinted in *Exc. Fas. Sci.* Vol. 4]

Darby, W. J. 1985. Some personal reflections on a half century of nutrition science: 1930s-1980s. *Annu. Rev. Nutr.* 5:1–24 [Reprinted in *Exc. Fas. Sci.* Vol. 3]

Davenport, H. W. 1985. The apology of a second-class man. *Annu. Rev. Physiol.* 47:1–14 [Reprinted in *Exc. Fas. Sci.* Vol. 3]

Dethier, V. G. 1990. Chemosensory physiology in an age of transition. *Annu. Rev. Neurosci.* 13:1–13 [Reprinted in *Exc. Fas. Sci.* Vol. 4]

Doisy, E. A. 1976. An autobiography. *Annu. Rev. Biochem.* 45:1–9

Du Bois, C. 1980. Some anthropological hindsights. *Annu. Rev. Anthropol.* 9:1–13

Du Bois, E. F. 1950. Fifty years of physiology in America—a letter to the editor. *Annu. Rev. Physiol.* 12:1–12 [Reprinted in *Exc. Fas. Sci.* Vol. 1]

Dupuis, C. 1974. Pierre Andre Lareille (1762-1833): the foremost entomologist of his time. *Annu. Rev. Entomol.* 19:1–13

Dupuis, C. 1984. Willi Hennig's impact on taxonomic thought. *Annu. Rev. Ecol. Syst.* 15:1–24

Philip, C. B., Rozeboom, L. E. 1973. Medico-veterinary entomology: a generation of progress. *Hist. Entomol.*, pp. 333–60

Eccles, J. C. 1978. My scientific odyssey. *Annu. Rev. Physiol.* 39:1–18 [Reprinted in *Exc. Fas. Sci.* Vol. 2]

Edsall, J. T. 1971. Some personal history and reflections from the life of a biochemist. *Annu. Rev. Biochem.* 40:1–28 [Reprinted in *Exc. Fas. Sci.* Vol. 2]

Eggan, F. 1974. Among the anthropologists. *Annu. Rev. Anthropol.* 3:1–19

Eliassen, A. 1982. Vihelm Berknes and his students. *Annu. Rev. Fluid Mech.* 14:1–12

Emerson, S. 1971. Alfred Henry Sturtevant. *Annu. Rev. Genet.* 5:1–4

Engelhardt, W. A. 1982. Life and science. *Annu. Rev. Biochem.* 51:1–19 [Reprinted in *Exc. Fas. Sci.* Vol. 3]

Erickson, R. O. 1988. Growth and development of a botanist. *Annu. Rev. Plant Physiol. Plant Mol. Biol.* 39:1–22 [Reprinted in *Exc. Fas. Sci.* Vol. 4]

Erlanger, J. 1964. A physiologist reminisces. *Annu. Rev. Physiol.* 26:1–14 [Reprinted in *Exc. Fas. Sci.* Vol. 1]

Estey, R. H. 1986. A. H. R. Buller: pioneer leader in plant pathology. *Annu. Rev. Phytopathol.* 24:17–25

Evenari, M. 1985. A cat has nine lives. *Annu. Rev. Plant Physiol. Plant Mol. Biol.* 36:1–25 [Reprinted in *Exc. Fas. Sci.* Vol. 3]

Eyring, H. 1977. Men, mines, and molecules. *Annu. Rev. Phys. Chem.* 28:1–13 [Reprinted in *Exc. Fas. Sci.* Vol. 3]

Feng, T. P. 1988. Looking back, looking forward. *Annu. Rev. Neurosci.* 11:1–12

Fenn, W. O. 1962. Born fifty years too soon. *Annu. Rev. Physiol.* 24:1–10 [Reprinted in *Exc. Fas. Sci.* Vol. 1]

Firth, R. 1975. An appraisal of modern social anthropology. *Annu. Rev. Anthropol.* 4:1–25

Fischer, H. O. L. 1960. Fifty years "Synthetiker" in the service of biochemistry. *Annu. Rev. Biochem.* 29:1–14 [Reprinted in *Exc. Fas. Sci.* Vol. 1]

Fish, S. 1970. The history of plant pathology in Australia. *Annu. Rev. Phytopathol.* 8:13–36

Flügge-Lotz, I. (with Wilhelm Flügge) 1973. Ludwig Prandtl in the 1930s: reminiscences. *Annu. Rev. Fluid Mech.* 5:1–8

Fortes, M. 1978. An anthropologist's apprenticeship. *Annu. Rev. Anthropol.* 7:1–30

Fraisse, P. 1984. Perception and estimation of time. *Annu. Rev. Psychol.* 35:1–36

Franz, J. M., Hagen, K. S. 1973. A history of biological control. *Hist. Entomol.*, pp. 433–476

French, C. S. 1979. Fifty years of photosynthesis. *Annu. Rev. Plant Physiol.* 30:1–26 [Reprinted in *Exc. Fas. Sci.* Vol. 3]

Fretwell, S. D. 1975. The impact of Robert Macarthur on ecology. *Annu. Rev. Ecol. Syst.* 6:1–13

Friedel, J. 1988. Dislocations and disclinations past and present: some personal views and reminiscences. *Annu. Rev. Mater. Sci.* 18:1–24 [Reprinted in *Exc. Fas. Sci.* Vol. 4]

Frumkin and N. M. Emanuel, A. N. 1968. Fifty years of Soviet physical chemistry. *Annu. Rev. Phys. Chem.* 19:1–30

Fulton, R. W. 1984. Pioneer leaders in plant pathology: James Johnson. *Annu. Rev. Phytopathol.* 22:27–34

Gaffron, H. 1969. Resistance to knowledge. *Annu. Rev. Plant Physiol.* 20:1–40

Gardner, M. W. 1977. Little-known plant pathologists: Ethelbert Dowlen. *Annu. Rev. Phytopathol.* 15:13–15

Garrett, S. D. 1972. On learning to become a plant pathologist. *Annu. Rev. Phytopathol.* 10:1–8

Garrett, S. D. 1981. Pioneer leaders in plant pathology: W. J. Dowson. *Annu. Rev. Phytopathol.* 19:29–34

Garrett, S. D. 1985. William Brown: pioneer leader in plant pathology. *Annu. Rev. Phytopathol.* 23:13–18

Gerard, R. W. 1952. The organization of science. *Annu. Rev. Physiol.* 14:1–12 [Reprinted in *Exc. Fas. Sci.* Vol. 1]

Gibson, Eleanor J. 1988. Exploratory behavior in the development of perceiving, acting, and the acquiring of knowledge. *Annu. Rev. Psychol.* 39:1–41

Gilluly, J. 1977. American geology since 1910—a personal appraisal. *Annu. Rev. Earth Planet. Sci.* 5:1–12

Ginzburg, V. L. 1990. Notes of an amateur astrophysicist. *Annu. Rev. Astron. Astrophys.* 28:1–36 [Reprinted in *Exc. Fas. Sci.* Vol. 4]

Goldsmith, J. R. 1991. Some Chicago georecollections. *Annu. Rev. Earth Planet. Sci.* 19:1–16 [Reprinted in *Exc. Fas. Sci.* Vol. 4]

Goldstein, S. 1969. Fluid mechanics in the first half of this century. *Annu. Rev. Fluid Mech.* 1:1–28

Good, N. E. 1986. Confessions of a habitual skeptic. *Annu. Rev. Plant Physiol. Plant Mol. Biol.* 37:1–22 [Reprinted in *Exc. Fas. Sci.* Vol. 3]

Goody, J. 1991. Towards a room with a view: a personal account of contributions to local knowledge, theory, and research in fieldwork and comparative studies. *Annu. Rev. Anthropol.* 20:1–23 [Reprinted in *Exc. Fas. Sci.* Vol. 4]

Granit, R. 1978. Discovery and understanding. *Annu. Rev. Physiol.* 34:1–12 [Reprinted in *Exc. Fas. Sci.* Vol. 2]

Green, G. J. (with T.Johnson, I. L. Conners) 1980. Pioneer leaders in plant pathology: J. H. Craigie. *Annu. Rev. Phytopathol.* 18:19–25

Greenberg, J. H. 1986. On being a linguistic anthropologist. *Annu. Rev. Anthropol.* 15:1–24 [Reprinted in *Exc. Fas. Sci.* Vol. 3]

Greenstein, J. L. 1984. An astronomical life. *Annu. Rev. Astron. Astrophys.* 22:1–35 [Reprinted in *Exc. Fas. Sci.* Vol. 3]

Gregory, P. H. 1977. Spores in air. *Annu. Rev. Phytopathol.* 15:1–11 [Reprinted in *Exc. Fas. Sci.* Vol. 3]

Griffin, J. B. 1985. An individual's participation in American archaeology. *Annu. Rev. Anthropol.* 14:1–23 [Reprinted in *Exc. Fas. Sci.* Vol. 3]

Grogan, R. G. 1987. The relation of art and science of plant pathology for disease control. *Annu. Rev. Phytopathol.* 25:1–8

Gueron, J. (with Michel Magat) 1971. A history of physical chemistry in France. *Annu. Rev. Phys. Chem.* 22:1–23

Gunsalus, I. C. 1984. Learning. *Annu. Rev. Microbiol.* 38:13–44 [Reprinted in *Exc. Fas. Sci.* Vol. 3]

Habermann, E. R. 1974. Rudolf Buchheim and the beginning of pharmacology as a science. *Annu. Rev. Pharmacol. Toxicol.* 14:1–8

Hales, A. L. 1986. Geophysics on three continents. *Annu. Rev. Earth Planet. Sci.* 14:1–20 [Reprinted in *Exc. Fas. Sci.* Vol. 3]

Hall, V. E. (with Ralph R. Sonnenschein) 1981. The Annual Review of Physiology: past and present. *Annu. Rev. Physiol.* 43:1–5

Hamburger, V. 1989. The journey of a neuroembryologist. *Annu. Rev. Neurosci.* 12:1–12 [Reprinted in *Exc. Fas. Sci.* Vol. 4]

Harpaz, Isaac. 1973. Early entomology in the Middle East. *Hist. Entomol.*, pp. 21–36

Harpaz, Isaac. 1984. Frederick Simon Bodenheimer (1897-1959): idealist, scholar, scientist. *Annu. Rev. Entomol.* 29:1–23

Hastings, A. Baird. 1970. A biochemist's anabasis. *Annu. Rev. Biochem.* 39:1–24

Heidelberger, M. 1978. A "pure" organic chemist's downward path. *Annu. Rev. Microbiol.* 31:1–12 [Reprinted in *Exc. Fas. Sci.* Vol. 2]

Heidelberger, M. 1979. A "pure" organic chemist's downward path: Chapter—the years at P. and S. *Annu. Rev. Biochem.* 48:1–21 [Reprinted in *Exc. Fas. Sci.* Vol. 3]

Heidelberger, M. 1967. Some contributions of immunochemistry to biochemistry and biology. *Annu. Rev. Biochem.* 36:1–12

Hendricks, S. B. 1970. The passing scene. *Annu. Rev. Plant Physiol.* 21:1–10

Herzberg, G. 1985. Molecular spectroscopy: a personal history. *Annu. Rev. Phys. Chem.* 36:1–30 [Reprinted in *Exc. Fas. Sci.* Vol. 3]

Hewitt, W. B. 1979. Conceptualizing in plant pathology. *Annu. Rev. Phytopathol.* 17:1–12

Heymans, C. 1963. A look at an old but still current problem. *Annu. Rev. Physiol.* 25:1–14 [Reprinted in *Exc. Fas. Sci.* Vol. 1]

Heymans, C. 1967. Pharmacology in old and modern medicine. *Annu. Rev. Pharmacol. Toxicol.* 7:1–14

Hildebrand, J. H. 1981. A history of solution theory. *Annu. Rev. Phys. Chem.* 32:1–23

Hilgard, E. R. 1980. Consciousness in contemporary psychology. *Annu. Rev. Psychol.* 31:1–26

Hill, A. V. 1958. The heat production of muscle and nerve. *Annu. Rev. Physiol.* 21:1–18 [Reprinted in *Exc. Fas. Sci.* Vol. 1]

Hill, R. 1975. Days of visual spectroscopy. *Annu. Rev. Plant Physiol.* 26:1–11

Hirota, E. 1991. From high-resolution spectroscopy to chemical reactions. *Annu. Rev. Phys. Chem.* 42:1–22 [Reprinted in *Exc. Fas. Sci.* Vol. 4]

Hirschfelder, J. O. 1983. My adventures in theoretical chemistry. *Annu. Rev. Phys. Chem.* 34:1–29 [Reprinted in *Exc. Fas. Sci.* Vol. 3]

Hirschfelder, J. O. 1983. Henry Eyring, 1901-1092. *Annu. Rev. Phys. Chem.* 34:xi-xvI [Reprinted in *Exc. Fas. Sci.* Vol. 3]

Hodgkin, A. L. 1983. Beginning: some reminiscences of my early life (1914-1947). *Annu. Rev. Physiol.* 45:1–16 [Reprinted in *Exc. Fas. Sci.* Vol. 3]

Holmes, F. O. 1968. Trends in the development of plant virology. *Annu. Rev. Phytopathol.* 6:41–62

Homans, G. C. 1986. Fifty years of sociology. *Annu. Rev. Sociol.* 12:xiii–xxx [Reprinted in *Exc. Fas. Sci.* Vol. 3]

Hopcroft, J. E. 1989. Reflections on computer science. *Annu. Rev. Comput. Sci.* 4:1–12 [Reprinted in *Exc. Fas. Sci.* Vol. 4]

Horsfall, J. G. 1975. Fungi and fungicides, the story of a nonconformist. *Annu. Rev. Phytopathol.* 13:1–13

Horsfall, J. G. 1979. Roland Thaxter. *Annu. Rev. Phytopathol.* 17:29–35

Horsfall, J. G. (with Stephen Wilhelm) 1982. Heinrich Anton de Bary: nach einhundertfunfzig Jahren. *Annu. Rev. Phytopathol.* 20:27–32

Horst, R. Kenneth 1984. Pioneer leaders in plant pathology: Cynthia Westcott, plant doctor. *Annu. Rev. Phytopathol.* 22:21–26

Houssay, B. A. 1956. Trends in physiology as seen from South America. *Annu. Rev. Physiol.* 18:1–12 [Reprinted in *Exc. Fas. Sci.* Vol. 1]

Hoyle, F. 1982. The universe: past and present reflections. *Annu. Rev. Astron. Astrophys.* 20:1–35

Hoytink, G. J. 1970. Physical chemistry in the Netherlands after Van't Hoff. *Annu. Rev. Phys. Chem.* 21:1–16

Humphrey, J. H. 1984. Serendipity in immunology. *Annu. Rev. Immunol.* 2:1–21 [Reprinted in *Exc. Fas. Sci.* Vol. 3]

Hungate, R. E. 1979. Evolution of a microbial ecologist. *Annu. Rev. Microbiol.* 33:1–20 [Reprinted in *Exc. Fas. Sci.* Vol. 3]

Hutt, P. B. 1984. Government regulation of the integrity of the food supply. *Annu. Rev. Nutr.* 4:1–20

Huxley, Sir Andrew. 1988. Prefatory chapter: muscular contraction. *Annu. Rev. Physiol.* 50:1–16 [Reprinted in *Exc. Fas. Sci.* Vol. 4]

Konish, M., Ito, Y. 1973. Early entomology in East Asia. *Hist. Entomol.*, pp. 1–21

Jasper, H. (with Theodore L. Sourkes) 1983. Nobel laureates in neuroscience: 1904-1981. *Annu. Rev. Neurosci.* 6:1–42

Jeffreys, Harold 1973. Developments in geophysics. *Annu. Rev. Earth Planet. Sci.* 1:1–13

Jonas, J. (with H. S. Gutowsky) 1980. NMR in chemistry - an evergreen. *Annu. Rev. Phys. Chem.* 31:1–27

Jones, R. T. 1977. Recollections from an earlier period in American aeronautics. *Annu. Rev. Fluid Mech.* 9:1–11 [Reprinted in *Exc. Fas. Sci.* Vol. 3]

Jones, D. P. 1973. Agricultural entomology. *Hist. Entomol.*, pp. 307–32

Jost, W. 1966. The first forty-five years of physical chemistry in Germany. *Annu. Rev. Phys. Chem.* 17:1–14

Jukes, T. H. 1990. Nutrition science from vitamins to molecular biology. *Annu. Rev. Nutr.* 10:1–20 [Reprinted in *Exc. Fas. Sci.* Vol. 4]

Kabat, E. A. 1983. Getting started 50 years ago—experiences, perspectives, and problems of the first 21 years. *Annu. Rev. Immunol.* 1:1–32 [Reprinted in *Exc. Fas. Sci.* Vol. 3]

Kabat, E. A. 1988. Before and after. *Annu. Rev. Immunol.* 6:1–24 [Reprinted in *Exc. Fas. Sci.* Vol. 4]

Kalckar, H. M. 1991. 50 years of biological research—from oxidative phosphorylation to energy requiring transport regulation. *Annu. Rev. Biochem.* 60:1–37 [Reprinted in *Exc. Fas. Sci.* Vol. 4]

Kamen, M. D. 1986. A cupful of luck, a pinch of sagacity. *Annu. Rev. Biochem.* 55:1–34 [Reprinted in *Exc. Fas. Sci.* Vol. 3]

Kamiya, N. 1989. My early career and the involvement of World War II. *Annu. Rev. Plant Physiol. Plant Mol. Biol.* 40:1–18 [Reprinted in *Exc. Fas. Sci.* Vol. 4]

Kato, G. 1970. The road a scientist followed: notes of Japanese physiology as I myself experienced it. *Annu. Rev. Physiol.* 32:1–20

Kelman, A. 1985. Plant pathology at the crossroads. *Annu. Rev. Phytopathol.* 23:1–11

Kent, G. C. 1979. Important little-known contributors to plant pathology: Mason Blanchard Thomas. *Annu. Rev. Phytopathol.* 17:21–28

Kerling, L. C. P. (with G. de Bruin-Brank & J. G. Ten Houten) 1986. Johanna Westerdijk: pioneer leader in plant pathology. *Annu. Rev. Phytopathol.* 24:33–41

Kern, H. 1985. Ernst Gaumann, 1893-1963: pioneer leader in plant pathology. *Annu. Rev. Phytopathol.* 23:19–22

Kety, S. S. 1979. The metamorphosis of a psychobiologist. *Annu. Rev. Neurosci.* 2:1–15 [Reprinted in *Exc. Fas. Sci.* Vol. 3]

Kingery, W. D. 1989. Ceramic materials science in society. *Annu. Rev. Mater. Sci.* 19:1–20 [Reprinted in *Exc. Fas. Sci.* Vol. 4]

Kiraly, Z. 1972. Main trends in the development of plant pathology in Hungary. *Annu. Rev. Phytopathol.* 10:9–20

Kleiber, M. 1967. Prefatory chapter: an old professor of animal husbandry ruminates. *Annu. Rev. Physiol.* 29:1–20 [Reprinted in *Exc. Fas. Sci.* Vol. 2]

Klein, G., Klein, E. 1989. How one thing has led to another. *Annu. Rev. Immunol.* 7:1–33 [Reprinted in *Exc. Fas. Sci.* Vol. 4]

Kornberg, A. 1989. Never a dull enzyme. *Annu. Rev. Biochem.* 58:1–30 [Reprinted in *Exc. Fas. Sci.* Vol. 4]

Kosterlitz, H. W. 1979. The best laid schemes o' mice an' men gang aft agley. *Annu. Rev. Pharmacol. Toxicol.* 19:1-12 [Reprinted in *Exc. Fas. Sci.* Vol. 3]

Kramer, P. J. 1973. Some reflections after 40 years in plant physiology. *Annu. Rev. Plant Physiol.* 24:1–24

Krogman, W. M. 1976. Fifty years of physical anthropology: the men, the material, the concepts, the methods. *Annu. Rev. Anthropol.* 5:1–14

Kuethe, A. M. 1988. The first turbulence measurements: a tribute to Hugh L. Dryden. *Annu. Rev. Fluid Mech.* 20:1–3

Kumagai, H. 1978. Pharmacology and medicine. *Annu. Rev. Pharmacol. Toxicol.* 10:277–284 [Reprinted in *Exc. Fas. Sci.* Vol. 2]

Lang, A. 1980. Some recollections and reflections. *Annu. Rev. Plant Physiol.* 31:1–28 [Reprinted in *Exc. Fas. Sci.* Vol. 3]

Lasagna, Louis 1985. Clinical pharmacology in the United States: a personal reminiscence. *Annu. Rev. Pharmacol. Toxicol.* 25:27–31

Leach, E. R. 1984. Glimpses of the unmentionable in the history of British social anthropology. *Annu. Rev. Anthropol.* 13:1–23 [Reprinted in *Exc. Fas. Sci.* Vol. 3]

Leake, C. D. 1976. How I am. *Annu. Rev. Pharmacol. Toxicol.* 16:1–14

Leben, Curt 1981. Pioneer leaders in plant pathology: G. W. Keitt. *Annu. Rev. Phytopathol.* 19:35–40

Lederberg, J. 1979. Edward Lawrie Tatum. *Annu. Rev. Genet.* 13:1–5 [Reprinted in *Exc. Fas. Sci.* Vol. 3]

Lederberg, J. 1987. Genetic recombination in bacteria - a discovery account. *Annu. Rev. Genet.* 21:23–46 [Reprinted in *Exc. Fas. Sci.* Vol. 3]

Leloir, L. F. 1983. Far away and long ago. *Annu. Rev. Biochem.* 52:1–15 [Reprinted in *Exc. Fas. Sci.* Vol. 3]

Lemberg, R. 1965. Chemist, biochemist, and seeker in three countries. *Annu. Rev. Biochem.* 34:1–20 [Reprinted in *Exc. Fas. Sci.* Vol. 1]

Libby, W. F. 1964. Berkeley radiochemistry. *Annu. Rev. Phys. Chem.* 15:7–12

Libby, W. F. 1964. Thirty years of atomic chemistry. *Annu. Rev. Phys. Chem.* 15:241–254 [Reprinted in *Exc. Fas. Sci.* Vol. 1]

Liljestrand, G. 1957. The increasing responsibility of the physiological sciences. *Annu. Rev. Physiol.* 19:1–12 [Reprinted in *Exc. Fas. Sci.* Vol. 1]

Lipmann, F. 1984. A long life in times of great upheaval. *Annu. Rev. Biochem.* 53:1–33 [Reprinted in *Exc. Fas. Sci.* Vol. 3]

Loewi, O. 1954. Reflections on the study of physiology. *Annu. Rev. Physiol.* 16:1–10 [Reprinted in *Exc. Fas. Sci.* Vol. 1]

Loitsianskii, L. G. 1970. The development of boundary-layer theory in the USSR. *Annu. Rev. Fluid Mech.* 2:1–14

Lowry, O. H. 1990. How to succeed in research without being a genius. *Annu. Rev. Biochem.* 59:1–27 [Reprinted in *Exc. Fas. Sci.* Vol. 4]

Luck, J. M. 1981. Confessions of a biochemist. *Annu. Rev. Biochem.* 50:1–22 [Reprinted in *Exc. Fas. Sci.* Vol. 3]

Lwoff, A. 1971. From protozoa to bacteria and viruses: fifty years with microbes. *Annu. Rev. Microbiol.* 25:1–26 [Reprinted in *Exc. Fas. Sci.* Vol. 2]

Mac Gillivray, H. 1968. A personal biography of Arthur Robertson Cushny, 1866-1926. *Annu. Rev. Pharmacol. Toxicol.* 08:1–24 [Reprinted in *Exc. Fas. Sci.* Vol. 2]

MacLeod, R. A. 1985. Marine microbiology far from the sea. *Annu. Rev. Microbiol.* 39:1–20 [Reprinted in *Exc. Fas. Sci.* Vol. 3]

Mann, F. C. 1955. To the physiologically inclined. *Annu. Rev. Physiol.* 17:1–16 [Reprinted in *Exc. Fas. Sci.* Vol. 1]

Maren, T. H. 1982. Great expectations. *Annu. Rev. Pharmacol. Toxicol.* 22:1–18 [Reprinted in *Exc. Fas. Sci.* Vol. 3]

Mark, H. F. (with Sheldon M. Atlas) 1976. Polymers as building materials. *Annu. Rev. Mater. Sci.* 6:1–32

Marrè, E. 1991. Short story of a plant physiologist and variations on the theme. *Annu. Rev. Plant Physiol. Plant Mol. Biol.* 42:1–20 [Reprinted in *Exc. Fas. Sci.* Vol. 4]

Matthews, R. E. F. 1987. The changing scene in plant physiology. *Annu. Rev. Phytopathol.* 25:11–23

Mayer, J. E. 1982. The way it was. *Annu. Rev. Phys. Chem.* 33:1–23 [Reprinted in *Exc. Fas. Sci.* Vol. 3]

McCrea, W. H. 1987. Clustering of astronomers. *Annu. Rev. Astron. Astrophys.* 25:1–22 [Reprinted in *Exc. Fas. Sci.* Vol. 3]

McElroy, W. D. 1978. From the precise to the ambiguous: light, bonding, and administration. *Annu. Rev. Microbiol.* 30:1–20 [Reprinted in *Exc. Fas. Sci.* Vol. 2]

McLean, Franklin C. 1960. Prefatory chapter: physiology and medicine: a transition period. *Annu. Rev. Physiol.* 22:1–16 [Reprinted in *Exc. Fas. Sci.* Vol. 1]

McCarty, M. 1980. Reminiscences of the early days of transformation. *Annu. Rev. Genet.* 14:1–15 [Reprinted in *Exc. Fas. Sci.* Vol. 3]

McCollum, E. V. 1953. My early experiences in the study of foods and nutrition. *Annu. Rev. Biochem.* 22:1–16 [Reprinted in *Exc. Fas. Sci.* Vol. 1]

McCallan, S. E. A. 1969. A perspective on plant pathology. *Annu. Rev. Phytopathol.* 7:1–12

Mead, Margaret 1973. Changing styles of anthropological work. *Annu. Rev. Anthropol.* 2:1–26

Mehl, R. F. 1975. A department and a research laboratory in a university. *Annu. Rev. Mater. Sci.* 5:1–26

Merton, R. K. 1987. Three fragments from a sociologist's notebooks: establishing the phenomenon, specified ignorance, and strategic research materials. *Annu. Rev. Sociol.* 13:1–28 [Reprinted in *Exc. Fas. Sci.* Vol. 3]

Mickel, C. 1973. John Ray: indefatigable student of nature. *Annu. Rev. Entomol.* 18:1–16

Miller, N. E. 1983. Behavioral medicine: symbiosis between laboratory and clinic. *Annu. Rev. Psychol.* 34:1–31

Millsaps, K. 1984. Karl Pohlhausen, as I remember him. *Annu. Rev. Fluid Mech.* 16:1–10

Mitchell, H. K. 1978. Ernst Hadorn. *Annu. Rev. Genet.* 12:1–3 [Reprinted in *Exc. Fas. Sci.* Vol. 3]

Mizushima, S. 1972. A history of physical chemistry in Japan. *Annu. Rev. Phys. Chem.* 23:1–14

Morgan, W. W. 1988. A morphological life. *Annu. Rev. Astron. Astrophys.* 26:1–9 [Reprinted in *Exc. Fas. Sci.* Vol. 4]

Morge, G. 1973. Entomology in the Western world in antiquity and in medieval times. *Hist. Entomol.*, pp. 37–80

Mrak, E. M. 1974. A microbiologist turned administrator—how it happened. *Annu. Rev. Microbiol.* 28:1–22

Mudd, S. 1969. Sequences in medical microbiology: some observations over fifty years. *Annu. Rev. Microbiol.* 23:1–28 [Reprinted in *Exc. Fas. Sci.* Vol. 2]

Mueller, Conrad G. 1979. Some origins of psychology as science. *Annu. Rev. Psychol.* 30:9–29

Mulliken, R. S. 1978. Chemical bonding. *Annu. Rev. Phys. Chem.* 29:1–30

Munk, M. M. 1981. My early aerodynamic research—thoughts and memories. *Annu. Rev. Fluid Mech.* 13:1–7 [Reprinted in *Exc. Fas. Sci.* Vol. 3]

Munk, W. H. 1980. Affairs of the sea. *Annu. Rev. Earth Planet. Sci.* 8:1–16 [Reprinted in *Exc. Fas. Sci.* Vol. 3]

Munro, H. N. 1986. Back to basics: an evolutionary odyssey with reflections on the nutrition research of tomorrow. *Annu. Rev. Nutr.* 6:1–12

Murray, R. G. E. 1988. A structured life. *Annu. Rev. Microbiol.* 42:1–34 [Reprinted in *Exc. Fas. Sci.* Vol. 4]

Muskett, A. E. 1967. Plant pathology and the plant pathologist. *Annu. Rev. Phytopathol.* 5:1–16

Nachmansohn, D. 1972. Biochemistry as part of my life. *Annu. Rev. Biochem.* 41:1–28 [Reprinted in *Exc. Fas. Sci.* Vol. 2]

Nanney, D. L. 1981. T. M. Sonneborn: an interpretation. *Annu. Rev. Genet.* 15:1–9 [Reprinted in *Exc. Fas. Sci.* Vol. 3]

Neel, J. V. 1983. Curt Stern, 1902-1981. *Annu. Rev. Genet.* 17:1–10 [Reprinted in *Exc. Fas. Sci.* Vol. 3]

Neergaard, P. 1986. Screening for plant health. *Annu. Rev. Phytopathol.* 24:1–16

Nelson, R. R. 1984. Pioneer leaders in plant pathology: E. C. Stakman. *Annu. Rev. Phytopathol.* 22:11–19

Newhall, A. G. 1980. Herbert Hice Whetzel: pioneer American plant pathologist. *Annu. Rev. Phytopathol.* 18:27–36

Nier, A. O. 1981. Some reminiscences of isotopes, geochronology, and mass spectrometry. *Annu. Rev. Earth Planet. Sci.* 9:1–17 [Reprinted in *Exc. Fas. Sci.* Vol. 3]

Nolla, J. A. B. 1976. Contributions to the history of plant pathology in South America, Central America, and Mexico. *Annu. Rev. Phytopathol.* 14:11–29

Norrish, R. G. W. 1969. Fifty years of physical chemistry in Great Britain. *Annu. Rev. Phys. Chem.* 20:1–24 [Reprinted in *Exc. Fas. Sci.* Vol. 2]

Northrop, J. H. 1961. Biochemists, biologists, and William of Occam. *Annu. Rev. Biochem.* 30:1–10 [Reprinted in *Exc. Fas. Sci.* Vol. 1]

Ochoa, S. 1980. The pursuit of a hobby. *Annu. Rev. Biochem.* 49:1–30 [Reprinted in *Exc. Fas. Sci.* Vol. 3]

Olson, G. B. (with M. Cohen) 1981. A perspective on martensitic nucleation. *Annu. Rev. Mater. Sci.* 11:1–30

Oort, J. H. 1981. Some notes on my life as an astronomer. *Annu. Rev. Astron. Astrophys.* 19:1–5

Opik, E. J. 1977. About dogma in science, and other recollections of an astronomer. *Annu. Rev. Astron. Astrophys.* 15:1–17 [Reprinted in *Exc. Fas. Sci.* Vol. 2]

Orlob, G. B. 1971. History of plant pathology in the Middle Ages. *Annu. Rev. Phytopathol.* 9:7–20

Osterhout, W. J. V. 1957. The use of aquatic plants in the study of some fundamental problems. *Annu. Rev. Plant Physiol.* 8:1–10

Oswatitsch, K. 1987. Ludwig Prandtl and his Kaiser-Wilhelm-Institut. *Annu. Rev. Fluid Mech.* 19:1–25

Ou, S. H. 1984. Exploring tropical rice diseases: a reminiscence. *Annu. Rev. Phytopathol.* 22:1–10 [Reprinted in *Exc. Fas. Sci.* Vol. 3]

Pappenheimer, J. R. 1987. A silver spoon. *Annu. Rev. Physiol.* 49:1–15 [Reprinted in *Exc. Fas. Sci.* Vol. 3]

Paton, W. D. M. 1986. On becoming and being a pharmacologist. *Annu. Rev. Pharmacol. Toxicol.* 26:1–22 [Reprinted in *Exc. Fas. Sci.* Vol. 3]

Pauling, L. C. 1965. Fifty years of physical chemistry in the California Institute of Technology. *Annu. Rev. Phys. Chem.* 16:1–14 [Reprinted in *Exc. Fas. Sci.* Vol. 1]

Pauling, L. 1986. Early days of molecular biology in the California Institute of Technology. *Annu. Rev. Biophys. Biophys. Chem.* 15:1–9 [Reprinted in *Exc. Fas. Sci.* Vol. 3]

Payne-Gaposchkin, C. 1978. The development of our knowledge of variable stars. *Annu. Rev. Astron. Astrophys.* 16:1–13

Peters, R. 1957. Forty-five years of biochemistry. *Annu. Rev. Biochem.* 26:1–16 [Reprinted in *Exc. Fas. Sci.* Vol. 1]

Phaff, H. J. 1986. My life with yeasts. *Annu. Rev. Microbiol.* 40:1–28 [Reprinted in *Exc. Fas. Sci.* Vol. 3]

Piaget, J. 1979. Relations between psychology and other sciences. *Annu. Rev. Psychol.* 30:1–8

Pittman, M. 1990. A life with biological products. *Annu. Rev. Microbiol.* 44:1–25 [Reprinted in *Exc. Fas. Sci.* Vol. 4]

Pitts, R. F. 1976. Why a physiologist? *Annu. Rev. Physiol.* 38:1–6

Pitzer, K. S. 1987. Of physical chemistry and other activities. *Annu. Rev. Phys. Chem.* 38:1–25 [Reprinted in *Exc. Fas. Sci.* Vol. 3]

Platt, B. S. 1956. Sir Edward Mellanby, G.B.E., K.C.B., M.D., F.R.C.P., F.R.S. (1884-1955), the man, research worker, and statesman. *Annu. Rev. Biochem.* 25:1–28 [Reprinted in *Exc. Fas. Sci.* Vol. 1]

Pontecorvo, G. 1968. Hermann Joseph Muller. *Annu. Rev. Genet.* 2:1–10

Popov, G. B., Waloff, N. 1990. Sir Boris Uvarov (1889–1970): the father of acridology. *Annu. Rev. Entomol.* 35:1–24 [Reprinted in *Exc. Fas. Sci.* Vol. 4]

Posnette, A. F. 1980. Recollections of a genetical plant pathologist. *Annu. Rev. Phytopathol.* 18:1–9 [Reprinted in *Exc. Fas. Sci.* Vol. 3]

Post, R. L. 1989. Seeds of sodium, potassium ATPase. *Annu. Rev. Physiol.* 51:1–15 [Reprinted in *Exc. Fas. Sci.* Vol. 4]

Pringsheim, E. G. 1970. Contributions toward the development of general microbiology. *Annu. Rev. Microbiol.* 24:1–16

Prosser, C. L. 1986. The making of a comparative physiologist. *Annu. Rev. Physiol.* 48:1–6 [Reprinted in *Exc. Fas. Sci.* Vol. 3]

Raffel, S. 1982. Fifty years of immunology. *Annu. Rev. Microbiol.* 36:1–26 [Reprinted in *Exc. Fas. Sci.* Vol. 3]

Raffel, S. 1973. Windsor Cooper Cutting (1907–1972). *Annu. Rev. Pharmacol. Toxicol.* 13:1–4

Ratner, S. 1977. A long view of nitrogen metabolism. *Annu. Rev. Biochem.* 46:1–24 [Reprinted in *Exc. Fas. Sci.* Vol. 3]

Raychaudhuri, S. P. (with J. P. Verma, T. K. Nariani, Beneeta Sen) 1972. The history of plant pathology in India. *Annu. Rev. Phytopathol.* 10:21–36

Reichenbach, H. 1983. Contributions of Ernst Mach to fluid mechanics. *Annu. Rev. Fluid Mech.* 15:1–28

Reisman, D. 1988. On discovering and teaching sociology: a memoir. *Annu. Rev. Sociol.* 14:1–24 [Reprinted in *Exc. Fas. Sci.* Vol. 4]

Remington, J. E. (with C. L. Remington) 1961. Darwin's contributions to entomology. *Annu. Rev. Entomol.* 6:1–12

Revelle, R. 1987. How I became an oceanographer and other sea stories. *Annu. Rev. Earth Planet. Sci.* 15:1–23 [Reprinted in *Exc. Fas. Sci.* Vol. 3]

Rheingold, H. L. 1985. Development as the acquisition of familiarity. *Annu. Rev. Psychol.* 36:1–17

Rhoades, M. M. 1984. The early years of maize genetics. *Annu. Rev. Genet.* 18:1–29 [Reprinted in *Exc. Fas. Sci.* Vol. 3]

Richard, G. 1973. The historical development of nineteenth and twentieth century studies of the behavior of insects. *Hist. Entomol.*, pp. 477–496

Richards, A. G. 1973. Anatomy and morphology. *Hist. Entomol.*, pp. 185–202

Riesman, David. 1988. On discovering and teaching sociology: a memoir. *Annu. Rev. Sociol.* 14:1–24

Riley, M. White. 1990. The influence of sociological lives: personal reflections. *Annu. Rev. Sociol.* 16:1–25 [Reprinted in *Exc. Fas. Sci.* Vol. 4]

Rodgers, J. 1985. Witnessing revolutions in the earth sciences. *Annu. Rev. Earth Planet. Sci.* 13:1–4 [Reprinted in *Exc. Fas. Sci.* Vol. 3]

Rohdendorf, B. B. 1973. The history of paleoentomology. *Hist. Entomol.*, pp. 119–154

Roman, H. 1986. The early days of yeast genetics: a personal narrative. *Annu. Rev. Genet.* 20:1–12 [Reprinted in *Exc. Fas. Sci.* Vol. 3]

Ross, H. H. 1973. Evolution and phylogeny. *Hist. Entomol.*, pp. 155–170

Rothlin, E. 1964. Outlines of a pharmacological career. *Annu. Rev. Pharmacol. Toxicol.* 4:9–32

Rott, N. 1985. Jakob Ackeret and the history of the Mach number. *Annu. Rev. Fluid Mech.* 17:1–9

Rott, N. 1990. Note on the history of the Reynolds number. *Annu. Rev. Fluid Mech.* 22:1–11 [Reprinted in *Exc. Fas. Sci.* Vol. 4]

Rouse, H. 1976. Hydraulics' latest golden age. *Annu. Rev. Fluid Mech.* 8:1–12

Rubey, W. W. 1974. Fifty years of the earth sciences—a renaissance. *Annu. Rev. Earth Planet. Sci.* 2:1–24

Russell, E. J. 1985. A history of mouse genetics. *Annu. Rev. Genet.* 19:1–28 [Reprinted in *Exc. Fas. Sci.* Vol. 3]

Russell, E. S. 1989. Sewall Wright's contributions to physiological genetics and to inbreeding theory and practice. *Annu. Rev. Genet.* 23:1–18 [Reprinted in *Exc. Fas. Sci.* Vol. 4]

Russell, L. M. 1978. Leland Ossian Howard: a historical review. *Annu. Rev. Entomol.* 23:1–15

Ruzicka, L. 1973. In the borderland between bioorganic chemistry and biochemistry. *Annu. Rev. Biochem.* 42:1–20

Sanger, F. 1988. Sequences, sequences, and sequences. *Annu. Rev. Biochem.* 57:1–28 [Reprinted in *Exc. Fas. Sci.* Vol. 4]

Schafer, H. 1985. On the problem of polar intermetallic compounds: the stimulation of E. Zintl's work for the modern chemistry of intermetallics. *Annu. Rev. Mater. Sci.* 15:1–41

Scharrer, B. 1987. Neurosecretion: beginnings and new directions in neuropeptide research. *Annu. Rev. Neurosci.* 10:1–17 [Reprinted in *Exc. Fas. Sci.* Vol. 3]

Schmidt, C. F. 1965. Pharmacology in a changing world. *Annu. Rev. Physiol.* 23:1–14 [Reprinted in *Exc. Fas. Sci.* Vol. 1]

Schmitt, F. O. 1985. Adventures in molecular biology. *Annu. Rev. Biophys. Biophys. Chem.* 14:1–22 [Reprinted in *Exc. Fas. Sci.* Vol. 3]

Scholander, P. F. 1978. Rhapsody in science. *Annu. Rev. Physiol.* 40:1–17 [Reprinted in *Exc. Fas. Sci.* Vol. 2]

Schwerdtfeger, R. 1973. Forest entomology. *Hist. Entomol.*, pp. 361–386

Scrimshaw, N. S. 1987. The phenomenon of famine. *Annu. Rev. Nutr.* 7:1–21

Sears, M. R., Sears, W. R. 1979. The Kármán years at Galcit. *Annu. Rev. Fluid Mech.* 11:1–10 [Reprinted in *Exc. Fas. Sci.* Vol. 3]

Segrè, E. 1981. Fifty years up and down a strenuous and scenic trail. *Annu. Rev. Nucl. Part. Sci.* 31:1–18 [Reprinted in *Exc. Fas. Sci.* Vol. 1]

Sela, M. 1987. A peripatetic and personal view of molecular immunology for one third of a century. *Annu. Rev. Immunol.* 05:1–19 [Reprinted in *Exc. Fas. Sci.* Vol. 3]

Sequeira, Luis 1988. On becoming a plant pathologist: the changing scene. *Annu. Rev. Phytopathol.* 26:1–13

Sewell, W. H. 1989. Some reflections on the golden age of interdisciplinary social psychology. *Annu. Rev. Sociol.* 15:1–16 [Reprinted in *Exc. Fas. Sci.* Vol. 4]

Sharp, R. P. 1988. Earth science field work: role and status. *Annu. Rev. Earth Planet. Sci.* 16:1–19 [Reprinted in *Exc. Fas. Sci.* Vol. 4]

Simpson, G. G. 1976. The compleat palaeontologist? *Annu. Rev. Earth Planet. Sci.* 4:1–13

Sloss, L. L. 1984. The greening of stratigraphy 1933–1983. *Annu. Rev. Earth Planet. Sci.* 12:1–10 [Reprinted in *Exc. Fas. Sci.* Vol. 3]

Smith, C. S. 1986. On material structure and human history. *Annu. Rev. Mater. Sci.* 16:1–11

Smith, E. H. 1976. The Comstocks and Cornell: in the people's service. *Annu. Rev. Entomol.* 21:1–25

Smith, H. 1989. The mounting interest in bacterial and viral pathogenicity. *Annu. Rev. Microbiol.* 43:1–22 [Reprinted in *Exc. Fas. Sci.* Vol. 4]

Snell, E. E. 1989. Nutrition research with lactic acid bacteria: a retrospective view. *Annu. Rev. Nutr.* 9:1–19 [Reprinted in *Exc. Fas. Sci.* Vol. 4]

Snyder, W. C. 1971. Plant pathology today. *Annu. Rev. Phytopathol.* 9:1–6

Sollman, T. 1965. Why an Annual Review of Pharmacology? *Annu. Rev. Pharmacol. Toxicol.* 1:453–460 [Reprinted in *Exc. Fas. Sci.* Vol. 1]

Sperry, R. W. 1981. Changing priorities. *Annu. Rev. Neurosci.* 4:1–15

Spitzer, L. Jr. 1989. Dreams, stars, and electrons. *Annu. Rev. Astron. Astrophys.* 27:1–17 [Reprinted in *Exc. Fas. Sci.* Vol. 4]

Sproull, R. L. 1987. The early history of the Material Research Laboratories. *Annu. Rev. Mater. Sci.* 17:1–12

Srb, A. M. 1990. G. W. Beadle. *Annu. Rev. Genet.* 24:1–4 [Reprinted in *Exc. Fas. Sci.* Vol. 4]

Stakman, E. C. 1964. Opportunity and obligation in plant pathology. *Annu. Rev. Phytopathol.* 2:1–12

Stanier, R. Y. 1980. The journey, not the arrival, matters. *Annu. Rev. Microbiol.* 34:1–48 [Reprinted in *Exc. Fas. Sci.* Vol. 3]

Stare, F. J. 1991. Nutrition research from respiration and vitamins to cholesterol and atherosclerosis. *Annu. Rev. Nutr.* 11:1–20 [Reprinted in *Exc. Fas. Sci.* Vol. 4]

Steward, F. C. 1971. Plant physiology: the changing problems, the continuing quest. *Annu. Rev. Plant Physiol.* 22:1–22

Stocking, C. R. 1984. Reminiscences and reflections. *Annu. Rev. Plant Physiol. Plant Mol. Biol.* 35:1–14 [Reprinted in *Exc. Fas. Sci.* Vol. 3]

Stockmayer, W. H. (with Bruno H. Zimm) 1984. When polymer science looked easy. *Annu. Rev. Phys. Chem.* 35:1–21 [Reprinted in *Exc. Fas. Sci.* Vol. 3]

Stout, J. W. 1986. The Journal of Chemical Physics: the first 50 years. *Annu. Rev. Phys. Chem.* 37:1–23 [Reprinted in *Exc. Fas. Sci.* Vol. 3]

Strömgren, B. 1983. Scientists I have known and some astronomical problems I have met. *Annu. Rev. Astron. Astrophys.* 21:1–11

Stuart, J. T. 1986. Keith Stewartson: his life and work. *Annu. Rev. Fluid Mech.* 18:1–14

Stumpf, S. E. 1981. The moral dimensions of the world's food supply. *Annu. Rev. Nutr.* 1:1–25

Sweeney, B. M. 1987. Living in the golden age of biology. *Annu. Rev. Plant Physiol. Plant Mol. Biol.* 38:1–9 [Reprinted in *Exc. Fas. Sci.* Vol. 3]

Swings, P. 1979. A few notes on my career as an astrophysicist. *Annu. Rev. Astron. Astrophys.* 17:1–7

Szentágothai, J. 1984. Downward causation? *Annu. Rev. Neurosci.* 7:1–11

Szent-Györgyi, A. 1959. Lost in the twentieth century. *Annu. Rev. Biochem.* 32:1–14 [Reprinted in *Exc. Fas. Sci.* Vol. 1]

Taliaferro, W. H. 1968. The lure of the unknown. *Annu. Rev. Microbiol.* 22:1–14

Talmage, D. W. 1986. The acceptance and rejection of immunological concepts. *Annu. Rev. Immunol.* 4:1–11 [Reprinted in *Exc. Fas. Sci.* Vol. 3]

Tamiya, H. 1966. Synchronous cultures of algae. *Annu. Rev. Plant Physiol.* 17:1–26

Tang, P. 1983. Aspirations, reality, and circumstances: the devious trail of a roaming plant physiologist. *Annu. Rev. Plant Physiol.* 34:1–19 [Reprinted in *Exc. Fas. Sci.* Vol. 3]

Tani, I. 1977. History of boundary-layer theory. *Annu. Rev. Fluid Mech.* 9:87–111

Tax, S. 1988. Pride and puzzlement: a retro-introspective record of 60 years of anthropology. *Annu. Rev. Anthropol.* 17:1–21 [Reprinted in *Exc. Fas. Sci.* Vol. 4]

Taylor, G. I. 1974. The interaction between experiment and theory in fluid mechanics. *Annu. Rev. Fluid Mech.* 6:1–16

Taylor, H. 1962. Fifty years of chemical kineticists. *Annu. Rev. Phys. Chem.* 13:1–18 [Reprinted in *Exc. Fas. Sci.* Vol. 1]

ten Houten, J. G. 1974. Plant pathology: changing agricultural methods and human society. *Annu. Rev. Phytopathol.* 12:1–11

Terroine, E. F. 1959. Fifty-five years of union between biochemistry and physiology. *Annu. Rev. Biochem.* 28:1–14 [Reprinted in *Exc. Fas. Sci.* Vol. 1]

Thimann, K. V. 1963. Plant growth substances: past, present and future. *Annu. Rev. Plant Physiol.* 14:1–18

Thomas, K. 1954. Fifty years of biochemistry in Germany. *Annu. Rev. Biochem.* 23:1–16 [Reprinted in *Exc. Fas. Sci.* Vol. 1]

Tiselius, A. 1968. Reflections from both sides of the counter. *Annu. Rev. Biochem.* 37:1–24 [Reprinted in *Exc. Fas. Sci.* Vol. 2]

Tiselius, A., Claesson, S. 1967. The Svedberg and fifty years of physical chemistry in Sweden. *Annu. Rev. Phys. Chem.* 18:1–18

Tomiyama, K. 1983. Research on the hypersensitive response. *Annu. Rev. Phytopathol.* 21:1–12 [Reprinted in *Exc. Fas. Sci.* Vol. 3]

Toussoun, T. A. 1986. William C. Snyder: pioneer leader in plant pathology. *Annu. Rev. Phytopathol.* 24:27–31

Truhlar, D. G. (with Robert E. Wyatt) 1976. History of H_3 kinetics. *Annu. Rev. Phys. Chem.* 27:1–43

Turnbull, David 1983. A commentary on the emergence and evolution of "materials science". *Annu. Rev. Mater. Sci.* 13:1–7

Tuxen, S. L. 1967. The entomologist, J. C. Fabricius. *Annu. Rev. Entomol.* 12:1–14

Tuxen, S. L. 1973. Entomology systematizes and describes: 1700–1815. *Hist. Entomol.*, pp. 95–118

Tyler, L. E. 1981. More stately mansions—psychology extends its boundaries. *Annu. Rev. Psychol.* 32:1–20

Uhlenbeck, G. E. 1979. Some notes on the relation between fluid mechanics and statistical physics. *Annu. Rev. Fluid Mech.* 12:1–9 [Reprinted in *Exc. Fas. Sci.* Vol. 3]

Ulrich, W. 1972. Hermann Burmeister, 1807 to 1892. *Annu. Rev. Entomol.* 17:1–20

Usinger, R. L. 1964. The role of Linnaeus in the advancement of entomology. *Annu. Rev. Entomol.* 9:1–16

Ussing, H. H. 1980. Life with tracers. *Annu. Rev. Physiol.* 42:1–16 [Reprinted in *Exc. Fas. Sci.* Vol. 3]

Uvnäs, B. 1984. From physiologist to pharmacologist—promotion or degradation? Fifty years in retrospect. *Annu. Rev. Pharmacol. Toxicol.* 24:1–18 [Reprinted in *Exc. Fas. Sci.* Vol. 3]

Van Allen, J. A. 1990. What is a space scientist? An autobiographical example. *Annu. Rev. Earth Planet. Sci.* 18:1–26 [Reprinted in *Exc. Fas. Sci.* Vol. 4]

Vanderplank, J. E. 1976. Four essays. *Annu. Rev. Phytopathol.* 14:1–10

van Niel, C. B. 1967. The education of a microbiologist: some reflections. *Annu. Rev. Microbiol.* 21:1–30 [Reprinted in *Exc. Fas. Sci.* Vol. 2]

van Niel, C. B. 1962. The present status of the comparative study of photosynthesis. *Annu. Rev. Plant Physiol.* 13:1–26

van Overbeek, J. 1976. Plant physiology and the human ecosystem. *Annu. Rev. Plant Physiol.* 27:1–17

Vennesland, B. 1981. Recollections and small confessions. *Annu. Rev. Plant Physiol.* 32:1–20 [Reprinted in *Exc. Fas. Sci.* Vol. 3]

Verhoogen, J. 1983. Personal notes and sundry comments. *Annu. Rev. Earth Planet. Sci.* 11:1–9 [Reprinted in *Exc. Fas. Sci.* Vol. 3]

Vickery, H. B. 1972. A chemist among plants. *Annu. Rev. Plant Physiol.* 23:1–28

Villat, H. 1972. As luck would have it—a few mathematical reflections. *Annu. Rev. Fluid Mech.* 4:1–65

Virtanen, A. I. 1961. Some aspects of amino acid synthesis in plants and related subjects. *Annu. Rev. Plant Physiol.* 12:1–12

Visscher, M. B. 1978. A half century in science and society. *Annu. Rev. Physiol.* 31:1–18 [Reprinted in *Exc. Fas. Sci.* Vol. 2]

von Békésy, G. 1978. Some biographical experiments from fifty years ago. *Annu. Rev. Physiol.* 36:1–16 [Reprinted in *Exc. Fas. Sci.* Vol. 2]

von Euler, U. S. 1978. Pieces in the puzzle. *Annu. Rev. Pharmacol. Toxicol.* 11:675–686 [Reprinted in *Exc. Fas. Sci.* Vol. 2]

von Muralt, A. 1984. A life with several facets. *Annu. Rev. Physiol.* 46:1–13 [Reprinted in *Exc. Fas. Sci.* Vol. 3]

Wadati, K. 1989. Born in a country of earthquakes. *Annu. Rev. Earth Planet. Sci.* 17:1–12 [Reprinted in *Exc. Fas. Sci.* Vol. 4]

Wagner, C. 1977. Point defects and their interaction. *Annu. Rev. Mater. Sci.* 7:1–22

Walker, J. C. 1975. Some highlights in plant pathology in the United States. *Annu. Rev. Phytopathol.* 13:15–29

Walker, J. C. 1979. Leaders in plant pathology: L. R. Jones. *Annu. Rev. Phytopathol.* 17:13–20

Walker, J. C. 1982. Pioneer leaders in plant pathology: Benjamin Minge Duggar. *Annu. Rev. Phytopathol.* 20:33–39

Walker, J. C. 1963. The future of plant pathology. *Annu. Rev. Phytopathol.* 1:1–4

Wallach, H. 1987. Perceiving a stable environment when one moves. *Annu. Rev. Psychol.* 38:1–27

Warburg, O. 1964. Prefatory chapter. *Annu. Rev. Biochem.* 33:1–14 [Reprinted in *Exc. Fas. Sci.* Vol. 1]

Wareing, P. F. 1982. A plant physiological odyssey. *Annu. Rev. Plant Physiol.* 33:1–26 [Reprinted in *Exc. Fas. Sci.* Vol. 3]

Washburn, S. L. 1983. Evolution of a teacher. *Annu. Rev. Anthropol.* 12:1–24 [Reprinted in *Exc. Fas. Sci.* Vol. 3]

Weissman, S. 1990. The way it was. *Annu. Rev. Phys. Chem.* 41:1–13 [Reprinted in *Exc. Fas. Sci.* Vol. 4]

Welch, A. D. 1985. Reminiscences in pharmacology: auld acquaintance ne'er forgot. *Annu. Rev. Pharmacol. Toxicol.* 25:1–26 [Reprinted in *Exc. Fas. Sci.* Vol. 3]

Welker, H. 1979. From solid state research to semiconductor electronics. *Annu. Rev. Mater. Sci.* 9:1–21

Went, F. W. 1974. Reflections and speculations. *Annu. Rev. Plant Physiol.* 25:1–26

Wheeler, J. A. 1989. Fission in 1939: the puzzle and the promise. *Annu. Rev. Nucl. Part. Sci.* 39:xiii–xxviii [Reprinted in *Exc. Fas. Sci.* Vol. 4]

Whipple, F. L. 1978. The Earth as part of the universe. *Annu. Rev. Earth Planet. Sci.* 6:1–8 [Reprinted in *Exc. Fas. Sci.* Vol. 3]

Whitford, A. E. 1986. A half-century of astronomy. *Annu. Rev. Astron. Astrophys.* 24:1–22 [Reprinted in *Exc. Fas. Sci.* Vol. 3]

Wiggers, C. 1951. Prefatory chapter: physiology from 1900–1920: incidents, accidents, and advances. *Annu. Rev. Physiol.* 13:1–20 [Reprinted in *Exc. Fas. Sci.* Vol. 1]

Wigglesworth, V. B. 1973. The history of insect physiology. *Hist. Entomol.*, pp. 203–228

Wilhelm, S. (with Helga Tietz) 1978. Julius Kuehn—his concept of plant pathology. *Annu. Rev. Phytopathol.* 16:343–358

Williams, Robley C. 1978. Spectroscopes, telescopes, microscopes. *Annu. Rev. Microbiol.* 32:1–18 [Reprinted in *Exc. Fas. Sci.* Vol. 3]

Wilson, M. K. 1975. The top twenty and the rest: big chemistry and little funding. *Annu. Rev. Phys. Chem.* 26:1–16

Wilson, E. W. 1979. Molecular spectroscopy. *Annu. Rev. Phys. Chem.* 30:1–27

Wilson, J. T. 1982. Early days in university geophysics. *Annu. Rev. Earth Planet. Sci.* 10:1–14 [Reprinted in *Exc. Fas. Sci.* Vol. 3]

Wilson, P. W. 1972. Training a microbiologist. *Annu. Rev. Microbiol.* 26:1–22

Wilson, E. B. (with John Ross) 1973. Physical chemistry in Cambridge, Massachusetts. *Annu. Rev. Phys. Chem.* 24:1–27

Wolfe, R. S. 1991. My kind of biology. *Annu. Rev. Microbiol.* 45:1–35 [Reprinted in *Exc. Fas. Sci.* Vol. 4]

Wolman, Abel 1986. Is there a public health function? *Annu. Rev. Publ. Health* 7:1–12

Wood, H. 1985. Then and now. *Annu. Rev. Biochem.* 54:1–41 [Reprinted in *Exc. Fas. Sci.* Vol. 3]

Wood, R. K. S. 1987. Physiological plant pathology comes of age. *Annu. Rev. Phytopathol.* 25:27–40

Woodruff, H. B. 1981. A soil microbiologist's odyssey. *Annu. Rev. Microbiol.* 35:1–28 [Reprinted in *Exc. Fas. Sci.* Vol. 3]

Wright, S. 1982. The shifting balance theory and macroevolution. *Annu. Rev. Genet.* 16:1–19

Wyman, J. (with Stanley J. Gill) 1987. Conversations with Jeffries Wyman. *Annu. Rev. Biophys. Biophys. Chem.* 16:1–23

Yokoyama, Tadao. 1973. The history of sericultural science in relation to industry. *Hist. Entomol.*, pp. 267–284

SUBJECT INDEX
VOLUME 4

The Leader for 65 years.

ANNUAL REVIEWS INC. A nonprofit publisher dedicated to serving the worldwide scientific community.

1995-1996 ANNUAL REVIEWS OF:

ANTHROPOLOGY
USA/other countries
- Volume 25 (avail. October 1996) ....$49/$54
- Volume 24 (1995) ....$47/$52

ASTRONOMY AND ASTROPHYSICS
- Volume 34 (avail. September 1996) ....$65/$70
- Volume 33 (1995) ....$60/$65

BIOCHEMISTRY
- Volume 65 (avail. July 1996) ....$59/$65
- Volume 64 (1995) ....$49/$55

BIOPHYSICS AND BIOMOLECULAR STRUCTURE
- Volume 25 (avail. June 1996) ....$67/$72
- Volume 24 (1995) ....$62/$67

CELL AND DEVELOPMENTAL BIOLOGY
- Volume 12 (avail. November 1996) ....$56/$61
- Volume 11 (1995) ....$49/$54

EARTH AND PLANETARY SCIENCES
- Volume 24 (avail. May 1996) ....$67/$72
- Volume 23 (1995) ....$62/$67

ECOLOGY AND SYSTEMATICS
- Volume 27 (avail. November 1996) .....$52/$57
- Volume 26 (1995) ....$47/$52

ENERGY AND THE ENVIRONMENT
- Volume 21 (avail. October 1996) ....$76/$81
- Volume 20 (1995) ....$71/$76

ENTOMOLOGY
- Volume 41 (avail. January 1996) ....$52/$57
- Volume 40 (1995) ....$47/$52

FLUID MECHANICS
- Volume 28 (avail. January 1996) ....$52/$57
- Volume 27 (1995) ....$47/$52

GENETICS
- Volume 30 (avail. December 1996) ....$52/$57
- Volume 29 (1995) ....$47/$52

IMMUNOLOGY
USA/other countries
- Volume 14 (avail. April 1996) ....$56/$61
- Volume 13 (1995) ....$48/$53

MATERIALS SCIENCE
- Volume 26 (avail. August 1996) ....$80/$85
- Volume 25 (1995) ....$75/$80

MEDICINE: Selected Topics in the Clincal Sciences
- Volume 47 (avail. April 1996) ....$52/$57
- Volume 46 (1995) ....$47/$52

MICROBIOLOGY
- Volume 50 (avail. October 1996) ....$53/$58
- Volume 49 (1995) ....$48/$53

NEUROSCIENCE
- Volume 19 (avail. March 1996) ....$52/$57
- Volume 18 (1995) ....$47/$52

NUCLEAR AND PARTICLE SCIENCE
- Volume 46 (avail. December 1996) ....$67/$72
- Volume 45 (1995) ....$62/$67

NUTRITION
- Volume 16 (avail. July 1996) ....$53/$58
- Volume 15 (1995) ....$48/$53

PHARMACOLOGY AND TOXICOLOGY
- Volume 36 (avail. April 1996) ....$52/$57
- Volume 35 (1995) ....$47/$52

PHYSICAL CHEMISTRY
- Volume 47 (avail. November 1996) ....$56/$61
- Volume 46 (1995) ....$51/$56

PHYSIOLOGY
- Volume 58 (avail. March 1996) ....$54/$59
- Volume 57 (1995) ....$49/$54

PHYTOPATHOLOGY *New in 1995! CD-ROM archive*
- Volume 34 (avail. September 1996) ....$54/$59
- Vol. 33 with 10 yrs on CD-ROM (1995). .$49/$54

PLANT PHYSIOLOGY & PLANT MOLECULAR BIOLOGY
USA/other countries
- Volume 47 (avail. June 1996) ....$52/$57
- Volume 46 (1995) ....$47/$52

PSYCHOLOGY
- Volume 47 (avail. February 1996) ....$48/$53
- Volume 46 (1995) ....$46/$51

PUBLIC HEALTH
- Volume 17 (avail. May 1996) ....$57/$62
- Volume 16 (1995) ....$52/$57

SOCIOLOGY
- Volume 22 (avail. August 1996) ....$54/$59
- Volume 21 (1995) ....$52/$57

Science Autobiography

THE EXCITEMENT AND FASCINATION OF SCIENCE:
- Volume 4 (1995) hardcover, ...$50/$55
  52 autobiographical, historical and philosophical essays by prominent scientists

ANNUAL REVIEWS INC.

BB96

4139 El Camino Way • P.O. Box 10139
Palo Alto, CA 94303-0139 • USA

Step 1 **Ordered by:**

Name ______________________________

Address ____________________________

____________________ Zip Code ___________

Call from USA or Canada
1.800.523.8635

Please Mention Priority Code **BB96** when placing orders by phone.

FAX orders 24 hours a day
1.415.424.0910

Today's Date __________ Day Phone: (____) ____________

Fax (____) ______________ e-mail ____________

Step 4 **Payment Method**

❏ Check or money order enclosed. Make checks payable to "Annual Reviews Inc."   *or charge*

❏ VISA   ❏ M/C   ❏ AMEX

Account Number

Mo __/__ Yr __/__
Expiration Date   Print name exactly as it appears on credit card.

Signature

Qty	Annual Review of	Vol.	Place on Standing Order? Save 10% now with payment		Price	Total
			❏ Yes, save 10%	❏ No	$	
			❏ Yes, save 10%	❏ No	$	
			❏ Yes, save 10%	❏ No	$	
			❏ Yes, save 10%	❏ No	$	
			❏ Yes, save 10%	❏ No	$	

Step 2
Enter Order

❏ **Student / Recent Graduate** (past three years) discount 30% off. Not applicable to standing orders. Proof of status enclosed.

❏ **California customers.** Add applicable California sales tax for your location.

❏ **Canadian customers.** Add 7% Canadian GST. **(Reg. # 121449029 RT)**

Step 3
Shipping and Handling

✔ **Handling Charges. Add $3 per volume. Applies to all orders.**

❏ **Standard shipping, US Mail 4th class bookrate** (surface) No extra charge.   N/C

❏ **Optional UPS Ground service, $3** extra per volume in 48 contiguous states only. UPS not available to PO boxes.

UPS Next Day Air ❏   UPS Second Day Air ❏   US Airmail ❏   Note option at left. We will calculate amount and add to your total.
Optional shipping to anywhere. Charged at actual cost and added to total. Prices vary by weight of volumes.

Total

Call Toll Free 1.800.523.8635 from USA or Canada 8am-4pm, M-F, Pacific Time. From elsewhere call 1.415.493.4400 ext. 1

Mail Orders, fill in form, send in attached envelope. *e* e-mail **service@annurev.org**

Orders may also be placed through booksellers or subscription agents or through our **Authorized Stockists**
From Europe, the UK, the Middle East, and Africa contact: **Gazelle Book Service Ltd.**, Fax 44 (0) 1524-63232
From India, Pakistan, Bangladesh or Sri Lanka contact: **SARAS Books**, Fax 91-11-941111.

ANNUAL REVIEWS INC. on the Web **http://www.annurev.org**